Advances in Intelligent Systems and Computing

Volume 1133

The series "Advances in Intelligent Systems and Computing" contains publications on theory, applications, and design methods of Intelligent Systems and Intelligent Computing. Virtually all disciplines such as engineering, natural sciences, computer and information science, ICT, economics, business, e-commerce, environment, healthcare, life science are covered. The list of topics spans all the areas of modern intelligent systems and computing such as: computational intelligence, soft computing including neural networks, fuzzy systems, evolutionary computing and the fusion of these paradigms, social intelligence, ambient intelligence, computational neuroscience, artificial life, virtual worlds and society, cognitive science and systems, Perception and Vision, DNA and immune based systems, self-organizing and adaptive systems, e-Learning and teaching, human-centered and human-centric computing, recommender systems, intelligent control, robotics and mechatronics including human-machine teaming, knowledge-based paradigms, learning paradigms, machine ethics, intelligent data analysis, knowledge management, intelligent agents, intelligent decision making and support, intelligent network security, trust management, interactive entertainment, Web intelligence and multimedia.

The publications within "Advances in Intelligent Systems and Computing" are primarily proceedings of important conferences, symposia and congresses. They cover significant recent developments in the field, both of a foundational and applicable character. An important characteristic feature of the series is the short publication time and world-wide distribution. This permits a rapid and broad dissemination of research results.

**** Indexing: The books of this series are submitted to ISI Proceedings, EI-Compendex, DBLP, SCOPUS, Google Scholar and Springerlink ****

More information about this series at http://www.springer.com/series/11156

Niranjan N. Chiplunkar · Takanori Fukao
Editors

Advances in Artificial Intelligence and Data Engineering

Select Proceedings of AIDE 2019

Springer

Editors
Niranjan N. Chiplunkar
NMAM Institute of Technology
Udupi, India

Takanori Fukao
Ritsumeikan University
Shiga, Japan

ISSN 2194-5357 ISSN 2194-5365 (electronic)
Advances in Intelligent Systems and Computing
ISBN 978-981-15-3513-0 ISBN 978-981-15-3514-7 (eBook)
https://doi.org/10.1007/978-981-15-3514-7

This Springer imprint is published by the registered company Springer Nature Singapore Pte Ltd.
The registered company address is: 152 Beach Road, #21-01/04 Gateway East, Singapore 189721, Singapore

Organization

Chief Patron

Shri. N. Vinaya Hegde, Chancellor, Nitte (Deemed to be University); President, Nitte Education Trust, Mangalore

General Chairs

Niranjan N. Chiplunkar, NMAM Institute of Technology, Nitte
I. R. Mithanthaya, NMAM Institute of Technology, Nitte
B. R. Srinivas Rao, NMAM Institute of Technology, Nitte
Sudesh Bekal, NMAM Institute of Technology, Nitte

Advisory Committee

N. R. Shetty, Chancellor, Central University, Karnataka, India
Omid Ansary, Penn State, Harrisburg, USA
B. S. Sonde, ASM Technologies Limited, India
L. M. Patnaik, IISc, Bengaluru, India
K. Chidananda Gowda, Kuvempu University, India
P. Nagabhushan, IIIT, Allahabad, India
D. S. Guru, University of Mysore, Mysuru, India
Narayan Choudhary, Central Institute of Indian Languages, Mysuru, India
H. N. Prakash, RIT, Hassan, India
N. V. Subba Reddy, MIT, Manipal, India
Mohan Hegde, NMAMIT, Nitte, India

M. Hanumanthappa, Bangalore University, Bengaluru, India
V. N. Manjunath Aradhya, JSSSTU, Mysuru, India
B. G. Satish, CSI Chairman, Bengaluru

Steering Committee

Shri. Vishal Hegde, Nitte (Deemed to be University), India
N. R. Shetty, Central University, Karnataka, India
B. S. Sonde, ASM Technologies Ltd., Bengaluru, India
G. Hemanth Kumar, University of Mysore, India
Omid Ansary, Penn State University, USA
Samson Ojawo, Ladoke Akintola University of Technology, Nigeria
Shripad T. Revankar, Purdue University, USA
Shuhaimi Mansor, Universiti Teknologi, Malaysia

Programme Chairs

K. R. Udaya Kumar Reddy, NMAM Institute of Technology, Nitte
D. S. Guru, University of Mysore, Mysuru

Organizing Chairs

K. R. Udaya Kumar Reddy, NMAM Institute of Technology, Nitte
Surendra Shetty, NMAM Institute of Technology, Nitte
B. H. Karthik Pai, NMAM Institute of Technology, Nitte

Conference Secretary

C. V. Aravinda, NMAMIT, Nitte

Track Chairs

Jyothi Shetty, NMAMIT, Nitte
P. S. Venugopala, NMAMIT, Nitte

Publication Chairs

Niranjan N. Chiplunkar, NMAMIT, Nitte, India
D. S. Guru, University of Mysore, Mysuru, India
Takanori Fukao, Ritsumeikan University, Japan

Publicity Chair

K. B. Sudeepa, NMAMIT, Nitte

Workshop Chair

D. Radhakrishna, NMAMIT, Nitte

Social Event Chair

Roshan Fernandes, NMAMIT, Nitte

Organizing Committee

Udaya Kumar K. Shenoy
Karuna Pandit
D. K. Sreekantha
Sharada U. Shenoy
Sarika Hegde
B. Ashwini
K. Raju
Mamatha Balipa
Pradeep Kanchan
Vasudeva Pai
T. Vijay Murari
K. N. Pallavi
P. R. Anisha
K. R. Raghunandan
Ramesh Shettigar
Shashank Shetty
R. Balasubramani
S. V. Aruna Kumar
K. Akshaya
Nikitha Sourabh
Tanzila Nargis
Saritha Shetty
B. Ravi
B. R. Puneeth
Balachandra Rao
Srikanth Bhat
Arahat Jain
Abhir Bhandary
Ranjan Kumar
M. S. Sannidhan
Pawan Hegde
R. P. Puneeth

Krishna Prasad Rao
Ankitha Nayak
V. Sanju
Devidas
Krishnaraj
M. Shruthi
Shilpa Karegoudar
Rashmi Naveen
Abhishek S. Rao
Prathyakshini
S. Sapna
Pallavi Shetty
Ramya
A. G. Mangala
R. Anand
Sampath Kini
Mahesh Kini
B. C. Keerthana
Sunil Kumar Aithal
Jason E. Martis
Asmita Poojary
Minu P. Abhraham
Spoorthi
Divya Jennifer D'Souza
Sandeep Kumar Hegde
Deepa
Savitha
Rajalaxmi Hegde
Rajashree
Shabari Shedthi
G. K. Shwetha
Chinmai Shetty
K. Shrikanth Bhat
Anusha
Alaka Ananth
Anantha Murthy

Preface

The editorial team is honoured to announce the publication of the conference proceedings of the International Conference on Artificial Intelligence and Data Engineering (AIDE 2019). The conference was organized by the Departments of Computer Science and Engineering, Information Science and Engineering and Master of Computer Applications of NMAM Institute of Technology, Nitte, India.

With the resurgence of machine intelligence, neural networks and huge data, the artificial intelligence and the data engineering have become extremely active areas of research. The focus of the conference was to understand the recent advancements in the artificial intelligence and data engineering era of computing systems that emulate the capability of human intelligent mind. The conference provided a forum for researchers, practitioners, undergraduate and postgraduate students to present and discuss the latest theoretical advances and real-world applications in all areas related to artificial intelligence and data engineering.

There were 360 papers submitted to AIDE 2019 from across globe, and they all underwent a rigorous review process. Upholding the quality requirements, around 35% of the papers received were selected for presentation.

The conference comprised of pre-conference workshop by Dr. Dinesh R. and Dr. Manjunath S. as resource persons from Samsung Electro Mechanics, Bengaluru, as well as keynote talks by Prof. (Dr.) Ernest Cachia, University of Malta, Malta; Dr. Vimala Balakrishnan, University of Malaya, Malaysia; Prof. (Dr.) D. S. Guru, University of Mysore, Mysuru, India; and Prof. (Dr.) Kavi Mahesh, Indian Institute of Information Technology, Dharwad, India.

We would like to gratefully acknowledge the support received from NMAM Institute of Technology, Nitte, in organizing this conference. We would also like to gratefully acknowledge the College of Science and Engineering, Ritsumeikan University, Japan, and the Department of Studies in Computer Science, University of Mysore, Mysuru, for their kind support in conducting this conference. We would like to acknowledge all the researchers, reviewers, speakers, members of various committees, international/national advisory committee members, general chairs, programme chairs, technical committee and organization committee for helping us attain the objectives of the conference. We are sure that this conference has given an excellent platform for all researchers to share their ideas to update themselves with the latest developments, knowing how other groups are applying the technology and exchange ideas with leading international experts in the fields of artificial intelligence, data engineering and its affiliated disciplines.

We hope that the conference book of abstracts will be inspiring to readers, which is ideal for researchers, practitioners, academicians and students from artificial intelligence community and data engineering community.

Udupi, India — Niranjan N. Chiplunkar
Shiga, Japan — Takanori Fukao
Mysuru, India — D. S. Guru
Udupi, India — K. R. Udaya Kumar Reddy
Udupi, India — Surendra Shetty
Udupi, India — B. H. Karthik Pai

Acknowledgements

The organization of a conference is always a stressful adventure because of all the very small things and all the very important issues that have to be planned and managed.

We would like first to express our deep gratitude to the NMAM Institute of Technology (NMAMIT), Nitte, in organizing this conference. We would like to gratefully acknowledge the College of Science and Engineering, Ritsumeikan University, Japan, and the Department of Studies in Computer Science, University of Mysore, Mysuru, for their kind support in conducting this conference. We would also like to extend our heartiest thanks to the Central Institute of Indian Languages (CIIL), Mysuru, Government of India, for sponsoring this conference under the platinum category.

Our gratitude to the Computer Society of India (CSI) and the Indian Society of Technical Education (ISTE) for extending their technical support to the conference.

We would like to address a warm thank to all the members of the scientific committee for their participation and expertise in the preparation of the conference.

Our sincere gratitude is extended immensely to all the eminent scientists (from international and national level) who agreed to deliver a keynote speech and took their precious time and joined us in sharing their invaluable inputs and expertise with all the delegates at the conference. We also thank all the experts who played the role of moderators and chairs of sessions for taking their time and giving valuable inputs to all author delegates.

We would like to acknowledge the entire organizing committee who worked intensively for months to manage the conference, such as website, call for proposal, answers, evaluations, to name a few.

We would like to acknowledge all the members of various committees, particularly international/national advisory committee members, general chairs, programme chairs and technical committee for their work.

Our heartiest thanks go to all the reviewers for reviewing the papers promptly and spending their valuable time and providing invaluable inputs to research papers despite their busy schedule.

Our heartiest greetings go to all the participants who proposed a paper as their valuable contribution and came to NMAM Institute of Technology, Nitte, to present it in making this event a grand success.

Finally, we immensely thank Springer Nature and their team for their high-quality support, coordination and cooperation.

Niranjan N. Chiplunkar
Takanori Fukao
D. S. Guru
K. R. Udaya Kumar Reddy
Surendra Shetty
B. H. Karthik Pai

About the Conference

We are very much pleased to present this special issue of abstracts which consists of the proceedings of the International Conference on Artificial Intelligence and Data Engineering 2019 (AIDE 2019) organized by NMAM Institute of Technology, Nitte, India, in collaboration with the College of Science and Engineering, Ritsumeikan University, Japan, and the Department of Studies in Computer Science, University of Mysore, Mysuru, India. The conference was held during 23–24 May 2019.

The AIDE 2019 conference is aimed at providing an international forum for discussing the latest theoretical advances and real-world applications in all areas related to artificial intelligence (AI) and data engineering (DE) and its affiliated disciplines. The addressed topics of interest are divided into four major groups: artificial intelligence, machine vision and robotics, ambient intelligence and data engineering.

The goal of AIDE 2019 was to bring researchers across the following disciplines: artificial intelligence, expert systems, machine learning, robotics, virtual reality, augmented reality, bioinformatics, intelligent systems, cognitive systems, computational intelligence, neural networks, evolutionary computation, speech processing, Internet of things, big data challenges, data mining, information retrieval and natural language processing to establish an effective channel of communication together with the AIDE community to discuss the latest theoretical work and experimental work, and to give AIDE attendees the opportunity to know how other groups are applying the technology and exchange ideas with leading international experts in these areas of interest. The book is ideal for students, researchers, practitioners and academicians from artificial intelligence community, data engineering community and combination of these to create an approach towards a global exchange of information on technological advances and new scientific innovations. Moreover, it is attractive in terms of being popular on artificial intelligence tools, data engineering tools, techniques and methods that are applicable to real-world problems.

Salient Features

- Coverage of all aspects of theoretical and applied AI and DE research.
- Coverage of invited talks on innovative applications of AI and DE.
- Student, researcher, practitioner and academician learning is supported by the cutting-edge technology along with the application of tools of both AI and DE and a combination of these.

The research papers of AIDE 2019 have been peer-reviewed by at least two reviewers drawn from the scientific committee, external reviewers and editorial board depending on the subject matter of the paper. Papers are selected on the basis of their scientific merit and relevance to the conference topics.

Keynote Speaker

Dr. Ernest Cachia
Dean and Head of Department,
Faculty of Information and Communication Technology,
University of Malta, Msida (Europe)

Title of Talk: "Staying Smart in a Smarter World"

Brief Biodata

Professor Ernest Cachia currently holds the position of Head of the Department of Computer Information Systems, and for 12 consecutive years (2007–2019) held the position of Dean of the Faculty of Information and Communication Technology. He is also the Chairman of the Institute of Aerospace Technologies at the University of Malta and Director of Malta Air Traffic Services (MATS). Furthermore, he is the University of Malta representative on the Malta Competition and Consumer Affairs—sitting on the EU CEN e-Skills e-Competencies Common Framework (e-CF) Standardisation Committee, he is a founding member of the Malta IT Agency e-Skills Alliance, founding member of the Malta Cloud Forum within the Malta Communications Authority, a core member of the e-Skills (Malta) Foundation and a founding member of (former) EuroCloud (Malta) organization. He acted as the Innovation and Technology Module Leader in the creation of a Joint Masters programme for Border Management, coordinated by Frontex, the European Border Guard and Border Management Agency. He is routinely engaged as a court-appointed expert in ICT-related cases and was nationally recognized by being nominated for the award of Worker of the Year 2011.

He holds a doctorate in software engineering from the University of Sheffield, UK. He has worked in various electronics and ICT companies, and private educational establishments, and since 1994 holds a tenured academic post at the University of Malta. He was involved in various EU projects and was the project leader for multi-million Euro ERDF project constructing and equipping the existing faculty of ICT building. He has managed teams of developers working on both EU and internal projects. Academically, he is currently responsible for the software engineering and software project management streams in both B.Sc. IT (Hons) software development and B.Sc. IT (Hons) computing and business degree programmes of the faculty of ICT at the University of Malta. He is also a member of the IT Services Board of Directors of the University of Malta.

His native languages are Maltese and English. However, he is also completely fluent in Italian and Russian, has working knowledge of French and Spanish and a basic understanding of Arabic. In 2006, he was knighted as a Knight of the Sovereign Order of Saint John of Jerusalem (The Hereditary Order) by the then Grandmaster of the Order himself—Michael Vella Haber.

Abstract

The modern ICT landscape is characterized by many issues that unless viewed holistically often lead to tunnel vision and isolated focus. The main research directions in today's ICT scenario can be seen as Bio-Info-Nano-Cogno (BINC). Based on this road map, specific domains attracting interest and advancement can be viewed as (1) big data (BD); (2) artificial intelligence (AI); and (3) Internet of Things (IoT). We will visit and reason about many such issues (and questions they give rise to) that many researchers often tend to tackle separately.

For instance, in a world that embraces, directly or indirectly, the proliferation, pervasiveness and permeation of smart devices, how can such smartness be controlled and exploited? Where are we now in terms of data generation and what does this mean to all aspects of development, social behaviour and business models? Is the Cloud "old news" or still very relevant? Why are we in the situation we are in and what exactly is big data and how does it drive the rest of the modern ICT landscape? Which are the current main business and social disruptors? Is there a new way to approach system functionality through data availability and value extraction? How must modern developers think in terms of innovation and creativity in a marketplace overflowing with data, and how should they appreciate the multi-faceted nature and fickle behaviour of data valorization? It is time to clearly define a new data-driven business model as well as new "job descriptions" such as that of a data scientist and a data analyst? Last but definitely not least—What is "smartness", where does it actually reside and how can it be controlled? This last question will be discussed through the types, and management, of smart devices, as well as, what I like to call, technology to handle technology.

We will analyse the evolution of technologies as well as today's disruptive technologies as seen through the adaptation of Kondratiev Economic Waves and the related "Hype Cycle". In a way, this mapping can help predict what is characteristically unpredictable. A technology evolutionary scheme is proposed that, together with the Kondratiev Wave approach, may help discern technological evolutionary patterns.

The notion of the "more the merrier" can seem attractive; however, unless properly managed and integrated into digital solutions, it can easily lead to false states of reliability and security. In this regard, we will dissect the Cloud, its services and related issues. We will look at the key issues to proper and managed Cloud adoption as well as Cloud and Computing Paradigm Drivers and Cloud and IoT interaction. Naturally, one cannot omit mentioning related risk and cybercrime implications together with blockchain reasoning.

Keynote Speaker

Dr. Vimala Balakrishnan
Senior Lecturer, and Data Scientist,
Faculty of Computer Science and Information Technology,
University of Malaya, Malaysia

Title of Talk: "Machine Learning and Data Analytics in Science"

Brief Biodata

Dr. Vimala Balakrishnan is Senior Lecturer and Data Scientist affiliated with the Faculty of Computer Science and Information Technology, University of Malaya, since 2010. She obtained her Ph.D. in the field of ergonomics from Multimedia University, whereas her masters and bachelor's degrees were from University of Science, Malaysia. Her main research interests are in data analytics and sentiment analysis, particularly related to social media. Her research domains include health care, education and social issues such as cyberbullying. She has published approximately 47 articles in top indexed journals and also serves as an associate editor to the *Malaysian Journal of Computer Science* and as Associate Member for the Global Science and Technology Forum. She is also a fellow for the Leadership in Innovation programme, a prestigious award by the Royal Academy of Engineering, UK, and a Fulbright Visiting Scholar.

Abstract

Her keynote topic during the Artificial Intelligence and Data Engineering 2019 would mainly focus on how artificial intelligence, particularly machine learning and data analytics, can be used not only in the field of science, but also in the field of social sciences. This, in fact, has been her research motivation for many years, as shown by her most recent works targeting detection mechanisms in cyberbullying, digital economy and electoral analysis, among others. Valuable data are available aplenty, especially on social media. These data can be leveraged to identify various patterns and insights, providing useful information to various stakeholders. The use of machine learning algorithms to develop detection mechanisms is gaining momentum not only in the Western world, but in the Asian countries as well. Dr. Balakrishnan will particularly highlight how and where AI and data engineering have been incorporated, by specifically focusing on her recent work on cyberbullying detection mechanism.

Keynote Speaker

D. S. Guru
Professor,
Department of Studies in Computer Science,
University of Mysore, India

Title of Talk: "Interval Valued Feature Selection: Filter Based Approaches"

Brief Biodata

D. S. Guru received his B.Sc., M.Sc. and Ph.D. in computer science and technology from the University of Mysore, Mysuru, India, in 1991, 1993 and 2000, respectively. He is currently Professor in the Department of Studies in Computer Science, University of Mysore, India. He was a fellow of BOYSCAST and a visiting research scientist at Michigan State University. He is the first ARP recipient from Vision Group of Science and Technology, Karnataka Government. He has successfully guided 17 Ph.D.s and currently supervising 6. He has authored 68 journals and 230 peer-reviewed conference papers at international and national levels. He is a co-author for three textbooks, editor of LNNS-43, LNNS-14 and LNEE-213. He is a founder Trustee cum Treasurer of Maharaja Education Trust, Mysore, which has established a couple of educational Institutions in and around Mysore. His area of research interest covers image retrieval, text mining, machine learning, object recognition, shape analysis, sign language recognition, biometrics and symbolic data analysis.

Abstract

This talk is intended to introduce an unconventional data analysis called symbolic data analysis [1], in the field of pattern recognition and its allied areas [2–6]. The talk initially presents various types of feature representation and subsequently highlights symbolic proximity measures in general and interval-valued proximity measures in particular [7, 8]. The problem of selectively choosing a subset of interval-valued features out of several available for compact representation of data would be addressed. Various models belonging to three categories, viz. models which transform interval data to crisp [9, 10], models which exploit conventional statistical measures directly on interval-valued data [11] and models which accomplish feature selection through feature clustering of interval-valued features [12, 13], shall be presented.

During the talk, necessary theory along with illustrative examples will be presented to highlight the abilities of symbolic data in preserving the reality in solving pattern recognition-related problems. Further, experimental results on some standard datasets shall also be discussed.

Keywords Symbolic data analysis, Interval-valued data, Interval proximity measures, Interval-valued feature selection.

References

1. Billard L, Diday E (2006) Symbolic data analysis: conceptual statistics and data mining. Wiley Publishers
2. Guru DS, Nagendraswamy HS (2007) Symbolic representation of two-dimensional shapes. Pattern Recogn Lett 28(1):144–155
3. Guru DS, Prakash HN (2009) Online signature verification and recognition: an approach based on symbolic representation. IEEE Trans Pattern Anal Mach Intell 31(6):1059–1073
4. Harish BS, Guru DS, Manjunath S, Dinesh R (2010) Cluster based symbolic representation and feature selection for text classification. In: Advanced data mining and applications. Lecture notes in computer science, vol 6441, pp 158–166
5. Guru DS, Kumar NV (2017) Interval valued feature selection for classification of logos. In: 17th international conference on intelligent systems, design and applications 2017, vol 736, Springer AISC, pp 154–165
6. Guru DS, Manjunatha KS, Manjunath S, Somashekara MT (2017) Interval valued symbolic representation of writer dependent features for online signature verification. Expert Sys Appl 80:232–243
7. Guru DS, Kiranagi BB (2005) Multivalued type dissimilarity measure and concept of mutual dissimilarity value for clustering symbolic patterns. Pattern Recogn 38(1):151–156
8. Guru DS, Kiranagi BB, Nagabhushan P (2004) Multivalued type proximity measure and concept of mutual similarity value useful for clustering symbolic patterns. Pattern Recog Lett 25(10):1203–1213
9. Guru DS, Kumar NV (2016) Novel feature ranking criteria for interval valued feature selection. In: IEEE ICACCI 2016, pp 149–155

10. Kumar NV, Guru DS (2017) A novel feature ranking criterion for interval valued feature selection for classification. In: 14th IAPR international conference on document analysis and recognition, IEEE, pp 71–76
11. Guru DS, Kumar NV (2018) Interval chi-square score (ICSS): feature selection of interval valued data. In: 18th international conference on intelligent systems, design and applications 2018, vol 941, Springer AISC, pp 686–698
12. Guru DS, Kumar NV (2017) Class specific feature selection for interval valued data through interval k-means clustering. vol 701, Springer CCIS, RTIP2R 2016, pp 228–239
13. Guru DS, Kumar NV, Suhil M (2017) Feature selection for interval valued data through interval k-means clustering. In: International journal of computer vision and image processing (IJCVIP), vol 7(2), IGI-global, pp 64–80

Keynote Speaker

Dr. Kavi Mahesh
Director,
Indian Institute of Information Technology,
Dharwad, India

Title of Talk: "AI and ML: Can Data Lead to Intelligence?"

Brief Biodata

Dr. Kavi Mahesh is Director of the Indian Institute of Information Technology, Dharwad. Previously, he was the Dean of Research at PES University, Director of the World-Bank-funded Research Centre for Knowledge Analytics and Ontological Engineering—KAnOE and Professor of computer science. His areas of interest are knowledge management, analytics, epistemology, ontology, classification studies, text processing and unstructured data management. He has three US patents and has published two books, 16 chapters and 80 papers which have received over 1500 citations. Notable among these are the textbooks—Theory of Computation: A Problem-Solving Approach (Wiley India, 2012) and Ten Steps to Maturity in Knowledge Management (Elsevier Pub. UK, 2006). He was earlier with Oracle Corporation, USA, and New Mexico State University and has consulted in the area of knowledge management with Infosys, Hewlett Packard, United Nations and www.EasyLib.com. He holds an M. Tech. in computer science from the Indian Institute of Technology, Bombay (1989), an MS (1991) and a Ph.D. (1995) in computer science from Georgia Institute of Technology, USA.

Abstract

Artificial Intelligence (AI) has come back in a big way in the last few years. However, making computing machines smarter, giving them human-like intelligence and enabling them to learn to do tasks better have been tried in a variety of ways in the previous several decades:

- Knowledge-based systems were developed based on the premise that intelligent behaviour requires significant amounts of knowledge; by providing well-designed and well-organized knowledge representations to machines, they were expected to carry out intelligent tasks.
- Rule-based expert systems were built to solve specific problems requiring domain expertise such as diagnosis or troubleshooting.
- Logical reasoning as the basis for rational thought was explored to a great extent with the development of higher-order, modal, temporal, uncertain, fuzzy, description, and other forms of logic.
- Neural networks were designed to automatically learn solutions to problems for which it was difficult to specify an algorithm.
- Cognitive models were built not only to make machines intelligent but also to serve as accurate models of how the human brain/mind appears to function.
- Other ideas were explored too, for example genetic algorithms and swarm intelligence.
- Of course, attempts were made to build robots to exhibit human-like physical motion and perceptual capabilities.

Today's enthusiasm about AI and machine learning (ML, including deep learning) must be seen in the broader context of these earlier developments. This is a new step, perhaps a very big and important step, in the ongoing journey to enhance the capabilities of computing machines, that is, to make them smarter and more efficient in solving problems of importance to us humans. Clearly, there is a lot of potential in the new AI equipped with deep ML in further easing our lives and reducing our dependence on unreliable and expensive human resources for carrying out mundane tasks in optimal ways.

Scepticism on whether machines can, in fact, be made intelligent is rarely to be seen today. Can machines think, Alan Turing asked way back in 1950. He provided a most practical answer to the question in the famous Turing Test wherein we are bound to accept that a machine can think if its behaviour is indistinguishable from that of a competent human being at the end of a controlled experiment. While no AI machine has thus far passed the Turing Test, its very idea has also been challenged by interesting counterarguments such as Searle's Chinese Room.

Leaving such philosophical issues aside for a minute, what has changed now that we all are so excited about the possibility of intelligent machines? More and more data has become available; more and more computing ability to crunch all the data is available everywhere at ever-reducing costs; and lots and lots of new pattern recognition algorithms, literally hundreds of them, have been developed. Even meta-algorithmic techniques such as ensembles are employed effectively to attack a vast variety of problems.

Yet, what grand challenge problems has AI solved? Unlike board games and quizzes where it has in fact done well against humans, machines do not yet understand natural languages, cannot yet replace a household maid or a medical nurse, or grow food for us. Yes, they are assisting human experts in many significant ways and are beginning to operate other machines as in driving a driverless car. In fact, whether AI has already even crashed two airplanes killing human lives is a topic of much speculation and debate at present.

Let us not be too eager to celebrate the successes of AI. Let AI first demonstrate a lot more of human-like decision-making capability during these exciting times in the second wave of development of machine intelligence.

Contents

Artificial Intelligence

NLP-Driven Ensemble-Based Automatic Subtitle Generation and Semantic Video Summarization Technique 3
V. B. Aswin, Mohammed Javed, Parag Parihar, K. Aswanth, C. R. Druval, Anupam Dagar, and C. V. Aravinda

A Generalized Model for Cardiovascular Disease Classification Using Machine Learning Techniques 15
Ankita Naik and Nitesh Naik

Classification of Road Accidents Using SVM and KNN 27
P. Joyce Beryl Princess, Salaja Silas, and Elijah Blessing Rajsingh

A Deep Convolutional Encoder-Decoder Architecture Approach for Sheep Weight Estimation 43
Nirav Alpesh Shah, Jaydeep Thik, Chintan Bhatt, and Aboul-Ella Hassanien

Supervised Machine Learning Model for Accent Recognition in English Speech Using Sequential MFCC Features 55
Dweepa Honnavalli and S. S. Shylaja

A Two-Level Approach to Color Space-Based Image Segmentation Using Genetic Algorithm and Feed-Forward Neural Network 67
B. S. Sathish, P. Ganesan, L. M. I. Leo Joseph, K. Palani, and R. Murugesan

Braille Cell Segmentation and Removal of Unwanted Dots Using Canny Edge Detector 79
Vishwanath Venkatesh Murthy, M. Hanumanthappa, and S. Vijayanand

Real-Time Detection of Distracted Drivers Using a Deep Neural Network and Multi-threading 89
Ajay Narayanan, V. Aiswaryaa, Aswesh T. Anand, and Nalinadevi Kadiresan

Analysing the Practicality of Drawing Inferences in Automation of Commonsense Reasoning 101
Chandan Hegde and K. Ashwini

Segmentation and Detection of Glioma Using Deep Learning 109
Navneeth Krishna, Mahammad Rumaan Khalander, Nandan Shetty, and S. N. Bharath Bhushan

Character Recognition of Tulu Script Using Convolutional Neural Network 121
Sachin Bhat and G. Seshikala

Exploring the Performance of EEG Signal Classifiers for Alcoholism 133
Nishitha Lakshmi, Rani Adhaduk, Nidarsh Nithyananda, S. Rashwin Nonda, and K. Pushpalatha

Type-2 Tetradecagonal Fuzzy Number 149
A. Rajkumar and C. Sagaya Nathan Stalin

Critical Path Problem Through Intuitionistic Triskaidecagonal Fuzzy Number Using Two Different Algorithms 159
N. Jose Parvin Praveena, C. Sagaya Nathan Stalin, and A. Rajkumar

Genetic-Neuro-Fuzzy Controller for Indirect Vector-Controlled Induction Motor Drive 169
B. T. Venu Gopal, H. R. Ramesh, and E. G. Shivakumar

Artificial Intelligence-Based Chatbot Framework with Authentication, Authorization, and Payment Features 179
Deena Deepika Cutinha, Niranjan N. Chiplunkar, Shazad Maved, and Arun Bhat

Disease Recognition in Sugarcane Crop Using Deep Learning 189
Hashmat Shadab Malik, Mahavir Dwivedi, S. N. Omkar, Tahir Javed, Abdul Bakey, Mohammad Raqib Pala, and Akshay Chakravarthy

Deep Learning-Based Car Damage Classification and Detection 207
Mahavir Dwivedi, Hashmat Shadab Malik, S. N. Omkar, Edgar Bosco Monis, Bharat Khanna, Satya Ranjan Samal, Ayush Tiwari, and Aditya Rathi

Sparse Reflectance Map-Based Fabric Characterization 223
Kayan K. Katrak, Rithvik Chandan, Sirisha Lanka, G. M. Chitra, and S. S. Shylaja

A Risk Assessment Model for Patients Suffering from Coronary Heart Disease Using a Novel Feature Selection Algorithm and Learning Classifiers 237
Sujata Joshi and Mydhili K. Nair

Toward Artificial Social Intelligence: A Semi-supervised, Split Decoder Approach to EQ in a Conversational Agent 251
Shruthi Shankar, V. Sruthi, Vibha Satyanarayana, and Bhaskarjyoti Das

Matrix Factorization for Recommendation System 267
T. Lekshmi Priya and Harikumar Sandhya

A Reinforcement Learning Approach to Inventory Management 281
Apoorva Gokhale, Chirag Trasikar, Ankit Shah, Arpita Hegde, and Sowmiya Raksha Naik

Human Resource Working Prediction Based on Logistic Regression 299
Anusha Hegde and G. Poornalatha

Kansei Knowledge-Based Human-Centric Digital Interface Design Using BP Neural Network 307
Huiliang Zhao, Jian Lyu, Xiang Liu, and Weixing Wang

DST-ML-EkNN: Data Space Transformation with Metric Learning and Elite k-Nearest Neighbor Cluster Formation for Classification of Imbalanced Datasets 319
Seba Susan and Amitesh Kumar

Classification Study and Prediction of Cervical Cancer 329
Kaushik Suresh

English Transliteration of Kannada Words with Anusvara and Visarga 349
Savitha Shetty, Saritha Shetty, Sarika Hegde, and Karuna Pandit

An Ensembled Scale-Space Model of Deep Convolutional Neural Networks for Sign Language Recognition 363
Neena Aloysius and M. Geetha

A Survey on Deep Learning-Based Automatic Text Summarization Models 377
P. G. Magdum and Sheetal Rathi

Automatic Multi-disease Diagnosis and Prescription System Using Bayesian Network Approach for Clinical Decision Making 393
P. Laxmi, Deepa Gupta, G. Radhakrishnan, J. Amudha, and Kshitij Sharma

Artificial Intelligence Techniques for Predicting Type 2 Diabetes 411
Ramyashree, P. S. Venugopala, Debmalya Barh, and B. Ashwini

Predictive Analysis of Malignant Disease in Woman Using Machine Learning Techniques 431
Akshaya, R. Pranam Betrabet, and C. V. Aravinda

Study on Automatic Speech Therapy System for Patients 439
Supriya B. Rao, Sarika Hegde, and Surendra Shetty

Data Engineering

The Design of Multiuser BGN Encryption with Customized Multiple Pollard's Lambda Search Instances to Solve ECDLP in Finite Time 457
Santosh Javheri and Uday Kulkarni

Internet Addiction Predictor: Applying Machine Learning in Psychology 471
S. N. Suma, Poornima Nataraja, and Manoj Kumar Sharma

An Approach Toward Stateless Chatbots with the Benefit of Tensorflow Over Spacy Pipeline 483
Chaithra, Roshan Fernandes, Anisha P. Rodrigues, and Venkatesh

Enhanced Processing of Input Data in Clustering Techniques of Data Mining Algorithms 497
K. Sampath Kini and B. H. Karthik Pai

A Comparative Analysis of MFIs in India Using ANOVA and Logistic Regression Model 503
M. G. Deepika and P. Sarika

Practical Analysis of Representative Models in Classifier: A Review 517
Angela Mathew and Sangeetha Jamal

Exponential Cipher Based on Residue Number System and Its Application to Image Security 529
Sagar Ramesh Pujar, Achal Ramanath Poonja, and Ganesh Aithal

Using Machine Learning and Data Analytics for Predicting Onset of Cardiovascular Diseases—An Analysis of Current State of Art 543
P. R. Mahalingam and J. Dheeba

Analysis of the Nearest Neighbor Classifiers: A Review 559
Yash Agarwal and G. Poornalatha

Analysis of Automated Log Template Generation Methodologies 571
Anoop Mudholkar, Varun Mokhashi, Deepak Nayak, Vaishnavi Annavarjula, and Mahesh Babu Jayaraman

Fraud Detection in Online Transactions Using Machine Learning Approaches—A Review 589
H. Dhanushri Nayak, Deekshita, L. Anvitha, Anusha Shetty, Divya Jennifer D'Souza, and Minu P. Abraham

Encryption and Decryption for Network Security Using Reverse Context-Free Grammar Productions 601
Aishwarya R. Parab and Teslin Jacob

A Survey on State-of-the-Art Applications of Variable Length Chromosome (VLC) Based GA 615
Ravi Domala and Upasna Singh

A Multi-level Access Technique for Privacy-Preserving Perturbation in Association Rule Mining 631
N. Komal Kumar and D. Vigneswari

LSB and RLE Based Approach for Increasing Payload and Security of Stego Images 647
Rupali Sanjay Pawar

Adaptive MoD Chatbot: Toward Providing Contextual Corporate Summarized Document as Suggestions and Reported Issue Ticket Routing 659
Shiva Prasad Nayak, Archana Rai, Kiran Vankataramanappa, Jalak Arvindkumar Pansuriya, and Joerg Singler

Classification of Text Documents 675
Pushpa B. Patil and Dakshayani M. Ijeri

Fine-Grained Sentiment Rating of Online Reviews with Deep-RNN 687
Ramesh Wadawadagi and Veerappa Pagi

Analysis of Strategic Market Management in Light of Stochastic Processes, Recurrence Relation, Abelian Group and Expectation 701
Prasun Chakrabarti, Tulika Chakrabarti, Siddhant Bane, Biswajit Satpathy, Indranil SenGupta, and Jonathan Andrew Ware

Peer-to-Peer Distributed Storage Using InterPlanetary File System . . . 711
A. Manoj Athreya, Ashwin A. Kumar, S. M. Nagarajath, H. L. Gururaj, V. Ravi Kumar, D. N. Sachin, and K. R. Rakesh

Knowledge Base Representation of Emails Using Ontology for Spam Filtering 723
V. Bindu and Ciza Thomas

Clinical Significance of Measles and Its Prediction Using Data Mining Techniques: A Systematic Review 737
Abhishek S. Rao, Demian Antony D'Mello, R. Anand, and Sneha Nayak

A Survey on Graphical Authentication System Resisting Shoulder Surfing Attack 761
S. Arun Kumar, R. Ramya, R. Rashika, and R. Renu

Analysis of Stock Market Fluctuations Incidental to Internet Trends 771
Vinayaka R. Kamath, Nikhil V. Revankar, and Gowri Srinivasa

Pseudo Random Number Generation Based on Genetic Algorithm Application 793
V. Pushpalatha, K. B. Sudeepa, and H. N. Mahendra

Analysis of an Enhanced Dual RSA Algorithm Using Pell's Equation to Hide Public Key Exponent and a Fake Modulus to Avoid Factorization Attack 809
K. R. Raghunandan, Rovita Robert Dsouza, N. Rakshith, Surendra Shetty, and Ganesh Aithal

A New Approach on Advanced Encryption Standards to Improve the Secrecy and Speed Using Nonlinear Output Feedback Mode 825
Dodmane Radhakrishna, Aithal Ganesh, and Shetty Surendra

Cyber-Bullying Detection: A Comparative Analysis of Twitter Data 841
Jyothi Shetty, K. N. Chaithali, Aditi M. Shetty, B. Varsha, and V. Puthran

An Optimal Wavelet Detailed-Coefficient Determination Using Time-Series Clustering 857
C. I. Johnpaul, Munaga V. N. K. Prasad, S. Nickolas, G. R. Gangadharan, and Marco Aiello

A Novel Data Hiding Technique with High Imperceptibility Using a 3-Input Majority Function and an Optimal Pixel Adjustment 873
P. V. Sabeen Govind, M. Y. Shiju Thomas, and M. V. Judy

Designing and Testing of Data Acquisition System for Satellite Using MIL-STD-1553 883
B. L. Lavanya and M. N. Srinivasa

Optimizing People Sourcing Through Semantic Matching of Job Description Documents and Candidate Profile Using Improved Topic Modelling Techniques 899
Lorick Jain, M. A. Harsha Vardhan, Ganesh Kathiresan, and Ananth Narayan

Mining Associations Rules Between Attribute Value Clusters 909
Shankar B. Naik

Machine Learning Approach to Stock Prediction and Analysis 919
Bhal Chandra Ram Tripathi, T. Satish Kumar, R. Krishna Prasad, and Visheshwar Pratap Singh

A Novel Approach for Error Analysis in Classified Big Data in Health Care 929
S. Kavitha, Mahesh S. Nayak, and M. Hanumanthappa

Multi-join Query Optimization Using Modified ACO with GA 937
Vikas Kumar and Mantosh Biswas

The Impact of Distance Measures in *K*-Means Clustering Algorithm for Natural Color Images 947
P. Ganesan, B. S. Sathish, L. M. I. Leo Joseph, K. M. Subramanian, and R. Murugesan

Designing an Adaptive Question Bank and Question Paper Generation Management System 965
Pankaj Dwivedi, R. Tapan Shankar, B. Meghana, H. Sushaini, B. R. Sudeep, and M. R. Pooja

Securing Media Information Using Hybrid Transposition Using Fisher Yates Algorithm and RSA Public Key Algorithm Using Pell's Cubic Equation 975
K. R. Raghunandan, Shirin Nivas Nireshwalya, Sharan Sudhir, M. Shreyank Bhat, and H. M. Tanvi

Analysis of Tuberculosis Disease Using Association Rule Mining 995
Ankita Mohapatra, Sangita Khare, and Deepa Gupta

Scalable Two-Phase Top-Down Specification for Big Data Anonymization Using Apache Pig 1009
Anushree Raj and Rio D'Souza

Segmentation of Lip Print Images Using Clustering and Thresholding Techniques 1023
S. Sandhya, Roshan Fernandes, S. Sapna, and Anisha P. Rodrigues

Filtering-Based Text Sentiment Analysis for Twitter Dataset 1035
Hiran Nandy and Rajeswari Sridhar

A Comparative Analysis of Clustering Quality Based on Internal Validation Indices for Dimensionally Reduced Social Media Data 1047
Shini Renjith, A. Sreekumar, and M. Jathavedan

Anomaly Detection for Big Data Using Efficient Techniques: A Review 1067
Divya Jennifer D'Souza and K. R. Uday Kumar Reddy

Data Science and Internet of Things for Enhanced Retail Experience 1081
Irfan Landge and Hannan Satopay

Machine Vision

An Experimental Study on the Effect of Noise in CCITT Group 4 Compressed Document Images 1101
A. Narayana Sukumara, Mohammed Javed, D. K. Sreekantha, P. Nagabhushan, and R. Amarnath

A Neck-Floor Distance Analysis-Based Fall Detection System Using Deep Camera 1113
Xiangbo Kong, Zelin Meng, Lin Meng, and Hiroyuki Tomiyama

An Introduction to Sparse Sampling on Audio Signal by Exploring Different Basis Matrices 1121
A. Electa Alice Jayarani, Mahabaleswara Ram Bhatt, and D. D. Geetha

Retrieval of Facial Sketches Using Linguistic Descriptors: An Approach Based on Hierarchical Classification of Facial Attributes 1131
S. Pallavi, M. S. Sannidhan, and Abhir Bhandary

Simplified SVD Feature Construction in Multiangle Images to Identify Plant Varieties and Weed Infestation 1151
K. Ramesh and Andrews Samraj

Old Handwritten Music Symbol Recognition Using Radon and Discrete Wavelet Transform 1165
Savitri Apparao Nawade, Rajmohan Pardeshi, Shivanand Rumma, and Mallikarjun Hangarge

Gender Recognition from Face Images Using SIFT Descriptors and Trainable Features 1173
Sneha Pai and Ramesha Shettigar

Multiscale Anisotropic Morlet Wavelet for Texture Classification of Interstitial Lung Diseases 1187
Manas Jyoti Das and Lipi B. Mahanta

A Review of Intelligent Smartphone-Based Object Detection Techniques for Visually Impaired People 1199
R. Devakunchari, Swapnil Tiwari, and Harsh Seth

Stereo Vision-Based Depth Estimation 1209
Zelin Meng, Xiangbo Kong, Lin Meng, and Hiroyuki Tomiyama

A Dynamic Programming Algorithm for Energy-Aware Routing of Delivery Drones 1217
Yusuke Funabashi, Atsuya Shibata, Shunsuke Negoro, Ittetsu Taniguchi, and Hiroyuki Tomiyama

Qualitative Approach of Empirical Mode Decomposition-Based Texture Analysis for Assessing and Classifying the Severity of Alzheimer's Disease in Brain MRI Images 1227
K. V. Sudheesh and L. Basavaraj

Facial Image Indexing Using Locally Extracted Sparse Vectors 1255
Vinayaka R. Kamath, M. Varun, and S. Aswath

Ambient Intelligence

Smart Agro-Ecological Zoning for Crop Suggestion and Prediction Using Machine Learning: An Comprehensive Review 1273
R. Chetan, D. V. Ashoka, and B. V. Ajay Prakash

Preparedness in the Aftermath of a Natural Disaster Using Multihop Ad hoc Networks—Drone-Based Approach 1281
Getzi Jeba Leelipushpam Paulraj, Immanuel Johnraja Jebadurai, and J. Jebaveerasingh

An IoT-Based Congestion Control Framework for Intelligent Traffic Management System 1287
Md. Ashifuddin Mondal and Zeenat Rehena

Link Prediction on Social Attribute Network Using Lévy Flight Firefly Optimization 1299
P. Srilatha, R. Manjula, and C. Pavan Kumar

Secure and Energy-Efficient Data Transmission 1311
H. V. Chaitra and G. K. RaviKumar

A Non-cooperative Game Theoretic Approach for Resource Allocation in D2D Communication 1323
Tanya Shrivastava, Sudhakar Pandey, Pavan Kumar Mishra, and Shrish Verma

IoT-Based Nursery Management System 1335
Mahendra S. Naik, Sreekantha Desai, K. V. S. S. S. S. Sairam, and S. N. Chaitra

Shortest Path Discovery for Area Coverage (SPDAC) Using Prediction-Based Clustering in WSN 1345
C. N. Abhilash, S. H. Manjula, R. Tanuja, and K. R. Venugopal

Smart Mirror Using Raspberry Pi for Intrusion Detection and Human Monitoring 1359
Raju A. Nadaf and Vasudha Bonal

A Home Security Camera System Based on Cloud and SNS 1375
Takuya Egashira, Lin Meng, and Hiroyuki Tomiyama

Design, Calibration, and Experimental Study of Low-Cost Resistivity-Based Soil Moisture Sensor for Detecting Moisture at Different Depths of a Soil 1383
S. Sunil Kumar, Ganesh Aithal, and P. Venkatramana Bhat

An IoT-Based Predictive Analytics for Estimation of Rainfall for Irrigation 1399
H. Shalini and C. V. Aravinda

Smart Watering System Using MQTT Protocol in IoT 1415
Mukambikeshwari and Asmita Poojary

Internet of Things (IoT) Enabling Technologies and Applications—A Study 1425
D. K. Sreekantha, Ashok Koujalagi, T. M. Girish, and K. V. S. S. S. S. Sairam

Evaluation of Standard Models of Content Placement in Cloud-Based Content Delivery Network 1443
Suman Jayakumar, S. Prakash, and C. B. Akki

IoT-Based Data Storage for Cloud Computing Applications 1455
Ankita Shukla, Priyatam Reddy Somagattu, Vishal Krishna Singh, and Mala Kalra

IoT-Based Heart Rate Monitoring System 1465
Jagadevi N. Kalshetty, P. Melwin Varghese, K. Karthik, Randhir Raj, and Nitin Yadav

About the Editors

Dr. Niranjan N. Chiplunkar is currently a Professor at the Department of Computer Science & Engineering and Principal of NMAM Institute of Technology, Nitte, Udupi, India. With more than 32 years of teaching experience, he has written one textbook on "VLSI CAD" published by PHI Learning, in 2011 and has edited two international and one national-level conference proceedings volume. He was selected for the "Bharatiya Vidya Bhavan National Award for Best Engineering College Principal" for the year 2014 by the ISTE New Delhi and for the "Excellent Achievement Award" by the Centre for International Cooperation on Computerization, Government of Japan, in 2002. Prof. Chiplunkar's major interests include CAD for VLSI, wireless sensor networks, and multicore architecture and programming. He is a Fellow of the Institution of Engineers (India), Senior Member of the IEEE, and a member of several other professional bodies, e.g. the Computer Society of India, ISSS, and ISTE. He has successfully completed two funded research projects—one from the AICTE, New Delhi, and another from the DST, Government of India, on "Network on Chip Architecture Design" and "Multicore Software Framework Development," respectively.

Dr. Takanori Fukao is currently a Professor at the Department of Electrical and Electronics Engineering, College of Science and Engineering, Ritsumeikan University, Japan. His primary research interests include perceptual information processing, intelligent robotics, automated driving, platooning, parking of automobiles, flight control for blimp robots and drones, automated driving of agricultural vehicles, active control including active suspension systems, and 3D model generation using motion stereo or stereo cameras. He heads the research laboratory Intelligent Vehicle Systems. He has published 47 research papers in reputed international journals, and three books with leading publishers. He has received several awards for his research and teaching achievements. Currently, he is an editor of IEEE Transactions on Intelligent Vehicles.

Artificial Intelligence

NLP-Driven Ensemble-Based Automatic Subtitle Generation and Semantic Video Summarization Technique

V. B. Aswin, Mohammed Javed, Parag Parihar, K. Aswanth, C. R. Druval, Anupam Dagar, and C. V. Aravinda

Abstract This paper proposes an automatic subtitle generation and semantic video summarization technique. The importance of automatic video summarization is vast in the present era of big data. Video summarization helps in efficient storage and also quick surfing of large collection of videos without losing the important ones. The summarization of the videos is done with the help of subtitles which is obtained using several text summarization algorithms. The proposed technique generates the subtitle for videos with/without subtitles using speech recognition and then applies NLP-based text summarization algorithms on the subtitles. The performance of subtitle generation and video summarization is boosted through ensemble method with two approaches such as intersection method and weight-based learning method. Experimental results reported show the satisfactory performance of the proposed method.

Keywords Summarization · NLP · Subtitle · Video · Text · Ensemble · Subtitles

---

V. B. Aswin (✉) · M. Javed · P. Parihar · K. Aswanth · C. R. Druval · A. Dagar
Department of IT, Indian Institute of Information Technology Allahabad, Allahabad, Prayagraj, India
e-mail: iit2016106@iiita.ac.in

M. Javed
e-mail: javed@iiita.ac.in

P. Parihar
e-mail: iit2016095@iiita.ac.in

K. Aswanth
e-mail: iit2016105@iiita.ac.in

C. R. Druval
e-mail: bim2016501@iiita.ac.in

A. Dagar
e-mail: iit2016128@iiita.ac.in

C. V. Aravinda
Department of CSE, NMAM Institute of Technology, Nitte, India
e-mail: aravinda.cv@nitte.edu.in

N. N. Chiplunkar and T. Fukao (eds.), *Advances in Artificial Intelligence and Data Engineering*, Advances in Intelligent Systems and Computing 1133,
https://doi.org/10.1007/978-981-15-3514-7_1

1 Introduction

A concept like video summarization has a huge scope in the modern era. Video repository websites like Google, Dailymotion, Vimeo, etc. are gaining popularity day by day. The popularity of these websites is enormous in the present scenario. A large number of videos are being uploaded as well as downloaded from these online video repository websites. For example, the total number of people who use YouTube is 1,300,000,000 [1]. 300 h of video are uploaded to YouTube every minute! Almost 5 billion videos are watched on YouTube every single day. YouTube gets over 30 million visitors per day [1]. In this scenario, for a concept like video summarization has a huge scope. The video summarization technique can be applied on the video thumbnail to attract more viewers. It can be developed to show only interesting and important parts of the video. It is not necessary for all the videos to come with a subtitle. It is very difficult to summarize the videos like security footage's as they do not have subtitles even after applying speech recognition. This reduces the domain of video summarization. But still, the summarization of video using subtitles is the most efficient and fastest way of doing it. If machine learning algorithms or histogram-based methods [2] were used to summarize videos, it would have taken a long time to train them which will increase the time of development. But dealing with subtitle which is obviously text is much more easy to deal with and faster which makes the video summarization easier and faster. But the main problem here is that most of the videos come without subtitles. In such cases, to rectify this problem, the technique of speech recognition which can be applied on the audio of the video and generate subtitles by formatting the text obtained after speech recognition. To extract different sentences from the video, there is a need to detect silence in the audio as it recognizes the end of the sentence from that. Once the subtitle is obtained, the video can be summarized with the generated subtitles. For summarization [3] of subtitles, natural language processing (NLP) algorithms [4] can be used which can be of various accuracy. Therefore, overall this paper proposes NLP-based subtitle generation and video summarization technique. The rest of the paper is organized as follows, Section 2 explains the proposed model, Section 3 explains experimental results, and Section 4 presents the summary of the report.

2 Proposed Model

For summarizing video using subtitles, the proposed method uses five text summarization algorithms which are NLP-based methods [5]. Further, there is an ensemble technique [6] using the text summarization [7] algorithms. The text summarization algorithms were used for filtering out key contents from subtitle file (.srt) [8], which will be taken as an input, implement the algorithm on the input, put the sentences in array, and rank them according to their importance using different domains, and will pick out the best sentences out of it, so as to form a concise subtitle keeping in mind where those subtitles were originally placed, according to subtitle id. Next is ensemble technique which combines the different algorithms together for a more perfect (intersection of all algorithms) abstract of the input. Also, we are training each algorithm in ensemble technique to give precise models. The flowchart in Fig. 1 describes the flow of video summarization using subtitles that have been implemented. The input video is fed in along with the subtitles, if the subtitle is not

Fig. 1 Flow control of video summarization

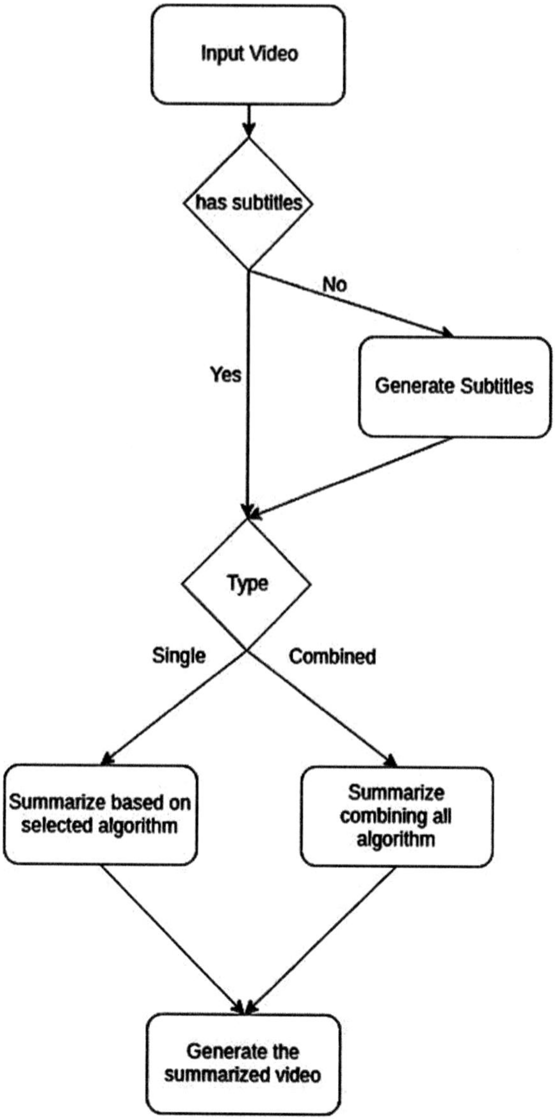

present, subtitle is generated using the subtitle generation algorithm and this subtitle along with the video is used to summarize depending upon whether single algorithm or the combined algorithm is to be used and the summarized video is the output.

2.1 NLP-Based Subtitle Generation

Since all videos may not have subtitles along with them and this method can be applied on videos which have subtitles. In case the user does not have a subtitle file, then the subtitles are generated first and then process the video using methods listed below. The subtitles (if not provided) are generated using speech recognition API of WIT.AI [9] which is used by Facebook for speech recognition. Basic idea on how the subtitle is generated is that chunks of audio are extracted from the video and apply speech recognition on them. To elaborate this, first the audio is extracted from the video file. Then max time interval is fixed for each subtitle for the video let it be 6 s which is in our case. Then the audio is scanned to detect silence in it; if the silence occurs before 6 s, the audio will be cut at that point. Also, a threshold is defined such that above this level the part of the audio is not treated as silence and vice versa. Also, to be noted that 1 s of extra silence is added at the starting and at the ending of the audio. So that words to be recognized by the speech recognizer is not missed out. Each time the words are recognized in a particular chunk of audio the whole sentence is formatted in the form of a subtitle file such that each of the sentences will be mentioned with the starting timestamp and the ending time stamp. Once the subtitle is available, it can be summarized to obtain the summarized video. There are five text summarization algorithms which have been used to summarize a video. These are:

Luhn 1958 came around as an emerging year in the field of summarization, when Hans Peter Luhn suggested an algorithm for thought summarization. This was a carrying a lot of weight achievement and a big head start in this sector, and followed his consider was an action in summarization area. Luhn received a rule of thumb to recognize salient sentences from the text by features well known as definition and definition frequency. This algorithm checks for words that have high occurrence frequency and the words are sorted based on decreasing frequency. The weight of a sentence is calculated by summing weights of each relevant word and these sentences are sorted in decreasing order based on the summed weight, and finally, P most relevant sentences are taken from the subtitles as the output [10].

Latent Semantic Analysis Latent semantic analysis (also known as LSA) is based on words and concepts. Because of the varying concepts within each word (like synonyms), LSA becomes a little tricky since each word maps to more than one concept. LSA works on the basis of title words and index words. Title words are those words which appear in all the sentences in LSA and index words are those which appear in more than two sentences which has the title word. A document is a bad of words, the order in which the words are arranged does not have any importance, whereas the count of the words plays an important role, and each word is supposed to have only one meaning. LSA follows a method which is based on the pattern of words. To understand the concept behind the word, clustering is used in LSA. The basic idea is to plot a XY graph including all the index words and title words based on their occurrences in each sentence. Then all the clusters in the graph are identified. Each of these clusters represents each concept and title in that sentence represents that particular concept. Hence, concept for each title word can be extracted. So in summarization, the technique was used in such a way that the sentence which is included in the most crowded cluster of the graph was taken. Using this concept of title words and index words, any text document can be summarized [11].

TextRank In the TextRank algorithm [12], it first takes the text which has to be split up and is converted into sentences and further into vectors. A similarity matrix is constructed from the vectors and a graph is created from this matrix. Using this graph, the sentence ranking is done. Based on the ranked sentences, summarized text is obtained. The probability of going from sentence A to sentence B is the similarity of two sentences. The text units which best define the sentence are identified. These text units are then added to the graph as vertices. The relations which connect texts are identified which are used to draw edges of the graph between vertices. A graph-based ranking algorithm is used until it converges. During the ranking algorithm, each vertex is assigned a value; this value is used for selection decisions. Finally, the vertices are sorted on the basis of their final score value and then the sentences are sorted based on the sorted vertices in the graph.

LexRank In the LexRank algorithm [13], first all the nouns and adjectives are separated out to form a document cluster. Now, the inverse document frequency (IDF) scores are found for each word in this cluster. For each word, let t_f be the frequency of that word in the cluster. The collection of all words which have $t_f * \text{IDF}$ score greater than a threshold value forms the centroid of the cluster. So, importance of a sentence will be higher if it contains more number of words which are present in the centroid. So, using this concept, P most relevant sentences are selected [13].

Edmundson Luhn method showed that the keywords can be extracted from the text as the most frequently occurring words excluding the stop words and then summarizing the text according to keywords that are obtained from the most occurring words. But since, this alone cannot properly produce summarized text; Edmundson [14] proposed another algorithm to summarize a text by adding three more methods to extract keywords, which are namely cue words, title and heading words, and structural indicators. The results produced by Edmundson method cleared the fact that the newly introduced factors played a dominant role in generating the best summarization of the text. Since this method works using the help of extra words called the stigma and bonus word, the summarization will be biased for different videos depending on these words given, so this method will not be used in the proposed methodology [14].

2.2 *Video Summarization*

From the algorithms explained in the previous section, Edmundson summarization will not be used because it is biased according to the bonus and stigma words given by the user. So, it cannot be used for a comparison.

Figure 2 is the summarization of the movie named Mrs. Doubtfire, which contains a frame of output using four algorithms (LexRank, Luhn, LSA, and TextRank). A video of Shah Rukh Khan's Ted Talk which was 17 min was taken and summarized. The subtitle file of the input video had 370 lines. The video was fed into the four abovementioned algorithms separately; the graph of subtitle text with it being relevant in the summarized video for the algorithms is described in Figs. 3, 4, 5, and 6.

Taking Intersection and Combining A very basic and simple approach is to combine them directly, meaning the idea is to run all the algorithms, keep their outputs side by side, and take a simple mathematical intersection. As it is very obvious, if a sentence is in all the combining algorithms, then it is must be of great importance; hence, it should be included in the output file. Hence, the user can choose from Luhn,

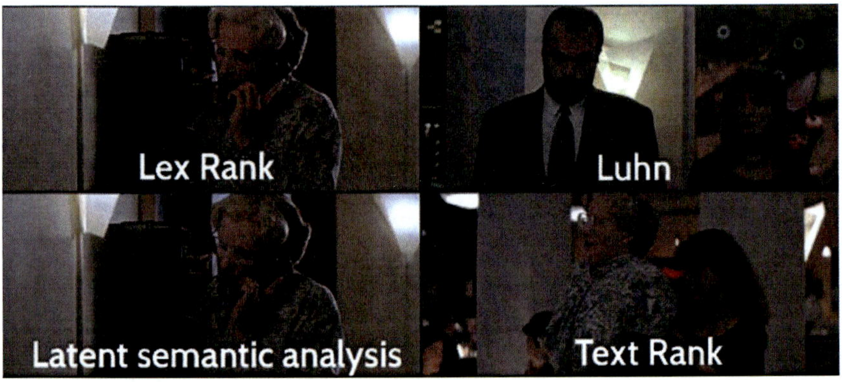

Fig. 2 Output of summarization using four algorithms

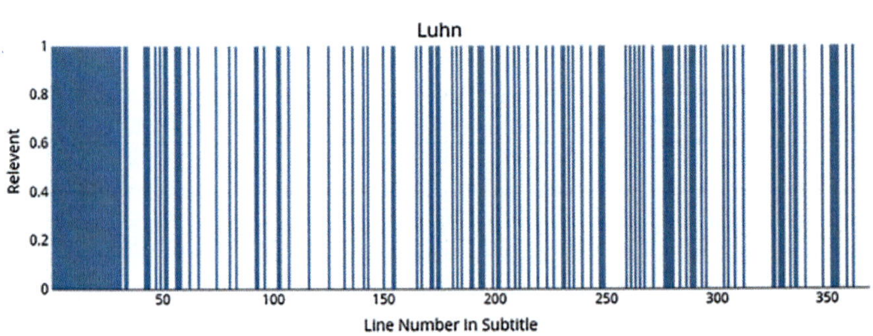

Fig. 3 Luhn algorithm

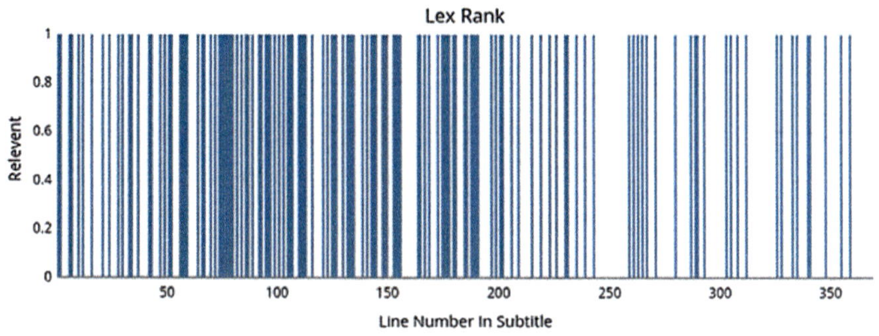

Fig. 4 LexRank algorithm

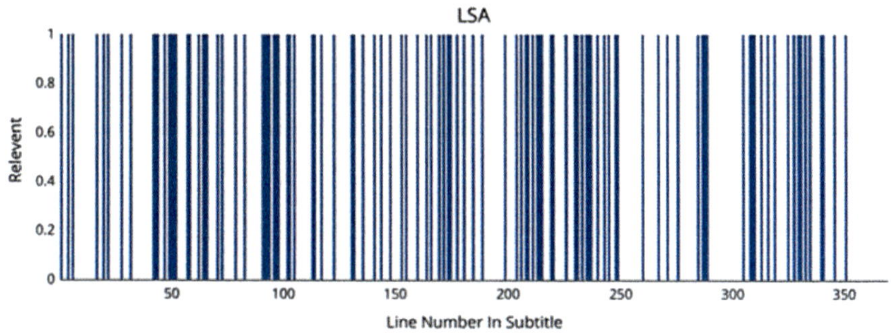

Fig. 5 LSA algorithm

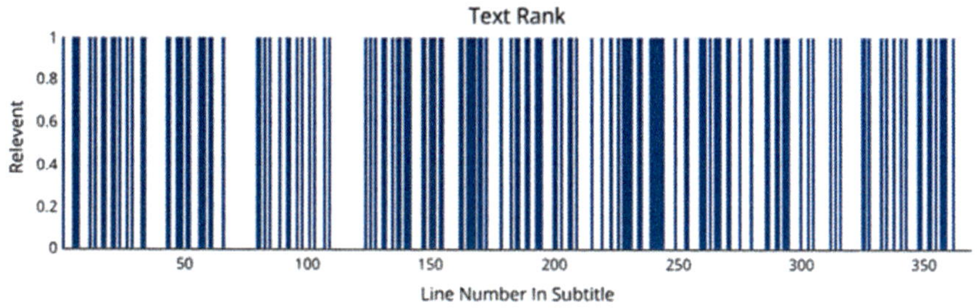

Fig. 6 TextRank algorithm

LSA, LexRank, TextRank, which all he wants to combine multiple correct, so that he can get the best results when all of them combined. Also, Edmundson cannot be concatenated in this process because it requires two extra files of bonus words and stigma words; hence, taking that while combining them was not ideally possible. This intersection gave very good results and implemented the ideology of ensemble properly.

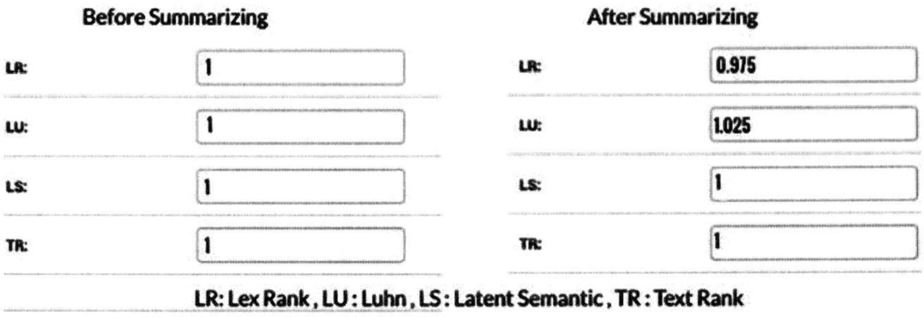

Fig. 7 Weights before and after summarization of first video

Weight-Based Learning Algorithms The problem with the previous method was that it gave equal powers to all the algorithms, but from the above explanations, it can be seen that not all algorithms behave properly, so here a method was devised, where each could get some weightage. The idea was simple, 'The one that performs the best, gets more.' As the name suggests, initial weights were taken for all algorithm and initialized all the algorithms with the same weights, i.e., WL, WLSA, WE, WLR, WTR = 1. Now, the check function will compare output of each of the algorithms and rank them accordingly on basis of their performance, so the ones that performed may get an increment in their weights. So, during future summarizations, they will get their scores according to their weights, and the sentences with higher score will be part of the output file. In this way, a clear view can be obtained of which algorithm gives the best output for the given input and the suggested algorithm can be used to obtain better results. Figure 7 is the weight allocation for different algorithms before and after summarization of Shah Rukh Khan's video. From the figure, it can be understood that the weight of LSA was increased and that of Lex was decreased. So for this video, LSA performed better and Lex performed the least. So, the weight of LexRank is increased by a unit and that of Luhn is decreased.

3 Experimental Results

Since there is no dataset for video summarization using subtitles and the four summarization techniques, a dataset of 40 videos was generated which were having different time lengths. These videos were given to these four algorithms and also the combined algorithm and the number of lines obtained in the summarized video was noted. The efficiency of each algorithm was decided based on the output of combined video which was the intersection of all these algorithms. So, the efficiency can be defined for an algorithm as the ratio of the number of subtitles in the combined video to that of the output of the particular algorithm.

$$\text{efficiency} = N_{\text{combined}}/N_{\text{algorithm}} \tag{1}$$

Table 1 Efficiency of intersection method

LexRank	LSA	Luhn	Text
37.1	40.6	38.0	37.7

Table 2 Efficiency of weighted ensemble method

LexRank	LSA	Luhn	Text
18.5	61.0	38.0	37.7

Table 3 Weights ensemble method

	LSA	Luhn	Text	LexRank
Initial	1	1	1	1
Finial	1.975	1	1	0.025

3.1 *Efficiency of Video Summarization*

By efficiency of an algorithm in summarization, it is meant how much part of the summarized algorithm is present in the video generated by the ensemble technique. As mentioned above, efficiency of each algorithm can be calculated using Formula (1), and on applying this on the intersection method, the following results were obtained as shown in Table 1.

On applying the dataset on the weighted ensemble technique, with initial weights set to 1 for all four algorithms, the following results were obtained as shown in Table 2 and the updated weights are shown in Table 3.

From Tables 1 and 2, it is observed that LSA performed better and LexRank performed the least. Since, for all videos, LSA performed best and Lex performed least their weights got affected, whereas the weight of Luhn and TextRank remained unchanged (Table 3). There is a huge difference in the efficiency value in weight-based and intersection method because in weight-based ensemble technique at each iteration, the weight of the better algorithm is increased and that of the worst algorithm is decreased. From these two methods, it is clear that LSA has a major contribution to the summarized video and Lex has the least contribution.

3.2 *Complexity*

Time Complexity

1. Single Summarization Algorithm: ***O(nk)***
   where n is the number of iterations until the summarization length is obtained and k is the number of sentences in the summarized subtitles.

2. Combined Summarization Algorithm: $\mathbf{O}\left(\sum_{i=1}^{\alpha} \boldsymbol{n}_i \boldsymbol{k}_i + \min(\boldsymbol{k}_i)\right)$
where α is the number of methods to be combined, n is the number of iterations until the summarization length is obtained, and k is the number of sentences in the summarized subtitles.

Space Complexity

1. Single Summarization Algorithm: $\boldsymbol{O}(\boldsymbol{r} + \boldsymbol{L} * \boldsymbol{r})$
where r is the total number of regions in the subtitle array; L is the average length of the sentences in the summarized subtitle.
2. Combined Summarization Algorithm: $\mathbf{O}\left(\sum_{k=1}^{\alpha} \boldsymbol{r} + \boldsymbol{L} * \boldsymbol{r}\right)$
where r is the total number of regions in the subtitle array; L is the average length of the sentences in the summarized subtitle.

4 Conclusion

A large number of videos are being generated and are increasing day by day. Hence, video summarization technique will be very helpful. Video summarization provided a faster way of browsing of large video collections and more efficient content indexing and access. The use of NLP Algorithms proved to be a very efficient way to form abstracts of videos. The case of no subtitles was by using subtitle generation method to convert speech to text, which turned out to be of great use in normal day-to-day usage. Many of the videos which are taken from phones, etc., do not contain subtitles; hence, there is future scope to work on this problem.

References

1. Mind blowing YouTube facts, figures and statistics. https://merchdope.com/youtube-stats/
2. Hannane R, Elboushaki A, Afdel K, Naghabhushan P, Javed M (2016) An efficient method for video shot boundary detection and keyframe extraction using SIFT-point distribution histogram
3. McQuillan S, Kriewel S, Goeuriot L, Jones G, Kelly L, Leveling J (2013) Report on summarization techniques
4. Pratibha Devihosur NR (2017) Automatic text summarization using natural language processing
5. Liu S (2017) Long text summarization using neural networks and rule-based approach
6. Dietterich TG (2000) Ensemble methods in machine learning
7. Dahale M (2014) Text summarization for compressed inverted indexes and snippets
8. Definition of subtitle. https://www.merriam-webster.com/dictionary/subtitle?
9. Constine J, Speech recognition using wit.ai. https://techcrunch.com/2015/01/05/facebook-wit-ai/
10. Nazari N, Mahdavi MA (2019) A survey on automatic text summarization. J AI Data Min 7:121–135
11. Landauer TK, Foltz PW, Laham D (1998) An introduction to latent semantic analysis. In: Discourse Processes, vol 2

12. Mihalcea R, Tarau P (2004) Textrank: bringing order into text
13. Erkan G, Radev DR (2004) Lexrank: graph-based lexical centrality as salience in text summarization. J Artif Int Res 22
14. Edmundson HP (1969) New methods in automatic extracting. J ACM 16

A Generalized Model for Cardiovascular Disease Classification Using Machine Learning Techniques

Ankita Naik and Nitesh Naik

Abstract The number of deaths caused due to cardiovascular ailments is increasing day by day. Prediction of such deadly disease is an unwieldy job for the medical practitioners as it needs adequate experience and knowledge. This study focuses on the real problem dealing with heart patient's data to make well-timed detection and prognosis of risk. The research revolves around building a prediction model to diagnose coronary heart disease. For this purpose, ensemble techniques are tried out along with eight standard classifiers. Two medical datasets are used. In this paper, standard methods are tried out to find the best technique that works for the two datasets chosen. The best results on two datasets are reported for each method.

Keywords Prediction · Ensemble · Cardiovascular ailment · Performance · Combination · Heart · Classification

1 Introduction

Cardiovascular ailments remain the key reason of mortality worldwide, and discovery of it at a prior stage will prevent the consequences that are caused further [1]. According to a survey conducted by World Health Organization, approximately 23 million people die from cardiovascular diseases which are equivalent to two-thirds of all deaths internationally [2]. Due to increasing temporality of cardiovascular disease patients, every year there is more rise in the number of mortality rate caused due to this particular disease [3]. Data mining plays a noteworthy part in the field of cardiovascular ailment prediction. The detection of unseen patterns from existing clinical data to identify the borderline between healthy and unhealthy individuals is present in the classification study [4].

Coronary heart disease is a type of disease wherein a waxy material called plaque accumulated within the coronary arteries. These arteries deliver oxygenated blood to your heart muscle [5]. When plaque collects (accumulates) in the arteries, the state is called atherosclerosis. The accumulation of plaque occurs over several years.

A. Naik (✉) · N. Naik
Computer Engineering Department, Goa College of Engineering, Farmagudi, Ponda, Goa, India

N. N. Chiplunkar and T. Fukao (eds.), *Advances in Artificial Intelligence and Data Engineering*, Advances in Intelligent Systems and Computing 1133,
https://doi.org/10.1007/978-981-15-3514-7_2

The plaque can solidify or break open over point in time. Solidified plaque causes thinning of the coronary arteries and lowers the flow of oxygenated blood to the heart. If the plaque breaks open, a blood clot can build on its surface. Complete blockage of blood flow through a coronary artery can occur due to a large blood clot. After some period of time, the accumulated plaque also hardens and narrows the coronary arteries [6]. The most important symptom of coronary heart disease is discomfort or chest pain, known as angina. This happens when your heart lacks blood or oxygen. Breathlessness is one more frequent symptom of coronary heart disease. The person might feel exhausted or frail. Discovery of such kind of disease at an early stage can facilitate prevention of it from getting worse. It can cause a heart attack or arrhythmias (irregular heartbeats) condition if it is not diagnosed [7, 8]. This research paper builds a prognosis model for classification of cardiovascular disease using ensemble approaches. Methods, namely bagging, boosting and stacking, are tried out on eight standard classifiers using two different medical datasets pertaining to the same disease. The classifiers used are J48, RIPPER, logistic regression, Bayesian net, SMO, Naïve Bayes, KNN and random forest. The best results on two datasets are reported for each method.

The rest of the paper is arranged as follows: Section 2 has the Related Works. Section 3 gives the dataset information that is used in the paper. Section 4 explains the methodology proposed in detail. The experimental results along with performance evaluation are discussed in Sect. 5. The conclusion is discussed in Sect. 6.

2 Related Works

Showman et al. [9] have applied different discretization approaches like equal width, chi merge, an equal frequency with diverse kinds of decision trees such as information gain, gini and gain ratio. They have computed the performance of the decision tree. Bashir et al. [4] have utilized the majority vote-based classifier to integrate different classifiers. In the experiment, ensemble methods gave fairly good results in comparison with single techniques. Chadha et al. [10] created a model by utilizing ANN, Naive Bayes and decision tree methods. They executed it by using Python and also utilized the C#. They showed that artificial neural networks proved to give better results. On the other hand, El-Bialy et al. [11] used an arrangement of the values associated with ML research carried on various datasets. Fast decision tree (FDT) and C4.5 tree (pruned) were used by them. Princy et al. [12] developed a model that used the KNN and ID3 algorithm for prognosis. Their system had two modules, namely the initial module and the prediction module. They showed an increase in accuracy upon addition of a certain number of attributes. Masethe et al. [1] offered a structure using standard data mining (DM) algorithms, namely CARTS, J48, Bayesian net, Naïve Bayes and REPTREE, for prognosis of heart attacks. The Bayesian net algorithm gave better results.

3 Dataset Information

Two coronary heart disease datasets were used. First dataset is taken from the UCI machine repository and the second one from the Kaggle repository. The first dataset, i.e., dataset1 has fourteen attributes. It has a total of 303 instances, of which 164 entries are of healthy patients and 139 having a heart ailment. [13] shows the attributes of the first dataset. The second dataset, i.e., Kaggle repository dataset [14], has fifteen attributes and a total of 4241 entries. [15] shows the attributes of the second dataset.

4 Proposed Method

4.1 Workflow of the Model

Figure 1 depicts the workflow used in the prediction of heart disease. The step-by-step process is discussed in detail below:

4.1.1 Preprocessing of Data

Preprocessing means changes that are made to the data before feeding the data as input to the algorithm. In this technique, the raw data is converted to clean data. Data obtained from many sources is termed to be raw data which is not suitable for analysis. In order to get better results, removal of outliers, noise and irregularities from data are necessary. Another characteristic is that the dataset must be cleaned in such a manner that many machine learning and data mining algorithms are run in one data set, and finest out of them is selected.

4.1.2 Data Splitting

The preprocessed data is then split into a training set which is 70% of original data and testing set which is 30% of the original data. Data having class labels are considered to be the training data without class labels is considered to be the testing data.

4.1.3 Ensemble Approach

The ability to integrate a miscellaneous set of single models together to improvise the steadiness and prognostic supremacy of the model is ensembling. The ensemble is a manner in which we combine the classifiers. Ensembles facilitate to integrate single classifiers together and are likely to enhance the accuracy of the classification.

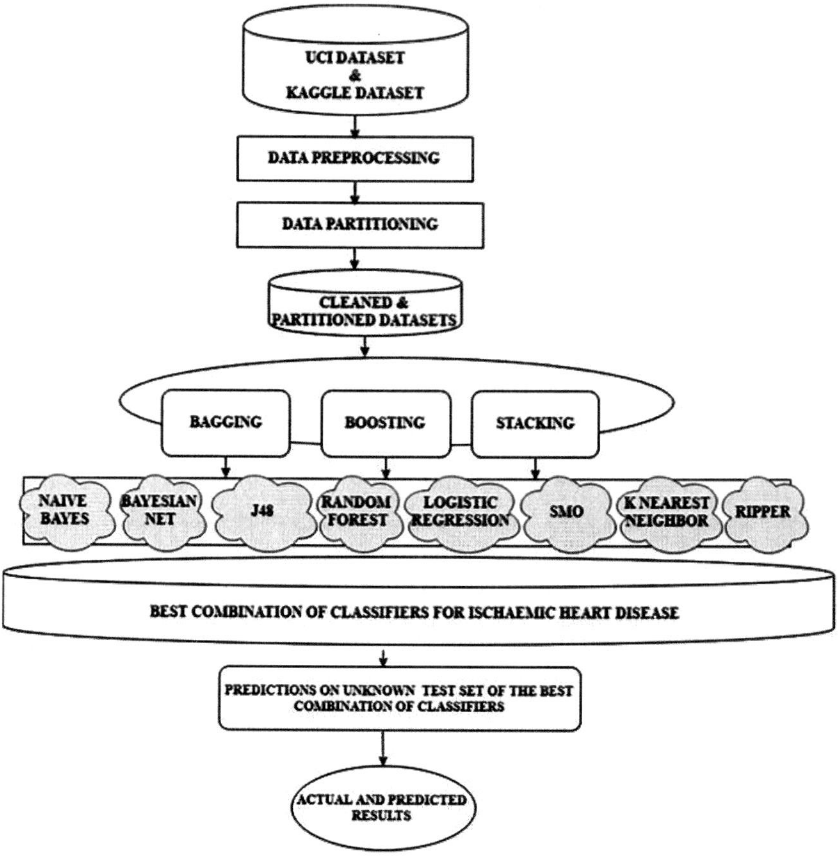

Fig. 1 Ensemble architecture

The most commonly used ensemble techniques are bagging, boosting and stacking [16].

4.1.4 Machine Learning Algorithms

The machine learning algorithms on which the ensembles are tried on are as follows: Naive Bayes Classifier (NB), K-nearest neighbor (KNN), logistic regression (LR), Bayesian net (BN), J48 (C4.5), random forest (RF), RIPPER and sequential minimal optimization (SMO). On basis of their accuracies for the cardiovascular disease datasets in general, the best combination of classifiers is found. For both the datasets, all the above described steps are repeated [17].

Table 1 Confusion matrix

Matrix	Predicted cases		
Actual cases		–	+
	–	TN	FP
	–	FN	TP

4.2 *Performance Evaluation*

Performance evaluation plays a vital role in developing a model. Best model from a set of several different models can be chosen with the assistance of evaluation. It also helps to locate how well the selected model will toil in the future. In machine learning and data mining, the performance evaluation of several models can be carried out based on computing the "confusion matrix." Table 1 depicts the standard confusion matrix, and the equations and principles are shown below.

The number of healthy patients correctly predicted as healthy is termed as the ***true positive rate (TP)***.

The number of unhealthy patients predicted to be healthy is termed as ***false positive rate (FP)***.

The number of unhealthy patients that were correctly classified as unhealthy is known as the ***false negative rate (FN)***.

The number of healthy patients that were incorrectly classified as unhealthy is termed as the ***true negative rate (TN)***.

Performance estimation of a model is done by computing the predictive accuracy of that model. It is defined as a number of correctly made predictions by the model. It is given by the equation:

$$\text{Accuracy} = \frac{\text{true positive cases} + \text{true negative cases}}{\text{true positive} + \text{true negative} + \text{false positive} + \text{false negative}} \quad (1)$$

K-fold cross-validation method was used to estimate the performance of each model. In k-fold cross-validation, the data is divided into k subdivisions and each subdivision is of the same size, i.e., if, for example, 100 samples are divided into 10 subsets, then each subset will contain 10 samples each. The building of models happens k times, every time excluding one particular subdivision from training and using it as the test set. This method is useful when a small amount of data is available in order to attain an unbiased approximation of performance of a model.

5 Experimental Results

The experimental results show the best technique that works for the two datasets chosen. The best results on two datasets are reported for each method. Analysis of the results was made using Java in order to know the accuracy of the different methods used on the two chosen datasets.

Figure 2 shows the classification accuracy when a single classifier is tried out on UCI and Kaggle datasets. When a single/individual classifier is applied for UCI dataset, the best classifier accuracy is produced by K-nearest neighbor and the second best classifier is produced by random forest.

For bagging technique applied as shown in Table 2, for the UCI dataset the best results were given by random forest (RF) and the second best results were given by K-nearest neighbor (KNN). In Table 3 for the Kaggle dataset, the classifier with the highest accuracy was K-nearest neighbor (KNN) and the second best classifier is random forest.

As can be seen in Table 4, on the UCI dataset when the boosting ensemble techniques were applied, the classifiers that gave best results were random forest (RF), J48 and K-nearest neighbor (KNN). In Table 5 when the boosting technique was

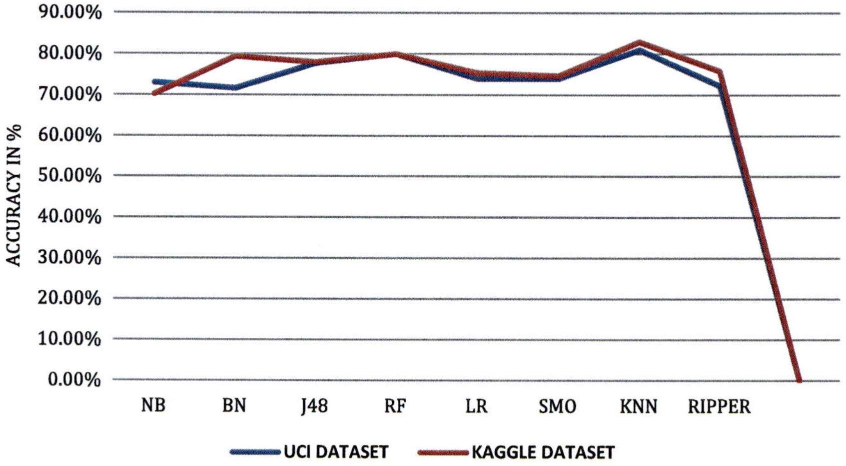

Fig. 2 Single classifier accuracy for UCI dataset and Kaggle dataset

Table 2 Performance estimates for bagging on the UCI dataset

Bagging	UCI dataset		
Classifiers	Correctly classified instances	Incorrectly classified instances	Predictive accuracy (in %)
NB	173	39	81.60
BN	176	36	83.01
J48	201	11	94.81
RF	209	3	98.58
LR	180	32	84.90
SMO	177	35	83.49
KNN	206	6	97.16
RIPPER	196	16	92.45

Table 3 Performance estimates for bagging on Kaggle dataset

Bagging	Kaggle dataset		
Classifiers	Correctly classified instances	Incorrectly classified instances	Predictive accuracy (in %)
NB	2391	577	80.55
BN	2398	570	80.79
J48	2773	195	93.42
RF	2871	97	96.73
LR	2532	436	85.31
SMO	2513	455	84.66
KNN	2896	72	97.57
RIPPER	2569	322	86.55

Table 4 Performance estimates for boosting on the UCI dataset

Boosting	UCI dataset		
Classifiers	Correctly classified instances	Incorrectly classified instances	Predictive accuracy (in %)
NB	175	37	82.54
BN	176	39	81.60
J48	208	4	98.11
RF	208	4	98.11
LR	178	34	83.96
SMO	178	34	83.96
KNN	208	4	98.11
RIPPER	203	9	95.75

Table 5 Performance estimates for boosting on Kaggle dataset

Boosting	Kaggle dataset		
Classifiers	Correctly classified instances	Incorrectly classified instances	Predictive accuracy (in %)
NB	2465	503	83.05
BN	2500	468	84.23
J48	2911	57	98.07
RF	2911	57	98.07
LR	2535	433	85.41
SMO	2513	455	84.66
KNN	2911	57	98.07
RIPPER	2544	424	85.71

Table 6 Performance estimates for stacking with two classifiers on the UCI dataset

Stacking	UCI dataset		
Classifiers	Correctly classified instances	Incorrectly classified instances	Predictive accuracy (in %)
NB+RF	183	29	86.32
NB+SMO	179	33	84.43
BN+RF	179	33	84.43
J48+RF	209	3	98.58
J48+RIPPER	179	33	84.43
LR+RF	195	17	91.98
LR+SMO	178	34	83.96
LR+RIPPER	178	34	83.96
RF+SMO	209	3	98.58
RF+KNN	209	3	98.58
RF+RIPPER	209	3	98.58
KNN+RIPPER	186	26	87.73

applied on the Kaggle dataset, the classifiers that gave the best accuracy were also random forest, J48 and K-nearest neighbor (KNN).

As can be seen in Table 6, on the UCI dataset when the stacking ensemble techniques were applied, the combinations that proved to be the best were J48 and random forest, random forest and SMO, random forest and K-nearest neighbor, and random forest and RIPPER.

In Table 7, when the stacking technique was applied on the Kaggle dataset, the combinations of classifiers that gave the best accuracy were J48 and random forest, random forest and SMO and random forest and K-nearest neighbor, respectively.

As can be seen in Table 8, on the UCI dataset when the stacking ensemble techniques combined with three classifiers were applied, the combinations that proved to be the best were random forest, SMO and KNN; random forest, KNN and RIPPER; J48, random forest and SMO; J48, RF and KNN; and J48, random forest and RIPPER.

In Table 9 when the stacking technique combined with three classifiers was applied on the Kaggle dataset, the combinations of classifiers that gave the best accuracy were random forest, SMO and KNN; random forest, KNN and RIPPER; J48, random forest and SMO; J48, RF, KNN; and J48, random forest and RIPPER, respectively.

As can be seen in Table 10, on the UCI dataset when the stacking ensemble techniques when combined with four classifiers were applied, the combinations that proved to be the best were J48, random forest, SMO and RIPPER.

In Table 11 when the stacking technique was applied on the Kaggle dataset, the combinations of classifiers that gave the best accuracy were J48 and random forest, random forest and SMO and random forest and K-nearest neighbor, respectively.

Table 7 Performance estimates for stacking with two classifiers on Kaggle dataset

Stacking	Kaggle dataset		
Classifiers	Correctly classified instances	Incorrectly classified instances	Predictive accuracy (in %)
NB+RF	2886	82	97.23
NB+SMO	2513	455	84.66
BN+RF	2932	36	98.78
J48+RF	2940	28	99
J48+RIPPER	2547	421	85.81
LR+RF	2616	352	88.1
LR+SMO	2522	446	84.97
LR+RIPPER	2531	437	85.27
RF+SMO	2940	28	99
RF+KNN	2940	28	99
RF+RIPPER	2935	33	98.88
KNN+RIPPER	2570	398	86.59

Table 8 Performance estimates for stacking with three classifiers on Kaggle dataset

Stacking	UCI dataset		
Classifiers	Correctly classified instances	Incorrectly classified instances	Predictive accuracy (in %)
J48+LR+RF	198	14	93.39
J48+LR+SMO	198	14	93.39
LR+RF+KNN	193	19	91.03
LR+RF+RIPPER	195	17	91.98
RF+SMO+KNN	206	6	97.16
RF+KNN+RIPPER	209	3	98.58
J48+RF+SMO	209	3	98.58
J48+RF+KNN	209	3	98.58
J48+RF+RIPPER	209	3	98.58

6 Conclusion

The main intention of this particular research study is to find the best technique that works for the two datasets chosen where standard methods are tried out for prognosis of cardiovascular ailments. The best results on two datasets are reported for each method. For this, eight machine learning classifiers were used . This was

Table 9 Performance estimates for stacking with three classifiers on Kaggle dataset

Stacking	Kaggle dataset		
Classifiers	Correctly classified instances	Incorrectly classified instances	Predictive accuracy (in %)
J48+LR+RF	2615	353	88.10
J48+LR+SMO	2615	353	88.10
LR+RF+KNN	2636	332	88.81
LR+RF+RIPPER	2611	357	87.97
RF+SMO+KNN	2888	80	97.30
RF+KNN+RIPPER	2888	80	97.30
J48+RF+SMO	2888	80	97.30
J48+RF+KNN	2888	80	97.30
J48+RF+RIPPER	2888	80	97.30

Table 10 Performance estimates for stacking with four classifiers on Kaggle dataset

Stacking	UCI Dataset		
Classifiers	Correctly classified instances	Incorrectly classified instances	Predictive accuracy (in %)
J48+RF+LR+SMO	197	15	92.92
J48+RF+LR+KNN	196	16	92.45
J48+RF+LR+RIPPER	198	14	93.39
J48+RF+SMO+KNN	207	5	97.64
J48+RF+SMO+RIPPER	209	3	98.58
RF+LR+SMO+KNN	188	24	88.67
RF+LR+SMO+RIPPER	193	19	91.03
RF+LR+KNN+RIPPER	192	20	90.56
RF+SMO+KNN+RIPPER	206	6	97.16

carried out on two datasets pertaining to coronary artery disease dataset having 14 attributes/303 records and the second dataset having 15 attributes/4241 records. These datasets are taken from two different sources. In general, according to the observations, the combinations that gave the best results for both datasets were the ones where random forest is one of the classifiers. The results obtained were nearly 98% accurate.

Table 11 Performance estimates for stacking with four classifiers on Kaggle dataset

Stacking	Kaggle dataset		
Classifiers	Correctly classified instances	Incorrectly classified instances	Predictive accuracy (in %)
J48+RF+LR+SMO	2615	353	88.10
J48+RF+LR+KNN	2643	325	89.04
J48+RF+LR+RIPPER	2616	352	88.14
J48+RF+SMO+KNN	2822	146	100
J48+RF+SMO+RIPPER	2822	146	100
RF+LR+SMO+KNN	2636	332	88.81
RF+LR+SMO+RIPPER	2611	357	87.97
RF+LR+KNN+RIPPER	2636	332	88.81
RF+SMO+KNN+RIPPER	2822	146	100

References

1. Masethe HD, Masethe MA (2014) Cardiovascular prediction of heart disease using classification algorithms. In: Proceedings of the World congress on engineering and computer science (WCECS), vol 2, 22–24 Oct 2014, San Francisco, USA
2. Xu S, Zhang Z, Wang D, Hu J, Duan X, (2017) Cardiovascular risk prediction method based on CFS subset evaluation and random forest classification framework. In: 2nd International conference on big data analysis (ICBDA), 10–12 Mar 2017. IEEE. https://doi.org/10.1109/ICBDA.2017.8078813
3. Shouman M, Turner T, Stocker R (2016) Using data mining techniques in heart disease diagnosis and treatment. In: 3rd 2012 Japan-Egypt conference on electronics, communications, and computers, 6–12 Mar 2012. IEEE. https://doi.org/10.1109/JEC-ECC.2012.6186978
4. Bashir S, Qamar U, Javed MY (2014) Prediction an ensemble based decision support framework for intelligent heart disease diagnosis. In: International conference on information society (i-Society 2014), 10–12 Nov 2014. https://doi.org/10.1109/i-Society.2014.7009056
5. National Heart, Lung, and Blood Institute What is coronary heart disease? Retrieved 20/8/2014 from http://www.nhlbi.nih.gov/health/healthtopics/topics/cad/
6. Coronary heart disease. Retrieved from https://www.nhlbi.nih.gov/health-topics/coronary-heart-disease
7. Coronary heart disease (CHD). Retrieved from https://familydoctor.org/condition/coronary-heart-disease-chd/
8. Gupta R, Mohan I, Narula J (2016) Trends in coronary heart disease epidemiology in India. Ann Glob Health 82(2). https://doi.org/10.1016/j.aogh.2016.04.002
9. Shouman M, Turner T, Stocker R (2011) Using decision tree for diagnosing heart disease patients. In: ACM, vol 121, 01–02 Dec 2011
10. Chadha R, Mayank S (2016) Prediction of heart disease using data mining techniques. CSI Trans ICT 4(2–4)
11. El-Bialy R, Salamay MA, Karam OH, Khalifa ME (2015) Feature analysis of coronary artery heart disease data sets. Procedia Computer Science 65. https://doi.org/10.1016/j.procs.2015.09.132 (International conference on communication, management and information technology (ICCMIT 2015), Oct 2015)
12. Princy T, Thomas J (2016) Human heart disease prediction system using data mining techniques. In: Circuit, power and computing technologies (ICCPCT), 2016. IEEE. https://doi.org/10.1109/ICCPCT.2016.7530265

13. UCI Machine Learning Repository. Arlington: The Association; 2006 [updated 1996 Dec 3; cited 2011 Feb 2. Available from http://archive.ics.uci.edu/ml/datasets/Heart+Disease
14. Meeshanthini V. Dogan, Isabella M. Grumbach, Jacob J. Michaelson, Robert A. Philiber (2018) Integrated genetic and epigenetic prediction of coronary heart disease in the Framingham heart study. PLoS ONE 13(1):e0190549. https://doi.org/10.1371/journal.pone.0190549
15. Framingham Heart Study: Available from https://www.kaggle.com/amanajmera1/framingham-heart-study-dataset/version/1
16. Skurichina M, Duin RPW (2002) Bagging, boosting and the random subspace method for linear classifiers. Pattern Anal Appl 5(2):121–135
17. Luo Y, Li Z, Guo H, Cao H, Song C, Guo X et al (2017) Predicting congenital heart defects: a comparison of three data mining methods. PLoS ONE 12(5):e0177811. https://doi.org/10.1371/journal.pone.0177811

Classification of Road Accidents Using SVM and KNN

P. Joyce Beryl Princess, Salaja Silas and Elijah Blessing Rajsingh

Abstract The rapid increase in automobiles in the recent past has caused an unrestrained escalation of road accidents. Due to road accidents, victims suffer from non-fatal injuries and incurring disabilities. The road accidents have tended many researchers to analyze the severity and type of the accident to enhance road safety measures and aid to speed up the post-crash support amenities. This paper attempts to classify the severity of the accidents by analyzing the accident images. The features of the accident images are extracted using algorithms such as histogram of oriented gradient (HOG), local binary pattern (LBP) and speeded up robust features (SURF). These features are given as input to k-nearest neighbor (KNN) and support vector machine (SVM) to classify the severity of the accidents. The performance of SVM and KNN classifiers with three feature extraction algorithms is assessed and compared. The classification results show that SVM classifier outperformed KNN. SVM with the HOG features shows better accuracy of 79.58% compared to LBP and SURF.

Keywords Features · Feature extraction · Local binary pattern · Histogram of oriented gradients · Speeded up robust features · Classification · Support vector machine · K-nearest neighbor

1 Introduction

The tremendous increase in automobiles leads to rising in road accidents. Every year millions of lives are lost all over the world due to road accidents [1]. Survivors suffer from grievous injury, slight injury and incurable disabilities due to road accidents [2]. Road accidents have been the major concern of many nations. The analysis on the statistical report [3, 4] and the investigation report [5] of the accidents as fatal and non-fatal helps to determine the appropriate emergency services, type of trauma facility and identify suitable treatment for the victims. The significant data required for the accident analysis is the accident image. The accident image can be captured

P. J. Beryl Princess · S. Silas (✉) · E. B. Rajsingh
Karunya Institute of Technology and Sciences, Coimbatore, India

N. N. Chiplunkar and T. Fukao (eds.), *Advances in Artificial Intelligence and Data Engineering*, Advances in Intelligent Systems and Computing 1133,
https://doi.org/10.1007/978-981-15-3514-7_3

by the spectators from the accident spot and by the surveillance cameras mounted on the roadways. Image processing contributes a key role in the analysis of the accident image. The two major steps required to understand the severity of the accidents are (1) feature extraction and (2) classification of the images. In this paper, the feature extraction algorithms such as HOG, SURF and LBP have been adopted to extract the features and are classified using SVM and KNN.

The overall paper is organized as follows. An overview of the existing system for analyzing the accident severity is briefed in Sect. 2. The feature extraction algorithms and classifiers are discussed in Sect. 3. The performance of the classifiers is shown in Sect. 4. The conclusion is given in Sect. 5.

2 Literature Review

An extensive literature survey has been carried out to understand the existing approaches used to determine the severity of the accident.

Mannera and Wunsch-Zieglerb [6] used multinomial logit model to statistically determine the circumstances for the severity of accidents. The model concluded stating that crashing with the objects by the side of the roads is very severe when compared to others. Zhu and Srinivasan [7] contributed a detailed analysis of the actual cause for the severity of the accidents. The major cause for the higher severity in road accidents is due to driver's attitudes such as emotional distraction and usage of alcohol during driving. Gao et al. [8] proposed a methodology to detect the road accident from the feed in social media using deep learning. Accident-related tweets were examined and evaluated using long short-term memory and deep belief networks.

De Ona et al. [9] analyzed the severity of the highway accidents using Bayesian networks classification model and latent class cluster. Latent class cluster-based Bayesian networks provided insights on the major cause of the severity of the accidents. Mujalli et al. [10] used Bayes classifier to forecast the level of severity of road accidents through available traffic accidents statistics. The Bayes classifier was able to identify killed and serious injuries efficiently than the slight injuries from the balanced set of accident data.

Lee et al. [11] employed structural equation modeling (SEM) to find the correlation among the accident severity of the accident and weather conditions. Garcia Cuenca and Puertas [12] compared naïve Bayes, gradient boosted trees and deep learning classification techniques to classify an accident data as serious or not serious accidents from Spanish traffic dataset. The comparison highlighted deep learning was the best classification technique. Zheng et al. [13] proposed road accident severity prediction using deep learning CNN. The proposed TASP-CNN model experimented on eight years period of accident data from Leeds City Council. The proposed model proved that it outperformed the models such as NBC, KNN, LR, DT and LSTM-RNN in the prediction of severity of the accident. Zhang and Li [14] performed a comparative analysis on crash injury severity using statistical models and machine learning techniques viz decision tree, k-nearest neighbor, random forest and support vector machine. The results proved that machine learning techniques have a greater prediction of road crash injury severity compared to statistical methods. Ditcharoen [15] reviewed the various factors influencing road traffic accidents severity. The most common factors analyzed causing the severity of road traffic injury were weather conditions, alcohol consumption, driver's fatigue and vehicle types. The logistic regression model was used in this study for analyzing the factors influencing the severity of the road traffic accidents. Caban et al. [16] analyzed data on road traffic accidents related to the type of accidents and have presented biomedical materials for the treatment of injuries caused due to the road accidents.

Most of the models have used the statistical report for the analysis of the severity of the accidents. However, incorporating the analysis of the accident images will be more efficient.

3 Classifiers for Accident Images

Classification techniques play a major role in the effective identification of the severity in an accident image. In this paper, the classifiers SVM and KNN are employed. Based on the literature survey, it is understood that SVM and KNN are applied for various pattern recognition and classification problems. While SVM has good generalization [17] with less possibility of overfitting [18], KNN is computationally less complex and effectively classifies less complex images [19].

Classification of accident images is relatively challenging due to the cluttered image background and complexity. Also, the performance of the classifiers is inclined to the robust and discriminant features extracted from the images. Therefore, feature extraction algorithms can be incorporated to increase the accuracy of classification [20, 21]. The purpose of the feature extraction is to determine the features that are unvarying to the rotation, scale and illumination changes.

Feature extraction algorithms and the classifier applied to identify the severity of the accident are briefed in this section.

3.1 Feature Extraction

Several feature extraction algorithms exist and can be broadly classified based on image features such as edge, texture and interest points. The algorithms have been analyzed and few algorithms have been identified. They are: (i) HOG (ii) LBP (iii) SURF. HOG is an edge feature-based feature descriptor, LBP is a texture-based algorithm and SURF is a feature detector that detects interest points from an image.

Histogram of Oriented Gradient (HOG). HOG is widely applied for object detection [22] in images and videos. It is resistant to illumination changes [20] and also efficient in critical environments such as cluttered background, occlusion [23] and noise [24]. In HOG, the images are divided into a smaller window and the window is further divided into cells of $n * n$ pixels. The magnitude and orientation are obtained by Eqs. (1) and (2)

$$\text{Magnitude} = \sqrt{g_x^2 + g_y^2} \tag{1}$$

$$\text{Orientation} = \arctan\left(\frac{g_y}{g_x}\right) \tag{2}$$

where g_x and g_y denote the horizontal and vertical gradient, respectively. The edge orientation and magnitude are calculated for each cell and accumulated as histogram bins of size M. The cells are grouped to form blocks and the weighted histograms are calculated and normalized for each block. The normalized histogram from all the blocks is combined together to form the final HOG descriptor.

Local Binary Pattern (LBP) Local binary pattern [25] describes the texture of an image. It is computationally simple, invariant to grayscale changes and has high discriminative power. Hence, it is broadly used for face analysis [26], gender classification [27] and recognition of humans [28]. The LBP, for a given (P, Q) dimensions, is attained by comparing the center pixel values with its corresponding neighboring pixel values. Q represents the radius of the neighborhood and P represents the number of neighbors. Initially, the threshold value of a pixel depends on its neighboring pixels and is calculated using Eq. (3)

$$S(g_i - g_c) = \begin{cases} 1, \text{ if } g_i \geq g_c \\ 0, \text{ if } g_i < g_c \end{cases} \tag{3}$$

where g_c denotes the center pixel and g_i denotes the neighbor pixel. The threshold values are encoded into LBP descriptor using Eq. (4)

$$\text{LBP}_{PQ}(x, y) = \sum_{i=0}^{P-1} S(g_i - g_c).2^i \tag{4}$$

Speeded Up Robust Features (SURF) SURF [29] feature detector and descriptor helps in extracting the local features used for tasks such as image classification, object detection and recognition [30]. The descriptors computed by SURF are invariant to scaling and rotational changes. The SURF descriptor is constructed by computing determinant of the Hessian matrix. The Hessian matrix depends on the integral images to find the interest points. Choosing a point $z = (x, y)$ in an image I, the determinant of the Hessian matrix can be computed with scale σ as given by Eq. (5)

$$H(z,\sigma) = \begin{vmatrix} L_{xx}(z,\sigma) & L_{xy}(z,\sigma) \\ L_{xy}(z,\sigma) & L_{yy}(z,\sigma) \end{vmatrix} \quad (5)$$

where $L_{xx}(z,\sigma)$, $L_{xy}(z,\sigma)$, $L_{yy}(z,\sigma)$ are the second-order derivatives. The blob regions in the image are deduced from the approximation of computed determinant of the Hessian matrix. Then non-maximal suppression is applied to locate interest points over different scales. It is further divided into subregions and Haar wavelet response is computed for each subregion. The computation results in a final feature descriptor with the length of 64.

3.2 Classifiers

Support Vector Machine (SVM) This supervised learning algorithm is broadly adopted in two-class classification with generalization performance [31]. In SVM, the input samples are mapped into the higher dimensional space and a decision plane (or hyperplane) is constructed between the input samples. In general, the hyperplane can be represented as given by Eq. (6)

$$w^{\mathrm{T}}.x + b = 0 \quad (6)$$

where w represents the weights, b represents the bias and x represents the input vector. For a given training set, SVM minimizes the training set error by maximizing the hyperplane between the different input categories. Applying various kernel functions viz., radial basis function (RBF), polynomial, sigmoid and linear improves the performance of the SVM classifier. Equation (7) represents the hyperplane after the inclusion of kernel functions.

$$f(x) = \sum_{i=1}^{N} a_i y_i \; R(x, x_i) \quad (7)$$

where R represents the kernel function, x_i is the training sample with class label y_i belongs to $\{-1, 1\}$ and $\sum_{i=1}^{N} a_i y_i$ is the Lagrange multiplier, where the sample corresponding to whose $ai \neq 0$ is called the support vector. RBF kernel function is employed to classify the accident images due to its heterogenous property. The RBF kernel for two samples x and $\mathrm{x_i}$ is represented by Eq. (8)

$$R(x, x_i) = \exp\left(\frac{-\|x - x_i\|^2}{2\gamma^2}\right) \quad (8)$$

where γ represents a hyperparameter to calculate the degree of bias and variance of the model.

k-Nearest Neighbor (KNN) KNN is widely used for classification based on neighborhood estimations [32]. In KNN, the test data are classified based on the k closest training samples present in the feature space. The selection of the number of nearest neighbors (or k) in KNN classifier is a crucial step, as it affects the classification accuracy. In this work, the number of nearest neighbors chosen is five and Euclidean distance as the distance metric. In general, the Euclidean distance is computed using Eq. (9)

$$D(x, y) = \sqrt{\sum_{j=1}^{n} (x_j - y_j)^2} \tag{9}$$

where D is the distance between test data and each training data, $x = \{x_1, x_2, \ldots.,\ x_n\}$ the features of test data and $y = \{y_1, y_2, \ldots, y_n\}$ the features of training data.

4 Results and Discussion

The experiment is carried out using MATLAB_R2018b software on 2.3 GHz Intel Core i5 processor with 8 GB RAM. The images for this work were collected from the public domain. In this experiment, 357 accident images are used and categorized as fatal and non-fatal accident images. The images such as victims crushed between vehicles, rollover vehicles, side-impact collisions, pile-up collisions and head-on collisions are considered as fatal accidents and a sample is shown in Fig. 1. The images such as slight collision, negligible damage in the vehicle body, cracked headlight and minor dent in the hood are considered as non-fatal accidents and a sample is shown in Fig. 2.

The images are converted into grayscale and resized to 256 × 256. The HOG, LBP and SURF features are extracted from the grayscale images. The HOG gradient

Fig. 1 Fatal accident image

Fig. 2 Non-fatal accident image

image of the fatal and non-fatal accident is depicted in Figs. 3 and 4. The prominent edges in the figures imply that HOG describes the shape and the appearance of the objects in the image. Hence, the essential features for the classification can be obtained. Figure 5 shows the feature variations between fatal and non-fatal accident images. The higher variation denotes that discriminant features are extracted which indicates the increase in the classification performance.

The LBP pattern image for fatal and non-fatal images is shown in Figs. 6 and 7 respectively. The LBP pattern of the non-fatal accident image is sharper than the fatal image. The feature variations between the fatal and non-fatal image are shown in Fig. 8. The variations between the histogram of fatal and non-fatal images are higher which implies the increase in the classification performance.

The SURF feature extraction with the feature points for the fatal and non-fatal accidents is shown in Fig. 9. In SURF, the feature points of the non- accidental area are also identified along with accidental area which declines the classification performance.

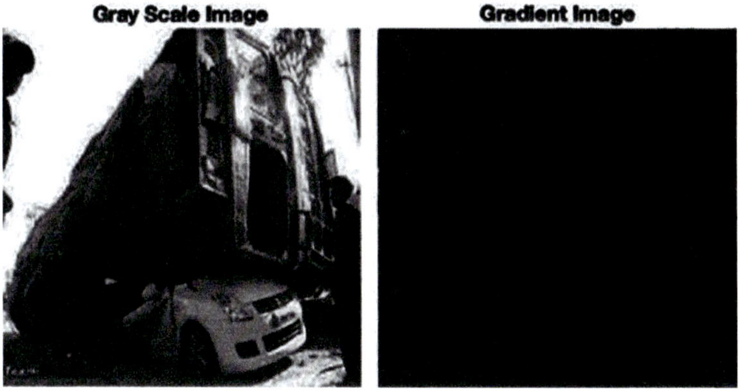

Fig. 3 Gradient image for fatal accident

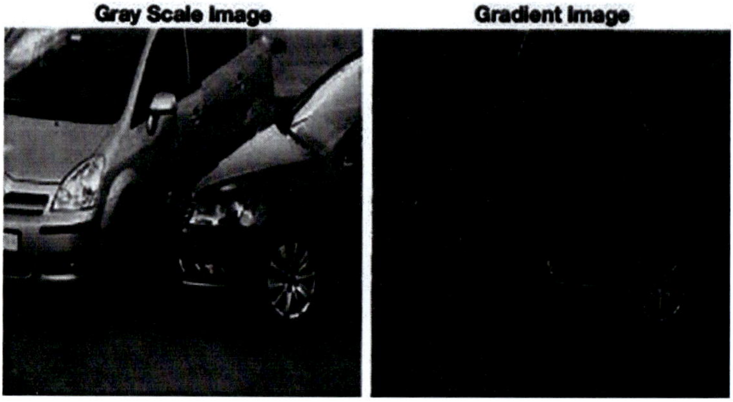

Fig. 4 Gradient image of non-fatal accident

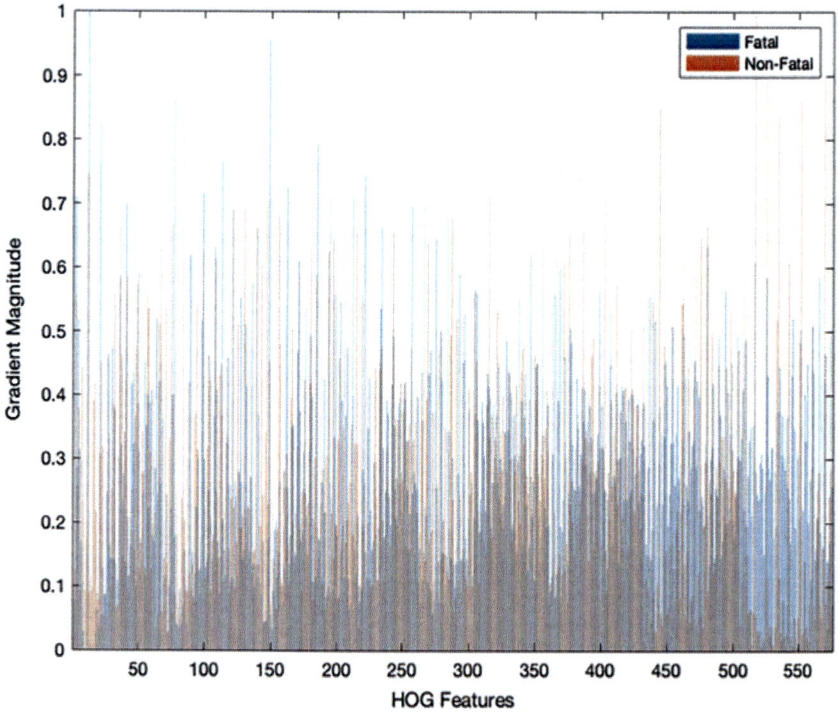

Fig. 5 The final HOG feature vector for a fatal and non-fatal accident images

Fig. 6 LBP pattern of the fatal accident

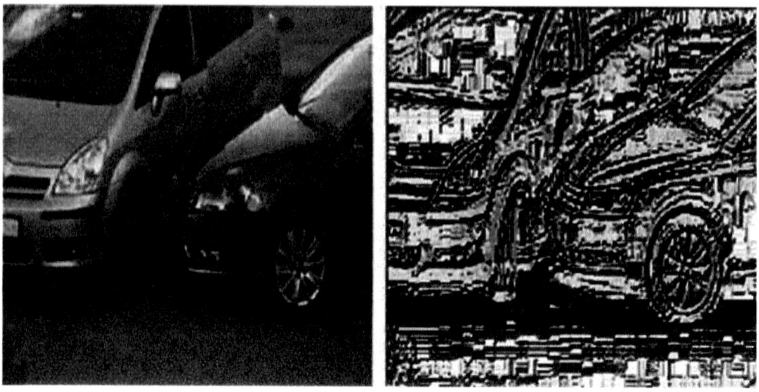

Fig. 7 LBP pattern of non-fatal accident

The 357 images are split into two groups: fatal (230) and non-fatal (127). These images are further split into training (60%) and testing (40%) dataset. The training images consist of 138 fatal images and 77 non-fatal images and testing images consist of 92 fatal images and 50 non-fatal images. The extracted HOG, LBP and SURF features are fed into SVM and KNN classifiers.

The classifier performance based on feature extraction techniques is examined by calculating the accuracy, sensitivity and specificity. Sensitivity is the proposition of true positive results to the actual positive samples. Specificity is the proportion of true negative outcomes to the actual negative samples. Accuracy is the proposition of correctly identified results to the overall samples. The aforementioned performance evaluations are mathematically expressed using Eqs. (10), (11) and (12)

$$\text{Accuracy} = \frac{\text{TN} + \text{TP}}{\text{TN} + \text{FN} + \text{TP} + \text{FP}} \tag{10}$$

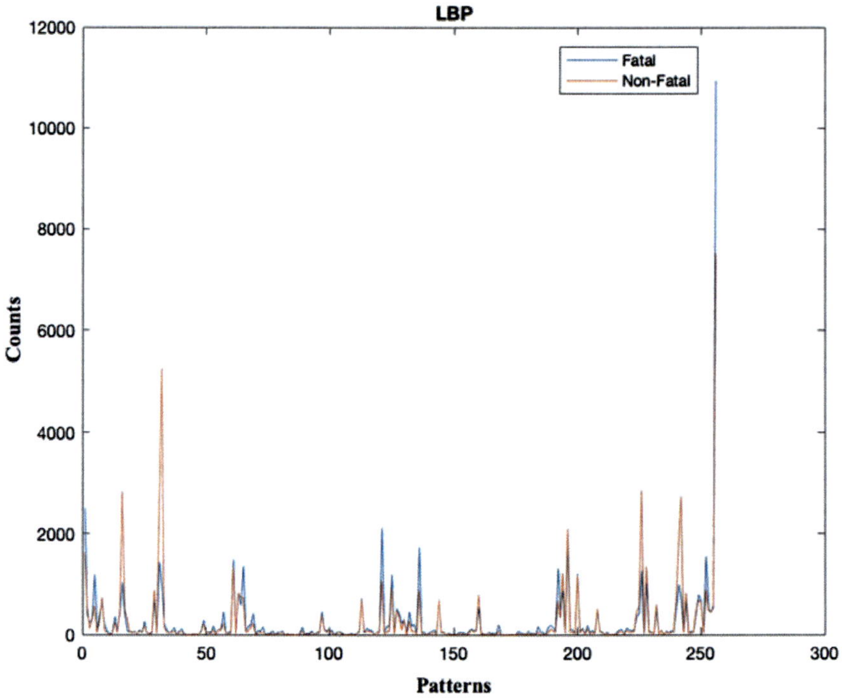

Fig. 8 Histogram of fatal and non-fatal accident image

Fig. 9 SURF feature points for fatal and non-fatal accident images

Table 1 Classification performance for SVM and KNN classifier with HOG, LBP and SURF feature extraction algorithms

Feature extraction algorithms	SVM classifier			KNN classifier		
	Accuracy (%)	Sensitivity (%)	Specificity (%)	Accuracy (%)	Sensitivity (%)	Specificity (%)
HOG	79.58	81.19	75.61	73.94	72.73	80.95
LBP	78.87	78.18	81.25	74.65	78.57	65.91
SURF	74.65	73.73	79.17	71.13	74.29	62.16

$$\text{Specificity} = \frac{\text{TN}}{\text{TN} + \text{FP}} \tag{11}$$

$$\text{Sensitivity} = \frac{\text{TP}}{\text{FN} + \text{TP}} \tag{12}$$

The accuracy of the classifier is influenced by the features extracted from the feature extraction algorithms. Table 1 depicts the classification performance of SVM and KNN classifiers with HOG, LBP and SURF feature extraction algorithms. From the results, it is evident that SVM with HOG features have attained the maximum accuracy of 79.58% compared with the LBP and SURF algorithms. SVM with LBP features has also attained 78.87% accuracy close to the accuracy of HOG. However, the sensitivity attained by LBP features is lower than the HOG features. The identification of fatal images by LBP is lesser by 3.01%. SVM with SURF features attained an accuracy of 74.65% which is lesser than HOG and LBP. The comparison of classification accuracy of SVM and KNN with different feature extraction algorithms is shown in Fig. 10. Compared to KNN classifier, the SVM classifier attained the highest classification accuracy in all the feature extraction techniques.

The comparison of sensitivity is shown in Fig. 11. The sensitivity of the SVM with HOG features is higher than KNN which implies the identification of fatal accidents by SVM with HOG features is higher. The comparison of specificity is shown in Fig. 12. It is evident from the figure that the specificity of LBP with SVM is higher which implies the identification of non-fatal accidents by SVM with LBP features is higher. However, identification of fatal accidents by SVM with LBP is lesser compared to SVM with HOG features. Hence, SVM classifies the accident images as fatal and non-fatal best with HOG features.

5 Conclusion

Road accidents are the prime cause for fatalities and incurable disabilities. Hence, the severity of the accidents must be identified to initiate necessary emergency services. In this work, the severity of the accident is assessed in real time based on the accident images. The feature extraction algorithms HOG, LBP and SURF are employed to

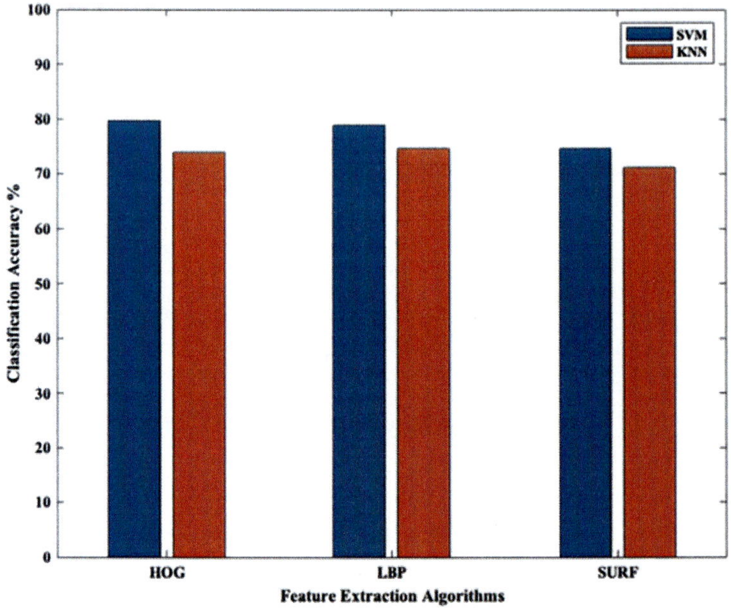

Fig. 10 Comparison of accuracy of SVM and KNN classifier

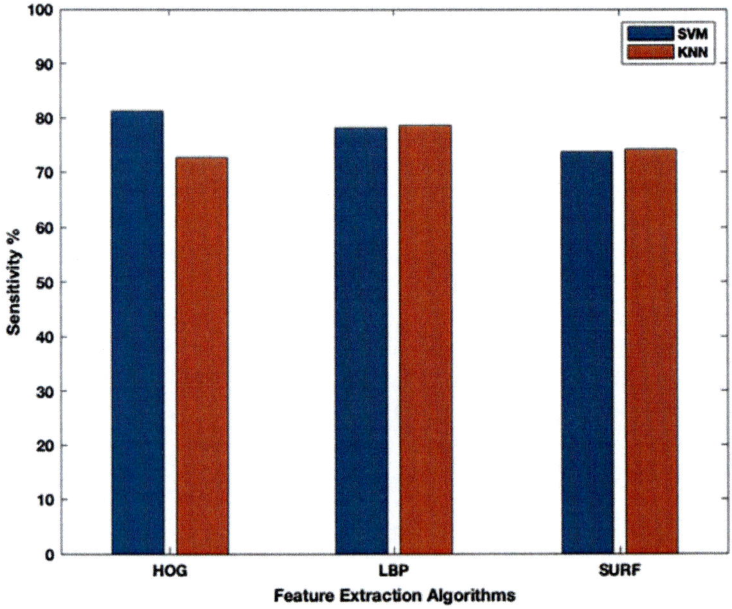

Fig. 11 Comparison of sensitivity

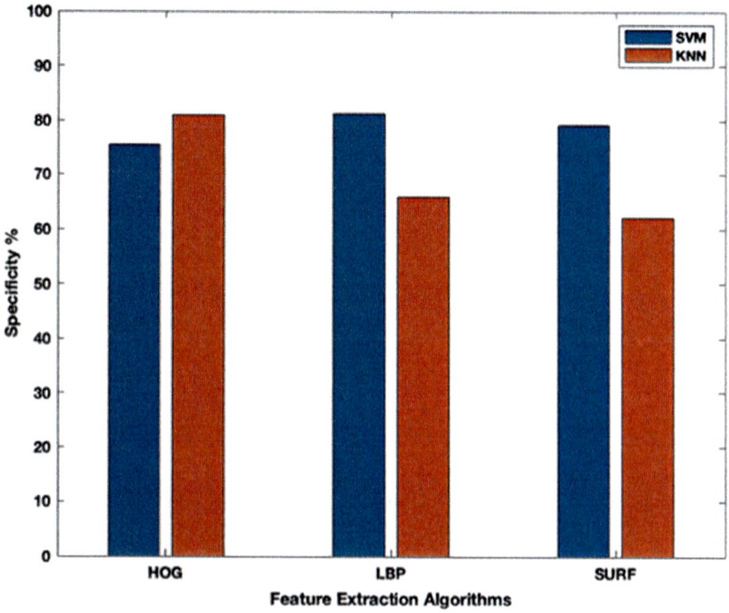

Fig. 12 Comparison of specificity

obtain the discriminant descriptors from the images. The extracted features in the form of feature vectors are fed into the classifier for the identification of the severity of the accident. There are classified as fatal or non-fatal accident using SVM and KNN classifier techniques. The classification performance implies that SVM classifier with the HOG features attained the highest classification accuracy of 79.58% compared with LBP and SURF.

Acknowledgements This work was sponsored by the Indian Council of Medical Research (ICMR) under the Health Systems Research Division, ad hoc project scheme (Project id: IRIS:2016-0395).

References

1. WHO. https://www.who.int/news-room/fact-sheets/detail/road-traffic-injuries
2. gov.in. https://data.gov.in/catalog/persons-injured-road-accidents
3. González MPS, Sotos FE, Ponce ÁT (2018) Data on the determinants of the risk of fatalities, serious injuries and light injuries in traffic accidents on interurban roads in Spain. Data Br 18:1941–1944. https://doi.org/10.1016/j.dib.2018.04.117
4. Chang DS, Tsai YC (2014) Investigating the long-term change of injury pattern on severity, accident types and sources of injury in Taiwan's manufacturing sector between 1996 and 2012. Saf Sci 68:231–242. https://doi.org/10.1016/j.ssci.2014.04.005

5. Griselda L, Juan DO, Joaquín A (2012) Using decision trees to extract decision rules from police reports on road accidents. Procedia Soc Behav Sci 53:106–114. https://doi.org/10.1016/j.sbspro.2012.09.864
6. Manner H, Wünsch-Ziegler L (2013) Analyzing the severity of accidents on the German Autobahn. Accid Anal Prev 57:40–48. https://doi.org/10.1016/j.aap.2013.03.022
7. Zhu X, Srinivasan S (2011) A comprehensive analysis of factors influencing the injury severity of large-truck crashes. Accid Anal Prev 43:49–57
8. Zhang Z, He Q, Gao J, Ni M (2018) A deep learning approach for detecting traffic accidents from social media data. Transp Res Part C Emerg Technol 86:580–596. https://doi.org/10.1016/j.trc.2017.11.027
9. De Oña J, López G, Mujalli R, Calvo FJ (2013) Analysis of traffic accidents on rural highways using Latent Class Clustering and Bayesian Networks. Accid Anal Prev 51:1–10. https://doi.org/10.1016/j.aap.2012.10.016
10. Mujalli RO, López G, Garach L (2016) Bayes classifiers for imbalanced traffic accidents datasets. Accid Anal Prev 88:37–51. https://doi.org/10.1016/j.aap.2015.12.003
11. Lee J, Chae J, Yoon T, Yang H (2018) Traffic accident severity analysis with rain-related factors using structural equation modeling—a case study of Seoul City. Accid Anal Prev 112:1–10. https://doi.org/10.1016/j.aap.2017.12.013
12. Garcí L, Puertas E, Andres JF (2018) Traffic accidents classification and injury severity prediction. In: 2018 3rd IEEE international conference intelligence transportation engineering, 52–57. https://doi.org/10.1109/ICITE.2018.8492545
13. Zheng M, Li T, Zhu RUI et al (2019) Traffic accident's severity prediction: a deep-learning approach-based cnn network. IEEE Access 7:39897–39910
14. Zhang J, Li Z (2018) Comparing prediction performance for crash injury severity among various machine learning and statistical methods. IEEE Access 6:60079–60087. https://doi.org/10.1109/ACCESS.2018.2874979
15. Ditcharoen A (2019) Road traffic accidents severity factors : a review paper. In: 2018 5th International Conference Bus Ind Res, 339–343
16. Caban J, Karpi R, Barta D (2018) Road Traffic accident injuries—causes and biomaterial related treatment. In: 2018 XI international science-technical conference automotive safety
17. Christopher J.C. Burges (Bell Laboratories LT.: A Tutorial on Support Vector Machines for Pattern Recognition. Data Min Knowl Discov 2, 121–167 (1998)
18. Chih-Wei Hsu, Chih-Chung Chang, Lin C-J, Chih-Wei Hsu, Chih-Chung Chang and C-JL, et al.: A Practical Guide to Support Vector Classification. BJU Int 101(1), 1396–1400 (2008). https://doi.org/10.1177/02632760022050997
19. Phyu TN (2009) Survey of classification techniques in data mining. Proc Int MultiConference Eng Comput Sci 1:18–20. https://doi.org/10.1136/bjsports2013092255
20. Tribaldos P, Serrano-cuerda J, L T, et al.: People Detection in Color and Infrared Video using HOG and Linear SVM. Nat Artif Comput Eng Med Appl IWINAC 2013 LNCS, vol. 7931 Springer, Berlin, Heidelb (2013)
21. Yan G, Yu M, Yu Y, Fan L (2016) Real-time vehicle detection using histograms of oriented gradients and AdaBoost classification. Opt Int J Light Electron Opt 127(19):7941–7951. https://doi.org/10.1016/j.ijleo.2016.05.092
22. Dalal N, Triggs B (2007) Histograms of oriented gradients for human detection. Comput Vis Adv Res Dev. https://doi.org/10.1109/CVPR.2005.177
23. Wang X, Han TX (2009) An HOG-LBP human detector with partial occlusion handling. In: 2009 IEEE 12th interantional conference on computer vision. pp 32–39
24. Zhuang J (2016) Compressive tracking based on HOG and extended haar-like feature. In: 2016 2nd International conference on computer and communications, vol 2. pp 326–331
25. Ojala T, Pietikäinen M, Harwood D (1996) A comparative study of texture measures with classification based on feature distributions. Pattern Recognit 29(1):51–59. https://doi.org/10.1016/0031-3203(95)00067-4
26. Ahonen T, Hadid A, Pietikäinen M (2004) Face recognition with local binary patterns: application on face recognition. 28:469–481. https://doi.org/10.1007/978-3-540-24670-1_36

27. Hadid A, Pietikäinen M (2009) Combining appearance and motion for face and gender recognition from videos. Pattern Recognit 42(11):2818–2827
28. Mu Y, Yan S, Liu Y, et al (2008) Discriminative local binary patterns for human detection in personal album. 26th IEEE conf comput vis pattern recognition, CVPR. https://doi.org/10.1109/CVPR.2008.4587800
29. Bay H, Ess A, Tuytelaars T, Van Gool L (2008) Speeded-Up Robust Features (SURF). Comput Vis Image Underst 110:346–359. https://doi.org/10.1016/j.cviu.2007.09.014
30. Asbach M, Hosten P, Unger M (2008) An evaluation of local features for face detection and localization. WIAMIS 2008—Proc 9th Int work image anal multimed interact serv, 32–35. https://doi.org/10.1109/WIAMIS.2008.58
31. Auria L, Moro RA (2008) Support vector machine (SVM) as a technique for solvency analysis, pp 1–16
32. Han Y, Virupakshappa K, Oruklu E (2015) Robust traffic sign recognition with feature extraction and k-NN classification methods. IEEE Int Conf Electro Inf Technol 2015:484–488. https://doi.org/10.1109/EIT.2015.7293386

A Deep Convolutional Encoder-Decoder Architecture Approach for Sheep Weight Estimation

Nirav Alpesh Shah, Jaydeep Thik, Chintan Bhatt, and Aboul-Ella Hassanien

Abstract A significant application has been found in making the task of automation easier. Computer vision is a very powerful method for understanding the contents from the images. Weighing sheep for the analysis of their health and growth is a very arduous task, and traditional methods of weighing using a balance are very time-consuming. Estimating sheep weights using computer vision algorithms has already been tried, but these methods are not efficient enough. They also require some extra instruments set-up. We tried to utilize this powerful technology to make the difficult task of estimating sheep weights quick and accurate. We bring a novel approach to image segmentation in order to calculate the area of the sheep directly. The various other features are fed to a neural network-based regression model for achieving better results for the task of estimating sheep's weight. The proposed approach has enabled us to minimize the complexities of carrying heavy equipment and errors through human involvement in measuring the weight of the sheep.

Keywords Segmentation · Weight estimation · Neural network · Regression

1 Introduction

Sheep's meat as a food source has always been used in high demand. The main factor which affects the proportion of meat of a sheep is the weight of the sheep. And thus it becomes a major factor for selecting the sheep. Estimating the weight of the

N. A. Shah · J. Thik · C. Bhatt (✉)
Charotar University of Science and Technology, Anand, India
e-mail: chintanbhatt.ce@charusat.ac.in

N. A. Shah
e-mail: 15ce128@charusat.edu.in

J. Thik
e-mail: 15ce140@charusat.edu.in

A.-E. Hassanien
Cairo University, Giza, Egypt
e-mail: aboitcairo@gmail.com

N. N. Chiplunkar and T. Fukao (eds.), *Advances in Artificial Intelligence and Data Engineering*, Advances in Intelligent Systems and Computing 1133,
https://doi.org/10.1007/978-981-15-3514-7_4

sheep through the naked eye is a difficult task and sometimes involves usage of time-consuming equipment like a weighing scale. So there needs to be a better and optimal way of finding the sheep weight through modern techniques. Computer vision has enhanced over the past few years and has also affected this area of predicting sheep's weight using some advanced computer vision and machine learning algorithms. There are already some algorithms available but they either had a constraint of high-end equipment or involved fix positions of cameras.

Menesatti et al. in [1] proposed a low-cost stereovision system to estimate the size and weight of live sheep. It uses two high-resolution web cameras connected to the laptop for this task. It involved the partial least square regression calculated on the linear distance between the images. The error was around 3.5 and 5.0% evaluated by the mean size error.

Kashiha et al. in [2] proposed an automatic weight estimation of individual pigs using image analysis and pattern recognition from the video images of pigs processed offline. The area calculated using ellipse fitting algorithm was used to estimate the weight of the sheep. Authors achieved 96.2% accuracy.

Prandana et al. in [3] proposed a beef cattle weight estimated approach by using digital image processing and authors used the largest area using active contour and extracted the number of pixels from the image to estimate cattle's weight. Front view image was used to detect chest dimensions to add extra features to estimate cattle's weight. Moreover, body size estimation of newborn lambs using image processing and its effect on the genetic gain of a simulated population.

Khojastehkey in [4] estimated using complete side view of newborn lamb achieved by cropping the image to estimate their size. This approach depended on the fact that all the lambs are black and the background is white. Therefore, the body was easily recognized by binarizing the image, so cattle size was measured by counting the white pixels. 89% accuracy was achieved in this research.

This paper proposed an approach based on computer vision and image segmentation algorithms for estimating sheep's weight. The main process of such image segmentation techniques is to correctly identify the objects in the images and label them precisely. The proposed approach extrapolated this basic idea of object detection and labeling to identify and segment out sheep's from the images by training it on a well-curated dataset. The labeled output was then used to extract out the area of the sheep in terms of a number of pixels covered. This extracted area along with other crucial parameters such as age and gender was used as features in a neural network-based regression model to finally estimate the weight of the sheep. The main aim of this paper is to have a real-time application which can be used by all type of users. None of the previous approaches can be used to do so because we cannot achieve real-time weight estimation as those approaches relied on video recordings and special camera settings. In this work, we are using only one side image of the sheep and the age and gender of the sheep to estimate sheep's weight. In this paper, dataset of 52 sheep of different physiological conditions and ages has been used. This research has also achieved higher accuracy than its previous approaches to sheep weight estimation.

The paper is organized as follows. In Sect. 2, the theoretical background is explained. In Sect. 3, the proposed weight estimation system is explained. In Sect. 4, results achieved through this model are discussed, In Sect. 5, conclusion is given.

2 Materials and Methods

2.1 The SegNet Architecture for Image Segmentation

Image segmentation is an application of computer vision that can be described as a process that partitions the digital image into segments where each segment represents a different class according to the color scheme. Image segmentation is used here to segment out the sheep from the background in order to estimate the pixel count to determine the surface area of the sheep.

The task of segmentation seems to be difficult here since the haystacks in the background sometimes merge with the texture of sheep wool. This problem has been successfully solved in a newer segmentation approach.

There are various image segmentation algorithms that are available and have been used to solve major segmentation problems in computer vision. The architecture is based on Convolution Neural Nets which is a deep learning approach to solve computer vision. Some of the architectures used are R-CNN, Fully Convolutional Networks [5], and SegNet [6].

Figure 1 illustrates the structure of SegNet [6]. There are no layers that are fully connected and therefore are only metaphysical. The decoder sets up its inputs using the billiard indicators transferred from its encoder to produce a separate map(s). The torsion is then carried out with a training able filter bank to intensify the characteristic map. The final decoder output maps are fed to the soft-max file for pixel-wise classification.

The newer approach is inspired from the SegNet [6] model that uses encoder-decoder architecture to carry out segmentation task at hand. One thing that is worth noting is that the ground truth in the images used in SegNet [6] architecture is three dimensional, i.e., the three dimensions correspond to the height, width and the number of classes in the image. It uses a per pixel soft-max activation at the last layer to determine which class the pixel belongs to.

The newer architecture completely eliminates the task of hand labeling the classes in the third dimension; in fact, the third dimension can represent a regular RGB channel. The newer architecture is an auto-encoder-based architecture with the last layer using a sigmoid activation to determine if a pixel belongs to the sheep or not. This enforces a complete end-to-end training of the segmentation model on the task at hand, thus making it more tasks-specific.

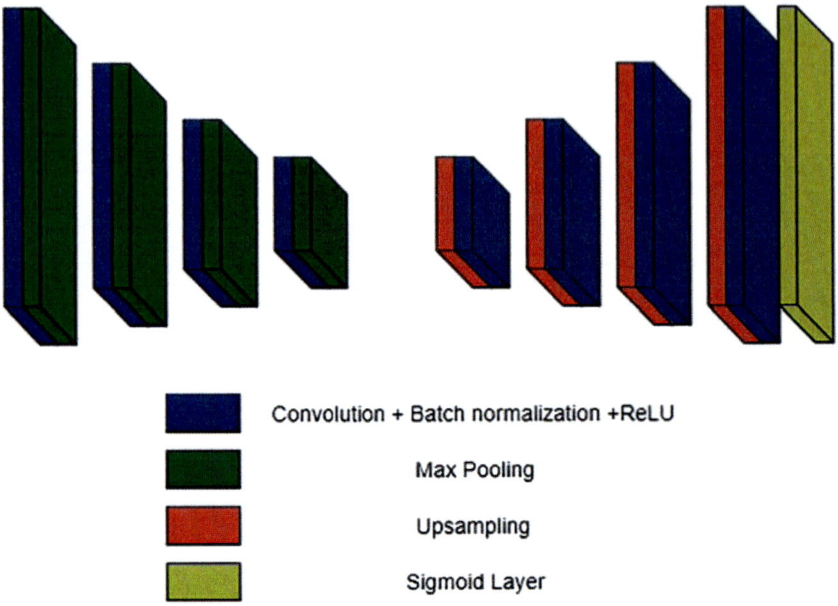

Fig. 1 SegNet architecture

The loss in the final output layer is a binary cross-entropy loss given by:

$$H(p,q) = -\sum_{x} p(x)\log(q(x)) \tag{1}$$

where $p(x)$ is the true underlying Bernoulli distribution and $q(x)$ is the estimated Bernoulli distribution by the learning model.

2.2 Artificial Neural Nets for Regression Task

Artificial neural networks are used for clustering through unsupervised learning or classification through supervised learning or regression. That means you can group unlabeled data, categorize labeled data or else predict the continuous values from the historical data.

The regression is mainly used to predict continuous data when a continuous input is provided to it. In artificial neural network, X input features are passed from the network's previous layer and are fed to the hidden layer. This X features will be multiplied with the corresponding weights and the sum of that is added to the bias.

$$(X * W) + b \tag{2}$$

This sum is then fed to the rectified linear unit, commonly known as ReLU. It is commonly used because it does not saturate on shallow gradients which is the benefit of using ReLU over sigmoid function. For each node, ReLU function outputs an activation 'a' which are then multiplied by the corresponding weights (W_2) between the hidden and the output layer (b_2 as the bias term) to generate the output. The output layer neuron is a linear function with no activation

$$\hat{y} = a * W_2 + b_2 \tag{3}$$

This is the ANN performing regression with a single output unit. This result is $\hat{y}$ also known as network's estimate dependent on the X which was supplied as input. This $\hat{y}$ is compared with the y which is also known as ground truth and adjust the weights and biases until the error between them is minimized. This is a simple working of an artificial neural network.

3 The Proposed Automatic Sheep Weight Estimation Approach

The main purpose of this paper is to have a real-time application that can help any person including buyer or consumer to estimate the sheep weight. And for thus no advanced webcam setting or video recording is required. Using this technique, weight can be estimated by a single photo of the side of the sheep taken from the mobile camera and also the age and gender of that sheep which helps to measure the weight of the sheep. The dataset used for this paper includes the images of 52 sheep from a high-definition mobile camera. The proposed weight estimation system consists of three phases: (1) preprocessing phase, (2) segmentation phase, and (3) weight estimation. The flowchart below shows the basic working of the proposed system which is further explained below in detail along with the involved steps with the characteristics for each phase (Fig. 2).

3.1 Preprocessing Phase

The images are cropped, and extra noise from the images is removed manually for training purpose only to accurately train the model while this preprocessing is not required for application purposes. Image can directly be used as input to predict the weight. Also, the images are annotated for training the encoder for segmentation. The sheep in the images are annotated by the orange color and also the original image is saved so as to train the segmentation model. Also, the age and gender which are available are organized in a text document separated by space. The annotated cropped image and the annotated image are shown in Fig. 3.

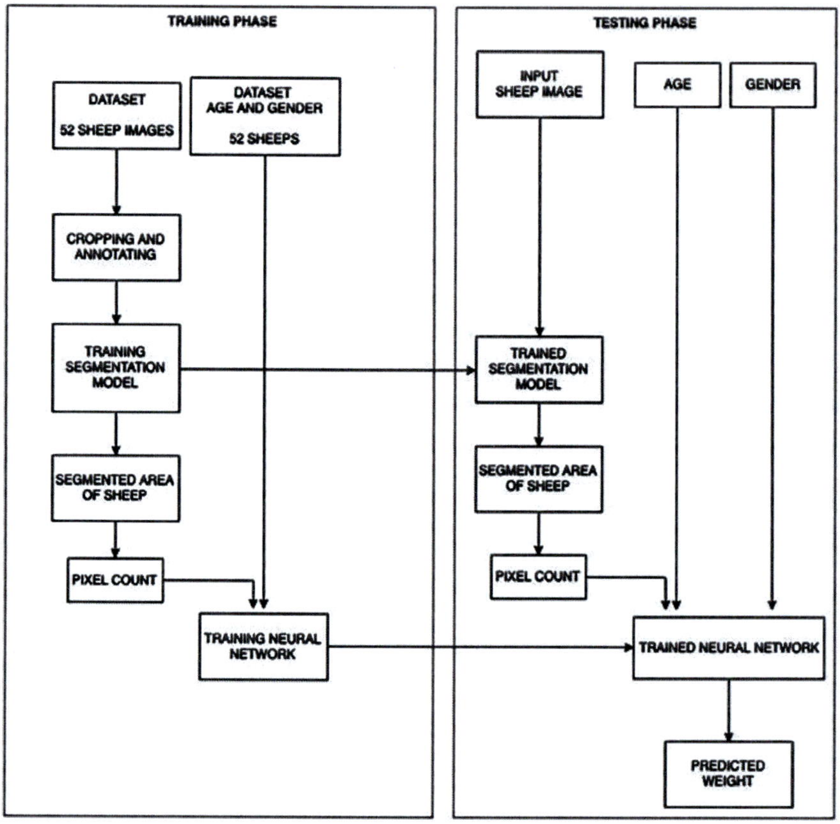

Fig. 2 General architecture of the proposed approach

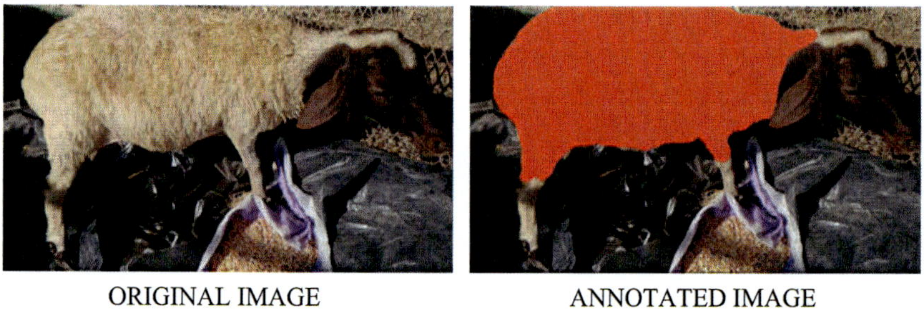

Fig. 3 Annotated images

3.2 Segmentation Phase

Segmentation is an inevitable part of any computer vision algorithm. There are many algorithms for segmenting and also various ways of doing so. Here we have used unsupervised image segmentation using an auto-encoder. The detailed discussion of the technique is already done in Sect. 2.1. Using this technique, we have segmented the images. The segmented images are further used to calculate the area of the sheep's image using pixel count technique. This area along with other features like sheep's age and gender is used to estimate the sheep's weight.

3.3 Sheep Weight Estimation Phase

Various attempts to predict the weight based on the extracted features using multiple linear regression failed to show up even a decent accuracy on the dataset, this suggested that the underlying function that maps the data to the output weights is not a linear one, and hence the use of neural networks is justified. ANN helped us in modeling a pretty decent approximation of the underlying function.

The structure of the neural net involves a three-neuron input layer followed by two hidden layers each with 10 and 5 hidden neurons, respectively. The last layer is a single neuron output layer with linear activation to predict the weights. Both the hidden layers have a ReLU activation unit. The ANN is using weight regularization to check to overfitting of the model, and the optimizer used is Adam [7] optimizer and mean squared error (MSE) as a loss.

The MSE is given as

$$\mathrm{MSE} = \frac{1}{m}\sum_{i=1}^{m}\left(y_{\mathrm{pred}} - y_i\right)^2 \tag{4}$$

where y_i is the observed actual value and y_{pred} is the predicted value

The structure of the ANN is as follows (Fig. 4).

4 Results and Discussion

The results of the segmentation are shown in Fig. 5, where the input is an RGB image and the output is the segmented sheep.

We used the data of 52 sheep which included the pixels extracted from the image, the age, and gender of the sheep. We used cross-validation to split the data into training and testing sets. We split the data into 80–20 ratio, i.e., 80% training data and 20% test data. The accuracy of the result was measured using r2 score; r2 score is measured using the below formula

```
Layer (type)                 Output Shape              Param #
=================================================================
dense_1 (Dense)              (None, 10)                40
_________________________________________________________________
dense_2 (Dense)              (None, 5)                 55
_________________________________________________________________
dense_3 (Dense)              (None, 1)                 6
=================================================================
Total params: 101
Trainable params: 101
Non-trainable params: 0
_________________________________________________________________
```

Fig. 4 Artificial neural network

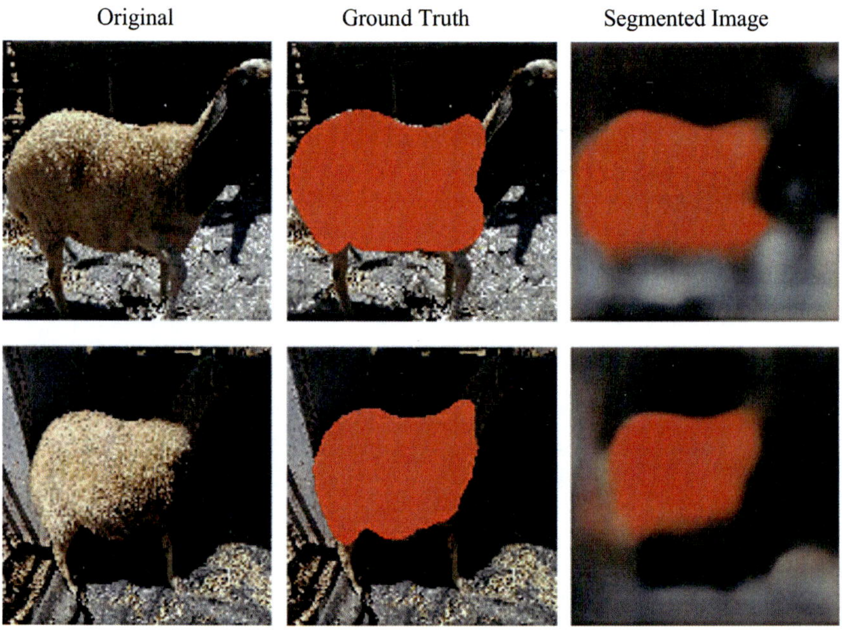

Fig. 5 Results of segmentation

$$R^2 = 1 - \frac{\sum(y_i - f_i)^2}{\sum(y_i - \bar{y})^2} \tag{5}$$

where y_i is the original value and f_i is the predicted value for ith sheep.

Using the above metrics, we have achieved a highest r2 score of 0.81 on the training set and 0.80 on the test set. The results were plotted against the original values on the graph for the training set and test set separately which is shown in Figs. 6 and 7.

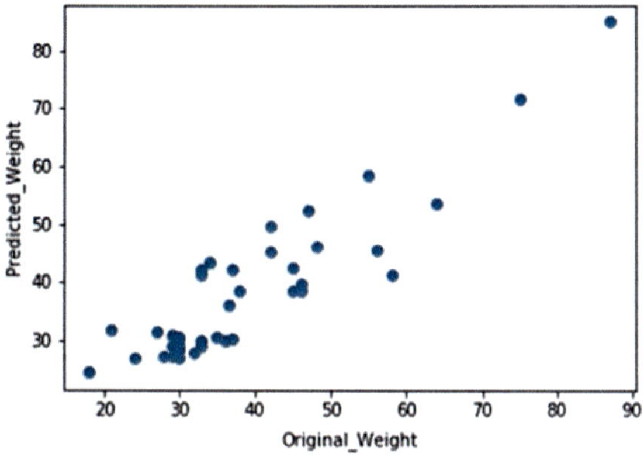

Fig. 6 Result of training data

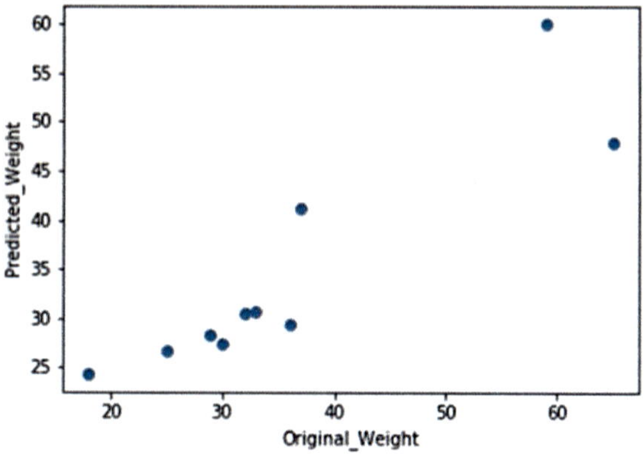

Fig. 7 Result of test data

Figure 6 represents the linear correspondence between the original weight and the predicted weight based on the training dataset. Figure 7 represents the linear correspondence between the original weight and the predicted weight based on the test dataset. These results show that the model has been trained accurately and does not overfit or underfit. Using this statistical analysis, we can assume that this trained model can be easily used with any other sheep's image for estimating its weight as well (Fig. 8).

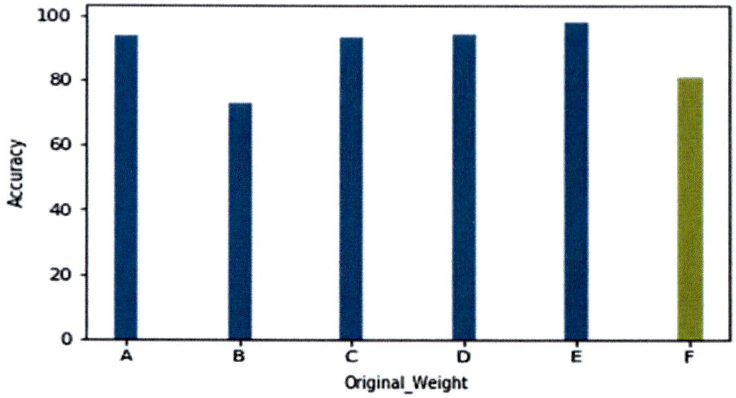

Fig. 8 Comparison of different approaches used

Label	Approach used
A	Ellipse fitting algorithm to calculate the area of sheep in the image and using it to estimate sheep's weight, in [2]
B	Linear regression model based on linear dimensions of sheep from image to estimate sheep's weight, in [3]
C	Binarizing the image in considering lambs are black with white background and calculate pixels to estimate sheep's weight, in [4]
D	Using height at withers (HW), heart girth (HG), body length (BL), head length (HL), head width (HDW), loin girth (LG), length of hindquarter (LHQ) and width of hindquarter (WHQ) as parameters for linear regression, in [8]
E	Phenotypic recording and image acquisition for extracting features and using best linear unbiased prediction (BLUP) method to predict weight, in [9]
F	Segmenting images using proposed segmentation technique to calculate the area of the sheep (pixel count) with gender and age as features to predict weight using a neural network

Note As you can see that each and every approach previously followed for this purpose did require some sort of physical equipment or any specific environment to get sheep's weight from image whereas the approach followed by us does not require any physical environment set-up or else any special equipment to capture image and predict the result. So the approach followed by us is mobile and can be independently used with a normal image taken from any camera source

5 Conclusion

Using this technique, we are able to estimate the sheep's weight with a simple image from a mobile device. This technique is different than the previous approaches as the previous approaches were either dependent on advanced system set-up or high-tech cameras. Also, the previous approaches did not take into consideration the age and gender of the sheep which our system takes care of and enables us to get better accuracy than the other similar approaches.

References

1. Menesatti P, Costa C, Antonucci F et al (2014) A low-cost stereovision system to estimate size and weight of live sheep. Comput Electron Agric 103:33–38. https://doi.org/10.1016/j.compag.2014.01.018
2. Kashiha M, Bahr C, Ott S et al (2014) Automatic weight estimation of individual pigs using image analysis. Comput Electron Agric 107:38–44. https://doi.org/10.1016/j.compag.2014.06.003
3. Pradana ZH, Hidayat B, Darana S (2016) Beef cattle weight determine by using digital image processing. In: 2016 International conference on control, electronics, renewable energy and communications (ICCEREC). https://doi.org/10.1109/iccerec.2016.7814955
4. Khojastehkey M, Aslaminejad AA, Shariati MM, Dianat R (2015) Body size estimation of new born lambs using image processing and its effect on the genetic gain of a simulated population. J Appl Anim Res 44:326–330. https://doi.org/10.1080/09712119.2015.1031789
5. Long J, Shelhamer E, Darrell T (2015) Fully convolutional networks for semantic segmentation. In: 2015 IEEE conference on computer vision and pattern recognition (CVPR). https://doi.org/10.1109/cvpr.2015.7298965
6. Badrinarayanan V, Kendall A, Cipolla R (2017) SegNet: a deep convolutional encoder-decoder architecture for image segmentation. IEEE Trans Pattern Anal Mach Intell 39:2481–2495. https://doi.org/10.1109/tpami.2016.2644615
7. Kingma DP, Ba JL (2017) Adam: a method for stochastic optimization
8. Sowande OS, Sobola OS (2007) Body measurements of west African dwarf sheep as parameters for estimation of live weight. Trop Anim Health Prod 40:433–439. https://doi.org/10.1007/s11250-007-9116-z
9. Girshick R, Donahue J, Darrell T, Malik J (2014) Rich feature hierarchies for accurate object detection and semantic segmentation. In: 2014 IEEE conference on computer vision and pattern recognition. https://doi.org/10.1109/cvpr.2014.81

Supervised Machine Learning Model for Accent Recognition in English Speech Using Sequential MFCC Features

Dweepa Honnavalli and S. S. Shylaja

Abstract Human–machine interfaces are rapidly evolving. They are moving from the traditional methods of input like keyboard and mouse to modern methods like gestures and voice. It is imperative to improve voice recognition and response since there is a growing market of technologies, world-wide, that use this interface. Majority of English speakers around the world have accents which are not exposed to speech recognition systems on a greater scale. In order to bridge the comprehension gap between these systems and the users, the systems need to be tuned according to the accent of the user. Accent classification is an important feature that can be used to increase the accuracy of comprehension of speech recognition systems. This paper recognizes Indian and American English speakers and distinguishes them based on their accents by constructing sequential MFCC features from the frames of the audio sample, oversampling the under-represented data and employing supervised learning techniques. The accuracies of these techniques reach a maximum of 95% with an average of 76%. Neural networks emerge as the top classifier and perform the best in terms of evaluation metrics. The results gleaned indicate that concatenating MFCC features sequentially and applying an apposite supervised learning technique on the data provide a good solution to the problem of detecting and classifying accents.

Keywords Accents · Speech · MFCC · Supervised machine learning

1 Introduction

Voice-interfaced gadgets are the new frontier of virtual assistants. Voice-controlled digital assistants like Apple's Siri, Google Assistant and Amazon's Alexa have been integrated into smartphones, smart speakers and computers. Surveys have indicated

D. Honnavalli (✉) · S. S. Shylaja
Department of Computer Science, PES University, Bangalore, India
e-mail: dweepa.prasad@gmail.com

S. S. Shylaja
e-mail: shylaja.sharath@pes.edu

N. N. Chiplunkar and T. Fukao (eds.), *Advances in Artificial Intelligence and Data Engineering*, Advances in Intelligent Systems and Computing 1133,
https://doi.org/10.1007/978-981-15-3514-7_5

that almost 50% of smart gadget owners in the USA use voice-enabled virtual assistants [1]. They save hours every day by helping users multitask and managing their activities and are the latest assistive technology for the visually impaired [2]. Virtual assistants are integrated into almost every smartphone and standalone smart speaker.

People all over the world have access to this technology but this technology is not helping everyone the way it should. 1.5 billion people around the world speak English either as a native language or a foreign language. Seventy-six per cent of this English-speaking population have accents that virtual assistants do not comprehend. According to a recent survey by Accenture, 1 in 3 online consumers in China, India, the USA, Brazil and Mexico will own a standalone digital voice assistant by the end of 2018 [3]. This means 950 million consumers based in India, Mexico, Brazil and China will use what is tuned to the accents of 108 million US consumers.

Automatic speech recognition systems (ASR) seem to work better with American speech rather than accented speech mainly because of the lack of inclusive training data [4]. The problem lies in the fact that there are so many different languages with many dialects and accents. Consider the English language, there are close to 100 accents and dialects and it is not feasible to collect enough annotated data for all accents and dialects. This limits the ASR's capability. With the explosion of data, ASRs have become more inclusive but there is always scope for improvement.

With the rise in the use of machine learning to solve real-world problems, detecting accents find application in many different domains. One of its use cases is in crime investigation. Evidence usually involves usable audio clips of 3–4 s. It is imperative to be able to detect the accent in these short, distorted clips to be able to recognize the speaker as well as the speech. If accents can be classified accurately with such short audio clips, investigators can gain insight into the identity of the criminal by identifying the criminal's ethnicity.

Classifying accents is a step towards more intelligent virtual assistants and expanded user-base and usability. This paper seeks to classify accents by concatenating sequential MFCC features which may aid in the ASR's comprehension of accented speech.

2 Background and Related Work

Automatic speech recognition systems (ASR) are ubiquitous and there is a pressing need for them to cover accented English as they form a large part of the population. To help with the ASR revolution, accents need to be understood and incorporated. There is a lot of active research on accented English in speech recognition and efforts to improve automatic speech recognition systems (ASRs) and make them more inclusive.

Research in this area has taken many paths depending on various points of impact such as the audio clip size, the type of accents the speakers have and how adept they are in the English language. Albert Chu et al. provided a comparative analysis of machine learning techniques in accent classification using self-produced 20-second audio clips and extracting various features to determine their influence. Their paper provides insights on the use of SVM and varying the number of features for feature description using PCA [5].

Each accent has a unique intonation that arises from the root language it is based on. ASRs can be tuned to comprehend accented speech better if they know what to look out for. Liu Wai Kat et al. presented a fast accent classification approach using phenome class models. The paper found that detecting accents and transforming the native accent pronunciation dictionary to that of the accented speech reduce the error rate of ASRs by 13.5% [6]. Accent classification can be used to switch to a more tuned ASR for better comprehension.

Information from audio signals can be extracted in many ways—by sampling, windowing, expressing the signal in the frequency domain or extracting perceptual features. The most common features extracted are linear predictive codes, perceptual linear predictions (PLP) and Mel-frequency cepstral coefficients (MFCC) [7]. As human voice is nonlinear in nature, linear predictive codes do not work as well as PLP and MFCC. PLP and MFCC are derived on the concept of logarithmically spaced filter banks, clubbed with the concept of the human auditory system and hence had a better response than LPC. Studies show that both (PLP and MFCC) parameterization techniques had provided almost comparable results. MFCC is chosen as the means of extracting information from the audio samples in this paper.

Hong Tang and Ali A. Ghorbani's paper on accent classification explores classifying accents based on features like word-final stop closure duration, word duration, intonation and F2-F3 contour. They used pairwise SVM, DAGSVM and HMM to obtain good accuracy results [8].

This paper takes inspiration from information garnered and proposes and compares methods to classify audio clips of 3–5 s containing accented English speech using concatenated MFCC perceptual features.

3 Methodology

3.1 Proposed Method

The proposed method depicted in Fig. 1 involves analysis of audio signals and extracting required features, following which the features are considered and respective frames are concatenated to form the input feature set to the classifier. The results of the classifiers are then validated and compared using different accuracy metrics.

Fig. 1 Block diagram of proposed method

3.2 Dataset

The data set is a collection of .wav files (3–5 s long) from VCTK-corpus. Speakers with American accents and Indian accents are recorded uttering the same content. There are 2301 audio samples in total which are divided into training (80%) and testing (20%) (Table 1).

3.3 Preprocessing

Table 2 indicates that if the current data is split into training and testing data sets, 'Indian Accent' would be underrepresented. To work around this problem, the data set is split into testing and training and the training set is oversampled on 'Indian Accent'. After choosing the oversampling algorithm [9], the ratio between the two outcomes in the training data is 1:1.

Table 1 Distribution of the dataset—I

Type of accent	Gender	Number of speakers	Total number of audio samples
Indian	Female	1	376
Indian	Male	2	656
American	Female	2	846
American	Male	1	423

Table 2 Distribution of the data set—II

Type of accent	Total number of audio samples
Indian	1032
American	1269

3.4 Feature Extraction

For each audio file in the data set, 20 Mel-frequency cepstral coefficients (MFCC) are calculated [10]. Mel-frequency cepstrum is the short-term power spectrum of a sound. Feature extraction is implemented using the Python library 'Librosa' [11].

The steps involved in calculating the MFCC cepstral features are:

1. Framing each signal into short frames of equal length with frame length as 2048 samples and hop length as 512 samples.
2. Calculating the periodogram estimate of the power spectrum for each frame.
3. Applying the Mel filterbank to the power spectra and summing the energy in each filter. The formula for converting from frequency to Mel scale is:

$$M(f) = 1125 \ln (1 + (f/700)) \tag{1}$$

4. Taking the logarithm of all filterbank energies.
5. Taking the discrete cosine transform of the log filterbank energies.
6. Selecting the required MFCC coefficients. In this case, 20 coefficients are selected.

These coefficients are used because they approximate the human auditory system's response closely. The coefficients of each frame of an audio file are concatenated to form an array of MFCCs.

The MFCC features extracted from an audio sample is outputted in the form of a matrix with 20 coefficients for each frame of the sample, i.e.

$$\text{MFCC} = \begin{bmatrix} c_0 f_0 & c_0 f_1 & c_0 f_2 & \cdots & c_0 f_m \\ \cdots & \cdots & \cdots & \cdots & \cdots \\ \cdots & \cdots & \cdots & \cdots & \cdots \\ c_{19} f_0 & c_{19} f_1 & c_{19} f_2 & \cdots & c_{19} f_m \end{bmatrix} \tag{2}$$

where $c_i f_0 \ldots c_i f_m$ are the values of coefficient i for frames $1, \ldots, m$.

The above matrix represents the MFCC coefficients for an audio sample with m frames. This $20 \times m$ matrix needs to be transformed into a format that is recognized by the machine learning model.

To be able to contrast the two accents based on their feature sets, MFCC coefficients are sequentially concatenated. This retains enough information required to identify the accent from the feature set.

The features are concatenated and flattened into a one-dimensional array of features for each frame. A sample vector of sequential MFCC features from a single audio sample is shown below.

$$M = \left[c_0 f_0 \ldots c_0 f_m \ldots\ldots\ldots c_{19} f_0 \ldots c_{19} f_m \right] \tag{3}$$

These features are used to train the model to distinguish a particular accent from others. The plot (Fig. 2) of MFCCs extracted from a speaker with American accent

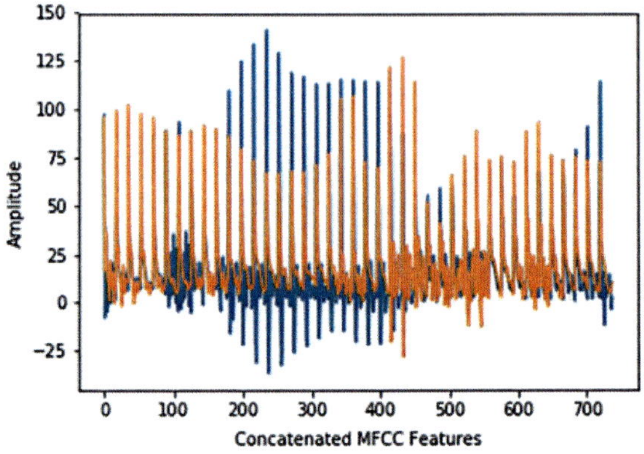

Fig. 2 Distinguishing plot of American and Indian accents

versus a speaker with Indian accent, both uttering the same sentence 'Please call Stella', shows there is a variation in speech. This variation helps the model to predict the class of accents.

3.5 *Supervised Learning*

To differentiate between American and Indian accent, the features are extracted and fed the feature data into a machine learning model to get the results.

K-Nearest Neighbour

K-nearest neighbours algorithm as a classifier calculates five nearest neighbours of each data point in the test and classifies it as either American or Indian based on which type of accent it is situated closest to. It uses Euclidean distance as the distance metric to calculate the nearest neighbours. It classifies based on the majority vote. Python's Sklearn package has a KNN classifier which does the above taking into consideration five of the nearest neighbours.

Support Vector Machine

Support vector machines are classification algorithms which find the decision boundaries between the classes. Since the data at hand is of higher dimension, we use a radial basis function kernel to achieve the desired functionality.

Equation for rbf (radial bias function) kernel:

$$K(X, Y) = \exp\left(-\gamma \frac{||XY||^2}{2\sigma^2}\right), \quad \gamma > 0 \tag{4}$$

Gaussian Mixture Model

Gaussian mixture model is a K-means clustering algorithm that groups the data points into two clusters on for each class: American and Indian. Sklearn's Gaussian mixture model is used to implement this.

Neural Networks
Multi-layer perceptron for binary classification is used with the number of features as the input layer of the neural network and one neuron as the output which states if the accent is Indian or American. The optimizer 'Adam' is used to update the weights after each iteration. The neural network was trained over 30 epochs. MLP is implemented using the Python library 'Keras' with Google's TensorFlow as the backend.
Logistic Regression
Logistic regression is suitable for analysis when the dependent variable is dichotomous. Math behind logistic regression:

$$l = \beta_0 + \beta_1 * x_1 \tag{5}$$

where l is log-odds, logistic regression was opted for the comparative analysis study because of various of reasons. The output is a probability that the given input point belongs to a certain class.

4 Results and Discussion

The results are evaluated based on various metrics like precision, recall, reject rate, accuracy score, area under the Receiver Operating Characteristic (ROC) curve (AUC) and K-fold validation results. From the test case and K-fold cross-validation results, we observe that neural networks, k-nearest neighbour and logistic regression perform the best in terms of overall accuracy, AUC and k-fold validation (Figs. 3 and 4).

4.1 Validation Accuracy

K-fold cross-validation is an evaluation metric in which the sample is partitioned into 'k' partitions. These partitions are of equal sizes and are partitioned at random. Among these k partitions, one of them is the validation or testing set and the rest

Model	Mean Validation Score	Std Dev of Validation Score
Neural Networks	0.95	0.02
KNN	0.91	0.06
Logistic Regression	0.95	0.03
SVM	0.36	0.01
GMM	0.39	0.10

Fig. 3 Validation results

Model	Precision	Recall	f-measure	Reject Rate	Accuracy
Neural Networks	0.96	0.94	0.95	0.97	0.95
KNN	0.9	0.92	0.91	0.9	0.91
Logistic Regression	0.94	0.96	0.95	0.95	0.95
SVM	1.0	0.02	0.04	1.0	0.54
GMM	0.43	1.0	0.60	0.0	0.43

Fig. 4 Test case results

serve as training data. The cross-validation process is repeated k times, such that all of the partitions are used as training data once. In general, 'k' remains an unfixed parameter, in our experimentation, k is taken to be as 10. The validation accuracies are tabulated in Fig. 2.

4.2 Test Case Accuracy Metrics

Metrics like precision, recall, reject rate and overall accuracy provide a holistic view on the performance of a model. The results presented apply to the data which is split into training and testing where the former constitutes 80% of the data and the latter 20%.

4.3 Analysis of Neural Networks, Logistic Regression and K-Nearest Neighbour

Neural Networks

Neural networks emerge as the top classifier for the job in terms of all the evaluation metrics. It has a high area under the curve (Fig. 5b) value which indicates that it

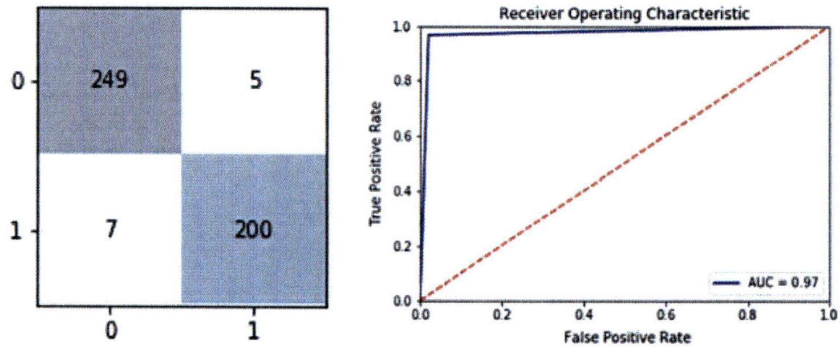

(a) Confusion Matrix (b) ROC Curve

Fig. 5 Neural networks

Table 3 Time taken for computation

Model	Time (s)
Neural networks	18.33
KNN	5.22
Logistic regression	0.58

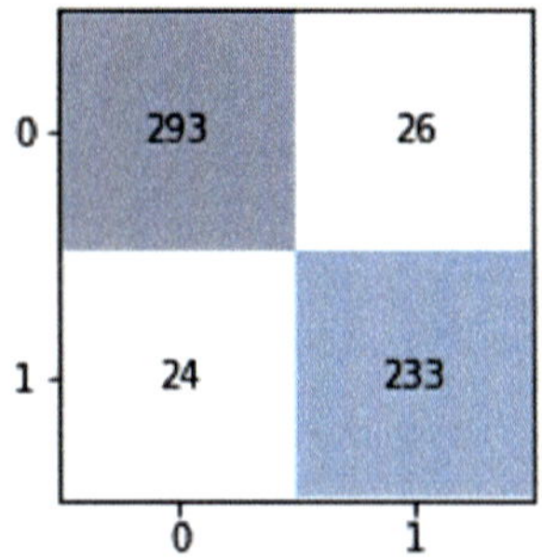

(a) Confusion Matrix

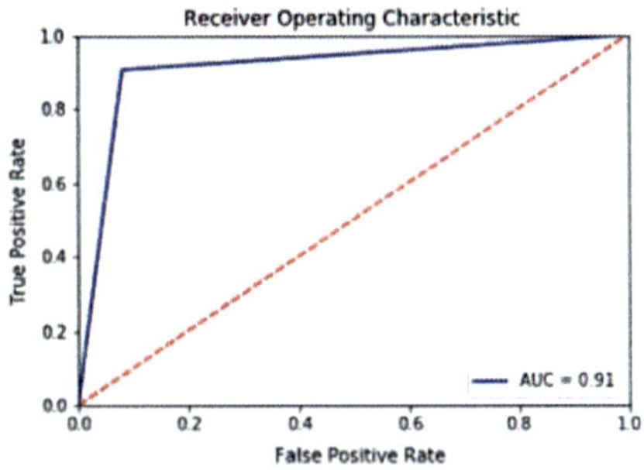

(b) ROC Curve

Fig. 6 K-Nearest Neighbour

is highly competent in separating the two classes. Confusion matrix describes high precision, recall and reject rate values which indicate the classifier is doing what is required by the problem. However, the only shortcoming is in terms of the time taken for computation which is very high and may prove undesirable for larger inputs, which is tabulated in Table 3.

K-Nearest Neighbour

The k-nearest neighbour classifier does not perform as well as neural networks or logistic regression classifiers, but it performs well, nonetheless. The results of neural network being slightly better than that of KNN can be attributed to the fact that as the number of features in the input data set increases KNN tends to perform worse. It is not far behind neural networks and logistic regression, and it does not take as much time to run as neural networks.

Logistic Regression

Logistic regression and neural network classifiers are very close to each other with respect to their evaluation metrics values. Logistic regression, like neural networks, perform very well in separating the two classes and predicting the classes with good accuracy. It has an edge over neural networks in the fact that it takes considerably less time to run on big inputs (Figs. 6 and 7).

4.4 Analysis of SVM and GMM

Support Vector Machine

SVM with RBF kernel, in this case, does a poor job of separating the two classes of accents. This can be concluded by looking at the AUC (Fig. 8b) which is 0.52

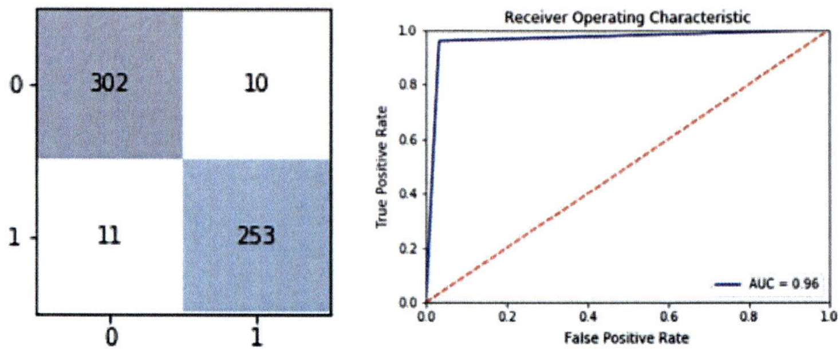

(a) Confusion Matrix (b) ROC Curve

Fig. 7 Logistic regression

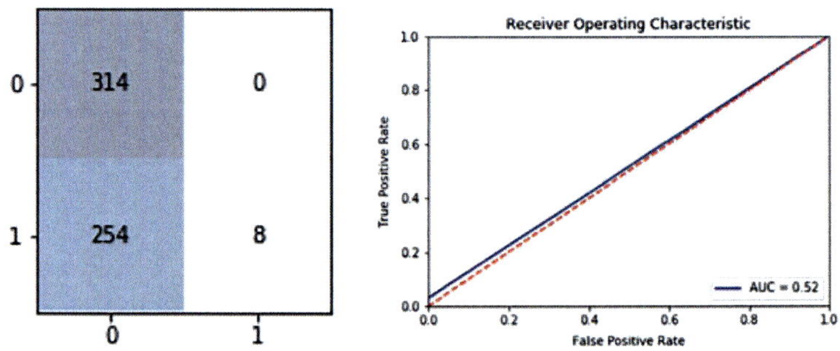

(a) Confusion Matrix (b) ROC Curve

Fig. 8 Support vector machine with rbf kernel

which is close to a classifier that predicts classes randomly. However, misleading conclusions may be made that SVM in indeed a good classifier due to its unusually high precision. On further observation, it is understood that it classifies the majority of the data samples as 'American'. This contributes to the unusually high precision, but starkly low recall and overall accuracy close to 50% which almost relates to a random classifier.

Gaussian Mixture Models

GMM, on the other hand, displays similar characteristics as the SVM model differing in the fact that it classifies the majority of the test case samples as 'Indian' which contributes to the high recall but low precision. The overall accuracy, AUC and k-fold cross-validation results (Fig. 9) lead to inferring that GMM does not work well for the problem at hand.

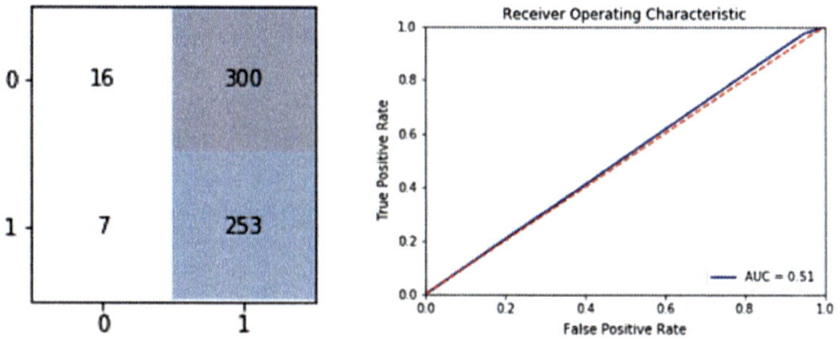

(a) Confusion Matrix (b) ROC Curve

Fig. 9 Gaussian mixture models

5 Conclusion and Future Work

Classifying accents based on their acoustic features using machine learning techniques provide good results. By using sequential MFCC features to construct the input, important information is retained that distinguishes the accents in a mathematical perspective. The models are calibrated with this input and validated using various metrics to arrive at a suitable solution to the problem at hand.

Accent classification is a preprocessing step to speech recognition. It can only help in fine-tuning speech recognition systems to detect accented speech better. Future work includes incorporating this into speech recognition systems to improve comprehension of accented English. This project could be extended to various accents and dialects, for example, regional Indian accents like Kannada, Malayali, Tamil and Bengali; British regional accents like Scottish, Welsh, etc. Classification using HMM has not been explored in this paper.

References

1. Center PR (2017) Voice assistants used by 46% of Americans, mostly on smartphones. https://pewrsr.ch/2l4wQnr
2. CBC (2018) Smart speakers make life easier for blind users, January. https://www.cbc.ca/radio/spark/380-phantom-traffic-jams-catfishing-scams-and-smart-speakers-1.4482967/smart-speakers-make-life-easier-for-blind-users-1.4482978
3. Accenture (2018) Accenture digital consumer survey. https://www.accenture.com/t20180105T221916Z__w__/us-en/_acnmedia/PDF-69/Accenture-2018-Digital-Consumer-Survey-Findings-Infographic.pdf
4. Ellis P (2017) Why virtual assistants can't understand accents, August. https://www.huffingtonpost.co.uk/philip-ellis/is-siri-racist-why-virtua_b_11423538.html?guccounter=2 (online)
5. Chu A, Lai P, Le D (2017) Accent classification of non-native english speakers. Available from World Wide Web: http://web.stanford.edu/class/cs224s/reports/Albert_Chu.pdf (online), cited 22 Nov 2018

6. Kat LW, Fung P (1999) Fast accent identification and accented speech recognition. In: 1999 IEEE international conference on acoustics, speech, and signal processing. Proceedings. ICASSP99 (Cat. No. 99CH36258), March, vol 1, pp 221–224. https://doi.org/10.1109/ICASSP.1999.758102
7. Dave N (2013) Feature extraction methods LPC, PLP and MFCC in speech recognition. Int J Adv Res Eng Technol 1. ISSN 2320-6802
8. Tang H, Ghorbani AA (2003) Accent classification using support vector machine and hidden Markov model. In: Canadian conference on AI
9. Lemaître G, Nogueira F, Aridas CK (2017) Imbalanced-learn: a python toolbox to tackle the curse of imbalanced datasets in machine learning. J Mach Learn Res 18(17):1–5. http://jmlr.org/papers/v18/16-365.html
10. Mel frequency cepstral coefficients. https://bit.ly/1mDpzDu
11. Librosa (2018) Audio and music processing in python, August. https://zenodo.org/record/1342708#.XDG9LS2B01I

A Two-Level Approach to Color Space-Based Image Segmentation Using Genetic Algorithm and Feed-Forward Neural Network

B. S. Sathish, P. Ganesan, L. M. I. Leo Joseph, K. Palani, and R. Murugesan

Abstract This work proposed an approach for the segmentation of high-resolution images such as satellite imagery stand on the supportive method of GA and FFNN. During this two-layer technique, the GA applies for the selection of the best individual (pixels in image). Based on the outcome generated by this process, feed-forward neural network is trained to carry out the detection and segmentation. Neural network is trained using an approach called Levenberg–Marquardt algorithm. To improve the quality of the segmentation process, the original test image is transformed into various color spaces and then segmentation is applied. In this work, bivariate image value actions are utilized to validate the excellence of the output (segmented) image based on the assessment of subsequent image pixels between input and output (segmented)

B. S. Sathish
Department of Electronics and Communication Engineering, Ramachandra College of Engineering, Eluru, Andhra Pradesh, India
e-mail: subramanyamsathish@yahoo.co.in

P. Ganesan (✉)
Department of Electronics and Communication Engineering, Vidya Jyothi Institute of Technology, Aziz Nagar, C.B. Road, Hyderabad, India
e-mail: gganeshnathan@gmail.com

L. M. I. Leo Joseph
Department of Electronics and Communication Engineering, S.R. Engineering College, Warangal, Telangana, India
e-mail: leojoseph@srecwarangal.ac.in

K. Palani
Department of Computer Science Engineering, Shadan College of Engineering and Technology, Hyderabad, India
e-mail: akpgda@gmail.com

R. Murugesan
Department of Electronics and Communication Engineering, Malla Reddy College of Engineering and Technology, Hyderabad, Secunderabad, India
e-mail: rmurugesan61@gmail.com

N. N. Chiplunkar and T. Fukao (eds.), *Advances in Artificial Intelligence and Data Engineering*, Advances in Intelligent Systems and Computing 1133,
https://doi.org/10.1007/978-981-15-3514-7_6

images. The investigational outcome illustrates the effectiveness of the two-layer process of GA and FFNN in support of the segmentation of high-resolution images. The experimental analysis based on the image quality measures exposed the crucial role of color space for the proposed work.

Keywords Segmentation · GA · FFNN · Color space · Image quality measure

1 Introduction

Segmentation is the procedure of detachment of an picture into a number of sub-imagery, sections, clusters with an akin aspect like color (for color images), texture (for rough surface images), or intensity (for grayscale images) [1]. This is a preliminary but a vital pace inside the picture analysis in the direction of congregate indispensable in sequence [2, 3]. The success of the image analysis mainly depends on this middle-level process [4]. In this work, for the segmentation of high-resolution images GA and FFNN are applied. Even though GA is the optimal choice intended for the optimization troubles, it obtains extra time for congregation [5]. At the same time, it is extremely to pick up the preeminent individual for a neural network (NN) [6, 7]. In our projected technique, GA is applied for the assortment of the best individual (pixels in image) and FFNN for the segmentation [8]. To get better the quality of the segmentation process, the original test image is transformed into various color spaces and then segmentation is applied. Color model is a numerical system to represent and explain the color system as three special color components [9]. In this work, to get better the quality of the segmentation process, the original test image is transformed into various color spaces before segmentation process. In this work, bivariate image value assessment is utilized to validate the excellence of the result (segmented) image based on the evaluation of consequent image pixels between input and output (segmented) images [10].

2 Color Space for Image Segmentation

Color model is a numerical system to represent and explain the color system as three special color components [11]. The conversion of RGB color model into YCbCr color model is mathematically shown in (1).

$$\begin{bmatrix} \mathrm{Y} \\ \mathrm{Cb} \\ \mathrm{Cr} \end{bmatrix} = \begin{bmatrix} 16 \\ 128 \\ 128 \end{bmatrix} + \begin{bmatrix} 0.257 & 0.504 & 0.098 \\ -0.148 & -0.291 & 0.439 \\ 0.439 & -0.368 & -0.071 \end{bmatrix} * \begin{bmatrix} R \\ G \\ B \end{bmatrix} \tag{1}$$

The hue, saturation, and intensity in HSI color space are defined [12] as:

$$H = \arctan\left(\frac{\sqrt{3}(G - B)}{(R - G) + (R - B)}\right) \tag{2}$$

$$I = \frac{(R + G + B)}{3} \tag{3}$$

$$S = 1 - \frac{\min(R, G, B)}{I} \tag{4}$$

The transformation of RGB image into HSV color space [13, 14] is as follows

$$H = \arccos\left(\frac{\frac{1}{2}(2R - G - B)}{\sqrt{(R - G)^2 - (R - B)(G - B)}}\right) \tag{5}$$

$$S = \frac{\max(R, G, B) - \min(R, G, B)}{\max(R, G, B)} \tag{6}$$

$$V = \max(R, G, B) \tag{7}$$

The conversion of an image from RGB to HSL color space is as follows

$$H = \arccos \frac{\frac{1}{2}(2R - G - B)}{\sqrt{(R - G)^2 - (R - B)(G - B)}} \tag{8}$$

$$L = \frac{\max(R, G, B) + \min(R, G, B)}{2} \tag{9}$$

$$S = \left\{ \frac{\max(R, G, B) - \min(R, G, B)}{\max(R, G, B) + \min(R, G, B)} \quad \text{for L} < 0.5 \right. \tag{10}$$

$$S = \left\{ \frac{\max(R, G, B) - \min(R, G, B)}{2 - \max(R, G, B) - \min(R, G, B)} \quad \text{for L} \geq 0.5 \right. \tag{11}$$

The transformation from RGB to $I_1I_2I_3$ color space is described as follows:

$$I_1 = \frac{1}{3}(R + G + B) \tag{12}$$

$$I_2 = \frac{1}{2}(R - B) \tag{13}$$

$$I_3 = \frac{1}{4}(2G - R - B) \tag{14}$$

RGB to CIEXYZ model is specified in (15) to (17).

$$X = 0.4124 * R + 0.3537 * G + 0.1804 * B \tag{15}$$

$$Y = 0.2126 * \mathrm{R} + 0.7151 * \mathrm{G} + 0.0722 * \mathrm{B} \tag{16}$$

$$Z = 0.0193 * \mathrm{R} + 0.1191 * \mathrm{G} + 0.9502 * \mathrm{B} \tag{17}$$

The main predicament of *XYZ* model is that the individual color colorimetric distances do not correlate with the perceived color differences. The conversion of *XYZ* to perceptual CIELab is as follows

$$L = 116 f\left(\frac{Y}{Y_n}\right) - 16 \tag{18}$$

$$a = 500\left[f\left(\frac{X}{X_n}\right) - f\left(\frac{Y}{Y_n}\right)\right] \tag{19}$$

$$b = 200\left[f\left(\frac{Y}{Y_n}\right) - f\left(\frac{Z}{Z_n}\right)\right] \tag{20}$$

X_n, $Y_{n,}$ and Z_n are the standardize tristimulus assessment of the reference. The conversion from CIELab to CIELch color space is defined as

$$L = L \tag{21}$$

$$C = \sqrt{a^2 + b^2} \tag{22}$$

$$H = 0 \text{ whether } a = 0$$

$$H = \frac{180}{\pi}\left\{\pi + \arctan\left(\frac{b}{a}\right)\right\} \tag{23}$$

3 Proposed Method for Image Segmentation

The genetic algorithm (GA) is a global search method based on the selection of the best individuals in a search space [15]. GA can be utilized for solving the complex optimization problems [16, 17]. The functional diagram of image segmentation system is shown in Fig. 1. Flow diagram of the two-layer approach of GA and FFNN for the segmentation of images is shown in Fig. 2.

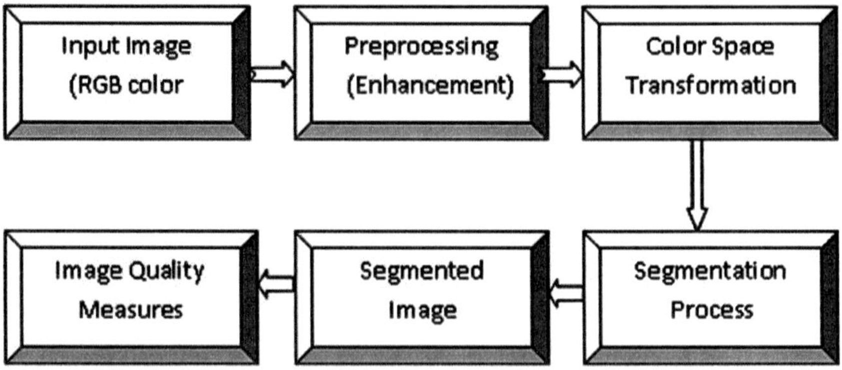

Fig. 1 Functional block diagram of image segmentation system

4 Experimental Results and Concert Analysis

Performance analysis of projected approach is explained as follows. Test illustration in various shade spaces exists and shown in Fig. 3. GA stage of this two-layer approach for the selection of the best individual (pixels in image) is explained in Fig. 4.

The segmentation using neural network stage is shown in Fig. 5. The feed-forward neural network generates the segmented output image. The performance of this two-layer feed-forward neural network is measured.

Figure 6 shows GA and FFNN support segmentation results for given input picture inside different shade spaces. Chart 1 illustrates input in CIELch color model and has best value of the execution time. This is followed by CIELab, HSI, HSV, and I_1, I_2, I_3 color space (Table 1).

In this work, bivariate image value estimation is utilized to validate the characteristic of output (segmented) image that supports on the assessment of related image pixels between input and output images. Table 2 illustrates the list of bivariate image quality measures utilized to validate the result.

Table 3 illustrates the analysis of color space based segmentation using the proposed two-layer approach based on image quality measures listed in chart 2.

5 Conclusion

The segmentation of high-resolution images in additive, user-oriented, and perceptually color spaces based on the proposed two-layer approach is explained. Experimental results demonstrated that input image in CIELch color space has the best value of execution time. This is followed by CIELab, HSI, HSV, and I_1, I_2, I_3 color space. The

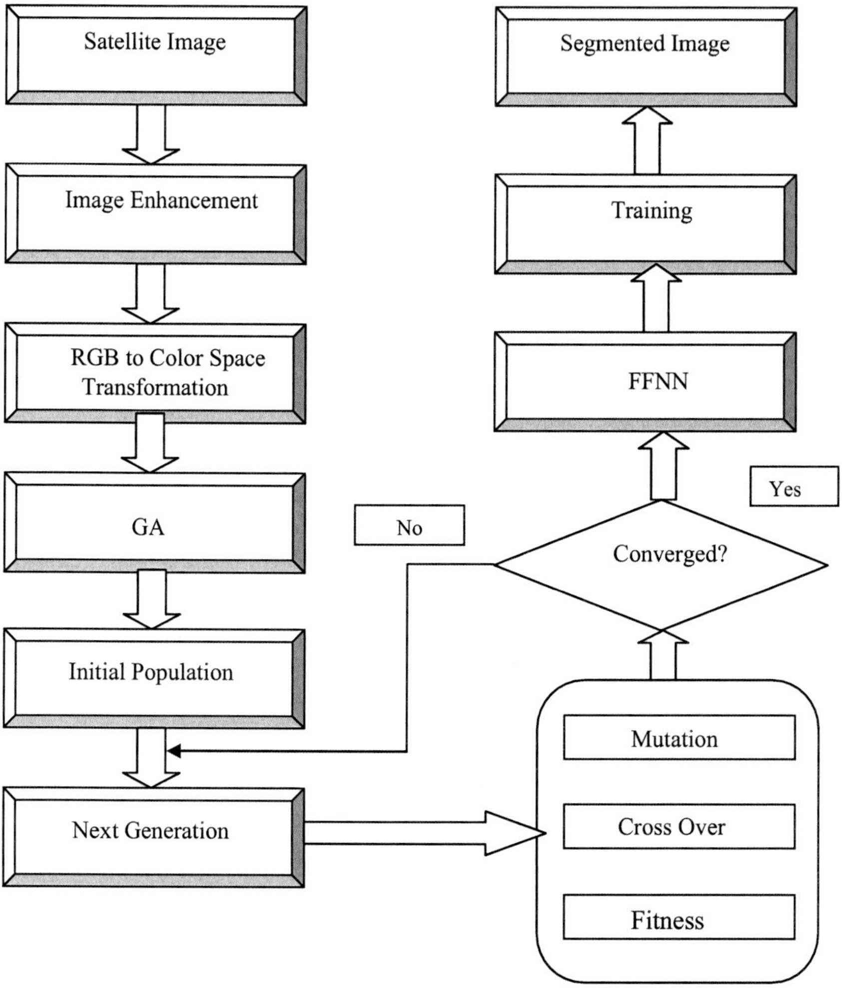

Fig. 2 Flow diagram of the two-layer approach of GA and FFNN for the segmentation of images

outcome of the proposed method is validated using image quality measures by comparing the related pixels of both output and input images. The experimental analysis based on the image quality measures exposed the crucial role of color space for the proposed work. The investigational upshot demonstrated the success of the planned two-layer advance for the segment of high-resolution images.

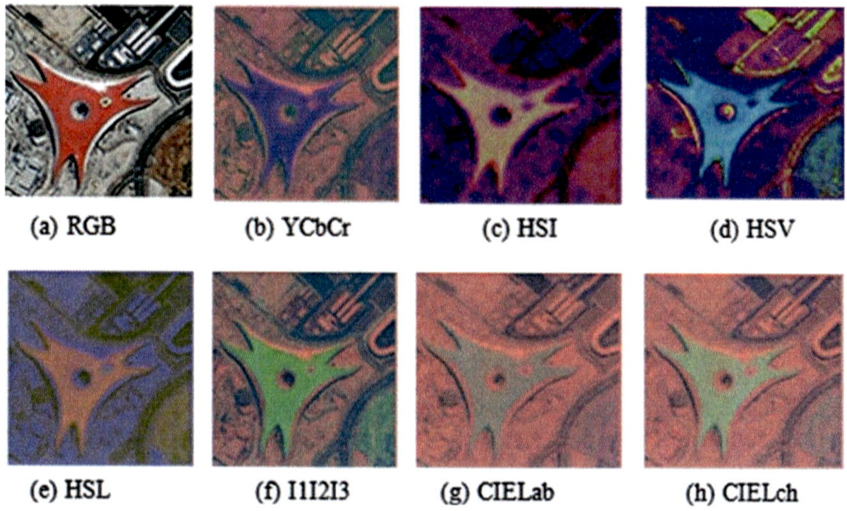

Fig. 3 Input image in different color spaces

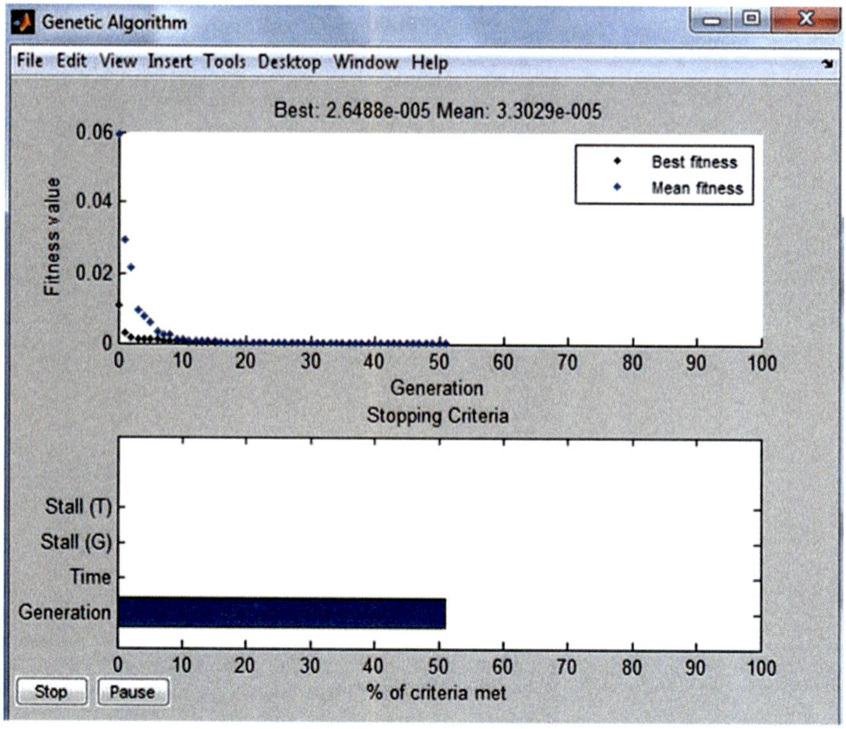

Fig. 4 Genetic algorithm stage of the cooperative approach

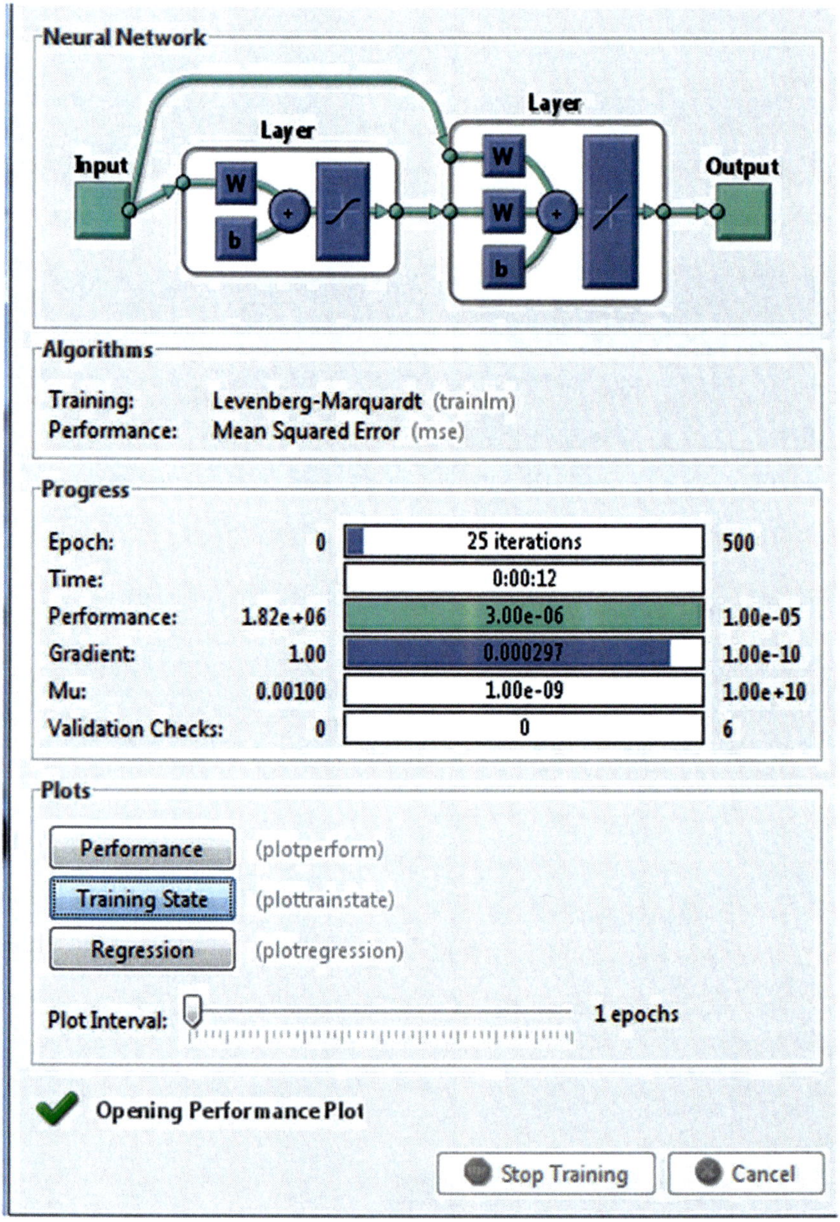

Fig. 5 Neural network stage of the cooperative approach

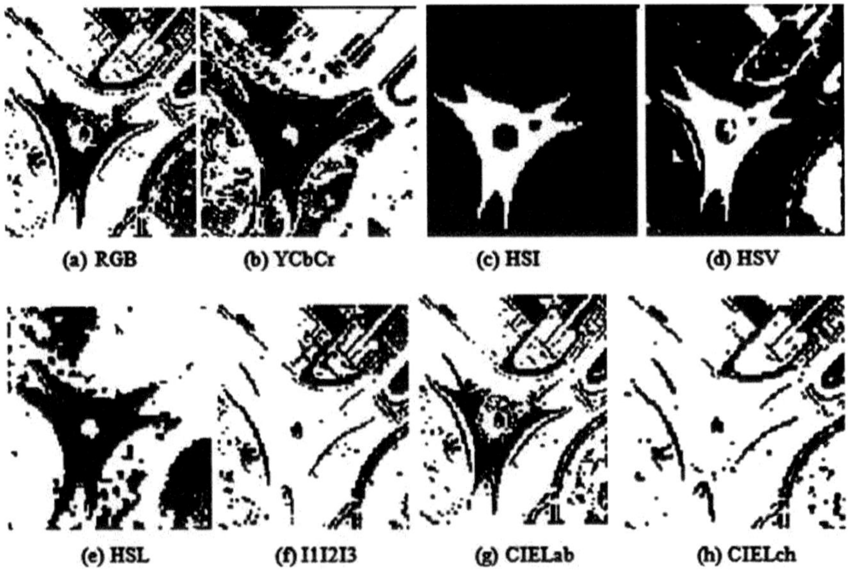

Fig. 6 GA and FFNN approach results for the given input image in different color spaces

Table 1 Cooperative approach result of input image in various color spaces

Color space	No. of. iteration	Performance (MSE)	Gradient	Execution time (in sec)
RGB	30	3.72e-06	0.000335	92.01
YCbCr	111	3.54e-06	0.000338	79.10
HSI	29	2.52e-06	0.000266	50.12
HSV	24	3.64e-06	0.000299	45.78
HSL	62	3.50e-06	0.000322	85.09
$I_1I_2I_3$	181	3.54e-06	0.000307	56.85
CIELab	24	3.39e-06	0.000366	36.17
CIELch	102	3.91e-06	0.000321	**30.12**

Bold indicates that the color space (CIELch) has lesser execution time as compared to others

Table 2 List of bivariate image quality measures utilized in this work

S. No.	Image quality measure	Formula
1.	MSE	$\frac{1}{MN}\sum_{j=1}^{M}\sum_{k=1}^{M}[f(j,k)-g(j,k)]^2$
2.	PSNR	$\frac{1}{MN}\sum_{j=1}^{M}\sum_{k=1}^{N}[f(j,k)-g(j,k)]^2/\sum_{j=1}^{M}\sum_{k=1}^{N}\mathrm{Max}[f(j,k)]^2$
3.	AD	$\frac{1}{MN}\sum_{j=1}^{M}\sum_{k=1}^{N}[f(j,k)-g(j,k)]$
4.	SC	$\frac{\sum_{j=1}^{m}\sum_{k=1}^{n}[f(j,k)]^2}{\sum_{j=1}^{m}\sum_{k=1}^{n}[g(j,k)]^2}$
5.	MD	$\mathrm{Max}\{f(j,k)-g(j,k)\}$
6.	NAE	$\sum_{j=1}^{M}\sum_{k=1}^{N}[f(j,k)-g(j,k)]/\sum_{j=1}^{M}\sum_{k=1}^{N}[f(j,k)]$
7.	LMSE	$\sum_{j=1}^{M-1}\sum_{k=1}^{N-1}[O\{f(j,k)\}-O\{g(j,k)\}]^2/ \sum_{j=1}^{M-1}\sum_{k=1}^{N-1}[O\{f(j,k)\}]^2)$
8.	NCC	$\sum_{j=1}^{M}\sum_{k=1}^{N}\left[F(j,k)\widehat{F}(j,k)\right]/\sum_{j=1}^{M}\sum_{k=1}^{N}[F(j,k)]^2$
9.	RMSE	$\sqrt{\frac{1}{MN}\sum_{j=1}^{M}\sum_{k=1}^{N}[f(j,k)-g(j,k)]^2}$
10.	NMSE	$\frac{\sum_{j=1}^{M}\sum_{k=1}^{N}[f(j,k)-g(j,k)]^2}{\sum_{j=1}^{M}\sum_{k=1}^{N}[f(j,k)]^2}$
11.	MB	$1-\left(\frac{\bar{g}}{\bar{f}}\right)$
12.	ERGAS	$100\frac{h}{l}\sqrt{\frac{1}{N}\sum_{k=1}^{N}\frac{\mathrm{RMSE}(B_k)^2}{\bar{f}_k^2}}$
13.	RASE	$\frac{100}{\bar{f}}\sqrt{\frac{1}{N}\sum_{k=1}^{N}\mathrm{RMSE}(B_k)^2}$
14.	SSIM	$\frac{(2\mu_f\mu_g+C_1)(2\sigma_{fg}+C_2)}{\left(\mu_f^2+\mu_g^2+C_1\right)\left(\sigma_f^2+\sigma_g^2+C_2\right)}$

Table 3 Concert analysis of color space support segmentation using the proposed two-layer approach

Quality measure	RGB	YCbCr	HSI	HSV	HSL	$I_1I_2I_3$	CIELab	CIELch
MSE	3211.6	8085.6	2075.7	759.55	812.98	1532.8	421.89	457.96
RMSE	56.671	89.92	35.56	27.56	28.51	39.14	20.54	21.4
PSNR	13.064	9.721	14.235	19.65	17.98	16.88	21.64	24.785
NCC	0.97888	0.951	0.8321	0.9819	0.9651	1.0712	0.9499	0.8976
AD	16.7913	21.652	12.762	7.784	2.7865	7.987	1.8308	3.246
SC	0.9534	0.8723	0.6995	5.894	0.9614	0.9765	1.0105	1.7634
MD	217	249	106	49	86	67	62	42
NAE	0.28402	0.5978	0.6456	0.2134	0.6799	0.7865	0.09462	0.0787
LMSE	16.311	28.542	12.564	21.543	13.876	9.7865	13.941	5.7865
Bias	0.3051	0.3845	0.1643	0.432	0.2019	0.1241	0.0069	0.0045
NMSE	0.6483	0.8451	0.561	0.4532	0.9601	0.7864	0.6397	0.8991
ERGAS	3339.1	6834.5	899.11	1478.09	1685.1	3411.8	1486.2	1672.5
SSIM	0.63851	0.1412	0.5781	0.8117	0.6972	0.6123	0.5673	0.8152
RASE	33.391	68.345	8.99	14.78	16.851	34.118	14.862	16.72

References

1. Kalist V, Ganesan P, Sathish BS, Jenitha JMM (2015) Possiblistic—fuzzy c-means clustering approach for the segmentation of satellite images in HSL color space. Procedia Comput Sci 57:49–56
2. Eskicioglu AM, Fisher PS (1995) Image quality measures and their performance. IEEE Trans Commun 43(12):2959–2965
3. Ganesan P, Palanivel K, Sathish BS, Kalist V, Shaik KB (2015) Performance of fuzzy based clustering algorithms for the segmentation of satellite images—a comparative study. In: IEEE seventh national conference on computing, communication and information systems (NCCCIS). Coimbatore, pp 23–27
4. Sajiv G, Ganesan P (2016) Comparative study of possiblistic fuzzy c-means clustering based image segmentation in RGB and CIELuv color space. Int J Pharm Technol 8(1):10899–10909
5. Gao B, Li X, Woo WL, Tian TY (2018) Physics-based image segmentation using first order statistical properties and genetic algorithm for inductive thermograph imaging. IEEE Trans Image Process 27(5):2160–2175
6. Zhang Y, Chandler DM, Mou X (2018) Quality assessment of screen content images via convolutional-neural-network-based synthetic/natural segmentation. IEEE Trans Image Process 27(10):5113–5128
7. Awad Mohamad (2010) An unsupervised artificial neural network method for satellite image segmentation. Int Arab J Inf Technol 7(2):199–205
8. Awad M, Chehdi K, Nasri A (2007) Multi component image segmentation using genetic algorithm and artificial neural network. Comput J Geosci Remote Sens Lett 4:571–575
9. Ganesan P, Rajini V (2014) YIQ color model based satellite image segmentation using modified FCM clustering and histogram equalization. In: International conference on advances in electrical engineering (ICAEE), pp 1–5
10. Avcıbas Ismail, Sankur Bulent, Sayood Khalid (2002) Statistical evaluation of image quality measures. J Electron Imaging 11(2):206–223
11. Ganesan P, Rajini V (2013) Value based semi automatic segmentation of satellite images using HSV color model, histogram equalization and modified FCM clustering algorithm. In: International conference on green computing, communication and conservation of energy (ICGCE), pp 77–82
12. Shaik KB, Ganesan P, Kalist V, Sathish BS (2015) Comparative study of skin color detection and segmentation in HSV and YCbCr color space. Procedia Comput Sci 57:41–48
13. Ganesan P, Rajini V (2014) Assessment of satellite image segmentation in RGB and HSV color model using image quality measures. In: International conference on advances in electrical engineering (ICAEE), pp 1–5
14. Zhang HZ, Xiang CB, Song JZ (2008) Application of improved adaptive genetic algorithm to image segmentation in real-time. Opt Precis Eng 4:333–336
15. Farmer ME, Shugars D(2006) Application of genetic algorithms for wrapper-based image segmentation and classification. In: IEEE congress on evolutionary computation, pp 1300–1307
16. Ibraheem Noor, Hasan Mokhtar, Khan Rafiqul, Mishra Pramod (2012) Understanding color models: a review. ARPN J Sci Technol 2(3):265–275
17. Sathish BS, Ganesan P, Shaik KB (2015) Color image segmentation based on genetic algorithm and histogram threshold. Int J Appl Eng Research 10(6):5205–5209

Braille Cell Segmentation and Removal of Unwanted Dots Using Canny Edge Detector

Vishwanath Venkatesh Murthy, M. Hanumanthappa, and S. Vijayanand

Abstract Braille cell segmentation is needed to detect, recognize and convert Braille cells of Braille document to natural language characters or word. The results can be used in the applications like converting Braille to text, embossing or printing Braille documents, distributing Braille documents over network or reproducing Braille documents on demand. Performance and quality of these applications get degraded due to noise introduced during scanning the Braille document. Noise may introduce due to scan resolution, poor lighting, introduction of skew or appearance of unwanted dots during scan. In line with the current work, skew correction and segmentation methods have been presented and published. Detecting the appearance of unwanted dots is a challenging task. The noise introduced during scan can sometimes be emitted as a dot in the scanned image. Also, deterioration of Braille plate can introduce some unwanted dots in the image while scanning. Poor lighting during scan also can be a cause of the unwanted dots. Identifying these unwanted dots is a major challenge. The relative position of the dot in Braille cell and their relativity with other dots helps to recognize Braille characters. Any unwanted dot makes it difficult in recognizing the character cell.

Keywords Braille · Canny edge · Gradient · Image segmentation · N-gram · OBR · Partial differential equation · Region-based method · Thresholding

V. V. Murthy (✉) · M. Hanumanthappa
Department of Computer Science and Applications, Bangalore University, Bengaluru, India
e-mail: vm.rnsit@gmail.com

M. Hanumanthappa
e-mail: hanu6572@hotmail.com

S. Vijayanand
Department of CSE, Rajrajeshwari Engineering College, Bengaluru, India

N. N. Chiplunkar and T. Fukao (eds.), *Advances in Artificial Intelligence and Data Engineering*, Advances in Intelligent Systems and Computing 1133,
https://doi.org/10.1007/978-981-15-3514-7_7

1 Introduction

Braille a writing language is generally used by the sightless people. The cell pattern of the Braille is made from collection of six elevated dots which exactly represents one cell that maps one text character or one word. [1] The plates used for Braille exist with either one or double-sided embossing. To avoid these plates from getting deteriorate, they are scanned, processed and transformed to natural text using OBR system. During scanning, the image produced may introduce noise due to uneven light, irregular pixel, image skew or slant, deprived dots, irregular spacing or uninvited dots [2].

The aim of segmentation is to divide an image into several parts/segments having similar features or attributes representing a meaningful cell pattern. Cell segmentation is again one of the decisive steps in OBR stages. Segmentation decays an image with sequence of cells representing characters or words in the form of meaningful pattern of sub-image. It divides the image into many meaningful regions which are having their own properties that can be analyzed and extracted as a desired target. [3] Extracted target is generally a cell with six dots representing a character or word of a natural language. The different techniques used for cell segmentation are threshold method, histogram segmentation, ANN-based method, compression and clustering-based method, edge-based method, region-based method and watershed-based method [3].

Thresholding divides image pixels using intensity level of an image which helps in distinguishing the foreground objects from background images. Different versions of thresholding segmentation are global thresholding, variable thresholding and multiple thresholding. [3] The edge-based thresholding has various methods like step edge that takes abrupt change in level of intensity, ramp edge considers gradual change in level of intensity, and spike edge accepts quick change in level of intensity, whereas roof edge has a steady variation in intensity. [1] Clustering methods use k-means method, where it selects random k-cluster centers and assigns the pixels to these canters that are closer to the cluster center. The algorithm performs these steps repetitively till convergence is achieved. Compression segmentation hypothesizes on optimal segmentation that the method minimizes the length of coding on the data using Huffman coding and lossy compression. Histogram segmentation methods do just a single pass over the pixels, and a histogram is calculated with all of these pixels. The histogram peaks and valleys are the basis to recognize the cluster centers. PDE-based method generates an initial curve for the lowest potential of a cost function which is non-trivial and provides smoothness constraints for the desired solution [4].

The main aim of this study was to present an efficient method that can correctly perform the cell segmentation process that will lead to further process of identifying the sequence of characters or words present in the document.

The research article is ordered in four segments. The Sect. 1 introduces the various stages of OBR system. Section 2 presents the literature survey on the current work that includes various methods of segmentation. In Sect. 3, the results of segmentation techniques applied using MATLAB 8.3 are presented. The Sect. 4 concludes the paper.

2 Current Work

Ravikumar [5] has segmented the cell dots in row and column order. The rows are numbered as rw1, rw2, …, rwn and columns into c1, c2, …, cn. Braille mesh of six points is prepared by calculating the upper edges, lower edges, left edges and right edges corresponding to the cell dot. The character of Braille is matched for the vigorous dots and its relative position. Each Braille cell is mapped using coordinates of the mesh box. The mesh box is formed with coordinates (rw1, c1) to (rw6, c4) that maps a single character. The mesh cell is copped as Ci, and the respective character is identified by generating bin value. The cropped and segmented cell pattern is shown in Fig. 1. Six centroid coordinate values are calculated to generate character box to extract the character.

Shahbazkia [6] after thresholding computed the center of mass of the point to determine the point's center. Further to isolate the lines of text, the lines without blob are marked. Author has used the linked list to store the coordinates and positions of the point center for each point in the line. Linked list is further parsed to recognize the cell and associated character.

Yin [7] has introduced a sub-block of relative size of 60 × 70 pixels, which locate the dots by dots' thresholding in the subsequent dispensation. The 3 × 2 array with six dots is represented by points (a, b, c, d, e, f) beforehand. The threshold value and the geometric center of each dots represent '8' as a mesh. Identified Braille cell is saved as binary string like 101111.

Minghu [8] has used rule base, sign base and knowledge base segmentation in which knowledge base carries out the process of disambiguation in Chinese text. Author has applied Viterbi search by constructing multi-level graph to identify the Chinese characters with high likelihood. Rule base generated using Chinese syntax was applied on word segmentation. Ambiguity is addressed by marking every word in token dictionary with pinyin in sequence. The N most likely character order is

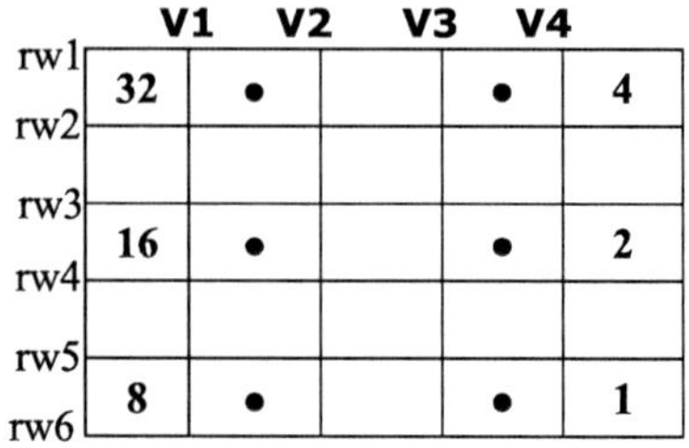

Fig. 1 Mesh with unique values given to dot positions

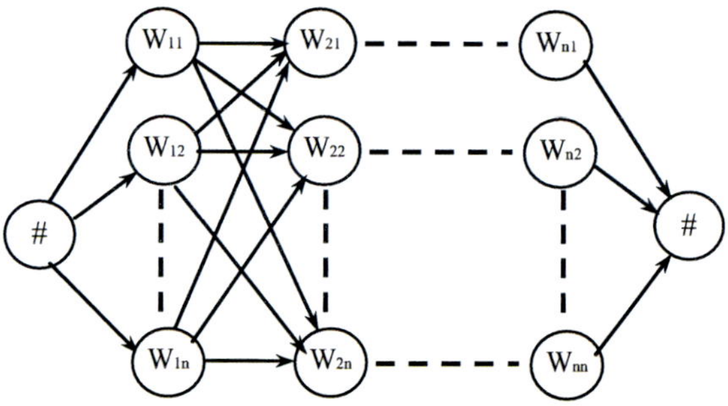

Fig. 2 Multi-level model of Chinese character set

retrieved by means of an N-Best algorithm. The experimental results proved that the translation of Braille codes to Chinese characters shows 94.38%.

Author has applied the Markov model on list of words $W = w_1, w_2, w_3 \ldots w_n$, to get conditional Eq. 1 [8].

$$P(w_1 w_2 \ldots w_n) \approx P(w_1) P\left(\frac{w_2}{w_1}\right) \ldots P\left(\frac{w_n}{w_{n-N} w_{n-N-1} \ldots w_{n-1}}\right) \quad (1)$$

where N-gram probability is defined by the general form—$P(w_n / w_{n-N} w_{n-N-1} \ldots w_{n-1})$. [8] Based on Eq. 1, ith word probability is calculated. Further smoothing bigram and unigram methods are used to develop multi-level pinyin model for Chinese alphabets as shown in Fig. 2.

Viterbi dynamic programming and N-best algorithms are used further to calculate sequence of characters with the highest likelihood. The precision level was far more than 99% for general text [8].

Smelyakov [9] has developed artificial neural network that recognizes Cyrillic alphabet images from Braille document. Neural network has been trained and applied on test database for recognizing scanned Cyrillic letters from Braille document. Backpropagation algorithm is used to estimate the gradient or ascent of loss function regarding the weights in ANN which is an input to optimization methods that help to minimize error function. Different training parameters used in ANN training are learning rate, minimum error, momentum factor, total of hidden layers, number of iteration which improved the identification results by 97.2% accuracy in average to all available letters of the training data set.

3 Braille Cell Segmentation and Removal of Unwanted Dots

Image preprocessing is performed on the scanned image to get high resolution and to eliminate noise. The preprocessing and segmentation process is implemented on MATLAB, and the results are presented. The algorithm used for Braille cell segmentation, detection and removal of unwanted dots is

1. *Read an image*
2. *Binarize the image*
3. *Remove shadow by applying morphological operations*
4. *Perform background correction*
5. *Apply canny edge detector to detect unwanted dots*
6. *Eliminate unwanted dots.*

Initially, in step 1, the image is read using the MATLAB 8.3 command of Eq. 2.

$$\text{colorImage} = \text{imread}\,('\text{Original.jpg}');\ \text{figure, imshow}\,(\text{colorImage});\ \text{title}\,('\text{Fig1 : original image}') \tag{2}$$

Figure 3 shows the original image which is displayed with the command image.show().

In second step, the color image is converted to binary image using the Eqs. 3 and 4.

$$\text{grayImage} = \text{rgb2gray(colorImage)}; \tag{3}$$

$$\text{BinaryImage} = \text{im2bw(grayImage, 0.7)} \tag{4}$$

Fig. 4 shows the resultant binary mage.

Step 3 is to remove the shadow of the dots and other noise introduced in the image. First the background correction is performed using morphological operation with command of Eq. 5.

$$\begin{aligned}&\text{se} = \text{strel}\,('\text{disk}', 20);\\&\text{background} = \text{imopen}\,(\text{BinaryImage, se});\end{aligned} \tag{5}$$

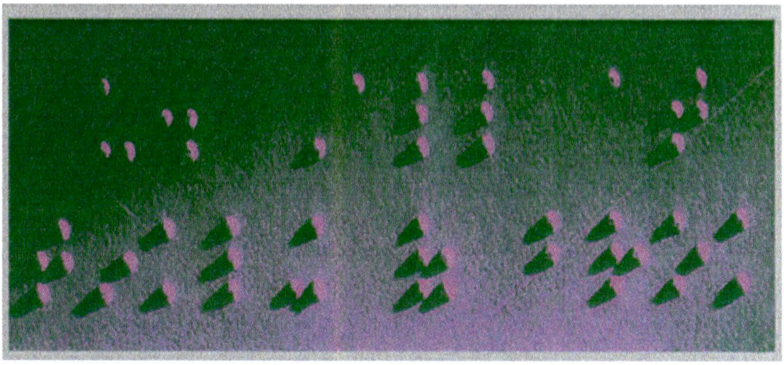

Fig. 3 Original image

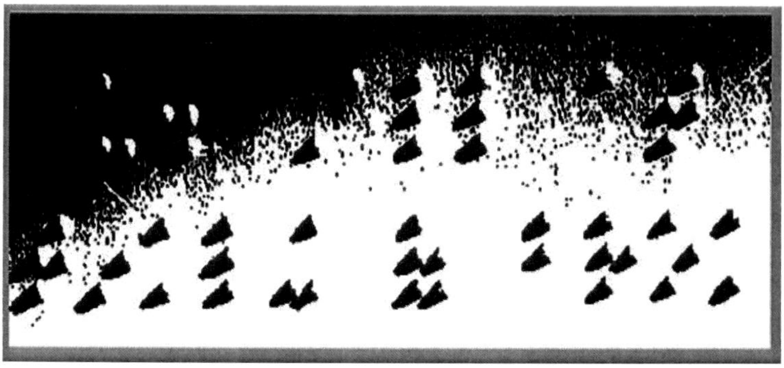

Fig. 4 Binarized image

imopen () performs morphological operation on the grayscale that includes erosion that follows the dilation and using the similar structuring component supplied by parameter 'se' on both operations. The method strel () creates the morphological disk-shaped structuring component with radius 20. The resultant image after morphological operations is shown in Fig. 5. Now, the background is uniform, and all the shadows are removed.

Background correction is performed by subtracting the uniform background from the existing binary image using Eq. 6 that produces the result as shown in Fig. 6.

$$\text{CorrectedImage} = \text{BinaryImage} - \text{background}; \tag{6}$$

After background correction is successfully performed, the next step-5, the unwanted dots are detected by applying the canny edge detector as shown in the Eq. 7.

$$\text{canny} = \text{edge}\left(\text{CorrectedImage}, '\text{canny}'\right); \tag{7}$$

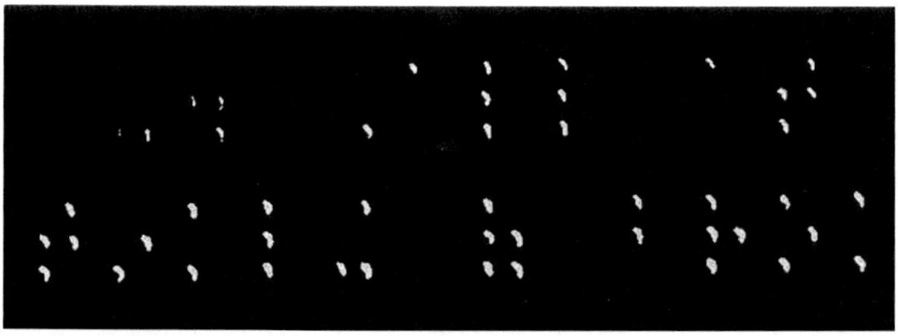

Fig. 5 Resultant Braille after morphological operation

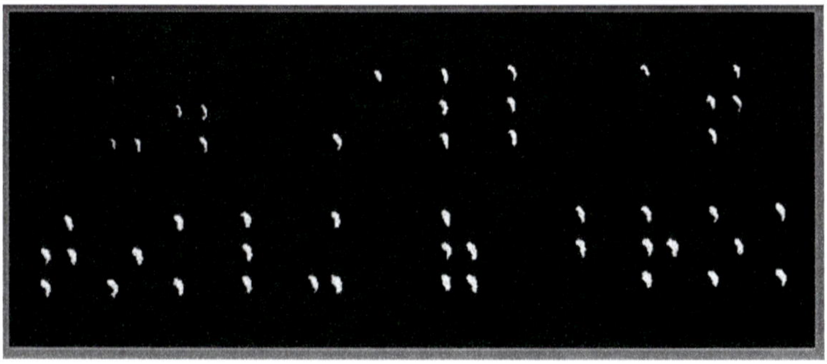

Fig. 6 Image after background correction

Canny edge detector is used to identify the cell boundary by dot edge detection. Canny edge detection helps to locate unwanted dots appearing in the image as highlighted by circular red ring in Fig. 7. There is a need to separate these dots from the rest of the original dots. The unwanted dots are successfully recognized in canny edge detection. These unwanted dots degrade the accuracy of Braille to text translation process if not removed.

In the last step-6, the removal of unwanted dots is performed using following set of code:

```
bw2 = imfill (canny,' 'holes');
bw3 = imopen (bw2, ones (1, 1));
bw4 = bwareaopen (bw3, 18);
bw4_perim = bwperim (bw4);
```

The parameter 'holes' in imfill function fills the dots that cannot be reached by from the edge of the Braille image. Again, the morphological operations are applied on array of ones (1, 1) using imopen () method. Now, the remaining small objects in the image are detected as unwanted dots and are eliminated using the method area opening. In the code, 'bwareaopen ()' is the area opening method that removes

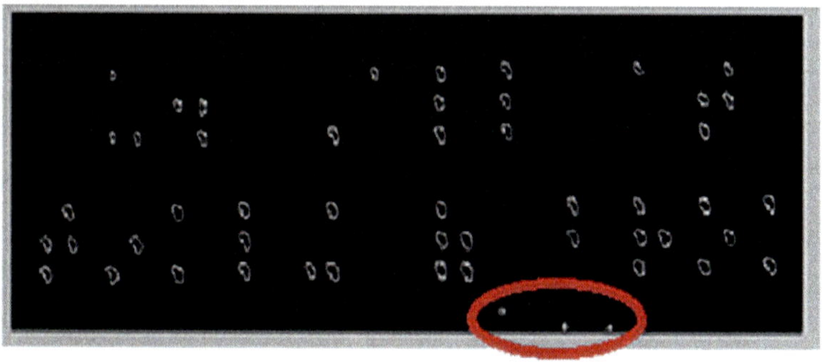

Fig. 7 Locating unwanted dots

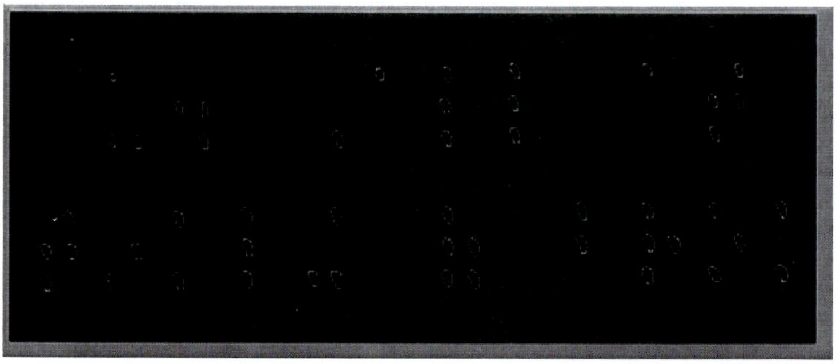

Fig. 8 Image after removal of unwanted dots

all small dots that has fewer than 18 pixels. Finally, the bwperim () method returns a binary image that has only the perimeter pixels of objects for the supplied input image. The resultant image is shown in Fig. 8.

Final image has eliminated all unwanted dots. This image can be further used for feature extraction process to locate individual Braille cells.

4 Conclusion and Future Work

In this paper, the morphological operations are used to segment and locate the unwanted dots. The dots are eliminated using area opening method. The implementation is done on six various Braille images on MATLAB 8.3. The 90% accuracy is observed in the removal of unwanted dots from these experiments considering the total count of number of dots as shown in Eq. 8. One of the experiments is presented in this paper.

$$\text{Accuracy} = (\text{number of unwanted dots removed})/(\text{total unwanted dots}) \quad (8)$$

In future work, we plan to take the results of this study and apply feature extraction process to locate individual Braille cell.

References

1. AlSalman A, El-Zaart A, Al-Salman S, Gumaei A (2012) A novel approach for braille images segmentation. In: 2012 International conference on multimedia computing and systems, Tangier, 2012, pp 190–195. https://doi.org/10.1109/icmcs.2012.6320146
2. Hanumanthappa M, Murthy VV (2016) Optical braille recognition and its correspondence in the conversion of braille script to text—a literature review. In: 2016 International conference on computation system and information technology for sustainable solutions (CSITSS), Bangalore, 2016, pp 297–301. https://doi.org/10.1109/csitss.2016.7779374

3. Rupanagudi SR, Huddar S, Bhat VG, Patil SS, Bhaskar MK (2014) Novel methodology for Kannada braille to speech translation using image processing on FPGA. In: 2014 International conference on advances in electrical engineering (ICAEE), Vellore, 2014, pp 1–6. https://doi.org/10.1109/icaee. 2014. 68384
4. Baird HS, Govindaraju V, Lopresti DP (2004) Document analysis systems for digital libraries: challenges and opportunities. In: Document analysis systems VI, Springer Berlin Heidelberg, Berlin, Heidelberg, pp 1–16
5. Kumar NR, Srinath S (2014) A novel and efficient algorithm to recognize any universally accepted braille characters: a case with Kannada language. In: 2014 Fifth international conference on signal and image processing, Bangalore, India, 2014, pp 292–296. https://doi.org/10.1109/icsip.2014.52
6. Shahbazkia HR, Silva TT, Guerreiro RM (2005) Automatic braille code translation system. In: 2005 Progress in pattern recognition, image analysis and applications, pp 233–241 Springer, Berlin Heidelberg. ISBN: 978-3-540-32242-9. https://doi.org/10.1007/11578079_25
7. Yin J, Wang L, Li J (2010) The research on paper-mediated braille automatic recognition method. In: 2010 Fifth international conference on frontier of computer science and technology, Changchun, Jilin Province, 2010, pp 619–624. https://doi.org/10.1109/fcst.2010.95
8. Minghu J, Xiaoyan Z, Ying X, Gang T, Baozong Y, Tang Xiaofang (2000) Segmentation of mandarin braille word and braille translation based on multi-knowledge. In: WCC 2000—ICSP 2000. 2000 5th international conference on signal processing proceedings. 16th world computer congress 2000, Beijing, China, 2000, vol 3 pp. 2070–2073. https://doi.org/10.1109/icosp.2000.893513
9. Smelyakov K, Chupryna A, Yeremenko D, Sakhon A, Polezhai V (2018) Braille character recognition based on neural networks. In: 2018 IEEE second international conference on data stream mining & processing (DSMP), Lviv 2018, pp 509–513. https://doi.org/10.1109/dsmp.2018.8478615

Real-Time Detection of Distracted Drivers Using a Deep Neural Network and Multi-threading

Ajay Narayanan, V. Aiswaryaa, Aswesh T. Anand, and Nalinadevi Kadiresan

Abstract Convolutional neural network (CNN) is a very popular deep learning architecture used for the problem of object detection and classification of images and videos. CNN is appropriate for image detection as it prunes the computational overheads as well as provides better performances. Visual Geometry Group (VGG) model is an instance of CNN architecture to solve the problem of object detection from images. VGG-16 model is one of the most potent classifiers that had won the second place in the ImageNet ILSVRC-2014 challenge. VGG also helps in reducing latency of detection of distracted drivers in real time (RT). The technique of transfer learning was applied using VGG-16, pre-trained on the ImageNet dataset intended to extract bottleneck feature that is further used to train a classifier. This technique thrives on the combined power of VGG convolutions and multi-layer perceptron. A key highlight of this paper is the coupling of transfer learning using VGG-16 with multi-threading to achieve a real-time prediction of distraction, from a video input. The proposed model performed notably well on real-time video, achieving an accuracy of 96%, and a video output of 26 frames per second which is comparable to the state-of-the-art algorithms, for real-time classification of objects.

Keywords Deep learning · CNN · VGG · Transfer learning

A. Narayanan (✉) · V. Aiswaryaa · A. T. Anand · N. Kadiresan
Department of Computer Science and Engineering, Amrita School of Engineering, Amrita Vishwa Vidyapeetham, Coimbatore, India
e-mail: cb.en.u4cse15203@cb.students.amrita.edu

V. Aiswaryaa
e-mail: cb.en.u4cse15202@cb.students.amrita.edu

A. T. Anand
e-mail: cb.en.u4cse15206@cb.students.amrita.edu

N. Kadiresan
e-mail: k_nalinadevi@cb.amrita.edu

N. N. Chiplunkar and T. Fukao (eds.), *Advances in Artificial Intelligence and Data Engineering*, Advances in Intelligent Systems and Computing 1133,
https://doi.org/10.1007/978-981-15-3514-7_8

1 Introduction

Many world organizations, engaged in studies on safe transportation, have identified distracted driving and distracted drivers as one of the foremost causes for road-related accidents. Distraction of drivers can be broadly classified into two types: cognitive and visual. Cognitive distraction refers to the distraction of the mind that cannot be identified through images. On the other hand, visual distraction refers to incidents, which can be easily identified through images. This research is focused on visual distraction. Common indicators of visual distraction, among drivers, are texting on the phone, talking on the phone or with co-passengers, setting eyes off the ongoing traffic, looking back, picking up a fallen article, eating, drinking, and adjusting the hair/makeup. Such distracted behavior, if detected and alerted in real time, may help save the life of the driver as well as the lives of others on the road. The magnitude of the problem of detecting distracted driving has impelled the research community to find efficient real-time solutions for the same. For the past decade or so, the focus of the research community has been on resorting to the concepts of machine learning and computer vision to solve this seemingly insurmountable problem. There have been promising measures that achieved near real-time (NRT) detection of distraction, using linear classification models such as support vector machine (SVM) [1, 2], ensemble methods like XGBoost. Recently, deep learning models like AlexNet, VGG, InceptionV3 have provided higher accuracy than traditional models. This paper presents an innovative utilization of the popular technique of transfer learning [3], for the real-time detection of distracted drivers. The bottleneck features are extracted from pre-trained VGG-16 [3] and fed into a fully connected layer, which is used as a classifier. The calibrated VGG-16 models were trained on the ImageNet, a publically available, visual image, large database that contains several hundred images from over 20,000 categories. This paper proposes a novel method of application of the deep neural net to a live video stream. The rest of this paper is organized as follows. Section 2 presents an overview on past research in this context and a comparison of the deep neural architectures. Section 3 describes the algorithms used. Section 4 describes proposed distracted driver detection architecture. Description of the dataset, model evaluation and validation, evaluation metrics, and result visualization are provided in Sect. 5. The analysis of the results is provided in Sect. 5 with closing remarks and conclusion in Sect. 6

2 Related Work

2.1 Real-Time Approaches

This section presents an overview of the various approaches made to solve the problem of real-time detection of driver distraction.

In [1, 2], support vector machines (SVMs) to detect distracted drivers in real time were implemented. An In-Vehicle Information System (IVIS) was used, while driving, to collect data which is obtained by taking images of eye movement.

In [4, 11], different training methods, model characteristics, and feature selection criteria of models like SVM, feed forward neural network (FFNN), layer recurrent neural network (LRNN), fuzzy inference system (FIS), and adaptive neuro-fuzzy inference system (ANFIS) have been compared.

[5, 6] makes a real-time prediction of distracted driving, by using the driver's gaze zone, which gives an insight into the precise behavior of the driver. Images, captured by cameras placed inside the vehicle, were used as the dataset. A face detector, based on Haar features, was combined with a MOSS tracker, which detects distraction, depending on where a driver is looking during driving. A convolutional neural network was trained to categorize the driver's gaze zone, from a given detected image of the driver's face.

2.2 Deep Learning Models

This section presents an overview of the various approaches pursued for the detection of driver distraction using deep learning models.

In [3], four models were compared for training the dataset, namely VGG16, pre-trained VGG16, VGG19, and pre-trained VGG19. Data is fed in as images and passed through the neural net architecture, for identification of the distraction.

[7] was used to select transfer learning over other models as this paper gives a detailed proof as to how transfer learning results in a better accuracy of detection and prediction.

[8] gave us a better understanding of how CNNs can be used as feature extractors. It compares the working of different CNN architectures.

3 Algorithms Used

3.1 VGG-16

VGG-16 is a 16-layer deep convolutional neural network. This model, publicly available through the Keras framework [9], has been pre-trained on the ImageNet, a huge image database consisting of about 14×10^6 images, belonging to over 1000 categories, and is a very good candidate for transfer learning.

3.2 Convolutional Neural Network

A neural network is an interconnection of neurons arranged in layers. Each neuron in the network performs a weighted sum or dot product operation on all its inputs and applies an activation function to produce a nonlinear output. The training of a neural net is achieved by updating the weights of the interconnections, in order to reduce a global loss function. A convolutional neural network is a neural network which has a 3D arrangement of its neurons whose inputs are images. The major building blocks of a CNN are explained below.

Input Layer—An input layer is an interface for the input images. The dataset considered consists of RGB images which are re-sized to 224 × 224 × 3.

Convolutional Layer—A convolutional layer consists of a set of filters or kernels of small size. These kernels are moved across the input image, and at each point, the dot product of the kernel weights and a small region beneath the kernel is taken as the output. The features are learnt by updating the weights of these trainable kernels.

Pooling Layer—The pooling layer applies down sampling along the spatial dimensions (width, height) to reduce over-fitting and to avoid computational redundancies.

3.3 Transfer Learning

Transfer learning is a deep learning technique that reuses the pre-trained model of one task as the starting point of another task. A pre-trained model is chosen, reused, and tuned according to the task requirement. Since the dataset used for training is rather small compared to standard dataset like ImageNet, the technique of transfer learning is implemented. This technique usually works better than a CNN, especially when the dataset used for training is small or computational resources are scarce. One of the techniques applied involves removing the last three fully connected layers of VGG model and using rest of the model as a feature extractor. The process of retraining the deeper and more specific parts of a convolutional network is called fine-tuning of the network and is being used extensively in this study. The major idea of using transfer learning is to use a pre-trained model which was trained on a very large database of about 14 million images that is similar to the dataset used for fine-tuning.

4 Proposed Architecture

The proposed architecture implements the technique of transfer learning using a pre-trained VGG-16 model. Transfer learning is applied in this context as it reduces the computational complexity of training the model. The ImageNet database was chosen due to its varied and widespread representation of classes of images. The

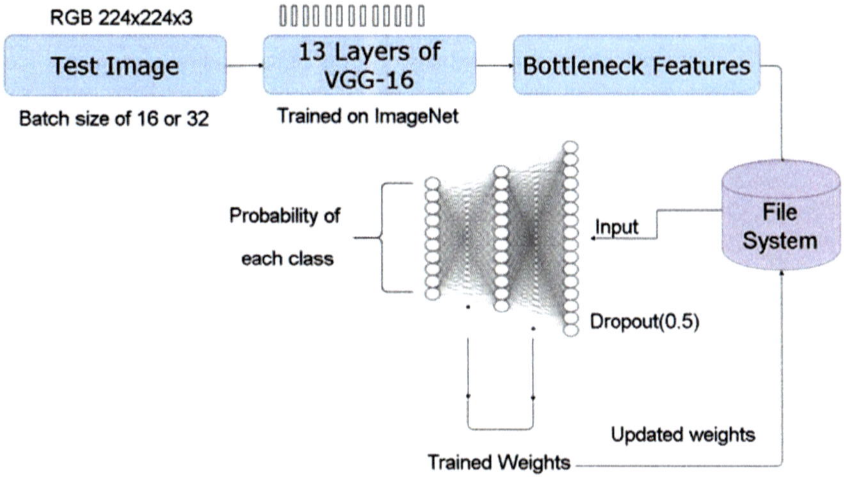

Fig. 1 Proposed system architecture

representative nature of the ImageNet database ensures that the pre-trained VGG-16 model would have learnt features, which would be useful for predicting driver distraction as well. The last fully connected layers of the pre-trained model were removed, as they would have features that are highly specific to the ImageNet dataset. The training and validation images were fed through the VGG-16, which extracts the bottleneck features, and the bottleneck features were saved on a local file. The bottleneck features hence extracted are a vector of numbers that represent the salient parts of the image. These bottleneck features were subsequently retrieved from the file system and passed as input to train on the newly created fully connected layers. The training and validation images were fed through the VGG-16, which extracts the bottleneck features, and the bottleneck features were saved on a local file. These bottleneck features were subsequently retrieved from the file system and passed as input to train on the newly created fully connected layers. Repeated application of this procedure, updated the weights of the fully connected layer, relevant to the detection of distracted driver. The weights of the VGG-16 model were only used to obtain the bottleneck features and were never retrained in the training process. The architecture of the proposed model is diagrammatically represented in Fig. 1.

4.1 Training Methodology

In the training phase, the input image set consisting of a batch size of 16 or 32 images is obtained from the training data. The images are first augmented by rotating about 180 degrees to suit the Indian driving context and adding Gaussian noise in order to generalize the training images and thereby improve the accuracy. The augmented images are pre-trained on 13 layers of the VGG model to extract the bottleneck features. The output from the VGG-16 network is obtained and is stored in the file system. The features are then passed to the fully connected layers of a neural network with a dropout of 0.5. The trained weights are then sent to the file system. A sigmoid function is applied to the output layer and is classified into one of the ten classes.

4.2 Real-Time Testing Methodology

The system was tested in real time by capturing a video of environment to be tested for distracted driver. Figure 2 shows the steps involved in the testing phase. The captured video split into image frames. Each frame was passed into a classification thread. Each classification thread performs testing. In this testing phase, the image frames of the video undergo the same process of feature extraction by 13 layers of the VGG-16 and then pass through the fully connected layers for classification. Trained weights required for classification were obtained from the file system, where it was previously stored. An argmax function was applied on the output layer to predict the output class. The intuition behind using the argmax function is to obtain the class with the highest probability of occurrence in the test image. On classification of each thread, it is passed into an output accumulator, which collects the output of the classified frames from the classification thread and appends those frames to produce a seamless output video feed. Each classification thread can be visualized as a VGG-16 network in its own sense. Each of these threads runs concurrently and even can be run on multiple processors for even better results. The process of multi-threading for this scenario is depicted in Fig. 3.

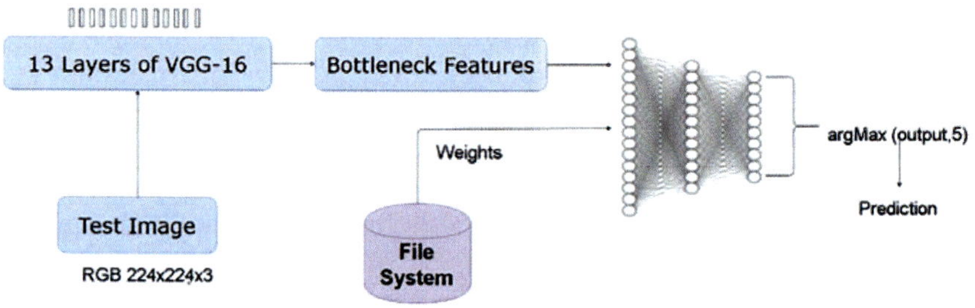

Fig. 2 Testing phase

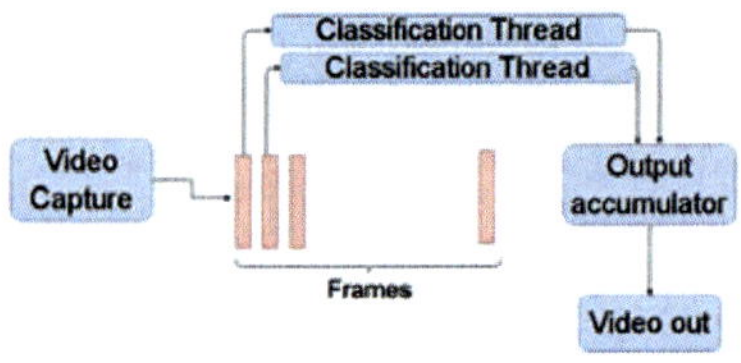

Fig. 3 Multi-threading for real-time prediction

Fig. 4 Distraction types

5 Performance Evaluation

5.1 Dataset

The dataset has been taken from the Kaggle Competition [10] conducted by State Farm to detect distracted drivers as in Fig. 4. The distraction of drivers from a live video was tested with the trained model and classified into one of the ten classes.

In order to train and evaluate the performance of the model, the dataset [7] containing 22,424 labeled images was split into 60% training data, 20% validation, and 20% testing data. The initial concern was whether the data would be sufficient enough to train the model; however, this was supplemented by data augmentation and transfer learning techniques.

5.2 Model Evaluation and Validation

To summarize the results, the VGG model seems to have some overhead cost in creating the bottleneck features, but since these bottleneck features can be stored offline and loaded at any time, the subsequent training of the fully connected layers takes significantly lesser time. This model in addition to producing high accuracy, it also trains on a regular laptop CPU with very reasonable training time.

The pre-trained VGG-16 model, retrained on the distracted driver dataset, is a state-of-the-art model for predicting the distraction of drivers from images. The VGG-16 acts as a feature extractor, responsible for extracting the salient and concise features in an input image. The 'features' hence extracted are a vector of numbers that represent the salient parts of the image. The features extracted by the VGG model aid the fully connected layer to easily classify the drivers into various output classes as defined in the dataset.

The testing was done with 5000 images used as the test subjects for the model evaluation. The following results were also a direct result of experiments done on this test set. The model was able to achieve 97% accuracy in prediction on test sets which it had never seen before.

5.3 *Accuracy of VGG-16 with Transfer Learning*

The change in accuracy can be observed in Fig. 5 as the model is trained through the different epochs. Accuracy in classification problems is the ratio between correct predictions and the total number of predictions made by an algorithm (1).

$$\text{Accuracy} = \frac{\text{TP} + \text{TN}}{\text{TP} + \text{FP} + \text{TN} + \text{FN}} \tag{1}$$

Both, the training and validation accuracies, progressively increase through the number of epochs until it reaches some threshold, indicating no further improvements are made. This causes the training to terminate in this case reaching an accuracy of 96% at epoch 8.

Fig. 5 Model accuracy

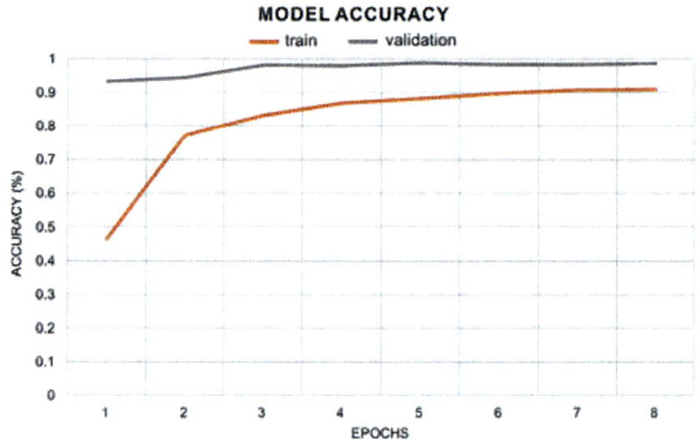

Fig. 6 Confusion matrix of VGG-16 with transfer learning

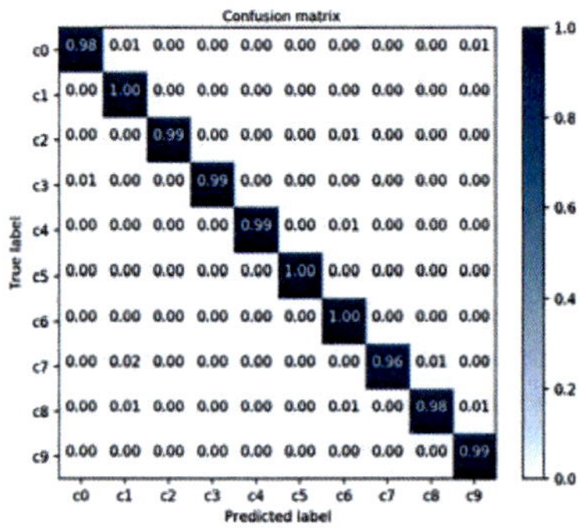

5.4 Confusion Matrix of VGG-16 with Transfer Learning

A confusion matrix as shown in Fig. 6 is usually visualized as a comparison between the true labels and the predicted labels. Each row of the matrix is the true label of the class, and each column represents a predicted table. The probability or the incorrect and correct predictions are then filled into the table. The diagonal of this matrix shows the probabilities of correct predictions, and all other values represent the misclassification errors made by the classification algorithm

Most of the predicted image classification labels have accuracy close to 100%. One, of the class c7, that is, 'reaching behind' stands out with 96% accuracy. It is relatively low, compared to the rest of the classes as seen in Fig. 6. The conclusion from this observation confusion matrix is that predicting whether a driver is reaching behind is the most misclassified label or most difficult category to predict. Furthermore, it can be seen from the confusion matrix that c7 is most commonly mislabeled as c1, meaning that reaching behind is typically mislabeled texting with the right hand. This is understandable since drivers that are typically reaching behind perform this action with the right hand raised. As such, the model will, at times, classify them as talking on the phone with the right hand if captured at a specific frame of motion.

5.5 Model Loss of VGG-16 with Transfer Learning

In Fig. 7, as noted, both the validation and training loss decrease as the model learns, until eventually terminating when the model begins to over-fit. The cross-entropy loss function (2) is applied to the probabilities that the deep learning model will predict for each class and hence will give the logarithmic loss based on the ground truth which is provided by the dataset used. Such a log-loss will provide a good measure of how good the model is at predicting the classes of unseen images. A high value close to 1 will mean that the model misclassifies a lot of images and is not able to capture the data properly.

Fig. 7 Model loss

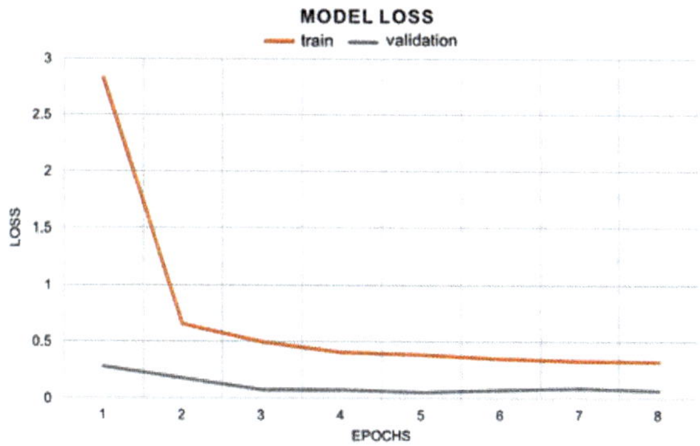

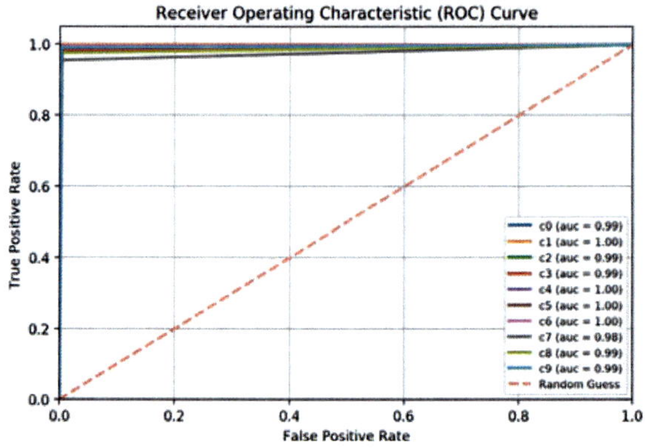

Fig. 8 ROC curve of VGG-16 with transfer learning

$$\sum_{c=0}^{M} y_0, c \log(p_o, c) \tag{2}$$

5.6 *Receiver Operating Curve of VGG-16 with Transfer Learning*

Finally, the performance of the model is analyzed using the ROC curve is shown in Fig. 8. A steeper and quicker increase to the top left corner indicates a close to optimal solution. Another characteristic to take note of in the graph is its sharp bend, which indicates that there is a very clear and near-perfect separation.

5.7 *Evaluation of Real-Time System*

The proposed multi-threading style application of the VGG-16 model on the individual frames, of the video, seems to do an excellent job, in achieving real-time detection of the class of distraction among the drivers. The results are validated by checking frames per second (FPS) rate of the output video compared to the input video stream. The input video stream, from the webcam, had an FPS of about 30. The output stream, by the proposed method, had a FPS of about 26. This lag of about 4 FPS is due to the lag in appending the output frames with the prediction from the VGG-16 network. The time taken for prediction of a single frame, by the VGG-16 network, is about 1.5 s. Since the latency of classification of a single frame is 1.5 s, fps produced by sequentially applying the VGG network to the video frame would be 1/1.5, a quantity less than 1. Therefore, by comparison to the achieved fps, a significant improvement has been achieved by separating the process of classification and that of output production. The latency of 1.5 s, in prediction, is manifested in the time taken, by the system, to respond to a change in the class of distractions, depicted by the test subject. This latency in the system's response time can further be reduced, by addition of a greater number of classification threads, by adding in the concept of locks to prevent deadlock situations. Overall, the system is able to achieve real-time classification of distraction type of the test subject.

6 Conclusion

This investigation was centered on the detection of distracted drivers, in near real time, using multi-threading on images, obtained from the State Farm Distracted Driver Detection competition on Kaggle [7]. By using a pre-trained VGG-16 network and retraining a new set of fully connected layers, the network was able to achieve 97% accuracy on test data. This network, coupled with multi-threading, yields a seamless classification of whether the driver is distracted, in real time. Evaluation of the confusion matrix revealed that the most mislabeled class, 'reaching behind,' was often confused with the driver 'talking on their phone with the right hand.' Overall, the model has proven to be effective at predicting distracted drivers and hopefully will aid in preventing further injuries and deaths, ensued from distracted driving. The latency in the classification of video frames can further be reduced by introducing locking mechanism and more parallelism through multi-threading

References

1. Liang Y, Reyes ML, Lee JD (2007) Real-time detection of driver cognitive distraction using support vector machines. IEEE Trans Intell Transp Syst 8(2):340–350
2. Ou C, Ouali C, Bedawi SM, Karray SM (2018) Driver behavior monitoring using tools of deep learning and fuzzy inferencing. In: 2018 IEEE international conference on fuzzy systems (FUZZ-IEEE)
3. Koesdwiady A, Bedawi SM, Ou C, Karray F (2017). End-to-end deep learning for driver distraction recognition. In: International conference image analysis and recognition, Springer, Cham, pp 11–18
4. Tango F, Botta M (2013) Real-time detection system of driver distraction using machine learning. IEEE Trans Intell Transp Syst 14(2):894–905
5. Choi I-H, Hong SK, Kim Y-G (2016) Real-time categorization of driver's gaze zone using the deep learning techniques. In: 2016 International conference on big data and smart computing (BigComp), Hong Kong, pp 143–148
6. Jiang Y-S, Warnell G, Stone P (2018): Inferring user intention using gaze in vehicles. In: Proceedings of the 2018 on international conference on multimodal interaction—ICMI '18
7. Damodaran N, Sowmya V, Govind D, Soman KP (2019). Scene classification using transfer learning. In: Recent advances in computer vision, Springer, Cham, pp 363–399
8. Mohan VS, Sowmya V, Soman KP (2018) Deep neural networks as feature extractors for classification of vehicles in aerial imagery. In: 2018 5th International conference on signal processing and integrated networks (SPIN), Noida, pp 105–110

9. Keras Documentation. https://keras.io/applications/#vgg16
10. State farm distracted driver detection. https://www.kaggle.com/c/state-farm-distracted-driver-detection/data
11. Atiquzzaman Md, Qi Y, Fries R (2018): Real-time detection of drivers' texting and eating behavior based on vehicle dynamics. In: Transportation research part F: traffic psychology and behaviour

Analysing the Practicality of Drawing Inferences in Automation of Commonsense Reasoning

Chandan Hegde and K. Ashwini

Abstract Commonsense reasoning is the simulation of human ability to make decisions during the situations that we encounter every day. It has been several decades since the introduction of this subfield of artificial intelligence but it has barely made some significant progress. The modern computing aids also have remained impotent in this regard due to the absence of a strong methodology towards commonsense reasoning development. Among several accountable reasons for the lack of progress, drawing inference out of commonsense knowledge base stands out. The study presented here emphasizes a detailed analysis of representation of reasoning uncertainties and feasible prospects of existing inference methods for reasoning. Also, the difficulties in deducing and systematizing commonsense reasoning and the substantial progress made in reasoning that influences the study have been discussed. Additionally, the paper discusses the possible impacts of an effective inference technique in commonsense reasoning.

Keywords Commonsense reasoning · Artificial intelligence · Knowledge base · Inference

1 Introduction

Commonsense reasoning (CR) is one among many subfields of artificial intelligence (AI) that deals with the simulation of human actions and decision making using commonsense knowledge. For example, making a computer to understand, whether an object can fly or not based on minimal information about the object. Introduced in the late 1980s, the field of CR is still in its early stages of development and has

C. Hegde (✉)
Surana College PG Departments, Bangalore, India
e-mail: chandanhegde1@gmail.com

Research Scholar at GAT (VTU), Bangalore, India

K. Ashwini
Global Academy of Technology, Bangalore, Karnataka, India
e-mail: dr.ashwini.k@gat.ac.in

N. N. Chiplunkar and T. Fukao (eds.), *Advances in Artificial Intelligence and Data Engineering*, Advances in Intelligent Systems and Computing 1133,
https://doi.org/10.1007/978-981-15-3514-7_9

plenty of room for significant contributions. Though implementing CR may seem interesting and possible with modern-day computer systems with extreme computing capacities, a practical model of the same could be far away from reality. This is mainly due to the limitations of computers related to reasoning and their inability for making assumptions voluntarily. While communicating, humans use certain set of knowledge which has been acquired through the years. We use assumptions about physical matters and actions related to them. For example, "a liquid is always wet" or "burning is related to heat" are assumptions made using the reasons behind a material's physical properties. The same cannot be derived for a computer system. That way, computers are very far away from having materialistic knowledge. Expert systems can be considered as "the first step" technologies towards achieving such a complete AI. Adding CR into the features of computing system itself would give rise to a new species of computers.

The most challenging part of automation of CR is to draw inferences from a set of knowledge bases. Reasoning with commonsense knowledge base differs from any other reasoning process in two ways. First, while most of the reasoning process involves deductive reasoning to arrive at conclusions, CR has very little use of such a method. Evidences play a vital role in deductive reasoning. CR works on the basis of observations rather than evidences. Often, these observations are subjected to periodic changes making them dynamic in nature. Therefore, a commonsense knowledge base, which acts as a primary source of information to draw inferences, must undergo constant updates. Second, to draw inferences, a machine must "understand" physical objects, time, space, basic rules of interaction, etc. Such type of information is referred to as common knowledge generally possessed by every human individual. Modern computing aids would have no immediate exertion in executing the first challenge of having an induced reasoning. It is the second part where one has to program a machine to understand the physical aspects related to everyday activity making it the most challenging part in automation of CR.

Inference drawing in association with a commonsense knowledge base is full of reasoning uncertainties. Several reasons account to this regard and one that stands out is the inductive nature of reasoning with respect to commonsense knowledge. A little can be done to avoid inductive reasoning in the process due to the lack of evidences in commonsense knowledge representation. It is inevitable that a conclusion derived from observations and experiences using induced reasoning will remain probable or uncertain. Unfortunately, uncertainty is not only restricted to the process of reasoning in commonsense knowledge but also exists in knowledge representation as well. The computer society has seen numerous developments in the field of knowledge representation. Yet a very few among them have made a positive impact on commonsense knowledge representation. Representation has a large influence on inference drawing. An efficient knowledge representation technique can be a catalyst in an effective reasoning process.

Overall, the challenges and perceptions related to commonsense reasoning study yield a lot of thought-provoking discussions. The modern computing aids have given several programming paradigms and tools to build effective knowledge representation techniques. Semantic networks are one among such tools which keep the hope of achieving commonsense reasoning alive. The subsequent sections of this paper discuss these matters in detail and analyse the amount of practicality in automation of commonsense reasoning.

2 What Constitutes a Commonsense Knowledge Base?

To be called as a commonsense knowledge base, a repository must have facts about activities, objects, and the outside world. These facts could be as simple as "sugar is sweet". To perform reasoning on such knowledge base, the information is needed to be stored in object-level [1]. Object-level information requires representing knowledge using an efficient data structure that can aid computer programs to constitute new information based on the available ones. In all these years, semantic networks have become the most practiced knowledge base constructors. There have been few more options in the form of linked lists, frames, and graphical representations [2], but semantic networks clearly outperform the rest. Irrespective of their structure or type, few key characteristics are expected from a commonsense knowledge base, which are as follows.

2.1 Atomicity

Reducing objects to their root level so that they become indivisible always simplifies data representation. This famous property known as atomicity of information works well with commonsense knowledge too. Functioning of inductive reasoning becomes better with simplified object notations. A lot of modern era data structures like dictionaries or JavaScript Object Notation (JSON) can feed a knowledge manager with ease if the information is stored at objet-level. Conclusions made using a lot of atomic objects are often more acceptable.

2.2 Inheritance

Inheritance brings abstraction into object-oriented approaches. Taxonomy in the form of semantic networks reflects inheritance at every stage. This could be a very important property for a commonsense knowledge base as inheritance has a huge effect on inductive reasoning. Sometimes, the decisions involving commonsense are largely influenced by inherited properties of objects. For example, the evaluation of the question "Can bats eat fruits?" using commonsense would involve a knowledge of a bat being an animal but not a wooden object. Though this fact is not explicitly mentioned in the phrase, the answer to that question must iterate the fact that bat is an animal.

Here, most of the general properties of animal would also apply to bat because of inheritance.

2.3 *Classification*

Classification is grouping of objects based on their properties. It is as important as inheritance. It reduces the amount of derivations one has to make during reasoning. Grouping has several advantages. It also reduces the complexity of prediction. Moreover, it provides better understanding of the world around us.

2.4 *Association*

Defining relation among objects is very crucial to apply commonsense knowledge. Associating one object with several using a well-defined relation leads to an effective inference method. Association also brings complexity in representation. While evaluating a commonsense knowledge, we expect objects and their properties to be dynamic. This may change the association between objects too. And in some cases, different objects of the same class may exhibit different associations among them.

3 Reasoning Uncertainties

Uncertainty in reasoning is not new. Every conclusion drawn from a set of knowledge bases need to be reiterated for its truth value [3]. Commonsense reasoning is no different in this regard. Every assumption made after considering substantial amount of facts, observations, and experiences need to be evaluated for its truthiness. And commonsense knowledge plays a very vital role in this evaluation process. Consider the following example. When an object is dropped to the floor, it is uncertain whether the object can take the impact or not. Apart from the basic physicality of the object, several other factors influence the result of this action. Height is one of the important factors that could decide the result right away. It is common sense that the height from which the object is dropped need to be considered before arriving at conclusions about the condition of the object after the impact. But, even after that, the conclusion remains uncertain.

One possible solution to reduce uncertainty with commonsense application is to use experience-based induction. Two things are very important in this process: number of positive experiences and probability of experiencing the same result in the near or distance future. Though it looks like a probabilistic model, the commonsense knowledge plays an important role in evaluating the expression. A hybrid approach comprising of experience-based learning with constantly updating knowledge base can certainly reduce the uncertainty in reasoning. When it comes to the representation of uncertainty, the probabilistic approach seems more practical [4] because the conclusion derived from commonsense knowledge are often subjected to a measure of belief. A probabilistic measure highlights the truthiness of a conclusion.

4 Drawing Inference for Commonsense Reasoning

4.1 The Inference Engine

There is an important question ahead of every individual trying to build a commonsense reasoning machine. How drawing inference differs in commonsense knowledge compared with any other inference methods? Generally, an inference engine works around two components: a well-defined knowledge base and an algorithm that acts as a knowledge manager. The knowledge base is usually a combination of curated structured data along with real-time data. A system ought to exhibit commonsense must have a third component in the form of commonsense knowledge (see Fig. 1) which may not have any direct association with objects of the knowledge base.

Commonsense knowledge in its simplest form is a materialistic knowledge about physical objects, their properties, laws of physics applied to such objects, and influence of time and space on actions of these objects. Drawing inference in commonsense reasoning would largely dependent on such a knowledge during conclusions as well as validation of conclusions. Taxonomy may not be of much help in representing this kind of knowledge. There are too many dynamic entities like active rules, real-time data, and observations in that representation.

Another certain fact about commonsense reasoning is that the inference cannot be drawn with the help of a single method. The reason why a monotonic inference method would not be ideal is that the validation of drawn conclusion with commonsense knowledge base may change the original course of inference and call a secondary inference to substantiate the truthiness of earlier conclusion. Such type of reasoning is known as plausible reasoning where the correctness of the result is not

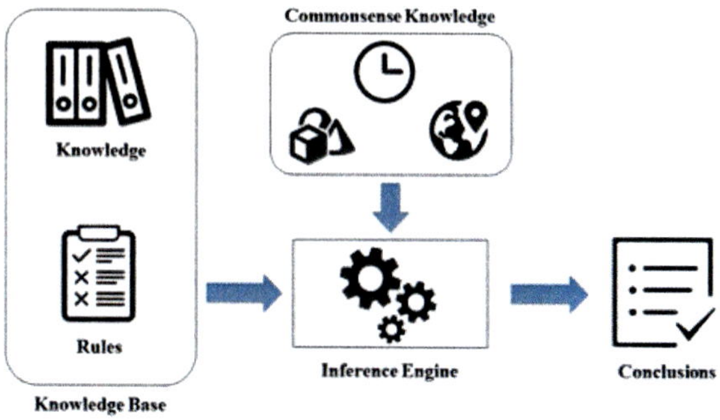

Fig. 1 An inference engine of CR machine backed-up by a separate commonsense knowledge

guaranteed [5]. While observation-based inference primarily concludes an action, probabilistic inference must validate the truthiness of that conclusion. Though different inference techniques can be experimented with commonsense reasoning, use of two aforementioned methods looks more realistic.

There are several methods for inference drawing being used in the field of expert systems. Not all of them ensembles the application of commonsense reasoning. A goal-driven approach or backward chaining for inference is not at all suitable in case of commonsense reasoning due to uncertain conclusion. On the other hand, reasoning cannot be data-driven too. A formulated hypothesis is required to trigger backward chaining whereas in-detail dataset can elicit forward chaining [6]. The idea of inference in commonsense reasoning follows a forward chaining with the primary knowledge base, arrives at certain conclusion, after which it reiterates the correctness using backward chaining with the secondary commonsense knowledge. It seems odd when a system uses two of the well-known yet contradictory methods to arrive at one conclusion. The reason why a single method does not seem to satisfy what commonsense reasoning simulates is simple: the absence of a strong association between objects of commonsense knowledge and the knowledge base. Thus, the indication is to incorporate multiple methods at different points of inference to keep both the knowledge bases separate yet parallel.

4.2 Systematizing Commonsense Reasoning

Computing power has reached a satisfactory level with cutting-edge technologies in modern-day computer systems. Yet, there has been no system that can give a satisfactory commonsense exhibition. There are few complexities ahead of systematizing commonsense reasoning:

Problem definition Before simulating commonsense, one has to understand physical objects, their materialistic properties, basic rules of interaction, etc. A human child learns about them by observing over a period of time hence developing his/her commonsense. From a computer's perspective, these are partially understood.

Plausible reasoning According to Ernest Davis—an active expert in the field of commonsense reasoning—the conclusions of commonsense reasoning are always expected to be plausible. It is very difficult to arrive at a truthfully acceptable conclusion even after providing considerable amount of input.

Size of data The amount of data on which the reasoning needs to be applied is of vast array. In most of the situations, a machine may have to consider ample amount of information before arriving at conclusions.

Knowledge representation Commonsense knowledge representation has no specific data structure or format. Moreover, abstraction is difficult to achieve when there is too much data describing a group of objects.

5 Impacts of an Effective Inference Technique

The most effective inference technique lies inside a human brain. Simulating what happens inside a biological matter which is barely understood would take all of computing powers associated with the most efficient reasoning methods. Recently, IBM Watson,a machine capable of working on unstructured data and give analytic solutions, proved that inference can be improvised [7]. An ideal inference technique must operate on a known or unknown datasets to derive conclusions that can be substantiated using real-time data. Also, it must incorporate a mistake-bound learning to enhance the accuracy of the result over a period of time. But performance is not the only concern in commonsense reasoning. Some applications of commonsense reasoning expect accuracy along with speed as the major factor. For instance, a driverless car expects input in a very short response time. An inference technique working with two distinct knowledge bases might want to finish the process of refining the conclusion within no time. If a machine-learning algorithm is expected to simulate commonsense then training the data can no longer be part of the process in some applications. In general, it is difficult to find a general solution to inference method for all the possible commonsense reasoning applications.

6 What is Expected from Commonsense Reasoning?

There have been a very few applications of commonsense reasoning in expert systems. Complexity with respect to practical simulation of commonsense has been addressed very few times and it is difficult to find notable works in that regard. Presently, commonsense reasoning operates under two subfields of artificial intelligence. One, knowledge-based systems and the other is machine-learning algorithms. There are a lot of problem statements in the field of artificial intelligence to which a touch of commonsense can yield a better result. When machine learning is applied over huge corpora, commonsense reasoning can boost-up the initial step of data aggregation and find quicker results than usual. For example, if a machine is asked to collect details of all the hotels with hot water swimming pools, if it had backed-up with commonsense knowledge suggested to avoid hotels from tropical regions where the temperature is likely to be more, searching would take lesser than the usual time. There are numerous such small-scale implementations that a commonsense knowledge can bring into the field of computer science. There is a wide spread area for commonsense to operate if it is made practical. Web mining, business analytics, deep learning, natural language processing, search engine optimization, pattern recognition, and e-governance are few in the line. Overall, commonsense reasoning is expected to refine the results rather than producing new ones. Accuracy of a computational device can hugely benefit from the use of commonsense reasoning.

7 Conclusion

It is clear that a different "modus operandi" is needed for machine-learning algorithms to tackle commonsense knowledge. Though the methods for inference are not less in number, there is a necessity of an effective technique which can cater the needs of an acceptable commonsense reasoning model. A benchmark in this regard is expected from the computer science society. Evolving into a system that can keep commonsense knowledge separate from the core system can help us build a better model. That way, a common repository with commonsense knowledge can be defined and reused with multiple systems. More work on existing semantic nets is needed to build a flexible knowledge base. Logical programming aids also need to be reiterated for the development of a strong commonsense reasoner.

References

1. Davis E (2014) Representation of commonsense knowledge. Morgan Kaufmann
2. Chandan H, Ashwini K (2017) Automation of commonsense reasoning: a study on feasible knowledge representation techniques. Int J Adv Res Comput Sci 8(9):601–604
3. Halpern JY (2017) Reasoning about uncertainty. MIT Press, Cambridge
4. Shachter RD, Kanal LN, Henrion M, Lemmer JF (eds) (2017) Uncertainty in artificial intelligence, vol 5, no 10. Elsevier, Amsterdam
5. Davis E, Marcus G (2015) Commonsense reasoning and commonsense knowledge in artificial intelligence. Commun ACM 58(9):92–103
6. Al-Ajlan A (2015) The comparison between forward and backward chaining. Int J Mach Learn Comput 5(2):106–113
7. Chen Y, Elenee-Argentinis JD, Weber G (2016) IBM Watson: how cognitive computing can be applied to big data challenges in life sciences research. Clin Ther 38(4):688–701

Segmentation and Detection of Glioma Using Deep Learning

Navneeth Krishna, Mahammad Rumaan Khalander, Nandan Shetty and S. N. Bharath Bhushan

Abstract The process of segmentation and detection of glioma plays a crucial role in medical image segmentation. With this paper, we demonstrate the capacity of deep learning algorithm-based U-Net model to handle varied spectrum of data and a comprehensive analysis of the performance of this deep learning approach with real-world data like BraTS 2018 dataset under consideration.

Keywords Deep learning · UNet · Medical image segmentation · Glioma

1 Introduction

Medical image processing has gained popularity in the recent times due to a dramatic increase in the computational capacity of modern day systems and also the recent increase in the demand for more sophisticated approaches in medical image processing. Also, due to a rapid increase in the number of diseases, there is a need for more accurate and error-free approaches in medical diagnosis to aid the doctors in the process of treatment of the diseases. One such disease which has gained popularity for being strenuous in its detection is glioma.

Glioma is a kind of brain tumor that may also occur in the spinal cord but more often found in the brain. Glioma can affect the normal functioning of the human brain and can finally lead to a painful death. The exact causes of glioma are although not

N. Krishna (✉) · M. R. Khalander · N. Shetty · S. N. Bharath Bhushan
Department of Computer Science and Engineering, Sahyadri College of Engineering and Management, Adyar, Mangaluru, Karnataka, India
e-mail: mnkb20@gmail.com

M. R. Khalander
e-mail: rumaankalander@gmail.com

N. Shetty
e-mail: nandanshetty97@gmail.com

S. N. Bharath Bhushan
e-mail: sn.bharath@gmail.com

N. N. Chiplunkar and T. Fukao (eds.), *Advances in Artificial Intelligence and Data Engineering*, Advances in Intelligent Systems and Computing 1133,
https://doi.org/10.1007/978-981-15-3514-7_10

known, its diagnosis and treatment are done by the professionals in the medical field. Today, it takes a radiologist with sufficient experience a lot of time to go through the symptoms and the MRI scans to diagnose and detect glioma. Moreover, this is a crude and manual task and is heavily subject to the expertise of the medical personnel. The current method of diagnosis of glioma involves taking images of the brain by making use of magnetic resonance imaging technique and then a manual inspection of the MRI slices by the medical professional. The doctor has to go through every slice to identify the presence of glioma and to localize the region of the tumor. Later, a comprehensive study of the characteristics and size of tumor is done to proceed with the treatment.

Magnetic resonance imaging is an imaging technology which is broadly used for the diagnosis of diseases confined to the inner human body. MRI works by exciting the water molecules subject to heavy magnetic field and mapping their activity using a sensor. There are four crucial types of MRI sequences that can be done in order to visualize the region of interest. This includes T1, T1-CE (Contrast Enhanced), T2 and fluid-attenuated inversion recovery (FLAIR). Figure 1 shows the different scans. Each of these scans differs by the way how they represent the internal parts of the human body like the cerebrospinal fluid (CSF), the gray matter and the cranium.

In the wake of the twenty-first century where the availability of faster and more precise computers available, the computers have become smarter. The fields of application of computers to solve problems have increased widely, and day by day, there

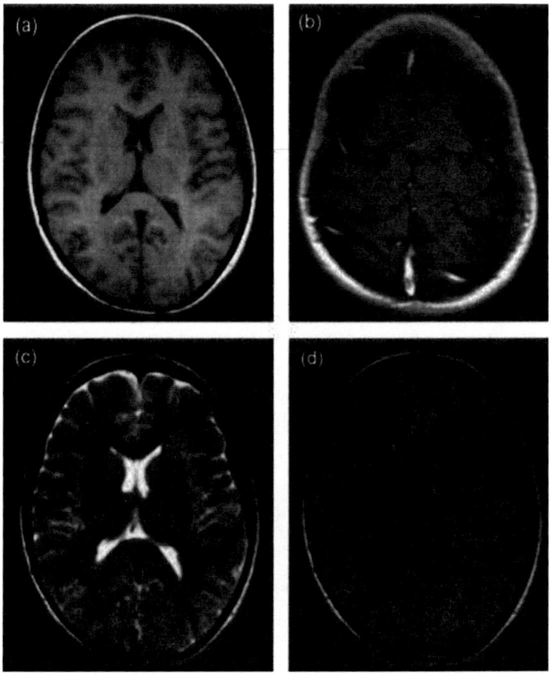

Fig. 1 **a** T1; **b** T1-CE (gadolinium); **c** T2; **d** FLAIR MRI sequences

is tremendous improvement in this field. At the heart of every computation lies the concept of computing algorithms which lay down simple and unambiguous steps to solve a given problem. Ever since this development, there have been significant improvement in the computer algorithms that were designed to make the computers smarter and be applicable to solve many problems in general. With the advancement in the field of machine learning, many such algorithms have taken birth and have been applied extensively in various field of computing. Another such field, which happens to be a subset of machine learning is deep learning. Aim of this field mimics the functioning of human brain in computers. There are various algorithms in deep learning that have proved their applicability in solving real-world problems [1].

1.1 Literature Survey

A literature survey was done on over twenty-five papers which had used various methods for segmenting the brain tumor. Some of the methods used a classical approach to the problem, while the others have used some advanced state-of-the art techniques to segment the brain tumor. Since these methods were applied on different datasets, we cannot generalize the performance of all the studies system and conclude on the overall drawbacks of the systems. However, we can come to a vague conclusion of the performance of various approaches and come to an understanding of the nature of various algorithms and their influence and performance in brain tumor segmentation.

The methodology in [2] has made use of convolutional neural network with densely connected layers and has made use of 3D voxel-level learning. This research has yielded a good result, and the segmentation was found to be successful although the paper does not mention any numerical result as such. The research [3] made use of fully connected CNN for segmentation of brain tumor with a hierarchical dice loss for loss metric. This has yielded better result and is also based on VGG16 network, which is a very popular network data for learning process.

The research methodology in [4] had made use of support vector machines in which the primary features of brain tumors were extracted and fed into an SVM. This approach yielded an accuracy of 94.12%. Also, the paper [5] made use of a collection of intelligent techniques where the near infrared imaging technology was made use along with seed growing and kernel-based fuzzy clustering was used. This method yielded an accuracy of 96.35%. The research titled An efficient brain tumor detection from MRI images using entropy measures [6] made use of probability matrix and gray level co-occurrence matrix method where the results were obtained from Havrda Charvat Entropy that was found to be better than the other entropy functions used for brain tumor detection.

Another method [7] made use of artificial neural networks with a preprocessing technique called Gabor filter. It extracted the texture features of the image. This method gave out 89.9% accuracy score. Moreover, another method as conducted in [8] which made use of K nearest neighbors along with fuzzy connected C-mean

(FCM) method yielded 95.80% accuracy, on the other hand, [9] which made use of K-means clustering algorithm yielded a better accuracy of 96.80%.

The brain tumor segmentation was also done using various other algorithms like in [10] which used watershed algorithm, CCL and multiparameter calculation. This method gave out good results in terms of entropy, eccentricity, area and perimeter values of the tumorous regions. Reference [11] shows a method of applying morphological operation on the image which is obtained by applying thresholding of the image. The tumorous edge was detected using the Sobel filter and achieved an accuracy of 95.6%. Method mentioned in [12] used conversion to grayscale, median filter application and high pass filter for edge detection. This gave out an accuracy of 97% for segmentation of brain tumor.

The paper [13] made use of displacement model to map the tumor growth and ended the segmentation by skull stripping. In various cases, it gave the accuracies of 93, 94, 95 and 87%.

Another approach made use of U-Net method [14] which is a CNN approach for biomedical segmentation of images using FCNN and got an accuracy of 77.5% for mere 20 images. Another method [15] used Berkeley wavelet transform (BWT) and SVM for brain tumor segmentation got an accuracy of 96.51%. The standard BRATS dataset in its original paper [16] gave an overall dice score of 0.8 as the base with Hausdroff distance metric for ranking and Ref. [17] made use of cascaded hierarchical model with an accuracy of 79.37% for 100 images. Another U-Net approach [18] achieved a dice score of 0.42. ImageNet classification with CNNs [19] made use of four-layer CNN with ReLU activation function got an accuracy of 78.1%.

Another approach made use of generative segmentation of brain tumor [20] with deep convolutional encoder–decoder network that scored a dice score of 0.94 and 0.82 for various datasets. Reference [21] made use of convolutional neural networks with input shifting and output interlacing and scored accuracy of 65.4% and [22] scored dice score of 86.38 with 3D U-net model. Another method as per [23] made use of type-specific sorting with LinkNet architecture got a dice score of 0.79, and Ref. [24] with V-Net CNN got a dice score of 0.89. Another method [25] with manual annotation of images and application of various method gave a range of dice score of 0.61 to 0.77.

2 Methodology

2.1 Model Architecture

In this paper, we have made use of a convolutional neural networks [1, 26] model for segmenting the glioma region (region of interest) from raw MRI images. The model is called U-Net model [27] and was introduced by Olaf Ronneberger et al. in 2015. The model makes use of a series of convolutions and de-convolutions (upscaling) to achieve better segmentation than the previous approaches.

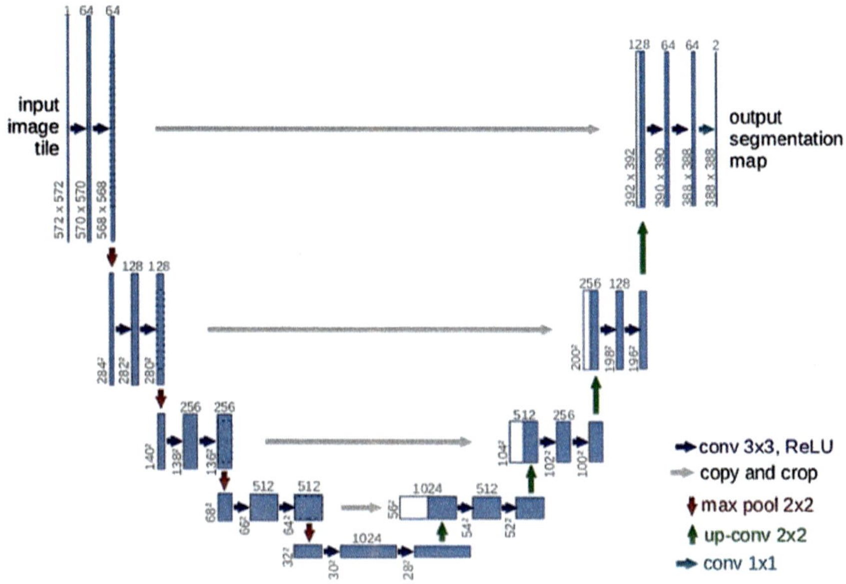

Fig. 2 U-Net model architecture

In Fig. 2, the blue cells refer to a multi-channel feature maps of the convolutional neural networks. The number above the cell represents the number of channels. The dimensions of each cell are also shown. The white cells show the copied feature map, and the arrows denote various operations.

The U-Net model consists of two paths, namely the reduction (contracting) path (to the left side of the architecture) and the dilating (expanding) path (to the right side). The contracting path is a basic convolutional neural network in which we repeatedly apply 3×3 unpadded convolutions followed by an application of the ReLU activation function.

ReLU:

$$\mathrm{ReLU}(x) = \max(0, x) \tag{1}$$

The ReLU activation function is a popular activation function which introduces maximum nonlinearity in the convolution layers. This activation function returns the maximum value of the weighted sum and 0 to maintain maximum possible nonlinearity in the network. ReLU layers are introduced in between the layers for the model to learn maximum variety in the input data. This step is followed by 2×2 max pooling for the purpose of down sampling. For every down-sampling step, we double the kernel (feature map) count. In the expanding path, we upsample the feature map which is followed by a 2×2 convolution which reduces the number of feature map to half (we make use of up-convolution technique to achieve this). Then, we concatenate the feature maps obtained with the cropped feature map from the contracting path followed by a two 3×3 convolution and a ReLU activation function. In the final layer, a 1 × 1 convolution is applied to map each 64 component feature vector to the number of target classes. Totally, there are 23 convolutional layers in this architecture. The entire network is trained by taking into consideration the cross-entropy as the loss function. The aim of the model is to reduce this loss function over the iterations and the epochs. The model is found to reduce this error as the iterations pass. However, the value of cross-entropy is found to be largely dependent on the type and variations in the training data.

Cross-Entropy:

$$-\sum_{c=1}^{M} y_{o,c} \log(p_{o,c}) \tag{2}$$

The last layer of the U-Net architecture makes use of the popular activation function Softmax. The reason for using this function is that since this segmentation is a two-class problem, the final layer of the network must be able to output a probabilistic value in order to decide the class of the corresponding pixel. Hence, the softmax activation function is used here.

Softmax:

$$p_k(X) = \frac{e^{a_k(X)}}{(\sum_{K'}^{K} e^{a_{k'}(x)})} \tag{3}$$

Dataset Description:

For the process of model evaluation, we have made use of BraTS 2018 dataset [16, 28]. The BraTS 2018 dataset contains two kinds of glioma MRI datasets. The high-grade glioma (HGG) and the low-grade glioma (LGG). In the dataset, there are 210 patients' data with HGG and 75 patients' data with LGG. Each of these datasets contains four different types of MRI sequences: T1, T1CE, T2 and FLAIR sequences. This is followed by the ground truth corresponding to that patient. To attain maximum variation in the dataset, our approach has randomly combined HGG and LGG and made use of the FLAIR sequence of the dataset only as it is quite popular in medical field. To greatly reduce the complexity of application, we have taken only a part of the BraTS 2018 training dataset and divided the dataset into training and testing data. Due to this, the heavy computational dependency of the model is greatly reduced and as the aim was to segment glioma and not to segment the various types of tumor like enhancing tumor or non-enhancing tumor. To further reduce the amount of computation, each FLAIR BraTS patient's data file is sliced at 86th slice and 66th slice. This has reduced the amount of computation required to convolve the entire 3D scan.

The model is implemented [29] using Tensorflow [30], an open-source deep learning library created primarily by Google. The inputs to the model are treated as tensors, and Tensorflow builds an executable graph of the entire model. Once the graph is build, the inputs are given to the graph in the form of tensors and the nodes which represent a computation unit, computes the required mathematical operation.

2.2 *Steps in the Approach*

The steps taken in our approach are as follows which include preprocess, training and testing.

Algorithm

Input: BraTS 2018 dataset directory containing all the images.

Output: Generated image with desired segmentation in accordance with the ground truth.

Method:

Preprocess:

- For every FLAIR BraTS file in the directory, read the data into a preprocessing program and slice the 86th slice and 66th slice and save in a directory.
- For every corresponding ground truth, slice the 86th slice and 66th slice and save in the same directory.
- Data augmentation is carried out to diversify the data and to avoid over-fitting.
- Split the data thus obtained into training and testing data.
- N4FieldBiasCorrection can be applied to further simplify the data and reduce the dimensions of the input images.

Training:

- Build the U-Net model as per the architecture with the required number of convolution layers.
- Pass the training image data from the training directory to the model with the corresponding ground truths for validation.
- Run the training process for desired number of epochs to achieve satisfactory performance.

Testing:

- Load the trained model and feed in the testing set of data.
- Run evaluation tests to obtain the corresponding results.

3 Results

The results of our approach are summarized below. The entire model was run on Google Colab with the native features available for any user. Google Colab allows the users to use up to 12 GB of Nvidia K80 GPU and Intel Core i7 processor in its cloud. The model was written using Python programming language and was run various times to evaluate its performance under rigorous conditions.

3.1 *Experimentation*

Table 1 represents the performance of the model with respect to various parameters considered during the training and testing procedure. Five trials were conducted with variable number of epochs for each trial. The number of iterations is also varied with respect to the experimentation conditions. The validation accuracy of the model can also be found in the table. The testing error is also noted.

Figure 3 shows the appearance of the segmented glioma region form the MRI images. The first column shows the raw MRI image which is fed into the model. The next column shows the ground truth (expert labeled glioma region), and the last column shows the segmentation of the glioma region as processed by our model.

Figure 4 shows only the ground truth and the segmentation of the MRI and the generated ground truth from the U-Net model. This is the testing output of the model.

Table 1 Result table of glioma segmentation using deep learning

Test	Epochs	Iterations	Validation accuracy	Testing error (%)
1	20	10	0.9769	9.05
2	20	10	0.9874	15.37
3	15	10	0.9835	14.13
4	15	10	0.9837	13.73
5	30	20	0.9843	13.87

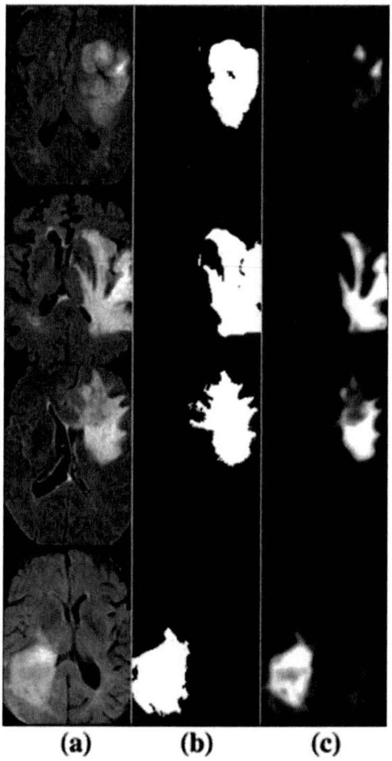

Fig. 3 **a** Original image; **b** ground truth; **c** segmented glioma

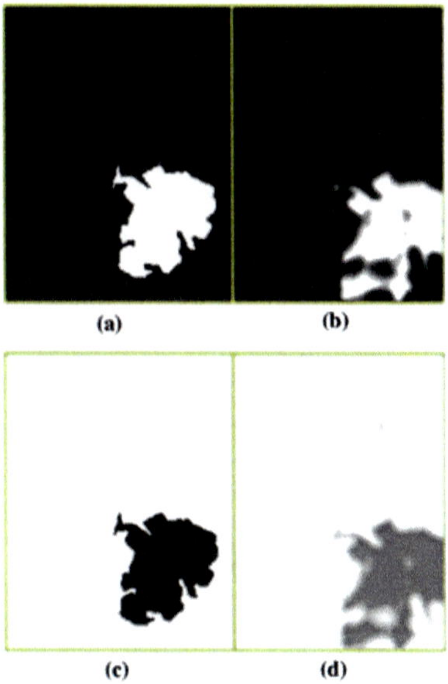

Fig. 4 **a** Ground truth (in white); **b** segmented ground truth (in white); **c** ground truth (inverted color); **d** segmented ground truth (inverted color)

Figure 5 shows the variation of accuracy with the iterations. It is clearly visible that the accuracy of the model increases as the number of iterations pass. The model is found to be performing at a very high accuracy after about 400 iterations. The model's learning is found to be almost constant even though the learning rate was varied as the training proceeded. The buildup of the accuracy over the iterations has been found to be on par with the state-of-art systems reaching very high levels of accuracy during the training process. The results come out to be interesting even though the model has not encountered the entire dataset.

Further improvement can be done with the predictions of the model by post-processing the obtained segmentation using some simple filters. This will enhance the appearance of the output image and give much more clearer borders for the desired segmentation.

4 Conclusion

The performance of the U-Net model for segmentation of glioma is thoroughly evaluated, and obtained results are inspected for superior performance of deep learning algorithms on real-time datasets like BraTS 2018. Even though the results are not top notch, due to highly restricted variations in the dataset used, the performance of the model was found to be significantly better than the previous approaches to this problem. This also concludes on the capacity and wide range of applicability of the U-Net architecture and the deep learning algorithms like CNNs to perform well in restricted conditions like variations in the data and limited availability of data, providing clear path to further improvements in this field.

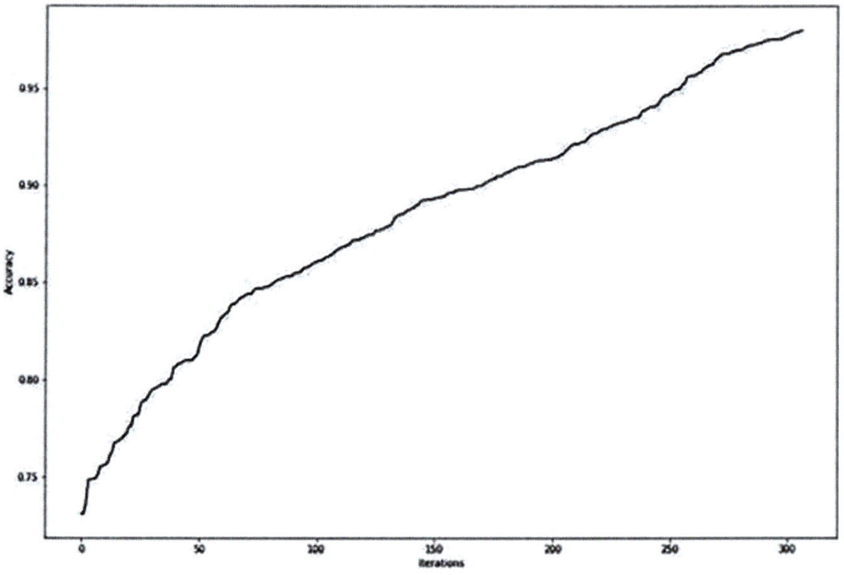

Fig. 5 Accuracy of the model over iterations

References

1. LeCun Y, Bengio Y, Hinton G (2015) Deep learning. Nature 521(7553):436. https://doi.org/10.1038/nature14539
2. Chen L et al (2018) MRI tumor segmentation with densely connected 3D CNN. Medical imaging 2018: image processing, vol 10574. International Society for Optics and Photonics. https://doi.org/10.1117/12.2293394
3. Ronneberger O, Fischer P, Brox T (2015) U-net: convolutional networks for biomedical image segmentation. In: International conference on medical image computing and computer-assisted intervention. Springer, Cham. https://doi.org/10.1007/978-3-319-24574-4_28
4. Telrandhe SR, Pimpalkar A, Kendhe A (2016) Detection of brain tumor from MRI images by using segmentation & SVM. In: 2016 world conference on futuristic trends in research and innovation for social welfare (startup conclave). IEEE. https://doi.org/10.1109/startup.2016.7583949
5. Jayalakshmi C, Sathiyasekar K (2016) Analysis of brain tumor using intelligent techniques. In: 2016 international conference on advanced communication control and computing technologies (ICACCCT). IEEE. https://doi.org/10.1109/icaccct.2016.7831598
6. Somwanshi D et al (2016) An efficient brain tumor detection from MRI images using entropy measures. In: 2016 international conference on recent advances and innovations in engineering (ICRAIE). IEEE. https://doi.org/10.1109/icraie.2016.7939554
7. Amsaveni V, Albert Singh N (2013) Detection of brain tumor using neural network. In: 2013 fourth international conference on computing, communications and networking technologies (ICCCNT). IEEE. https://doi.org/10.1109/icccnt.2013.6726524
8. Aswathy SU, Glan Deva Dhas G, Kumar SS (2014) A survey on detection of brain tumor from MRI brain images. In: 2014 international conference on control, instrumentation, communication and computational technologies (ICCICCT). IEEE. https://doi.org/10.1109/iccicct.2014.6993081

9. Kharrat A et al (2009) Detection of brain tumor in medical images. In: 2009 3rd international conference on signals, circuits and systems (SCS). IEEE. https://doi.org/10.1109/icscs.2009.5412577
10. Dhage P, Phegade MR, Shah SK (2015) Watershed segmentation brain tumor detection. In: 2015 international conference on pervasive computing (ICPC). IEEE. https://doi.org/10.1109/pervasive.2015.7086967
11. Hunnur MSS, Raut A, Kulkarni S (2017) Implementation of image processing for detection of brain tumors. In: 2017 international conference on computing methodologies and communication (ICCMC). IEEE. https://doi.org/10.1109/iccmc.2017.8282559
12. Aswathy SU, Glan Deva Dhas G, Kumar SS (2014) A survey on detection of brain tumor from MRI brain images. In: 2014 international conference on control, instrumentation, communication and computational technologies (ICCICCT). IEEE. https://doi.org/10.1109/iccicct.2014.6993081
13. Bauer S, Nolte L-P, Reyes M (2011) Segmentation of brain tumor images based on atlas-registration combined with a Markov-Random-Field lesion growth model. In: 2011 IEEE international symposium on biomedical imaging: from nano to macro. IEEE. https://doi.org/10.1109/isbi.2011.5872808
14. Dhage P, Phegade MR, Shah SK (2015) Watershed segmentation brain tumor detection. In: 2015 international conference on pervasive computing (ICPC). IEEE. https://doi.org/10.1109/pervasive.2015.7086967
15. Bahadure NB, Ray AK, Thethi HP (2017) Image analysis for MRI based brain tumor detection and feature extraction using biologically inspired BWT and SVM. Int J Biomed Imaging 2017. https://doi.org/10.1155/2017/9749108
16. Menze BH et al (2015) The multimodal brain tumor image segmentation benchmark (BRATS). IEEE Trans Med Imaging 34(10):1993–2024
17. Bauer S, Nolte L-P, Reyes M (2011) Segmentation of brain tumor images based on atlas-registration combined with a Markov-Random-Field lesion growth model. In: 2011 IEEE international symposium on biomedical imaging: from nano to macro. IEEE. https://doi.org/10.1109/isbi.2011.5872808
18. Dong H et al (2017) Automatic brain tumor detection and segmentation using U-Net based fully convolutional networks. In: Annual conference on medical image understanding and analysis. Springer, Cham. https://doi.org/10.1007/978-3-319-60964-5_44
19. Krizhevsky A, Sutskever I, Hinton GE (2012) Imagenet classification with deep convolutional neural networks. Advances in neural information processing systems. https://doi.org/10.1145/3065386
20. Menze BH et al (2015) The multimodal brain tumor image segmentation benchmark (BRATS). IEEE Trans Med Imaging 34(10):1993–2024
21. Seyedhosseini M, Sajjadi M, Tasdizen T (2013) Image segmentation with cascaded hierarchical models and logistic disjunctive normal networks. In: Proceedings of the IEEE international conference on computer vision. https://doi.org/10.1109/iccv.2013.269
22. Erden B, Gamboa N, Wood S (2018) 3D convolutional neural network for brain tumor segmentation
23. Krizhevsky A, Sutskever I, Hinton GE (2012) Imagenet classification with deep convolutional neural networks. Advances in neural information processing systems. https://doi.org/10.1145/3065386
24. Kim, JU, Kim HG, Ro YM (2017) Iterative deep convolutional encoder-decoder network for medical image segmentation. arXiv preprint arXiv:1708.03431. https://doi.org/10.1109/embc.2017.8036917
25. Long J, Shelhamer E, Darrell T (2015) Fully convolutional networks for semantic segmentation. In: Proceedings of the IEEE conference on computer vision and pattern recognition. https://doi.org/10.1109/cvpr.2015.7298965
26. LeCun Y et al (1998) Gradient-based learning applied to document recognition. Proc IEEE 86(11):2278–2324. https://doi.org/10.1109/5.726791

27. Sobhaninia Z et al (2018) Brain tumor segmentation using deep learning by type specific sorting of images. arXiv preprint arXiv:1809.07786
28. Bakas S et al (2017) Advancing the cancer genome atlas glioma MRI collections with expert segmentation labels and radiomic features. Sci Data 4:170117. https://doi.org/10.1038/sdata.2017.117
29. Akeret J et al (2017) Radio frequency interference mitigation using deep convolutional neural networks. Astron Comput 18:35–39. https://doi.org/10.1016/j.ascom.2017.01.002
30. Bakas S et al (2017) Advancing the cancer genome atlas glioma MRI collections with expert segmentation labels and radiomic features. Sci Data 4:170117. https://doi.org/10.1038/sdata.2017.117

Character Recognition of Tulu Script Using Convolutional Neural Network

Sachin Bhat and G. Seshikala

Abstract Handwriting classification and identification is one of the most interesting issues in the current research because of its variety of applications. It has leveraged its potential in reducing the manual work of converting the documents containing handwritten characters to machine-readable texts. The deep convolutional neural networks (DCNNs) are successfully implemented for the recognition of characters in various languages. This paper proposes a DCNN-based architecture for the classification of Tulu language characters. Tulu is one of the five Dravidian groups of languages used by around 50 Lakh people in the states of Karnataka and Kerala. This model is mainly developed to assist the character recognition of Tulu documents. A total of 90,000 characters including both vowels and consonants have been included in the dataset. This architecture is showing a satisfactory test accuracy of 92.41% for the classification of 45 handwritten characters.

Keywords Computer vision · Character recognition · Tulu language · Convolutional neural network

1 Introduction

Optical character recognition (OCR) is the process of converting the documents or images containing printed or handwritten characters into machine-readable format. In recent decades, it has leveraged its potential in reducing the tedious manual work of digitizing the images of printed or handwritten text. It has become one of the most interesting research fields in these days helping to recognize the characters from images. Different traditional approaches are used in OCR so far like support vector machines (SVM), template matching, K-nearest neighbor, hidden Markov model and so on. With the advancement of technology and increased computational

S. Bhat (✉) · G. Seshikala
Reva University, Bengaluru, India
e-mail: sachinbhat88@gmail.com

S. Bhat
Shri Madhwa Vadiraja Institute of Technology and Management, Bantakal, Udupi, India

N. N. Chiplunkar and T. Fukao (eds.), *Advances in Artificial Intelligence and Data Engineering*, Advances in Intelligent Systems and Computing 1133,
https://doi.org/10.1007/978-981-15-3514-7_11

power, deep learning techniques are acquiring the limelight from a couple of years. But, OCR of handwritten text is still a complex and gainsaying task to researchers as the model should deal with the challenges in identifying the characters from a variety of handwriting styles. This is especially true for the Indian languages which have a large number of character sets and the presence of compound characters. OCR with high accuracy is reported in English and other western languages which have a minimal number of characters and less structural complexity. But character recognition of Indian scripts is relatively intense because of its complex structure and identical nature of characters.

CNN is a famous deep learning approach which has been successfully used in different classification problems. In the domains of pattern classification, face recognition, natural language processing, CNN has a great ability to identify patterns in 2D data. It was first proposed by Fukushima [1] in the 80s. Due to the difficult process of training, it was not used for several years. It became famous only after LeCun who got good results in the identification of digits. In later years, several new architectures using CNN were proposed which recorded excellent performance and surpassed the traditional methods.

Remaining part of the paper has been arranged in the below-mentioned order. Section 2 describes the Tulu language script. Section 3 reviews the related work carried out on character recognition using neural networks. In Sect. 4, we introduce the dataset and the method in detail. Experiments are conducted, and evaluation of results is shown in Sect. 5. Section 6 will be the conclusion.

2 Tulu Script

Tulu is one of the five Dravidian families of languages in Southern India. There are more than 5 million people who speak Tulu in coastal Karnataka and northern Kerala. Tulu has derived its script from *Grantha Lipi*, having evolved from *Tamil Brahmi*. This script is also called as *Arya-zhuttu* or *Tigalari*. It is a well-known fact that most of the ancient documents in Kerala as well as in *Karavali* (costal) and *Malenadu* (hilly) regions of Karnataka are in *Tigalari* script. British linguist Robert Caldwell academically established the Dravidian languages in his monumental work on the grammar of South Indian languages. In this, Robert refers Tulu as one of the most developed Indian languages. Unfortunately, German missionaries started publishing the Tulu literature in Kannada script as printing presses had been established in Kannada in Mangalore. Disuse of Tulu due to the unavailability of printing technology gradually led to the extinction of this script over the period of time. However, in these days, there is regenerated interest in the language as manifested by the reality that many institutions and universities in and outside India are encouraging more research of Tulu language. Also, its near extinct script has sired new exuberance amongst the linguists as well.

Tulu is written as a sequence of syllables called *Aksharas*. Set of basic *Aksharas* of Tulu contains 13 *Swara* (vowels) and 34 *Vyanjana* (consonants). For our dataset, we have taken 45 *Aksharas* excluding *Anuswara* (Am) and *Visarga* (Aha) which do not have an independent existence of their own. They are pronounced with other vowels or consonants.

Handwritten data is collected from 40 native *Tuluvas* between the ages of 20 to 40. All writers have minimum knowledge of Tulu script. Train and test sets are divided based on collected images. Every person is asked to write 50 samples of each alphabet. Datasheets are scanned using Canon E560 scanner. Binarization and noise removal are done on the manuscript [2, 3]. Handwritten characters are extracted from scanned datasheets using machine learning OCR tool. Extracted characters are manually tagged and grouped to different classes.

3 Literature Review

In this part, we briefly depict some of the deep learning-based OCR methods used by earlier researchers. In the area of text classification, there are several works reported with respect to the benchmark online databases. [4] deals with character and word recognition for Chars74K [5] and ICDAR2003 [6] word datasets using compact CNN architecture. He [7] built a deep recurrent network for character classification in natural images. In this, CNN was used to create an ordered sequence from the word and long short-term memory (LSTM) to recognize these sequences. Shi et al. [8] offered a combined model of CNN with RNN unitedly achieved better results in recognition. Mhiri in [9] proposed a new approach for offline word recognition using DCNN without any explicit segmentation for IAM database [10].

A few remarkable works are also available for the character recognition of modern day Indian scripts. Most of the works in this category are either on Devanagari or Bengali handwritten characters. Two online databases are available for Bangla language namely Banglalekha [11] and CMATERdb [12] with 90,000 and 15,000 characters, respectively. Rabby [13] used a lightweight CNN-based model called BornoNet for classifying these handwritten characters. Alom in [14] deals with the implementation and performance analysis of Bengali character recognition using different DCNN models including FractalNet [15], ResNet [16], VGG [17], AlexNet [18] and DenseNet [19]. Pixel-level reconstruction and recognition of basic Bengali numerals and alphabets are addressed in [20]. This uses CNN pretrained on ImageNet database and a deep belief network to reconstruct the noisy pixels through transfer learning. [21] introduces a database of 17,000 Telugu characters including *vattu* and *gunithas* and exploits simple CNN to classify them. The only available Devanagari OCR using CNN is given by Avadhesh [22]. This OCR employs an ACNN model for classifying the alphabetic characters in an image. The limitation with this is that it is sensitive to the quality of image which can be accounted to the relative simplicity of the proposed model.

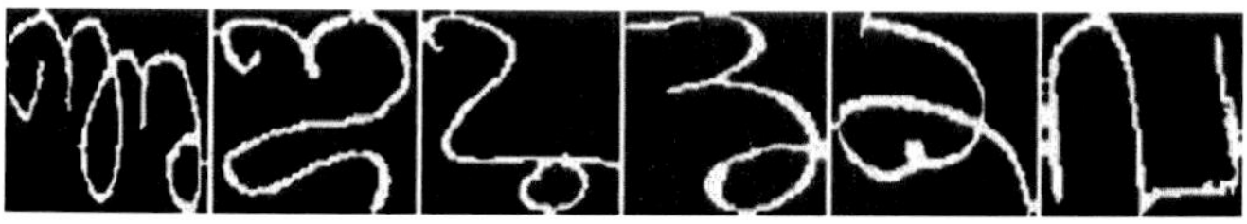

Fig. 1 Randomly chosen character images

4 Proposed Methodology

Our proposed model is based on the famous VGGNet [17] architecture. In this section, first we have given the description of our dataset and its preparation. After that, a brief review of VGG architecture and its evolution is provided. Then, we describe the adaptations done in the proposed model to show the best results on our customized data set.

4.1 Dataset and Preparation

A total of 90,000 characters belonging to 45 different classes are included in the dataset. Train, test and validation sets are taken in a proportion of 60:20:20. That means 54,000 training images, 18,000 test images and 18,000 validation images. All images in the dataset are resized to 32 × 32 pixels. Some of the random characters taken from six different classes of the dataset are shown in Fig. 1.

4.2 VGGNet: A Brief Review

It was introduced by Vision Geometry Group for Large-Scale Visual Recognition Challenge (ILSVRC) on ImageNet database in 2015. ImageNet is a Google database comprising of more than 1.2 million images. The basic idea of VGG is to use small convolutional filters in every layer to achieve deep network architecture. VGG16 model contains a stack of 16 weight layers where 13 convolutional layers are followed by three fully connected (FC) layers. Few other variants of VGG are also present like VGG19, the deepest with 19 weight layers. In each convolutional layer, it uses a receptive field of 3 × 3. Deep learning models prior to these used bigger receptive fields like 7 × 7 or 11 × 11 compared to VGG. Filters of small size help to expand the depth of the network with more discrimination. It has five convolution blocks, whose width starts from 64 and increments by 2 after every block till 512. Two or three convolutional layers are piled together with a pooling layer in each block . FC layers are generally used at the last stages of the CNN which connects to the output layer and constructs the desired number of outputs. In VGG, first two FC layers have 4096 channels and the third one with 1000 nodes fitting to the 1000 different classes in ImageNet dataset.

4.3 Proposed Model

Alike VGGNet, our model too has various sequential small-sized convolutional and FC layers. But, four basic changes have been made in the network architecture setting. We have used 32 × 32 grayscale images in input layer instead of 224 × 224. As characters used here are much smaller compared to objects in the pictures, smaller-sized images make it able to get the trade-off between the size of data and training time.

After several rounds of experiments, we have observed that test results remain same even after eliminating the last two convolutional layers. Hence, they are removed from the architecture. Lesser layers are used than VGG founded on the empirical analysis on the customized dataset.

Features extracted from the convolutional layers are sent to dense layers which are trained according to our data set. We have removed the top fully connected layers and replaced by our own customized layers by making last four layers trainable. Customized layers include two fully connected dense layers, a flatten layer and dropout layer with dropout ratio 0.5. Also, we have reduced the channels from 4096 to 1024. This will fruitfully cut down the depth of the entire network and makes the training quicker without diminishing the overall performance.

The output layer is also replaced with a new softmax layer of 45 classes relevant to our problem. We just train the weights of these layers and try to recognize the characters. The architecture of this model is represented in Fig. 2. Transfer learning-based approach is adopted here where basic VGGNet deep learning model is trained on ImageNet dataset. The theme behind transfer learning is that it is economic and effective to use deep learning models trained on huge image datasets and "transfer" this learning ability to new classification assumption rather than train a deep CNN classifier from scratch.

32 × 32 gray images are fetched as input to this 13 layer network. Adam optimizer [23] which has a combining advantage of Adagard and RMSProp is used as an alternative to classical stochastic gradient descent. Adam updates network weights iteratively based on training data (1). Adam paper says "it makes use of average of the second moments of the gradients i.e. uncentred variance rather than adapting the parameter learning rates established on the average first moment." Most of the recent NLP and deep learning applications are using Adam due to its straightforwardness in implementation and less memory requirement. Back propagation model with a learning rate of 0.01 and momentum of 0.1 is applied.

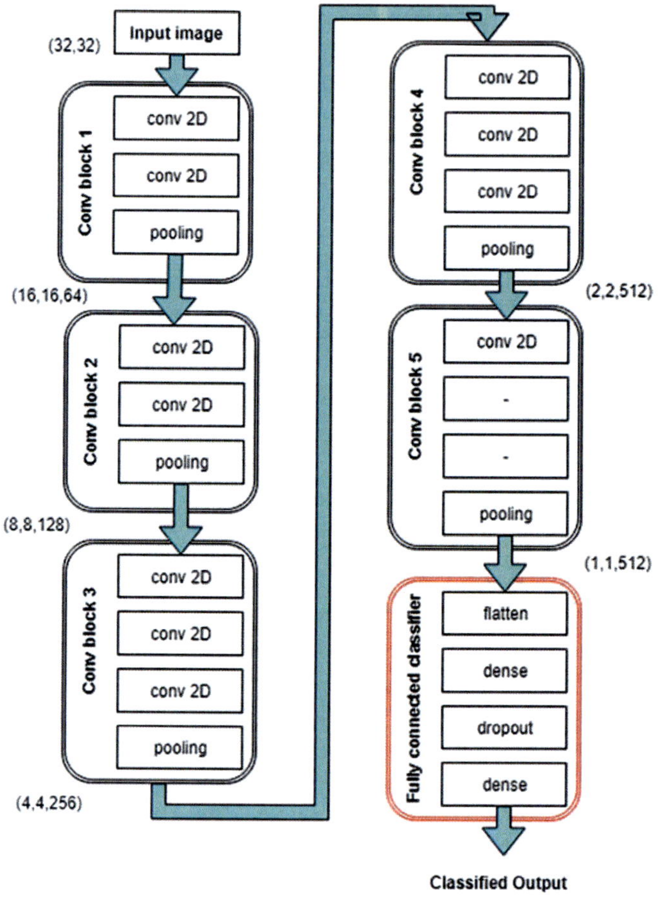

Fig. 2 Proposed architecture based on VGGNet

$$\theta_{t+1} = \theta_t - \frac{\eta}{\sqrt{\widehat{v_t}} + \varepsilon}\widehat{m_t} \tag{1}$$

In the first block, 2 two layers of 64 filters of size 3×3 in each layer are used with zero padding of size $(k-1)/2$ where k is size of filter and ReLU activation function [24] which is given by

$$\text{ReLU}(X) = \text{MAX}(0, X) \tag{2}$$

The same process is continued in the subsequent layers with 128, 256 and 512 filters, respectively. Outputs of these convolutional layers are converted to a monodimensional tensor using a flatten layer. Then, a FC dense layer of 1024 hidden nodes succeeded by a dropout layer is used. Dropout layer [25] with a dropout ratio of

Table 1 Architecture summary of the proposed model

Layer type	Model	Parameters
Input	32 × 32 image	–
Conv	Conv-64	1792
	Conv-64	36,928
Pool	2 × 2 Maxpool	0
Conv	Conv-128	73,856
	Conv-128	147,584
Pool	2 × 2 Maxpool	0
Conv	Conv-256	295,168
	Conv-256	590,080
	Conv-256	590,080
Pool	2 × 2 Maxpool	0
Conv	Conv-512	1,180,160
	Conv-512	2,359,808
	Conv-512	2,359,808
Pool	2 × 2 Maxpool	0
Conv	Conv-512	2,359,808
Pool	Maxpool	0
Fully connected	Flatten-512	0
Dense	Dense-1024	525,312
	Dropout-1024	0
	Dense-45	47,150
Output	Softmax	45
Total params: 10,567,534		
Trainable params: 10,567,534		
Non-trainable params: 0		

0.5 randomly mutes 50% of nodes. This helps in creating a sparse network with a decreased possibility of overfitting and makes the weights to spread over the input features. At last, a final output layer has a softmax classifier [1] with 45 classes as shown below

$$\sigma(z)_j = \frac{e^{z_j}}{\sum_{k-1}^{K} e^{z_k}} \text{for } j = 1, 2 \ldots k \tag{3}$$

The proposed model is trained with different train and validation sets for 15 epochs with a batch size of 500 (Table 1).

5 Evaluation

The dataset contained 90,000 images. Train, test and validation sets are taken on 60, 20 and 20% basis. Architecture has been trained with 54,000 images. 18,000 images are used in test set and remaining 18,000 in validation. For the customized handwritten Tulu database, after 15 epochs, model shows an accuracy of 92.41% on the validation dataset. This is shown in Fig. 3. It can be observed that it is neither under fitting nor overfitting as train accuracy follows same path as validation accuracy. Performance of the model is evaluated by plotting confusion matrix, precision matrix and recall matrix. Figure 4 shows the visualization of misclassified points using confusion matrix.

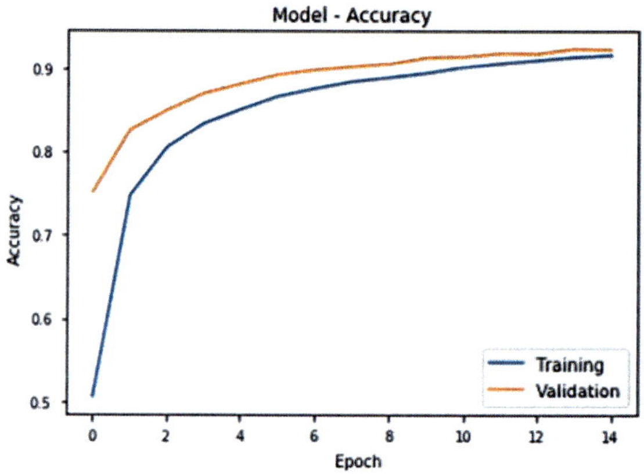

Fig. 3 Accuracy graph on train and validation set

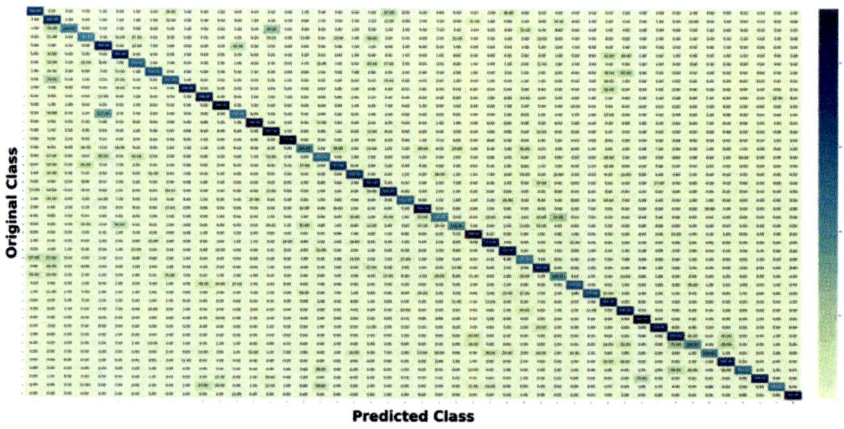

Fig. 4 Misclassification visualization using confusion matrix

Precision is a ratio of total true positives to the predicted positive instances, and recall is a ratio of total true positives to the total number of positive instances. Total predicted positive examples are the sum of true positives (TP) and false positives (TP), whereas total positive examples are the sum of true positives and false negatives. These are given in Figs. 5 and 6.

This model is executed in Python3 supported by Google Colab GPU. It is compared with some of the traditional machine learning algorithms. The test accuracies and score times are calculated for each model and are compared with the proposed method as shown in Table 2. Classifiers considered for comparison are K-nearest-

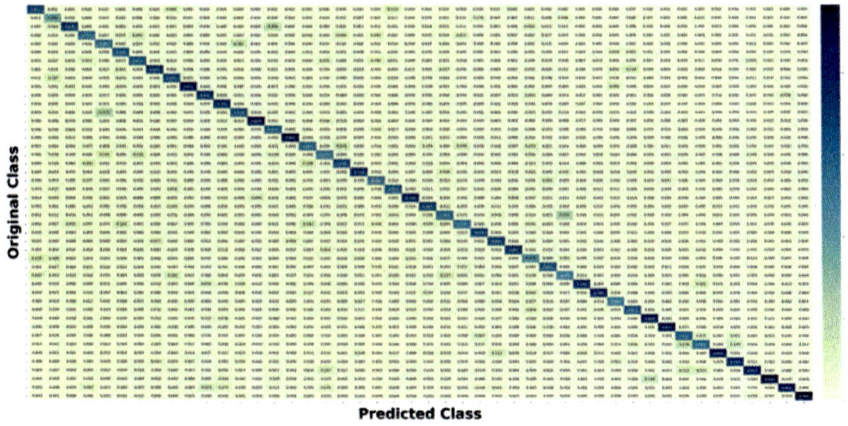

Fig. 5 Visualization of precision matrix

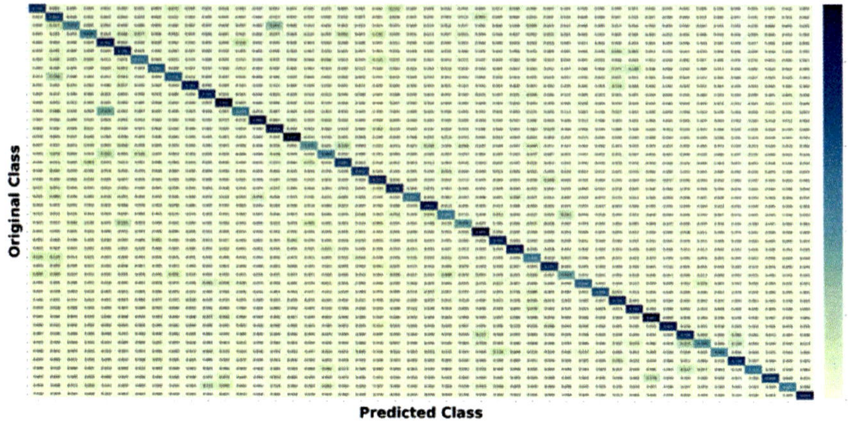

Fig. 6 Visualization of recall matrix

Table 2 Result comparison with different models

Algorithm	Accuracy in %	Score time in seconds
Proposed	92.41	26.1177
KNN	88.02	36.4137
Ridge classifier	42.33	1.9267
Naive Bayes	58.70	0.5952
Logistic regression	69.00	0.2366
Extra tree classifier	56.85	1.8726
SVM	65.90	3.3007
Decision tree	57.21	5.6463
Random forest	80.48	1.3523

neighbor classifier for $k = 5$, Ridge classifier, Naive Bayes, logistic regression, extra tree classifier, support vector machine(SVM), decision tree with depth 45 and random forest with depth 45.

6 Conclusion

Handwritten text recognition is a challenging task because of its inter- and intra-class variation of character patterns. Lack of benchmark datasets is one of the major problems encountered while addressing the issue of document analysis for all the Indian language scripts. This paper deals with the creation of handwritten Tulu character database and its classification using DCNN approach. Forty-five classes of characters including vowels and consonants have been considered for the dataset creation. Ninety thousand samples have been created based on the orthographic shape of the characters. VGGNet-based DCNN model is designed which achieved a recognition accuracy of 92.41%. This is evaluated against many machine learning algorithms, and their accuracies with execution times are noted down. This is an inception model for the recognition of Tulu script. Future work will concentrate on the expansion of dataset by including the compound characters and improving the system accuracy by introducing new deep learning models.

References

1. Caldwell R (1856) Comparative grammar of dravidian or South Indian family of languages, Trübner & Co., London
2. Bhat Sachin, Seshikala G (2019) Preprocessing of historical manuscripts using phase congruency Features and gaussian mixture model. Far East J Electron Commu 19(1):47–67
3. Kingma DP, J Ba (2014) Adam: a method for stochastic optimization. arXiv preprint arXiv: 1412.6980

4. Zhao H, Hu Y, Zhang J (2017) Character recognition via a compact convolutional neural network. In: 2017 International conference on digital image computing: techniques and applications (DICTA). IEEE
5. de Campos, TE (2012) The Chars74K dataset: character recognition in natural images. University of Surrey. Guildford, Surrey, UK
6. Lucas SM et al (2003) ICDAR 2003 robust reading competitions. Null. IEEE
7. He P, Huang W, Qiao Y, Loy CC, Tang X (2016) Detecting oriented text in natural images by linking segments. In: AAAI conference on artificial intelligence (AAAI)
8. Shi B, Bai X, Yao C (2016) An end-to-end trainable neural network for image-based sequence recognition and its application to scene text recognition. IEEE Trans Pattern Anal Mach Intell
9. Mhiri M, Desrosiers C, Cheriet M (2018) Convolutional pyramid of bidirectional character sequences for the recognition of handwritten words. Pattern Recogn Lett 111:87–93
10. Marti U-V, Bunke Horst (2002) The IAM-database: an english sentence database for offline handwriting recognition. Int J Doc Anal Recogn 5(1):39–46
11. Biswas Mithun et al (2017) Banglalekha-isolated: a multi-purpose comprehensive dataset of handwritten Bangla isolated characters. Data in brief 12:103–107
12. Sarkar R, Das N, Basu S, Kundu M, Nasipuri M, Basu DK (2012) Cmaterdb1: a database of unconstrained handwritten Bangla and Bangla—English mixed script document image. Int J Doc Anal Recognit (IJDAR) 15(1):71–83
13. Rabby ASA et al (2018) BornoNet: bangla handwritten characters recognition using convolutional neural network. Procedia Comput Sci 143:528–535
14. Alom MZ et al (2017) Handwritten bangla digit recognition using deep learning. arXiv preprint arXiv:1705.02680
15. Larsson G, Maire M, Shakhnarovich G (2016) Fractalnet: ultra-deep neural networks without residuals. arXiv preprint arXiv:1605.07648
16. He K et al (2016) Deep residual learning for image recognition. In: Proceedings of the IEEE conference on computer vision and pattern recognition
17. Sachin B, Seshikala G (2018) Preprocessing and binarization of inscription images using phase based features. In: 2018 Second international conference on advances in electronics, computers and communications (ICAECC). IEEE
18. Krizhevsky A, Sutskever I, Hinton GE (2016) Imagenet classification with deep convolutional neural networks. In: Advances in neural information processing systems, pp 1097–1105; Huang G et al (2012) Densely connected convolutional networks. arXiv preprint arXiv:1608.06993
19. Karki M et al (2018) Pixel-level reconstruction and classification for noisy handwritten bangla characters. arXiv preprint arXiv:1806.08037
20. Prakash KC et al (2018) Optical character recognition (OCR) for Telugu: database, algorithm and application. In: 2018 25th IEEE international conference on image processing (ICIP). IEEE
21. Avadesh M, Goyal N (2018) Optical character recognition for sanskrit using convolution neural networks. In: 2018 13th IAPR international workshop on document analysis systems (DAS). IEEE
22. Simonyan K, Zisserman A (2014) Very deep convolutional networks for large-scale image recognition. arXiv preprint arXiv:1409.1556
23. Agostinelli F et al (2014) Learning activation functions to improve deep neural networks. arXiv preprint arXiv:1412.6830
24. Srivastava Nitish et al (2014) Dropout: a simple way to prevent neural networks from overfitting. J Mach Learn Res 15(1):1929–1958
25. Zang F, Zhang J (2011) Softmax discriminant classifier. In: 2011 Third international conference on multimedia information networking and security (MINES). IEEE

Exploring the Performance of EEG Signal Classifiers for Alcoholism

Nishitha Lakshmi, Rani Adhaduk, Nidarsh Nithyananda, S. Rashwin Nonda and K. Pushpalatha

Abstract Alcoholism is a tendency to continually rely on alcohol. Unchecked ingestion leads to gradual deteriorating mental health of the abusers. To study the changes in brain activity, electroencephalography (EEG) is one of the acute and low-cost methods. In this study, different sampling rates are experimented on the input EEG signals. The most favorable sampling rate is applied to extract features using the statistical parameters such as mean, median, variance and standard deviation. The best result obtained is an accuracy of 96.84% with support vector machine (SVM) classifier for a sampling rate of 16 by combining all the features extracted using aforementioned statistical parameters.

Keywords EEG signals · Alcoholism · Artificial neural network · Cosine similarity · K-nearest neighbors · Statistical parameters · Support vector machine

Nomenclature

ANN	Artificial Neural Networks
BCI	Brain-Computer Interface
EEG	Electroencephalography
KNN	K-Nearest Neighbors

N. Lakshmi (✉) · R. Adhaduk · N. Nithyananda · S. Rashwin Nonda · K. Pushpalatha
Department of Computer Science and Engineering, Sahyadri College of Engineering and Management, Mangaluru, Karnataka, India
e-mail: nishithalakshmi1@gmail.com

R. Adhaduk
e-mail: rani.adhaduk@gmail.com

N. Nithyananda
e-mail: nidarsh39@gmail.com

S. Rashwin Nonda
e-mail: rashwinnonda@gmail.com

K. Pushpalatha
e-mail: pushpalatha.cs@sahyadri.edu.in

N. N. Chiplunkar and T. Fukao (eds.), *Advances in Artificial Intelligence and Data Engineering*, Advances in Intelligent Systems and Computing 1133,
https://doi.org/10.1007/978-981-15-3514-7_12

RAM	Random-Access Memory
RBF	Radial Basis Function
SVM	Support Vector Machine
SWT	Stationary Wavelet Transform

1 Introduction

Alcohol is usually considered a 'social beverage.' Unfortunately, the abuse of alcohol has caused societal rot. Alcoholism dulls brain activity and causes muddled physical reactions. Innumerable cases are recorded of liver cirrhosis deaths, and many of the road traffic crash deaths caused by uncontrolled alcohol consumption. The term healthy society cites a condition of complete well-being of every individual at all facets of the existing society, which includes spiritual, emotional and physical. This state of serenity of a society is disrupted when crime and diseases run rampant. As humans are social beings, these transgressions cause hindrance in the progress of a dynamic society and negatively affect mental health. Brain signal study is ideal to obtain the insights about mental health and activities as they are the blueprint of all nervous activity. Electroencephalography (EEG) is one of the most effective brain signal recording systems that helps to study the brain activities.

EEG is a neuro-electricity activity that collects information in bulk that represents the psychological and physiological state of the human body by a conductive medium. Out of norm, EEG signals represent irregular brain activities. By analyzing these EEG signals, we gain an understanding about occurrences of abnormal brain activity which plays an important role in diagnosis of different mental disorders like epilepsy, schizophrenia, addiction. EEG signals are also employed in criminal psychology studies where abnormal brain waves are significant in analyzing the violent crime patterns, lie detection and deception. In this paper, we provide performance analysis on various classification algorithms on our data set to diagnose abnormal brain activity and classify input EEG signals into control or alcoholic groups.

The paper is organized as follows. Section 2 tackles any work that is published on the analysis of EEG signals. Section 3 illustrates our proposed work in this field. Section 4 demonstrates experiments that were undertaken and the results it yielded. And Sect. 5 deals with the concluding remarks for the system that we have proposed.

2 Related Work

Over the past few years, EEG has gained popularity in brain–computer interface (BCI) applications. EEG has been successfully used for diagnosis and treatment of mental abnormalities, brain-neuro-degenerative diseases and criminology studies. EEG can also be used to diagnose numerous neurological disorders such as dementia, brain tumors [17], Parkinson's disease, Alzheimer's and countless others. Apart from these, EEG is also used to detect traumatic disorders and chronic diseases [21]. EEG is an effective tool for detecting weariness onset over a long drive [28].

With the advancement of technologies, the crime scenes also require a newer method wherein investigators can use modern ways to look for clues which might not be visible to the traditional method of collecting evidence. Based on the results that were published by the researchers on this topic, application of EEG significantly improves the accuracy of criminal identification to over 70%. This proves that EEG can also be used in the forensics discipline [7].

In [30], the authors exhibited discoveries of a few EEG variations from the norm in various sorts of criminal conduct in contrast with healthy controls and thus summarize, the most elevated event of EEG anomalies was found in the gathering of impulsive criminals.

Yasmeen and Karki [27] suggested a model which analyzes EEG signal in seizure detection utilizing wavelet transform with statistical parameters. They took two different data sets, each having a different sampling rate. One had 128 Hz, whilst the other had 1024 Hz. Using discrete wavelet transformation, feature extraction was done, and a multi-layered neural network was used to differentiate between normal brain signals and brain signals indicating seizures.

It is proven true that the emotions are affected by the alcohol intake and influence adverse effects on human abilities to think, act and behave. These progressions or harms appear in the brainwave recording of an EEG. The target behind this research [18] was to demonstrate the unfavorable impacts of alcohol on the brain. The EEG signals were denoised utilizing independent component analysis and classified using probability neural network to successfully recognize the brain signals influenced by alcohol.

Motive behind the research conducted by authors [6] was to differentiate between truth and deception by extracting EEG features that most suited the differences between the two stages. The recordings that were obtained were alpha waves from two midline and four frontal electrodes. In this study, multi-layer perception was used for classification to differentiate between truth and deception EEG classes.

From the study [4] by Bayram et al., an experiment was conducted to find the most efficient of the four kernel functions of support vector machine to classify whether the subjects were willing to undertake a task at hand by studying their EEG signals. The results obtained were found to be comparable to the recent works.

In [26], authors presented a new method with stationary wavelet transform (SWT) and artificial neural network (ANN) for the detection of brain tumor. First EEG signals were preprocessed and fed as input to SWT where low and high signals were separated. The features extracted from the processed signal were fed to the ANN aimed at the categorization of the test EEG signal as normal or abnormal. The outcomes demonstrated that the accuracy of the proposed technique utilizing SWT and ANN was superior to the current strategies.

Support vector machine (SVM) [4, 5, 10, 23, 29], artificial neural networks [9, 11, 19, 22], k-nearest neighbors (KNN) [3, 14] and many other machine learning methods [1, 2, 8, 9, 13, 20, 22, 25] are a wide assortment of classification methods for seizure prediction and other disorders.

The experimental study confirmed the association among electroencephalogram and memory recall in human. EEG coherence was calculated, and the memory recall was used to check the correlation. The outcome [12] of this investigation demonstrated that the coherent connections are related to recall of memory in widespread gamma networks and frontal delta.

Kaundanya et al. [15] have proposed an algorithm for classifying emotions where the EEG data is preprocessed then followed by feature extraction. The obtained feature values were then classified using k-nearest neighbors with different neighbors. The results showed that the classifier with $k = 3$ neighbors gives high accuracy among all others.

Therefore, from perusal of the related works, we perceive that EEG recordings contain significant information regarding mental activities and are ideal in diagnosis of mental and chronic disorders. We also studied various feature classification algorithms suitable for EEG signals, which are employed in our experimental study. The goal of this paper is to obtain the most efficient classifier by comparing the accuracy rates resulted by experimental procedures undertaken.

3 Proposed Study

To analyze the performance of the classifiers, following phases are employed, i.e., sampling, feature extraction with statistical parameters and evaluation of the classification algorithms. Sampling allows the reduction in dimension of our large data set. The sampled data set with the optimal sampling rate is extracted for features using statistical parameters. These extracted features are fed as training data for various classifiers, and their performance is analyzed.

3.1 EEG Data

Brain cells communicate with each other by sending messages as electrical impulses. Transportation of these impulses from one neuron to another causes ionic drift between the two, which is recorded by the electrodes in electroencephalogram. Based on the frequency of EEG signals resulting due to conscious and sub-conscious brain activities, EEG signals are divided into four categorical waveforms, i.e., alpha, beta, theta and delta. Delta(<4 Hz) is found in the adults under slow-wave sleep. Theta (4–7 Hz) associated with weariness in teenagers and adults. Alpha (8–15 Hz) waveform represents relaxed state in the adults and is associated with inhibition control. Beta (16–31 Hz) is associated with state of high computation and alertness [16]. These bandwidths are significant during the irregular EEG signal study for many applications in fields like medical, criminal, research,etc. EEG signals are highly sensitive and are equipped with extremely sophisticated temporal resolution which are in the order of milliseconds. Hence, we choose EEG signals for our study to analyze the abnormal brain activity.

3.2 Feature Extraction

Feature extraction methods distill the sampled data for its characteristic attributes. Features are the representative parameters of input sampled patterns that facilitate differentiating between the samples of input feature patterns.

To analyze a representative subset of data points and identify the feature patterns, we employ the sampling method. In the domain of statistics, the process of sampling selects a subset of data from a large group of observations (called population) to represent the whole population. Sampling also brings about reduction of the dimension of our data set, making it simpler for ease of interpretation.

We are implementing four of the statistical parameters mean, median, standard deviation and variance to extract the prominent features from our EEG sampled signals.

Mean: Mean is a scale of central tendency, which ascertains the feature around which central grouping occurs. We calculate the mean value of a sample using Eq. 1.

$$\mu = \frac{1}{n}\sum_{i=1}^{n} x_i \tag{1}$$

where x_i is sample value and n is the number of samples.

Median: Median is an estimate lying at the center of a frequency distribution of data samples.

Variance: Variance gives the spread or scale of distribution of data samples. It is a measure of squared deviation of the data samples from its average values. Variance of a sample is calculated using Eq. 2.

$$\text{Variance}(x_1, \ldots, x_n) = \frac{1}{n}\sum_{i=1}^{n}(x_i - \mu_x)^2 \tag{2}$$

μ_x denotes the mean of n sample values, x_i is the sample value, and n is the number of samples.

Standard Deviation: Standard deviation is a quantification measure of dispersed data samples. It is also described as the root of variance. We calculate standard deviation of a sample using Eq. 3.

$$\sigma(x_1, \ldots, x_n) = \sqrt{\text{Variance}} \tag{3}$$

3.3 Classification

We aim to compare the performance of classifiers SVM, ANN, KNN and cosine similarity, thus obtaining the most favorable classifier on our classifier. Extracted features are fed into classifiers to classify the abnormal EEG signals and finally sort if the input EEG signals belong to the control group or alcoholic group.

3.3.1 Support Vector Machine

Collection of affiliated supervised learning techniques proposed for classification. A maximal separating hyperplane is constructed by inserting a vector in the higher dimensional place by mapping. As a result, an output map of the sorted data with the margins between the two as far apart as possible using kernel functions. The function of kernel is to take data as input and transform it into the required form. SVM has many kernel functions, and the ones used in our study are as follows:
Sigmoid Kernel:

$$K(x_i, x_j) = \tanh(\gamma x_i^T x_j + r) \tag{4}$$

RBF Kernel:

$$K(x_i, x_j) = exp(-\gamma |x_i - x_j|^2), \quad \gamma > 0 \tag{5}$$

Linear kernel:

$$K(x_i, x_j) = x_i^T x_j \tag{6}$$

Polynomial Kernel:

$$K(x_i, x_j) = (\gamma x_i^{\mathrm{T}} x_j + r)^d, \quad \gamma > 0 \tag{7}$$

where x_i and x_j are training vectors, x^{T} is the transpose of training vector, $k(x_i, x_j)$ is a kernel function, and γ, r, d are the kernel parameters.

3.3.2 K-Nearest Neighbors

A non-parametric approach, which classifies the training data point according to majority of its nearest neighbors. Performance of KNN depends on the number of nearest neighbor values, i.e., $k = 2, 3, 4, 5$. To classify the neighbor, we made use of Minkowski metric distance as given in Eq. 8.

$$D(X, Y) = \left(\sum_{i=1}^{n} \mid x_i - y_i \mid^p \right)^{1/p} \tag{8}$$

where $X = (x_1, x_2, x_3, x_4, \ldots, x_n)$, $Y = (y_1, y_2, y_3, y_4, \ldots, y_n)$ and p is the parameter.

3.3.3 Artificial Neural Network

A framework of machine learning algorithms motivated by the working of biological neural networks. It consists of single or more layers to perform complex tasks. Each layer is made up of artificial neurons (nodes) and node connectors called edges. ANN encompasses three layers, namely the input, hidden and output layer. The inputs to the artificial neural network are stored in the artificial neurons of the input layer along with its respective weights, which is assigned based on its relative importance. The output of each artificial neurons is obtained only when a nonlinear function called activation function is triggered by the weighted sum of its inputs, provides a result exceeding the threshold value.

Activation functions utilized in this experiment are

ReLU: Known as rectified linear unit, it inputs a real-valued data and calculates a threshold at zero (replaces negative values with zero) using Eq. 9.

$$f(x) = \max(0, x) \tag{9}$$

where x is a real sample value.

tanh: The real-valued input is transformed into the range within limits $[-1, 1]$ by using Eq. 10.

$$\tanh(x) = \frac{2}{(1 + \exp(-2x))} - 1 \tag{10}$$

where x is a real sample value.

Sigmoid: takes a real-valued input and transforms it into the range $[0, -1]$ by using Eq. 11.

$$\sigma(x) = \frac{1}{1 + \exp(-x)} \tag{11}$$

where x is a real sample value.

3.3.4 Cosine Similarity

A similarity measure which is obtained by observing the cosine angle between the feature vectors. The outcome of the cosine similarity is bound in $[0, 1]$ where similarity score of 1 denotes vectors having same orientation and similarity score of 0 denoting vectors being relatively oriented at 90°. Cosine similarity then gives a useful measure of how similar two signals are likely to be in terms of their label, i.e., alcohol or control using Eq. 12.

$$\cos\varphi = \frac{A \cdot B}{|A| \cdot |B|} \tag{12}$$

A and B are sample vectors of EEG data, $|A|$ and $|B|$ represent the cardinality of two vectors.

4 Experimental Results

We experimented using HP Pavilion 15-au007tx with the configurations of RAM—8GB DDR4, CPU—Intel i5 6th generation. The coding for the experiment was done in Python2 language.

We have led our analyses by taking the data set from an open source, contributed by Henri Begleiter at the Neurodynamics Laboratory at the State University of New York Health Center at Brooklyn [24]. Placement of the electrodes was done in the internationally accepted 10–20 method. Data set comprises EEG signal recordings from 64 electrodes from each subject which were sampled at the rate of 256 Hz for a second with 30 trials each. Each subject belonged to either one of the groups: control or alcoholic. Training and testing data comprise 468 trials and 480 trials, respectively. Training and testing trials were randomly combined in the ratio of 80:20.

4.1 Performance Evaluation of Classifiers Without Sampling

The motive behind our experiment is to analyze the performance of SVM, ANN, KNN and cosine similarity classification techniques on our data set. As each electrode on a subject's scalp records 256 readings in a second, 64 electrodes record 16,834 values. To begin with, we experimented the classifiers without sampling the EEG data, i.e., with 16,834 feature values. Initially, we experiment with the SVM classifier and analyze the performance of four of its kernel functions. From our analysis, we see that the linear kernel function on SVM is more favorable on our data set with an accuracy of 84.2%. Performance of ANN is analyzed with various activation functions. It is evident that logistic function is the most effective activation function of the four functions with the accuracy of 83.15%. When experimented with 2, 3, 4 and 5 neighbors of KNN classifier, it is evident that three nearest neighbors are suitable for our data set, giving an accuracy of 73.68%. The data set is lastly analyzed for the cosine similarity method, which gives an accuracy of 68.73%. Hence, we conclude that the best performance on our experimental input data is given by linear kernel of SVM with an accuracy of 84.21%.

4.2 Performance Evaluation of Classifiers with Different Sampling Rates

To study the effects of sampling rate on classification algorithms for our data set, we conducted the classification with differing sampling rates of 4, 8, 16 and 32. The sampled data is then extracted for features using mean.

We commence the experiment on our data set with sampling rate of 4. Therefore, from 64 electrodes, we get $84 \times 64 = 5376$ values. We experiment with the SVM classifier and analyze the performance of four of its kernel functions. It is found that linear kernel function on SVM is more favorable on our data set with an accuracy of 86.84%. Performance of ANN is analyzed with various activation functions. It is evident that logistic function is the most effective activation function of the four functions with the accuracy of 85.70%, Fig. 1. When experimented with 2, 3, 4 and 5 neighbors of KNN classifier, it is evident that three nearest neighbors are suitable for our data set, giving an accuracy of 65.26%. The data set is lastly analyzed for the cosine similarity method, which gives an accuracy of 66.60%. Hence, we conclude that the best performance on our experimental input data is given by SVM using linear kernel function, resulting in an accuracy of 86.84%.

When sampled at the rate of 8 for each electrode, we get ($32 \times 64 = 2048$) values. From our experimental studies of SVM and its four kernel functions, it is found that linear kernel function on SVM is more favorable on our data set with an accuracy of 88.42%. Performance of ANN is analyzed with various activation functions. It is evident that logistic function is the most effective activation function of the four functions with the accuracy of 86.31%, Fig. 1. When experimented with 2, 3, 4 and 5 neighbors of KNN classifier, it is evident that three nearest neighbors are suitable for our data set, giving an accuracy of 65.26%. The data set is lastly analyzed for the cosine similarity method, which gives an accuracy of 66.49%. Hence, we conclude that the best performance on our experimental input data is given by linear kernel of SVM with an accuracy of 88.42%.

With sampling rate of 16 on an electrode, we get one feature from every 16 feature samples. Hence, from 64 electrodes, we get $16 \times 64 = 1024$ values, where the input data of size (948×1024) is fed to the classifiers SVM, ANN, KNN and cosine similarity. Following results are obtained. We experiment with the SVM classifier

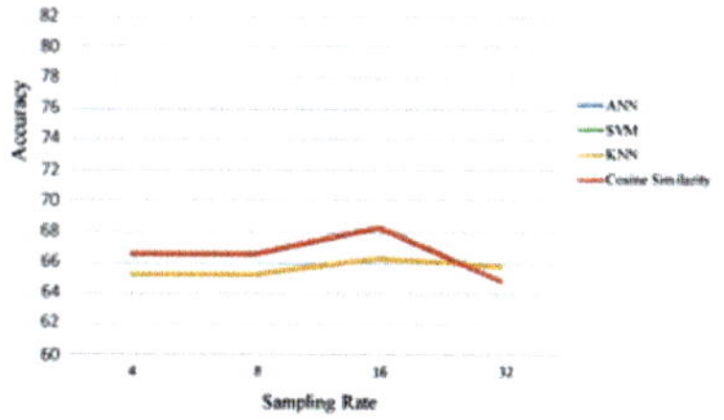

Fig. 1 Performance analysis of classifiers with different sampling rates

and analyze the performance of four of its kernel functions. Therefore, from our analysis, we see that the linear kernel function on SVM gives an accuracy of 89.47%. Performance of ANN is analyzed with various activation functions, and logistic function gives an accuracy of 86.84%, Fig. 1. When experimented with 2, 3, 4 and 5 neighbors of KNN classifier, it is evident that three nearest neighbors are suitable for our data set, giving an accuracy of 66.31%. The data set is lastly analyzed for the cosine similarity method, which gives an accuracy of 68.21%. Hence, we conclude that the best performance on our experimental input data is given by linear kernel of SVM with an accuracy of 89.47%.

When sampled at the rate of 32 for each electrode, we get ($8 \times 64 = 512$) values. From our experimental studies of SVM and its four kernel functions, it is found that linear kernel function on SVM is more favorable on our data set with an accuracy of 85.78%. Performance of ANN is analyzed with various activation functions. It is evident that logistic function is the most effective activation function of the four functions with the accuracy of 86.84%, Fig. 1. When experimented with 2, 3, 4 and 5 neighbors of KNN classifier, it is evident that three nearest neighbors are suitable for our data set, giving an accuracy of 65.95%. The dataset is lastly analyzed for the cosine similarity method, which gives an accuracy of 64.90%. Hence, we conclude that the best performance on our experimental input data is given by ANN with logistic function resulting with an accuracy of 86.84%.

From the comparison of these obtained results, it is proven that the sampling rate of 16 is ideal for our data set. Therefore, the proceeding experiments in our paper are done by employing the sample rate of 16 values.

4.3 Performance Analysis of Classifiers with Statistical Parameters

Data set with samples of 16 is extracted for prominent features with mean, median, standard deviation and variance. The performance of each classifiers is recorded and evaluated for every statistical parameter.

Mean: When mean is the parameter used, the highest accuracy of all classifiers is observed in Table 1. For SVM, linear function gives a score of 89.47%. Accuracy of ANN with activation function logistic gives an accuracy of 86.84%. With KNN, an accuracy of 66.31% is observed. Accuracy of cosine similarity is 68.21%. Thus, it is evident from Table 1 that when mean is the statistical parameter used SVM with linear function is favorable.

Median: When median is the statistical parameter used, the performance of all classifiers is observed from Table 2. It is noted that SVM with linear kernel function gives maximum accuracy of 87.89%. Accuracy of ANN with identity activation function gives a maximum score of 89.47%. With KNN, maximum accuracy of 66.31% is observed. When cosine similarity is applied, an accuracy of 65.17% is recorded. Thus, in this instance, identity function of ANN is the ideal candidate for our data set.

Table 1 Accuracy values of classifiers with mean

Classifier	Kernel/activation function	Accuracy
SVM	Linear	89.47
	RBF	62.45
	Sigmoid	52.11
	Polynomial	73.68
ANN	Logistic	86.84
	Relu	85.78
	Tanh	85.26
	Identity	84.36
KNN	$K = 2$	64.09
	$K = 3$	66.31
	$K = 4$	59.47
	$K = 5$	62.1
Cosine similarity	–	68.21

Table 2 Accuracy values of classifiers with median

Classifier	Kernel/activation function	Accuracy
SVM	Linear	87.89
	RBF	49.47
	Sigmoid	52.10
	Polynomial	81.052
ANN	Logistic	88.42
	ReLU	86.84
	tanh	84.21
	Identity	89.47
KNN	$K = 2$	58.42
	$K = 3$	66.31
	$K = 4$	61.57
	$K = 5$	66.31
Cosine similarity	–	65.17

Standard Deviation: When standard deviation is the parameter used, the highest accuracy of all classifiers is observed in Table 3. For SVM, linear function gives a maximum score of 94.73%. Accuracy of ANN with activation function logistic gives maximum accuracy of 94.73%. With KNN, maximum accuracy of 86.84% is observed. Accuracy of cosine similarity is 78.49%. Thus, it is evident from Table 3 that when standard deviation is the statistical parameter used, ANN with logistic function and SVM with linear function are favorable.

Table 3 Accuracy values of classifiers with standard deviation

Classifier	Kernel/activation function	Accuracy
SVM	Linear	94.73
	RBF	82.10
	Sigmoid	46.84
	Polynomial	91.57
ANN	Logistic	94.73
	ReLU	94.21
	tanh	93.15
	Identity	94.21
KNN	$K = 2$	78.94
	$K = 3$	86.84
	$K = 4$	81.57
	$K = 5$	84.21
Cosine similarity	–	78.49

Table 4 Accuracy values of classifiers with variance

Classifier	Kernel/activation function	Accuracy
SVM	Linear	90
	RBF	47.89
	Sigmoid	47.36
	Polynomial	85.26
ANN	Logistic	88.94
	ReLU	89.47
	tanh	90
	Identity	89.47
KNN	$K = 2$	73.68
	$K = 3$	81.05
	$K = 4$	76.84
	$K = 5$	76.84
Cosine similarity	–	77.17

Variance: When variance is the parameter used, accuracy of classifiers is calculated. From Table 4, it is evident that SVM with linear kernel function gives the maximum accuracy of 90%. Accuracy of ANN with activation, tanh function is noted to give a maximum score of 90%. With KNN, maximum accuracy of 81.05% is observed. Accuracy of cosine similarity is 77.17%. Thus, it is evident that when variance is the statistical parameter used, SVM with linear function and ANN with tanh function are the ideal candidate.

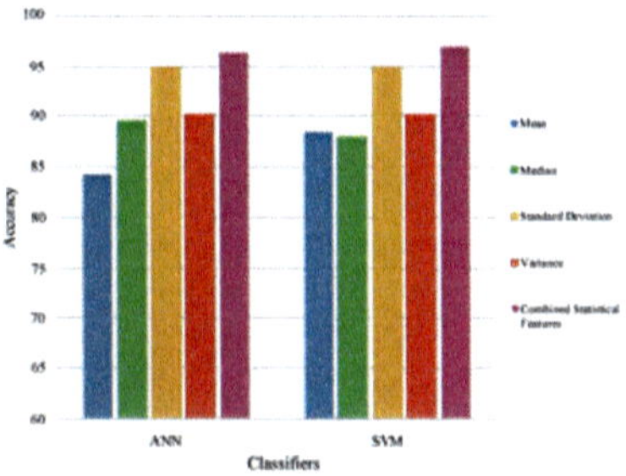

Fig. 2 Performance analysis of classifiers with statistical parameters

Therefore, from the comparison of performances of all the statistical parameters for feature extraction on our data set, it is evident that using standard deviation is ideal.

4.4 Performance Analysis of Classifiers with Combination of Statistical Parameters

The performance analysis of SVM, ANN, KNN and cosine similarity demonstrates that performances of SVM with linear function and ANN with logistic function are more favorable. Hence, we proceed with analyzing the performance of SVM and ANN classifiers where features are the combination of all statistical parameters, mean, median, standard deviation and variance. When applied on SVM with linear kernel function, we get an accuracy score of 96.84%. Similarly, when experimented on ANN with logistic activation function, it results in an accuracy of 96.31% as observed in Fig. 2. Hence, we conclude that the best performance on our experimental input data is given by the classifiers when features are extracted with the combination of statistical parameters.

4.5 Comparison of Computational Time Taken by EEG Classifiers

From the previous experiments conducted, it is evident that with combined statistical parameters used for feature extraction, SVM gives the highest accuracy score of 96.84%, and the second highest score, i.e., 96.31% is given by ANN with logistic function. Hence, we compare the computational time for these classifiers. From Table 5, it is observed that SVM takes the minimal time of 1.92 s. Therefore, we conclude that SVM classifier is the most efficiently suited classifier for our data set.

Table 5 Time taken by classifiers in seconds

Classifier	Raw data	Mean	Median	Standard deviation	Variance	Combined statistical features
ANN	277.90	5.11	13.56	20.59	18.67	84.77
SVM	11.55	0.736	0.77	0.58	0.61	1.92

5 Conclusion

Alcoholism is an addiction risen with a need for reliance on a substance to make crucial decisions and to cope with events. Alcoholism predominantly dulls brain activity and triggers numerous societal illnesses. The aim of this paper is to analyze EEG signals and classify them as belonging to control or alcoholic group, efficiently. We conducted performance analysis of SVM, ANN, KNN, cosine similarity classifiers on our data set. Experiments are commenced by sampling phase. Sampling is performed to reduce the dimension of our data set. Performance of the classifiers is checked for different sampling rates of 4, 8, 16 and 32. The sampling rate of 16 showed comparatively promising results. Features are extracted using the statistical parameters mean, median, standard deviation and variance. When all the extracted statistical features are combined as input data and fed to SVM classifier with linear kernel function, the highest accuracy of 96.84% was obtained.

References

1. Adeli H, Zhou Z, Dadmehr N (2003) Analysis of EEG records in an epileptic patient using wavelet transform. J Neurosci Methods 123(1):69–87. https://doi.org/10.1016/S0165-0270(02)00340-0
2. Ahmadi A, Shalchyan V, Mohammad RD (2017) A new method for epileptic seizure classification in EEG using adapted wavelet packets. In: 2017 electric electronics, computer science, biomedical engineerings' meeting (EBBT). IEEE. https://doi.org/10.1109/EBBT.2017.7956756
3. Bablani A, Edla DR, Dodia S (2018) Classification of EEG data using k-nearest neighbor approach for concealed information test. Procedia Comput Sci 143:242–249. https://doi.org/10.1016/j.procs.2018.10.392
4. Bayram K, Ayyuce Sercan M, Kizrak Bolat B (2013) Classification of EEG signals by using support vector machines. In: 2013 IEEE INISTA. IEEE. https://doi.org/10.1109/INISTA.2013.6577636
5. Bhuvaneswari P, Satheesh Kumar J (2013) Support vector machine technique for EEG signals. Int J Comput Appl 63(13)
6. Cakmak R, Zeki AM (2015) Neuro signal based lie detection. In: 2015 IEEE international symposium on robotics and intelligent sensors (IRIS). IEEE. https://doi.org/10.1109/IRIS.2015.7451606

7. Chan H-T et al (2017) Applying EEG in criminal identification research. In: 2017 international conference on applied system innovation (ICASI). IEEE. https://doi.org/10.1109/ICASI.2017.7988484
8. Gandhi T et al (2010) Expert model for detection of epileptic activity in EEG signature. Expert Syst Appl 37(4):3513–3520. https://doi.org/10.1016/j.eswa.2009.10.036
9. Ghosh-Dastidar S, Adeli H, Dadmehr N (2008) Principal component analysis-enhanced cosine radial basis function neural network for robust epilepsy and seizure detection. IEEE Trans Biomed Eng 55(2):512–518. https://doi.org/10.1109/TBME.2007.905490
10. Guler I, Ubeyli ED (2007) Multiclass support vector machines for EEG-signals classification. IEEE Trans Inf Technol Biomed 11(2):117–126. https://doi.org/10.1109/TITB.2006.879600
11. Guler NF, Ubeyli ED, Guler I (2005) Recurrent neural networks employing Lyapunov exponents for EEG signals classification. Expert Syst Appl 29(3):506–514
12. Hanouneh S et al (2015) Functional connectivity of EEG regional delta and inter-regional gamma activity correlates with memory recall. In: 2015 IEEE international conference on control system, computing and engineering (ICCSCE). IEEE. https://doi.org/10.1109/ICCSCE.2015.7482237
13. Harikumar R, Sunil Kumar P (2015) Dimensionality reduction techniques for processing epileptic encephalographic signals. Biomed Pharmacol J 8(1):103–106. https://doi.org/10.13005/bpj/587
14. Huang J et al (2018) An improved kNN based on class contribution and feature weighting. In: 2018 10th international conference on measuring technology and mechatronics automation (ICMTMA). IEEE. https://doi.org/10.1109/ICMTMA.2018.00083
15. Kaundanya VL, Patil A, Panat A (2015) Performance of k-NN classifier for emotion detection using EEG signals. In: 2015 international conference on communications and signal processing (ICCSP). IEEE. https://doi.org/10.1109/ICCSP.2015.7322687
16. Kirmizi-Alsan E et al (2006) Comparative analysis of event-related potentials during Go/NoGo and CPT: decomposition of electrophysiological markers of response inhibition and sustained attention. Brain Res 1104(1):114–128. https://doi.org/10.1016/j.brainres.2006.03.010
17. Murugesan M, Sukanesh R (2009) Towards detection of brain tumor in electroencephalogram signals using support vector machines. Int J Comput Theory Eng 1(5):622
18. Rachman NT, Tjandrasa H, Fatichah C (2016) Alcoholism classification based on EEG data using independent component analysis (ICA), wavelet de-noising and probabilistic neural network (PNN). In: 2016 international seminar on intelligent technology and its applications (ISITIA). IEEE. https://doi.org/10.1109/ISITIA.2016.7828626
19. Rout N (2014) Analysis and classification technique based on ANN for EEG signals. IJCSIT 5(4):5103–5105
20. Shahid A et al (2013) Epileptic seizure detection using the singular values of EEG signals. In: 2013 ICME international conference on complex medical engineering. IEEE. https://doi.org/10.1109/ICCME.2013.6548330
21. Siuly S, Li Y, Zhang Y (2016) Significance of EEG signals in medical and health research. EEG signal analysis and classification. Springer, Cham, pp 23–41. https://doi.org/10.1007/978-3-319-47653-7_2
22. Srinivasan V, Eswaran C, Sriraam N (2007) Approximate entropy-based epileptic EEG detection using artificial neural networks. IEEE Trans Inf Technol Biomed 11(3):288–295. https://doi.org/10.1109/TITB.2006.884369
23. Subasi A, Ismail Gursoy M (2010) EEG signal classification using PCA, ICA, LDA and support vector machines. Expert Syst Appl 37(12):8659–8666. https://doi.org/10.1016/j.eswa.2010.06.065
24. State University of New York Health Center, Neurodynamics Laboratory (1999) UCI machine learning repository, 13 October 1999 (online). Available at: https://archive.ics.uci.edu/ml/datasets/EEG+Database
25. Supriya S et al (2016) Weighted visibility graph with complex network features in the detection of epilepsy. IEEE Access 4:6554–6566. https://doi.org/10.1109/ACCESS.2016.2612242
26. Thiyagarajan M (2019) Brain tumour detection via EEG signals. Indian J Appl Res 9:213–215

27. Yasmeen S, Karki MV (2017) Neural network classification of EEG signal for the detection of seizure. In: 2017 2nd IEEE international conference on recent trends in electronics, information & communication technology (RTEICT). IEEE. https://doi.org/10.1109/RTEICT.2017.8256658
28. Yeo MVM et al (2009) Can SVM be used for automatic EEG detection of drowsiness during car driving? Saf Sci 47(1):115–124. https://doi.org/10.1016/j.ssci.2008.01.007
29. Zavar M et al (2011) Evolutionary model selection in a wavelet-based support vector machine for automated seizure detection. Expert Syst Appl 38(9):10751–10758. https://doi.org/10.1016/j.eswa.2011.01.087
30. Zukov I, Ptacek R, Fischer S (2008) EEG abnormalities in different types of criminal behavior. Activitas Nervosa Superior 50(4):110–113. https://doi.org/10.1007/BF03379552

Type-2 Tetradecagonal Fuzzy Number

A. Rajkumar and C. Sagaya Nathan Stalin

Abstract This paper deals with a project construction of normal and crashing activity using type-2 tetradecagonal fuzzy number. Our aim is to obtain the total cost and completion time of the project. The working of the algorithm has been illustrated by numerical example.

Keywords Triskaidecagonal fuzzy number · Intuitionistic Triskaidecagonal fuzzy number · Time–cost trade-off

1 Introduction

The concept of type-2 fuzzy set is the extension of ordinary fuzzy set. Type-2 fuzzy sets can give a better improvement in certain kinds of uncertainty arise in performance measure. The solution may be viewed in three-dimensional where each type-2 fuzzy set itself is fuzzy [1]. The concept of time–cost relationship deals with normal and crashing activity. The slope of crashing activity will increase the outcome in terms of cost that activity per unit period. This paper includes with definition, algorithm and illustration calculated with time–cost relationship method.

A. Rajkumar (✉) · C. Sagaya Nathan Stalin
Hindustan Institute of Technology and Science, Chennai 603103, India
e-mail: arajkumar@hindustanuniv.ac.in

C. Sagaya Nathan Stalin
Loyola College, Vettavalam, India

N. N. Chiplunkar and T. Fukao (eds.), *Advances in Artificial Intelligence and Data Engineering*, Advances in Intelligent Systems and Computing 1133,
https://doi.org/10.1007/978-981-15-3514-7_13

2 Definitions

2.1 Type-2 Trapezoidal Fuzzy Number [2]

Let $a = (a_1, a_2, a_3, a_4)$ be a trapezoidal fuzzy number. A normal type-2 trapezoidal fuzzy number $\widetilde{A} = \{(x, \mu_A^1(X), \mu_A^2(X), \mu_A^3(X), \mu_A^4(X)); X \in R\}$ and $\mu_A^1(X) \leq \mu_A^2(X) \leq \mu_A^3(X) \leq \mu_A^4(X)$ for all $X \in R$. Denote $\widetilde{A} = (\widetilde{A_1}, \widetilde{A_2}, \widetilde{A_3}, \widetilde{A_4})$, where $\tilde{A} = ((\widetilde{a_1^\beta}, \widetilde{a_2^\beta}, a_3^\beta, \widetilde{a_4^\beta}), (\widetilde{a_1^\delta}, a_2^\delta, \widetilde{a_3^\delta}, \widetilde{a_4^\delta}), (\widetilde{a_1^\mathrm{T}}, \widetilde{a_2^\mathrm{T}}, \widetilde{a_3^\mathrm{T}}, \widetilde{a_4^\mathrm{T}})(\widetilde{a_1^\sigma}, \widetilde{a_2^\sigma}, \widetilde{a_3^\sigma}, \widetilde{a_4^\sigma}))$

2.2 Type-2 Triskaidecagonal Fuzzy Number [3]

A type-2 Triskaidecagonal fuzzy number

$$\widetilde{A} = \begin{pmatrix} x, \mu_A^1(X), \mu_A^2(X), \mu_A^3(X), \mu_A^4(X), \mu_A^5(X), \mu_A^6(X), \mu_A^7(X), \mu_A^8(X), \\ \mu_A^9(X), \mu_A^{10}(X), \mu_A^{11}(X), \mu_A^{12}(X), \mu_A^{13}(X) \end{pmatrix}; X \in R$$

$$\{(\mu_A^1(X) \leq \mu_A^2(X) \leq \mu_A^3(X) \leq \mu_A^4(X) \leq \mu_A^5(X) \leq \mu_A^6(X) \leq \mu_A^7(X) \leq \mu_A^8(X) \\ \leq \mu_A^9(X) \leq \mu_A^{10}(X) \leq \mu_A^{11}(X) \leq \mu_A^{12}(X) \leq \mu_A^{13}(X))\text{for all } X \in R$$

Denote

$$\widetilde{A} = (\widetilde{A_1}, \widetilde{A_2}, \widetilde{A_3}, \widetilde{A_4}, \widetilde{A_5}, \widetilde{A_6}, \widetilde{A_7}, \widetilde{A_8}, \widetilde{A_9}, \widetilde{A_{10}}, \widetilde{A_{11}}, \widetilde{A_{12}}, \widetilde{A_{13}})$$

where

$\widetilde{A_1} = (\widetilde{A_1^\alpha}, \widetilde{A_1^\beta}, \widetilde{A_1^\sigma}, \widetilde{A_1^\delta}, \widetilde{A_1^\mathrm{T}}, \widetilde{A_1^\Omega}, \widetilde{A_1^{\acute{\rho}}}, \widetilde{A_1^{\acute{u}}}, \widetilde{A_1^{\breve{e}}}, \widetilde{A_1^{\acute{\omega}}}, \widetilde{A_1^\Upsilon}, \widetilde{A_1^\chi}, \widetilde{A_1^\Phi})$,
$\widetilde{A_2} = (\widetilde{A_2^\alpha}, \widetilde{A_2^\beta}, \widetilde{A_2^\sigma}, \widetilde{A_2^\delta}, \widetilde{A_2^\mathrm{T}}, \widetilde{A_2^\Omega}, \widetilde{A_2^{\acute{\rho}}}, \widetilde{A_2^{\acute{u}}}, \widetilde{A_2^{\breve{e}}}, \widetilde{A_2^{\acute{\omega}}}, \widetilde{A_2^\Upsilon}, \widetilde{A_2^\chi}, \widetilde{A_2^\Phi}), \ldots,$
$\widetilde{A_{13}} = (\widetilde{A_{13}^\alpha}, \widetilde{A_{13}^\beta}, \widetilde{A_{13}^\sigma}, \widetilde{A_{13}^\delta}, \widetilde{A_{13}^\mathrm{T}}, \widetilde{A_{13}^\Omega}, \widetilde{A_{13}^{\acute{\rho}}}, \widetilde{A_{13}^{\acute{u}}}, \widetilde{A_{13}^{\breve{e}}}, \widetilde{A_{13}^{\acute{\omega}}}, \widetilde{A_{13}^\Upsilon}, \widetilde{A_{13}^\chi}, \widetilde{A_{13}^\Phi})$

2.3 Type-2 Tetradecagonal Fuzzy Number

A type-2 Tetradecagonal fuzzy number

$$\widetilde{A} = \begin{pmatrix} x, \mu_A^1(X), \mu_A^2(X), \mu_A^3(X), \mu_A^4(X), \mu_A^5(X), \mu_A^6(X), \mu_A^7(X), \mu_A^8(X), \\ \mu_A^9(X), \mu_A^{10}(X), \mu_A^{11}(X), \mu_A^{12}(X), \mu_A^{13}(X), \mu_A^{14}(X) \end{pmatrix}; X \in R$$

$$\{(\mu_A^1(X) \leq \mu_A^2(X) \leq \mu_A^3(X) \leq \mu_A^4(X) \leq \mu_A^5(X) \leq \mu_A^6(X) \leq \mu_A^7(X) \leq \mu_A^8(X) \\ \leq \mu_A^9(X) \leq \mu_A^{10}(X) \leq \mu_A^{11}(X) \leq \mu_A^{12}(X) \leq \mu_A^{13}(X) \leq \mu_A^{14}(X))\text{for all } X \in R$$

Denote

$$\widetilde{A} = \left(\widetilde{A_1}, \widetilde{A_2}, \widetilde{A_3}, \widetilde{A_4}, \widetilde{A_5}, \widetilde{A_6}, \widetilde{A_7}, \widetilde{A_8}, \widetilde{A_9}, \widetilde{A_{10}}, \widetilde{A_{11}}, \widetilde{A_{12}}, \widetilde{A_{13}}, \widetilde{A_{14}}\right)$$

where

$$\widetilde{A_1} = (\widetilde{A_1^{\alpha}}, \widetilde{A_1^{\beta}}, \widetilde{A_1^{\sigma}}, \widetilde{A_1^{\delta}}, \widetilde{A_1^{\text{Ţ}}}, \widetilde{A_1^{\Omega}}, \widetilde{A_1^{\acute{\rho}}}, \widetilde{A_1^{\acute{u}}}, \widetilde{A_1^{\check{e}}}, \widetilde{A_1^{\acute{\omega}}}, \widetilde{A_1^{\Upsilon}}, \widetilde{A_1^{\chi}}, \widetilde{A_1^{\phi}}),$$
$$\widetilde{A_2} = (\widetilde{A_2^{\alpha}}, \widetilde{A_2^{\beta}}, \widetilde{A_2^{\sigma}}, \widetilde{A_2^{\delta}}, \widetilde{A_2^{\text{Ţ}}}, \widetilde{A_2^{\Omega}}, \widetilde{A_2^{\acute{\rho}}}, \widetilde{A_2^{\acute{u}}}, \widetilde{A_2^{\check{e}}}, \widetilde{A_2^{\acute{\omega}}}, \widetilde{A_2^{\Upsilon}}, \widetilde{A_2^{\chi}}, \widetilde{A_2^{\phi}}), \ldots,$$
$$\widetilde{A_{14}} = (\widetilde{A_{14}^{\alpha}}, \widetilde{A_{14}^{\beta}}, \widetilde{A_{14}^{\sigma}}, \widetilde{A_{14}^{\delta}}, \widetilde{A_{14}^{\text{Ţ}}}, \widetilde{A_{14}^{\Omega}}, \widetilde{A_{14}^{\acute{\rho}}}, \widetilde{A_{14}^{\acute{u}}}, \widetilde{A_{14}^{\check{e}}}, \widetilde{A_{14}^{\acute{\omega}}}, \widetilde{A_{14}^{\Upsilon}}, \widetilde{A_{14}^{\chi}}, \widetilde{A_{14}^{\phi}})$$

3 Algorithm to Estimate the Critical Path in Time–Cost Trade-off Problem [4]

A new procedure is implemented to check out the fuzzy complete result in time–cost trade-off problems testing with type-2 tetradecagonal fuzzy number, computing a network to find out the critical path by any normal time duration and computing the normal total cost based on project completion, in order to find the project cost and cost slope for each activity by the following formula

$$\text{Project cost} = \text{Direct Cost} + (\text{Indirect cost} * \text{Project duration})$$
$$\text{Cost Slope} = (\text{Crash cost} - \text{Normal cost})/(\text{Normal time} - \text{Crash time}).$$

The new crashing networks process is continued until the minimum cost slope has crashed up to a desired time. This process is stopped where further crashing is not possible.

3.1 Numerical Result [5, 6]

A construction of independent house with the first floor around 1000 ft^2 has listed down with various activities involved. Type-2 Tetradecagonal fuzzy number applied in this defined problem in order to get the total cost and the project duration. The main stages are

(1) Stage-1: Planning and design; (2) Stage-2: Permit application; (3) Stage-3: Tendering and foundation; (4) Stage-4: Construction process; (5) Stage-5: Interior and exterior work; (6) Stage-6: Detailed finishing and evaluation; (7) Stage-7: Clean up and final touches (Tables 1, 2 and 3; Fig. 1).

Step-1

See Fig. 2.

Step-2

See Fig. 3.

Table 1 Details of project time

Activity number	Activity	Normal time (N_t)	Crash time (C_t)
A	$1 \to 2$	(18, 20, 22, 24, 26, 28, 32, 38, 42, 44, 46, 48, 50, 52), (16, 18, 20, 22, 24, 26, 32, 38, 44, 46, 48, 50, 52, 54), …, (2, 3, 4, 5, 6, 7, 32, 38, 60, 64, 65, 66, 68, 70)	(14, 16, 18, 20, 23, 25, 28. 30, 32, 34, 36, 40, 42, 44), (13, 14, 16, 18, 20, 22, 28, 30, 33, 36, 40, 42, 44, 46), …, (1, 2, 3, 4, 5, 6, 28, 30, 48, 50, 53, 55, 58, 59)
B	$2 \to 3$	(25, 27, 30, 32, 33, 35, 43, 44, 46, 48, 49, 52, 53), (24, 26, 29, 30, 32, 34, 43, 45, 46, 49, 50, 53, 56), …, (4, 5, 6, 8, 9, 11, 43, 65, 67, 70, 72, 77, 80)	(18, 20, 22, 24, 26, 28, 32, 38, 42, 44, 46, 48, 50, 52), (16, 18, 20, 22, 24, 26, 32, 38, 44, 46, 48, 50, 52, 54), …, (2, 3, 4, 5, 6, 7, 32, 38, 60, 64, 65, 66, 68, 70)
C	$3 \to 4$	(18, 20, 22, 24, 25, 28, 32, 38, 42, 44, 46, 48, 50), (17, 19, 21, 23, 24, 25, 32, 38, 43, 46, 48, 50, 51), …, (2, 3, 4, 5, 6, 7, 32, 58, 59, 61, 63, 67, 70)	(14, 16, 18, 20, 23, 25, 28, 30, 32, 34, 36, 40, 42, 44), (13, 14, 16, 18, 20, 22, 28, 30, 33, 36, 40, 42, 44, 46),…, (1, 2, 3, 4, 5, 6, 28, 30, 48, 50, 53, 55, 58, 59)
D	$4 \to 5$	(18, 20, 22, 24, 26, 28, 32, 38, 42, 44, 46, 48, 50, 52), (16, 18, 20, 22, 24, 26, 32, 38, 44, 46, 48, 50, 52, 54), …, (2, 3, 4, 5, 6, 7, 32, 38, 60, 64, 65, 66, 68, 70)	(14, 16, 18, 20, 23, 25, 28, 30, 32, 34, 36, 40, 42, 44), (13, 14, 16, 18, 20, 22, 28, 30, 33, 36, 40, 42, 44, 46), …, (1, 2, 3, 4, 5, 6, 28, 30, 48, 50, 53, 55, 58, 59)
E	$4 \to 6$	(25, 27, 30, 32, 33, 35, 43, 44, 46, 48, 49, 52, 53), (24, 26, 29, 30, 32, 34, 43, 45, 46, 49, 50, 53, 56), …, (4, 5, 6, 8, 9, 11, 43, 65, 67, 70, 72, 77, 80)	(14, 16, 18, 20, 23, 25, 28, 30, 32, 34, 36, 40, 42, 44), (13, 14, 16, 18, 20, 22, 28, 30, 33, 36, 40, 42, 44, 46), …, (1, 2, 3, 4, 5, 6, 28, 30, 48, 50, 53, 55, 58, 59)
F	$5 \to 6$	(24, 26, 28, 29, 32, 34, 36, 38, 40, 43, 44, 46, 48), (22, 24, 26, 27, 30, 33, 36, 40, 43, 44, 46, 47, 50), …, (2, 4, 5, 7, 8, 10, 36, 61, 63, 65, 66, 70, 71)	(14, 16, 18, 20, 23, 25, 28, 30, 32, 34, 36, 40, 42, 44), (13, 14, 16, 18, 20, 22, 28, 30, 33, 36, 40, 42, 44, 46), …, (1, 2, 3, 4, 5, 6, 28, 30, 48, 50, 53, 55, 58, 59)
G	$6 \to 7$	(26, 28, 30, 32, 34, 36, 37, 41, 42, 44, 46, 48, 50, 52), (24, 26, 28, 30, 32, 34, 37, 41, 44, 46, 48, 50, 52, 54), …, (4, 6, 8, 9, 11, 12, 37, 41, 64, 66, 68, 71, 73, 76)	(17, 19, 21, 13, 25, 26, 30, 34, 37, 38, 40, 44, 46, 48), (16, 17, 19, 21, 23, 25, 30, 34, 38, 40, 42, 45, 48, 50), …, (1, 2, 3, 5, 6, 7, 30, 34, 53, 57, 59, 61, 64, 66)

Step-3

See Fig. 4.

Step-4

See Fig. 5 and Table 4.

Table 2 Details of the project cost

Activity number	Activity	Normal cost (N_c)	Crash cost (C_c)
A	$1 \to 2$	(140,000, 150,000, 160,000, 170,000, 180,000, 190,000, 200,000, 220,000, 240,000, 250,000, 260,000, 270,000, 280,000, 290,000), …, (10,000, 20,000, 30,000, 40,000, 50,000, 60,000, 200,000, 220,000, 370,000, 380,000, 390,000, 400,000, 410,000, 420,000)	(250,000, 270,000, 290,000, 310,000, 330,000, 340,000, 390,000, 420,000, 450,000, 470,000, 490,000, 510,000, 53,000, 550,000), …, (40,000, 50,000, 60,000, 80,000, 90,000, 110,000, 390,000, 420,000, 670,000, 690,000, 720,000, 740,000, 760,000, 780,000)
B	$2 \to 3$	(250,000, 270,000, 290,000, 310,000, 330,000, 340,000, 390,000, 420,000, 450,000, 470,000, 490,000, 510,000, 530,000, 550,000), …, (40,000, 50,000, 60,000, 80,000, 90,000, 110,000, 390,000, 420,000, 670,000, 690,000, 720,000, 740,000, 760,000, 780,000)	(440,000, 460,000, 480,000, 500,000, 520,000, 540,000, 580,000, 660,000, 690,000, 700,000, 720,000, 740,000, 760,000), …, (200,000, 220,000, 240,000, 260,000, 280,000, 300,000, 580,000, 900,000, 920,000, 940,000, 960,000, 980,000, 990,000)
C	$3 \to 4$	(180,000, 200,000, 220,000, 240,000, 260,000, 280,000, 320,000, 380,000, 420,000, 440,000, 460,000, 480,000, 500,000, 520,000), …, (20,000, 30,000, 40,000, 50,000, 60,000, 70,000, 320,000, 380,000, 600,000, 640,000, 650,000, 660,000, 680,000, 700,000)	(260,000, 280,000, 290,000, 320,000, 330,000, 340,000, 420,000, 440,000, 460,000, 480,000, 500,000, 510,000, 530,000, 580,000), …, (40,000, 50,000, 80,000, 90,000, 130,000, 100,000, 120,000, 520,000, 440,000, 690,000, 710,000, 720,000, 740,000, 760,000, 780,000)
D	$4 \to 5$	(140,000, 150,000, 160,000, 170,000, 180,000, 190,000, 200,000, 220,000, 240,000, 250,000, 260,000, 270,000, 280,000, 290,000), …, (10,000, 20,000, 30,000, 40,000, 50,000, 60,000, 200,000, 220,000, 370,000, 380,000, 390,000, 400,000, 410,000, 420,000)	(500,000, 530,000, 540,000, 570,000, 590,000, 600,000, 610,000, 630,000, 650,000, 670,000, 700,000, 720,000, 730,000, 750,000), …, (270,000, 300,000, 310,000, 330,000, 340,000, 350,000, 610,000, 630,000, 910,000, 910,000, 940,000, 950,000, 960,000, 970,000)

(continued)

Table 2 (continued)

Activity number	Activity	Normal cost (N_c)	Crash cost (C_c)
E	$4 \to 6$	(140,000, 150,000, 160,000, 170,000, 180,000, 200,000, 220,000, 260,000, 280,000, 300,000, 320,000, 340,000, 350,000, 360,000), …, (10,000, 20,000, 30,000, 40,000, 50,000, 60,000, 220,000, 260,000, 410,000, 440,000, 450,000, 470,000, 480,000, 490,000)	(300,000, 320,000, 340,000, 360,000, 380,000, 400,000, 420,000, 440,000, 460,000, 480,000, 500,000, 520,000, 540,000), …, (60,000, 80,000, 100,000, 120,000, 140,000, 160,000, 420,000, 680,000, 700,000, 720,000, 740,000, 760,000, 780,000)
F	$5 \to 6$	(180,000, 200,000, 220,000, 240,000, 260,000, 280,000, 320,000, 380,000, 420,000, 440,000, 460,000, 480,000, 500,000, 520,000), …, (20,000, 30,000, 40,000, 50,000, 60,000, 70,000, 320,000, 380,000, 600,000, 640,000, 650,000, 660,000, 680,000, 700,000)	(450,000, 480,000, 500,000, 520,000, 540,000, 560,000, 580,000, 600,000, 620,000, 640,000, 660,000, 680,000, 700,000, 720,000), …, (250,000, 260,000, 270,000, 300,000, 310,000, 320,000, 580,000, 600,000, 830,000, 870,000, 890,000, 900,000, 930,000, 940,000)
G	$6 \to 7$	(250,000, 270,000, 290,000, 310,000, 330,000, 340,000, 390,000, 420,000, 450,000, 470,000, 490,000, 510,000, 530,000, 550,000), …, (40,000, 50,000, 60,000, 80,000, 90,000, 110,000, 390,000, 420,000, 670,000, 690,000, 720,000, 740,000, 760,000, 780,000)	(300,000, 320,000, 340,000, 360,000, 380,000, 400,000, 420,000, 440,000, 460,000, 480,000, 500,000, 520,000, 540,000), …, (60,000, 80,000, 100,000, 120,000, 140,000, 160,000, 420,000, 680,000, 70,720,000, 740,000, 760,000, 780,000)

Table 3 Details of the slope cost

Activity number	Activity	N_t	C_t	N_c	C_c	$\Delta t = N_t - C_t$	$\Delta c = C_c - N_c$	$\Delta c/\Delta t$
A	$1 \to 2$	35	28.71429	214,285.7	400,000	6.28571	185,714.3	29,545.47
B	$2 \to 3$	40	35	400,000	598,461.5	5	198,461.5	39,692.3
C	$3 \to 4$	34	28.71429	350,000	410,000	5.28571	60,000	11,351.36
D	$4 \to 5$	35	28.71429	214,285.7	627,857.1	6.28571	413,571.4	65,795.49
E	$4 \to 6$	40	28.71429	245,000	420,000	11.28571	175,000	15,506.33
F	$5 \to 6$	36	28.71429	350,000	589,285.7	7.28571	239,285.7	239,278.41
G	$6 \to 7$	39	32	400,000	420,000	7	20,000	2857.14

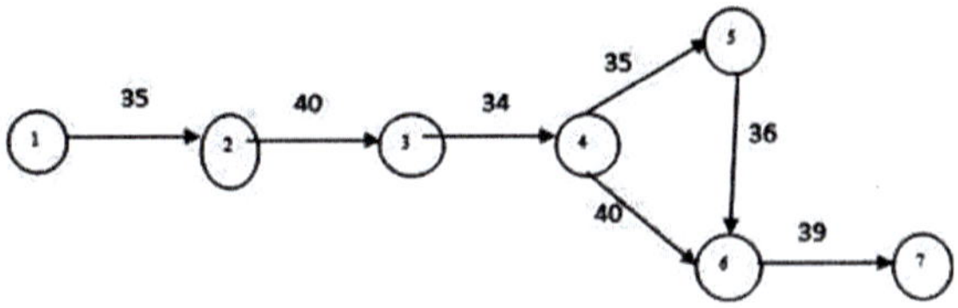

Fig. 1 Critical path

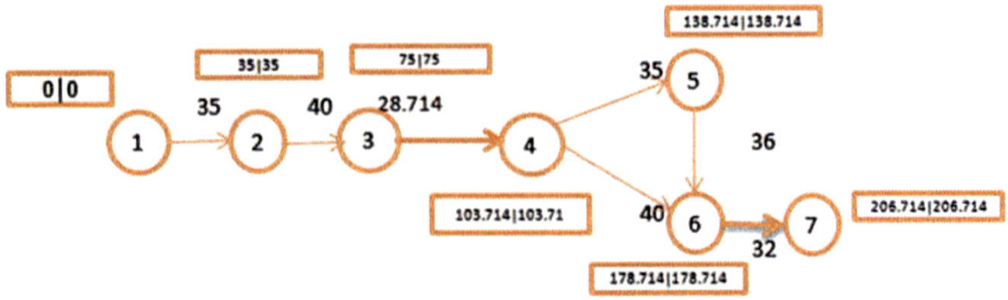

Fig. 2 Critical path

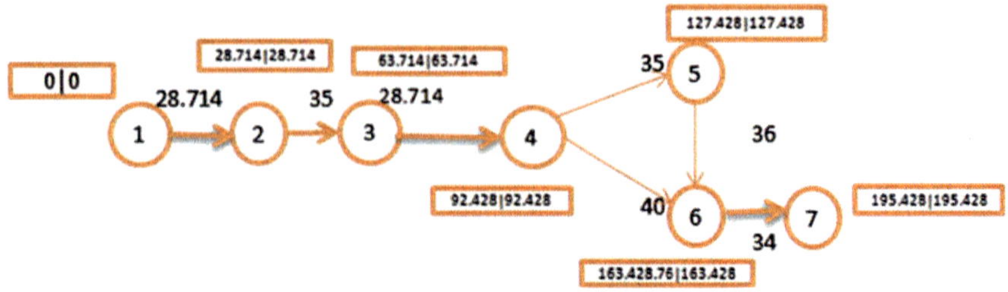

Fig. 3 Critical path

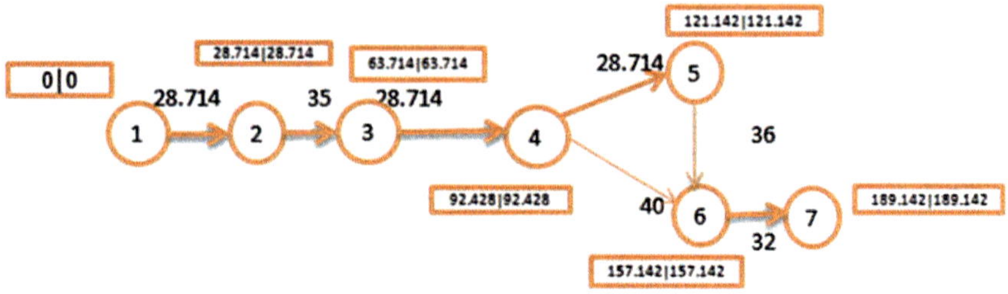

Fig. 4 Critical path

Critical path: 1 → 2→3 → 4→5 → 6→7
Project duration: 219 days
Direct cost = Rs. 2,173,571.399 and indirect cost = Rs. 0

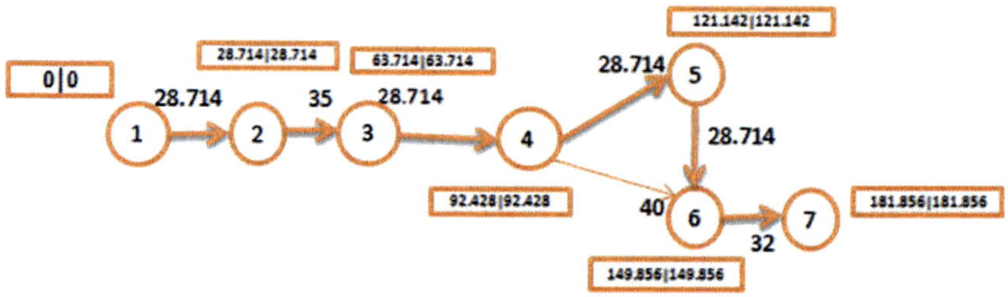

Fig. 5 Critical path

Table 4 Details of the slope cost

Step size	Critical path	Project duration (days)	Total cost (Rs)
Step-1	1 → 2→3 → 4→5 → 6→7	206.71	2,187,779.899
Step-2	1 → 2→3 → 4→5 → 6→7	195.428	2,257,017.669
Step-3	1 → 2→3 → 4→5 → 6→7	189.142	2,322,813.159
Step-4	1 → 2→3 → 4→5 → 6→7	181.856	2,562,091.569

Total cost = Rs. 2,173,571.399.

4 Conclusion

This method is very easy to understand and will help the decision maker to determine the total cost and completion time of the project. This method is most significant in the use of the practical test in industries.

References

1. Helen R, Sumathi R (2017) Time cost trade-off problem using type -2 trapezoidal fuzzy numbers. Inf Sci 96(1):1–6
2. Rajkumar A, Helen D (2016) A new arithmetic operations of triskaidecagonal fuzzy number using alpha cut. In: Advances in intelligent system computing, Soft computing theories and applications, vol 583. Springer, Singapore, pp 77–78
3. Rajkumar A, Helen D (2016) A new arithmetic operations in inverse of triskaidecagonal fuzzy number using alpha cut. In: Advances in intelligent system computing, Soft computing theories and applications, vol 583. Springer, Singapore, pp 79–86
4. Rajkumar A, Helen D (2017) Tree trigger success of door bell using fuzzy number. Int J Pure Appl Math 114(5):71–77

5. Rajkumar A, Helen D (2017) A project construction of normal and crashing activity using type-2 triskaidecagonal fuzzy number. Int J Pure Appl Math 117(11):237–244
6. Rehab R, Carr RI (1986) Time cost trade off among related activities. J Constr Eng Manag 115:475–486

Critical Path Problem Through Intuitionistic Triskaidecagonal Fuzzy Number Using Two Different Algorithms

N. Jose Parvin Praveena, C. Sagaya Nathan Stalin, and A. Rajkumar

Abstract This paper deals to find critical path through Intuitionistic Triskaidecagonal fuzzy numbers and Triskaidecagonal fuzzy numbers by two distinct methods. The formulae for α-cut ranking method and Euclidean-ranking method are derived. The Triskaidecagonal fuzzy numbers are defuzzied by the magnitude measure which is derived in this paper.

Keywords Triskaidecagonal fuzzy number · Intuitionistic triskaidecagonal fuzzy number · Critical path · Magnitude measure · α-cut ranking · Euclidean ranking

1 Introduction

In numerous genuine cases, the choice on information human judgments with inclinations is regularly unclear, so the customary methods for utilizing fresh qualities are lacking additionally utilizing fuzzy numbers, for example, triangular and trapezoidal are not appropriate in where the vulnerabilities emerge in thirteen distinct focuses in such cases the Triskaidecagonal fuzzy number can be utilized to tackle the issues [1]. In 2015, decagonal and dodecagonal have been produced. It has been additionally stretched out into Triskaidecagonal fuzzy number. Triskaidecagonal, another type of fuzzy number, has been inspected under unverifiable etymological condition which would be less difficult to represent with Triskaidecagonal fuzzy semantic-scale esteems [2]. Alpha-cut strategy is the standard technique that performs diverse number juggling activities including expansion, subtraction, increase and division. Accordingly, Triskaidecagonal, another type of fuzzy number, had been inspected under semantic-scale esteems in which number-crunching task is to be performed

N. Jose Parvin Praveena (✉)
St. Joseph's College of Engineering, Semmenchery, Chennai 600119, India
e-mail: jose30102003@gmail.com

C. Sagaya Nathan Stalin
Hindustan Institute of Technology and Science, Loyola College, Vettavalam, India

C. Sagaya Nathan Stalin · A. Rajkumar
Hindustan Institute of Technology and Science, Chennai 603103, India

N. N. Chiplunkar and T. Fukao (eds.), *Advances in Artificial Intelligence and Data Engineering*, Advances in Intelligent Systems and Computing 1133,
https://doi.org/10.1007/978-981-15-3514-7_14

by standard alpha-cut strategy with numerical precedents. This will be connected in biomedical, numerical examination, activity look into and so forth. The α-cut ranking method, Euclidean-ranking method and magnitude measure are derived in Sect. 2. In Sect. 3, new JOSE algorithm was introduced through Intuitionistic Triskaidecagonal fuzzy numbers to find the critical path. In Sect. 4, the critical path is obtained using Triskaidecagonal fuzzy numbers. Section 5 concludes this paper.

2 Definitions

2.1 The Membership Function and Non-membership Function of the Intuitionistic Triskaidecagonal Fuzzy Numbers

$$\mu_{\tilde{R}_{\mathrm{TD}}}(x)=\begin{cases}0 & x\le l_1\\ \frac{1}{6}\frac{x-l_1}{l_2-l_1} & l_1\le x\le l_2\\ \frac{1}{6}+\frac{1}{6}\frac{x-l_2}{l_3-l_2} & l_2\le x\le l_3\\ \frac{2}{6}+\frac{1}{6}\frac{x-l_3}{l_4-l_3} & l_3\le x\le l_4\\ \frac{3}{6}+\frac{1}{6}\frac{x-l_4}{l_5-l_4} & l_4\le x\le l_5\\ \frac{4}{6}+\frac{1}{6}\frac{x-l_5}{l_6-l_5} & l_5\le x\le l_6\\ \frac{5}{6}+\frac{1}{6}\frac{x-l_6}{l_7-l_6} & l_6\le x\le l_7\\ 1-\frac{1}{6}\frac{x-l_7}{l_8-l_7} & l_7\le x\le l_8\\ \frac{5}{6}-\frac{1}{6}\frac{x-l_8}{l_9-l_8} & l_8\le x\le l_9\\ \frac{4}{6}-\frac{1}{6}\frac{x-l_9}{l_{10}-l_9} & l_9\le x\le l_{10}\\ \frac{3}{6}-\frac{1}{6}\frac{x-l_{10}}{l_{11}-l_{10}} & l_{10}\le x\le \lambda_{11}\\ \frac{2}{6}-\frac{1}{6}\frac{x-l_{11}}{l_{12}-l_{11}} & l_{11}\le x\le l_{12}\\ \frac{1}{6}\frac{l_{13}-x}{l_{13}-l_{12}} & l_{12}\le x\le \lambda_{13}\\ 0 & x\ge \lambda_{13}\end{cases}$$

$$\gamma_{\tilde{R}_{\mathrm{ITD}}}(x)=\begin{cases} 1 & x\le m_1\\ 1-\frac{1}{6}\frac{x-m_1}{m_2-m_1} & m_1\le x\le m_2\\ \frac{5}{6}-\frac{1}{6}\frac{x-m_2}{m_3-m_2} & m_2\le x\le m_3\\ \frac{4}{6}-\frac{1}{6}\frac{x-m_3}{m_4-m_3} & m_3\le x\le m_4\\ \frac{3}{6}-\frac{1}{6}\frac{x-m_4}{m_5-m_4} & m_4\le x\le m_5\\ \frac{2}{6}-\frac{1}{6}\frac{x-m_5}{m_6-m_5} & m_5\le x\le m_6\\ \frac{1}{6}\frac{x-m_6}{m_7-m_6} & m_6\le x\le m_7\\ \frac{1}{6}\frac{x-m_7}{m_8-m_7} & m_7\le x\le m_8\\ \frac{1}{6}+\frac{1}{6}\frac{x-m_8}{m_9-m_8} & m_8\le x\le m_9\\ \frac{2}{6}+\frac{1}{6}\frac{x-m_9}{m_{10}-m_9} & m_9\le x\le m_{10}\\ \frac{3}{6}+\frac{1}{6}\frac{x-m_{10}}{m_{11}-m_{10}} & m_{10}\le x\le m_{11}\\ \frac{4}{6}+\frac{1}{6}\frac{x-m_{11}}{m_{12}-m_{11}} & m_{11}\le x\le m_{12}\\ \frac{5}{6}+\frac{1}{6}\frac{m_{13}-x}{m_{13}-m_{12}} & m_{12}\le x\le m_{13} \end{cases}$$

2.2 *Defuzzification of Triskaidecagonal Fuzzy Number*

The α-cut of the fuzzy set $\tilde{H}$ is described as

$$\tilde{H}=\{y\varepsilon Y/\mu(y)\ge\alpha\}$$

where $\alpha\varepsilon[0,1]$

α-cut of Triskaidecagonal fuzzy number and is as follows:

$$\tilde{\mathrm{T}}\mathrm{D}_\alpha=\begin{cases} [6\alpha(l_2-l_1)+l_1,\ -6\alpha(l_{13}-l_{12})+l_{13}] & \alpha\in(0,1/6)\\ [6\alpha(l_3-l_2)-l_3+2l_2,\ -6\alpha(l_{12}-l_{11})+2l_{12}-l_{11}] & \alpha\in(1/6,2/6)\\ [6\alpha(l_4-l_3)-2l_4+3l_3,\ -6\alpha(l_{11}-l_{10})-2l_{10}+3l_{11}] & \alpha\in(2/6,3/6)\\ [6\alpha(l_5-l_4)-3l_5+4l_4,\ -6\alpha(l_{10}-l_9)-3l_{12}+4\theta_{13}] & \alpha\in(3/6,4/6)\\ [6\alpha(l_6-l_5)-4l_6+5l_5,\ -6\alpha(l_9-l_8)-4l_8+5l_9] & \alpha\in(4/6,5/6)\\ [6\alpha(l_7-l_6)-5l_7+6l_6,\ -6\alpha(l_8-l_7)-5l_7+6l_8] & \alpha\in(5/6,1) \end{cases} \quad (1)$$

Magnitude measure of $\tilde{F}$ is as follows:

$$M\left(\tilde{F}\right)=\frac{1}{2}\int_0^1\begin{Bmatrix} [6\alpha(l_2-l_1)+l_1,\ -6\alpha(l_{13}-l_{12})+l_{13}]+\\ [6\alpha(l_3-l_2)-l_3+2l_2,\ -6\alpha(l_{12}-l_{11})+2l_{12}-l_{11}]+\\ [6\alpha(l_4-l_3)-2l_4+3l_3,\ -6\alpha(l_{11}-l_{10})-2l_{10}+3l_{11}]+\\ [6\alpha(l_5-l_4)-3l_5+4l_4,\ -6\alpha(l_{10}-l_9)-3l_{12}+4\theta_{13}]+\\ [6\alpha(l_6-l_5)-4l_6+5l_5,\ -6\alpha(l_9-l_8)-4l_8+5l_9]+\\ [6\alpha(l_7-l_6)-5l_7+6l_6,\ -6\alpha(l_8-l_7)-5l_7+6l_8] \end{Bmatrix}\alpha d\alpha$$

$$= \frac{1}{12}\begin{pmatrix} -9l_1 + 6l_2 + 6l_3 + 6l_4 + 6l_5 + 6l_6 - 30l_7 \\ +6l_8 + 6l_9 + 6l_{10} + 6l_{11} + 6l_{12} - 9l_{13} \end{pmatrix} \tag{2}$$

2.3 α-Cut Ranking Technique for Intuitionistic Triskaidecagonal Fuzzy Number [3]

For membership function,

$$R\left(\tilde{F}_1\right) = \int_0^1 \begin{Bmatrix} [6\alpha(l_2 - l_1) + l_1, -6\alpha(l_{13} - l_{12}) + l_{13}] + \\ [6\alpha(l_3 - l_2) - l_3 + 2l_2, -6\alpha(l_{12} - l_{11}) + 2l_{12} - l_{11}] + \\ [6\alpha(l_4 - l_3) - 2l_4 + 3l_3, -6\alpha(l_{11} - l_{10}) - 2l_{10} + 3l_{11}] + \\ [6\alpha(l_5 - l_4) - 3l_5 + 4l_4, -6\alpha(l_{10} - l_9) - 3l_{12} + 4\theta_{13}] \\ [6\alpha(l_6 - l_5) - 4l_6 + 5l_5, -6\alpha(l_9 - l_8) - 4l_8 + 5l_9] + \\ [6\alpha(l_7 - l_6) - 5l_7 + 6l_6, -6\alpha(l_8 - l_7) - 5l_7 + 6l_8] \end{Bmatrix} \alpha d\alpha$$

$$= \frac{1}{6}\begin{pmatrix} -9l_1 + 6l_2 + 6l_3 + 6l_4 + 6l_5 + 6l_6 - 30l_7 \\ +6l_8 + 6l_9 + 6l_{10} + 6l_{11} + 6l_{12} - 9l_{13} \end{pmatrix}$$

For non-membership function,

$$R\left(\tilde{F}_1\right) = \int_0^1 \left\{ \begin{bmatrix} -6\alpha(m_2 - m_1) + 6m_2 - 5m_1 + 6\alpha(m_{12} - m_{13}) \\ +6m_{13} - 5m_{12} - 6\alpha(m_3 - m_2) + 5m_3 - 4m_2 \\ +6\alpha(m_{12} - m_{11} - 4m_{12} + 5m_{11} - 6\alpha(m_4 - m_3) \\ +4m_4 - 3m_3 + 6\alpha(m_{11} - m_{10}) - 3m_{11} + 4m_{10} \\ -6\alpha(m_5 - m_4) + 3m_5 - 2m_4 + 6\alpha(m_{10} - m_9) - 2m_{10} \\ +3m_9 - 6\alpha(m_6 - m_5) + 2m_6 - m_5 + 6\alpha(m_9 - m_8) \\ -m_9 + 2m_8 + 6\alpha(m_7 - m_6) + m_6 + 6\alpha(m_8 - m_7) + m_7 \end{bmatrix} \right. \alpha d\alpha$$

$$= \frac{1}{6}\begin{pmatrix} -3m_1 + 6m_2 + 6m_3 + 6m_4 + 6m_5 + 6l_6 - 15m_6 \\ +3m_7 + 6m_8 + 6m_9 + 6m_{10} + 6m_{11} - 3m_{12} + 6m_{13} \end{pmatrix}$$

Therefore, α-cut ranking for Intuitionistic Triskaidecagonal fuzzy number is as follows:

$$R\left(\tilde{E}_1\right) = \left(\frac{1}{6}\begin{pmatrix} -9l_1 + 6l_2 + 6l_3 + 6l_4 + 6l_5 + 6l_6 - 30l_7 \\ +6l_8 + 6l_9 + 6l_{10} + 6l_{11} + 6l_{12} - 9l_{13} \end{pmatrix}, \frac{1}{6}\begin{pmatrix} -3m_1 + 6m_2 + 6m_3 + 6m_4 + 6m_5 + 6l_6 - 15m_6 \\ +3m_7 + 6m_8 + 6m_9 + 6m_{10} + 6m_{11} - 3m_{12} + 6m_{13} \end{pmatrix}\right) \tag{3}$$

2.4 Euclidean-Ranking Technique for Intuitionistic Decagonal Fuzzy Number [4]

Let

$$P_i = \big((\eta_{1,1}, \eta_{2,1}, \eta_{3,1}, \eta_{4,1}, \eta_{5,1}, \eta_{,6,1}, \eta_{7,1}, \eta_{8,1}, \eta_{9,1}, \eta_{10,1}, \eta_{11,1}, \eta_{12,1}, \eta_{13,1}),\\ (\eta'_{1,1}, \eta'_{,2,1}, \eta'_{3,1}, \eta'_{4,1}, \eta'_{5,1}, \eta'_{6,1}, \eta'_{7,1}, \eta'_{8,1}, \eta'_{9,1}, \eta'_{10,1}, \eta'_{11,1}, \eta'_{12,1}, \eta'_{13,1})\big)$$

be the ith fuzzy path length and

$$\begin{aligned} P_{\max} &= \big((\beta_{,11}, \beta_{2,1}, \beta_{3,1}, \beta_{4,1}, \beta_{5,1}, \beta_{6,1}, \beta_{7,1}, \beta_{,8,1}, \beta_{9,1}, \beta_{10,1}, \beta_{11,1}, \beta_{12,1}, \beta_{13,1}),\\ &\quad (\beta'_{1,1}, \beta'_{2,1}, \beta'_{3,1}, \beta'_{4,1}, \beta'_{5,1}, \beta'_{6,1}, \beta'_{7,1}, \beta'_{8,1}, \beta'_{9,1}, \beta'_{10,1}, \beta'_{11,1}, \beta'_{12,1}, \beta'_{13,1})\big)\\ &= (P_{\mathrm{MAX}}, p_{\max}) \end{aligned}$$

be the lengthiest length.

The Euclidean ranking is given by

$$\mathrm{ER}(P_i) = \left(\frac{\sqrt{\begin{array}{r} (\beta_{1,1} - \eta_{1,1})^2 + (\beta_{2,1} - \eta_{2,1})^2 + (\beta_{3,1} - \eta_{3,1})^2 + (\beta_{4,1} - \eta_{4,1})^2 \\ +(\beta_{5,1} - \eta_{5,1})^2 + (\beta_{6,1} - \eta_{6,1})^2 + (\beta_{7,1} - \eta_{7,1})^2 + (\beta_{8,1} - \eta_{8,1})^2 \\ +(\beta_{9,1} - \eta_{9,1})^2 + (\beta_{10,1} - \eta_{10,1})^2 + (\beta_{11,1} - \eta_{11,1})^2 + (\beta_{12,1} - \eta_{12,1})^2 \\ +(\beta_{13,1} - \eta_{13,1})^2 \end{array}}}{\sqrt{\begin{array}{r} (\beta'_{,11} - \eta'_{1,1})^2 + (\beta'_{2,1} - \eta'_{2,1})^2 + (\beta'_{3,1} - \eta'_{3,1})^2 + (\beta'_{4,1} - \eta'_{4,1})^2 \\ +(\beta'_{5,1} - \eta'_{5,1})^2 + (\beta'_{6,1} - \eta'_{6,1})^2 + (\beta'_{7,1} - \eta'_{7,1})^2 + (\beta'_{8,1} - \eta'_{8,1})^2 \\ +(\beta'_{9,1} - \eta'_{9,1})^2 + (\beta'_{10,1} - \eta'_{10,1})^2 + (\beta'_{11,1} - \eta'_{11,1})^2 + (\beta'_{12,1} - \eta'_{12,1})^2 \\ +(\beta'_{13,1} - \eta'_{13,1})^2 \end{array}}} \right) \tag{4}$$

3 New JOSE Algorithm for Critical Path [5]

Step: 1 Build an acyclic system. V and E are represented for vertex and edge set.
Step: 2 Possible paths $T_i, i = 1, \ldots, n$, path lengths $M_i, i = 1, \ldots, n$ are determined.
Step: 3 The longest fuzzy path length is calculated and it is denoted by $L_{\max}$.
Step: 4 The α-cut ranking and the Euclidean ranking are calculated for all the path lengths $M_i, i = 1$ to n. In α-cut ranking, the membership function is highest and the non-membership function is lowest. Such a path is called the critical path. In Euclidean ranking, both membership and non-membership values are lowest (Fig. 1; Tables 1 and 2).

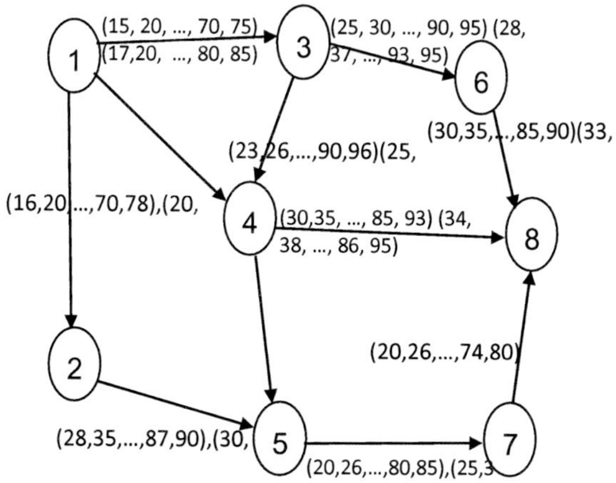

Fig. 1 Acyclic network

3.1 *Numerical Result*

Consider **an acyclic system**

Calculate $P_{\max}$ and P_{MAX}

$$P_{\text{MAX}} = (103, 133, 165, 191, 14, 240, 265, 296, 324, 351, 378, 402, 426)$$
$$P_{\max} = (122, 146, 182, 207, 230, 255, 282, 309, 337, 377, 402, 424, 452)$$

By the algorithm, it is identified that 1-3-4-5-7-8 is the fuzzy critical path.

4 Technique for the Fuzzy Critical Path Using Triskaidecagonal Fuzzy Numbers [6, 7]

Let

$$e_{mn} = \left((\varsigma_{1,1}, \varsigma_{2,1}, \varsigma_{3,1}, \varsigma_{4,1}, \varsigma_{5,1}, \varsigma_{6,1}, \varsigma_{7,1}, \varsigma_{8,1}, \varsigma_{9,1}, \varsigma_{10,1}, \varsigma_{11,1}, \varsigma_{12,1}, \varsigma_{13,1})\right)$$

represents the Triskaidecagonal fuzzy number and

$$h^*(m) = \max(e^*_{mn} + g^*(n)/m, n \in E)$$
$$h^*(u) = 0$$

where $h^*(m)$ is the distance of the lengthiest path. e_{mn} is defuzzified by using the following formula (Fig. 2).

Table 1 Pathways and pathway lengths

Pathways	Pathway length
1-3-6-8	((70, 85, 105, 120, 141, 159, 175, 190, 205, 220, 233, 245, 260), (78, 95, 119, 130, 151, 165, 186, 197, 216, 238, 253, 263, 275))
1-4-8	((50, 60, 70, 80, 90, 101, 108, 116, 134, 147, 157, 171, 183), (59, 68, 80, 90, 100, 107, 115, 125, 141, 154, 160, 171, 185))
1-2-5-7-8	((84, 107, 124, 149, 172, 190, 211, 230, 250, 271, 292, 311, 333), (100, 120, 141, 167, 190, 211, 227, 244, 260, 280, 299, 320, 347))
1-4-5-7-8	(85, 112, 135, 157, 176, 196, 215, 237, 262, 286, 307, 328, 345), (105, 127, 153, 177, 194, 215, 234, 251, 275, 303, 318, 337, 360)
1-3-4-8	((68, 81, 100, 114, 128, 145, 158, 175, 196, 212, 228, 245, 264), (76, 87, 109, 120, 136, 147, 163, 183, 203, 228, 244, 258, 277))
1-3-4-5-7-8	((103, 133, 165, 191, 14, 240, 265, 296, 324, 351, 378, 402, 426), (122, 146, 182, 207, 230, 255, 282, 309, 337, 377, 402, 424, 452))

Table 2 Euclidean ranking and α-cut ranking

Paths	$R(P_i)$	$\mathrm{ER}(P_i)$
1-3-6-8	(333, −1184)	(384, 405)
1-4-8	(237, −778)	(600, 646)
1-2-5-7-8	(416, −1424)	(229, 251)
1-4-5-7-8	(476, −1517)	(193, 208)
1-3-4-8	(336, −1134)	(407, 439)
1-3-4-5-7-8	(576, −1873)	(0, 0)

Fig. 2 Acyclic network

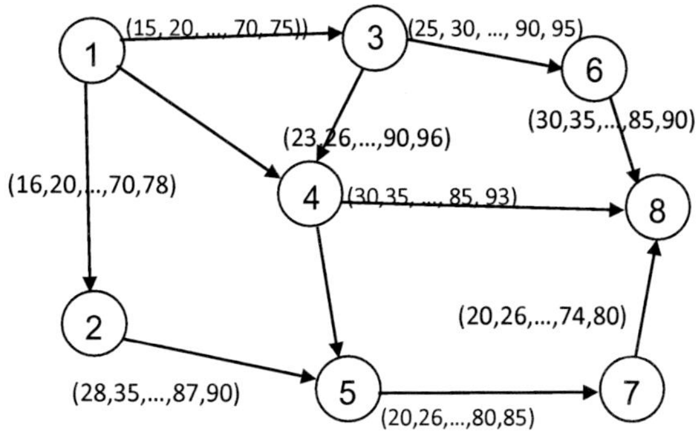

$$e^*_{mnj} = \frac{1}{12}\begin{pmatrix} -9\varsigma_{1,1} + 6\varsigma_{2,1} + 6\varsigma_{3,1} + 6\varsigma_{4,1} + 6\varsigma_{,15} + 6\varsigma_{6,1} \\ -30\varsigma_{7,1} + 6\varsigma_{8,1} + 6\varsigma_{9,1} + 6\varsigma_{10,1} + 6\varsigma_{11,1} + 6\varsigma_{12,1} - 9\varsigma_{13,1} \end{pmatrix} \quad (6)$$

Estimate the defuzzified edge weights for each edge e_{ij} which is represented by [8] e^*_{ij}

$$e^*_{12} = 40, e^*_{13} = 45, e^*_{14} = 62, e^*_{25} = 66, e^*_{45} = 74.25, e^*_{34} = 66.75,$$
$$e^*_{36} = 61.5 e^*_{68} = 60, e^*_{48} = 56.25 e^*_{57} = 51.75, e^*_{78} = 50$$

The calculation is as follows:

$$\begin{aligned}
f^*(8) &= 0, f^*(7) = e^*_{78} + f^*(8) = 50 + 0 = 50, \\
f^*(6) &= e^*_{68} + f^*(8) = 60 + 0 = 60, \\
f^*(5) &= e^*_{57} + f^*(7) = 51.75 + 50 = 101.75 \\
f^*(4) &= \max(e^*_{48} + f^*(8), e^*_{45} + f^*(5)) \\
&= \max(56.25 + 0, 74.25 + 101.75) \\
&= \max(56.25, 176) = 176 \\
f^*(3) &= \max(e^*_{34} + f^*(4), e^*_{36} + f^*(6)) = \max(66.75 + 176, 61.5 + 60) \\
&= \max(242.75, 121.5) = 242.75 \\
f^*(2) &= e^*_{25} + f^*(5) = 66 + 101.75 = 167.75,
\end{aligned}$$

$$
\begin{aligned}
f^*(1) &= \max(e_{13}^* + f^*(3), e_{14}^* + f^*(4), e_{12}^* + f^*(2) \\
&= \max(45 + 242.75, 62 + 176, 40 + 167.75) \\
&= \max(287.75, 238, 207.75) = 287.75 \\
f^*(1) &= e_{13}^* + f^*(3) = e_{13}^* + e_{34}^* + f^*(4) = e_{13}^* + e_{34}^* + e_{45}^* + f^*(5) \\
&= e_{13}^* + e_{34}^* + e_{45}^* + e_{57}^* + f^*(7) = e_{13}^* + e_{34}^* + e_{45}^* + e_{57}^* + e_{78}^* + f^*(8) \\
&= e_{13}^* + e_{34}^* + e_{45}^* + e_{57}^* + e_{78}^*
\end{aligned}
$$

Therefore, **1-3-4-5-7-8** is the fuzzy critical path.

5 Conclusion

The proposed model has been designed to find the critical path using 13 parameters. The model has been illustrated with α-cut ranking method, Euclidean-ranking method and dynamic encoding recursion of the critical path towards the Intuitionistic Triskaidecagonal and Triskaidecagonal fuzzy numbers. The path **1-3-4-5-7-8** is the fuzzy critical path for this acyclic network.

References

1. Dubois D, Prade H (1980) Fuzzy sets and systems. Academic Press, New York, NY
2. Okada S, Soper T (2000) A shortest path problem on a network with fuzzy arc lengths. Fuzzy Sets Syst 109(1):129–140
3. Okada S, Gen M (1994) Fuzzy shortest path problem. Comput Ind Eng 27(1–4):465–468
4. Nagoor Gani A, Mohammed Jabarulla M (2010) On searching intuitionistic fuzzy shortest path in a network. Appl Math Sci 4(69–72):3447–3454
5. Okada S, Gen M (1993) Order relation between intervals and its application to shortest path problem. Comput Ind Eng 25(1):147–150
6. Hernandes F, Lamata MT, Verdegay JL (2007) The shortest path problem on networks with fuzzy parameters. Fuzzy Sets Syst 158(14):1561–1570
7. Jayagowri P, Geetharamani G (2012) On solving network problems using new algorithm with intuitionistic fuzzy arc length. In: Proceedings of the international conference on mathematics in engineering & business management
8. Yu D (2013) Intuitionistic trapezoidal fuzzy information aggregation methods and their applications to teaching quality evaluation. J Inf Comput Sci 10(6):861–869

Genetic-Neuro-Fuzzy Controller for Indirect Vector-Controlled Induction Motor Drive

B. T. Venu Gopal, H. R. Ramesh, and E. G. Shivakumar

Abstract In this work, genetic-neuro-fuzzy controller (GA-NFC) for speed control of induction motor, which has the advantages of fuzzy logic control (FLC), genetic algorithm (GA), and artificial neural networks (ANN), is presented. Genetic algorithms are used to tune the membership functions of neuro-fuzzy controller and advance the neuro-fuzzy controller (NFC). To execute this, normalization parameters and membership functions are translated into binary bit string in order to be optimized for fitness function. Multipoint crossover, binary encoding method, and roulette wheel selection techniques are used to improve the efficiency of existing genetic algorithm. Neural networks are good at recognizing patterns and fuzzy logic is good at taking decisions. In fuzzy logic, experts should write the fuzzy rules, but in case of NFC, computer writes rules by itself. Input to proposed NFC is only speed error, but conventional NFCs use both speed error and its derivative as inputs. GA-based NFC controller for a field-oriented/vector control of induction motor is drive-simulated using MATLAB/Simulink. Simulation results indicate a great development in shortening settling time, decreased speed, and torque ripples. GA-NFC offers significant speed accuracy over a conventional NFC.

Keywords GA-NFC · NFC · Artificial neural network (ANN) · Vector control · Induction motor

1 Introduction

Almost 85% of the motors used in industry are induction motors (IM's) because of their simple structure, solid quality, and unchallenged execution. In late 1970s,

B. T. Venu Gopal (✉) · H. R. Ramesh · E. G. Shivakumar
Department of Electrical Engineering, UVCE, Bangalore University, Bengaluru, Karnataka, India
e-mail: btvgopal@gmail.com

H. R. Ramesh
e-mail: hrramesh74@gmail.com

E. G. Shivakumar
e-mail: shivaettigi@gmail.com

N. N. Chiplunkar and T. Fukao (eds.), *Advances in Artificial Intelligence and Data Engineering*, Advances in Intelligent Systems and Computing 1133,
https://doi.org/10.1007/978-981-15-3514-7_15

Hasse and Blaschke proposed vector control method, which is delivered upon the field-oriented control, which decouples the torque and flux control in an induction motor. Thusly, it impacts the control of IM drives like separately excited dc motor while keeping up the benefits of ac over dc motors and which is suitable for high-performance adjustable speed drives. With the happening to late power semiconductor advancements and differing pricise control techniques, a useful control system in context on vector control thought can be associated in real-time applications. Today, field-oriented control-based IM drives have included most extreme of the situations that were situated by dc drives. Among different ac motors, induction motors involve practically 90% of the modern drives because of its strong construction; be that as it may, the control of IM is difficult due to its parameter change with working conditions and also nonlinear nature of induction motor drive [1].

Artificial intelligent controller is the only best way to control the induction motor drive. Figure 1 shows the different combinations of different AI techniques. From last 2 decades, researchers and engineers are working to apply AI techniques for induction motor control [1]. When compared to Pi, PID controllers, AI controllers has many benefits. Because for implementing AI controllers, mathematical model of the system is not required and it offers high performance [2, 3]. Due to limitations of both fuzzy logic and neural networks, a genetic-neuro-fuzzy controller as an AIC is considered in this paper. If we take a simple fuzzy logic controller, it needs more manual adjustment of membership functions by trial and error method to get superior performance. Then again, it is incredibly hard to make a sequential of preparing big data for neural networks that can operate on all the working modes. Neuro-fuzzy controllers (NFCs) with genetic algorithm tuning, which overcome limitations of fuzzy logic controllers and neural system controllers, have been used for motor drive applications [1, 4].

Fuzzy controller structure is a controller structure built on fuzzy reasoning—a intelligent system that evaluates input values in terms of logical variables which take on values between zeros and ones. Fuzzy logic provides the linguistic methods control version in automatic control schemes. In fuzzy control structure, the action of controller is mainly constructed on rules of fuzzy concept, which are produced by fuzzy set principle. Fuzzification, decision making, and defuzzification are the main steps involved in the fuzzy logic controller. Fuzzification is the method of changing the crisp value into fuzzy value [5].

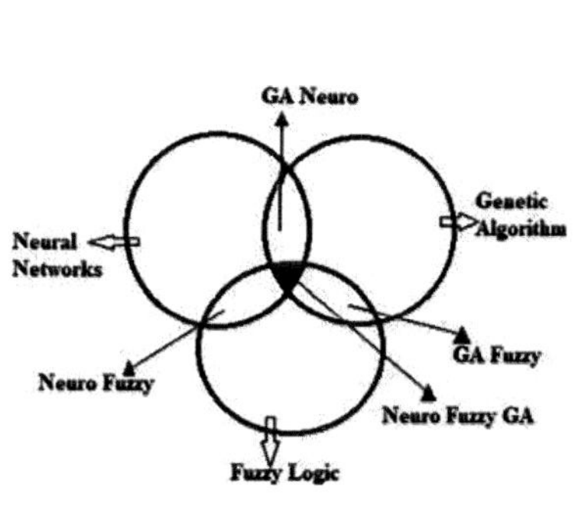

Fig. 1 Integration of neural networks, fuzzy logic, and genetic algorithm technologies

An artificial neural network is a computational model which is derived from natural neural systems. It consists of an inter-related collecting of artificial neurons and actions data using a connectionist method to deal with calculation. By and large, an NN is a adaptable system that develops its arrangement in terms of outside or interior data that interchanges through the system along with the learning step. Neural networks are used to represent the complicated connections of information source and to get the designs in information. NN is a man-made brainpower method that is used for producing and preparing informational collection and testing the connected information [6].

2 Methodology

For high-performance applications, vector control is the most suitable technique to control the induction motor drive. Vector control process can be explained by using Fig. 2. The vector control technique consists of 2 basic ideas. The first is the flux and other is torque creating current components. An induction motor can be presented by utilizing two quadrature current components instead of three-phase currents fed to the motor. These 2 currents known as direct (id) and quadrature (iq) are in charge of creating flux and torque separately in the motor. By definition, the (iq) current is in-phase with the stator flux and (id) is at right angels [7]. The second key feature is the

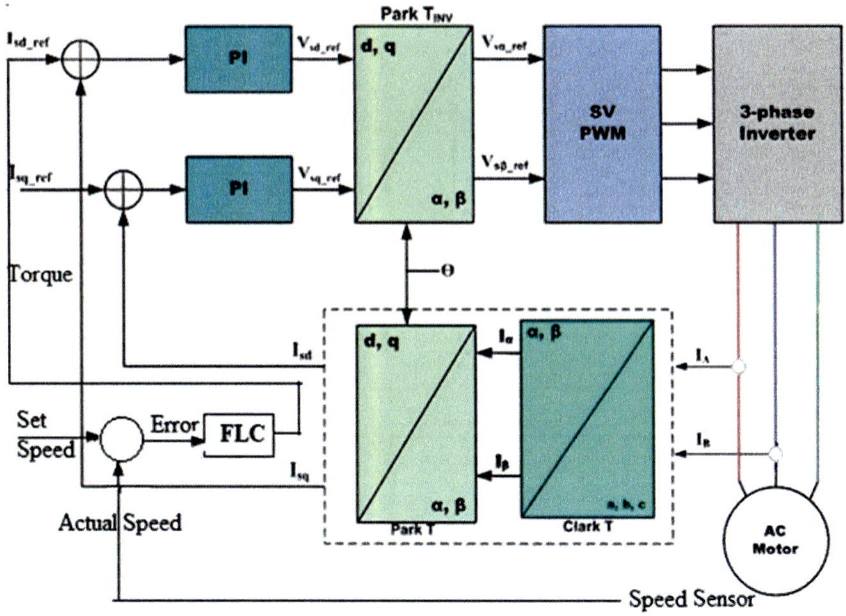

Fig. 2 Induction motor indirect vector control method

use of reference frame. In this technique, an amount of sinusoidal quantity is changed to a constant value in rotating frame which rotates at same frequency. When we get constant value from sinusoidal values by using Clarke and Parke transformations, at the end that quantity can be controlled by using traditional PI controllers. In rotating reference frame, the equation for the electromagnetic torque of induction motor is similar as that of the electromagnetic torque of the separately excited DC motor [8]. NFC block can be simulated in MATLAB either by typing anfisedit command or by programming. Since we want to interlink NFC with GA, we have selected programming method.

3 Neuro-Fuzzy Optimization by Genetic Algorithm

Adaptive Neuro-Fuzzy Inference System: A neuro-fuzzy structure is a fuzzy structure that rehearses an inferred learning algorithm or simulated by the hypothesis of neural systems to decide its variables (fuzzy set and fuzzy rules) by taking care of data models. NFC is the course of action of fuzzy inference system (FIS) and NN. The fuzzy logic is actuated allowing the fuzzy standard and NN is practiced dependent on the preparation information [6]. Standard genetic or genetic searching algorithms are utilized for numerical parameter advancement and depend on the standards of the natural selection process. Later, genetic algorithm is integrated with neuro-fuzzy to form genetic-neuro-fuzzy hybrid as shown in Fig. 3.

GA ideas can be adjusted into the different forms that are reasonable for the execution.

1. To produce the primary random populace consists of individuals whose characteristics were coded by the string of zeros and ones.

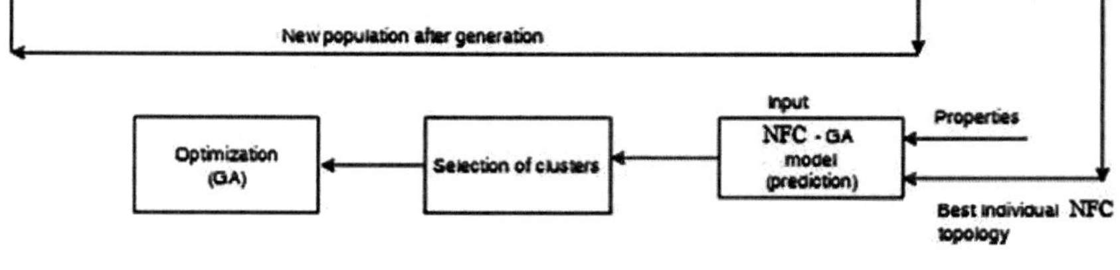

Fig. 3 Block diagram of genetic-neuro-fuzzy hybrids

Table 1 GA-based controller parameters

Possibility over crossover	0.845
Possibility over mutation	0.0082
Number of generations	100
Individual generations	25
Length of generator	72 bit

2. The each selective individual is given a fitness function and produces fitness value for each series inside the population dependent on the performance standards.
3. On a probability basics, choose the sets of parents, to bread off-springs, where off-springs with a greater fitness value will be more preferred than those with a lesser value in reproduction.
4. By separating the binary coding of respective parent into at least two fragments and after that the crossover links to provide additional different child string that has hereditary portion of its coding from individual parent.
5. With a less probability, the mutation inverse bits are coded.
6. Break when a permissible generation is accomplished or look objective is accomplished, or else jump to step 2 [9].

GA parameters used for simulation of GA-NFC controller are mentioned in Table 1.

4 Development of Simulink Model

The proposed GA-NFC control algorithm is executed in MATLAB. At that point, the speed execution of GA-NFC method was tried with induction motor 10 HP/400 V. The speed execution of the GA-NFC method was compared with neuro-fuzzy controller. From the created model, torque, current, the speed, and voltages were analyzed. The Simulink model of GA-NFC controller is as shown in Fig. 4 and Simulink block of induction motor vector control system is shown in Fig. 5, respectively.

5 Results

The comparison performance of GA-NFC and NFC are described. When we change speed reference from 100 to 150 rad/s and 150 to 60 rad/s, speed response is shown in Fig. 7. From the comparative analysis, the suggested GA-NFC offers smooth speed control performance when compared to other controllers. Overshoot is significantly less in case of genetic-neuro-fuzzy controller. GA-NFC produces very less torque ripples compared to other controllers but takes more search space and CPU time

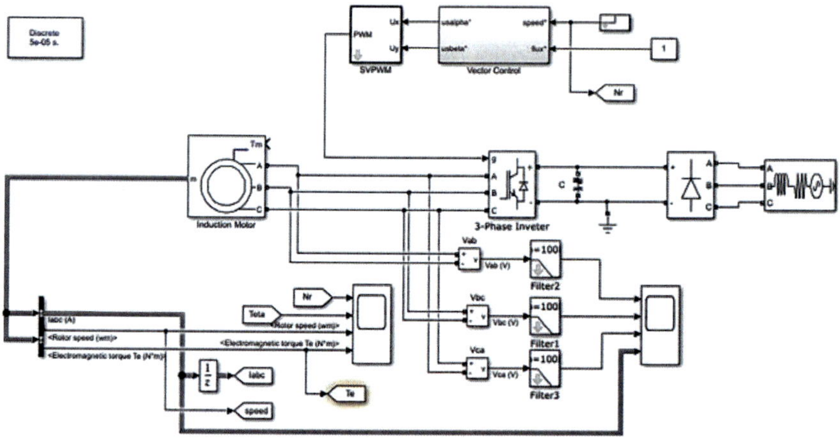

Fig. 4 Genetic algorithm-based neuro-fuzzy-controller simulation model

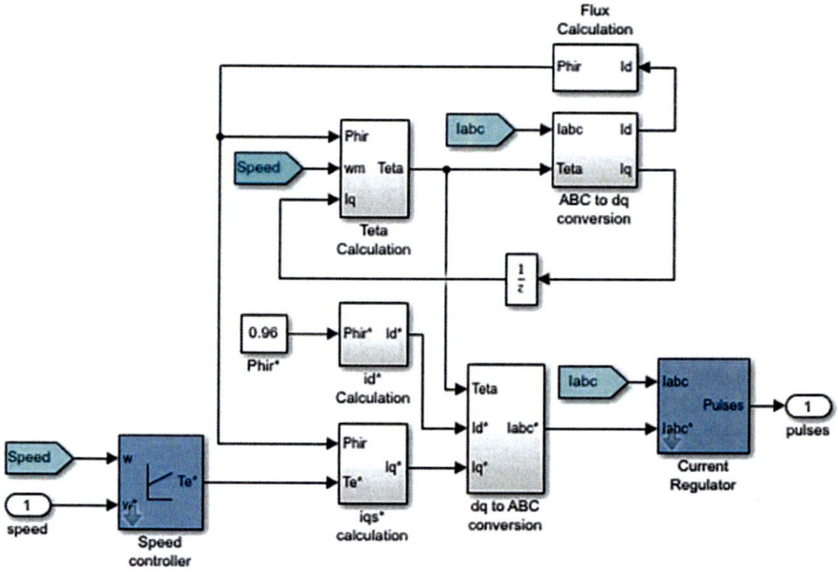

Fig. 5 Simulink block of induction motor vector control system

for training. The performance of current, voltage, torque, and speed of genetic-neuro-fuzzy controller and neuro-fuzzy controller is shown in Figs. 6, 7, 8, and 9.

Speed and Torque ripples are reduced significantly in case of GA-NFC controller compared to NFC controller. This will help to improve the speed accuracy and reliability of the motor drive. From Table 2 readings, we can observe that GA-NFC

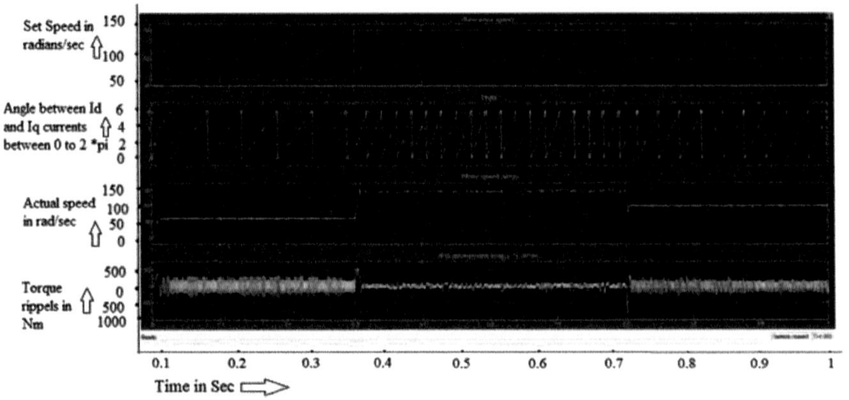

Fig. 6 Set speed, angle theta (in radians), rotor speed (wm), and electromagnetic torque (Nm) of GA-NFC controller

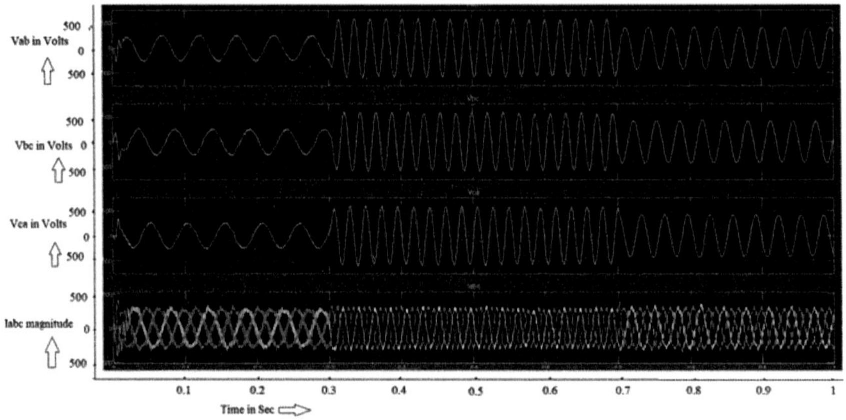

Fig. 7 Line-to-line voltages Vab, Vbc, Vca, and phase currents IABC of GA-NFC controller

gives great speed precession compared to NFC. Since motor drive is trying to operate at rated speed (150 rad/s), there is no slip during 0.3–0.7 s, so we can find lot of speed ripples when we try to operate motor drive at rated speed.

6 Conclusions

In this paper, another new advanced method for the design of GA-NFC controller for the induction motor drive is suggested. From research discoveries, we can conclude that GA-NFC creates significantly decreased torque and speed ripples compared with NFC controller, thus reducing losses in IM drives. To reduce the calculation burden,

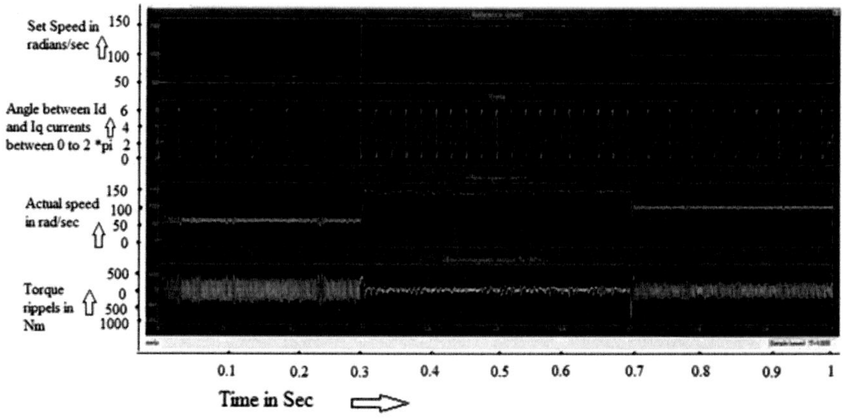

Fig. 8 Set speed, angle theta (in radians), rotor speed (wm), and electromagnetic torque (Nm) of NFC controller

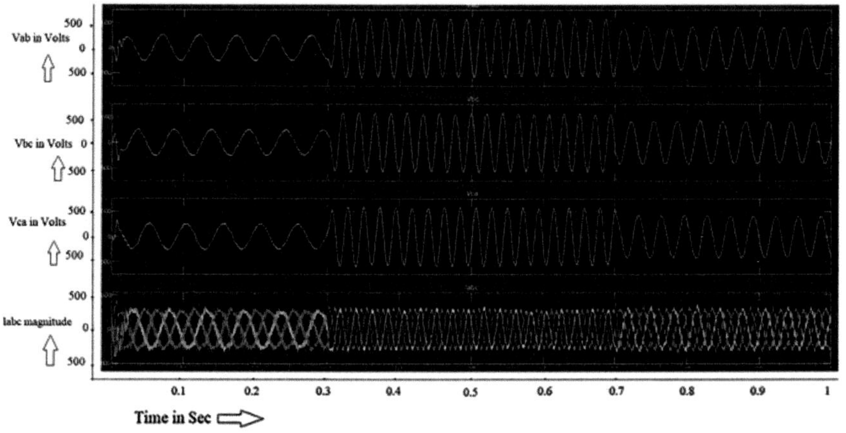

Fig. 9 Line-to-line voltages Vab, Vbc, Vca, and phase currents Iabc of NFC controller

Table 2 Result analysis

Controller	Settling time (in sec)	Rise time (in sec)	Peak time (in sec)	Steady-state error (rad/sec)
NFC	0.0200	0.015	0.017	−1.80
GA-NFC	0.0125	0.010	0.011	+0.02

we have taken only speed error as input to the controller, instead of taking both speed error and its derivative, since induction motor is a very nonlinear device, so it produces more torque ripples which in turn affects the reliability of the induction motor drive. If we increase one parameter, another parameter decreases. Since induction motors are rigid and reliable, it can withstand such torque ripples. GA-NFC needs substantial search space and more CPU time. Results demonstrated that GA-NFC was capable of reproducing the ideal speed accuracy and significantly less speed ripples. This method is very successful in identification and simple in execution. GA-NFC controller must be preferred where speed accuracy is very much important.

7 Future Scope

Having accomplishment in the utilization of GA, we should consider enhancing the convergence rate by utilizing hybrid GA. In future, the speed control system can be implemented to control five-phase or multiphase motor with a similar drive and also advanced deep learning method. The experimental setup might be implemented by using ARM Cortex M4/FPGA Spartan 7 to enhance the effectiveness of the induction motor drive. Nowadays, switched reluctance motors are attracting huge attention due to its advantages over induction motor. Similar control technique can be implemented to control the switched reluctance motors.

Appendix

Parameters of Induction Motor:

Nominal power	10 HP	Stator resistance	0.19 Ω
Line to line voltage	400 volts	Stator inductance	0.21 mH
Frequency	50 Hz	Rotor resistance	0.39 Ω
Inertia	0.0226 kg m^2	Rotor inductance	0.6 mH
Friction	0.01 Nms	Mutual inductance	4 mH
Number of poles	2		

Parameters of PI speed controller:

Proportional Constant (Kp)	9.69
Integral Constant (Ki)	148.64

References

1. Jalluri SR, Ram BVS A neuro fuzzy controller for induction machines drives. J Theor Appl Inf Technol 102–108
2. Cheon K, Kim J, Hamadache M, Lee D (2015) On replacing PID controller with deep learning controller for DC motor system. J Autom Control Eng 3(6):452–456
3. Keerti Rai S, Seksena BL, Thakur AN (2018) A comparative performance analysis for loss minimization of induction motor drive based on soft computing techniques. Int J Appl Eng Res 13(1):210–225. ISSN: 0973-4562
4. Venu Gopal BT, Shivakumar EG (2019) Design and simulation of neuro-fuzzy controller for indirect vector-controlled induction motor drive. In: Nagabhushan P et al (eds) Data analytics and learning, lecture notes in networks and systems, vol 43. Springer, Singapore, pp 155–167
5. Jhon Y, Langari R (2016) Fuzzy logic, intelligence, control and information. Pearson
6. Srikanth G, Madhusudhan Rao G Adaptive neuro fuzzy based maximum torque control of three phase induction motor. Helix 8(2):3067–3071
7. Douiri MR, Cherkaoui M, Essadki A (2012) Genetic algorithms based fuzzy speed controllers for indirect field oriented control of induction motor drive. Int J Circuits Syst Signal Process 6(1):21–28
8. Selvi S, Gopinath S (2015) Vector control of induction motor drive using optimized GA technique. In: IEEE sponsored 9th international conference on intelligent systems and control (ISCO)
9. Rushi Kumar K, Sridhar S (2015) A genetic algorithm based neuro fuzzy controller for the speed control of induction motor. Int J Adv Res Electr Electron Instrum Eng 4(9):7837–7846

Artificial Intelligence-Based Chatbot Framework with Authentication, Authorization, and Payment Features

Deena Deepika Cutinha, Niranjan N. Chiplunkar, Shazad Maved, and Arun Bhat

Abstract Chatbot is a computer program that can simulate a human conversation via a chat application such as Facebook messenger, WeChat, Slack, etc., which permits businesses to deliver interactive experiences that help in client engagement. We create an artificial intelligence-based chatbot-building framework which is a subscription-based chatbot builder platform with authentication, authorization along with other major modules. It provides better intelligence for conversations compared to present flow-based chatbot builders. Brands can sign up for the platform and build their own unique chatbot based on their business needs and deploy to any messenger platforms such as Facebook, WhatsApp, Telegram, WeChat, etc.

Keywords Chatbot · Rasa stack · Authentication · Authorization · Zoho subscription · Razor Pay

1 Introduction

Chatbot is an interface where people interact with the machine. It is a text-based conversational agent which perhaps has ability to perform actions in response to user's conversation. It runs inside a popular messaging application, such as Facebook Messenger, Slack, or SMS. It answers user's question, rather than merely directing to a website.

D. D. Cutinha (✉) · N. N. Chiplunkar
NMAM Institute of Technology, Nitte, India
e-mail: deena.cutinha17@gmail.com

N. N. Chiplunkar
e-mail: nchiplunkar@nitte.edu.in

S. Maved · A. Bhat
Mindstack Technologies, Mangalore, India
e-mail: shazad@mindstack.in

A. Bhat
e-mail: arun@mindstack.in

N. N. Chiplunkar and T. Fukao (eds.), *Advances in Artificial Intelligence and Data Engineering*, Advances in Intelligent Systems and Computing 1133,
https://doi.org/10.1007/978-981-15-3514-7_16

We create an artificial intelligence-based chatbot by making use of Rasa [1]. Rasa has got better intelligence and conversation compared to the present flow-based chatbot builders. Rasa stack consists of two components: Rasa NLU and Rasa Core. NLU stands for natural language understanding which understands the user inputs in the form of text or speech. NLU is a subset of natural language processing which understands human language; NLU is tasked with communicating with untrained texts and understanding their intent. Rasa NLU is used to identify intents and entities in the given text. Intent depicts the intention of a user and entities are used to make responses more customized. The second element, Rasa Core, takes structured input in the form of intents and entities and chooses which action the bot should take using probabilistic model.

With authentication and authorization, brands can sign up for the platform and build their own unique chatbots according to their business needs and deploy to any messenger platforms. The framework used for backend working is Ruby on Rails. Rails are Web-application framework which has everything required to create Web applications according to the Model–View–Controller (MVC) pattern.

2 Literature Survey

ELIZA is a natural language processing computer program created from 1964 to 1966. Chatbots have resulted in the creation of a variety of technologies and have taken a variety of approaches. Given an input sentence, ELIZA identifies keywords and matches those keywords against a set of pre-programmed rules to generate appropriate responses [2].

Watson, designed by IBM could be a question responsive computer system designed to use advanced linguistic communication processing, information retrieval, knowledge representation, automated reasoning, and machine learning technologies to the field of open domain question answering. Watson uses IBM's DeepQA computer code and the Apache Unstructured info Management Architecture (UIMA) framework. It runs on the SUSE Linux Enterprise Server 11 operating system using Apache Hadoop framework to provide distributed computing [3, 4].

ALICE was created in early 1980s and led to the development of artificial intelligence mark-up language (AIML) [5]. AIML is used for pattern-matching rule that links user-submitted words and phrases with topic categories. It is extensible markup language (XML) based, and supports most chatbot platforms and services. As intelligent agents, they are autonomous, reactive, proactive, and social [6].

In [7], the authors have built a chatbot which incorporates a number of rules and a rules engine for controlling message delivery to users. Rules stored in the rules database, or the rules engine itself, incorporate different sorts of rules, including "when" and "if-then" type rules. In the view of these rules, rules engine decides the state of relevant conditions.

The idea behind rule-based bot-building platforms is that a bot contains a knowledge base with documents; in turn it consists of patterns and templates. When the bot receives a text which matches the pattern, it sends the message stored in the template as a response back to the user. The pattern is a regular expression. This can be very time consuming and memory expensive.

2.1 Why Rasa?

There are many bot-building frameworks such as Dialogflow bot framework, Microsoft bot framework. Both Dialogflow bot framework and Microsoft Bot Framework have pre-built custom language understanding models. These frameworks seem to be great tools if we do not have existing chat logs that we can use as a training data. For instance, we want to develop a chatbot for a business and this chatbot will be receiving potentially sensitive or confidential information from its users. In such case, we may be more comfortable keeping all the components of our chatbot in the house.

This is where Rasa platform comes in picture. Rasa NLU is used to identify intents and entities in the given text. Intent depicts the intention of a user and entities are used to make responses more customized. The second part, Rasa Core, takes structured input in the form of intents and entities and chooses which action the bot should take using a probabilistic model.

Systems are no longer dependent on deterministic responses from rules-based pattern matching. Systems require supervised learning which should have large training sets, unsupervised learning, and hybrid intelligence where humans participate in the training process over time.

3 Artificial Intelligence-Based Chatbot Framework

The idea behind this work is to setup a platform, which helps people to build their own conversational agents according to their business requirements. Initially, customer/user must register himself on the website. Upon authentication, customer/user can select the plan and based on that can create chatbots and can be easily integrated into the websites or any other applications.

3.1 Authentication and Authorization

Authentication
Authentication is a process of verifying the identity of a user. Devise [7] is a flexible authentication solution used and it supports Rack-based MVC solution for Rails.

Here, multiple models can be signed in at the same time based on modularity concept. To setup authentication, devise should be installed and we need to add protect_from_forgery in application controller and before_action: authenticate_user! in all other controllers. Authentication is very important in every application. It is helpful to know whether the current user has ability to perform the requested action.

Once devise is installed, it will create some helpers and these can be used to check user's authentication. We check for authentication before user logs in.

Authorization

Cancancan [8] is a flexible authorization solution for Rails. Authorization is used to determine whether user has privileges to access system resources, data, and application features. Once the authentication is performed, authorization can be done. During authorization, system verifies an authenticated user and either grants or refuses the access. This is used to create different permissions. The Ability category is where all user permissions are outlined.

The can method needs two arguments. The first one is the action where the permission is set and the second one is the class of object on which the permission is set. For example, Can: update, User. Using can and cannot methods, we check users permission. The authorize method in the controller will raise an exception if the user does not have permission.

3.2 Zoho Subscriptions and Razor Pay

Zoho subscription

Zoho subscriptions are subscription billing platform, used to handle the entire customer subscription life cycle. We have used Razor Pay for the payment processing and it offers customers alternative payment options to reduce time of payment. When user does the payment, we create customer, subscription, and invoice for the same user in Zoho and based on the billing frequency, Zoho subscriptions send out invoices before due dates.

Zoho accounting software, consists of many other features such as mail, accounting books, and human resource software. It is an all in one package. Easy to use and configure and cheaper compared to all other subscription-based platforms. Payment can be processed with the selected interval.

Initially, product and plan entries are made on Zoho application. A product refers to the service that would be offered to a customer. Each product will have completely different plans and addons related to it. An association between product and plans is one to many. The product name in Zoho should be unique and each product will generate a unique product id. Updating or deletion of the product can be done based on the product id. A plan contains plan features which includes billing and pricing information. Plan has got many features with an association of one to many.

Application uses the nested attributes to save attributes of a record through its connected parent. To create a new product and plan, a POST request is sent to Zoho subscriptions through an API. To create a product, a unique name is must and to create a plan specify name, plan code, price, interval and product id along with the post request. Product id is required to associate the plan with a particular product. Plan has a name of our choice to be displayed in the interface and invoices. Customer is charged an amount over an interval for the subscription based on the price. Interval indicates the number of cycles between each billing. Plan is associated with role. Once the user is registered, he can purchase a plan and based on that role gets changed.

A Zoho subscription enables us to charge customers at specified intervals for a plan of their choice. Customer object needs to be passed to the respective API for the creation of the new customer. Customer_id of that customer is passed using POST request to Zoho subscriptions through an API for the creation of the new subscription. Some of the features supported by Zoho in subscription are creating, updating, deleting, retrieving, listing, and marking a customer as active or inactive. Subscription can be either canceled immediately or at the end of the current term based on the value of "cancel_at_end." If "cancel_at_end" is set to true then the "status" of the subscription is changed to non_renewing and if it is false, the "status" would be canceled. Invoices along with the subscription are created at the time of purchase. Invoices describe how much a customer owes for the subscription.

Invoices are created for recurring charges, one-time charges including any prorated adjustments. After the payment, the user gets the role based on the selected plan and he will be allowed to create a chatbot, which is saved in database. The user can have many chabots with one to many associations.

Razor Pay

Razor Pay has been used for supporting the payment process. Razor Pay is the payments solution which allows businesses to accept and process payments with its product suite. Checkout Form is the primary way to integrate Razor Pay with our front end and accept payments online. It provides a well designed and tested checkout form for the customers. On successful payment, customer/user name, selected plan, authenticity token, payment id, and status of the payment would be saved.

On successful payment, Zoho customer and Zoho subscription entries would be created. A customer object allows to keep track all customer-related information. To create a customer in Zoho, name and email are the required mandatory fields. On successful payment, user gets the role based on the selected plan and will be allowed to create a chatbot; the user can have many chabots with the association of one to many.

Razor Pay is an Indian payment solution for businesses to accept and process the payments. Easy to use and configure and they have options of accepting recurring payments. Payment happens from Razor Pay and gets accounted in Zoho. During payment, we create customer and subscription along with the invoice. Customer object stores all customer-related information such as name, mail, and other attributes where as subscription stores all plan details of selected plan.

The flow of customer subscription process is shown in Fig. 1.

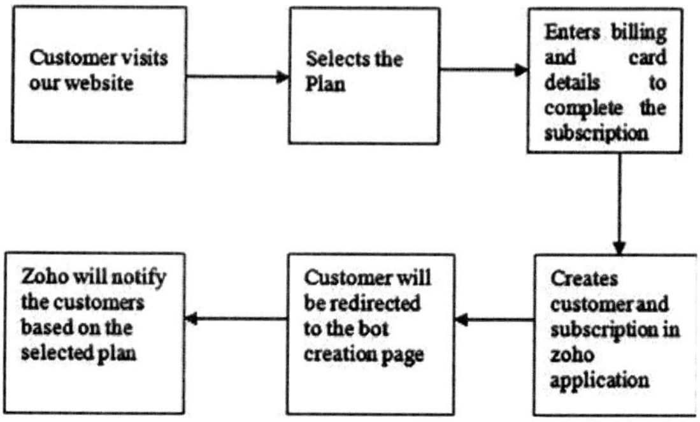

Fig. 1 Flow diagram of customer subscription life cycle

3.3 Creating AI-Based Chatbot Framework

Chatbot may be described as software that uses artificial intelligence to chat with people. These softwares are used to perform tasks such as quickly responding to users, informing them, helping to purchase products, and providing better service to customers. Chatbot architecture consists of three main parts. Request Interpretation, Response Retrieval, and Message Generation. Request Interpretation is the process of classifying the intents and extracting the entities in a user text. Response Retrieval receives the current state of the action and chooses which action to take next. Message Generation is the response of the bot, back to the user based on the context.

Rasa is an open source and it is based on artificial intelligence. Rasa NLU has various pipelines and these can be used to process user messages. We are utilizing the pre-characterized spacy_sklearn pipeline. The spacy_sklearn pipeline utilizes pre-prepared word vectors from either GloVe or fastText.

In order to create chatbot framework, it accomplished in three steps viz., design phase, forms creation, and training phase.

In design phase, user selects the theme, avatar, color, position, and other required attributes.

Second step is to create forms. Form indicates the intension of a user. For example, a greeting form indicates one of the basic functions of communication and triggers positive conversations. User can create any number of forms with different intension. Each form has form elements which in turn consist of entities and its associated values. Entities are pieces of information which needs to extract from messages.

Last step is to give training data which is important to develop chatbots. It should include texts to be interpreted and the structured data, i.e., intent and entities. Once the training data is ready, post request is sent to rasa to parse the text and identify intents and entities.

Once the chatbot is trained, user gets a chat widget and an embedded id which can be inserted inside any Web application. The architectures and retrieval process of chatbots take advantage of advances in machine learning to give advanced "information retrieval" processes, in which responses are generated based on analysis of the results of Web searches.

Example bot
Food bot is used to search for nearby food shops with different cuisines or with location. Steps involved are

Step 1: Setting up the design phase

Here, user selects appropriate font style, color, avatar, and position of the bot.

Step 2: Creating forms

A Food_Bot has three forms,

- greet
- food_fancy
- good_bye.

Food fancy is used to search and browse recipes. Search using cuisine or diet and find nearby stores. Greeting and good-bye are used as basic functions of communication. Along with forms, we create actions, which are nothing but the things bot should do in response for a user query.

Here, forms are nothing but the intents which indicate the intention of user. Each form is created with intention, and Rasa NLU will predict the intent on a user text and send an appropriate result back to the user.

Step 3: Preparing the Training Data

The training data includes texts to be interpreted and the structured data. For example, in the form food_fancy, we create location and cuisine as the entities. For example, on a search query "show me Southindian food shops," Rasa will identity which is the intent and entities in the user text. The intent is associated with the text. Such as greet good-bye or food_fancy. The entities are specific parts of the text which need to be identified. The first job of Rasa NLU is to assign any given sentence to one of the intent categories: greet food_fancy, or good-bye. The second task is to attach words like "southindian" and "mangalore" as values to cuisine and location entities. Along with intents and entities, rasa will predict the confidence level.

Training data consists of user messages, the intents to which that particular user text can be classified to and the entities which can be extracted from the user text. The user's text will be classified into any one of the above intents. For example, there are many ways that the user can ask about the restaurants,

- I am searching for restaurants nearby!
- Indian restaurants in mangalore.

Rasa NLU will classify the user text to any of the intents. The words like cuisine, location, etc., will be the entities. The model is, hence, built which can do the intent classification and entity extraction.

For example, taking a sentence like,

"I am looking for a south indian restaurant in mangalore" and NLU returns structured data like,

{
"intent": "food_fancy",
"entities": {
"cuisine" : "mangalore",
"location" : "south Indian"
}
}

User text is processed by a sequence of components. There are components for entity extraction, for intent classification, preprocessing, and others. Each component processes the input and creates an output. The output can be used by any component that comes after a particular component in the pipeline.

Step 4: Talking To our Bot

Once the bot is trained, we get a chat widget where we can have a conversation using natural language. On each message, bot returns the structured data and based on the confidence level, it will trigger intent and returns the action back to the user.

For example, for the sentence "I am looking for Chinese food" the output is:

{
"text": "I am looking for Indian food",
"entities": [

{"start": 18, "end": 23, "value": " Indian ", "entity": "cuisine", "extractor": "ner_crf", "confidence": 0.864}
],

"intent": {"confidence": 0.6485910906220309, "name": "food_fancy"},
"intent_ranking": [
{"confidence": 0.6485910906220309, "name": "food_fancy"},
{"confidence": 0.1416153159565678, "name": "goodbye"}
]
}

Based on confidence level, bot sends an action back to the user. Actions are the utterances and activities of the bot. Actions are just the things bot should do in response to user input. For example, an action can be a utter messages, or it can make an external API call, or it can query a database. There are three kinds of actions.

In Food Bot example, action is the Get Food Results which is invoked if the user asks for the availability of food shops with particular cuisine and location. Slots can be used to store data given by the user such as location, cuisine, etc. Slots are nothing but the bot's memory which can be user-defined variables. During a conversation, system keeps track of slots. For example, to book a restaurant, we should keep track of location, cuisine, etc., so all such information will come under slots.

4 Conclusions

This section concludes that an artificial intelligence-based chatbot-building framework with authentication, authorization along with other major modules has been built. The explained framework has got features and capabilities that give good service and solutions than the other chatbot builders. The integrated payment options make it very easy for the customer to finish the purchase process.

On the other hand, the details of the customers and subscription can be managed easily using the application. People can communicate with the chatbot and the business keeps track of the intents, entities, and visitors of the chatbot. The architectures and retrieval process of chatbots take advantage of advances technologies of machine learning concept to give advanced "information retrieval" processes, in which responses are generated based on analysis of the results of Web searches.

References

1. Bocklisch T, Faulker J, Pawlowski N, Nichol A (2017) Rasa: open source language understanding and dialogue management. Preprint at arXiv:1712.05181
2. Weizenbaum J (1966) ELIZA—a computer program for the study of natural language communication between man and machine. Commun ACM 9(1):36–45
3. Jackson J (2011) IBM watson vanquishes human jeopardy foes. PC World. IDG News. Retrieved Feb 17
4. Takahashi D (2011) IBM researcher explains what Watson gets right and wrong. VentureBeat. Retrieved Feb 18
5. Marietto MDGB, de Aguiar RV, Barbosa GDO, Botelho WT, Pimentel E, França RDS, da Silva VL (2013) Artificial intelligence markup language: a brief tutorial
6. Neves AM, Barros FA, Hodges C (2006) Iaiml: a mechanism to treat intentionality in aiml chatterbots. In: 18th IEEE international conference tools with artificial intelligence (ICTAI'06), pp 225–231
7. https://github.com/plataformatec/devise
8. https://github.com/nathanvda/cocoon

Disease Recognition in Sugarcane Crop Using Deep Learning

Hashmat Shadab Malik, Mahavir Dwivedi, S. N. Omkar, Tahir Javed, Abdul Bakey, Mohammad Raqib Pala, and Akshay Chakravarthy

Abstract Crop diseases recognition is one of the considerable concerns faced by the agricultural industry. However, recent progress in visual computing with improved computational hardware has cleared way for automated disease recognition. Results on publicly available datasets using convolutional neural network (CNN) architectures have demonstrated its viability. To investigate how current state-of-the-art classification models would perform in uncontrolled conditions, as would be faced on site, we acquired a dataset of five diseases of sugarcane plant taken from fields across different regions of Karnataka, India, captured by camera devices under different resolutions and lighting conditions. Models trained on our sugarcane dataset achieved a top accuracy of 93.40% (on test set) and 76.40% on images collected from different trusted online sources, demonstrating the robustness of this approach in identifying complex patterns and variations found in realistic scenarios. Furthermore, to accurately localize the infected regions, we used two different types of object-detection algorithms—YOLO and Faster R-CNN. Both networks were evaluated on our dataset, achieving a top mean average precision score of 58.13% on the test set. Taking everything into account, the approach of using CNN's on a considerably diverse dataset would pave the way for automated disease recognition systems.

H. S. Malik (✉) · M. Dwivedi · S. N. Omkar · T. Javed · A. Bakey · M. R. Pala · A. Chakravarthy
Indian Institute of Science, Bangalore, India
e-mail: hashmat.shadab.malik@gmail.com

M. Dwivedi
e-mail: mahaviredx@gmail.com

S. N. Omkar
e-mail: omkar@aero.iisc.ernet.in

T. Javed
e-mail: tahirjmakhdoomi@gmail.com

A. Bakey
e-mail: mirbakey@gmail.com

M. R. Pala
e-mail: raqib0027@gmail.com

N. N. Chiplunkar and T. Fukao (eds.), *Advances in Artificial Intelligence and Data Engineering*, Advances in Intelligent Systems and Computing 1133,
https://doi.org/10.1007/978-981-15-3514-7_17

Keywords Computer vision · Deep learning · Object detection · Crop diseases · Automated plant pathology

1 Introduction

Food is one of the mainstays of human existence. Although there has been a boon in the yield of food, yet, the security of food is challenged by various factors. These include increased temperature and plant diseases [1] in addition to other factors. Plant diseases affect the agricultural yield of the country as well as the livelihoods of farmers who contribute 80% of the total agriculture productions [2]. Fortunately, plant diseases can be handled if we have a timely and accurate diagnosis.

The past plant disease diagnosis system was based on visually observing the symptoms in the sample and identifying the disease. This task required human expertise, restricting its domain and making it difficult for farmers to avail. Also, a large variation in the symptoms sometimes even made it difficult for the experts to identify the disease. An artificial intelligence-based system would offer valuable assistance in the diagnosis of the disease through image-based observation. Advances in computer vision provide an opportunity to enhance plant disease diagnosis and extend the field of computer vision for precision agriculture. Especially, if the system is affordable, it would be practically possible for even the farmer to make a timely diagnosis of the disease and act accordingly, apart from making the work of experienced pathologists precise and accurate.

Deep learning for computer vision, particularly object classification and detection has made a significant progress in recent years. The large-scale visual recognition challenge (ILSVRC) [3] based on ImageNet dataset [4] has been used as standard for various computer vision problems. In 2012 AlexNet, a CNN [5]-based deep learning classification model achieved 16.4% top-5 error rate on the ImageNet dataset [6] beating all other state-of-the-art classification models which were non-CNN based. Subsequent progress in deep convolution networks has brought the error rate down to 3.57% [6–9] illustrating the technical feasibility of the deep learning-based approach for plant disease detection and the reliability of the results. The training of these networks takes quite a bit of time but once trained, these networks can make real-time classification of images, suiting consumer applications on smartphones. Smartphones in particular provide a novel approach in the deployment of these systems partly because of their computational power, advanced cameras, high-resolution displays, and partly because of their affordability. This will make it technically feasible for the farmers to diagnose the plant disease from a mere picture of a plant leaf. In fact, it is established that around 6 billion smartphones would be available by 2020 [10].

We focused our work on sugarcane. It is a kind of crop that has experienced a considerable increase in terms of production especially in India. India has become the largest producer and consumer of sugar in the world. This has become possible because of 45 million sugarcane farmers and a large agricultural force which constitute about 7.5% of the total rural population [11]. This indicates the vitality of sugarcane both in the economy of the country and the sustenance of a large number of sugarcane cultivators, majority of them being small scale. Unfortunately, the onset of diseases still remains a threat in the large-scale production of this crop mostly because of the lack of pathological facilities. Moreover, majority of the cultivation takes place in rural areas and cultivators fail to have an accurate, timely diagnosis of the diseases eventually leading to damage to the crop. Fortunately, it turns out that this problem can be solved if we have an effective means of timely diagnosis of the disease.

In this paper, we present an effective and reliable way of diagnosing major sugarcane diseases which directly affect the crop. We present a classification report on five diseases (or their absence) using 2940 images with a convolutional neural network approach. We based our work on the physical characteristics and symptoms of each disease. We measure the performance of the models VGG-19, Resnet-34, and Resnet-50 based on their ability to predict the correct disease (or their absence). Going a step further, we localize the exact diseased spots in the images of the leaves for four classes, thus distinguishing the infected areas from the rest of the leaf. For this purpose, we used two powerful detection models Faster R-CNN [12] and YOLOv3 [13]. Our results are a big step toward an automated plant disease diagnosis system.

This paper is organized as follows. In Sect. 1, we briefly review related work in the field of image processing and specifically, in crop disease recognition. The sugarcane dataset is introduced in Sect. 3. Sections 4 and 5 present the approach followed for classification and detection task, respectively, as well as the achieved results. Conclusion and future work are given in Sect. 6.

2 Related Work

In recent years, CNNs [5] have shown a tremendous advancement and are being applied in different domains including crop disease detection. Mohanty et al. [14] used two deep learning architectures GoogleNet [8] and AlexNet [6] on Plant-Village-Dataset [15] to identify 26 diseases among 14 crops, achieving a peak accuracy of 99.35% at test time. Working on the same dataset [16] shows a test accuracy of 90.4% using a VGG-16 [17] model. Another work [18] also uses a deep learning system to identify 13 types of diseases in five crops using images from Internet achieving an accuracy up to 96.3%.

2.1 Plant-Village Dataset

The Plant-Village-Dataset [15] contains 54,306 images distributed among 38 classes. Disease names are used as class labels. The dataset in some cases has more than one image of a leaf which vary in orientation. All the images have been taken in a controlled environment in the laboratory.

3 Sugarcane Dataset

Although all the works mentioned above show prolific results, the problem with these works is that either the images were downloaded from Internet [18] or the images were taken in controlled environment in laboratory [14, 16], questioning their applicability in real world where we may encounter numerous variations in images. Mohanty et al. [14] have also mentioned that the accuracy of their work drops substantially to 31.4% when testing is done on images taken under different conditions from the one under which training images were taken (laboratory conditions). To encounter all these, we obtain a more realistic dataset of sugarcane for real-world applicability.

The dataset contains 2940 images of sugarcane leaves belonging to six different classes (consisting of five diseases and one healthy). These include major diseases that affect the crop in India. All the images were taken in natural environment with numerous variations. The images were taken at various cultivation fields including University of Agricultural Sciences, Mandya, Bangalore, and nearby farms belonging to farmers. All the images were taken using phone cameras at various angles, orientations, backgrounds accounting for most of the variations that can appear for images taken in real world. The dataset was collected with the company of experienced pathologists (Sect. 6). For localizing the infected spots on the leaves (object detection) corresponding to four diseases, we manually annotated the dataset. Most of the images in the dataset contain multiple infected spots of varying patterns. All these spots were individually annotated using different patches accordingly. The distribution of the images into different classes is detailed in Table 1. Figure 1 shows the classes used for classification and detection and their corresponding distribution (Fig. 2).

Table 1 Distribution of images into different classes

S. No	Class	Count
1	*Cercospora* leaf spot	346
2	*Helminthosporium* leaf spot	410
3	Rust	382
4	Red rot	454
5	Yellow leaf disease	420
6	Healthy	928

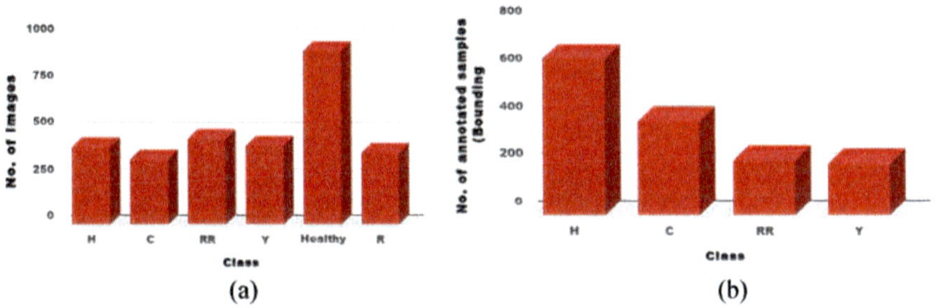

Fig. 1 Distribution of images into various classes used for classification 1 (**a**) and detection 1 (**b**). The letters represent diseases, 'H' denotes *Helminthosporium* leaf spot, 'RR' Red Rot, 'C' *Cercospora*, 'R' Rust, and 'Y' denotes Yellow leaf disease

Fig. 2 Example of leaf images from our dataset, representing every class. (1) *Helminthosporium* leaf spot (2) Red rot (3) *Cercospora* leaf spot (4) Rust (5) Yellow leaf disease (6) Healthy

4 Classification

4.1 Approach

We used three popular architectures Resnet-50 [9], Resnet-34 [9], VGG-19 [17] and assessed their applicability for the said classification problem. All these architectures are CNN [5] based and are discussed below.

VGG-19 [17] follows a simple architecture, a set of stacked convolution layers followed by fully connected layers. Concretely, it contains 16 convolution layers and three fully connected layers, finally followed by a Softmax layer. It only uses 3×3 convolutions with both stride and padding of 1. Set of stacked convolution layers also contain MaxPool layers which use a 2×2 filter and a stride of 2. It contains around 143.6 million parameters. It is appealing because of its uniform structure, yet achieving high accuracy, though it comes at the cost of slow training. It uses ReLu activation throughout. This architecture ended as runner-up at the ILSVRC [3] (2014) competition.

ResNets [9] introduce skip connections (or shortcuts) to fit the input from the previous layer to the next layer without any modification of the input. This architecture emerged as the winner of ILSVRC [3] (2015) in image classification, detection, and localization. In addition to using the architecture of a set of stacked convolution layers, finally followed by a fully connected layer, it uses skip connections or shortcuts. It does not rely on stacked layers to obtain the underlying mapping rather a residual it lets these layers fit a residual mapping. If the desired underlying mapping is denoted as $H(x)$, it lets the stacked nonlinear layers fit another mapping $F(x)$ related to $H(x)$ as:

$$F(x) = H(x) - x \tag{1}$$

This relies on the fact that it is easier to optimize the residual mapping than to optimize the original, unreferenced mapping. Even in the extreme case, if an identity mapping were optimal, it would be easier to make the residual function zero rather than to an identity function by a stack of nonlinear layers. ResNets also use 3×3 convolutions. ResNet-34 and ResNet-50 are two variants of this architecture, which are same except for the number of layers, former having 34 while latter has 50. Both these variants have considerably less parameters than VGG-19, ResNet-34 having 21.8 million and ResNet-50 having 25.6 million parameters.

We evaluate the performance of our models in two ways, by training the entire model in one case, and only the fully connected part in other case. Transfer learning was used in both cases starting from pretrained weights on the ImageNet dataset. Here also, we note that weights obtained from training the fully connected part only were used as the starting point for training of entire network. To sum up, we have a total of six experimental configurations depending on following parameters:

Deep learning architecture:

- VGG-19
- ResNet-34
- ResNet-50.

Training mechanism:

- Training only fully connected layer
- Training all layers.

The learning rate (LR) range test showed 0.01 was a reasonable choice for base-learning rate for the fully connected layer in all cases. Then, using differential learning rate, the initial convolution layers used a base-learning rate 0.0001 and the later ones used 0.001. This also ensured a fair enough comparison between the results as rest of the hyper parameters was same in all configurations. To get best accuracies possible on these datasets, following techniques were used while training the networks, solver being [19]stochastic gradient descent with restarts (SGDR) in all cases:

Differential Learning Rate [20] We first found the optimal base-learning rate for the fully connected layers of the network (training only fully connected layers) by plotting loss versus epochs for a few epochs as shown in Fig. 3 and choosing a value in the optimal LR (learning rate) range. $1e{-}2$ turned out to be a reasonable value in all cases. The base-learning rates of the network are set in a way such that it increases from initial to later convolution layers and from later convolution layers to fully connected part, by order one in both cases. Using different learning rates for different parts of the network allows us to control the rate at which weights change for each part of our network during training.

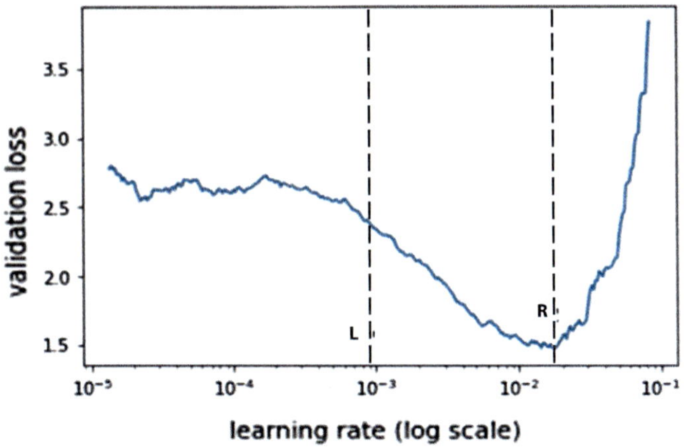

Fig. 3 Learning rate range test for Resnet-34 (shows 1e−2 to be a reasonable value)

Cyclic Learning Rate [19] Starting from its base value, learning rate cosine annealed its way to zero, completing one cycle in one epoch. Cycling the learning rate allows the network to get out of spiky minima and enter a more robust one. While annealing, the learning rate allows it to traverse quickly from the initial parameters to a range of good parameters and perform a more thorough search of the weight space as it approaches the minima, accounting for substantial increase in performance of the network.

Test Time Augmentation At test time, we performed a 1.1× zoom of the original image and took five random crops of it. The network was evaluated on these augmented images and the network generalized exceptionally well as depicted by test accuracies in Table 1.

Solver: Stochastic gradient descent with restarts (SGDR) [19]
Learning rate policy: Cyclic learning rate, cosine annealing from base-learning rate to zero (completes one cycle in one epoch).

All the above trainings were done using fastai, which is an open-source library for deep learning.

4.2 Results

PlantVillage-Dataset Since this dataset is publicly available, we also did classification work on it, training three well-known CNN architectures viz VGG-19 [17], Resnet-34 [9], Resnet-50 [9] achieving exemplary accuracies up to 99.84% at test time. All the results are shown in Table 2. The results are taken from the best epoch. Here, we note that weights obtained from training the fully connected part only were used as the starting point for training of entire network. Loss plots for all networks are given in Fig. 4.

Sugarcane Dataset The overall accuracy we obtained on our dataset ranged from 83.20% (*VGG-19: Training only FC layers*) to 93.2% (*Resnet-50: Training all layers and then testing with augmentation*). All the networks were trained using 80:20 train–test split.

Table 2 Accuracies for different networks

Model	Testing accuracy (training FC layers only, for 5–10 epochs) (%)	Testing accuracy (training all layers, 10–15 epochs) (%)	Testing accuracy (training all layers, testing with augmentation) (%)
VGG-19	96.66	99.55	99.72
ResNet-34	97.13	99.62	99.68
ResNet-50	97.41	99.81	99.84

Fig. 4 Loss plots for different architectures on Plant-Village-Dataset

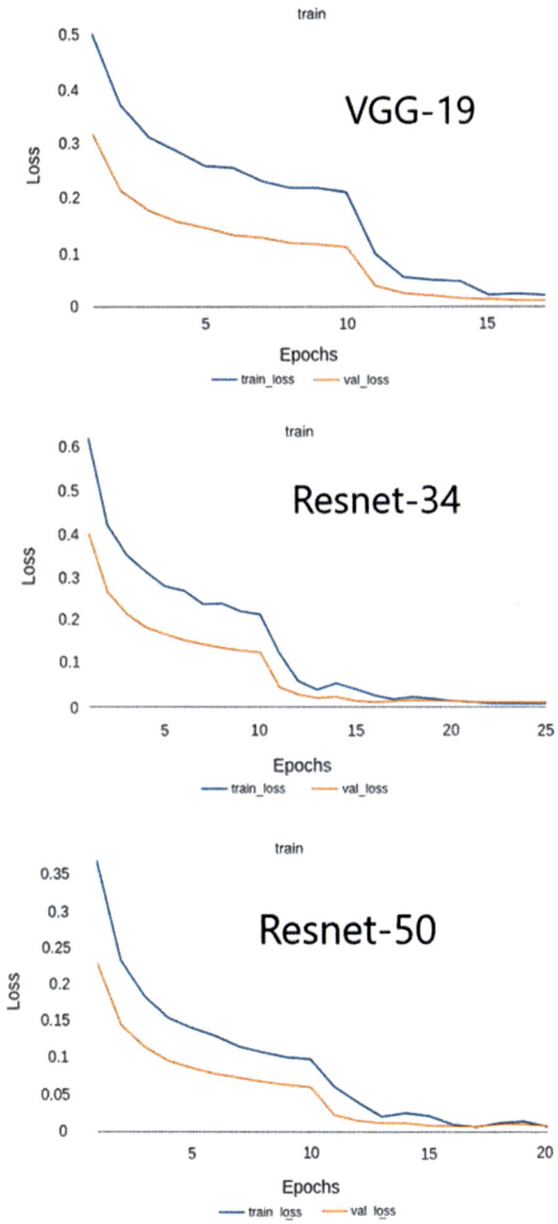

We used mean F_1 score, mean recall [21], mean precision [21], and overall accuracy as the metrics for evaluation of our models. All these metrics were calculated on the entire test set. These metrics are briefly discussed here:

Mean Precision Precision is the measure of ability of classifier not to label a negative sample as positive. Mathematically,

$$\text{Precision(for particular class)} = \frac{\text{True Positives(for that class)}}{\text{True Positives(for that class)} + \text{False Positives(for that class)}} \tag{2}$$

Averaging this precision score over all classes (in our case 6) gives us the mean precision.

Mean Recall Recall measures the ability of the classifier to find all the positive samples. Mathematically,

$$\text{Recall(for particular class)} = \frac{\text{True Positives(for that class)}}{\text{True Positives(for that class)} + \text{False Positives(for that class)}} \tag{3}$$

Averaging this recall score over all classes gives the mean Recall.

Mean F_1 Score F_1 score considers both precision and recall. Mathematically,

$$F1\text{score(for particular class)} = \frac{2 * \text{Recall(for that class)} * \text{Precision(for that class)}}{\text{Recall(for that class)} + \text{Precision(for that class)}} \tag{4}$$

Averaging these scores over all classes gives the mean F_1 score.

Overall Accuracy It is a simple measure for evaluating the performance of a model. It is the ratio of the total number of items correctly classified to the total number of items present in the test set.

Table 3 shows these metrics on the test set across all our experimental configurations (taken from best epoch). All the networks ran for a total of ten epochs when training only the fully connected layers and 15 epochs when training all the layers. To check overfitting, we obtained a small dataset of 52 images from trusted online sources belonging to these six classes and tested our best predicting model Resnet-50 on these images and we obtained an overall accuracy of 76.4% illustrating the generalization of our model. The results achieved are significantly better than [14]; they achieved an accuracy of 31.4% on a similar test setup (Fig. 5).

We note that these promising results were obtained on a rather small dataset containing only 2940 images, suggesting results will only get better if a reasonably large dataset is used. Loss plots for the networks are shown in Fig. 6.

Table 3 Mean metrics (precision, recall, and F_1) and overall accuracies across various experimental configurations

Model	Mean precision	Mean recall	Mean F_1-score	Overall testing accuracy (training all layers, testing with augmentation)	Testing accuracy (training all layers, testing without augmentation)	Testing accuracy (training only FC layers)
VGG-19	0.9044	0.9087	0.9026	0.9200	0.9120	0.8320
ResNet-34	0.9095	0.9107	0.9066	0.9240	0.9240	0.8360
ResNet-50	0.9260	0.9200	0.9213	0.9320	0.9280	0.8600

(a) Example image of a leaf from our test dataset suffering from Helminthosporium Leaf Spot

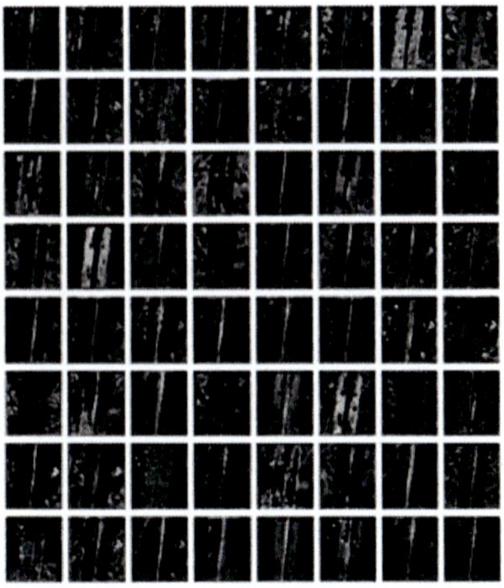

(b) Visualization of activations in the first convolution layer of Resnet-50 architecture trained using 80:20 split when doing a forward pass on the image in (a)

Fig. 5 Visualization of activations in the initial layer of Resnet-50 architecture depicting that the model has efficiently learnt to activate against diseased spots on the example leaf

5 Detection

5.1 Approach

To accurately recognize infected regions in the images, two state-of-the art detections networks, YOLOv3 and Faster R-CNN, will be used for evaluation. The above-mentioned models are significantly swift than models which preceded it, such as R-CNN [22] and Fast R-CNN [23]. [22, 23] used a more time-consuming method to find the regions on which CNN will be passed separately for classifying the label,

Fig. 6 Loss plots for different architectures on our dataset (decrease in validation loss depicts the learning of the networks)

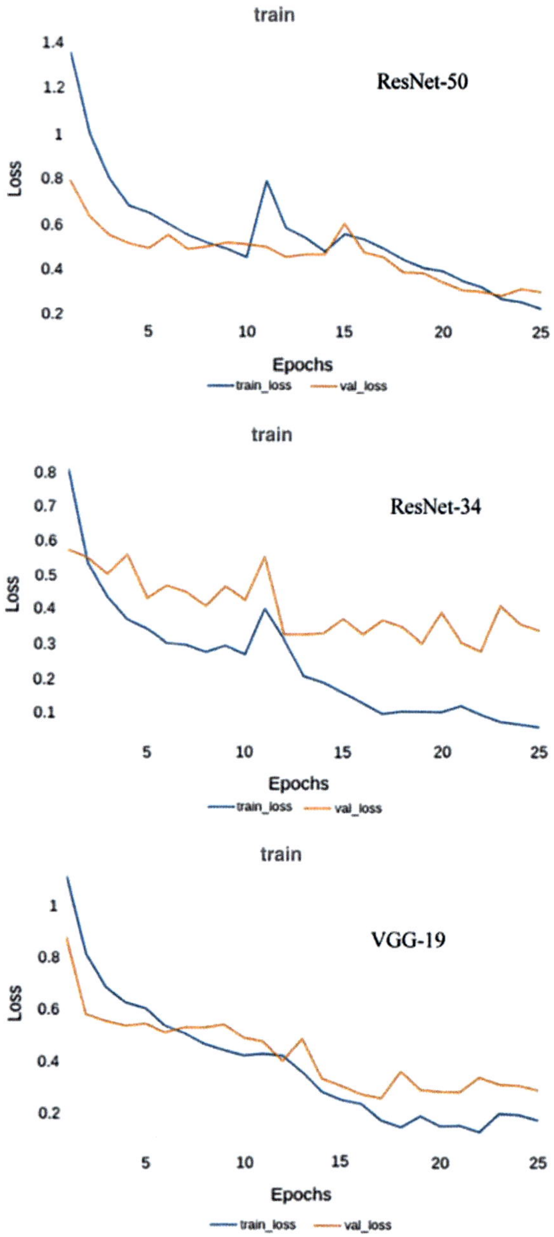

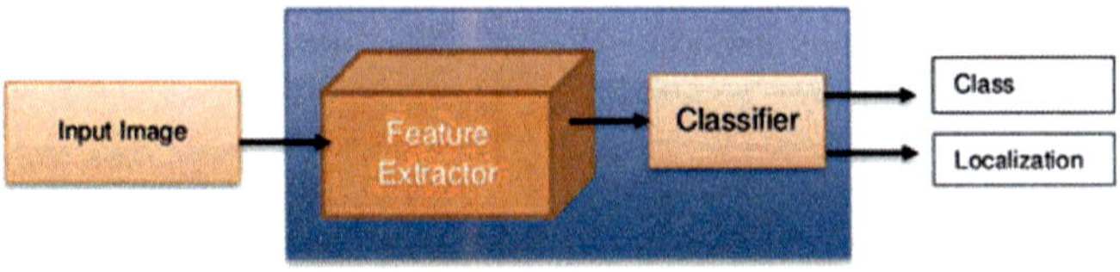

Fig. 7 Pictorial representation of detection process

known as selective search [24]. A few thousand regions of interest were generated by [24], which were then separately passed to the network for classification. This method made [22, 23] ill-suited for real-time inference. In case of Faster R-CNN, region proposals are predicted on a convolutional feature map using a region proposal network (RPN) [12] after passing image through a CNN. After that an ROI layer helps in finding the class suitable to a proposed region. Region proposals are generated in significantly less time and it takes around 2.3 s to make an inference [25]. Faster R-CNN with a VGG-16 backbone [12] when evaluated on the PASCAL VOC dataset [26] achieved a mAP of 0.76 at 5 frames/second. YOLOv3 does all in one pass through the network. Instead of different modules for predicting regions and localizing the object, YOLOv3 [13] does all this in a single CNN. YOLO [27] performs real-time detection at 45 fps, yet achieving a comparable mAP of 63.4% [28] when trained and tested on the above-mentioned dataset [26]. The feature extractor for YOLOv3 is a 106-layer CNN block, a variant of Darknet. For our work, we used Faster-RCNN with VGG-16 as its backbone and YOLOv3.

We evaluated the implementation of these architectures on our dataset by training the entire model starting from pretrained weights for the convolutional block on ImageNet dataset [3] in both the models. Faster R-CNN was trained on 600 × 1000 resolution images for 15 epochs and tested on same resolution images. YOLOv3 was trained on 416 × 416 resolution images for 6000 iterations and tested on 416 × 416 and 608 × 608 resolution images. Faster R-CNN was trained using [29] implementation while YOLOv3 was trained using [30] implementation (Fig. 7).

5.2 *Results*

The most common metrics used in examining the performance of these models is mean average precision (mAP) [28]. Metrics are briefly discussed below:

***Mean Average Precision (mAP)* [28]** It evaluates across different recall values, the average of maximum precisions. If the Intersection over Union (IoU) of the prediction which matches the ground truth label is ≥0.5, it is considered as true positive. To calculate mAP, we first calculate the maximum precision values AP_r for all classes individually at 11 recall values viz 0, 0.1, 0.2, …, 0.9, and 1.0. The maximum precision (AP_r) at any recall value $\tilde{r}$ is the highest precision value for any recall $\geq \tilde{r}$.

Then, we calculate average precision (AP), again for all classes individually, which is the average of these 11 precision values.

$$\begin{aligned} \mathrm{AP} &= \frac{1}{11} \sum_{r \in \{0.0, \ldots, 1.0\}} \mathrm{AP}_r \\ &= \frac{1}{11} \sum_{r \in \{0.0, \ldots, 1.0\}} P_{\text{interp}}(r) \end{aligned} \tag{5}$$

where

$$p_{\text{interp}}(r) = \max_{\tilde{r} \geq r} p(\tilde{r})$$

mAP is the average of these APs over all the classes. As mentioned, we used an IoU threshold of 0.5 for both the models. We also used other metrics for assessing the performance of our models including precision (2), recall (3), and F_1 score (4). All the scores are on the entire test set.

YOLOv3 trained across a set consisting of images of 416 × 416 resolution for 6000 iterations, produced mAP score of 51.73% on the validation set. When tested across a set of 608 × 608 images, score was improved by 2.38 percent to 54.11%, showing that although trained on a lower resolution images, evaluation at a relatively higher resolution leads to better results in this case. We believe it is due increase in accuracy of detecting small-scale regions when inference is done on a higher resolution image. Results on the model are detailed in Tables 4 and 5, mean precision and mean recall are evaluated on a withheld validation set at different confidence thresholds to find a suitable trade-off between the two. Faster R-CNN trained on 600 × 1000 resolution images for 15 epochs produced a mAP score of 58.30%, performing a bit better than YOLOv3 in terms of accuracy as expected. However, significant time reduction in YOLOv3 during inference makes it more suitable for implementation in automated disease recognition systems (Fig. 8).

Figure 9b shows the results produced by Faster R-CNN when image in Fig. 9a was passed through it. The predicted image contains more predictions (bounding

Table 4 Results produced by YOLOv3 on 416 × 416 resolution images at different thresholds

Threshold	Precision	Recall	F_1-score	Avg. IOU (%)
0.00	0.22	0.60	0.32	14.57
0.25	0.58	0.46	0.51	39.24
0.40	0.64	0.43	0.52	43.46
0.50	0.67	0.43	0.52	45.53
0.60	0.69	0.42	0.52	46.79
0.70	0.70	0.40	0.51	47.75

Table 5 Result on YOLOv3 (608 × 608)

Threshold	Precision	Recall	F_1-score	Avg. IOU (%)
0.00	0.15	0.65	0.24	9.96
0.25	0.45	0.47	0.46	30.23
0.40	0.49	0.44	0.46	33.25
0.50	0.54	0.41	0.47	36.39
0.60	0.56	0.40	0.47	38.49
0.70	0.61	0.39	0.48	41.55

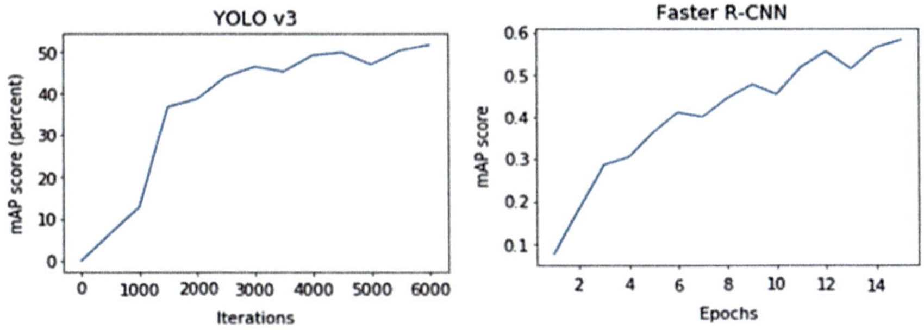

Fig. 8 mAP score plots of the models used for detection

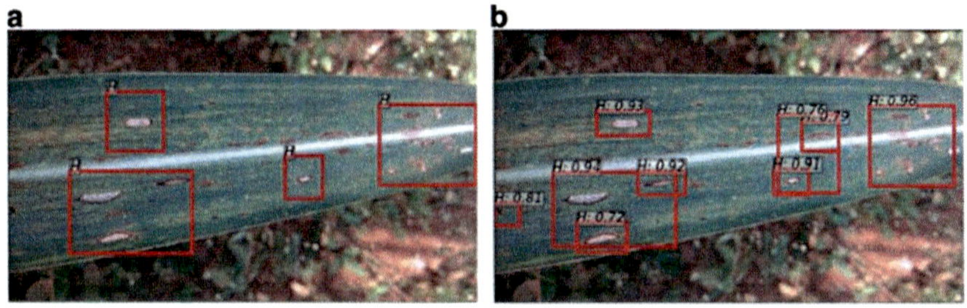

Fig. 9 Visualizations of the predictions made by faster R-CNN (**b**) when an image from our test set (**a**) was passed through it

boxes) of diseased spots than in the input-annotated image, Fig. 9a. As seen, the model is bounding even those diseased spots that were not annotated in the input image. This shows that the model has learnt the diseased spots thoroughly. This difference between the number of predictions of diseased spots made by the model and in the input image, although reduces the mAP score, simultaneously illustrates the accuracy of the model in predicting the diseased spots. Moreover, since the images were taken in realistic conditions, they showed wide variations in the patterns of diseased spots leading to variable-sized bounding boxes in the training images as in Fig. 9a, sometimes encompassing the disease specifically and sometimes a bunch

Fig. 10 Visualizations of predictions made by YOLOv3 on held out test set

of them altogether. The model learns all these variations and predicts all types of bounding boxes as in Fig. 9b. Since the mAP score [28] is dependent on the mapping of bounding boxes between the input and the predicted image, this does reduce the mAP score statistically; however, the real time results Fig. 9b depict the robustness of the model. The training set contains a lot of these types of images. Keeping all this in view and the small dataset, the scores are quite reasonable. Given a larger dataset and better annotations of training data with a even distribution of objects at different scales, the statistics are only expected to improve (Fig. 10).

6 Conclusion

The conventional approaches of finding anomalies in plants by supervision of expert pathologists are a difficult and time-consuming task. The availability of experts and the time taken in the process can delay immediate identification of diseases. CNN's have been found to be very robust in finding visually observable patterns in images and the growth of computational hardware has made it viable to utilize them. Using CNN's for finding visible anomalies in plants will result in faster identification and quicker interventions to subdue the effects of diseases on the plants. Our work focused on evaluating current state-of-art classification architectures on a publicly available dataset of plant diseases comprising of 54,306 images of different plants and exhibiting the contrast in test accuracy when the trained model is tested on images taken in controlled and uncontrolled conditions, respectively. Performance decreases manifold when evaluated on images from different sources. To circumvent this, we introduced sugarcane dataset, which has been collected by taking into account the issues which would be faced in identification of diseases on site. CNN models reached an accuracy of 93.20% on this dataset. The results of our model on images collected from different sources showed a significant improvement compared to performances reported in literature (e.g., Mohanty et al. 2016 [13] 31% accuracy for a problem with 38 plant disease classes), we achieved an accuracy of 76.4% with six plant disease classes. These results show huge potential of these deep learning models and allude to the fact that for robust plant disease identification system, a more diverse set of training data from different areas and under different conditions is needed.

Furthermore, it is apparent localizing the infected region will be the next step in progression from coarse to refine inference. Both models, Faster R-CNN and YOLOv3 were trained on a subset of sugarcane dataset. In context of the diversity and uncontrolled conditions of the dataset, both frameworks showed promising results in successfully detecting five different diseases in sugarcane. Larger dataset for both tasks, we believe would improve the results further.

Acknowledgements We thank College of Agriculture, Mandya, Bangalore, for helping us in the collection of sugarcane dataset and providing expertise for identifying the different types of diseases that the sugarcane plants were suffering from while collecting the images of their leaves.

References

1. Strange RN, Scott PR (2005) Plant disease threat to global food security. Phytopathology 43
2. UNEP (2013) Smallholders, food security and the environment
3. Russakovsky O et al (2015) ImageNet large scale visual recognition challenge. Int J Comput Vis (IJCV) 115(3):211–252
4. Deng J et al (2009) ImageNet: a large-scale hierarchical image database in computer vision and pattern recognition. In: 2009 IEEE conference on (IEEE), pp 248–255
5. LeCun Y, Bottou L, Bengio Y, Haffner P (1998) Gradient-based learning applied to document recognition
6. Krizhevsky A, Sutskever I, Hinton GE (2012) Imagenet classification with deep convolutional neural networks. In: Advances in neural information processing systems, pp 1097–1105
7. Zeiler MD, Fergus R (2014) Visualising and understanding convolutional networks. In: Computer vision-ECCV 2014, pp 818–833
8. Szegedy C et al (2015) Going deeper with convolutions. In: Proceedings of the IEEE conference on computer vision and pattern recognition, pp 1–9
9. He K, Zhang X, Ren S, Sun J (2015) Deep residual learning for image recognition. Preprint at arXiv:1512.03385
10. CNBC news channel. https://www.cnbc.com/2017/01/17/6-billion-smartphones-will-be-in-circulation-in-2020-ihs-report.html
11. Nandhini TSD, Padmavathy V (2017) A study on sugarcane production in India. Int J Adv Res Bot 3(2):13–17. http://dx.doi.org/10.20431/2455-4316.0302003

12. Ren S, He K, Girshick R, Sun J (2016) Faster R-CNN: towards real-time object detection with region proposal networks. IEEE Trans Pattern Anal Mach Intell 39:1137–1149
13. Redmon J, Farhadi A (2018) Yolov3: an incremental improvement. Preprint at arXiv:1804.02767
14. Mohanty SP, Hughes DP, Salathé M (2016) Using deep learning for image-based plant disease detection. Front Plant Sci 7. Article: 1419. http://dx.doi.org/10.3389/fpls.2016.01419
15. Hughes DP, Salathé M (2015) An open access repository of images on plant health to enable the development of mobile disease diagnostics. Preprint at coRR abs/1511.08060
16. Wang G, Sun Y, Wang J (2017) Automatic image-based plant disease severity estimation using deep learning. Comput Intell Neurosci 2017:2917536. https://doi.org/10.1155/2017/2917536
17. Simonyan K, Zisserman A (2014) Very deep convolutional networks for large-scale image recognition. Preprint at arXiv:1409.1556v6
18. Sladojevic S, Arsenovic M, Anderla A, Culibrk D, Stefanovic D (2016) Deep neural networks based recognition of plant diseases by leaf image classification. Comput Intell Neurosci 2016:3289801. https://doi.org/10.1155/2016/3289801
19. Smith LN (2015) Cyclic learning rates for training neural networks. Preprint at arXiv:1506.01186v6
20. Fastai deep learning course lesson 1. https://course.fast.ai/videos/?lesson=1
21. https://www.biostat.wisc.edu/~page/rocpr.pdf
22. Girshick R, Donahue J, Darrell T, Malik J (2014) Rich feature hierarchies for accurate object detection and semantic segmentation. Preprint at arXiv:1311.2524
23. Girshick R (2015) Fast R-CNN. Preprint at arXiv:1504.08083
24. Uijlings J, Sande K, Gevers T, Smeulders A (2013) Selective search for object recognition
25. https://webcourse.cs.technion.ac.il/236815/Spring2016/ho/WCFiles/RCNN_X3_6pp.pdf
26. Everingham M, Eslami S, Gool L, Williams C, Winn J, Zisserman A (2014) The PASCAL visual object classes (VOC) challenge: a retrospective. Int J Comput Vis 111:98–136 (2015). https://doi.org/10.1007/s11263-014-0733-5
27. Redmon J, Divvala S, Girshick R, Farhadi A (2016) You only look once: unified, real-time object detection. In: 2016 IEEE conference on computer vision and pattern recognition (CVPR), pp 779–788
28. Beitzel SM, Jensen EC, Frieder O (2009) MAP. In: Liu L, Özsu MT (eds) Encyclopedia of database systems. Springer, Boston, MA
29. https://github.com/chenyuntc/simple-faster-rcnn-pytorch
30. https://github.com/AlexeyAB/darknet

Deep Learning-Based Car Damage Classification and Detection

Mahavir Dwivedi, Hashmat Shadab Malik, S. N. Omkar, Edgar Bosco Monis, Bharat Khanna, Satya Ranjan Samal, Ayush Tiwari, and Aditya Rathi

Abstract In this paper, we worked on the problem of vehicle damage classification/detection which can be used by insurance companies to automate the process of vehicle insurance claims in a quick fashion. The recent advances in computer vision largely due to the adoption of fast, scalable and end-to-end trainable convolutional neural networks make it technically feasible to recognize vehicle damages using deep convolutional networks. We manually collected and annotated images from various online sources containing different types of vehicle damages. Due to the relatively small size of our dataset, we used models pre-trained on a large and diverse dataset to avoid overfitting and learn more general features. Using CNN models pre-trained on ImageNet dataset and using several other techniques to improve the performance of the system, we were able to achieve top accuracy of 96.39%, significantly better than the current results in this work. Furthermore, to detect the region of damage, we used state-of-the-art YOLO object detector and achieving a maximum map score

M. Dwivedi (✉) · H. S. Malik · S. N. Omkar · E. B. Monis · B. Khanna · S. R. Samal · A. Tiwari · A. Rathi
Indian Institute of Science, Bangalore, India
e-mail: mahaviredx@gmail.com

H. S. Malik
e-mail: hashmat.shadab.malik@gmail.com

S. N. Omkar
e-mail: omkar@aero.iisc.ernet.in

E. B. Monis
e-mail: edgarmonis@gmail.com

B. Khanna
e-mail: khannabharat001@gmail.com

S. R. Samal
e-mail: samalsatya990@gmail.com

A. Tiwari
e-mail: tiwariayush1997oct@gmail.com

A. Rathi
e-mail: rathiaditya53@gmail.com

N. N. Chiplunkar and T. Fukao (eds.), *Advances in Artificial Intelligence and Data Engineering*, Advances in Intelligent Systems and Computing 1133,
https://doi.org/10.1007/978-981-15-3514-7_18

of 77.78% on the held-out test set, demonstrating that the model was able to successfully recognize different vehicle damages. In addition to this, we also propose a pipeline for a more robust identification of the damage in vehicles by combining the tasks of classification and detection. Overall, these results pave the way for further research in this problem domain, and we believe that collection of a more diverse dataset would be sufficient to implement an automated vehicle damage identification system in the near future.

Keywords Car damage classification/detection · Pre-trained CNN models · YOLO object detector

1 Introduction

In recent times, car insurance companies deal with more frequent insurance claims. Manually validating on large scale of claims cannot meet the speed requirement for express claim process anymore resulting claim leakage [1]. Claim leakage is simply defined as lost money through claims management failures that ultimately result from inefficiency in the current manual and automated processes. Claim amount primarily relies on the type of damage and damaged part of the car, so we need an automated system for the whole process of car insurance claim as used by some start-ups [2] which can efficiently classify and detect damage and helps to minimize the claim leakage.

The revolution because of Krizhevsky and others leaves the field of computer vision open for further research. Convolutional neural networks have shown superior accuracy on image classification tasks, as shown on the ImageNet Large Scale Visual Recognition Challenge (ILSVRC) and are dominating most of the problems faced in computer vision. CNNs have also been applied on the problem of object detection, such as R-CNN [3], that are significantly better than the conventional feature-based detectors in terms of accuracy. In R-CNN, selective search [4] is used to propose regions which are more likely to have an object. These proposed regions are then resized to pass through a CNN with SVM classifiers to assign the type of object in the region. However, as for each proposal generated by selective search algorithm, we have to perform a forward pass through the CNN, and it is quite slow. Further improvements in the architecture came in Fast R-CNN [5] which instead of passing proposed region through the CNNs processes the whole image to extract a feature map. Feature map is used for object proposals with the help of a region of interest (ROI) pooling layer. In addition, Faster R-CNN [6] improves the architecture further by proposing region proposal networks (RPNs) that share convolutional layers with object detection networks. Architectures following the approach of [3] for object detection use regions to localize objects within the image and tend to look at the only those regions of image which have a higher chance of containing an object. In contrast to this technique, YOLO framework [7] uses the whole image simultaneously for detection. Furthermore, it is significantly faster than R-CNN family of architectures.

YOLOv3 [8] is currently the best among different available models in the YOLO family; it is improved by making the network bigger and adding residual networks by adding shortcut connections.

Taking this into consideration, we used CNNs for classification/detection of vehicle damages. For this problem statement, we created our own dataset by collecting images from Internet and manually annotating them. We focused on seven commonly observed car damage types such as smashed, scratch, bumper dent, door dent, head lamp broken, glass shatter and tail lamp broken. We started with basic CNN model, pre-training a CNN using autoencoder followed by fine-tuning image classification models pre-trained on the ImageNet dataset. We observed that transfer learning performs the best. To increase the robustness of the system, we propose an architectural pipeline that will first extract the damaged region of the vehicle from the image and then pass it to the trained classifier network. As described earlier, for detection purposes, we also use YOLOv3 [8].

2 Related Works

Several damage detection approaches have been proposed and applied to car body damage detection. Srimal et al. [9] proposed to use 3D CAD models to handle automatic vehicle damage detection via photograph. To identify damage in a car, they use a library of undamaged 3D CAD models of vehicles as ground truth information. In principle, image edges which are not present in the 3D CAD model projection can be considered to be vehicle damage.

An Anti-fraud System for Car Insurance Claim Based on Visual Evidence—they proposed an approach to generate robust deep features by locating the damages accurately and used YOLO to locate the damage regions. Gontscharov et al. [10] try to solve vehicle body damage by using multi-sensor data fusion. Keyence Vision [11] proposed an industrial solution for car damage by hail, by applying a high-resolution multi-camera vision system.

Cha et al. [12] adopt an image-based deep learning to detect crack damages in concrete; the methodology used is acquiring images with the help of camera and then the preprocessing stage where the acquired images undergo scaling and segmentation, and finally to get the shape of crack, feature extraction is done, while [13] adopted a phase-based optical flow and unscented Kalman filters. A. Mohan and S. Poobal study and review crack detection using image processing. Based on the analysis, they conclude that more number of researchers have used the camera-type image for the analysis with a better segmentation algorithm [14].

3 Dataset Description

As far as we know, there is no publicly available dataset for car damage classification; without a large and diverse dataset, it becomes rather difficult to apply standard computer vision techniques; therefore, we created our own dataset by collecting images using Web crawling as done by Patil et al. [15]. We manually filtered and categorized images into seven commonly observed damage types as shown in Table 1. We also collected images belonging to no damage class. Some sample images of our dataset are shown in Fig. 1 where each column from left to right represents different damage types, bumper dent, scratches, door dent, glass shatter, head lamp broken, tail lamp broken and smashed, respectively. For the purpose of detection, we manually annotated different types of damaged regions.

Table 1 Dataset description

Classes	Train size	Aug. train size	Test size
Bumper dent	150	750	30
Scratch	112	560	22
Door dent	146	730	25
Glass shatter	104	520	25
Head lamp broken	107	535	20
Tail lamp broken	39	195	11
Smashed	256	1280	30
No damage	947	4735	225

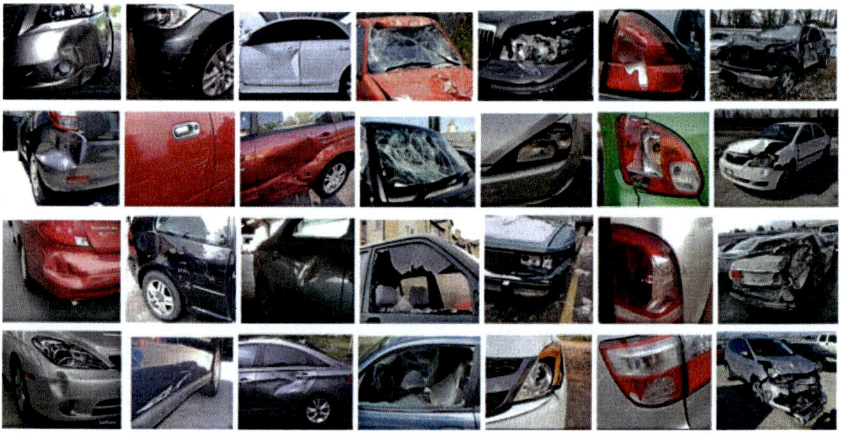

Fig. 1 Sample images of different damage types

3.1 Data Augmentation

Some issues like overfitting may occur as model is trained with small dataset, so to overcome this problem and improve our model, we used data augmentation to increase dataset size. We enlarged the dataset to approx. $4\times$ by applying rotation of $20°$, shear range of 0.2, zoom range of 0.2 and horizontal flip. For classification, we split the data randomly in 4:1 ratio, where 80% was used for training and 20% was used for testing.

4 Transfer Learning

To overcome the problem of overfitting on small datasets, instead of training the CNN models from scratch, we can use transfer learning which has shown a significant improvement on classification problems when the dataset available is scarce [16, 17]. So, it is more common to train a CNN on an already available large dataset (e.g., ImageNet, which contains 1.2 million images with 1000 categories) and then transfer this knowledge to the target task with the intuition that some knowledge may be common between the source domain and target domain. Learning features from a large and diverse input data distribution has been found to be more effective than using features from a data distribution specific to just the target class. Instead of just pre-training our model on a large car dataset, it is better to learn features on a more diverse dataset [15]. After training on ImageNet dataset, we retrain the classifier on top of the CNN on our dataset. We also fine-tune all the layers of the CNN while keeping in mind that the earlier layers learn more generic features that are common in all classification tasks [18].

4.1 AlexNet

AlexNet [19], was designed by Alex Krizhevsky, is one of the deep ConvNets designed to deal with complex scene classification task on ImageNet data. The network had a very similar architecture to LeNet [20] but was deeper, bigger and featured convolutional layers stacked on top of each other. It contains eight layers; the first five are convolutional layers, some of them followed by max-pooling layers, and the last three are fully connected layers. To reduce overfitting, AlexNet uses another technique called dropout that was introduced by G. E. Hinton in a paper [21] in 2012.

4.2 VGG19

The outstanding accomplishment achieved by submission to the ILSVRC2013 [30] took advantage of smaller receptive window size and smaller stride of the initial convolutional layer. VGG-19 [18] focuses on another important aspect of ConvNet architecture design—its depth. They kept other parameters of the architecture constant and steadily increase the deepness of the network by adding more convolutional layers, which is convenient due to the fact that they used very small (3×3) convolution filters in all layers. As an outcome, they came up with significantly more accurate ConvNet architecture.

4.3 Inception V3

The Inception V3 [22] or GoogleNet is an architecture with a large set of 42 layers and was created by Google. With such a large set of layers, error rate is reduced effectively. All other architectures preceding to the Inception performed the convolution on the channel and spatial-wise domain together. Inception is a multiple step process where a pattern of convolutional layer is repeated multiple times, and in each step, the architecture has set of convolutions layers through which the input is passed which ultimately leads to better learning rate.

By performing the one-on-one convolution, the inception block is doing cross-channel correlations, ignoring the spatial dimensions. This is followed by cross-spatial and cross-channel correlations via the 3×3 and 5×5 filters. All of this then goes through dimension reduction to end up in 1×1 convolutions.

4.4 MobileNets

As consumer technology becomes thinner and lighter, interest in lightweight neural networks for mobile applications has gained traction [23–25]. MobileNets [26] are based on the concept of factorized convolutions, where a standard convolution is split into a depthwise convolution and a pointwise convolution, to reduce the number of parameters, originally introduced in [27]. Considering M as the number of input channels, N as the number of output channels and D_k as the spatial dimension of the kernel which is assumed to be square, the reduction in computation is [26]:

$$\frac{1}{n^2} + \frac{1}{D_k^2}$$

Table 2 Test accuracy of different pre-trained (on ImageNet) models

Model	Parameters (m)	Acc. (without Aug.)	Acc. (with Aug.)
AlexNet	60	82.71	89.89
VGG19	144	93.2	94.9
Inception V3	5	66.26	74.18
ResNet50	25.6	89.58	90.26
MobileNets V1.0	4.2	69.6	70.8

4.5 *ResNet50*

ResNet works by using microarchitecture modules to create a "Network of Networks" architecture. In recent times, deep modules have increased the accuracy multiple times. This network introduces residual learning first introduced by He et al. in their 2015 paper [28]. Residual learning aims to try to learn residuals instead of trying to learn features. We used ResNet50 pre-trained on the ImageNet dataset for feature extraction.

Keeping the convolution layer freezed, we trained the densely connected classifier on the augmented data. In order to make the model generalize well with our dataset, we fine-tuned it. In which, we unfreezed a few of the top layers of the freezed model used for feature extraction and jointly trained the fully connected classifier and these top layers. Validation accuracy increased in all the models. Accuracy of the models with and without data augmentation is shown in Table 2.

5 Further Improvements

To further enhance the accuracy and speed up the training process, we implemented various techniques that have demonstrated significant improvements than the conventional techniques.

Finding the optimal learning rate region is utmost important as it drastically affects the performance and speed of the network. We used the technique developed by Smith [29]; a good learning rate bound is found by increasing the learning rate until the loss stops decreasing. We then choose an optimal learning rate by estimating "reasonable bounds." Further, also varying the learning rate cyclically between reasonable boundary values helps in achieving improved classification accuracy and often in fewer iterations. This can be understood by keeping in mind the intuition that as we get closer to the optimal weights, we might want to take smaller steps. To get a more accurate and stable model, we must find weight space that is robust to small changes to the weight and in turn generalize well to the unseen data. And to prevent from getting stuck in a "spikey" region of weight space, the cyclic restarts of learning rate can help to jump to a different weight landscape and avoid getting stuck in a region which will not generalize well.

The initial layers of the pre-trained models can be thought of learning more general-purpose features [30]. So, while fine-tuning our models, we would be expecting the initial layers to go through less changes in order to accommodate our dataset. Keeping this in mind, we will use different learning rates for different layers: increasing the learning rate with the depth of the network. We have differentiated our models into three blocks, and then, accordingly, we chose a learning rate for each block. It is referred to as differential learning rates [31].

While data augmentation during training time significantly improves the performance of the model, we can also use it during inference. Instead of making prediction on a single image, we will be making predictions on a number of augmented versions of the original image and taking the average prediction as the final answer.

All the networks ran for a total of 10 epochs when training only the fully connected layers and 15 epochs when training all the layers (Fig. 2; Table 3).

The confusion matrix of ResNet34 in Fig. 3 shows that it predicted the classes correctly to large extent. Some of the misclassifications are merely due to the large similarity between the classes such as head lamp broken, tail lamp broken and glass shatter because all three involve broken glasses. The uncertainty in tail lamp broken is also because of the small percentage of images in this category. Moreover, smashed

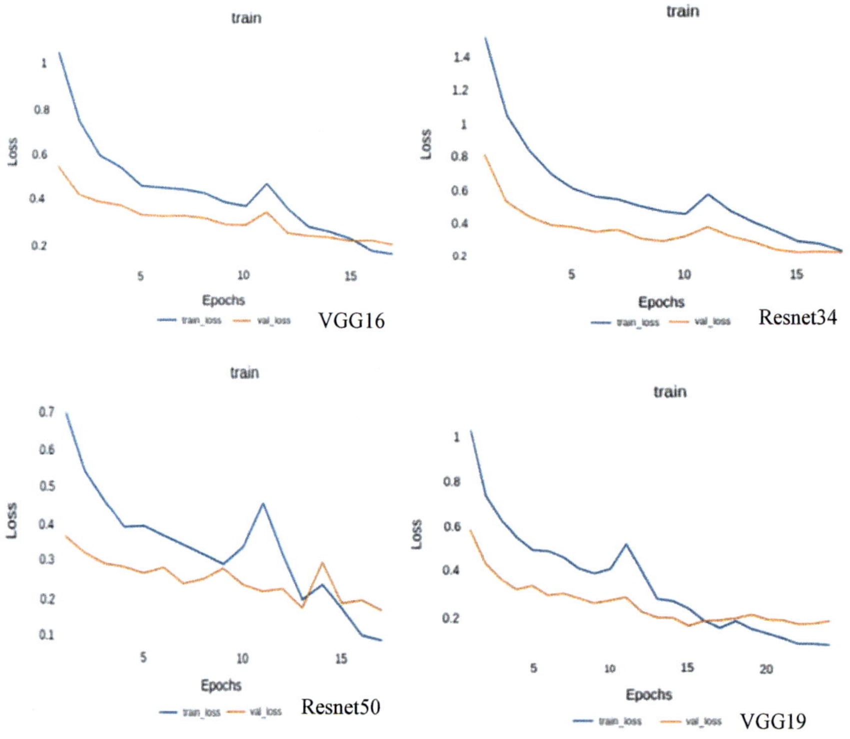

Fig. 2 Training and validation loss plots of various models

Table 3 Test accuracy using these techniques

Model	Acc (training FC layers only)	Acc (training all layers with differential learning rate annealing)	Precision	Recall	*F*-beta score	Acc (using test time augmentation) (%)
VGG16	90.97% (10 epochs)	93.81% (15 epochs)	0.907	0.886	0.891	94.84
VGG19	90.46% (10 epochs)	95.36% (15 epochs)	0.922	0.908	0.914	95.87
ResNet34	90.20% (10 epochs)	93.29% (15 epochs)	0.857	0.830	0.837	93.88
ResNet50	91.75% (10 epochs)	96.13% (15 epochs)	0.928	0.922	0.922	96.39

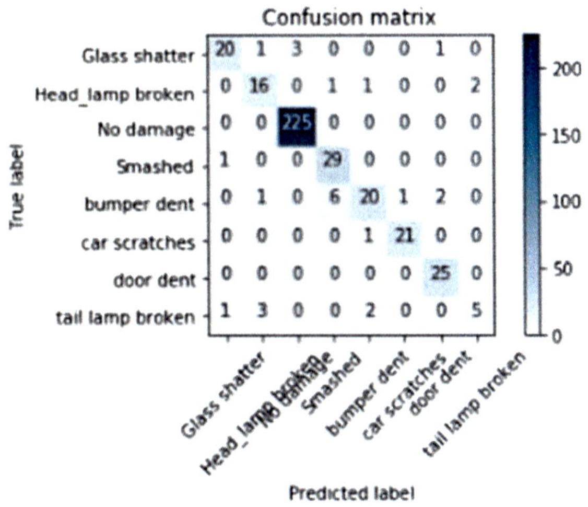

Fig. 3 Confusion matrix for ResNet34

cars contain dents, so few images which belong to bumper dent were misclassified as smashed. Zoomed viewpoints of images involving dents may also have led to misclassification of bumper dent as door dent. Out of the 388 testing images, 27 images are misclassified and 361 images are classified correctly.

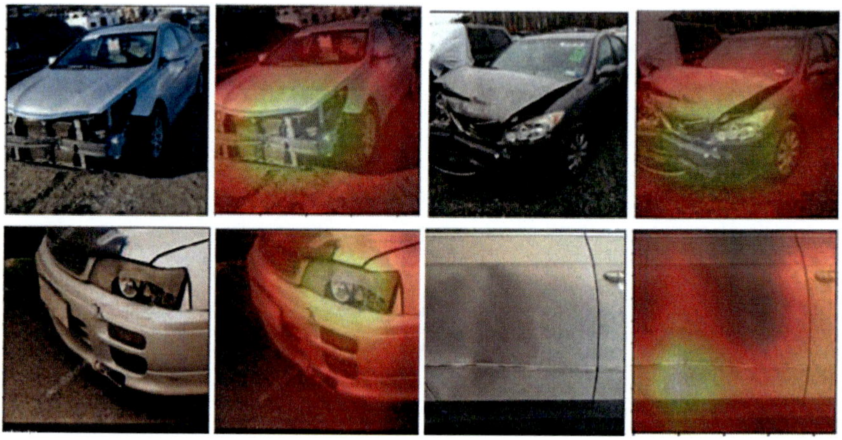

Fig. 4 Examples of CAMs, few images of damaged cars and their class activation maps with them. We observe that the damaged regions of the images are often highlighted. Apart from this, similar observations are recorded in other samples

6 Class Activation Mapping

Class activation mapping (CAM) technique developed in the 2015 [32] allows one to visualize which pixels of the original image are responsible for predicting the corresponding class, highlighting the discriminative object parts detected by the CNN. Moreover, we can be more certain that the CNN is working properly and gains some intuitions where the CNNs are failing as it localizes the main object in the image. It enables classification-trained CNN models to learn to perform object localization, without using any bounding box annotations. These results show that our model is generalizing well; even though the dataset is small, it is not learning the idiosyncrasies of our dataset rather learning to find more general features. We generated class activation maps (CAM) using global average pooling (GAP) in CNNs. A class activation map for a particular category indicates the discriminative image regions used by the CNN to identify that category (Fig. 4).

7 Damage Detection Using YOLOv3

YOLOv3 consists of a 106-layer architecture [8]. It improves on version 2 by using residual connections and upsampling (YOLOv3: An Incremental Improvement). It uses Darknet architecture. Darknet-53 uses fewer floating-point operations and is in general 1.5× faster than ResNet-101 and 2× faster than ResNet-152 even though it achieves similar performance. We first tested the object detection capabilities of Yolov3 by only labeling damaged areas of the image. After achieving satisfactory results (Table 6), we split the data into six categories: (1) bumper dent (BD), (2) car scratches (CR), (3) door dent (DD), (4) glass shatter (GS), (5) lamp broken (LB), (6) smashed (SM). As observed, the average precision of the "Car scratches" and "Lamp Broken" classes is comparatively lower because of fewer images in the training dataset. Results of YOLOv3 were validated across a set consisting of images of 416 × 416 resolution. The MAP score obtained was 74.23%. When tested across a set of 608 × 608 images, the precision was improved by 3.5–77.78% which shows that the model accuracy can be improved during inference by passing a higher resolution image through the network trained on relatively lower resolution; this increases the precision and makes it possible to detect small objects (Fig. 5; Tables 4 and 5).

AP scores: **BD**: 77.16% **CR**: 59.21% **DD**: 88.21% **GS**: 71.07% **LB**: 64.56% **SM**: 85.19%.

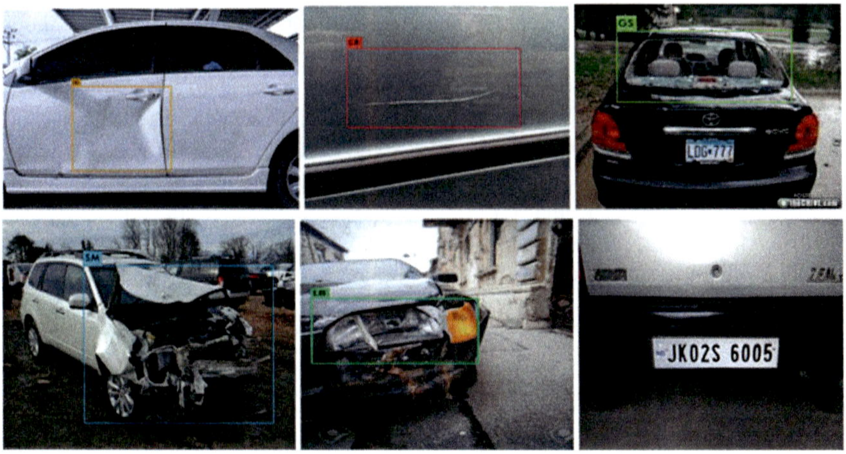

Fig. 5 Sample images of damage detection via YOLOv3

Table 4 Results on YOLOv3 (416 × 416) map score: 0.7423

Threshold	Precision	Recall	F_1-score	Avg. IOU (%)
0.00	0.59	0.81	0.68	43.22
0.25	0.82	0.73	0.77	61.25
0.40	0.84	0.71	0.77	62.99
0.50	0.86	0.70	0.77	64.18
0.70	0.88	0.69	0.77	65.70

Table 5 Result on YOLOv3 (608 × 608) map score 0.7778

Threshold	Precision	Recall	F_1-score	Avg. IOU (%)
0.40	0.70	0.80	0.74	50.21

8 Pipeline for Insurance Claim Process

To make the system more robust, we developed a two-step process to combine object detection and classification. The basic strategy was to use YOLOv3 framework to detect the damaged region and classifying that region using a CNN model trained on the damage dataset. Initially, the end user will upload an image to the server using an insurance claim app. Then, the image will be passed through an object detector; in our case, it is YOLOv3. The detector will be able to detect the region of the vehicle which is damaged. After localizing the damaged region, the proposed region will go through a CNN trained on the damage dataset to classify the type of damage. Furthermore, images aggregated over time can be used to increase the size of the dataset and in turn, increase the accuracy of the system by training on a more diverse dataset. The current accuracy on detecting damages using YOLOv3 on our dataset is depicted in Table 6. Taking into consideration, the relatively small size of our dataset, the results are very encouraging, and our model is able to learn features that are common in all the different type of damages. We believe that increasing the dataset size would be enough to increase the accuracy even further. So, as we keep storing images uploaded by customers and then training our system on them, it would improve the performance of the pipeline in detecting regions of damage as it will lead to a larger dataset which eventually allows the network to learn more significant features which generalize well to inter- and intra-class variations of the dataset (Fig. 6).

Table 6 Results on damage detection (map score: 0.663)

Threshold	Precision	Recall	F_1-score	Avg. IOU (%)
0.00	0.65	0.80	0.72	45.76
0.25	0.81	0.78	0.80	57.20
0.40	0.82	0.77	0.79	57.66
0.50	0.82	0.77	0.79	57.66
0.70	0.84	0.75	0.79	59.53

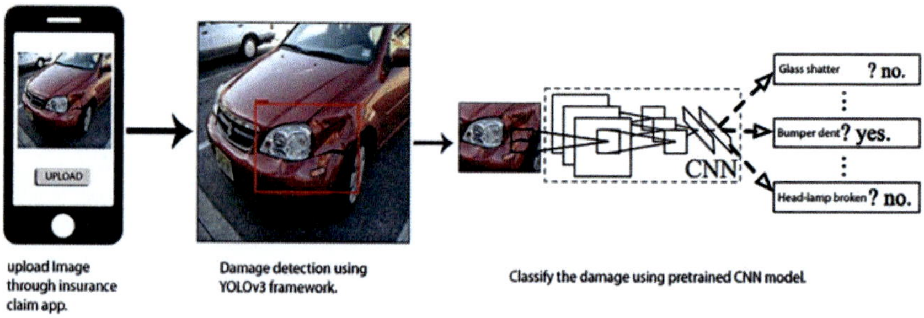

Fig. 6 Pipeline overview

9 Hardware Specifications

The advancement in the computation power is one of the factors which has resulted in the success of deep learning techniques. For training our model, we used a Nvidia GTX 1080Ti. This machine had 8 GB of graphics card memory and 24 GB of system RAM memory.

Once the system was trained for required number of epochs and the model started achieving a benchmark accuracy over the test dataset, we moved it for inference. We optimized our pipeline in such a way that our system is deployable in real-life scenario like insurance companies using it over laptops and handheld mobile phones for vehicle damage detection. The inference model runs at 10 fps on a laptop computer having Nvidia 940 MX graphics card with 2 GB of graphics card memory and 8 GB of system memory. Also, we ported the model to TFlite so that it can run on a handheld mobile phone such as OnePlus3T.

10 Conclusion

Stating the two most prudent quotes of our time by Andrew Ng, "AI is the new electricity," and Clive Humby "Data is the new oil," based on this motivation, we here amalgamate both data and AI to provide a novel approach for automating the vehicle damage insurance claims. In this paper, we demonstrate a mechanism to classify/detect damage in vehicles. For this, we manually collected versatile dataset from the Internet through running Web crawler on various search engines like Google and Bing and used deep learning models for the damage classification task. Combining transfer learning with cyclic learning rates for training neural networks, we were able to outperform the current state of the art in vehicle damage classification by a significant margin. We are also successful in detecting the damaged part of the vehicle using YOLO framework. Even though our dataset is comparatively small by the standard of other deep learning datasets, several quantitative evaluations show the power and potential of the proposed approach. We have also proposed a practical system application which can fully automate the insurance claim process and can profit many car insurance companies. We believe a larger dataset would improve the results and make this system ready for much more dynamic real-life scenarios.

References

1. http://www.ey.com/publication/vwluassets/ey-doesyour-firm-need-a-claims-leakage-study/ey-does-yourfirm-need-a-claim-leakage-study.pdf
2. https://tractable.ai/

3. Girshick R, Donahue J, Darrell T, Malik J (2014) Rich feature hierarchies for accurate object detection and semantic segmentation. In: The IEEE conference on computer vision and pattern recognition (CVPR)
4. Uijlings JRR, van de Sande KEA, Gevers T, Smeulders AWM (2013) Selective search for object recognition. Int J Comput Vision 104:154–171
5. Girshick R (2015) Fast R-CNN. In: The IEEE international conference on computer vision (ICCV)
6. Ren S, He K, Girshick R, Sun J (2017) Faster R-CNN: towards real-time object detection with region proposal networks. IEEE Trans Pattern Anal Mach Intell 39:1137–1149
7. Redmon J, Divvala S, Girshick R, Farhadi A (2016) You only look once: Unified, real-time object detection. In: The IEEE conference on computer vision and pattern recognition (CVPR)
8. Redmon J, Farhadi A (2018) Yolov3: an incremental improvement
9. Jayawardena S (2013) Image based automatic vehicle damage detection. Ph.D. thesis, College of Engineering and Computer Science (CECS)
10. Gontscharov S, Baumgartel H, Kneifel A, Krieger K-L (2014) Algorithm development for minor damage identification in vehicle bodies using adaptive sensor data processing. Procedia Technol 15:586–594. 2014. 2nd international conference on system-integrated intelligence: challenges for product and production engineering
11. Multi-camera vision system inspects cars for dents caused by hail
12. Cha Y-J, Choi W, Büyüköztürk O (2017) Deep learning-based crack damage detection using convolutional neural networks. Comput Aided Civ Infrastruct Eng 32(5):361–378
13. Cha Y-J, Chen J, Büyüköztürk O (2017) Output-only computer vision based damage detection using phase-based optical ow and unscented kalman filters. Eng Struct 132:300–313
14. Mohan A, Poobal S (2018) Crack detection using image processing: a critical review and analysis. Alex Eng J 57(2):787–798
15. Patil K, Kulkarni M, Sriraman A, Karande S (2017) Deep learning based car damage classification. In: 2017 16th IEEE international conference on machine learning and applications (ICMLA), pp 50–54
16. Yosinski J, Clune J, Bengio Y, Lipson H (2014) How transferable are features in deep neural networks? In: Ghahramani Z, Welling M, Cortes C, Lawrence ND, Weinberger KQ (eds) Advances in neural information processing systems, vol 27. Curran Associates, Inc., Red Hook, pp 3320–3328
17. Oquab M, Bottou L, Laptev I, Sivic J (2014) Learning and transferring mid-level image representations using convolutional neural networks. In: The IEEE conference on computer vision and pattern recognition (CVPR)
18. Simonyan K, Zisserman A (2014) Very deep convolutional networks for large-scale image recognition. arXiv:1409.1556
19. Krizhevsky A, Sutskever I, Hinton GE (2012) ImageNet classification with deep convolutional neural networks. In: Pereira F, Burges CJC, Bottou L, Weinberger KQ (eds) Advances in neural information processing systems, vol 25. Curran Associates, Inc., Red Hook, pp 1097–1105
20. Lecun Y, Bottou L, Bengio Y, Haffner P (1998) Gradient-based learning applied to document recognition. Proc IEEE 86:2278–2324
21. Hinton GE, Srivastava N, Krizhevsky A, Sutskever I, Salakhutdinov R (2012) Improving neural networks by preventing co-adaptation of feature detectors. arXiv:1207.0580
22. Szegedy C, Vanhoucke V, Ioffe S, Shlens J, Wojna Z (2016) Rethinking the inception architecture for computer vision. In: The IEEE conference on computer vision and pattern recognition (CVPR)
23. Jin J, Dundar A, Culurciello E (2014) Flattened convolutional neural networks for feedforward acceleration. arXiv:1412.5474
24. Hubara I, Courbariaux M, Soudry D, El-Yaniv R, Bengio Y (2018) Quantized neural networks: training neural networks with low precision weights and activations. J Mach Learn Res 18(187):1–30
25. Rastegari M, Ordonez V, Redmon J, Farhadi A (2016) XNOR-Net: ImageNet classification using binary convolutional neural networks. In: Leibe B, Matas J, Sebe N, Welling M (eds) Computer vision ECCV 2016. Springer, Zurich, pp 525–542

26. Howard AG, Zhu M, Chen B, Kalenichenko D, Wang W, Weyand T, Andreetto M, Adam H (2017) Mobilenets: efficient convolutional neural networks for mobile vision applications. arXiv:1704.04861
27. Sifre L (2014) Rigid-motion scattering for image classification. Ph.D. thesis
28. He K, Zhang X, Ren S, Sun J (2016) Deep residual learning for image recognition. In: The IEEE conference on computer vision and pattern recognition (CVPR)
29. Smith LN (2017) Cyclical learning rates for training neural networks. In: 2017 IEEE winter conference on applications of computer vision (WACV), pp 464–472
30. Zeiler MD, Fergus R (2014) Visualizing and understanding convolutional networks. In: Fleet D, Pajdla T, Schiele B, Tuytelaars T (eds) Computer vision ECCV 2014. Springer, Zurich, pp 818–833
31. Fastai deep learning course lesson 1. https://course.fast.ai/videos/?lesson=1
32. Zhou B, Khosla A, Lapedriza A, Oliva A, Torralba A (2016) Learning deep features for discriminative localization. In: The IEEE conference on computer vision and pattern recognition (CVPR)

Sparse Reflectance Map-Based Fabric Characterization

Kayan K. Katrak, Rithvik Chandan, Sirisha Lanka, G. M. Chitra, and S. S. Shylaja

Abstract The research in the field of fabric characterization is reaching its zenith due to the increasing e-commerce activity and ever-growing digitalization of fabric information. With the increase in variety and heterogeneity of fabric classes, fabric characterization has become a very challenging task. Many approaches have been implemented to solve this problem, and the most common solutions are based on texture analysis. Since many fabrics, especially man-made fabrics have untextured or similar textured surfaces, it poses a problem to distinguish between them. Considering these complexities, an interesting way to solve this problem is to leverage the fabric's reflection information. In this paper, the problem has been addressed using reflection property along with an SVM algorithm. Instead of deriving a complete reflectance model with an elaborate laboratory set-up, a model was developed where all that is needed is an image of the fabric taken using a regular commercial camera with an HDR feature, which captures the appearance in a single image under unknown natural illumination. This method was evaluated with a synthetic as well as a real-life data set, and it achieves an accuracy of 76.63% as compared to the human accuracy of 53.87%.

Keywords Fabric classification · Silk · Denim · Linen · Leather · Support vector machine · High dynamic range · Reflectance maps

---

K. K. Katrak · R. Chandan · S. Lanka · G. M. Chitra · S. S. Shylaja (✉)
PES University, Bangalore, India
e-mail: shylaja.sharath@pes.edu

K. K. Katrak
e-mail: kayankkatrak@gmail.com

R. Chandan
e-mail: rithvikchan1@gmail.com

S. Lanka
e-mail: siri181.lanka@gmail.com

G. M. Chitra
e-mail: chitragm8@gmail.com

N. N. Chiplunkar and T. Fukao (eds.), *Advances in Artificial Intelligence and Data Engineering*, Advances in Intelligent Systems and Computing 1133,
https://doi.org/10.1007/978-981-15-3514-7_19

1 Introduction

Global textile industry has crossed trillion USD and is growing fast due to urbanization, increasing population and higher disposable income. With e-commerce powered by high-quality digital catalogues, the ability to shop anytime and anywhere is becoming the norm for shopping clothes. Considering this, the ability for designers and manufactures to provide different fabric choices is becoming a critical need. Beyond e-commerce applications, the fabric characterization is expected to help in the field of robotics as well. This paves a way for a new field of study known as fabric characterization. While humans may make errors in differentiating between various classes of fabric, models that employ computer vision techniques execute this function much more accurately. Previously, the classification methods have mainly focused on handling texture perception and description, synthesis, material recognition and segmentation [1–3]. However, with materials often having similar textures or no macroscopic texture at all, it is identified that the reflection property provides better insight.

Humans can get a rough estimate of the material class from its RGB bands, even if this comes about through the spectral reflectance of the material and its illumination. But, the task at hand is to accomplish this using computer vision and image processing techniques. A spectral reflectance measurement can yield information about the material from which any fabric sample is made, since light that is not reflected from the sample is absorbed due to its chemical and physical compositions, otherwise is scattered or transmitted. This amount of light that is scattered or reflected is what determines a body's reflectance property.

Since obtaining the reflectance property involves complex systems such as gonio-reflectometers and spectrometers, measuring pixel intensities is a much more efficient method. This information comprises luminance and chrominance attributes.

In this paper, reflection properties extracted from a sparse reflectance map of an HDR image are used, along with a grid search algorithm to match efficient kernels generating features for SVM classification. The aim is to categorize four different classes of fabric—denim, silk, linen and leather without considering additional features such as texture and weave patterns.

This method is assessed on two data sets: a commercially available MERL BRDF data set and our own People's Education Society-Variable Sparse-mapping of Textiles on Reflection (PES-VaSTR) data set collected from real-life samples (Fig. 1).

2 Related Work

Fabric characterization has been a primal problem and has been implemented using multiple techniques.

Texture analysis has been one of the oldest methods for fabric classification [4, 5]. Song et al. [5] acquire the texture of a fabric sample by using a sensor that imitates

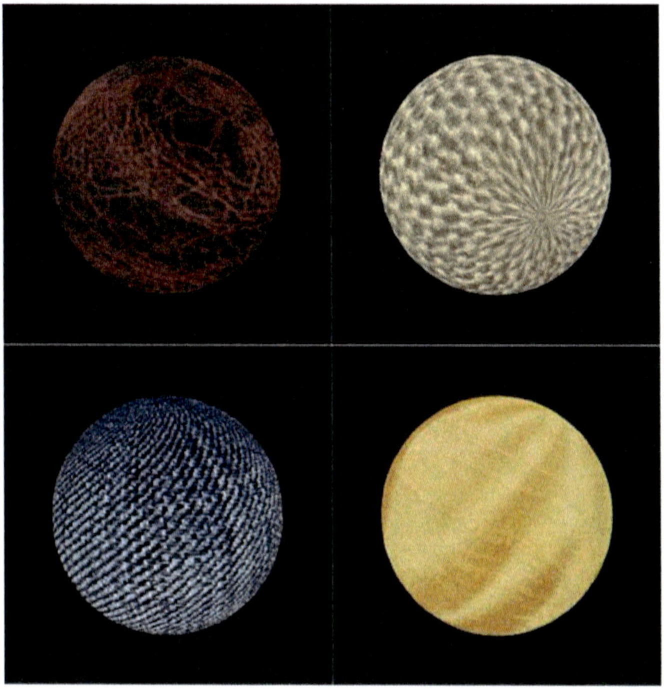

Fig. 1 Sample images of the sparse reflectance maps for (clockwise) leather, linen, silk and denim

human touch developed using thin polyvinylidene fluoride (PVDF). Notwithstanding, the impediment of this surface sensor framework is that a test sample ought to be physically present before the set-up.

Zhang et al. [6] use wavelet transform to extract surface textures of cashmere and merino wool fibres. The fibre height, shape, brightness and interval are measured to classify the different animal fibres. It has a limited application and is only restricted to synthetic fibres.

With the advancement of technology as well as the introduction of man-made fibres, characterizing fabrics based on reflectance property is a more favourable approach [7, 8]. The work of Nielsen et al. [9] is a good example of this, and it addresses the problem by reconstructing measured BRDF into complete reflectance maps. The bidirectional reflectance distribution function (BRDF) uses four variables to obtain the reflectance of an opaque surface object requiring an extensive laboratory set-up to study the samples.

Georgoulis et al. [10] have used reflection information to characterize four classes: metal, paint, plastic and fabric. This paper implements the Gaussian Process Latent Variable Model (GPLVM) to distinguish the four classes. The reflectance values of these classes are spread out widely making it easy to classify between them.

3 Method

The problem of fabric characterization is tackled by taking real-life images on an HDR camera and converting them into corresponding spherical maps by applying image processing techniques to maintain uniformity among the images. The spherical map is further processed, and the sparse reflectance information of the sample is obtained by calculating pixel intensities of the map. This entire procedure of converting the image to a spherical sparse reflectance map is known as "Spherification".

Consequently, a multi SVM model [3, 11] is trained for efficient recognition of the fabric material. This method of characterization proves to be extremely cost-effective as (i) it does not require extensive laboratory set-ups, (ii) it uses a commercial camera with HDR feature to capture images of samples, (iii) and the samples are observed under natural lighting (Fig. 2).

3.1 *Spherification*

Images taken in high dynamic range mode are passed through a spherification model which generates a spherical mapping of the image. A sphere shape is chosen as the properties of reflection are best studied on simulated curved surfaces. A sphere provides us with a uniform measurement for reflection due to its multidirectional symmetry. The raw data set has also been converted into such spheres to maintain uniformity in the number of pixels in each data input. This simulation is obtained

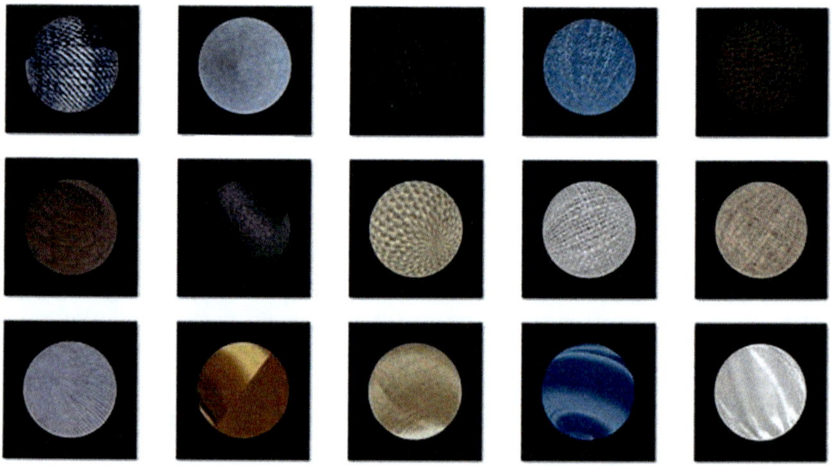

Fig. 2 Reflectance maps from the PES-VaSTR data set extracted from HDR images using natural lighting

by geometrical manipulations in order to obtain the relative x- and y-coordinates of each pixel (Fig. 3).

Given an image, the central coordinates of the image x_c, y_c are determined. Using this, the relative radius of x- and y-coordinates (pixel values) from the centre is calculated by using the following formula (Figs. 4 and 5).

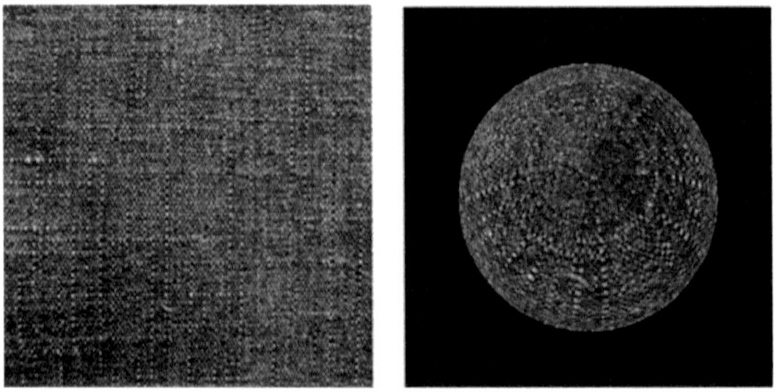

Fig. 3 HDR image of a denim fabric along with the spherical mapping of HDR image taken from a commercial camera

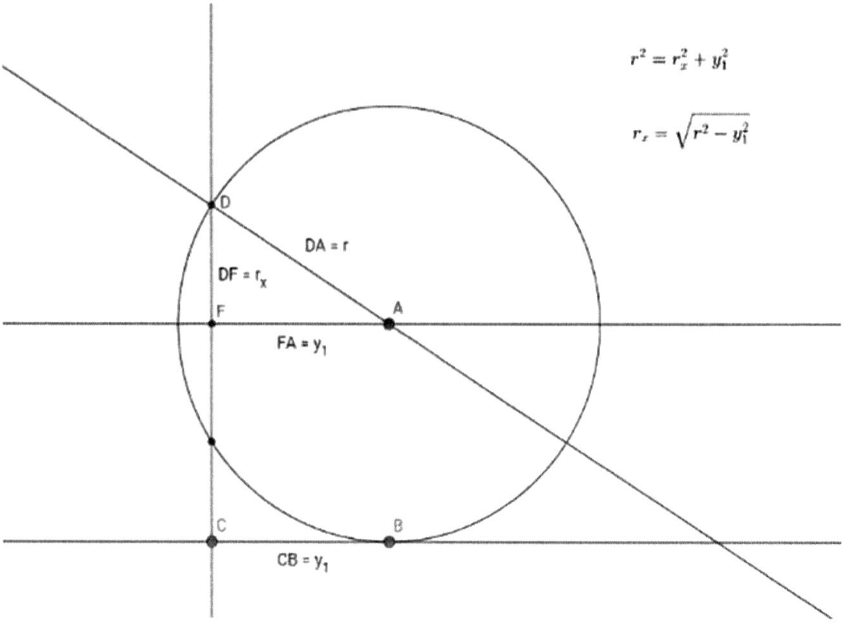

Fig. 4 To find the effective radius with which one can calculate resultant x-location based on the displacement from the centre of the sphere in y. *Courtesy* [12], https://stackoverflow.com

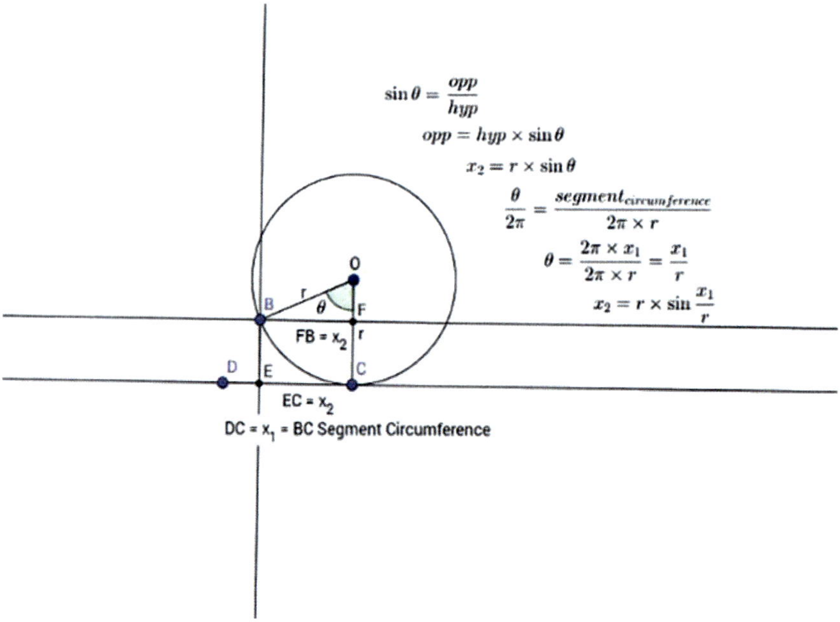

Fig. 5 Concepts involved to obtain simulated/new pixel coordinates *y*. *Courtesy* [12] https://stackoverflow.com

$$\sqrt{\text{rad}^2 - y_1^2} \text{ and } \sqrt{\text{rad}^2 - x_1^2} \tag{1}$$

With the relative radius, the new position of x and y is determined. Let x_1 be the position of a pixel. The new position x_2 is determined by getting the resultant pixel location as if seen from the top of the sphere on which the image has been pasted. This works in one dimension should the radius remain the same in the other dimension. However for a sphere this effective radius changes since the circle drawn out on the inside of the sphere becomes smaller the further away you are from the centre.

3.2 Model

For the classification of the spherical mappings into the different fabric classes, the image is rescaled to a fixed standard, followed by smoothening into a rundown of crude pixel intensities. These pixel intensities represent the luminance of the sample.

A feature vector is a vector containing multiple elements or features. Examples of such features are colour components, length, area, circularity, etc. We have chosen reflection property, i.e. pixel intensities as our primary feature. The feature vector strategy is a very gullible function that basically takes an input image and rescales it to a fixed width and stature and afterwards levels the RGB pixel intensity into a solitary rundown of numbers. This implies contraction to 64 × 64 pixels and giving it three channels for red, green, and blue parts separately. Our yield "feature vector" will be a listing of 64 × 64 × 3 = 12,288 numbers.

This list is then passed into the SVM-SVC algorithm which finds out a hyper-plane in multidimensional space that separates out classes.

Given training vectors $x_i \in \mathbb{R}$, $i = 1, \ldots, n$, in two classes, and a vector $y \in \{-1, 1\}^n$, SVC solves the following primal problem:

$$\min_{w,b,\zeta} \frac{1}{2} w^T w + C \sum_{i=1}^{n} \zeta_i \text{ subject to } y_i \left(w^T \Phi(x_i) + b \right) \geq 1 - \zeta_i, \zeta_i \geq 0, i = 1, \ldots, n \tag{2}$$

where e is the vector of all ones, $C > 0$ is the upper bound, Q is an n by n positive semidefinite matrix, $Q_{ij} = y_i y_j K(x_i, x_j)$, where $K(x_i, x_j) = \varnothing(x_i)^T \varnothing(x_j)$ is the kernel.

Here, training vectors are implicitly mapped onto a higher-(maybe infinite) dimensional space by the function $\varnothing$. The decision function being

$$\operatorname{sgn}\left(\sum_{i=1}^{n} y_i \alpha_i K(x_i, x_j) + \rho \right) \tag{3}$$

Polynomial is the kernel used for our classification. A degree-3 polynomial is chosen, and the polynomial kernel is defined as follows:

$$K(x, y) = (x^T y + c)^d \tag{4}$$

where x and y are vectors in the information space, for example, vectors of features processed from training or test samples and $c \geq 0$ is a free parameter exchanging off the impact of higher-order versus lower-order variables in the polynomial.

Kernel K relates to an inner product in a feature space dependent on some mapping φ:

$$K(x, y) = \langle \varphi(x), \varphi(y) \rangle \tag{5}$$

The penalty parameter c is for the soft margin cost function, which controls the impact of every individual support vector; this procedure includes an error penalty for stability. The value of c was chosen by conducting a grid search until the value which gave the highest accuracy was found. The grid search yielded an optimum value of 947 for the c parameter which gave the most stable as well as high accuracy (Fig. 6).

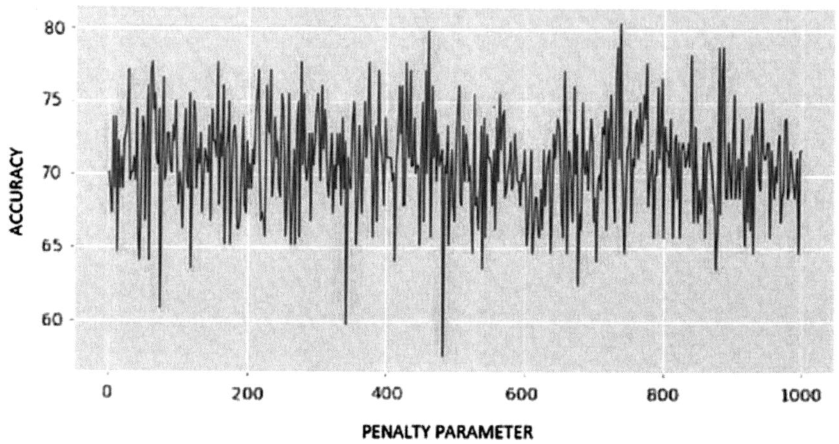

Fig. 6 Above graph represents how our accuracy varies with a change in the penalty parameters

4 Results

4.1 Data Set and Evaluation Protocol

To know if something truly works, it must be checked and cross-checked multiple times. The technique was cross-checked utilizing engineered and real-life samples seen under natural illumination. For the engineered data set, the MERL BRDF data set was utilized which involves 94 reflectance samples (36 plastics, 20 paints, 13 fabrics and 25 metals) as per the work of Nielsen et al. [9]. All these examples were rendered on a synthetic sphere positioned in 20 diverse HDR condition maps, which gives us 1880 reflectance maps altogether [10]. Be that as it may, for our investigations, a raw data set was created from real-life images. It consists of an average of 75 HDR images for each of the four classes and performed the spherification function on every image thrice to include randomness in the data set which improves its size and quality, amounting to a total of 919 images.

An image for testing is then passed and manually cropped by the user to remove unwanted noise from the experimental image and then resized using the spherification function in order to have a uniform size as the ones in our real-life database (768 × 768 pixels). This was run ten times on a single experimental input image, and the predictions were recorded to boost our accuracy.

Upon validation, the model's kernel parameters as well as its penalty parameters were fine-tuned. It was observed that giving low penalty parameters resulted in low accuracy. On the other hand, keeping it too high also leads to the same result, finally settling for "poly" as our kernel parameter and 947 as our penalty parameter. For every experiment conducted, an 80:20 random split for training and validation was performed. This resulted in obtaining accuracy as low as 62% and jumping to a maximum of 76.63%. However, since the training and validation are based on a random shuffle of the data set, the model has also achieved an accuracy of 80.83% statistically.

4.2 Experiments on Synthetic Database

To get professional evaluation on the performance of our model, it was tested with a synthetic database. For this purpose, the "MERL BRDF" database which contains reflectance functions of 100 distinct materials was picked and stored as a densely measured BRDF. For each experiment conducted on the model, 1504 reflectance maps were used for training followed by 376 reflectance maps for validation. The training and validation databases were not fixed for every experiment and kept changing due to the random functions applied in the model. The MERL BRDF database was created under monitored laboratory lighting conditions which gives us perfect reflectance maps. The model achieved an accuracy of 95% with this data set.

4.3 Experiments on Our Database

To accomplish a good result, training and validation were done with our own data set which comprises real-life HDR images taken by a commercial camera or smartphone. The database is then split, 735 images for training followed by 184 images for validation and gave our experimental images for testing. However, the number of images for each class getting fed as training data may not be the same due to our method of randomly splitting the data. For this database, the same kernel and penalty parameters were maintained. In general, a lower accuracy is achieved when compared with the synthetic data set. This is due to the extensive laboratory set-up being used to build the synthetic database, whereas our database is created using commercial HDR cameras.

In addition to our experiment, an online survey was conducted to measure human accuracy of differentiating various fabric classes. The confusion matrices for the characterization of the four classes from the human survey and the model are given in Figs. 7 and 8.

The diagonal elements in the matrices give the accuracy of the correct prediction. It can be inferred that the values in the model's matrix have higher or comparable values as to that of the human's confusion matrix. This model especially outranks human's ability to detect linen. It can also be observed that this model has a lower confusion rate compared to humans. For example, as in the case of linen and denim, humans predict linen instead of denim 42 out of 100 times, however the model just does it seven out of 100 times clearly outperforming the human visualization. These results are tabulated in Table 1 (Fig. 9).

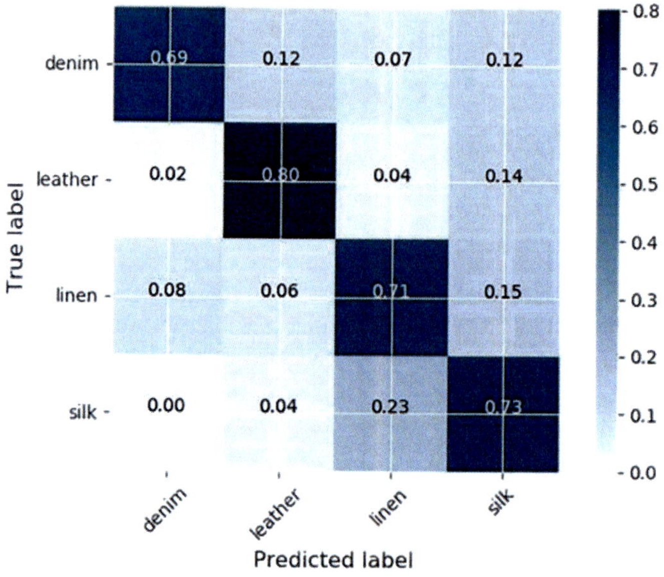

Fig. 7 Confusion matrix of our model with normalization

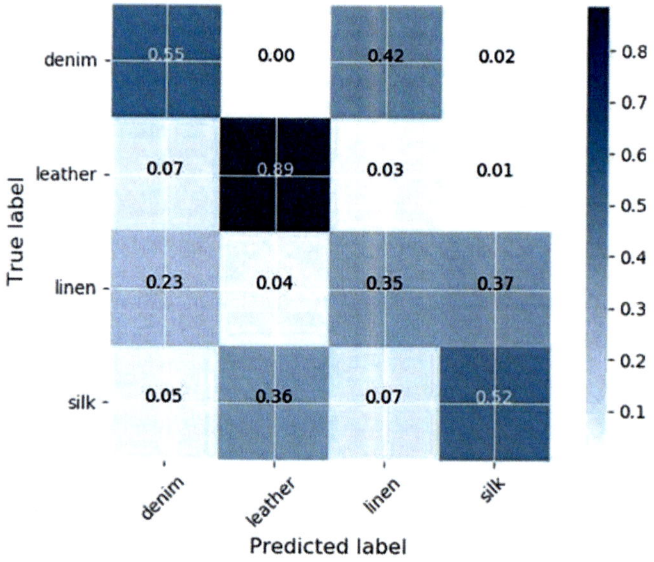

Fig. 8 Confusion matrix for humans with normalization

Table 1 Accuracy of prediction of the four classes by the model and human beings

Prediction of class	Our model (%)	Humans (%)
Denim	69	55
Leather	80	89
Linen	71	35
Silk	73	52

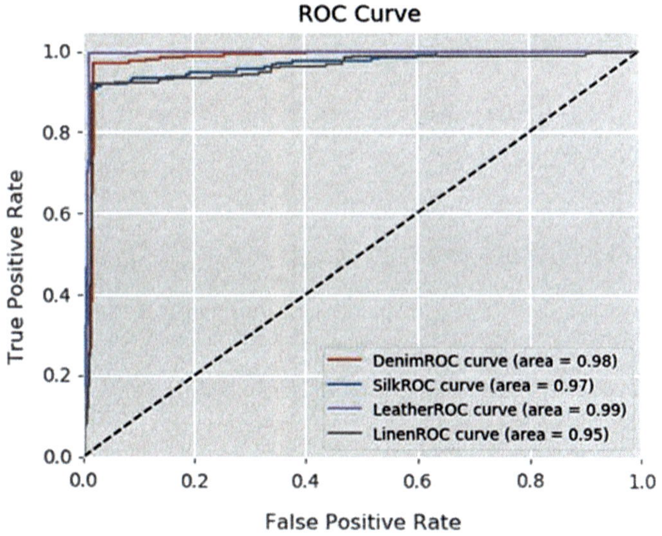

Fig. 9 ROC curve with x-axis as False Positive rate and y-axis as True Positive rate

Upon constructing a ROC curve which has the False Positive rate on its x-axis and the True Positive rate on its y-axis, we observe that the areas below each curve are very high, thereby indicating that the model is good at distinguishing between the classes and gives us accurate results.

5 Conclusions

Fabric characterization is a challenging task when it comes to material characterization especially when no information about the texture of the material is known. In this paper, an attempt was made to solve the problem of classifying fabric materials with the help of image processing and machine learning techniques. The materials are characterized based on their reflectance properties without the consideration of features like textures and weave patterns. The four classes dealt in this paper are denim, silk, linen and leather. Upon conducting a survey to measure human accuracy, it has been concluded that our SVM model outperforms human accuracy by 22.75%. With the use of sparse reflectance maps, an accuracy of 76.63% was achieved.

The future scope of this paper would be the inclusion of more fabric classes especially cotton as it is one of the most prevalent fabric materials used in the industry and to obtain a complete reflectance map to help improve our data set in turn improving our accuracy. The improvement of the model's accuracy with an increase in the number of classes in the data set is another primary focus.

Acknowledgements This project was carried out under "Centre for Data Science and Applied Machine Learning", PES University, India.

References

1. Ben Salem Y, Nasri S (2009) Texture classification of woven fabric based on a GLCM method and using multiclass support vector machine. In: 2009 6th international multi-conference on systems, signals and devices. Djerba, pp 1–8. https://doi.org/10.1109/ssd.2009.4956737
2. Zhang J, Marszalek M, Lazebnik S, Schmid C (2006) Local features and kernels for classification of texture and object categories: a comprehensive study. In: Proceedings of the 2006 conference on computer vision and pattern recognition workshop (CVPRW'06). IEEE Computer Society, Washington, DC, p 13. https://doi.org/10.1007/s11263-006-9794-4
3. Yang Y, Li J, Yang Y (2015) The research of the fast SVM classifier method. In: 2015 12th international computer conference on wavelet active media technology and information processing (ICCWAMTIP). IEEE, pp 121–124
4. Ghosh A, Guha T, Bhar RB, Das S (2011) Pattern classification of fabric defects using support vector machines. Int J Cloth Sci Technol 23(2/3):142–151. https://doi.org/10.1108/09556221111107333
5. Song A, Han Y, H H, Li Jianqing (2014) A novel texture sensor for fabric texture measurement and classification. Instrum Meas IEEE Trans On 63:1739–1747. https://doi.org/10.1109/TIM.2013.2293812
6. Zhang J, Palmer S, Wang X (2010) Identification of animal fibers with wavelet texture analysis. In: WCE 2010: proceedings of the world congress on engineering 2010. Newswood Limited/International Association of Engineers, Hong Kong, pp 742–747
7. Liu C, Gu J (2014) Discriminative illumination: per-pixel classification of raw materials based on optimal projections of spectral BRDF. IEEE Trans Pattern Anal Mach Intell 36:86–98. https://doi.org/10.1109/TPAMI.2013.110
8. Chandraker M, Ramamoorthi R (2011) What an image reveals about material reflectance. In: Proceedings of the 2011 international conference on computer vision (ICCV'11). IEEE Computer Society, Washington, DC, pp 1076–1083. https://doi.org/10.1109/iccv.2011.6126354
9. Nielsen JB, Jensen HW, Ramamoorthi R (2015) On optimal, minimal BRDF sampling for reflectance acquisition. ACM Trans Graph 34(6):11, Article 186 (October 2015). https://doi.org/10.1145/2816795.2818085
10. Georgoulis S, Vanweddingen V, Proesmans M, Gool LV (2017) Material classification under natural illumination using reflectance maps. In: 2017 IEEE winter conference on applications of computer vision (WACV), Santa Rosa, CA, pp 244–253. https://doi.org/10.1109/wacv.2017.34

11. Ahuja Y, Yadav SK (2012) Multiclass classification and support vector machine. Glob J Comput Sci Technol Interdiscip 12(11):14–20
12. http://www.stackoverflow.com

A Risk Assessment Model for Patients Suffering from Coronary Heart Disease Using a Novel Feature Selection Algorithm and Learning Classifiers

Sujata Joshi and Mydhili K. Nair

Abstract The aim of this research is to develop an efficient risk assessment model to assess the risk in patients suffering from coronary heart disease. The proposed technique classifies patients as having low risk or high risk of coronary heart disease. CVD dataset from Cleveland database is used to develop the model. The various parameters considered are cholesterol, blood pressure, electrocardiogram and echocardiogram tests among others as well. The patients suffering from coronary heart diseases are labeled as lower or higher risk. Feature selection is one of the critical tasks in developing predictive models. It reduces the computational cost by removing insignificant features. This leads to a simpler, accurate and comprehensible model. Here, two feature selection techniques namely Mean selection (MS) and a novel feature selection technique, VAS-CHD, a variance-based attribute selection for coronary heart disease diagnosis is implemented to capture the right features for the purpose of prediction. The decision tree has resulted in accuracy of 80.5% and lower False Negatives for features obtained from MS. The multilayer perceptron has resulted in 73.7% accuracy and lower False Negatives for features obtained from VAS-CHD.

Keywords Feature selection · Variance · Classification · Heart disease · Decision tree · Multilayer perceptron

1 Introduction

According to the reports of WHO, the leading cause of global deaths is cardiovascular diseases (CVDs). The reports reveal that in 2015 about 17.7 million deaths occurred due to CVDs, representing about one-third of the deaths in the world. Among these deaths, 7.4 million deaths were due to coronary heart disease which is the most

S. Joshi (✉)
Nitte Meenakshi Institute of Technology, Bangalore, India
e-mail: sujata.joshi@nmit.ac.in

M. K. Nair
M. S. Ramaiah Institute of Technology, Bangalore, India

N. N. Chiplunkar and T. Fukao (eds.), *Advances in Artificial Intelligence and Data Engineering*, Advances in Intelligent Systems and Computing 1133,
https://doi.org/10.1007/978-981-15-3514-7_20

prevalent type of heart-related disease and 6.7 million deaths were reported to be because of cardiovascular stroke [1]. With this rise of global deaths, it is opined that by 2030, about 23.6 million deaths may occur due to CVDs. It is also reported that South East Asia will see a steep increase in the number of deaths due to non-communicable diseases in general and CVDs in particular. This is attributed to changing lifestyle and food habits, work culture and stress [1].

Most of the CVDs prevalent in India are coronary heart diseases (CHD) stroke and hypertensive heart diseases. In people suffering from CHD, the coronary arteries, which are blood vessels supplying oxygen and blood to the heart, are narrowed. This happens when a substance called plaque accumulates on the walls of the arteries, thus narrowing the arteries and resulting in reduced flow of blood to the heart. Due to this, sufficient blood does not flow to the heart which may result in angina, shortness of breath, heart attack or heart failure. Diagnosis of CHD is done through blood tests, electrocardiogram, echocardiogram, stress test, CT scan and coronary catheterization. Although the risk of CHD increases with age, there are other factors as well, which contribute to the disease. Some of these factors are high blood pressure, smoking, obesity, high alcohol intake, diabetes, genetics and family history, lack of exercises and stress [2–4].

Hence, there is a need for early detection and management of CHD for people who are at high risk due to the presence of risk factors. This research is motivated by these factors to deal with the problem of heart disease diagnosis in our country.

2 Literature Review

It is reportedly observed that the number of people suffering from heart diseases has steadily increased in the recent past. Applications in information technology and data mining have been playing a key role in almost all sectors in general and healthcare sector in particular [5]. In the healthcare sector, the data mining applications are developed for diagnosis of diseases, spread of epidemics, adverse drug detection and analysis [6]. The researchers in [7] have developed a classifier for assessing the risk in patients who have been suffering from congestive heart failure. Here, the parameter used to develop the classifier is long-term heart rate variability. A model is developed to assess the risk factors associated with coronary heart diseases by the researchers in [8]. The model is developed using the algorithm C4.5. The most important risk factors are smoking and hypertension among others. A study was taken up by the researchers in [9] to assess the heart failure severity and to predict the mortality due to the presence of adverse events. To identify the early stage of CVD, the researchers have used ECG signals to categorize the patient as normal or abnormal using fuzzy logic and artificial neural network [10]. In another work by the authors, ECG data is analyzed to improve the risk of cardiovascular disease. Here, ECG features and other features were employed to detect abnormality [11]. As reported in the work by the researchers in [12], a risk model to predict mortality due to heart failure is developed using an improved random forest. Here, the model uses a new method to split data to improve accuracy as well as identify more accurate predictors. The researchers in [13] have developed a decision tree for assessing the risk in patients suffering from long-term heart variability. This work analyzes the heart rate variability on nominal 24 h recordings and selects the best subset of features with the lowest misclassification error. In another work, the risk factors are assessed for coronary heart events using classification trees [7].

Many researchers have been working on feature selection techniques in the area of data mining in health care. An attribute selection measure based on the distance for the decision tree induction is proposed in [14]. This method produces smaller trees for the dataset which has attributes having different number of values. The researchers have developed a feature selection method using neuro-fuzzy technique and multi-objective genetic algorithm in [15]. Here, the concept of dominance is used for multi-objective feature selection and fast subset evaluation. It is applied to small and high-dimensional regression problems. In [16], a framework for local dimension reduction is proposed which is based on a partial least squares method. It focuses on extracting features from high-dimensional multicategory microarray data to determine the biomarkers. Cancer biomarkers are identified using feature selection in a micro-RNA of gene expression, where the gene expression levels are monitored in [17]. The authors have proposed a combined approach using semi supervised support vector machine to find biomarkers related to cancer. In [18], the authors have proposed an efficient and effective group incremental feature selection algorithm using rough set theory. For a gradually increasing dataset, knowledge can be discovered using an incremental approach where computation can be directly carried out from original data. It uses information entropy to determine the significance of features and includes the features based on their significance level. This algorithm finds the feasible feature subset much faster compared to heuristic algorithms. In [19], the researchers have proposed a novel bare bones particle swarm optimization (BBPSO)-based feature selection which uses an adaptive chaotic jump strategy for better extraction. The proposed BBPSO is compared with eight evolutionary computation (EC)-based wrapper methods and two filters can select the most discriminative features from the entire feature set and achieve significantly better classification performance than other comparative methods.

3 Methods

3.1 Data

In our study, we have used the CVD dataset from the UCI [20] Learning Repository. It contains 14 attributes which are listed in Table 1.

The records include 96 female and 207 male patients of ages ranging from 29 to 77 years old. The other attributes consist of categorical attributes, binary attributes

Table 1 Dataset description

Attribute	Data type	Description
'age'	real	Age of the patient in years
'gender'	string	Gender {m: male; f: female}
'chest pain'	string	Type of Chest pain {'anginal','non-anginal'}
'blood pressure'	real	Systolic and diastolic blood pressure
'cholesterol'	real	Serum cholesterol
'restecg'	string	Readings of Electrocardiography {'normal','abnormality in ST-T wave', 'left ventricular hypertrophy'}
'thal_ach'	real	Heartbeat rate at maximum
'fast blood sugar'	binary	Fasting blood sugar > 120 mg/dl {1 = yes; 0 = no}
'exang'	binary	Angina induced due to exercise {1 = true; 0 = false}
'old_peak'	real	ST wave reading at rest
'slope_ST'	string	ST slope during peak exercise
'colored arteries'	real	Number of major heart vessels colored during fluoroscopy
'thal'	string	{'normal'; 'fix_defect'; 'rev_defect'}
'diag'	{0,1,2,3,4}	Diagnosis of heart disease
'risk'	string	{low, high}

and continuous attributes. The dataset is a typical mix of the attributes of different data types. The class attribute 'diag' has five values from 0 to 4 indicating

0- <50% narrowing in the vessels
1- >50% narrowing in one vessel
2- >50% narrowing in two vessels
3- >50% narrowing in three vessels
4- >50% narrowing in four vessels.

A new class attribute 'risk' is created and has the values {low, high} depending on the narrowing of the vessels.

The research question to be answered is: Is it possible to *predict the risk level of patient based on physical examinations, blood test reports and various tests like electrocardiograph test and angiography accurately?*

3.2 *Learning Framework*

Since the objective is to develop the most appropriate model for the prediction of risk in patients suffering from coronary heart diseases, we assessed the most popular machine learning algorithms for their suitability to the task. Two feature selection techniques mean selection (MS) and VAS-CHD are implemented to select the relevant features for the purpose of prediction. We have then applied individual data mining algorithms to the dataset with different feature sets, both using MS and VAS-CHD along with other known feature selection techniques.

3.3 Feature Selection

As we know, the selection of appropriate features is one of the critical tasks in data mining as selecting the right features results in a comprehensible model. To select the most relevant features, two feature selection algorithms MS and VAS-CHD are proposed and implemented using the concept of entropy.

In information theory, the Shannon entropy measure is generally used to measure the impurity of a collection of data [21, 22]. It is given by

$$\text{Entropy}(S) = \sum_{i=0}^{c-1} p(i|S)\log_2 p(i|S) \tag{1}$$

where $p(i|S)$ is the record set belonging to class i, and c denotes the number of classes [21].

The effectiveness of a feature in classifying the data can be measured by information gain which is given by Han et al. [21].

$$\text{Gain}(S, A) = \text{Entropy}(S) - \sum_{j=1} \frac{|Sv|}{|S|}\text{Entropy}(Sv) \tag{2}$$

The computed gain for all the features can also be used to eliminate all those features which have low gain as compared to others. Hence, we can use feature selection with ranking for elimination of low ranked features.

Mean selection (MS)
The algorithm MS is presented in Fig. 1.

In addition to MS feature selection technique described above, we have also applied other standard feature selection techniques based on subset evaluation, chi square, gain ratio and information gain with search methods best first search (BFS), exhaustive search (ES), genetic search (GS), greedy stepwise search (GRS), forward stepwise search (FWS), random search (RS) and ranker (RK). The summarized results are presented in Table 2.

Most of the algorithms have selected seven features, whereas MS has selected six features. The features obtained from MS are

v1={cp, fbs, exang, oldpeak,,ca,thal}

```
Mean Selection(MS)

Input: Dataset D with attributes F <f1…fm>
Output: A reduced feature set F`

For each attribute <f1…fm> ,
Compute the Gain.  Let the computed gain be <g1…gm>

Compute mean : MV=(1/m) ∑_{i=1}^{m} g

For each attribute f € F
If g(f) >= MV
  F`=F`U f
```

Fig. 1 Mean selection algorithm for selection of features

However, the domain expert suggested that cholesterol and blood pressure are important indicators of heart disease. None of the techniques listed in Table 3 selected the features chol and restbps. In this regard, a novel algorithm, VAS-CHD, a variance-based attribute selection for coronary heart disease diagnosis is proposed and implemented to capture the discriminating and right features for the purpose of prediction. The algorithm developed is given in Fig. 2.

Variance-based Attribute Selection (VAS)

Variance is a statistical way to measure variability. It measures how far a set of data is spread out. Since variance measures the variability from an average or mean, it can be employed to find the most discriminating features from a dataset. It is the average of the squared differences from the mean and is given by

$$\text{Variance } V = \frac{1}{N}\sum_{i=1}^{N}(x_i - \bar{x})^2 \tag{3}$$

Standard deviation is then computed as $\sqrt{V}$. Since we are looking at the most discriminating features, we select a feature if its deviation from mean is not in the range of [−SD, +SD]. Otherwise, it is discarded. The algorithm is presented in Fig. 2.

The algorithm VAS-CHD based on variance is successful in selecting these features chol and trestbps along with other features. The features obtained from VAS-CHD are

v2={trestbps, chol, fbs, thalach, slope, ca}

Table 2 Attributes selected using different techniques

Attribute	Subset evaluation						Chi square	Gain ratio	
	BFS	ES	GS	GRS	FWS	RS	RK	RK	MS
'age'							x	x	
'gender'							x	x	
'type_of_chest pain'	x	x	x	x	x	x	x	x	x
'blood pressure'							x	x	
'cholesterol'							x	x	
'restecg'							x	x	
'thal_ach'	x	x	x	x	x	x	x	x	
'fast_blood_sugar'	x	x	x	x	x	x	x	x	x
'exang'	x	x	x	x	x	x	x	x	x
'old_peak'	x	x	x	x	x	x	x	x	x
'slope_ST'							x	x	
'colored_arteries'	x	x	x	x	x	x	x	x	x
'thal'	x	x	x	x	x	x	x	x	x
# attributes	7	7	7	7	7	7	13	13	6

Table 3 Evaluation metrics

Measure	Description
Accuracy	$\frac{TP}{TP+TN+FP+FN}$
Error rate	$\frac{TP}{TP+TN+FP+FN}$
Sensitivity	$\frac{TP}{TP+FN}$
Specificity	$\frac{TN}{TN+FP}$
Area under the curve	$\frac{1}{2}\left(\frac{TP}{TP+FN}+\frac{TN}{TN+FP}\right)$

TP is the number of low risk patients correctly classified; *TN* is the number of high risk patients correctly classified; *FP* is the number of low risk patients incorrectly classified as high risk patients; *FN* is the number of high risk patients incorrectly classified as low risk patients

```
Variance based Attribute Selection(VAS_CHD)

Input: Dataset D with attributes F <f1…fm>
Output: A  new feature set F` with discriminating features.

Compute the Gain ratio for each attribute <f1…fm>.Let the
computed gain be  <g1…gm>
Compute mean : M=(1/m) Σ_{i=1}^{m} g

For each attribute f € F
    Compute deviation from mean DM_i  =>  (g_i-M)
    Compute  square of deviation from mean  SM=> (g_i-M)^2

Compute variance =>  V=(1/m) Σ_{i=1}^{m}(g_i-M)^2
Compute Standard deviation SD= √V
For each attribute f € F
   If DM_i is not range[-SD,+SD]
     F`=F`U f
```

Fig. 2 Variance-based attribute selection algorithm for selecting features

3.4 Classification Algorithms

The machine learning algorithms, namely decision tree (DT), Naïve Bayes classifier (NB), adaptive boosting (AB), *k*-nearest neighbor (*k*-NN) and multilayer perceptron (MLP), were considered for the classification task. The selected features obtained from MS and VAS-CHD algorithms were used to develop the models.

Feature set v1={ ca, thal, exang, thalach,cp, oldpeak *} obtained from the algorithm MS.*

Feature set v2={ ca, thalach, slope, fbs, chol, trestbps} obtained from the algorithm VAS-CHD.
The predictive models were developed individually for both feature sets v1 and v2.

4 Model Evaluation and Results

In our work, the dataset was split into 2/3 which was used for training the models and 1/3 which was used for testing. The measures for the evaluation of models for a two-class problem are given in Table 3 [23].

The results of the predictive models for feature set *v*1 are depicted in Table 4. Also the graphical representation of the performance measures is presented in Fig. 3.

The results of the predictive models for feature set *v*2 are shown in Table 5. Also the graphical representation of performance of different classifiers is presented in Fig. 4.

Although accuracy and error rate show the overall performance of the classifier and sensitivity and specificity show the correct positive predictions and negative

Table 4 Model evaluation results for feature set *v*1

Model	Accuracy	Error rate	Sensitivity	Specificity	AUC
DT	0.805	0.195	0.806	0.807	0.861
NB	0.864	0.135	0.864	0.864	0.907
k-NN	0.825	0.174	0.825	0.826	0.913
Adaboost	0.854	0.145	0.854	0.856	0.916
MLP	0.815	0.184	0.816	0.817	0.89

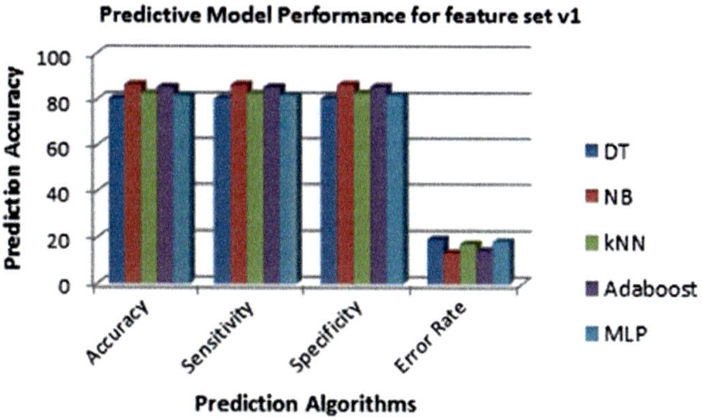

Fig. 3 Performance of different prediction algorithms for feature set v1

Table 5 Model results for feature set $v2$

Model	Accuracy	Error rate	Sensitivity	Specificity	AUC
DT	0.708	0.292	0.709	0.711	0.775
NB	0.766	0.233	0.767	0.767	0.83
k-NN	0.747	0.252	0.748	0.749	0.809
Adaboost	0.737	0.262	0.738	0.797	0.804
MLP	0.737	0.262	0.738	0.742	0.816

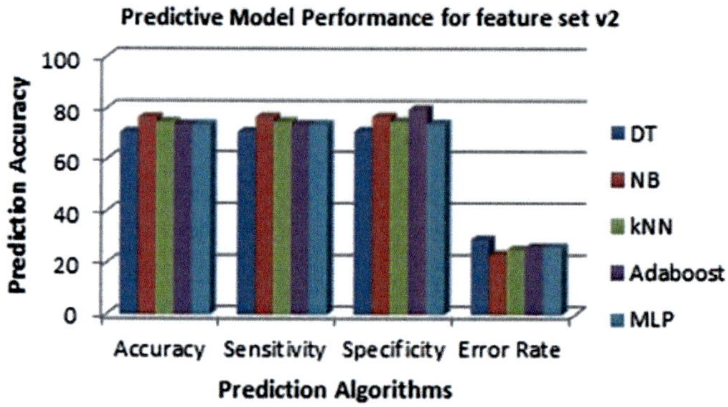

Fig. 4 Performance of different prediction algorithms for feature set v2

predictions, respectively, FN indicates the number of patients with high risk, who were classified incorrectly as having low risk, which is a matter of concern. This misclassification may lead to casualties to the patient because they are treated as low risk patients but in fact, they are high risk patients. Hence, we aim at reducing the number of such cases and choose a classifier which satisfies this objective. The analysis is done for both feature sets $v1$ and $v2$. For feature set $v1$, decision tree and MLP have lower values for FN as compared to other algorithms as shown in Fig. 5.

Fig. 5 False Negatives and prediction algorithms for feature set $v1$

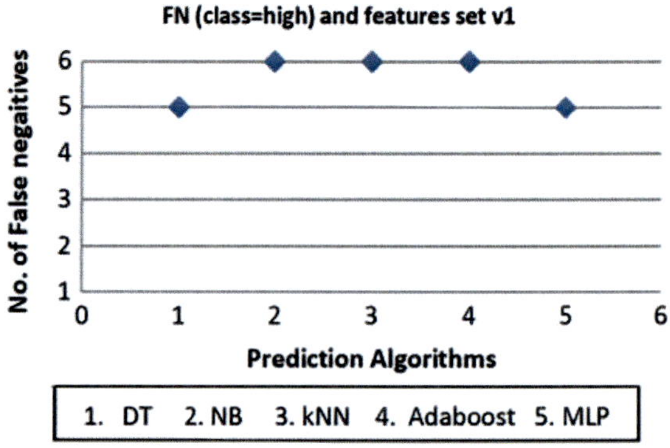

Hence, we can either choose MLP or decision tree for the purpose of classification.

The resulting decision tree model for feature set $v1$ is shown in Fig. 6.

For feature set $v2$, MLP has the lowest value of FN as compared to other algorithms as shown in Fig. 7. Hence, we choose MLP as a classifier.

The resulting model using MLP for feature set $v2$ is shown in Fig. 8.

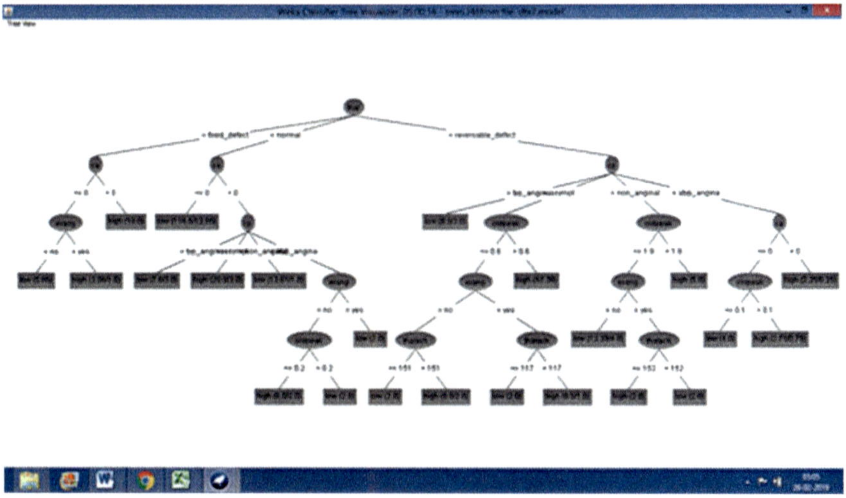

Fig. 6 Decision tree for feature set $v1$

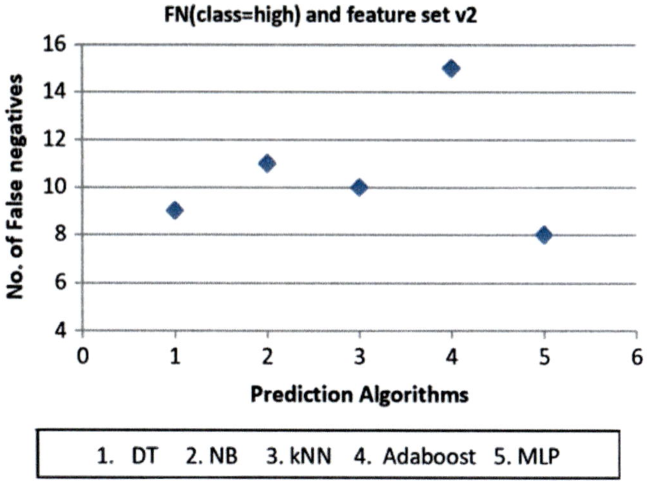

Fig. 7 False Negatives and prediction algorithms for feature set $v2$

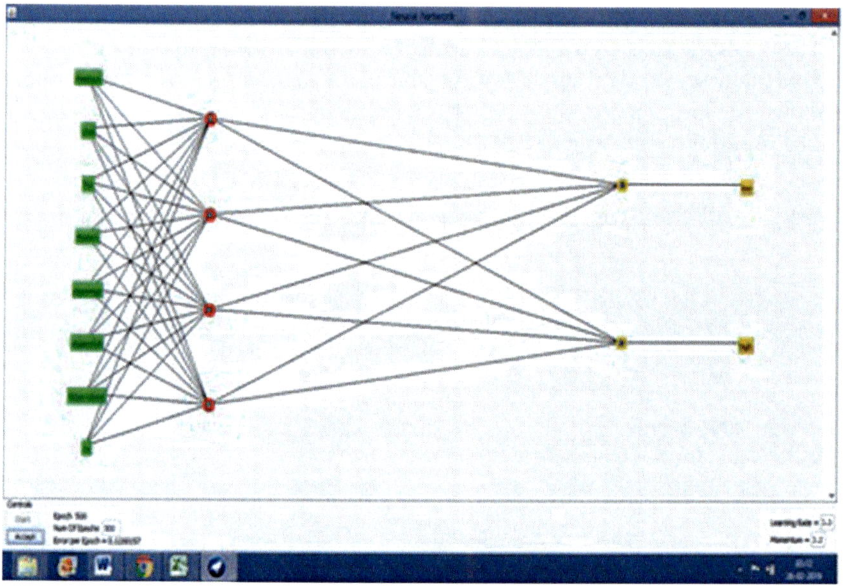

Fig. 8 MLP model for feature set *v*2

5 Conclusion

In this research, we have developed risk assessment models for patients suffering from coronary heart diseases. We have obtained feature sets from two different algorithms for the development of models. Different prediction algorithms are explored for their applicability to the selected features. Experimental results show modest accuracies in predicting the risk status targeting in the reduction of False Negatives. Future work involves further analysis and interpretation with the help of a domain expert.

References

1. Prabhakaran D, Jeemon P, Roy A (2016) Cardiovascular diseases in India: current epidemiology and future directions. Circulation 133(16):1605–1620
2. Heart disease—general info and peer reviewed studies. Online available http://www.aristoloft.com
3. https://www.who.int/cardiovascular_diseases/en/cvd_atlas_01_types.pdf
4. Joshi S, Nair MK (2015) Prediction of heart disease using classification based data mining techniques. Proc Comput Intell Data Min India 2:503–511
5. Tomar D, Agarwal S (2013) A survey of data mining approaches for healthcare. Int J Bio Sci Bio Technol 5(5):241–256
6. Obenshain MK (2004) Application of data mining techniques to healthcare data. Infect Control Hosp Epidemiol 25(8):690–695

7. Melillo P, De Luca N, Bracale M, Pecchia L (2013) Classification tree for risk assessment in patients suffering from congestive heart failure via long-term heart rate variability. IEEE J Biomed Health Inform 17(3):727–733
8. Karaolis MA, Moutiris JA, Demetra H, Constantinos S (2010) Assessment of risk factors of coronary heart events on data mining with decision tree. IEEE Trans Inf Technol Biomed 14(3):559–566
9. Tripoliti EE, Papadopoulos TG, Karanasiou GS, Naka KK, Fotiadis DI (2017) Heart failure: diagnosis, severity estimation and prediction of adverse events through machine learning techniques. Comput Struct Biotechnol J 15:26–47
10. Kamaruddin NH, Murugappan M, Omar MI (2012) Early prediction of cardiovascular diseases using ECG signal: review. IEEE Conf Res Dev
11. Shen Y, Yang Y, Parish S, Chen Z, Clarke R, Clifton DA (2016) Risk prediction for cardiovascular disease using ECG data in the China kadoorie biobank. In: Engineering in medicine and biology society (EMBC), pp 2419–2422
12. Miao F, Cai YP, Zhang YX, Fan XM, Li Y (2018) Predictive modeling of hospital mortality for patients with heart failure by using an improved random survival forest. IEEE Access
13. Karaolis MA, Moutiris JA, Hadjipanayi D, Pattichis CS (2010) Assessment of risk factors of coronary heart events on data mining with decision tree. IEEE Trans Inf Technol BioMedicine 14(3):559–566
14. De Mántaras RL (1991) A distance-based attribute selection measure for decision tree induction. Mach Learn 6(1):81–92
15. Emmanouilidis C, Hunter A, Macintyre J, Cox C (2001) A multi-objective genetic algorithm approach to feature selection in neural and fuzzy modeling. Evol Optim 3(1):1–26
16. You W, Yang Z, Yuan M, Ji G (2014) Totalpls: local dimension reduction for multicategory microarray data. IEEE Trans Hum Mach Syst 44(1):125–138
17. Chakraborty D, Maulik U (2014) Identifying cancer biomarkers from microarray data using feature selection and semisupervised learning. IEEE J Transl Eng Health Med 2:1–11
18. Liang J, Wang F, Dang C (2014) A group incremental approach to feature selection applying rough set technique. IEEE Trans Knowl Data Eng 26(2):294
19. Qiu C (2018) Bare bones particle swarm optimization with adaptive chaotic jump for feature selection in classification. Int J Comput Intell Syst 11(1):1–14
20. UCI machine learning repository. http://archive.ics.uci.edu/ml/. Accessed 17 Oct 2017
21. Han J, Kamber M, Pei J (2014) Data mining: concepts and techniques. 3rd edn. Morgan Kaufmann, Burlington
22. Tan P-N, Steinbach M, Kumar V (2005) Introduction to data mining. 4th edn. Addison Wesley, Boston
23. Cios KJ, William Moore G (2002) Uniqueness of medical data mining. J Artif Intell Med 26(1):1–24

Toward Artificial Social Intelligence: A Semi-supervised, Split Decoder Approach to EQ in a Conversational Agent

Shruthi Shankar, V. Sruthi, Vibha Satyanarayana and Bhaskarjyoti Das

Abstract This paper explores the idea of making text-based conversational agents more human-like by engineering emotional capabilities in them. The emotional quotient is still missing in the current generation of conversational chatbots. Since this is a broad problem, this research focuses on a key building block of such agents, i.e., ability to respond with a particular emotion given an input text embedded with a certain emotion. This work achieves this research goal by using a few innovative strategies, i.e., a layering of the system architecture, semi-supervised learning to workaround lack of labeled data, an innovative agent model based on a classical decision-maker coupled with a deep learning-based split decoder architecture, greedy training of the individual components of the model, an innovative approach for model evaluation based on emotion intensity, and finally, selection of final weights of the components of the model based on experimental results measured in terms of perplexity. The results obtained are extremely promising.

Keywords Artificial social intelligence · Conversational agent · Encoder–decoder · Split decoder · Greedy training · Semi-supervised learning · Emotional intelligence

---

S. Shankar · V. Sruthi · V. Satyanarayana · B. Das (✉)
Department of Computer Science, PES University, Bangalore, India
e-mail: bhaskarjyoti01@gmail.com

S. Shankar
e-mail: shruthi.shankar2512@gmail.com

V. Sruthi
e-mail: vsruthi98@gmail.com

V. Satyanarayana
e-mail: vib.satya@gmail.com

N. N. Chiplunkar and T. Fukao (eds.), *Advances in Artificial Intelligence and Data Engineering*, Advances in Intelligent Systems and Computing 1133,
https://doi.org/10.1007/978-981-15-3514-7_21

1 Introduction

It is clear that AI systems' ability to perform certain tasks, like classification and prediction, has surpassed that of humans. This category of AI, known as Weak AI, has made its way into modern day-to-day life. It consists of tasks that are extremely specific and are well-defined in a narrow domain. But in order for AI systems to really integrate into the real world and exist alongside humans as well as other systems without conflict, there is a pressing necessity to move toward "Strong/General AI," i.e., AI that is at least as intelligent as human and is apparently self-aware. It needs to exhibit socioculturally appropriate and acceptable behavior. These aspects are vague and are not determined by any specific set of rules. They cannot be taught easily in a straightforward manner even to a human, let alone a machine, as they vary with different geographical places, contexts, and cultures. It is the responsibility of humans to teach these AI systems so that they behave in an acceptable manner. This phenomenon of AI systems with the capabilities of sociocultural adeptness and behavioral intelligence is referred to as "artificial social intelligence." This is a multi-disciplinary field that requires inputs from linguists, psychologists, researchers as well as programmers.

Artificial social intelligence (ASI) is an umbrella term that includes a variety of features/characteristics. Today, the main factor that differentiates a human from a robot is the emotional quotient (EQ). According to the four branch models [1], emotional intelligence involves four aspects:

1. Perceiving emotions
2. Reasoning with emotion
3. Understanding emotion
4. Managing emotion.

Integrating these aspects into an AI system is not an easy task. It involves understanding of complex issues like what emotions mean. It also requires understanding artificial psychology [2], what it means, and how it can be mapped to human psychology.

Since ASI is such a broad term, in order to be able to implement it and get meaningful results, it is necessary to make some choices and define the scope of the problem statement at each stage. In our work, it is narrowed down to ASI in textual conversational agents, with a focus of implementing EQ into it.

2 Related Work

The first thing to understand is what exactly constitutes the social aspect of an AI system. Once this is established, one can look at the currently available state of the art in socially aware conversational agents. This gives an idea on what systems currently exist and how much functionality they have. Then it is necessary to delve deeper into the implementation aspect and look into the different possible architectures to build a conversational agent. Finally, one has to examine how emotion is extracted, perceived, and embedded in the response delivered by such agents.

According to Dautenhahn [3], AI has only focused on isolated non-social aspects in the past and in order to enable the social integration of robots into the lives of humans, they require both domain-specific knowledge and effective human-like communication skills. For example, a "nurse robot" that must assist and aid humans has to have medical knowledge and must treat humans as individuals instead of anonymous patients.

Shum et al. [4] have presented a brief survey of state-of-the-art conversational agents, out of which the chatbot of interest to us is Xiaoice, a social bot with emotional capability. The authors argue that the Turing test is not a proper metric for evaluating the success of emotional engagement and define an evaluation metric—conversational turns per session (CPS), which is the average number of conversation turns between the chatbot and the user in a conversational session.

As far as architecture is concerned, a widely used approach to the machine translation problem is to build the "sequence-to-sequence" model by Sutskever et al. [5] which was used by Qiu et al. [6] in combination with an information retrieval model.

Another technique is to apply reinforcement learning, which is effective for neural machine translation (NMT) but difficult to train due to its instability. Wu et al. [7] have conducted a systematic study on how to train better NMT models using reinforcement learning (how to set up the reward and the baseline reward). They introduced a new method leveraging large-scale monolingual data from both source and target sides.

The use of transfer learning in this domain is shown by Mo et al. [8] where they built a personalized task-oriented dialogue system which, when compared to non-personalized systems, helped the user complete dialogue tasks better and faster by learning about the users' preferences and habits. This in turn positively impacted the conversation.

As a significant step toward integrating socially acceptable behavior by performing emotion engineering, Hu et al. [9] solved the issue of implementing a form of social aspect into the conversational agent by including tonal information. This was done by identifying the underlying tone (a set of pre-defined tones) in a message, and learning how to respond to the different types of tones through training.

Li et al. [10] addressed the challenge of consistently providing data-driven systems with the coherent "persona" needed in modeling human-like behavior. After evaluating the performance and seeing that their model performed better than the basic seq2seq architecture, the authors concluded that personas can be encoded to generate a more personalized response. However, mood and emotion were left as future work.

The main challenge in introducing emotion factor is in obtaining quality emotion-labeled data and creating the final response with grammar and expression of emotions.

In an attempt to compare different techniques for emotion detection, Hirat and Mittal [11] first focus on keyword-based techniques where text is broken into tokens and emotion words are extracted. Once the intensity of the emotion word is analyzed, the sentence is checked for negation, and finally, an emotion class is given as output.

An example of this can be seen in Hinrichs and Le [12] where the authors use the emotion lexicon EmoLex [13] and also mark negative sentence parts with a separate token.

Learning-based methods detect emotions based on previously trained classifiers. The lexical affinity method again detected emotion based on keywords, especially the probabilistic affinity for a particular emotion to arbitrary words. Hybrid methods that try to combine these first have a rule-based approach to extract semantics related to emotions, then a lexicon to extract attributes. They are combined to form emotion association rules, and classifiers are trained on those rules.

For emotion synthesis, Al Masum and Ishizuka [14] followed the OCC emotion model by Ortony et al. [15]. The OCC model defines emotions as a set of "valanced" reactions to events, agents, or objects, and according to it, these valanced reactions are necessary to differentiate between emotions and non-emotions. SenseNet [16] is also used, whose main idea is to form a network of senses from the sentences that were given as input; assign a numerical value to every lexical unit based on the corresponding lexical sense affinity; and output a sense-valence for each lexical unit after assessing the value of the sense.

Zhou et al. [17] built an emotional chatting machine, an open domain, end-to-end framework trained on a manually annotated corpus capable of recognizing emotions and giving appropriate responses. The model takes corpus of the form "post, response, emotion of response" and the decoder of the seq2seq architecture utilizes the emotional category embedding, internal memory, and external memory to generate a response.

3 Our Approach

In order to build the entire end-to-end system, there are multiple ways in which each part of the system can be built. This paper presents a layered view, which lets one think about the design in an organized manner where the layers are modularized, and a set of decisions have to made at each step to choose the appropriate technique for that particular layer. This approach is used to aid the decision-making process of choosing from a plethora of techniques at each stage.

The entire system is composed of three layers:

1. **Agent Architecture**—This layer forms the base of the entire system, where one goes through multiple processes to decide on the basic, fundamental architecture of the baseline chatbot. It involves choosing the different deep learning techniques that can be used to build a conversational bot, choosing the right frameworks to build them, finding the right datasets, choosing between sequence-to-sequence architecture and reinforcement learning-based techniques, transfer learning, and fine-tuning the model parameters.
2. **Emotion Engineering**—This layer deals with choosing from multiple ways to engineer emotion and integrate it with the base agent. This is the experimental

layer, where there are a lot of methods under research. This layer is the main focus of this paper. Emotion engineering can be done using keyword-based methods like using the NRC Lexicon, SenseNet, etc. It can also be done using methods such as topic analysis, emotion ontology, learning-based methods, using lexical resources such as WordNet to calculate emotion weights, using the OCC model, using emotion embedding.

3. **Socially Acceptable Behavior**—This layer is all about the agent behaving "appropriately." It needs to know how to respond with an appropriate emotion for that situation. This layer is about managing emotion over a conversational context. It also involves a few other additional aspects like sarcasm detection, hate speech detection, detection of foul language, and handling of controversial subjects.

4 Implementation Details

4.1 Learning Methodology

Greedy training Typically, transfer learning utilizes the learned model from one work in another work, whereas layer-wise greedy training [18] is a technique used in deep neural network. In our work, a greedy training approach is followed which can be termed transfer learning within the same project. The model was trained over multiple iterations with different datasets or configurations, and at each iteration, principle of transfer learning was applied. This way, at each iteration, the model learned a little more and built on top of the knowledge it gained previously.

Semi-supervised learning Emotion annotated dataset is hard to obtain for conversational agent, and we also had a very limited quantity of such training data available. However, there is no lack of training dataset for general-purpose conversational agent. Typically, semi-supervised learning is an approach used to workaround the lack of annotated training data. This approach does not exceed the accuracy offered by the supervised kernel but addresses the training dataset availability issue. In this work, a similar method is used to generate emotion annotated training dataset required for extensive deep learning.

4.2 Dataset and Preprocessing

The training of this model required conversational datasets that were labeled with emotion, which are not easily available. A combination of multiple datasets was used for the purpose of training this architecture. The datasets that were used are:

1. Cornell movie dataset [19]
2. A Twitter Conversations dataset

3. The subtitles of the television show “Friends” [20]
4. The EmotionLines dataset [21] (a set of 14,000 conversational lines that are classified into one of the six Ekman emotions or neutral)
5. The rDany conversation dataset [22] (which is a conversation between a human pretending to be a bot and a human).

Multiple steps were performed in order to make all the above-mentioned datasets suitable for our work.

Common phrases removal The Cornell movie dataset is the largest dataset that was used, consisting of about three lakh lines. Firstly, the dataset was cleaned. Secondly, apart from some of the regular data cleaning steps that are performed on text (e.g., removal of special symbols like @), a few additional sets of text had to be removed. This was required as some phrases like “I don’t know” and “I’m sorry” were plenty in the dataset, making it quite imbalanced. Due to this, the base agent had initially performed poorly as it almost always replied with these phrases. The model became biased, and hence, some downsampling had to be performed. About 75% of them, along with a few more common phrases were removed from the dataset, and the base model was retrained.

Removal of non-English words The other datasets, i.e., the twitter conversational dataset, the subtitles of the show Friends, and the rDany conversation dataset contained a lot of text that were not English. These datasets were cleaned by removing all languages other than English.

Generating annotated data by semi-supervised learning As explained above, due to the unavailability of labeled/annotated datasets, a semi-supervised learning approach was used. All the datasets were first annotated and labeled using an emotion classifier by Colneriĉ and Demsar [23]. These datasets, which are now emotion-labeled, are then used to train the pre-trained model (that was trained on the Cornell movie dataset).

4.3 Architectural Choices

In each layer of the above-mentioned layered architecture, an architectural choice was made.

1. The **Agent architecture** that is chosen is a sequence-to-sequence architecture by Sutskever et al. [5] that is commonly used for neural machine translation problems. It is composed of an encoder and a decoder. The input to the model is the text given by the user, which is encoded and sent into the decoder. The encoder understands the input and creates an internal representation of it in a fixed-length lower dimension, also known as the context vector. The decoder takes this context vector and outputs a corresponding sequence, one word at a time. The basic blocks of both the encoder and the decoder are recurrent neural networks composed of long short-term memory (LSTM) cells. The sentence to vector conversion and

vice versa can be done by the encoder and decoder, respectively, can be done in two ways—either by mapping words to dictionary numbers or by using Glove embeddings.
2. Three distinct choices were made in the **Emotion Engineering** layer. Apart from semi-supervised learning and transfer learning techniques mentioned in the methodology section is an **innovative split decoder** architecture. This was done by modifying the base architecture itself and integrating it into the conversational agent. This modification is explained further in the next section.
3. The third layer of **Socially Acceptable Behavior** has not been the focus of this phase of our work though one aspect of the agent responding with the appropriate emotion for the given input is also taken care of by the modified architecture in the previous layer. Additionally, some naive approaches of including other socially acceptable behavior like using appropriate emoticons in the reply and foul language detection are implemented.

4.4 Final Architecture of the Conversational Agent

In our work, an innovative "split decoder" architecture has been used in place of the conventional sequence-to-sequence model. The final architecture uses a classical machine learning approach to predict the emotion in the response given the input text and then uses a deep learning-enabled model to generate the appropriate response with embedded emotion. It consists of

- An Encoder
- A Decision-Maker
- Six Decoders, one for each emotion category.

The encoder of the encoder–decoder architecture is retained but the decoder is split into the decision-maker and the individual emotion decoder. Each emotion decoder is trained to respond only with one particular emotion. Once the encoder encodes the input text, it is passed on to the decision-maker, which uses the input emotion to make a decision on what the output emotion should be and then sends it to that particular decoder (Fig. 1).
Decision-maker The way emotional intelligence is integrated into this architecture is by enabling the agent to respond with a particular emotion. The decision-maker is the agent that decides the output emotion that the agent must respond with. This can be done by using some pre-defined hard-coded set of rules that declare what the response emotion should be given the input emotion. Another way to do this is to use a machine learning-based classifier which can learn from all the data and predict what the output emotion should be given the emotion of the input and the input text itself. The following steps were performed:

1. Due to the lack of conversational data that comes annotated with emotions, the diverse datasets were first annotated using the emotion classifier by Colneriĉ and Demsar [23].

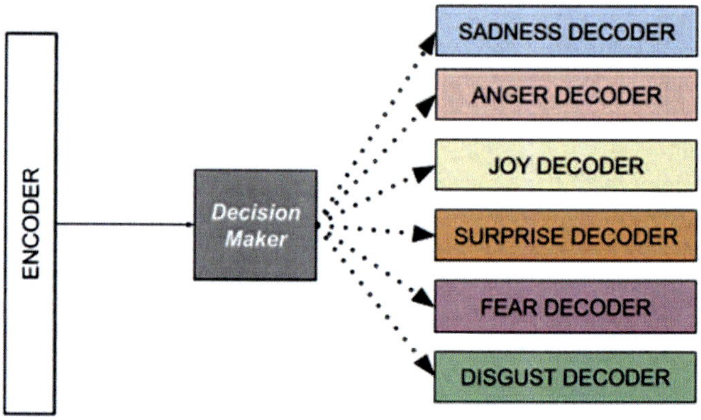

Fig. 1 Split decoder architecture

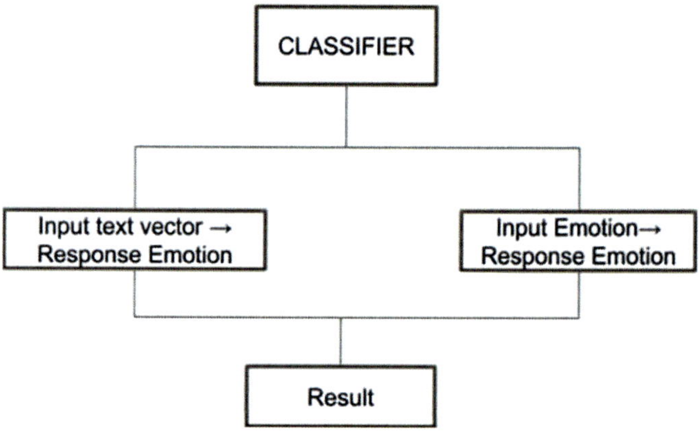

Fig. 2 Ensemble classifier

2. Two models were built, one to predict the output emotion given the input text, and another to predict output emotion given input emotion label. For this, random forest classifiers were used since tree-based models are more immune to class imbalance in the data. Since the output emotion is dependent on the input text and the input emotion, an ensemble model was created using these two, which were combined to give the final result. The accuracy obtained by this model is about 65%, which can be improved by a large margin, provided a larger set of training data is made available. The ensemble performed better than the stand-alone classifier.
   The input sentences were vectorized using Glove embedding. Again, since this is a semi-supervised learning approach, the accuracy of the decision-maker's model depends on the accuracy of the classifier used to label the datasets (Figs. 2 and 3).

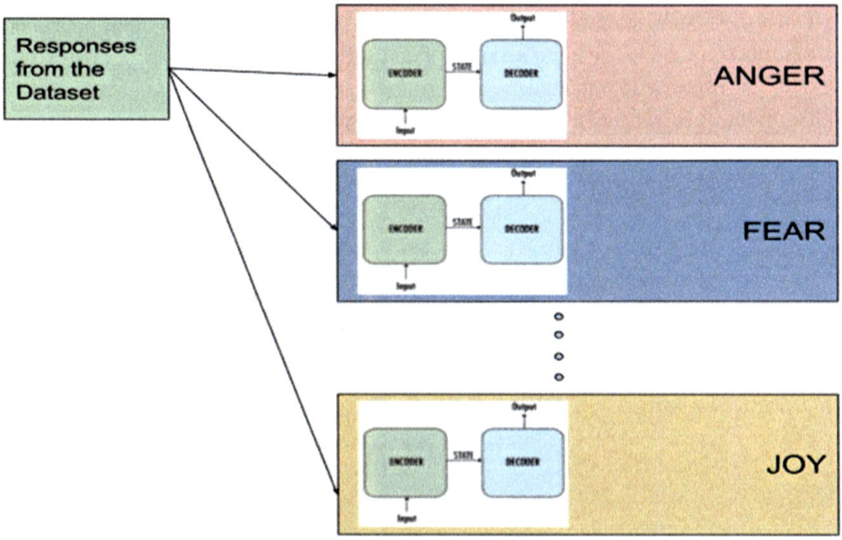

Fig. 3 Dividing the annotated dataset according to the emotion of the responses while training. These divided responses are then used to train each encoder–decoder pair

Greedy training of the split decoder As mentioned previously, this system was trained over multiple iterations where in each iteration, transfer learning was applied on the trained models of the previous iteration, ensuring that something new was learned each time. The following steps were performed:

1. The initial crude model was built and trained on the Cornell movie dataset alone, which had no emotional engineering integrated into it at all, by training it over a standard single encoder–decoder architecture.
2. This model was used as a base and was further trained with emotions. Six different models were trained, one for each emotion. In order to do this, the datasets were segregated based on the response emotion. This way, the encoder of each model can accept sentences of any emotion, but the decoder will be trained on only one particular emotion.
3. The weights of the decoder from each of these models were taken and used in the proposed architecture. It was then retrained separately on the combined dataset to ensure that the encoder learns a larger vocabulary of all emotions.

Final Pipeline The final pipeline can be summarized as:

1. It begins with classifying the input text into one of the six categories of Ekman emotions—joy, anger, sadness, fear, surprise, and neutral.
2. Then both the text and its emotion are fed to a model that will decide what the emotion of the response should be. This was initially done in a rule-based manner and was later changed to use a prediction model.

3. Once the response emotion is decided on, the decoder trained for that particular emotion is picked. The input text is encoded and then sent to this particular decoder and the response is obtained.
4. As mentioned in the subsequent evaluation section, the weights chosen for the final conversational bot are decided based on experimental results, i.e., the weights that delivered lower perplexity values were selected (Figs. 4 and 5).

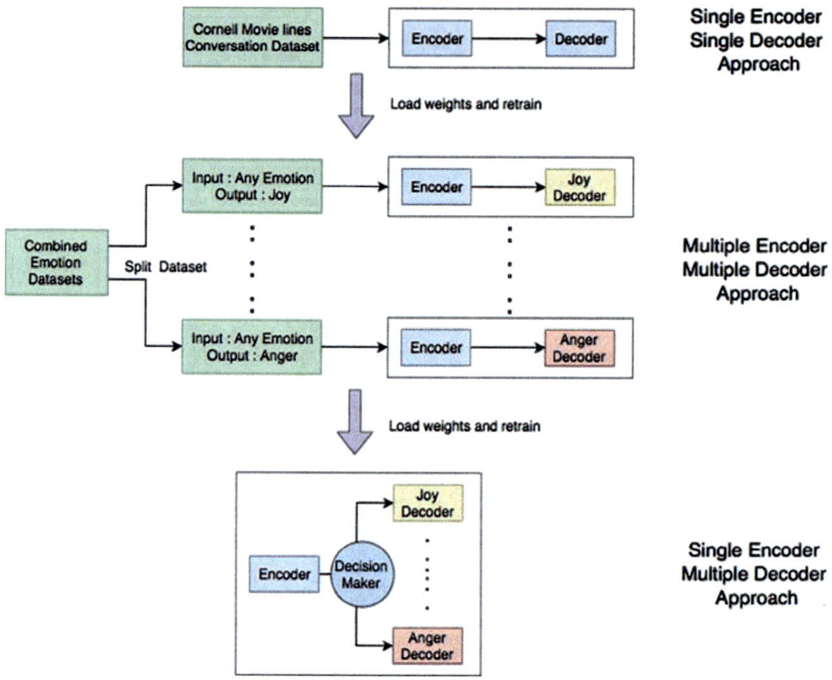

Fig. 4 Greedy training over multiple iterations

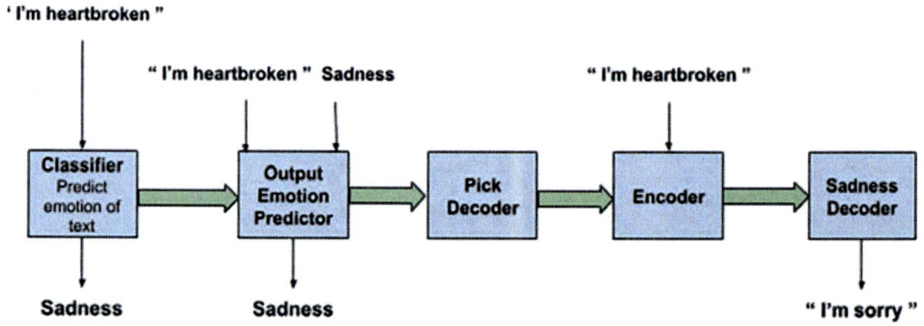

Fig. 5 Entire pipeline for the sample input text "I'm heartbroken"

4.5 Assumptions

Since an external classifier [23] was used in order to get more emotion annotated data, making use of a semi-supervised learning approach, the biggest assumption made is that this is the ground truth. All further training is done using this data. The performance of the decision-maker also heavily relies on this. Hence, any error on its part will be propagated along.

One of the datasets, the rDany conversation set that was obtained from Kaggle is a dataset of conversations between a human and a human pretending to be a robot. Since this dataset was also used to train the agent, the assumption made is that the rDany human–bot has replied appropriately to all the input text. In order to evaluate the model created, the measure used is perplexity. This compares the response of the model to the response in the dataset. Therefore, the assumption made here is that the dataset contains the correct and appropriate response to a given input text.

5 Evaluation

The evaluation of conversational agents with emotion is challenging. There is no definite metric, and for the most part, needs to be performed by humans. There are several approaches. The most widely used mathematical approaches to chatbot evaluation are perplexity and BLEU score [24]. Since most neural machine translation models are evaluated using these scores, we followed the same method.

Perplexity is the inverse probability of the test set, normalized by the number of words. The unigram model of the corpus was calculated. This gave the unigram probability of each word. Perplexity was then calculated using the formula given below:

$$\mathrm{PP}(W) = P(w_1 w_2 w_3 ... w_n) = \sqrt[n]{\frac{1}{P(w_1 w_2 w_3 ... w_n)}} \tag{1}$$

$$\mathrm{PP}(W) = \sqrt[n]{\prod_{i=1}^{n} \frac{1}{P(w_i)}} \tag{2}$$

As seen from the definition, a lower perplexity value is desired.

Another common metric used in language models is the bilingual evaluation understudy score or the BLEU score, as mentioned previously. This method is mainly used for machine translation models, but can also be used for text generation models. It is calculated by an ngram comparison of the expected text and the response text. A perfect match gives a score of 1.0 and an imperfect match results in a 0.0. Although this method is not a very accurate way to evaluate a text generation system since the generated text and the reference text provided need to have an exact match, it provides a quick and easy way to quantitatively observe the conversational agent's responses.

Both methods provide some indication of performance but are not entirely reliable since they expect responses very similar to the training data. The only reliable way to determine the bot's performance is by human judgment. In order to enhance this process and evaluate the performance of this particular agent with emotion, another novel method is used. For this, emotion intensity levels are predicted for each sentence generated. Since the dataset also had the actual response, the emotion intensity levels are predicted for that as well. These two can be compared to check how different the emotion intensities are for the actual response and the generated response. This method provides a quantifiable way to compare the emotion levels of the generated and actual text.

6 Results

In the sample conversation between the user and the agent as shown in Fig. 6, the column "Forced Emotion" signifies what the emotion was intended to be. For the actual responses generated by the agent, the emotion intensity levels are predicted to analyze how close the generated emotion is to the actual emotion. In the case of sadness, the generated and forced emotions are the same. But in the cases where the agent was expected to reply with joy, it seemed to enforce surprise. With more training with larger annotated dataset, this can improve further.

The results obtained from each version of the conversational bot was evaluated using perplexity. For a given set of input sentences, the perplexity values were calculated on the agent's response. This was done independently for each emotion. This was also done for the emotionless chatbot in order to compare versions. The weights

Speaker	Text	Forced Emotion	Actual Emotion Intensities					
			Joy	Anger	Sadness	Fear	Surprise	Disgust
User	Hello		63 %	1.3 %	1 %	7.3 %	25.6 %	0.2 %
Agent	cool	joy	29 %	5.8 %	12.7 %	12.3 %	37.1 %	3.1 %
User	how are you		29.7 %	9.4 %	9.9 %	23.4 %	19.9 %	7.7 %
Agent	i 'm good ! and yourself ?	joy	11.3 %	0.1 %	1.5 %	4.5 %	87.7 %	0.06 %
User	I'm heartbroken		1 %	0.6 %	94.3 %	2.2 %	94.3 %	0.3 %
Agent	i miss you	sadness	13.5 %	0.46 %	80.3 %	0.7 %	4.8 %	0.08 %
User	i miss you too		2.04 %	0.7 %	95.5 %	0.3 %	1.2 %	0.06 %
Agent	that sucks	sadness	13.3 %	1.4 %	74.8 %	1.4 %	8.4 %	0.4 %
User	Are you awesome?		64.1 %	1.1 %	1.7 %	2 %	30.5 %	0.4 %
Agent	i 'm hype	joy	6.1 %	0.9 %	9.8 %	14.1 %	67.9 %	0.9 %
User	Are you emotional?		9.4 %	35.2 %	29 %	21.7 %	1.6 %	2.8 %
Agent	yeah	neutral	52.8 %	1.8 %	22.6 %	4.1 %	17.3 %	1.2 %
User	Why are you emotional?		5.4 %	26.1 %	55.1 %	7.3 %	25.6 %	0.2 %
Agent	:(	sadness	2 %	0.5 %	89.3 %	4.9 %	2.9 %	0.1 %

Fig. 6 Sample conversation showing intended emotion versus actual emotion intensity in the generated response

Input	Agent's Response	Expected Response	Perplexity	BLEU
hello	cool	hi	415.4948801260333	0
how are you?	i' m good and yourself ?	im good	11.764290427828872	0.408248290463863
Im heartbroken	i miss you	im sorry	10.848009802912609	0
I miss you too	That sucks	That sucks	149.00257853223732	1
Are you emotional?	yeah	-	362.48380617687246	-
Why are you emotional?	:(	-	10.0	-

Fig. 7 Perplexity and BLEU scores of the agent's conversation

that provided lower perplexity values were selected. The final chatbot was created using these selected weights for each emotion.

Figure 7 shows that BLEU score is not a reliable metric. "-" indicates that the input sentence was not part of the dataset. As it can be seen, the agent's responses are acceptable, yet the BLEU score is low since it does not match the response in the dataset. Hence, human judgment was the only reliable way to evaluate the bot.

Another observation made was that the agent trained without emotional intelligence could not construct syntactically correct and meaningful sentences. The bot trained on a larger dataset performed poorly proving that the addition of data from different sources only added noise.

7 Conclusion

Building an EQ enabled chatbot is an unsolved research problem and has many dimensions. The current generation of conversational bots, both rule-based and corpus-based, using both information retrieval and neural architectures focus mostly on facts and conversational goals. They are not successful in implementing emotions like a true human. This work focuses on a specific building block of an EQ enabled chatbot, i.e., addressing the aspect of ensuring a response embedded with an intended emotion given an input text with a certain emotion. The results obtained so far exemplify this intended behavior of the conversational bot.

In terms of engineering an intended emotion in the response by a conversational agent, this work succeeds in making specific contributions, i.e., coming up with a layered architecture of the end to end system delivering appropriate emotional quotient in conversation, workaround the lack of emotion-labeled conversational datasets by using semi-supervised learning, successful usage of greedy training approach in deep neural networks used in conversational agents, an innovative split decoder architecture that uses a decision-maker as the forcing function to choose the appropriate emotion decoder, an innovative evaluation strategy using emotion intensity as both BLEU score and perplexity have issues in human-intensive natural language generation problems such as this.

Although this conversational bot is trained to respond with appropriate emotions, what it lacks is the ability to generate proper sentences and keep up with the conversation like a human does. Lack of modeling of conversational context is an open problem of the current generation of conversational agent. In the future, the bot needs to learn how to keep track of context in the conversation, as well as be able to generate more meaningful sentences with more training. We also intend to do some small modifications to the architecture like adding beam search decoders for incremental improvement in performance.

Furthermore, as mentioned previously, emotion engineering is not the only aspect of artificial social intelligence. Implementing other socially acceptable behavior, for example, handling of controversial subjects, detecting sarcasm, enforcing a consistent personality, etc., is also necessary in order to ensure a wholesome and pleasant experience with a conversational bot.

References

1. Mayer JD, Geher G (1996) Emotional intelligence and the identification of emotion. Intelligence 22(2):89–113
2. Crowder J, Friess S (2012) Artificial psychology: the psychology of AI. In: Proceedings of the 3rd annual international multi-conference on informatics and cybernetics, Orlando, FL
3. Dautenhahn K (1995) Getting to know each other-artificial social intelligence for autonomous robots. Robot Auton Syst 16(2–4):333–356
4. Shum H-Y, He X-d, Li D (2018) From Eliza to Xiaoice: challenges and opportunities with social chatbots. Front Inf Technol Electron Eng 19(1):10–26
5. Sutskever I, Vinyals O, Le QV (2014) Sequence to sequence learning with neural networks. In: Advances in neural information processing systems, pp 3104–3112
6. Qiu M, Li F-L, Wang S, Gao X, Chen Y, Zhao W, Chen H, Huang J, Chu W (2017) AliMe chat: a sequence to sequence and rerank based chatbot engine. In: Proceedings of the 55th annual meeting of the association for computational linguistics, vol 2: short papers, pp 498–503
7. Wu L, Tian F, Qin T, Lai J, Liu T-Y (2018) A study of reinforcement learning for neural machine translation. arXiv preprint arXiv:1808.08866
8. Mo K, Li S, Zhang Y, Li J, Yang Q (2016). Personalizing a dialogue system with transfer reinforcement learning. arXiv preprint arXiv:1610.02891
9. Hu T, Xu A, Liu Z, You Q, Guo Y, Sinha V, Luo J, Akkiraju R (2018) Touch your heart: a tone-aware chatbot for customer care on social media. In: Proceedings of the 2018 CHI conference on human factors in computing systems. ACM, p 415
10. Li J, Galley M, Brockett C, Spithourakis GP, Gao J, Dolan B (2016) A persona-based neural conversation model. arXiv preprint arXiv:1603.06155
11. Hirat R, Mittal N (2015) A survey on emotion detection techniques using text in blogposts. Int Bull Math Res 2(1):180–187
12. Hinrichs H, Le N-T (2018) Which text-mining technique would detect most accurate user frustration in chats with conversational agents?
13. Mohammad SM, Turney PD (2013) Crowdsourcing a word–emotion association lexicon. Comput Intell 29(3):436–465

14. Al Masum SM, Ishizuka M (2006) Integrating natural language understanding and a cognitive approach to textual emotion recognition for generating human-like responses
15. Ortony A, Clore G, Collins A (1988) The cognitive structure of emotions. Cambridge University Press, New York
16. Al Masum SM, Prendinger H, Ishizuka M (2007) SenseNet: a linguistic tool to visualize numerical-valence based sentiment of textual data. In: Proceedings of the international conference on natural language processing (ICON), pp 147–152
17. Zhou H, Huang M, Zhang T, Zhu X, Liu B (2017) Emotional chatting machine: emotional conversation generation with internal and external memory. arXiv preprint arXiv:1704.01074
18. Bengio Y, Lamblin P, Popovici D, Larochelle H (2007) Greedy layer-wise training of deep networks. In: Advances in neural information processing systems, pp 153–160
19. Danescu-Niculescu-Mizil C, Lee L (2011) Chameleons in imagined conversations: a new approach to understanding coordination of linguistic style in dialogs. In: Proceedings of the workshop on cognitive modeling and computational linguistics, ACL 2011
20. Zahiri SM, Choi JD (2017) Emotion detection on tv show transcripts with sequence-based convolutional neural networks. arXiv preprint arXiv:1708.04299
21. Chen S-Y, Hsu C-C, Kuo C-C, Ku L-W et al (2018) EmotionLines: an emotion corpus of multi-party conversations. arXiv preprint arXiv:1802.08379
22. Caraballo G (2017) rDany chat
23. Colneriĉ N, Demsar J (2018) Emotion recognition on twitter: comparative study and training a unison model. IEEE Trans Affect Comput
24. Novikova J, Dušek O, Curry AC, Rieser V (2017) Why we need new evaluation metrics for NLG. arXiv preprint arXiv:1707.06875

Matrix Factorization for Recommendation System

T Lekshmi Priya and Harikumar Sandhya

Abstract Recommendation systems (RS) aim at prediction of user preferences for a given set of items. Conventionally, RS uses collaborative filtering, content-based filtering, or hybrid of both the approaches for generating recommendation lists. Each of these approaches suffers from one or the other problems such as cold-start, sparsity, scalability, processing streaming data, and low latency. Furthermore, with high frequency of data updates, the user preferences change too, that demands for latest recommendation list for each user based on recent activity, capturing the concept shift and eliminating the stale item preferences. Hence, we propose a solution approach based on matrix factorization, that is robust, handles sparse and streaming data. The user preferences are represented in a matrix form and are decomposed into smaller matrices for ease of interpretation and information retrieval. Four matrix factorization techniques, namely NMF, NMFALS, CUR, and SVD, have been used for empirical analysis, and it is found that SVD outperforms NMF, NMFALS, and CUR in terms of time and recommendation accuracy.

Keywords Recommendation system (RS) · Nonnegative matrix factorization · Nonnegative matrix factorization with alternating least squares · CUR decomposition · Singular value decomposition

1 Introduction

Recommendation systems (RS) are used for information retrieval to predict the user preferences of items based on the historic preferences of the user. The main motivation behind RS is to increase customer retention and satisfaction thereby improving

T. Lekshmi Priya (✉) · H. Sandhya
Amrita School of Engineering, Amrita Vishwa Vidyapeetham, Amritapuri, India
e-mail: 4lekshmipriya@gmail.com

H. Sandhya
e-mail: sandhyaharikumar@am.amrita.edu

N. N. Chiplunkar and T. Fukao (eds.), *Advances in Artificial Intelligence and Data Engineering*, Advances in Intelligent Systems and Computing 1133,
https://doi.org/10.1007/978-981-15-3514-7_22

sales. RS has been dominantly utilized by Amazon, Netflix, LinkedIn, and many such organizations to help users to search for their relevant products, movies, music, job, etc.

For an instance, YouTube has 1.3 billion users and 300 hours of videos uploaded to YouTube every minute, 30 million viewers who watch 5 billion videos per day. The conventional RS provides less accurate predictions for voluminous, sparse, and streaming data. The other critical problem is the cold-start problem [1, 2] that occurs due to a new item or new user entering the system. For example, RS uses rating of the users on different videos that they watched. Based on this rating, RS recommends videos to the user. However, for a new video uploaded requires some initial rating for recommending to users, and similarly, a new user requires to acquaint himself to the system for a new video to be recommended to the user. This situation is a cold-start problem. Sparsity [3–5] is another problem that arises due to insufficient number of video views and user rating the videos. Streaming is the continuous data created ceaselessly by a large number of information sources. It is a vital method for information expansion. In today's era of online business, users are more interested in new items and companies are interested in personalizing user's interests thereby recommending the items to them. The challenges are how to identify these videos, how long should a recommendation list be maintained for a user, how to capture the change in user's concept, and how to change the current recommendation list eliminating the stale items. Different researches [6–9] are going on in the field to overcome these challenges. In Fig. 1, recommendation list provided by YouTube for a particular user based on his/her history is shown as an example.

There are different approaches for solving these problems such as content-based filtering (CB), collaborative filtering (CF), and hybrid approaches. In content-based filtering, based on the item profile items that are similar to an already rated item is found. Here, capturing of user preferences is difficult and may affect the privacy of the user. In collaborative filtering based on the rating given by the user, similar users

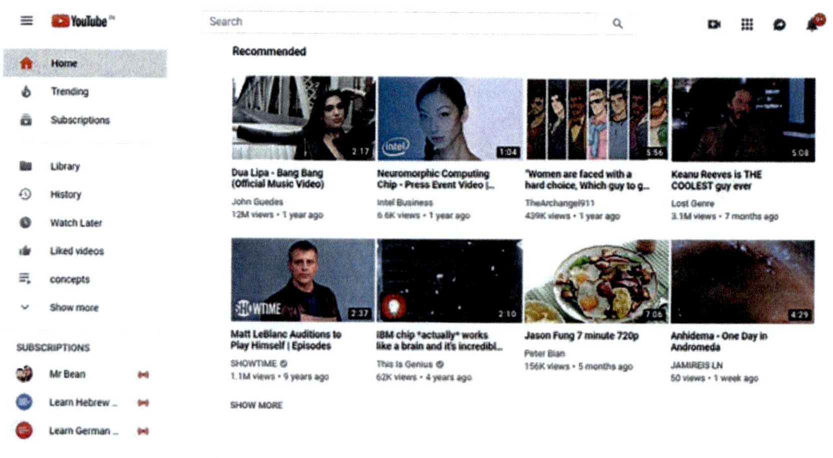

Fig. 1 YouTube recommendation list for a particular user

are found and items which are rated by a similar user are recommended for watching. This can overcome problems of content-based filtering but cold-start problems exist in these kinds of systems. In hybrid approach, a combination of this method is applied for overcoming issues in individual approaches and providing a better prediction. The recent research on RS comes up with methods like clustering, complex network analysis, and matrix factorization.

Matrix factorization (MF) is a model-based collaborative filtering approach which partition the user-item rating matrix into factors for understanding and evaluation. In matrix factorization, a high-dimensional data matrix is represented as a product of two or more low-dimensional matrix. There are different algorithms for matrix factorization. The popular MF techniques are nonnegative matrix factorization (NMF), NMF with alternating least squares (NMFALS), singular value decomposition (SVD), and CUR. In NMF [10, 11], a matrix A is factorized into two matrices $W_{m\times k}$ and $H_{k\times n}$, where $k \le \min(m, n)$ with the property that factorized matrices do not have any negative elements. In the procedure, A is factorized as W and H such that $A \approx WH$. NMFALS [12–14] is similar to that of NMF except that it uses least squares approach for convergence. In this algorithm for every iteration, it first fixes W and solves for H, and then it fixes H and solves for W. In SVD, an $m \times n$ matrix A is a factorized to the form $U\Sigma V^T$, where the matrices are $U_{m\times m}$ containing left singular vectors, $\Sigma_{m\times n}$ real diagonal matrix, and $V_{n\times n}$ containing right singular vectors of matrix A. The diagonal entries σ_i of S are known as the singular values of A. In CUR Approximation matrix is factorized to set of three matrices with the end goal that when all three multiplied together, the product closely approximate to the given matrix. This approximation is similar to that of low-rank approximation of SCD except that; Instead of taking transformed matrix of A, original columns and rows of A are chosen as factors.

In our work, we compared different matrix factorization techniques with different comparing strategies to compare the performance of these factorization techniques. Rating matrix is a sparse matrix and most of the rating value will be unknown. In our method, we propose a method for predicting the unknown rating from the factorized matrix. In our work, we use SVD for factorizing and when a user query comes; with the factor, V form V^T we were able to fill the unknown values in user query thereby recommending movies to a user.

2 Related Works

There are different approaches for solving these problems in RS. One of the approaches is content-based filtering [1] in this approach based on the item profile and find items which are similar to an already rated item. In these kinds of approaches, capturing of user preferences is difficult and may affect the privacy of the user. In collaborative filtering [2, 3] based on the rating given by user find similar users and items which are watched by the similar user can be recommended for watching. This can overcome problems of content-based filtering but cold-start problem exists in these kinds of system. In hybrid approach [4], a combination of these method is applied for overcoming issues in individual approaches and providing a better prediction. The recent research on RS come up with methods like clustering, complex network analysis, and matrix factorization. There are different problems which restrict RS from providing the best feed for users, and they are sparsity, cold-start, and streaming data. Sparsity [5] is tackled in this paper with collaborative filtering technique with different phases.

RS is used in different fields from online shopping to research paper recommendation. Bobadilla et al. [3] in their paper, they proposed a new RS for e-learning. The problem in e-learning recommendation is that it is difficult to understand whether one user is qualified than other here classifying users in a different class is not a good approach so they used a weighting value for each user. The equations used for similarity function which use memory-based CF. Also in [6, 7], they considered online shopping and feedback is used to understand the behavior of users and their sentiments were analyzed for better recommendations. Keunho Choi and Yongmoo Suh in [8] their work, they proposed different approaches for finding the similarity to improve accuracy new similarity calculation methods are used. In Wesomender [9] approach, they use both CF and CN. Content based find similarity depending on novelty, user history and trust worthiness of news. For adaptive and hybrid there is a part that assesses the current dataset and picks the best collaborative filtering algorithm which has been implemented. From this best algorithm, CF value is calculated. From the two values, recommendations are made. Here also main aim is to improve efficiency.

NMF is introduced in [10], and it uses multiplicative update rule for finding factors of the matrix. In [11], they used NMF with clustering in recommendation system for a better performance; they initially clustered users and items, and then this information is used in matrix factorization for the recommendation. The accuracy of NMF is improved by using alternating least squares; hence in [12], they used ALS in a multiplicative update of nonnegative matrix factorization. In [13], they used dimensionality reduction in hyperspectral images. In [14, 15], they used SVD for dimensionality reduction. In [16], they used linear time CUR algorithm for factorizing the matrix. In this procedure matrix, C is created by randomly sampling columns of matrix A, matrix R is created by independently sampling rows of matrix A and matrix U, and the intersection of C and R properly rescales such that $A = \text{CUR}$.

3 Methodology

In this work, we used four different techniques for matrix factorization, namely NMF, NMF with alternating least squares (NMFALS), singular value decomposition (SVD), and CUR.

Nonnegative Matrix Factorization It is an algorithm in which matrix $V_{m\times n}$ is factorized into two matrices $W_{m\times k}$ and $H_{k\times n}$ where $k \leq \min(m, n)$, with the property that all three matrices do not have negative elements in them which help to understand and evaluate the matrix.

The algorithm uses random initialization for W and H, and optimum value for W and H is not unique. There is no guarantee that we can recover the original matrix, so we will approximate it as best as we can. Now, suppose that A is composed of m rows, $a_1, a_2, a_3, \ldots, a_m$, W is composed of k rows, $w_1 w_2, \ldots, w_k$, and H is composed of m rows $h_1, h_2, \ldots, h_m$. Each row in A can be considered a data point. Frobenius norm is used for checking convergence. Formalizing this, we obtain the following objective:

$$\text{minimize} \ \parallel A - WH \parallel_f^2 \ \ \text{wrt}\ W, H\text{s.t.}\ W, H \geq 0 \tag{1}$$

A commonly used method of optimization is the multiplicative update method. In this method, W and H are each updated iteratively as follows:

$$H \leftarrow H \odot \frac{W^T A}{W^T W H} \tag{2}$$

$$W \leftarrow W \odot \frac{A H^T}{W H H^T} \tag{3}$$

The algorithm for NMF can be basically stated as follows.

Algorithm 1 Multiplicative Update rule for NMF

Input: $A \in \mathbb{R}^{\mathrm{m} \times \mathrm{n}}$, $1 \leq k \leq min(m, n)$
Output: $W \in \mathbb{R}^{\mathrm{m} \times \mathrm{k}}$, $H \in \mathbb{R}^{\mathrm{k} \times \mathrm{n}}$

Initialization: Initialize W and H with different random values, initialize $S_{threashold}$ to a small value
while $S_{threashold} \leq \parallel X - WH \parallel_f$ **do**
 $H \leftarrow H \odot \frac{W^T A}{W^T W H}$
 $W \leftarrow W \odot \frac{A H^T}{W H H^T}$
end while

NMF Alternating Least Squares Method It is similar to that of NMF ordinary least squares (OLS) are used to find optimal factors for W and H. In this, initially they fix W and optimize for H alone and repeat the process with fixing hand optimizing for W. Alternating least squares is used which is an iterative optimization process with two-step. In ALS, the cost function in each step can either decrease or stay unchanged but never increase hence provide a unique solution and guarantees minimal MSE. Alternating between the two steps guarantees reduction of the cost function, until convergence.

Algorithm 2 Basic ALS algorithm for NMF

Input: $A \in \mathbb{R}^{m \times n}$, $1 \leq k \leq min(m, n)$
Output: $W \in \mathbb{R}^{m \times k}$, $H \in \mathbb{R}^{k \times n}$

```
Initialization: W ← random(m, k), initialize ε to a small value
while ‖ A − W^(k) H^(k) ‖_f − ‖ A − W^(k+1) H^(k+1) ‖_f ≥ ε do
    H ← [W^+ A]_+
    W ← [A H^+]_+
end while
```

Singular Value Decomposition The SVD of a matrix A is the factorization of matrix $A_{m \times n}$ into 3 matrices such that $A = U \Sigma V^T$ where the columns of U and V are orthonormal and the matrix Σ is positive real-valued diagonal matrix. In SVD, an $m \times n$ matrix A is a factorized to the form $U \Sigma V^T$, where the matrices are $U_{m \times m}$ containing left singular vectors, $\Sigma_{m \times n}$ real diagonal matrix, and $V_{n \times n}$ containing right singular vectors of matrix A. The diagonal entries σ_i of S are known as the singular values of A, and also, this factorization is an example of low-rank approximation. The algorithm for SVD is as follows.

Algorithm 3 SVD algorithm

Input: $A \in \mathbb{R}^{m \times n}$
Output: $A = U \Sigma V^T$ where $U \in \mathbb{R}^{m \times n}$, $V \in \mathbb{R}^{n \times n}$, $\Sigma \in \mathbb{R}^{n \times n}$

```
Find eigenvalues of A^T A, from characteristic equation find eigenvalues λ_i
For diagonal matrix Σ calculate σ_i = √λ_i
For V, each row v_i is given by eigenvector for corresponding eigenvalue
For U, each column U_i is the eigenvectors of A A^T
return U, Σ, V^T
```

CUR Decomposition CUR matrix decomposition is the same as that of low-rank approximation of SVD. In CUR matrix, $A_{m \times n}$ is factorized to three matrices $C_{m \times K}$, $U_{k \times k}$, and $R_{k \times n}$ where $k \leq \min(m, n)$. In CUR, factorization approximations are less accurate as compared to SVD, but the advantage of CUR over SVD is no transformation of original matrix takes place. Original rows and columns are used in factorized matrix. Formally, a CUR matrix approximation of a matrix A is three matrices C, U, and R such that C and R are produced using columns of A and rows of A, respectively. Hence, approximation of rating matrix A is found. The algorithm for CUR is given below in this procedure matrix C is created by randomly sampling k columns of matrix A, matrix R is created by independently sampling k rows of matrix A, and matrix U is the pseudo-inverse of the intersection of C and R and properly rescales such that $A =$ CUR.

Algorithm 4 Linear Time CUR

Input: $A \in \mathbb{R}^{n \times d}$
Output: $C \in \mathbb{R}^{n \times c}$, $U \in \mathbb{R}^{c \times r}$, $R \in \mathbb{R}^{r \times d}$

for $t \leftarrow 1$ to c **do**
  pic $j \in \{1, \ldots d\}$ with probability $p_j = \parallel A_{:,j} \parallel^2 / \parallel A \parallel_F^2$
  Set $C_{:}, t = A_{:}, j/\sqrt{cp_j}$
end for
Set $k = min(k, rank(C^T C))$
for $t \leftarrow 1$ to r **do**
  pic $j \in \{1, \ldots n\}$ with probability $q_i = \parallel A_{i,:} \parallel^2 / \parallel A \parallel_F^2$
  Set $R_{t,:} = A_{i,:}/\sqrt{rq_i}$
  Set $\Psi_{t,:} = C_{i,:}/\sqrt{rq_i}$
end for
$U = ([C^t C]_k)^{-1\Psi^T}$
return C,U,R

3.1 Proposed Method

In our system, we compared four different techniques for matrix factorization, namely NMF, NMFALS SVD, and CUR. Our rating matrix $A_{m \times n}$ contains m users and n items and is factorized using given factorization techniques. User query $Q_{1 \times n}$ is given as the input. The query contains known rating for the user, and our point is to anticipate the unknown rating for an item. In NMF and NMFALS, the original rating matrix $A_{m \times n}$ is factorized into two thin matrices $W_{m \times k}$ and $H_{k \times n}$. To find the rating given by user$_i$ to item$_j$, we can directly multiply corresponding rows and columns of W and H.

$$R_{i,j} = \sum_k W_{i,k} * H_{k,j} \tag{4}$$

In real scenario, we need to find unknown rating, and for that, we need to find similar users as in collaborative filtering; so we multiply factorized matrices and retrieve matrix A. From this, we find the most similar user by using cosine similarity and the rating given for the particular item by similar user is chosen for unknown rating.

$$Q' = Q * V \tag{5}$$

$$Q'' = Q' * V^T \tag{6}$$

In the case of SVD and CUR, the rating matrix $A_{m \times n}$ is factorized to three matrices $U_{m \times k}$, $S_{k \times k}$, and $V_{k \times n}^T$. To find the unknown rating, we take user vector $Q_{1 \times n}$ and multiply it with $V_{n \times k}$. This new vector $Q'_{1 \times k}$ is again multiplied with $V_{k \times n}^T$ to form predicted vector $Q''_{1 \times n}$. After this operation, unknown values will be filled with predictions.

Algorithm 5 Proposed method for recommendation

Input: Rating matrix $A \in \mathbb{R}^{m \times n}$, user query $Q \in \mathbb{R}^{1 \times n}$, R_{thresh} minimum rating for recommending an item.
Output: rating updated for user $Q'' \in \mathbb{R}^{1 \times n}$

```
Factorize the rating matrix A using SVD, A = USV^T
User vector Q_{1xn} is multiplied with V_{nxk}
Q' = Q * V
New vector Q'_{1xk} is again multiplied with V^T_{kxn} to form predicted vector Q''_{1xn}
Q'' = Q' * V^T
for i ← 1 to n do
    if Q''_i ≥ R_thresh then
        Recommend the item_i to the user
    end if
end for
```

4 Experimental Analysis and Results

To store this huge data, we are using matrix factorization. There are different matrix factorization techniques available, and we had done a comparative study of different matrix factorization algorithms to find best-suited method for recommendation system.

4.1 Comparison Methods

There are different matrix factorization methods available to find the best-suited algorithm evaluations are done in different dataset. For evaluation, we used four matrix factorization algorithms: NMF, NMFALS, CUR, and SVD.

To check the accuracy of algorithms in retrieving original matrix after factorization is tested by using six datasets, namely rating dataset, Glass dataset, Diabetes dataset, Leukemia, B cell and pendigits dataset. Rating dataset is of size 80×1000, Glass dataset is of size 214×9, Diabetes data is 569×8, Leukemia dataset is 7129×29, B cell data 33×4026, and pendigits is of size 7494×17.

4.2 Performance

To compare the performance of matrix factorization in retrieving the original matrix from factorized matrix, Frobenius norm is used.

4.3 Accuracy

For this evaluation, we chose MovieLens dataset and BBC sports dataset [17, 18] and in that four different rating data of varying size, and they are of size 80×1000, 943×1682, 4613×737, and 9635×2225. For evaluation, we used 80, 20 split, 80% rows are chosen for training, and 20% rows are chosen for testing. From test data, we chose each user and deliberately select a random item whose rating is already known and we try to find the value from factorized matrices. To find that value in NMF and NMFALS we multiplied the factors W, H and formed the matrix A' and by applying cosine similarity we find the most similar user and used rating given by that user. In the case of SVD and CUR the chosen user and factorized matrix V, we find the specific rating. In this evaluation, SVD gives a better accurate prediction for unknown data and is shown in Fig. 2.

Frobenius Norm Plot of Frobenius norm versus rank is used for evaluation. In this, size of matrix kept the same as the original size of respective matrices. From the plots, it can be inferred that as rank increases closeness of factorized matrix with original matrix increases. On all the data were taken for study, NMF and CUR perform poor as compared to SVD and NMFALS. The performance of NMFALS is comparable with SVD. The corresponding plots are given in Fig. 3.

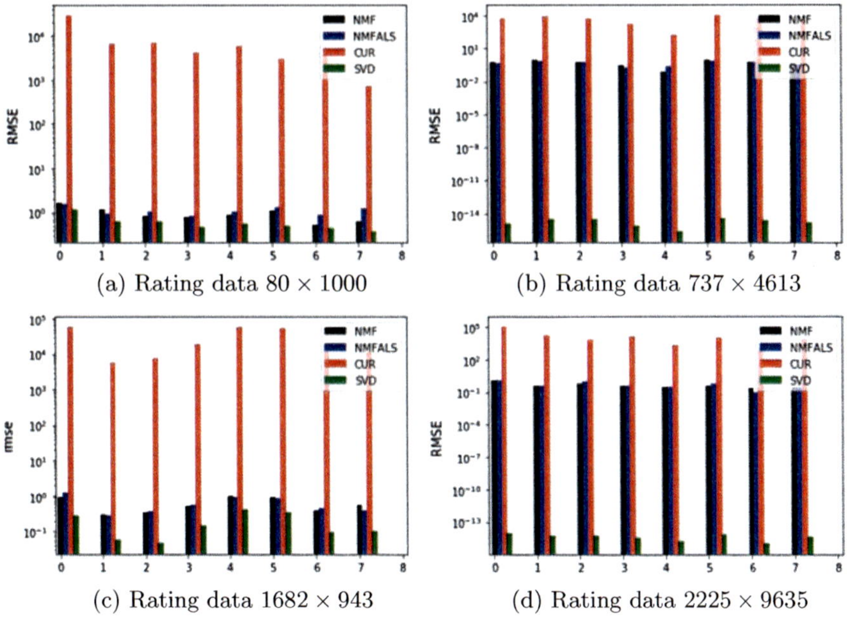

(a) Rating data 80×1000

(b) Rating data 737×4613

(c) Rating data 1682×943

(d) Rating data 2225×9635

Fig. 2 RMSE of number of users: for each user in test data evaluating the accuracy of prediction using different algorithms. X-axis gives query given by each user, and Y-axis gives log scale value of RMSE value of each query

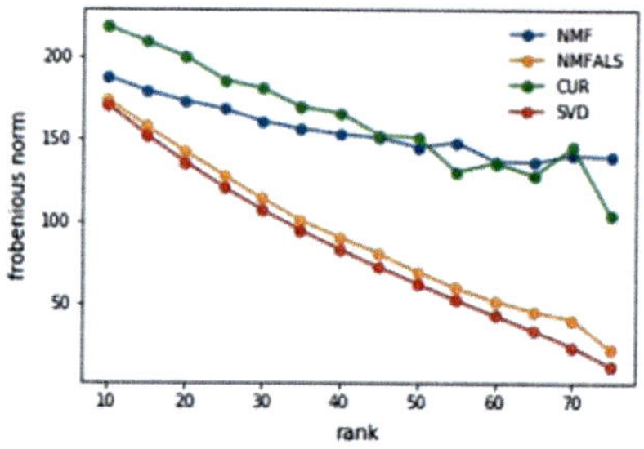

(a) Rating data

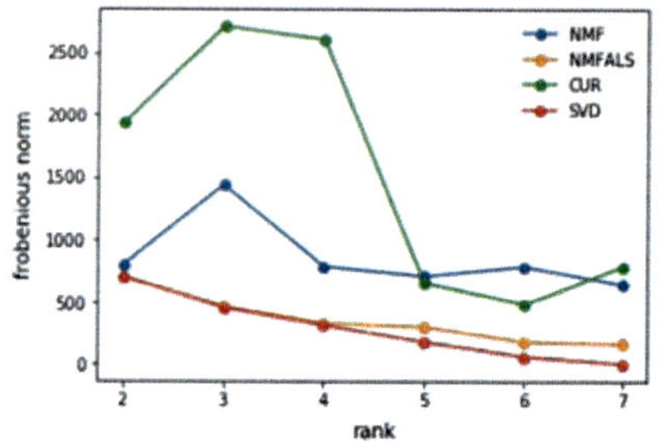

(b) Diabetes data

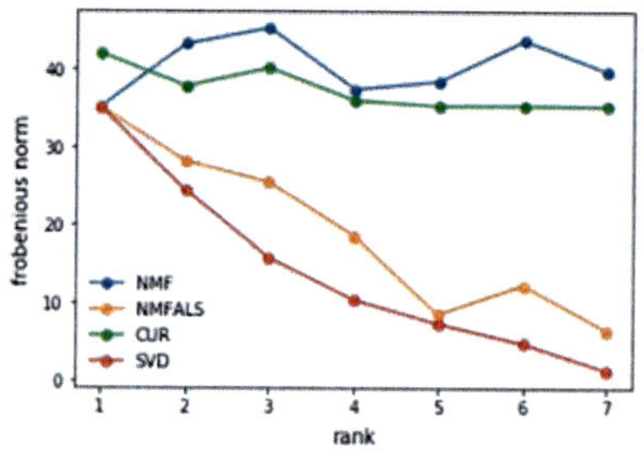

(c) Glass data

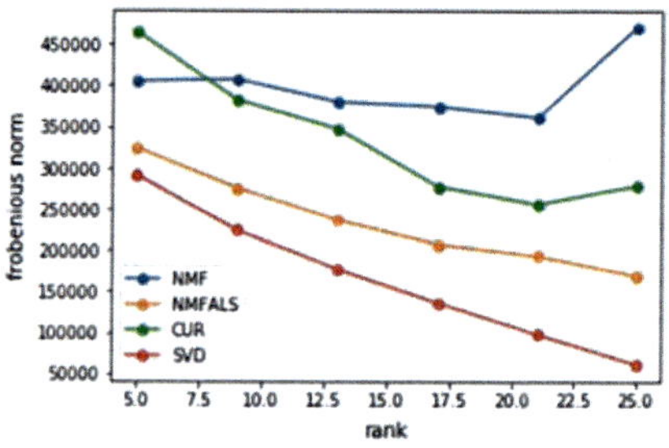

(d) Leukemia data

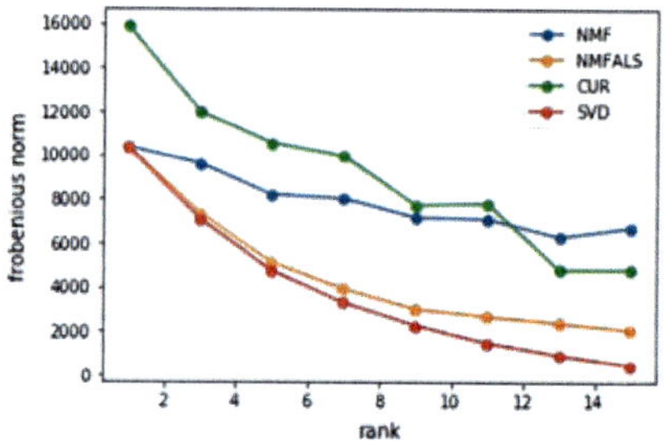

(e) Pendigits

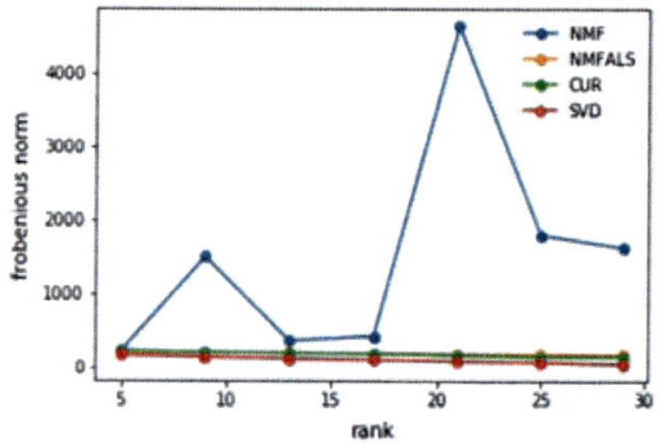

(f) B-Cell data

Fig. 3 Variation of Frobenius norm with change in rank and keeping size constant

Time In the plot Time is compared by varying the size where as rank is kept constant, there is no visible variation in time as compared to variation in rank. NMFALS shows that it is taking longer to find factorized matrices for actual matrix. All other algorithms take similar time for factorization in that; also, NMF takes slightly high time, whereas CUR and SVD are taking comparable time. The reason for NMFALS to take more time is because of time is taken for convergence. The plot for this evaluation is given in Fig. 4.

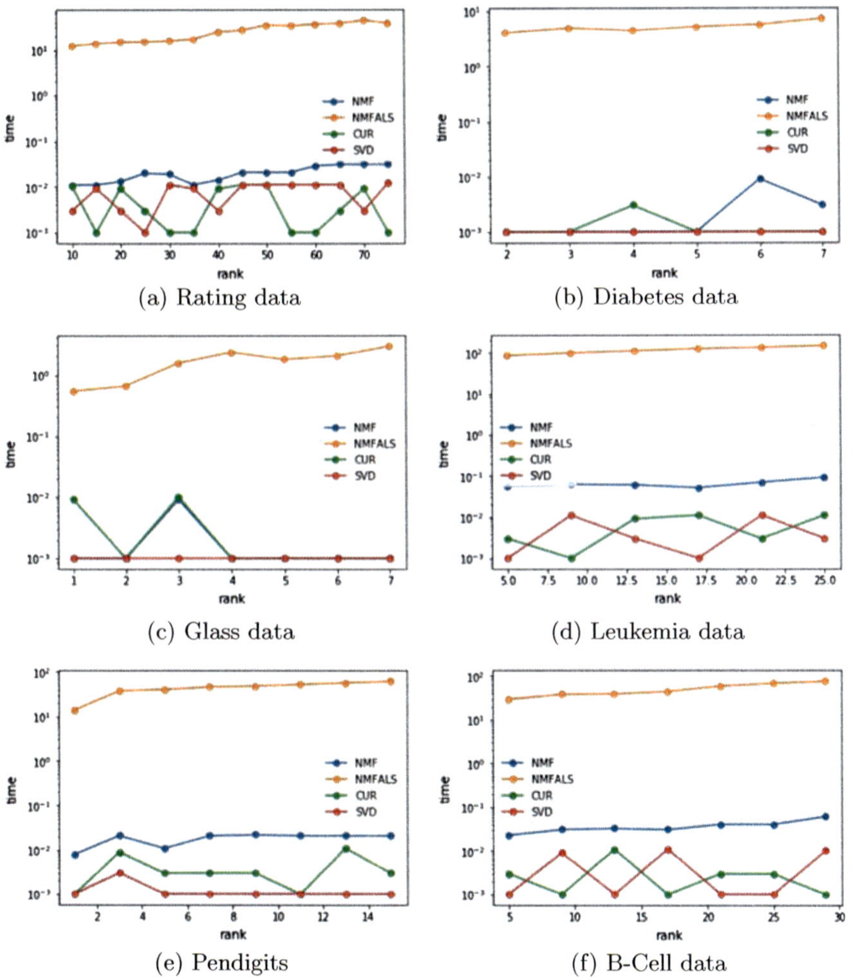

(a) Rating data (b) Diabetes data

(c) Glass data (d) Leukemia data

(e) Pendigits (f) B-Cell data

Fig. 4 Variation of time with change in rank while keeping size constant. X-axis is rank, and Y-axis is time for factorization in log scale

4.4 *Scalability*

It is used to check the performance of algorithm when the size of matrix increases. To check this, rank kept the same and we increase the number of rows in the algorithm. **Frobenius Norm** In the plot Frobenius norm is compared by varying the size where as keeping rank constant. In this plot, as size increases Frobenius norm also increases, i.e., retrieval of actual data from factorized data is not accurate, and it is shown in Fig. 5. In this, SVD and NMFALS perform well as compared to NMF and CUR. For evaluation purposes, the rank given to every algorithm is five.

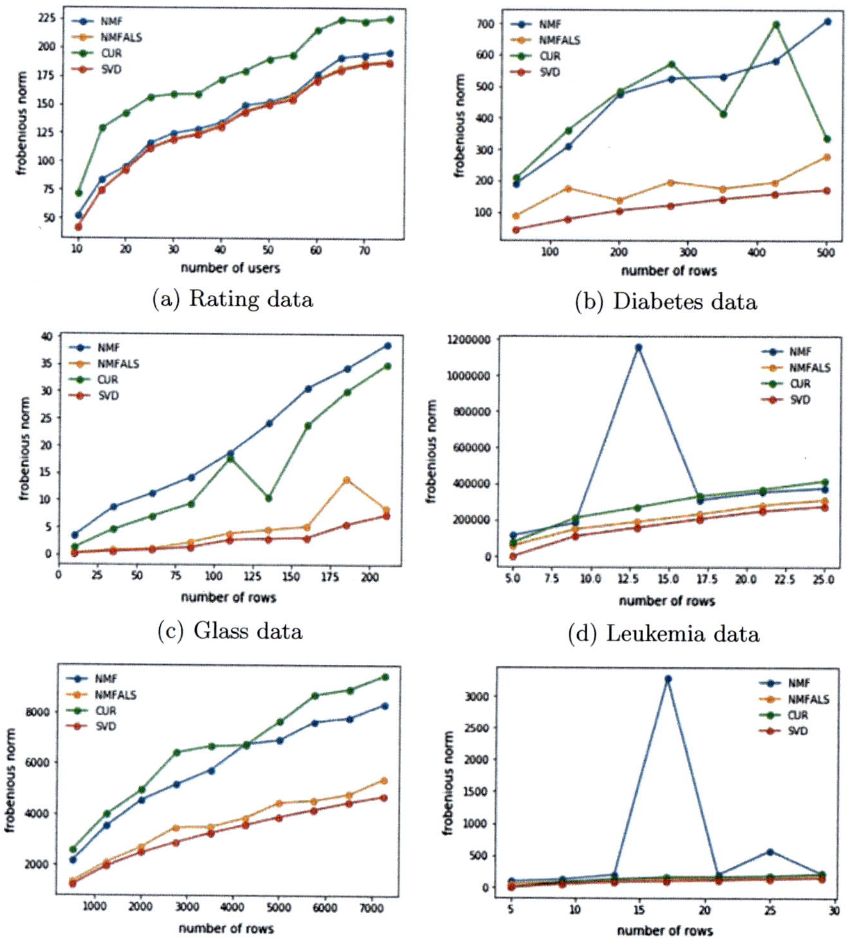

(a) Rating data (b) Diabetes data

(c) Glass data (d) Leukemia data

(e) Pendigits (f) B-Cell data

Fig. 5 Variation of Frobenius norm with change in size while keeping rank constant

Time In the plot of time versus size keeping rank constant. In this plot, as the size of the matrix there is also a small increase in time taken for factorizing. NMFALS shows that it is taking longer to find factorized matrices as compared to SVD, NMF, CUR. All other algorithms take similar time for factorization and in that time taken for SVD is less as compared to CUR and NMF, and the graph is given in the plot in Fig. 6.

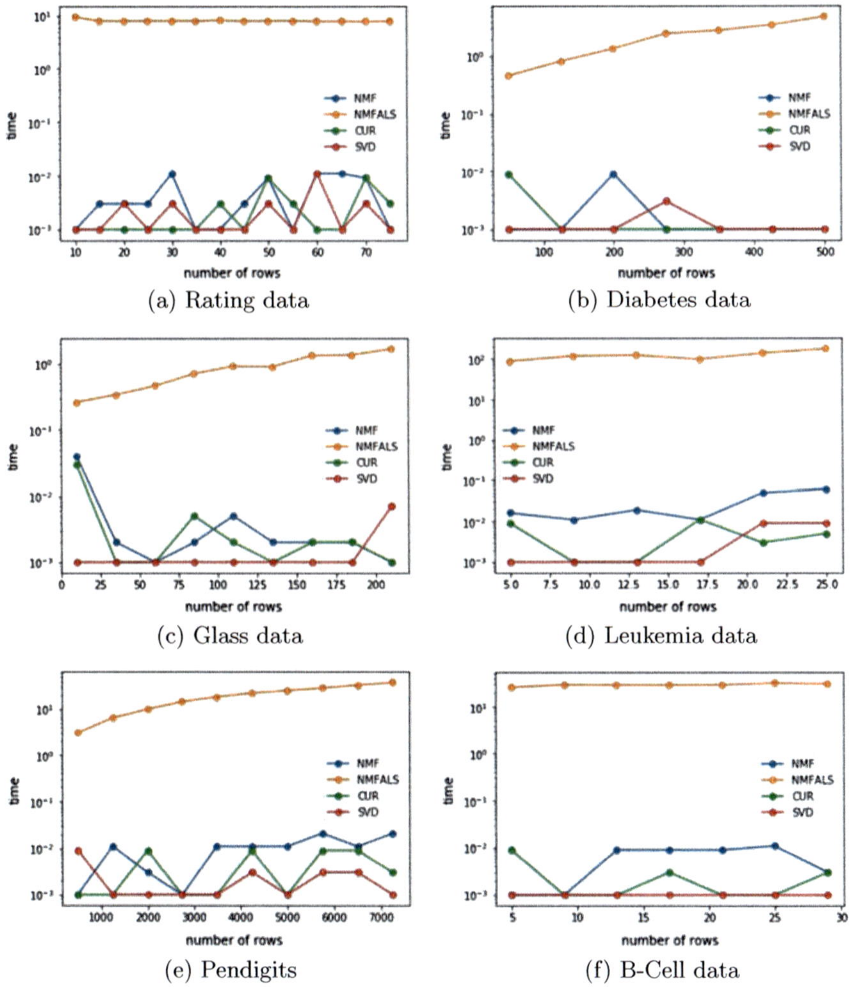

Fig. 6 Variation of time with change in size while keeping rank constant. X-axis is size of rating matrix, and Y-axis is time for factorization in log scale

5 Conclusion

In the evaluation study of matrix factorization method, we compared four different methods, namely NMF, NMFALS, CUR, and SVD. NMF takes time similar to that of SVD and CUR but it fails to retrieve actual data from factorized data and fails to retrieve accurate predictions for unknown users. In the case of CUR, time taken for factorizing is better or comparable with SVD but its performance in the prediction and retrieval of actual data is poor. NMFALS shows comparable performance in the prediction and retrieval of actual data by compromising in terms of time. So based on the evaluation, SVD outperforms in every evaluation method in different data. Hence by the conclusion, we chose to use SVD as the factorization method for the project.

References

1. Erkin Z et al (2012) Privacy-preserving content-based recommender system. ACM
2. Ahn HJ (2008) A new similarity measure for collaborative filtering to alleviate the new user cold-starting problem. Inf Sci 178(1):37–51. https://doi.org/10.1016/j.ins.2007.07.024
3. Bobadilla J, Serradilla F, Hernando A (2009) Collaborative filtering adapted to recommender systems of e-learning. Knowl-Based Syst 22(4):261–265. https://doi.org/10.1016/j.knosys.2009.01.008
4. De Campos LM et al (2010) Combining content-based and collaborative recommendations: a hybrid approach based on Bayesian networks. Int J Approx Reason 51(7):785–799
5. Al-Bakri NF, Hashim SH (2018) Reducing data sparsity in recommender systems. Al-Nahrain J Sci 21(2):138–147
6. Kumar KV, Reddy RR, Balasubramanian R, Sridhar M, Sridharan K, Venkataraman D (2015) Automated recommendation system with feedback analysis. Int J Appl Eng Res 10:22201–22210
7. Kiran MVK, Vinodhini RE, Archanaa R, Vimalkumar K (2017) User specific product recommendation and rating system by performing sentiment analysis on product reviews. In: 2017 4th international conference on advanced computing and communication systems (ICACCS)
8. Choi K, Suh Y (2013) A new similarity function for selecting neighbors for each target item in collaborative filtering. Knowl-Based Syst 37:146–153. https://doi.org/10.1016/j.knosys.2012.07.019
9. Montes-García A et al (2013) Towards a journalist-based news recommendation system: the Wesomender approach. Expert Syst Appl 40(17):6735–6741
10. Lee DD, Sebastian Seung H (2001) Algorithms for non-negative matrix factorization. Advances in neural information processing systems
11. Wang X, Zhang J (2014) Using incremental clustering technique in collaborative filtering data update. In: Proceedings of the 2014 IEEE 15th international conference on information reuse and integration (IEEE IRI 2014). IEEE
12. Liu H, Li X, Zheng X (2013) Solving non-negative matrix factorization by alternating least squares with a modified strategy. Data Min Knowl Discov 26(3):435–451. https://doi.org/10.1007/s10618-012-0265-y
13. Reshma R, Sowmya V, Soman KP (2016) Dimensionality reduction using band selection technique for kernel based hyperspectral image classification. In: 6th international conference on advances in computing and communications, ICACC-2016, Rajagiri School of Engineering and Technology, vol 93, pp 396–402
14. Pan M, Yang Y, Mi Z (2016) Research on an extended SVD recommendation algorithm based on user's neighbor model. In: 2016 7th IEEE international conference on software engineering and service science (ICSESS). IEEE. https://doi.org/10.1137/S0097539704442702
15. Kogbetliantz EG (1955) Solution of linear equations by diagonalization of coefficients matrix. Q Appl Math 13(2):123–132
16. Drineas P, Kannan R, Mahoney MW (2006) Fast Monte Carlo algorithms for matrices III: computing a compressed approximate matrix decomposition. SIAM J Comput 36(1):184–206
17. Maxwell Harper F, Konstan JA (2015) The MovieLens datasets: history and context. ACM Trans Interact Intell Syst (TiiS) 5(4):Article 19, 19pp. https://doi.org/10.1145/2827872
18. Greene D, Cunningham P (2006) Practical solutions to the problem of diagonal dominance in kernel document clustering. In: Proceedings of the 23rd international conference on Machine learning. ACM

A Reinforcement Learning Approach to Inventory Management

Apoorva Gokhale, Chirag Trasikar, Ankit Shah, Arpita Hegde, and Sowmiya Raksha Naik

Abstract This paper presents our approach for the control of a centralized distributed inventory management system using reinforcement learning (RL). We propose the application of policy-based reinforcement learning algorithms to tackle this problem in an effective manner. We have formulated the problem as a Markov decision process (MDP) and have created an environment that keeps track of multiple products across multiple warehouses returning a reward signal that directly corresponds to the total revenue across all warehouses at every time step. In this environment, we have applied various policy-based reinforcement learning algorithms such as Advantage Actor-Critic, Trust Region Policy Optimization and Proximal Policy Optimization to decide the amount of each product to be stocked in every warehouse. The performance of these algorithms in maximizing average revenue over time has been evaluated considering various statistical distributions from which we sample demand per time step per episode of training. We also compare these approaches to an existing approach involving a fixed replenishment scheme. In conclusion, we elaborate upon the results of our evaluation and the scope for future work on the topic.

Keywords Reinforcement learning · Inventory management · Markov decision process · Policy gradients · Agent based systems

A. Gokhale (✉) · C. Trasikar · A. Shah · A. Hegde · S. R. Naik
CE & IT Department, Veermata Jijabai Technological Institute, Mumbai, India
e-mail: apgokhale_b15@it.vjti.ac.in

C. Trasikar
e-mail: cstrasikar_b15@it.vjti.ac.in

A. Shah
e-mail: acshah_b15@it.vjti.ac.in

A. Hegde
e-mail: ahegde_b15@it.vjti.ac.in

N. N. Chiplunkar and T. Fukao (eds.), *Advances in Artificial Intelligence and Data Engineering*, Advances in Intelligent Systems and Computing 1133,
https://doi.org/10.1007/978-981-15-3514-7_23

1 Introduction

The ever increasing demand among people for a luxurious lifestyle has resulted in a corresponding increase in the variety and number of products available in the market. The improper management of such large inventories of goods leads to greater costs for producers and retailers causing cost inefficiency which gets passed on to the consumers, as a consequence of which the purchasing power of people is marginalized. Realizing the importance of curbing this inefficiency due to poor management of inventory, we decided to address the problem of inventory management.

Due to the high variability of customer demands, traditional statistical methods are rendered ineffective. We require modern approaches such as supervised machine learning or reinforcement learning [1] to tackle the problem of inventory management, which have shown greater adaptability and generalization power. Reinforcement learning, specifically, is independent of explicit labels, hence requiring little or no prior expert knowledge. Moreover, reinforcement learning is only bounded by the objective function and is hence free of any labelling bias and associated bounds to the maximum achievable returns.

1.1 The Inventory Management Problem

Most companies have warehouses distributed across their region of operations. These warehouses are dedicated to the storage of raw materials or finished goods, to supply to local manufacturing units or retailers, respectively. The amount of each of the goods stored in a warehouse should be sufficient in order to meet the demands for those goods. At the same time, storing goods in the warehouses also has associated costs, such as rent for real estate and any cost involved in keeping the goods' value from degrading. Due to shortening of product life cycles and the rapid rise in production and consumption levels, the inventory management problem is virtually omnipresent in modern industry, and it is becoming increasingly necessary to tackle uncertainties in demand and fulfilment from the consumer and by the supplier, respectively, while making decisions. Raw material and finished products form the core of the business, and a shortage of inventory is hence extremely detrimental since it may lead to unfulfilled customer demand. At the same time, a large inventory becomes a liability that has a heightened risk of damage, spoilage or even theft. It is also highly susceptible to shifts in demand. If not sold in time, surplus inventory may have to be disposed of at clearance prices or discarded, leading to a loss of revenue for the manufacturer. Thus, it is important to know when to restock the inventory, the amount of stock to purchase, at what price to sell and when.

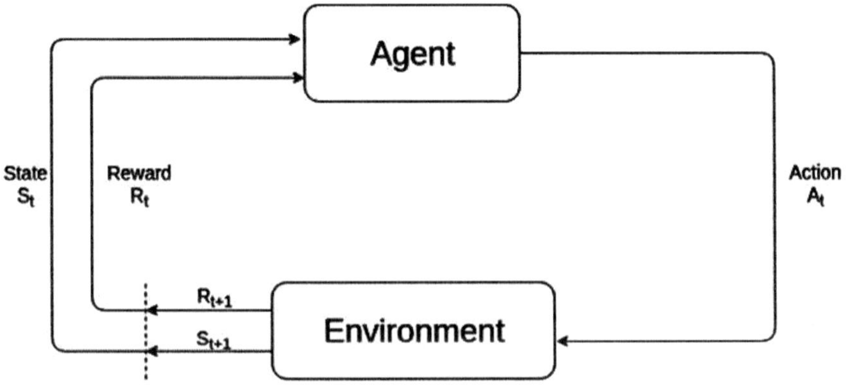

Fig. 1 RL framework (Sutton, Barto) [1]

1.2 Reinforcement Learning

Reinforcement learning [1] (RL) involves an agent (the algorithm) that acts upon the environment and receives feedback from it in the form of a reward signal and corresponding change in the state of the environment. In order to apply reinforcement learning algorithms on it, a problem needs to be modelled as a Markov decision process (MDP). Reinforcement learning algorithms are capable of optimizing an objective through tuning parameters that lead to effective decisions, thereby maximizing an expected reward signal. No prior expert knowledge about the task at hand is provided to the algorithm (Fig. 1).

Reinforcement learning algorithms can have either tabular or approximation approaches. When the state space of the environment is very large and/or continuous, tabular methods are not feasible. Approximation approaches are preferred in such scenarios. Further, the problem of continuous and/or large action spaces can be tackled by making use of policy-based approaches that directly optimize the decisions to be made instead of computing an intermediate value function for a state or state-action pair. We hence model the inventory management problem as a Markov decision process and use various policy gradient-based reinforcement learning algorithms to optimize it, as described in Sect. 3.

2 Literature Review

Stockheim et al. [2] present a reinforcement learning approach for the problem of decentralized supply chain management, involving multiple independent scheduling agents learning an optimal strategy for acceptance of jobs offered in order to maximize yield. The optimizer schedules the requested jobs, which are arriving in a non-deterministic manner from the customers, into a production queue if the job yields sufficient returns. The reinforcement learning agent accepts offers according to the job price, the delivery due date, the penalty cost associated in case of timeout and the output of the scheduling component.

In [3], the problem of optimizing profit for a single perishable product in an inventory is solved using Q-Learning and SARSA. Various other parameters, e.g. lead time, product lifetime and demand variance are considered. It concludes that SARSA performs better in terms of cost performance compared to the other algorithms.

An estimation is made to find the reward of actions not taken (rumination) called RSARSA [4]. It encourages exploration but achieves poor long-term learning quality. Other variants of the algorithm consider the probability of future action to be taken.

Supply chain management (SCM) is a superset of inventory management involving the coordination among various functional units such as manufacturing, inventory and retail. A Semi Markov Average Reward Technique (SMART) technique is used to model the system where the rewards may, or may not be obtained at each step due to lead time for transportation of goods [5].

In [6], Jia Yuan Yu has taken a backward induction-based approach to revenue maximization. The reward function allowed for partial sales, i.e. even in case the demand is not completely fulfilled, the goods stocked in the inventory are all sold to partially satisfy the demand, and the profit made being capped to the number of goods sold. The reward at the 'final' state has been taken to be the value after selling the goods remaining at salvage price. The value of the prior states is then derived by applying the Bellman equation assuming fixed transition probabilities. Only a single product in the inventory has been considered.

C. O. Kim et al. have performed demand forecasting in [7], with both centralized and decentralized approaches. The decision that the agent has to take is setting the value of the factor that determines the 'Safety Lead Time' which is an early estimate of after how long the inventory will be exhausted, and the 'Safety Factor' which determines the amount of surplus stock to store in the inventory, as a forecast of upcoming demand for the single product under consideration.

In [8], modelling for an inventory management environment as a Semi Markov Decision Process (SMDP) has been proposed. They use the Semi Markov Average Reward Technique (SMART) to maximize the returns from the environment. The average reward or gain is as follows: $P = (\text{Total Reward})/(\text{Total Time})$, where the total reward is formulated as the difference between the total selling price and the total cost price. The total selling price can be calculated from the number of products sold in total episode time. The cost price can be calculated in a similar manner. The system state variable is given by the vector $(\text{IP}_1, \text{IP}_2, \text{IP}_3)$, which describes the Global Supply Chain Inventory Position as the Inventory Position IP_i at every stage i. The Inventory Position IP_i at a given stage depends on Schedule Receipts (SR_i), On hand Inventory (OH_i) and back-orders (BO_i) as $\text{IP}_i = \text{OH}_i + \text{SR}_i - \text{BO}_i$. Each must take an action that ranges from ordering nothing up to a maximum amount equal to the stock point capacity plus the current back-order plus the estimated consumption during the transportation lead time minus the stock on hand.

In [9], Oroojlooyjadid et al. apply deep learning to the newsvendor problem. The newsvendor problem assumes that the agent purchases the goods at the beginning of a time period and sells them during the period. If the agent runs out of goods, then

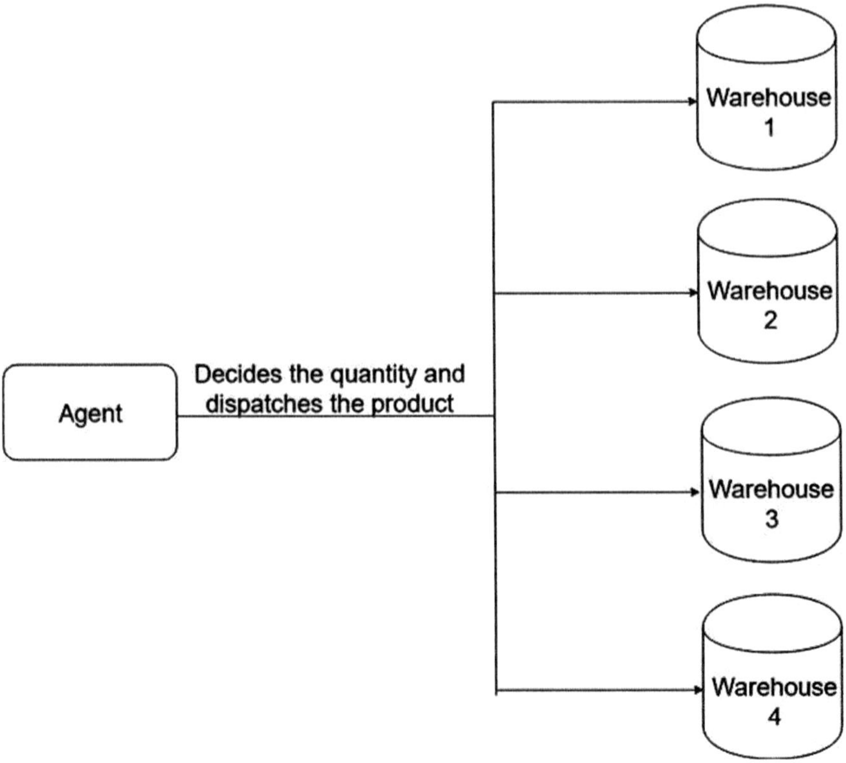

Fig. 2 Interaction of agent with the environment

a shortage cost is incurred by it. Whereas, if the agent is left with excess goods at the end of a day, then it has to discard the excess goods, incurring a holding cost in the process. Therefore, the agent is motivated to order an optimal amount of goods to minimize the aforementioned costs which is given by

$$\min_{y} C(y) = E_d[c_p(d - y)^+ + c_h(y - d)^+]$$

where d is the demand generated randomly, y is the quantity ordered, c_p and c_h are the shortage and holding costs per unit, respectively, and $(a)^+ := \max\{0, a\}$. We use a modified version of this function as our reward function.

In [10], the authors consider the problem of supply chain management as a multi-arm non-contextual bandit problem and apply a policy gradient approach to find a robust policy. Using historical data, the authors build a simulator to demonstrate their approach. We use a similar simulator to demonstrate our approach which samples demands from a variety of statistical distributions instead of generating demands from historical data, and we consider the inventory management problem to be a full reinforcement learning problem instead of a non-contextual bandit problem.

3 Our Approach

3.1 The Inventory Environment as an MDP

We have developed the environment as that of an inventory, where there are N warehouses at different locations, each capable of storing M products. There is a central authority, our agent, who is responsible for deciding how much quantity of each product is to be sent to each warehouse for storing, based on the demand for that product in the warehouse's locality. Our environment represents the inventory status and goods transactions across all the warehouses. We have modelled it as a Markov decision process, and hence, have accordingly defined the state, action, reward and transition representations. The environment is implemented such that it provides an interface like those in OpenAI Gym [11]. Contrary to the existing approaches in the literature, our environment fully considers multiple products across warehouses. The state is represented as a 2-D matrix in which the (i, j)th element represents the amount of the jth product in ith warehouse. The amount of product can take values from zero up to the inventory limit for the warehouse for that product (Fig. 2).

The action represents the order given to replenish stock. It is a matrix of dimensions similar to the state matrix, but here the (i, j)th element represents the amount of the ith product ordered for storing at the jth warehouse. This representation results in an enormous state and action space. The time interval between the time of placing the order and the time when the order arrives at the warehouse is the lead time and is set to 10 days.

At every time step, the demand for every product at every warehouse is deducted automatically if it can be completely fulfilled, and the corresponding sales are added to the revenue. If the demand is not completely fulfilled, it is counted as a back-order, and no deduction from the inventory takes place for that product at that warehouse. Thus, the transition from one state of the inventory to another is caused by the deduction of the fulfilled demands from the previous state, and addition of the orders in the action taken Lead Time time steps ago. Also, there is an upper limit to the amount of any particular product that can be stored at a warehouse. Any quantity above that will not be stored in that warehouse, but the ordering cost of that amount will be included in the reward calculation.

The reward is modelled to directly reflect the total revenue generated due to the transactions taking place in all the warehouses. Our reward definition incorporates the following costs and incomes:
The income from successful sales:

$$\sum(\text{selling\_price} * \text{demand\_satisfied}) \tag{1}$$

where

$$\text{selling\_price} = (1 + \text{profit\_factor}) * \text{cost\_price} \tag{2}$$

The cost of holding inventory in the warehouses:

$$\sum (\text{holding\_cost} * \text{quantity}) \tag{3}$$

The cost of ordering the goods at the current time step:

$$\sum (\text{cost\_price} * \text{quantity\_ordered}) \tag{4}$$

The back-order cost, to account for lost sales due to unfulfilled demands:

$$\sum (\text{cost\_price} * \text{demand\_not\_satisfied}) \tag{5}$$

Thus, the complete reward formula is as follows:

$$\text{Reward}\, R(t) = (1) - (3) - (4) - (5) \tag{6}$$

We have also normalized the reward value returned by dividing it by the product of the number of warehouses and the number of products per warehouse. This is done to remove the large variability of the reward caused due to changing the number of warehouses and products.

We have modelled different statistical distributions (Poisson, normal and Gamma distributions) to generate consumer demands. The demand for every product at every warehouse is assumed to follow a particular probability distribution function and is sampled from that distribution at every time step.

3.2 *The Algorithms Implemented*

We have evaluated the performance of three policy-based algorithms to solve our inventory management environment, with the objective of maximizing the reward over a fixed number of time steps and compared the results obtained from each of them. We have also evaluated the performance of a naive, heuristic-based approach, which is one of the existing approaches to tackle this problem. The large state space makes it infeasible to apply tabular reinforcement learning methods to solve our environment. Moreover, the value function approximation algorithms such as Deep Q-Network [12] and Least-Squares Policy Iteration [13] cannot be applied here owing to the large, continuous action space. Hence, we have applied policy-based algorithms, namely Actor-Critic [14], Trust Region Policy Optimization [15] and Proximal Policy Optimization [16].

Since all of the algorithms evaluated are policy gradient methods that follow the Actor-Critic structure, they involve two function approximators: the policy network or actor, which takes the state descriptor as input and outputs an action, and the value network or critic, which takes the state descriptor as input and outputs the value, or the ‘goodness’ of the state.

The action space that we have modelled is a matrix of integer values. So, we have implemented the actor such that it outputs a mean and standard deviation for each product–warehouse pair. Each element of the action matrix, signifying the amount of each product to be ordered for each warehouse, is sampled from a normal distribution fit to the corresponding mean and standard deviation as output by the actor. Each mean is computed by passing the state through a deep neural network that has two hidden layers having 64 rectified linear units each. Each standard deviation is computed separately from the mean, as a variable that is directly optimized by the gradient of the actor loss with respect to it.

The critic, or the value network, is also a neural network having a single hidden layer with 32 rectified linear units. It takes in the state matrix as input, and outputs a single value.

We have used this architecture for applying and evaluating all of the following algorithms:

Algorithm 1: Advantage Actor-Critic [14]

Input: policy parameters (initial) θ_0, value function parameters (initial) ϕ_0;

for k = *0, 1, 2 ...* **do**

Collect trajectories $D_k = \{\tau_i\}$ by running the policy $\pi_k = \pi(\theta_k)$ on environment;

Calculate rewards $\hat{R}_t$;

Calculate advantage estimate, $\hat{A}_t$ (with any method of advantage estimation) depending on current value function V_{ϕ_k};

Estimate policy gradient as

$$\hat{g}_k = \frac{1}{|D_k|} \sum_{\tau \epsilon D_k} \sum_{t=0}^{T} \nabla_\theta \log \pi_\theta(a_t|s_t)|_{\theta_k} \hat{A}_t;$$

Calculate policy update, using standard gradient ascent,

$$\theta_{k+1} = \theta_k + \alpha_k \hat{g}_k,$$

Fit value function regressing on mean-squared error:

$$\phi_{k+1} = \arg\min_{\phi} \frac{1}{|D_k|T} \sum_{\tau \epsilon D_k} \sum_{t=0}^{T} (V_\phi(s_t) - \hat{R}_t)^2$$

Using gradient descent.

end

Heuristic-Based Approach: As an existing approach adopted by popular inventory automation tools [17], we consider a threshold-based fixed policy and evaluate its performance on our environment. This agent follows the rule that whenever the inventory level of a product at a particular warehouse falls to a threshold value, the agent issues an order for a fixed quantity of the product to be stocked at that warehouse. Since for a particular value of the threshold inventory level, this approach follows a fixed policy, and it is incapable of improving with experience. So, we only evaluate the agent over 1000 epochs, to ensure that a large enough number of samples are drawn from the demand distribution, and a majority of possible demands are encountered by the agent.

Algorithm 2: Trust Region Policy Optimization [15]

Input: policy parameters (initial) θ_0, value function parameters (initial) ϕ_0;
Hyperparamters: Limit for KL-Divergence δ, coefficient for backtracking α, max backtracking steps K;
for $k = 0, 1, 2 \ldots$ **do**

Collect trajectories $D_k = \{\tau_i\}$ by running policy $\pi_k = \pi(\theta_k)$ on environment;

Calculate rewards $\hat{R}_t$;

Calculate advantage estimate, $\hat{A}_t$ (with any method of advantage estimation) depending on current value function V_{ϕ_k};

Estimate policy gradient as

$$\hat{g}_k = \frac{1}{|D_k|} \sum_{\tau \epsilon D_k} \sum_{t=0}^{T} \nabla_\theta \log \pi_\theta(a_t|s_t)|_{\theta_k} \hat{A}_t;$$

Use conjugate gradient algorithm to compute

$$\hat{x_k} = \hat{H_k}^{-1} \hat{g_k}$$

where $\hat{H_k}$ is sample average KL-Divergence's Hessian;
Update policy using backtracking line search,

$$\theta_{k+1} = \theta_k + \alpha^j \sqrt{\frac{2\delta}{\hat{x_k}^T \hat{H}_k \hat{x_k}}}$$

where j ϵ {0, 1, 2…K} is the minimum value improving sample loss & satisfying the KL-Divergence constraint;
Fit value function regressing on mean-squared error:

$$\phi_{k+1} = \arg\min_{\phi} \frac{1}{|D_k|T} \sum_{\tau \epsilon D_k} \sum_{t=0}^{T} (V_\phi(s_t) - \hat{R}_t)^2$$

Using gradient descent.

end

Advantage Actor-Critic (A2C): Advantage Actor-Critic, as described by Sutton et al. [14] is an policy-based following the idea of end to end policy optimization. The actor's probability of taking an action is pushed towards actions that perform well with respect to a baseline, which corresponds to the TD-error approximated by the critic. This performs better than making use of a constant advantage to quantify the value of an action.

Trust Region Policy Optimization (TRPO): It is an on-policy algorithm only applicable to environments with continuous action spaces. In normal policy gradients, taking a step too large can degrade the policy performance beyond recovery, making the use of large step sizes dangerous. Whereas with TRPO, the largest step possible to improve performance by updating the policy is taken, meeting a constraint on KL-divergence between the stochastic policies represented by the parameters of the old and the new actors, which indicates how different the two policies are allowed to be from each other.

Proximal Policy Optimization (PPO): In order to limit the change made to the current policy, so that we achieve improvement on our baseline, at the same time, do not overstep in changing the policy by too much, which would lead to a collapse in performance due to subsequent bad on-policy samples, we make use of PPO [16].

There are two alternative ways in which PPO imposes a constraint on the change in the policy: PPO-penalty and PPO-clip.

Algorithm 3: Proximal Policy Optimization [16]

Input: policy parameters (initial) θ_0, value function parameters (initial) ϕ_0;

for $k = 0, 1, 2 \ldots$ **do**

Collect trajectories $D_k = \{\tau_i\}$ by running policy $\pi_k = \pi(\theta_k)$ on environment;

Calculate rewards $\hat{R}_t$;

Calculate advantage estimates, $\hat{A}_t$ (with any method of advantage estimation) depending on current value function V_{ϕ_k};

Perform policy update that maximizes the PPO-Clip objective:

$$\theta_{k+1} = \arg\max_{\theta} \frac{1}{|D_k|T} \sum_{\tau \epsilon D_k} \sum_{t=0}^{T} \min(\frac{\pi_\theta(a_t|s_t)}{\pi_{\theta_k}(a_t|s_t)} A^{\pi_{\theta_k}}(s_t, a_t),$$

$$g(\epsilon, A^{\pi_{\theta_k}}(s_t, a_t)))$$

using stochastic gradient descent with Adam;

Fit value function regressing on mean-squared error:

$$\phi_{k+1} = \arg\min_{\phi} \frac{1}{|D_k|T} \sum_{\tau \epsilon D_k} \sum_{t=0}^{T} (V_\phi(s_t) - \hat{R}_t)^2$$

using gradient descent.

end

PPO-penalty performs an update meeting a constraint on the KL-divergence between the previous and current policies. The constraint is incorporated as a penalty in the objective, which has a penalty coefficient that can be changed over the course of training.

PPO-clip, on the other hand, does not make use of any KL-divergence-based constraint at all. Instead, it relies on a clipped objective function that removes the incentive for the new policy changing by too much from the old policy. We have made use of PPO-clip here.

4 Results

For evaluating the performance of the various algorithms, we run them on our environment and performed a comparative study of the trend of average rewards per epoch achieved by the algorithms throughout the training procedure for 15,000 epochs each. We have considered four separate warehouses, stocking 100 distinct products each. Accordingly, the dimensions of the state, action and demand matrices of the environment are 4×100. The results have been visualized as the following line plots of episode number versus the average reward per episode received by the agent (Figs. 3, 4, 5, 6, 7, 8, 9, 10, 11, 12, 13 and 14).

We observe the following:

- The policy-based optimization approaches that come under Actor-Critic methods, namely A2C, PPO and TRPO are successful at improving their performance on the inventory environment, for a variety of demand distributions. This can be inferred from the steady improvement in the average reward per epoch achieved by these methods for most of the distributions considered.
- The definition of the environment, mainly the reward function, is successful in motivating the agent(s) to forecast, and hence, maximize the fulfilment of demand,

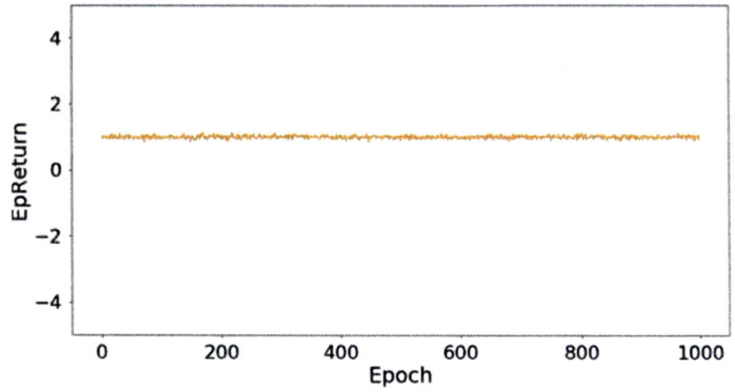

Fig. 3 Average reward per episode for normal distributed demand using threshold heuristic method

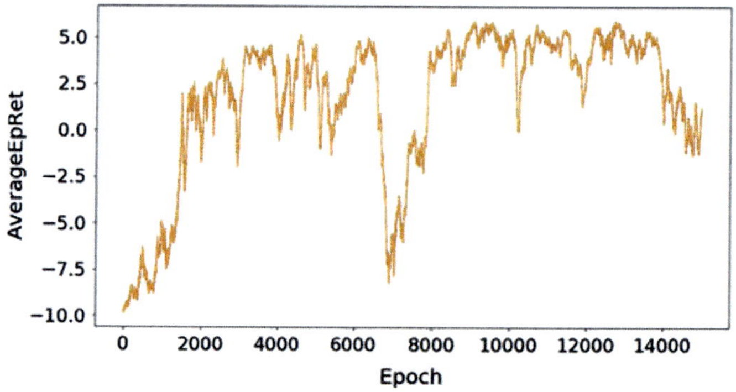

Fig. 4 Average reward per episode for normal distributed demand using A2C

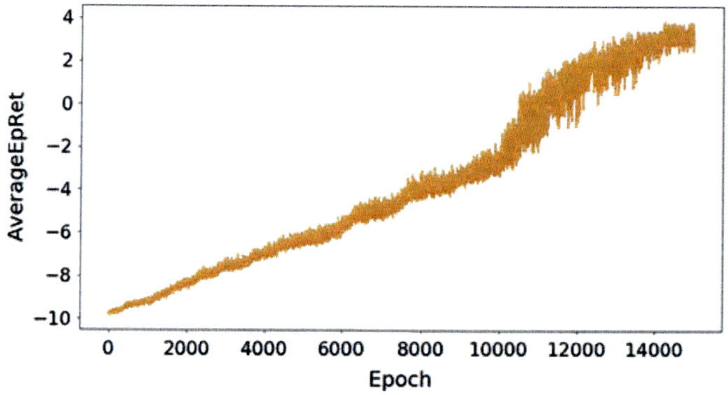

Fig. 5 Average reward per episode for normal distributed demand using TRPO

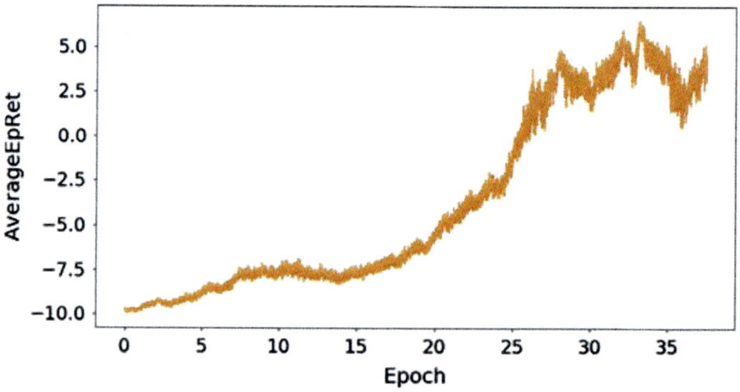

Fig. 6 Average reward per episode for normal distributed demand using PPO

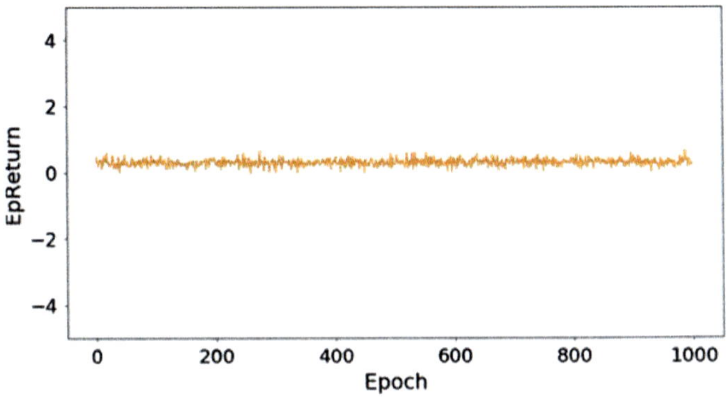

Fig. 7 Average reward per episode for Poisson distributed demand using threshold heuristic method

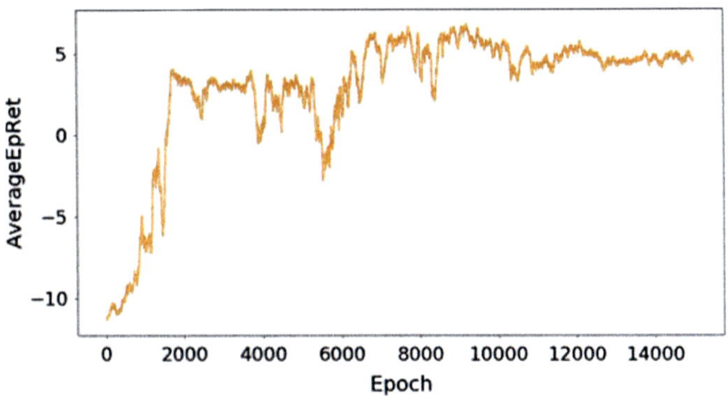

Fig. 8 Average reward per episode for Poisson distributed demand using A2C

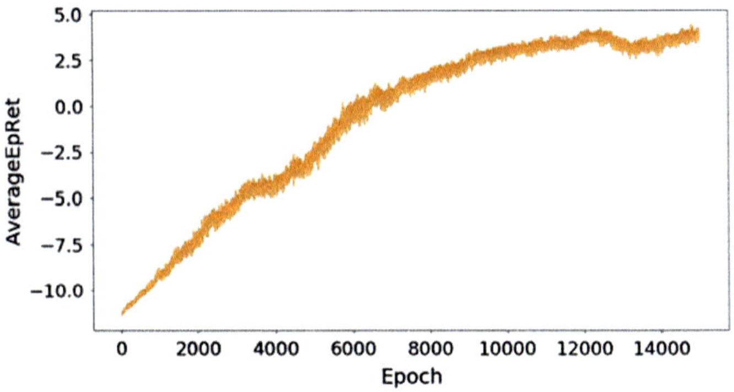

Fig. 9 Average reward per episode for Poisson distributed demand using TRPO

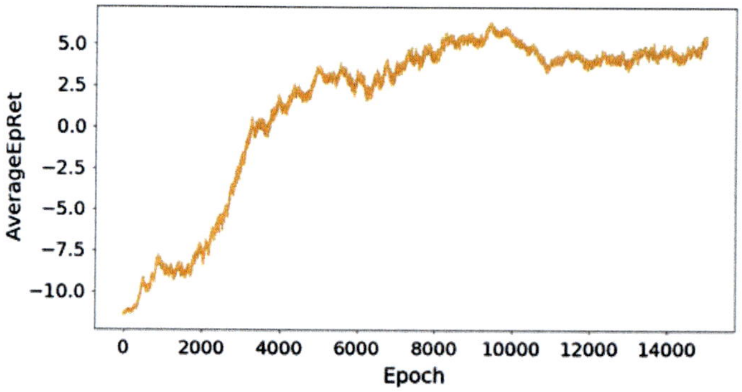

Fig. 10 Average reward per episode for Poisson distributed demand using PPO

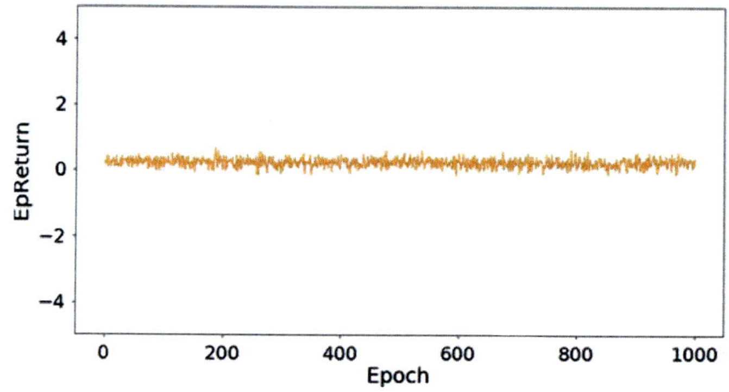

Fig. 11 Average reward per episode for Gamma distributed demand using threshold heuristic method

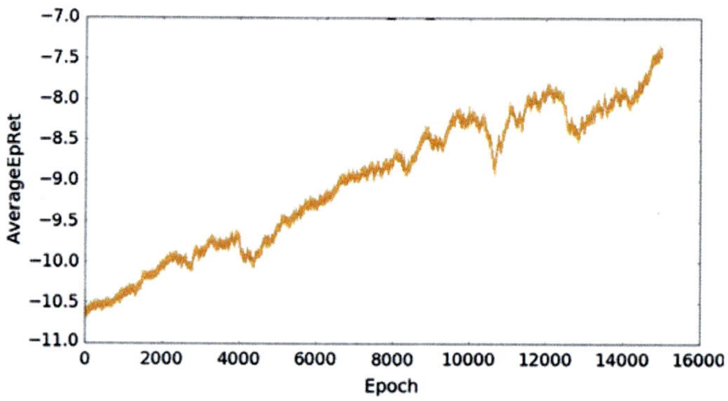

Fig. 12 Average reward per episode for Gamma distributed demand using A2C

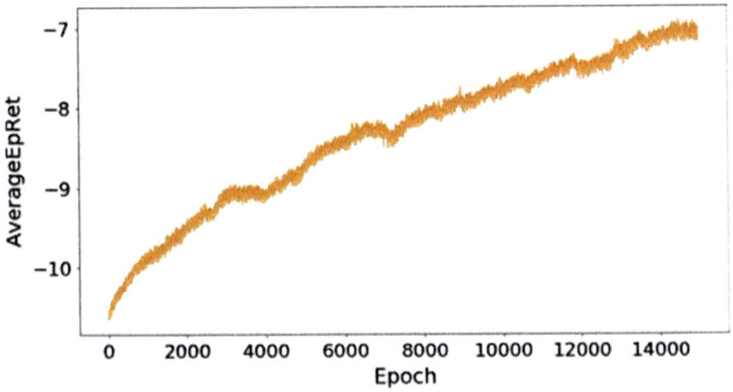

Fig. 13 Average reward per episode for Gamma distributed demand using TRPO

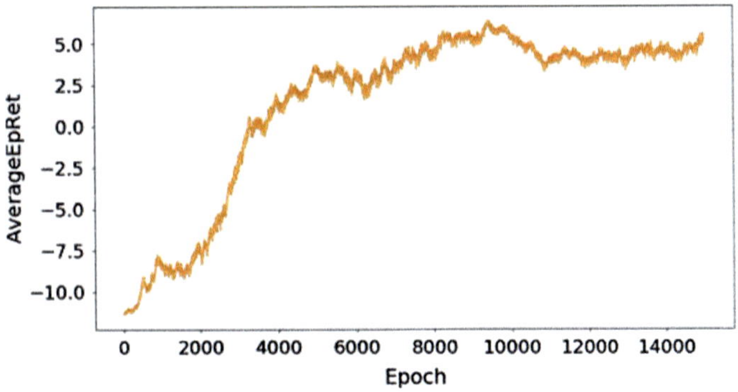

Fig. 14 Average reward for Gamma distribution demand using PPO

despite the significant lead time involved in the delivery of the products after an order is placed for them. Moreover, the reward normalization has proved to be critical in assisting proper convergence of the agents to good ordering policies. In contrast, returning rewards without normalization leads to slower convergence, due to the corresponding large value function baselines and gradients.

- Due to the varying complexities of various approaches, the time taken by each to converge is different. We observe that A2C is the fastest to converge to good policy, while TRPO is the slowest.
- Also, we observe that various approaches show different variability in the average rewards per episode. A2C shows large variability in the average returns over multiple episodes. This is due to the unchecked large improvement steps that the agent is allowed to take to change its policy by changing the weights of the actor network. Meanwhile, PPO shows much less variance, which we attribute to the bounds imposed on policy improvement by the clipping function. The policy is not allowed to improve drastically on achieving good incentives, hence preventing it

Table 1 Average returns ($\times 10^4$) of algorithms over different demand distributions after 15k episodes

	Normal	Poisson	Gamma
A2C	4.36	4.66	−7.32
TRPO	3.85	4.21	−7.07
PPO	4.09	5.48	5.49
Heuristic	1.21	0.91	0.24

from over-correcting its policy, which could lead to bad on-policy samples, causing the policy to collapse. At the same time, we observe a trade-off between the rate of policy improvement and variability in average rewards per episode (Table 1).

5 Conclusion and Future Work

Our environment, modelling multiple products across multiple inventories, is capable of being solved by policy-based RL methods. Effectively, the algorithms are capable of taking decisions that maximize the overall revenue generated over a span of time from all the warehouses. The algorithms, specifically PPO, have shown to be robust to a variety of demand distributions. In comparison with existing approaches, we find that RL-based approaches significantly outperform the heuristic-based traditional ordering policy. Also, they are capable of adapting to the demand distribution with experience, unlike the heuristic-based approach which will require manual intervention to tune the hyperparameters. As prospective future work, we plan to evaluate the performance of these methods on demand distributions that change over an episode. Another aspect to evaluate is the ability of these approaches to handle stochastic lead times in product delivery to the warehouses. Yet, another direction of prospective work in the future is to incorporate perishable commodities as products in the inventory.

References

1. Sutton R, Barto A (1995) Reinforcement learning: an introduction. MIT Press, Cambridge, MA. http://www.incompleteideas.net/book/the-book-2nd.html
2. Stockheim T, Schwind M, Koenig W (2003) A reinforcement learning approach for supply chain management
3. Kara A, Dogan I (2018) Reinforcement learning approaches for specifying ordering policies of perishable inventory systems. Expert Syst Appl 91:150–158
4. Katanyukul T, Chong EKP (2014) Intelligent inventory control via ruminative reinforcement learning. https://doi.org/10.1007/11823285_121

5. Pontrandolfo P, Gosavi A, Okogbaa OG, Das TK (2002) Global supply chain management: a reinforcement learning approach. Int J Prod Res 40:1299–1317
6. Yuan Yu J (2017) Quality assurance in supply chain management lesson 4, INSE6300 supply chain management. https://users.encs.concordia.ca/~jiayuan/scm17/reinforcement.pdf
7. Kim CO, Jun J, Baek JK, Smith RL, Kim YD (2005) Adaptive inventory control models for supply chain management. Int J Adv Manuf Technol 26:1184–1192. https://doi.org/10.1007/s00170-004-2069-8
8. Giannoccaro I, Pontrandolfo P: Inventory management in supply chains: a reinforcement learning approach. Int J Prod Econ 78:153–161. https://doi.org/10.1016/S0925-5273(00)00156-0
9. Oroojlooyjadid A, Snyder LV, Takac M: Applying deep learning to the newsvendor problem. arXiv:1607.02177
10. Mehta D, Yamparala D (2014) Policy gradient reinforcement learning for solving supply-chain management problems. https://doi.org/10.1145/2662117.2662129
11. Brockman G, Cheung V, Pettersson L, Schneider J, Schulman J, Tang J, Zaremba W (2016) OpenAI Gym. arXiv:1606.01540 [cs.LG]
12. Mnih V, Kavukcuoglu K, Silver D, Rusu A, Veness J, Bellemare MG, Graves A, Riedmiller M, Fidjeland A, Ostrovski G, Petersen S, Beattie C, Sadik A, Antonoglou I, King H, Kumaran D, Wierstra D, Legg S, Hassabis D (2015) Human level control through deep reinforcement learning. Nature 518:529–533. https://doi.org/10.1038/nature14236
13. Lagoudakis M, Parr R (2003) Least-squares policy iteration
14. Sutton RS, McAllester D, Singh S, Mansour Y (2000) Policy gradient methods for reinforcement learning with function approximation. http://dl.acm.org/citation.cfm?id=3009657.3009806
15. Schulman J, Levine S, Moritz P, Jordan M, Abbeel P (2015) Trust region policy optimization. arXiv:1502.05477
16. Schulman J, Wolski F, Dhariwal P, Radford A, Klimov O (2017) Proximal policy optimization algorithms. arXiv:1707.06347
17. Adobe Magento eCommerce Software: threshold-based inventory replenishment. https://docs.magento.com/m1/ce/user_guide/catalog/inventory-out-of-stock-threshold.html
18. National Programme for Technology Enhanced Learning, Reinforcement Learning, Ravindran B. https://onlinecourses.nptel.ac.in/noc16_cs09
19. Deep RL Bootcamp lecture series. Berkeley, CA (2017). https://sites.google.com/view/deep-rl-bootcamp/lectures
20. Analytics Vidhya website. https://www.analyticsvidhya.com/blog/2018/09/reinforcement-learning-model-based-planning-dynamic-programming

Human Resource Working Prediction Based on Logistic Regression

Anusha Hegde and G. Poornalatha

Abstract A promising organization depends on the competitiveness and professional development of its employees. As an organization reaches new levels, the pressure on employees to achieve goals is in its peak. The work activity of the employee is highly related to the growth of the company. While setting these strategies, the business insights should recognize achievable target for the human force and the factors affecting the employee in achieving the given targets. The targets and deadlines cannot be met if the employees are not reporting to work and there is no suitable plan to overcome the loss. Hence, it is required to analyse and understand the working as well as the absence pattern of employee to minimize the possible loss to the company. In the present work, logistic regression is used to analyse these kinds of pattern to predict the absence of employees which enables the employer to take necessary actions and meet the deadlines in time.

Keywords Logistic regression · Employee absenteeism · Performance · Prediction · Working pattern · Confusion matrix

1 Introduction

1.1 Absenteeism

The future of business or an organization is the most discussed topic in recent articles and papers. The current millennium has given way to many exciting and dynamic growth opportunities for the development of humans through technology and innovation. The environment for a business is complex and competitive in which there

A. Hegde · G. Poornalatha (✉)
Department of Information & Communication Technology, Manipal Institute of Technology, Manipal Academy of Higher Education, Manipal, Karnataka 576104, India
e-mail: poornalatha.g@manipal.edu

A. Hegde
e-mail: anusha.hegde@manipal.edu

N. N. Chiplunkar and T. Fukao (eds.), *Advances in Artificial Intelligence and Data Engineering*, Advances in Intelligent Systems and Computing 1133,
https://doi.org/10.1007/978-981-15-3514-7_24

is accelerating rate of changes. The companies always measure the customer loyalty and satisfaction as a means of setting strategies. For customer satisfaction to be achieved, the key point is the performance of the employees.

The performance of the employees is mainly dependent on factors like health, mindset, family, commute time, etc. Absenteeism of employees is loss for the company in terms of productivity, i.e. reduction in revenue. If there are no substitutes for certain tasks, then the effect on productivity is at risk. Thus, if prior information regarding the likeliness of an employee being absent is known, the organization can take up relevant measures to cope with the situation.

The main cause of absenteeism is the lack of a benefit policy. If the employee is not contented with the benefits provided by the company, he/she might not complete tasks assigned to him/her in time by creating some constant delays for no valid reason.

Absenteeism due to known causes like weddings, vacations, births and deaths is supported by law. Hence, absence with permission during these kinds of situation are justifiable to the employer. Absenteeism due to ignored causes includes worker and their dependent health problems or any random diverse factor.

Recognizing the employees who may not be available to carry out the scheduled work can eliminate these problems. So, if an employee does not turn up for work, then there is a possibility of that work being done by another employee or not done at all. Hence, it is important to identify the absenteeism of an employee, to avoid the possible problems supposed to be faced by the company.

The rest of the paper is organized as follows. The related works are discussed in Sect. 2, and the methodology used is presented in Sect. 3. Section 4 presents the results obtained, and paper conclusion is discussed in Sect. 5.

1.2 Logistic Regression

Logistic regression is a classification technique that can be primarily used for binary classification with an outcome of either 0 or 1. Logistic regression assesses the relationship between the label (dependent variable) and the selected features (independent variables), by evaluating probabilities using its underlying logistic function. To make a prediction, these values are mapped into binary numbers by sigmoid function (logistic function). The sigmoid function takes real-valued number between 0 and 1 and transforms to either 0 or 1 based on certain threshold.

Logistic regression is a statistical method that helps in performing categorization according to the rules of probability. The aim of the logistic regression is to present a scientific model that determines the relationship between dependent and independent variables by using less number of variables. It is a regression model that examines the relationship between discrete and continuous (independent) variables and those which have binary result variables (dependent variables). The logistic regression model formulated in this work associates the probability of the outcome to a series of potential predictor variables as shown in the Eq. (1).

$$\log\left(\frac{1}{1-p}\right) = \beta 0 + \beta 1X1 + \beta 2X2 + \cdots + \beta nXn \quad (1)$$

where p is the probability of the absenteeism, β_0 is an intercept term, β_1 to β_n are coefficients associated with each variable X_1–X_n.

Logistic regression is a widely used technique due to its efficient nature. The interpretation of the results is easy and does not require too many computational resources. It performs better when certain attributes which are not related to the output variable are removed. The removal of highly correlated attributes contributes to better performance [1]. Hence, it is used in the present work.

2 Related Work

Performance was equated to work attendance in the past. Showing up for work during sickness is termed as presenteeism, and it can be traced to loss of productivity just like absenteeism. In the light of these findings, organizations are reconsidering the approaches for daily work attendance. Changes in organizational techniques are required to prevent absenteeism turning to presenteeism [2].

Organizations are concerned about improvement in efficiency and performance by neglecting the likelihood of the problem of absence. Absenteeism hinders the growth of the company, and it may be described as an unorganized conflict where employees are showing their discontent through absenteeism. To achieve a true competitive edge in the market, managers are becoming aware of the importance to make the most use of existing resources. Absenteeism is directly related to technology management as it causes inefficiency in several ways [3].

Employee work is affected by the type of supervision; a boss who creates pressure for employees at work will not be able to get support by his workers during the hour of need. Greater work satisfaction is necessary to ensure that employees are motivated rather than the work which would require just employee's presence [4]. Absenteeism results in ill effects on the individual, his co-workers and the organization.

In a school scenario, the performance of the students is adversely affected by teacher absenteeism. In addition to this, it puts pressure on the colleagues who have to perform the duties of the teacher who is absent. Also, it may create a scenario wherein the absent teacher extends his/her leave just to avoid the more workload that awaits on return. There could be unavoidable reasons like sickness for which organizations must come with better solutions. Employee absenteeism generally implies job dissatisfaction. To create a solution for elimination of this problem, it is important to understand the root causes and trends of employee work pattern [5].

The performance appraisal of a company always considers the employee attendance criteria. Employee attendance pattern is used to study the relationship between employee absence and employee characteristics. Two years data of employee absence pattern was used to predict the special characteristics of frequent absent employees using decision tree. The characteristics which were most absent are female employee with three kids aged between 33 and 39 [6].

Career management is part of human resource management by which organizations try to recruit employees considering their career interests. Organization determines a person's career objective based on their potential and guides them towards achieving organization objectives. Employee absence pattern is a major criteria to be considered for achieving organization objectives [7].

The absenteeism prediction is being done using an efficient technique of logistic regression. Employee attributes like social drinking and smoking are considered in the model built as it leads to employee absenteeism.

3 Methodology

The dataset consists of 740 records, and each record has 13 independent attributes and one outcome variable. The dataset was scanned for missing attributes, and there was no record with missing attributes found. The outcome variable initially had the value as number of hours of absence which was converted as whether an employee is absent for one full day or not. Here, the criteria considered are if an employee has nine or more hours of absence in work, then that record is labelled with value 1 else the record is assigned a label of 0. Usually, the work duration per day is 9 h across many organizations, and hence, the same number of hours is considered.

The correlation between the 13 attributes in Table 1 was measured. Attribute 1—Transportation expense is positively correlated to attribute 10—Son and attribute 11—Pet. Attribute 6—Distance from residence to work and attribute 7—Service time

Table 1 Selected attributes

S. No.	Attribute name	Type
1	Transportation expense	Numeric
2	Disciplinary failure	Binomial
3	Education	Categorical
4	Reason for absence	Categorical
5	Day of the week	Categorical
6	Distance from residence to work	Numeric
7	Service time	Numeric
8	Age	Numeric
9	Workload average/day	Numeric
10	Son	Numeric
11	Pet	Numeric
12	Social drinker	Binomial
13	Social smoker	Binomial

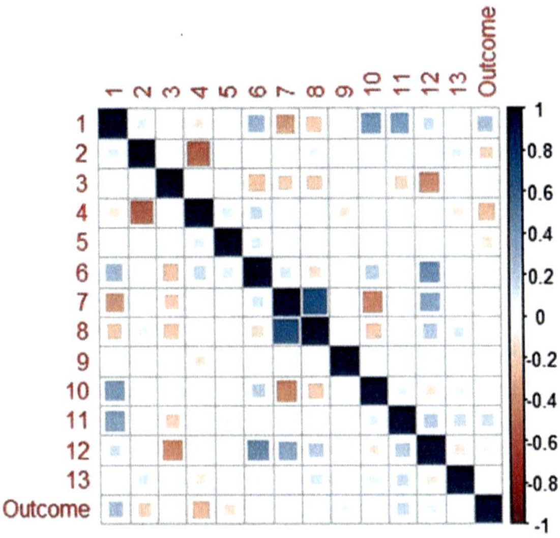

Fig. 1 Correlation between the selected attributes

is positively correlated to attribute 12—Social drinker. The most positive correlation was found between the attribute 7—Service time and attribute 6—Age. Attribute 2—Disciplinary failure was seen to have positive correlation between multiple attributes, but the attribute 3—Education had negative correlation between multiple attributes.

The correlation between the attributes signifies the importance of their consideration in the analysis. Initially, the logistic model was built using glm function in R with four attributes and the AIC score was 972. The dataset consisted of 20 attributes, and as we added one more attribute to the model, the AIC score was improving. With each step, the highest influencing factor was recorded (Fig. 1).

The logistic Model-I was built considering the following eleven attributes: Transportation expense, disciplinary failure, education, reason for absence, day of the week, distance from residence to work, service time, age, workload average/day, son, pet. For this model, the AIC score was observed as 736, and four highly affecting parameters were recorded as transportation expense, reason for absence, day of the week and son.

For the previous model, two more attributes, social drinker and social smoker, were considered, and a new logistic Model-II was created. The AIC score for this model was recorded as 724. In this model, it was observed that social drinker also affects the outcome in a significant way as the p-value is less than 0.0001 for this attribute.

The relative quality of the models based on same dataset can be estimated by Akaike information criterion (AIC). AIC evaluates the quality of each model, and hence, it serves as a means for model selection. AIC is based on information theory. During the process of statistical modelling, there will be loss of some data information, i.e. model will never be exact representation of the dataset. AIC estimates the information lost by a model. So, less information loss implies high-quality model. While estimating the information lost, there is a trade-off between simplicity of the model and goodness of fit of the model.

For validation, the dataset was divided into training data consisting of 629 [85%] records and testing data consisting of 111 [15%] records. The training model was built using 13 attributes and validated with the testing dataset. The predicted probabilities are interpreted as 1 and 0 based on the mean values. The values greater than mean values are mapped to 1, and less than mean values are mapped to 0.

4 Results

The two models are evaluated based on the confusion matrix obtained after validation. Confusion matrix is used to summarize or evaluate the performance of a binary classification task or model. The confusion matrix calculates the number of correct and incorrect predictions which is further condensed with the number of count values and classify into each classes. It ultimately shows the path in which classification model is confused while making predictions.

The Model-I has eleven attributes and AIC score of 736. The Model-I gives the confusion matrix of Table 2 and an accuracy of 74.8%. The Model-II has thirteen attributes and AIC score of 724. The Model-II gives the confusion matrix as in Table 3 and an accuracy of 81.9%.

The residuals versus leverage graph of the two models in Fig. 2 depicts that there are considerably few points which are residuals in the Model-II.

The results of Model-I and Model-II are recorded in Table 4. It is seen that Model-II provides better results for all of the values in consideration.

Sensitivity can be defined as the proportion of observed absenteeism that was predicted to be true. In other words, of all the cases that were labelled as prone to be absent, how many were correctly classified true by the two models. Clearly, Model-II is giving a better sensitivity value of 83%.

Table 2 Confusion matrix of Model-I

Model-I	1	0
1	52	10
0	18	31

Table 3 Confusion matrix of Model-II

Model-II	1	0
1	59	8
0	12	32

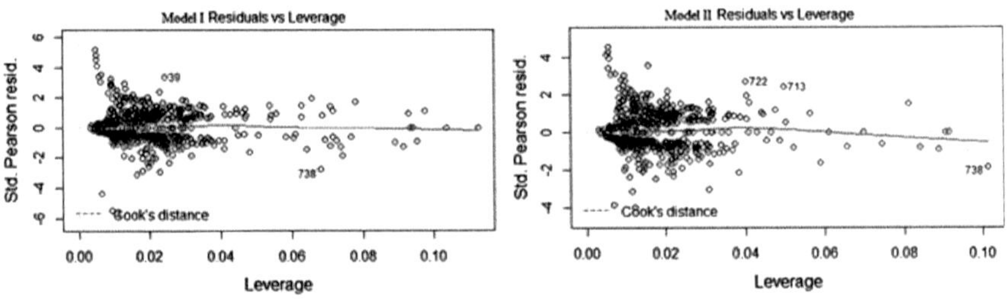

Fig. 2 Residuals versus leverage comparison

Table 4 Results statistics of Model-I and Model-II

Results	Model-I	Model-II
Accuracy	0.7477	0.8198
95% CI	(0.6565, 0.8254)	(0.7355, 0.8863)
No information rate	0.6306	0.6396
P-value [Acc>NIR]	0.005948	2.551e−05
Kappa	0.4796	0.6175
Mcnemar's test *P*-value	0.185877	0.5023
Sensitivity	0.7429	0.8310
Specificity	0.7561	0.8000
Pos pred value	0.8387	0.8806
Neg pred value	0.6327	0.7273
Prevalence	0.6306	0.6396
Detection rate	0.4685	0.5315
Detection prevalence	0.5586	0.6036
Balanced accuracy	0.7495	0.8155

Specificity can be defined as the proportion of observed negatives that were predicted to be negatives. In other words, of all the transactions that were legitimate, what percentage was correctly predicted by the model. Model-II is giving higher specificity than Model-I.

5 Conclusion

The logistic regression for absenteeism at work prediction can be an excellent tool to aid decision making. The deadlines can be maintained with planned measures, like holiday distribution, and production maintenance measures such as notices of overtime, work in weekend or holiday, and hiring of temporary workers. With the advantage of timely planning, summoning of employees for temporary contract can be done in advance.

References

1. Kleinbaum DG, Klein M (2010) Introduction to logistic regression. In: Logistic regression. statistics for biology and health. Springer, New York. https://doi.org/10.1007/978-1-4419-1742-3_1
2. Gosselin E, Lemyre L, Corneil W (2013) Presenteeism and absenteeism: Differentiated understanding of related phenomena. J Occup Health Psychol 18:75–86. https://doi.org/10.1037/a0030932
3. Ramdass K (2017) Absenteeism—a major inhibitor of productivity in South Africa: a clothing industry perspective. In: Portland international conference on management of engineering and technology, Portland, OR, pp 1–7. https://doi.org/10.23919/picmet.2017.8125404
4. Martiniano A, Ferreira RP, Sassi RJ, Affonso C (2012) Application of a neuro fuzzy network in prediction of absenteeism at work. In: 7th Iberian conference on information systems and technologies, Madrid, pp 1–4
5. Badubi RM (2017) A Critical Risk Analysis of Absenteeism in the Work Place. J Int Bus Res Mark 2:32–36. https://doi.org/10.18775/jibrm.1849-8558.2015.26.3004
6. Qomariyah NN, Sucahyo YG (2014) Employees' Attendance Patterns Prediction using Classification Algorithm Case Study: A Private Company in Indonesia. Int J Comput Commun Instrum Eng 1:68–72
7. Mallafi H, Widyantoro DH (2016) Prediction modelling in career management. In: International conference on computational intelligence and cybernetics, Makassar, pp 17–21. https://doi.org/10.1109/cyberneticscom.2016.7892560

Kansei Knowledge-Based Human-Centric Digital Interface Design Using BP Neural Network

Huiliang Zhao, Jian Lyu, Xiang Liu, and Weixing Wang

Abstract Digital interface has increasingly replaced the traditional human–computer hardware interface and become the main carrier of human–computer interaction in information intelligent system. How to design and develop an effective digital interface is a new problem faced by enterprises and designers. Aiming at the practical problems of cognitive difficulties such as overload and mismatch in the field of digital interface design of complex information systems, this paper proposed a method for human-centric digital interface design based on Kansei knowledge. It was done to study the Kansei knowledge of digital interface to determine the Kansei images that affects the interface, identify the key elements of interface design including interface layout style, main color style, font style, and core component expression, and then construct a nonlinear mapping and mathematical prediction model between the Kansei images and elements of interface design based on BP neural network. Finally, the feasibility of this method was verified, which can effectively match the user's specific perceptual cognitive needs of complex digital interface.

Keywords Digital interface · Kansei knowledge · BP neural network

1 Introduction

With the rapid development of computer and communication technology, human beings have entered the information age. Digital interface has increasingly replaced the traditional human–computer hardware interface and become the main carrier of human–computer interaction in information intelligent system [1]. The traditional

H. Zhao (✉)
School of Fine Arts, Guizhou Minzu University, 550025 Guiyang, China
e-mail: fightingzhl@163.com

J. Lyu · X. Liu
Key Laboratory of Advanced Manufacturing Technology, Ministry of Education, Guizhou University, 550025 Guiyang, China

W. Wang
School of Mechanical Engineering, Guiyang University, 550005 Guiyang, China

N. N. Chiplunkar and T. Fukao (eds.), *Advances in Artificial Intelligence and Data Engineering*, Advances in Intelligent Systems and Computing 1133,
https://doi.org/10.1007/978-981-15-3514-7_25

instrument control system is gradually transiting to the highly integrated digital control system. The operation mode based on the computer digital interface can effectively improve the human–machine experience and reduce the production cost of enterprises by integrating the information of the central console and increasing the human–machine communication. It has been widely used in aerospace, nuclear power monitoring, intelligent transportation, and other complex man–machine interface interaction systems. How to design and develop an effective digital interface is a new problem faced by enterprises and designers, which has become the most important part of the sustainable development of many enterprises. How to systematically design the human–computer interaction information interface in complex systems is a key issue in the current academic and engineering fields, and also a blank spot in the cross-research between design and engineering fields.

In the information age, the digital interface design must solve the problem of transforming the computer's behavior, thinking, and information expression into acceptable ways, so that the digital interface can adapt to the user's thinking characteristics and behavior, which is the main idea of human-centric digital interface design [2]. In the design of digital interface, we should focus on users, analyze the information organization structure of the interface according to users' cognitive mechanism, optimize the design of the interface according to users' cognitive characteristics, and improve the usability and user experience of the interface. Therefore, in the context of the information age, the research on the design method and theory of digital interface are new contents of user-centered research, and it is necessary to conduct in-depth research. From the point of view of satisfying users' perceptual cognitive needs, this paper uses users' Kansei knowledge to seek the optimal design strategy of complex information system digital interface.

2 Related Works

For the design method of digital interface, Norman first introduced the concept of mental model into the field of design, and now, it has been developed into a new user research method of interface [3]. Mendel et al. [4] point out that the display, interaction, and consistency design of interface information in information search tasks will affect the internal and external cognitive loads. Chang et al. [5] studied the effects of continuous presentation and simultaneous presentation at different densities on learners'visual ability and working memory load and found that information density played a decisive role. In terms of design methods, Van Merriënboer et al. [6] point out that building a good cognitive framework based on cognitive load theory can reduce the external load and manage the internal load, so as to better guide people's cognition. Chalmers et al. [7] synthesized schema theory, cognitive load theory, and memory theory and pointed out that the wrong design in man–machine interface design would lead to the lack of user's memory and understanding of interface information, and proposed that schema should be established from the aspects of interface layout, color, spatial display, and consistency. Oviatt S, et al. [8] argue that design should be more intuitive and easy to learn, and that user-centered design should be used as a starting point to reduce the cognitive load of the interface. He also pointed out that the user's natural behavior model can make the interface more intuitive, easy to learn and reduce human errors; by minimizing the user's cognitive load, the human-centric design can maintain the same degree of coordination of the interface while minimizing the psychological resources. In terms of coding principles, Sweller et al. [9] analyzed the internal, external, and correlative cognitive loads in arithmetic from the perspective of element interactivity, believing that element interactivity affects cognitive loads by changing working memory resources.

Kansei engineering is a product development method used to investigate the human feelings and to discover quantitative relationships between the affective responses and design features [10]. By using Kansei engineering, a lot of research has been done to improve product design. For instances, Fung, et al. [11] used this method to conduct a set of custom designs. Llinares and Page [12] designed a questionnaire that uses this method to measure subjective consumer perceptions that affect property purchase decisions. Shieh et al. [13] used this method to explore the relationship between the shape and color of toothbrushes.

3 Proposed Method

According to psychological research, the correct rate of users' behavior under high cognitive load and positive emotional experience is significantly improved. Garbarino and Eden found that when cognitive tasks became more difficult and cognitive efforts increased, users' emotions became more negative [14]. Fredrickson et al. [15] put forward the Broaden-and-build theory through the integration of early research on positive emotions. This theory hypothesis holds that positive emotions broaden the instantaneous thinking–behavior sequence of individuals, and then promote individuals to form a lasting personal resource, including physical, intellectual, and psychological, which will affect the future behavior of individuals, and bring about a sustained positive impact. This also lays a theoretical foundation for this paper to put forward a cut-in point to satisfy users' perceptual perception to reduce users' cognitive difficulties in complex system digital interface.

User's cognitive process of digitized man–machine interface is a series of psychological processes that make people to know things from the surface to the inside through feeling, perception, memory, imagination, thinking, and emotion. It is the user's acceptance and processing of information. It mainly plays the role of information transmission through the layout of the interface itself, color, shape, and other formal factors, thus producing the recognition of the interface. In this paper, the research results of cognitive psychology and the characteristics of perceptual knowledge are closely linked, and the perceptual cognition in the design of digital human–machine interface is studied.

In the complex human-centric digital interface design based on Kansei knowledge, the most important step is to carry out the representation of Kansei images and the

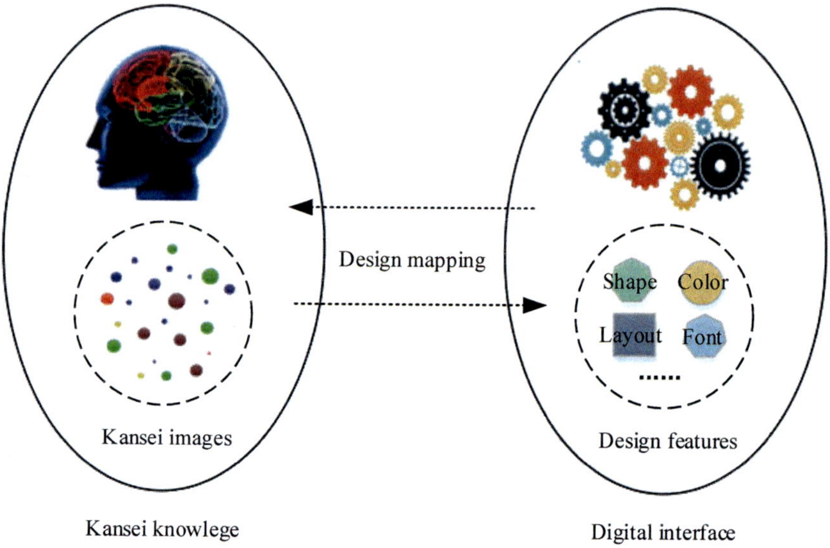

Fig. 1 Framework of research content

analysis of interface design elements. By means of quantitative analysis, the original abstract perceptual information and the actual design elements can be interpreted to form the relationship between Kansei images and interface design features (see Fig. 1). It can provide a support for the designing of digitized interface of engineering vehicles based on perceptual perception.

3.1 Design Features of Digital Interface

The design features of digital interface refer to the integral formal relationship formed by the combination of multiple components, forming a unique interface recognition mode, and it is the expression vector of vehicle information attributes. Unlike three-dimensional modeling features, which can be presented clearly in line, surface or body, the design features of digital interface are synthetically represented by the formal relations of various features, and its meaning is higher than the single expression of each part. Therefore, the characterization of design features of digital interface is quite complex. This complex design features create the style of the interface and can be recognized significantly. Design features mainly include product layout, shape (icon, character), color, and so on. The combination, collocation, interrelationship, and importance of features form the perceptual information of the whole digital interface, and also constitute the design feature framework of the digital interface.

3.2 Kansei Knowledge

Kansei engineering has been widely studied and applied in academia and business circles because it provides an implementation framework for acquiring users' perceptual needs and translating them into a specific product design elements. Kansei knowledge is a logical-reasoning structure that quantifies the relationship between perceptual image and product design elements. It can not only directly guide product development but also realize rapid automation and intelligent design by further establishing expert system. As a whole, perceptual knowledge is used to manage users' perceptual demand information, which not only affects the description of design tasks but also is closely related to the user-decoding process of design results. It is the basis of communication between design subjects. For the study of the design process and design evaluation of digitized human–machine interface for engineering vehicles, the standardized expression and quantification of perceptual knowledge has important research and practical significance and is the basis for the study of the design process of digitized human–machine interface for engineering vehicles.

On the basis of psychological semantics, through the relevant methods of semantics, this paper carries the information of culture, society, and products expressed in the interface design and builds the visual relationship between the interface and the social and life levels. It describes and expresses the features of digitized man–machine interface by language. This description and expression are either intuitive or abstract, but it can basically and comprehensively reflect the external form of the interface and the sensory information among the audience. The flow of perceptual knowledge among design subjects, interfaces, and users constitutes the process of design, use, demand investigation, and so on.

The basic assumption of the executability of perceptual knowledge is that the specific design attributes of a specific user interface will cause the user's specific perceptual response. From the perspective of decision making, the attributes of perceptual information and the attributes of interface design features can be regarded as decision-making attributes and conditional attributes, respectively. Therefore, a common sensory engineering problem can be transformed into a multi-criteria decision-making problem.

3.3 Design Mapping

Artificial intelligence is a new technology to extend theory, method, technology, and application system for simulating, extending, and expanding human intelligence. Artificial intelligence is also a branch of computer science, which attempts to understand the essence of intelligence, produce a new kind of human intelligence and can be a similar way to respond to intelligent machines.

Due to the different complexity of the content of each digitized human–machine interface, the interaction and influence between interface elements are obvious. Traditional linear methods such as multiple regression and quantification theory are difficult to deal with the complex nonlinear relationship among the design elements of digital interface and Kansei images. Considering the strong nonlinear mapping ability of BP neural network and its simple structure and easy realization, this paper proposes to use BP neural network to study the design of digital human-centric interface based on Kansei knowledge.

4 Case Study

4.1 Design Features Analysis of Interface

Based on the questionnaires of interface designers, industrial control interface and literature analysis, combined with actual design requirement of the digital interface of a remote control engineering vehicle, this paper synthetically considers the constraints of balance, proportion, conciseness, and responsiveness, and finally determines the interface design elements that mainly affect the perceptual image of digitized human–machine interface, including interface layout style, main color style, font style, and core component expression. Each design element consists of different number of interface design categories, as shown in Table 1. The 'main color style' is measured by the HSB color mode based on people's psychological feeling of color, which refers to the main color value under the condition that the background color (H: 0, S: 0, B: 0) of the digitized man–machine interface of the remote control engineering vehicle is determined. The research object of core component expression is determined by the characteristics of remote control engineering vehicle and the weight of corresponding components including battery pack style, body movement style, body azimuth style, and tachometer style.

4.2 Kansei Knowledge Characterization of Digital Interface

On the basis of Mohammed's research method [16], 75 adjectives describing Kansei images of digital interface are collected through enterprise investigation, interviews with operators and designers, and literature analysis. Factor analysis is used to input them into SPSS statistical software. Principal component analysis (PCA) and maximum variance rotation factor are used to analyze the results. Four factors with characteristic value greater than 1 are selected, and their cumulative total interpretable variance is obtained 95.540%. According to the result of factor analysis, the Kansei image of digital interface design can be explained by four factors, each of which has its representative meaning. In order to further select representative Kansei vocabulary

Table 1 Digital interface design elements

Project		Category			
		1	2	3	4
Interface layout style X_1		 C11	 C12	 C13	
Main color style X_2		 C21	 C22	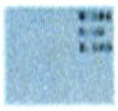 C23	 C24
Font style X_3		Sharp C31	Smooth C32		
Core component expression X_{4-7}	Battery pack style X_4	Sector C41	Rectangle C42	Cylindrical C43	Square column C44
	Body movement style X_5	Flat C51	3D C51		
	Body azimuth style X_6	SphereC61	Calibration loop C62	Radar scanning C63	
	Tachometer style X_7	Circular C71	Square C72	Sector C73	

and understand the similarity among them, the factor load of each pair of vocabulary is analyzed by cluster analysis, and 22 perceptual vocabularies are grouped into four groups. Finally, the Kansei images of digital interface are determined as innovative Y_1, readable Y_2, sophisticated Y_3, and scientific Y_4. Based on the semantic difference method (SD), 50 subjects (20 designers, 15 operators, 8 technical developers, and 7 business managers) were selected to conduct a survey and determine the vocabulary evaluation values of each perceptual image. Take sophisticated Y_3 as an example, the Kansei image evaluation of 'sophisticated' is shown in Table 2. The sample of digital interface of remote control engineering vehicle comes from the design concepts provided by many professional designers in this field.

4.3 Building Mathematical Prediction Model

Combined with the experimental data in Table 2, the complex nonlinear relationship between the operator perceptual image evaluation value and the design elements of the digitized interface is simulated by using the neural network tool of MATLAB. At the same time, the relationship between the design elements and the perceptual image is established by using the 7-5-4-1 four-layer BP network based on the empirical common and multiple comparative experiments. The network structure is shown in

Table 2 Kansei image evaluation for 'Sophisticated'

Sample	Y_3	X_1	X_2	X_3	X_4	X_5	X_6	X_7
1	5.42	3	4	2	4	2	3	3
2	2.86	1	1	1	2	1	2	1
3	5.11	2	2	2	3	2	3	3
4	4.75	1	1	1	4	2	3	3
5	5.36	3	4	2	3	2	2	3
…	…	…	…	…	…	…	…	…
23	4.11	2	1	1	2	2	3	2
24	3.78	3	4	2	4	1	3	2
25	4.37	3	3	2	1	2	3	2
26	4.38	1	2	1	3	2	1	2
27	4.35	1	4	1	4	2	2	1
28	4.81	2	3	1	3	2	3	3
29	3.07	1	1	2	2	1	2	1
30	4.43	3	4	1	4	2	2	1

Fig. 2. The input layer is the key design element of digital interface, and the output layer is the Kansei image evaluation value.

The first 27 samples in Table 1 are input into the established network, and the setting error is 0.01. Trangdm with momentum gradient descent method is used to start training and achieve the training objectives, as shown in Fig. 3. Thus, the trained

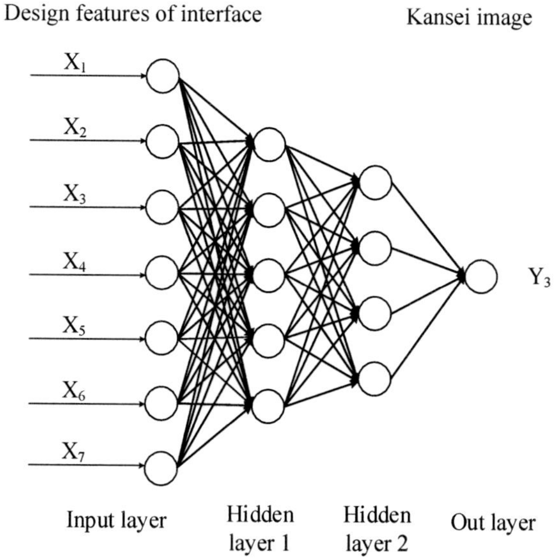

Fig. 2 Established BP neural network structure

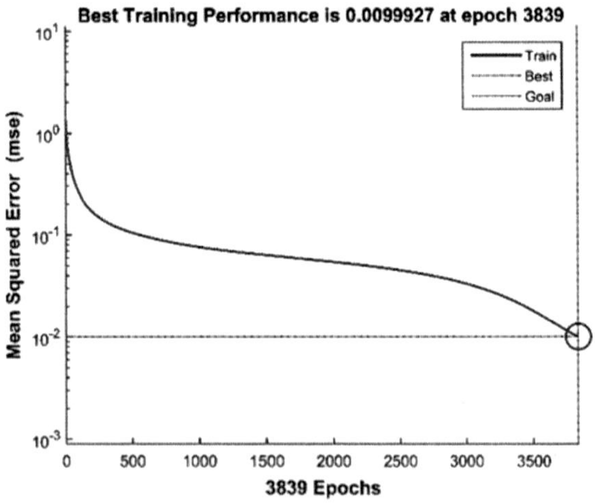

Fig. 3 MATLAB network training

BP neural network establishes a nonlinear connection between input layer (interface design elements) and output (perceptual vocabulary evaluation values) and has the ability to process input information of similar samples by itself.

4.4 Analysis of Experimental Results

The last three samples in Table 2 are used to test the network performance. By comparison, it was found that the predicted values of the output and the evaluation values of the questionnaire basically coincided with the two groups of data, as shown in Table 3. It can be seen that the established BP neural network achieves the correct mapping from design elements to perceptual image evaluation values. The mean square error MSE value is 0.0114, and the prediction accuracy is good, which verifies the validity of the BP neural network model.

For 'sophisticated' digital interface, there are $3 * 4 * 2 * 4 * 2 * 3 * 3 = 1728$ different design schemes under seven design elements. After BP network model operation, the maximum value of perceptual evaluation is 5.512 and the minimum value is 2.341. The corresponding combination numbers of design elements are

Table 3 Comparison between evaluation value and predicted value for selected Kansei image

Evaluation value		Tested sample		
		1	2	3
Sophisticated	Evaluated value	4.81	3.07	4.43
	BP predicted	4.73	2.89	4.50
	Absolute error	0.08	0.14	0.09

342,323 and 2,111,122, respectively. The two groups of numbers correspond to the digitized interface design elements table of Table 1, and the optimal and worst combination of design elements about "'precision' in theory is obtained.

Similarly, we can get the forecasting models of such Kansei images as 'innovative,' 'readable,' and 'scientific.' According to the above research, we can find that using BP neural network to design perceptual image of digital interface can guide the design process, help designers design products that meet users' perceptual needs, and scientifically construct design decision support database, which can be used for design decision under multi-objective image preference conditions.

5 Conclusion

With the rapid development of computer and communication technology, digital interface has increasingly replaced the traditional human–computer hardware interface and become the main carrier of human–computer interaction in information intelligent system. From the point of view of satisfying users' perceptual cognitive needs, this paper uses users' Kansei knowledge to seek the optimal design strategy of complex information system digital interface.

It was done to study the Kansei knowledge of digital interface to determine the Kansei images that affects the interface, identify the key elements of interface design including interface layout style, main color style, font style, and core component expression, and then construct a nonlinear mapping and mathematical prediction model between the Kansei images and elements of interface design based on BP neural network. Taking 'precision' as an example, the mathematical model of perceptual image prediction and the combination of the best and worst design elements based on BP neural network are established. The test results show that the method is feasible, and it has certain theoretical guiding significance for assisting the design and development of digitized interface of the product.

References

1. Dai Y, Xue C, Guo Q (2018) A study for correlation identification in human-computer interface based on HSB color model. In: Human interface and the management of information. interaction, visualization, and analytics. Springer International Publishing, pp 477–489. https://doi.org/10.1007/978-3-319-92043-6_40
2. Wu X, Chen Y, Zhou F (2016) An interface analysis method of complex information system by introducing error factors. In: Engineering psychology and cognitive ergonomics. Springer International Publishing, pp 116–124. https://doi.org/10.1007/978-3-319-40030-3_13
3. François M, Osiurak F, Fort A, Crave P, Navarro J (2016) Automotive HMI design and participatory user involvement: review and perspectives. Ergonomics, pp 541–552. https://doi.org/10.1080/00140139.2016.1188218

4. Mendel J, Pak R (2009) The effect of interface consistency and cognitive load on user performance in an information search task. In: Proceedings of the human factors and ergonomics society annual meeting, pp 1684–1688. https://doi.org/10.1177/154193120905302206
5. Chang T-W, Kinshuk, Chen N-S, Yu P-T (2012) The effects of presentation method and information density on visual search ability and working memory load. Comput Educ, pp 721–731. https://doi.org/10.1016/j.compedu.2011.09.022
6. Van Merriënboer JJG, Sweller J (2010) Cognitive load theory in health professional education: design principles and strategies. Med Educ, pp 85–93. https://doi.org/10.1111/j.1365-2923.2009.03498.x
7. Chalmers PA (2003) The role of cognitive theory in human–computer interface. Comput Hum Behav, pp 593–607. https://doi.org/10.1016/s0747-5632(02)00086-9
8. Oviatt S (2006) Human-centered design meets cognitive load theory. In: Proceedings of the 14th annual ACM international conference on multimedia—MULTIMEDIA '06. ACM Press. https://doi.org/10.1145/1180639.1180831
9. Sweller J (2010) Element interactivity and intrinsic, extraneous, and germane cognitive load. Educ Psychol Rev, pp 123–138. https://doi.org/10.1007/s10648-010-9128-5
10. Nagamachi, M (1989) Kansei engineering approach to automotive. J Soc Automot Eng Jpn, pp 94–100
11. Fung CKY, Kwong CK, Chan KY, Jiang H (2013) A guided search genetic algorithm using mined rules for optimal affective product design. Eng Optim, pp 1094–1108. https://doi.org/10.1080/0305215x.2013.823196
12. Llinares C, Page AF (2011) Kano's model in Kansei Engineering to evaluate subjective real estate consumer preferences. Int J Ind Ergon, pp 233–246. https://doi.org/10.1016/j.ergon.2011.01.011
13. Shieh M-D, Yeh Y-E, Huang C-L (2015) Eliciting design knowledge from affective responses using rough sets and Kansei engineering system. J Ambient Intell Humaniz Comput, pp 107–120. https://doi.org/10.1007/s12652-015-0307-6
14. Garbarino EC, Edell JA (1997) Cognitive effort, affect, and choice. J Consum Res, pp 147–158. https://doi.org/10.1086/209500
15. Fredrickson BL, Branigan C (2005) Positive emotions broaden the scope of attention and thought-action repertoires. Cognit Emot, pp 313–332. https://doi.org/10.1080/02699930441000238
16. Mohamed MSS, Shamsul BMT, Rahman R, Jalil NAA, Said AM (2015) Determination of salient variables related to automotive navigation user interface research survey for Malaysian consumers. Adv Sci Lett, pp 2089–2091. https://doi.org/10.1166/asl.2015.6217

DST-ML-EkNN: Data Space Transformation with Metric Learning and Elite k-Nearest Neighbor Cluster Formation for Classification of Imbalanced Datasets

Seba Susan and Amitesh Kumar

Abstract Most of the real-world datasets suffer from the problem of imbalanced class representation, with some classes having more than sufficient samples, while some other classes are heavily underrepresented. The imbalance in the class distributions of the majority and minority samples renders conventional classifiers like the k-Nearest Neighbor (kNN) classifier ineffective due to the heavy bias toward the majority class. In this paper, we propose to counter the class-imbalance problem by data space transformation of the training set by distance metric learning prior to an enhanced classification phase. The classification phase comprises of partitioning the set of k-Nearest Neighbors of each transformed test sample into two clusters based on their distances from the two extreme members in the set. A majority voting of the training samples within the 'Elite' cluster that is closest to the transformed test sample indicates the label of the test sample. Our proposed method is called Data Space Transformation with Metric Learning and Elite k-Nearest Neighbor cluster formation (DST-ML-EkNN). Extensive experiments on benchmark datasets using a variety of metric learning methods, with comparisons to the state of the art, establish the supremacy of our learning approach for imbalanced datasets.

Keywords Imbalanced datasets · Imbalanced learning · Distance metric learning · Data space transformation · kNN classifier

1 Introduction

Learning from imbalanced datasets is a matter of great concern for the current crop of real-world datasets that are compiled from multiple sources. The imbalanced datasets are characterized by uneven class distributions in which minority class samples are underrepresented as compared to the more than sufficient majority samples.

S. Susan (✉) · A. Kumar
Department of Information Technology, Delhi Technological University, Shahbad Daulatpur, Delhi 110042, India
e-mail: seba_406@yahoo.in

N. N. Chiplunkar and T. Fukao (eds.), *Advances in Artificial Intelligence and Data Engineering*, Advances in Intelligent Systems and Computing 1133,
https://doi.org/10.1007/978-981-15-3514-7_26

In such a scenario, classical machine learning techniques like the nearest neighbor classifier [1] fail to perform since the data space is insufficiently represented especially near the boundary lines of the two classes [2]. Nonlinear classifiers like Support Vector Machines (SVMs) are also found to fail under such circumstances [3]. The false negative rate is very high for the target (minority) class due to the skewed class boundary that favors the majority class. Sampling is a typical solution for solving data imbalance problems with both approaches of undersampling the majority class and oversampling the minority class found amply in literature [4–8]. While oversampling the minority class, problems of duplicity of instances and false alarm rates arise, which is mitigated to some extent by advanced oversampling techniques such as MAHAKIL in [9]. Methods that incorporate both undersampling of the majority class and oversampling of the minority class are now garnering interest among researchers working on the issue of imbalanced learning [10–13]. Sampling techniques have the disadvantage that they remove or duplicate samples without any change in the actual input space. Distance metric learning (ML) techniques [14] induce data space transformation, that makes high accuracies by distance-based classifiers feasible. The distance- or similarity-based classifiers are still popular in the field of Computer Vision despite of advances in deep learning [15, 16]. Sometimes a nonlinear transformation of the data space helps in constructing improved classifiers for real-world datasets [17–19].

In this paper, we address the class-imbalance problem through data space transformation by metric learning (ML) followed by 'Elite' cluster formation of the k-Nearest Neighbors (kNN) of the transformed test sample. The latter step is performed in order to remove impostors from the polling process in the kNN classification. The organization of this paper is as follows. The metric learning process and the subsequent data space transformation achieved are discussed in Sect. 2, the proposed methodology of utilizing the transformed data space to find the 'Elite' closest neighbors and the label of the test sample is explained in Sect. 3, the experimentation is analyzed in Sect. 4, and the overall conclusions are drawn in Sect. 5.

2 Data Space Transformation by Metric Learning for *Training* and *Testing* Sets

The need for a transformed input data space for distance-based classification was emphasized in [20] that iteratively applied Large Margin Nearest Neighbor (LMNN) till convergence was achieved. The advantage claimed here with respect to other metric learning techniques is a more stable neighborhood for the minority class samples. Sampling techniques that are conventionally employed for imbalanced learning lead to loss of information and are not basically designed for distance-based classifiers [11, 13]. Other alternatives to input space transformation also exist. In [3], conformal transformation is applied in the feature space as per the spatial distribution of the support vectors. The minority support vectors are given higher spatial resolution by this transformation. LMNN was introduced in [21] as a metric learning technique that seeks to learn the Mahalanobis distance metric from labeled samples of the input. A transformation of the input space is achieved in the process that improves the efficiency of distance-based classifiers in the transformed input space.

Here, we review some basic formulae of input space transformation taking LMNN as an example since most of the other ML techniques easily relate to it. We have, in our work, tested and tried several ML techniques that justify our scheme of DST-ML-EkNN. The loss function for LMNN is the combination of the push and pull functions which is a function of the transformation function $\boldsymbol{L}$ as shown below.

$$\varepsilon(\boldsymbol{L}) = (1-\mu)\varepsilon_{\boldsymbol{push}}(\boldsymbol{L}) + \mu\varepsilon_{\boldsymbol{pull}}(\boldsymbol{L}) \tag{1}$$

The push function pushes away impostors from the target sample, while the pull functions attracts similar neighbors. The parameter μ is a value between 0 and 1 that balances these two acts. For more details on the push and pull functions, the reader is referred to [21]. The transformation matrix $\boldsymbol{L}$ is used to compute transformed distances between two vectors in the original space as

$$\boldsymbol{D}(\mathbf{x}_j, \mathbf{x}_k) = \left\| \boldsymbol{L}(\mathbf{x}_j - \mathbf{x}_k) \right\|_2^2 \tag{2}$$

where the samples $(\mathbf{x}_j, \mathbf{x}_k)$ belong to the training set $\boldsymbol{X}$. The transformed training set $\boldsymbol{X}'$ is given by

$$\boldsymbol{X}' = \boldsymbol{L} * \boldsymbol{X} \tag{3}$$

3 Cluster Formation of k-Nearest Neighbors in the Transformed Data Space

The transformation of the data space described in Sect. 2 brings similar valued samples closer to each other. These close-knit groups may contain members of both categories, no matter how efficient the transformation is. The nearest neighbor classifier is a typical choice for the classification in the transformed space [3]. Our next task is to improvise the k-Nearest Neighbor classifier, inspired by the ambiguity which surrounds the selection of k since it is usually data-specific. We exclude outliers which may have crept in the list of k-Nearest Neighbors by forming a two-cluster formation based on distances from the extreme members of the set as explained below.

Let $\boldsymbol{T}$ be the test sample and $\boldsymbol{T}'$ be the transformed test sample obtained by

$$\boldsymbol{T}' = \boldsymbol{L} * \boldsymbol{T} \tag{4}$$

Here, $\boldsymbol{L}$ is derived from (2) in the training phase. Let $\boldsymbol{S}$ be the set of k-Nearest Neighbors from the transformed *Training* set. Let $\mathbf{s_{max}}$ and $\mathbf{s_{min}}$ be the two extreme neighbors within $\boldsymbol{S}$ such that $\mathbf{s_{max}}$ is at a maximum distance from the transformed test sample and $\mathbf{s_{min}}$ is at a minimum distance from the transformed test sample. Two clusters of $\boldsymbol{S}$ are formed in a single step by comparing distance of each member **s** in $\boldsymbol{S}$ with the two extreme members $\mathbf{s_{max}}$ and $\mathbf{s_{min}}$. The two clusters formed are called $\mathbf{S_{max}}$ and $\mathbf{S_{min}}$, respectively. $\mathbf{S_{min}}$, we term as the 'Elite' cluster containing the closest neighbors of the transformed test sample excluding outliers if any.

$$s \in S_{\min}, |s - s_{\min}| \leq |s - s_{\max}| \tag{5}$$

and

$$s \in S_{\max}, |s - s_{\min}| > |s - s_{\max}| \tag{6}$$

A majority voting of class labels within the Elite cluster $\mathbf{S_{min}}$ provides the required class label of the test sample.

$$\boldsymbol{CLASS}(\boldsymbol{T}') = \underset{\forall s}{\arg\max}\, \boldsymbol{count}(\boldsymbol{CLASS}(s)) \tag{7}$$

Here, ***CLASS*** indicates the class label. The algorithm for the proposed method that we call as Data Space Transformation with Metric Learning and Elite k-Nearest Neighbor cluster formation (DST-ML-EkNN), is summarized below.

Algorithm DST-ML-EkNN
Input: *Training set*, *Test* sample
Parameter: ML method specific
Output: Class label of *Test* sample ***T***
1: Select ML method, say NCA
2: Perform metric learning on *Training set*,
3: Obtain L: Transformation matrix
4: Transform the *Test* input ***T*** by Eq. (4)
5: Find the Elite set of k-Nearest Neighbors in the *Training set* for the transformed *Test* sample as per Eqs. (5–7)
6: Majority voting in the Elite set returns the class label of the *Test* sample ***T***
7: **return** solution

The procedure of data space transformation is demonstrated for the *Wilt* dataset used in our experiments in Fig. 1 plotted for the Mean_NIR (mean NIR value) versus SD_Pan (standard deviation of Pan band) features of *Wilt*. More details can be found in the UCI repository Web site at [22]. The training set comprising of the even-numbered samples only is plotted in Fig. 1a. The red dot markers denote the majority samples (4265 in number), and the blue cone markers denote the minority samples (74 in number). The single green square marker in Fig. 1b denotes a test instance of the minority class. The training distribution with the single test sample embedded is

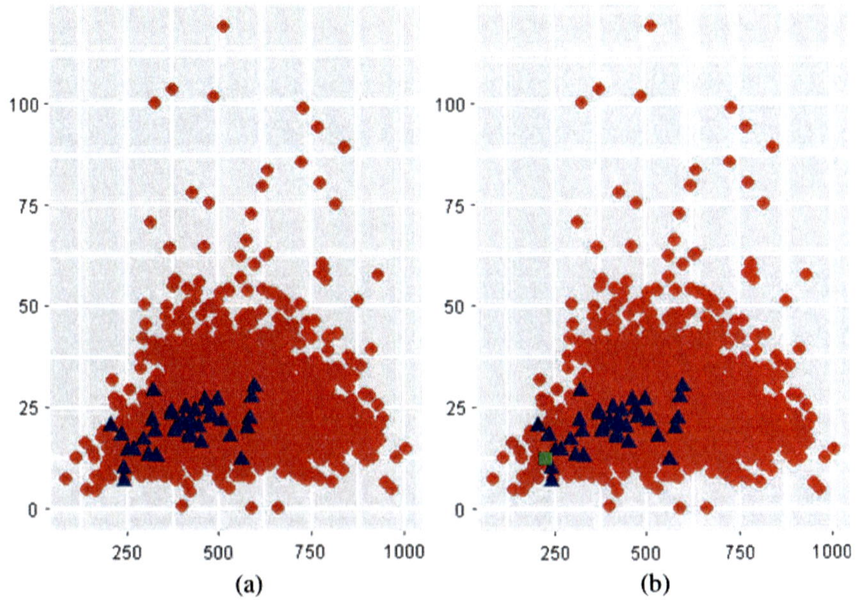

Fig. 1 Original distribution of training set (*Wilt* dataset—even-numbered samples). **a** Original training distribution without pure test sample embedded. **b** Original training distribution with test sample embedded (green square marker)

transformed by the metric learning method neighborhood component analysis (NCA) [23] and plotted in Fig. 2. The minority instances in the transformed distribution are shown grouped close together at the outer fringe of the majority samples. The (transformed) test instance is now surrounded by more minority instances than in the original distribution Fig. 1b.

4 Experimentation and Discussion

The experiments are conducted in JAVA programming platform and the IDE: Eclipse and Python software (for metric learning modules) on an Intel Core i3-4005U CPU with a clock frequency of 1.70 GHz. All our codes just take a few seconds to execute. The seven benchmark datasets used for the experimentation are shown in Table 1 along with their state of imbalance. These datasets are taken from the UCI repository available online [22] with the exception of *Redaktor* dataset which is one of the imbalanced datasets in [9].

They are each split into odd- and even-numbered samples (on a 50% basis). The even-numbered samples are applied for training initially and the odd samples for testing. As discussed in Sect. 2, only the training samples are subject to metric learning (ML) and the transformation matrix derived is applied to the test samples as

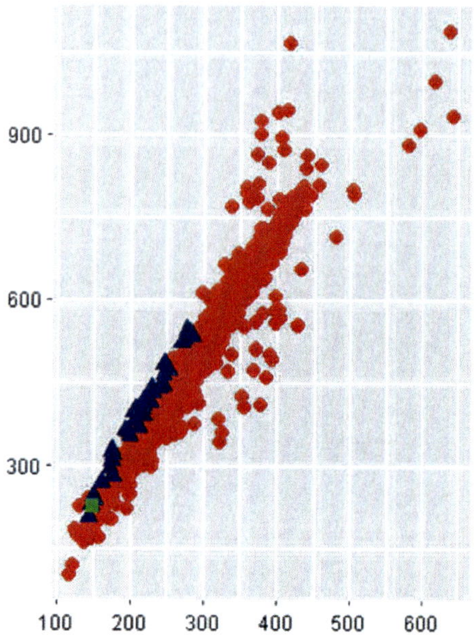

Fig. 2 Transformed training distribution of *Wilt* dataset with the (transformed) test sample embedded

Table 1 Datasets used for the experimentation along with their state of imbalance

Dataset	Minority	Majority	Imbalance ratio (= Majority/Minority)
Abalone	42	688	16.38
Contraceptive method	333	1140	3.42
Diabetes	268	500	1.865
Redaktor	27	149	5.51
Statlog (vehicle)	218	628	2.88
Wilt	74	4265	57.63
Wireless indoor	500	1500	3

well. The classification results are evaluated using F-score computed from precision and recall values as

$$F - score = 2(precision * recall)/(precision + recall) \tag{8}$$

where precision and recall are computed from the number of true positives (TP), false positives (FP), and false negatives (FN) of the predicted class labels as shown below.

$$precision = TP/(TP + FP) \tag{9}$$

Table 2 *F*-score for the proposed DST-ML-EkNN for different ML schemes

Dataset	MLKR	NCA	LMNN	ITML	LSML	SDML	MMC
Abalone	(0.619, 0.54)	(0.525, 0.594)	(0.552, 0.566)	(0.521, 0.672)	(0.375, 0.333)	(0.399, 0.269)	(0.399, 0.484)
Contraceptive method	(0.466, 0.40)	(0.478, 0.487)	(0.465, 0.456)	(0.442, 0.468)	(0.432, 0.446)	(0.459, 0.472),	(0.456, 0.467)
Diabetes	(0.668, 0.677)	(0.619, 0.667)	(0.670, 0.634)	(0.636, 0.683)	(0.493, 0.532)	(0.603, 0.518)	(0.652, 0.690)
Redaktor	(0.578, 0.226)	(0.705, 0.210)	(0.545, 0.327)	(0.545, 0.5)	(0.529, 0.454)	(0.310, 0.214)	(0.731, 0.310)
Statlog (Vehicle)	(0.869, 0.851)	(0.918, 0.944)	(0.935, 0.979)	(0.942, 0.947)	(0.593, 0.585)	(0.740, 0.831)	(0.863, 0.861)
Wilt	(0.821, 0.770)	(0.818, 0.742)	(0.885, 0.822)	(0.702, 0.520)	(0.711, 0.692)	(0.592, 0.304)	(0.821, 0.75)
Wireless indoor	(0.993, 0.996)	(0.995, 0.998)	(0.989, 0.993)	(0.971, 0.974)	(0.815, 0.810)	(0.990, 0.994)	(0.983, 0.972)

(V, CV): V Validation, *CV* Cross-validation

$$\boldsymbol{recall} = \boldsymbol{TP}/(\boldsymbol{TP} + \boldsymbol{FN}) \tag{10}$$

The classification results so obtained are termed as the validation (V) results. The training and testing sets are now swapped, and the results obtained are termed the cross-validation (CV) results. The k-Nearest Neighbor classifier is used for our method, and the effect of changing k is empirically studied. The variation of k in k-Nearest Neighbor cluster formation does not produce any effect when k is varied from $k = 3$ to $k = 19$. Our results are shown for $k = 7$ in Table 2. This invariance to the choice of k is due to the two-cluster formation of k-Nearest Neighbors prior to polling and considering only the Elite cluster.

We compare different metric learning (ML) techniques such as metric learning for kernel regression (MLKR) [24], neighborhood component analysis (NCA) [23], large margin nearest neighbor (LMNN) [21], information-theoretic metric learning (ITML) [25], Mahalanobis metric learning for clustering (MMC) [26], least squares metric learning (LSML) [27], and sparse discriminant metric learning (SDML) [28] for the application to our algorithm. The *F*-scores obtained by these techniques are summarized in Table 2 for our method. As observed, NCA, MLKR, and LMNN are the ML techniques that give the best performance in terms of high values of *F*-scores for our method DST-ML-EkNN. Table 3 compares the performance (for $k = 7$ in k-Nearest Neighbor classifier) of some state-of-the-art methods for imbalanced learning, namely the undersampling method—Sample Subset Optimization—optimized by Particle Swarm Optimization (SSO-PSO) [7], an adaptive oversampling technique ADASYN [6], and an intelligent variant of SMOTE called Borderline-SMOTE [5]. As observed by comparison, our methods in Table 2 show higher *F*-scores as compared to the existing methods in Table 3 for imbalanced learning, especially for the

Table 3 F-scores for other comparison methods

Dataset	SSO-PSO [7]	ADASYN [6]	Borderline-SMOTE [5]
Abalone	(0.491, 0.533)	(0.586, 0.698)	(0.410, 0.409)
Contraceptive method	(0.615, 0.657)	(0.589, 0.612)	(0.554, 0.578)
Diabetes	(0.786, 0.751)	(0.722, 0.738)	(0.768, 0.742)
Redaktor	(0.232, 0.269)	(0.539, 0.479)	(0.12, 0.133)
Statlog (Vehicle)	(0.818, 0.827)	(0.880, 0.880)	(0.838, 0.827)
Wilt	(0.224, 0.164)	(0.764, 0.693)	(0.678, 0.654)
Wireless indoor	(0.995, 0.998)	(0.995, 0.994)	(0.996, 0.998)

datasets: *Abalone*, *Redaktor*, *Statlog (Vehicle)*, *Wilt* and *Wireless indoor*. Our scores are specifically high for the severely imbalanced cases of *Wilt* for which the ratio of imbalance is very high (refer Table 1 for state of imbalance). The proposed technique of data space transformation of the original data distribution by metric learning coupled with the enhanced classification by kNN cluster formation forms an effective learning technique for imbalanced datasets. Our approach is devoid of problems, such as loss of information and duplicity of redundant information, that accompany sampling which is the popular remedy for imbalanced learning. Our method lays the foundation for further investigations into exploring data space transformation for imbalanced datasets with enhanced distance classifiers. Exploring hybrid methods that combine sampling with data space transformation techniques form the future scope of our work.

5 Conclusions

A new learning technique for overcoming the class-imbalance problem prevalent in real-world datasets is proposed in this paper that investigates distance metric learning (on the training set) for transforming the entire data space followed by enhanced kNN classification. The latter is composed of the application of the k-Nearest Neighbor classifier in a manner that excludes outlier neighbors and allows polling only for the 'Elite' k-Nearest Neighbors that are closest to the transformed test sample. Experimental results indicate high F-scores for our method, especially for severely imbalanced datasets, as compared to the state of the art on imbalanced learning. Distance metric learning by NCA and MLKR is observed to be the best data space transformation techniques for our application. We call our proposed method as Data Space Transformation with Metric Learning and Elite k-Nearest Neighbor cluster formation (DST-ML-EkNN). The future scope of our work involves exploring hybridizing data space transformation with sampling techniques.

References

1. Kriminger E, Principe JC, Lakshminarayan C (2012) Nearest neighbor distributions for imbalanced classification. In: The 2012 international joint conference on neural networks (IJCNN), pp 1–5. IEEE, New York
2. Chawla NV (2009) Data mining for imbalanced datasets: an overview. In: Data mining and knowledge discovery handbook, pp 875–886. Springer, Boston, MA
3. Wu G, Chang EY (2003) Adaptive feature-space conformal transformation for imbalanced-data learning. In: Proceedings of the 20th international conference on machine learning (ICML-03), Chawla, pp 816–823
4. Chawla NV, Bowyer KW, Hall LO, Kegelmeyer WP (2002) SMOTE: synthetic minority oversampling technique. J Artif Intell Res 16:321–357
5. Han H, Wang WY, Mao BH (2005) Borderline-SMOTE: a new over-sampling method in imbalanced data sets learning. In: International conference on intelligent computing. Springer, Berlin, pp 878–887
6. He H, Bai Y, Garcia EA, Li S (2008) ADASYN: adaptive synthetic sampling approach for imbalanced learning. In: 2008 IEEE international joint conference on neural networks (IEEE world congress on computational intelligence). IEEE, New York, pp 1322–1328
7. Yang P, Yoo PD, Fernando J, Zhou BB, Zhang Z, Zomaya AY (2014) Sample subset optimization techniques for imbalanced and ensemble learning problems in bioinformatics applications. IEEE Trans Cybern 44(3):445–455
8. Lin WC, Tsai CF, Hu YH, Jhang JS (2017) Clustering-based undersampling in class-imbalanced data. Inf Sci 409:17–26
9. Bennin KE, Keung J, Phannachitta P, Monden A, Mensah S (2018) Mahakil: diversity based oversampling approach to alleviate the class imbalance issue in software defect prediction. IEEE Trans Softw Eng 44(6):534–550
10. Ramentol E, Caballero Y, Bello R, Herrera F (2012) SMOTE-RSB*: a hybrid preprocessing approach based on oversampling and undersampling for high imbalanced data-sets using SMOTE and rough sets theory. Knowl Inf Syst 33(2):245–265
11. Susan S, Kumar A (2019) SSOMaj-SMOTE-SSOMin: three-step intelligent pruning of majority and minority samples for learning from imbalanced datasets. Appl Soft Comput 78:141–149
12. de Morais RF, Vasconcelos GC (2019) Boosting the performance of over-sampling algorithms through under-sampling the minority class. Neurocomputing
13. Susan S, Kumar A (2018) Hybrid of intelligent minority oversampling and PSO-based intelligent majority undersampling for learning from imbalanced datasets. In: International conference on intelligent systems design and applications. Springer, Cham, pp 760–769
14. Yang L, Jin R (2006) Distance metric learning: a comprehensive survey. Mich State Univ 2(2):4
15. Susan S, Kakkar G (2015) Decoding facial expressions using a new normalized similarity index. In: 2015 annual IEEE India conference (INDICON). IEEE, New York, pp 1–6
16. Susan S, Hanmandlu M (2013) Difference theoretic feature set for scale-, illumination-and rotation-invariant texture classification. IET Image Proc 7(8):725–732
17. Susan S, Chandna S (2013) Object recognition from color images by fuzzy classification of gabor wavelet features. In: 2013 5th international conference and computational intelligence and communication networks. IEEE, New York, pp 301–305
18. Susan S, Sharma S (2012) A fuzzy nearest neighbor classifier for speaker identification. In: 2012 fourth international conference on computational intelligence and communication networks. IEEE, New York, pp 842–845
19. Susan S, Kadyan P (2013) A supervised fuzzy eye pair detection algorithm. In: 2013 5th international conference and computational intelligence and communication networks. IEEE, New York, pp 306–310
20. Wang N, Zhao X, Jiang Y, Gao Y, BNRist KLISS (2018) Iterative metric learning for imbalance data classification. IJCAI, pp 2805–2811
21. Weinberger KQ, Saul LK (2009) Distance metric learning for large margin nearest neighbor classification. J Mach Learn Res 10:207–244

22. Frank A, Asuncion A (2010) UCI machine learning repository http://archive.ics.uci.edu/ml. University of California, School of Information and Computer Science, Irvine, CA
23. Goldberger J, Hinton GE, Roweis ST, Salakhutdinov RR (2005) Neighbourhood components analysis. In: Advances in neural information processing systems, pp 513–520
24. Weinberger KQ, Tesauro G (2007) Metric learning for kernel regression. In: Artificial intelligence and statistics, pp 612–619
25. Davis JV, Kulis B, Jain P, Sra S, Dhillon IS (2007) Information-theoretic metric learning. In: Proceedings of the 24th international conference on machine learning. ACM, pp 209–216
26. Bar-Hillel A, Hertz T, Shental N, Weinshall D (2003) Learning distance functions using equivalence relations. In: Proceedings of the 20th international conference on machine learning (ICML-03), pp 11–18
27. Xing EP, Jordan MI, Russell SJ, Ng AY (2003) Distance metric learning with application to clustering with side-information. In: Advances in neural information processing systems, pp 521–528
28. Qi GJ, Tang J, Zha ZJ, Chua TS, Zhang HJ (2009) An efficient sparse metric learning in high-dimensional space via l 1-penalized log-determinant regularization. In: Proceedings of the 26th annual international conference on machine learning. ACM, pp 841–848

Classification Study and Prediction of Cervical Cancer

Kaushik Suresh

Abstract The cancer disease has created a grave threat to the life of the people over more than a decade. As time increases, the problems and cases related tend to grow at a higher scale with respect to the different lifestyles lead by people. One such cancer, which plays a dominant role in the death of many women, is cervical cancer. This study helps us in classifying the presence of cervical cancer among subjects using machine learning classification approaches, such as logistic regression, decision tree, k nearest neighbor (KNN), random forest and support vector machine by constructing effective ensemble models and overcoming the biased nature of the class distribution in the dataset in order to identify the best classifier model to identify the presence of cervical cancer using the behavioral data of the subjects. Models are constructed based on the different types of input factors, such as lifestyle, habits, medical history of the patients and sexual practice of the patient. Among all the models constructed using different machine learning classification algorithms KNN algorithm provides the best performance in classifying the patients under two categories, such as affected by cervical cancer and not affected by cervical cancer for multiple target variables, like Hinselmann test, Schiller, cytology and biopsy test and discussing the important features in the dataset responsible for the cause of cervical cancer by identifying the variable importance using random forest approach.

Keywords Cervical cancer · Classification · Class imbalance · Ensemble models · Sampling

1 Introduction

One out of every six deaths that occur is due to cancer, where the count stands with 171.2 per 100,000 die out of cancer, where the death rate of women counts for 145.4 in every 100,000 deaths recorded. Women suffer from different types of cancer like breast, colon, endometrial, lung, cervical, skin and ovarian cancers. Cervical

K. Suresh (✉)
PSG College of Technology, Coimbatore, India
e-mail: Kaushikkash147@gmail.com

N. N. Chiplunkar and T. Fukao (eds.), *Advances in Artificial Intelligence and Data Engineering*, Advances in Intelligent Systems and Computing 1133,
https://doi.org/10.1007/978-981-15-3514-7_27

cancer is one of the rare types of cancer, which causes the death of 2.3 women for every 100,000 women. A recent study indicates nearly about 67.1% of women survive 5 years or more after being diagnosed for cervical cancer [1]. Unaware of the presence of cervical cancer until the growth of cancer has reached the advanced stage and due to the lack of awareness among women is found to be a common aspect for the death of the most number of women. The main types of cervical cancer are squamous cell carcinoma and adenocarcinoma [2]. One of most causative agent for cervical cancer is Human Papilloma Virus (HPV) resulting in long-lasting infections, which spread cancer throughout the reproductive parts (uterus, cervix, vagina, ovaries) and chances of experiencing a kidney failure is also possible due to the presence of cancer cells which blocks the urinary bladder and affecting the entire pelvic region. The International Federation of Gynecology and Obstetrics (FIGO) staging system is used most often for cancers of the female reproductive organs, including cervical cancer (staging from I to IVb) [3]. Medical database plays a major role as it is associated with people's lives, medical history and other related health information resulting in the data to be huge, complex and unstructured [4].

Classification plays an important role in data mining and machine learning. A large number of classification algorithms have been implied to study the medical application related to the medical data to identify substantial patterns for the betterment and prediction of the health related to the patients. Class imbalance is one of the major challenges faced during the classification of the medical data; therefore in many cases, a skewed distribution is observed as the number of instances in the majority and the minority class is not equally balanced. In cases of class imbalance, the primary interest lies in the minority class. The cost of misclassifying a minority class is more expensive than the misclassification of the majority class [5, 6]. Various methods are introduced in order to handle the class imbalance [7, 8]. Several issues related to the learning with skewed class distributions have been briefly studied in [9]. Some of the imbalance resolving methods are such as data-level methods [9], algorithm level methods, cost-sensitive [4, 10] and ensembles [11]. A survey paper [12] summarizes the current research works on high-class imbalance issues in big data investigating the research works conducted from 2010 to 2018. Methods to overcome the imbalance are by either adding minority class data called as oversampling or removal of the majority class data called as undersampling. Both undersampling and oversampling are performed until there is a balance between the majority class and minority class. Undersampling techniques involve less training data with less time and computation but suffer from information loss as the majority class data records are removed in order to balance with the minority class. If the deletion of the instances in the majority class is large, it could possibly change the distribution of the majority class which is representative of the given domain [12]. Unlike undersampling techniques oversampling provides better accuracy with no information loss but the addition of new data records might end up increasing the noise and chances to overfit the model.

Oversampling a non-heuristic approach, which balances the class distribution, consists of a wide range of sampling techniques. One such commonly and frequently used oversampling technique is Synthetic Minority Oversampling Technique

(SMOTE), where oversampling of the minority class is done by generating synthetic minority examples [13, 14]. SMOTE identifies the new minority class examples by interpolating between several minority class examples that lie close to each other [15]. Methods like Random Undersampling (RUS) and Random Oversampling (ROS) are relatively simpler methods compared to SMOTE approach [16]. An oversampling algorithm based on preliminary classification (OSPC) proposed in [17], where the useful information of the majority class is mostly saved by performing a preliminary classification on the test data. Therefore, the test data which is predicted to belong to the minority class is reclassified in order to obtain an improvement in the classification of the minority class. Classification performance of the majority and minority class is argued to be better by using OSPC compared to undersampling techniques and SMOTE [18]. The heuristic approach of undersampling requires shorter training time but with the chances of removing the potentially useful data. OSS randomly obtains one majority class example with all the other minority class examples in order to eliminate the examples from the majority class that is found distant from the decision boundary [19]. Wilson's Edited Nearest Neighbor Rule (ENN) removes any given example, whose class label differs from the class of at least two of its three nearest neighbors [20]. Comparison and combination of various sampling strategies have been discussed in detail in [21]. Studies [21–24] show that under extreme skewed conditions of the data both undersampling and oversampling are often combined in order to provide an improved generalization of the learner. Ensemble learning helps in providing better performance in the process of classification by combining several models. A significant amount of increase in performance has been observed by using ensemble-based models for classification compared to the single model approach [25]. Usage of ensemble idea in supervised learning has been studied since the late 1970's, where [26] proposed combining two linear regression models. Later the partitioning of the input space by two or more classifiers was proposed by Dasarathy and Sheel [27]. A hybrid ensemble approach for the classification of the medical data is proposed in [25]. In [11] support vector machine (SVM)-based ensemble approach has been used to address the issue of predicting rare class using various classifier combinational strategies like majority voting, sum-rule and hierarchical SVMs. Binary classification problems have been studied using ensemble classifier, has been discussed in order to sample several subsets from the majority class and training a learner using each of the subsets and combine the outputs of the learners followed by an approach based on training the learners sequentially, where the correctly classified majority class samples are removed for further consideration [28]. Performance of various different ensemble methods, while using tree-based classifiers as base learners, has been studied under the imbalanced data conditions in [29].

The data which is used to build the classification models is obtained from UCI Repository [30] collected at Hospital Universitario de Caracas' in Caracas, Venezuela initially used in [31] and later studied in [4, 32, 33]. The data depicts the factors of a patient's lifestyle and their past medical data and certain specified habits which are considered as an causative agent for cancer has been studied in form of mixture of both categorical and numerical data factors. The dataset consists of 858 patient's data with

32 features and four classes (Colposcopy using acetic acid—Hinselmann, colposcopy using Lugol's iodine—Schiller, cytology and biopsy). The dataset consists of some NA values as some of the patients considered in the study refused to answer due to personal reasons. And those missing NA values have been replaced by sample mean. The data suffers from class imbalance, where class 0 is the majority class and class 1 is the minority class. The paper focusses on the classification of the dataset in two different ways. The first approach is by building traditional machine learning classification model to predict the presence of cervical cancer with the imbalanced dataset with a split of training dataset (75%) and testing dataset (25%). The second approach is by resolving the imbalance in class distribution by balancing the majority and minority class by constructing sampled datasets with the equal amount of majority and minority class and the construction of classification models using $n - 1$ parallel ensemble architecture.

Classification algorithms like logistic regression, decision tree, random forest, support vector machines and KNN are used. The sampling process involved in this study to overcome the imbalance in the cervical cancer dataset does not involve the addition of minority class data records or the removal of the majority class data records, therefore leaving no information loss or the increase of noise. Multiple copies of the same minority class training data are considered with the randomly generated non-repetitive majority class data which is expressed in detail in Sect. 2. The sampled data obtained is further classified by different algorithms in an ensemble fashion to provide better performance in the prediction of cervical cancer among patients. In [32], the cervical cancer data has been studied using diverse classification techniques and points out the advantage of feature selection approaches for the prediction of cervical cancer claiming that decision tree provides higher performance. Similarly, in [33] three SVM-based approaches such as standard SVM, SVM-RFE (recursive feature elimination) and SVM-PCA (principal component analysis) are used to classify cervical cancer. The standard SVM outperforms SVM-RFE and SVM-PCA but SVM needed 30 out of 32 features to gain a higher accuracy. SVM-RFE and PCA were able to achieve the classification with a decent accuracy by reducing the number of features from 30 to 8. Fatlawi [4] studied cervical cancer data using cost-sensitive classification with the decision tree. A comparison of results obtained in [4, 32, 33] is made with the results obtained from this study.

The paper is organized as follows: Sect. 2 deals with the class imbalance in the cervical cancer dataset and the method to overcome the imbalance and briefly discusses the use of performance measures considered in the study to evaluate the classification models. Section 3 deals with the $n - 1$ ensemble architecture of the classification models, which are trained for the prediction of cervical cancer among patients using the sampled datasets with an equal balance of majority and minority data. Section 4 deals with the classification algorithms used in this study and 2×2 confusion matrix obtained from each classification algorithm for all the four target variables (screening tests for cervical cancer) in the cervical cancer dataset for the normal and the proposed approach. Section 5 deals with the results obtained from the normal single classification models trained with the initial imbalanced dataset and their inability to classify the patients, who are affected by cervical cancer and discusses implementing an $n - 1$ ensemble parallel architecture of the classification algorithms trained with the balanced sampled datasets to overcome the imbalanced situation. Section 5 also discusses the variables, which plays a major role in the occurrence of cervical cancer in women. Finally, we conclude in Sect. 6.

2 Class Imbalance and Performance Measures

2.1 Imbalanced Cervical Cancer Data

The dataset holds record of 858 patients with 32 different features (20 categorical and 12 numerical features) and with four cervical cancer screening tests (Hinselmann, Schiller, cytology and biopsy) as target variables holding the value 0 and 1, where 0 denotes the absence of cervical cancer in patients and 1 denotes the presence of cervical cancer. Among all the four target variables the number of patient's records with class 0 is high compared to the number of patients with class 1 leaving to the fact that class 0 as the majority class and class 1 as the minority class. Therefore, in such cases, when the degree of the class imbalance is high for the majority class, the classifier built without resolving the class imbalance shows a higher overall rate of accuracy as the model likely predicts most of the data from majority class [34]. In the case of medical data of the patients, adding or removing data might end up in less inference and is not accurate every time. Therefore, in this paper to construct a model with unbiased nature giving equal importance to both majority and minor class data for the better prediction of cervical cancer is carried out by placing the minority class training data as constant with randomly selected different subsets of majority class training data. So in this study balance of the cervical cancer dataset is achieved without removal, addition or duplication of data records.

2.2 Random Sampling of Majority Class Based on the Size of Minority Class

For each target variable (Hinselmann, Schiller, cytology and biopsy) is being sampled into the sample dataset (s_n) with the size S. The size of the sample dataset (s_n) depends on the OLC (Occurrence of the Least Class in the dataset) value. N denotes the number of classes in the dataset ($N = 2$) and OLC is an integer value which holds the number of instances present with the minority class (class 1). The OLC values tend to differ for each target variable. The number of training datasets built depends on the OLC Value. In sample dataset (s_n), where $S/2$ data points are occupied by the randomly non-repetitive subset of majority class, data records and the other half of the dataset will consist of 50% of minority class data, which is made constant in every sampled dataset (s_n). The rest 50% of the minority class is kept for the testing purpose in the test data s_n.

$$\text{Size of the Training Dataset}\,(S) = N * (\text{OLC} * 0.5) \tag{1}$$

$$S_n = \begin{cases} \frac{S}{2}, & \text{Random selected non-repetitive majority class} \\ \frac{S}{2}, & \text{Minority class training data} \end{cases}$$

Equation (1) is the product of the number of classes in the classification study where ($N = 2$) with half of the value of the minority class as the other half is considered as the test data. In this case, the size of each sample set (S_n) generated is always twice the size of the minority class training dataset as the majority class is been reduced in order to balance with the minority class. Random selection of data points with class 0 for each target variable is made. **f** is the fixed percentage value of dataset denoting the population of the training dataset which ranges from 0 to 100%. The **f** value has been set to 50% in this study.

$$F_{\text{train}} = (\text{nrows}(D) * f \tag{2}$$

$$n = \frac{F_{\text{train}}}{\text{OLC}} \tag{3}$$

In our case of scenario, **f** value is fixed to 50% and nrows(D) is 858 by substituting the values in Eq. (2). Obtain $\mathbf{F_{train}}$ value is obtained as 429. The $\text{OLC}_{\text{Hinselmann}}$ value for Hinselmann target variables is 36. Substituting the OLC value of the Hinselmann target variable in Eq. (3) the number of training sample sets (n) for the target variable Hinselmann is identified by dividing the OLC value of Hinselmann and F_{train} value, where the n value is 11. The values of the other target variables Schiller, cytology and biopsy are shown in Table 1. Sample dataset s_1 to s_{n-1} is considered for the training purposes and the s_n data is used for the testing as shown in Fig. 1a–c.

Table 1 Number of models to be built for each target variable

Target variable	OLC value	n (No of training sample sets)	Models built ($n - 1$)
Hinselmann	36	11	10
Schiller	75	5	4
Cytology	56	7	6
Biopsy	56	7	6

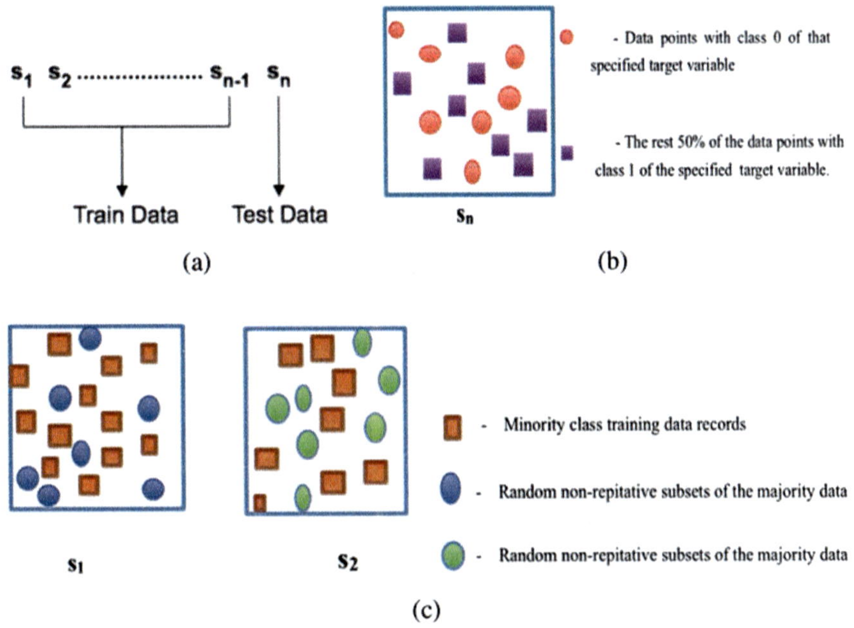

Fig. 1 **a** Split among the sampled dataset (s_1 to s_{n-1}) into training and testing data generated for each target variable T_i, $i = 1{:}4$. **b** Test data to evaluate the models which are constructed using the sample datasets from s_1 to s_{n-1}. **C** Sample datasets s_1 and s_2 with size S, consisting of $S/2$ randomly generated non-repetition of majority class data along with $S/2$ minority class data, which is considered constant among the training data

2.3 Performance Measures

The performance measure of the models M_i, $i = 1$ to $n - 1$ for each target variable T_i, $i = 1$–4 is identified by using True Positive (TP), False Positive (FP), True Negative (TN) and False Negative (FN) values forming a 2×2 confusion matrix as shown in Table 2. TP denotes the correct classification of the model by predicting the absence of cervical cancer in patients; TN denotes the correct classification of the number of patients affected by cervical cancer. FP denotes the incorrect classification, therefore predicting the patients with the absence of cervical cancer as being affected by cervical cancer. FN is opposite to FP, where the classifier wrongly predicts the patients with cervical cancer as not affected by cervical cancer.

Table 2 2×2 confusion matrix for binary classification

		Prediction	
		0	1
Actual	0	TP	FN
	1	FP	TN

$$\text{Accuracy} = \frac{\text{TP} + \text{TN}}{\text{TP} + \text{TN} + \text{FP} + \text{FN}} \tag{4}$$

$$\text{Precision} = \frac{\text{TP}}{\text{TP} + \text{FP}} \tag{5}$$

$$\text{Recall} = \frac{\text{TP}}{\text{TP} + \text{FN}} \tag{6}$$

$$\text{FPR} = \frac{\text{FP}}{\text{FP} + \text{TN}} \tag{7}$$

$$\text{TPR} = \frac{\text{TP}}{\text{TP} + \text{FN}} \tag{8}$$

$$F\text{-measure} = \frac{(2 \times \text{Recall} \times \text{Precision})}{(\text{Recall} + \text{precision})} \tag{9}$$

One of the commonly used approach to identify the performance measure of a model is the accuracy as it determines the overall performance of the classifier model in predicting the classes. But accuracy suffers from a disadvantage of classifying the minority class as the accuracy measure place more weight on the majority class compared to the minority class [18]. So certain various evaluation metrics based on confusion matrix have been implied to better understand the models, which are used for the prediction of cervical cancer among patients. Apart from accuracy measure certain other metrics like precision, recall, Truth Positive rate (TPR), False Positive rate (FPR) and F-measure is used Eqs. (4–9). Recall helps us in identifying the model's ability to classify the absence of cervical cancer among patients. Low precision and recall are observed in the minority class compared to the majority class. TPR stands for the percentage of correctly classified un-affected cervical patients shown in Eq. (8) and FPR denotes the percentage of misclassified cervical cancer patients represented in Eq. (7). F-measure is the harmonic mean obtained from the combination of precision and recall, which is one of the suitable and popular metrics used in case of imbalanced data [35]. Increase in the precision and recall value leads to the increase in the F-measure value Eq. 9

3 $n - 1$ Ensemble Model Architecture

Ensemble methods are extensively used to handle class imbalance problems as they combine the result of many classifiers as their base learners are usually of diversity in principle or induced with various class distributions [18]. The idea of ensemble classifier in [36] has been adopted based on the fact that different classifier produces different opinions and therefore those results are combined in order to achieve higher performance as they deal great importance in the classification process. In [37] under conditions like class imbalance better classification, results are obtained

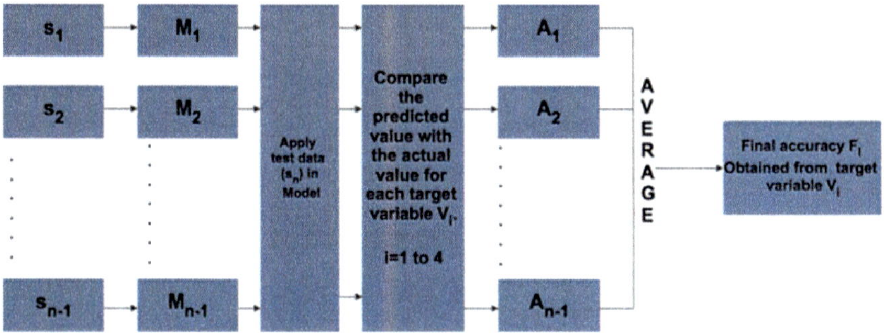

Fig. 2 $n - 1$ ensemble classification architecture with the balanced sample datasets (s_1 to s_{n-1}) for training and s_n for testing

due to ensemble-based techniques. Models $M = \{M_1, \mathrm{M}_2, \ldots, M_{n-1}\}$ are designed and implemented in a speculated parallel way of computation in an ensemble fashion with the help of the sample datasets obtained s_1 to s_{n-1}, $\mathbf{T_i}$ (i = 1–4), denotes the four target variable present in the dataset. Thus (n − 1) models are built for each target variable T_i (Hinselmann test, Schiller test, cytology test and biopsy test). The number of models (n) built for each target variable differs as shown in Table 1. Based on the OLC value of the target variable, the number of sample datasets and models to be built is identified. For all the four target variables, sample dataset s_1 to s_{n-1} is considered for training the ensemble models, which therefore results in the $n - 1$ ensemble model architecture and the s_n dataset is used for testing purposes as shown in Fig. 2. Important characteristics of using the sampled datasets with different randomly selected data points from the majority class along with the minority class training data, which is maintained as constant from s_1 to s_{n-1}, are to ensure that the models possess a strong prediction in the presence of cervical cancer and resolves the issue of the model turning out into a biased model.

4 Classification Algorithms

Classification algorithms like logistic regression, support vector machine, decision tree, random forest and KNN are used for the classification of the patients based on their features. A comparative classification of study is being carried out between the imbalanced dataset and the balanced dataset. Initially, the classifier is constructed with the initial imbalance cervical dataset with 75% (643 patients) of the data as training and the rest 25% (215 patients) of the data as testing data. Similarly, ensemble models of the classifier based on the figure are constructed for the balanced dataset and compared with the results obtained from the traditional approach without resolving the class imbalance problem. 2 × 2 confusion matrices are obtained for each classification algorithm considered in the study with respect to the target variable is obtained as shown in Fig. 3 for both balanced data and imbalanced dataset.

	0	1
0	201	4
1	7	3

Hinselman

	0	1
0	165	31
1	13	6

Schiller

	0	1
0	204	0
1	9	2

Citology

	0	1
0	201	0
1	12	2

Biopsy

(a)

	0	1
0	17	1
1	1	17

Hinselman

	0	1
0	34	3
1	1	36

Schiller

	0	1
0	26	2
1	1	27

Citology

	0	1
0	26	2
1	1	27

Biopsy

(b)

	0	1
0	206	9
1	0	0

Hinselman

	0	1
0	193	22
1	0	0

Schiller

	0	1
0	204	10
1	1	0

Citology

	0	1
0	200	15
1	0	0

Biopsy

(c)

	0	1
0	17	1
1	0	18

Hinselman

	0	1
0	35	2
1	1	36

Schiller

	0	1
0	26	2
1	2	26

Citology

	0	1
0	27	1
1	1	27

Biopsy

(d)

Fig. 3 **a** 2×2 confusion matrix of logistic regression using A1 approach for all four target variables. **b** Confusion matrix of logistic regression using A2 approach. **c** Confusion matrix obtained by using decision tree with A1 approach. **d** Confusion matrix obtained by applying a decision tree with A2 approach for all four target variables

The performance measure of the classifier models is determined based on the TP, TN, FP and FN values. A1 denotes the traditional classification models trained with 75% of the initial imbalanced data and 25% as the testing data. A2 denotes the ensemble models built as shown in Fig. 2 with the balanced data obtained from the random sampling of majority class data without repetition and the minority class data as constant data among the subsets generated from s_1 to s_{n-1} as shown in Fig. 1.

Logistic regression is one of the most commonly used classification algorithms. The algorithm uses maximum likelihood estimation (MLE) to evaluate the model parameters to maximize classification accuracy. But in case of imbalanced data scenarios, logistic regression does not consider the difference between the majority and minority class, therefore not considering the fact that the minority class is more valuable than the majority class in conditions like class imbalance, which makes the model incorrectly classify the minority class [38]. The classification performance of the decision tree is based on the degree of balance in the class distribution, which makes the decision tree strongly sensitive to imbalance situations [39]. Random forest algorithm works as a large collective decorrelated decision tree considering each decision as the subset and classifying them accordingly [40]. KNN is one of the most widely used techniques in the classification process [41]. The instance base classifier KNN is simple and intuitive but also ends up incorrectly classifying the minority class data as the samples of the minority class are sparsely present in the dataset [42].

One of the important parameters in the KNN algorithm is the k value [43], in fact, there is no accurate value for k and it depends on the data distribution and space of the problem. KNN algorithms mainly focus on the improvement of classification accuracy in terms of balanced data. But fail to achieve the same kind of accuracy in terms of imbalanced data. In [43] an improvised KNN algorithm based on the sample mean has been proposed in order to handle the imbalanced data. The classification performance of the KNN algorithm depends on the how optimal the number of neighbors (k value) is chosen, therefore choosing of the optimal k value is almost impossible for a variety of problems; as the classification performance differs in great deal accordance to the k value, therefore, an ensemble KNN classifier is used each time with different k values ranging from one to the square of the size of the training data [44]. Many research studied related to the determination k value has been done, where [45] choose $k = 3$, $k = 1–10$ in [46], $k = 1–50$ in [47] and $k =$ square of (number of samples) [48, 49]. Larger k values present less sensitive to noise and make smoother between the boundaries of class. Mostly the optimal k value for most of the datasets lies between 3 and 10 [50]. Since the cervical cancer dataset being handled in this study is too crucial and the inappropriate choice of the k value might end up in less accuracy of the model; therefore, k values are considered from the range of 1 to the square root of the number of patients data considered in the study ($\sqrt{858} = 29.29$ considered as 30). Higher accuracy is achieved when $k = 10$ and the accuracy eventually decreased with increase in k value.

Logistic regression and the other classification algorithms considered in the study are applied to both imbalanced and balanced data. The 2×2 confusion matrix obtained from the imbalanced data (A1 approach) clearly shows that the models fail in predicting the minority class as the model tends to be biased over the majority class projecting a higher FPR rate, where the model fails in identifying the patients with cervical cancer. But in case of A2 approach shows a lower FPR rate compared to the FPR rate obtained from the traditional model approach classifiers trained with imbalanced data. The F-measure and accuracy also tend to be higher compared to the traditional approach models.

Figures 3a, b and 4a, b are the 2×2 confusion matrix obtained from logistic regression and decision tree using both imbalanced data and balanced data. Based on Figs. 3a, c and 4a, it clearly shows that the FP and TN values are very less compared to the values of the TP and FN; therefore, this is due to the fact that the traditional classification model approach trained with the initial imbalanced data turns out to be a biased model toward class 0 due to the presence of more number of class 0 patient's records (absence of cervical cancer in patients), which lead to the inability of the model to classify the class 1 records (presence of cervical cancer patients). Classifiers like logistic regression and decision tree tend to bias toward the majority class and tend to predict only the majority class as they consider the features related to the minority class as noise and ignore it [51]. On the other hand, in respective Figs. 3b, d and 4b the confusion matrix show us that the $n - 1$ level ensemble classifier models trained with the sample subsets obtained by random selection of majority data with constant minority class data points were not only able to identify the patients with cervical cancer diseases but also provided a higher accuracy, precision, F-measure values and lower FPR rates compared to the traditional classifiers approach models trained with imbalanced data.

(a)

	0	1
0	206	0
1	3	6

Hinselman

	0	1
0	195	0
1	9	11

Schiller

	0	1
0	204	0
1	4	7

Citology

	0	1
0	203	0
1	6	6

Biopsy

(b)

	0	1
0	18	0
1	0	18

Hinselman

	0	1
0	37	0
1	0	37

Schiller

	0	1
0	28	0
1	0	28

Citology

	0	1
0	28	0
1	1	27

Biopsy

Fig. 4 **a** 2×2 confusion matrix of KNN algorithm over the initial imbalanced dataset (A1 approach). **b** Confusion matrix of all four target variables using KNN over the balanced dataset obtained by A2 approach to classify the presence or absence of cervical cancer

5 Results and Discussion

Based on the results obtained from the classification of the presence or absence of cervical cancer in balanced and imbalanced dataset using classification algorithms such as logistic regression, SVM, decision tree, random forest and KNN clearly show that the A2 approach built with $n - 1$ ensemble classifiers trained with the balanced data outperforms the A1 approach trained with imbalanced data; therefore, A1 tends to be a biased model due to its inability to classifying the patients with cervical cancer and A2 overcomes this disadvantage of A1 by providing an unbiased model helping us to better classify the patients in order to identify the presence or absence of cervical cancer. KNN with A2 approach shows a better accuracy, precision, recall and F-measure values compared to the other classification models as shown in Table 3. Performance measures of A1 and A2 approach over classification algorithms like logistic regression (LR), decision tree (DT), support vector machine (SVM), random forest (RF) and k nearest neighbor (KNN) for the classification of cervical cancer among patients is shown in Fig. 5a, b.

Compared to the other classification models the performance results obtained from logistic regression using A2 approach is less. Decision tree, random forest and support vector machine values are quite comprising compared to logistic regression during the classification of the presence of cervical cancer in patients but less compared to the values obtained from the KNN models using A2 approach. KNN models provide a better prediction as they hold higher performance measure values under all four metrics considered in the study. Accuracy proves to be a less informative inference in case of imbalanced data scenarios; therefore, measures like F-measure are also used in order to evaluate the complete performance of each and every model. The tree-based classifiers hold a god FPR rate in both A1 and A2 approaches thus ensuring that the tree-based classifiers work well in imbalanced data situations. Logistic regression holds the highest FPR rates followed by the SVM compared to the other classification algorithms. SVM, therefore, has been claimed less prone to the class imbalance problem than other classification algorithms [41]. Considering all the inferences obtained from the study in classifying the presence of cervical cancer or not in patients, KNN model provides a better classification model. The result obtained from the A2 approach is compared with [32, 33] and [4] as shown in Table 4. Compared to the results obtained from [4] a higher TPR (sensitivity), precision and F-measure have been observed in [32] by considering the decision tree as the recommended classifier. Table 4 clearly shows the performance of the study carried out with the other published works. The KNN algorithm with A2 approach provides a higher F-measure value compared to the cost-sensitive decision tree in [4], decision tree in [3] and SVM in [33] providing a better classification for the identification of cervical cancer in patients.

Table 3 Performance measures of classification algorithms with balanced (A2) and imbalanced dataset (A1)

	Logistic regression		Decision tree		kNN		Random forest		SVM	
	A1	A2	A1	A2	A1	A2	A1	A2	A1	A2
Accuracy	91.15	94.57	93.14	95.62	97.43	99.55	94.88	98.01	94.06	97.31
Precision	94.86	96.04	99.87	96.62	97.33	99.13	99.8	97.72	95.61	96.70
Recall	95.55	93	93.47	94.57	100	99.16	94.29	97.21	93.81	95.79
FPR	76.10	5.58	0.25	3.35	41.09	0.92	0	4.5	53.27	9.82
F-measure	84.63	94.48	96.55	95.57	98.62	99.16	96.96	97.46	94.82	97

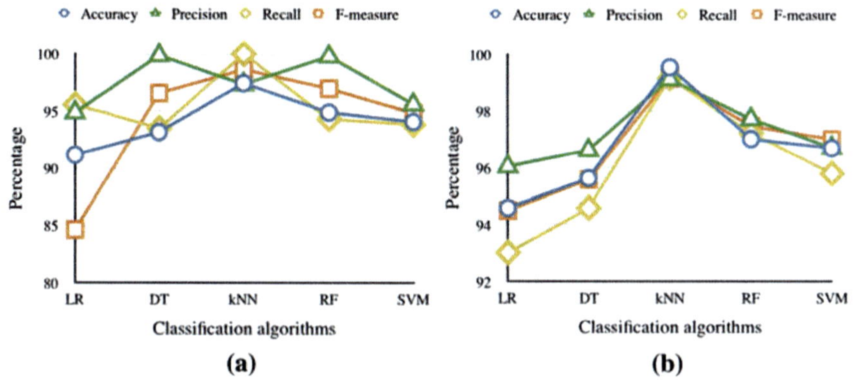

Fig. 5 **a** Performance measures obtained from the traditional classification approach with imbalanced data. **b** Performance measures of $n-1$ ensemble classifier models constructed with balanced data

Table 4 Comparison of results obtained in this study with the other published works [4, 32, 33]

	[32]	[4]	[33]	Obtained result	Obtained result	Obtained result
Algorithm	DT	Cost-sensitive DT	SVM	SVM (A2)	DT (A2)	KNN (A2)
Accuracy	97.515	–	92.6075	97.31	95.629	99.55
Precision	100	42.9	99.68	96.70	96.62	99.13
Recall	95	42.9	89.25	95.79	94.57	99.16
F-measure	97	30.6	90.98	97	95.57	99.16

5.1 Variable Importance

The important features among the 32 set of features in the cervical cancer dataset, which plays a dominant role in the occurrence of cervical cancer, are identified using a random forest approach. Based on the performance measures, results obtained for the various considered classification algorithms in this study random forest have been choose to identify the variable, though KNN algorithm was argued to outperform the other classification algorithms in the previous sections and random KNN [52] for variable importance is fast and a more stable alternative approach compared to random forest in certain scenarios like when the data is very large in dimension consisting of thousands of variables with multiple class. Therefore, based on the cervical data studied in this paper deals with less number of features with binary class in nature, random forest approach is used to study the important aspects responsible for the cause of cervical cancer in patients as the feature selection is based on Gini importance and providing an advantage of double benefit in the process by achieving dimensionality reduction and elimination of noise [53]. Mean Decrease in Gini is

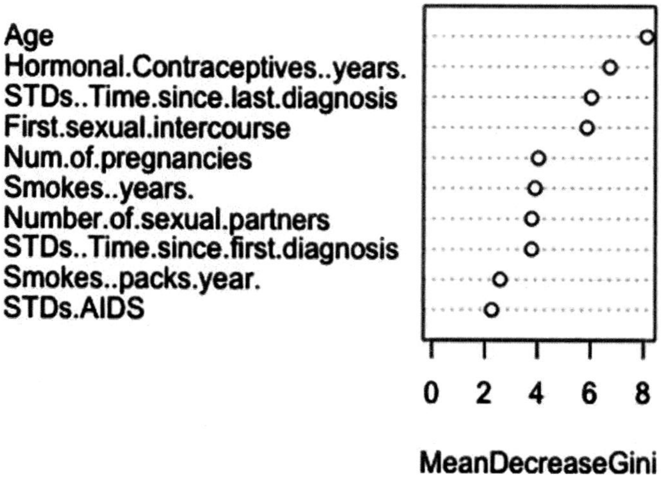

Fig. 6 Important factors responsible for the cause of cervical cancer using random forest

a measure of variable importance based on the Gini impurity index used for the calculation of splits during training [53]. Therefore, adding up the Gini decrease for each variable among all the trees in the forest delivers a consistent faster approach in identifying the variable importance. The following factors responsible for the cause of cervical cancer are shown in Fig. 6.

6 Conclusion

Medical data of patients is some crucial data, the cervical cancer data studied in this paper suffered from class imbalance. Further construction of models based on these types of imbalanced data would end up as a biased model toward the majority class, therefore, the normal traditional classification approach algorithms work in a more generalized way unaware of the importance of the data. Clearly, the traditional approach of classifier models trained with 75% of the imbalanced training data provided results based on the majority class but not the minority class; therefore, the models failed to predict patients who are affected by cervical cancer. This study tries to provide a balance between the majority class and the minority class by random sampling of the majority class and considering the minority class as a constant data followed by a $n - 1$ ensemble architecture for the better prediction of the presence of cervical cancer as the number of instances considered in the sampled dataset gives a fair chance to both the majority and minority data, therefore providing a better unbiased model. KNN algorithm provides better performance in terms of accuracy, precision, recall, TPR and FPR rate compared to the other classification algorithms. Among all the 32 variables in the cervical cancer dataset certain important features, such as age, usage of hormonal contraceptives for years, time since the last and first diagnosis taken after being affected sexually transmitted diseases, age of the patient during the first sexual intercourse, number of pregnancies the patient has undergone, number of years the patient is used to the habit of smoking and the number of cigarettes the patient smokes every year and number of sexual partners the patient had and if the patient is diagnosed with Acquired Immune Deficiency Syndrome (AIDS).

References

1. Cervical cancer—cancer stat facts, https://seer.cancer.gov/statfacts/html/cervix.html
2. National Cancer Institute—cervical cancer, https://www.cancer.gov/types/cervical
3. Cervical cancer stages—American Cancer Society, https://www.cancer.org/cancer/cervical-cancer/detection-diagnosis-staging/staged.html
4. Fatlawi HK (2017) Enhanced classification model for cervical cancer dataset based on cost sensitive classifier
5. Margineantu D (2000) When does imbalanced data require more than cost-sensitive learning? In: Proceedings of the AAAI'2000 workshop on learning from imbalanced datasets, pp 47–50, Austin TX
6. Weiss G, Provost F (2003) Learning when training data are costly: the effect of class distribution on tree induction. J Artif Intell Res 19:315–354
7. Galar M., Fransico.: A review on ensembles for the class imbalance problem: bagging, boosting and hybrid-based approaches. IEEE Trans Syst Man Cybern Part C Appl Rev 42 (2012)
8. Kotsiantis S, Kanellopoulos D, Pintelas P (2006) Handling unbalanced datasets: A review. GESTS Int Trans Comput Sci Eng 30(1):25–36
9. Song Q, Zhang J, Chi Q (2010) Assistant detection of skewed data streams classification in cloud security. IEEE Trans
10. Liu X, Zhou Z (2006) The Influence of class imbalance on cost-sensitive learning: an empirical study. In: Sixth international conference on data mining (ICDM'06), Hong Kongpp 970–974
11. Yan R, Liu R, Jin R, Hauptmann A (2003) On predicting rare classes with SVM ensembles in scene classification. In: IEEE international conference on acoustics, speech, and signal processing, ICASSP '03, Hong Kong, pp III-21
12. Leevy JL, Khoshgoftaar TM, Bauder RA, Seliya N (2018) A survey on addressing high-class imbalance in big data. J Big Data 5(1)
13. Chawla NV, Bowyer KW, Hall LO, Kegelmeyer WP (2002) SMOTE: synthetic minority over-sampling technique. J Artif Intell Res 16:321–357
14. Seiffert C, Khoshgoftaar TM (2010) RUSBoost: a hybrid approach to alleviating class imbalance. IEEE Trans Syst Man Cybern Part A 40(1):185–197
15. Fernandez A, Rio S, Chawla N, Herrera F (2017) An insight into imbalanced Big Data classification: outcomes and challenges. Complex Intell Syst 3:105–120
16. Rodriguez D, Herraiz I, Harrison R, Dolado J, Riquelme J (2014) Preliminary comparison of techniques for dealing with imbalance in software defect prediction. In: Proceedings of the 18th international conference on evaluation and assessment in software engineering, Article no. 43
17. Han H, Wang L, Wen M, Wang WY (2006) Over-sampling algorithm based on preliminary classification in imbalanced datasets learning. J Comput Alloc 26(8):1894–1897 (in Chinese)
18. Guo X, Yin Y, Dong C, Yang G, Zhou G (2008) On the class imbalance problem. In: Fourth international conference on natural computation, Jinan, pp 192–201
19. Kubat M, Matwin S (1997) Addressing the curse of imbalanced training sets: one sided selection. In: Proceedings of the fourteenth international conference on machine learning, Nashville, Tennesse, Morgan Kaufmann, pp 179–186

20. Wilson DL (1972) Asymptotic properties of nearest neighbor rules using edited data. IEEE Trans Syst Man Cybernet 2:408–420
21. Batista G, Prati M, Monard M (2004) A study of the behavior of several methods for balancing machine learning training data. SIGKDD Explor 6(1):20–29
22. Laurikkala J (2001) Improving identification of difficult small classes by balancing class distribution. Technical Report, A-2001-2, University of Tampere
23. Kotsiantis S, Pintelas P (2003) Mixture of expert agents for handling imbalanced datasets. Ann Math Comput TeleInform 1:46–55
24. Han H, Wang WY, Mao BH (2005) Borderline-SMOTE: a new over-sampling method in imbalanced datasets learning. In: Proceedings of the international conference on intelligent computing, Part I, LNCS 3644, pp 878–887
25. Verma B, Hassan SZ (2011) Hybrid ensemble approach for classification. Appl Intell 34(2):258–278
26. Tukey JW (1977) Exploratory data analysis. Addison-Wesley, Reading
27. Dasarathy BV, Sheela BV (1979) Composite classifier system design: concepts and methodology. IEEE 67(5):708–713
28. Liu X-Y, Wu J, Zhou Z-H (2009) Exploratory undersampling for class-imbalance learning. IEEE Trans Syst Man Cybern Part B (Cybernetics) 39(2):539–550
29. Matteo RE, Valentini G (2012) Ensemble methods: a review. In: Advances in machine learning and data mining for astronomy. Chapman & Hall, pp 563–594
30. Cervical cancer (Risk Factors) Data Set, https://archive.ics.uci.edu/ml/datasets/Cervical+cancer+%28Risk+Factors%29
31. Fernandes K, Cardoso JS, Fernandes J (2017) Transfer learning with partial observability applied to cervical cancer screening. In: Iberian conference on pattern recognition and image analysis. Springer, Berlin
32. Alwesabi Y, Choudhury A, Won D (2018) Classification of cervical cancer dataset
33. Wu W, Zhou H (2017) Data-driven diagnosis of cervical cancer with support vector machine-based approaches. IEEE Access 5:25189–25195
34. Bauder RA, Khoshgoftaar TM (2018) The effects of varying class distribution on learner behavior for medicare fraud detection with imbalanced Big Data. Health Inf Sci Syst 6:9
35. Estabrooks, Japkowicz N (2001) A mixture-of-experts framework for learning from unbalanced datasets. In: Proceedings of the 2001 intelligent data analysis conference, pp 34–43
36. Daia W, Brisimia TS, Adamsb WG, Melac T, Saligramaa V, Paschalidisa IC (2015) Prediction of hospitalization due to heart diseases by supervised learning methods. Int J Med Informatics 84:189–197
37. Galar M, Fernandez A, Barrenechea E, Bustince H, Herrera F (2012) A review on ensembles for the class imbalance problem: bagging-, boosting-, and hybrid-based approaches. IEEE Trans Syst Man Cybern C Appl 42(4):463–484
38. Dong Y, Guo H, Zhi W, Fan M (2014) Class imbalance oriented logistic regression. In: International conference on cyber-enabled distributed computing and knowledge discovery. IEEE, pp 187–192
39. Lang J, Sun J (2014) Sensitivity of decision tree algorithm to class-imbalanced bank credit risk early warning. In: Seventh international joint conference on computational sciences and optimization
40. Anbarasi MS, Janani V (2017) Ensemble classifier with random forest algorithm to deal with imbalanced healthcare data. In: International conference on information, communication & embedded systems
41. Shouman M, Turner T, Stocker R (2012) Applying k-nearest neighbour in diagnosing heart disease patients. Int J Inf Educ Technol 2(3)
42. Sun Y, Wong AKC, Kamel MS (2009) Classification of imbalanced data: a review. Int J Pattern Recognit Artif Intell 23(4):687–719
43. Miao Z, Tang Y, Sun L, He Y, Xie S (2014) An improved KNN algorithm for imbalanced data based on local mean. J Comput Inf Syst 10(12):5139–5146

44. Hassanat AH, Abbadi AM, Altarawneh GA, Alhasanat AA (2014) Solving the problem of the K parameter in the KNN classifier using an ensemble learning approach. Int J Comput Sci Inf Secur 12(8)
45. Yang W, Wang K, Zuo W (2012) Fast neighborhood component analysis. Neurocomputing 83:31–37
46. Sierra B, Lazkano E, Irigoien I, Jauregi E, Mendialdua I (2011) *K* nearest neighbor equality: giving equal chance to all existing classes. Inform. Sci 181:5158–5168
47. Wang J, Neskovic P, Cooper LN (2007) Improving nearest neighbor rule with a simple adaptive distance measure. Pattern Recognit Lett 28:207–213
48. Mitra P, Murthy CA, Pal SK (2002) Unsupervised feature selection using feature similarity. IEEE Trans Pattern Anal Mach Intell 24:301–312
49. Bhattacharya G, Ghosh K, Chowdhur AS (2012) An affinity-based new local distance function and similarity measure for kNN algorithm. Pattern Recognit Lett 33:356–363
50. Ma C-M, Yang W-S, Cheng B-W (2014) How the parameters of K-nearest neighbor algorithm impact on the best classification accuracy: in case of parkinson dataset. J Appl Sci 14(2):171–176
51. Liu XY, Li QQ, Zhou ZH (2013) Learning imbalanced multi-class data with optimal dichotomy weights. In: IEEE 13th international conference on data mining (ICDM). IEEE, pp 478–487
52. Shengqiao L, Harner EJ, Adjeroh DA (2011) Random KNN feature selection—a fast and stable alternative to Random Forests. BMC Bioinform
53. Breiman L (2001) Random forest

English Transliteration of Kannada Words with Anusvara and Visarga

Savitha Shetty, Saritha Shetty, Sarika Hegde, and Karuna Pandit

Abstract Kannada is one of the frequently spoken languages of state Karnataka. It is descendents of Brahmi and one among the five Dravidian languages. Resources in Kannada are used by only those people who can read Kannada. Kannada language is spoken by most of the people in southern Indian state of Karnataka and in Andhra Pradesh, Maharashtra, Kerala, Tamil Nadu and Goa. Type of writing system in Kannada is called alphasyllabary and very less amount of work is done in this area. Kannada literature remained limited only for those who know it, and we improved the access to Kannada literature by proposing a method for Kannada to English transliteration. Our input language is Kannada and the output language is English. Here, input word is divided into transliteration units and then mapped to output language. We proposed an approach which takes Kannada words as input and produces exact transliterations from Kannada to English with Anusvara ("*O*") and Visarga (":") and gives results with 100% accuracy including numbers.

Keywords Kannada language · Anusvara · Visarga · Natural language · Processing · Transliterations

S. Shetty (✉) · S. Shetty · S. Hegde · K. Pandit
NMAM Institute of Technology, Nitte, Udupi, Karnataka 574110, India
e-mail: shettysavi1@nitte.edu.in

S. Shetty
e-mail: shettysaritha1@nitte.edu.in

S. Hegde
e-mail: sarika.hegde@nitte.edu.in

K. Pandit
e-mail: karunapandit@nitte.edu.in

N. N. Chiplunkar and T. Fukao (eds.), *Advances in Artificial Intelligence and Data Engineering*, Advances in Intelligent Systems and Computing 1133,
https://doi.org/10.1007/978-981-15-3514-7_28

1 Introduction

Processing of natural language is making computers to understand human language as it is when it is spoken by the people. It is nothing but the automatic manipulation of natural language like speech and text by the software. Human language is highly ambiguous, and there are specific rules that govern the language. There are approximately about 6500 languages.

Natural language processing is Artificial Intelligent method of communicating with an intelligent system using natural language. It is used for computers to process or understand natural language in order to perform a task that is useful. Fully understanding and representing meaning of a language are a difficult task. Few of the advantages of natural language processing are automatic summarization, where it produces readable summary of the part of text. Through natural language processing, we can learn about customer habits, preferences, tendencies, etc. regarding a product purchase. Development of natural language applications is difficult because computers work with unambiguous and highly structured languages like programming languages. Natural languages are ambiguous and linguistic structure always depends upon complex variables, dialects and social context. Unicode approach is used for mapping of Kannada words to English [1]. Section 2 explains the related work, proposed system is written in Sect. 3, Sect. 4 describes Kannada language, Sect. 5 describes methodology of the proposed system, and Sect. 6 describes results and discussions.

2 Related Work

Transliteration is method of mapping the words considering input language script into another. Transcription is mapping phonemes from one language into another. Transliteration is very much useful in translation of language and information retrieval of multiple languages. Model of transliteration should be designed in such a way that structure of sounds of the language should be preserved to the possible extent [2]. Various transliteration schemes are available for languages like European, English languages, and few attempts have been done for some of the Indian languages but still the work is in initial stage [3]. Less amount of work is done in Kannada language transliteration.

Surana and Singh [4] apply various techniques based on word origin, but they did not use training data on target side, but better algorithms are to be used on the source side. Their method explains transliteration as CLIR problem. Ganapathiraju et al. [5] proposed a mapping approach 'Om' for Indic scripts which uses the fact that all languages used in India are having similar sound structures. They developed 'Microsoft WinWord Text Editor,' which is helpful to people who can speak but cannot read their original language. They developed translator framework for non-native speakers.

Vijaya and Ajith et al. [6] proposed an approach for transliteration which is modeled as classification problem for transliteration from English to Tamil using WEKA environment. They used training data from a parallel corpus and achieved English to Tamil mapping with 84.82% accuracy for English names. Prakash et al. [7] proposed a transliteration approach which does transliteration from Kannada to English which gives 99% accuracy, and also user-friendly transliterator was developed for Kannada. But proposed approach gives inaccurate results when input includes 'Anusvara' ('O' or 'um') and 'Visarga' (':' or 'aha') in Kannada. Occurrence of various sound structures for same letters is dependent on the context and sometimes the system fails to produce correct output.

Mallamma and Hanumanthappa [8] proposed a model for information retrieval in case of Kannada and Telugu languages. It uses rule-based method for mapping of words. Prakash et al. [9] proposed a method which removes ambiguity of period symbol in Kannada. Period symbol can be used both as boundary and as abbreviation marker. This method divides the input data into sentences without intermediate tools. Mallamma and Hanumanthappa [10] proposed an approach where query is expressed in users own language and information retrieval system matches input query to the appropriate documents in the target language. Mallamma and Hanumanthappa [11] proposed a model for English to Kannada/Telugu machine translation and language identification; it uses simple rule-based and dictionary-based methods.

3 Proposed System

From literature survey, it is found that our work is not previously proposed and not implemented. We have developed the system in which exact transliteration is done from Kannada to English considering Anusvara and Visarga considering the following points.

- Exact Kannada to English transliterations considering Anusvara.
- Exact Kannada to English transliterations considering Visarga.

Kannada paragraph is given as input to the system and mapping is done to get exact transliterations in English (see Fig. 1). Unicode approach is used for mapping of Kannada words to English by taking vowels as anchor points to identify a syllable.

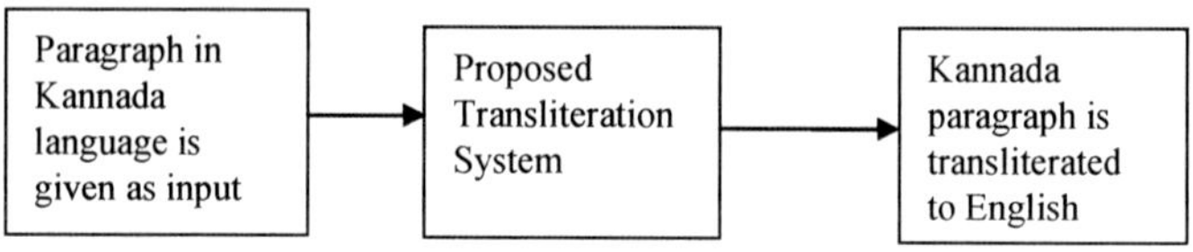

Fig. 1 System showing mapping from Kannada input to English output

4 Kannada Language

Kannada script is evolved from Kadamba script of the fifth century, and now Kannada is written using Kannada script. Kannada is descendents of Brahmi and it is one among the five Dravidian languages [12]. It is 25th more spoken language in the whole world. Indian languages like Tulu, Kodava are written using Kannada script [8]. Kannada language is spoken by most of the people in southern Indian state of Karnataka and in Andhra Pradesh, Maharashtra, Kerala, Tamil Nadu and Goa. Type of writing system in Kannada is called alphasyllabary. Here, all consonants have built-in vowel, vowels are written with diacritics.

Kannada language is collection of both basic and compound (ottaksharas) characters which results in equal and unequal sized characters [14]. Kannada script consists of akshara. Here, the consonant with the following vowel is called diacritics. Akshara without vowel-diacritics forms a consonant having a vowel (schwa) [15]. For the root word suffixes are added to form word in Kannada language [16]. Kannada literature remained limited only for those who know it. We improved the access to Kannada literature by proposing a method for Kannada to English transliteration. It is written generally from left side to right. When consonants are present along with the intervening vowels in the CCV form, then the second consonant is denoted as special conjunct below the first. Kannada language has totally 49 phonemic letters, which are divided into three groups. We have thirteen letters named as 'Swaragalu,' thirty-four letters named as 'Vyanjanagalu' and two letters as 'Yogavaahakagalu' [17].

In transliteration process, every word in the source script is transcribed into another language. It helps in machine translation and retrieval of information in different languages [18]. It is useful when we want to express the concepts in one language by using the script of another language [19]. Transliteration generally depends on the context. For example, the English (source language) grapheme 'a' can be transliterated to Kannada (target language) graphemes like 'a,' 'aa,' 'ei,' etc. based on its context. This is because vowels in English may represent short vowels or long vowels or combination of vowels in Kannada during the process of transliteration [20].

5 Methodology

5.1 *Proposed Transliteration Approach*

Unicode approach is used to convert the Kannada words in the input file into English words (see Fig. 2). Suppose 'C' is consonant and 'V' is vowel, then we find C*V. Vowels are used as anchor points to identify syllable, each syllable ends with the vowel. Kannada script syllables are represented by using UTF-8 encoding format. Unicode Transformation Format encoded texts are given as—input to our system. We perform phonetic mapping for these codes to obtain the output. At the end, we

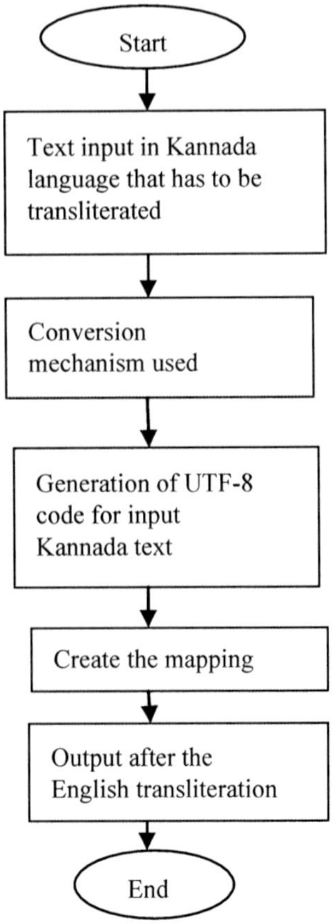

Fig. 2 System workflow of our proposed approach

obtain the mapped English words. We have focused on transliteration of Kannada alphabets which includes Anusvara and Visarga.

Each and every written symbol in Kannada represents one syllable but in English it is called phoneme. Writing system in Kannada is referred as abugida with consonants having an inherent vowel. Each sound has a distinct letter so each word is pronounced as it is spelt. We have short one (ಕ್, known as hrasva), which does not use any of vowel and long one (ಕ, known as deergha) which is used along with first vowel (ಅ). During transliteration, multiple characters in a script map to single character in other script, which makes transliteration a difficult task. We have different ways to write a source word into target word. For example, person with name 'saMyukta' has multiple ways to present it among different transliterations in Kannada language like ' ಸಂಯುಕ್ತ,' ' ಸಮ್ಯುಕ್ತ.' The right word is chosen depending on the user's perception [21].

Glyphs are added with other vowels and consonants to form consonant conjuncts (see Table 1). If we consider the input which has Anusvara ('M' or '0') and Visarga ('Ha' or ':') in the words of Kannada. Expressing these letters in English language

Table 1 Some of consonant conjuncts in Kannada for the letter 'ಕ'

ಕ್ಕ (kka)	ಕ್ಗ (kga)	ಕ್ಚ (kcha)	ಕ್ಜ (kja)	ಕ್ನ (kna)	ಕ್ತ (kta)	ಕ್ದ (kda)	ಕ್ಪ (kpa)	ಕ್ಬ (kba)
ಕ್ಮ (kma)	ಕ್ಯ (kya)	ಕ್ರ (kra)	ಕ್ಲ (kla)	ಕ್ವ (kva)	ಕ್ಶ (ksha)	ಕ್ಷ (kSha)	ಕ್ಸ (ksa)	ಕ್ಲ (kla)

correctly is impossible in some cases. The sound structure of 'Anusvara' depends on its use, it can be pronounced as 'M' in 'kaMpu' having Anusvara (0), or as 'N' in 'chaNdana' having Anusvara (0). Visarga also has conflicting sound structures, it uses 'H' for the word 'duHkha' which contains (:), 'Ha' in case of 'namaHa' having (:). Occurrence of these two sound structures is dependent on the context, and hence, the system sometimes fails to produce the correct output.

Table 2 shows transliterations considering Anusvara for Kannada vowels, Table 3 shows mapping for velars, Table 4 shows mapping for palatals, Table 5 shows mapping for retroflex, Table 6 shows mapping for dentals, and Table 7 shows mapping for labials. We have considered all Anusvara word contexts in Kannada and found that 'M' appears only when any of the four structured Vyanjanaas as in Table 8 and eight unstructured Vyanjanaas as in Table 9 appear. And also these Vyanjanaas appear as the second letter in the word.

Table 2 Transliterations considering Anusvara for Kannada vowels

Vowel	Kannada word	Current English transliterations available	Our transliteration method
ಅ	ಅಂಗಳ	aMgala	aNgala
ಆ	ಆಂತರಿಕ	aaMtarika	aaNtarika
ಇ	ಇಂಗಿತ	iMgita	iNgita
ಉ	ಉಂಗುರ	uMgura	uNgura
ಎ	ಎಂತ	eMta	eNta
ಒ	ಒಂಟೆ	oMTe	oNTe
ಅಂ	ಅಂಗಿ	aMgi	aNgi

Table 3 Transliterations considering Anusvara for Kannada consonants (velars)

Velars	Kannada word	Current English transliterations available	Our transliteration method
ಕ	ಕಂದ	kaMda	kaNda
ಖ	ಖಂಡ	khaMDa	khaNDa
ಗ	ಗಂಟೆ	gaMTe	gaNTe
ಘ	ಘಂಟೆ	ghaMTe	ghaNTe

Table 4 Transliterations considering Anusvara for Kannada consonants (palatals)

Palatals	Kannada word	Current English transliterations available	Our transliteration method
ಚ	ಚೆಂಡು	cheMDu	cheNDu
ಛ	ಛಂದಸ್ಸು	CaMdassu	CaNdassu
ಜ	ಜಂಗಮವಾಣಿ	jaMgamavaaNi	jaNgamavaaNi
ಝ	ಝಂಡಾ	JaMDaa	JaNDaa

Table 5 Transliterations considering Anusvara for Kannada consonants (retroflex)

Retroflex	Kannada word	Current English transliterations available	Our transliteration method
ಡ	ಡಂಗೂರ	DaMgoora	DaNgoora

Table 6 Transliterations considering Anusvara for Kannada consonants (dentals)

Dentals	Kannada word	Current English transliterations available	Our transliteration method
ತ	ತಂಟೆ	taMTe	taNTe
ದ	ದಂಗ	daMga	daNga
ಥ	ಇಂಥ	iMtha	iNtha
ಧ	ಧಂಗ	dhaMga	dhaNga
ನ	ನಂಟು	naMTu	naNTu

Table 7 Transliterations considering Anusvara for Kannada consonants (labials)

Labials	Kannada word	Current English transliterations available	Our transliteration method
ಪ	ಪಂದಳ	paMdaLa	paNdaLa
ಫ	ಫಂದಳ	phaMdaLa	PhaNdaLa
ಬ	ಬಂದರು	baMdaru	baNdaru
ಭ	ಭಂಗ	bhaMga	bhaNga
ಮ	ಮಂಗ	maMga	maNga

Considered structured Vyanjanaas are 'ಪ','ಫ','ಬ','ಭ'. For these structured consonants we get 'M' if they appear as second letter of the word. We checked structured Vyanjanaas with almost all words, few examples like ' ಪಂಪ' (paMpa), ' ಪಂಫ' (paMpha), ' ಕಂಬ' (kaMba), and ' ಕುಂಭ(kuMbha) as shown in Table 8. Considered unstructured Vyanjanaas are ' ಯ', ' ರ ', ' ವ ' , ' ಶ ', ' ಷ ', ' ಸ ', ' ಹ ', ' ಳ '.

Table 8 Transliterations considering Anusvara for Kannada words ending with letter 'ಪ' 'ಫ' 'ಬ' 'ಭ'

	Kannada word	Current English transliterations available	Our transliteration method
ಪ	ಪಂಪ	paMpa	paMpa
ಫ	ಪಂಫ	paMpha	paMpha
ಬ	ಕಂಬ	kaMba	kaMba
ಭ	ಕುಂಭ	kuMbha	kuMbha

Table 9 Transliterations considering Anusvara for Kannada unstructured consonants

	Kannada word	Current English transliterations available	Our transliteration method
ಯ	ಸಂಯುಕ್ತ	saMyukta	saMyukta
ರ	ಸಂರಕ್ಷಣೆ	saMrakShaNe	saMrakShaNe
ವ	ಹೋಂವರ್ಕ್	hOMwark	hOMwark
ಶ	ಅಂಶ	aMsha	aMsha
ಷ	ಅಂಷ	aMSha	aMSha
ಸ	ಸಂಸಾರ	saMsaara	saMsaara
ಹ	ಸಂಹಾರ	saMhaara	saMhaara
ಕ್ಷ	ಅಕ್ಷಾಂಶ	akShaaMsha	akShaaMsha

For these unstructured consonants we get 'M' if they appear as second letter of the word. We checked unstructured Vyanjanaas with almost all words, few examples like 'ಸಂಯುಕ್ತ,' 'ಸಂರಕ್ಷಣೆ,' 'ಹೋಂವರ್ಕ್', etc. as shown in Table 9.

Few of the letters in unstructured consonants like 'ಲ', 'ಳ', 'ಜ್ಞ'do not form a word if it appears as second letter in the word with Anusvara, so it is not considered. The letter 'ಮ' in structured consonant does not form a word if it appears as second letter of the word with Anusvara, so it is not considered. It also works for the words like 'ಪಂಪಾವತಿ' and 'ಕುಂಭಮೇಳ,' etc. If we have a word with Anusvara at the end like 'ಮಾಂ,' it gives correct output. For all the other vowels and consonants, they have 'N.'

For Kannada words which includes Visarga if Visarga appears in between the letters, then the mapping considers 'H' and if Visarga appears at the end of the word, mapping considers 'Ha' as shown in Table 10.

Table 10 Transliterations considering Visarga for Kannada words

Kannada word	Current English transliterations available	Our transliteration method
ದುಃಖ	duHkha	duHkha
ಅಂತಃಕರಣ	antaHkaraNa	antaHkaraNa
ನಮಃ	namaH	namaHa
ಸಂಶಯಃ	saMshayaH	saMshayaHa

6 Results and Discussions

Proposed Kannada transliteration system takes Kannada paragraph as input from a file or as a string and produces transliterated English as output. Previous models which were designed for transliterations from Kannada to English are not giving correct result if we have Kannada words with Anusvara and Visarga. Our approach gives correct output for all Kannada words with Anusvara and Visarga occurring in the middle or end of the word.

We have tested our approach considering a file having 5000 Kannada words with Anusvara and Visarga considering all the possibilities. System gives correct output for all possible combinations of words with Anusvara and Visarga. If there are English words in the Kannada input, it is displayed as it is. If there are Kannada numbers from '೦-೯' in the Kannada input, it is represented by the digits in English as '0–9.' System accuracy is calculated using Formula (1).

$$\text{Accuracy} = \frac{\text{Number of words correctly transliterated}}{\text{Total number of words}} \quad (1)$$

Initially, we have taken 1000 words in the input file and tested our system. Then, we increased the number of words to 5000 and accuracy remains 100% (see Fig. 3).

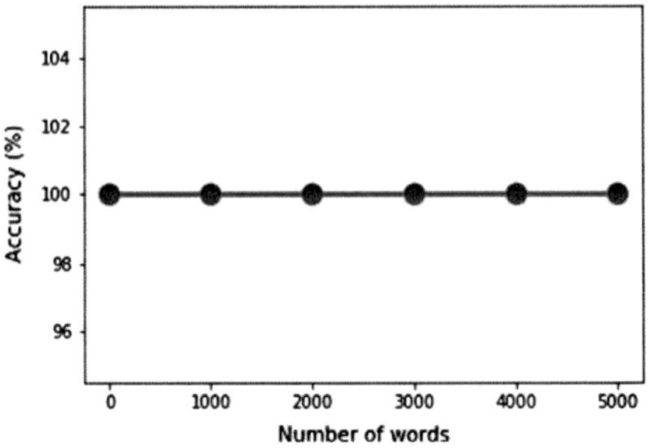

Fig. 3 Graph showing increase in the number of words but accuracy remains the same

We tested our system by giving the input from the Web site 'Kendasampige' and famous Kannada newspapers like 'Udayavaani,' 'Vijayavaani' and 'Prajaavaani,' etc. as shown in Tables 13 and 14. To test Kannada words with Visarga we used Hindu sacred text 'Bhagavad Gita' chapter 10 and 12 as input which is Sanskrit shloka written using Kannada. Mapping is shown for few lines of chapter as shown in Tables 11 and 12. Purpose of taking this input is, it includes Sanskrit shlokas written using Kannada script, which consists of a greater number of Visargas (':'). Our system gives 100% accurate results for all possible combinations of Kannada words with Anusvara and Visarga and also other words.

Table 11 12th Chapter of Bhagavad Gita as input to the system

ಅಥದ್ವಾದಶೋಧ್ಯಾಯಃ
ಅರ್ಜುನಉವಾಚ
ಏವಂಸತತಯುಕ್ತಾಯೇಭಕ್ತಾಸ್ತ್ವಾಂಪರ್ಯುಪಾಸತೇ
ಯೇಚಾಪ್ಯಕ್ಷರಮವ್ಯಕ್ತಂತೇಷಾಂಕೇಯೋಗವಿತ್ತಮಾಃ೧
ಶ್ರೀಭಗವಾನುವಾಚ
ಮಯ್ಯಾವೇಶ್ಯಮನೋಯೇಮಾಂನಿತ್ಯಯುಕ್ತಾಉಪಾಸತೇ
ಶ್ರದ್ಧಯಾಪರಯೋಪೇತಾಸ್ತೇಮೇಯುಕ್ತತಮಾಮತಾಃ೨
ಯೇತ್ವಕ್ಷರಮನಿರ್ದೇಶ್ಯಮವ್ಯಕ್ತಂಪರ್ಯುಪಾಸತೇ
ಸರ್ವತ್ರಗಮಚಿಂತ್ಯಂಚಕೂಟಸ್ಥಮಚಲಂಧ್ರುವಮ್೩
ಸಂನಿಯಮ್ಯೇಂದ್ರಿಯಗ್ರಾಮಂಸರ್ವತ್ರಸಮಬುದ್ಧಯಃ
ತೇಪ್ರಾಪ್ನುವಂತಿಮಾಮೇವಸರ್ವಭೂತಹಿತೇರತಾಃ೪
ಕ್ಲೇಶೋಧಿಕತರಸ್ತೇಷಾಮವ್ಯಕ್ತಾಸಕ್ತಚೇತಸಾಮ್
ಅವ್ಯಕ್ತಾಹಿಗತಿರ್ದುಃಖಂದೇಹವದ್ಭಿರವಾಪ್ಯತೇ೫
ಯೇತುಸರ್ವಾಣಿಕರ್ಮಾಣಿಮಯಿಸಂನ್ಯಸ್ಯಮತ್ಪರಾಃ
ಅನನ್ಯೇನೈವಯೋಗೇನಮಾಂಧ್ಯಾಯಂತಉಪಾಸತೇ೬
ತೇಷಾಮಹಂಸಮುದ್ಧರ್ತಾಮೃತ್ಯುಸಂಸಾರಸಾಗರಾತ್
ಭವಾಮಿನಚಿರಾತ್ಪಾರ್ಥಮಯ್ಯಾವೇಶಿತಚೇತಸಾಮ್೭
ಮಯ್ಯೇವಮನಆಧತ್ಸ್ವಮಯಿಬುದ್ಧಿಂನಿವೇಶಯ
ನಿವಸಿಷ್ಯಸಿಮಯ್ಯೇವಅತಊರ್ಧ್ವಂನಸಂಶಯಃ೮
ಅಥಚಿತ್ತಂಸಮಾಧಾತುಂನಶಕ್ನೋಷಿಮಯಿಸ್ಥಿರಮ್
ಅಭ್ಯಾಸಯೋಗೇನತತೋಮಾಮಿಚ್ಛಾಪ್ತುಂಧನಂಜಯ೯
ಅಭ್ಯಾಸೇಪ್ಯಸಮರ್ಥೋಸಿಮತ್ಕರ್ಮಪರಮೋಭವ
ಮದರ್ಥಮಪಿಕರ್ಮಾಣಿಕುರ್ವನ್ಸಿದ್ಧಿಮವಾಪ್ಸ್ಯಸಿ೧೦

Table 12 English output after doing the mapping

['atha dvaadashoodhyaayaHa arjuna uvaacha \nEvaM satatayuktaa yE bhaktaastvaaM paryupaasatE \nyE chaapyakSharamavyaktaM tEShaaM kE yoogavittamaaHa 1 \nshrIbhagavaanuvaacha \nmayyaavEshya manoo yE maaM nityayuktaa upaasatE \nshraddhayaa parayoopEtaastE mE yuktatamaa mataaHa 2 \nyE tvakSharamanirdEshyamavyaktaM paryupaasatE \nsarvatragamachiNtyaM cha kUTasthamachalaM dhruvam 3 \nsaNniyamyENdriyagraamaM sarvatra samabuddhayaHa \ntE praapnuvaNti maamEva sarvabhUtahitE rataaHa 4 \nklEshoodhikatarastEShaamavyaktaasaktachEtasaam\navyaktaa hi gatirduHkhaM dEhavadbhiravaapyatE 5 \nyE tu sarvaaNi karmaaNi mayi saNnyasya matparaaHa \nananyEnaiva yoogEna maaM dhyaayaNta upaasatE 6\ntEShaamahaM samuddhartaa mRityusaMsaarasaagaraat \nbhavaamina chiraatpaartha mayyaavEshitachEtasaam 7\nmayyEva mana AAdhatsva mayi buddhiM nivEshaya \nnivasiShyasi mayyEva ata oordhvaM na saMshayaHa 8\natha chittaM samaadhaatuM na shaknooShi mayi sTiram \nabhyaasayoogEna tatoo maamichChaaptuM dhanaNjaya 9\nabhyaasEpyasamarToosi matkarmaparamoo bhava \nmadarthamapi karmaaNi kurvansiddhimavaapsyasi 10']

Table 13 Kannada newspaper input including Anusvara and Visarga

ಇಂಗಿತ ಉಂಗುರ ಅಂತಃಕರಣ ಸಂಶಯಃ ನಮಃ ದುಃಖ
ಆಂತರಿಕ ಅಕ್ಷಾಂಶ ಸಂಹಾರ ಸಂಸಾರ ಅಂಷ ಅಂಶ
ಹೋಂವರ್ಕ್ ಸಂರಕ್ಷಣೆ ಸಂಯುಕ್ತ ಕುಂಭ ಪಂಫ ಪಂಪ
ಬಂದರು ಫಂದಳ ದಂಗ ತಂಟೆ ಡಂಗೂರ ಝುಂಡಾ
ಜಂಗವಾಣಿ ಛಂದಸ್ಸು ಘಂಟೆ ಚೆಂಡು ಖಂಡ ಒಂಟೆ ಕಂದ ಕಂಬ

Table 14 English output after processing the Kannada newspaper input including Anusvara and Visarga

['eNgita uNgura aNtaHkaraNa saMshayaHa namaHa duHkha aaNtarika akShaaMsha saMhaara saMsaara aMSha aMsha hooMv
ark saMrakShaNe saMyukta kuMbha paMpha paMpa baNdaru phaNdaLa daNga taNTe DaNgUra jhaNDaa jaNgamavaaNi ChaNdassu ghaNTe cheNDu khaNDa oNTe kaNda kaMba']

7 Conclusion

We presented a transliteration system which takes input consisting of Kannada sentences which may include punctuations, numbers, etc. We focused on Kannada words which have two Yogavaahakaas, Anusvara and Visarga. Unicode mapping approach is used for Kannada to English transliteration. Testing is done by taking 12th Chapter of Bhagavad Gita and few Kannada newspapers. The purpose of taking Bhagavad Gita as input is since it includes more number of Anusvara and Visarga words. Overall accuracy of the system is 100% and it performs exact transliterations from Kannada to English including Yogavaahaka's.

References

1. Madhavi GT, Kishore P, Lavanya P (2005) A simple approach for building transliteration editors for Indian languages. J Zhejiang Univ Sci A 6(11):1354–1361. https://doi.org/10.1007/BF02841675
2. Ajith VP, Antony PJ, Soman KP (2010 Mar) Kernel method for English to Kannada transliteration. In: 2010 international conference on recent trends in information, telecommunication and computing. IEEE, pp 336–338. https://doi.org/10.1109/itc.2010.85
3. Antony PJ, Soman KP (2011) Machine transliteration for indian languages: a literature survey. Int J Sci Eng Res 2(12)
4. Singh AK, Surana H (2008) A more discerning and adaptable multilingual transliteration mechanism for Indian languages. In: Proceedings of the third international joint conference on natural language processing: Volume-I
5. Raj R, Madhavi G, Mini B, Balakrishnan N (2005) Om: one tool for many (Indian) languages. J Zhejiang Univ Sci A 6(11):1348–1353. https://doi.org/10.1631/jzus.2005.A1348
6. Soman KP, Ajith VP, Vijaya MS, Shivapratap G (2009) English to tamil transliteration using weka. Int J Recent Trends Eng 1(1):498
7. Prakash A, Ashok M, Swathi R, Sahana C, Kamath S, iKAN a Kannada transliteration tool for assisted linguistic learning
8. Mallamma VR, Hanumanthappa M (2011) Kannada and Telugu native languages to English cross language information retrieval. Int J Comput Sci Inf Technol 2(5):1876–1880
9. Parakh M, Rajesha N, Ramya M (2011) Sentence boundary disambiguation in Kannada texts. Language in India, www.languageinindia.com, Special Volume: Problems of Parsing in Indian Languages, 17–19
10. Reddy MV, Hanumanthappa M, CLIR: English to Kannada/Telugu
11. Reddy MV, Hanumanthappa M (2013) Indic language machine translation tool: English to kannada/telugu. In: Multimedia processing, communication and computing applications. Springer, New Delhi, pp 35–49
12. Sukumar M, Prasad MM (2013) 2D-LDA based online handwritten Kannada character recognition. Int J Comput Sci Telecommun 4(1):14–18
13. Vijaya PA, Padma MC (2008) Language identification of Kannada, Hindi and English text words through visual discriminating features. Int J Comput Intell Syst 1(2):116–126. https://doi.org/10.1080/18756891.2008.9727609
14. Nag S (2007) Early reading in Kannada: the pace of acquisition of orthographic knowledge and phonemic awareness. J Res Reading 30(1):7–22. https://doi.org/10.1111/j.1467-9817.2006.00329.x
15. Kumar R, Shambhavi BR (2012) Kannada part of speech tagging with probabilistic classifiers. Int J Comput Appl 975:888

16. Rashmi S, Reddy M, Jyothi NM, Hanumanthappa M (2014) Phonetic dictionary for natural language processing: Kannada. Int J Eng Res Appl 4(7):01–04
17. Soman KP, Antony PJ (2011) Machine transliteration for Indian languages: a literature survey. Int J Sci Eng Res 2(12)
18. Soman KP, Loganathan R, Sasidharan S (2009) English to Malayalam transliteration using sequence labeling approach. Int J Recent Trends Eng 1(2):170
19. Kaur J, Josan GS (2011) Punjabi to Hindi statistical machine transliteration. Int J Inf Technol Knowl Manag 4(2):459–463
20. Reddy MV, Hanumanthappa M (2011) English to Kannada/Telugu name transliteration in Clir: a statistical approach. Int J Mach Intell 3(4)

An Ensembled Scale-Space Model of Deep Convolutional Neural Networks for Sign Language Recognition

Neena Aloysius and M. Geetha

Abstract A sign language translator is a utilitarian in facilitating communication between the deaf community and the hearing majority. This paper proffers an innovative specialized convolutional neural network (CNN) model, Sign-Net, to recognize hand gesture signs by incorporating scale-space theory to deep learning framework. The proposed model is an ensemble of CNNs—a low resolution network (LRN) and a high resolution network (HRN). This architecture of the proposed model allows the ensemble to work at different spatial resolutions and at varying depths of CNN. The Sign-Net model was assessed with static signs of American Sign Language—alphabets and digits. Since there exists no sign dataset for deep learning, the ensemble performance is evaluated on the synthetic dataset which we have collected for this task. Assessment of the synthetic dataset by Sign-Net reported an impressive accuracy of 74.5%, notably superior to the other existing models.

Keywords Deep learning · Convolutional neural networks · Ensemble · Sign language · Classification

1 Introduction

Sign language recognition (SLR) facilitates smooth communication among the deaf community. Since sign language is not a worldwide language, very few normal people know this language. Obviously, this adds further complications to the interaction between the deaf and the hearing majority. The alternative form of written communication is cumbersome, since the deaf community is normally less proficient in writing a spoken language. Besides, written communication is much slower than face-to-face conversations as well as unbiased [1].

N. Aloysius (✉) · M. Geetha
Department of Computer Science and Engineering, Amrita School of Engineering,
Amrita Vishwa Vidyapeetham, Amritapuri, Kollam, India
e-mail: neenaloysius@am.amrita.edu

N. N. Chiplunkar and T. Fukao (eds.), *Advances in Artificial Intelligence and Data Engineering*, Advances in Intelligent Systems and Computing 1133,
https://doi.org/10.1007/978-981-15-3514-7_29

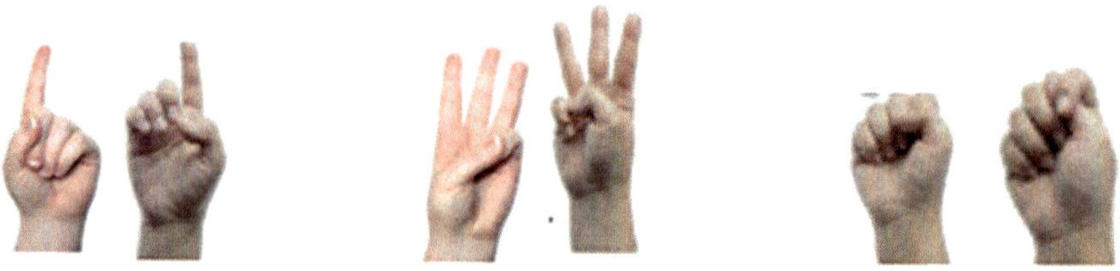

Fig. 1 Challenging classes: (1, d), (6, w), (s, t)

Despite much research effort in this area for the past few years [2], the recognition of sign language has been a daunting exercise. Interpretation of sign language requires simultaneous understanding, in real time, of multi-modal data such as facial expression, body posture, hand pose and hand movement. With a small set of specific gestures, recognition of gestures is much easier, however, there are at least thousands of words synonymous to similar hand poses. Furthermore, with varying signers and different viewpoints, even same signs are interpreted or recognized differently.

In this work, we deal with deciphering static signs of the American Sign Language (ASL). Accurate recognition of static signs is a matter of concern even in continuous sign language recognition. Visually similar yet different signs make this an arduous task. For instance, certain signs can be distinguished only by the thumb positions Fig. 1). Other variations are introduced with different camera-viewpoints and different signers.

We introduce Sign-Net, a novel ensemble of CNNs, specialized to recognize static gesture sign images for fingerspelled ASL characters. Conventional CNN architectures like AlexNet yield outstanding performance, for high detail images, as in case of the ImageNet dataset. However, the accuracy is compromised for low detail images, showing less texture and color features for gesture and character images. The proposed method is an augmented version of AlexNet, incorporating the concepts of scale-space theory. An ensemble of scaled models of CNNs is predicted to return better results.

The specific contributions of this study are as follows: a trained, ensembled, scale-space model of CNN, which is a combination of a deep high resolution network (HRN) and a shallow low resolution network (LRN) for the discernment of fingerspelling in sign language. The ensembled system considers 24 English alphabets (excluding j and z, which have dynamic signs, so not included) and digits 0–9, thus a total of 34 classes. While the conventional LeNet architecture was applied to LRN, the HRN in ensemble is a new 13-layer deep CNN network. The popularity of LeNet and AlexNet architectures has prompted us to fix the resolutions for LRN and HRN as 28×28 and 227×227, respectively. From a detailed study, it can be conjectured that the proposed model would be applicable to any classification task that requires learning low-level details of any image, with distinct boundary shapes. Another major contribution is the dataset, which we have created as part of this project, having more than 1000 images per class and additional test images from different signers not included in training set. Till now there are no deep-net available for signs.

2 Related Work

The feature extraction technique, inherent to CNN can be put to beneficial use to reduce the load of preprocessing done on raw images; consequently, the overall complexity of the system is reduced. Initially, CNN was widely used for tasks associated with object recognition. Nowadays, it is being examined in other domains as well [3, 4]. The hand gesture recognition system proposed by Nagi et al. [5] is the first implementation of a big and deep CNN for this task. Nagi's team developed a real-time gesture recognition system, for mobile robots, with an error rate of 3.23% much lower than contemporary methods like PHOG, FFT, skeletonization, spatial pyramid (BoW), etc. Ma et al. [6] devised an unsupervised feature learning technique for extraction of the edge features, from a depth image database of eight different gestures, using an auto-encoding neural network; the features were then fed to a classifier, ensuing in a recognition rate of 84% on their synthetic dataset.

Besides ASL, other recognition systems have been developed, specific to the language of a region, or a country. An Italian Sign Language recognition system was proposed by Pigou et al. [7] which employed CNN. Although they achieved an accuracy of 95.68% for 20 Italian gestures, their test data had signers in training or validation set. Though Liwicki et al. [8] have also dealt with the fingerspelling task, their team followed a non-CNN-based methodology. Liwicki's team successfully modeled a British Sign Language recognition system, to decode fingerspelled words from a video. The alphabets were recognized using histogram of gradients (HOG) descriptors; words were then recognized using hidden Markov models (HMM); the dataset used refers to a single signer. Srujana et al. [9] put forth an ASL image database, which was exercised by six contemporaneous methods of gaging human pose; they used the pose information for recognition of sign language, merely considering upper body features. Their results showed good pose estimation accuracy for the ASL dataset.

The problem we deal within sign language recognition is the identification of fingerspelled signs. Isaac et al. [10] proposed a neural network system, applied to wavelet features, for the translation of ASL fingerspelled signs. Their team's feed-forward neural network was trained using Levenberg–Marquardt training. Their system recognized static ASL signs with an accuracy of 99.9%. Nevertheless, the number of signers and the size of the dataset were not explicitly documented. Such information is of crucial significance for the determination of the efficacy of any SLR system. A similar recognition system was proposed by Pugeault et al. [11] that used Gabor filters and random forest. ASL's 24 fingerspelling signs (excluding j and z), for alphabets, were recognized by their system; their dataset, collected from five subjects, reported a recognition accuracy of 69% using only depth, 73% using only color and 75% using both color and depth.

A study, similar to ours, was done by Tripathi et al. [12] for recognizing static ASL signs of alphabets and digits. Their proposed system was developed using the famous AlexNet architecture, trained on 31,000 depth maps, collected using a depth sensor: Creative Senz3D camera of resolution 320×240. They reported an accuracy of 85% for the subjects not in training or validation. The CNN model, put forth by Ameen and Vadera [13], also used depth maps for recognition of fingerspelled ASL signs. Similarly, depth maps were deployed in the recognition system of Wenjin et al. [14]; besides, they introduced the concepts of multi-view augmentation (to multiply training data) and inference fusion (for better production). Their experimental results, on the ASL benchmark dataset and the NTU digit dataset, show significant improvement over previous methods.

Oyebade et al. [15] were successful in developing two sign recognition systems—one using CNN and the other using a stacked denoising autoencoder. The deep learning framework was applied on Thomas Moeslund's gesture recognition database, reporting recognition accuracies of 91.33% (for CNN) and 92.83% (for autoencoder). Our previous probes [16, 17] pertinent to dynamic sign language recognition have introduced a new method of stroke based representation of signs to decode sign language using key maximum curvature points and 3D chain codes. Exclusive consideration of the keystrokes rendered the training phase less expensive compared to other prevailing methods.

A real-time ASL fingerspelling recognition system was proposed by Kuznetsova et al. [18] that used multi-layered random forest. Merits of their technique were divulged by evaluation of the algorithm on publicly available synthetic dataset, comprised of 24 ASL signs, and a new dataset, collected by using Intel Creative Gesture Camera. Recently, Dong et al. [19] presented an ASL recognition system that used multi-layered random forest (MLRF), trained on rotation, translation and scale-invariant features derived from depth images. They reported 70% accuracy for new signers and 90% accuracy for the subjects, with whom the system was trained. In contrast, this research pursued a path, divergent from those previously reported, in several aspects. From the literature survey, it is clear that not much work has been done on the problem of recognition of fingerspelled sign language using CNN to translate signs into text. In fact, reportedly, Tripathi et al. [12] and Wenjin et al. [14] pursued this task of recognition, using CNN; the former reused the extant AlexNet model and retrained on depth maps of fingerspelled signs, while the latter conceived a new CNN network, trained on depth maps. The focus of this paper has been to augment these methods by modeling a new ensemble working at different spatial resolutions on raw images.

3 Proposed Ensemble Architecture

Sign-Net, the newly advocated system architecture, can be deemed as a combination of two different CNN models ensembled to work at different spatial resolutions. The ensemble is constituted of a high resolution network (HRN) and a low resolution network (LRN). While the LeNet architecture is reused for LRN, the HRN is a new 13-layer deep model. The advantage of working with two different spatial scales was found to be multi-fold: reduced training time of the ensemble framework since LeNet is a shallow network; learning at two different scales—global features by LRN and minor features by HRN, with their combination expected to perform even better. Moreover, the recommended ensemble model has incorporated the concept of scale-space theory into deep learning networks. Scale-space theory [20] is a framework for multi-scale signal representation for handling image structures at different scales by representing image as a one-parameter family of scaled and smoothed images. The scale-space representation was used for the suppression of fine-scale structures. Each of the pipelines of CNN in the ensemble represents and learns data at a different scale space.

Figure 2 shows the high-level diagram of the proposed recognition system. As its input, the system takes the fingerspelled or static signs of alphabets and digits; as outputs, the system returns the corresponding label for the associated gesture. Figure 3 shows all the components of the ensemble CNN. The size and depth of filters at each stage, as well as the number of such filters are denoted in this figure. The following sections specify all the steps involved in the working of ensemble.

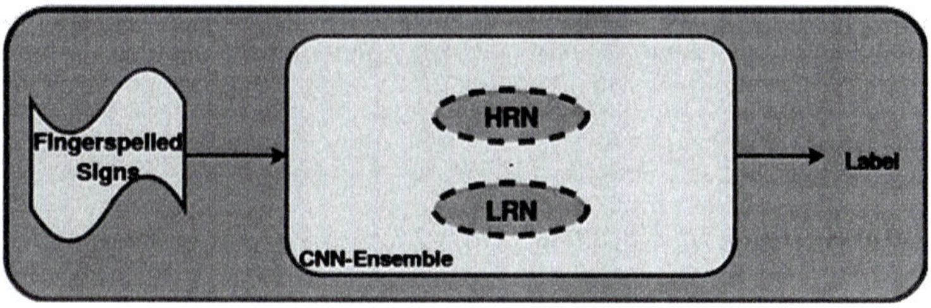

Fig. 2 Proposed recognition system

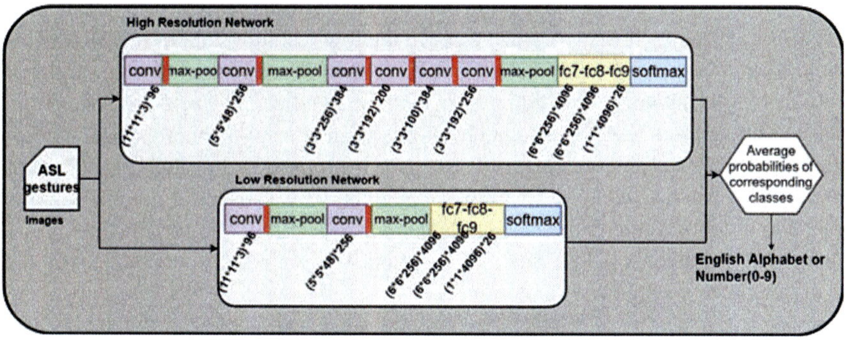

Fig. 3 Sign-Net

3.1 High Resolution Network (HRN)

The creative CNN architecture is a 13-layer deep network having 6 convolutions, 3 max-pooling, 3 fully connected and 1 softmax layer. The red markings in Fig. 3 indicate the activation function, rectified linear units (ReLu). The network takes images of static ASL signs, of size 227 × 227, as input. The weights of AlexNet model, trained on ImageNet, were used as filter weights for the first three convolution layers of the proposed model. This was based on the understanding that the initial layers of any trained CNN model will be learning the edges and other general features, which can be conveniently reused for other tasks. Filter weights of other layers were initialized randomly. The network was trained using backpropagation algorithm for 20 epochs and the softmax loss layer is updated using stochastic gradient descent for a batch size of 50. A weight decay of 0.001 is used for maintaining the objective function value within a reasonable range. Another new CNN architecture of 14 layers is also trained for the same task to study performance variations with depth. As a perceptive investigation, this research considered 12-layer AlexNet, 13-layer new HRN model and the 14-layer CNN model. The outcomes are detailed in the results section.

3.2 Low Resolution Network (LRN)

LeNet was chosen as the low resolution network since it takes as input, images of dimension 28 × 28. The architecture was formed of alternating convolution and max-pooling layers followed by fully connected and softmax layer. The original raw images of dataset were scaled down to form the low resolution input for LRN. This scaling suppressed the fine details of image and in-turn led to the learning of global features by this network. The red markings in Fig. 3 indicate the activation function, rectified linear units (ReLu). The network was trained using backpropagation algorithm for 10 epochs and the softmax loss layer is updated using stochastic gradient descent for a batch size of 50. A weight decay of 0.001 was used for maintaining the objective function value within a reasonable range. Learning rate for all convolution layers was assigned as 0.002. Since the network was trained from scratch, weights of all layers were initialized randomly.

3.3 Sign-Net—The CNN Ensemble Model

One approach to speeding up the networks was to reduce the number of layers and neurons in each layer, but this may consistently lower the performance. As a trade-off between the computation complexity (owing to number of layers) and performance, the shallow LRN and deeper HRN were integrated. Ensemble was designed to work at varying spatial scales by combining the high resolution CNN and low resolution CNN. Even though the HRN had better accuracy than the LRN, their combination as an ensemble was much more powerful in doing the classification task. The resultant ensemble achieved improved performance than the individual constituent networks. Compared to the conventional CNN ensembles which are formed of same networks trained with different parameter settings or varying training data, the proposed model is a synthesis of two different CNNs of varying depths that works at different spatial resolutions, thereby incorporating the scale-space theory into CNNs. The ensemble training algorithm is given below.

Algorithm 1 Ensemble_Training

Input: Images of signs of alphabets or digits (v)
Output: Trained models
1: **procedure** ENSEMBLE_TRAINING(v)
2: Read input image dataset v.
3: Train_New_Net(v)
4: Train_LeNet(v)
5: Return trained models.
6: **end procedure**

3.4 *Ensemble Classifier*

The ensemble classifier is score-averaging softmax classifier. Softmax function normalizes the 34-dimensional output vector v according to the equation

$$p(v)_j = \frac{e^{v_j}}{\sum_{k=1}^{34} e^{v_k}} \tag{1}$$

where e^{v_j} and e^{v_k} represent vector values. The results of softmax normalization of individual classes, of both networks, were averaged, to derive the final predictions, which is maximum of all the average scores and the corresponding label will be the final ensemble output, i.e.,

$$\text{score} = \max(p(v)_j) \quad \text{and} \quad \text{class} = j \tag{2}$$

For conventional ensembles prediction by combining the results of two or more networks have proved to be better than using the individual ones. Similarly, it worked well for the new ensemble model as well.

Algorithm 2

```
Input: Image of a sign(img)
Output: Class or Sign Label
procedure ENSEMBLE_CLASSIFIER(img)
    lenet ← Load LRN
    newnet ← Load HRN
    Scores1[] ← Classify with lenet(img)
    Scores2[] ← Classify with newnet(img)
    for i = 1 to 34 do
        Scores3[i] ← Scores1[i] + Scores2[i]
    end for
    Final_Score ← max(Scores3)
    Return corresponding label.
end procedure
```

4 Results and Performance Evaluation

The new ensemble model was trained on a newly collected dataset of fingerspelled signs (see Sect. 4.1) using random initialization weights for LRN and HRN, except for first three layers of HRN, wherein weights were transfer learned from AlexNet trained on ImageNet. Also, an ensemble of LRN and AlexNet as HRN was evaluated for performance comparison. Different CNN models as well as ensemble combinations were compared to arrive at an optimal design. Figure 4 shows performance variation of HRNs—12 layer (AlexNet), 13 layer and 14 layer CNNs. It is evident that the accuracies decrease with increase in network depth, as long as training data remains constant. As the depth of network increases, it requires even more data for generating a good model. This fact has led us to choose the 13-layer deep model, which shows decent performance, rather than adding more layers or increasing the size of dataset.

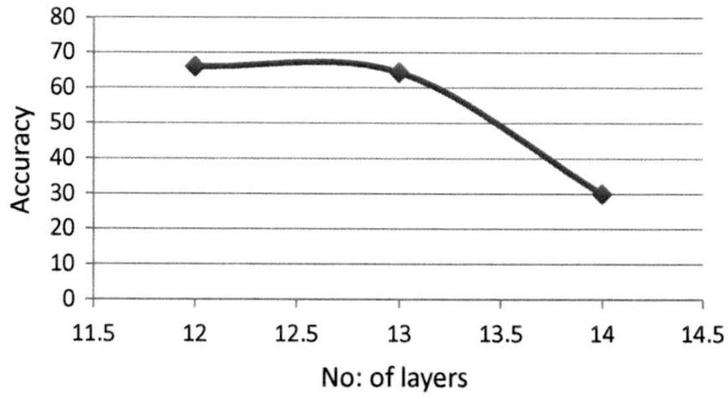

Fig. 4 Performance variation of HRNs with depth

4.1 Dataset

The newly collected dataset, used for this work, had a total of 12,138 static images of 34 signs—300 images per class for training, 50 images per class for validation during training and 238 test images, i.e., 7 images per class. Sample images from the dataset is given in Fig. 5. The testing and training samples were from different signers. The training data was collected from four different signers with maximum variations. Testing data corresponded to fingerspelled signs from four different signers. The images were captured in a home environment without any special lighting, using mobile camera: Samsung Galaxy S9 Plus. All data were collected in a black background so that learning happened only about the required features, and thus irrelevant background details were eliminated. This obviated the need for explicit hand segmentation.

4.2 Performance Comparison of All Models

We conducted a comparative evaluation of our Sign-Net model with the following models: (1) LRN, (2) HRN, (3) ensemble of LRN and HRN, (4) ensemble of LRN and AlexNet and (5) CNN architecture used by Tripathi et al. [12] (retrained AlexNet). Testing was performed on the same dataset and results are presented in Table 1. Accuracies or recognition rates for all the models were calculated according to the definition:

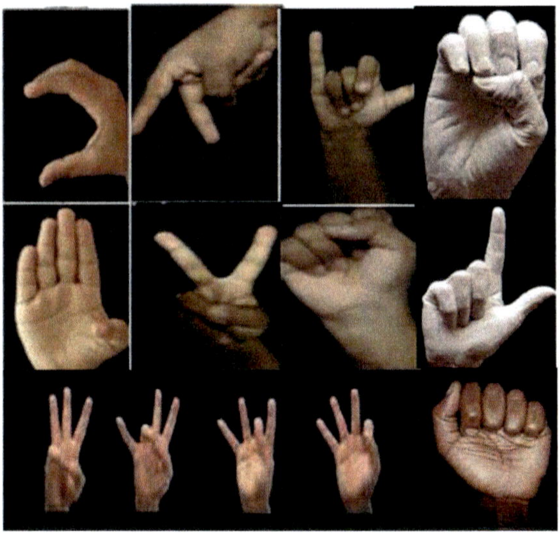

Fig. 5 Sample images from dataset

Table 1 Recognition rates of all trained models

Model	Accuracy (%)
Low resolution network (LeNet)	60.32
High resolution network (new model)	64.4
Sign-Net (proposed ensemble)	74.5
LRN + AlexNet	66.4
AlexNet [12]	66

$$\text{accuracy} = \frac{\text{Number of correctly classified images}}{\text{Total number of images}} \times 100 \tag{3}$$

The LRN obtained an accuracy of 60.32 and 64.4% for HRN and the combination score-averaging ensemble had a better recognition rate of 74.5%. Model [12], which refers to retrained AlexNet, showed a classification rate of 66%. The ensemble of LRN and AlexNet as HRN has reported an accuracy of only 66.4% much less than the proposed model. Based on these findings and the fact that the performance drops with depth, HRN was fixed as the new 13-layer architecture. These results show that the scale-space model has been highly effective in the deep learning framework for the task of image classification. The proposed model's performance characteristics were noticeably superior to the existing ones [12, 14]. The details given in Table 1 and Fig. 6 show that scale-space motivated approach yields better results than training an CNN on a single-scale image.

Low resolution images lead to faster training time than high resolution ones, but in terms of performance, it is still an open question whether high resolution images lead to better performance than the low resolution ones. A larger scale means a lower resolution image while a finer scale implies a high resolution image. Usage of low resolution images will help the neural net capture more representations of global

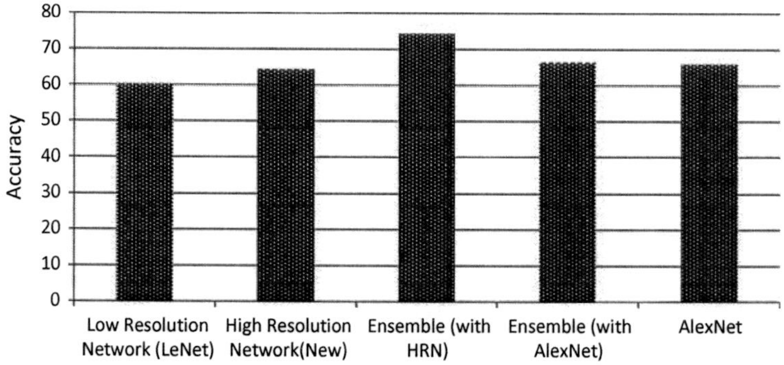

Fig. 6 Performance comparison of all trained models. The proposed ensemble shows superior performance compared to retrained AlexNet as well as the ensemble formed with AlexNet as HRN

features but the finer features can be lost; the upside of using high resolution images is speedier training and capture of fine details, but loss of the global features. There is a trade-off between the two, as it is important to be aware of the context in which CNN is applied. With classic computer vision problems, it is better to work at multiple resolutions. It is beneficial to work at multiple scales, so as to capture a variety of features that can only be best captured at particular scale levels. The efficiency of the proposed scale-space model could be further appraised by an analysis of the class-wise accuracies of challenging classes, i.e., signs corresponding to 1, d, s, t, 6 and w. The class-wise accuracies for these classes are reported in Table 2.

Based on the results in Table 2, the average recognition rates of all the models for the challenging classes are shown in Fig. 7. For these classes as well the new ensemble model has outperformed other models. The new HRN has also excelled over LRN, AlexNet and ensemble of LRN and AlexNet as HRN.

Table 2 Class-wise accuracies of challenging classes—1, d, 6, w, s, t.

Class	Sign-Net (proposed) (%)	LRN + AlexNet (%)	LRN (%)	HRN (%)	AlexNet (%)
1	87.5	75	75	37.5	75
d	57.14	0	0	57.14	0
6	87.5	100	50	75	87.5
w	85.71	42.85	42.85	85.71	42.85
s	83.33	83.33	33.33	66.67	83.33
t	85.71	57.14	57.14	71.43	28.57

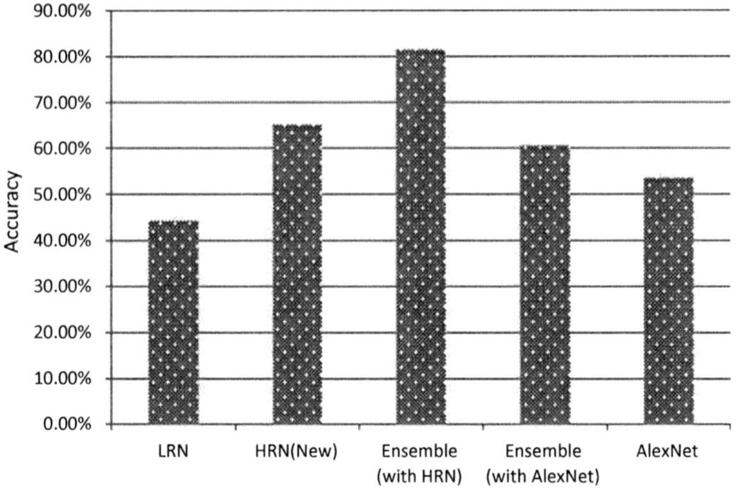

Fig. 7 Performance comparison of all trained models for challenging classes

Table 3 Recognition rates of all LRN, HRN and the proposed ensemble on MNIST dataset

Model	Accuracy(%)
Low resolution network (LeNet)	97
High resolution network (new model)	96
The ensemble (proposed)	98.5

4.3 Evaluation

Even if the scale-space motivated approach as outlined above seems more computationally intensive, the upside is more accurate results than merely running the ConvNet on a single resolution image, as shown by our empirical observations. The tabulated results show that the HRN performs better than the LRN in our task, which may not be the same in all cases. The ensemble, a combination of HRN and LRN performed better than either of the two by itself. Sign-Net also outperformed the famous AlexNet architecture when retrained for recognition of sign language.

Additionally, the performance of our new ensemble model for other applications was also analyzed. For this purpose, the datasets chosen were Caltech-101, CIFAR-10 and MNIST. While the performance on CIFAR and Caltech were very low, the ensemble showed good results for MNIST (Table 3). These outcomes lead to an understanding that the proposed ensemble model works best for applications that require learning of outer shapes of the object under consideration, rather than its finer details. CIFAR and Caltech type datasets require learning of minute details, for which the proposed ensemble fails.

5 Conclusion

The new CNN ensemble, Sign-Net, composed of different CNN models, has been shown to work better at multiple spatial resolutions, which enable capture of multiple input features. The ensemble showed better performance compared to its component individual networks, by incorporation of the scale-space model. Training time or computational complexity is reduced to a large extend by usage of the shallow LeNet architecture. This ensemble model is particularly suited well for the recognition of fingerspelled signs; the model can be reused for other tasks that requires learning of outer shape of objects. As a future work, we would like to appraise the application of fingerspelled word recognition that involve video processing.

References

1. Williams E (1977) Experimental comparisons of face-to-face and mediated communication: a review. Psychol Bull 84:963
2. Cooper H, Holt B, Bowden R (2011) Sign language recognition. In: Visual analysis of humans. Springer, Berlin, pp 539–562
3. Nithin DK, Sivakumar PB (2015) Generic feature learning in computer vision. Procedia Comput Sci 58:202–209
4. Ramachandran R, Rajeev D, Krishnan S, Subathra P (2015) Deep learning—an overview. Int J Appl Eng Res 25433–25448
5. Nagi J, Ducatelle F, Di Caro GA, Ciresan D, Meier U, Giusti A, Nagi F, Schmidhuber J, Gambardella LM (2011) Max-pooling convolutional neural networks for vision-based hand gesture recognition. In: Signal and image processing applications (ICSIPA). IEEE, pp 342–347
6. Ma L, Huang W (2016) A static hand gesture recognition method based on the depth information. In: Intelligent human-machine systems and cybernetics (IHMSC), pp 136–139
7. Pigou L, Dieleman S, Kindermans PJ, Schrauwen B (2014) Sign language recognition using convolutional neural networks. In: Workshop at the European conference on computer vision. Springer, Berlin, pp 572–578
8. Liwicki S, Everingham M (2009) Automatic recognition of fingerspelled words in British sign language. In: Computer vision and pattern recognition workshops. IEEE, pp 50–57
9. Gattupalli S, Ghaderi A, Athitsos V (2016) Evaluation of deep learning based pose estimation for sign language recognition. In: Proceedings of the 9th ACM international conference on pervasive technologies related to assistive environments. ACM, p 12
10. Isaacs J, Foo S (2004) Hand pose estimation for American sign language recognition. In: Proceedings of the thirty-sixth southeastern symposium. IEEE, pp 132–136
11. Pugeault N, Bowden R (2011) Spelling it out: real-time ASL fingerspelling recognition. In: Computer vision workshops (ICCV workshops). IEEE, pp 1114–1119
12. Kang B, Tripathi S, Nguyen TQ (2015) Real-time sign language fingerspelling recognition using convolutional neural networks from depth map. In: 3rd IAPR Asian conference on pattern recognition. IEEE, pp 136–140
13. Ameen S, Vadera S (2017) A convolutional neural network to classify American sign language fingerspelling from depth and colour images. Expert Syst 34:e12197
14. Tao W, Leu MC, Yin Z (2018) American sign language alphabet recognition using convolutional neural networks with multiview augmentation and inference fusion. Eng Appl Artif Intell 76:202–213
15. Oyedotun OK, Khashman A (2016) Deep learning in vision-based static hand gesture recognition. Neural Comput Appl 1–11
16. Geetha M, Kaimal MR (2018) A 3d stroke based representation of sign language signs using key maximum curvature points and 3d chain codes. Multimedia Tools Appl 1–34
17. Geetha M, Aswathi PV (2013) Dynamic gesture recognition of Indian sign language considering local motion of hand using spatial location of key maximum curvature points. In: Intelligent computational systems (RAICS). IEEE, pp 86–91
18. Kuznetsova A, Leal-Taixe L, Rosenhahn B (2013) Real-time sign language recognition using a consumer depth camera. In: Proceedings of the IEEE international conference on computer vision workshops, pp 83–90
19. Dong C, Leu MC, Yin Z (2015) American sign language alphabet recognition using microsoft kinect. In: Proceedings of the IEEE conference on computer vision and pattern recognition workshops, pp 44–52
20. Lindeberg T (1994) Scale-space theory: a basic tool for analyzing structures at different scales. J Appl Stat 21:225–270

A Survey on Deep Learning-Based Automatic Text Summarization Models

P. G. Magdum and Sheetal Rathi

Abstract Text summarization is referred to the process of rewriting a particular content into its brief version by understanding it. It is a serious need to obtain concise and relevant information from huge information. Automatic text summarization is fast-growing research area in natural language processing from last few years. It has improved from simple heuristics to neural network-based text summarization techniques. The extractive and abstractive text summarization generate the condensed text which saves time in reading the entire document. The abstractive approach is more complex as it needs natural language processing and neural network. Due to the availability of huge unsupervised data, traditional approach fails to provide accuracy in text summarization. The deep learning-based text summarization model gives a good performance as compared to the conventional techniques. In this paper, we have reviewed the recent work on text summarization based on deep learning techniques.

Keywords Text summarization · Natural language processing · Deep learning

1 Introduction

There is a large amount of text material available on the World Wide Web. So, searching relevant documents and absorbing relevant information become a time-consuming task. There is a big need to provide an improved system to extract relevant information quickly and efficiently. There is a massive growth in the online availability of data. Such data is available from Web pages, e-mails, news, e-books, science journals, learning content, and social media [1]. As the data grows larger in size, the redundancy may increase in text and become difficulty to produce concise information. The research on automatic text summarization has gained a lot of

---

P. G. Magdum (✉) · S. Rathi
Thakur College of Engineering and Technology, Kandivali(E), Mumbai, India
e-mail: magadumpg@rmcet.com

S. Rathi
e-mail: sheetal.rathi@thakureducation.org

N. N. Chiplunkar and T. Fukao (eds.), *Advances in Artificial Intelligence and Data Engineering*, Advances in Intelligent Systems and Computing 1133,
https://doi.org/10.1007/978-981-15-3514-7_30

importance. Some models help in finding extractive or abstractive summarized part of text content and trying to handle the overloading problem [2].

Automatic summarization is a very beneficial technique to facilitate users to refer a large number of documents efficiently. There is big improvement in automatic text summarization, and so many methods have been developed in this domain of research [3]. The text summary produced by different text summarization approaches consists of relevant information in text document, and also it occupies less space as compared to original text document [4]. There are several issues yet to be solved such as handling grammaticality of summarized text, redundancy, semantic similarity, long sentences, and consistency in generated summary. For such issues, the techniques used for text summarization are not able to produce required accuracy. Another main problem of large available information on the Internet is unlabeled data where machine learning algorithms have limitation to generate up to mark summary [5]. Recent advancements in artificial intelligence, like deep learning techniques, would help to gain better accuracy in summary generation of text documents [6]. The summarization of document improves decision making. So better to use deep learning techniques which could tackle such unlabeled data. So here the main objective of this survey paper is a review of some recent research papers based on text summarization supervised and unsupervised deep learning.

The rest of the paper is organized as follows: the need of text summarization, types of text summarization, classification of evaluation methods, review of five recent research papers of text summarization based on supervised deep learning and five unsupervised deep learning. The performance of each text summarization model is tabularized which is based on ROUGE tool. After that by considering overall aspects, advantages, disadvantages, and how we can improve the accuracy of text summarization models are discussed in the discussion section. Finally, the paper is concluded with a remark on the overall review of different papers.

2 The Need of Text Summarization

Nowadays, there is a huge increase in online publication of data; large Internet users and social media need of text summarization have emerged. The needs of text summarization in different fields are as follows:

- Sentiment analysis from customer's response on social media.
- Text summarization reduces reading time.
- For finding significant reference from historical book.
- To generate news headline.
- To retrieve important content blogs.
- Summaries make the selection process easier.

Table 1 Types of text summarization

S. No.	Types of summarization	Description
01	Single-document and multi-document	In single-document summarization, to generate summary, only one document is processed at a time In multi-document summarization, to generate summary, multiple documents are considered
02	Extractive and abstractive	The extractive summary is some relevant sentences from the source document The abstractive summary contains different words and phrases from the source document
03	Supervised and unsupervised	In supervised system, labeled data is used to generate significant text from the input documents. The large size of labeled data is required for training supervised techniques [8] The unsupervised systems use unlabeled data to generate summary from the input documents [8]
04	Web-based	The Web-based summary is relevant content in Web page
05	Personalized	A personalized summary is a piece of significant information that the user wants

3 Types of Text Summarization

The text summarization classification [7] according to application is shown in Table 1.

4 Evaluation of Summarization Systems

In text summarization, the main critical task is an evaluation of generated summary by different algorithms. There are intrinsic and extrinsic methods used for evaluation [9]. The supervised and unsupervised deep learning systems require dataset having large size for training. Nowadays, commonly used datasets are DUC2001, DUC2002, DUC2003, DUC2004, DUC2005, DUC2006, and DUC2007; also CNN and Daily Mail news datasets developed as good source for training and testing purposes; also many authors utilized Gigaword dataset for abstractive as well as extractive summarization. Figure 1 shows classification of evaluation methods used to evaluate summarized text.

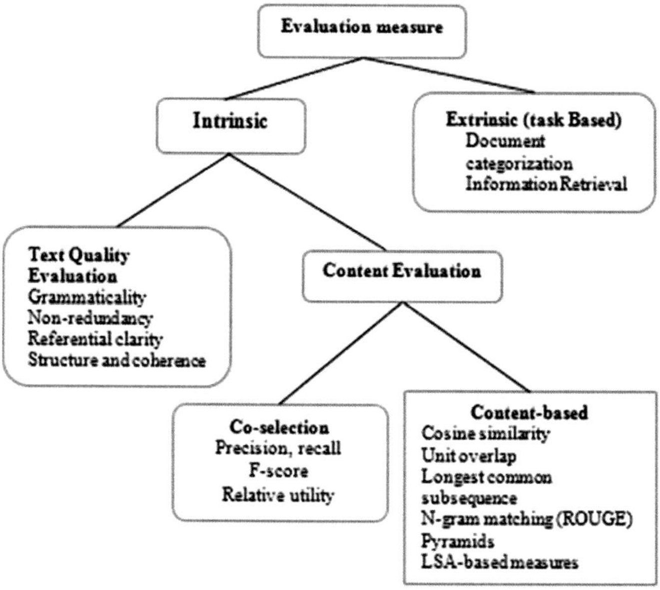

Fig. 1 Classification of evaluation methods [9]

ROUGE

The Recall-Oriented understudy for Gisting Evaluation (ROUGE) tool is used to evaluate output of text summarization models. It provides measures that can be used to evaluate generated summaries or GIST against one or more standard summaries [10]. It combines precision, recall, and *F*-measure in the measures.

The mathematical formula for precision (*P*) is as follows:

$$\text{Precision}\,(P) = \frac{\text{Number of relevant text sentences retrieved}}{\text{Number of retrieved text sentences}} \tag{1}$$

Recall (*R*) is given as follows:

$$\text{Recall}\,(R) = \frac{\text{Number of relevant text sentences retrieved}}{\text{Number of relevant text sentences}} \tag{2}$$

F-measure (*F*1 score) is given as follows:

$$F1 = \frac{\text{Precision} * \text{Recall}}{\text{Precision} + \text{Recall}} \tag{3}$$

Various versions of ROUGE are as follows:

ROUGE-N

$$\text{ROUGE - N} = \frac{\sum_{\text{Sen}\in(\text{Reference Summaries})} * \sum_{\text{gram}\in(S)} \text{Count}_{\text{mat}}(\text{gramsn})}{\sum_{\text{sen}\in(\text{Reference Summaries})} * \sum_{\text{gram}\in(S)} \text{Count}_{\text{gramsn}}} \quad (4)$$

where sen is a sentence, n is n-grams, gram_n and $\text{Count}_{\text{mat}}$ are the maximum number of n-grams co-occurring in candidate summary and a reference summary.

ROUGE-L: It computes longest common subsequence of documents.
ROUGE-W: It is the weighted longest common subsequence of documents.
ROUGE-S: It is the skip-bigrams co-occurrences statistics metrics.
ROUGE-SU: It measures skip grams plus unigram-based co-occurrences metrics.

ROUGE-N, ROUGE-S, and ROUGE-L are granularities of texts being compared between the system summaries and reference summaries.

5 Deep Learning-Based Summarization Techniques

The introduction of deep learning, the researcher get a great way for improvement in automatic text summarization. The neural network-based techniques render good performance as compared to traditional techniques with less human interference if the training data is rich. The deep learning-based text summarization models use supervised or unsupervised learning techniques. There is a small difference in model construction for text summarization in abstractive and extractive text summarization. In abstractive summarization model, text processing and encoder–decoder model and attention mechanism are included, while in extractive model text processing, feature extraction using deep learning techniques are included.

In this section, extractive and abstractive neural network-based model overviewed in details.

Generally, all neural network-based models follow the following steps:

- Sentence words to continuous vectors are called word embedding; use the neural network to obtain pre-learned lookup tables such as Word2vec, CW vectors, and GloVe.
- Sentences are encoded as continuous vectors using word embedding CNN or RNN which can be used as encoders for extracting sentence features.
- In the case of unsupervised learning, some models use RBM, autoencoder, and deep belief network as an encoder.
- In some model, a decoder uses LSTM, GRU, recurrent generator decoder, pointer generation network [11], etc.
- Some may use the attention model.

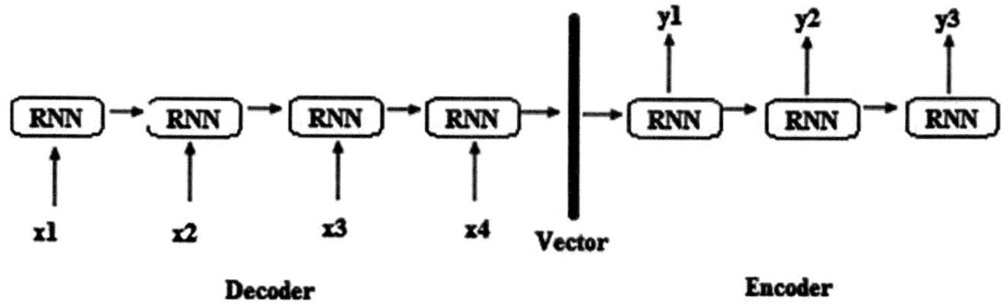

Fig. 2 Encoder–decoder model for text summarization

- Such input to the model for the extractive or abstractive summary.

The commonly used components in abstractive text summarization are:

Word Embedding
Word embedding is a popular way of representation of document vocabulary. It is used to convert input sentences into vector form required for further processing using deep learning techniques. Word2Vec and GloVe are popular techniques used to learn word.

Encoder–Decoder Model
Encoder–decoder model is sequence to sequence mapping model which uses recurrent neural network (RNN). The model trains source and target sentence and also handles variable length and output sequences of input and output sentences. Figure 2 shows encoder–decoder model for text translation.

6 Deep Learning Models in Text Summarization

6.1 Extractive Text Summarization Using Deep Learning

Extractive text summarization is selection-based method which includes representation of sentences and taking in consideration the most important sentences as summary by ranking. It uses supervised deep learning techniques for extractions of high-scored sentences. The extraction of summary in extractive text summarization follows path as shown in Fig. 3.

The following are some extractive summarization papers reviewed with some criteria:

NN-SE [12]
This is single-document summarization model which uses hierarchical document encoder and attention-based extractor. The proposed hierarchical framework system

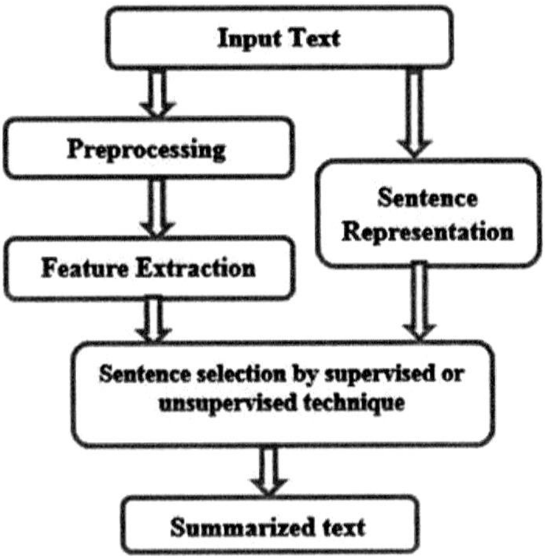

Fig. 3 Extractive text summarization

makes use of CNN and RNN to generate sentences and document representation, respectively. It uses dataset DUC 2002 and daily mail for evaluation.

Sentence representation

Initially, the sentences and words are separated out and form their respective sets. The given document NN-SE does not use exactly the same step in text processing like stop word removal, part of speech, stemming, but it uses hierarchical document reader to represent sentences. The reader is constructed using convolutional neural network (CNN).

Sentence Selection

The output summary is selected by attention-based extractor from the respective sets generated by hierarchical document encoder. The sentence and word extractor is attention mechanism which is constructed using recurrent neural network and long-short term memory.

R2N2_ILP [3]

The R2N2_ILP is the framework constructed using recursive neural network. It ranks the sentences of multi-document summarization. It also assigns additional ranking to sentences by constructing parse tree. The ranking scores obtained for word and sentences help to form more accurate sentence selection process. It uses dataset DUC 2002 and DUC2004 for evaluation.

Sentence representation

The input sentence is converted into binary tree. These parsing trees provide the meaningful representation words to sentences.

The score formula is as follows:

$$s(n) = R1(n), \quad n \in \mathrm{Nw} \tag{5}$$

$$s(n) = \alpha R1(n) + (1 - \alpha)R2(n), \quad n \in N - \mathrm{Nw} \tag{6}$$

where $R1$ is ROUGE-1 score, $R2$ is ROUGE-2 score, $N = (n)$ is whole non-terminal nodes and Nw is set of pre-terminals nodes, α is coefficient normally set to 0.5.
Sentence Selection
This model provides two sentence selection methods, and these are greedy-based sentence selection and ILP-based sentence selection. In greedy-based sentence selection method, the sentence with maximal salience value is added in summary. It uses tf-idf cosine similarity with threshold value Tsim = 0.6. The second method, ILP-based sentence selection, defines one objective function which combines the weights of sentences and weights of words. The summary is generated by considering these weights and redundancy of words.

$$\text{Max} = \lambda \sum_{i \in \text{Na}} \text{liWS}(i)\text{xi} + (1 - \lambda) \sum_{j \in \text{Nw}} \text{liWC}(j)\text{cj} \tag{7}$$

where xi indicates whether sentence i is summary, cj indicates the appearance of word j, li is length of sentence, λ is tradeoff between sentence and word, WS is weight of sentences and WC is weight of words.

SummaRuNNer [13]
SummaRuNNer is a RNN-based sequence classifier and renders visualization of its output. It presents training mechanism which allows training of extractive model using abstractive summaries. It uses dataset DUC 2002 and CNN/daily mail for evaluation.
Sentence representation
SummaRuNNer uses two-layer bidirectional recurrent neural network for sentences representation. In the two layers of RNN, both layers are of bidirectional GRU. The first layer of bidirectional RNN accepts sentences in the form of word embedding and produces a set of hidden states. The hidden states are calculated in form of averaged vector. The second layer of RNN takes the sentence representation as input produced by first layer and converts it into document representation by nonlinear transformation.
Sentence selection
The RNN decoder performs function of sentence selection. The decoder is with softmax function layer to generate word at time step. The soft generation of significant word at each time step is given by feedforward layer followed by softmax layer. So, probability at each time step is as follows:

$$\text{fk} = \tanh(w'\text{fh hk} + w'\text{fxXk} + w'\text{fcS} - 1 + b'f \tag{8}$$

$$\text{Pv}(w)k = \text{softmax}\big(w'\text{vfk} + b'v\big) \tag{9}$$

Bilingual Automatic Text Summarization Using Unsupervised Deep Learning [14]
The described model reduces the input bilingual document into its condensed text by using unsupervised deep learning technique, i.e., Boltzmann restricted machine [RBM]. This model improves much accuracy of automatic text summarization.
Text preprocessing
The text preprocessing phase includes techniques like sentence segmentation, tokenization, stop word removal and POS tagging. This will help in further processing of text.
Feature extraction
In the feature extraction phase, the preprocessed texts are represented by feature vector. The different features are extracted on each sentence to generate feature matrix which is useful for further calculation of sentence score. The features calculated are sentence position, sentence TF-ISF, sentence centroid similarity features, sentence similarity, numeric token, sentence length, proper noun, named entity, Bi-gram and tri-gram key features. On the basis of extracted different features, 2D sentence matrix is constructed.
Sentence selection
Sentence score of the document is calculated by using the sentence matrix. The sentence score is given by

$$\text{sen\_ score}_i = \sum_{1}^{1} 1\,\mathrm{j}\,(\mathrm{fi}) \tag{10}$$

where sen_score$_i$ is sentence score of ith sentence and fi is jth feature value of i. All the sentences are arranged in descending order according to sentence score, and top sentences are considered for summary.

Text Summarization Using Unsupervised Deep Learning [15]
Introduces deep autoencoder which introduces feature that improves query-based single-document summarization.
Sentence representation
The sentence representation by using local word representation is constructed using vocabulary of input document. It uses ensemble noisy autoencoder model which is used in sentence representation and sentence ranking
Sentence selection
The sentence selection is done using sentence ranking. The model ranks the sentences of documents based on important and relevant features of text documents. The encoder produces the concept vector. The cosine similarity metric is used on concept space to select sentences. The cosine similarity is calculated as follows:

Table 2 Performance comparison of extractive model

Model	Sentence representation	Sentence selection	Dataset	Performance evaluation using ROUGE tool		
				ROUGE-1	ROUGE-2	ROUGE-L
NN-SE [12]	RNN and CNN	Attention mechanism + RRN + LSTM	DUC 2002, Daily mail	48.5	21.5	43.5
R2N2_ILP [3]	Parse tree and ranking	Greedy-based and ILP sentence selection method	DUC 2001, 2002, 2004	37.92	8.88	–
SummaRuNNer [13]	Bidirectional RNN decoder having two layers of GRU	Softmax and feedforward layer	Daily mail	46.6	23.1	43.3
Bilingual automatic text summarization using unsupervised deep learning [14]	RBM	Sentence score	HinDoc, EngDoc	Recall = 0.8331	Precision = 0.8724	$F1 =$ 0.8523
Text summarization using unsupervised deep learning [15]	Autoencoder and RBM	Ensemble noisy Autoencoder	SKE email	–	11.2	–

$$\cos(\mathrm{Cq}, \mathrm{Cs}) = \frac{\mathrm{Cq}.\mathrm{Cs}}{\|\mathrm{Cq}\|\,\|\mathrm{Cs}\|} \tag{11}$$

where Cq is query and Cs is sentence.

Comparative Performance of Abstractive Model
All the above abstractive models described are compared against the points: Sentence representation, sentence selection, dataset and performance on different datasets are shown in Table 2.

6.1.1 Abstractive Text Summarization Using Deep Learning

Extractive text summarization is based on capturing the meaning of document and generates abstractive summarization by considering meaning representation. The

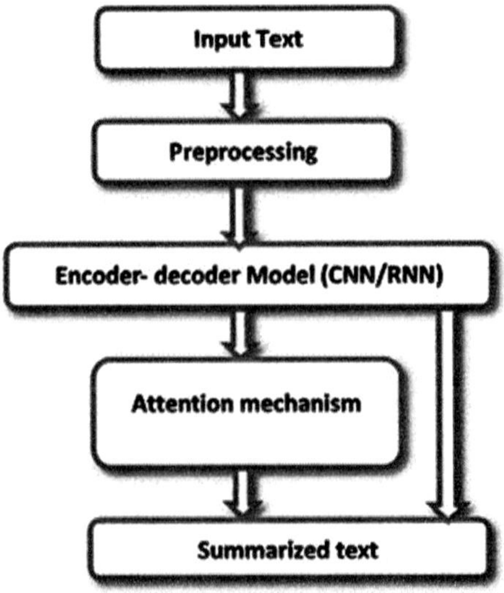

Fig. 4 Abstractive text summarization

neural network first represents whole document by the encoder and then generates the word sequence by decoder. The extraction of summary in extractive text summarization follows the path as shown in Fig. 4.

The techniques used in the abstractive summarization using deep learning are compared with their performance as follows.

6.2 Attention-Based Summarization [16]

Proposes neural attention model combined with generation algorithm to generate accurate abstractive summary

Encoder

The meaning representation of text document is obtained by using three encoder structures. The main function of encoder is to perform word to vector conversion. The following are three instantiates of the encoder.

- Bag-of-words encoder: The ABS model uses bag of words of source sentence having some definite size.
- Convolutional Encoder: some modeling issues of bag of words are resolved by using convolutional encoder. It allows local interaction between words. In each layer of convolutional operations, feature vector is extracted and the count of feature vectors is reduced by max pooling.
- Attention-based encoder: This encoder generates document representation.

Decoder

The model uses neural network language model (NNLM) as decoder. The word generation is done by estimating probability distribution.

DEATS [17] It introduces dual encoder system, having encoder of two types, namely primary encoder and secondary encoder which help in the generation of summary for long sequences and reduce repetitions.

Encoder
The model has dual encoding mechanisms. The purpose of primary encoder is feature extraction; the first encoder computes the significance of each word from text sentence, and the secondary mechanism calculates weight for each word in input sentence. The primary encoder uses bidirectional GRU-based RNN, and secondary encoder uses unidirectional GRU-based RNN.

Primary encoder equation

$$C^{\mathrm{p}} = \tanh\left(W_{\mathrm{p}}\frac{1}{N}\sum_{i=1}^{N} h_t^{\mathrm{p}} + b_{\mathrm{p}}\right) \tag{12}$$

where C^{p} is Word representation, W_{p}, b_{p} are parameter matrices and N is length of input sequence.
Decoder
The decoder is with GRU and attention mechanism which generates fixed-length output sequence at each stage. The context vector produces meaningful and relevant output.

NLP_ONE [18] It introduces joint attention model to avoid redundant word on output sequence for better accuracy.

Encoder
The encoder with bidirectional neural network + LSTM produces the feature representation for input sequence. The encoder based on LSTM compares set of input sequence vectors $X = (x_1, x_2, \ldots, x_t)$ to the set of LSTM state vectors

$$H^e = \{h_1^e, h_2^e, \ldots, h_t^e\} \tag{13}$$

$$h_t^e = \mathcal{F}(x_t, h_{t-1}^e) \tag{14}$$

where $\mathcal{F}$ is dynamic function of bidirectional LSTM.
Decoder
The decoder with LSTM generates a target sequence $Y = \{y_1, y_2, \ldots, y_t'\}$.

The target sequence is produced using context vector C into a set of decoder state vectors.

$$\mathcal{H}^{\mathrm{d}} = \{h_1^{\mathrm{d}}, h_{21}^{\mathrm{d}}, \ldots, h_{t'}^{\mathrm{d}},\}$$

The prediction model is given by

$$h_{t'}^{\mathrm{d}} = f(y_{t'-1}, h_{t'-1}^{\mathrm{d}}, C) \tag{15}$$

$$P(y_{t'}|y_{<t'}X) = g\left(y_{t'-1}, h^{\mathrm{d}}_{t'-1}, C\right) \tag{16}$$

RAS-Elman [19] This paper proposes the convolutional attention-based conditional model with RNN to generate abstractive text summary. It generates news headline of news article.

Encoder
The encoder in this model is called as attentive encoder. It computes context vector for each time step *t*. The encoder uses convolutional network to encode input. The encoder output is as follows:

$$c_t = \sum_{j=1}^{M} \alpha j, t - 1xj \tag{17}$$

where c_t is context vector, xj is jth word of input sentence, αj is weight.
Decoder
The decoder is also called recurrence decoder. It uses recurrent neural network (RNN). The proposed conditional model is represented by attentive decoder.

ATSDL [6] It proposes the combination of LSTM and CNN to generate better quality summery

Encoder
It is used to compute document representation based on CNN and LSTM. The encoder uses single-layer bidirectional LSTM.
Decoder
The decoder uses sequence to sequence model to generate summary.

Comparative Performance of Abstractive Model
All the above abstractive models described are compared against the points: Encoder model, decoder model, dataset and performance on different datasets are shown in Table 3.

7 Discussions

There is great improvement in the performance of text summarization using deep learning techniques. The techniques like RNN provide solution to critical problem where many other basic techniques fail. Introducing joint attention mechanism with deep techniques reduces inaccuracies and redundancy of word. Deep learning model improves readability of abstractive and extractive text summarization.

Even though there is much research on text summarization using deep learning, still some problems are facing by deep learning model.

Table 3 Performance comparison of abstractive models

Model	Encoder model	Decoder model	Dataset	Performance evaluation using ROUGE tool		
				ROUGE-1	ROUGE-2	ROUGE-L
Attention-based summarization [16]	Attention-based encoder bag-of-word encoder Convolution encoder	NNLM decoder	DUC 2002 DUC 2004, Giga word	28.18	8.49	23.81
DEATS [17]	Primary encoder: GRU-RNN Secondary encoder: Unidirectional GRU-RNN	GRU + attention model	CNN/Daily Mail and DUC 2004	29.91	9.61	25.95
NLP_ONE [18]	Bidirectional LSTM-based encode	Unidirectional LSTM-based decoder	NLPCC 2017 shared task3	34.94	21.17	30.66
RAS-Elman [19]	RNN attention	Elman RNN and LSTM	Gigaword DUC 2004	28.97	8.26	24.06
ATSDL [6]	CNN-LSTM	Sequence to sequence model	CNN and daily mail	34.9	17.8	–

- Since we are using supervised learning, it consumes much time in labeling the input data, which needs a lot of human efforts for labeling data manually.
- Models are not sufficiently robust for accepting long sentences.
- There is yet the issue of representation of semantic meaning of sentence.
- Improvement in grammar of generated text is needed
- Proper and large dataset is very important for training of model.

Due to the availability of a huge amount of information online from different sources, it is difficult to create the labeled dataset to train supervised learning deep learning model.

To overcome such a problem, we need to use unsupervised learning model which accepts unsupervised or unlabeled data. The unsupervised learning model tries to find out natural structure of data. The techniques search for related patterns and structures in the information and put into respective clusters. The formed clusters are used for further processing.

Yet, need to improve the unsupervised model to handle such vast information. Also, the use of reinforcement learning techniques may improve significant result in automatic text summarization.

8 Conclusion

The growth of information due to the Internet and other means has created a great need to develop efficient and accurate summarization systems which will be fulfilled by such a system. The abstractive and extractive text summarization using supervised deep learning techniques perform well for large size dataset, but still, there are many problems like large training time, grammaticality, semantic similarity.

The development of more focused and efficient unsupervised deep learning model for abstractive and extractive text summarization will produce better and grammatically powerful text summarization model. It will also reduce much time-consuming in preparation of supervised dataset.

References

1. http://home.iitk.ac.in/~soumye/cs396a/pres.pdf
2. Afsharizadeh M, Ebrahimpour-Komleh H, Bagheri A (2018) Query-oriented text summarization using sentence extraction technique. In: 2018 4th international conference on web research (ICWR). https://doi.org/10.1109/icwr.2018.8387248
3. Cao Z, Wei F, Li S, Li W, Zhou M, Wang H (2015) Learning summary prior representation for extractive summarization. In: ACL
4. Jo T (2018). Text summarization. In: Text mining, pp 271–294. https://doi.org/10.1007/978-3-319-91815-0_13
5. Spärck Jones K (2007) Automatic summarizing: the state of the art. Inf Process Manag 43(6):1449–1481. https://doi.org/10.1016/j.ipm.2007.03.009
6. Song S, Huang H, Ruan T (2018) Abstractive text summarization using LSTM-CNN based deep learning. Multimed Tools Appl. https://doi.org/10.1007/s11042-018-5749-3
7. Gambhir M, Gupta V (2016) Recent automatic text summarization techniques: a survey. Artif Intell Rev 47(1):1–66. https://doi.org/10.1007/s10462-016-9475-9
8. https://towardsdatascience.com/automated-text-classification-using-machine-learning-3df4f4f9570b
9. Steinberger J, Ježek K (2009) Evaluation measures for text summarization. Comput Inf 28:1001–1026 (V 2009-Mar-2)
10. Lin C-Y (2004). ROUGE: a package for automatic evaluation of summaries. In: Proceedings of workshop on text summarization branches out, post-conference workshop of ACL, Barcelona, Spain
11. See A, Liu PJ, Manning CD (2017) Get to the point: summarization with pointer-generator networks. In: Proceedings of the 55th annual meeting of the Association for Computational Linguistics, ACL, Vancouver, Canada, July 30–Aug 4, Volume 1: Long Papers, pp 1073–1083
12. Cheng J, Lapata M (2016) Neural summarization by extracting sentences and words. In: Proceedings of the 54th annual meeting of the Association for Computational Linguistics (Volume 1: Long Papers). Berlin, Germany. Association for Computational Linguistics, pp 484–494
13. Nallapati R, Zhai F, Zhou B (2017) SummaRuNNer: a recurrent neural network based sequence model for extractive summarization of documents. In: Proceedings of the thirty-first AAAI conference on artificial intelligence, San Francisco, California, USA, Feb 4–9, pp 3075–3081
14. Singh SP, Kumar A, Mangal A, Singhal S (2016) Bilingual automatic text summarization using unsupervised deep learning. In: International conference on electrical, electronics, and optimization techniques (ICEEOT), pp 1195–1200. https://doi.org/10.1109/iceeot.2016.7754874

15. Yousefi-Azar M, Hamey L (2017) Text summarization using unsupervised deep learning. Expert Syst Appl 68:93–105. https://doi.org/10.1016/j.eswa.2016.10.017
16. Rush AM, Chopra S, Weston J (2015) A neural attention model for abstractive sentence summarization. In: Proceedings of the 2015 conference on empirical methods in natural language processing, EMNLP, Lisbon, Portugal, Sept 17–21, 2015, pp 379–389
17. Yao K, Zhang L, Du D, Luo T, Tao L, Wu Y (2018) Dual encoding for abstractive text summarization. IEEE Trans Cybern, pp 1–12. https://doi.org/10.1109/tcyb.2018.2876317
18. Hou L, Hu P, Bei C (2018) Abstractive document summarization via neural model with joint attention. Lect Notes Comput Sci, 329–338. https://doi.org/10.1007/978-3-319-73618-1_28
19. Chopra S, Auli M, Rush AM (2016) Abstractive sentence summarization with attentive recurrent neural networks. In: The 2016 conference of the North American Chapter of the association for computational linguistics: human language technologies, San Diego California, USA, June 12–17, 2016, pp 93–98

Automatic Multi-disease Diagnosis and Prescription System Using Bayesian Network Approach for Clinical Decision Making

P. Laxmi, Deepa Gupta, G. Radhakrishnan, J. Amudha, and Kshitij Sharma

Abstract A clinical decision support system (CDSS) is used as an aid in decision-making processes of health care providers in their day-to-day activities. This research attempts diagnosis of multiple diseases based on symptoms provided by patients. The work also recommends laboratory tests related to the predicted diseases and medications based on their results. The methodology adopted for implementation of CDSS is the Bayesian network approach. The modeling of the Bayesian network structure was undertaken in consultation with experts from medical domain. Clinical data has been used for estimation of network parameters such as conditional probability tables thereby bringing in machine learning into Bayesian methodology. The model developed is a learning model wherein the system input is saved for future training of the model. The results indicate that Bayesian approach is suitable for implementing a CDSS for multiple disease diagnosis. The proposed work will be useful towards increasing physicians throughput.

Keywords Bayesian network · Bayesian learning · Expert systems

P. Laxmi (✉) · D. Gupta · G. Radhakrishnan · J. Amudha
Department of Computer Science & Engineering, Amrita School of Engineering, Amrita Vishwa Vidyapeetham, Bengaluru, India
e-mail: laxmip20@gmail.com

D. Gupta
e-mail: g_deepa@blr.amrita.edu

G. Radhakrishnan
e-mail: g_radhakrishnan@blr.amrita.edu

J. Amudha
e-mail: j_amudha@blr.amrita.edu

K. Sharma
Paralaxiom Technologies Private Limited, Bengaluru, India
e-mail: kshitij.sharma@paralaxiom.com

N. N. Chiplunkar and T. Fukao (eds.), *Advances in Artificial Intelligence and Data Engineering*, Advances in Intelligent Systems and Computing 1133,
https://doi.org/10.1007/978-981-15-3514-7_31

1 Introduction

Health care industry has a direct bearing on the personal satisfaction in lives of people in every nation. Modern health care is an extremely competitive sector that continuously strives to enhance people's chances of survival. Infusing large portions of its shares of profits into R&D, modern health care industry is a major economic influencer.

In this aspect, clinical decision support systems (CDSS) prove to be a vital tool in health care. Decision support systems (DSS) are developed using mathematical tools and techniques for enabling managements to arrive at decisions under uncertainty. Use of CDSS on aspects of clinical process was understood in late 50s leading to implementation and usage in early 60s [1].

Rapid development in CDSS can be attributed to artificial intelligence and quite recently, machine learning. The initial era of CDSS saw successful systems like MYCIN, QMR, and Protege. CDSS can be broadly categorized into knowledge-based and non-knowledge-based systems [2]. The former has three major components which includes a knowledge base,an inference engine, and a method to communicate with the user. Collation and population of knowledge in such systems use data from experts in medical field. In non-knowledge-based systems, the concept of knowledge base is to learn from clinical data.

The step of knowledge acquisition is the foundation of any CDSS. The correct representation of knowledge in knowledge base is a crucial step [3]. Knowledge representation can be logical wherein inference involves a simple lookup of known facts or procedural where representation takes form of 'If-Else' rules,logic or probabilities. Commonly used methodology for procedural systems is fuzzy logic and Bayes rules.Knowledge representation can also be graph/network-based as commonly seen in methodologies like Bayesian network,decision tree and artificial neural network. The representation can also be structural where in knowledge is packed into different levels according to its importance.

This paper focuses on the design details of a clinical decision support system for disease diagnosis of some commonly occurring Indian diseases. A Bayesian network methodology is used for subject implementation. The data set required has been generated with the help of medical domain experts. The system also recommends laboratory tests related to the disease(s) diagnosed by the model. Further, medication is also recommended to the patients based on their age. The proposed system is used to aid the doctors in decision making and is not intended for the individual use of a patient without supervision from the medical expert. The remainder of this paper is organized as follows: Section 2 discusses about the literature survey done in the related medical domain. Section 3 gives a brief overview of Bayesian network. Section 4 discusses about architecture and implementation details of the proposed work. Section 5 enlists the experimental results and Sect. 6 discusses the conclusion, limitations and future work proposed.

2 Related Works

Literature discusses CDSS built using many methodologies. Selection of a particular methodology in implementing a CDSS is subject to factors like method of knowledge acquisition used, the type of disease diagnosis, etc.
Rule-based CDSS which comes under knowledge-based systems, are in use since inception stages. In such systems are a set of 'If-Else' rules obtained from experts forms knowledge base. A rule engine which interprets subject rules is the main component of such systems. The limitation in using such systems lies primarily in the heterogeneity of clinical data.Mapping of complex clinical data to rule expression variables is extremely difficult. Zhang et al. propose a shared ontology for interaction between rule engine and external systems in implementing a CDSS [4]. The work involved creation of an ontology from local vocabularies used in external clinical systems. This ontology becomes a part of the CDSS and is used by the rule editor and rule engine for rules generation and inference. Scalable multi-disease diagnosis can also be easily developed using a rule-based system [5]. Chen et al. proposes a rule-based CDSS for diagnosing hematological disorder [6]. 'If-Else' rule engine is used for analyzing the CBC count of the patient and disease diagnosis is made based on CBC result.

The symptoms which the patient describes could be very vague and thus, uncertain. Fuzzy logic-based CDSS are mostly used to handle such uncertainties. These fall under knowledge-based systems. Malmir et al. proposes a fuzzy logic-based CDSS which tries to capture all uncertain inputs provided by patients [7]. Knowledge base for a disease for the proposed work is collected from experts. Fuzzy rules are derived out of this knowledge base. Symptom inputs from users are taken using linguistic and numeric inputs which helps in capturing uncertainty. Inference mechanism used in the work is Mamdani inference which works on rules and provides correct output. Fuzzy systems were also used for determining obstructive pulmonary disease in people exposed to chemicals [8]. Taha et al. have developed a model where variables affecting the disease were initially determined through clinical guidelines and rules were generated and stored in database. Mamdani method was used for inference and was observed to give very good results. Recent development in fuzzy systems is its integration with neural network (ANFIS). Shaker et al. have done a comparison of two soft computing techniques fuzzy analytical hierarchy process(FAHP) and ANFIS in detection of liver fibrosis from diagnostic data [9]. ANFIS approach utilizes information gain to determine the most relevant features. Clustering was applied initially and the ANFIS model was trained on clustered data using Sugeno model. The feasibility of using an ANFIS or FAHP model was confirmed by the results.

Ontologies represent semantic relationships between concepts. Ontology-based CDSS prove successful in incorporating volumes of available data on diseases, symptoms, medicines and other ancillaries. Shen et al. propose a work where ontology is used for implementing a CDSS which works without need of any health care provider [10]. The patient can perform disease diagnosis by entering symptoms. The work also mentions medications to be provided on the basis of disease diagnosis. It uses existing ontologies integrated into the system along with newly created ontology. Multiple clinical process guidelines(CPG) can also be integrated into one functional CDSS with ontology [11]. Alexandre et al. proposed the system to integrate HT and T2D CPG into a CDSS. The concepts were designed for patient variables and clinical management variables. The ontological reasoning brings flexibility in CDSS and also offers the ability to deal with patients suffering from multiple disorders.

Machine learning(ML) has been recently introduced in implementing CDSS. These non-knowledge-based systems learn patterns from clinical data. These systems do not require expert involvement.Decision tree methodology which is a ML algorithm is used in implementing a diagnostic system for acute respiratory infection [12]. Support vector machine which is an extremely popular supervised machine learning algorithm is also used in diagnosing diseases [13]. Self-organizing neuro fuzzy logic is a ML-based technique which is used in CDSS [14]. Baig et al. propose a CDSS for diagnosis of physiological events. Adaptive network-based fuzzy inference system is used in the proposed work for combining both fuzzy logic and neural network.

Highly useful ML techniques are also used in detection of correlation of ICD-Codes and chronic diseases. The system designed by Deepa et al. has applied machine learning techniques like InfoGain and Adaboost to the Centre for Medicare Services (CMS) data set to find reduced set of diagnostic code for a chronic disease [15]. To determine relationships among factors determining a disease has been found extremely useful. A study of cardiovascular disease using association mining techniques has been performed by Sangita et al. [16]. Data mining techniques were used for finding relationship between a diseases and its effects on other parts of body [17]. Classification techniques like decision tree, multi-layer perceptron, Naive Bayes, support vector machine have been used for finding parameters that greatly affect breast cancer diagnosis [18].

Probabilistic approaches are also widely used in CDSS. A three-layer multi-symptom Naive Bayes methodology is used by Jiang et al. [19]. Bayesian network methodology, a knowledge-based systems rely on knowledge base populated using expert data. The approach can also be considered as a ML approach as parameter estimation can also be done from clinical data. Bayesian approach has been used in detection of Alzheimers, Dementia and Mild Cognitive Impairment [20]. A 'three-layer' approach comprising demographic details, clinical manifestation and symptoms is used for disease detection. Network creation of the proposed work has been created with help from experts. Parameter estimation is done from data set. A similar Bayesian approach has been used for Dengue detection [21]. In this case, network was created using expert help and the parameter estimation was inferred from data. Bayesian approach is also used for breast cancer detection [22]. Constantinou et al. have designed a framework for designing a CDSS in the most appropriate way that involves data acquisition from patients and doctors [23]. The framework proposes eight important steps for successful implementation of a Bayesian network methodology in CDSS.

Bayesian network methodology was selected for implementation of the proposed work as it has been successfully used in CDSS as mentioned in literature. It is well suited for representing uncertainty and causality of clinical domain. Its modular representation of data is another added advantage. It also provides an efficient and principled approach for avoiding over-fitting data in conjunction with Bayesian statistical methods.

Most of the work in literature mentions a single disease modeling using Bayesian approach. Very few works have been done in multi-disease modeling this methodology. The proposed work attempts multi-disease modeling for commonly occurring Indian diseases. A two-layer approach having symptoms and diseases was selected for the implementation of CDSS. The work also includes Laboratory Test recommendations and medicine prescriptions. A brief overview of basics of Bayesian network is mentioned in the following section.

3 Bayesian Network Basics

Bayesian network is a representation of knowledge in the form of nodes and links. The graphical model of Bayesian network is in the form of a directed acyclic graph. The Bayesian network represents data in a compact form by eliminating the need of storing the entire joint probabilities of the network. Instead, only the conditional probability tables (CPT) are created for nodes by taking into account parent nodes probabilities. The proposed work has a Cause-Evidence structure as mentioned in Fig. 1. The model tries to predict probability of cause nodes depending on the evidence.

The Bayesian network methodology has three integral components as mentioned:
Bayesian Network creation: The network creation can be done either manually with help of experts or by learning from data. The methodology adopted for network creation for this work involves manual creation of network with expert help.
Bayesian Parameter creation: The parameter estimation involved in Bayesian network involves creation of conditional probability tables for every network node either manually or through learning from data.

Fig. 1 Structure of Bayesian Network

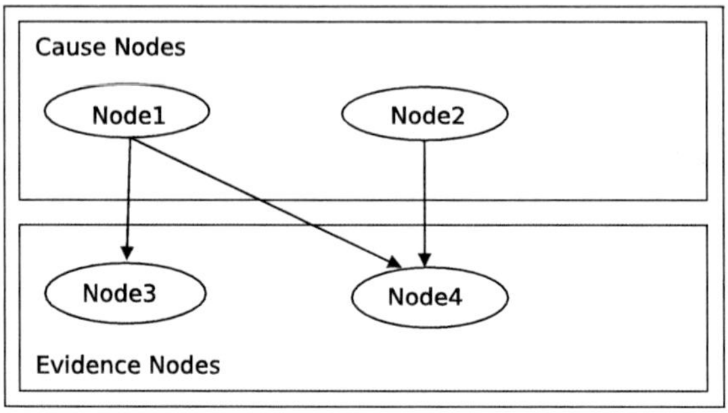

The proposed model falls in the category 'Known structure, Unknown parameters'. The proposed work uses maximum likelihood estimator which is one of the two most commonly used data learning algorithms for parameter learning. Maximum Likelihood Estimator: The maximum likelihood estimator algorithm tries to find optimum parameters that maximize the likelihood of seeing all data instances. The likelihood of parameter estimate for network structure G with estimates θ is given by Eq. 1.

$$L(\theta/D) \Rightarrow \prod_{i=1}^{N} Pr_{\theta}(d_i) \tag{1}$$

Here, probability of each instance of data set is computed and multiplied. The final parameter estimates chosen by MLE are the estimates that maximize the likelihood function as mentioned in Eq. 2.

$$\theta^* \Rightarrow \arg\max_{\theta}(\theta/D) \tag{2}$$

Bayesian Network Inference: Inference refers to questions asked to the Bayesian network. Evidence refers to the instantiation of some variables that are provided to the network to compute probabilities of some network nodes. The proposed work uses an exact inference algorithm called variable elimination algorithm (VE) for inference. VE is the simplest form of inference. VE uses CPTs as factors and works by eliminating variables one by one. Variables which are not a part of evidence are summed up and the marginal distributions over the remaining variables are computed by merging all rows that agree on those remaining variables.

4 The Proposed CDSS with Laboratory Test and Medicine Recommendations

The high-level architecture of the proposed work is as depicted in Fig. 2. The architecture is divided into six major components depending on the functionality of the CDSS as mentioned below:

- Data collection module
- Probable disease identification module
- Laboratory tests recommendation module
- Disease confirmation module
- Medication module
- Learning module.

The dataflow diagram of the proposed system combining all of the above-mentioned modules is as depicted in Fig. 3. The initial step in the proposed system is to create the Bayesian model and store it in memory. The details of the Bayesian network creation are mentioned in Sect. 4.1. The input to the system comprises of 'symptom input' and 'demographic input' from users. A total of 28 symptoms have

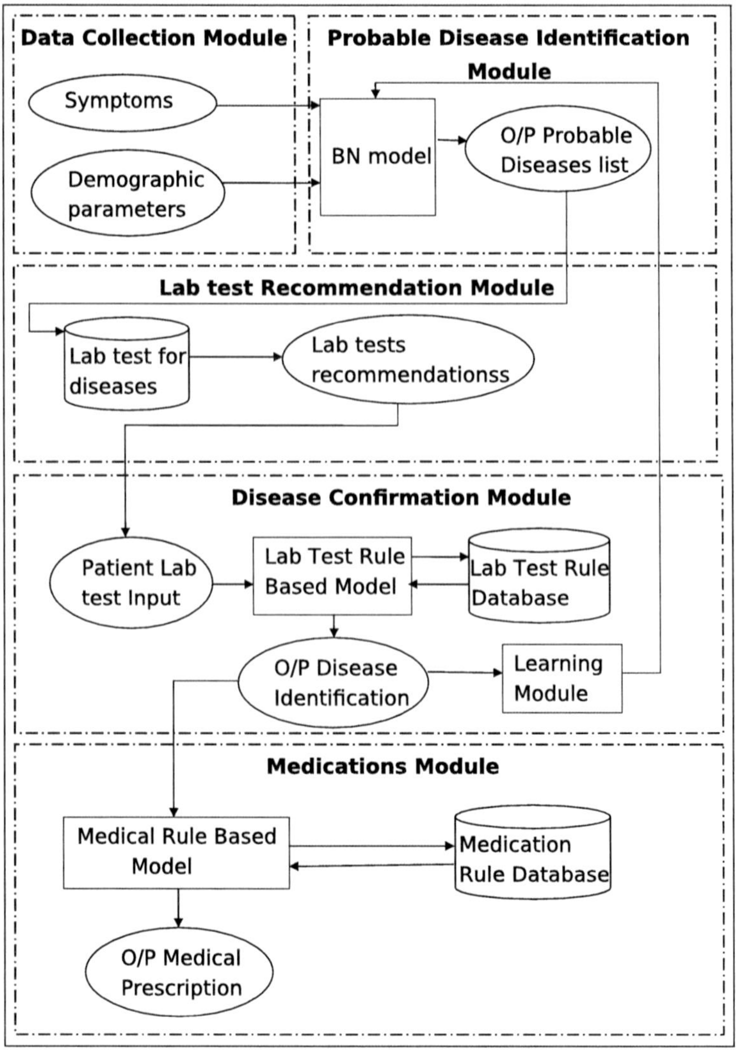

Fig. 2 High-level architecture of the proposed CDSS

been identified for modeling eight diseases. The input data is pre-processed for mapping the symptom states to Bayesian modeled network states. Evidence vector for inference is created based on entered symptoms and passed on to inference engine. The inference engine queries the built-in Bayesian model for getting a list of probable diseases. Ranking function takes these diseases as input and produces a sorted list based on probabilities. This data is passed on to visualization module for generating a pie chart. The top three most probable diseases are selected from the sorted data

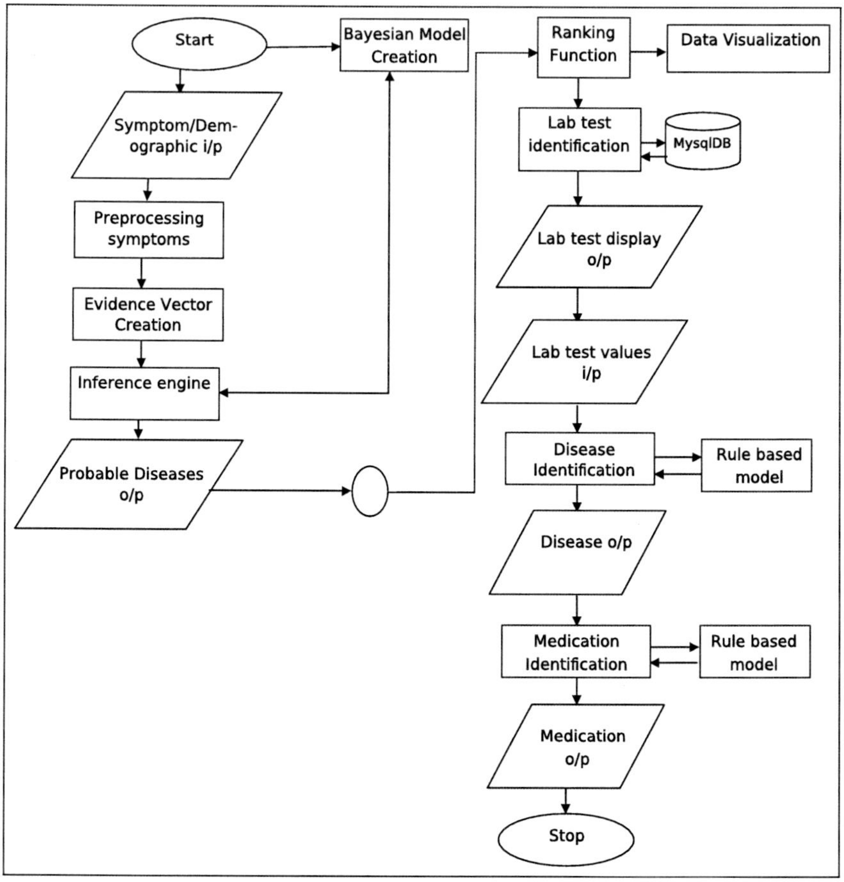

Fig. 3 Data flow diagram of the proposed CDSS

and passed on to the Laboratory test identification function. Database is queried to select the laboratory test data related to inferred disease(s).

The user inputs the laboratory test values to the system. The disease identification function takes these inputs and queries the rule-based model for laboratory test result analysis. The output of the identification function is the ‘top three’ diagnosed diseases. The data is saved into database to enable future learning. Depending on the disease(s) diagnosed, prescription generator accesses the database and prescribes medication. A discussion on different modules mentioned in Fig. 2 are provided in subsequent subsections.

Table 1 Symptom-disease table in the proposed CDSS architecture

Disease	Symptom
Malaria	Rigour, chills, cold, fever, headache, shivering
Typhoid	Chills, diarrhoea, abdominal pain, fever, headache, vomiting
Dengue	Joint pain, muscle pain, diarrhoea, fever, rash, headache, eye pain, nausea, vomitting
Conjunctivitis	Itchy eyes, discharge eyes, sensitive eyes, redness eyes, eye pain
Gastritis	Abdominal pain, vomitting, bloating, nausea, indigestion, heartburn
Viral fever	Cold, fever, rash, wet cough, chest congestion, sore throat
Tonsillitis	Fever, sore throat, wetcough, neck heavy, body ache
Gastroenteritis	Fever, diarrhoea, abdominal pain, vomitting, indigestion, heartburn, blood stool

4.1 Data Collection Module

The data collection module refers to creation of knowledge base of CDSS. The first phase of data collection refers to collection of symptoms related to diseases. A total of eight diseases and 28 symptoms were chosen for initial implementation. The symptoms chosen were a set of common symptoms experienced by a patient. The symptom collection of data required for Bayesian network creation is explained in Sect. 4.1. The symptom-disease table for all diseases modeled is as shown in Table 1.

4.2 Probable Disease Identification Module

Bayesian network methodology is used for implementing probable diseases module. Bayesian network basics are as discussed in Sect. 3. Disease modeling is divided into two phases, namely training and testing phases. The first involves network creation and parameter learning phase. The second implements the Bayesian inference. The Bayesian architecture is as shown in Fig. 4.

In the proposed work, the knowledge base is populated with data collected from experts as well as from the clinical dataset.

The approaches used for testing and training phases in the proposed work are as discussed.

- Bayesian Network Creation: The data collected from experts as symptom-disease relations is used to create the network structure manually as mentioned in Sect. 4.1. A 'two-layer' approach in creation of Bayesian network has been adopted. The first layer has the disease nodes and the second has the symptom nodes. The network design, thus, created for the eight diseases identified for the modeling is as depicted in Fig. 5.

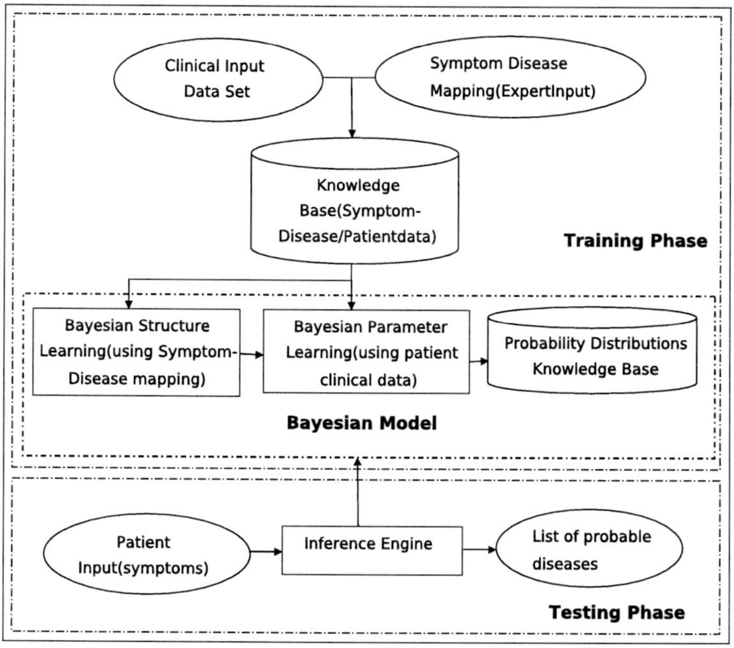

Fig. 4 Bayesian network training and testing architecture of the proposed CDSS

- Bayesian Parameter Learning: Clinical input data is added to the knowledge base for parameter learning. This involves creation of local conditional probability tables using maximum likelihood estimator algorithm.
- Bayesian Inference: The input to the inference engine is evidence which is basically the symptoms of a patient. Inference engine computes the posterior probability of evidence entered and predicts the output of the disease using variable elimination algorithm.

Symptoms entered by users form the input to the Probable Disease Identification module. Output is a list of probable diseases ranked in order of increasing probability. 'Top three' probable diseases are chosen from the output list as the final output.

4.3 *Lab Tests Recommendation Module*

Laboratory tests recommendation module is designed as a database dependent module and is acquired from expert interaction. Output of probable disease identification module is given as an input to this module. The laboratory tests relevant to those diseases are retrieved and displayed to users. The laboratory tests were populated from the data provided by doctors who were a part of this design. This module uses

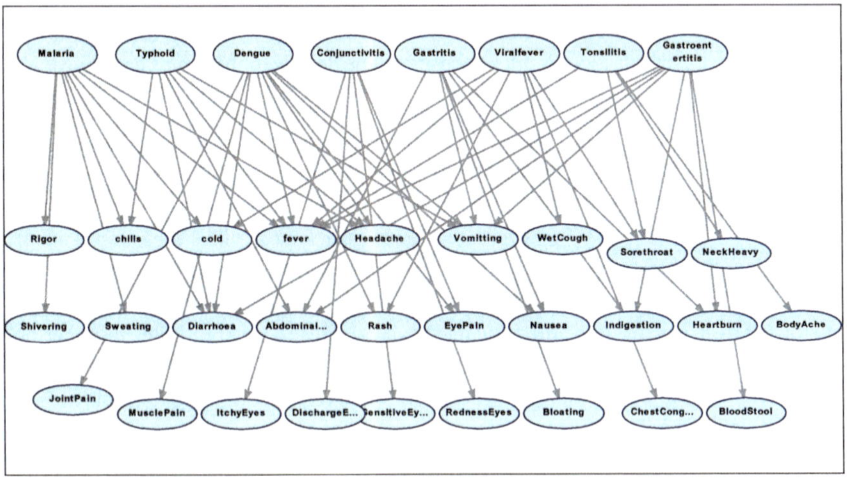

Fig. 5 Bayesian Network created for the proposed CDSS

Table 2 Database table details designed for the laboratory tests and medication modules

Table name	Attributes	Purpose
disease_info	disease_id, disease_name	Stores all disease names
disease_test_tbl	disease_id, test_id, test_name	Stores all tests related to the diseases
test_param_range	id, test_id,parameter, range_from, range_to, inference, inferval	Stores all test range values related to tests
disease_medication	id, disease_id, medicine, start_age, end_age, dosage_description	Stores all age-based medication details for diseases

disease_info and disease_test_tbl for recommending tests related to diseases. The details of the tables are mentioned in Table 2. The patient need not undergo all the laboratory tests recommended for him. Doctor will be recommending the tests for the patient based on the output of the phase.

4.4 Disease Confirmation Module

The disease confirmation module uses rule-based methodology. Every laboratory test will have some set of associated parameters with the range details that are stored in the database. The user entered parameter values are checked using 'If-Else' rules. Based on the values entered, the system confirms the presence or absence of a disease.

This module uses disease_info and disease_test_tbl and test_param_range tables for confirming diseases based on input test results. The details of the tables are mentioned in Table 2.

4.5 Medication Module

The medications module is designed on 'age dependent' logic. Medications are recommended based on the output of disease confirmation module. The medication data is gathered from doctors and authentic medical websites like webmd. The medication-related information is stored in the database. The database is queried for the confirmed disease and medicine information is retrieved based on 'age' of patients. This module uses disease_info and disease_medication tables for accessing medications. Details of the tables are mentioned in Table 2.

4.6 Learning Module

The learning module is in place to make sure that new observations get stored in the database. The new data thus gets integrated with existing data and the entire set is used in future for recalculation of Bayesian probabilities. As the data set size increases the probability calculations are predicted to become more accurate.

The system is implemented in Python language. MySql database used is for back-end operations.

5 Experimental Results and Discussion

Bayesian approach for the implementation of CDSS has been tried on a sample data set having 185 patient records. The patient data set was constructed with the help of doctors involved in design of the proposed work. The approach for data collection is as mentioned in Sect. 4.1.

The procedure adopted for testing the proposed CDSS is as follows:

- Create the symptom vector along with actual type of disease diagnosed for the symptoms.
- Input the symptom vector to the prediction model of CDSS.
- Record the probabilities of the diseases predicted by the CDSS.
- The actual disease type should match the top three probable diseases predicted by CDSS.

The entire system input and output cycles of the proposed CDSS are as discussed. The symptom input to the system is provided as mentioned in Fig. 6. The input is provided numerically which is mapped to the symptoms modeled in CDSS.

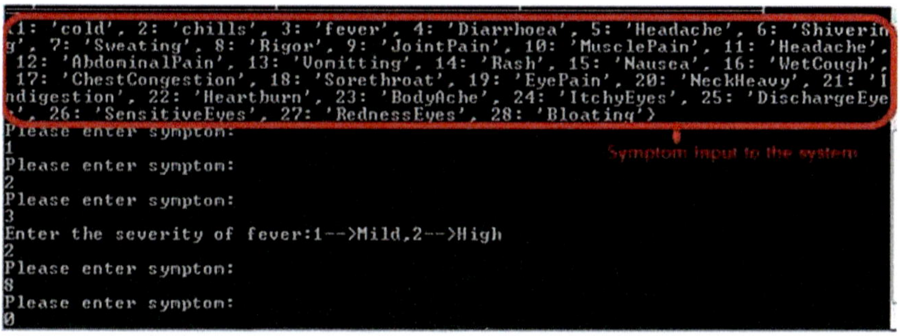

Fig. 6 Symptom input to the proposed CDSS

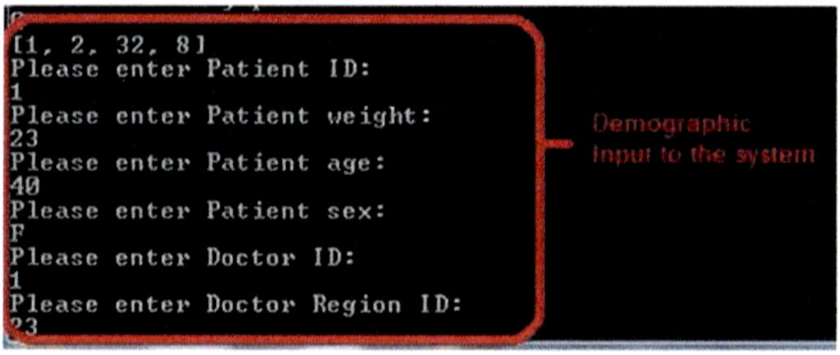

Fig. 7 Demographic input to the proposed CDSS

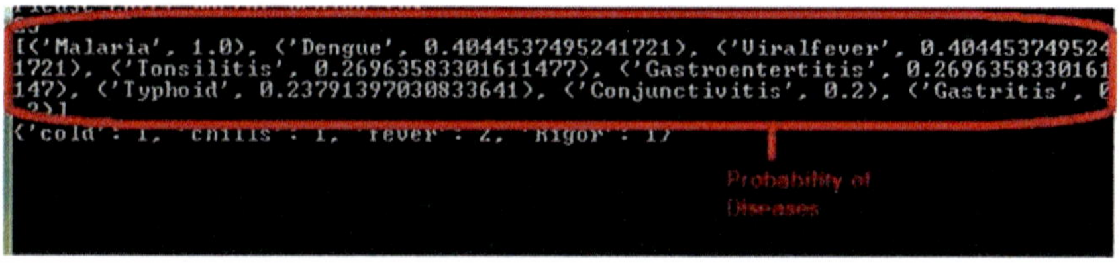

Fig. 8 Diseases probability output of the proposed CDSS

The system also takes demographic parameters as input as shown in Fig. 7. Currently, demographic parameters are not taken as a part of modeling and can be used in future. The Bayesian model is applied on input symptom data and probabilities of the diseases are displayed accordingly. The model details are provided in Sect. 4.2. Results are displayed in decreasing order of probabilities as depicted in Fig. 8.

The system also provides a pie-chart visualization of the diseases modeled in the system as shown in Fig. 9. The three top-ranked diseases will be displayed along with recommended laboratory tests as shown in Fig. 10. The laboratory test modeling details are provided in Sect. 4.3. The user will select the laboratory test from the menu and provide relevant laboratory test values as input as in Fig. 11. The test is confirmed based on the entered values. The details of modeling are mentioned in

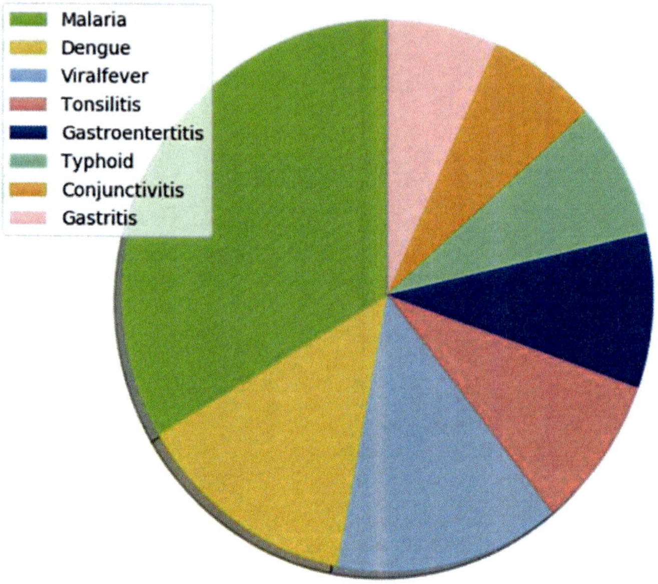

Fig. 9 Visualization of diseases probability of the proposed CDSS

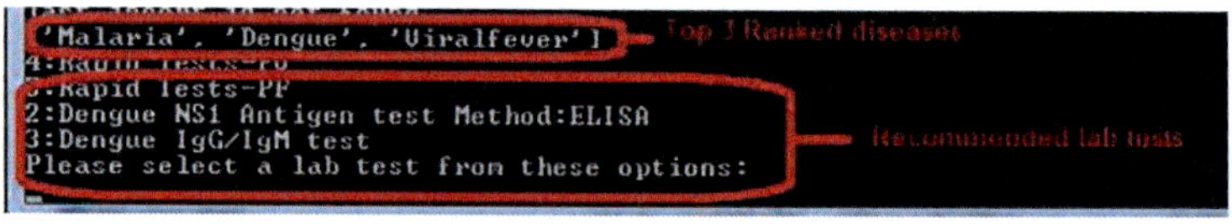

Fig. 10 Laboratory tests display of the proposed CDSS

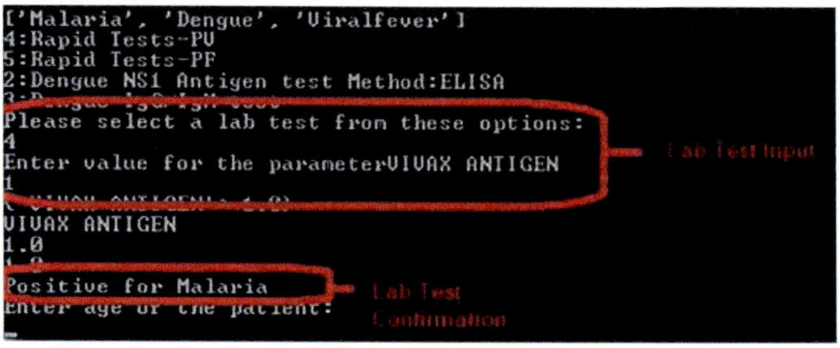

Fig. 11 Laboratory tests confirmation of the proposed CDSS

Sect. 4.4. Medical prescription is displayed based on test confirmation of the disease. Currently, medicines are advised based on the 'age'. The results are as shown in Fig. 12.

The system is tested with the help of doctors involved in sample data set creation. Each of them has tested the system for three diseases by providing symptom inputs.

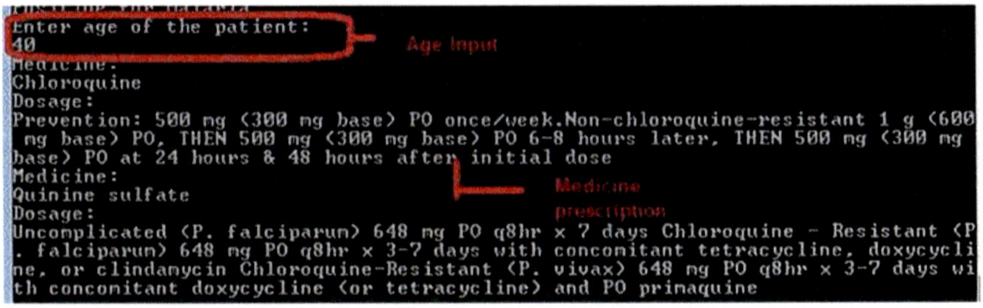

Fig. 12 Age-related medicine prescriptions of the proposed CDSS

	Precision	Recall
Malaria	0.6	1
Typhoid	1	0.75
Dengue	0.5	1
Conjunctivitis	1	0.5
Gastritis	1	1
ViralFever	1	1
Tonsillitis	1	1
Gastroentertitis	1	0.5

Fig. 13 Precision and recall of the proposed CDSS

Results were recorded in terms of precision and recall as mentioned Fig. 13. Prescription and laboratory tests were populated in the database. The testing for these modules has not been undertaken in this phase of the proposed work. The system needs to be tested with more patient data by deployment in the clinics. Learning module as mentioned in Sect. 4.5 will enable patient data to be saved for further analysis.

6 Conclusions and Future Work

The work proposes use of Bayesian network in modeling a clinical decision support system. Bayesian ML approach has been used for inferring parameters of the network, which brings in the notion of learning. The novelty of the work is that in addition to predicting diseases, the work tries to suggest laboratory tests, infers diseases from the values entered for laboratory tests and also recommends age-based medical prescriptions for commonly occurring diseases in India. A rule-based methodology is used for modeling laboratory tests and medical prescriptions.

The limitation of the system is the manual creation of the Bayesian network. Scalable system design requires automatic creation of the Bayesian network from patient data. The model is intended for learning from past experiences and deployment in clinics for assessment and data collection. The system needs to be extensively tested using clinical data. With the incorporation of data, the predicted results shall see an improvement.

The future enhancement to this CDSS will involve extension of its model to undertake diagnosis of multiple diseases. Automatic creation of Bayesian network from clinical data to make the system scalable is also a proposed enhancement. Enhancements in GUI with NLP techniques will be proposed for increased usability.

References

1. Farooq K, Khan BS, Niazi MA, Leslie SJ, Hussain A (2017) Clinical decision support systems: a visual survey. arXiv e-prints, arXiv:1708.09734
2. Dinevski D, Bele U, Sarenac T, Rajkoviĉ U, Sustersic O (2011) Clinical decision support systems. IntechOpen. https://doi.org/10.5772/25399
3. Kong G, Xu D, Yang J (2008) Clinical decision support systems: a review on knowledge representation and inference under uncertainties. Int J Comput Intell Syst 1(2):159–167. https://doi.org/10.2991/jnmp.2008.1.2.6
4. Zhang Y, Li H, Duan H, Shang Q (2016) An integration profile of rule engines for clinical decision support systems. In: International conference on progress in informatics and computing (PIC), pp 762–766. https://doi.org/10.1109/PIC.2016.7949601
5. Chu H, Yang Y, Li Q, Xu Y, Wei H (2016) A scalable clinical intelligent decision support system. In: Proceedings of the 14th international conference on inclusive smart cities and digital health, vol 9677, pp 159–165. https://doi.org/10.1007/978-3-319-39601-9_14
6. Chen YY, Goh KN, Chong K (2013) Rule based clinical decision support system for hematological disorder. In: IEEE 4th international conference on software engineering and service science, pp 43–48. https://doi.org/10.1109/ICSESS.2013.6615252
7. Malmir B, Amini M, Chang SI (2017) A medical decision support system for disease diagnosis under uncertainty. Expert Syst Appl 88:95–108. https://doi.org/10.1016/j.eswa.2017.06.031
8. Samad-Soltani T, Ghanei M, Langarizadeh M (2015) Development of a fuzzy decision support system to determine the severity of obstructive pulmonary in chemical injured victims. Acta Inform Med 23(3):138–141. https://doi.org/10.5455/aim.2015.23.138-141
9. El-Sappagh S, Ali F, Ali A, Hendawi A, Badria FA, Su DY (2018) Clinical decision support system for liver fibrosis prediction in hepatitis patients: a case comparison of two soft computing techniques. IEEE Access 6:52911–52929. https://doi.org/10.1109/ACCESS.2018.2868802
10. Shen Y, Yuan K, Chen D, Colloc J, Yang M, Li Y, Lei K (2018) An ontology-driven clinical decision support system (IDDAP) for infectious disease diagnosis and antibiotic prescription. Artif Intell Med 86:20–32. https://doi.org/10.1016/j.artmed.2018.01.003
11. Galopin A, Bouaud J, Pereira S, Seroussi B (2015) An ontology-based clinical decision support system for the management of patients with multiple chronic disorders. In: Studies in health technology and informatics, vol 216, pp 275–279. https://doi.org/10.3233/978-1-61499-564-7-275
12. Subiyanto, Mulwinda A, Andriani D (2017) Intelligent diagnosis system for acute respiratory infection in infants. In: 3rd international conference on science in information technology (ICSITech), pp 558–562. https://doi.org/10.1109/ICSITech.2017.8257175
13. Ahmad A, Tundjungsari V, Widianti D, Amalia P, Rachmawati UA (2017) Diagnostic decision support system of chronic kidney disease using support vector machine. In: Second international conference on informatics and computing (ICIC). IEEE, pp 1–4. https://doi.org/10.1109/IAC.2017.8280576

14. Baig MM, Hosseini HG, Lindên M (2016) Machine learning-based clinical decision support system for early diagnosis from real-time physiological data. In: 2016 IEEE region 10 conference (TENCON), pp 2943–2946. https://doi.org/10.1109/TENCON.2016.7848584
15. Gupta D, Aggarwal A, Khare S (2016) A method to predict diagnostic codes for chronic diseases using machine learning techniques. In: Fifth IEEE international conference on computing communication and automation (ICCA), pp 281–287. https://doi.org/10.1109/CCAA.2016.7813730
16. Khare S, Gupta D (2016) Association rule analysis in cardiovascular disease. In: Second international conference on cognitive computing and information processing (CCIP), SJCE, Mysuru, India. IEEE, pp 1–6. https://doi.org/10.1109/CCIP.2016.7802881
17. Dominic V, Aggarwal A, Gupta D, Khare S (2015) Investigation of chronic disease correlation using data mining techniques. In: 2nd international conference on recent advances in engineering and computational sciences (RAECS), pp 1-6. https://doi.org/10.1109/RAECS.2015.7453329
18. Shastri SS, Nair PC, Gupta D, Nayar RC, Rao R, Ram A (2017) Breast cancer diagnosis and prognosis using machine learning techniques. In: Intelligent systems technologies and applications, pp 327–344. https://doi.org/10.1007/978-3-319-68385-0_28
19. Jiang Y, Qiu B, Xu C, Li C (2017) The research of clinical decision support system based on three-layer knowledge base model. J Healthc Eng 2017:6535286. https://doi.org/10.1155/2017/6535286
20. Seixas FL, Zadrozny B, Laks J, Conci A, Saade CM (2014) A Bayesian network decision model for supporting the diagnosis of dementia, Alzheimer's disease and mild cognitive impairment. Comput Biol Med 51:140–158. https://doi.org/10.1016/j.compbiomed.2014.04.010
21. Sa-ngamuang C, Haddawy P, Luvira V, Piyaphanee W, Iamsirithaworn S, Lawpoolsri S (2018) Accuracy of dengue clinical diagnosis with and without NS1 antigen rapid test: comparison between human and Bayesian network model decision. PLoS Negl Trop Dis 12(6):e0006573. https://doi.org/10.1371/journal.pntd.0006573
22. Liu S, Zeng J, Gong H, Yang H, Zhai J, Cao Y (2016) Quantitative analysis of breast cancer diagnosis using a probabilistic modelling approach. Comput Biol Med 92:168–175. https://doi.org/10.1016/j.compbiomed.2017.11.014
23. Constantinou AC, Fenton N, Marsh W, Radlinski L (2016) From complex questionnaire and interviewing data to intelligent Bayesian network models for medical decision support. Artif Intell Med 67:75–93. https://doi.org/10.1016/j.artmed.2016.01.002

Artificial Intelligence Techniques for Predicting Type 2 Diabetes

Ramyashree, P. S. Venugopala, Debmalya Barh, and B. Ashwini

Abstract Diabetes is the most common disease experienced recently. Type 1 diabetes, type 2 diabetes, and gestational diabetes are the most common types of diabetes. The aim is to predict the type 2 diabetes with various parameters. "Diabetes risk score or test system" is designed with the various risk factors like age, waist circumference, physical activity, family history, and BMI using artificial intelligence technique and to design a universally acceptable diabetes prediction system that predicts the possibility of diabetes risk. This process is carried out using the various parameters of the patient's lifestyle and without using the data from medical test results. The individuals who are interested to know about their risk score can use this diabetes risk score system.

Keywords Type 2 diabetes · Artificial intelligence · Risk score · Age · Waist circumference · BMI

Ramyashree (✉) · P. S. Venugopala · B. Ashwini
NMAM Institute of Technology, Nitte, Udupi, Karnataka 574110, India
e-mail: ramyashreebhat1994@gmail.com

P. S. Venugopala
e-mail: venugopalaps@nitte.edu.in

B. Ashwini
e-mail: ashwinib@nitte.edu.in

D. Barh
Nitte University Centre for Science Education and Research, Paneer Rd, Kotekar, Mangalore, Karnataka, India
e-mail: dr.barh@gmail.com

N. N. Chiplunkar and T. Fukao (eds.), *Advances in Artificial Intelligence and Data Engineering*, Advances in Intelligent Systems and Computing 1133,
https://doi.org/10.1007/978-981-15-3514-7_32

1 Introduction

In the present scenario, diabetes is one of the common diseases. Type 1 diabetes (T1D) and type 2 diabetes (T2D) are the most common types. T1D occurs when pancreas does not produce insulin. The work presented in this paper is based on type 2 diabetes (T2D). In type 2 diabetes, cells cannot utilize glucose proficiently for strength. This happens when the cells end up unfeeling to insulin and the glucose slowly gets excessively high. There are various reasons for causing the T2D that are obesity, lack of physical activity, stress, genetics, and eating or beverages with sugar. Risk factors may include history of family having the diabetes, sedentary lifestyle, obesity. The major symptoms of T2Ds are excess thirst, dark skin under armpits, chin, or groin, blurry vision, etc. [1–4].

Count of people having type 2 diabetes is gradually increasing. It is a vital factor for death. Several researchers have carried out experiments on type 2 diabetes and proved that the prevention for this disease can be done by lifestyle modification [5–8].

The aim of this work is to develop the diabetes risk score system with the most used artificial intelligence (AI) methodologies. In the present scenario, AI can be applied in variety of the research areas because of its widespread application domains. In this paper, a system is introduced that will predict the T2D based on the different parameters and also it will inform the patients the most affected parameter for T2D based on the expert system [9, 10]. The user can use this system to know his T2D risk score.

2 Literature Survey

The diabetes risk score system is developed by several researchers.

Mohan v [1] developed the Indian Diabetes Risk Score with the help of Madras Diabetes Research Foundation (MDRF-IDRS), to recognize the undiscovered T2DM in the dataset. While developing the MDRF-IDRS, they took samples of 26,001 from 155 wards. Logistic regression method was used to build IDRS score with a maximum limit set as 100. IDRS score with <30 is considered as low risk, 30–50 is considered as medium risk, and $\geq$60 is considered as high risk. Receiver operating curve is used to detect the optimum value ($\geq$60).

Lindström and Tuomilehto [2] developed the practical tool to predict the type 2 diabetes risk for France population. According to this paper, "A sample of approximately 4500 patients is taken, while developing the system, various categorical variables such as age, gender, food, family history, waist, physical activity, BP, high blood glucose, and BMI are considered." In this tool, risk score with $\leq$7 is considered as the low risk, 7–14 is considered as the moderate risk, 15–20 is considered as the high risk, and >20 is considered as very high risk.

Katulanda et al. [3] developed the diabetes risk score for Sri Lanka (SLDRISK). To develop the SLDRISK, samples from 4276 patients were collected. To identify the variables, univariate regression analysis is applied. To derive the risk score, the β-coefficient values are identified using the analysis called logistic regression. For finding the optimal cutoff value, sensitivity, and specificity, ROC analysis is done. They also validated the SLDRISK with IDRS and CRS. When compared, it is concluded that sensitivity is 77.9% and specificity is 65.6% which is higher than IDRS and CRS.

Al-Lawati and Tuomilehto [4] proposed the diabetes risk score in Oman. In this paper, the authors have developed the diabetes risk score system for identifying the diabetes mellitus with samples 4881. In this paper, logistic regression method is used with the different parameters. Analysis of the samples concluded that when age, WC, and BMI are increased, the probability of getting T2D is more. Age and the family history are the strongest predictors, whereas BMI, BP, and WC are the moderate parameters. The Oman risk score system is validated with the Nizwa survey which contains the same parameter with 1432 data in which 145 had diabetes.

Griffin et al. [5] developed the diabetes risk score system for Cambridge. Here, the system is developed only for the age group 40–64 such that 1077 data are collected. Regression model is used for calculation. Results that is 72% specificity, 77% of sensitivity and ROC is up to 80% is identified.

Zhou et al. [6], developed the risk score for T2D mellitus for the Chinese population. Here, they took the samples around 5453 based on the lifestyle and other features like gender, age, physical activity, family history, waist circumference, history of dyslipedemia, diastolic blood pressure, and BMI. A cutoff BMI value of 17 was taken, with sensitivity 67.9% and specificity 67.8%. A system was developed and validated such that it is compared with the American Diabetes Association score (0.636), Inter99 score (0.669), Oman score (0.675). While comparing, it is concluded that using Area Under Curves, the Chinese diabetes risk score is 0.723 which is higher compared to other systems.

Glümer et al. [7] proposed the Danish diabetes risk score. Samples of 6784 patients were collected to create the system. Danish risk score was developed based on the above parameters. In this paper, authors built up a straightforward down to earth risk score based on the survey. This chance score recognizes 75.9% of people with already undiscovered composes type 2 diabetes. Besides, the hazard score has a specificity of 72% and reductions the extent of people in the populace that need ensuing testing to 29%.

Chen et al. [8] developed the risk assessment tool for Australian called AUSDRISK. According to the survey by 2025, people with diabetes will reach 2 million. So for preventing this, lifestyle should be improved. Here, 6060 samples are collected. Based on this, prediction and risk score manipulation are done.

Bang and Edwards [9] created and approved the patient self-evaluation diabetes screening score for US grown-ups. In NHANES (in ARIC/CHS), 30 (40%) of people underwent for diabetes screening and yielded affectability of 79 (72%), specificity of 67 (62%), constructive prescient estimation of 10 (10%), and probability proportion constructive of 2.39 (1.89). Conversely, the examination scores yielded affectability of 44–100%, particularity of 10–73%, positive prescient estimation of 5–8%, and probability proportion positive of 1.11–1.98. This new diabetes screening score, basic and effectively executed, appears to show enhancements upon the current strategies. Future examinations are expected to assess it indiverse populaces in certifiable settings.

Rigla and García-Sáez [11] proposed the artificial intelligence (AI) technique and the uses that help to detect the diabetes. AI techniques are widely used in variety of applications. Based on the varieties of the abilities, such as learning, reasoning, it can be applied in predicting the diabetes risk score. Here, the various AI techniques such as data mining, fuzzification, defuzzification, SVM, heuristic approach, hybrid systems, naive Bayes, supervised learning, and unsupervised learnings are explained.

Observations: Different countries diabetes risk score systems are studied, and various parameters considered in these systems are observed as shown in Table 1.

3 Methodology

Machine learning is the logical field, managing the manners by which machines gain the fact from expertise [11]. To implement machine algorithms, Python apparatuses and modules are used. Here, in this case, matplotlib, numpy and pyplots for plotting yield results additionally bolster machine learning algorithms like classification, logistic regression, decision tree, random forest, linear, and different algorithms were utilized. Here, the accuracy, confusion matrix, sensitivity, and specificity are calculated using the machine learning algorithm [12]. Specificity or the true negative rate is defined as the level of patients who are accurately distinguished as being healthy. (1-Specificity) is the level of patients who are mistakenly recognized as being having the disease. Sensitivity or the true positive rate is defined as the percentage of patients who are correctly identified as being having the disease. In machine learning grouping models, one basic proportion of model exactness is AUC or area under the curve. ROC represents Receiver Operating trademark which can be drawn as sensitivity versus 1-Specificity [13].

The motivation behind this work is to detect the type 2 diabetes of individuals who are interested to know about their risk score. Therefore, diabetes risk score system is designed, without any laboratory tests.

Its design steps are as follows:

1. Different diabetes risk score systems are studied to understand the parameters that are being used in risk estimation.
2. Built a dataset using the parameters that are being used for the prediction.
3. Applied the suitable machine learning algorithm and find the score on this designed data set (with suitable parameter).

Table 1 Comparison of different systems

Country	Simple size	Age	Gender	Food	Family history	Waist circumference	Physical activity	BP	High blood glucose	BMI	Smoking
India	26,001	✓	✓		✓	✓	✓				
China	5453	✓	✓		✓	✓	✓	✓	✓	✓	
Sri Lanka	4276	✓	✓		✓	✓	✓	✓		✓	
Oman	4881	✓	✓		✓	✓		✓		✓	✓
Cambridge	1077	✓	✓		✓	✓		✓	✓	✓	
France	~4500	✓	✓	✓	✓	✓	✓	✓	✓	✓	
UK	6186	✓	✓		✓	✓		✓		✓	
Danish	6784	✓	✓	✓	✓	✓	✓	✓	✓	✓	✓
Australia	6060	✓	✓	✓	✓	✓	✓	✓	✓		✓
Brazil	1224	✓	✓		✓	✓		✓		✓	✓
US	5258	✓	✓		✓	✓	✓	✓		✓	

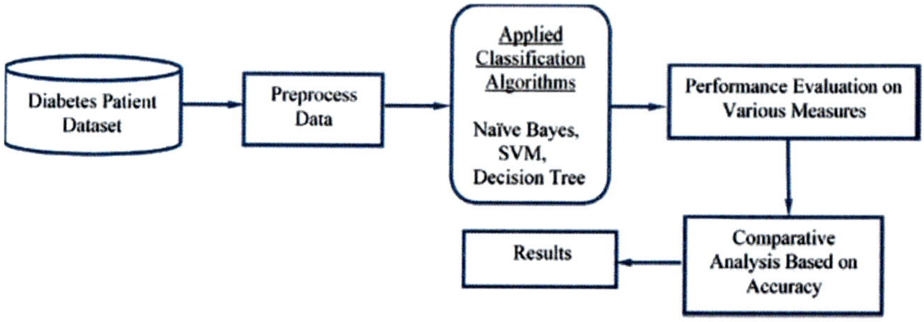

Fig. 1 Proposed model diagram

4. The built dataset should represent all the scoring systems that are existing and should be able to represent people from around the world.
5. Add more features to the designed system and again calculate the score, compare with the original system. (How the original system can be fine-tuned if we add some other feature).
6. Validate the data using the existing diabetes risk score system.

Proposed model is represented in Fig. 1.

3.1 Dataset Selection

To develop a uniform T2D risk scoring system for Asians, we used scoring values form IDRS (India) [1], Chinese score system [6], SLDRISK (Sri Lanka) [3], and Omanees score system [4]. The details of the scores of these four systems are given in Table 2.

Table 1 represents the β-coefficients value of Asian system by considering five parameters, namely age, waist circumference, physical activity, family history, and BMI. To design a uniform T2D risk scoring system for Asians among the different parameters, waist circumference has a significant role. The protection from insulin increments as the individual progresses toward becoming overweight. The risk factor to diabetes likewise an increment if an individual has a family ancestry that is, if a parent or sibling of the subject has/had diabetes. As the age expands, the risk of diabetes also increments on the grounds that the physical action reduces yet type II diabetes is likewise been observed to be expanding in youthful populace [14, 15]. Using all these constraints, we have identified the strong parameters which effect more for T2D, and based on that parameter, we have developed the diabetes risk score system. The important parameters that are identified are age, gender, waist circumference, physical activity, family history, and BMI. The score such that the dimension of hazard within the sight of some hazard factors is called the risk score. Risk score can be easily calculated using β-coefficient value. It can be most used in the developing countries. β-coefficients calculations are explained in Sect. 3.2.

Table 2 Beta scores of Indian, Chinese, Sri Lanka, and Omanees system

Variable	India	China	Sri Lanka	Oman
Age				
<35	0	0	0	0
35–49	0.84	0.845	0.95	1.8
≥50	1.47	1.357	1.61	2.3
60–69				
≥70				
Waist circumference				
Female <80 Male <90	0	0	0	0
Female 80–89 Male 90–99	0.44	0.952	1	0.38
Female ≥90 Male ≥100	0.81	1.493		
>109		2.271		
Physical activity				
Vigorous	0	0	0	
Mild	1.13		0.17	
No	1.45	0.352	0.32	
Family history				
Two non-diabetic	0	0	0	0
Either parent	0.54	0.656	0.52	1.9
Both parent	0.83			
BMI				
<25		0	0	0
25–29		0.679	0.52	0.54
30–34		0.948	0.72	0.69
≥35		1.418		
		1.784		

Risk score of probability using β-coefficients value can be calculated using the following formula:

$$P\,(\text{diabetes}) = \frac{e^{(\beta_0+\beta_1 x_1+\beta_2 x_2+\cdots)}}{1+e^{(\beta_0+\beta_1 x_1+\beta_2 x_2+\cdots)}} \quad (1)$$

where independent risk factor is $x_1, x_2 \ldots$etc., and intercept β_0, regression coefficient β_1, β_2,…etc., are used.

3.2 Data Preparation and Imputation

For Age, Waist Circumference, and BMI

IDRS is used as template to reverse calculation and create an imputed dataset. As the aim is to provide individual age-specific personalized T2D risk score, calculated beta coefficient for each year instead of making an age group. To achieve this, we took IDRS as reference and created the imputed data set. In IDRS, the β-coefficients of age groups <35 are 0, 35–49 are 0.84, and $\geq$50 are 1.47. Created a continuous dataset for individual ages from 21 to 80 using these β-coefficient values. To do so, the lowest value considered is −0.4 for 21–34 years, and highest value for 21–34 years is calculated based on the next value of the category. So, here highest β-coefficients value is determined as 0.2; similar technique is done for the age category 35–49 and 50–80, respectively. So, the value obtained for the age 35 is 0.699, age 49 is 1.1, age 50 is 1.2, and age 80 is 1.64. Once the above step is completed, then impute the data according to the average value between the ranges. While doing this, β value is increasing according to the individual specific age. Similarly, the calculation is done for China, Sri Lanka, and Oman. Similar approach was taken to personalize BMI and waist circumference.

For Physical activity

Considered the three categories of physical activity according to IDRS. Those categories are vigorous—exercise with β-coefficient is 0, no exercise with β-coefficients is 1.45, and mild exercise with β-coefficients is 1.13. In view of the inquiries shaped by the International Physical Activity (IPAQ), physical action was separated as low, moderate, and high [16]. Here, the lively physical exercises are alluded to the exercises that require hard physical exertion and influence us to inhale a lot harder than typical. Such physical exercises resemble hard work, burrowing, high impact exercise, and quick bicycling. Moderate exercises allude to exercises that require moderate physical exertion and influence us to inhale fairly harder than ordinary.

For Family history

Since the family history is another important parameter for predicting the type 2 diabetes. We have considered the three categories of family history according to IDRS. Those categories are two non-diabetic—with β-coefficient is 0, either parent with β-coefficients is 0.54, and both parent with β-coefficients is 0.83. All these five categories are included while creating the dataset.

3.3 Computation for Data Imputation

Once the β-coefficients are calculated as explained in the data computation part, in the next stage imputing the data is very much essential. Here, Python library scikit-learn is used and also there is a python module dedicated to permutations and combinations

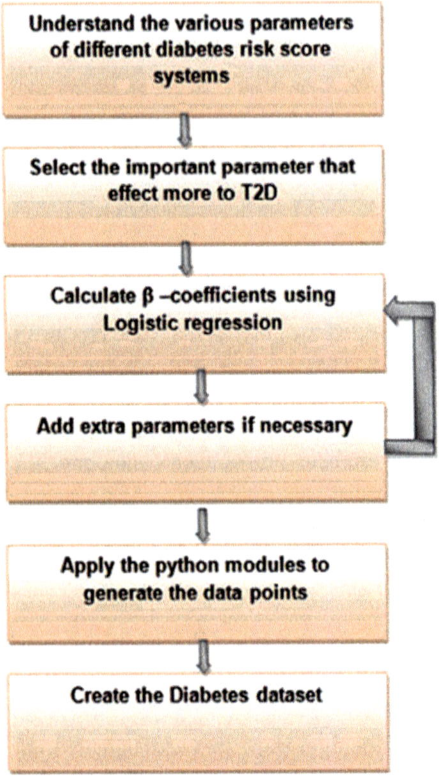

Fig. 2 Steps involved in creating the dataset

called itertools. One of the greatest corners of the Python 3 standard library: Itertools. Product (): This tool computes the cartesian product of input timetables. This module implements a number of iterator building blocks in a form suitable for Python. This is the efficient tool that can be used for variety of combinations. Initially, we took the four parameters beta coefficient value, namely age, waist, physical activity, and family history, and BMI are stored it in the list. Once all the values are stored in the particular list, then product (*) is used with itertools module. This works like nested for loop, and total samples obtained for these combination are 514,384. Similar approach was taken to create the dataset of India (IDRS) with BMI, China, SLDRISK, and Oman. The steps that are used while creating the dataset is shown in Fig. 2.

3.4 Description of the Dataset

Dataset is created based on the certain important attributes for India, China, Sri Lanka, and Oman which does not contain any missing values. All variables were

Table 3 Attributes

Attribute number	Attribute
1	Age
2	Waist
3	Physical activity
4	Family history
5	BMI
6	Outcome

categorized by age (21–34 years versus 35–49 and ≥50 years), waist circumference (men <90, 90–99, ≥100 cm versus women <80, 80–89, >90 cm), body mass index (BMI) (weight in kg divided by height in m^2) (BMI <25 versus 25–29 versus 30–34 and ≥35), family history of diabetes (two non-diabetic parents versus either parent having diabetes and both parent having diabetes), physical activity (vigorous exercise versus mild and no exercise is considered). Based on the outcome of the diabetes (that is 0/1), training data should be classified such that 0 indicates no diabetes and 1 indicates diabetes. Table 3 represents the attribute of the created dataset.

3.5 Data Modeling and Algorithms Used for Prediction

Different algorithms are used, namely multiple logistic regression (MLR), Gaussian Bayes (GB), random forest (RF), decision tree (DT) [17] applied to the imputed Indian, Chinese, SLDRS, and Oman diabetic data sets. The data were grouped into training (70%) and test sets (30%) comprising 50% of T2D. Three combinations of parameters, such as (Combination 1)—age, gender, physical activity, family history, waist circumference, (Combination 2)—age, gender, physical activity, family history, waist circumference, BMI, and (Combination 3)—age, gender, physical activity, family history, and BMI were used in predicting the efficacy (specificity and sensitivity) of each algorithm using ROC and AUC with 95% of CIs. Further, the outcomes of each algorithm were compared among each other, and the best model is selected. In another approach, we used a consensus algorithm [18] to get the average of scores from the entire algorithm. Similarly, we developed the consensus-based Asian score as described in Fig. 3. The essential issue to decide positioning accord is an issue to join a few rankings those are chosen by at least two decision maker

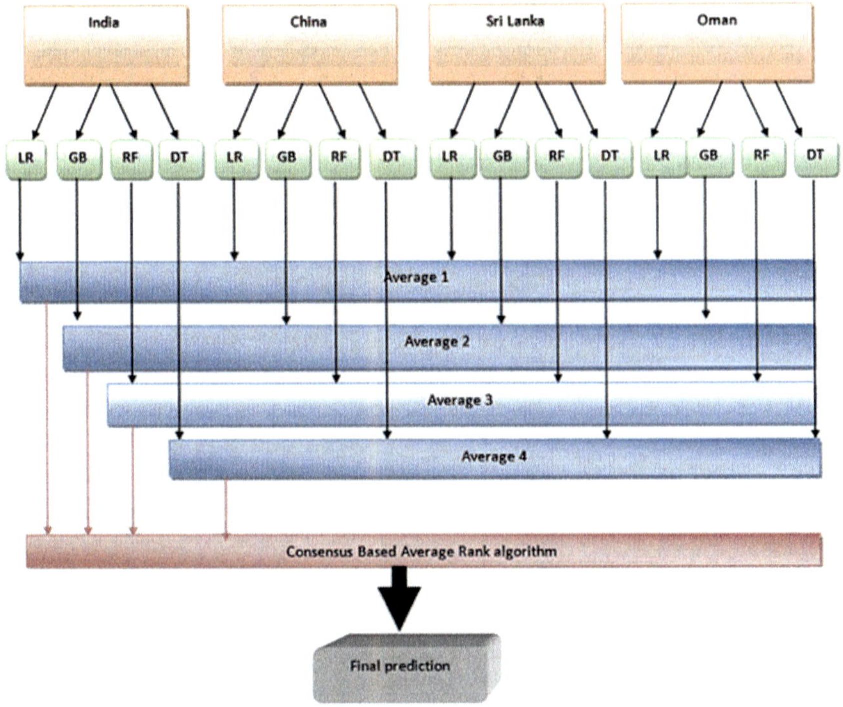

Fig. 3 Workflow of the consensus-based for final prediction

(DM) into positioning agreement. For the different Asian countries initially applied the machine learning algorithms. The average value of each method is identified, then final value is predicted using consensus-based average rank algorithm as shown in Fig. 3.

4 Results and Discussion

Three combinations of parameters are used for prediction using different algorithms for different Asian countries. Accuracy, precision, sensitivity, and specificity are calculated, and the values are shown in Tables 4, 5, and 6. Sensitivity and specificity rate is calculated which is needed to draw the ROC for prediction. Three combinations of parameters were used for prediction. Using the logistic regression model predicts that precision is almost up to 87% of the time. On the off chance that there are patients who have diabetes in the test set and logistic regression model can recognize it 91% of the time. Specificity calculation using logistic regression model is almost 71.17%. So similarly, these values are identified using four different algorithms for Asian countries.

Table 4 Combination 1 (without BMI)

	Accuracy	Precision	Sensitivity	Specificity
India				
Logistic regression	85.05	87.40	91.39	71.17
Gaussian Bayes model	78.4	78	79.0	77.0
Random forest	71.1	69.5	71.9	69.3
Decision tree	72.7	67	67.9	67.8
China				
Logistic regression	93.98	92.97	72.17	70.81
Gaussian Bayes model	95.30	96	96.0	95.0
Random forest	96	95	95.0	92.0
Decision tree	89.2	85.5	70.2	65.8
Sri Lanka				
Logistic regression	94.37	94.28	74.19	61.53
Gaussian Bayes model	94.50	89	94	92
Random forest	94.0	88.2	94.8	91.7
Decision tree	92.0	77.7	70.3	65.6
Oman				
Logistic regression	92.55	90.19	73.18	69.72
Gaussian Bayes model	92.62	86.7	93.0	89.9
Random forest	93.0	86	93.0	89.0
Decision tree	92.0	79.0	79.0	66.1

While developing the multiple logistic regression models for Indian system, initially X and Y values are defined. Such that X is the matrix which contains the attributes from the dataset and Y is the vector based on that prediction can be done. Once X and Y values are defined, then split the X and Y values into corresponding training set and testing set. Here, using the sklearn, splitting is done such that random state value is set as zero. Once the classification of training set and testing set is over in the next stage on the training set, we have to train a logistic regression model and fit the model on the X_train and Y_train. Once the model is fit, then prediction based on the testing set that is X_test should be carried out and accuracy should be calculated. Such that for the Indian data set, the accuracy for Combination 1 is 85.05 and similarly for Combination 2 is 97.78. True negative rate is determined using logistic regression model; for all the three combinations, it is almost up to 71.17%, 63.07%, and 94.5%, respectively. True positive rate is determined using logistic regression model; for all the three combinations, it is almost up to 91.39%, 99.18%, and 96.2%, respectively. So similarly, these values are calculated using remaining algorithm, namely Gaussian Bayes, random forest, and decision tree. Similar technique is applied to Asian countries to design the Asian diabetes risk score system.

Table 5 Combination 2 (with considering BMI)

	Accuracy	Precision	Sensitivity	Specificity
India				
Logistic regression	97.78	98.51	99.18	63.07
Gaussian Bayes model	96.06	92.00	96.12	94.3
Random forest	96.0	92.0	96.0	94.0
Decision tree	93.7	98.1	74.8	72.9
China				
Logistic regression	96.98	98.14	98.64	70.94
gaussian bayes model	94.7	89.23	94.80	91.0
Random forest	94.0	89.0	94.0	91.0
Decision tree	94.0	90.01	75.7	56.0
Sri Lanka				
Logistic regression	97.08	98.18	98.74	69.38
Gaussian Bayes model	97.01	97.0	97.0	97.0
Random forest	99.9	99.8	98.7	98
Decision tree	96.7	97.7	66.2	64.0
Oman				
Logistic regression	97.175	98.141	98.54	80.18
Gaussian Bayes model	95.90	96.28	96.0	95.0
Random forest	96.0	96.0	96.0	95.0
Decision tree	92.14	72.47	76.05	78.20

There might be unpredictable and unknown connections between the factors in the dataset. It is critical to find and evaluate how many factors in the dataset are needy upon one another. This information can enable to more readily set up the information to meet the desires for machine learning calculations [19]. Factors inside a dataset can be connected for bunches of reasons. A relationship could be sure, which implies the two elements move a comparative way, or negative, inferring that when one variable's regard fabricates, the other elements' characteristics decrease [20]. Association can similarly be neural or zero, inferring that the variables are insignificant. Relation between different features for the Indian system is as shown in Fig. 4. Plotting of the graph is done according to the pair using the correlation feature as shown in Fig. 5. Similar approach is applied to all the remaining T2D systems of Asian countries and analyzed.

According to the survey "Using Receiver Operating Characteristic (ROC) curve, the cutoff value for the risk score can be identified. The ROC curves were plotted for the diabetes risk score, the sensitivity was plotted on the *y*-axis, and the false positive rate (1_specificity) was plotted on the *x*-axis. The more precise segregating the test, the more extreme the upward part of the ROC bend and the higher the zone under the bend (AUC). Optimum value is considered as the high risk score for the diabetes that

Table 6 Combination 3 (without considering waist circumference)

	Accuracy	Precision	Sensitivity	Specificity
India				
Logistic regression	96.0	95.8	96.2	94.5
Gaussian Bayes model	96.24	96.0	96.0	94.0
Random forest	99.0	99.0	99.2	98.0
Decision tree	92.5	95.0	84.0	78.0
China				
Logistic regression	96.24	97.51	98.52	60.74
Gaussian Bayes model	95.50	95.25	96.0	94.0
Random forest	98.0	98.0	98.0	98.0
Decision tree	96.0	96.0	96.0	94.0
Sri Lanka				
Logistic regression	96.52	97.78	98.46	64.07
Gaussian Bayes model	95.47	92.5	91.0	89.5
Random forest	98.0	97.8	96.4	92.2
Decision tree	80.0	93.0	84.0	79.0
Oman				
Logistic regression	85.8	91.5	85.8	81.0
Gaussian Bayes model	94.48	94.0	94.0	93.0
random forest	98.0	98.0	98.0	98.0
Decision tree	95.0	92.0	90.15	89.7

are detected based on the receiver operating curves. The ROC curves using different combinations for Indian system are as shown in Fig. 6. So from this, it is observed that logistic regression performs better when compared to all other algorithms. Since the AUC is constantly used to identify how well the test is performed between the two gatherings like if the value of AUC increases which indicates, better the test. So Indian system is validated using AUC, and its value is 0.98. Similarly, China, Sri Lanka, and Oman systems are also validated, and the result obtained are 0.98, 0.97, and 0.94, respectively. Similar approach is applied for remaining countries.

Similarly, about 5, 14,384 samples are used for analysis. Optimization of diabetes data using KNN classifier, decision tree, random forest, support vector machine, and Gaussian Bayes model is compared as shown in Table 7.

By considering the simple five parameters, system was developed. Quickly, the data for these factors were gotten by five inquiries, and scores acquired for these elements are as shown in Table 8.

From this work, it is seen that execution of the framework is better utilizing logistic relapse contrasted with other AI calculations for both Asian nations. The corresponding bar chart of accuracy calculation using logistic regression for Asian

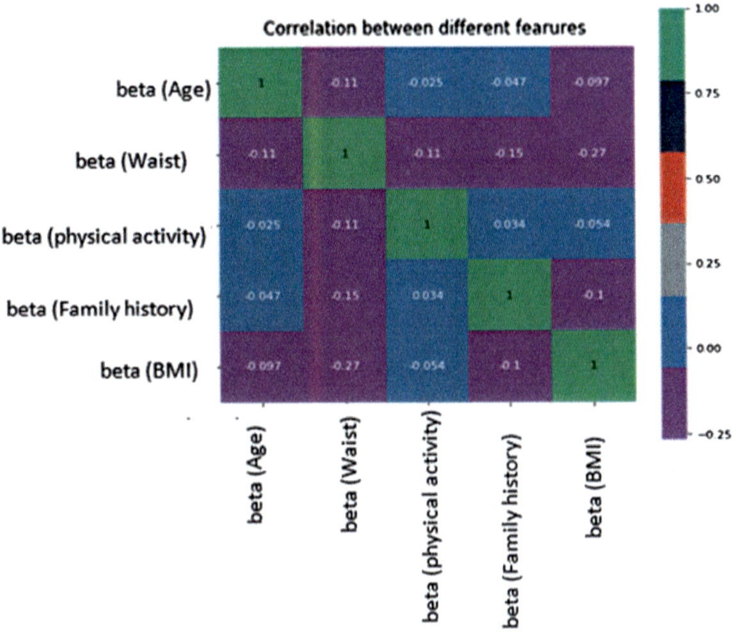

Fig. 4 Correlation between the features for Indian system

countries is as shown in Fig. 7. For India, China, Sri Lanka, and Oman, the accuracy is 97.78%, 96.98%, 97.08%, and 97.175%, respectively.

5 Conclusion

Simple diabetes risk assessment tool is developed and validated. Diabetes risk score system is developed which can be universally acceptable using the five different parameters, namely age, waist circumference, family history, physical activity, and BMI.

Several risk score tools are developed, but predicting the correct risk score without losing simplicity is really the challenging task. The proposed system is compared with the existing risk score system for the accuracy and performance. It can also be applied to the different ethnic groups. With the reference of Table 6, it is concluded that score with <35 is considered as low risk, 35–69 is considered as medium risk, 70–95 is considered as the high risk, and finally $\geq$95 is considered as very high risk. To detect the optimum value ($\geq$95), the receiver operating curve (ROC) was used, and these values were validated using area under curves (AUC). This tool also provides the information about which factor mostly effects the type 2 diabetes .

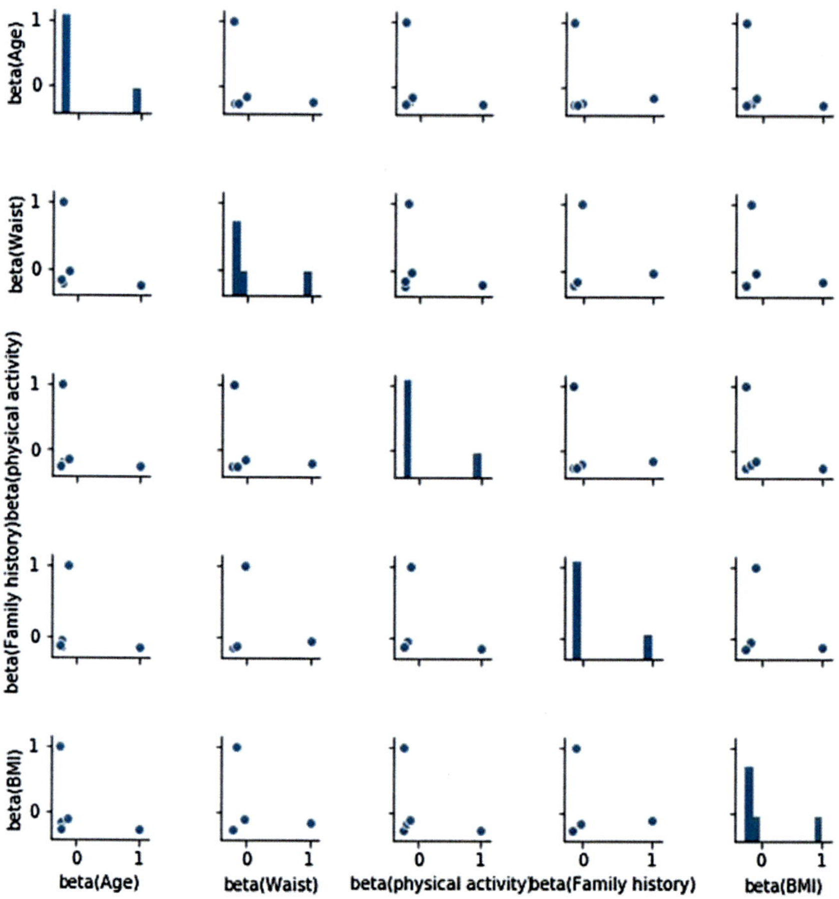

Fig. 5 Pair plot according to correlation feature values

As a future work, the system can be developed which enables the experts to provide the precautionary measure for this T2D.

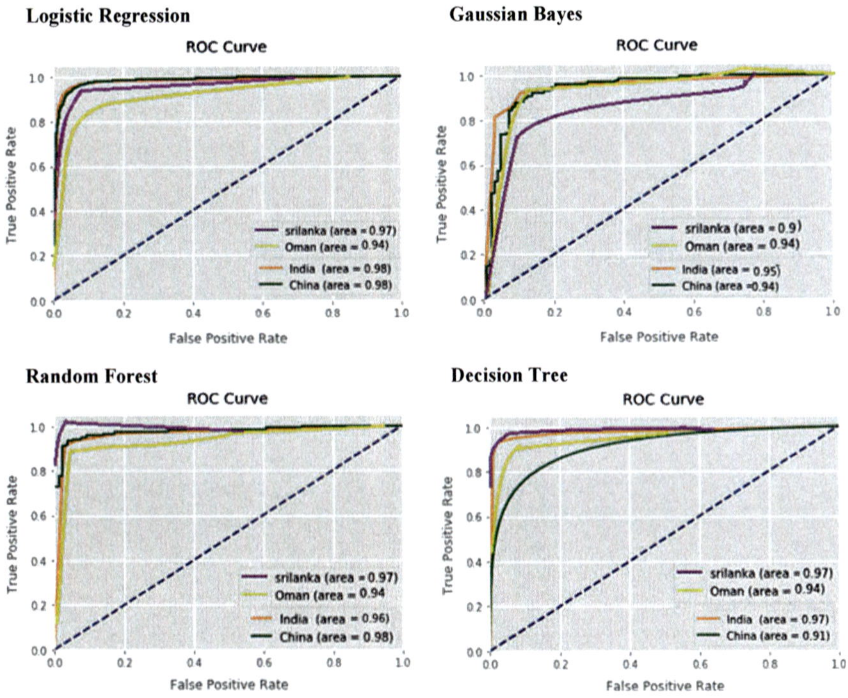

Fig. 6 Representation of ROC for Asian countries

Table 7 Analysis of optimization of different machine learning algorithms

Machine learning algorithm	Accuracy-India	Accuracy-China	Accuracy-Sri Lanka	Accuracy-Oman
KNN classifier	Training set: 1.00 Test set: 1.00	Training set: 1.00 Test set: 1.00	Training set: 1.00 Test set: 1.00	Training set: 1.00 Test set: 1.00
Decision tree	Training set: 1.00 Test set: 1.00	Training set: 1.00 Test set: 1.00	Training set: 1.00 Test set: 1.00	Training set: 1.00 Test set: 1.00
Random forest	Training set: 1.00 Test set: 1.00	Training set: 1.00 Test set: 1.00	Training set: 0.94 Test set: 0.94	Training set: 0.93 Test set: 0.93
Support vector machine	Training set: 0.99 Test set: 0.99	Training set: 0.95 Test set: 0.95	Training set: 0.965 Test set: 0.965	Training set: 0.95 Test set: 0.95
Gaussian Bayes model	Training set: 0.9611 Test set: 0.9608	Training set: 0.9404 Test set: 0.9414	Training set: 0.9443 Test set: 0.9450	Training set: 0.9261 Test set: 0.9262

Table 8 Asian countries diabetes risk score

Parameters	Risk score
Age	
<35 35–49 ≥50	0 22 34
Waist circumference	
Female <80 cm, male <90 cm Female 80–89 cm, male 90–99 cm Female ≥90 cm, male ≥100 cm	0 11 20
Physical activity	
Vigorous exercise Mild exercise No exercise	0 13 18
Family history	
Two non-diabetic parents Either parent having diabetes Both parent having diabetes	0 18 29
BMI	
<25 25–29 30–34 ≥35	0 11 16 29
Maximum score	130
Score ≥95: very high risk, 70–95: high risk, 35–69: medium risk, <35: low risk	

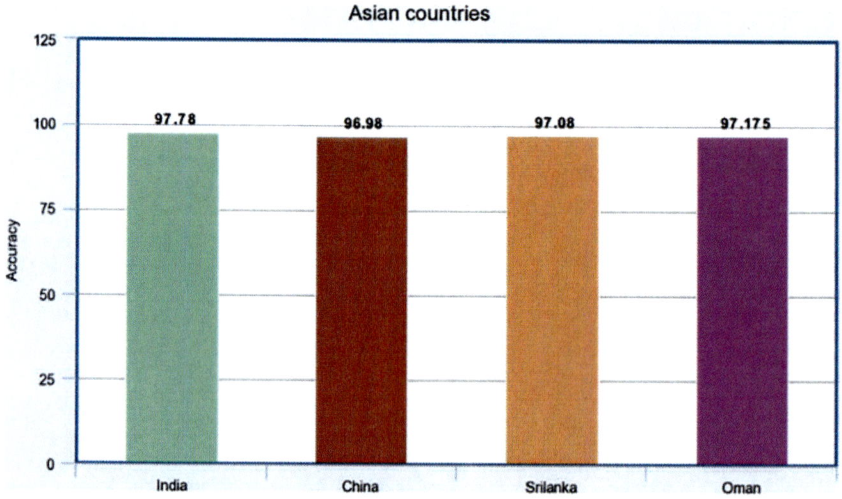

Fig. 7 Accuracy using logistic regression for Asian countries

References

1. Mohan V, Deepa R, Deepa M, Somannavar S, Datta M (2005 Sept) A simplified Indian diabetes risk score for screening for undiagnosed diabetic subjects. J Assoc Phys India 53:759–763
2. Lindström J, Tuomilehto J (2003 Mar) The diabetes risk score: a practical tool to predict type 2 diabetes risk. Diabetes Care 26(3):725–731
3. Katulanda P, Hill NR, Stratton I, Sheriff R, De Silva SD, Matthews DR (2016 July 25) Development and validation of a diabetes risk score for screening undiagnosed diabetes in Sri Lanka (SLDRISK). BMC Endocr Disord 16(1):42. https://doi.org/10.1186/s12902-016-0124-8
4. Al-Lawati JA, Tuomilehto J (2007) Diabetes risk score in Oman: a tool to identify prevalent type 2 diabetes among Arabs of the Middle East. 77(3):438–444. Epub 2007 Feb 15
5. Griffin SJ, Little PS, Hales CN, Kinmonth AL, Wareham NJ (2000 May–June) Diabetes risk score: towards earlier detection of type 2 diabetes in general practice. Diabetes Metab Res Rev 16(3):164–171
6. Zhou H, Li Y, Liu X, Xu F, Li L, Yang K, Qian X, Liu R, Bie R, Wang C (2017 Feb 17) Development and evaluation of a risk score for type 2 diabetes mellitus among middle-aged Chinese rural population based on the RuralDiab Study. Sci Rep 7:42685. https://doi.org/10.1038/srep42685.5
7. Glümer C, Carstensen B, Sandbaek A, Lauritzen T, Jørgensen T, Borch-Johnsen K (2004 Mar) A Danish diabetes risk score for targeted screening: the Inter99 study. Diabetes Care 27(3):727–733
8. Chen L, Magliano DJ, Balkau B, Colagiuri S, Zimmet PZ, Tonkin AM, Mitchell P, Phillips PJ, Shaw JE (2010 Feb 15) AUSDRISK: an Australian Type 2 Diabetes Risk Assessment Tool based on demographic, lifestyle and simple anthropometric measures. Med J Aust 192(4):197–202
9. Bang H, Edwards AM, Bomback AS, Ballantyne CM, Brillon D, Callahan MA, Teutsch SM, Mushlin AI, Kern LM, A patient self-assessment diabetes screening score: development, validation, and comparison to other diabetes risk assessment scores. 10.1059/0003-4819-151-11-200912010-00005
10. Schmidt MI, Duncan BB, Bang H et al (2005) Identifying individuals at high risk for diabetes: the Atherosclerosis Risk in Communities study. Diabetes Care 28:2013–2018
11. Rigla M, García-Sáez G, Pons B, Hernando ME (2018) Artificial intelligence methodologies and their application to diabetes. J Diabetes Sci Technol 12(2):303–310. © 2017 Diabetes Technology Society Reprints and permissions: sagepub.com/journalsPermissions.nav. https://doi.org/10.1177/1932296817710475
12. Kim MJ, Lim NK, Choi SJ, Park HY (2015) Hypertension is an independent risk factor for type 2 diabetes: the Korean genome and epidemiology study. Hypertens Res 38:783–789
13. Gupta N, Rawal A, Narasimhan VL, Shiwani S (2013 May–June) Accuracy, sensitivity and specificity measurement of various classification techniques on healthcare data. IOSR J Comput Eng (IOSR-JCE) 11(5):70–73. e-ISSN: 2278-0661, p-ISSN: 2278-8727
14. Gray LJ, Taub NA, Khunti K, Gardiner E, Hiles S, Webb DR, Srinivasan BT, Davies MJ (2010 Aug) The Leicester Risk Assessment score for detecting undiagnosed Type 2 diabetes and impaired glucose regulation for use in a multiethnic UK setting. Diabet Med 27(8):887–895. https://doi.org/10.1111/j.1464-5491.2010.03037.x
15. Sharma KM, Ranjani H, Nguyen H, Shetty S, Datta M, Narayan KM et al (2011) Indian Diabetes Risk Score helps to distinguish type 2 from non-type 2 diabetes mellitus (GDRC-3). J Diabetes Sci Technol 5:419–425
16. Pires de Sousa AG, Pereira AC, Marquezine GF, Marques do Nascimento-Neto R, Freitas SN, de C Nicolato RL, Machado-Coelho GL, Rodrigues SL, Mill JG, Krieger JE (2009) Derivation and external validation of a simple prediction model for the diagnosis of type 2 diabetes mellitus in the Brazilian urban population. Eur J Epidemiol 24(2):101–109. https://doi.org/10.1007/s10654-009-9314-2. Epub 2009 Feb 4
17. Contreras I, Vehi J (2018) Artificial intelligence for diabetes management and decision support. J Med Internet Res 20(5):e10775. https://doi.org/10.2196/10775

18. Nalluri JJ, Barh D, Azevedo V, Ghosh P (2017 Jan 03) *miRsig*: a consensus-based network inference methodology to identify pan-cancer miRNA-miRNA interaction signatures. www.nature.com/scientificreports
19. Shanbhogue VV, Vidyasagar S, Madken M, Varma M, Prashant CK, Seth P et al (2010) Indian Diabetic Risk Score and its utility in steroid induced diabetes. J Assoc Phys India 58:202
20. Shera AS, Jawad F, Maqsood A (2007) Prevalence of diabetes in Pakistan. Diabetes Res Clin Pract 76(2):219–222

Predictive Analysis of Malignant Disease in Woman Using Machine Learning Techniques

Akshaya, R. Pranam Betrabet, and C. V. Aravinda

Abstract Healthcare organization is suffering from unsustainable costs and lack of data utilization. Thus, it is important to find solution which will minimize needless costs and improve the quality of health care. If we separate the death cause between preventable death and other death in India, preventable death stands in the third position due to cancer and heart attack. So, there is a huge problem present in modern house care, including high cost, high waste, and low quality of health. In order to tackle this problem, machine learning can help in analyzing better care and low-cost health care. In this research, we are investigating the feasibility of employing the machine learning algorithms say SVM, naive Bayes, logistic regression, and KNN for identifying the malignant or benign type of breast cancer found in women.

Keywords Human cancer type · Cancer data set · Machine learning · Predictive model

1 Introduction

Cancer is considered as an evolving health concern of the present population. The population of age 35–65 is now being part of this deadly tumor. The social and the economic cost implications of tumor are enormous to the society. Families with this kind of illness will suffer a lot due to direct cost involved in reduced productivity, and it adds indirect cost to the society.

It is also important that clinical services should be more effectively equipped in order to serve the necessary level of care for the illness through healthcare delivery system at the primary level as well as at the secondary level. Nowadays, it is observed that there is an increase in the occurrence of cancer all over the world [1], and this high rate of cancer is chiefly attributed to population growth, urbanization, lifestyle

Akshaya (✉) · R. Pranam Betrabet · C. V. Aravinda
NMAM Institute of Technology, Nitte, Karkala, India
e-mail: akshaya@nitte.edu.in

C. V. Aravinda
e-mail: aravinda.cv@nitte.edu.in

N. N. Chiplunkar and T. Fukao (eds.), *Advances in Artificial Intelligence and Data Engineering*, Advances in Intelligent Systems and Computing 1133,
https://doi.org/10.1007/978-981-15-3514-7_33

changes, industrialization, and ecological factors like heavy metals and chemicals. The survey says that the two-third of cancer-affected patients obtains the curative amenities at the last stage when the tumor has already reached an advanced level. Thus, survivals of the cancer-affected patients are very less.

Breast cancer is considered as most hazardous kind of malignant tumor among women across the world. These kinds of cancer begin in the cells of duct and lobe. Rate of survival can be high if the disease is diagnosed at the early stage. In the year 2007, it was reported that around 40,598 women died in the USA due to breast cancer tumor [2]. Thus, it is very crucial to predict these types of cancer at an early stage.

2 Related Works

In this section, the description of previous research related to this area is being discussed in brief. Table 1 shows an overview of related works.

Joseph et al. [9] explained about different machine learning algorithms, its benefits, assumptions, and limitations. Further, this paper also gave a clear picture of predicting methods which is used to show the types of cancer, and the algorithm they chose for the research, clinical reports, choice of training data, and the performance obtained.

M.H. Tafish et al. [1] in this research developed a model which helped them in resolving the trouble of crucially identifying the degree of risk for the cancer disease. This model applied KNN, ANN, and SVM classification techniques on the breast

Table 1 Machine learning techniques for diagnosis of human type cancer

S. No.	Reference	ML technique used	Data set	Accuracy
1	Shrivastava et al. [3]	J48 (Breast Cancer)	UCI machine learning repository	98.14%
2	Venkatesan et al. [4]	Decision Tree (Breast Cancer)	Swami Vivekananda Diagnostic Centre Hospital	97%
3	Sivakami [5]	DT + SVM (Breast Cancer)	UCI machine learning repository	91%
4	Shajahaan et al. [6]	CART Naïve Bayes (Breast Cancer)	University of Wisconsin hospitals	92.42% 97.42%
5	Safiyari et al. [7]	Ensembling learning methods (Lung Cancer)	SEER Data set	Not survived—78.21% Survived—21.79%
6	Xu et al. [8]	SVM (Oral Cancer)	Clinical, genomic	75%

cancer data set collected by Gaza Strip hospital. This model in turn predicts the severity of breast cancer and obtained 77% of prediction accuracy.

Umesh et al.[10] this research investigated the feasibility of using association rule to identify the breast cancer in SEER data set. Three arbitrary sets of 574 records of data set were used in predicting the recurrence of cancer and then found out the error rate. WEKA tool investigated the behavior of model used. After examination of execution, it is seen that model could predict the cancer up to 87.72%.

3 Experiments and Results

This section through the empirical observation measures the performance of different machine learning models on breast cancer data set. Description of data set, machine learning algorithms, methodology used in experimentation and experimental results is presented below in this section.

3.1 Data Concerned to Breast Cancer

Data set consisted of total 569 samples with 32 attributes. Attributes are shown in Table 2, and detailed description of attributes is given in [11]. Attributes used for prediction are computed from a digitized image of fine-needle aspiration (FNA) of breast mass. 357 out of the 569 samples had benign kind of tumor and rest of the 212 samples of peoples had malignant kind of tumor. These samples were obtained from Wisconsin data set. Figure 1 specifies the same. This data set is preprocessed and stored in CSV format.

Table 2 Attributes selected for the prediction of cancer

fractaldimension_mean	radius_worst	smoothness_se	area_worst
fractaldimension_worst	radius_mean	compactness_se	smoothness_worst
fractal_dimension_se	radius_se	texture_worst	compactness_worst
symmetry_mean	texture_se	perimeter_mean	concavity_worst
smoothness_mean	area_se	concavity_se	concave points_worst
compactness_mean	area_mean	perimeter_worst	symmetry_worst
concavity_mean	perimeter_se	symmetry_se	Class
concave points_mean	texture_mean	concave points_se	ID

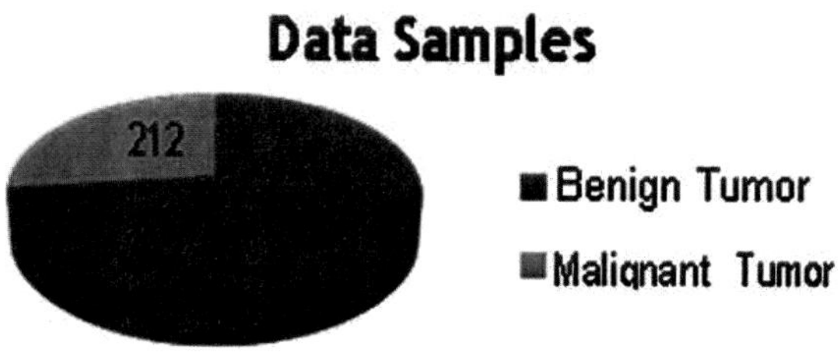

Fig. 1 Data samples containing benign and malignant tumors

3.2 Prediction Algorithms

Four machine learning algorithms were in consideration in this research, namely KNN [12], naive Bayes [13], logistic regression [14], and SVM [1]. Each algorithm was experimented for ten folds, and the average accuracy was calculated for the prediction.

KNN classification is a prominent technique for classifying an unseen instance. It is done by characterizing the occasions nearest to it. KNN algorithm works by discovering K instances that are near to the unseen instance [12]. This can be done using Euclidean, Manhattan distance, etc. At last, the algorithm chooses the class for the unseen instance by taking the most basic class in the closest K instances.

Naive Bayes is one of the techniques for constructing classifiers. It based on Bayes' theorem which is a probabilistic classifier. Naive Bayes classifiers assume that for a given class variable, the estimation of a specific component is free from the estimation of some other elements. Bayes' theorem:

$$P(C|X) = P(X|C) * P(C)/P(X) \tag{1}$$

where C is the class; $P(X)$ is constant for all classes; and X is the data tuple [13]. In spite of the fact that it expects an impossible condition that attribute values are restrictively free, it performs shockingly well on huge data sets where this condition is assumed and holds good.

Logistic regression is a predictive analysis that is used to predict the outcome of a categorically dependent variable based on one or more independent variables [14]. It is used to describe data and to explain the relationship between one-dependent binary variable and one or more ordinal, nominal, interval, ratio level, or independent variables. From the previous research, it is observed that logistic regression is being used mainly in medical research especially for the correlation of dichotomous outcomes with the predictor variables that may include different physiological data.

SVM is considered as supervised machine learning technique, and it will classify the data into two different classes over a hyperplane. The vectors or the cases which define the hyperplane are called support vectors. SVM is considered as a classification method and is based on statistical learning theory. SVM plays out the same role like C4.5 with the exception of that; it will not use decision trees at all [1]. SVM endeavors

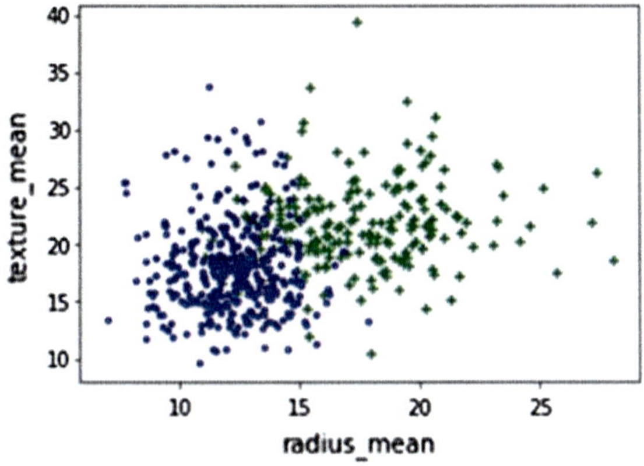

Fig. 2 SVM classifier

to amplify the edge to diminish any opportunity of misclassification. Figure 2 shows the SVM classifier model.

3.3 *Methodology*

In order to evaluate the above-mentioned prediction algorithms, data set is being partitioned into ten folds and different folds were assigned to training and testing sets. In each run, we calculated the accuracy. Then, the average of overall k runs was reported and this was the prediction accuracy of testing. Flow diagram of complete procedure is shown in Fig. 3.

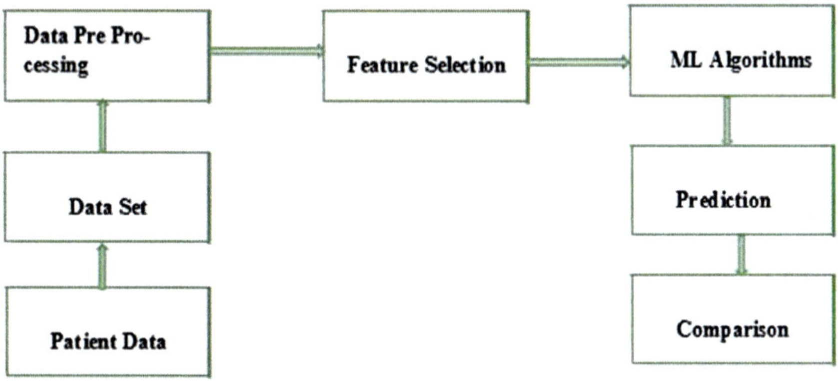

Fig. 3 Flow diagram of complete procedure

At first, KNN was chosen with $n = 3$ and we got the average accuracy of 92.56%. Secondly, the data set was tested with naive Bayes method which produced the accuracy of 54.32%. For the third time, the data set was experimented with logistic regression algorithm and obtained the average accuracy of 94.26%. And finally, the data set was experimented with SVM method, and the accuracy obtained was 95.67%. Confusion matrix for all the four algorithms with ten folds and the accuracy obtained are shown in Tables 3, 4, 5, and 6. Execution time to train the data was also computed and shown in the respective tables of all the four models. It is found that logistic regression trains the model at very least time compared to other models and Fig. 4 shows the results of accuracy for the four machine learning algorithms; confusion matrix is a table that can be used to evaluate the performance of classification models. By using confusion matrix, we can even visualize the performance of any given algorithm. Class-wise summing up the number of correct and incorrect predictions is the main fundamentals of confusion matrix. Using the confusion matrix, the classification rate or the accuracy is calculated using the relation (1) given below:

$$\text{Accuracy} = (\text{TP} + \text{TN})/(\text{TP} + \text{TN} + \text{FP} + \text{FN}) \quad (2)$$

Table 3 Confusion matrix for SVM

	Predicted no	Predicted yes	Total
Actual no	59	5	64
Actual yes	2	105	107
Total	61	110	171
Accuracy	95.67%		
Execution time	1.3218 s		

Table 4 Confusion matrix for logistic regression

	Predicted no	Predicted yes	Total
Actual no	57	6	63
Actual yes	4	104	108
Total	61	110	171
Accuracy	94.26%		
Execution time	0.0105 s		

Table 5 Confusion matrix for KNN

	Predicted no	Predicted yes	Total
Actual no	54	8	62
Actual yes	4	102	106
Total	58	110	168
Accuracy	92.56%		
Execution time	1.2746 s		

Table 6 Confusion matrix for naive Bayes

	Predicted no	Predicted yes	Total
Actual no	20	47	67
Actual yes	31	73	104
Total	51	120	171
Accuracy	54.32%		
Execution time	0.1536 s		

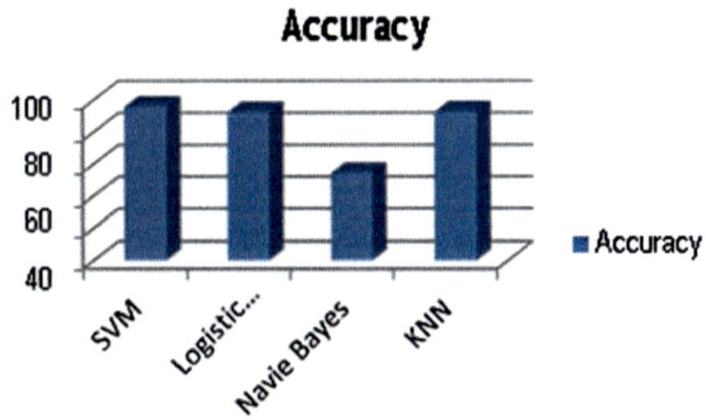

Fig. 4 Accuracy obtained for the classification and prediction model

4 Conclusions and Future Work

In this research, we investigated the feasibility of employing the four machine learning algorithms—SVM, naive Bayes, logistic regression, and KNN—for identifying the malignant or benign type of breast cancer found in women. These experiments were conducted on real-time data set which was preprocessed as an initial stage. Missing value was filled with mean value of the column. To obtain the accuracy, we have chosen ten folds' average accuracy. Confusion matrix was built for all the four prediction and classification algorithms. From our observation, we can say that SVM classification method gives the better prediction accuracy. Furthermore, the actual and the predicted values obtained in the process of classification are also shown. In any case, these results which are acquired mainly depend on Wisconsin data set. As the future work, we plan to collect the time series data from the hospitals and apply machine learning algorithms. The aim of the problem is to merge feature learning with the machine learning algorithms employed in this research work.

References

1. Tafish MH, El-Halees AM (2018) Breast cancer severity degree prediction using data mining techniques in the Gaza Strip. In: International conference on promising electronic technologies (ICPET), Deir El-Balah, 2018, pp 124–128
2. NCI cancer fact sheets (2008), available: http://www.cancer.gov/cancertopics/types/breast

3. Shrivastava (2013) An overview on data mining approach on breast cancer data. Int J Adv Comput Res 3:256–262
4. Venkatesan E (2015) Performance analysis of decision tree algorithms for breast cancer classification. Indian J Sci Technol 8:1–8
5. Sivakami K (2015) Mining big data: breast cancer prediction using DTSVM hybrid model. Int J Sci Eng Appl Sci 1:418–429
6. Shajahaan S (2013) Application of Data Mining techniques to model breast cancer data. Int J Emerg Technol Adv Eng 3:362–369
7. Safiyari A, Javidan R (2017) Predicting lung cancer survivability using ensemble learning methods. In: Intelligent systems conference, IEEE
8. Xu X, Zhang Y, Zou L, Wang M (2012) Agene signature breast cancer prognosis using SVM, IEEE, 928–931
9. Cruz JA, Wishart DS (2006) Applications of machine learning in cancer prediction and prognosis. David S Wishart, 2-21 Athabasca Hall, University of Alberta, Edmonton, AB Canada.
10. Umesh DR, Ramachandra B (2015) Association rule mining based predicting breast cancer recurrence on SEER breast cancer data. In: International conference on emerging research in electronics, computer science and technology, IEEE, pp 376–380
11. https://www.kaggle.com/uciml/breast-cancer-wisconsin-data
12. Shah C, Jivani AG (2013) Comparison of data mining classification algorithms for breast cancer prediction. In: 4th ICCCNT
13. Rashmi GD, Lekha A, Bawane N (2015) Analysis of efficiency of classification and prediction algorithm for breast cancer data set. In: International conference on emerging research in electronics, computer science and technology, IEEE, pp 108–113
14. Zhou X, Liu K-Y, Wong ST (2004) Cancer classification and prediction using logistic regression with Bayesian gene selection. J Biomed Inf Sci Direct 37(4):249–259

Study on Automatic Speech Therapy System for Patients

Supriya B. Rao, Sarika Hegde, and Surendra Shetty

Abstract Many children face difficulty with communication at an early age of their life. Some children catch up gradually but few might continue to have problems. The children are brought to the voice clinic because of some kind of dysphonia. The diagnosis must be done as soon as possible, and if problem persists, speech therapy will be recommended by the doctor. The main purpose of automatic speech therapy system (STS) is to help children improve their communication skills. The proposed work in this paper is to study on various voice disorder problems and the research works carried out in developing automatic STS.

Keywords Children · Speech therapy · Voice disorder

1 Introduction

Voice disorder is caused when the normal speech flow is corrupted. Due to this, the child feels difficult to communicate with others. It can be categorized into speech disorders, language disorders, swallowing disorders, social communication disorders and cognitive–communication disorders [1]. Stuttering is one of the very common speech disorder problems found in children. In order to detect stuttering, the children must repeat a given word in front of a speech therapist. This manual method can be automated using automatic STS. The system will check if the pronounced word has a stuttering problem or not. The stuttered word of one child might be different than

S. B. Rao (✉) · S. Hegde · S. Shetty
Department of CSE, NMAM Institute of Technology, Nitte, Udupi, Karnataka 574110, India
e-mail: supriyabrao@gmail.com

S. Hegde
e-mail: sarika.hegde@nitte.edu.in

S. Shetty
e-mail: hsshetty@nitte.edu.in

N. N. Chiplunkar and T. Fukao (eds.), *Advances in Artificial Intelligence and Data Engineering*, Advances in Intelligent Systems and Computing 1133,
https://doi.org/10.1007/978-981-15-3514-7_34

the other child. The stuttering behaviour includes repetition, blocking, interjection or prolongation of the word. Repetition can occur at different levels of speech. There can be a repetition of sound (e.g. A-A-A Are you alright?), repetition of a syllable (e.g. "Can I eat Ma-Ma-Ma Mango?"), repetition of a word (e.g. "He–He-He likes ice cream"), repetition of phrase includes repeating one or more word (e.g. "Shall I buy-some-buy-some chocolates") and repetition of the whole sentence ("This is a book. This is a book"). Repetition can occur in any part of the sentence. But most of the times, the repetitions occur at the beginning of the sentence. It can also occur in between or end of a sentence. Blocking usually occurs at the beginning of the sentence. When children try to speak, there might be some disturbance in the airflow due to which there will be a blockage and sound may not come out easily. When a stuttered word is stretched out, there will be a prolongation (e.g. What is your name?). "Fillers" or the interjections are the words that a child may use regularly (e.g. "um") [2].

Following are the benefits obtained by the STS [3].

1. Patients need not visit hospital or speech therapist regularly. Due to this travel, cost can be reduced. It saves time and the money of patients who are economically backward.
2. Patient can get speech therapy by making use of mobile/computer/robot/virtual reality.
3. Patient can also be reminded for speech therapy through mobile phones.
4. The doctors can also give appointments when necessary at their fingertips using mobile.

1.1 Existing STS

Speech-language pathologist (SLP) provides treatment to the children with communication, language and swallowing disorder. They also give counselling for the children as well as the parents or caregivers to educate about the problem. Children with language impairment will not be able to produce a sound that has quality. SLP provides treatment to overcome the stuttering problem and helps in enhancing the communication skill. The SLP has to go through some assessment to find how stuttering problem is affecting children at school or activities or family interactions. The SLP provides children with pictures or objects, and children are supposed to identify and produce sounds. To obtain speech-language therapy, the patient either visits a hospital or speech therapy institute or individual practitioners as explained by De Silva [4] (Fig. 1).

Automated STS Most of the children today make use of computers, smartphones, tablets and other technologies that change their interactions and learning preferences. Technology helps children to read and write easily. The STS is a virtual therapist that helps children in improving their communication problem. The devices can help

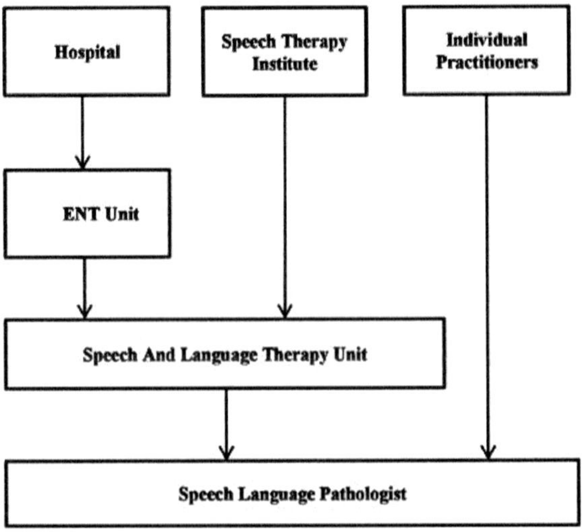

Fig. 1 Accessibility of speech-language pathologist [4]

in providing speech therapy for children with a stuttering problem. Speech therapy devices motivate children easily than paper-based assessment method [5]. STS can also be inexpensive and also provides more accessibility. STS provides many tasks and exercises for children to practise which are widely accepted by SLP. Hence, by practising regularly, the communication skill gets improved, and children will be able to express their thoughts efficiently during natural conversation.

2 Methodology

Speech signals are processed, and then the feature extraction of the signal is carried out. Many methods can be used for feature extraction. The pronounced quality is measured, and the classification of signals is done using a suitable algorithm [6, 7] (Fig. 2).

3 Related Work

We have surveyed papers related to speech therapy and categorized them based on the type of interactive system used for speech therapy as computer-based, mobile, robots and augmented reality/virtual reality. The summary of the mentioned categories is shown in Tables 1, 2, 3 and 4, respectively.

The below papers are closely related to our studies.

[Grzybowska 6] The application helps in self-assessment of the practised sound without the presence of SLP. Here, HFCC is applied for speech signals. The HFCC

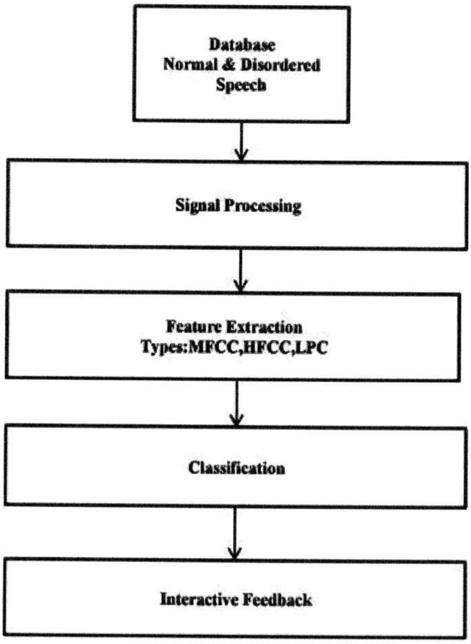

Fig. 2 Speech therapy system

is the modification of MFCC. For measuring the similarity between the signals, DTW algorithm was used. The system provides computer-based speech therapy. The recording of the Polish words is done by making use of microphones which transforms the acoustic pressure into signals. In the pre-acoustic stage, the ratio of signal is improved. With reference to the magnitude of lower frequency, higher frequency is increased. In the signal extraction stage, the threshold is set to differentiate between the voiced frames and the unvoiced frames. The feature vector is obtained in the next stage. The DTW algorithm finds the lowest distance path using a distance matrix. Two sequences are more similar if the distance is low. Using the K-nearest neighbours algorithm, recorded signals will be classified.

[Tan 2] The Malay speech therapy assistance tools (MSTAT) system assists the speech therapist in diagnosing the language disorder. It provides training to the children suffering from the utterance problem. In order to evaluate the stuttering problem in children, the HMM technique is used. The normal and disordered voice samples are taken to train the HMM model. The feature is extracted from the speech signals, and the size of the data is reduced by making use of MFCC. Samples of 20 normal speeches and 15 stuttered speeches are used. By taking ten samples of normal and ten samples of stuttered speeches, the speech model will be generated. A threshold is set. To test the model, five normal speech samples and five stuttered speech samples are used. If the obtained score is less than the threshold, then the resultant speech is diagnosed as the stuttered speech.

Table 1 Summary of research works on computer-based speech therapy with microphone

S. no.	Title	Description
1	Computer-assisted HFCC-based learning system for people with speech sound disorders [6]	• The application helps in self-assessment of the practised sound without the presence SLP • Here, the human factor cepstral coefficient (HFCC) which is the modification of mel frequency cepstral coefficient (MFCC) is applied for speech signals. For measuring the similarity between the signals, dynamic time warping (DTW) algorithm was used
2	Virtual clinicians for the treatment of aphasia and speech disorders [8]	• The virtual clinician helps aphasia [9] patients in improving communication that is essential for everyday life • The text is passed to the avatar which is then converted into the speech • The avatar provides basic animation and displays visual emotion • Once the in-clinic therapy ended, the patient was given with the software suite that can be used at home so that the patient could continue the therapy after they had ended the in-clinic treatment
3	Robust scoring of voice exercises in computer-based speech therapy systems [10]	• The proposed model is helpful in providing speech therapy. It is mainly used for increasing or decreasing in the pitch variations • The model is trained using the support vector machine (SVM) and double cross validation. These learned models are useful in providing children with feedback based on their performance • The exercised voices are automatically classified which is helpful in deciding if the exercises are proper. The learned models provide high accuracy, and false negative rate is low
4	Application of Malay Speech Technology in Malay speech therapy assistance tools [2]	• The proposed system helps speech therapist in identifying children suffering from the stuttering problem • In this work, speech recognition system is the main engine that makes use of hidden Markov model (HMM) technique which evaluates speech disorder like stuttering. MFCC is used for feature extraction

(continued)

Table 1 (continued)

S. no.	Title	Description
5	Improving the intelligibility of dysarthria speech towards enhancing the effectiveness of speech therapy [11]	• Speech therapy tools provide auditory feedback with delay which helps patients improve their speech intelligibility by relearning which in turn helps in boosting their confidence • Linear predictive coding (LPC) coefficients mapping the frequency warping transformation are used to improve disordered speech. DTW is used for similarity measurement between two speech utterances
6	An automated speech-language therapy tool with interactive virtual agent and peer-to-peer feedback [12]	• Computer-only treatment (COT) is the speech-language therapy tool which is used in the proposed work • The virtual speech therapist interacts with the patient and also provides feedback which helps in improving the communication skill of the patient
7	A machine learning-based system for the automatic evaluation of aphasia speech [7]	• The machine learning system is trained using the speech samples like articulation or prosody and phonology • The large database of aphasia patient and the interviewer dialogue has been used • The pre-processing is done, and the patient's dialogue part is extracted • After training the system with the speech samples of aphasia patient, feature extraction is carried out • The words are represented as bag-of-words using bag-of-audio-words (BoAW) approach • The classifier is trained in order to differentiate between the severe levels of aphasia
8	Towards a multimodal emotion recognition framework to be integrated in a computer-based speech therapy system [13]	• Multimodal emotion recognition framework has been used for improving computer-based STS • Pre-processing is done to obtain the feature vector • Different classifier systems like HMM and artificial neural network (ANN) have been used • Multimodal decision level fusion provides context-sensitive emotional state as a result

(continued)

Table 1 (continued)

S. no.	Title	Description
9	Speech therapy software on an Open Web Platform [14]	• The Open Web Platform is used where the patients can avail speech therapy through the ordinary Web browser • From Open Web Platform, speech therapy software packages have been developed which can be accessed easily • The Web browsers have all the necessary APIs in order to develop a speech therapy application • The speech therapist can keep track on the patient's progress outside the clinic as well
10	Application of visual speech synthesis in therapy of auditory verbal hallucinations [15]	• The animated head talks to the patient using synthetic voice • The synthetic voice is generated by making use of text-to-speech system which is available in the market • The opinion was gathered from the patient who underwent avatar-based speech therapy • They reported that they changed their opinion towards the hallucinations and were also satisfied with the audio video content that was presented during the speech therapy
11	Vowel and diphthong tutors for language therapy [16]	• An ICATIANI application was developed which helps in the acquisition of speech for the children with a hearing problem • An animated character was used. The name of the character was "Baldi", he smiles when children answers properly, and he makes sad face when children provide wrong answer and asked them to try again • Speech recognition was used to recognize the pronounced sound • Applications were developed using the CSLU toolkit

(continued)

Table 1 (continued)

S. no.	Title	Description
12	Speech therapy system to Kannada language [17]	• Kannada is one among the Indian languages. The system transforms the entered Kannada text into the spoken waveform • The proposed work helps in providing the communication technique for the patient suffering from speech and language problem • The main advantage of Kannada speech therapy system is it helps in the conversion of entered text into a speech waveform • The proposed work can be modified and used for various Indian languages
13	An interactive speech therapy session using linear predictive coding in MATLAB and Arduino [18]	• The concept of linear predictive coding (LPC) is used for speech recognition. The hardware interface used here is Uno board • The proposed work provides an interactive speech therapy session. It determines if the patient's pronunciation is correct or not • If the patient's pronunciation is correct, then LED displays green light, and suppose if the patient's pronunciation is wrong, then red light is displayed • Feedback provided by the system is friendly and accurate
14	A versatile assistive device for ADHD with speech therapy using embedded system [19]	• The Lab View soft core is used for analysing speech. UNO board is used to give feedback • The input speech is processed by which a number of frames are obtained and energy frames are calculated by summing up the frames obtained • The energy level of input is calculated and compared with different threshold levels to obtain the voice status • The threshold is compared, if the voice matches the first threshold level, then the child must increase the pitch, if it matches with the second threshold, then the child's voice is either too low or high, and if the voice matches with the third threshold, then the child's voice is very high and the pitch must be lowered

(continued)

Table 1 (continued)

S. no.	Title	Description
15	Automated and self-learning sentence generation methodology for Stammer patients by means of natural language generation [4]	• The proposed work provides automated speech therapy for the children subjected to stammer disorder • Markov model along with the natural language generation (NLG) suits the best for self-learning • Simple poems are used as trained data for training the NLG model
16	E-Inclusion technologies for the speech handicapped [20]	• An application called Vocaliza is used which is a freeware that helps in providing speech therapy for speech impaired and handicapped patients • Automatic speech recognition system (ASR) is used for decoding the utterances, and text-to-speech helps in showing the correct pronunciation of the word or sentence • Systems are enabled by the speaker adaption techniques to get better performance when a specific speaker is using it • Improved communication is evaluated by using the utterance verification technique which is embedded within the application
17	Measuring the performance of children with speech and language disorders using a serious game [21]	• A serious game is provided for the children suffering from speech and hearing problem • By playing this serious game, children will be able to improve the communication impairment • The serious game has three different stages. In the first stage, the diagnosis is done to see if speech therapy is necessary for the child. The second stage includes playing the game by which the child is trained through the game. The third stage is used for testing and evaluating child advancement

Table 2 Summary of research works on mobile-based speech therapy

1	A mobile multimedia application inspired by a spaced repetition algorithm for assistance with speech and language therapy [5]	• The proposed work aims at providing speech-language therapy using the Web application • Spaced repetition algorithm helps in learning the keywords by arranging reviews of the cards for which the user finding difficult to appear at more frequent intervals
2	Speech therapy mobile application for speech and language impairment children [22]	• Mobile speech application is developed to give speech therapy for the speech-language impaired children of age 3–6 • The application is designed appropriately which helps in playing an important role in providing speech therapy to children • Children found it easy using the application since they already had a gaming experience
3	Speech therapy and assessment [3]	• The mobile and media technology is used to provide speech therapy for children suffering from cleft problem [23] • In order to compare the speech pattern between normal speech and the cleft patient speech, some basic algorithms have been developed • The proposed system motivates the children by providing games with singalong • By making use of VOIP-based video conference, the patient can communicate remotely with the speech therapist • Speech therapist can advise patient if any changes are required in the speech therapy drills • Speech therapy drills are provided using mobile technology
4	The table to tablet (T2T) therapy software development approach [24]	• The proposed project provides a tablet to tablet speech therapy software • The developed software aims at providing a valid solution for the speech therapist to work with speech impaired children • The set of games have been provided as the children's homework

[Kohlschein 7] Aphasia is a language disorder caused by the damage of related networks in the brain. The proposed work is aimed at providing speech therapy for aphasia patient. The machine learning system is trained using speech samples like articulation or prosody and phonology. The large database of aphasia patient and the interviewer dialogue has been used. The pre-processing of the speech signal is

Table 3 Summary of research works on robot-based speech therapy

1	Noisy speech training in MFCC-based speech recognition with noise suppression towards robot-assisted autism therapy [25]	• The robot is used to give speech therapy for children suffering from autism [26]. Even in a noisy environment, the robot should have the ability to recognize speech • Here, the noisy speech found in the environment is used for training in MFCC-based speech recognition system • The results show that the performance of MFCC with noise suppression is better than MFCC with clean speech
2	A robotic assistant to support the development of communication skills of children with disabilities [27]	• The speech therapy is provided by robot assistant and has the ability to manage information and various functionalities • Robot stores the database of the entire task, results and also the behaviour of the children • Different guidelines are provided based on individual preferences • Multimedia resources are used for speech therapy sessions. It keeps track of patient's progress • Robot costumes can be changed based on the type of disability • The face and voice recognitions are done, and it provides reaction to the patient using arm and head movements, sounds or songs and animations
3	RAMSES: A robotic assistant and a mobile support environment for speech and language therapy [28]	• RAMASES is an intelligent environment where the mobiles and robotic assistant are used to give speech therapy • The robot provides different activities like hand gestures, reproduces songs, stories and also helps in searching the words or sentences from the Internet by making use of voice commands. It also detects patient's face • On the Android devices, the control software is installed and controlled remotely by the graphical user interface

done, and the patient's dialogue part is extracted. OPENSMILE toolkit is used for language disease classification. After training the system with the speech samples of aphasia patient, feature extraction is carried out. The words are represented as bag-of-words using BoAW approach. Since the large database is already put together, the additional linguistic features are tested for classification of aphasia. The classifier is trained in order to differentiate between the severe levels of aphasia.

Table 4 Summary of research works on virtual reality-based speech therapy

S. no.	Title	Description
1	Language therapy of aphasia supported by augmented reality applications [29]	• The augmented reality provides speech therapy which is very interactive. The user interface design is very simple, and the patients can be trained at home itself • The augmented reality software recognizes the objects present at patient home • The user interface is natural which is very helpful for the elderly people to meet high alertness • The repetitive training of objects is well supported by augmented reality • It also helps in reducing the financial burden on the patient and provides speech therapy for the patient suffering from aphasia
2	A game-based assistive tool for rehabilitation of dysphonic patients [30]	• In the proposed work, the virtual reality-based application has been developed in order to provide speech therapy for patients suffering from dysphonia [31] • A serious game is developed, and the microphone arrays are used as an input device • The patient plays the game under the guidance of a speech therapist. The patient undergoes voice training simultaneously when playing the game • The recorded voice is used to check the patient's progress • Emotion recognition has been done which helps in better decision making

[Diogo 10] The proposed model is used for voice exercising that is used in the speech therapy sessions. It helps in strengthening the vowel and increasing or decreasing pitch exercises. The robust model makes use of SVM and double cross validation. The main advantage of this system is that it provides the children with feedback based on their performance. The classification models are used to decide if the exercise is correct or not. The classification models like sustained vowel, increasing pitch and decreasing pitch are learned using the SVM. The feature extraction is done using OPENSMILE tool. The learned model provides high accuracy. The false negative rate is very low.

[De Silva 4] Automated speech therapy is provided for the children with stammer disorder. NLG along with the HMM is used for automatic sentence generation. The NLG model is trained using data like simple poems. The trained data is organized in the syntactic structure by Markov chain. Markov chains build the new syntactic structure whenever the new data train is provided. The accuracy is high if the syntactic structure is large. The output obtained will be the text generation. The natural language algorithm traverses randomly in order to generate the new sentence. The second input for the algorithm is obtained from the table that has repeated words in the database. The sentence with stuttered words is the result. The patients have to stutter the same word once again which is put in the different sentence.

4 Conclusions

We have presented the need for speech therapy and also various interactive systems used for speech therapy. The STS acts as a virtual therapist and provides treatment to children with a voice disorder. STS recognizes the sentence pronounced by the patient. It identifies if the sentence pronounced is correct or stuttered and provides feedback to the patient. Many significant research works have been carried out in this area. However, more focus is needed for developing practical solutions which will be interactive and useful for the patients to overcome the communication problems.

References

1. Patel Rita R et al (2018) Recommended protocols for instrumental assessment of voice: American speech-language-hearing association expert panel to develop a protocol for instrumental assessment of vocal function. Am J Speech-Lang Pathol 27(3):887–905
2. Tan T-S et al (2007) Application of malay speech technology in malay speech therapy assistance tools. In: International conference on intelligent and advanced systems, ICIAS 2007. IEEE
3. Dhaky K, Bulsara M, Sethna B (2011) Speech therapy and assessment (via multimedia devices for cleft lip and palate patients). In: 2011 IEEE global humanitarian technology conference (GHTC). IEEE
4. De Silva Y (2016) Automated and self-learning sentence generation methodology for stammer patients by means of natural language generation smart speech therapist for stammer. In: 2016 International conference on signal processing, communication, power and embedded system (SCOPES). IEEE
5. Folksman D et al (2013) A mobile multimedia application inspired by a spaced repetition algorithm for assistance with speech and language therapy. In: 2013 Sixth international conference on developments in eSystems engineering (DeSE). IEEE
6. Grzybowska J, Klaczynski M (2014) Computer-assisted HFCC-based learning system for people with speech sound disorders. In: 2014 XXII annual pacific voice conference (PVC). IEEE
7. Kohlschein C et al (2017) A machine learning based system for the automatic evaluation of aphasia speech. In: 2017 IEEE 19th international conference on e-health networking, applications and services (Healthcom). IEEE
8. Teodoro G et al (2013) Virtual clinicians for the treatment of aphasia and speech disorders. In: 2013 International conference on virtual rehabilitation (ICVR). IEEE
9. Holland AL et al (1996) Treatment efficacy: aphasia. J Speech Lang Hear Res 39(5):S27–S36
10. Diogo M et al (2016) Robust scoring of voice exercises in computer-based speech therapy systems. In: 2016 24th European Signal Processing Conference (EUSIPCO). IEEE

11. Kumar SA, Kumar CS (2016) Improving the intelligibility of dysarthric speech towards enhancing the effectiveness of speech therapy. In: 2016 International conference on advances in computing, communications and informatics (ICACCI). IEEE
12. Das M, Saha A (2017) An automated speech-language therapy tool with interactive virtual agent and peer-to-peer feedback. In: 2017 4th International conference on advances in electrical engineering (ICAEE). IEEE
13. Schipor O-A, Pentiuc S-G, Schipor M-D (2011) Towards a multimodal emotion recognition framework to be integrated in a computer based speech therapy system. In: 2011 6th Conference on speech technology and human-computer dialogue (SpeD). IEEE
14. Awad SS, Piechocki C (2014) Speech therapy software on an open web platform. In: 2014 10th International computer engineering conference (ICENCO). IEEE
15. Sorokosz K (2018) Application of visual speech synthesis in therapy of auditory verbal hallucinations. In: 2018 International interdisciplinary PhD workshop (IIPhDW). IEEE
16. Kirschning I et al (2005) Vowel & diphthong tutors for language therapy. In: Sixth Mexican international conference on computer science, 2005. ENC 2005. IEEE
17. Udayashankara V, Havalgi S (2016) Speech therapy system to Kannada language. In: 2016 Second international conference on cognitive computing and information processing (CCIP). IEEE
18. Vijayalakshmi R, Priya S (2016) An interactive speech therapy session using linear predictive coding in Matlab and Arduino. In: 2016 International conference on advanced communication control and computing technologies (ICACCCT). IEEE
19. Chinnaaiah MC et al (2016) A versatile assistive device for ADHD with speech therapy using embedded system. In: International conference on research advances in integrated navigation systems (RAINS). IEEE
20. Vaquero C et al (2008) E-inclusion technologies for the speech handicapped. In: IEEE International conference on acoustics, speech and signal processing, 2008. ICASSP 2008. IEEE
21. Nasiri N, Shirmohammadi S (2017) Measuring performance of children with speech and language disorders using a serious game. In: 2017 IEEE International symposium on medical measurements and applications (MeMeA). IEEE
22. Tommy CA, Minoi J-L (2016) Speech therapy mobile application for speech and language impairment children. In: 2016 IEEE EMBS conference on biomedical engineering and sciences (IECBES). IEEE
23. Lewis CW et al (2017) The primary care pediatrician and the care of children with cleft lip and/or cleft palate. Pediatrics 139(5)
24. Jesus LMT et al (2015) The table to tablet (T2T) therapy software development approach. In: 2015 10th Iberian conference on information systems and technologies (CISTI). IEEE
25. Attawibulkul S, Kaewkamnerdpong B, Miyanaga Y (2017) Noisy speech training in MFCC-based speech recognition with noise suppression toward robot assisted autism therapy. In: 2017 10th Biomedical engineering international conference (BMEiCON). IEEE
26. Prelock Patricia (2001) Understanding autism spectrum disorders: The role of speech-language pathologists and audiologists in service delivery. The ASHA Leader 6(17):4–7
27. Ochoa-Guaraca M et al (2016) A robotic assistant to support the development of communication skills of children with disabilities. In: 2016 IEEE 11th Colombian computing conference (CCC). IEEE
28. Robles-Bykbaev VE et al (2015) RAMSES: a robotic assistant and a mobile support environment for speech and language therapy. In: 2015 Fifth international conference on innovative computing technology (INTECH). IEEE
29. Antkowiak D et al (2016) Language therapy of aphasia supported by augmented reality applications. In: 2016 IEEE 18th international conference on e-Health networking, applications and services (Healthcom). IEEE

30. Lv Z et al (2015) A game based assistive tool for rehabilitation of dysphonic patients. In: 2015 3rd IEEE VR international workshop on virtual and augmented assistive technology (VAAT). IEEE
31. Carding Paul N et al (2006) The prevalence of childhood dysphonia: a cross-sectional study. J Voice 20(4):623–630

Data Engineering

The Design of Multiuser BGN Encryption with Customized Multiple Pollard's Lambda Search Instances to Solve ECDLP in Finite Time

Santosh Javheri and Uday Kulkarni

Abstract Data privacy breach is an important concern for data owners, in today's online world. Data science applications, smart city applications and smart grid applications generate precious knowledge through analyzing user's data. But many a times, it is experienced by the users that their sensitive personal data is exposed by these applications, without their consent. Due to the serious issues of data privacy preservation problems, public cloud-based data centers have failed in achieving popularity in third-party computation. To solve this privacy preservation problem, data can be collected and uploaded in unidentifiable format, such a way that meaningful computation can be performed easily. Homomorphic encryption is one of the favorite techniques, wherein direct computation on encrypted data is permitted and its results are equivalent to their plaintext format. BGN 'doubly' homomorphic encryption can execute infinite additions and single multiplication over encrypted data values at a time. BGN is based on the elliptic curve. Constant size of ciphertext is an attractive feature of BGN encryption. User has to solve elliptic curve discrete logarithm problem (ECDLP) to receive the desired results of addition and multiplication in plaintext format. Solving ECDLP is most time-consuming operation. Hence, BGN encryption can be used in small message space only. Customize Pollard's lambda method can increase the message space of BGN encryption and solve the problem. Multiple instances of Pollard's lambda attack can be executed with different range of intervals to solve ECDLP in finite time. This customization will make narrow search intervals to Pollard's lambda to solve even 78-bit ECDLP for 88-bit prime field in finite times.

Keywords Homomorphic encryption · Elliptic curve · Elliptic curve discrete logarithm problem (ECDLP) · Pohlig-Hellman · Pollard's Lambda · BGN cryptosystem

---

S. Javheri (✉)
JSPM's Rajarshi Shahu College of Engineering, Pune, India
e-mail: sbjavheri@gmail.com

U. Kulkarni
SGGS Institute of Engineering and Technology, Nanded, India

N. N. Chiplunkar and T. Fukao (eds.), *Advances in Artificial Intelligence and Data Engineering*, Advances in Intelligent Systems and Computing 1133,
https://doi.org/10.1007/978-981-15-3514-7_35

1 Introduction

The true benefits of information technology and data science will be acknowledged in more meaningful way if it is able to preserve data owners' privacy [1] during execution of joint computation and bringing out the knowledge through data science processes. Smart city [2] and smart grid [3] projects collect users' real-time data for their operations, enhancement of performance and accuracy. Collaborative learning is used to incorporate datasets of multiple parties in the learning process and to enhance the accuracy of machine learning models [4]. Cloud data centers [5] are highly in demanding position in public environment for uploading and performing third-party computations or for storage purposes. These examples illustrate data owners' privacy can be breached easily without his/her consent. Even sometimes data owner may come to know about breach from some other sources [6].

To avoid any misuse of data, all data should be transformed in ciphertext format. Subsequently, algebraic computations or data science operations can be performed. The obtained results will be similar to the operations derived on plaintext data. Secure multiparty computation (SMC) [7] and data randomization [8] are two main techniques used to execute algebraic operations on transformed data. Wherein, results can be expected similar to plaintext, if performed on same algebraic operations. SMC incorporates encryption techniques, whereas 'data randomization' uses the addition of the 'noise' in plaintext data. SMC methods offer higher level of security as compared to data randomization, whereas data randomization techniques are more efficient [9]. High resource demanding nature of SMC techniques makes them less practical [4]. Similarly, homomorphic encryption executes SMC or data randomization operations on encrypted data [4]. It operates in finite time with data owners' privacy preservation.

Homomorphic encryption allows computation of certain algebraic operations on encrypted data, by maintaining characteristics of functions and formats of encrypted data. It is a map associating two algebraic structures, which preserves algebraic structure [10, 11]. These abilities have made homomorphic encryption more popular in execution of SMC type operations. Users can encrypt plaintext data into ciphertext and handover to other users to execute further functions. The results will be similar in case of execution of similar type function on plaintext data. Reliability on third-party clouds and distributed computations can be increased by using homomorphic encryption [5, 12, 13].

Boneh-Goh-Nissim (BGN) [14], homomorphic encryption, can execute infinite number of additions and single multiplication. BGN encryption operates on elliptic curve of finite groups $\mathbb{G}$ and $\mathbb{G}_1$ of composite order, and a bilinear map $e\colon \mathbb{G} * \mathbb{G} \rightarrow \mathbb{G}_1$. Users require solving discrete logarithm on elliptic points, to obtain final answer in plaintext form, which is computationally hard. In recent years, many lattice-based and coding-based encryption techniques known as fully homomorphic encryption are developed. These techniques theoretically solve the arbitrary number of additions and multiplications, but practically demand very high resources. Therefore, they are less useful, also the security of these schemes is not proven [15]. BGN encryption offers robust semantic security. Constant ciphertext size is desirable characteristics of all elliptic curve cryptographic systems. BGN operates on elliptic curve; therefore, it generates constant size ciphertext [11]. BGN encryption is more suitable where frequent execution of addition operations and few multiplications are involved [15].

A parallelize approach to execute multiple instances of Pollard's lambda method [16] is introduced for algorithms with customize bounds. This customization incorporates with Pohlig-Hellman [17] ECDLP attack for factorizing composite order discrete logarithm problem in prime order subgroups. Pohlig-Hellman factorizes large composite number in multiple prime factors. In present research, ECDLP method generates exactly two prime factors every time. Pollard's lambda algorithm solves ECDLP in range of integer values. Complete range of interval is divided into multiple parts; starting points will be calculated in close range. Therefore, it takes less time to find discrete logarithm. After obtaining the results, parent computing process stops all other instances of Pollard's lambda, so computing resources CPU and memory will be free for further computations. Usually, Pollard's lambda method can solve 30–40-bit ECDLP in minutes or hours [4]. Range partition of Pollard's lambda search, in BGN multiplication, can successfully solve 78-bit ECDLP on single contemporary computer. These results are promising with Pollard's lambda space efficiency.

The remaining section of paper elaborates related work in Sect. 2. Mathematical background of elliptic curve is explained in Sect. 3. Multiuser BGN encryption with customize Pollard's lambda algorithm is explained in Sect. 4. Discussion on experimental setup and results are presented in Sect. 5. Modification and conclusion are discussed in Sect. 6.

2 Related Work

Koblitz [18] and Miller [19] from their independent research work suggested new directions in cryptography, known as elliptic curve cryptography (ECC). Security of this technique is depending on discrete logarithm problem on finite multiplicative group created by elliptic curve points.

'Doubly' homomorphic encryption designed by Boneh-Goh-Nissim (BGN) is pioneer contribution based on elliptic curve cyclic groups and bilinear map. BGN technique is useful in private information retrieval (PIR), election protocol applications wherein users perform number of additions and few multiplications. The difficulty in solving of ECDLP results in execution of application in small message space.

In the practical implementation of BGN encryption, Freeman [20] created BGN type cryptosystem through the prime order elliptic set of groups. Eom et al. [21] introduced a composite order BGN system with multiple prime order factoring to increase message space, by assuming existence of minimum required security on Decisional Diffie-Hellman (DDH). Yuan et al. [22] have performed splitting of large numbers into smaller chunks, with the power of cloud computing and parallel decryption. They have solved 90-bit ECDLP successfully using Pollard's lambda method for multiuser

BGN encryption. Herbert et al. [15] proposed a modified BGN cryptosystem. They have enhanced multiplication operation on encrypted data by simultaneously multiplying four numbers in small message space. Javheri and Kulkarni [22] suggested the multiprocessing approach with Pohlig-Hellman and customized interval baby-step giant-step (BSGS) [23]. This approach is useful to solve 78-bit ECDLP, for BGN encryption, with secure multiparty multiplication on the single contemporary computer in finite time.

3 Methodology

3.1 Elliptic Curve Groups

An elliptic curve E can be defined over prime field $\mathbb{F}_p$ created by prime integer p. Weierstrass Eq. (1) is used to define the elliptic curve as

$$E: y^2 = x^3 + ax + b, \quad \text{where } a, b \in \mathbb{F}_p \tag{1}$$

An abelian cyclic group $(\mathbb{G}, +)$ is formed with a set of points $E(\mathbb{F}_p)$ and addition rule. The point '∞' is the only point at '*line of infinity*', represents '*identity element*', where elliptic curve remains parallel to 'y' axis and no new points are obtained. The number of points in $E(\mathbb{F}_p)$ is called an *order* of curve. Supersingular elliptic curve contains $p + 1$ elliptic point, wherein non-supersingular elliptic curve contains any number of points satisfying equation of curve only. Elliptic curve cryptographic systems are constructed using abelian group $(\mathbb{G}, +)$ [24, 25]. The elliptic curve generated by $y^2 = x^3 - x$ is shown in Fig. 1.

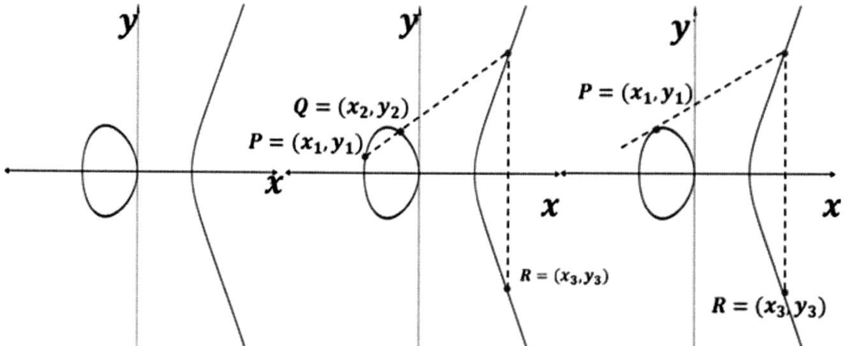

Fig. 1 Elliptic curve and geometric operations on elliptic points

Geometric rule *chord-and-tangent* is used for the addition of two points in $E(\mathbb{F}_p)$ and obtains the third point in $E(\mathbb{F}_p)$. Algebraic operations *addition* and *double* on elliptic curve are shown in Fig. 1. Let $P = (x_1, y_1)$ and $Q = (x_2, y_2)$ are two distinct points on the elliptic curve. The *sum* $R = P + Q$, (3) received by drawing a line passing through P and Q and finds a point of intersection on the elliptic curve and find reflection of this point on x-axis as R. For *double* of P point (4), draw a tangent line to point P, find its intersection on the elliptic curve; point R is a reflection of this point on the x-axis.

$$\textbf{Group law for } E/\mathbb{F}_p : y^2 = x^3 + ax + b \quad \textbf{char}(\mathbb{F}_p) \neq 2, 3 \tag{2}$$

Let $P = (x_1, y_1)$, $Q = (x_2, y_2)$ where $P, Q \in E(\mathbb{F}_p)$, then

- *Identity*: $P + \infty = \infty + P = P$
- *Negative*: $-P = (x, -y), P + (-P) = \infty$
- *Point Addition*: $P + Q = R$, where $R = (x_3, y_3)$

$$x_3 = \left(\frac{y_2 - y_1}{x_2 - x_1}\right)^2 - x_1 - x_2 \text{ and, } y_3 = \left(\frac{y_2 - y_1}{x_2 - x_1}\right) - (x_1 - x_3) - y_1 \tag{3}$$

- *Point doubling*: If $P \neq -P$, then $2P = (x_3, y_3)$

$$x_3 = \left(\frac{3x_1^2 - a}{2y_1}\right)^2 - 2x_1 \text{ and, } y_3 = \left(\frac{3x_1^2 - a}{2y_1}\right)^2 - (x_1 - x_3) - y_1 \tag{4}$$

Elliptic curve points can be represented in *affine* coordinates or *projective* coordinates.

- *Affine* coordinates: The set of *affine* points in prime field $\mathbb{F}_p$ denoted with x and y coordinates as:

$$\mathbb{A}(\mathbb{F}_p) = \{(x, y) : x, y \in \mathbb{F}_p\}$$

- *Projective* coordinates: The set of *projective* points in prime field $\mathbb{F}_p$ denoted with x, y and z coordinates as:

$$\mathbb{P}(\mathbb{F}_p)^0 = \{(X : Y : Z) : (X, Y, Z) \in \mathbb{F}_p, Z = 0\}$$

Affine coordinates system is suitable for additions and point doubling operations, but point inversion operation is time consuming. Point inversion is easy in projective coordinates system, but to execute addition and point doubling operation, projective points transform to affine coordinates.

3.2 Elliptic Curve Discrete Logarithm Problem (ECDLP)

Discrete logarithm problem is an area of interest after the introduction of key-agreement protocol introduced by Diffie-Hellman [26] in 1976. The hardness of elliptic curve discrete logarithm problem is essential for security of all elliptic curve encryption schemes.

Elliptic curve E is defined over finite field $\mathbb{F}_p$. A point $\boldsymbol{P} \in \mathbb{F}_p$, of order n, point $\boldsymbol{Q} \in \langle P \rangle$ where $\langle P \rangle$ is cyclic subgroup generated by P, find the integer $l \in [0, n-1]$, such that $Q = lP$. The integer l is called as discrete logarithm of Q with base P. Robustness of elliptic curve cryptosystems depends on these carefully selected parameters. Other side of this is large computations required to solve ECDLP, resulting in encryption scheme operating only on small message space.

Researchers invented generic attacks to solve discrete logarithm problem which are successfully applied for elliptic curve discrete logarithm problem. Pohlig-Hellman [17] generic attack divided discrete logarithm problem into prime order subgroups to efficiently compute discrete logarithm. Pollard [16] in his remarkable work introduced Pollard's rho and Pollard's lambda attacks to solve discrete logarithm problem in finite time with efficient use of memory space. Pollard's rho attack executes only on prime order groups, whereas Pollard's lambda operates on both prime and composite order groups. After the invention of ECC, researchers have extensively studied techniques to solve ECDLP in finite time. Van Oorschot and Wiener [27] efficiently parallelize Pollard's rho attack and Pollard's lambda attack.

3.3 Pollard's Lambda Algorithm

To increase the message space, it is essential to solve ECDLP in finite time. Pollard's lambda or Pollard's kangaroo [16] method finds discrete logarithm in interval length b. It executes with two starting points and select two random walks, finally, they meet each other on a common point, and subsequently further computation will get executed to find a discrete log. Execution of both random walks is similar to the Greek letter 'λ'; therefore, it is known as Pollard's lambda method. Started with two random points, there are chances of Pollard's lambda algorithm to miss the collision and fails in solving discrete logarithm problem. Even two selected points are apart from each other with sufficiently large distance, and it may take more time requirement for Pollard's lambda algorithm to solve discrete logarithm problem.

4 Multiuser BGN Cryptosystem Range Partition Pollard's Lambda Algorithm

BGN encryption can perform an arbitrary additions and single multiplication of encrypted data in the row with Okamoto-Uchiyama [28], Pallier [29]. BGN type encryption is pairing-based cryptosystem.

In present research work, researcher has attempted successfully the 'optimization' of time, memory use and computing speed in the application of multiuser BGN cryptographic system designed for secure addition and multiplication. Secure sum operations can execute among s users encrypted data values collectively. Due to the bilinear map in symmetric groups, the secure multiplication operation can execute on two encrypted data values in same time.

To generate BGN parameters, select q_1 and q_2—two large prime numbers. Wherein q_1 acts as system master secret key, which splits as q_{1s} among 's' users for secure decryption. BGN is constructed on supersingular elliptic curve $y^2 = x^3 + 1 \bmod p$, where p is prime. Two cyclic groups, $\mathbb{G}_1$ of N that is composite order, where $N = q_1 * q_2$ with bilinear map $\mathbf{e}\colon \mathbb{G} \times \mathbb{G} \rightarrow \mathbb{G}_1$ creates for cryptographic operations. The system generates two random generator denoted as g, u for encryption and calculate h parameter. Each s user encrypts his message m_s to C_{T_s} as his ciphertext.

4.1 Secure Sum Operation

Secure sum operation performs addition of all ciphertext. Each user s sends his ciphertext C_{T_s} to joint addition operation to generate $C_{T_{\mathrm{Sum}}}$ as the joint sum. Each user securely decrypts joint sum by his secret key q_{1s}. These individuals' secure decrypted values and again send for cumulative sum $C_{T_{\mathrm{FinalSum}}}$. By solving ECDLP for $C_{T_{\mathrm{FinalSum}}}$ with base g_1^q, and then process return final sum in plaintext form, which is equal to the sum of individual 's' users message 'm_s'.

4.2 Secure Multiplication Operation

Secure multiplication operation allows multiplication of two plaintext messages m_1 and m_2 in their ciphertext form C_{T_1} and C_{T_2} from cyclic group $\mathbb{G}$. Bilinear map h_1 is created by solving ECDLP for ciphertexts C_{T_1} and C_{T_2} with base 'g' and $\mathbb{G}_1$. This map creates pairing of two ciphertexts from group $\mathbb{G}$ to group $\mathbb{G}_1$, which is the most time-consuming operation. Addition of ciphertext, return as intermediate multiplication $C_{T_{\mathrm{Prod}}}$ from cyclic group $\mathbb{G}_1$. This intermediate result $C_{T_{\mathrm{Prod}}}$ is securely decrypted by each user s using its secret key q_{1s}, and transfer to cumulative addition $C_{T_{\mathrm{FinalProd}}}$ is the final multiplication of ciphertexts, by solving ECDLP for $C_{T_{\mathrm{FinalProd}}}$ with base g_1^q and $\mathbb{G}_1$. This is the final multiplication of C_{T_1} and C_{T_2} in plaintext form is equal to the multiplication of messages m_1 and m_2.

4.3 Reduce the Complexity of ECDLP

To solve ECDLP in finite time, researcher uses Pohlig-Hellman [17] attack, which effectively reduces the discrete logarithm computation of $l = \log_P Q$ in prime order subgroups of $\langle P \rangle$. Pollard's kangaroo [16] is used to solve discrete logarithms of both prime and composite order subgroups. Pollard's kangaroo attack finds discrete logarithm in integer range. Standard techniques are used Hasse-interval or complete range of elliptic curve order. These both techniques give a wide range to search discrete logarithm value for given elliptic points.

A customized range approach keeps interval in the limited range of integer values. In all computations, the user requires to solve ECDLP to obtain the final result. Solving ECDLP needs to perform 'mod' operation on the intermediate result with the order of the base point. Therefore, researcher keeps range for Pollard's kangaroo in between 0 and N, where N is the order of base point P. By using parallel instances of Pollard's lambda method, then again go for partition in the range of different intervals of equal size. For each instance of Pollard's lambda, final range is very small as compared to original range of zero to order of elliptic curve. It helps to reduce time requirement in finding discrete logarithm of specific intervals. Discrete logarithm value will be found by any one instance of Pollard's lambda. Then stop all other instances after finding discrete logarithm, so that CPU and memory resources occupied by these instances immediately will get free for next computations. This customization helps to solve 78-bit ECDLP in 88-bit prime field in finite time on the single contemporary computer. Algorithm given below gives details of proposed multiuser BGN cryptosystem with Pollard's lambda range partition.

4.4 Multiuser BGN Cryptosystem Algorithm with Pollard's Lambda Range Partition for ECDLP

BGN parameter generation

- Select q_1 and q_2 τ-bit prime numbers
- Compute $N = q_1 * q_2$
- Create elliptic curve $y^2 = x^3 + 1 \bmod p$ where prime $p = lN - 1$, l is small integer, wherein $p = 2 \bmod 3$
- Generate two cyclic groups $\mathbb{G}$ and $\mathbb{G}_1$order N and bilinear map $e\colon \mathbb{G} * \mathbb{G} \rightarrow \mathbb{G}_1$.
- Select two random generators $g, u \in \mathbb{G}$, calculate $h = u^{q_2}$.

- Publish $P_{bk} = (N, \mathbb{G}, \mathbb{G}_1, e, g, h)$ as public key.
- Each user s receives his/her secrete key q_{1s} from the system by randomly splitting system master key q_1 as:
  For Z users $q_1 = (q_{11} + q_{12} + \cdots + q_{1s})\ mod\ N$, wherein q_{1s} randomly distributed among $s \in [1, Z]$. System module passes q_1 master key.

Encryption
User s selects message m_s, where $0 \le m_s \le q_2$ encrypt his message m_s, by selecting arbitrary integer $r_s \in (0, 1, 2, 3, \ldots, N)$ and produce ciphertext C_{T_s} as: $g^{m_s} h^{r_s}$.

Secure Sum

- Secure sum module performs addition of all ciphertext as:$C_{T_{\text{sum}}} = \prod_{s=1}^{Z} C_{T_s}$
- Each user s calculates his $C_{T_{S\text{sum}}} = \left(C_{T_{\text{sum}}}\right)^{q_{1s}}$
- Secure sum module calculate:

$$C_{T_{\text{FinalSum}}} = \prod_{s=1}^{i=Z} C_{T_{S\text{sum}}} \tag{5}$$

- Solve ECDLP to receive

$$P_{T_{\text{sum}}} = g^{q_1 g} \text{discrete\_log}\, C_{T_{\text{FinalSum}}} \tag{6}$$

Secure Product

- Let us consider two ciphetexts $C_{T_{1s}}$ and $C_{T_{2s}}$ Calculate discrete logarithm of each ciphertext as: $\text{DLC}_{T_s} = \mathbb{G}\,\text{discrete\_log}\, C_{T_s}$
- Construct bilinear map $h_1 = e(g, h)$ and $e(C_{T_{1s}}, C_{T_{2s}})$.
- Calculate $C_{T_{\text{Prod}}} = \left((\mathbb{G}_1)^{\sum_{s=1}^{Z} \text{DLT}_s}\right)^{rh_1}$ where r is $0 \le r \le N$.
- Each s user calculates $C_{T_{S\text{Prod}}} = \left(C_{T_{\text{sum}}}\right)^{q_{1s}}$.
- Secure product calculate:

$$C_{T_{\text{FinalProd}}} = \prod_{s=1}^{i=Z} C_{T_{S\text{Prod}}} \tag{7}$$

- Final multiplication of ciphertexts is obtained by calculating discrete logarithm as

$$P_{T_{\text{prod}}} = \mathbb{G}_1^{q_1 g n^2} \text{discrete\_log}\, C_{T_{\text{FinalProd}}}, \quad \text{where } n \stackrel{R}{\leftarrow} \mathbb{Z}_N \text{ and } g = \mathbb{G}^n \tag{8}$$

Solve ECDLP using Pohlig-Hellman and Range Partition Pollard's Lambda Search

- Pohlig-Hellman attack factorizes composite order P_{Order} of base point P for the discrete logarithm of Q in prime order subgroup.

$$P_{\text{Order}} = p_1 e^1 * p_2 e^2 * p_r e^r \quad (9)$$

- Each factor in Eq. (9), $p_r e^r$ independently solve by Pollard's lambda attack by splitting $p_r e^r$ value in J partitions with each partition in the intervals 0 to $p_r e^r / J$ to respective running instances of Pollard's lambda method.
- This customization reduces interval range of Pollard's lambda method, returns discrete logarithm of the subgroup in less time. At the end, using Chinese remainder theorem (CRT), Pohlig-Hellman finds the unique solution as the final discrete logarithm.

5 Results and Discussion

5.1 *Experimental Setup*

Multiuser BGN cryptosystem with range partition approach for Pollard's lambda algorithm implemented it using SageMath 8.3 under Ubuntu 16 environment. Intel Xeon machines with ten cores and 16 GB RAM is used for experiments. Prime numbers q_1 and q_2 are used in the interval between 32-bit to 43-bit size to create prime fields of 64-bit to 88-bit. BGN cryptosystems allow message space $\leq q_2$. Therefore, message space half of q_2 is required to keep our multiplication results less than q_2.

5.2 *Result Analysis*

Pollard's lambda method is a space-efficient attack for ECDLP as like Pollard's rho version. The main strength of Pollard's lambda algorithm is to solve ECDLP for composite order numbers. But to solve ECDLP, time requirement of Pollard's lambda is very high. This is due to the wide interval bound to find ECDLP. Pollard's lambda method, while solving discrete logarithm for 30–40 bit numbers, takes time in minutes to hours [13].

Customization of Pollard's lambda method is found successful in solving ECDLP problem for BGN addition from 32-bit to 43-bit numbers in minutes. Multiplication operation requires to solve two discrete logarithms of nearly equal to prime field

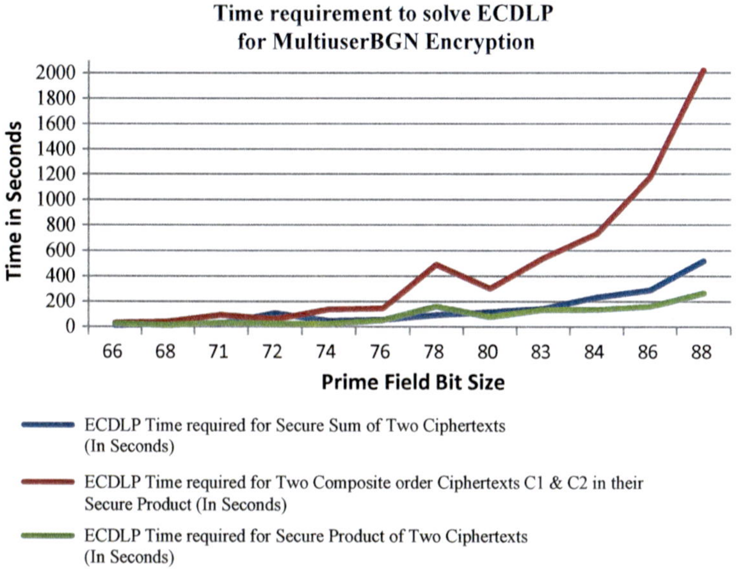

Fig. 2 Time requirement to solve ECDLP (66-bits to 88-bits prime field)

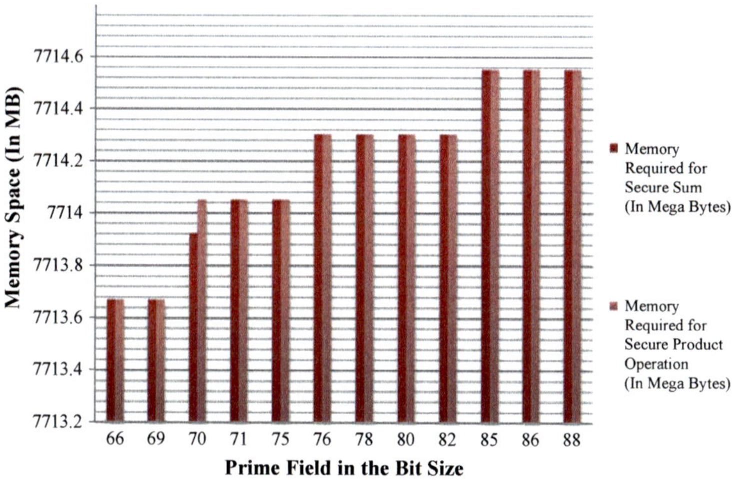

Fig. 3 Memory usage to solve ECDLP (66-bits to 88-bits prime field)

bit size. These discrete logarithms solving process takes time, but again two nearly 78-bit sizes discrete logarithm solves in minutes on a single contemporary computer. Outcomes are stated in Table 1. Figure 2 shows time required to solve ECDLP for 66-bit to 88-bit size prime field and Fig. 3 shows memory usage in solving ECDLP for a single 3.6 GHz Xeon computer with 16 GB RAM.

Table 1 Time and memory required in solving ECDLP for multiuser BGN system

Prime numbers q_1 and q_2 (bit size)	Elliptic curve prime field $E(\mathbb{F}_p) = 6 * (q_1 * q_2) - 1$ (bit size)	Time required to solve ECDLP for secure sum of two ciphertexts (s)	ECDLP solved for ciphertext C_1 (bit size)	ECDLP solved for ciphertext C_2 (bit size)	Time required to solve ECDLP for composite order ciphertexts C_1 and C_2 for their secure product (In s)	Time required to solve ECDLP for secure product of two ciphertexts (s)	Memory required for secure sum (MB)	Memory required for secure product operation (MB)
32	66	11.54	63	62	30.71	24.57	7713.67	7713.67
33	69	19.87	66	66	40.51	11.78	7713.67	7713.67
34	70	23.03	64	67	92.89	30.62	7713.92	7714.05
35	71	106.58	65	66	58.96	21.44	7714.05	7714.05
36	75	46.29	69	70	135.16	22.43	7714.05	7714.05
37	76	54.50	69	69	143.76	51.72	7714.30	7714.30
38	78	90.79	72	73	488.21	159.41	7714.30	7714.30
39	80	115.10	73	73	301.26	75.69	7714.30	7714.30
40	82	143.89	72	72	537.71	133.38	7714.30	7714.30
41	85	232.60	75	76	728.53	134.70	7714.55	7714.55
42	86	288.56	74	75	1174.38	160.12	7714.55	7714.55
43	88	517.42	77	78	2019.92	265.64	7714.55	7714.55

6 Conclusion and Scope for Future Study

The success of BGN type elliptic curve cryptosystems depends on solving ECDLP in finite time. Since invention, researchers try to give solutions for ECDLP problem for BGN encryption. This research work proposes a customized approach to Pollard's lambda attack in solving ECDLP for BGN homomorphic encryption. On a single contemporary computer, researcher introduced solution, which returns results of secure sum and secure addition for multiuser BGN cryptosystem in finite time with Pollard's lambda attack. Our customization improves the performance of ECDLP up to solving ECDLP of 78-bit in the 88-bit prime field. As a future work, with advanced network or cloud computing, it can split ECDLP problem among multiple nodes. This will help to solve higher size ECDLP problem on larger key size.

References

1. Cranor L, Rabin T, Shmatikov V, Vadhan S, Weitzner D (2016) Towards a privacy research roadmap for the computing community. White paper, Computing Community Consortium. https://arxiv.org/abs/1604.03160
2. Zhang Q, Yang LT, Chen Z (2015) Privacy preserving deep computation model on cloud for big data feature learning. IEEE Trans Comput 65(5):1351–1362. https://doi.org/10.1109/TC.2015.2470255
3. Abdulatif A, Kumarage H, Khalil I, Atiquzzaman M, Yi Xun (2017) Privacy-preserving cloud-based billing with lightweight homomorphic encryption for sensor-enabled smart grid infrastructure. IET Wirel Sens Syst 7(6):182–190. https://doi.org/10.1049/iet-wss.2017.0061
4. Yuan J, Yu S (2014) Privacy preserving back-propagation neural network learning made practical with cloud computing. IEEE Trans Parallel Distrib Syst 25(1):212–221. https://doi.org/10.1109/TPDS.2013.18
5. Naehrig M, Lauter K, Vaikuntanathan V (2011) Can homomorphic encryption be practical? In: CCSW'11 Proceedings of the 3rd ACM workshop on cloud computing security. ACM, pp 113–124. https://doi.org/10.1145/2046660.2046682
6. Xu L, Jiang C, Wang J, Yuan J, Ren Y (2014) Information security in big data: privacy and data mining. IEEE Access 2:1151–1178. https://doi.org/10.1109/ACCESS.2014.2362522
7. Yao AC (1982) Protocols for secure computations. In: Proceeding 23rd annual symposium foundations of computer science (SFCS'82). IEEE Computer Society, Washington DC, USA, pp 160–164. https://doi.org/10.1109/sfcs.1982.69
8. Agrawal R, Srikant R (2000) Privacy-preserving data mining. In: SIGMOD'00 Proceedings of the 2000 ACM SIGMOD international conference on Management of data, vol 29, no 2. ACM, pp 439–450. https://doi.org/10.1145/342009.335438
9. Vaidya J, Clifton C (2004) Privacy-preserving data mining: why, how, and when? IEEE Secur Priv 2(6):19–27. https://doi.org/10.1109/msp.2004.108
10. Acar A, Aksu H, Uluagac AS, Conti M (2018) A survey on homomorphic encryption schemes: theory and implementation. ACM Comput Surv (CSUR) 51(4):A:1–A:35. https://doi.org/10.1145/3214303
11. Yi Xun, Paulet Russell, Bertino Elisa (2014) Homomorphic encryption and applications. Springer, Heidelberg
12. Vaidya J, Shafiq B, Fan W, Mehmood D, Lorenzi D (2014) A random decision tree framework for privacy-preserving data mining. IEEE Trans Dependable Secure Comput 11(5):399–411. https://doi.org/10.1109/TDSC.2013.43

13. Bansal A, Chen T, Zhong S (2011) Privacy preserving back-propagation neural network learning over arbitrarily partitioned data. Neural Comput Appl 20(1):143–150. https://doi.org/10.1007/s00521-010-0346-z
14. Boneh D, Goh E, Nissim K (2005) Evaluating 2-DNF formulas on ciphertexts. In: TCC'05 Proceedings of the second international conference on theory of cryptography. Springer, Berlin, Heidelberg, pp 325–341. https://doi.org/10.1007/978-3-540-30576-7_18
15. Herbert V, Biswas B, Fontaine C (2018) Design and implementation of low-depth pairing-based homomorphic encryption scheme. J Cryptographic Eng 2018:1–17. https://doi.org/10.1007/s13389-018-0192-y
16. Pollard JM (1978) Monte Carlo methods for index computation (mod p). Math Comput 32(143):918. https://doi.org/10.1090/s0025-5718-1978-0491431-9
17. Pohlig S, Hellman M (1978) An improved algorithm for computing logarithms over gf(p) and its cryptographic significance. IEEE Trans Inf Theor IT 24(1):106–110. https://doi.org/10.1109/tit.1978.1055817
18. Koblitz N (1987) Elliptic curve cryptosystems. Math Comput 48(177):203–209. https://doi.org/10.1090/S0025-5718-1987-0866109-5
19. Miller V (1985) Use of elliptic curves in cryptography In: Williams HC (ed) Advances in cryptology—CRYPTO'85 Proceedings CRYPTO 1985, vol 218. Springer, Berlin, Heidelberg, pp 6–10. https://doi.org/10.1007/3-540-39799-x_31
20. Freeman DM (2010) Converting pairing-based cryptosystems from composite-order groups to prime-order groups. In: Gilbert H (ed) Advances in cryptology–EUROCRYPT 2010, vol 6110. Springer, Berlin, Heidelberg, pp 44–61. doi: 10.1007/978-3-642-13190-5_3
21. Eom SK, Lee HS, Lim S (2016) Message expansion of homomorphic encryption using product pairing. Electron Telecommun Res Inst (ETRI) J 38(1):123–132. https://doi.org/10.4218/etrij.16.0115.0630
22. Javheri SB, Kulkarni UV (2018) Design multiuser BGN encryption with customizing Pohlig-Hellman algorithm to solve ECDLP in the prime order subgroups. J Appl Sci Comput V(XII):650–657. doi: 16.10089.JASC.2018.V5I12.453459.1500179
23. Shanks D (1969) Class number, a theory of factorization and genera. In: Lewis DJ (ed) Symposium of pure mathematics, vol 20. State University of New York, Stony Brook, New York, pp 415–440. https://doi.org/10.1090/pspum/020
24. Hankerson D, Menezes A, Vanstone S (2004) Guide to elliptic curve cryptography. Springer, New York
25. Anuradha N, Katre SA (1999) Number of points on the projective curves $aY^{l} = bX^{l} + cZ^{l}$ and $aY^{2l} = bX^{2l} + cZ^{2l}$ defined over finite fields, l an odd prime. J Number Theor 77(2):288–313. https://doi.org/10.1006/jnth.1999.2382
26. Diffie Whitefield, Hellman ME (1976) New directions in cryptography. IEEE Trans Inf Theor 22(6):644–654. https://doi.org/10.1109/TIT.1976.1055638
27. Van Oorschot PC, Wiener MJ (1999) Parallel collision search with cryptanalytic applications. J Cryptology 12(1):1–28. https://doi.org/10.1007/PL00003816
28. Okamoto T, Uchiyama S (1998) A new public-key cryptosystem as secure as factoring. In: Nyberg K (ed) Advances in cryptology—EUROCRYPT'98, vol 1403. Springer, Berlin, Heidelberg, pp 308–318. https://doi.org/10.1007/bfb0054135
29. Paillier P (1999) Public-key cryptosystems based on composite degree residuosity classes In: Stern J (ed) Advances in cryptology—EUROCRYPT'99, vol 1592. Springer, Berlin, Heidelberg, pp 223–238. https://doi.org/10.1007/3-540-48910-x_16

Internet Addiction Predictor: Applying Machine Learning in Psychology

S. N. Suma, Poornima Nataraja, and Manoj Kumar Sharma

Abstract This work is an effort to exploit the unique ability of supervised machine learning model to explore how causal systems have an influence on Internet addiction disorder occurrence. The paper aims to pursue the possibilities of predicting Internet addiction based on a set of predictor variables. Here, the predictor variable set is selected such that there exists a strong relationship between the parameters considered to have an influence toward problematic Internet usage. Healthcare sector data always poses a challenge to the researchers in terms of unbalanced size of class representative samples available for the study. This kind of unbalanced dataset does affect the efficiency of the machine learning model to learn unbiased and to predict the unseen/test data accurately. Further, to this challenge, we propose to make a study on understanding the effect of an unbalanced dataset on the efficiency and performance of the machine learning model.

Keywords Supervised learning · Random forest model · Internet addiction · Unbalanced data · Upsample · SMOTE · Performance

1 Introduction

Internet technology is now an inevitable part of every person's life. The easy access and utility across multiple fields have made this as one of the most central components in various activities that humans perform. The ubiquity of Internet technology has its applications in the fields of education, research, health care, entertainment, commerce

S. N. Suma (✉)
Anand Diagnostic Laboratory (A Neuberg Associate), Bengaluru, India
e-mail: snsuma112@gmail.com

P. Nataraja
Dayananda Sagar College of Engineering, Bengaluru, India
e-mail: poorni.na@gmail.com

M. K. Sharma
SHUT Clinic, NIMHANS, Bengaluru, India
e-mail: shutclinic@gmail.com

N. N. Chiplunkar and T. Fukao (eds.), *Advances in Artificial Intelligence and Data Engineering*, Advances in Intelligent Systems and Computing 1133,
https://doi.org/10.1007/978-981-15-3514-7_36

and communication. However, Internet addiction has become an undesired offshoot of technological revolution on the Web and a well-known problem today. Excessive use of social media is leading to an increase in the number of people suffering from depression, anxiety, stress and other psychological issues which were not prevalent in the previous generations. Children and young adults are getting hooked to these devices and falling prey to many ailments which are uncommon for their age.

The prevalence of Internet addiction among the youth across India and globe is high due to the explosive growth in e-technology, e-commerce, e-learning, e-banking, etc. Excessive usage of Internet can manifest into emotional and physical problems, which can cause problematic psychological and physiological repercussions leading to clinical and professional interventions.

2 Aim of the Study

This study proposes to build a classification model which can predict the plausibility of the subject being Internet-addicted or not. The study is based on multiple associated features like stress, anxiety and depression levels, number of siblings, browsing in Incognito mode, etc. During our preliminary investigation, we observed that the performance of the classification model was inconsistent with the training and test dataset, thus leading us to explore techniques for balancing the dataset to build an efficient and consistent high-performance supervised learning model.

3 Literature Survey

Internet addiction disorder is radically increasing across the globe, and the studies by psychologists and psychiatrists have shown that Internet addiction is positively related to anxiety, depression and greater levels of loneliness [1–3]. Poor emotional skills and poorer social adaptation [4], lower level of self-esteem [5], decreased social adoption and emotional skills [4], considerable decrease in self-esteem [5] and psychological well-being [6] have been observed in association with Internet addiction. Sharma A [7] did a study on university students and the findings show that Internet addiction is strongly related to psychological well-being of the students.

In India, there are statistical studies carried out on Indian university students which show noticeable positive relation between Internet addiction and depression, stress, anxiety [8, 9]. Extended period of Internet usage per week, constant exposure to online contents, depression, stress and anxiety were noticed to be independent predictors of Internet addiction by Gupta [10].

Machine learning can be a powerful tool in preventive health care, especially in psychological disorders like Internet addiction, online gaming addiction, etc. The scholars Singh and Babbar [11] have used Bayesian network in their study to predict the probability of Internet addiction disorder occurrence. In many of the studies, depression has been one of the major predictors of Internet addiction. Studies show that Internet addiction is positively correlated to anxiety, depression and stress. Impulsive traits, compulsive traits and symptomology have been used to build different machine learning models for problematic Internet usage [12].

4 Methodology

4.1 *Procedure*

The proposed study was conducted among the college students pursuing graduation program in Bangalore. The students who have been using Internet for at least the past one year and willing to participate in the study were chosen. The selected 107 participants were appraised of the purpose behind the study. Psychometric questionnaires along with demographic details questions were administered to them.

The required data was collected by administering the following psychometric questionnaires—Depression, Anxiety and Stress Scale (DASS 21) by Fernando Gomez and Young's Internet Addiction Test. Demographic information was collected from each student such as age, gender, number of siblings, education qualification, annual income of the family and additional information like what all devices are used for accessing Internet? Do they share their device with family members? Do they browse in Incognito mode?

Young's scale of Internet addiction is used for screening Internet addiction disorder. This is one of the most reliable scales used for evaluating levels of Internet addiction [13]. Young's scale of Internet addiction is a 20-item questionnaire whose responses are based on Likert scale. Participants must respond to each question with a number between 1 and 5 of the five-point Likert scale continuum. Alavi et al. [14] in their study have proved the reliability and validity of this test with their following findings—internal consistency (Cronbach's $\alpha = 0.88$) and test–retest reliability ($r = 0.82$) [14]. Participants were classified based on sum of item scores. A score of 21–49 was classified as not addicted and 50–100 as addicted.

DASS 21 is a self-reporting questionnaire consisting of 21 items and is designed to assess the severity of the symptoms of anxiety, depression and stress. In our present study, DASS 21 scores are used as predictors.

4.2 *Supervised Learning*

Supervised learning is a type of machine learning where the model learns to classify or regress from the training data which has a set X of input variables and the output variable Y. The cognitive algorithm learns to map these input variables (X) to the desired output (Y). The goal is to build a model which, given an unknown dataset, can predict Y using the independent input variables. There are a variety of supervised algorithms, viz. decision tree, ensembled random forest, logistic regression, neural networks, Bayesian networks, K-nearest neighbor (KNN) and support vector machines, etc., which have been developed to achieve this task. In our present study, we have built a supervised classification model called random forest model (RFM). **Supervised Classifier: RFM** "Random forest is an ensemble of unpruned classification or regression trees, induced from bootstrap samples of the training data, using random feature selection in the tree induction process" [15]. The algorithm can be used for both numeric as well as categorical data. The algorithm can be used for both regression as well as classification. The algorithm uses aggregation of the majority votes for classification and average of the predictions for regression. The random forest algorithm is very simple to understand and also user friendly with only two hyperparameters which can be used to tune the algorithm, viz. total number of decision trees in the forest (*ntree*) and the number of randomly sampled variables at each split (*mtry*).

The performance of random forest algorithm is substantially higher than the single decision tree algorithms like C4.5, CART, etc., because of its robustness to noise. Even when the numbers of trees are more in the forest, the algorithm does not overfit. It can handle more features and can provide the importance of features in the prediction, which makes feature engineering very easy and simple. The pseudocode for RFM is as follows:

Pseudocode for RFM creation

1. Select k features from the input feature set X having m members such that $k < m$.
2. Using the best split for the randomly selected k features, find the root node.
3. Split the trees using the root node and create left and right children.
4. Repeat steps 1–3 until no more possible splits are possible.
5. Repeat steps 1–4, for the specified number of trees to be built.

Pseudocode for RFM prediction

1. Read test data features, use the rules decision trees to predict. Also, store the outcome.
2. Compute votes for each predicted outcome.
3. Predict the class of the test data as the class with majority votes from the forest of decision trees.

4.3 Data Set

The dataset consists of 107 instances. The data consists of both male and female gender (female: 30% and male: 70%) of the age group between 17 and 25 years. 57 of them share their device with others, and 50 of them do not share their devices. 39 subjects sometimes activated Incognito mode on browser, 13 used it often, 45 subjects never used it and 10 of them always used it while using the browser.

Of the 107 instances, 81 are found to be not addicted, 26 are addicted users. Table 1 depicts the summary of the total addicted and not-addicted users, and Table 2 depicts the scores and interpretation of depression, anxiety and stress for the subjects.

5 Results and Discussions

The results of our work mainly demonstrate the applicability of machine learning models in the field of psychology. The findings set the hope and encourage us to conduct more extensive work in the same area, which can lead to the development of a predictive tool for plausible Internet addiction. The outcome of the study also proves the need of a balanced dataset in building predictive models. Additionally, it evaluates the performance of the machine learning models built with unbalanced data as well as with data balanced using different data balancing methods.

The dataset was split into two parts in the ratio 80:20. Eighty percent of the dataset was used as training set and twenty percent as test set. The training dataset serves to train and build the machine learning model, whereas the test dataset serves to evaluate and validate the learnt model. The composition of the categories of Internet addiction in the training and test datasets is as shown in Table 3.

Initially, RFM was built using all the predictor variables. R programming language code has been written to build and test the model. After training the model using random forest, importance of variables was studied using accuracy-based importance measure and Gini-based importance measure. The mean decrease accuracy is the

Table 1 Number of addicted and not-addicted users

	Not addicted	Addicted	Total
Number of subjects	81	26	107

Table 2 Scores and interpretations of depression, anxiety and stress for the subjects

Interpretation	Depression	Anxiety	Stress
Normal	45	41	66
Mild	16	4	10
Moderate	27	21	17
Severe	8	10	7
Extremely severe	11	31	7

Table 3 Training and test dataset

	Addicted	Not addicted
Training dataset	21	64
Test dataset	5	17
Total	26	81

measure of how much the accuracy decreases when the variable is excluded from the model. The mean decrease Gini is the measure of the decrease of Gini impurity when a variable is chosen to split a node. Higher the value, higher the contribution to decrease the overall mean squared error of the model. Variables with high importance are natural drivers of the classification, and they will have a significant impact on the model's accuracy. Hence, the top eight important variables were chosen to build a model that is simpler and faster to fit and predict. Table 4 depicts the importance of variables for the respective classes, mean decrease accuracy and mean decrease Gini value of each of the variables.

Once again RFM was built using the six important predictor variables shown in Fig. 1. The RFM was built with 500 trees and two variables tried at each split. The out-of-the-bag (OOB) estimation of error rate was 18.92%. The confusion matrix of the training dataset is shown in Table 5.

Testing dataset which is unknown to the trained model was predicted, and the predictions were compared to the actual classification. The results are given in Table 6.

The confusion matrix of both test and training dataset shows that the model's accuracy in learning to predict not-addicted users is good with a class error of 6% and 13% in training and test dataset respectively, whereas the model's accuracy in learning to classify addicted users is poor and the class error is 57% in both training and testing dataset. This can be attributed to the bias in learning due to unbalanced training dataset. Learning problems of this type are characterized by the uneven proportion of cases that are available for each class of the problem. The purpose of

Table 4 Variables importance table

Variables	Addicted	Not addicted	Mean decrease accuracy	Mean decrease Gini
Siblings	1.478	0.673	1.406	2.164
Gender	2.197	−1.144	0.134	0.530
Annual income	5.914	−3.381	0.864	2.655
Device sharing	9.289	−0.632	5.562	1.547
Incognito	3.764	1.852	3.411	2.476
Depression	9.908	7.428	11.345	6.031
Anxiety	6.019	5.355	7.203	5.267
Stress	12.132	7.661	12.511	5.068

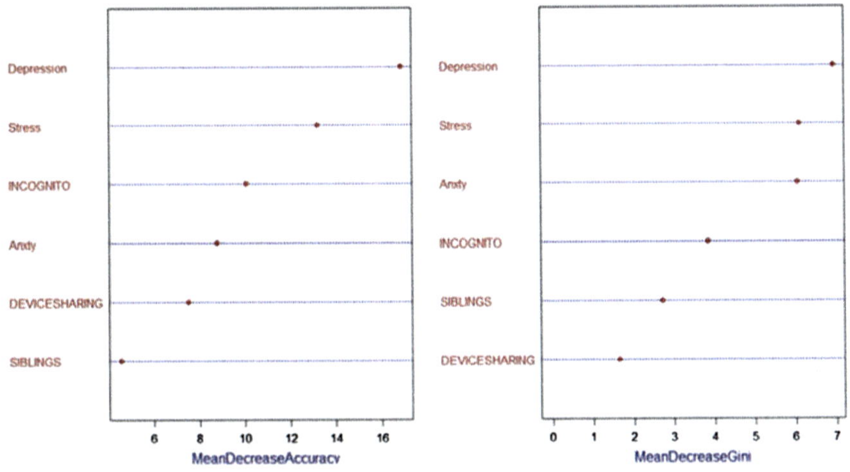

Fig. 1 Graphical representation of variables importance

Table 5 Confusion matrix for training dataset

	Addicted	Not addicted	Class. error
Addicted	8	11	0.5789
Not addicted	4	62	0.0606

Table 6 Confusion matrix for test dataset

	Addicted	Not addicted	Class. error
Addicted	3	2	0.5714
Not addicted	4	13	0.1333

RFM is to reduce the overall error rate. Hence, the focus of the model will be more towards accuracy of majority class prediction. However, this focus often tends to yield poor accuracy in minority class. The brute force way of solving the problem of unbalanced dataset is to balance the size of each of the different classes in the training dataset. This can be achieved by collecting more samples or by using any of the data balancing methods like oversampling the minority class, downsampling the majority class or combination of both.

The problem of unbalanced dataset was addressed by applying the data stratification or data balancing methods available in R programming language, viz. upsample and SMOTE. Upsample function samples with replacement of minority class to make the class distributions equal. SMOTE function generates a new "SMOTE" dataset that addresses the class unbalance problem by artificially generating more samples of minority class and randomly selecting lesser number of majority class representatives. The downsampling of majority class method creates the training set by randomly selecting the records from majority class equal to the size of minority class resulting in loss of information. Since the minority class representation is small

Table 7 Confusion matrix for training dataset predicted by SMOTE model

	Addicted	Not addicted	Class. error
Addicted	49	11	0.0833
Not addicted	5	75	0.0625

Table 8 Confusion matrix for test dataset predicted by SMOTE model

	Addicted	Not addicted	Class. error
Addicted	7	1	0.000
Not addicted	0	14	0.0667

Table 9 Confusion matrix for training dataset predicted by upsample model

	Addicted	Not addicted	Class. error
Addicted	64	2	0.0303
Not addicted	7	59	0.1060

Table 10 Confusion matrix for test dataset predicted by upsample model

	Addicted	Not addicted	Class. error
Addicted	3	3	0.5714
Not addicted	4	12	0.2000

in our study, we choose not to use this method. RFMs were built with the balanced training dataset using the above-mentioned methods to compare the classification performance of the models.

SMOTE method. This method was applied to the training dataset as in Table 3. The training sample after SMOTE consists of 60 addicted and 80 not-addicted cases. The model was trained with this balanced dataset. The RFM was built with 500 trees and two variables tried at each split, and the OOB estimation of error rate was 11.43%. Tables 7 and 8 depict the confusion matrix of the training and test datasets respectively.

Upsample method was applied to the same training set as in Table 3. The training set after upsampling consists of 60 addicted and 60 not-addicted cases. The model was trained with this balanced dataset. The RFM was built with 500 trees and two variables tried at each split, and the OOB estimation of error rate was 6.82%. The confusion matrix of the training dataset and testing data set is depicted in Tables 9 and 10 respectively.

5.1 Performance Measurement

The most widely used metrics for evaluating performance of classification models are all functions of the confusion matrix. The metrics are precision, recall, true positive rate, true negative rate, G-mean, weighted accuracy and F-score. For any

Table 11 Definition of confusion matrix

	Positive class (prediction)	Negative class (prediction)
Positive class (actual)	True positive (TP)	False negative (FN)
Negative class (actual)	False positive (FP)	True negative (TN)

classification model, there is always a trade-off between true negative rate and true positive rate. Similar trade-off is true for precision and recall. Weighted accuracy is often used in situations where the data is highly imbalanced and the need is to predict the minority class with high prediction accuracy while maintaining the accuracy of majority class reasonably well. F-score is the measure of accuracy of the test, as the harmonic mean of recall and precision and G-mean is the geometric mean of recall and precision.

In this study, equal weights have been used for both true negative rate and true positive rate, i.e., beta is set to 0.5. The confusion matrix and the performance metric for comparison are defined in Table 11 and the subsequent formulae:

$$\text{True Negtive Rate (TNR)} = \frac{\text{TN}}{\text{TN} + \text{FP}}$$

$$\text{True Positive Rate (TPR)} = \frac{\text{TP}}{\text{TP} + \text{FN}}$$

$$G - \text{mean} = (\text{TNR} \times \text{TPR})^{1/2}$$

$$\text{Weighted Accuracy (WA)} = \beta \times \text{TPR} + (1 - \beta) \times \text{TNR}$$

$$\text{Precision} = \frac{\text{TP}}{\text{TP} + \text{FP}}$$

$$\text{Recall} = \frac{\text{TP}}{\text{TP} + \text{FN}}$$

$$\text{F-Score} = \frac{2 \times \text{Precision} \times \text{Recall}}{\text{Precision} + \text{Recall}}$$

Comparison of performance of the RFMs on the training set and the test set is depicted in Tables 12 and 13 respectively.

Though performance on training dataset is one of the metrics to evaluate the performance of the model, nothing conclusive can be said just based on this. This is because an overfit model may show higher performance accuracy on training data but may show lower performance while predicting the test data. Hence, it is always advisable to consider the model's performance on the test dataset as the gold standard for comparing the performance of multiple models. We can infer the same

Table 12 Performance comparison on training dataset

Data balancing methods	Precision	Recall	TPR	TNR	G-mean	F-score	WA
Unbalanced	66.66	72.72	72.72	93.93	82.64	69.55	83.32
SMOTE	90.74	81.66	81.66	93.75	87.49	85.96	87.70
Upsample	90.14	96.96	96.96	89.39	93.09	93.42	93.17

Table 13 Performance comparison on test dataset

Data balancing methods	Precision	Recall	TPR	TNR	G-mean	F-score	WA
Unbalanced	42.85	60.00	60.00	76.47	67.73	49.99	68.23
SMOTE	100	87.50	87.50	100	93.54	93.33	93.75
Upsample	42.85	50.00	50.00	75.00	61.23	46.15	62.5

from Tables 12 and 13, where the performance of the unbalanced and upsample models is higher for the training dataset and is not consistent with the testing dataset.

From the comparison Table 13, it is evident that it not just important to balance the unbalanced data, but it is also important to choose the right data balancing technique that gives the best performance for the given problem. The model trained using data balancing using SMOTE method shows consistency in its performance on the training as well as test dataset and is proved to be the best model among the three.

6 Conclusion

The evolution of technology is mostly intended towards the improvement of lifestyle and efficiency of humankind. However, it is also known to have a long-lasting psychological and physiological implications. These go mostly unnoticed during the initial phases of development and are prominent and observable only at the later stage when an intervention is necessitated.

The outcome of the study proves the applicability of machine learning in psychology. An improved and more robust model built and validated using a larger dataset can be used by the psychologists and other practitioners as preventive healthcare tool to identify plausible Internet addiction in an early stage and guide the users for healthy use of technology as well as management of psychological issues.

The study also helped us to note that like any other supervised learning model, the performance of RFM suffers because of unbalanced training data set. The pilot study helped us in understanding, implementation and evaluation of data balancing methods for building the supervised machine learning models for psychological data. The RFM built using the data set balanced by SMOTE method has proved to be the best performing model amongst different models built. We propose to improve accuracy and reliability of the results obtained by increasing the sample size and also by comparing the RFM with other suitable supervised learning models.

References

1. Young KS (1998) Caught in the net: how to recognize the signs of internet addiction and a winning strategy for recovery, 196
2. Moody EJ (2001) Internet use and its replationship to lonines. CyberPsychology Behav 4:393–401
3. Yen J-Y (2007) The comorbid psychiatric symptoms of internet addiction: attention deficit and hyperactivity disorder (ADHD), depression, social phobia, and hostility. J Adolesc Health 41(1):93–98
4. Engelberg E, Sjöberg L (2004) Internet use, social skills and adjustment. Cyberpsychology Behav 7:41–47
5. Kraut R, Patterson M (1998) Internet paradox. A social technology that reduces social involvement and psychological well-being? Am Psychol 53:1017–1031
6. Yoo YS, Cho O (2014) Associations between overuse of the internet and mental health in adolescents. Nurs Health Sci 16:193–200. https://doi.org/10.1111/nhs.12086
7. Sharma A (2018) Internet addiction and psychological well-being among college students: a cross-sectional study from Central India. J Family Med Primary Care 7(1):147–151
8. Goel D, Subramanyam A, Kamath R (2013) A study on the prevalence of Internet addiction and its association with psychopathology in Indian adolescents. Indian J Psychiatry 55:140–143. https://doi.org/10.4103/0019-5545.111451
9. Kawa SA (2015) A study of internet addiction and depression among university students. Int J Behav Res Psychol 3(4):105–108
10. Gupta A, Khan A (2018) Internet addiction and its mental health correlates among undergraduate college students of a university in North India. J Family Med Primary Care 7(4):721–727
11. Singh A, Babbar S (2018) Detecting internet addiction disorder using Bayesian networks. In: Communications in computer and information science, vol 799. https://doi.org/10.1007/978-981-10-8527-7_8
12. Ioannidis K, Chamberlain SR (2016) Problematic internet use (PIU): associations with the impulsive-compulsive spectrum An application of machine learning in psychiatry. J Psychiatr Res 83:94–102
13. Young KS, Rogers R (1998) The relationship between depression and internet addiction. Cyber Psychol Behav 1:25
14. Alavi SS, Eslami M, Meracy M, Najafi M, Jannatifard F, Rezapour H (2010) Psychometric properties of young Internet addiction test. Int J Behav Sci 4:183–189
15. Breiman L (2001) Mach Learn 45(1):5–32. https://doi.org/10.1023/A:1010933404324

An Approach Toward Stateless Chatbots with the Benefit of Tensorflow Over Spacy Pipeline

Chaithra, Roshan Fernandes, Anisha P. Rodrigues, and Venkatesh

Abstract Conversational bots are assuming control over the business space. Consistently, individuals send in excess of a billion messages to organizations and associations through numerous messaging applications. The most usual method for building chatbots is the state machine approach which usually involves creating distinctive states and dependent on some logic invoking actions. But, with the increase in the number of states, lot of rules are required to be added, with additional logic, and hence create a delicate code which is difficult to keep up and maintain. In this work, we have shown how the stateless approach of building a chatbot using Rasa Core eliminated the need of complex state machine approach, as it makes use of machine learning-based dialog management. Along with this, we trained the model separately with the two pipelines "spacy" and "Tensorflow embedding". We evaluated the two pipelines with cross-validation and without cross-validation for intent classification. The result showed that, with more training examples, Tensorflow embedding shows good accuracy for intent classification.

Keywords Natural language understanding · Tensorflow embedding · Spacy

1 Introduction

State-full bots are more modern than bots that carry on like computerized telephone menus. This influences the utilization of sate machines, where we make use of diverse

Chaithra (✉) · R. Fernandes · A. P. Rodrigues
NMAM Institute of Technology, Nitte, India
e-mail: chaithrasa02@gmail.com

R. Fernandes
e-mail: roshan_nmamit@nitte.edu.in

A. P. Rodrigues
e-mail: anishapr@nitte.edu.in

Venkatesh
Alva's Institute of Engineering and Technology, Moodbidri, India
e-mail: venkateshbhat2007@gmail.com

N. N. Chiplunkar and T. Fukao (eds.), *Advances in Artificial Intelligence and Data Engineering*, Advances in Intelligent Systems and Computing 1133,
https://doi.org/10.1007/978-981-15-3514-7_37

states, rules, and activities. However, as the quantity of states builds, levels of nested rules also get increased, which makes it harder to keep up and manage. In any case, Rasa platform gave a best approach of building stateless bots, eliminating scaling issues.

Rasa Core performs dialog management, which implies monitoring a conversation and choosing how to continue. Rasa Core's basic role is to construct logical and layered discussions. To have a genuine conversation, we need some memory and expand on things. Rasa Core gives us a chance to do that scalably.

Rasa [1] Core handles start to finish conversation stream. All things considered, when we send any sentence to chatbot, it first goes through Rasa natural language understanding (NLU), which is a natural language processing tool. This NLU interpreter classifies the intent and extracts the entities out of that sentence. Now, RASA Core provides the guidance further and makes the responses.

RASA Core handles end-to-end chat flow. When we send any sentence to chatbot, it first passes through any NLU interpreter. This NLU interpreter classifies the intent and extracts the entity out of that sentence. Now, RASA core comes into role. It gives the direction further and creates the user response.

2 Literature Survey

The authors in research paper [2] developed a student bot project. Few artificial algorithms are used to analyze user's message. As they have made use of state-full approach, there were only linear conversation routes, and complicated branching paths are not handled.

In research paper [3], the authors have built a chatbot which incorporates a number of rules and a rules engine for controlling message delivery to users. Rules stored in the rules database, or the rules engine itself, incorporate different sorts of rules, including "when" and "if-then" type rules. In the view of these rules, rules engine decides the state of relevant conditions. This can be very time consuming, and it takes lot of memory.

The authors in research paper [4] developed a set of active probing techniques to detect stateless botnet conversations. According to their idea, there will be a clear command-response pattern with the typical interaction of botnet. Hence, stateless bot can show deterministic character to dialog replays, whereas state-full bot interaction will behave in a nondeterministic manner.

The authors in research paper [5] eliminated the need of checking the validity of runtime inputs by using XML documents and consequently attempted to wipe out longer development cycles. But, instant messaging bots in their development had a necessity of creating a state transition diagram, which defines states, state transitions, and methods to be carried out upon each state.

As these state-full bots create a lot of development cycles, a little improvement is shown in research article [6] where they have created the conversational agent giving a stateless reactive dialog layer, with a first level in their rule hierarchy. Anyway, to give access to the content of the XML-based story database as well as to the state of the interaction, they extended the pattern processing with a rule engine, which was time consuming.

A chatbot was developed in research paper [7] using server-less platform. Stateless functions were developed to perform actions. Each invocation of stateless function is independent of the previous runs. Similarly, in research article [8], a social chatbot is been built with a “Storing” component taking the message object at the last stage of computation. The current output of the bot is saved into the DB. It also stores outputs of each component. “History” and “Storing” components allow for stateless execution of sessions keeping user context. It also provides the ability to scale on any number of AWS Lambda calls.

The authors in research paper [9] have shown how dialog management can support a more robust handling of context in closed domain (state-full context). But, they failed in providing a broader definition of context.

Researchers and developers building chatbots have realized that the vast majority of the state-full bots [10] require state machines which are not versatile. As the quantity of states builds, levels of nested rules also get increased. It may also require the logic for transition from one state to another. Hence with such methodologies, maintaining and updating of the code are made difficult.

The next strategy to overcome total stateliness is to utilize the reinforcement learning. Quite a bit of this has concentrated on utilizing reinforcement learning (RL) to fabricate frameworks which can learn complex tasks, simply by attempting again and again and accepting a reward on success. The straightforwardness of reinforcement learning is both quality and a shortcoming. System does not have to comprehend anything about its own behavior, just which activities result in a reward.

Then again, systems may settle on several choices such as, previously accepting a reward, which of those decisions aided and which requires a great deal of involvement. Reinforcement learning can turn into an important technology for the eventual fate of conversational AI, yet it is not a best methodology for the designers who need to construct a functioning dialog system. This is because reinforcement learning is data hungry for which we require a large number of conversations to adapt even simple patterns.

But, keeping track of conversations is done well with Rasa Core which is a great platform to perform dialog management. It supports the interactive learning, by making use of Rasa NLU for natural language understanding and Rasa Core.

3 Eliminating the State Machine Approach

Most bots available today are stateless. Rasa platform has given a way to build stateless bots, where scaling is not a problem. State-full bots are more sophisticated.

Consider an example of building a bot for buying a car as shown in Fig. 1. We have two states “car_details” and “payments_details”. If the customer or user asks “I want to book a car”, the control jumps to “car_details” state. Now, if user asks

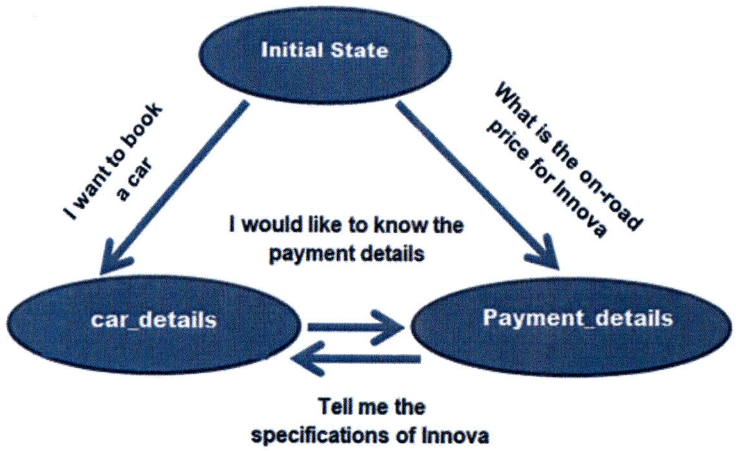

Fig. 1 A typical state machine flow

"ok I would like to pay" or "I would like to know the payment details", it jumps to the "payment_details" state. So here, we have some rules for switching between the states. Now, the user and bot conversation can be as follows:

User: "What are the different payment options?"
Bot: "You can go with debit cards, or NEFT option.
User: "Ok I would like to know the benefits".
Bot: "Through debit cards there is a discount of Rs. 1000, and through NEFT you can save upto 2k.

Now, the user can say "Wow, I would like to go with the second option" or "I would like to choose to pay through NEFT"; the control can jump to "choose" state or intent and can continue. But, if the user goes away from the happy path and asks "I want to go with the third option" when third option is not at all available, the logic can get more complicated.

Adding more levels of nested logic or adding even more rules to get in and out of the states can lead to complicated code which will be difficult to maintain. But, Rasa platform gave a best stateless approach, where scaling is not at all an issue.

4 Advantage with Stateless Bots Using Rasa Platform

With Rasa Core, bot responses can be specified manually. We call these as "actions". Greeting the user can be one kind of action, and another action can be to call an API, or query a database. Then, we can predict which action to take based on the history of a conversation after training a probabilistic model.

In interactive learning mode, we give step-by-step feedback on what our bot chose to do. It is sort of like reinforcement learning, however with feedback on each and every progression, as opposed to exactly toward the finish of the chat.

At the point, when our bot picks the wrong action, we reveal to it what the correct one would have been. The model updates itself promptly, so we are less inclined to experience a similar slipup again, and once we complete, the discussion gets logged to a document and taken as a training data.

Thus, this settles an edge case by giving step-by-step feedback. As opposed to a single reward toward the end, we are training the system substantially more specifically what is good and bad.

5 Few Terminologies in Rasa Core for Building a Bot

Interpreter: The main job of Rasa NLU interpreter is to parse the messages. The structured output indicating the intents and entities is the format to which message will be converted. This helps in carrying out natural language understanding.

Domain: The domain file is created in yml format. Domain file includes intents, entities, templates, slots, and actions.

Intents: Intention of the chatter text.

Entities: Information extracted from user's message.

Templates: These are the strings that indicate bot responses. The format for specifying a template string is with the format utter_<intent>. This indicates the utter actions for the bot.

Actions: This shows the utterances and activities of the bot. One action is just the utter messages from the bot; the other is some customized actions. Customized actions are referenced in domain file with some logic defined.

Slots: Slots are nothing but the bot's memory which can be user-defined variables. During a conversation, system keeps track of slots. For example, to book a flight, we should keep track of to_location, from_location, time, etc., so all such information will come under slots.

Stories: This will decide which action to be taken next. A probabilistic model is generated, in accordance with these stories, which helps bot in taking the next action.

Subsequently, the message arriving will be converted to structured output indicating the intents and entities. The conversation state is monitored by the tracker. It gets the information about the new message. At that point, dependent on domain and stories, the dialog model gets generated. Policy chooses about the next action. The tracker then logs up the picked action, hence sending the response back to the user.

Dialog Model: Training of model takes place based on the stories, in view of which the policy will make the action. The strategies by which stories can be created are supervised and reinforcement learning.

Supervised learning will make the narratives by hand, keeping in touch with them straightforwardly in a record. Though it is easy to create story manually, however, if there should be an occurrence of complex use cases, it is hard to face all situations.

In reinforcement learning, step-by-step feedback is given by the user for chosen action of the policy. Hence, this can also be called as interactive learning. Whenever an action is chosen by the policy, the user is been asked for the correctness of the chosen action; if it is wrong, action can be corrected and stories are stored to train the model once more. Hence, this aid in getting in the edge cases which are hard to write manually.

6 Designing the Flight Bot

6.1 Creating the Training Data

The following section tells about the initial stage of creating the training data for stateless bots using Rasa platform. It contains the user messages, the intents to which that particular user text can be classified to and the entities which can be extracted from the user text. Designing a Flight_Search_Bot, the intents might initially have

- greet
- inform_flight_details
- good_bye.

The user's text will be classified into any one of the above intents. For example, there are many ways that the user can ask about the flights,

- Could I know the availability of flights at 9.00 A.M?
- Name of the flights available to the Netherlands.

Rasa NLU will classify the user text to any of the intents. The words like time, location, etc., will be the entities. The model is hence built which can do the intent classification and entity extraction.

6.2 Domain File Creation

The following section tells about the universe that our bot behaves in. Domain includes slots, intents, entities, actions. It can also include the utter messages of the bot. As an example, the domain file for Flight_Search_Bot has the following yml definition,

```
slots:
  Time:
    type: time
To_Location:
   type: text
From_location:
  type: text

intents:
 - greeting
- inform_flight_details
 - good_bye

entities:
 - To_location
- From_location
- Time

templates:
  utter_greeting:
    - 'Hey Dude, what's up?'
  utter_good_bye:
    - 'See you Later'
    - 'Bye Take Care :('
  utter_ask_to_location:
    - 'your landing point please?'
utter_ask_from_location:
  -'Your boarding location?'
utter_ask_time:
  -'At what time you want to book your flight?'

actions:
 - utter_greeting
 - utter_good_bye
 - utter_ask_to_location:
- utter_ask_from_location:
- utter_ask_time:
 - actions.GetFlightResults
```

Slots can be used to store data given by the user (e.g., address, phone) and data gathered about the outside world.

Actions are what bot does or say in response to user input. For example, an action can be an utter message, or it can make an external API call, or it can query a database. There are three kinds of actions in Rasa Core:

- Default actions (action_listen, action_restart, action_default_fallback)
- Utter actions which are the utter messages by the bot
- The arbitrary code which is called the custom actions.

We can refer to the custom actions with their name. In the Flight_Search_Bot example, custom action is the GetFlightResults which is invoked if the user asks for the availability of flights with particular time and location.

To do slot filling, the bot collects few pieces of information from the user such as time, to_location, from_location in case of Flight_Search_Bot. In Flight_Search_Bot, if user asks for flight availability details, we need to know the to_location, from_location, and the time. When the user does not provide the required details, we will ask them for it.

6.3 Machine Learning Model Definition

The following section tells about choosing a pipeline to create a model based on training data. In Rasa NLU, various pipelines can be utilized to process user messages. In case of Flight_Search_Bot, we are utilizing the pre-characterized spacy_sklearn pipeline, alongside nlp_spacy, tokenizer_spacy.

Tensorflow_embedding and spacy_sklearn are the two most important pipelines. The spacy_sklearn pipeline utilizes pre-prepared word vectors from either GloVe or fastText, whereas Tensorflow embedding pipeline does not utilize any pre-prepared word vectors, however fits these specifically for your dataset.

For few training examples, spacy_sklearn would be better. Spacy_sklearn pipeline has a benefit, for example, if we have a training text like: "I want to buy bananas" and Rasa is asked to predict the intent for "get fruits", and our model already knows that the words "bananas" and "fruits" are very similar. This comes to benefit if there is less training data.

By using Tensorflow_embedding pipeline, the benefit is that word vectors are modified for our domain. For example, the word "switch" is closely related to "shift" or "change" in general English, however altogether different to "wire". In electrical domain, "switch" and "wire" are firmly related. We would like our model to catch that.

6.4 Machine Learning Model Training

Once the training data and the configuration file with the pipeline are ready, the model will be trained with the command rasa_nlu.train as,

!python -m rasa_nlu.train -c config.yml --data data.json -o models --fixed_model_name weathernlu --project current --verbose

The second option is to create a Python file, in which we provide the path for training data file, configuration file, and model path. To utilize the new model in Python, we can pass a user text to its parse () method of the interpreter object.

6.5 Writing Stories

Once the model is trained, stories can be provided by giving out the intents and corresponding actions, which will help the bot to show the general pathway. "stories" for Flight_Search_Bot can be provided in the following alternative ways.

```
## story 1
* inform_flight_details {"to_location": "Italy", "from_location": "India", time: "9.00"}
  - action_ get_flight_results
## story2
* greeting
  - utter_greeting
* inform_flight_details
  - utter_ask_to_location
* inform_flight_details{"to_location": "Italy"}
  - utter_ask_from_location
* inform_flight_details{"to_location": "Italy"}
  - utter_ask_time
* inform_flight_details{"time ": "9.00"}
  - action_ get_flight_results
```

6.6 Interactive Learning Using Rasa Core

Interactive learning means giving step-by-step feedback to our bot while we converse.

Load up an Existing Bot

With the basic bot built up with stories, it can be improved by giving out feedback on its mistakes. The following section tells about how the stories can be improved by making use of interactive learning. The following command gives us the result of interactive learning,

```
$ python -m rasa_core.train \
  --online -o models/dialogue \
  -d domain.yml -s stories.md \
  --endpoints endpoints.yml
```

With the above command, bot will be loaded with the interactive mode. In case of Flight_Search_Bot, it is as follows:

Bot loaded. Type a message and press enter.

? Next chatter input:
-hey good morning
? Is the NLU classification for 'hey good morning' with intent 'greeting' correct?
-Yes

Chat History:

#	Bot	User
1	action_listen	
2		hey good morning
3.	intent: 'greeting'	

? The bot felt to run 'utter_greeting', right? (Y/n)

As we can see, for the message "hey good morning", the bot took the proper action utter_greeting, so we type Y. Until the wrong action is chosen, this can be continued.

Providing feedback on errors

The following section tells about how the bot errors can be corrected by making use of the feedback given by the user. Considering the Flight_Search_Bot example, if we ask "how many flights available to India?", the bot should suggest action_get_flight_results and then action_listen

Chat History:

#	Bot	User
1	action_listen	
2		'how many flights available to India?'
3	intent: inform_flight_details Action: action_get_flight_results	
4	Next action: action_listen	
5		'Bye bot'
6	intent: good_bye	
7	next action: action_listen	

? The bot felt to run 'action_listen', right? No

As the bot chose the wrong action, we type no. Hence on typing no, it asks for the feedback from the user as follows:

? What can be the next action of the bot
1. action_restart
2. utter_greeting
3. action_get_flight_details
4. utter_good_bye
5. utter_default
6. action_listen

Now, the bot should actually do utter_goodbye, so we select that action. Conversation with the bot can now be stored as stories to a file. Hence, depicting the dialog with machine learning is easier with Rasa Core because it allows giving the feedback on errors and hence can get improved with every conversation.

7 Results of Evaluating the Model Against Tensorflow and Spacy Pipelines

The default method used is spacy sklearn, where it considers the sentences as a sum of word vectors. Based on that representation the classifier can be trained. Though it is a good approach, it has few limitations. One disadvantage is that, with huge number of vectors that will be never used, as most of the conversational AI manages a restricted space.

But the Tensorflow method does nearly the correct inverse. Using of pre-prepared word vectors is eliminated, and it can be supported with any language. For both intents and words, it can learn embedding at the same time. Tensorflow embedding method will rank the closeness between user text and all of the intents, rather than training a classifier. Hence, learning is carried out specifically for own domain rather than getting stuck with out-of-case pre-defined word vectors.

We built a company_query_bot, where the training data classified to five intents "jobs", "services", "technology", "greeting" and "good_bye". The description for each intent is given below,

Greeting_form—greet messages from the user
Jobs—various job-related training data from the user to the company
Services—various services-related training texts
Technology—questions asked regarding the technologies
Good_bye—getting off messages or texts from the user

We trained the model separately with the two pipelines "spacy" and "Tensorflow". We evaluated the two pipelines with cross-validation and without cross-validation for intent classification.

Table 1 Result of evaluating Tensorflow over spacy pipeline on 35 training examples

	F1-score (%)	Precision (%)	Accuracy (%)
Spacy	79	87	80
Tensorflow embedding	67	73	67

Table 2 Result of evaluating Tensorflow over spacy pipeline on 100 training examples

	F1-score (%)	Precision (%)	Accuracy (%)
Spacy	77	78	80
Tensorflow embedding	73	77	78

Table 3 Result of evaluating Tensorflow over spacy pipeline on 400 training examples

	F1-score (%)	Precision (%)	Accuracy (%)
Spacy	75	76.8	73.2
Tensorflow embedding	81.6	83.1	89.8

7.1 With Cross-Validation

Initially, for small dataset (35, 100 labeled utterances), spacy showed good results, but as the dataset is increased in number, the Tensorflow embedding pipeline showed good accuracy comparatively. Table 1 and Table 2 shows the result of precision, F1-score, and accuracy for 35, 100 labeled utterances, when evaluating with cross-validation with three folds using spacy and Tensorflow embedding as shown in Table 1 and Table 2. As we have increased the training data to 400 labeled utterances, drastically, the accuracy has been increased with Tensorflow embedding pipeline as shown in Table 3.

7.2 Without Cross-Validation

Even 'without cross-validation' evaluation leads the same result as 'with cross-validation'. Here, we have taken 40 training data in JSON file for evaluation against a model generated with 350 training examples. Table 4 shows the result of precision, F1-score, and accuracy, when evaluating without cross-validation using spacy and Tensorflow embedding. The result shows that Tensorflow embedding pipeline gives better accuracy.

Table 4 Result of evaluating Tensorflow over spacy pipeline on 40 test data and 350 training data

	F1-score (%)	Precision (%)	Accuracy (%)
Spacy	71.7	72	72.8
Tensorflow embedding	83.7	83.1	84.2

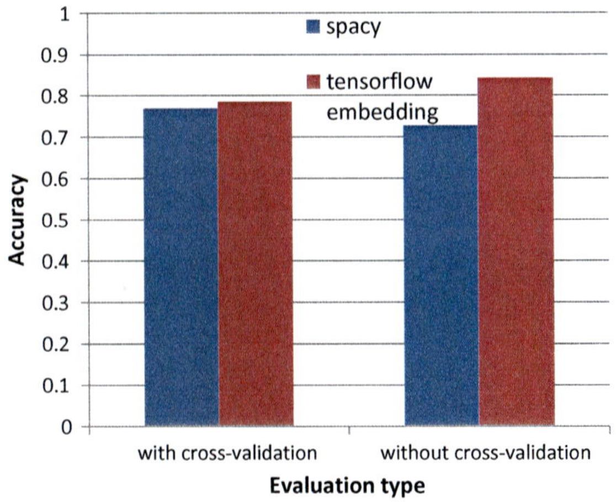

Fig. 2 Graph showing the accuracy for spacy and Tensorflow pipelines with cross-validation and without cross-validation

We have plotted a graph to check the accuracy with cross-validation (by taking the average) and without cross-validation for spacy and Tensorflow pipelines. The graph is shown in Fig. 2.

The result shows that Tensorflow embedding pipeline gives better accuracy on huge amount of training data on both with and without cross-validation evaluation, compared to that of spacy.

8 Conclusions

Vast majority of the state-full bots require state machines which are not versatile. As the quantity of states builds, levels of nested rules also get increased. It may also require it to add an extra state to the state machine. To have a genuine conversation, we need some memory and expand on things. Rasa stack gives us a chance to do that in a very scalable way and hence eliminated the need for state-full bot approach. With the stateless approach, the results also show the advantage of using the Tensorflow embedding over spacy pipeline. With Tensorflow embedding, using of pre-prepared word vectors is eliminated, and it shows better accuracy for intent classification with the presence of large amount of training data.

References

1. Bocklisch T, Faulker J, Pawlowski N, Nichol A (2017) Rasa: open source language understanding and dialogue management. ArXiv Prepr. ArXiv:171205181
2. Tiwari A, Talekar R, Patil SM (2017) College information chat bot system. Int J Eng Res Gen Sci 5(2)
3. Digate CJ, Herot CF, Ketudat T, Kopikis AM (2007) Rules based real-time communication system. US Patent 7,184,524
4. Gu G, Yegneswaran V, Porras P, Stoll J, Lee W (2009) Active botnet probing to identify obscure command and control channels. In: Annual computer security applications conference, Honolulu, HI, pp 241–253
5. Zhou N, Shu C, Meliksetian DS (2008) Method and system for instant messaging Bots specification using state transition methodology and XML. US Patent 7,454,469
6. Tarau P, Figa E (2004) Knowledge-based conversational agents and virtual storytelling. In: Proceedings of the 2004 ACM symposium on applied computing, Nicosia, Cyprus, 14–17 Mar 2004. https://doi.org/10.1145/967900.967913
7. Yan M, Castro P, Cheng P, Ishakian V (2016) Building a chatbot with serverless computing. https://doi.org/10.1145/3007203.3007217
8. Adewale O, Beatson A, Buniatyan D, Ge J, Khodak M, Lee H, Prasad N, Saunshi N, Seff A, Singh K, Suo D, Zhang C, Arora S (2017) Pixie: a social chatbot. In: Alexa prize proceedings
9. Bianchini A, Tarasconi F, Ventaglio R, Guadalupi M (2017) Gimme the usual-how handling of pragmatics improves chatbots. In: CLiC-it
10. https://medium.com/bots-for-business/how-to-build-a-state-full-bot-a2703ff2d57b

Enhanced Processing of Input Data in Clustering Techniques of Data Mining Algorithms

K. Sampath Kini and B. H. Karthik Pai

Abstract Techniques of data mining and its applications have become significant in almost all domains. Each technique has its own significance in a given problem context. Likewise, clustering technique is in use to group elements with similar properties together. Objective of this research work is to apply multithreading approach on grouping the elements into various clusters. This multithreading technique determines the target cluster for multiple elements simultaneously. Thus, this approach improves overall performance of clustering technique which involves many iterations required for finding out target cluster. Implementation approach in this work involves, partitioning the input data elements and associating each of these partitions to multiple threads. Each thread is responsible for picking up the input element from the respective partition and determining the destination cluster. This implementation technique uses master–slave design. Master component makes various partitions of input data elements, creates instances of slave components, and invoke the services offered by slaves. Each slave runs in separate thread of computation and performs the task of determining target cluster of the given input element from the corresponding partition. Slaves will terminate its task once it finishes the determination of target cluster for the data elements present in the respective partition. We have seen that the above-designed technique does well when it comes to consumption of computational time. Partitioning of input data elements and number of data points in a partition is decided based on number of cores available for multithreading. It consumes less computational time as each partition is processed simultaneously on different available cores. Thereby, it eliminates the large computational time needed for existing techniques. We did an experiment sample dataset from the data mining open-source library source spfm. The size of the datasets we have used is closed to 50 MB. Our test environment created maximum of four simultaneous threads for the processing of entire input data points. When observed with the current approaches,

K. Sampath Kini (✉) · B. H. Karthik Pai
NMAM Institute of Technology, Nitte, Karnataka, India
e-mail: sampath@nitte.edu.in

B. H. Karthik Pai
e-mail: karthikpai@nitte.edu.in

N. N. Chiplunkar and T. Fukao (eds.), *Advances in Artificial Intelligence and Data Engineering*, Advances in Intelligent Systems and Computing 1133,
https://doi.org/10.1007/978-981-15-3514-7_38

experimental results tell that our approach provides a performance gain of 30% on an average.

Keywords Clustering · Multithreading · Mining · Partitioning · Performance

1 Introduction

This research paper provides a new approach that can be used for the processing of large set of input data elements to determine the target cluster. Cluster, in this context means grouping similar objects together [1]. Every clustering technique has its own strengths and liabilities [2]. Most of the techniques would scan through all the input data points one after another, determine target cluster, and recalculate centroid of the cluster after every input data point is processed. This approach causes performance problem when quite a large number of iterations are involved in determining target cluster for each input data element. This new approach splits the large set of input data points into different multiple partitions and process each partition in a separate thread of execution for determination of target cluster. This sub-partitioning design would help to eliminate large computational time needs of clustering techniques such as k-means data mining algorithm.

The objective of this paper is implemented in two stages. (1) Partitioning of large input dataset based on number of cores available for multithreaded computation. (2) Perform target cluster determination on each of the sub-partition in a different thread of execution. Comparison of computational time is carried out for this approach.

2 Methodology

In this section, methods followed and algorithms applied are captured. Methodology for finding out target cluster for each input data element involves various underlying subtasks. Subtasks include making subpartitions of large input data points, selecting each partition, and assign it to different slave instances for separate thread of execution, each slave instance determining target cluster for all the data points in its cluster. The steps followed in our approach are depicted via a flowchart (see Fig. 1). Two types of components are identified for the implementation of this technique. One of them is ClusterMaster and it will provide the services of master. ClusterMaster will have the responsibility of splitting the input data points, assigning the partitioned data points to various slaves and declaring the final result once all slave instances have finished their execution in separate threads. The other component is ClusterSlave and it will do the task of determining target cluster of the data point in its partition. Runtime interaction between ClusterMaster and ClusterSlave is depicted via interaction diagram (see Fig. 2).

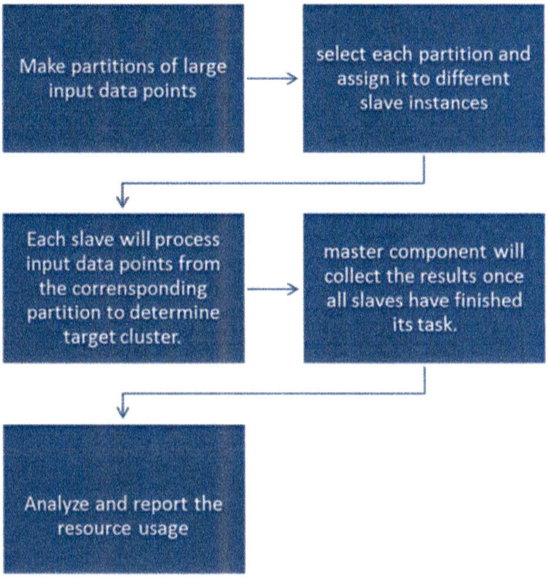

Fig. 1 Steps for processing input elements in multiple threads

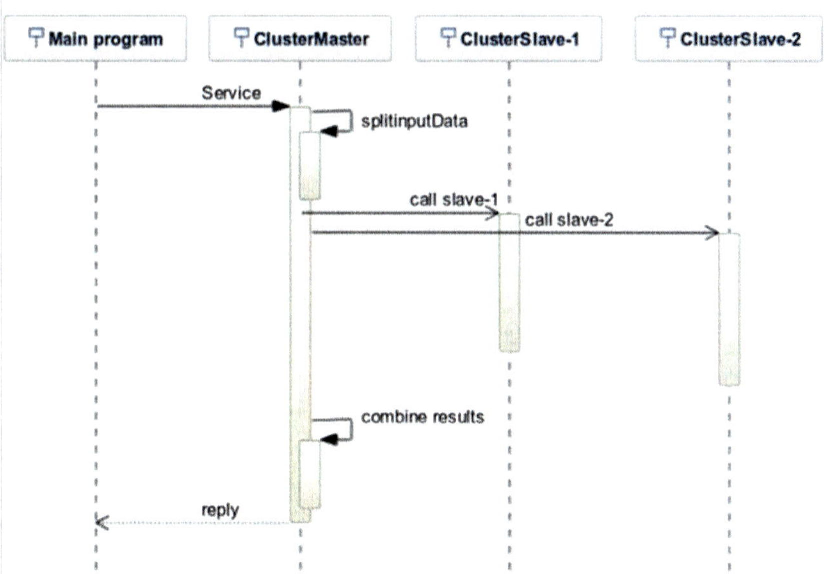

Fig. 2 Runtime interaction between ClusterMaster and ClusterSave

2.1 Design of Algorithms

Now, we will discuss the algorithms that are followed in our approach.

Algorithm 1: P-DP

This algorithm is used for creating N number of partitions from the large input data elements.

Input: D, a dataset of input data points; m, number of core processors available in the system

Output: P, set of partitions; P contains partition details in terms of start and end indices of partitioned dataset.

```
r ← |D|    /* no: of data points in entire input data set */
q ← r/m    /* q represents total number of elements in a partition */
x ← m      /* x: no of partitions equals no of cores */
sindex ← 0
For i = 1 to x
begin
   eindex← sindex + q
     /* sindex and eindex represents both start and end indices of the partition */
   add {sindex,eindex} to P
        sindex ← eindex
     End
     return P;
```

Algorithm 2: CL-PTDB

This algorithm is used for invoking k-means [3] algorithm on each partition belonging to separate instance of slave. The function k-means can be the existing algorithm of clustering process.

Input: D, a dataset of input data points; P, set of partitions;

Output: CL, a set of clusters

```
For each partition Pq Є P do
begin
   Tr← create new Thread
   Kmeans(Tr, Pq, CL);    /* CL represents set of clusters as output */
End
WaitAll() /* wait till all the threads have finished its job */
return CL;
```

Now, with an example, we will show the process of clustering technique using multiple threads. Let the input data elements be as shown in Table 1. Let us define two clusters with centroids (1.0, 1.0) and (5.0, 7.0), respectively. Let us assume that system has four cores available for multithreaded execution. Since there is total of 12 data points present in this example, four different partitions with three data points in each partitions are made. Each partition is used as input data for various ClusterSlave instances that computes target cluster using technique such as k-means.

Table 1 Sample input data points, various partitions, and the target cluster

Input data points	Partition ID/thread ID	Target cluster
(1.0, 1.0)	1	1
(1.5, 2.0)		1
(3.0, 4.0)		2
(5.0,7.0)	2	2
(3.5, 5.0)		2
(1.5,1.0)		1
(3.5, 4.5)	3	2
(1.1, 2.0)		1
(5.2, 7.0)		2
(1.4, 2.0)	4	1
(1.5, 2.6)		1
(3.0, 4.0)		2

3 Results of Experiments

We conducted experimentation on sample datasets provided by open-source data mining library spmf [4]. This multithreading implementation for processing input data points is run using 5000, 10,000, and 15,000 data points, respectively. Recorded computational time for all the three sets of data points is shown through graph (See Fig. 3). We conducted experiments on following system environment.
Processor type: Intel(R) Pentium(R) CPU N3540 @2.16GHZ
RAM size: 4 GB
Number of cores: 4

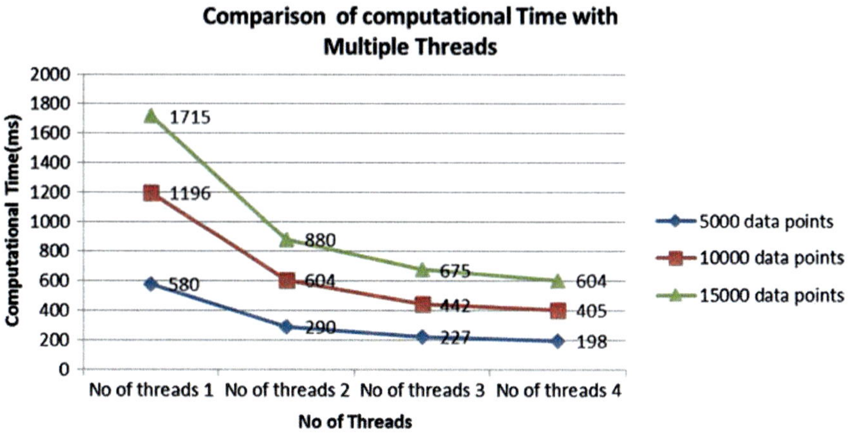

Fig. 3 Computational time with varying number of threads and input data points using k-means clustering

System: windows 8.1, 64 bit, x-64 based processor
Java/JVM: 1.8.

4 Conclusions

We have come to experimental conclusions from our findings.

- Existing clustering techniques such as k-means are suffering from performance issues in terms of computational time with many iterations involved.
- The multithreading-based technique explained in this paper eliminates large computational time needs by clustering techniques.
- The multithreading-based technique also enables performance gain of 30% on an average.
- This technique cannot be applied if the system does not support multithreading execution in simultaneous cores.

References

1. Berkhin P (2016) A survey of clustering data mining techniques. Recent advances in clustering. Springer, Berlin, pp 25–71
2. Xu D, Tian Y (2015) A comprehensive survey of clustering algorithms. Ann Data Sci 2(2):165–193
3. K-means clustering. https://en.wikipedia.org/wiki/K-means_clustering
4. Open source library spmf. http://www.philippe-fournier-viger.com/spmf/

A Comparative Analysis of MFIs in India Using ANOVA and Logistic Regression Model

M. G. Deepika and P. Sarika

Abstract The microfinance industry in India is undergoing radical changes with changes in regulations and competition from the banking sector. With the RBI providing licences for opening up of Small Finance Banks and Payment Banks and commercial banks planning to penetrate into microlending leading to increased competition to microfinance institutions (MFIs), sustainability of MFIs is questioned. The not-for-profit NGO MFIs have largely converted themselves into for-profit entities. Given the current situation in the MFI industry, the paper tries to analyse as to what determines the performance and sustainability of MFIs in India. What are the reasons for not-for-profit MFIs to move out of business and was their performance significantly different from that of the for-profit sector? We examine if there is a significant difference in the financial, social and organizational indicators of for-profit MFIs and not-for-profit MFIs and sustainable and non-sustainable MFIs through the application of statistical procedures, the analysis of variance (ANOVA), a parametric statistical tool for comparing group means, and a limited dependent model, logistic regression.

Keywords Microfinance · Sustainability · Legal status · Logistic regression · ANOVA · India

1 Introduction

The microfinance industry in India is undergoing radical changes with the changes in regulations and the competition from the banking sector. With the RBI providing licences for the opening up of Small Finance Banks and Payment Banks and the commercial banks planning to penetrate into microlending, the sustainability of microfinance institutions is questioned. For MFIs to grow, there is a pressing need for

M. G. Deepika (✉) · P. Sarika
Department of Management, Amrita University, Bangalore, India
e-mail: mgdeepika@gmail.com

P. Sarika
e-mail: sarikaperumpilly@gmail.com

N. N. Chiplunkar and T. Fukao (eds.), *Advances in Artificial Intelligence and Data Engineering*, Advances in Intelligent Systems and Computing 1133,
https://doi.org/10.1007/978-981-15-3514-7_39

capital which can come with the support of venture capital, equities or angel investors who invest seeking profits in return. The not-for-profit NGO MFIs who were earlier playing a larger role in microfinance sector by depending on donor funding are now waning away and have now become less relevant. Given the current situation in the MFI industry, the paper tries to analyse as to what determines the performance and the sustainability of MFIs in India. What should the investors look for to assure good returns? What are the reasons for the not-for-profit MFIs to move out of the business and was their performance significantly different from that of for-profit MFIs in India? Next section provides a brief review of the literature relating to the performance of MFIs, sections three and four deal with the objectives, database and methods used in the study, section five deals with the results and discussion, and section six provides the conclusions of the study.

2 Performance of MFIs: A Review

It was earlier debated if the MFIs could strike a balance between realizing financial sustainability and attaining societal goals. The performance and sustainability of MFIs are usually gauged by financial sustainability indicators. The MIX market defines financial sustainability as MFIs having the operational self-sufficiency ratio (OSS) of 110% or more. Financial sustainability is seen as the capacity of MFIs to cover all expenses through the revenues and also generate a given surplus without the subsidies. The OSS ratio which is seen as a proxy for sustainability of MFIs could be defined as total financial revenue divided by the sum of financial and operating expenses [1].

Ayayi and Sene (2010) see financial sustainability as a function of a given set of independent variables like portfolio at risk, interest rate, operating expense ratio, productivity ratios like number of loans per staff member, number of borrowers per and loans per loan officer. Other set of independent variables were also considered in the initial model like percentage of women borrowers, the average loan balance, the number of staff members and the age of MFIs. Results of their study indicate that quality credit portfolio along with reasonably high interest rates and sound management practices are prerequisites for financial sustainability of MFIs. It is also seen that the outreach of clients in MFI programs and the age of MFIs generally have positive but smaller impact on achievement of financial sustainability [2]. Cull et al. (2007) find that the age of institution and size have notable impact on financial performance. It is also seen that the sustainable individual lenders provide smaller loans, especially to women, suggesting that attaining profits and outreach to the poor can be achieved simultaneously [3].

In the earlier studies on the performance of microfinance institutions, financial performance is typically measured by a set of profitability and viability indicators [3–6]. Return on equity (ROE) is typically used as a measure of financial performance in private microfinance companies [7]. Hermes and Hudon (2018) while reviewing the existing literature on determinants of MFIs' social and financial performance found that while some studies used traditional financial performance indicators like return on equity (ROE) and return on assets (ROA), others used more microfinance specific indicators like OSS and financial self-sufficiency to evaluate financial performance [8]. In cooperatives and NGOs, the larger objective is said to be to attain social performance than achieve higher profits. As the MFIs are of varied structures, it is more appropriate to adopt a profitability indicator which can be common for all categories of MFIs. Unlike ROE, ROA measures the profitability irrespective of the financial structure of the institution. This allows for the comparison of private MFIs like NBFCs with that of not-for-profit NGOs and cooperatives [7, 9].

OSS is a widely used measure of sustainability. We see in the earlier literature that comparison is made across social, commercial and organizational dimensions to demonstrate the differences between different institutional forms. Social performance is measured through outreach, which is measured through the number of active borrowers and depth of reach, using the average loan size which also speaks of the intensity of transaction. Efficiency of organizations is measured by operating expense ratio (ORR), cost of credit by borrower (CPB), the profit margin ratio and the portfolio quality of MFIs by the indicator portfolio at risk greater than thirty (PAR > 30) [3, 5, 10].

We also come across a few studies which looked into the performance of MFIs by governance mechanism or lending methodology [3–5, 7, 10]. When many not-for-profit MFIs in the last decade had converted their status into for-profit organizations, it was assumed that the conversion was due to the fact that the financial performance of the latter is better than not-for-profit MFIs. This prompted some of the studies to investigate the performance difference by the legal status of MFIs, like between the NGOs and private companies. In doing so, [7] had used a multidimensional approach to performance of MFIs across countries. It was seen that the comparison of performance indicators for two groups did not show any notable differences in profitability indicators between NGOs and private MFIs.

Though the not-for-profit NGO MFIs and cooperative MFIs are generally assumed to have better social performance and less financial performance than their for-profit counterparts, the studies have found mixed results. While some studies found this assumption true [3, 11–15], other studies like Mersland and Strøm [4] and Louis and Baesens [16] found that institutional type of MFIs (not-for-profit NGO MFIs, for-profit shareholder companies such as commercial banks and NBFIs and credit and savings cooperatives) has no significant role in determining their financial performance. Some studies examining the performance of cooperative MFIs found to be more cost efficient [7, 17–19]. Ibrahim et al. (2018) using OSS, return on assets (ROA) and profit margin (PM) as indicators of financial performance of different forms of MFIs found that the credit union and cooperatives, NBFIs and NGO MFIs' financial performance is better than their for-profit counterparts [20]. While it was seen that for-profit MFIs are more socially efficient and productive than not-for-profit MFIs, the commercial approach of microfinance did not seem varying with the social mission of MFIs [7]. Some other studies had also shown that the portfolio quality is better with cooperatives and NGOs due to the decentralized decision-making process in these firms as compared to for-profit MFIs [21]. This was attributed to the fact that

NGOs had less risk of adverse selection. The operational efficiency was also seen to be higher in the case of not-for-profit sector due to the low running costs which was attributed to lower payment to the staff who work in the not-for-profit sector with a social motive as compared to the for-profit ones [11].

Given the availability of latest data on financial, social and organizational indicators of MFIs with mix market, the current study separates the data on Indian MFIs from the MFIs of other countries and tries to do a comparative analysis of for-profit and not-for-profit MFIs for the above indicators using the analysis of variance. We then separate the sustainable Indian MFIs from non-sustainable MFIs and analyse the factors contributing to sustainability with a given set of independent variables using the widely used limited dependent model, the logistic regression.

3 Objectives of the Study

1. To examine if there is a notable difference in the performance of MFIs based on their legal status using ANOVA.
2. To examine the factors contributing to sustainability of MFIs using a logistic regression model.

4 Database and Methodology

The study is based on the secondary sources of information. We depend on the data available with mix market on MFIs in India. Performance is seen under three heads—financial, organizational and social using different available indicators across the two categories of MFIs. Return on assets, operational self-sufficiency ratio, profit margins, portfolio at risk greater than 30 days, yield on gross portfolio are the financial indicators chosen for comparison of financial performance of MFIs; average loan per borrower, number of active borrowers and percentage of female borrowers are social indicators; operational efficiency ratio, cost per borrower, number of loans per staff member, number of borrowers per loan officer, number of loans per loan officer are organizational performance indicators.

A comparison is made of the financial, organizational and social (outreach) indicators of MFIs across different legal structures of 104 MFIs in India for different years varying from 2000 to 2013. MFIs are listed under three different legal statuses in the mix market data, the non-banking financial institutions (NBFIs), the NGOs and cooperatives. While the number of cooperatives covered is very less in number, we limit the analysis to comparison of NBFIs representing the for-profit sector and NGO MFIs representing the not-for-profit sector. We compare the above group of MFIs through the analysis of variance (ANOVA) to analyse the differences in performances of MFIs under for-profit and not-for-profit statuses.

Analysis of variance is a statistical method used to analyse if the group means are significantly different. This is an extension of the t-test when applied for more than two groups. It is more or less similar to the multiple two-sample t-test. ANOVA is seen to be a more conservative test as compared to the t-test. In the general application of ANOVA, the null hypothesis is that the given samples are drawn from the same population, i.e. the groups are the simple random sample of the same population. In ANOVA, the factors are treated as dependent variables (variables chosen for analysis shown in Table 1), and the classified groups are independent variables. Rejection of the null hypothesis would mean the difference in the observations between the groups is unlikely to be due to the random choice. In our study, the results of ANOVA would indicate as to whether the legal status of the firm would influence the select factors representing MFI performance.

To meet the second objective of the study, we run a logistic regression model. Here, we try to examine if the predictor variables chosen for analysis (financial, social and organizational indicators) influence the sustainability of MFIs which is a categorical variable. Sustainability of MFIs is measured through their OSS ratio. We classify MFIs into two groups sustainable (represented as 1 with OSS ratio above 110%) and non-sustainable MFIs (represented as 0 with OSS ratio less than 110 percent) based on the definition provided by the mix market.

Logistic regression is a multivariate statistical dependency model. It is a method for analysing the set of data when we have more than one independent variable and dependent variable is binary or dichotomous in nature which is coded as 0 and 1. The purpose behind this method is to arrive at best-fitting model to describe the relation between the dependent variable which is a dichotomous characteristic of interest and is also a response or outcome variable and a set of independent variables.

In the case of the logit model, the probability function reads the following:

$$\begin{aligned} P_i &= E(Y = 1|X_i) \\ &= \text{for } X \text{ independent variable (when the dichotomous value of } Y \text{ is 0 and 1)} \end{aligned} \tag{1}$$

Or what is the expected value that your $Y = 1$ given your X. The subscript i represents the ith observation.

The logistic distribution function can be written as

$$P_i = E\left(Y = \frac{1}{X_i}\right) = \frac{1}{1 + e^{-(\beta_1 + \beta_2 X_i)}} \tag{2}$$

We can write

$$Z_i = -(\beta_1 + \beta_2 X_i) \tag{3}$$

As the denominator is always positive and greater than 1, P_i will always be less than 1 and greater than 0.

Table 1 Variables covered for analysis and their definitions

Variables	Definitions	What does it indicate
Financial indicators		
Return on assets (ROA)	Net income/total assets	Capacity of the MFI to utilize the assets to get more returns
Return on equity (ROE)	Net income/average shareholder's equity	The capacity of a firm to get profits from shareholders investments in the company
Equity	Assets–liabilities	Represents capital invested by shareholders (owners)
Debt-to-equity ratio (DER)	Short-term debt + long-term debt/shareholder's equity	Explains the relative proportion of shareholders' equity and debt used to finance a company's assets
Profit margin (PM)	Net operating income/financial revenue	Higher the percentage, higher is the performance of the organization
Yield on gross portfolio in real (YGP nominal)	Interest and fee income during period/average outstanding loan portfolio	It tells us how much the MFI will receive as interest payments from its clients during the year
Operational self-sufficiency (OSS)	Total financial revenue/(financial expense + loan loss provision expense + operating expense)	Ability of MFIs to cover the costs through its revenues
Capital to asset ratio (CAR)	Solvency ratio	CAR is expressed as a percentage of a bank's risk-weighted credit exposures
Social indicators		
Number of active borrowers (NAB)	Number of active borrowers of an MFI	This indicates the breadth of outreach
Average loan balance (ALB)	Average loan balance per borrower/GNI per capita	A close proxy for the extent of outreach indicating that the smaller loan size would lead to reaching down the poverty spectrum and therefore higher depth of outreach
Percentage of female borrowers (PFB)	Number of female borrowers to total borrowers	Higher the percentage of female borrowers higher the MFI meeting the social mission of reaching to vulnerable sections of the society

(continued)

Table 1 (continued)

Variables	Definitions	What does it indicate
Organizational indicators		
Operating expense ratio (OER)	Operating cost/average gross portfolio	This is an indicator of operational efficiency of the organization which calculates the cost incurred by an institution to provide credit. The lower the ratio the more efficient the organization is
Cost per borrower (CPB)	Operating costs/average number of active borrowers	The ratio gives a measurement of the efficiency of the institution by showing the average cost to serve a borrower over a year
Number of loans per staff member (LSM)	Number of loans per staff member	Higher the number, higher the efficiency of staff
Portfolio at risk greater than 30 days (PAR > 30)	Outstanding balance on arrears over 30 days + total gross outstanding refinanced portfolio/total gross portfolio	This shows the portfolio on risk that may not be paid. The threshold is <10% given that financial guarantees in microfinance are not always sufficient
Institutional indicators		
Total assets (TA)	Total assets	Shows the size of MFIs
Age of MFIs	Seen as new, young and matured denoting 1, 2 and 3	Shows the age and maturity of MFIs

We can interpret the probabilities of the occurrence of the event using the odds ratio.

The odds ratio can be defined as

$$\frac{P_i}{1 - P_i} \tag{4}$$

This is the ratio of probability of Y equal to the expected outcome (1) to the probability of Y not equal to 1.

So if the odds ratio = 1, then the odds are the same whether you will get $Y = 1$ or 0. In which case $P_i = 0.5$. If the odds are greater than 1, then the odds favour the chances of $Y = 1$.

The estimation of the logit function can be explained as follows

$$P_i = E\left(Y = \frac{1}{X_i}\right) = \frac{1}{1 + \mathrm{e}^{-(\beta_1 + \beta_2 X_i)}} \tag{5}$$

where

$$Z_i = -(\beta_1 + \beta_2 X_i) \tag{6}$$

The odds ratio expressed earlier can now be rewritten as

$$\frac{pi}{1 - pi} = \frac{1 + \mathrm{e}^{zi}}{1 + \mathrm{e}^{-zi}} = \mathrm{e}^{zi} \tag{7}$$

Taking log on both sides:

$$\ln \frac{P_i}{1 - P_i} = \ln\left(\frac{1 + \mathrm{e}^{Z}}{1 + \mathrm{e}^{-Z}}\right) = \ln(\mathrm{e}^{Z}) = Z_i = \beta_1 + \beta_2 X_i \tag{8}$$

Now, the equation can be estimated using ordinary least squares method. With more than one independent variable, we can rewrite the equation as

$$L_i = \left(\frac{P_i}{1 - P_i}\right) = \beta_1 + \beta_2 X_1 + \beta_3 X_2 + \beta_4 X_3 + \mu_i \tag{9}$$

5 Results and Discussions

5.1 *Results of ANOVA*

The results of ANOVA (Table 2) on analysing the difference in performance between not-for-profit and for-profit MFIs reveal that not-for-profit MFIs have been performing better as seen through many indicators like ROA as compared with for-profit MFIs, lesser PAR > 30 and higher OER ratio. There is no significant difference seen with sustainability indicator, the OSS ratio being same across two categories of firms. The outreach measure is better for the for-profit MFIs. This again supports the earlier cross country studies that not-for-profit MFIs due to their smaller loan size, better decentralized management and less of asymmetry in information have performed better in terms of recoveries of loans leading to lesser NPAs. As we see in earlier studies, these MFIs due to lower salaries to the staff have lower operational costs and better operational efficiency indicators. It is therefore not the efficiency parameter that has made the not-for-profit MFIs to switch over to for-profit organizations. The limitation with these MFIs however lies with the shortage of capital and less product portfolio as most NGOs had depended on donor money which is now dwindling. As NGOs are less regulated, they are unable to raise capital to run their operations which also would have prompted some of them to transform into for-profit entities.

Table 2 Results of one way ANOVA for difference in performance among for-profit NBFC MFIs and not-for-profit NGO MFIs

	Type	Number	Mean	Standard deviation	*F*	Sig.
ROA	Non-profit	340	0.86	8.85	3.84	0.05
	Profit	543	−0.56	11.37		
ROE	Non-profit	334	28.35	211.82	0.18	0.68
	Profit	541	56.14	1197.77		
Equity	Non-profit	437	1,161,774	4,026,350	36.87	0
	Profit	619	12,438,508	38,670,642		
DER	Non-profit	420	85.56	1054.86	3	0.08
	Profit	605	10.56	124.67		
OSS (%)	Non-profit	407	106.82	33.79	1.74	0.19
	Profit	604	109.65	33.23		
PM (%)	Non-profit	406	−12.67	1.8	0.45	0.5
	Profit	603	−21	234.45		
Profit/Loss	Non-profit	407	249,382	1,012,778	0.25	0.62
	Profit	601	−228,502	19,127,131		
PAR > 30 (%)	Non-profit	256	4.43	14.6	2.31	0.13
	Profit	457	6.93	24.02		
CAR	Non-profit	436	15.65	18.79	47.2	0
	Profit	619	26.42	28.69		
YGP (%)	Non-profit	278	22.48	10.04	2.35	0.13
	Profit	489	23.48	7.83		
ALB (in USD)	Non-profit	434	130.78	94.48	26.7	0
	Profit	607	161.77	96.07		
NAB	Non-profit	442	79,850.26	326,438.2	44.28	0
	Profit	619	368,087.7	867,738.7		
PFB (%)	Non-profit	387	95.29	13.58	0.34	0.55
	Profit	552	94.71	15.66		
OER (%)	Non-profit	345	13.68	14.02	2.92	0.09
	Profit	543	17.4	38.86		
CPB	Non-profit	331	14.49	11.62	15.29	0
	Profit	532	23.84	42.54		
LSM	Non-profit	296	20,259.38	242,969.7	2.31	0.13
	Profit	533	316.54	203.14		
Assets	Non-profit	447	10,771,409	42,328,841	37.88	0
	Profit	619	69,519,048	1.99E+08		
Age	Non-profit	455	2.48	0.71	24.94	0
	Profit	628	2.24	0.82		

5.2 Results of Logistic Regression

As discussed under methodology section, we run the logistic regression model to analyse the factors influencing the sustainability of MFIs. While we run the initial model with all the variables listed in Table 1 as predictor variables, we omit many variables due to the problem of multicollinearity. The reduced final model consists of sustainability (1 = sustainable, with OSS ratio greater than 110) and (0 = non-sustainable with OSS ratio less than 110) as dependent variable and eight predictor variables. The predictor variables included in the final model are debt equity ratio (DER), yield on gross portfolio (YGP), capital adequacy ratio (CAR), number of active borrowers (NAB), percentage of female borrowers (PFB), portfolio at risk greater than 30 (PAR > 30), assets and operational efficiency ratio (OER). The results of the logistic regression analysis using the SPSS software are displayed in Tables 3, 4, 5, 6 and 7.

The overall model summary in Table 3 shows that the model is overall a good fit as the Nagelkerke R square is at 0.391 and the Hosmer and Lemeshow test is significant at one percent level (Table 4). Table 6 shows that 72% of the cases are rightly classified which again speaks of the overall goodness of fit of the model. Table 7 shows the beta values and their significance for the predictor variables included in the final model along with the odds ratio in the last column. The beta coefficients of the predictor variables show almost expected results. Four important variables YGP, CAR, PAR > 30, OER have emerged significant of which YGP and CAR are positively related with dependent variable, and odds ratio greater than 1 also indicates that higher the values of these variables higher the probability of $Y = 1$. OER and PAR > 30 are negatively associated with odds ratio less than one showing inverse relationship with the dependent variable. The rest of the predictors included in the model have not emerged significant. The overall results of the logistic regression analysis indicate that MFIs should charge a reasonable rate of interest, should have capital adequacy, should minimize risky assets and have good operational efficiency to attain sustainability.

Table 3 Model summary

Step	−2 Log likelihood	Cox and Snell R square	Nagelkerke R square
1	597.161	0.292	0.391

Table 4 Hosmer and Lemeshow test

Step	Chi-square	df	Sig.
1	19.743	8	0.011

Table 5 Dependent variable encoding

Original value	Internal value
Non-sustainable	0
Sustainable	1

Table 6 Classification table

	Observed		Predicted		
			SI		Percentage correct
			Non-sustainable	Sustainable	
Step 1	SI	Non-sustainable	155	104	59.8
		Sustainable	58	263	81.9
	Overall percentage				72.1

Table 7 Variables in the equation

		B	S.E.	Wald	Sig.	Exp(B)
Step 1	DER	−0.003	0.003	1.375	0.241	0.997
	YGPnom	0.152	0.023	44.863	0.000	1.164
	CAR	0.017	0.007	6.104	0.013	1.017
	NAB	0.000	0.000	0.272	0.602	1.000
	PFB	−0.001	0.009	0.004	0.952	0.999
	PAR30	−0.027	0.009	9.832	0.002	0.973
	Assets	0.000	0.000	2.070	0.150	1.000
	OER	−0.287	0.033	75.724	0.000	0.751
	Constant	−0.143	0.921	0.024	0.876	0.867

6 Conclusions

Given the current situation in the microfinance industry, where the MFI sector is undergoing radical changes with the competition from the banking sector, the paper tries to analyse as to what would determine the performance and sustainability of MFIs in India. What are the reasons for the not-for-profit MFIs to move out of the business and was their performance significantly different from that of the for-profit sector? We examine if there is a significant difference in the financial, social and organization indicators by categorizing MFIs into for-profit and not-for-profit MFIs and sustainable and non-sustainable MFIs. For the first objective, we run ANOVA and examine if the legal status of the firm influences the financial, social and organizational indicators of the firm. To examine the factors influencing the sustainability of MFIs, we run the logistic regression model. The results of ANOVA reveal that not-for-profit MFIs have been performing better as seen through indicators like better ROA, lesser PAR > 30 and higher OER ratio. No significant difference was seen through the sustainability indicator, the OSS ratio. The limitation with the not-for-profit MFIs however lies with the shortage of capital and less product portfolio as most NGOs had depended on donor money which is now dwindling. As NGOs are less regulated, they are unable to raise capital to run their operations which led to their transformation into for-profit entities. In order to analyse the factors influencing the sustainability of MFIs, we run the logistic regression model in which the odds ratio greater than 1 indicates that higher the values of these variables higher the probability of $Y = 1$. YGP and CAR are positively related with dependent variable and odds ratio, and OER and PAR > 30 are negatively associated with odds ratio less than one showing inverse relationship with the dependent variable. Thus, the overall results of the logistic regression analysis indicate that the existing MFIs should charge a reasonable rate of interest, minimize their risk and maintain sufficient capital adequacy in order to be sustainable.

References

1. The Mix Market. Glossary. [Online]. https://www.themix.org/resources/glossary. Accessed 4 Feb 2019
2. Ayayi AG, Sene M (2010) What drives microfinance institution' s financial sustainability. J Developing Areas 44(1):303–324. http://www.jstor.org/stable/41428207 (Published by : College of Business, Tennessee State University)
3. Cull R, Demirgu¨çKunt A, Morduch J (2007) Financial performance and outreach: a global analysis of leading microbanks. Econ J 117(517):F107–F133. Retrieved from http://siteresources.worldbank.org/DEC/Resources/Financial_Performance_and_Outreach.pdf on 2 May 2019
4. Mersland R, Øystein Strøm R (2009) Performance and governance in microfinance institutions. J Bank Finance 33(4):662–669. https://doi.org/10.1016/j.jbankfin.2008.11.009
5. Hartarska V (2005) Governance and performance of microfinance institutions in Central and Eastern Europe and the Newly Independent States. World Dev 33(10):1627–1643. https://doi.org/10.1016/j.worlddev.2005.06.001
6. Lafourcade A, Isern J, Mwangi P, Brown M (2005) Overview of the outreach and financial performance of microfinance institutions in Africa, microfinance information eXchange (MIX). Retrieved from https://www.griequity.com/resources/industryandissues/financeandmicrofinance/Africa_Data_Study.pdf on 2 May 2019
7. Tchakoute-Tchuigoua H (2010) Is there a difference in performance by the legal status of microfinance institutions? Q Rev Econ Finan 50(4):436–442. https://doi.org/10.1016/j.qref.2010.07.003
8. Hermes N, Hudon M (2018) Determinants of the performance of microfinance institutions: a systematic review. J Econ Surv 32(5):1483–1513. https://doi.org/10.1111/joes.12290
9. Bruett T (ed) (2005) Measuring performance of microfinance institutions: a framework for reporting, analysis, and monitoring. The SEEP Network, Washington DC. Retrieved from http://www.findevgateway.org/sites/default/files/mfg-en-toolkit-measuring-performance-of-microfinance-institutions-a-framework-for-reporting-analysis-and-monitoring-2006.pdf on 2 May 2019

10. Hartarska V, Nadolnyak D (2007) Do regulated microfinance institutions achieve better sustainability and outreach? Cross-country evidence. Appl Econ 39(10):1207–1222. https://doi.org/10.1080/00036840500461840
11. Gutiérrez-Nieto B, Serrano-Cinca C, Mar Molinero C (2007) Microfinance institutions and efficiency. Omega 35(2):131–142. https://doi.org/10.1016/j.omega.2005.04.001
12. Gutiérrez-Nieto B, Serrano-Cinca C, Mar Molinero C (2009) Social efficiency in microfinance institutions. J Oper Res Soc 60(1):104–119. https://doi.org/10.1057/palgrave.jors.2602527
13. Servin R, Lensink R, van den Berg M (2012) Ownership and technical efficiency of microfinance institutions: empirical evidence from Latin America. J Bank Finance 36(7):2136–2144. https://doi.org/10.1016/j.jbankfin.2012.03.018
14. D'Espallier B, Hudon M, Szafarz A (2013) Unsubsidized microfinance institutions. Econ Lett 120(2):174–176. https://doi.org/10.1016/j.econlet.2013.04.021
15. Gutierrez-Goiria J, San-Jose L, Retolaza JL (2017) Social efficiency in microfinance institutions: identifying how to improve it. J Int Dev 29(2):259–280. https://doi.org/10.1002/jid.3239
16. Louis P, Baesens B (2013) Do for-profit microfinance institutions achieve better financial efficiency and social impact? A generalised estimating equations panel data approach. J Dev Effectiveness 5(3):359–380. https://doi.org/10.1080/19439342.2013.822015
17. Aboagye AQQ (2009) A baseline study of Ghanian microfinance institutions. J Afr Bus 10(2):163–181. https://doi.org/10.1080/15228910903187734
18. Abate GT, Borzaga C, Getnet K (2014) Cost-efficiency and outreach of microfinance institutions: trade-offs and the role of ownership. J Int Dev 26(6):923–932. https://doi.org/10.1002/jid.2981
19. Marwa N, Aziakpono M (2015) Financial sustainability of Tanzanian saving and credit cooperatives. Int J Soc Econ 42(10):870–887. https://doi.org/10.1108/IJSE-06-2014-0127
20. Ibrahim Y, Ahmed I, Mohd M (2018) The influence of institutional characteristics on financial performance of microfinance institutions in the OIC countries. Econ Sociol 11(2):19–35. https://doi.org/10.14254/2071-789X.2018/11-2/2
21. Jansson T, Rosales R, Westley GD (2004) Principles and practices for regulating and supervising microfinance. Retrieved from https://publications.iadb.org/en/publication/11768/principles-and-practices-regulating-and-supervising-microfinance on 2 May 2019

Practical Analysis of Representative Models in Classifier: A Review

Angela Mathew and Sangeetha Jamal

Abstract Handling large amount of information with the advancement of technologies is beyond the scope of conventional database technologies. As a solution to the above task, document management systems and machine learning techniques are used. The widely used supervised machine learning technique employing representative models are reviewed in this paper. A classifier creates a model for prediction. The fundamental concern in the classifier is to create an efficient document representation for representative modeling. The model represents documents in an algebraic vector form. The classical models for document representation are numerical vector representational model which gives the frequency of occurrence of the feature as the vector element. Feature extraction and feature selection are done for dimension reduction techniques. In this paper, we have reviewed some existing methods for document representation. The practical analysis and comparison on various techniques for document representation are presented here.

Keywords Classifier · Document representation · Feature extraction · Representation models · Supervised learning · Text mining

1 Introduction

With the advancement of technology, there is an exponential increase in the data collected for analysis and review. Text analytic deals with analyzing text in the form of data which is in an unstructured form to deduce useful information. It uses techniques such as extraction and classification to extract the information hidden in the raw text data.

A. Mathew (✉) · S. Jamal
Rajagiri School of Engineering and Technology, Kochi 682031, India
e-mail: angelathaikattil@gmail.com

S. Jamal
e-mail: sangeethaj@rajagiritech.edu.in

N. N. Chiplunkar and T. Fukao (eds.), *Advances in Artificial Intelligence and Data Engineering*, Advances in Intelligent Systems and Computing 1133,
https://doi.org/10.1007/978-981-15-3514-7_40

Machine learning analyses data in the form of documents, i.e., text, image, digital data, etc., to automate slow learning processes, pattern analysis, data classification, clustering, decision making, etc. In real-life applications, most of the data will be in the form of text documents, images, digital data, etc. Analyzing and learning these documents are of great importance in the present day scenario. It can be supervised, unsupervised or semi-supervised learning methods. Learning techniques use a mathematical model for predicting. Thus, the processing of text documents as a string of words as such will not be of any help for integrating with learning algorithms. So, the document data are converted to numerical form.

Classification is a process of classifying documents to different categories for application like sentiment analysis, routing, spam filtering, etc. In supervised machine learning technique for classification, the learning process is carried out with the help of a trained model. Here, we train the model with large number of documents belonging to different classes. The document term metrics for the training model represents the whole set of training data numerically. The rows of the metrics represent the documents and columns of the metrics represent the words considered as features. The elements in the metrics represent the occurrence of feature in the document. For the test document, a feature vector has to be created with number of columns corresponding to the number of features in the document term metrics and the elements representing the occurrence of each feature within the test document. A classifier algorithm compares the feature vector of the test document with the model to predict the category of the test document.

The document term metrics suffer from high dimensionality, sparse elements, inability to identify semantic meanings of the text, existence of polysemic words, non-consideration of syntax, etc. Text preprocessing is done to address some of the deficiencies mentioned above. However, these improvement techniques are not sufficient to give accurate results. Additional techniques such as feature selection and feature extraction models are used to further improve the model.

Document classification is a major problem concerning text mining. Several papers on the latest models for document classification by supervised machine learning specifically on the representational models for predictive analysis are reviewed.

The paper is organized as follows, Sect. 2 gives a brief on the various methods used for text document representation. Section 3 gives the various representational models and Sect. 4 reviews the various works in the related field. A comparative study on the various models is done in Sects. 5 and 6. And Sect. 7 summarizes the work done.

2 Types of Representation

2.1 Word-Based

Words are considered as the independent feature. Here, documents are converted to a vector of words. The conventional bag-of-words (BoW) model is an example of this method [1]. It ignores the structure of data and semantic meaning behind the document. Despite all the disadvantages, it is still a simple and efficient representational form.

2.2 *Phrase-Based*

Phrases are the features in phrase-based method. Here, a cluster of n words considered as a phrase is used to represent documents. N-gram model for representation is an example for this model. It overcomes the ignorance of syntactic similarity in the classical word-based representation models. But it is still ignorant of the relation and order of words within the same cluster [2]. These models are efficient and accurate for modeling documents to identify the polarity of the data. Using this model for the classification of large text documents leads to a considerable fall in the accuracy of the system.

2.3 *Synset-Based*

This model uses only a specific synonym of words in the word-based model using word sense disambiguation. This can be done using WordNet ontology which identifies similar words within the same context. This has shown a considerable effect on analyzing the concept behind the data of the word-based models [3].

2.4 *Graph-Based*

Here, a graphical representation of a document is done. This is an improved phrase-based model representation, which disregards the similarity of words within the same cluster [4]. A document graph is created which associated neighboring pair of n-grams where the edges represent the frequency of occurrence. The closeness of graph represents the document similarity.

2.5 *Hybrid Model*

Combinations of different models like word-based with phrase-based are used for text document representations. The combined model is used particularly for domain-based classification problems which give more accurate results [5].

3 Representation Models

3.1 Boolean Modeling

Here, documents are represented as set of words. The common AND, NOT, OR operations are done on this representation of the documents to retrieve similar vectors. This is a simple information retrieval (IR) model for document retrieval and classification. This modeling technique disregards every other detail behind the text like semantic and syntactic meanings. But it uses the conventional Boolean operations on the text representations for effective information retrieval. It is a simple IR technique for classification. This model is used along with WordNet and decision trees for efficient document classification [6].

3.2 Vector Space Modeling

A document is represented as a vector with each element representing features extracted from the text document and the elements represent the importance of the feature in that document. The document is hence defined in a space of vocabulary features. The commonly used weighing method is occurrence of the word in the document. But an effective representation is using term frequency-inverse document frequency (TF-IDF) method. In a vector space model, the words are considered to be mutually independent in the vector space. This will in effect neglect the syntactic information within the text. Also, identifying a semantic relation behind the text document data is difficult. This model compares this vector space representation models to identify document similarity for classification models [7]. Even with so many disadvantages, this method shows great advantages over other models by its simple representation [8]. Latent semantic analysis (LSA) which identifies the latent topics behind the text data is based on a linear mathematical model on the classical bag-of-words (BoW) representation is an example of a vector space modeling [7–10]. Implements a vector space model for document classification.

3.3 Probabilistic Topic Modeling

These modeling techniques identify the latent topics behind the text data. The modeling method represents documents within a latent topic space identified by a probabilistic method. It helps to identify the structure and semantics of the raw text data collected. It works on statistical models which take the classical BoW model for basic representation and do statistical calculations to identify the latent topics behind the unstructured data. Unsupervised machine learning is done on the probabilistic values of the raw text to analyze the latent topics. Probabilistic latent semantic analysis (pLSA), latent Dirichlet allocation (LDA), etc., are the main topic modeling techniques [11]. The main idea is adopted from a vector-based model where a statistical calculation on the model identifies the latent topics behind the text. This model is based on documents represented as mixture of topics and topics as a probabilistic distribution of words [12]. It cannot capture complex topics and word sense within the documents. But it shows improved accuracy in identifying topics for document classification.

4 Related Works

4.1 *Bag-of-Words (BoW)*

This is a vector space model where the elements represent the occurrence or frequency of features within the document. A document term matrix is formed considering the distinct features in all the documents for predictive modeling. This matrix is considered for document classification model. The recent work on this model proposes a word enrichment model for BoW model [1]. Here, the model is used specifically for short text classification. The documents that were taken for training, lack all the features. Thus, for all the low frequent terms, k-nearest neighbors are identified from the dense word vector model and are included to the document term matrix as new features with the same element value as the identified word. The model ignores the relation between terms in the dictionary and semantics behind the text. A hybrid approach is always used along with these models for a better prediction [9]. A model which identifies a vector form from the given vocabulary of words can be used to determine the word embedding of each word in the test set. A WordNet model can be used for multiple document classification [13]. Reference [14] proposes a word embedding model on BoW to find similarity among documents as such to find similarity.

4.2 *Latent Semantic Analysis (LSA)*

This method uses linear model to identify latent topics within raw text data. Singular value decomposition is done on the document term matrix created on classical BoW model to identify the distribution of topics over the document and words over the topics. Decomposition is done based on document term matrix using the formula.

$$A = USV^T \tag{1}$$

The matrix A represents the document term matrix with dimensionality $m \times n$, where m is the number of documents and n is the number of features. U and V are orthogonal matrix and S is a diagonal matrix. The matrix U is obtained by identifying the eigenvector of vector AA^T. Matrix V is obtained by identifying the eigenvector of matrix $A^T A$. The matrix S is a diagonal matrix where the diagonal elements represent the root of eigenvalues in descending order. The vector U is identifying similar documents over the latent topic in the matrix and V is identifying similar words over the latent topic in the matrix. The decomposition of matrix A given by

$$A_{m\times n} = U_{m\times m} S_{m\times n} V^T_{n\times n} \tag{2}$$

is done and further select K topics from the diagonal matrix for approximating k latent topic. This will in effect reduce the dimensionality by making the vector

$$A_{m\times n} = U_{m\times k} S_{k\times k} V^T_{k\times n} \tag{3}$$

Reference [7] gives a method for faster information retrieval using LSA and clustering. Here, a k-means clustering is done on the K value approximated decomposed matrix and the central points of the clusters are identified. When a document has to be retrieved the similarity can be done on the cluster points and further matching can be done on the most similar cluster documents. This can improve the speed of retrieval in case of classifying documents in classification models. This model generation will be complex with large data set hence it is used for a smaller data set. Reference [9] proposes a dimensionality reduction technique by using fuzzy clusters for improved performances in LSA for information retrieval.

4.2.1 Probabilistic Latent Semantic Analysis (pLSA)

This is a topic modeling method where probabilistic values are used to identify the latent topics within the raw model. The same decomposition as in the case of LSA is done using probabilistic values. Here, $P(W/D)$ is the probability of word W given a document D in the matrix and is equated to

$$P(W/D) = P(Z/D) \,.\, P(W/Z) \tag{4}$$

where Z is the latent topic, $P(Z/D)$ gives the probability of a topic Z given a document D and $P(W/Z)$ is the probability of a word W given a topic Z. The values of decomposition are identified through unsupervised learning algorithms. Paper [5] suggests a method that uses multichannel pLSA to classify biomedical signals. Here, similar to a text document the signals are considered with the time-specific signals as distinct documents and the value of signal at each points as the words within each signals. An expectation–maximization algorithm is used for identifying the decomposition components in pLSA. A similar method can be used for classifying text documents as in case of [7]. Here, online texts are classified based on semantic relation obtained by pLSA and TF-IDF feature weighting method. This is used for multilanguage text document clustering for faster information retrieval. It has its application in biomedical fields as given by [15] classifying protein, DNA, cell, gene, etc. It uses a probabilistic model which takes large number of parameters which grows linearly with number of training data set hence may lead to over fitting.

4.3 Latent Dirichlet Allocation (LDA)

Latent Dirichlet allocation is a biased framework of pLSA. Here, Dirichlet priori values are used to analyze the multinomial distribution values and using this distribution biased probabilities are found for word distribution within a document based on the latent topics. For a topic k, the word distribution ϕ given by

$$\phi \sim \text{Dirichlet}(\beta) \tag{5}$$

where k ranges from 1 to K is the approximated number of topics and β is the Dirichlet priory for tag distribution. For a document D_i draw topic distribution δ

$$\delta \sim \text{Dirichlet}(\alpha) \tag{6}$$

where i ranges from 1 to M are the distinct documents. And for each word draw, W

$$W \text{ Multinomial}(\phi \,.\, \delta) \tag{7}$$

Reference [12] suggests a method based on joint LDA where the distribution δ is calculated based on two characteristic features of the document collected. It suggests a method to classify multimedia tags. The words being the tag content and documents being the multimedia content are used to create the matrix representation. Reference [12] suggests that user interest distribution and objects topic distribution together contribute in identifying the δ value of the main model. Reference [5] suggests a hybrid model for document classification using LDA and GA (genetic algorithm). This LDA-GA model uses genetic algorithms to efficiently allocate weights to the LDA latent topics. A genetic evolution algorithm is used to identify optimization values for ϕ and a new ϕ' matrix is created for LDA model. Reference [16] proposes this algorithm to identify the subjective and objective words within documents. In this reference, the distribution values are created for the objective and subjective topics and are applied to the classical LDA model.

4.4 Word Movers Distance (WMD)

It is a method to identify concept similarity among documents. According to [2], document distance is identified on vector space to measure the similarity between two documents. A new metric to measure the distance between documents is defined by identifying word to word relation among words from two documents. WMD measures how words are separated from one document to another. Cosine similarity measurement is used to find similarity among words. A cumulative sum of similarity measurement is identified to find the document similarity. Paper [17] proposes a similar method which finds the distance between topics within document for classification purposes.

4.5 Fuzzy Bag-of-Words (FBoW)

The model is identified as a method which can replace the classical mapping models. The model uses a fuzzy mapping instead of a hard mapping as in case of classical BoW model. The fuzzy representation is based on word similarity as given by a word embedding model [15]. The model determines similarity based on cosine similarity between features determined using the word embedding. Comparison with other models on real-world database shows an improved performance of this representational model. But these models are expensive when considering larger data set for classification with other domain-specific data sets.

5 Practical Analysis

A comparative study for document classification is done by providing test data to various representative models described above. The real-world data sets 20newsgroups (20NG) and Reuters (RET) are considered for comparison. The statistics on the data sets are shown in Table 1.

20NG data set has comparatively equal distribution of test set and train set over 20 categories. Each having sufficient training and test set. Data set is considered for all the 20 categories and 6 distinct categories. Full train and test data from the specified categories are considered for comparison.

- In 20NG data set, all the train and test document set are considered.
- In 20NG(6), 6 distinct classes are considered. They are comp.graphics, rec.autos, sci.crypt, talk.politics.misc, soc.religion.christian, misc.forsale.

Table 1 Database statistics

Data set	Train samples	Test samples	Words	Features
20NG	11,314	7532	15,41,522	77,348
20NG(6)	3422	2279	47,992	30,697
RET(4)	966	342	88,651	5282
RET(20)	123	79	11,541	2284

RET data set a collection of articles on 90 categories. Here, the distribution of train set and test set is drastically varying. The set contains a minimum of 1 and maximum of 2877 train set and minimum of 1 and maximum of 1087 documents as subsets in different classes within the set. Here, 4 dissimilar data set and 20 random categories are considered for comparison.

- In RET(4), 4 distinct classes are considered. They are grain,interest,reserves, dlr.
- In RET(20), 20 random classes are considered. They are nkr, palladium, rye, nzdlr, dfl, naphtha, rand, cpu, potato, propane, coconut, jet, platinum, oat, nickel, dmk, fuel, hog, tin, orange.

The data sets undergo the preprocessing steps and the most common 2000 features are considered for comparison. Support vector machine (SVM) is considered as the classifier algorithm [8].

6 Results

The comparison is done on data sets for various representation models such as BoW, LSA, LDA, WMD and FBoW. Table 2 shows the results obtained on the different data sets.

Graphical depiction of the accuracy obtained by analysis is given by Fig. 1. The figure gives a comparison of accuracy given by various models for the data sets considered.

The data sets show a drastic performance using the fuzzy model [15] which gives the maximum accuracy when compared with classical BoW, linear mathematical dimensionality reduction model and statistical model in case of 20NG and 20NG(6). Soft mapping outperforms the conventional hard mapping methods to build an effective fuzzy representational model. Fuzzy model delegates the root causes of the classical BoW model which is suitable for classification of large domain-specific models which are used in real-life classification problems. It also accounts for reducing the dimensionality and sparse elements within the matrix representational model. Fuzzy mapping can be integrated with other dimensionality reductions techniques for improved accuracy [15].

Table 2 Accuracy comparison of representative models on different data sets

Data set	Accuracy				
	BoW	LSA	LDA	WMD	FBoW
20NG	0.664233	0.731944	0.713426	0.721240	0.751593
20NG(6)	0.866608	0.904782	0.901324	0.921024	0.973457
RET(4)	0.967836	0.950292	0.957638	0.642138	0.883041
RET(20)	0.607594	0.645570	0.594549	0.432798	0.088608

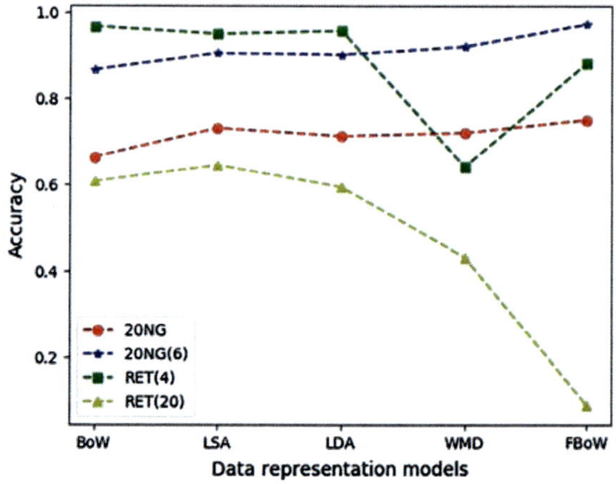

Fig. 1 Comparison of various data representation models on different data sets

BoW model gives maximum accuracy in case of RET(4). And LSA model gives maximum accuracy in case of RET(20). The results show that the most accurate model FBoW for 20NG data set performs poorly in case of heterogeneous samples like RET. The degradation occurs as there are no sufficient features in the trained data to remove the ambiguity introduced when a fuzzy membership is used.

7 Summary

A practical analysis on different document representational models is done. The features to be analyzed like document type as comments used for sentiment analysis, short text from a social media element for clustering and recommendation systems, or raw form of data from digital media for classification, etc., requires different document representation methods. There is no representational model which best suits every data sets. The main objective is to form a simple and reliable representational model suitable for real-life data sets. Here, the study on data sets is done. Fuzzy mapping models show improved accuracy based on its performance on data sets with sufficient train data. But the classical BoW model and LSA model perform better in case of RET data set which resembles practical application data sets used in everyday processes.

References

1. Heap B, Bain M, Wobcke W, Krzywicki A, Schmeidl S (2017) Word vector enrichment of low frequency words in the bag-of-words model for short text multi-class classification problems. In: Computational and language and machine learning ACM classes I.2.7;I.2.6. PMID arXiv: 1709.05778v1. https://arxiv.org/pdf/1709.05778.pdf

2. Huo Z-L, Wu J-F, Lu Y, Li C-Z (2018) A topic-based cross-language retrieval model with PLSA and TF-IDF. In: IEEE 3rd international conference on big data analysis. https://doi.org/10.1109/ICBDA.2018.8367704
3. Yang L, Chen X, Liu Z, Sun M (2017) Improving word representations with document labels. ACM Trans Audio Speech Lang Process 25:863–870. https://doi.org/10.1109/TASLP.2017.2658019
4. Aisopos F, Tzannetos D, Violos J, Varvarigou T (2016) Using n-gram graphs for sentiment analysis: an extended study on Twitter. In: IEEE second international conference on big data computing service and applications. https://doi.org/10.1109/BigDataService.2016.13
5. Hsu C-I, Chiu C (2017) A hybrid Latent Dirichlet Allocation approach for topic classification. In: IEEE international conference on innovations in intelligent systems and applications (INISTA). https://doi.org/10.1109/INISTA.2017.8001177
6. Maiti R (2018) Determining the best classifier for predicting the value of a Boolean field on a blood donor database using genetic algorithms. Machine Learning (stat.ML); Machine Learning (cs.LG), arXiv: 1802.07756. https://doi.org/10.5281/zenodo.1336304
7. Wenli C (2016) Application research on latent semantic analysis for information retrieval. In: Eighth international conference on measuring technology and mechatronics automation. https://doi.org/10.1109/ICMTMA.2016.37
8. Ge L, Moh T-S (2018) Improving text classification with word embedding. In: IEEE international conference on big data (BIGDATA). https://doi.org/10.1109/BigData.2017.8258123
9. Karami A (2017) Taming wild high dimensional text data with a fuzzy lash. In: Article on machine learning, computation and language (cs.CL). Information retrieval (cs.IR). https://doi.org/10.1109/ICDMW.2017.73
10. Vences R, Gómez J, Menéndez V (2016) A document recommendation system using a document. Similarity ontology. IEEE Lat Am Trans 14(7):3329–3334. https://doi.org/10.1109/TLA.2016.7587638
11. Wang J, She M (2016) Probabilistic latent semantic analysis for multichannel biomedical signal clustering. IEEE Sig Process Lett 23:1821–1824. https://doi.org/10.1109/LSP.2016.2623801
12. Yao J, Wang Y, Zhang Y, Sun J, Zhou J (2017) Joint latent Dirichlet allocation for social tags. IEEE Trans Multimedia 20(1):224–237. https://doi.org/10.1109/TMM.2017.2716829
13. Regmi S, Bal BK, Kultsova M (2017) Analyzing facts and opinions in Nepali subjective texts. In: 8th international conference on information, intelligence, systems and applications (IISA). https://doi.org/10.1109/IISA.2017.8316445
14. Liu M, Lang B, Gu Z, Zeeshan A (2017) Measuring similarity of academic articles with semantic profile and joint word embedding. Tsinghua Sci Technol 22:619–632. https://doi.org/10.23919/TST.2017.8195345
15. Li L, Li H (2018) A novel document distance based on concept vector space. In: 17th IEEE international conference on communication technology. https://doi.org/10.1109/ICCT.2017.8359982
16. Wang H, Wu F, Lu W, Yang Y, Li X, Li X, Zhuang Y (2017) Identifying objective and subjective words via topic modeling. IEEE Trans Neural Netw Learn Syst 29:718–790. https://doi.org/10.1109/TNNLS.2016.2626379
17. Wang B, Ou Z, Tan Z (2017) Learning trans-dimensional random fields with applications to language modeling. IEEE Trans Pattern Anal Mach Intell 40:876–890. https://doi.org/10.1109/TPAMI.2017.2696536

Exponential Cipher Based on Residue Number System and Its Application to Image Security

Sagar Ramesh Pujar, Achal Ramanath Poonja, and Ganesh Aithal

Abstract Residue number system (RNS) which was inspired by Chinese remainder theorem (CRT) provides a method for using groups of smaller numbers to represent large integers. Although cryptographic algorithms based on decimal arithmetic have their advantages, direct sum of smaller rings properties-based algorithms offers better benefits. This application of RNS number systems successfully utilized in an image security. An exponential cipher based on RNS is discussed in this paper. RNS-based exponential cipher systems exhibit properties that increase computation speed, reduce the time complexity and makes the system immune to known-plaintext attacks, side-channel attacks and algebraic attacks. In addition, the characteristics of exponential cipher systems, the method of key generation and RNS-based encryption and decryption are discussed in this paper.

Keywords CRT · RNS · Stream cipher · Key sequence · Exponential cipher

1 Introduction

Demand on information security has widely increased due to the sensitivity of the exchanged information over public communication channels. One of the primary goals of the cryptographic systems (cryptosystems) is to help communicators exchange their information securely. This goal is achieved by cryptographic applications and protocols. Ciphertext is an encryption method of converting any message

S. R. Pujar (✉) · A. R. Poonja
Mangalore Institute of Technology and Engineering, Badaga Mijar, Moodbidri, Mangalore, Karnataka, India
e-mail: pujar.sagar@gmail.com

A. R. Poonja
e-mail: achalpoonja@gmail.com

G. Aithal
Shri Madhwa Vadiraja Institute of Technology and Management, Bantakal, Udupi, Karnataka, India
e-mail: ganeshaithal@gmail.com

N. N. Chiplunkar and T. Fukao (eds.), *Advances in Artificial Intelligence and Data Engineering*, Advances in Intelligent Systems and Computing 1133,
https://doi.org/10.1007/978-981-15-3514-7_41

to an inscrutable form. The negation of ciphertext into the original messages is decryption, which helps obtain plaintext from ciphertext.

Cryptographic systems can be generally classified into two ways. (i) Symmetric key systems that use an individual key for encryption and decryption at the sender and the receiver, respectively. (ii) Asymmetric key systems use a pair of different keys for encryption and decryption. For example, RSA [1], elliptic curve cryptography [2], etc. These public-key systems utilize a public key–private key pair, where public key is known to everyone and private key is confidential to the receiver [3–5].

Block and stream ciphers are two variety of symmetric cryptosystems. Operation of block ciphers on bigger blocks of plaintext data by transforming them into secure ciphers; operate on individual plaintext with a time-varying characters is known as stream ciphers [5, 6]. This classification cannot be claimed as complete, and hence, block ciphers are one of the variants of stream ciphers. Block ciphers can be converted to stream ciphers using operations such as output feedback mode, cipher feedback mode (CFB) and counter mode. The security affects the stream cipher security, with the corresponding block cipher [3, 4, 6].

The systems which provide a model of dividing data into blocks and encrypting and decrypting them block by block is called as block cipher, whereas stream ciphers differ considerably as encryption, and decryption is processed pixel by pixel, or byte by byte or word by word or character by character. Block ciphers and stream ciphers still find applications in the modern world. In general, stream cipher follows a methodology of adding bit by bit modulo 2 binary random sequences called the key sequence to the binary message for encryption. Stream cipher systems provide real-time encryption and decryption capabilities and thus find application in business and military sectors [3–5, 7–9].

The uncertainty properties of the sequence called the key sequence affect the stream cipher security. Therefore, generating the key sequence is a critical component of the stream cipher system. Binary bit sequences with prudent statistical properties generating using a random bit generator have got vital usage in applications of cryptography. The cryptographic research community still faces a challenge when it comes to design of secure and efficient pseudorandom sequence generators [10].

2 Literature Review

Residue number systems is an extension of the CRT grants the usage of smaller integers to represent large integers. This can be proved in residue number system that the direct sum of commutative rings formulated by relatively prime products, which is obtained by the set of all integers from 1 to M based on residue number system, is component-wise modular addition and multiplication of the same. Direct sum of smaller commutative rings-based encryption and decryption algorithms are constructed in parallel which is more robust than decimal arithmetic. Additive, affine and multiplicative stream cipher systems utilize the representation of integers using

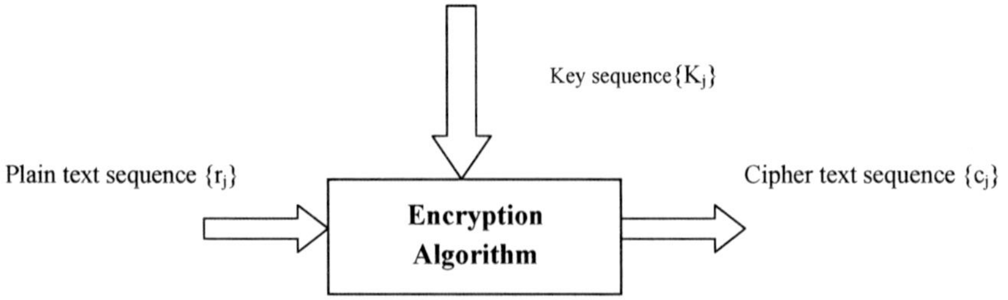

Fig. 1 Schematic representation of a basic cipher system

RNS [11] is possible which is called as RNS-based cipher system. These RNS-based cipher systems have enhancements to the speed of the encryption/decryption process, reduction in the time complexity and immunity for the system against various attacks like algebraic attacks, side-channel attacks and known-plaintext attacks.

RSA, Diffie–Helmann, elliptic curve likewise cryptographic applications have a requirement of modulo multiplication and modulo exponentiation of large numbers. By the usage of RNS, the operation (X^Y) mod N can be performed in parallel and faster [12].

In stream cipher system, sequence of integers which is random in nature as key sequence $\{K_j\}$ is used to encrypt the plaintext sequence $\{r_j\}$, to get sequence of ciphertext $\{c_j\}$. The schematic representation of a basic cipher system is shown in Fig. 1. Here, $\{K_j\}$, $\{r_j\}$ and $\{c_j\}$ are from Z_m, ring of residue class integer modulo m.

Generally, an additive integer modulo m operation is subjected over the input text sequence or the plaintext sequence and the key sequence elements. This is consequently called as additive cipher systems. Here, encryption algorithm is modular addition key elements with plaintext. Additive systems utilize a single key sequence for encryption. Similar to modular addition, modular multiplication and modular exponential based cipher systems are also possible. This paper explores the conditions and characteristics of exponential cipher system.

Mathematical Base of the Paper: Assume M to be any composite integer, expressible in terms of product of its k factors. These factors are to be pair-wise relatively prime.

$$M = \prod_{i=1}^{k} m_i \tag{1}$$

Indicating, GCD of any two factors in general is equal to one. Let r be any integer which is plaintext, is represented as, $0 \le r \le M$, Then, the k-tuple can be represented as $r_1 \equiv r \bmod m_1$, system of congruence, $r_2 \equiv r \bmod m_2$, $r_3 \equiv r \bmod m_3$, $r_4 \equiv r \bmod m_4 \ldots.\ r_k \equiv r \bmod m_k$,

This satisfies congruencies as shown in Eq. 2.

$$r = \{r_{k-1}, r_{k-2}, \ldots r_1, r_0\} \tag{2}$$

This is called as the Chinese remainder theorem.

The digits r_i, $i = 2, 3, \ldots, k$, are such that $0 \leq r_i \leq m_i$. This lets r to be expressed in terms of unique k-tuple using CRT. {'r_i'} represented in terms of $\{k_j\}$ digits is RNS representation based on CRT.

Conversion of RNS Representation to Fixed Radix: Suppose a positive integer r, $0 \leq r \leq M$, has RNS representation $r \rightarrow (r_k\, r_{k-1}\, r_{k-2} \ldots r_2\, r_1)$ then r is given by,

$$(r_k \times w_k + r_{k-1} \times w_{k-1} + r_{k-2} \times w_{k-2} + \cdots + r_2 \times w_2 + r_1 \times w_1) \bmod M \tag{2.a}$$

w_js can be obtained as follows. We define M_j equal to M/m_j, which is relatively prime to m_j, $1 \leq j \leq k$. Now, w_j is defined as [3].

$$w_j = \left[\left\{M_j^{-1} \bmod m_j\right\} M_j\right] \bmod M \tag{3}$$

and

$$r = \left[\sum r_i\, w_i\right] \bmod M \tag{4}$$

After brief mathematical introduction, further paper moves to the methodology in the next section.

3 Methodology

Given a finite composite, integer M can be expressed as distinct RNS representation as shown in the Eq. (4). Based on this, equation is possible to develop a stream cipher system. This proposed scheme uses any integer, $0 \leq r < M$ as a k-tuple representation based on RNS. Encryption is essentially a key dependent transformation of k-tuples from plaintext to ciphertext in k-dimensional space. The use of proper decryption key maps the k-tuple ciphertext back to its original plaintext k-tuple. The scheme can be regarded as mapping of one-dimensional point to a k-dimensional space where encryption and decryption are defined. The scheme is illustrated in Fig. 2.

Plaintext $\{r_i\}$ is represented in RNS to get k-tuple $(r_{k-1}, r_{k-2}, \ldots, r_1, r_0)$ denoted by $\{r_i\}$. After encryption using key $\{K_j\}$, the ciphertext is $(c_{k-1}, c_{k-2}, \ldots, c_1, c_0)$ denoted by $\{c_i\}$. With decryption key $\{K_{i'}\}$, the cipher $\{c_i\}$ text is mapped back to original plaintext $(r_{k-1}, r_{k-2}, \ldots, r_1, r_0)$, $\{r_i\}$. In the proposed work, conversion of r, $0 \leq r < M$ into RNS representation, encryption/decryption and reconversion to r are all regarded as appropriate mappings. After RNS representations of the plaintext, encryption and decryption algorithms are performed. These operations carry free

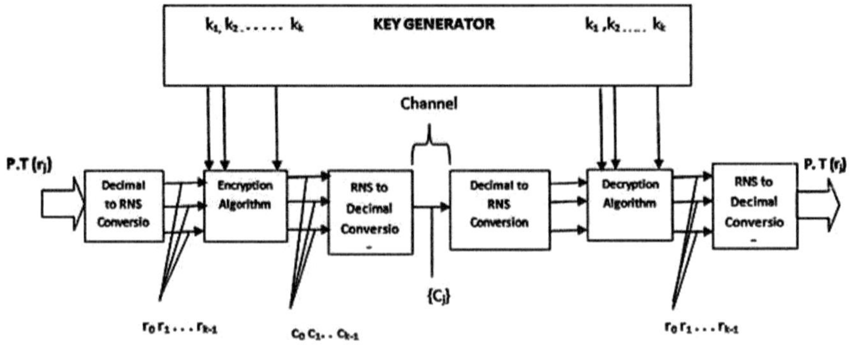

Fig. 2 Proposed scheme uses any integer, $0 \leq r < M$ as a k-tuple representation based on RNS

arithmetic operations. No arithmetic computation in Z_M is carried out; however, all the operations are carried out using RNS representations.

The conditions to be satisfied by random key sequences for exponential cipher and generation of key sequences are investigated. The cipher system compares these with cipher system.

This scheme employs integers from Z_m. The exponential cipher can be defined when m is not of the form p^e. Let $\phi(m)$ denotes the number of integers relatively prime to m. Such integers are called as units in Z_m and are used in multiplicative and exponential cipher. As stated earlier, the set of all such integers is called reduced residue class of integer modulo m. This set is denoted by U_m and can be shown to be a commutative group under multiplication modulo m. Let b be an integer which is relatively prime to $\phi(m)$. Then, there exists an integer d which is multiplicative inverse of b mod $\phi(m)$ that is bd mod $\phi(m) \equiv 1$. There are $\phi(\phi(m))$, number of integers which are relatively prime to $\phi(m)$ and having multiplicative inverse modulo $\phi(m)$. Then, as in the case of multiplicative cipher, key sequence $\{b_i\}$ of integers relatively prime to $\phi(m)$ can be generated. Let $\{r_i\}$ be the plaintext integer sequence, $\{b_i\}$ be the key sequence, and then, corresponding ciphertext is given by

$$\{c_i\} = \{r_i\}^{\{b_i\}} \bmod m \tag{5}$$

Decryption algorithm follows from Euler's generalization of Fermat's theorem for the generation of key to get plaintext back again.

$$\{r_i\} = \{c_i\}^{\{d_i\}} \bmod m \tag{6}$$

where d_i is multiplicative inverse of b_i modulo $\phi(m)$. As mentioned earlier, the scheme can be defined for the case where m is not of the form p^e *or* p^e as factor, (p is a prime integer, and e is the integer exponent) as factor.

Key generation based on FSR: As it was said in the previous section that d_i is multiplicative inverse of b_i modulo $\phi(m)$. where *di* is the key for the receiving end, b_i is the key for the sending end and m is modular function. In case of the considered

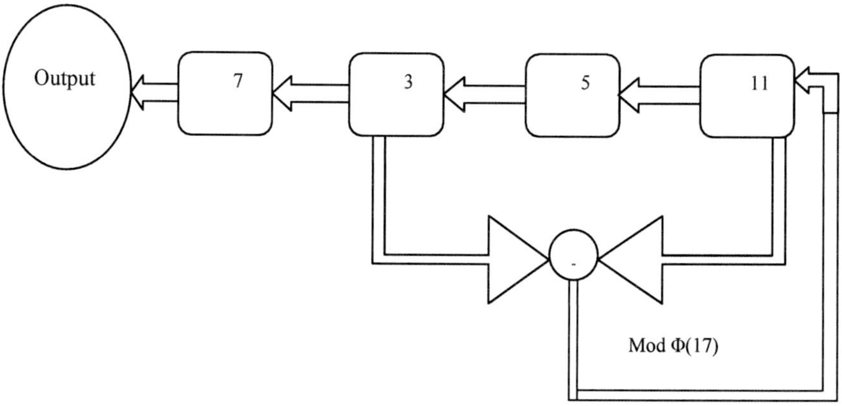

Fig. 3 The schematics diagram of the key generator of the third tuple of the RNS converted from the pixel of a gray scale image of the Lena image

example in the next section, m becomes either of three that is m_0, m_1 or m_2 that is 3, 5 or 17.

The key is generated for the three separate encryption processes, first one is the generation for exponential modulo 3, where the keys are to be between zero and $\Phi(m_0)$ where m_0 is 3. Therefore, the generation limit is from 0 to 2 where 2 is $\Phi(m_0)$ or $\Phi(3)$ since the only one integer is present in this group that is 1, we cannot have encryption for this tuple of the residue number system. Next group of key is from 0 to $\Phi(5)$ that is 0 to 4. In this case only two elements of the set, that is, 1 and 3 will have multiplicative inverse. Figure 3 gives the schematic for the generation of the key for the exponential cipher part of modulo 17 tuple, with the initial contents 7, 3, 5 and 11 are shown. These numbers will have inverse since the number is relatively prime to $\Phi(17)$. In this case, the groups of the numbers which can be used for the generation of the key are $\{0 \text{ to } \Phi(17)\} = \{0 \text{ to } 16\}$ which is relatively prime to $\Phi(17)$.

Operation $\odot$ indicates content of the register to the power of content of another register; also, it should be added with the modular functions mentioned. This work has taken three relatively primes which are actually factors of the highest pixel value, 255, i.e., 3, 5 and 17. Generations of keys, for each of these factors, are based on linear-feedback shift registers. It is capable of generating keys which are equal to the number of pixels in the input picture. So as one-time pad criteria can be obtained. The condition for the key generation is, as said earlier:

As mentioned in Fig. 3, initially four registers are taken, the value in second register is operated with $\odot$ the value in the fourth register, and result is calculated by applying modular function with Φ(m). In this work, value of m is taken as prime as 3, 5 and 17 so as they are suitable to represent in residue number system.

Application for image encryption: Let us consider $M = 255$ and its factors $m_1 = 3$, $m_2 = 5$ and $m_3 = 17$. Since all three factors are pair-wise relatively prime, this can be used as a factor for residue number system. In case of a monochrome image, each pixel will be represented in terms of 0 to 255 gray levels, and these pixels can be represented in terms of residue numbers based on the tuple 3, 5 and 17. In this paper, the standard monochrome image Lena is used for the encryption.

Example A pixel value of 173 can be represented in residue number of 3, 5 and 17 factors as three tuple as follows.

(173 mod 3, 173 mod 5, 173 mod 17) = (2, 3, 3) in case if an arbitrary three tuple is (1, 3, 7) taken as key for each of plaintext tuple. This pixel can be then encrypted as defined from the Eq. (1) given in the previous section as based on encryption algorithm as

$$\mathrm{CT} = (21 \bmod 3), (33 \bmod 5), (37 \bmod 17) = (2, 2, 11)$$

This can be reconverted into pixel of limits 0 to 255 as
where 3 is number of tuple of the residue number system.
$A_i = M/m_i$ and a_i^{-1} is A_i multiplicative inverse modulo mi

$$r = \left(\sum_{i=1}^{k} A_i\, \mathrm{a}_{\mathrm{i}}^{-1} m_i \right) \bmod 255 \tag{7}$$

$$\mathrm{A}_1 = 255/3 = 85 \quad a^{-1} = (85^{-1} \bmod 3) = 1$$
$$\mathrm{A}_2 = 255/5 = 51 \quad a_2^{-1} = (51^{-1} \bmod 5) = 1$$
$$\mathrm{A}_3 = 255/17 = 15 \quad a_2^{-1} = (15^{-1} \bmod 17) = 8$$

Therefore, (2, 2, 11) in terms of decimal equivalent to

$$((85 \times 1 \times 2) + (51 \times 1 \times 2) + (15 \times 8 \times 11)) \bmod 255 = 62.$$

This 62 is reconverted at the receiving end into an equivalent RNS number as

$$(62 \bmod 3, 62 \bmod 5, 62 \bmod 17) = (2, 2, 11)$$

Decryption keys are (1, 3, 7) which are multiplicative inverse mod Φ(3) Φ(5) and Φ(17) of key assumed during sending end, respectively.
Therefore, as per decryption algorithm,

$$\mathrm{PT} = \left(\left(2^1 \bmod 3\right), \left(2^3 \bmod 5\right), \left(11^7 \bmod 17\right) = (2, 3, 3)\right.$$

That is (2, 3, 3) in terms of decimal equivalent to

$$((85 \times 1 \times 2) + (51 \times 1 \times 3) + (15 \times 8 \times 3)) \bmod 255 = 173$$

The above equation gives the plain text back again. In the next section, the results obtained are analyzed.

4 Performance Evaluation

Performance of the schemes can be evaluated by conversion of the ciphertext elements into decimal representation by using RNS to decimal conversion methodologies. The following parameters taken into account for investigation of performance of encryption algorithms are as follows.

1. Visual analysis,
2. Histogram of number of occurrences of both plaintext and ciphertext,
3. Entropy of number of occurrences of both plaintext and ciphertext,
4. Efficiency of encryption by taking into account the plaintext and the ciphertext,
5. Observation of the avalanche effect is also done by using two different key sequences generated as a result of alteration of initial values in one bit position and using this to encrypt the same plaintext [11].

In the experiment, standard monochrome image, 'Lena' is taken as an example. The size of each pixel is eight bits, and this is converted into an equivalent RNS of three tuple. This three tuple of RNS is based on $M = 255$ followed by $m_1 = 3, m_2 = 5$ and $m_3 = 15$ ($m_1 \times m_2 \times m_3 = 255 = 3 \times 5 \times 17$). Each tuple is encrypted based on the algorithm mentioned in above section. The following results are obtained.

5 Result and Observation

Further, in a stream cipher system if the period of the key sequence is larger than the length of plaintext, the cipher system approaches **one-time pad**, which is theoretically secure [13]. Side-channel attack is also reduced by making single plaintext into multiple components [14].
Visual Analysis: The following Figs. 4 and 5 indicate insignificant residues of input image in the cipher image after encryption. It indicates a better encryption process, even though there is small residue. Hence, the proposed encryption scheme is secure and efficient.
Histogram: The histograms of the number of occurrences of input image and cipher image are plotted versus pixel values. The histograms of the number of occurrences of encrypted form of the original image and the original image are represented in Figs. 6 and 7, respectively. Visually, the histogram of the encrypted image can be observed to be uniformly distributed, unlike the original image. So, the encrypted image dissolves any opportunity of carrying out statistical attacks on the image encryption procedure proposed. These properties elicit the proposed image encryption provides high immunity against statistical attacks. The concept of residue number system helps split the information of the pixel into k parts. There is enhancement to the pace of encryption and decryption by parallel encryption and decryption.
Entropy: Entropy is evaluated by the following equation,

Fig. 4 Plain image of Lena

Fig. 5 Cipher image Lena mentioned in Fig. 4

$$\text{Entropy} = \operatorname*{Lim}_{N\to\infty} \sum_{i=0}^{M-1} p_i \log_2 \frac{1}{p_i} \tag{8}$$

The evaluated entropy is compared with A5/1 and additive system and is shown in Fig. 8. It has been observed that comparatively it is better.

Encryption Efficiency: Entropy values are calculated by using the Eq. (8). This is also compared with the same set of algorithm given in the previous test, and results are encouraging, which is shown in Fig. 9.

Avalanche effect: For the one bit different in the initial values of two key sequences, ten values of ciphertext are plotted, with plaintext value is shown in Fig. 10. The initial

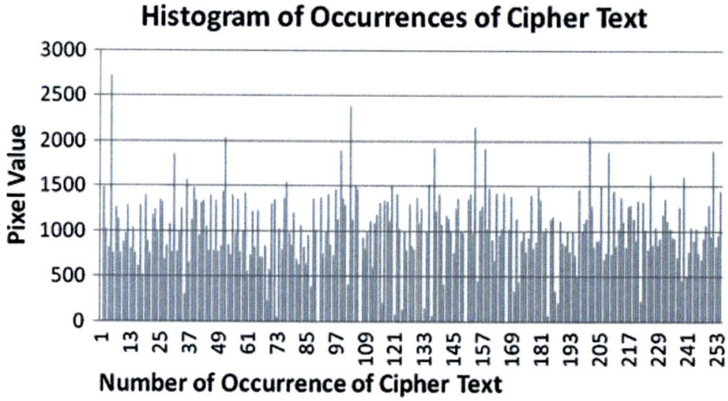

Fig. 6 Histogram of number of occurrences of ciphertext with its pixel value

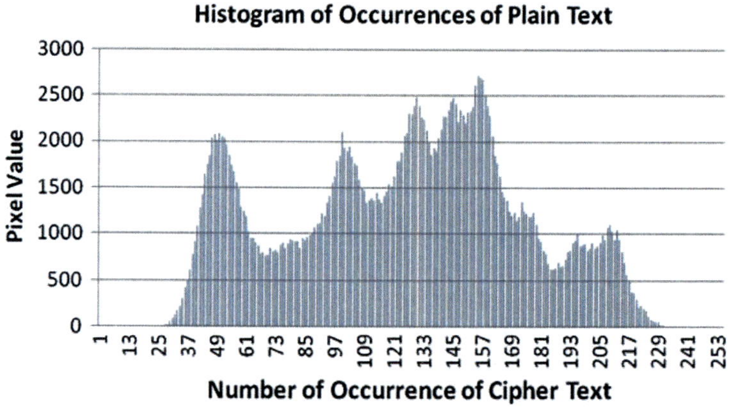

Fig. 7 Histogram of number of occurrences of plaintext with its pixel value

values and encrypting the same plaintext with the two new key sequences obtained can be used to produce an avalanche effect of this algorithm. Eavesdroppers work would become more complex as small changes in the initial value of key impart disparity in the ciphertext.

6 Immunity Against Attacks

As said earlier the immunity for the side channel, algebraic and known-plaintext attacks are briefed below.

Side-Channel Attack: In binary stream cipher system, timing, radiation and power analysis become simple since there are only four combinations of operations. That

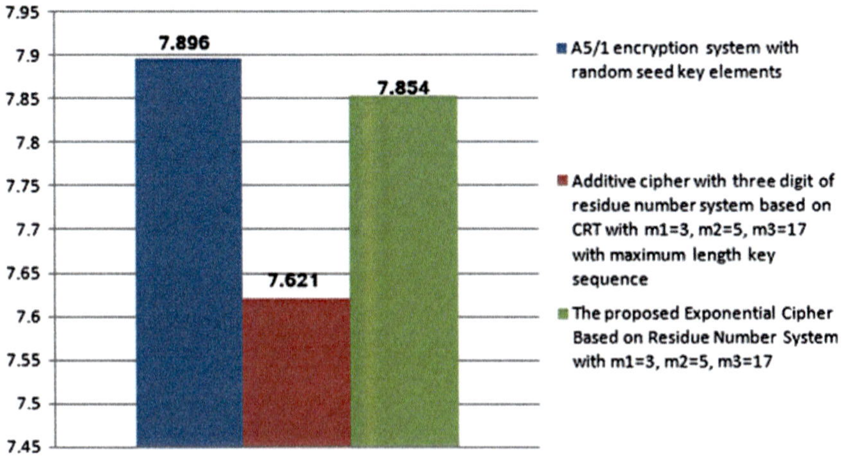

Fig. 8 Entropy comparison of two previously proposed systems with proposed exponential cipher-based residue number system [15]

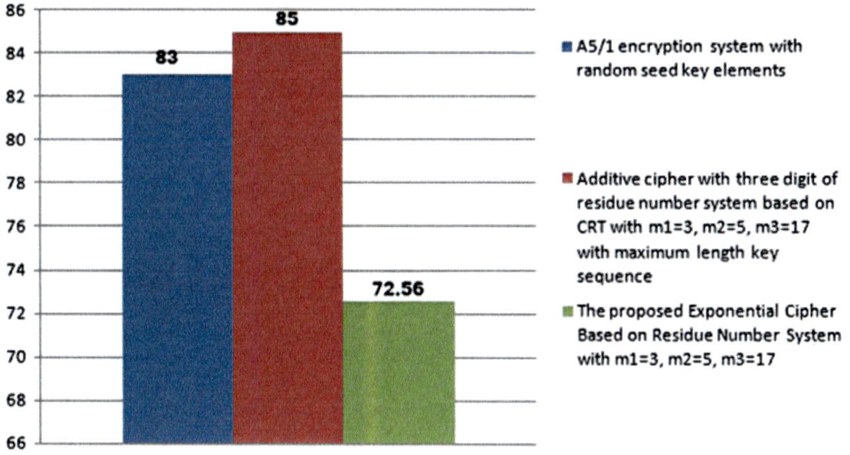

Fig. 9 Encryption efficiency comparison of two previously proposed systems with proposed exponential cipher-based residue number system

is 0 is operated on 0, 0 is operated on 1, 1 is operated on 0 and 1 is operated on 1. Because of these four possibilities, it becomes simple for the eavesdropper to predict the result based on timing, radiation and power.

In the above-proposed example, three bit streams are working in parallel since direct sum of three commutative rings is used for encryption process. There are three arithmetic exponential operations in parallel. Hence, the prediction from side-channel attack becomes difficult as the parallel operation increases complexity of the prediction. Hence, the side-channel attack can be foiled.

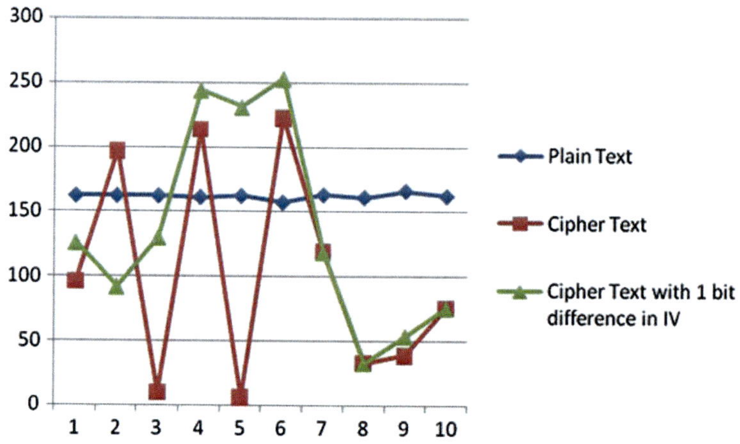

Fig. 10 Avalanche effect with one bit difference in IV is mentioned in Fig. 3

Algebraic attack: The exponential and modular operation lead to nonlinear algebraic operations. Hence, algebraic attack can be foiled.
Known-plaintext Attack: Known-plaintext attack can also be eliminated by the parallel operation by splitting the number into *k* components.

7 Conclusion

In the proposed system, it is possible to generate maximum length, key sequence, by choosing large numbers of shift registers. Hence, one-time padding is possible, and this enables the system as theoretically secure since key length is maximum. The exponential cipher gives a non-linear output; it is obvious that statistical attack can be foiled. Also, this cipher system possesses nonlinearity hence algebraic attack. A single plaintext is made into multiple components, and hence, the side-channel attack is also reduced. The visual effect of the ciphertext gives absolutely no additional information about the plaintext, hence able to define strong diffusion and confusion. Known-plaintext attack is also thwarted in this case. The analysis of entropy and encryption efficiency shows the result from the proposed system is comparatively better than previously proposed systems.

References

1. Rivest RL, Shamir A, Adleman L (1978) A method for obtaining digital signatures and public-key cryptosystems. Commun ACM 21(2):120–126
2. Koblitz N (1997) Elliptic curve cryptosystems

3. Menezes A, Van Oorschot P, Vanstone S (1997) Handbook of applied cryptography, 1st edn. CRC Press, Boca Raton
4. Schneier B (1996) Applied cryptography, 2nd edn. Wiley, ISBN: 0471128457
5. Rueppel RA (1992) Stream ciphers. In: The science of information integrity. IEEE press, New York
6. Stallings W (2006) Cryptography and network security, principles and practice, 5th edn. Pearson Education Inc
7. Zenner E (2004) Cryptanalysis of LFSR-based pseudorandom generators—a survey. University of Mannhoim (Germany)
8. Golomb S (1967) Shift register sequences. Aegean Park Press (1982) reprint, Laguna Hills, California
9. Robshaw MJB (1995) Stream ciphers. RSA Laboratories Technical Report TR-701, Version 2.0
10. Ramesh S, Haribhat KN, Murali R (2010) On linear complexity of binary sequences generated using matrix recurrence relation defined over Z4. Int J Distrib Parallel Syst (IJDPS) 1(2)
11. Aithal G, Hari Bhat KN, Sripathi U (2010) Implementation of stream cipher system based on representation of integers in residue number system. IEEE, pp 211–213
12. Anand Mohan PV (2016) Residue number systems. Springer Nature
13. Shannon C (1949) Communication theory and secrecy system. Bell Syst Tech J 28(4):656–715
14. Garrett P (2000) Introduction to cryptography, notes
15. Sudeepa KB, Aithal G (2016) Generation of maximum length non-binary key sequence and its application for stream cipher based on residue number system. J Comput Sci. 31:379–386. https://doi.org/10.1016/j.jocs.2016

Using Machine Learning and Data Analytics for Predicting Onset of Cardiovascular Diseases—An Analysis of Current State of Art

P. R. Mahalingam and J. Dheeba

Abstract Cardiovascular diseases are becoming one of the largest causes of natural fatalities around the world today. The major reason for this is attributed to the unhealthy lifestyle trends followed by developed nations and lack of proper diagnostics in developing nations. In this paper, we observe how different diagnostic methods can contribute to building an automated decision support system that will help experts to predict the onset of heart diseases. As with any data analytics problem, we initially try to classify the data based on annotated classes and explore a set of algorithms in that domain. Then, we observe peculiarities of heart disease data, and find that some level of clustering will help increase the accuracy of the algorithm. This leads to the formation of an ensemble, which clubs together suitable clustering and classification algorithms, and we run tests on various disease datasets to evaluate the performance of the algorithms. From the analysis, we can conclude that Bayesian models perform well when supported by clustering algorithms, and they can be generally applied over a range of disease data. Analysis was also done by considering the time taken for making the decision, and found that Bayesian algorithms are competent enough to give a good level of accuracy in reasonable amount of time for cardiovascular diseases.

Keywords Machine learning · Cardiovascular diseases · Data analytics

1 Introduction

One of the most common health problems faced by the world today is heart failure, where the heart fails to pump enough blood to vital organs [1]. This is of high importance since the issue is compounded by lifestyle conditions like obesity. While lifestyle disorders are of concern in developed nations, lack of appropriate diagnosis

P. R. Mahalingam (✉) · J. Dheeba
School of Computer Engineering, Vellore Institute of Technology, Vellore, India
e-mail: prmahalingam@gmail.com

J. Dheeba
e-mail: dheeba.j@vit.ac.in

N. N. Chiplunkar and T. Fukao (eds.), *Advances in Artificial Intelligence and Data Engineering*, Advances in Intelligent Systems and Computing 1133,
https://doi.org/10.1007/978-981-15-3514-7_42

measures is of concern in developing nations. Conventional diagnosis models use the person's lifestyle and other medical history to measure the possibility that the person has a heart problem. But this is time consuming and may turn out to be fatal. Here, we try to perform a survey of different methods that have been aimed at predicting onset of cardiovascular diseases based on easily accessible clinical data.

According to Mayo Clinic, "Heart disease describes a range of conditions that affect your heart. Diseases under the heart disease umbrella include blood vessel diseases, such as coronary artery disease; heart rhythm problems (arrhythmias); and heart defects you're born with (congenital heart defects), among others". They mention that diagnosis of heart diseases may use the following.

1. Blood tests like cholesterol, high sensitivity C-reactive protein, lipoprotein (a), plasma ceramides, natriuretic peptides, etc.
2. Chest X-rays for general heart size and pericardial effusion
3. Electrocardiogram (ECG)—measuring heart rhythm using electric signals. We look at the initial P-wave, intermediate QRS complex, and the closing T-wave
4. Holter monitoring—continuous ECG over 24–72 h
5. Echocardiogram—ultrasound imaging
6. Stress test—heart rate analysis
7. Cardiac catheterization—blood flow and pressure analysis using special dyes
8. Cardiac computerized tomography (CT) scan
9. Cardiac magnetic resonance imaging (MRI).

Diagnosis involves a range of methods, ranging from simple tests to complex imaging.

The rest of the paper is organized as follows. Section 2 is concerned with different data sources and models associated with medical information. Section 3 deals with classification and prediction. Section 4 deals with dynamics of heart disease data, while Sect. 5 deals with clustering of dynamic information. A brief comparison of the current state of art is discussed in Sect. 6 with a variety of datasets.

2 Medical Data Collection and Management—Background and Current State of Art

Medical data is generally measured using two methods [2]—invasive and non-invasive. Invasive methods work by taking samples from the body and analyzing them for different parameters. Non-invasive ones take data from the body by listening to various senses like heartbeat, blood pressure, etc. In general, data like age, gender, cuisine, location, diabetic status, blood pressure, genetic information, ECG signals, etc., are used for studying heart diseases. Hence, we have a mix of possible input formats including numerical data, nominal data, textual data, medical rule data, medical images, etc.

Wright et al. [3] stressed the possibility of using data mining methods to identify associations between diagnostics, medication, and disorders. They proposed frequent itemset mining, association rule mining, and interestingness to mine relations between medications, results, and diagnosis. Diagnosis can be done in a non-invasive fashion and still be fed into computer-assisted decision systems, as described by Tsipouras et al. [2]. Sometimes, for accurate measurements, physicians may use biomarkers as detailed by McRae et al. [4], or genetic parameters as described by Atkov et al. [5]. Missing values in datasets are sometimes filled using standard algorithms, as in the case of Purushottam et al. [6] who proposed using AllPossible-MV algorithm.

Pattern discovery and predictive analysis can be done on any massive dataset. The primary concern is dimensionality of the dataset, which can be managed by methods like Monte Carlo sampling. This was described in detail by Mullins et al. [7] who took the case of data from 667,000 patients. Mining from clinical notes has also been explored by Chen et al. [8], but we predominantly take into consideration more structured information. Dictionary-based keyword spotting, machine learning, and rule-based methods are used to identify contextual information from text. During that, we may have to deal with conflicting information also. Temporal information will have to be supported by suitable annotations like "before," "during," and "after" to notify relative ordering of events, as demonstrated by Yang et al. [9]. Suitable semantic models can be used to improve the accuracy of conventional semantic models using contextual information. Semantic models may include pointwise mutual information, latent semantic analysis, and tensor space models. This was explored in detail by urban [10]. Data aggregation may be used to generate contextual relations in temporal data. A cause-effect model can be built which can evaluate preconditions and generate conclusions, similar to the model proposed by Yang et al. [9]. Khalifa et al. [11] proposed adding lexical analysis and tagging, along with synonym analysis to improve accuracy of such input methods.

While research on cardiovascular diseases is best done on real data, there are some datasets available that may be of help. Some of them which have been used in the survey are

1. MIT-BIH [12]
2. Invasive Cardiology department of University Hospital of Inoannina [2]
3. Cleveland heart disease dataset from UCI machine learning repository [13, 14]
4. Fantasia and St. Petersburg Institute for Cardiology Technics [15]
5. Clinical notes data from https://www.i2b2.org/NLP/DataSets [16]
6. Tehran heart disease data [17]
7. NewYork-Presbyterian Hospital Outpatient Dataset [18]
8. Hungarian heart disease dataset from UCI machine learning repository [19].

While clinical notes are widely available, their unstructured nature makes them less dependable. Hence, we go for more structured diagnostic information to study disease onset. Sometimes, the clinical notes themselves can be converted to structured form in order to supplement the dataset. Expert advice also has its importance in such decision systems.

3 Data Classification and Prediction

When the primary purpose to perform a prediction of whether an event is going to occur, we have to resort to classification and prediction models. In the case of cardiovascular diseases, the model should read clinical parameters as input, and give an output which gives indications of whether a heart disease is present (and if possible, the level of disease). We generally categorize the models into three—regression models (which give a numeric output that can be interpreted accordingly), learning models (which give outputs based on the input annotations), and classification models (which place the data item into one or more of predefined classes based on the supplied parameters).

3.1 Regression Models

Regression involves creating mathematical expressions to predict a value as a function of input parameters. It was used in the work by Verma et al. [20] as multinominal logistic regression to achieve an accuracy of 84% on live data. Logistic regression and linear regression are applicable in healthcare predictions, but we can note an obvious preference toward logistic regression because it performs better on multiple types of data, and the relation is much more relatable to real-world data. McRae et al. [4] proposed a scoring system for prediction based on lasso logistic regression. A number of parameters were interpreted to create an in vitro diagnostic multivariate index assay. The main feature of this method was that it was hardware-oriented, and can be implemented on a chip. Austin et al. [21] noted that logistic regression gave an accuracy of 77% over a dataset in which conventional models like bagging, boosting, and random forests gave an average accuracy of 70%. At the same time, Chaurasia et al. [19] noted that bagging works well enough when used on datasets of limited variability. Wu et al. [22] used logistic regression to classify diabetes data, but had a level of preprocessing where initial data points were set with some level of intelligence.

3.2 Learning Models

Artificial neural networks have been widely studied in order to predict diseases, including heart diseases. Neural networks are explored as a "universal" method that can map any set of inputs to a defined set of outputs. But when we consider learning, we have to consider the fact that the process is primarily supervised in nature, since we need to know in advance what to look for in the input. Samuel et al. [1] used neural networks as a learning model. Since the accuracy of neural models depends on weight matrices, they used fuzzy analytical hierarchy process to set the initial weights to achieve an accuracy of more than 90%. Basic neural network models were also used by Atkov et al. [5], Chaitrali et al. [23], and Rajeswari et al. [24] to get an accuracy of more than 90%. Similarly, Arabasadi et al. [25] and Amin et al. [26] used genetic algorithms to generate initial weight matrices, which may suffer from the side effect of slow convergence but can generate a good vector (observed accuracy of about 94%) given enough time. Studies by Verma et al. [20] used learning models as a second stage to classify a given set of clustered data by exploring the use of perceptrons and rule induction. A diagnosis system developed by Das et al. [27] extends this by creating an ensemble of neural network models and combines the results to get an accuracy of 89%, and Mustaqeem et al. [28] performed a statistical evaluation for attribute selection followed by learning-based methods like support vector machines and perceptrons.

Acharya et al. [12] discussed the application of deep learning models in arrhythmia to achieve an accuracy of about 90%. But one of the most convincing uses of deep learning models was proposed by Tan et al. [29]. They proposed a combination of convolutional neural networks (CNN) and long short-term memory (LSTM) to achieve a classification accuracy of 99% on ECG signals. While LSTM works well on time series signals, CNN compensates for the speed by reducing data points. It may be further be improved by using region of interest (ROI) to create smaller windows of ECG signals, as done by Lee et al. [30]. Acharya et al. [15] also used CNN with four convolution layers, four maxpooling layers, and three fully connected layers to achieve accuracy of 95%.

Shah et al. [31] used support vector machines based on radial basis function to classify data. Recurrent fuzzy neural networks were used by Uyar et al. [13] to achieve accuracy of over 97% on Cleveland UCI dataset. They used it as a binary classifier and progressively scaled down the input (which contained 13 neurons) through a single hidden layer (containing seven neurons) to reach a single-neuron output. Least square SVM and Levenberg–Marquardt ANN were used by Polar et al. [32] to achieve an accuracy of 83.7% on mixed-form data.

3.3 *Classification Models*

A clinical decision support system based on fuzzy rules was proposed by Anooj [33]. Fuzzy inference is done using Mamdani system. Classification is done by regulating the type of fuzzy membership function. Mustaqeem et al. [28] explored random forests to classify data after statistical preprocessing. Binary classification was also explored by Nahar et al. [34].

Decision trees also help in classification if there are a number of parameters, and they can be used to hierarchically build a decision system, as depicted by Tsipouras et al. [2], and Karaolis et al. [35]. In their study, C4.5 was used to build a decision tree, which was then processed to generate rules for classification. Fuzziness was also added in [9], and then optimized, to improve accuracy. Information gain, gini index, likelihood ratio chi-squared statistics, gain ratio, and distance measure were used as decision tree induction parameters in Karaolis et al. [35]. Bhatla et al. [36] also performed a weak evaluation of decision trees, and found them to be effective.

Association rules are also proposed by Nahar et al. [37], who presented a study that observed gender-based variations in certain symptoms. They suggested predictive apriori and Tertius algorithms as possible candidates for rule mining. But when we generate association rules, a level of validation is required by experts from the medical domain. Validation is easy since rules are easier to understand when compared to other data structures. Similarly, Karaolis et al. [38, 39] mentioned that algorithms like C4.5, k-means, decision trees, and apriori methods may be used to generate association rules. Karystianis et al. [16] found that rule-based system and conditional random field classifier can be used for heart disease models from clinical notes, and that it takes into consideration temporal information to a good extent.

El-Bialy et al. [40] proposed using fast decision tree and Pruned C4.5 for fast classification, and it gave an accuracy of more than 78%. Classification is made better using information gain parameters and gain ratio. Purushottam et al. [6] also used tree-based models and later improved on them using Hill climbing methods to achieve an accuracy of 86.7%, but the method is highly dependent on the dataset itself. Chaurasia [14] suggested the use of CART instead of conventional tree-based methods since the information gain on certain datasets is low, and methods like ID3 may underperform in that case. This was further supported in the case of diabetes by Breault et al. [41] who pointed out that CART can improve accuracy by managing the depth of trees. Pivovarov et al. [18] extend the conceptual framework into graphs, which makes diagnosis more efficient. They call it the unsupervised Phenome model. Kumari et al. [42] proposed augmenting decision tree models with a pruning method like RIPPER to achieve better accuracy and performance.

Sen et al. [43] proposed a system in which parameters are fuzzified and then classified into two layers based on how critical they are. A neurofuzzy system then takes the parameters and performs further prediction. The layers of parameters are weighed appropriately. Hachesu et al. [17] proposed an ensemble which combines SVM and C5.0 to obtain an accuracy of 98% on a specific dataset.

4 Dynamics of Heart Disease Data

4.1 Variations in Data

Heart rate variability (HRV) is a parameter that can be used to identify heart diseases, as mentioned by Lee et al. [30]. Thresholding may be required when we have high variability in data. This is demonstrated by Pathan et al. [44] in the case of skin lesions. As the number of parameters increases, the possibility of variation also increases. Tomography images may need additional signal processing, as done by Polat et al. [32]. When we consider the case of heart diseases, the diagnostic information from Mayo Clinic which we mentioned in Sect. 2 lists a number of diagnostic measures.

Since body parameters cannot be exactly located to a number, we will have to consider a range of values for each parameter (called the tolerance limit) as mentioned in Pathan et al. [44]. Using classifiers for that may turn out to be suboptimal since minor variations in diagnostic data will be possibly classified as a disorder state. In order to avoid that, we need to perform some level of clustering to support the classifier, and it is most recommended to cluster the data into "bags" and send the bags to classification.

4.2 Feature Selection

Considering the large number of tests that are done to diagnose the disorder, a lot of data is generated. But not all of them will be of interest to us. Verma et al. [20] proposed correlation-based feature selection to pick essential features among the available attributes. This method will be helpful especially when there are indirect correlations among a number of parameters. But this can be time consuming if the number of attributes is high. Nahar et al. [34] also propose a computerized feature selection method to reduce the complexity of further operations. Acharya et al. [45] proposed a projection model to select data from processed ECG signals, called Locality Preserving Projection, which considers variations in the signals and picks up variances in the image. Alizadehsani et al. [46] also advocated feature selections for improving the performance of data mining algorithms in the cardiovascular domain. Shah et al. [31] propose probabilistic principal component analysis for feature selection. Polay et al. [32] used a kernel function to generate an F-score and assist in selecting valuable features. Vivekanandan et al. [47] presented a new algorithm that uses differential evolution to select and optimize features before classification or clustering. This method was shown to achieve an accuracy of 83%.

5 Clustering of Data

Even though classification and prediction models give a good estimate, sometimes, it is better to avoid exact outputs, especially when handling medical data. This is since exact outputs may give a false impression of confidence or panic (the processes are not 100% accurate). Hence, we normally try to group cases together based on the parameters, and consider known outcomes of some cases to understand the overall behavior. This happens in clustering models. We consider two types of models here—nearest neighbor model and optimization models.

5.1 Nearest Neighbor Methods

Acharya et al. [45] performed k-nearest neighbor (kNN) to cluster data selected through Locality Preserving Projection to yield an accuracy of 98.5%. kNN can also be augmented with genetic search methods by pruning irrelevant and redundant attributes, as demonstrated by Jabbar et al. [48]. It showed an improvement of 5% over genetic algorithms and kNN separately.

5.2 Optimization Models

The study by Verma et al. [20] highlighted the use of optimization models as a method to preprocess input data before classification. They use correlation-based feature selection, particle swarm optimization, and k-means clustering to generate data annotations, which help in further classification. Fuzzification was shown to improve prediction by up to 15% by Tsipouras et al. [2]. But all fuzzy parameters have to be optimized by global optimization so that the curve fits the data to the best extent. Wavelet transforms may also be used to optimize the information provided by taking small intervals of ECG signals, which is proposed by Acharya et al. [45]. Expectation–maximization was used by Sufi et al. [49] to cluster compressed ECG signals to get an abnormality detection rate of 97%. Uyar et al. [13] used genetic algorithms to help to train a neural network and achieved an accuracy of over 97% on Cleveland UCI dataset. Wu et al. [22] also use k-means clustering to set initial seed points for further regression modeling.

Roy et al. [50] used scale-space filtering to make data more separable, and use fuzzy c-means clustering to achieve up to 98% accuracy on UCI Iris dataset. But it may not give that level of accuracy in cases like heart diseases since there will be lot of possible combinations that will predict the occurrence. To avoid local optima, more randomized searches like PCA or genetic algorithms may be needed. Framingham risk score (FRS) is one risk parameter, we can use to evaluate conditions, which was used by Amin et al. [26]. If temporal information is needed, we may explore methods like frequent itemset, as explored by Gotz et al. [51]. Ilayaraja et al. [52] improved this further using pinser search to remove redundancy.

6 Performance Evaluation

We have selected datasets related to disease diagnosis to measure the performance of different algorithms. The following datasets have been selected.

1. Breast cancer (Generated by Matjaz Zwitter and Milan Soklic, Physicians, Institute of Oncology, University Medical Center, Ljubljana, Yugoslavia)

2. Diabetes (Generated by National Institute of Diabetes and Digestive and Kidney Diseases)
3. Heart disease (statlog data from Kaggle)
4. Hypothyroidism (Supplied by the Garavan Institute and J. Ross, Quinlan, New South Wales Institute, Syndney, Australia)
5. Heart disease (Cleveland data from UCI)
6. Arrythmia (from Bilkent University, Department of Computer Engineering and Information Science, Ankara, Turkey).

We have evaluated the algorithms based on accuracy (in terms of accuracy percentage/root mean squared error) and runtime. Algorithms were selected based on the type of data. Testbed was set on RapidMiner Studio. The results are given in Tables 1 and 2.

In order to consolidate the results, we generate a score for each combination of dataset and algorithm. Put a score in descending order of accuracy or ascending order of RMS error. Generate the score as accuracy score/runtime, and scale it by multiplying with 100. The score is such that it is higher if accuracy is high, and reduces as runtime increases. This generates the score as in Table 3, and is visualised in Fig. 1.

We can see that Naive Bayes performs well for the first four datasets (which are well suited for classification), while the generalized linear model works well for the last two datasets (which are more suited for clustering). But the way in which datasets are built influence the performance to a huge extent. This was observed by Nahar et al. [8] in the case of UCI Cleveland heart dataset which became unbalanced when applied to binary classifiers.

Table 1 Accuracy of algorithms

	Accuracy (%)				RMS error	
Algorithm	Breast cancer	Diabetes	Heart statlog	Hypothyroid	Cleveland data	Arrythmia
Naive Bayes	68.4	76	83.3	97.2		
Generalized linear model	73.7	74.7	83.3	92.6	0.91	3.836
Logistic regression	71.9	75.3	77.8			
Decision tree	78.9	70.1	77.8	92.3	1.156	3.984
Random forest	71.9	73.4	74.1	92.3	0.927	3.719
Gradient boosted trees	77.2	76.6	83.3	96.9	0.883	4.475

Table 2 Runtime of algorithms (ms)

Algorithm	Breast cancer	Diabetes	Heart statlog	Hypothyroid	Cleveland data	Arrythmia
Naive Bayes	51	32	23	160		
Generalized linear model	259	377	353	833	38	5000
Logistic regression	656	283	202			
Decision tree	202	368	212	587	107	9000
Random forest	2000	5000	2000	6000	2000	30,000
Gradient boosted trees	27,000	31,000	19,000	56,000	8000	54,000

Table 3 Accuracy—runtime ratio

Algorithm	Breast cancer	Diabetes	Heart statlog	Hypo thyroid	Cleveland data	Arrythmia
Naive Bayes	1.96	15.63	17.39	3.13		
Generalized linear model	1.16	0.8	1.13	0.24	7.89	0.08
Logistic regression	0.3	1.41	0.99			
Decision tree	0.5	0.5	0.3	0.03	1.68	0.03
Random forest	2.48	0.27	0.94	0.17	0.93	0.03
Gradient boosted trees	0.1	0.04	0.05	0.02	0.1	0.02

The arrhythmia dataset was considered to check if the performance of Naive Bayes can be extended to a classification model. The ensemble was created by discretization (where each class is assigned a name), and then classified using a Naive Bayes classifier with kernel density estimation [53]. The ensemble gave a classification accuracy of 99.78%. When regular Naive Bayes was used with discretization, the classification accuracy was 92.82%. The discretization operation creates an impression of clustering (but since classes have been preassigned, no explicit clustering is needed). The

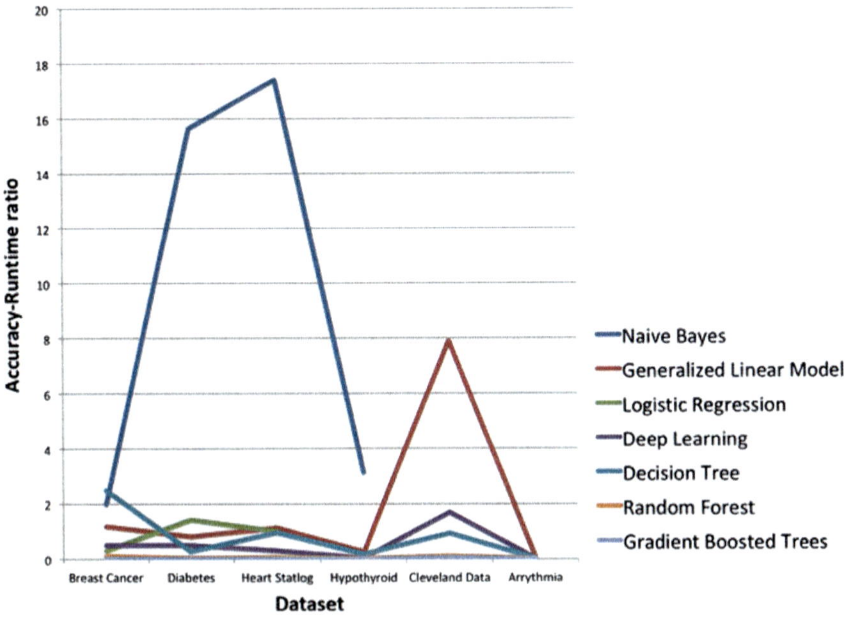

Fig. 1 Comparison of accuracy—runtime ratio

process was validated usingten-fold cross validation. The data contained 16 classes of output as given in Table 4.

Even though the results we obtained are conclusive to some extent, it is not generalized. Hence, this will serve as a direction on how data can be further processed. As the study was conducted on normalized datasets, the accuracy is quiet high and is not the exact representatives of real-time data.

The algorithm behavior is as observed in the case of earlier models, primarily because Bayesian and regression models work better in sequential (continuous) data, while decision models work better on categorical data.

7 Discussion and Conclusion

In this paper, we have discussed how heart diseases may occur, how they are diagnosed, and how we may be able to computerize a decision support system for diagnosis. We have considered a number of available methods that have been explored by researchers, and how they may or may not work. For comparative purposes, we have selected datasets for a variety of diseases (including those outside the domain of cardiovascular diseases). This was done so that we explore different types and combinations of data and reach a general conclusion on which types of algorithms work well in different circumstances.

Table 4 Classes of output in arrhythmia dataset

Class code	Class	Number of instances
1	Normal	245
2	Ischemic changes (coronary artery disease)	44
3	Old anterior myocardial infarction	15
4	Old inferior myocardial infarction	15
5	Sinus tachycardia	13
6	Sinus bradycardia	25
7	Ventricular premature contraction (PVC)	3
8	Supraventricular premature contraction	2
9	Left bundle branch block	9
10	Right bundle branch block	50
11	One-degree atrioventricular block	0
12	Two-degree AV block	0
13	Three-degree AV block	0
14	Left ventricular hypertrophy	4
15	Atrial fibrillation or flutter	5
16	Others	22

We have closed the main discussion with a comparative evaluation which gives Naive Bayes and generalized linear model an edge over the others, in terms of diagnosis time and accuracy. Accuracy may be higher for some other models, but the time taken to build those may be too much to fall within the tradeoff margins. The findings correlate to the discussion in Sect. 4.1 where a cluster-classify ensemble is suggested. In situations where generalized linear model performed well, we were able to observe a large number of numerical data, which enables the creation of proper mathematical models. But when we consider the actual accuracy, we can see that the performance is not up to the mark, since the model primarily performs binary classification, which does not work well under dynamic data. Hence, we consider Bayesian models as one of the most suitable for classification, especially in the case of disease data. This was further confirmed in the study since it performed well on heart disease data (which is our domain of interest) compared to other methods. The method works well since a good amount of clustering was done beforehand, and dynamics were accounted for in the clusters.

As mentioned, the dataset considered in the study may not be counted as exact representations of real-time data, since some level of normalization and outlier elimination would have been done during creation itself. Hence, in the future, we will have to explore hybrid algorithms to improve the accuracy while keeping diagnostic time within limits when applied on real-time data.

References

1. Samuel OW, Asogbon GM, Sangaiah AK, Fang P, Li G (2017) An integrated decision support system based on ANN and Fuzzy_AHP for heart failure risk prediction. Expert Syst Appl 68:163–172
2. Tsipouras MG et al (2008) Automated diagnosis of coronary artery disease based on data mining and fuzzy modeling. IEEE Trans Info Tech Med 12:447–458
3. Wright A, Chen ES, Maloney FL (2010) An automated technique for identifying associations between medications, laboratory results and problems. J Biomed Inform 43:891–901
4. McRae MP et al (2016) Cardiac ScoreCard: a diagnostic multivariate index assay system for predicting a spectrum of cardiovascular disease. Expert Syst Appl 54:136–147
5. Atkov OY et al (2012) Coronary heart disease diagnosis by artificial neural networks including genetic polymorphisms and clinical parameters. J Cardiol 59:190–194
6. Purushottam K, Saxena R Sharma (2016) Efficient heart disease prediction system. Procedia Comput Sci 85:962–969
7. Mullins IM et al (2006) Data mining and clinical data repositories: Insights from a 667,000 patient data set. Comput Biol Med 36:1351–1377
8. Chen Q et al (2015) An automatic system to identify heart disease risk factors in clinical texts over time. J Biomed Inform 58:S158–S163
9. Yang H, Garibaldi JM (2015) A hybrid model for automatic identification of risk factors for heart disease. J Biomed Inform 58:S171–S182
10. Urbain J (2015) Mining heart disease risk factors in clinical text with named entity recognition and distributional semantic models. J Biomed Inform 58:S143–S149
11. Khalifa A, Meystre S (2015) Adapting existing natural language processing resources for cardiovascular risk factors identification in clinical notes. J Biomed Inform 58:S128–S132
12. Acharya UR et al (2017) A deep convolutional neural network model to classify heartbeats. Comput Biol Med 89:389–396
13. Uyar K, Ilhan A (2017) Diagnosis of heart disease using genetic algorithm based trained recurrent fuzzy neural networks. Procedia Comput Sci 120:588–593
14. Chaurasia V (2013) Early prediction of heart diseases using data mining. Caribb J Sci Technol 1:208–217
15. Acharya UR, Fujita H, Lih OS, Adam M, Tan JH, Chua CK (2017) Automated detection of coronary artery disease using different durations of ECG segments with convolutional neural network. Knowl-Based Syst 132:62–71
16. Karystianis G, Dehghan A, Kovacevic A, Keane JA, Nenadic G (2015) Using local lexicalized rules to identify heart disease risk factors in clinical notes. J Biomed Inform 58:S183–S188
17. Hachesu PR, Ahmadi M, Alizadeh S, Sadoughi F (2013) Use of data mining techniques to determine and predict length of stay of cardiac patients. Healthc Inform Res 19:121–129
18. Pivovarov R, Perotte AJ, Grave E, Angiolillo J, Wiggins CH, Elhadad N (2015) Learning probabilistic phenotypes from heterogeneous EHR data. J Biomed Inform 58:156–165
19. Chaurasia V, Pal S (2013) Data mining approach to detect heart diseases. Int J Adv Comput Sci Inf Technol 2:56–66

20. Verma L, Srivastava S, Negi PC (2016) A hybrid data mining model to predict coronary artery disease cases using non-invasive clinical data. J Med Syst 40:1
21. Austin PC, Tu JV, Ho JE, Levy D, Lee DS (2013) Using methods from the data-mining and machine-learning literature for disease classification and prediction: a case study examining classification of heart failure subtypes. J Clin Epidemiol 66:398–407
22. Wu H, Yang S, Huang Z, He J, Wang X (2018) Type 2 diabetes mellitus prediction model based on data mining. Inform Med Unlocked 10:100–107
23. Chaitrali DS, Sulabha AS (2012) A data mining approach for prediction of heart disease using neural networks. Int J Comput Eng Technol 3:30–40
24. Rajeswari K, Vaithiyanathan V, Neelakantan TR (2012) Feature selection in ischemic heart disease identification using feed forward neural networks. Procedia Eng 41:1818–1823
25. Arabasadi Z, Alizadehsani R, Roshanzamir M, Moosaei H, Yarifard AA (2017) Computer aided decision making for heart disease detection using hybrid neural network-Genetic algorithm. Comput Methods Programs Biomed 141:19–26
26. Amin SU, Agarwal K, Beg R (2013) Genetic neural network based data mining in prediction of heart disease using risk factors. In: ICT 2013—Proceeding of 2013 IEEE conference on information and communication technologies, pp 1227–1231
27. Das R, Turkoglu I, Sengur A (2009) Effective diagnosis of heart disease through neural networks ensembles. Expert Syst Appl 36:7675–7680
28. Mustaqeem A, Anwar SM, Khan AR, Majid M (2017) A statistical analysis based recommender model for heart disease patients. Int J Med Inform 108:134–145
29. Tan JH et al (2018) Application of stacked convolutional and long short-term memory network for accurate identification of CAD ECG signals. Comput Biol Med 94:19–26
30. Lee HG, Noh KY, Ryu KH (2008) A data mining approach for coronary heart disease prediction using HRV features and carotid arterial wall thickness. In: 2008 International conference on biomedical engineering and informatics, pp 200–206
31. Shah SMS, Batool S, Khan I, Ashraf MU, Abbas SH, Hussain SA (2017) Feature extraction through parallel probabilistic principal component analysis for heart disease diagnosis. Phys A Stat Mech Appl 482:796–807
32. Polat K, Güneş S (2009) A new feature selection method on classification of medical datasets: Kernel F-score feature selection. Expert Syst Appl 36:10367–10373
33. Anooj PK (2012) Clinical decision support system: risk level prediction of heart disease using weighted fuzzy rules. J King Saud Univ Comput Inf Sci 24:27–40
34. Nahar J, Imam T, Tickle KS, Chen YPP (2013) Computational intelligence for heart disease diagnosis: a medical knowledge driven approach. Expert Syst Appl 40:96–104
35. Karaolis M, Moutiris J, Hadjipanayi D, Pattichis CS (2010) Assessment of the risk factors of coronary heart events based on data mining with decision trees. IEEE Trans Inf Technol Biomed 14:559–566
36. Bhatla N, Jyoti K (2012) A novel approach for heart disease diagnosis using data mining and fuzzy logic. Int J Comput Appl 54:975–8887
37. Nahar J, Imam T, Tickle KS, Chen Y-PP (2013) Association rule mining to detect factors which contribute to heart disease in males and females. Expert Syst Appl 40:1086–1093
38. Karaolis M, Moutiris JA, Papaconstantinou L, Pattichis CS (2009) Association rule analysis for the assessment of the risk of coronary heart events. In: Conference Proceeding of the IEEE engineering in medicine and biology society, pp 6238–6241
39. Karaolis M, Moutiris JA, Pattichis CS (2006) Assessment of the risk of coronary heart event based on data mining. Comp A J Comp Educ, pp 1–5
40. El-Bialy R, Salamay MA, Karam OH, Khalifa ME (2015) Feature analysis of coronary artery heart disease data sets. Procedia Comput Sci 65:459–468
41. Breault JL, Goodall CR, Fos PJ (2002) Data mining a diabetic data warehouse. Artif Intell Med 26:37–54
42. Kumari M, Godara S (2011) Comparative study of data mining classification methods in cardiovascular disease prediction. Int J Comput Sci Trends Technol 2:304–308

43. Sen AK, Patel SB, Shukla DP (2013) A data mining technique for prediction of coronary heart disease using neuro-fuzzy integrated approach two level. Int J Eng Comput Sci 2:1663–1671
44. Pathan S, Prabhu KG, Siddalingaswamy PC (2018) Techniques and algorithms for computer aided diagnosis of pigmented skin lesions—a review. Biomed Signal Process Control 39:237–262
45. Acharya UR et al (2017) Automated characterization and classification of coronary artery disease and myocardial infarction by decomposition of ECG signals: a comparative study. Inf Sci (Ny) 377:17–29
46. Alizadehsani R et al (2013) A data mining approach for diagnosis of coronary artery disease. Comput Methods Programs Biomed 111:52–61
47. Vivekanandan T, Sriman Narayana Iyenga NC (2017) Optimal feature selection using a modified differential evolution algorithm and its effectiveness for prediction of heart disease. Comput Biol Med 90:125–136
48. Jabbar MA, Deekshatulu BL, Chandra P (2013) Classification of heart disease using K-nearest neighbor and genetic algorithm. Procedia Technol 10:85–94
49. Sufi F, Khalil I (2011) Diagnosis of cardiovascular abnormalities from compressed ECG: a data mining-based approach. IEEE Trans Inf Technol Biomed 15:33–39
50. Roy P, Mandal JK (2013) A novel selective scale space based fuzzy C-means model for spatial clustering. Procedia Technol 10:596–603
51. Gotz D, Wang F, Perer A (2014) A methodology for interactive mining and visual analysis of clinical event patterns using electronic health record data. J Biomed Inform 48:148–159
52. Ilayaraja M, Meyyappan T (2015) Efficient data mining method to predict the risk of heart diseases through frequent itemsets. Procedia Comput Sci 70:586–592
53. Minnier Jessica et al (2015) Risk classification with an adaptive Naive Bayes Kernel machine model. J Am Stat Assoc 110:393–404

Analysis of the Nearest Neighbor Classifiers: A Review

Yash Agarwal and G. Poornalatha

Abstract We are living in a data age and with the expansion of 'Internet of Things' platform, there is an upsurge in devices connected to the Internet. Everything from smart sensors to smartphones and tablets, systems installed in manufacturing units, hospitals, vehicles, etc. is generating data. Such developments in the technological world have escalated the generation of data and require an analysis to be performed on the raw data to identify patterns. The data mining techniques are deployed extensively to extract information and they yield far-reaching effects on the trade and the lives of the people concerned. The accuracy and effectiveness of data mining techniques in providing better outcomes and cost-effective methods in various domains have been established. Usually, in supervised learning, density estimation is used by instance-based learning classifiers like k-nearest neighbor (kNN). In this paper, the regular kNN classifier is compared with the various classifiers conceptually and the ARSkNN that uses mass estimation has been proved to be commensurate to kNN in accuracy and has reduced computation time drastically on datasets chosen for this analysis. Tenfold cross-validation is used for testing.

Keywords Data mining · Machine learning · Data analysis · Classification · Nearest neighbor · K-fold cross-validation

1 Introduction

Technology is the backbone of today's modern world and its outreach has spanned across various domains, such as health care, travel, education, entertainment, and communication. Slowly, but steadily it has permeated the spheres of our daily lives. Owing to its increasing applications and aggregation of a vast volume of data from

Y. Agarwal · G. Poornalatha (✉)
Department of Information & Communication Technology, Manipal Institute of Technology, Manipal Academy of Higher Education, Manipal, Karnataka 576104, India
e-mail: poornalatha.g@manipal.edu

Y. Agarwal
e-mail: y4shagarwal@gmail.com

N. N. Chiplunkar and T. Fukao (eds.), *Advances in Artificial Intelligence and Data Engineering*, Advances in Intelligent Systems and Computing 1133,
https://doi.org/10.1007/978-981-15-3514-7_43

a variety of sources, the technology is in need of data mining algorithms that can produce accurate results and can effectively reduce the time complexity. In recent years, there has been an upsurge in processing power of computers which are available at economical prices, coupled with the need to analyze massive datasets, and has compelled researchers to come up with ways and new techniques that serve the need of the hour [1–6].

Data mining is an intersection of various subjects, such as statistics, machine learning, artificial intelligence, and database systems. These techniques are used to handle and organize homogeneous data into a well-structured dataset which makes sense. Various methods have been developed over time in machine learning, enabling intelligent systems to perform predictive data mining tasks. Learning from data can be supervised, unsupervised, or reinforcement, depending on the samples provided in the data.

Classification techniques are used in machine learning to classify data such that the data points of the same category are similar to each other. Classification uses a supervised learning approach which says that training dataset that needs to be classified should have associated labeled class. Unlike regression techniques which require continuous values, classification methods are robust to both categorical and numeric values. Nearest neighbors are one of the most fundamental and straightforward classifiers. The k-nearest neighbors estimate the nearest neighbors by calculating the closeness or the distance of the k-nearest values of the test sample. The accuracy and classification of the k-nearest neighbor algorithm depend on the similarity and distance measure [7–10] chosen. In the standard kNN, distance metric used to measure the closeness between two points is Euclidean distance. kNN is a non-parametric classifying method which was introduced by Hodges and Fix [11].

A dataset can be binary or multi-class depending on the dimensions of the data. HTRU2 dataset being used in this analysis is binary. If the test candidate is not a pulsar, it will be grouped with class 0 and if it is a pulsar, belong to class 1. In this type of dataset, the binary classification method is used in which two classes are predefined. An odd value of k is chosen in a binary classification to keep an even number of data points from getting selected from both the classes.

This paper has been branched into six sections. Section 2 presents the literature review, which talks about previous works and improved nearest neighbor algorithms. Section 3 describes the experimental setup and gives an introduction to the dataset details. Section 4 gives a detailed description of the experiment results. Section 5 discusses about the results. Section 6 formally concludes the paper, followed by the references at the end.

2 Literature Review

In case of data classification, kNNs are widely used by researchers' community to classify the samples. It is the simplest and one of the most prominent supervised learning classifier. Major drawbacks of using general NN approach are that it stores the entire training set, then compares the test sample with every training sample. Basic NN algorithm also has an undesirable property of storing noisy instances (error in the input instance or output class) degrading the accuracy rate. This mechanism proves to be expensive on a large dataset, and many researchers have tried to remove the unnecessary data of the training set to relieve this limitation [12–14].

Prototype selection is a process that gravitates on selecting a smaller set of patterns from the training sample. This influences the computation time and alleviates the memory requirement obstacle, and if followed accurately, it can result in higher accuracy percentages.

In 2000, a comprehensive survey and analysis of various machine learning algorithms which are used for reducing the size of the training set were done by Wilson and Matinez [15]. They further suggested various instance reduction techniques which are described and analyzed as given in Table 1.

The HTRU dataset has been tested with a variety of algorithms like (C4.5, Multilayer Perceptron, Support Vector Machines, Naive Bayes) by Lyon RJ [19]. The aim of these tests and grouping is to reduce the false positives and to maximize the predictions of true positives.

3 Experimental Setup

In this experiment, two datasets have been used from the UCI online repository. WEKA Experimenter was used to analyze and classify datasets with algorithms kNN and ARSkNN. Tenfold cross-validation was used 10 times (refer Fig. 1) on a machine with Core i7 processor, 2.4 GHz clock speed and 8 GB RAM, to get the average accuracy percentage of the algorithms.

3.1 Dataset Details

In this analysis, both the algorithms have been tested on two datasets, namely HTRU2 and Chess (king-Rook vs. king) as detailed in Table 2.

HTRU2: HTRU2 dataset, shown in Fig. 2, characterizes a sample of pulsar candidates collected during the High Time Resolution Universe Survey (South). Pulsars are highly magnetized, rotating neutron starts which emit electromagnetic radiations toward earth [20, 21].

Promising candidates are selected using a purpose-built tree-based machine learning classifier, the Gaussian Hellinger Very Fast Decision Tree, and a new set of features have been chosen for depicting the pulsar candidates [19].

Table 1 Analysis and comparison of existing classifiers

kNN (k-nearest neighbor) [16]	Key idea	Uses the basic nearest neighbor rule to predict class of the queried sample
	Advantages	• Reduction in training time • simple and straightforward to implement • High accuracy generalizations
	Disadvantages	• Increases storage requirements • Classification requires time as queried sample is compared with training set • Dependent on the value of k
WkNN (weighted k- nearest neighbor) [2]	Key idea	Training instances are allotted weights depending on the distance between instance and queried sample and is effective on large datasets
	Advantages	• Overcomes standard kNN's limitations of assuming the weight of all samples to be equal • Considers the whole training set for predictions instead of just points which lie under the region of interest depending on k
	Disadvantages	• Calculation of weights causes an increase in complexity • Algorithm tends to work slower
CNN (condensed nearest neighbor)	Key idea	Excludes data points similar to each other and are not contributing any new information
	Advantages	• Reduction of size in training set • Improvement in computation time for classification • Effective in case of memory constraints
	Disadvantages	• Does not guarantee a minimal set • It is order dependent, which means it is undesirable to pick points near to the decision boundary • If training set has less instances, prediction is prone to errors
MCNN (modified condensed nearest neighbor)	Key idea	It is an improved CNN technique
	Advantages	Order independent, so it gives the same set, independent of processing order of the data

(continued)

Table 1 (continued)

	Disadvantages	• Works well for Gaussian distribution, but is unlikely to select boundary points in other scenarios • In case of many classes, it requires lots of iterations to converge
FCNN (fast condensed nearest neighbor)	Key Idea	Selects data points closer to the decision boundary
	Advantages	• It is order independent • Less quadratic complexity compared to CNN • Results do not depend on the order in which the data is processed
	Disadvantages	• Highly iterative method
RNN (reduced nearest neighbor)	Key idea	It works in a decremented manner; initially result set and the train set are equal. Then, the instances which do not affect the classification of instances in the training set are eliminated from the result set
	Advantages	• Always produces a subset of CNN • Improves search time • Memory requirement is less compared to the tradition kNN
	Disadvantages	• Expensive than CNN • Complexity is more
DROP1 (decremental reduction optimization procedure 1)	Key Idea	Resulting set(S) and training set(T) are equal initially. Instance P is isolated from S if associates of P can be categorized accurately exclusive of P
	Advantages	• Eliminates instances and reduces training set size • Accuracy not degraded by noisy instances • Serves as the baseline for other DROP algorithms to be compared
	Disadvantages	• Lower rate of accuracy • The consistency of S is checked instead of T • Information from eliminated instances(P) cannot be used for classification
DROP2 (decremental reduction optimization procedure 2)	Key idea	Reported better results than DROP1 and eliminates instances when enough associates of P can be classified correctly without P

(continued)

Table 1 (continued)

	Advantages	• Decreases the training set size • Removes instances • Less storage requirement • Checks consistency of T rather than S
	Disadvantages	• Removes points located near centroid first instead of borderline, which neglects the noisy instances • For a huge dataset, adequate amount of points are not removed
DROP3 (decremental reduction optimization procedure 3)	Key idea	It uses a noise-filtering pass, before sorting S. The misclassified instance from the nearest neighbors is removed
	Advantages	• Reduces the dataset size and eliminates instances • Storage requirement is less than DROP 1, 4, 5 • Firstly, the data points that lie beyond the decision boundary are eliminated • Reports higher accuracy that kNN when noisy instances are present
	Disadvantages	• The possibility of eliminating too high a number of instances
KdNN (Kd tree nearest neighbor) [17]	Key idea	Divides the training set into two halves and organizes multidimensional data
	Advantages	• It is quick and easy to use • Forms a perfectly balanced tree
	Disadvantages	• More computation is required • Follows insensitive search method
ARSkNN [18, 28]	Key idea	It uses massism as a similarity measure
	Advantages	• Computation time required is just a fraction of what it takes in kNN • Has higher accuracy compared to other kNN classifiers
	Disadvantages	• Requires large amount of RAM with 100 sTrees in modeling stage

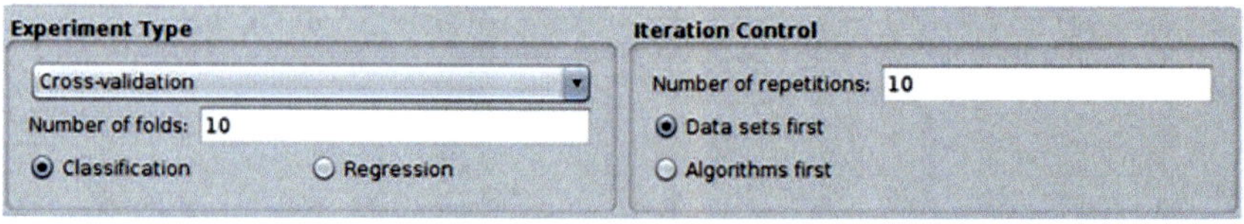

Fig. 1 Ten times—ten cross-validation in WEKA

Table 2 Datasets

Dataset name [Ref.]	#Instances	#Attributes	#Classes	Domain
HTRU2 [10]	17,898	9	2	Distinguishing authentic pulsars with non-pulsar radiations
Chess (king-Rook vs. king) [11]	28,056	6	17	Predicting endgame results

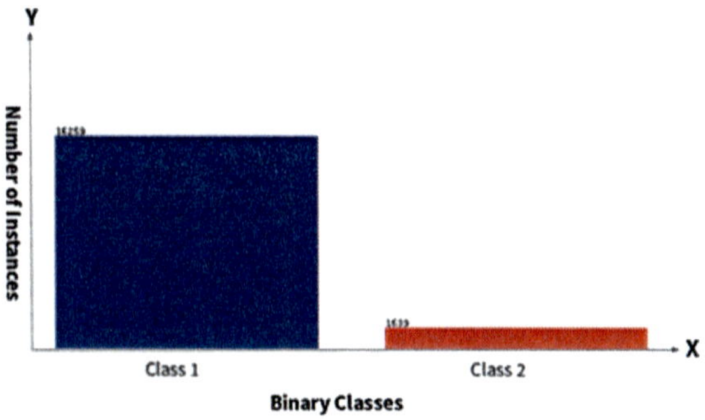

Fig. 2 Class distribution of HTRU2 dataset (16,259 = non-pulsar, 1639 = pulsar)

The features have been chosen to

(i) maximize the partition among candidates resulting from noise and plausible astrophysical origin,
(ii) be an impartial from independent analysis.

Chess:
Chess endgame is a multi-class dataset as shown in Fig. 3 and has a complex domain with enumerable set of positional values. This database stores the game-theoretic values of legal positions in a chess game. The game-theoretic values stored also denote whether the position played were a win, loss or a draw for either side. Game-theoretic values of the optimum number of moves in which the endgame results were obtained are also stored.

4 Empirical Evaluation

In this experiment, datasets were trained and tested on both the classifiers using tenfold cross-validation [29–31]. While testing for accuracy, the number of iterations (for $k = n$) was set to 10. After the cycle of ten iterations was completed, average accuracy percentage of the results was calculated to extract the most optimum and probable accuracy percentage which proves that the results produced in Tables 1 and 2 are reliable.

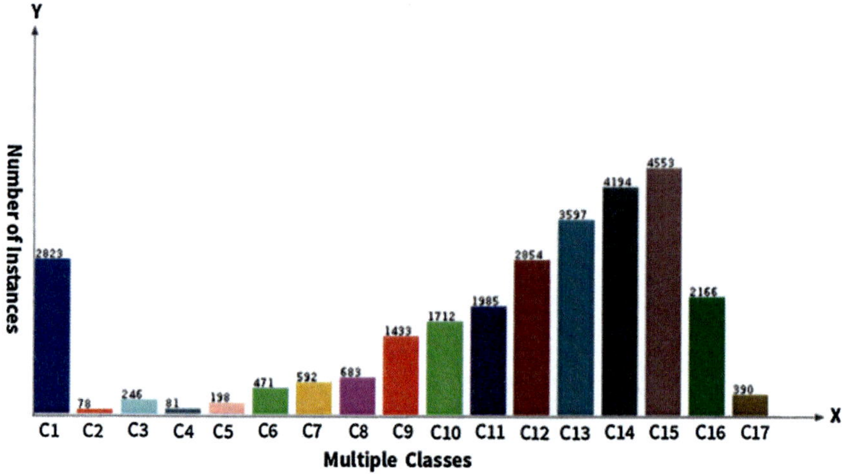

Fig. 3 Class distribution of chess dataset

Subsequently, results were predicted for similarity trees (sTrees = 10, 50, 100) as can be seen in Tables 2 and 3. The highlighted values denote the best results. ARSkNN was successful in significantly raising the accuracy while computation time has dropped by order of magnitude of 2. The major drawback of radical kNN is the amount of computation power required to store the training samples in memory until a new unlabeled data sample is classified [22]. This drastic shrinkage in computation time can solve these long-standing problems of lazy learners [23] and also trim the computation costs.

The ARSkNN classification technique, when implemented on this dataset, has achieved results comparable to the kNN classifier.

The algorithms were implemented on both the dataset (binary and multi-class) to check the effectiveness compared to each other. HTRU2 is a binary dataset which has continuous attribute variables and a class variable denoted by 0(negative) and 1(positive). For (k = 1, 3) ARSkNN achieved the average accuracy percentage of 97.92% and 97.83%, respectively.

The performance of ARSkNN in the data HTRU2 when k is fixed at 5 and 10 although comparable was not able to surpass the kNN results. When increasing the value of k, there is the risk of over-smoothing of the large region and outliers

Table 3 Average accuracy percentage for HTRU2

	k = 1 (%)	k = 3 (%)	k = 5 (%)	k = 10 (%)
kNN	97.10	97.74	97.84	97.83
ARSkNN with 10 sTrees	97.67	97.76	97.72	97.66
ARSkNN with 50 sTrees	97.91	97.82	97.79	97.75
ARSkNN with 100 sTrees	97.92	97.83	97.79	97.76

Table 4 Average accuracy percentage for chess

	$k = 1$ (%)	$k = 3$ (%)	$k = 5$ (%)	$k = 10$ (%)
kNN	56.18	61.12	58.07	54.59
ARSkNN with 10 sTrees	58.16	57.60	55.49	51.11
ARSkNN with 50 sTrees	67.21	63.76	60.60	55.04
ARSkNN with 100 sTrees	69.35	65.03	61.54	55.73

from various classes. This influences the predicted results of the classifier and thus degrades the performance.

It is evident in Table 4 that ARSkNN has produced better results for all values of k and a significant rise of accuracy is observed when the algorithm ran with 100 sTress for $k = 1$.

As shown in Table 5 computation time taken by ARSkNN is astonishingly less when compared with the IBK. For, $k = 1$ IBK completes the classification in approximately 1.15 s while ARSkNN is completing the same classification in barely 0.03 s. With such results, it is evident that high-dimensional datasets can be classified with ease which would not have been possible earlier.

As shown in Table 6, ARSkNN has provided results similar to Table 5 for the values of $k = 1, 3, 5$ and 10, respectively, ARSkNN has shown consistency in reducing the computation time by a significant amount. For instance, if the value of $k = 1$ is taken, ARSkNN is successful in reducing the computation time by 138 times.

Table 5 Average runtime—HTRU2

	$k = 1$ (s)	$k = 3$ (s)	$k = 5$ (s)	$k = 10$ (s)
kNN	1.15	1.17	1.13	1.26
ARSkNN with 10 sTrees	0.03	0.03	0.03	0.02
ARSkNN with 50 sTrees	0.20	0.21	0.19	0.18
ARSkNN with 100 sTrees	0.35	0.38	0.37	0.38

Table 6 Average runtime—Chess

	$k = 1$ (s)	$k = 3$ (s)	$k = 5$ (s)	$k = 10$ (s)
IBk	5.53	4.07	3.33	3.70
ARSkNN with 10 sTrees	0.04	0.03	0.03	0.03
ARSkNN with 50 sTrees	0.23	0.23	0.19	0.16
ARSkNN with 100 sTrees	0.40	0.42	0.44	0.38

Fig. 4 ARSkNN algorithm

Algorithm 1: ARSkNN (y, D, k)

Input:y – Query instance, D – Dataset which has $\{(x_1, c_1), (x_2, c_2), \ldots, (x_n, c_n)\}$, k - number of nearest neighbor,

Output:c_y – Class of query instance y

Let A $\leftarrow$ { }

fori = 1 $\rightarrow$ n do

Mass $\leftarrow$ Mass (x_i, y, F, e)

A $\leftarrow$ A ∪ $\{x_i, c_i, \text{Mass}\}$

end for

Sort in ascending order, the pairs in U using third components

c_y $\leftarrow$ the most frequent class in [Select the first k instances from U]

returnc_y

5 Discussion

kNN has been a conventional and a widely accepted data mining technique by many researchers and many improved versions of kNN, like center-based nearest neighbor classifier (CBNNC) [24], meaningful nearest neighbor (MNN) [25], probably correct k-nearest neighbor (PCkN) [26], have been proposed. They are directly or indirectly calculating similarity/dissimilarity among the probed illustration and the class-label samples using distance metrics.

ARSkNN is a novel algorithm which implements massism, which is a mass-based similarity measure proposed by Ting et al. [27] in 2010. Mass is a unary function but massism is a binary function based on mass, which calculates mass similarities between queried instance and training instance instead of calculating distances between them.

ARSkNN works in a two-stage process as shown in Algorithm 1. First is a modeling stage (preprocessing) in which a similarity forest(sForest) with n number of similarity trees(sTress) are made from a dataset. The complexity of this stage is O(ntlog(d)) which can be improved to O((n + d)t) using indexing techniques. Stage 2 of this algorithm is called Class Assignment in which query instance calculates similarity compared to the nearest data points using mass estimation. The data points that have less mass compared to the query instance will be more similar. Then, by voting of the kNNs, class of the query instance will be predicted (Fig. 4) [28].

6 Conclusion

From the study and analysis of the comparison between the data mining algorithms, it is certain that ARSkNN outperforms the standard kNN and produces more accurate results. It has been consistent in predicting results in a significantly lesser time than a kNNs classifier would take. This major reduction in time is primarily because of the concept called massism that is used in ARSkNN. This experiment also confirms that implementing different similarity or dissimilarity measures can yield different outcomes.

References

1. Audibert JY, Tsybakov AB (2007) Fast learning rates for plug—in classifiers under the margin condition. Ann Stat 35:608–633. https://doi.org/10.1214/009053606000001217
2. Bailey T, Jain A (1978) A note on distance-weighted k-nearest neighbor rules. IEEE Trans Syst Man Cybern 8:311–313. https://doi.org/10.1109/TSMC.1978.4309958
3. Baoli L, Shiwen Y, Qin L (2003) An improved k-nearest neighbor algorithm for text categorization. https://pdfs.semanticscholar.org/490a/b325ba480f6fb71cdbb5f87ff4cb70918686.pdf
4. Bauer ME, Burk TE, Ek AR, Coppin PR, Lime SD, Walsh TA, Walters DK, Befort W, Heinzen DF (1994) Satellite inventory of Minnesota's forest resources. Photogram. Eng. Remote Sens. 60(3):287–298
5. Bax E (2000) Validation of nearest neighbor classifiers. IEEE Trans Inform Theor 46:2746–2752. https://doi.org/10.1109/18.887892
6. Imandoust SB, Bolandraftar M (2013) Application of K-nearest neighbor (KNN) approach for predicting "economic events: theoretical background. Int J Eng Res Appl 3(5):605–610
7. Weinberger KQ, Lawrence KS (2009) Distance metric learning for large margin nearest neighbor classification. J Mach Learn Res 10:207–244
8. Shalev-Shwartz S, Singer Y, Ng AY (2004) Online and batch learning of pseudo-metrics. In: Twenty-first International conference on machine learning. ACM, New York, NY, USA, vol 94. https://doi.org/10.1145/1015330.1015376
9. Baoli L, Qin L, Shiwen Y (2004) An adaptive k-nearest neighbor text categorization strategy. ACM Trans Asian Lang Inf Process (TALIP) 3(4):215–226
10. Chen YS, Hung YP, Yen TF, Fuh CS (2007) Fast and versatile algorithm for nearest neighbor search based on a lower bound tree. Pattern Recogn 40(2):360–375
11. Fix E, Hodges J (1951) Discriminatory analysis, non parametric discrimination: consistency properties. Technical report, vol 4. USA, School of aviation medicine Randolph field texas
12. Hart P (1968) The condensed nearest neighbor rule. IEEE Trans Inf Theory 14(3):515–516. https://doi.org/10.1109/TIT.1968.1054155
13. Gate G (1972) The reduced nearest neighbor rule. IEEE Trans Inf Theory 18(3):431–433. https://doi.org/10.1109/TIT.1972.1054809
14. Alpaydin E (1997) Voting over multiple condensed nearest neighbors. Artif Intell Rev 11:115–132. https://doi.org/10.1023/A:1006563312922
15. Wilson D, Martinez T (2000) Reduction techniques for instance-based learning algorithms. Mach Learn 38(3):257–286. https://doi.org/10.1023/A:1007626913721
16. Aha DW, Kibler D, Albert M (1991) Instance-based learning algorithms. Mach Learn 6(1):37–66. https://doi.org/10.1007/BF00153759
17. Sproull RF (1991) Refinements to Nearest neighbor searching. Tech Rep Int Comput Sci ACM 18(9):507–517
18. Kumar A, Bhatnagar R, Srivastava S (2018) ARSkNN: an efficient k-nearest neighbor classification technique using mass based similarity measure. J Intell Fuzzy Syst 35(4):1–12. https://doi.org/10.3233/JIFS-169701
19. Lyon RJ, Stappers BW, Cooper S, Brooke JM, Knowles JD (2016) Fifty years of pulsar candidate selection: from simple filters to a new principled real-time classification approach. Mon Not R Astron Soc 459(1):1104–1123

20. Keith MJ (2010) The high time resolution universe pulsar survey—I. System configuration and initial discoveries. Mon Not R Astron Soc 409(2):619–627. https://doi.org/10.1111/j.1365-2966.2010.17325.x
21. Lorimer DR, Kramer M (2004) Handbook of pulsar astronomy. Cambridge observing handbooks for research astronomers. Cambridge University Press, Cambridge, vol 4
22. Archana S, Elangovan K (2014) Survey of classification techniques in data mining. Int J Comput Sci Mob Appl 2(2):65–71
23. Aha DW (1997) Lazy learning. Kluwer, Norwell
24. Gao QB, Wang ZZ (2007) Center-based nearest neighbor classifier. Pattern Recogn 40(1):346–349
25. Omercevic D, Drbohlav O, Leonardis A (2007) High-dimensional feature matching: employing the concept of meaningful nearest neighbors. In: IEEE eleventh international conference on computer vision, pp 1–8
26. Toyama J, Kudo M, Imai H (2010) Probably correct k-nearest neighbor search in high dimensions. Pattern Recogn 43(4):1361–1372
27. Ting KM, Zhou GT, Liu FT, Tan SC (2010) Mass estimation and its applications. In: Sixteenth ACM SIGKDD international conference on knowledge discovery and data mining, pp 989–998
28. Kumar A, Bhatnagar R, Srivastava S (2018) Analysis of credit risk prediction using ARSkNN, pp 644–652. https://doi.org/10.1007/978-3-319-74690-6_63
29. Cristoph B, Rob JH, Bonsoo K (2017) A note on the validity of cross-validation for evaluating autoregressive time series prediction. Comput Stat Data Anal 120:70–83. https://doi.org/10.1016/j.csda.2017.11.003
30. Ji-Hyun K (2009) Estimating classification error rate: repeated cross-validation, repeated hold-out and bootstrap. Comput Stat Data Anal 53:3735–3745
31. Gaoxia J, Wengian W (2017) Error estimation based on variance analysis of k-fold cross validation. Pattern Recogn 69:94–106

Analysis of Automated Log Template Generation Methodologies

Anoop Mudholkar, Varun Mokhashi, Deepak Nayak, Vaishnavi Annavarjula, and Mahesh Babu Jayaraman

Abstract Decision making and advanced diagnostics over log messages from operations point of view are an important and challenging area giving that these log messages do not adhere to a format, variable in length and essentially remains unstructured. Log analysis is a key function that deals with analyzing these messages to produce insights that help operate, understand, debug and manage the services, applications and/or functions to (i) *detect failures* (ii) *detect consistency issues* (iii) *deduce anomalous behaviors* and (iv) *continuous monitoring with prediction*. An infrastructure management practice relies on huge amounts of log messages collected from the devices such as servers, routers and switches in a data center, telecom, datacom networks or IT operations. Entities emit log messages revealing state of the running system for management perspective. The growing scale and complexity of infrastructure make it unrealistic and impractical to analyze log messages with manual or subject matter experts or even with expert systems. Toward this emerged automated methods for log parsing, which carefully study and extract features of interest from these messages and produces message templates for applications to easily discover constituent properties. These methods are largely based on clustering which is an unsupervised machine learning approach. Different parsers make use of clustering in different ways, and in this work, we study these techniques and compare its template generation capabilities. The generated templates so formed

A. Mudholkar (✉) · V. Mokhashi · D. Nayak
School of Information Sciences, MAHE, Manipal, India
e-mail: anoopmudholkar1994@gmail.com

V. Mokhashi
e-mail: varunmkhs@gmail.com

D. Nayak
e-mail: meetdeepaknayak03@gmail.com

V. Annavarjula
Blekinge Institute of Technology, Karlskrona, Sweden
e-mail: vaan15@student.bth.se

M. B. Jayaraman
Ericsson Research, Bangalore, India
e-mail: mahesh.babu@ericsson.com

N. N. Chiplunkar and T. Fukao (eds.), *Advances in Artificial Intelligence and Data Engineering*, Advances in Intelligent Systems and Computing 1133,
https://doi.org/10.1007/978-981-15-3514-7_44

face challenges such as human readability, machine interpretability and inconsistent structures. In this paper, we present an analysis of existing log template generation methodologies quantiatively and qualitatively. We conclude with challenges that are still prevailing identifying the functional weaknesses using examples when templatizing the log message history.

Keywords Log analysis · Template generation · Unsupervised learning · Clustering · Machine learning

1 Introduction

Logging frameworks help developers inject logging-related features through libraries, application programming interface(API) or software development kit(SDK) generated in megabytes to gigabytes of dumps and hence I/O intensive. Choice of one over the other can significantly impact the performance of applications. Well-known methods include syslog [1], Windows EventLog [2], Log4j [3], SLF4J [4] and LOGBack [5].

Log message structures

- Traditional unstructured log
- Structured log

Structured logging gave greater control to developers by enabling to encode structure using SDK, parameter identification mechanisms and log message decorator frameworks. Libraries like message template [6], slog, NLog, etc, provide single-line, multi-line, JavaScript object notation (JSON) formats. Structured logging benefited by reducing the need for regular expression-based matching during log parsing and leading to solve leakage problems. Free-form logging prevails and still continues to be in use due to its convenience and model-free nature.

Traditional Log parsing feature extraction techniques

- Regular expression-based parsing
- Rule-based parsing and handling

Rigorous domain-agnostic processing methods like text analytics techniques were used, i.e., like a counters accounting for number of times a certain log message, variable portion and set of messages that are repeated. These were inspired from bag of words model, n-gram model and also using techniques like transactionized log batching.

There are varieties of logs viz. Hadoop, Apache, sshd and other flavors include network, cloud infrastructure, standard-specific and product-specific logs. It becomes complex for the log parsers to perform domain-based feature extraction. Key-value notation-based logging results in increase of the size of logs. Number of templates, rules or expressions that needs to be maintained in context of parsing adds significant bookkeeping requirements. Hence, this remains a challenge for backward compatibility, freedom for log structure and flexibility during development.

1.1 Log Parsing Frameworks

Notable log parsing frameworks include liblognorm [7], logstash's grok filter [8] which essentially extracts structured data from unstructured logs using regular expression and tagging-based solutions. With large scale and converged infrastructure deployments, the problem of processing and analyzing log messages manually becomes intractable. Traditional methods do not meet the functional, scaling and flexibility requirements.

Hence evolved the new trend of machine learning-based automated log parsing techniques. viz., Iterative Partitioning Log Mining (IPLOM)[9], log key extraction (LKE) [10], message template extraction (MTE) [11], LogSig [12], simple logfile clustering tool (SLCT) [13] and The LogCluster [14] (Table 1).

Other works like Semantic Query Federation and Linked Data paradigm are being adopted to ensure better detection over log data [15]. This is an interestingly new dimension for log analysis but still dependent on domain knowledge to codify the semantics. Unsupervised log signature extraction [16] by Thaler et al. shows domain-agnostic method using auto-encoder neural network as an alternative to IPLoM and LogCluster. However, the extracted features are not human interpretable and remain an internal property within a black box model.

Table 1 Different log parsers

Parsing method	Key techniques used
IPLoM	Feature extraction, Tokenization, Cluster formation, Templatization
LKE	Log keys clustering, Empirical rules, Log key templates
MTE	Predefined ASCII tokens, Constant generation, Templatization
LogSig	Unsupervised clustering, Word pair generation, Templatization
SLCT	Word vocabulary dictionary, Support value, Clustering
LogCluster	Input support value, Line patterns, Clustering

1.2 Log Management Frameworks

Log analysis is carried out either in centralized or decentralized manner. Decentralized approach essentially distributes the log message processing functions to edges or jobs to shards and aggregates the processed results at a different level. A typical log management framework involves collecting, processing and analyzing processes. Storage and cluster infrastructure are other important areas that these log management frameworks dwell with.

Most commonly used log management frameworks in the industry are ELK Stack, Splunk, LogPacker, LogRhythm, LogScape, Loggly, etc. ELK Stack is the most flexible open-source framework which also recently has machine learning capabilities for log analysis. A brief summary of log management framework features is discussed below.

1.2.1 Log Collection

Large-scale systems continuously generate logs to record system states and runtime information, each comprising a time-stamp and a log message indicating what has happened. These information can be used for multiple purposes like anomaly & network fraud detection, etc. Storage, indexing, buffering and fast-retrieval capabilities are key system features enabled by the underlying platform.

1.2.2 Log Processing

Logs are dominantly unstructured, which contain free-form text. Processing it means cleaning, parsing, aggregating, filtering, converting and transforming such that applicable, relevant and important features are extracted into a structured and organized information.

1.2.3 Log Analysis & Visualization

Log processing yields a database like structured information. This helps carry out generation of insights, derive key performance indicators through further processing and search/query over structured log information base. Analysis and visualization capabilities enable just-in-time generation of results (reports, summary and alerts)and visualization (dashboards, graphical and interactive) for enhanced user experience.

Our observation is that these automated methods produce templates which are still not intuitive to human reading, supportive of machine interpretability and method consistency aspects. Machine interpretability in this context refers to the ability of the templating algorithm to identify the data type and form a schema of the logs. A detailed quantitative assessment is discussed in work by Pinjia et al. [17] which had more of use-case centric measures from respective demonstrator application such as anomaly detection [18] or from problem diagnostics [11] perspective or detect conditions [19]. No existing work has compared the quality of generated templates, i.e., feature extraction for generating quality templates. This log template generation problem is also referred to as signature or log key extraction in the literature.

The log management frameworks are rule-based engines and are most appropriate to derive insights and perform analysis of the logs but are not useful in forming log templates.

In this paper, we show examples of these functional weaknesses and impress upon the need for newer techniques. We present a brief study of algorithms in Sect. 2, and we discuss the quantitative and qualitative assessment of known log template generation methods and techniques in Sect. 3 including the comparative analysis. Present conclusions in Sect. 4.

2 Related Work

We are interested in the aspect of log parsing that extracts a group of event templates which are derived using clustering, partitioning and grouping techniques, whereby logs can be mapped to template structures as part of subsequent analysis. More specifically, each log message can be parsed into an event template (constant part) with some specific parameters (variable part). A detailed analysis of domain-agnostic automatic feature extraction techniques is discussed below.

2.1 IPLoM

Iterative Partitioning Log Mining (IPLoM) [9] is a widely referred feature extraction techniques for log messages. First step groups messages by number of tokens into a cluster. Second step involves grouping by token position creating sub-clusters. Third step we search for bijective relationships between the set of unique tokens by searching for mapping relationships creating partitions. Final step is to extract the event template which involves verifying cardinality of the unique tokens. If the cardinality of the unique token values in a position is 1, then that token position becomes constant part in the template. Otherwise, variable token position is replaced with symbol '*'. The results are very close to a regular expression structure differentiating constant and variable portions, which we believe is the key reason behind its broad adoption.

2.2 LKE

Log key extraction (LKE) [10] makes use of log key to deal with the unstructured data. Log key is defined as the common content among all log messages. Assuming that log messages printed by same statement are highly similar, clustering techniques are used to group log messages printed by the same statement together. Parameters in the form of URIs, IP address, special symbols may cause distortions while clustering, they are handled by writing empirical rules in the form of regular expressions and produce log key templates.

2.3 MTE

Message template extractor (MTE) component is part of FDiag—a diagnostic tool [11]. This method uses predefined ASCII token expression identifying the constant parts to arrive at templates. First it splits the raw logs based on respective individual days the messages were generated, then extract constant from the message bodies based on ASCII code message segment, merge the constants, find unique constants and represent it as a template, and by computing the daily frequency of occurrence of the constant template, identify the filtered templates. These are constant part-based templates, a structure very specific to the FDiag system used for detecting correlation between any two log messages. This method enables construct episodes that will further support debugging and however may not be friendly to allow generation of templates.

2.4 LogSig

LogSig [12] is dependent on unsupervised clustering to derive the templates. It takes 'k'-clusters as input. First, it generates word pairs by extracting words preserving the order as part of the structure. It further iterates to determine clusters based on number of word pairs present and its frequencies. This step is repeated until all logs are identified with its cluster home. The formed k-clusters are then assessed for the most frequently occurring word pair order and is chosen as the log message template in that cluster. The key disadvantage to this method is to specify a 'k' that is appropriate for a given dataset. This method exhibits template fragmentation issue where multiple templates reported when 'k' is high. However, reducing 'k' either is not supportive because the collapsed template structure is not representative of all the variants.

2.5 SLCT

Simple logfile clustering tool (SLCT) [13] in its first step builds a dictionary of word vocabulary that contains word frequency with respect to its position. In the next step, it makes a second pass over the logs and constructs cluster candidates based on the most frequent words/phrases calling these as log templates, and each log template is denominated as cluster candidate. Further, based on user configured minimum support value—'N', it filters candidate clusters to arrive at final log templates. The candidate clusters that have less than 'N' logs are binned as outliers class.

2.6 LogCluster

The LogCluster [14] algorithm addresses the shortcomings of SLCT. With the assumption that there are 'n' lines and each line is a sequence of 'k' words, LogCluster takes support threshold 's' as input and divides log lines into clusters each with 's' line. Then, it mines for line patterns with words and wildcards where multiple log lines match to the pattern in that cluster. LogCluster finds patterns with support 's' or higher by identifying words which occur in at-least 's' event lines. But unlike IPLoM and SLCT, LogCluster considers each word agnostic to position in the log line.

3 Analysis of Log Parsers

In this section, we carry out a detailed analysis of these template generation methods. The primary motive is to identify opportunities for improvement by identifying the prevailing algorithmic challenges that enhances applications and its usefulness the generated templates. We try to build on top of the detailed analysis by Pinje et al. [17] which enables log parser selection by appropriate parsing techniques and reiterated reuse of existing algorithms. It evaluated four log parsers using five different log message datasets with six insightful findings. However, the previous works did not functionally validate the readability of templates and consistency issues with template generation methods.

In this paper, we present an evaluation based on readability and consistency criteria detailed further by selecting two candidate datasets.

From the parsing algorithms mentioned in Sect. 2, we have eliminated LKE, MTE and LogCluster due to the nature of generated templates. LKE depends on predefined rule-based preprocessing that removes content beyond prefix of the message which we felt is doing too much change to the original message and making this method farther to produce readable templates. MTE generates constant templates which are seen as message keys rather than log template. LogCluster requires user to input

Table 2 HDFS dataset

PacketResponder 1 for block blk_38865049064139660 terminating
BLOCK* NameSystem.addStoredBlock: blockMap updated: 10.251.73.220:50010 is added to blk_7128370237687728475 size 67108864
Received block blk_3587508140051953248 of size 67108864 from /10.251.42.84

Table 3 Dartmouth dataset

Interface Dot11Radio0, Station 001360469694 001360469694 Reassociated KEY_MGMT[NONE]
Station 001360469694 Roamed to 004096ec1414
Interface Dot11Radio0, Deauthenticating Station 001360469694 Reason: Disassociated because sending station is leaving (or has left) BSS

parameters such as support threshold 's' which in-turn requires domain knowledge and understanding of dataset, which we felt is very expensive step.

3.1 Candidate Datasets

This analysis makes use of (i) HDFS Dataset [20] referred as Dataset 1, a subset of messages from Darkstar data based on Amazon EC2 cluster shown in Table 2 and (ii) Dartmouth Dataset [21] referred as Dataset 2, a subset of syslog messages from the wireless network of Dartmouth college is shown in Table 3. These input datasets have been pruned of the prefix portion for simplicity in respective table views.

3.2 Quantitative Analysis

We compare the parsers based on the templates generated for datasets namely HDFS(1k,2k,4k) and Dartmouth(5k,10k,20k) variants. The numbers in Table 4 have a lot to say with respect to functionality of each parser, in terms of number of templates, consistency and duplication.

3.2.1 Number of Templates Generated

From Figs. 1 and 2, we can infer that the number of templates generated increases with the increase in number of log lines. While it is possible that there may be increase in log templates, however, the differences in numbers between different parsers are large, and it forced us to explore these variations individually.

Table 4 Dataset to #log templates generated

Dataset	Number of logs	IPLoM	Logsig	SLCT
Dataset 1	1k	12	31,27	16
Dataset 1	2k	16	32,36	27
Dataset 1	4k (doubled 2k)	16	35,38	161
Dataset 2	5k	19	58,69	78
Dataset 2	10k	19	44,43	164
Dataset 2	20k	19	62,67	275

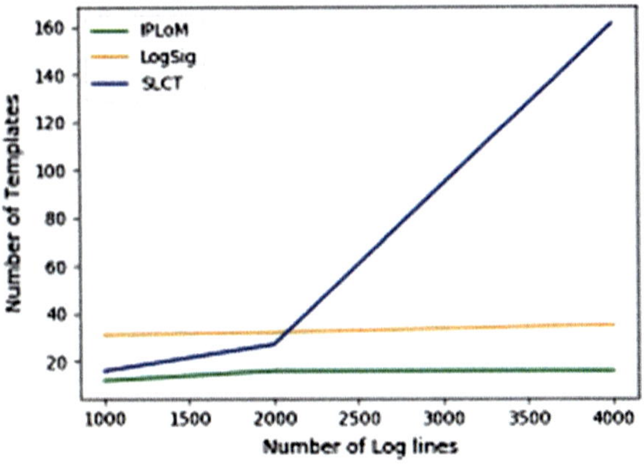

Fig. 1 HDFS

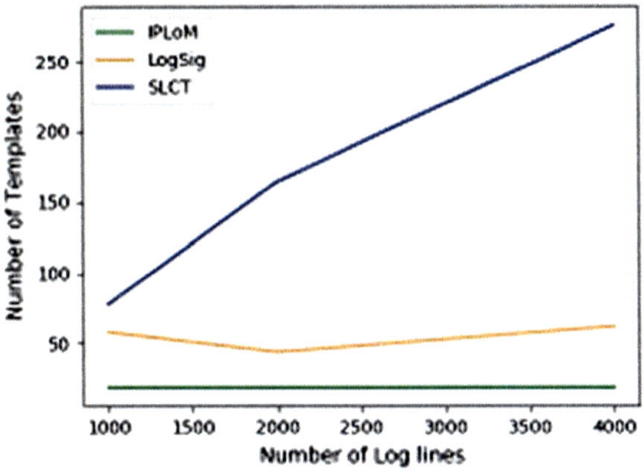

Fig. 2 Syslog

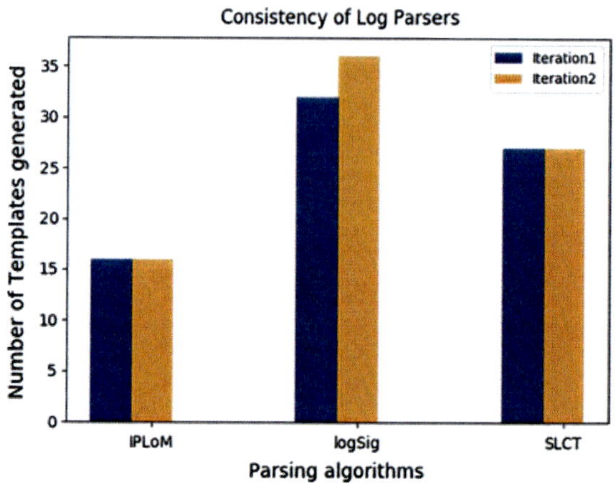

Fig. 3 Consistency of log parsers

IPLoM Number of templates generated is small, and it is almost constant throughout for all the test cases.
*LogSig*The numbers generated are not reliable since this method has consistency issue which is discussed in following section.
SLCT The curve is steeply increasing in number of templates. This was because SLCT uses a frequent word clustering approach which uses frequently occurring words to form the clusters. When more number of messages are given as input, it results in generating more number of templates which is not a desired behavior.

Across these parsers, except for IPLoM, there is observable difference in the output and clearly indicating large variations that is not indicative of characteristics of the generated templates.

3.2.2 Consistency of the Parsers

In Fig. 3, we observe a major issue with respect to consistency of LogSig. Although the templates for IPLoM, SLCT are consistent when run for more than single iteration, LogSig is inconsistent where the number of templates generated varies with every iteration.

3.2.3 Duplication of Log Lines

In Fig. 4, we can observe how parsers react to duplicate log lines. Duplicates may occur when same log is registered multiple times either with different time-stamp or as repeated messages.

To look deeper, we took HDFS 2k dataset and doubled it to new dataset called HDFS 4K. The expectation is that the number of log templates generated is same as

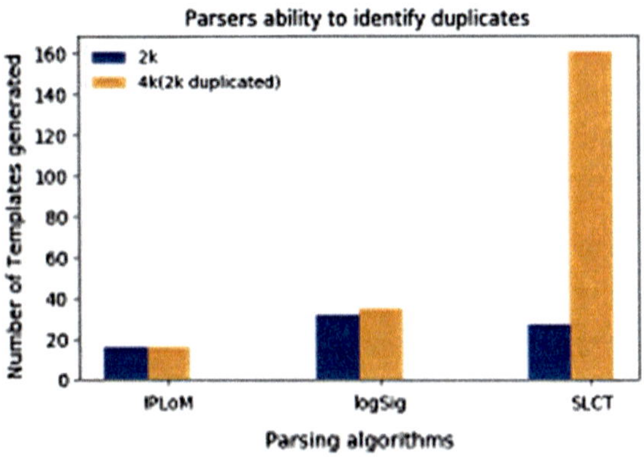

Fig. 4 Duplication response of Log Parsers

that of 2k variant. This is because doubling the dataset will not add a new template structurally.

IPLoM Generates the same number of templates which highlights that this parser can handle repeating messages.

LogSig There were some duplicates that were insignificant. However, since this parser already exhibits inconsistency with number of generated templates, we skipped duplication check.

SLCT With 4K variant of the dataset, it only contributed to the internal support levels increasing, i.e., there are more messages that are frequent with 4K variant compared to that of 2k and hence resulting in lot more templates which was not expected. This is a major drawback of SLCT parser.

Quantitative Analysis Summary: Neither the number of templates generated nor the consistency of parsers during iterations or the duplication of log lines leads us to a conclusive understanding of the log parsers. This motivated and led us to perform qualitative analysis which is described in the following section.

3.3 *Qualitative Analysis*

Qualitative analysis in this context refers to analyzing the templates generated by each of the parsers considered in detail. We have made an attempt to examine the templates with respect to certain characteristics such as fragmentation of templates, handling of delimiters and detection of variables, etc. This section shows an example set of templates that were generated and a discussion with its pros and cons.

3.3.1 Variable Detection

Variable Detection refers to the ability of the parser to identify variable segments in the message and represent them appropriately in the template either as wildcard or with an identifier name. Most of the log template generation methods place a wildcard when it identifies a variable, and this again leads to problems with human and machine interpretation.

In Fig. 5, we can observe variation with respect to variable detection in templates generated from different parsers.
IPLoM did excellent in identifying the variables (e.g., interface id, station id) and replaced these with wildcard to arrive at suitable templates.
SLCT results were a bit confusing, not all station identifiers were identified as variables as shown in the example.
LogSig did not recognize the variable portions expansively, and this method produced multiple templates for same family of log messages as shown. Also, another observation is that, this parser does not place wildcard when deduced as variable which is a significant drawback when considering only the template structure aspect.

An ideal parser should identify the variable part and replace it with respective identifiers and its deduced type rather than wildcards. Also, an ideal parser with variable detection would name this part potentially with pseudo names if not realistic like a domain expert would do, e.g., as shown below:

Interface (Interface ID:AplhaNumeric) Station (Station ID:AlphaNumeric) Associated KEYMGMT[NONE]

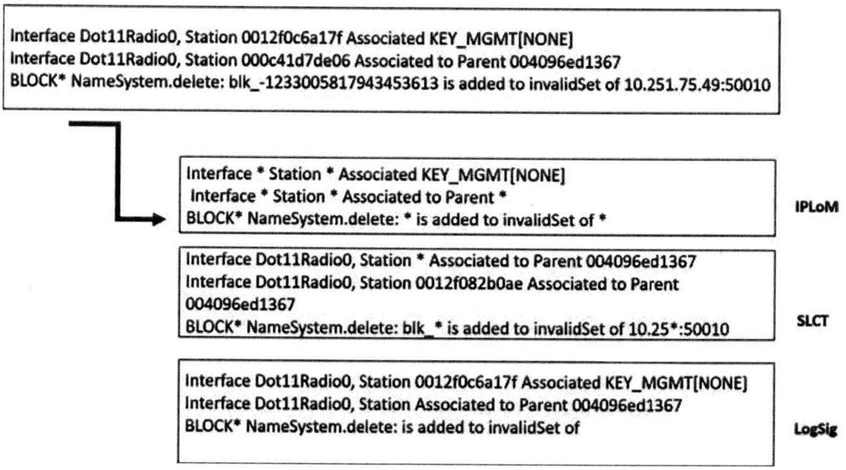

Fig. 5 Variable detection

3.3.2 Delimiters

Log messages are subtly embedded with set of delimiters to indicate separation between segments of the message or to separate the constant and variable parts whichever order they occur. Log parsers are expected to identify these correctly and templatize the messages accordingly.

In Fig. 6, we can observe different interpretation of DELIMITERS by respective parsers.

IPLoM has interpreted two variable parts separated by delimiter correctly and placed wildcard for these portions in the generated template fittingly.

SLCT comparatively has done a better job by identifying the constant prefixes within the variable part and replaced the remaining portion with wildcards.

LogSig parser observably seems to drop portions of messages in the generated templates. This is because this parser does not place wildcard, and hence, the generated template is not reflective of possible variables and its delimiters that separate them.

3.3.3 Fragmentation

Fragmentation can be referred to as generation of multiple templates for similar log lines. These generated templates may be representing the log message line either completely or partially or in duplicates. Partial template case arises when parser generated structure is not fully representative of the log message line or when it is compounded with other problems as discussed elsewhere in this paper. Fragmentation

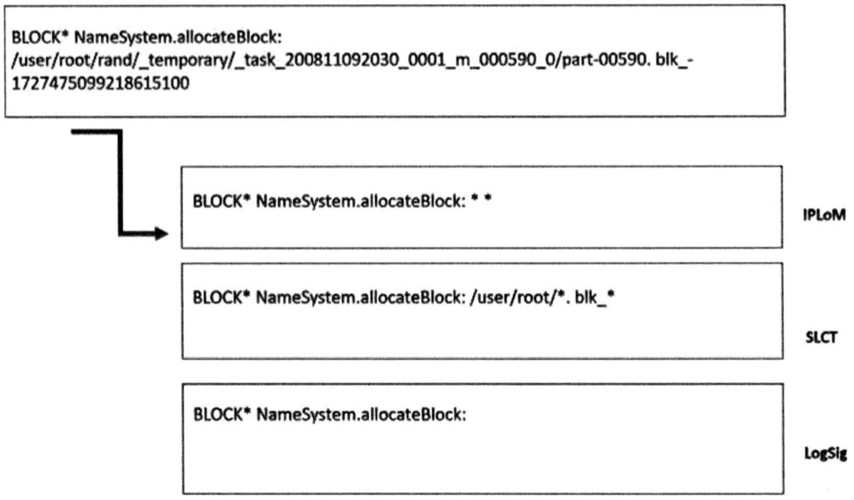

Fig. 6 Delimiters

further leads to the main challenge mentioned earlier, human readability and machine interpretability because multiple templates of same kind lead to ambiguity.

In Fig. 7, we can observe FRAGMENTATION problem with almost all the parsers. *IPLoM* in general produces very good templates by detecting and replacing variables with wildcards. However, we observe scarce conditions where a variable part in a message that typically flips between fixed set of values (e.g., associated and disassociated) is not identified as variables, which is discussed in detail in Enumeration detection problem section as in Fig. 8.

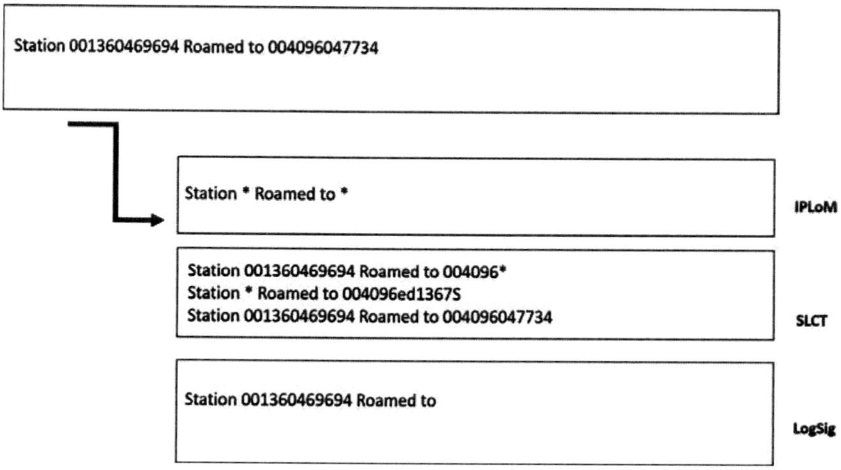

Fig. 7 Fragmentation

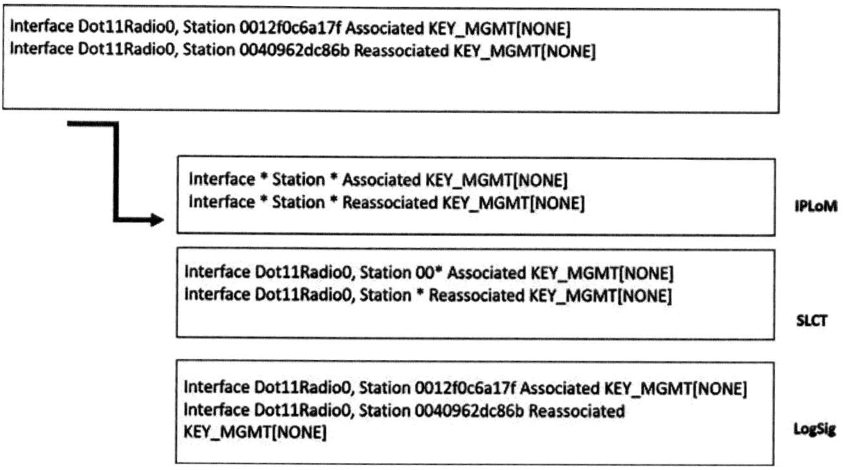

Fig. 8 Enumeration

SLCT produces templates with wildcards for parts of the variable portions in the message which leads to multiple templates. This fragmentation occurs because this parser algorithm is inherently depending on support threshold which fragments templates in countless circumstances.
LogSig on the other hand is skipping certain segments of the log line which is more severe problem when compared to SLCT. This issue is similar to delimiters issue too.

3.3.4 Enumeration

Enumeration refers to identifying variants that are flipping between certain set of values. Unlike an IP address or a station id type of variable, enumeration type can be associated to when its value swings between examples like 'up, down,' 'associated, disassociated,' etc. These shifts in values are consistently within a globally derivable dataset for the given input log message history. Not dealing with enumerated value type only leads to fragmentation, i.e., multiple independent templates. Similar to fragmentation, enumeration also leads to problems with respect to consistent human and machine interpretation.

In Fig. 8, we can observe ENUMERATION problem. The words *Associated* and *Reassociated* in the input log message history are not translated to variable in any of the log parser methods. This is because these values being the only option at that position and also not varying unlike a variable value these words easily cross the support levels measured within the algorithm and hence considered as constants. Subsequently, these methods end up creating two or multiple different templates for the log message history.

3.3.5 Preprocessing

Preprocessing refers to cleaning the raw log messages. Identifying and removing segments of a log line like time-stamp, host-name, port number and facility in case of syslog & time-stamp and prefix pattern in case of HDFS logs, most of the parsers have the functionality for preprocessing inbuilt, but human expertise is needed to identify these segments that need removal, before applying the log parsing algorithm. Here, we wanted to evaluate how the existing log parsers perform when cleaning is not undertaken. The primary motive is to establish and measure how much these algorithms are domain agnostic and sensitive to those structures and how generalized these algorithms are to automatically learn such features.

Figure 9 shows templates created without performing the expected PREPROCESSING. Two key observations were made across all parsers. First, the number of templates created increased significantly, and secondly, part of the raw log lines where preprocessing is normally expected was not templatized properly. IPLoM algorithm is the one that gave the best results very close match as when generated after cleaning. However, this resulted in too many wildcard replacements

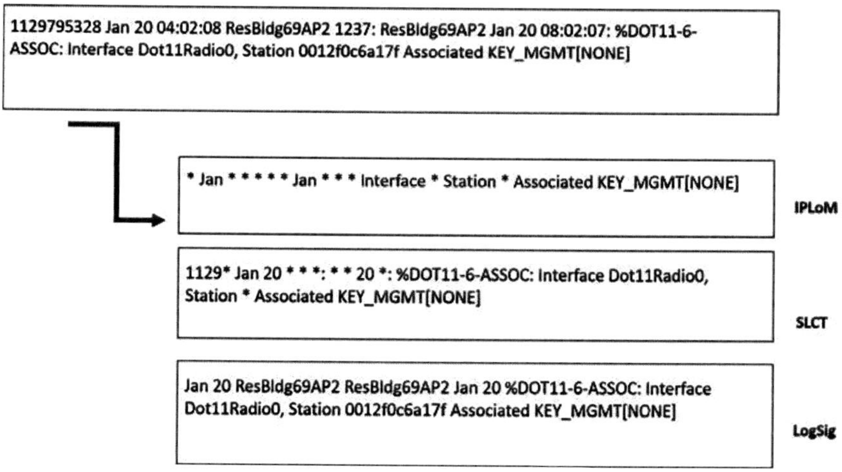

Fig. 9 Preprocessing

and fragmentation of templates because month-related parts in the log message like *Jan*, *Feb*, etc., were not deduced as variables.

4 Conclusion and Future Work

We highlighted results from additional quantitative measurements (over and above the already reported performance figures in earlier work [17]) for the extracted templates. It was conclusive and evident that the generated templates were not easily human readable and also machine interpretable. Also, in some cases, our measurements indicated sufficient variability and inconsistencies in number of generated templates. In this paper, we presented a discussion of observations, viz. what makes these algorithms unique and its respective challenges from the structure of generated templates.

Open challenges that remains in log template generation

- Fragmented templates
- Enumerations not captured
- Inappropriate delimiter handling
- Data types not inferred

In summary, there is a need for new algorithms that are domain agnostic, preprocessing free and ability to extract features without human or subject matter expert intervention for all possible formats. Toward this, in our future work, we are further studying tree-based clustering, text analytics and pattern mining algorithms to overcome these challenges.

References

1. Syslog (2018) Syslog—Wikipedia, the free encyclopedia (Online). https://en.wikipedia.org/w/index.php?title=Syslog
2. Microsof (2018) Windows events: simple logging facade for Java (Online). https://docs.microsoft.com/en-in/windows/desktop/Events/windows-events
3. Apache (1999–2018) Apache Log4j 2: a next generation logging framework for Java applications (Online). https://logging.apache.org/log4j/2.x/
4. SLF4J (2004–2017) SLF4J: Simple logging facade for Java (Online). https://www.slf4j.org/
5. LOGBack (2018) LOGBack: proposed successor framework for log4j (Online). https://logback.qos.ch/
6. Message Templates (2018) Message Templates: a language neutral specification for capturing and rendering structured log events (Online). https://messagetemplates.org/
7. liblognorm (2018) liblognorm: making sense out of syslog data into a well understood interim structured format (Online). http://www.liblognorm.com/news/liblognorm-2-0-5-released/
8. Elastic (2018) Grok: a pattern based log parsing framework to create structured data from unstructured (Online). https://www.elastic.co/guide/en/logstash/current/plugins-filters-grok.html
9. Makanju A, Zincir-Heywood AN, Milios EE (2012) A lightweight algorithm for message type extraction in system application logs. IEEE Trans Knowl Data Eng 24(11):1921–1936
10. Fu Q, Lou J-G, Wang Y, Li J (2009) Execution anomaly detection in distributed systems through unstructured log analysis. In: Ninth IEEE international conference on data mining, 2009. ICDM'09. IEEE, pp 149–158
11. Chuah E, Kuo S, Hiew P, Tjhi W-C, Lee G, Hammond J, Michalewicz MT, Hung T, Browne JC (2010) Diagnosing the root-causes of failures from cluster log files. In: 2010 international conference on high performance computing (HiPC). IEEE, pp 1–10
12. Tang L, Li T, Perng C-S (2011) Logsig: generating system events from raw textual logs. In: Proceedings of the 20th ACM international conference on information and knowledge management. ACM, pp 785–794
13. Vaarandi R (2003) Data clustering algorithm for mining patterns from event logs. In: 3rd IEEE workshop on IP operations & management, 2003 (IPOM 2003). IEEE, pp 119–126
14. Vaarandi R, Pihelgas M (2015) Logcluster—a data clustering and pattern mining algorithm for event logs. In: 2015 11th international conference on network and service management (CNSM). IEEE, pp 1–7
15. Kurniawan K. Semantic query federation for scalable security log analysis
16. Thaler S, Menkovski V, Petkovic M (2017) Unsupervised signature extraction from forensic logs. In: Joint European conference on machine learning and knowledge discovery in databases. Springer, Berlin, pp 305–316
17. He P, Zhu J, He S, Li J, Lyu MR (2016) An evaluation study on log parsing and its use in log mining. In: DSN'16: Proceedings of the 46th annual IEEE/IFIP international conference on dependable systems and networks
18. Ahirwar DK, Saxena SK, Sisodia M (2012) Anomaly detection by Naïve Bayes & RBF network. Int J Adv Res Comput Sci Electron Eng (IJARCSEE) 1(1):14–18

19. Xu W, Huang L, Fox A, Patterson DA, Jordan MI (2010) Detecting large-scale system problems by mining console logs. In: Proceedings of the 27th international conference on machine learning (ICML-10). Citeseer, pp 37–46
20. Xu W, Huang L, Fox A, Patterson D, Jordan MI (2009) Detecting large-scale system problems by mining console logs. In: Proceedings of the ACM SIGOPS 22nd symposium on operating systems principles. ACM, pp 117–132
21. Kotz D, Henderson T, Abyzov I, Yeo J (2009) CRAWDAD dataset dartmouth/campus (v. 2009-09-09), Sept 2009. Downloaded from https://crawdad.org/dartmouth/campus/20090909

Fraud Detection in Online Transactions Using Machine Learning Approaches—A Review

H. Dhanushri Nayak, Deekshita, L. Anvitha, Anusha Shetty, Divya Jennifer D'Souza, and Minu P. Abraham

Abstract Anomaly is referred to as an object that does not follow the footprints of usual data object or the data that contains pattern that does not fit to a well-defined normal behavior. Cybercrime is a pervasive threat for today's Internet-dependent society. Machine learning is becoming increasingly important for fraud detection. This means machine learning can analyze a large amount of data to identify the pattern associated with the fraud. Machine learning provides speed, scale and efficiency. In this paper, we are giving a machine learning model that will detect the fraud and give a known difference between fraud and genuine transactions. We use machine learning algorithms for efficient fraud detection in online transaction and represent those using graphs. Graph exhibits interdependencies between data in an effective way.

Keywords Machine learning · Anomaly detection · Synthetic minority over-sampling technique · Naïve Bayesian · Support vector machine · Random forest

H. D. Nayak (✉) · Deekshita · L. Anvitha · A. Shetty · D. J. D'Souza · M. P. Abraham
NMAM Institute of Technology, Nitte 574110 , Karnataka, India
e-mail: nayakdhanushri@gmail.com

Deekshita
e-mail: deekshitaaithal@gmail.com

L. Anvitha
e-mail: anvithabl@gmail.com

A. Shetty
e-mail: anushashetty97@gmail.com

D. J. D'Souza
e-mail: jenniferdsouza87@nitte.edu.in

M. P. Abraham
e-mail: minupjuly12@nitte.edu.in

N. N. Chiplunkar and T. Fukao (eds.), *Advances in Artificial Intelligence and Data Engineering*, Advances in Intelligent Systems and Computing 1133,
https://doi.org/10.1007/978-981-15-3514-7_45

1 Introduction

Security is the major issue that has come into consideration in recent times. Detecting frauds in an online transaction can be done using any type of techniques. Credit card frauds have been significant and have caused many banks to increase their security systems. Proper and secured authentication needs to be given to any kind of systems. The implementation can be done using unsupervised or supervised machine learning techniques. By considering all the possibilities, the usage of supervised learning might be considered better. The main issue in implementing is that for supervised learning where the need of a dataset is necessary.

Due to this problem in the systems: Banks are facing tremendous number of issues and taking the issue of fraud detection seriously, thus making the transactions to be more sophisticated. Even if any frauds are detected, it needs to be detected as fast as possible. That is the reason why the system transaction needs to have a high-speed security and authentication. From the earlier studies, they have implemented using unsupervised and semi-supervised machine learning techniques. Usage of graphs [1] has proven it to be more tedious and time consuming. By using supervised learning, these issues may be resolved.

Machine learning is the process of training the machine on the basis of how we want it to work. Machine learning is being used in various fields, such as speech recognition, detection, social network filtering. There are basically three types of machine learning techniques: supervised, unsupervised and semi-supervised techniques. In supervised machine learning techniques, there is a desired output for every input. Some of the supervised techniques are Naïve Bayesian, random forests, support vector machine and K-nearest neighbors. In unsupervised machine learning techniques, there is a set of input but no proper desired output. It mainly focuses on grouping together based on the classes and structure of the data. Some of the techniques are K-means clustering. Semi-supervised machine learning techniques have an incomplete training data with no input labels.

In this paper, we focus on the concept used for anomaly detection in banking transactions. This paper is divided into five sections. Section 2 gives the literature survey of the previous methods that have been used for fraud detection using machine learning. Section 3 gives the work proposed by us which contains the problem statement, details about the dataset used and finally the experimental procedure. Section 4 gives the results of the experiment in the tabular column and also plotted in the form of bar graph. Section 5 gives the conclusion of the paper and possible future work that can be done.

2 Related Works

Some of the studies have given a real-world fraud detection system (FDS) [2] which consists of five layers: (1) the terminal: which provides the security checks by providing the PIN codes to the user before transaction; (2) transaction blocking rule: which consists of if…else statements and mainly does not let any transactions happen if the site is not secure also at this stage the featured data that includes all the personal data of the person is collected and compared with the current transaction to check if it is a fraud or genuine transaction; (3) scoring rules: here the if…else statements operate on the featured vector and give a score to each transaction taking place. At the end, it concludes the higher the score more is the fraud transaction. (4) Data-driven model: This proposed model has to find the frauds in the data fed to it. (5) Investigators: This level is mainly to cross-check if the results given by the model are correct or wrong. Basically, it checks for the alerts [3] in transactions happening in the day. The challenges faced in fraud detection are the unbalanced dataset and size of the dataset, non-availability of real dataset [4], finding the appropriate evaluation parameters.

Some use the method of bagging ensemble classifier for decision tree [5]. It was proposed by Breiman [6]. It says about how the machine learning algorithms can improve in terms of accuracy, stability and performance.

There can be many algorithms in order to detect a fraud in online transaction, such as the artificial neural network, sequence alignment algorithm [7], meta-learning agents and fuzzy systems [8, 9]. Hidden Markov Models are also used for the detection of fraud [9, 10]. The parameters that are used for fraud detections are none other than the time taken for training the model, cost, true positive and false positive values [9].

The frauds can be detected using the semi-supervised data too [11], in this initially generation of rules using the a priori of association takes place then using these rules the legitimate transactions are discarded. With the left out data, all the data that has no anomalies is eliminated. From the reference [12] the model proposed by the paper says that in order to come up with the final methods for preventing the fraud the training to data is being traced by the rules replicating them each time. Finally leading to the set of rules that can be used to detect and prevent frauds. It includes online behaviors, fingerprints, voice recognition, face image, etc.

This paper demonstrates how to implement various supervised machine learning techniques. The dataset is taken from the past usage of online transactions in different places. All the needs are to be satisfied by the trained dataset.

3 Proposed Work

3.1 Problem Statement

The credit card fraud detection includes modeling the past credit card transaction with the knowledge of the ones that turned out to be fraud. This model is then used to learn if the transaction happening is fraudulent or not. Our aim here is to identify 100% of the fraudulent transaction while minimizing the incorrect frauds. In addition our classifier model must be able to detect the frauds happening during generalization; else the work would be practically useless.

Since most of the people do not spend time to understand how and where the fraud has occurred. Our methods will help to detect the fraud and prevent it from happening. The proposed idea aims at developing efficient techniques in detecting fraud using a synthetic credit card transaction dataset.

3.2 Dataset

The dataset used for this experiment a synthetic dataset that contains the details of all the European credit card holders and their transactional details. The dataset has 284,807 transactional details in which 492 transactions are fraudulent and the remaining are genuine transactions. It has 31 columns in which we shall focus on only time of the transaction and amount that is transacted. The rest of the columns tells about the details of the customer. Therefore, finally predicting if the transaction is fraudulent or not. Figure 1 shows dataset.

Since the number of non-fraud data is more than the fraud data and the data has to be normalized in order to be trained properly. This is done using synthetic minority over-sampling (SMOTE) technique which adds equal number of 0 s (non-fraud) and 1 s (fraud) to the dataset. It helps in getting the model trained faster and understanding the prediction of the model that is which transaction is fraudulent and which is genuine.

Once the synthetic data is trained and tested we derive the real-time data from it. In real-time data, we consider the fields' time of transaction and the amount transacted by the customer. This later predicts if the transaction is fraud or not fraud that is 0 or 1, respectively. The data is been appended onto the dataset at few interval of time and later training using the real-time data to find out if there is any increase in accuracy. This is done several times.

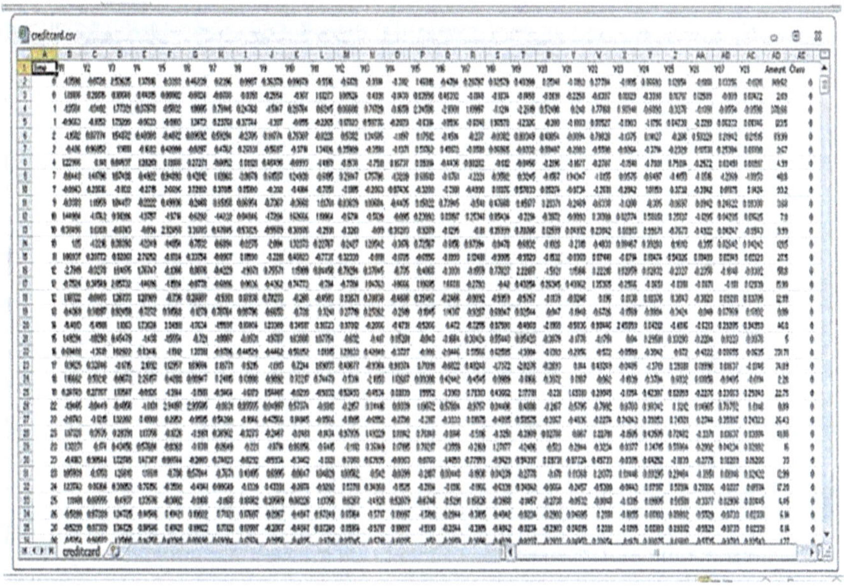

Fig. 1 Clear view on the dataset used for the experiment

3.3 *Experiment Procedure*

The main three steps for predicting the fraud using machine learning are:

Step 1: Extract the required data from the dataset.

Since there is abundant amount of data it will easy to access the required data from the dataset and collect it and store it.

Step 2: Build the model.

The model is build using the algorithms of Naïve Bayesian, support vector machine and random forest.

- Naïve Bayesian Algorithm: Includes the conditional probability.

$$P(H|E) = (P(E|H) * P(H))/P(E) \tag{1}$$

where

- Probability of hypothesis class denoted by P(H).
- Probability of evidence class denoted by P(E).
- Probability of evidence given hypothesis denoted by P(H|E).
- Probability of hypothesis given evidence denoted by P(E|H).

- Support vector machine (SVM) consists of the two classes, one with the fraud data and another with the non-fraud data. Therefore, when the new set of transaction arrives, it must be classified in such a way by plotting the graph and checking if the point belongs to the fraud class or non-fraud class.
- Random forest: The best split is done on the basis of randomly selected features that are obtained. The nodes are split until no further split is possible.

Step 3: Learning parameters.

We take the true positive, false positive, the training set values in order to check with the actual data available. Check if the transaction is fraud or genuine.

Step 4: Compute errors.

We use different matrices to see if any error has occurred, such as we can check if area under the receiver operating characteristics (ROC) curve is equal to 1 then we say that the model is best suitable. If the area is 0.5, then we can say that the model is as good as the random. The model—Fig. 2—gives the clear idea of the steps used for detecting fraud in online transaction.

In every machine learning concept, the dataset is divided into training set which has the large amount of data and testing set which has lesser amount of data say 70% and 30%, respectively, in this experiment. The training set is used by the model to train itself by mapping the input to its respective output. The testing set of the dataset is used to test or predict the model.

Figure 3 shows the anomaly detection in transactions of a credit card. It starts with the preprocessing of the data and splitting the data as train data and test data and finally working on the model to detect the anomalies in the transaction.

4 Results

Results are obtained by finding the accuracy using the confusion matrix. The formula for accuracy is as follows.

$$\text{Accuracy} = (\text{TP} + \text{TN})/(\text{TP} + \text{FP} + \text{TN} + \text{FN}) \tag{2}$$

where FN = False Negative, FP = False Positive, TN = True Negative and TP = True Positive

From using Python and R programming in an 8 GB RAM with i5 CORE processor, we infer the following results for the machine learning methods Naïve Bayesian, SVM and random forest.

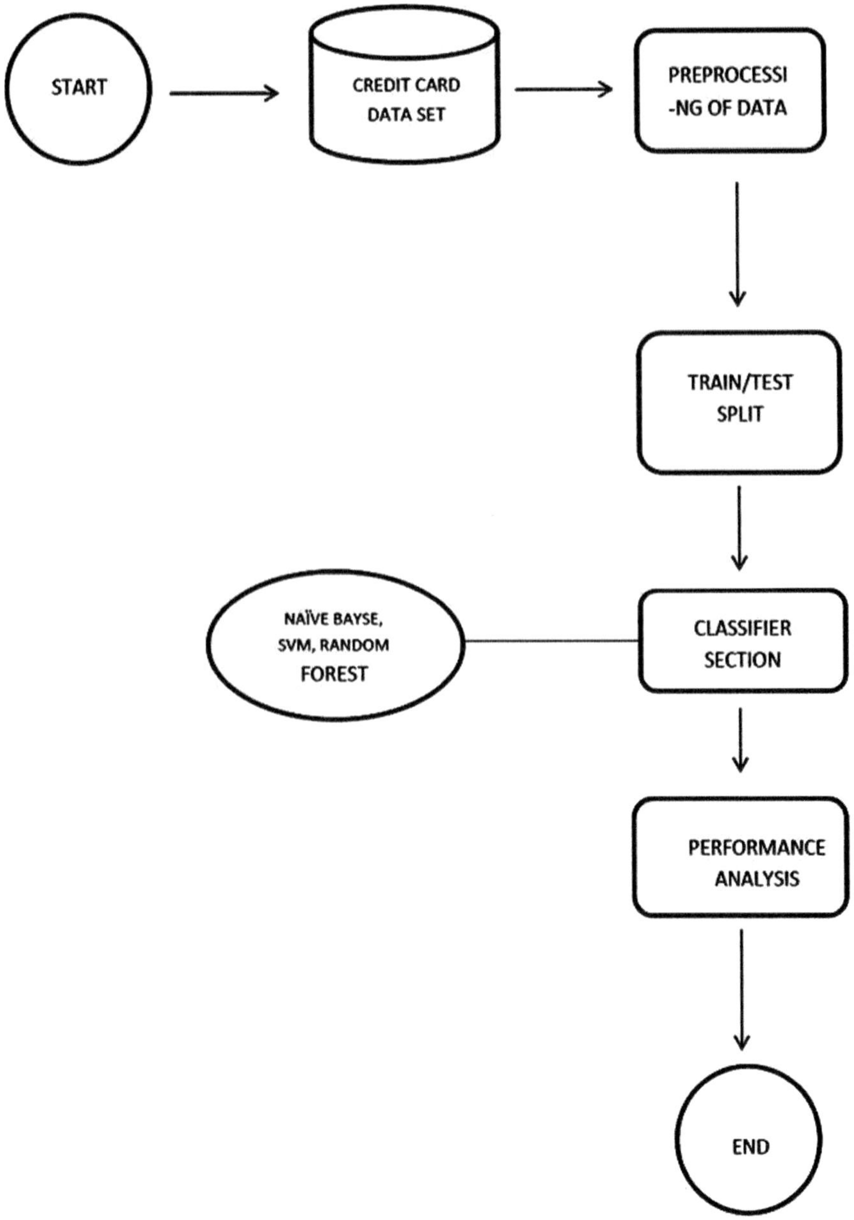

Fig. 2 Model for fraud detection

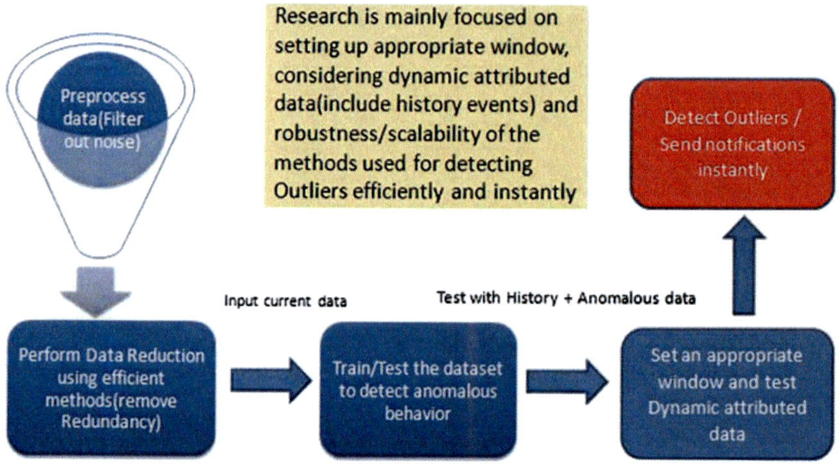

Fig. 3 The process of detecting anomalies in a transaction

Table 1 The accuracy obtained from Python programming

Methods	Accuracy (%)
Naïve Bayesian	99.25
Random forest	98.99

4.1 Results of Python Programming

We see that Naïve Bayesian and random forest have approximately the same accuracy. But being more specific, Naïve Bayesian has the highest accuracy. Table 1 gives the results obtained from Python programming, and Fig. 4 gives the bar graph of the accuracy in percentage.

4.2 Results of R Programming

We get to see the results from R programming from Table 2 contains the accuracy of all the methods used say Naïve Bayesian, SVM and random forest. These accuracies have been plotted in the form of a bar graph in Fig. 5. We infer that random forest has the least accuracy whereas Naïve Bayesian and SVM have approximately the same accuracy but, being more precise Naïve Bayesian has the highest accuracy of 97.6%.

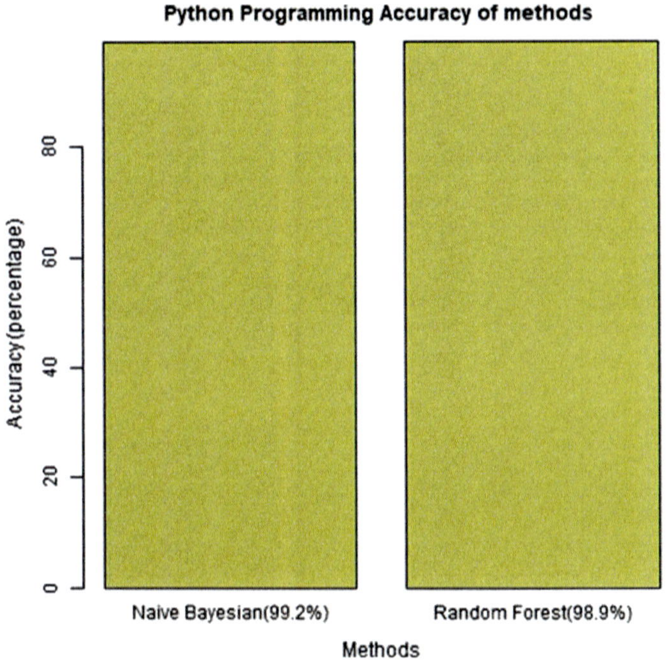

Fig. 4 Bar graph showing the accuracy in python programming

Table 2 The accuracy obtained from R programming

Methods	Accuracy (%)
Naïve Bayesian	97.66
Random forest	89.13
Support vector machine	96.69

5 Conclusion

We can infer that Naïve Bayesian has the highest accuracy in both R and Python programming. Therefore, the machine learning model can be trained using the Naïve Bayesian algorithm. Also we infer that Naïve Bayesian is easier to train and understand when compared to SVM and random forest. We also see that random forest is understandable but since the nodes get randomized each time it is not highly preferred. SVM on the other hand can be used for all classification of data say structured, unstructured and semi-structured data but it is difficult to understand and interpret the final model.

Furthermore, the future work that can be done is the detection of fraudulent and non-fraudulent data can be shown where the fraud is taking place. The geographic location can be notified to the user whose credit card detection has been done. Also,

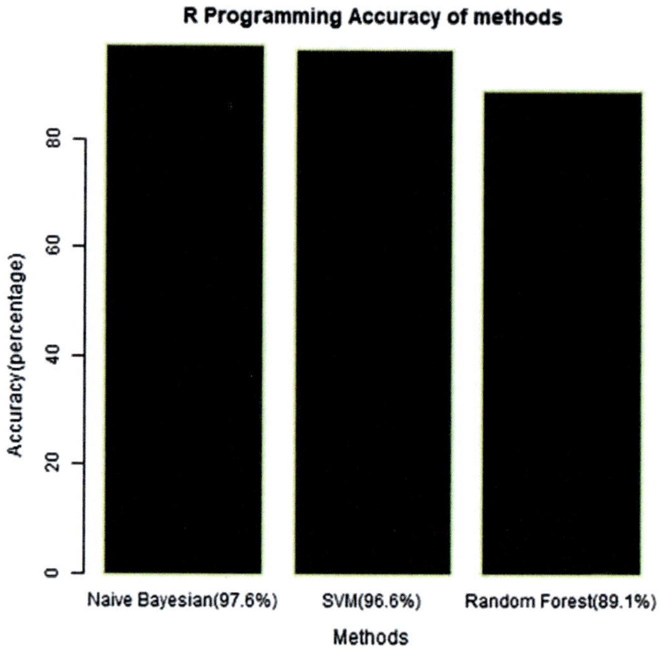

Fig. 5 Bar graph showing accuracy in R programming

the users when use their credit card the notification to them can be done by making it more reliable and user-friendly.

References

1. Akoglu L, Tong H, Koutra D (2015) Graph based anomaly detection and description: a survey. Data Min Knowl Disc 29(3):626–688
2. Dal Pozzolo A, Boracchi G, Caelen O, Alippi C, Bontempi G (2017) Credit card fraud detection: realistic modelling and a novel learnig stratergy. In: IEEE Transaction 2017
3. Krivko M (2010) A hybrid model for plastic card fraud detection systems. Expert Syst Appl 37(8):6070–6076
4. Qibei L, Chunhua J (2011) Research on credit card fraud detection model based on class weighted support vector machine. J Convergence Inf Technol 6(1):62–68
5. Zareapoor M, Shamsolmoali P (2015) Application of credit card fraud detection: based on bagging ensemble classifier. ICCC
6. Breiman L (2001) Random forests. Mach Learn 45(1):5–32
7. Kundu A, Panigrahi S, Sural S, Majumdar AK BLAST-SSAHA hybridization for credit card fraud detection. IEEE Trans Dependable Secure Comput 6(4):309–315
8. Bentley PJ, Kim J, Jung G-H, Choi J-U (2000) Fuzzy darwinian detection of credit card fraud. 14th annual fall symposium of the Korean information processing
9. Raj SBE, Portia A (2011) Analysis on credit card fraud detection methods. In: International conference paper on computer, communication and electrical technology

10. Srivastava A, Kundu A, Sural S, Majumdar A (2008) Credit card fraud detection using hidden Markov model. IEEE Trans Depend Secure Comput 5(1):37–48
11. Farvaresh H, Sepehri MM (2011) A data mining framework for detecting subscription fraud in telecommunication. Eng Appl Artif Intell 24(1):182–194
12. Suman MB, Bansal M (2014) Survey paper on credit card fraud detection. Int J Adv Res Comput Eng Technol (IJARCET) 3(3):827–832

Encryption and Decryption for Network Security Using Reverse Context-Free Grammar Productions

Aishwarya R. Parab and Teslin Jacob

Abstract An enhanced symmetric key cryptographic algorithm is presented in this paper. This new technique uses context-free grammar as it presents a unique property of cryptography which states that one can generate strings from a given grammar; but one cannot classify a grammar if provided with only the strings generated by it. The planned idea consists of four modules: encode, encrypt, decrypt, and decode. Firstly, the file is encoded by making use of the encoding techniques to obtain an intermediate text. The intermediate text is then encrypted to get the ciphertext file using context-free grammar along with the secret key. The secret key is generated by using random number generation algorithm. At the receiver's side, the ciphertext file is then decrypted by making use of context-free grammar followed by decoding techniques to obtain the original file.

Keywords Context-free grammar · Cryptosystem · Decryption · Encryption · Symmetric

1 Introduction

In today's world of data communications, data security is a demanding issue that touches numerous areas such as safe transmission channel, powerful encryption methods, and trusted third party to keep the database. Most of the encryption methods solely maintain information security. The information obtained can be used for malicious purpose by any unauthorized user. Hence, it is obligatory to use effectual encryption and decryption strategies to reinforce information security [1]. Cryptographic mechanisms which are considered extremely dead set against cryptanalysis are often termed as cryptographically strong or strong cryptography [2, 3].

A. R. Parab (✉) · T. Jacob
Department of Computer Science and Engineering, Goa College of Engineering, Ponda, Goa, India
e-mail: aishwaryaparab94@gmail.com

T. Jacob
e-mail: teslinjacob@gec.ac.in

N. N. Chiplunkar and T. Fukao (eds.), *Advances in Artificial Intelligence and Data Engineering*, Advances in Intelligent Systems and Computing 1133,
https://doi.org/10.1007/978-981-15-3514-7_46

Sensitive information transmitted over the channel has focused on the requirement for quick and safe electronic network to attain the purpose for reliability, secrecy, and no duplication of the data exchanged. Cryptography provides a technique to secure and authenticate the data transmitted on protected channel. It allows us to store confidential data and/or send it across insecure network so that it is not read by any unauthorized people.

Cryptography is referred to as encryption that is the method of converting plaintext to incomprehensible ciphertext. The reverse of it is decryption, moving from incomprehensible ciphertext to plaintext. Cipher is said to be a duo of algorithms formed by encryption and its reverse decryption. The algorithm and the key control the exhaustive operation of a cipher. It is an important factor for an explicit message exchange context. Traditionally, ciphers were frequently been used for encoding or decoding, with no supplementary actions such as validation and reliability check. New technologies and new applications bring in new threats, and with the ever-increasing growth of data communication, the need for security and privacy has become a necessity [4–6].

2 Grammar

2.1 Context-Free Grammar

Most of the cryptographic algorithms to provide security against adversaries make use of one-way functions and are yet valuable for authoritative parties. A one-way function can be defined as: It is easy to find $f(x)$ given x. However, it is hard to find x given $f(x)$.

A context-free grammar [7] G is said to be a quadruple if it consists of a finite set of grammar rules (N, T, P, S), where

- N is a set of non-terminal symbols.
- T is a set of terminals where $N \cap T = NULL$.
- P is a set of rules, P: $N \rightarrow (N \cup T)^*$.
- S is the start symbol.

Context-free grammars generally operate on a set of rules. The rules are generated from the intermediate text. Given G (context-free grammar) and a string x, there exists a polynomial time algorithm to decide if x belongs to L(G) [8]. Thus, authentication mostly occurs in polynomial time.

EM Gold [9] states that it is unfeasible to classify the context-free languages if the input data consists of only the text generated by it. There is no way to learn the grammar G using only the string from the language L(G) generated by G, because reduced grammar may fail at the new input string. Pitt and Warmuth [7] in their paper have explained that context-free grammars are difficult to predict accurately just as certain cryptographic predicates to calculate. Therefore, the foundation layer of security of the algorithm is due to this property of context-free grammars and makes it appropriate for planning of a robust cryptosystem.

3 Current System

An often made clear-cut attack on the encrypted data is merely an aim to decrypt the data with each probable key. Majority of those tries will not succeed, but one may, in which the message will be able to be decrypted.

By employing an amalgamation of subtle arithmetic and computational command, most of the encryption algorithms can be defeated. The outcome is that a lot of messages which are encrypted and can be decoded without the knowledge of the key. A skillful cryptanalyst will generally decode encrypted message without the knowledge of the encryption algorithm.

AES in particular is a symmetric-key, iterative block cipher and it encrypts/decrypts 128 bit (16 bytes) blocks of data and makes use of 128, 192, and 256 bits of keys. In contrast to symmetric-key ciphers that make use of the same keys, public-key ciphers use a pair of keys for encryption and decryption. The input and output of block cipher encryption have the same number of bits. Iterative ciphers make use of a round functions which repetitively perform permutations and substitutions on the data.

Data encryption standard (DES) is the predecessor of the advanced encryption standard (AES). The AES algorithm [10] is based on permutations and substitutions. A permutation is rearrangement of data, substitutions trade one component of data with a different one. Permutations and substitutions are performed via numerous different techniques.

The AES encryption algorithm [11] starts by adding the 16-byte input array into a 4×4 byte matrix named State. The encryption algorithm performs a beginning processing step which is called AddRoundKey. AddRoundKey performs a byte-by-byte XOR operation on the State matrix using the first four rows of the key, and XORs input State[r,c] with round keys table w[c,r].

In the coming future, the new AES will undoubtedly replace DES and becomes the actual standard for encrypting all forms of electronic information [12]. The data encrypted by the AES algorithm is so strong that any cryptanalysis attack cannot decrypt the AES ciphertext without making use of a brute-force search all through 256-bit possible keys.

4 Proposed System

This paper introduces a new technique for encrypting and decrypting files. The proposed algorithm is a theoretical study based on the context-free grammar. The security of the system is provided by context-free grammar. The algorithm makes cryptanalysis even tougher owing to the utilization of "random number generator" function that additionally helps in the encryption rounds and generates the keys to encrypt the plaintext [13]. This makes algorithm even safer and removes the overhead of deciding a predetermined key by the end user. The effectiveness of the ciphertext generated is stated by key choice technique and the encoding method [14].

The cryptographic property of context-free grammars states that it is easy to generate and authenticate strings from a given grammar; but it is hard to classify a grammar if provided with the strings generated by it [15]. The primary aim is to develop a context-free grammar-based cryptosystem that will encrypt the data to defend it from different security attacks. This cryptosystem will make use of a symmetric key algorithm which will have a secret key. The input file will be transformed into an intermediate ciphertext which is then sent to the receiver over a transmission channel for decryption.

As shown in Fig. 1, the algorithm takes text characters as input and produces a secret key as the output. The random number generator is used to generate the secret key [16, 17].

As shown in Fig. 2, the encryption algorithm takes input a plaintext file and a secret key and produces out as the ciphertext file. It begins by adding the secret key into the plaintext file. It then performs Base64 encoding on the input file. The next step is to generate (4 × 4) matrices [18] . On applying the transposition techniques on

Fig. 1 Key generation algorithm

```
    Procedure Key Generation ()
    Input: text characters
    Output: Secret key
    Begin
    Enter text
    Secret key generation
End
```

Fig. 2 Encryption algorithm

```
Procedure Encryption ()
Input: Plaintext file, secret key
Output: Ciphertext file
Begin
Add the key to plaintext file
Base64 Encode
Generate matrices
Generate Reverse productions
Generate ASCII and Binary
Generate Ciphertext
End
```

Fig. 3 Decryption algorithm

```
Procedure Decryption ()
Input: Ciphertext file, Secret Key
Output: Plaintext file
Begin
Generate ASCII and Binary
Generate reverse productions
Generate matrices
Base64 Decode
Key extraction
Key matching
If secret key == key in text file
Display plaintext
Else
Display garbage value
End
```

the matrices, a string will be obtained. The next step is to generate reverse context-free grammar productions on the string followed by generating the ASCII and binary values of the characters. The user finally gets a ciphertext file.

As shown in Fig. 3, the input at the decryption side is the ciphertext file generated by the encryption algorithm and the secret key. The decryption process is exactly opposite of the encryption process. It begins by generating the ASCII and the binary characters of the contents in the file. It then produces reverse context-free grammar productions followed by the generation of matrices. It then performs Base64 decoding on the file. The next step is key matching, wherein the key which is input by the receiver is matched with the one in file. If the key matches, then the message is displayed to the user.

4.1 Advantages of Proposed System

Time complexity is one of the major advantages. The proposed algorithm is faster as compared to AES as it does not make use of a round function. AES uses a round function, wherein it performs permutation and substitution methods on the data which is input by user repetitively. Since AES works on a fixed block size and takes approximately the same size independent of the input; thus, its complexity is O(1).

Another advantage of the proposed cryptosystem is irreversibility. The use of context-free grammar makes the algorithm even more robust because of its cryptographic property. Even if the attacker gets the context-free grammar productions, it will be very difficult to obtain the grammar, and this provides added security to the algorithm.

The use of random number generator for the generation of secret key is another advantage of the proposed algorithm. Therefore, a user will not have to define a fixed key each time and it wants to encrypt a message. The proposed algorithm is a theoretical study; the experimental results will be compared on implementing the new technique.

4.2 Comparison with the Existing System

Most of the existing cryptosystems take an input of 128-bit block of plaintext [19] including AES and make use of Feistel network for the encryption process. The algorithm proposed above does not make use of a round function. More the number of rounds, more secure the system but also more inefficient and slow encryption and decryption. AES makes use of block cipher which requires more memory. A small error in one symbol may corrupt the entire block. The proposed algorithm makes use of stream cipher which is more efficient than block cipher.

5 Design

The design of the proposed algorithm is illustrated in this section in detail. It consists of all the processes occurring in the encryption and the decryption algorithms (Fig. 4).

Along with the file, the secret key is added. The next step is to add the key into the plaintext file using mathematical operations. The next step is to generate matrices of the data into the file. After this step, cryptographic transposition techniques will be applied in order to jumble up the data.

The next module is the encrypt module in which the first step is to generate the productions using context-free grammar (CFG). After generating productions, the next step is to compute the ASCII values and binary values of the characters. The resulting data is the ciphertext. On receiving the ciphertext, the receiver then decrypts it (Fig. 5).

The decryption algorithm consists of two modules: the decrypt and the decode modules. It is the turn round of the encryption algorithm. The receiver inputs the same secret key along with the ciphertext file in order to get the plaintext.

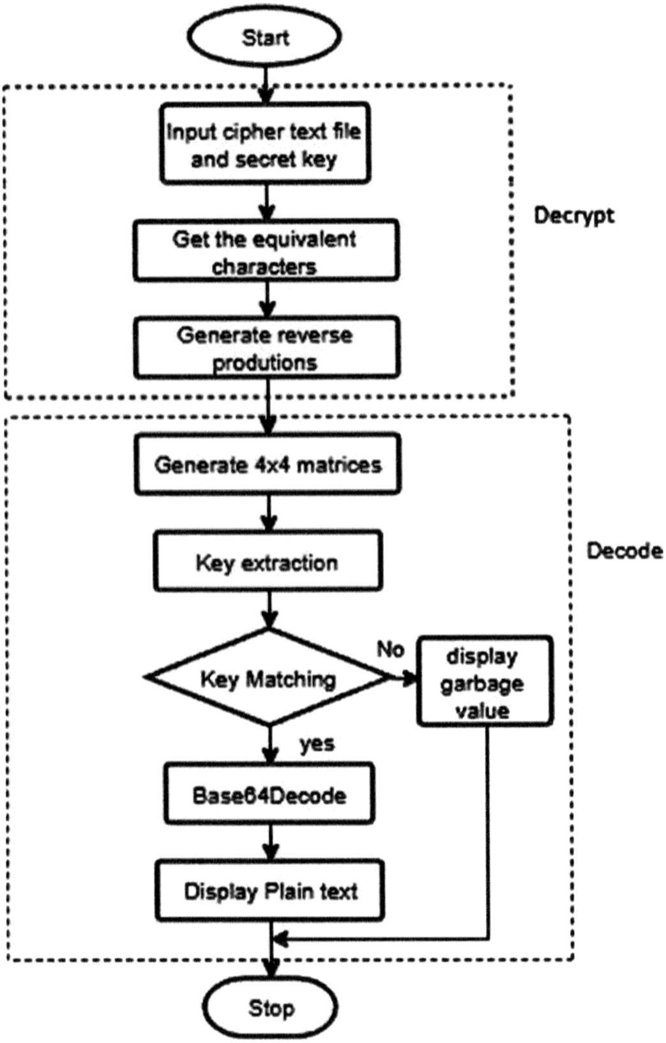

Fig. 4 Flowchart for decryption

6 Security Attacks

6.1 Brute Force Attack

This thoroughly attempts each probable key. It is mostly used in a ciphertext-only attack or known-plaintext attack. Given a limited key length and adequate time, a brute-force attack always leads to success. As we make use of context-free grammar, it is very effortless to generate productions but very difficult to get the grammar back.

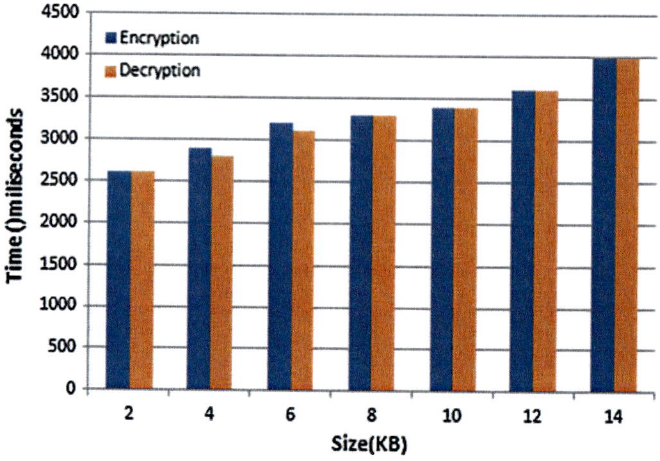

Fig. 5 Graph of time (ms) versus size (KB)

Also, here we use of 128-bit key; therefore, 2128 key combinations are needed which is a large number for any attacker.

6.2 *Ciphertext-Only Attack*

In CFG-based cryptosystem, it is not possible to get the original data/plaintext if only ciphertext is available, as there are two levels of security in it. If you try all the possible keys also, you will get the text which has no meaning.

6.3 *Known-Plaintext Attack*

This attack aims at discovering the secret key along with the cryptographic algorithm which is used to decode any new message. Given that we make use of random key generation algorithm, and the key is going to be unique each time; hence, it will be difficult for the attacker to determine the key.

6.4 *Attack by Breaking the Ciphertext into Strings*

Since in CFG-based cryptosystem each production is of the same length/size and since it uses random key generation, it is difficult to get the plaintext.

Fig. 6 Flowchart for encryption

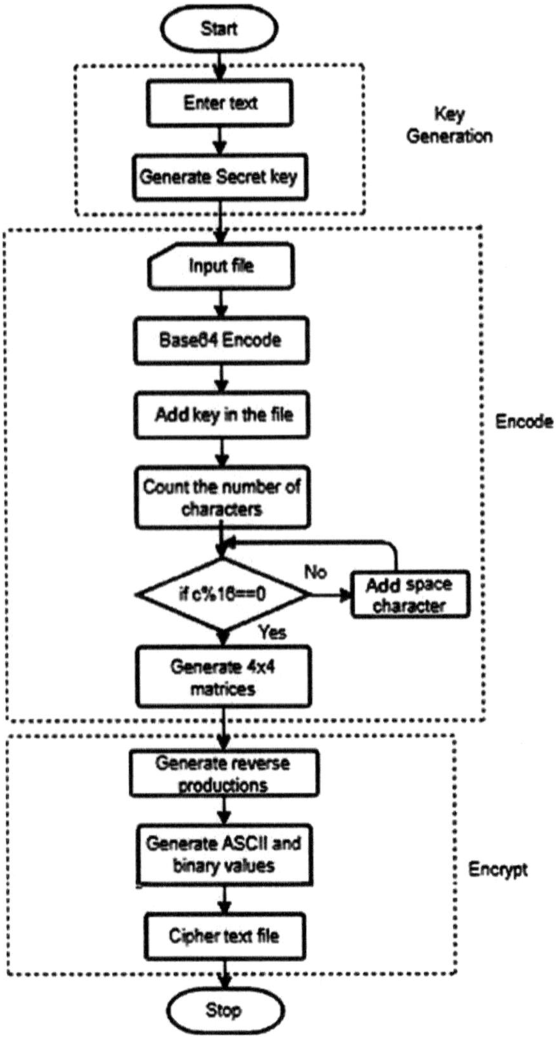

7 Results and Discussion

A computer program is written in C# language for the proposed cryptographic algorithm shown in Fig. 6. The system is intended to take a file (e.g., .txt, .doc, .pnj, .jpeg, .pdf) as input of any size and produce a corresponding encrypted file. It makes use of symmetric keys of 128 bits. The algorithm has been experimented and tested for keys of 64- and 128-bit lengths which are randomly generated.

The demonstration and the implementation steps for the algorithm execution are shown in the next subsection with an example, and basic randomness tests are included in the subsequent part.

7.1 Algorithm Implementation

The subsequent example will show the execution of the algorithm stated above intended for a .txt file. Consider a file containing the data "The quick brown fox jumps over the lazy dog" which needs to be encrypted.

The first module is **KEY GENERATION** module, wherein the user will enter the characters for the secret key to be generated. Suppose the user has entered the characters "pass". The secret key is the combination of user-entered text along with the randomly generated numbers. The output of the key generation module will be—"pass040043214" which is the secret key.

The key is divided into four parts. The numbers are randomly generated which will be used in the further processing of the algorithm.

The next module is **ENCODE** which consists of the following processes:

- User input text file.
- Key embedded into the text file.
- Character count in text file.
- Generation of matrices (4 × 4).

On adding key into the file, the output is:
The pquicak brsown sfox jumps over the lazy dog

Next step is the counting of the characters.
Let "c" be the count
c = 47 the following computations take place.
c mod 16
47 mod 16 = 15
16 − 15 = 1

Thus, 1 padding character is added into the text file.

In the next step, the matrices are generated. The matrices will be generated from each character into the file and matrix transposition techniques will occur.

Therefore, we get the following string:

T skpqcbhai eusrofjsnxmowou pvehad ygrezotl

The third module is the encryption algorithm is **ENCRYPT** module. In this, the first step is to generate reverse context-free grammar productions.

The above string is divided into eight characters, and for each string of eight characters, reverse context-free grammar productions are generated [20].

Once we eliminate the non-terminals, we get the following string-

"reauqkhsoT pic b uxomnovwpf jsso z eyrg e hlatd"

The algorithm then generated ASCII values and binary values for each character in the file.
1110010110010111000011110101111000111010111101000111001111011111
0101000100000111000011010011100011010000011000100100000010000011

1010111110001101111110110111011101101111110110111011111110000110
0110010000011010101110011111001111011110100000111101001000001100
1011111001111001011001110100000110010101000001101000110110011000
0111101001100100

And finally, the ciphertext is generated:

ËÃëâÖÑæ©AàÓÇAÅ@ëðÞÚÜÞïáÍ@ÕçæAõ@ËóåÏÊAÐØÂèÈ

The above obtained ciphertext file will then be transmitted over a secure channel to the receiver. The file is decrypted using the same secret key by the receiver.

The algorithm at the decryption side consists of two modules: **DECRYPT** and **DECODE**. It is the reverse of the encryption algorithm with one exception. It consists of key matching at the end of the algorithm. The message is displayed to the receiver only if the key is matched.

7.2 Testing Results

The algorithm has been tested for different types of file, and it produces the output at a desirable time complexity. Tests were conducted to compute the encryption and decryption time for different file sizes (KB). The encryption and decryption time is almost same for all the different file sizes. The test results for the same are shown in Table 1.

The size of the file is taken in KB (kilobytes), and for each file size, the encryption and decryption time (ms) is recorded. As seen in the above results, the time taken for encryption as well as decryption of different files is almost equivalent. As the size increases, the encryption/decryption time also increases.

A graph has been developed of time (ms) versus size (KB). The time for encryption and decryption varies for text files of different sizes.

Table 1 Time comparison for encryption and decryption

Size (KB)	Time (ms)	
	Encryption	Decryption
2	2600	2600
4	2900	2600
6	3200	2900
8	3300	2500
10	3000	2700
12	3600	3000
14	2800	2800

8 Conclusion

A powerful cryptosystem based on context-free grammar is projected in this paper. The system discussed does not rely on any other cryptographic protocol and provides security without the need of an additional encryption layer like SSL Secure Sockets Layer (SSL). The prominent features of the proposed algorithm are: three-step protocols, easy to use, do not depend on any other cryptographic protocol, and no large overheads. The paper displays and analyzes the convention with incentive to its durability against malevolent assaults.

The cryptosystem is described based on context-free grammars uses its interesting issues, which until now have only been used in designing programming languages. It also makes use of random number generation algorithm for secret key generation. Tests were then conducted to determine that; given a string from a language, it is hard to create another string that belongs to the same language. As the size of the file increases, percentage of accepted strings generated after breaking the string decreases. Hence, the chances of guessing the key and the data in the file become nearly impossible.

Acknowledgements I would like to take this opportunity to express my profound gratitude and deep regard to my Prof. Teslin Jacob, Computer Engineering Department, Goa College of Engineering, for his guidance and valuable feedback and constant encouragement throughout the project.

References

1. Kaushik S, Singhal A (2012, December) Network security using cryptographic techniques. Int J Adv Res Comput Sci Softw Eng 2(12)
2. Abhilash G, Sudhakar KN, Mungara Jitendranath (2012) Advanced symmetric key cryptography using extended MSA method: BLZ symmetric key algorithm. IJCSET 2(7):1321–1326
3. Gupta V, Singh G, Gupta R (2011) A hyper modern cryptography algorithm to improved data security: HMCA. Int J Comput Sci Commun Netw 1(3):258–263
4. Stallings W (2006) Cryptography and network security, 4th edn. William Stallings PEA. ISBN 978-81-7758-774-6
5. AbdulWahab HB, AK AbdulWahab, Nedhal AA (2012) Proposed new algorithm to generate cryptography session keys based on CFG and Huffman code. Eur J Sci Res. ISSN 1450-216X 78(4):546–545
6. Ayushi, Dey SK, Nandy T (2010) A symmetric key cryptographic algorithm. Int J Comput Appl 1(15)
7. Leonard PM, Warmuth MK (1988) Reductions among prediction problems: on the difficulty of predicting automata. In: 3rd structure in complexity theory, pp 60–69
8. Lewis HR, Papadimitriou CH (1998) Elements of the theory of computation. Prentice-Hall, Upper Saddle River
9. Mark GE (1967) Language identification in the limit. Inf Control 10(5):447–474
10. Jain R (2017) Advanced encryption standard (AES). Washington University in Saint Louis, St. Louis
11. Rao S, Mahto D, Khan DA (2017, January) A survey on advanced encryption standard. Int J Sci Res (IJSR)

12. Adki VK, Hatkar SS (2016, June) A survey on cryptography techniques. 6(6)
13. L'Ecuyer P (2007) Random number generation. Departement d'Informatique et de Recherche Opérationnelle, Université deMontréal, Canada
14. Maaita AA, Al_Sewadi HAA (2015) Deterministic random number generator algorithm for cryptosystem keys. Int J Comput Electr Autom Control Inf Eng 9(4)
15. Singh A, dos Santos ALM (2011) Grammar based off line generation of disposable credit card numbers. In: Proceedings of 17th ACM symposium on applied computing SAC
16. Rastogi R, Mittal S, Shekhar S (2015) Linear algorithm for imbricate cryptography using pseudo random number generator. In: IEEE 2nd international conference on computing for sustainable Global development
17. Moosavi SR, Nigussie E, Virtanen S, Isoaho J (2017) Cryptographic key generation using ECG signal. In: 14th IEEE annual consumer communications & networking conference (CCNC)
18. Chatterjee D, Nath J, Dasgupta S, Nath A (2011) A new symmetric key cryptography algorithm using extended MSA method: DJSA symmetric key algorithm. In: International conference on communication systems and network technologies
19. Udit K, Gupta N (2015, August) Modified symmetric key algorithm for improving data security. MIT Int J Comput Sci Inf Technol 5(2)
20. Singh A, dos Santos ALM (2011) Context free grammar for the generation of a one time authentication identity. In: IEEE international conference for advancement of artificial intelligence

A Survey on State-of-the-Art Applications of Variable Length Chromosome (VLC) Based GA

Ravi Domala and Upasna Singh

Abstract Variable length chromosome (VLC)-based genetic algorithms (GA) were applied for various graph-based optimization problems, viz. positional navigation system (PNS), tour planning problem (TPP), motion of robot (MoR), etc., which have obtained efficient results. In these problems, the domain space of the problem variable is known a priori, i.e., number of decision variables to solve the problem is known at the beginning. However, there exist problems wherein the domain space of the problem variable is not known a priori, i.e., number of decision variables required for solving the problem is not known at the begining. The problems with this nature include optimization of complex topologies, viz. automatically finding the number of layers and number of neurons in each hidden layer (HL) of a neural network (NN), optimal architecture of deep autoencoders (DAE), convolutional neural networks (CNN), automatic generation of sufficient number of test cases for a program, etc. Currently, the architectures for these deep neural network (DNN) models are designed manually with human expertise, trial and error approach. These models are data specific and hence consume quantum amount of time in designing the architectures.The traditional GAs can be used to obtain optimal architectures for these models. However, they produce sub-optimal solutions as the size of the chromosome is fixed at the beginning and does not change over the generations. To solve this type of problem efficiently, we need to explore the domain space of the problem variable dynamically and then find the optimal solution starting from simple domain space to complex one. So, we need adaptive approaches wherein the length of chromosome, i.e., number of decision variables, can be increased dynamically. The VLC-based GAs have the capability to explore the domain space of the problem variable dynamically and hence most suitable approaches to solve the problems of this nature. As on date, to the best of our knowledge, less or no research is ventured in solving these optimization problems using VLC-based GA. Therefore, we felt that a thorough survey and critical review on the capabilities of VLC-based GAs may help in understanding their capability to solve these state-of-the-art problems mentioned above. Keeping this view in mind, we did a thorough survey on VLC-based

R. Domala (✉) · U. Singh
Defence Institute of Advanced Technology (DIAT), Pune 411025, India
e-mail: ravidrdo158819@gmail.com

N. N. Chiplunkar and T. Fukao (eds.), *Advances in Artificial Intelligence and Data Engineering*, Advances in Intelligent Systems and Computing 1133,
https://doi.org/10.1007/978-981-15-3514-7_47

GAs and brought out the state-of-the-art applications of it. We also provided various approaches to solve the problem of optimal archiecture design for DNNs.

Keywords ANN · CNN · GA · LCS · PGA · AGA · SAGA · SAMUEL · VLC · MGA · VIV

1 Introduction

Genetic algorithms (GAs) are a type of stochastic search technique built on the basic principles of natural selection and genetics [1]. These algorithms find the solutions to the given problem stochastically and do not need any extra information during evolution of the solution. These algorithms combine fittest individuals among population of individuals which are structured yet allow the randomized information exchange to form a search algorithm similar to human-like search. These algorithms perform much better than the random and exhaustive search algorithms [2], given no other information about the problem of interest. Genetic algorithms (GA) work with coding of parameters, viz. binary coding, ordered coding, tree coding, etc., but not parameters of the problem. To implement the process of natural selection and genetic evolution, we need to consider two aspects, viz. encoding of the problem variable and then a performance measure which can be used to distinguish between the good and bad chromosomes, respectively, for natural selection. Encode process maps problem variable to the corresponding genotype. Generally, GAs encode the problem variable(s) of a given search problem into fixed-length individual strings with each string consisting of alphabets, numerical values, etc. The individual strings which satisfy the constraints are feasible solutions to the problem. The feasible/candidate solutions are known as chromosomes. Each chromosome is composed of alphabets wherein each alphabet acts as gene. Each gene is composed group of values known as alleles. Having represented the candidate or feasible solutions as chromosome, the next step involves the performance measure which enables the selection of good individuals from the weak individuals. The performance of each individuals/chromosome is measured using fitness function which in turn be used by the GA to stochastically guide the search process. Various GAs are proposed depending the type of the problem to be solved and the encoding scheme used. Various types of GA include simple genetic algorithms (SGA), steady-state genetic algorithms (SSGA) and messy genetic algorithms (mGA), respectively.

1.1 Simple Genetic Algorithm (SGA)

Simple genetic algorithm follows five major steps, viz. following major steps while solving an optimization problem. (1) Represent the problem variable domain as fixed-length chromosome with constant population size, crossover probability (CP) and the mutation probability (MP). (2) Fitness function is used as a performance measure of individual chromosomes. (3) Selection of the fittest individual. (4) Perform crossover and mutation with their respective probabilities. (5) Size of the population is kept constant over all the generations.

In this SGA, the offsprings generated using crossover and mutation operations are added to the population of the next generation irrespective of their fitness, whereas the parent chromosomes will be discarded from the population of the next generation.

1.2 Steady-State Genetic Algorithms (SSGA)

In this algorithm, the steps followed are same as that of SGA. However, only two best chromosomes are selected among the two offsprings and the two parent chromosomes that generated these offsprings. Therefore, this ensures the guidance of the search process stochastically.

1.3 Messy Genetic Algorithm (mGA)

Problems solved using SGA and SSGA may suffer from sub-optimal points. The sub-optimal solutions are obtained due to lack of underutilization of full search space of the problem variable. To address problem of local optimization of deceptive problems, David E. Goldberg et al. proposed a new genetic algorithm called messy genetic algorithm (mGA) [3]. Messy GAs are able to process variable length chromosomes and have the capability to work with both under- or over- specifications with respect to problem being solved. Analogous to the nature (wherein nature has evolved its complex life forms of genotype starting from simple forms), the messy GAs as well solve problems by starting from simple, well-tested building blocks to form complex and more complex individuals which in turn cover all features of a problem space. This approach is different from the usual fixed-length coding of the genetic algorithm.

Organization of the papers is as follows: In Sect. 2, we bring out the details of the literature survey done on VLC-based GA. Section 3 describes the problems which are solved using VLC-based GA. In Sect. 4, we provide various state-of-the-art applications of VLC-based GAs and approaches to solve them. We also explain in detail how VLC-based GA can be applied to derive optimal architecture for deep neural networks (DNN). Finally, we conclude with Sect. 5.

2 Literature Survey of Variable Length Chromosomes (VLCs)

Nature has formed its genotypes by processing simple life forms to complex one. Inspired from nature, the VLC is obtained by combining simple well-tested short chromosomes to form to more complex chromosomes which will cover all the features of the problem under consideration. Simple GA solves optimization or search problems. However, they are not suitable for solving complex optimization problems which include topology optimization, path optimization and optimal architecture selection; therefore, we use VLC-based GA to solve these problems effectively. Many variants of VLC-based GAs have been developed over the years. We discuss various VLC-based GAs that are found in the literature along with their applications.

2.1 *The Messy GA*

The first and well-known VLC-based GA is the messy GA [3]. The messy GA was developed to remove the fixed bit position dependency criteria in which simple GA operates. The position independent chromosome representation of Messy GA allows the evolution in both the bit locations and bit values (gene values) while going through the generations. Objective of the messy GA is to establish a tight link between the bits of individuals as well as good bit values of individuals simultaneously so that they are not disturbed by the random crossover phenomenon. In messy GA, the chromosome representation has a generic structure wherein each gene is represented using two values, viz. position of the bit and value associated with that bit. Therefore, an ordered pair <position, value> is referred as a messy gene. Prior to the evaluation of the fitness of the individuals, the first step involves extraction of the bit values from the chromosome and then followed by reordering based on their corresponding positions. Goldberg et al. [3] called this phenomenon as a "*moving-locus scheme*" which means bits within a genotype are flexible and hence can move across the chromosome and generate better building blocks.

2.2 *Implementation of Messy GA*

To implement mGA, first we need a representation of mGA chromosomes and then define the operators on mGA chromosomes. Below sections cover the representation and operations on mGA.

2.3 Representation of Messy GA Chromosomes

As discussed above, in messy GA, the chromosome is represented as name value pair, viz. (Name, value). For example, a five-bit chromosome string "11101" is represented as follows.

((1,1) (2,0) (3,1) (4,0) (5,1))

In mGAs, since variable length strings are allowed, interpretations must be found for strings with too few or too many bits. For example, the strings ((l l) (2 0) (3,1)) and ((l 1) (2 0) (3 1) (4,0) (5,1)(1 0)) are both valid mGA strings in a five-bit problem despite the lack of a fourth and fifth bit in the first and despite the extra first bit in the second.

Operations on mGA chromosomes The messy GAs support two types of operations, viz. crossover and mutation. However, the cross operations is a cut and splice unlike single or two point crossover in SGA. The operator cut and splice produces VLC. The cut operation separates the chromosomes into two parts, whereas the splice operation combines the parts of two different chromosomes. Examples of cut and splice are given below.

As it can be seen from Fig. 1, the two chromosomes are cut initially, and then, splice operation was used to combine parts 1–4 and 3–2, thereby creating two children ch1 and ch2. It is also possible to not create splice for 1 and 4 in which case Cr1 is directly included in new population. For Cr2, an inversion operation is applied, i.e., child is created by splicing 4–3. The reproduction can sometimes produce children wherein the children may contain different values for a given name or position which is called over-specification. It is also possible that there may be an absence of bit values for a given name called under-specification. The individuals suffering from either over-specification or under-specification must be resolved before a full pledged solution can be sent to the fitness function for evaluation. There are various approaches to resolve the over-specification and under-specification issues. This algorithm may not be suitable a good choice for solving problems open-ended search space.

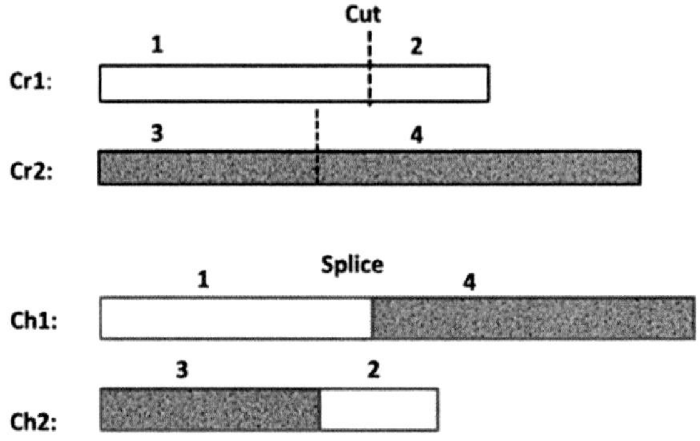

Fig. 1 Cut and splice operation

2.4 The Virtual Virus (VIV)

Another genetic algorithm which uses variable length chromosomes is the virtual virus or VIV [2]. The goal of this algorithm is to bridge the gap between GA and natural systems. The gap is due to the lack of understanding the representation of genetic information, i.e., the encoding mechanism. Authors of the paper brought out various key features of biological genetic representation that are different from standard GA and are mentioned below.

- Length of the biological genomes varies during evolution.
- The biological genome may contain non-coded regions.
- The biological genes are independent of their locations.
- The biological genomes have overlapping of frames.
- The biological genome may contain either duplicate or competing genes.
- Genomes contain a multi-character alphabet, viz. A, T, C and G which indicates nucleotides.

The above features are incorporated into the VIV algorithm to some extent.

2.5 The SAMUEL's System

The SAMUEL's system [4] is a type of learning classifier system (LCS). SAMUEL uses a VLC-based GA approach as its learning module. The learning module learns the rules for making sequential decisions. In this system, each individual/chromosome consists of a finite number of rules known as activity plan which will be used to guide the SAMUEL's system to perform its tasks. Each plan is composed of a variable number of rules wherein each rule acts as genes and many such plans will be constructed and fed to the performance module of SAMUEL for their execution. The rules in an activity plan or individual are expressed using if then else representation instead of binary strings. An action is taken for each matched condition(s). In SAMUEL's system, rules are analogous to genes within a chromosome. Two operations are usually performed in the system which are crossover and mutation. The crossover operation occurs between rules of the chromosomes, thereby allowing the good rules to combine and produce better solution. Mutation operation allows the creation of new rule(s). This type of GA is suitable for obtaining better decision-making strategies.

2.6 Propositional GA (PGA)

The PGA is a multi-character-based GA [5]. The characters are chosen from the fixed alphabet. The size of the biological alphabet is four. It is based on the fundamental principle of DNA level of information encoding wherein each character is used to represent a specific variable whose solution to be found. The goal is to look for a desired character pattern or sequence whose encoding produces a desired value. It evolves the resolution of encoded information dynamically. As it provides one-to-one correspondence between the characters appearing in the genome with that of variable whose solution to be found, it is used to compute the value of the variable. Computation of variable values using PGA1 algorithm is given by Eq. 1.

$$\text{PGA1}(x) = \#\text{Characters of variable } x \text{ on genome/Chromosome Length} \quad (1)$$

where x is the ith variable in the solution.

The sum of the values of all variables is equal to 1. Hence, this algorithm is a right choice for solving resource allocation problems. An example of PGA1 representation is shown below.

Chromosome: FBBDAFECEFADBAAABFAD

Chromosome Length: 20

Computation of values corresponding to each variable/ character is listed in Table 1. As discussed in the previous paragraphs, the PGA representation of chromosome uses characters of alphabets as basic units in place of binary bits or genes. The authors [5] said that the PGA representation relies on the fact that the content of the encoded information is important than the ordering of encoded information. The main advantage of this algorithm is that the process of evolution of the genome at the character level unlike the gene level wherein each gene may be composed of multiple bits. In this algorithm, the size of the alphabet must be determined at the beginning based on the number of variables required for a complete solution of the problem variable. Therefore, large collection of alphabet set is required for a problem having large number of variables. The main drawback of this representation is that problem with large number of variable necessitates the need for large alphabet size. This representation (PGA) handles both over- and under-specifications problems effectively. Th PGA considers all the assigned characters of the genome while calculating the proportions for each variable. Chromosome under-specification is accomplished by setting the missing character value in the chromosome to 0. Another advantage of this representation is that there is no need of extra information such as tags, viz. start and end to be maintained in the genotype.

Table 1 Variables name and corresponding value

Variable name	Value
A	0.3
B	0.2
C	0.05
D	0.15
E	0.1
F	0.2

2.7 The Species Adaptation Genetic Algorithm (SAGA)

Harvey proposed "Species Adaptation Genetic Algorithm—SAGA" [6] wherein the author argues that the random crossover operation on variable length (VL) representations may produce individuals having widely varying lengths as population members. This results in a genetic search through a large hyperspace of the problem variable. This genetic search through large hyperspace is mainly due to large variations in the size of individual genomes. Therefore, the hyperspace for the solution becomes uncorrelated landscape if the size variation of the individuals within a population is large. The best strategy is that GA must first converge to a population of individuals with appropriate size and then narrow down its search operation for the optimal solution. Harvey proposes that the genome size within the population is to be increased gradually over time instead of taking large jumps. The gradual increase in genome size indicates the addition of one gene per generation wherein large jumps indicate addition of multiple genes per generation. A genetic search through the correlated landscapes is achieved by gradual increase in the size of the genotype over the generations. Therefore, this systematic way of increasing the genome size enables the GA to search through a series of smaller to larger and large hypercubes on its way. The incremental size growth of the genome is accomplished using the SAGA Cross [6]. This is a good choice for obtaining optimal topology or architectures. However, the converge time is longer if the problem consists of large number of variables.

2.8 Adaptive Genetic Algorithm (AGA)

The authors, Kim and de Weck [7], proposed an AGA wherein the VLC is used to enable the adaptive search in the increase search domains, i.e., the search domain space is increased over the period of time. The proposed approach finds the solution in two stages, viz. stage 1 and stage 2, respectively. During stage 1, it finds the optimal solution to the problem by exploring small design space using short chromosomes. The optimal solution obtained in stage 1 is then carried forward to stage 2 by means of transfer learning. In stage 2, the sizes of the chromosomes are increased, and then, stage 1 optimal solution is used for initialization of these new individuals. Hence, the more accurate solutions are generated in the subsequent stages. The length of the chromosome is increased in such a way that the population diversity is maintained. Author also applied the algorithm to solve the short cantilever problem and obtained optimal load structure for cantilever.

This algorithm is suitable for exploring the problem variable domain dynamically. Therefore, it is a very good choice for obtaining the optimal architecture for ANN, deep networks, etc., provided an appropriate increase in genotype is established based on domain knowledge. In the paper, author doubled the chromosome length during generation and also during fine correction. Also, the author did not mention the criteria for doubling length of the chromosome length. Therefore, this approach cannot be used directly for exploring the problem variable domain dynamically.

2.9 VLC-Based GA Approach for Fixed Domain Space Applications

Various real-world problems solved by VLC-based GA are available in the literature. For all these problems, problem variable space is known a priori. The problem variable space is known a priori means that the dimension of decision variable used genome is known at the beginning of the genotype formation. The problem of this nature is listed in Table 2.

Authors Pawar et al. [14] proposed VLC-based GA methodology for intrusion detection system in network. Here, authors represented the chromosome as a series of rules, i.e., each rule acts as gene in chromosome representation. Population of chromosome is used for generating a combination of new rules. Authors Arora et al. [15] applied VLC-based GA for generating test cases for state-based representation of software module.

From the literature survey on VLC-based GAs, we found that most of the problems solved using VLC-based GAs are graph-based problems wherein the dimensionality

Table 2 VLC-based GA solution to fixed domain space applications

Application	Optimization algorithm	Remarks
PNS	The VLC-based GA was used to produce optimal trip with multiple destination points, relative importance and restrictions on arrival/staying time as constraints	Reference [8]
Transportation system	The VLC-based GA was used for obtaining optimal transportation path for transporting goods	Reference [9]
Tour planning problem	Solved the tour planning in complex and large urban areas with time-dependency constraint	Reference [10]
Project scheduling	To obtain optimal schedule of a project given resource constraints	Reference [11]
Motion of robot	The VLC-based GA was used to find an obstacle free path for robot movement in a given environment	Reference [12]
Fast robot path planning	To find a fast path for robot movement using better initialization strategy in a genetic algorithm-based robot path planning problem	Reference [13, 16]

of problem search space is known a priori. However, to the best of our knowledge as on the date, there is no research ventured into the problem wherein the dimension of search space is not known a priori which includes optimal architecture of NN, AEs and CNNs.

3 State-of-the-Art Applications of VLC-Based GA

In this section, we discuss two aspects, viz. state-of-the-art applications of VLC-based GA and how VLC-based GA can solve them.

3.1 State-of-the-Art Applications

The state-of-the-art applications of VLC-based GA include automatically designing network architectures, automatic generation of software test cases, design of optimal hardware circuit, design of optimal software functionality, automatic generation of new malware.

Some of the state-of-the-art applications which can be solved effectively using VLC-based GA approach are listed Table 3. The description for each of the applications is given below.

3.1.1 Automatic Design of Deep Neural Network Architectures

This application targeted toward the automatic design of optimal NNA for multi-layer neural networks such as ANN and deep neural networks (DNN) such as auto-encoder–decoder networks, CNNs and recurrent neural networks (RNNs). Training [17] and finding architecture in these networks time consuming. Currently, the suitable the architecture for these networks are designed manually with trial and error phenomenon.

Table 3 State-of-the-art applications

S. No.	Application
1	Automatic design of deep neural network architectures
2	Automatic generation of software test cases
3	Automatic design of hardware circuit efficiently
4	Optimal software functionality design
5	Automatic generation of new malware

3.1.2 Automatic Generation of Software Test Cases

This application involves the automatic generation of test cases for a given software module based on functional acceptance level.

3.1.3 Automatic Design of Hardware Circuit

This involves designing of an optimal hardware circuit by considering the space and number of components as constraints.

3.1.4 Optimal Software Functionality Design

This includes the design of software under memory, functionality and acceptance constraints.

3.1.5 Automatic Generation of New Malware

In this application, the VLC-based GA is useful for producing new malware by combining the different existing malware features. It also has the capability to increase the number of features dynamically.

The above-said applications are the current state-of-the-art applications. Currently, the solutions to these applications are obtained by human experts manually using trial and error methodology. Therefore, we believe that the VLC-based GAs are the most suitable optimization method for the above-said applications.

4 VLC-Based GA Approaches

In the following approaches, we propose to solve the optimal architecture by considering "number of layers and neurons in each layer" as "optimization parameters."

4.1 Problem Formulation

We formulate the automatic designing of optimal NNA as an optimization problem consisting of number of HLs and number of neurons in each HL as optimization parameters. The mathematical representation is shown below.

$$\text{Optimize}\, z = \text{Min}\ \{f_i(X, y)\}_{i=1}^{N} \tag{2}$$

where y: number of HL in NN.
X: number of neurons in each of the 'y' HLs (Vector).

4.2 VLC-Based GA with Adaptive Domain Space Refinement for Optimal Architectures Design

In this section, we describe automatically designing of multi-layer feed-forward neural network (MLFFNN) architecture using VLC-based GA. In this approach, we consider the problem of automatically designing NNA as 1xN-dimensional space search problem. The dimension corresponds to number of decision variables, i.e, NHL and number of neurons in each HL of NNA. To obtain optimal NNA, we start generating NNA in simple domain space (small dimension), i.e., 1D or 2D or up to 5D (which indicates a NNA with one or 2 or 5 HLs and associated neurons), etc., randomly. We then train these NNAs using back-propagation (BP) algorithm and obtain optimal architecture in small domain space and then progressively expand the domain space to high and very high dimension till we obtain a NNA architecture with desired accuracy. The increase in search domain space shall be achieved using high mutation rate. The mutation operation is used to add layers to the NNA which automatically increases the length of the chromosome, thereby automatically searching for the optimal architecture in the next high-dimensional space.

The proposed architecture of automatically designing NNA is shown in Fig. 2.

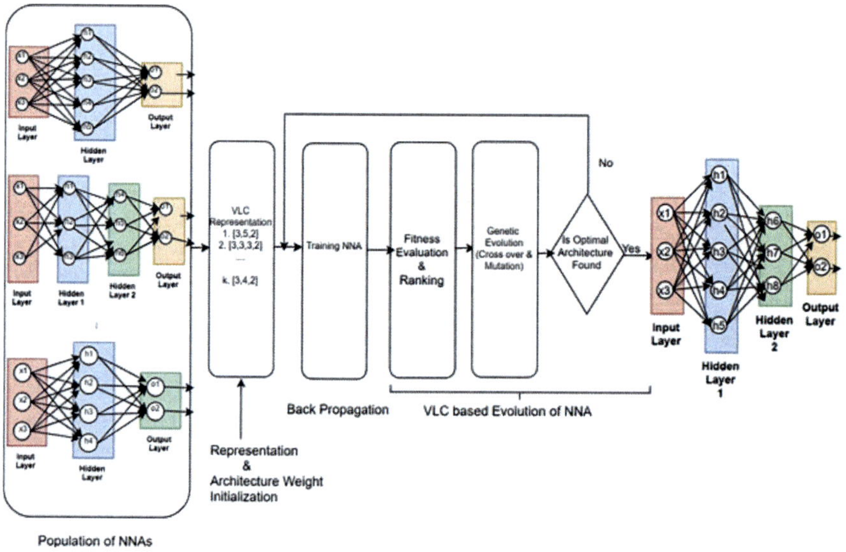

Fig. 2 Architecture of automatic designing of NNA using VLC-based GA

The components of the proposed approach are given below.

- Representation of NNA
- Encoding of NNA
- Creation of initial population of NNA
- Training the NNA
- Fitness evaluation and ranking
- Genetic evolution.

4.2.1 Representation of NNA

The structure of the VLC for representation of NNA is shown in Fig. 3.
where k: Number of HLs
gene1: neurons in HL1
gene2: neurons in HL2
genek: neurons in HLk.

4.2.2 Encoding Method

Integer-based encoding method is the most suitable one for NNA. The chromosome representation (encoding) of neural network is shown in Fig. 4.
Chromosome: [3, 3, 3, 2]

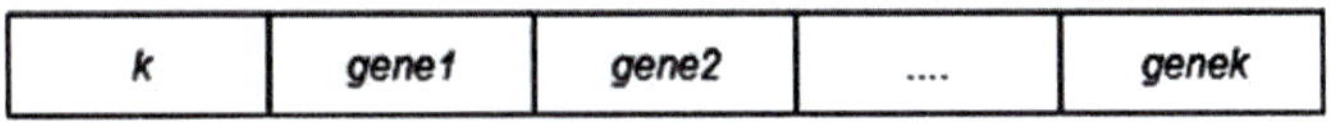

Fig. 3 Structure of VLC

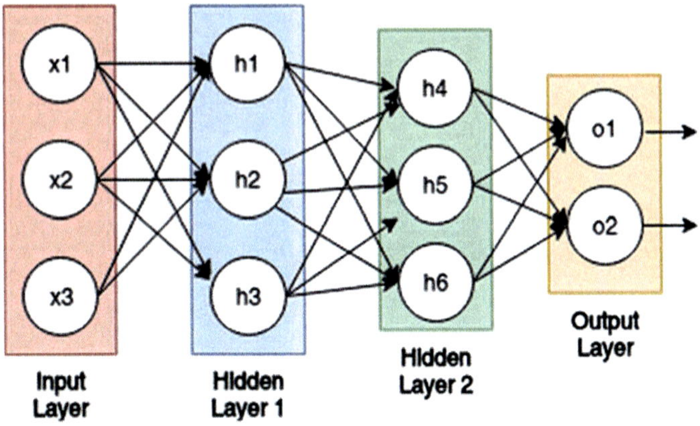

Fig. 4 Multi-layer perception

4.2.3 Creation of Initial Population of NNA

Initially, we create a population of NNA randomly in the simple domain space. This random creation ensures the diversity among the chromosomes or individual NNA.

4.2.4 Training the NNA

This step involves training the neural networks using back-propagation algorithm for a given data sample set. Various network training parameters, viz. learning rate, number of epochs and batch size, etc., can be experimented.

4.2.5 Fitness Evaluation and Ranking

The objective of fitness evaluation is to retain the NNA with good fitness and remove the worst. We then order the NNA based on their fitness starting from highest to least fit.

4.2.6 Genetic Evolution

This step involves two operations to be performed on NNAs, viz. crossover and mutation.

Crossover: It takes genes from two parent NNAs and stochastically produces new architectures which are assumed to be better than the parents in terms of fitness value.

Mutation: It is done to increase the population diversity. In our approach, mutation is used to add new HL to the NNA, i.e., increase the chromosome length.

The algorithm for automatically designing of NNA using VLC-based GA is given below.

*Therefore, we claim that state-of-the-art applications mentioned in Table 3 can be solved efficiently by formulating them as optimization problem mathematically and then applying VLC-based GA.

4.3 Multistage Graph-Based Approach

It is based on construction and exploration of multistage graph wherein initially, a simple multistage graph is to be constructed in simple design space of neurons and layers. Subsequently, the domain space of the multistage graph shall be expanded layer-wise till complex in order to obtain the optimal architecture followed by BP-based training. In this graph, the links indicate the number of neurons in each layer, whereas the nodes of the graph indicate layers of the neural network. Assign weights

Algorithm for Automatically Designing Optimal NNA

1. Representation of NNA as VLC Chromosome with its HLs and neurons as genes or decision variables;
2. Generate Random population of NNA ;
3. Initialization of NNAs;

while *Is Optimal Architecture Not found* **do**
  Training NNAs ;
  Fitness Evaluation of NNAs;
  Ranking NNAs based fitness;
  Evolution of NNAs using Cross over and Mutation;
end
Result: Optimally Designed NN Architecture

Algorithm 1: Automatically Designing of NN Architecture using VLC based GA

to input to the next stage, kth stage to $k+1$th stage and till all the stages using He initialization [14]. Then, apply any graph-based path optimization using VLC-based GA in order find an optimal solution.

4.4 Network Pruning Approaches

This approach is already experimented using VLC-based GA. However, the time required for converge is slow as it initially finds an architecture in large domain space and then prunes for optimal one in smaller domain space.

5 Conclusion

We explored various VLC-based GAs that are available in the literature and their features and found that the VLC-based GAs are applied mostly for graph-based applications wherein the problem variable domain space is known a priori. However, there exist problems wherein the domain space of the problem variable is not known a priori and to be explored dynamically. We also found that the VLC-based GA has the capability of dynamic exploration domain space. We also found state-of-the-art real-world applications which include design of optimal architecture for DNN models, state-based test case generation and new malware generation, etc. In these applications, the problem variable domain space needs to be explored dynamically. Therefore, the VLC-based GA approaches are the most suitable for these problems. To the best of our knowledge, less research is ventured into this direction. In this regard, we proposed few approaches, provided a detailed procedure on how to obtain optimal archiecture for MLFFNN using VLC-based GA, and we also provided a hint where the research effort can put in order to solve the said problems efficiently.

References

1. Goldberg DE (1989) Genetic algorithms in search, optimization, and machine learning. Addison-Wesley, Reading
2. Burke DS, De Jong KA, Grefenstette JJ, Ramsey CL, Wu AS (1998) Putting more genetics into genetic algorithms. Evol Comput 6(4):387–410
3. Goldberg DE, Korb B, Deb K (1989) Messy genetic algorithms: motivation, analysis, and first results. Complex Syst 3(5):493–530. https://doi.org/10.1162/evco.1997.5.1.61
4. Schultz A, Grefenstette J (1992) Using a genetic algorithm to learn behaviors for autonomous vehicles. In: Guidance, navigation and control conference, p 4463
5. Wu AS, Garibay I (2002) The proportional genetic algorithm: gene expression in a genetic algorithm. Genet Program Evolvable Mach 3(2):157–192
6. Harvey I (1992) Species adaptation genetic algorithms: a basis for a continuing SAGA. In: Toward a practice of autonomous systems: Proceedings of the first European conference on artificial life, pp 346–354
7. Kim IY, de Weck O (2004) Variable chromosome length genetic algorithm for structural topology design optimization. In: 45th AIAA/ASME/ASCE/AHS/ASC structures, structural dynamics & materials conference, Apr, p 1911
8. Maruyama A, Shibata N, Murata Y, Yasumoto K, Ito M (2004) P-tour: a personal navigation system with travel schedule planning and route guidance based on schedule. IPSJ J 45(12):2678–2687
9. Ojha A, Das B, Mondal S, Maiti M (2010) A solid transportation problem for an item with fixed charge, vechicle cost and price discounted varying charge using genetic algorithm. Appl Soft Comput 10(1):100–110
10. Abbaspour RA, Samadzadegan F (2011) Time-dependent personal tour planning and scheduling in metropolises. Expert Syst Appl 38(10):12439–12452
11. Mendes JJDM, Gonçalves JF, Resende MG (2009) A random key based genetic algorithm for the resource constrained project scheduling problem. Comput Oper Res 36(1):92–109
12. Alajlan M, Koubâa A, Châari I, Bennaceur H, Ammar A (2013). Global path planning for mobile robots in large-scale grid environments using genetic algorithms. In: 2013 international conference on individual and collective behaviors in robotics (ICBR), Dec. IEEE, pp 1–8
13. Lee J, Kim DW (2016) An effective initialization method for genetic algorithm-based robot path planning using a directed acyclic graph. Inf Sci 332:1–18. https://doi.org/10.1016/j.ins.2015.11.004
14. Pawar SN, Bichkar RS (2015) Genetic algorithm with variable length chromosomes for network intrusion detection. Int J Autom Comput 12(3):337–342. https://doi.org/10.1007/s11633-014-0870-x
15. Arora A, Sinha M (2014) State based test case generation using VCL-GA. In: International conference on issues and challenges in intelligent computing techniques (ICICT). IEEE
16. Qiongbing Z, Lixin D (2016) A new crossover mechanism for genetic algorithms with variable-length chromosomes for path optimization problems. Expert Syst Appl 60:183–189. https://doi.org/10.1016/j.eswa.2016.04.005
17. Glorot X, Bengio Y (2010) Understanding the difficulty of training deep feedforward neural networks. In: Proceedings of the thirteenth international conference on artificial intelligence and statistics, pp 249–256

A Multi-level Access Technique for Privacy-Preserving Perturbation in Association Rule Mining

N. Komal Kumar and D. Vigneswari

Abstract Data generated by the digital media is enormous day by day. Privacy-preserving perturbation techniques such as additive, multiplicative, geometric, differential privacy provide reliable and safe data publicizing, where sensitive data stored in the database cannot be inferred from the published data. Existing privacy-preserving perturbation techniques focus on publishing a single copy of data assuming that the user has the same privacy level of data, but the user has data at different sensitive levels that can be shared to improve the business. Even though the user data can be stored in the selective levels of data, it increases the storage barrier which incurs high cost to the company. In this paper, a multi-level access technique for privacy-preserving perturbation in association rule mining is proposed, which generates multiple copies of data at different privacy levels. We evaluated the proposed technique with the existing privacy-preserving technique with a real-world dataset, and our result shows that the proposed technique has more number of hidden rules and less number of lost rules.

Keywords Multi-level · Privacy · Perturbation · Sensitive · Preservation · Association

1 Introduction

Recent advancement in data analytics made possible to extract a data from published data. Disclosure of data related to an individual without prior permission leads to privacy risks. An individual has data at different levels of privacy [1], which requires relative mechanisms for sharing. The privacy risks always arise with the association between sensitive and non-sensitive attributes [2], such as providing sharing social

N. Komal Kumar (✉)
St. Peter's Institute of Higher Education and Research, Avadi, Chennai, India
e-mail: komalkumarnapa@gmail.com

D. Vigneswari
KCG College of Technology, Karapakkam, Chennai, India
e-mail: vigneswari121192@gmail.com

N. N. Chiplunkar and T. Fukao (eds.), *Advances in Artificial Intelligence and Data Engineering*, Advances in Intelligent Systems and Computing 1133,
https://doi.org/10.1007/978-981-15-3514-7_48

security number/Aadhar, medicine purchased in a pharmacy or products purchased online. For example, an individual sharing his/her Aadhar number for purchasing online via third-party wallets is subjected to disclosure of Aadhar number which is impervious. With this, the enemy can extract the insights of a particular individual where the privacy is compromised. Privacy-preserving perturbation techniques [3] aim at publishing sensitive information by perturbing raw data set. There are several techniques which provide only a single copy of a data with a fixed level of privacy. But when the user needs to share data at variable levels of privacy, the existing privacy-preserving perturbation techniques are inefficient. In this paper, sequential generation of multiple copies of the perturbed data with masking key at various privacy levels is proposed. For example, when the user purchases goods via e-commerce, he/she uses banking transactions for payment which lead to the sharing of credential data. Another example is a company employee sharing business sales to a data analyst to forecast the amount of sales which is most sensitive as the data analyst may compromise the credentials to the competing business company resulting in deflation of the company sales.

Privacy-preserving data perturbation techniques are designed to prevent inference problem [1]. Data owners use privacy-preserving perturbation technique to the data before storing them, which can be later accessed by the user to extract information. The stored data is subjected to a fixed level of privacy, i.e., the information provided to the user without unfair bias. Figure 1 shows the sharing of data to the users with same level of privacy; before sharing the data to the third-party users, the data is perturbed by the owner with perturbation techniques and the perturbed data is shared to all the users without any sensitivity levels, thus reducing the usability of the data for some user. Figure 2 shows the traditional perturbation technique where the data is perturbed according to the data sharing sensitivity levels. If the owner needs to share data to a reliable user, owner can perturb the data to lower levels and store it

Fig. 1 Sharing data with same privacy level

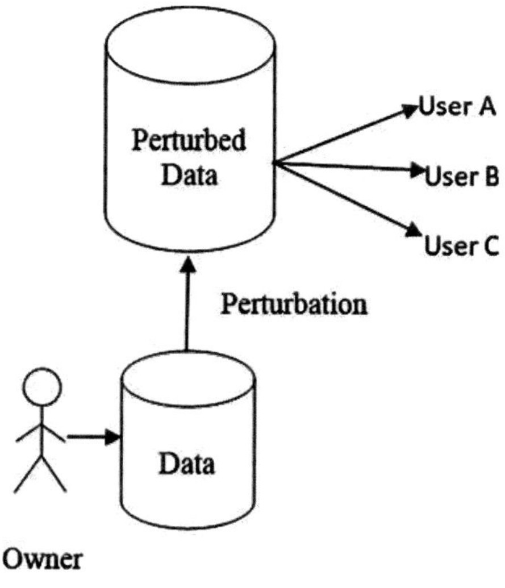

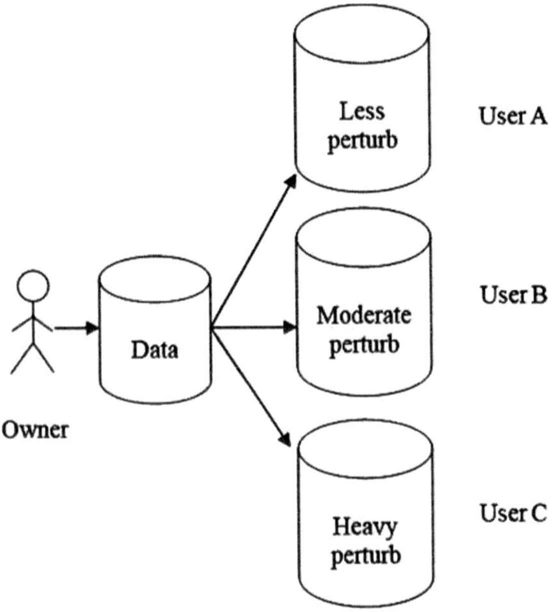

Fig. 2 Traditional perturbation

in separate data storage; if the user is a moderate, owner can perturb to moderate level; and if the user is heavy, owner can perturb data at heavy level. The problem with the traditional method is reliability and storage. When the owner updates set of data, the same has to be updated to all the privacy level data which reduces the data reliability. Storing separate copies of data increases the cost to the company. Rather than storing separate copies of the data, the proposed method uses multi-level perturbation technique where the data is perturbed only when the user needs. The proposed technique perturbs the data according to the user trustworthiness.

The rest of the paper is organized as follows: The related work is discussed in Sect. 2. The proposed multi-level access technique for privacy-preserving perturbation in association rule mining is discussed in Sect. 3. We present the experimental evaluation and discussion of the proposed technique in Sect. 4 and finally conclude in Sect. 5.

2 Related Work

The inference problem and its disclosure of information have been studied abundantly. Sweeney [4] and Samarati [5] proposed *k*-diversity approach which focuses on publishing microdata without revealing information of the individual records,and such release of anonymity approach led to some new techniques such as *t*-closeness [6] and *l*-diversity [7] techniques. Lee [8] proposed an index-based privacy-preserving technique for context-aware computing environments where the privacy-preserving constraints are based on the users' recommendation rather than system-adopted restrictions. Rottondi [9] concerns the openness of the automated meter readings (AMRs) of energy such as water, electricity and gas and proposed an infrastructure and a communication protocol where the user interacts in different spatiotemporal regions without disclosing information to the other nodes in the infrastructure. Zwattendorfer [10] discussed the migration of the components of eID system to a public cloud with cryptographic algorithms such as proxy re-encryption and censored signatures which have improved the privacy-preserving access of data records. Xia et al. [11] proposed a content-based image retrieval technique with encryption techniques without revealing the sensitive information in the cloud server. Ravi et al. [12] proposed synthetic data perturbation techniques such as additive, multiplicative, random for privacy preservation in association rule mining which provided optimal privacy to the datasets. Later by understanding the techniques related to perturbation, Ravi et al. [13] proposed a non-synthetic data perturbation technique which minimizes information loss that occurs in synthetic data perturbation, where the non-sensitive information is injected as a noise which also maintains the mean vectors of data. The comparison of perturbation techniques for privacy-preserving is studied in [3]; the study also proves that the non-synthetic perturbation technique is advisable for data in numeric form which maintains the mean vector of data thus reducing information loss. Vigneswari [14] proposed a privacy-preserving synthetic logarithmic perturbation technique for incremental dataset for cloud environment. Synthetic perturbation technique produces effective results than the traditional systems. Based on the encryption techniques, there had been many works focusing on the efficient privacy-preserving approaches. Recent work had focused on mutual and batch authentication schemes for efficient and reliability; however, the existing schemes do not support the multi-level access scheme to publish and access the dataset. This paper extends the work presented in [14] with a suitable multi-level access scheme which provides safe and reliably technique for data publication and retrieval.

3 Multi-level Access for Privacy-Preserving Perturbation in Association Rule Mining

Traditional privacy-preservation perturbation techniques suffer from reliability and data storage. Rather than storing data at various privacy levels, the proposed method perturbs data only when it is needed using non-synthetic data perturbation technique [13]. The equation for perturbation is shown below

$$Z_i = RC_i + (1 - R)X\mathrm{NC}_i + \mathrm{sqrt}\big((1 - R^2)(1 - X^2)N\big) \tag{1}$$

where

Z_i is the perturbed data for index $i = 0, 1, 2, 3 \ldots$ n
R is the resemblance parameter
C_i Confidential data
NC_i = Non-confidential data
X is the correlation between C and NC
N—Normally distributed with 0 mean and unit variance.
Algorithm 1: Key Allocation

Input: User U, User Trustworthiness T_U-{0, 1, 2), counter c, index I, number of
Output: Lkey K_L, Mkey K_M, Hkey K_H
begin
Initialize c=0, i=0 for all n
  while $c<n_i$ do
check T_U for all n
    if T_U=0, then allocate K_L to U
    end if
      if T_U=1, then allocate K_M to U
      end if
          if T_U=2, allocate K_H to U
        end if
  end while
end

The key allocation algorithm gets the user data and user trustworthiness; the user trustworthiness is based on the number of trusted access to the data, based on the trustworthiness of the user; the key is allocated to the appropriate user, and the multi-level masking algorithm takes the role in masking the data according to the resemblance parameter subjected to the level of trustworthiness.

Algorithm 2: Multi-level Masking

Input: User U, Key K, counter c, index i, Resemblance parameter R, where R->0 to 1, Confidential data C_i for all i=0, 1, 2, 3…n, Non- Confidential data NC_i for all i=0, 1, 2, 3…n, X- Correlation between C and NC, N- Normally distributed with 0 mean and unit variance

Output: Perturbed data Z_i for all i=0, 1, 2, 3…n
begin
Initialize c=0, i=0 for all n
  while $c<n$ do
check K for all n
    if $K=K_L$, then **set** R=0.9 in equ(1)
      calculate Z_i for all C_i and NC_i where i=0, 1, 2, 3…n
      return Z_i for all i=0, 1, 2, 3…n
    end if
    if $K=K_M$, then **set** R=0.5 in equ(1)
      calculate Z_i for all C_i and NC_i where i=0, 1, 2, 3…n
      return Z_i for all i=0, 1, 2, 3…n
    end if
    if $K=K_H$, then **set** R=0.2 in equ(1)
      calculate Z_i for all C_i and NC_i where i=0, 1, 2, 3…n
      return Z_i for all i=0, 1, 2, 3…n
    end if
end while
end

The proposed multi-level privacy-preservation perturbation technique is shown in Fig. 3. When the user tries to access data, first the user sends a request to the owner; second, the owner checks the credentials of the user and their reliability, and based on their trustworthiness, the owner sends the appropriate key. The scenario falls in three cases.

Case 1: When the user is most trustworthy, owner allocates Lkey, which takes the resemblance parameter value as 0.9; the less perturbed data is obtained and shared to the user.
Case 2: When the user is more trustworthy, owner allocates Mkey, which takes the resemblance parameter value as 0.5; the moderate perturbed data is obtained and shared to the user.
Case 3: When the user is less trustworthy, owner allocates Hkey, which takes the resemblance parameter value as 0.2; the heavy perturbed data is obtained and shared to the user.

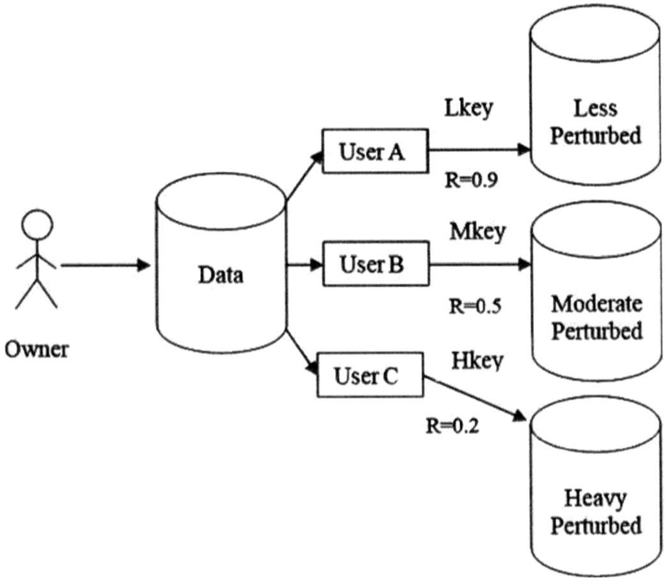

Fig. 3 Multi-level privacy-preservation perturbation technique

4 Experimental Evaluation and Discussion

In this section, we present the experimental study on the performance of the proposed multi-level access technique. We first briefly describe the experimental setup of the evaluation. Our experiment setup was implemented in Java with an Intel Core i7 CPU, with 16 GB RAM laptop computer. The datasets used in the experiment were measured of birth and death from [13]. The dataset contained twenty-one quantitative and six categorical attributes, and only seven quantitative and three categorical attributes were used for the analysis. Experiments are conducted for 1000 transactions, EHR dataset from [14], which contained eight attributes of 556 patients and science clips [15] dataset. The experimental evaluations are organized into three stages, namely perturbing data with synthetic data perturbation techniques such as synthetic multiplicative and synthetic logarithmic [16] and non-synthetic data perturbation technique with three granularity levels ($R = 0.2, 0.5, 0.9$), secondly performing associations rule mining with Apriori algorithm, and finally evaluating the protection offered by the perturbation techniques with number of lost and hidden rules. The privacy-preserving multi-level perturbation can be also applied to medical dataset [17–19] which performs machine learning classifications to overcome privacy-related issues. Table 1 shows the number of rules, number of lost rules and number of hidden rules generated for measure of birth and death dataset for synthetic multiplicative and synthetic logarithmic perturbation techniques. Table 2 shows the number of rules, number of lost rules and number of hidden rules generated for EHR dataset for synthetic multiplicative and synthetic logarithmic perturbation techniques.

Table 1 Rules generated for measure of birth and death dataset

Algorithm	Confidence	Number of rules generated	Number of lost rules	Number of hidden rules
Synthetic multiplicative	20	8	8	0
	30	4	4	0
	40	4	4	0
	50	2	2	1
	60	1	1	1
Synthetic logarithmic	20	8	8	0
	30	4	4	0
	40	4	4	0
	50	2	2	1
	60	1	1	1

Table 2 Rules generated for EHR dataset

Algorithm	Confidence	Number of rules generated	Number of lost rules	Number of hidden rules
Synthetic multiplicative	20	9	7	0
	30	6	6	0
	40	5	5	0
	50	1	3	1
	60	1	1	1
Synthetic logarithmic	20	7	7	0
	30	5	6	0
	40	3	5	0
	50	3	3	1
	60	2	1	1

Table 3 shows the number of rules, number of lost rules and number of hidden rules generated for science clips dataset for synthetic multiplicative and synthetic logarithmic perturbation techniques. Table 4 shows the number of rules, number of lost rules and number of hidden rules generated for measure of birth and death dataset for non-synthetic perturbation technique with varying *R*. Table 5 shows the number of rules, number of lost rules and number of hidden rules generated for EHR dataset for non-synthetic perturbation technique with varying *R*. Table 6 shows the number of rules, number of lost rules and number of hidden rules generated for science clips dataset for non-synthetic perturbation technique with varying *R*.

The number of lost rules for multiplicative perturbation with varying confidence is shown in Fig. 4, the number of hidden rules for multiplicative perturbation with

Table 3 Rules generated for science clips dataset

Algorithm	Confidence	Number of rules generated	Number of lost rules	Number of hidden rules
Synthetic multiplicative	20	10	8	0
	30	8	5	0
	40	7	4	0
	50	5	2	1
	60	2	2	1
Synthetic logarithmic	20	8	8	0
	30	6	7	0
	40	4	5	0
	50	2	3	1
	60	1	1	1

Table 4 Non-synthetic—rules generated for measure of birth and death dataset

R value	Confidence	Number of rules generated	Number of lost rules	Number of hidden rules
$R = 0.2$	20	10	4	2
	30	6	3	3
	40	5	2	3
	50	5	2	4
	60	3	2	5
$R = 0.5$	20	12	3	2
	30	8	2	2
	40	7	1	3
	50	6	1	1
	60	4	0	2
$R = 0.9$	20	6	2	0
	30	3	1	0
	40	3	0	0
	50	3	0	1
	60	1	0	2

varying confidence is shown in Fig. 5, the number of lost rules for logarithmic perturbation with varying confidence is shown in Fig. 6, and the number of hidden rules for logarithmic perturbation with varying confidence is shown in Fig. 7, respectively.

The number of rules generated for non-synthetic perturbation when $R = 0.2$ for varying confidence is shown in Fig. 8, the number of rules generated for non-synthetic perturbation when $R = 0.5$ for varying confidence is shown in Fig. 9, and the number

Table 5 Non-synthetic—rules generated for EHR dataset

R value	Confidence	Number of rules generated	Number of lost rules	Number of hidden rules
$R = 0.2$	20	11	5	2
	30	7	4	3
	40	6	3	3
	50	6	2	3
	60	4	2	5
$R = 0.5$	20	2	1	0
	30	10	3	0
	40	6	2	2
	50	5	1	1
	60	3	0	2
$R = 0.9$	20	9	1	0
	30	4	1	0
	40	4	0	0
	50	3	0	2
	60	1	0	1

Table 6 Non-synthetic—rules generated for science clips dataset

R value	Confidence	Number of rules generated	Number of lost rules	Number of hidden rules
$R = 0.2$	20	10	4	2
	30	5	3	2
	40	4	2	3
	50	3	2	1
	60	1	2	2
$R = 0.5$	20	1	3	2
	30	9	2	2
	40	8	1	0
	50	6	1	1
	60	3	0	2
$R = 0.9$	20	1	2	0
	30	1	1	0
	40	1	1	0
	50	0	1	2
	60	0	0	1

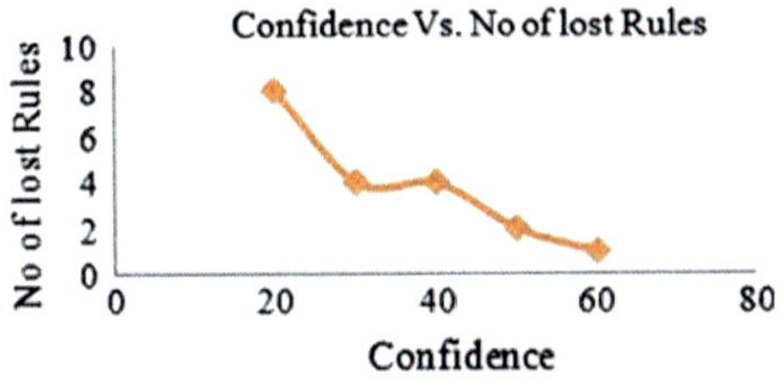

Fig. 4 Synthetic multiplicative—minimum confidence versus number of lost rules (measure of birth and death dataset)

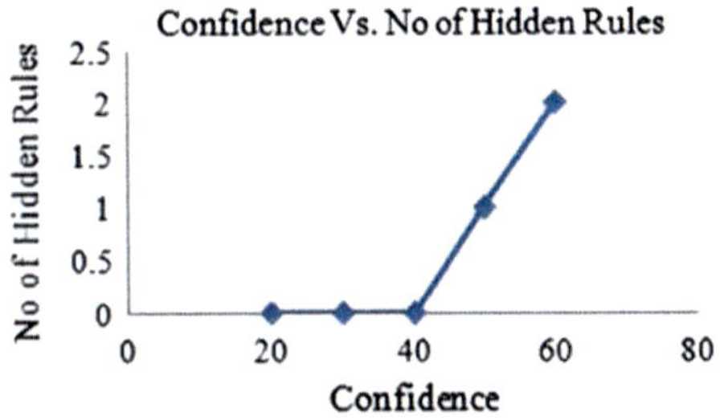

Fig. 5 Synthetic multiplicative—minimum confidence versus number of hidden rules (measure of birth and death dataset)

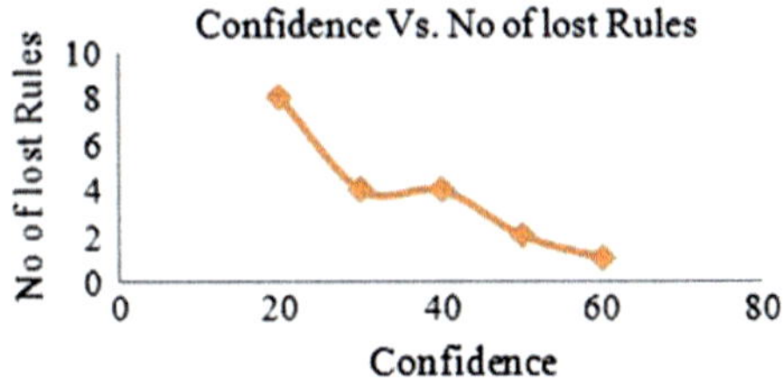

Fig. 6 Synthetic logarithmic—minimum confidence versus number of lost rules (measure of birth and death dataset)

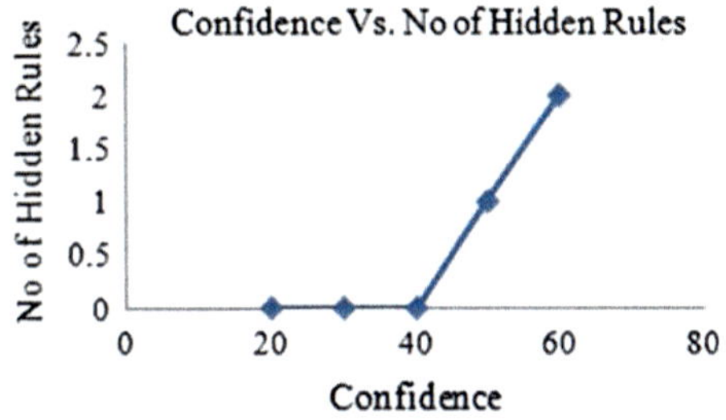

Fig. 7 Synthetic logarithmic—minimum confidence versus number of hidden rules (measure of birth and death dataset)

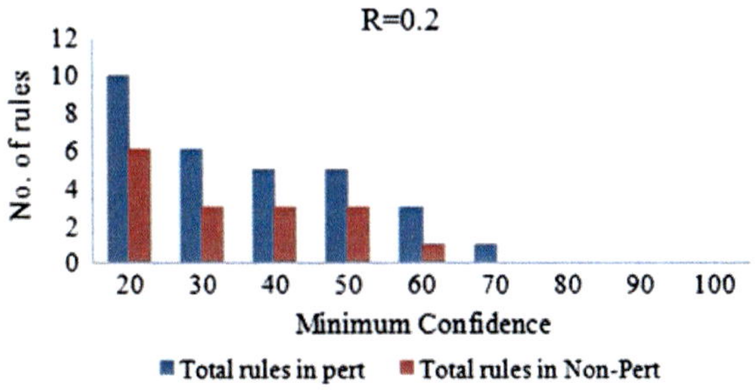

Fig. 8 Non-synthetic—minimum confidence versus number of rules when $R = 0.2$ (measure of birth and death dataset)

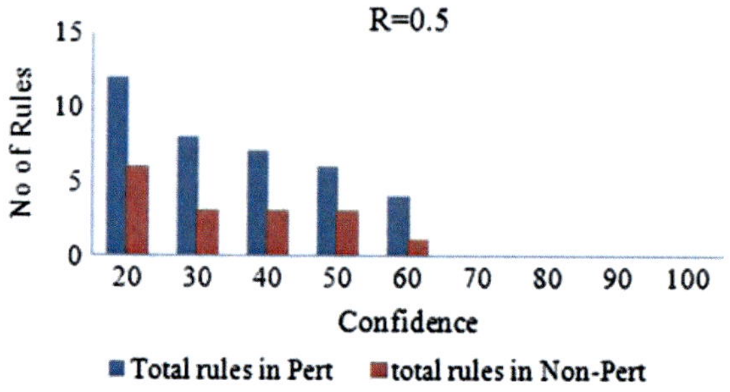

Fig. 9 Non-synthetic—minimum confidence versus number of rules when $R = 0.5$ (measure of birth and death dataset)

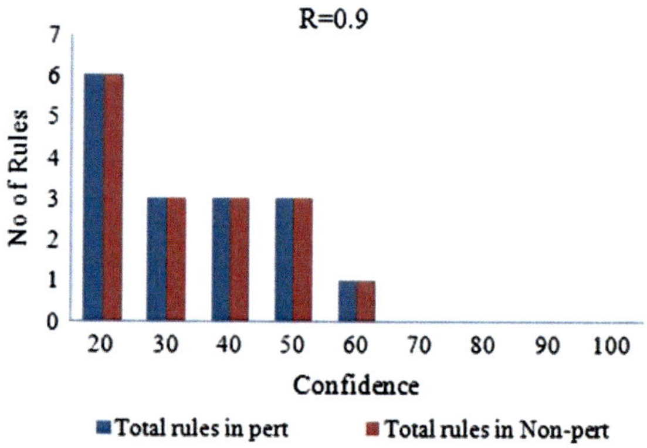

Fig. 10 Non-synthetic—minimum confidence versus number of rules when $R = 0.9$ (measure of birth and death dataset)

of rules generated for non-synthetic perturbation when $R = 0.9$ for varying confidence is shown in Fig. 10, respectively.

The number of lost rules for non-synthetic perturbation when $R = 0.2$ for varying confidence is shown in Fig. 11, and the number of hidden rules for non-synthetic perturbation when $R = 0.2$ for varying confidence is shown in Fig. 12.

The number of rules for synthetic multiplicative, synthetic logarithmic and non-synthetic perturbation for varying confidence is shown in Fig. 13, the number of lost rules for synthetic multiplicative, synthetic logarithmic and non-synthetic perturbation for varying confidence is shown in Fig. 14, and the number of hidden rules

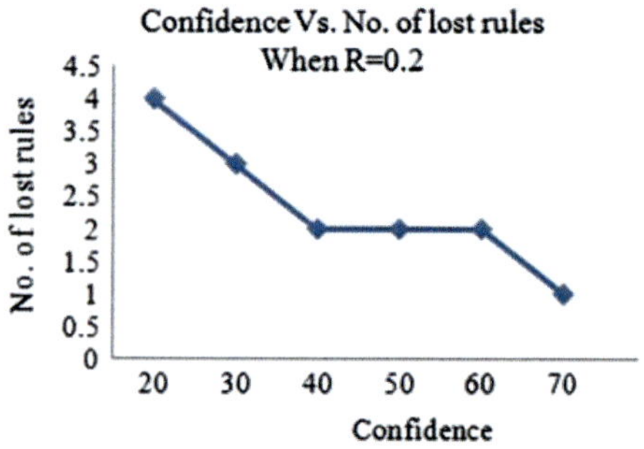

Fig. 11 Non-synthetic—minimum confidence versus number of lost rules when $R = 0.2$ (measure of birth and Death dataset)

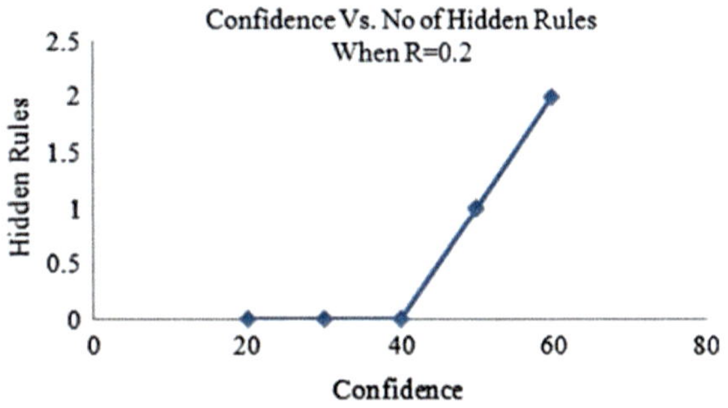

Fig. 12 Non-synthetic—minimum confidence versus number of hidden rules when $R = 0.2$ (measure of birth and death dataset)

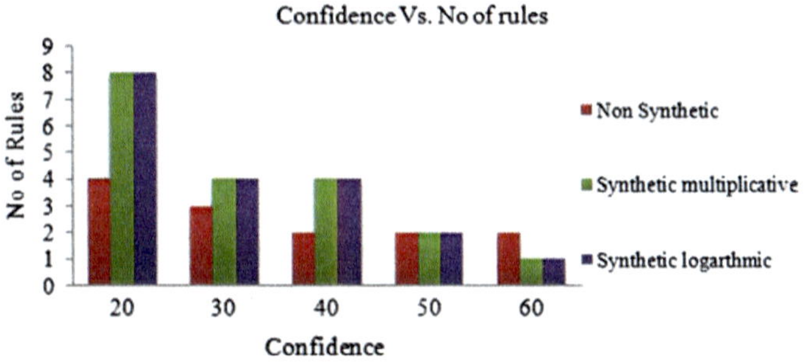

Fig. 13 Minimum confidence versus number of rules

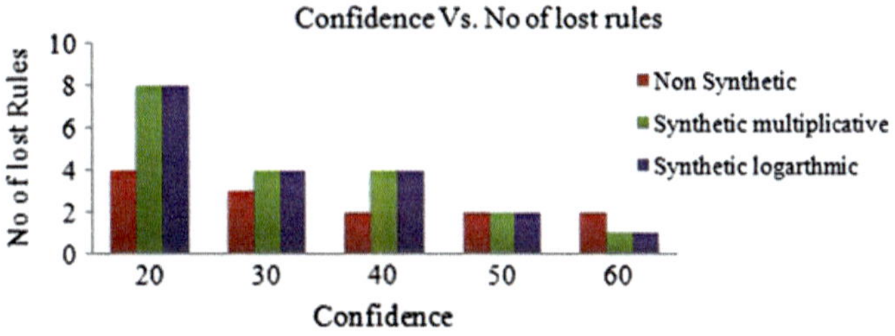

Fig. 14 Minimum confidence versus number of lost rules

for synthetic multiplicative, synthetic logarithmic and non-synthetic perturbation for varying confidence is shown in Fig. 15, respectively.

Perturbation techniques such as synthetic and non-synthetic were applied to the three datasets, and set of rule were generated. Rules were generated for varying confidence. Synthetic perturbation techniques produce many unfavorable rules, making

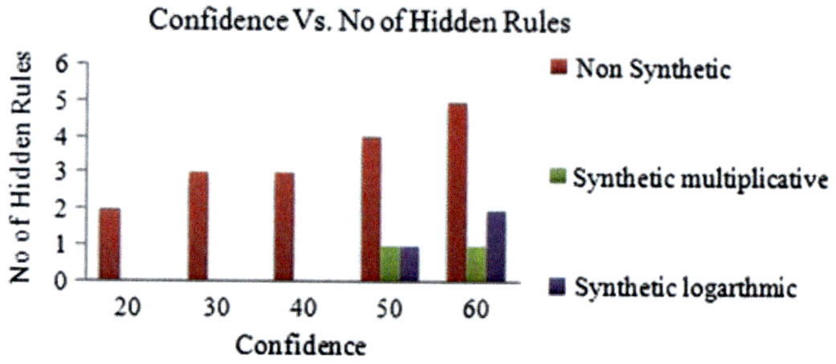

Fig. 15 Minimum confidence versus number of hidden rules

data less utilizable. But, non-synthetic data perturbation technique produces favorable, maintaining the utility of data. Synthetic perturbation algorithm produces lesser number of hidden rules whereas non-synthetic data perturbation technique hides many rules, providing greater privacy.

5 Conclusion and Future Outlook

Privacy-preserving data access techniques are censorious for protecting individual's data published in cloud environments. Existing solutions have focused on producing a single copy of the perturbed data assuming that user has the same level of privileges, but when the owner needs a variable level of privileges to the users, the existing system is ineffective. In case, when the organization have different levels of privacy, such an approach needs multiple copies of the data to be stored separately which increases storage cost. In this paper, we develop a multi-level access technique which employs variable levels of perturbation subjected to the trustworthiness of the user. In order to support the cost, the proposed technique requires only one copy of the data; it only perturbs the data only when required. We evaluate the proposed technique with the existing synthetic perturbation techniques on real-world dataset. The experiments demonstrate that the proposed techniques are scalable and effectively and efficiently supports multi-level access with only one copy of data. The proposed work can be extended by applying genetic algorithm for finding the trustworthiness of the user to provide better utility.

Acknowledgements The authors are grateful to the managements' of St Peter's Institute of Higher Education and Research, Avadi, Chennai, and KCG College of Technology, Karappakam, Chennai, that greatly assist research.

References

1. Chow R, Golle P, Staddon J (2008) Detecting privacy leaks using corpus-based association rules. In: Proceedings of the 14th ACM SIGKDD international conference on Knowledge discovery and data mining (KDD '08). Association for Computing Machinery, New York, NY, USA, 893–901. https://doi.org/10.1145/1401890.1401997
2. Gkoulalas-Divanis A, Verykios VS (2010) Introduction. In: Association rule hiding for data mining. Advances in database systems, vol 41. Springer, Boston. https://doi.org/10.1007/978-1-4419-6569-1_1
3. Vigneswari, Komal Kumar N, Bharath Kumar GV, Vamsi Krishna M (2018) Performance comparison of privacy preserving perturbation algorithms in association rule mining. Adv Eng Res 142:1–6. https://doi.org/10.2991/pecteam-18.2018.1
4. Sweeney L (2002) K-anonymity: a model for protecting privacy. Int J Uncertainty Fuzziness Knowl Based Syst 10(05):557–570. https://doi.org/10.1142/S0218488502001648
5. Samarati P (2001) Protecting respondents identities in microdata release. IEEE Trans Knowl Data Eng 13(6):1010–1027. https://doi.org/10.1109/69.971193
6. Li N, Li T, Venkatasubramanian S (2007) T-closeness: privacy beyond k-anonymity and l-diversity. In: 2007 IEEE 23rd International Conference on Data Engineering, pp. 106–115. IEEE. https://doi.org/10.1109/icde.2007.367856
7. Machanavajjhala A, Kifer D, Gehrke J, Venkitasubramaniam M (2007) l-diversity: privacy beyond k-anonymity. ACM Trans Knowl Disc Data (TKDD) 1(1):3–es. https://doi.org/10.1145/1217299.1217302
8. Lee Y, Kwon O (2010) An index-based privacy preserving service trigger in context-aware computing environments. Expert Syst Appl 37(7):5192–5200. https://doi.org/10.1016/j.eswa.2009.12.072
9. Rottondi C, Verticale G, Capone A (2013) Privacy-preserving smart metering with multiple data consumers. Comput Netw 57(7):1699–1713. https://doi.org/10.1016/j.comnet.2013.02.018
10. Zwattendorfer B, Slamanig D (2016) The Austrian eID ecosystem in the public cloud: how to obtain privacy while preserving practicality. J Inf Secur Appl 27:35–53. https://doi.org/10.1016/j.jisa.2015.11.004
11. Xia Z, Xiong NN, Vasilakos AV, Sun X (2017) EPCBIR: an efficient and privacy-preserving content-based image retrieval scheme in cloud computing. Inf Sci 387:195–204. https://doi.org/10.1016/j.ins.2016.12.030
12. Ravi T, Prasanna Kumar R, Komal Kumar N, Ragu G (2015) Synthetic data perturbation technique for privacy preservation in association rule mining. J Appl Sci Res 11(10):55–59
13. Ravi T, Prasanna Kumar R, Komal Kumar N (2014) A non-synthetic data perturbation technique for privacy preservation in association rule mining. Int J Appl Eng Res 9(24):24311–24320
14. Vigneswari D, Komal Kumar N, Tulasi L (2018) A privacy-preserving technique for incremental data set on cloud by synthetic data perturbation. Int J Eng Technol 7(3.34):331–334. https://doi.org/10.14419/ijet.v7i3.34.19219
15. https://www.kaggle.com/cdc/scienceclips/version/64#science-clips.csv
16. Ravi T, Prasanna Kumar R (2015) data perturbation techniques for privacy preservation in association rule mining. Aust J Basic Appl Sci 9(20):220–227
17. Komal Kumar N, Lakshmi Tulasi R, Vigneswari D (2019) An ensemble multi-model technique for predicting chronic kidney disease. Int J Electr Comput Eng 9(2):1321–1326. https://doi.org/10.11591/ijece.v9i2.pp1321-1326
18. Komal Kumar N, Vigneswari D, Kavya M, Ramya K, Lakshmi Druthi T (2018) Predicting non-small cell lung cancer: a machine learning paradigm. J Comput Theor Nanosci 15(6/7):2055–2058. https://doi.org/10.1166/jctn.2018.7406
19. Komal Kumar N, Vigneswari D, Vamsi Krishna M, Phanindra Reddy GV (2019) An optimized random forest classifier for diabetes mellitus. In: Abraham A, Dutta P, Mandal J, Bhattacharya A, Dutta S (eds) Emerging technologies in data mining and information security advances in intelligent systems and computing, vol 813, pp 765–773. https://doi.org/10.1007/978-981-13-1498-8_67

LSB and RLE Based Approach for Increasing Payload and Security of Stego Images

Rupali Sanjay Pawar

Abstract Data security is a major issue for Internet communication. This problem is solved by using various cryptography and steganography techniques. The challenges for different steganography methods are security, payload (embedding capacity) and robustness. This paper suggests spatial domain technique where the conventional least significant bit (LSB) substitution method is used with run-length encoding (RLE) for improving security and payload. The methods mentioned in literature give embedding capacity 20–50%, [1] whereas the work presented here shows embedding capacity up to 88% for three-bit evaluation (Table 4). At the same time, it maintains low distortion of the vessel image which is shown by calculating MSE and PSNR at each stage (Table 1). Also, there is an insignificant effect on various moments before and after embedding (Table 2).

Keywords Steganography · LSB · RLE · Payload · Security

1 Introduction

Steganography is an art of communicating secretly [2]. It refers to covert transmission.

R. Sanjay Pawar (✉)
Cummins College of Engineering for Women, Pune 411052, India
e-mail: rupali.pawar@cumminscollege.in

N. N. Chiplunkar and T. Fukao (eds.), *Advances in Artificial Intelligence and Data Engineering*, Advances in Intelligent Systems and Computing 1133,
https://doi.org/10.1007/978-981-15-3514-7_49

There is a very interesting story about steganography. In 400 BC, an ancient Greek named Hestaieus was fomenting revolt against Persian king and wanted to deliver a secrete message. He shaved the head of his bondsman and tattooed a secrete message on it. When his hair grew back in, he was sent on mission. At other end, the recipient shaved his head again to read the alert.

Steganography essentially blends of two words in Greek: "steganos" meaning "cover" or "conceal" and "graphein" meaning "writing," which means "covered writing" [3]. The cover objects can be either of image, audio or video. The images are most popular for many reasons—(1) Images are used widely on the Internet (2) They can be used as carrier objects (3) Spatial redundancy is more observed in images. The original image after embedding the message is referred as stego image. The message is hidden in such a way that stego image resembles original image.

Often steganography is mistaken with cryptography as both deal with secret transmission of information [3]. The difference is the later fetches the attention of the intruder whereas the former does not [4].

Different applications of steganography include—(1) Covert transmission. (2) Security against data alteration. (3) Protection against unauthentic access for digital data distribution [5].

The four basic traits of data hiding include (i) transparency—less distortion of vessel image (ii) payload-embedding capacity (iii) robustness—ability to handle various attacks (iv) security—access control only for authorized users. Different applications of steganography require trade-offs between these four aspects.

In this paper, a new approach is discussed which increases embedding capacity as well as security of stego images maintaining transparency and robustness.

2 Related Work

Image steganography approaches are mainly classified into two types: spatial domain and transformed domain approaches, each one having its own pros and cons. The former one gives better imperceptivity with higher payload capacity, whereas the latter one has excellent robustness properties. Also, the complexity of transformed domain techniques is much more than spatial domain techniques [6] (Fig. 1).

The basic spatial domain method given in the literature is LSB substitution [7] (Fig. 2). This method is modified many times mainly to enhance security [8]. E. Kawaguchi introduced technique of Bit Plane Complexity Segmentation (BPCS) which gives high embedding capacity with increased computational complexity [1]. The author claims information hiding capacity as high as 50% of the original image data.

The main reason for the success of LSB substitution technique is that the modification in LSB bits will have negligible impact on visual perception as compared to MSB bits. The MSB bits hold coarse image information, whereas LSB bits hold

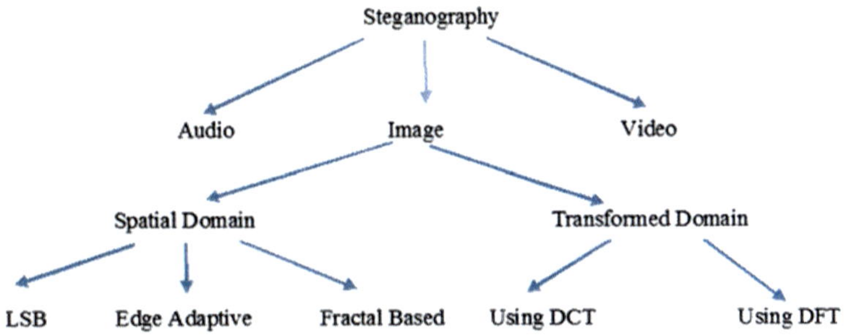

Fig. 1 Classification of steganography approaches

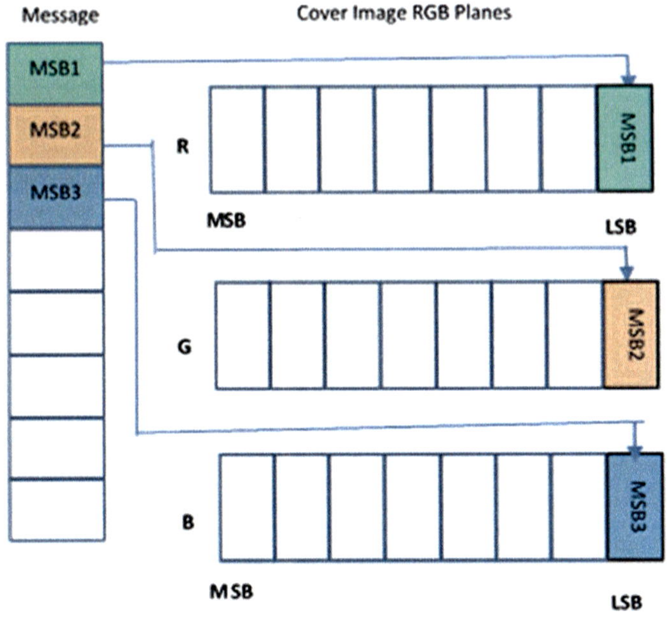

Fig. 2 LSB method

detail information. That means even though LSB bits are modified, there is very less image distortion produced [7]. It aids the purpose of information hiding.

3 Proposed Work

3.1 The RLE Process

In this paper, the attempt is made to achieve both security as well as increased payload together. In this, LSB substitution method is expanded in a simple way. The difficulty in calculating complex segments is avoided here, rather concept of data compression is used here which results in increased payload as well as security.

The work is essentially a blend of LSB and run-length encoding (RLE) techniques. RLE is lossless image compression technique. The data to be hidden is first encoded using RLE algorithm which results into compaction and data security as well. The ability to embed data is thus enhanced. Data compaction increases data embedding capability.

The RLE procedure is explained in Fig. 3. The first three bits in the original stream are compared with the group of next three bits in the sequence until the mismatch is found. The match count which is considered as the "run length" is coded in binary format to form compressed stream. This compressed stream is suitably decoded to obtain original stream as shown in Fig. 3.

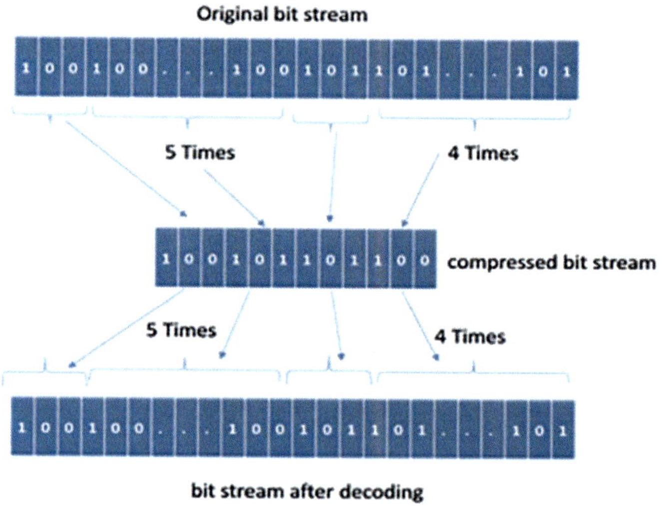

Fig. 3 RLE process

3.2 The RLE Algorithm

```
================================================================
p=[ ];

while i <end
if(i ==1)
    for i =1:3
        z1(i)=x(i);
    end

p=[p z1];

i =i+1;
end

for j =1:3
    z2(j)=x(i);
    i =i+1;
end

t= strcmp(z1,z2);

if(t==1)
Runlength = Runlength +1;
end
================================================================
```

The work presented in the paper is carried out in four steps as explained in Figs. 4, 5, 6 and 7.

Run-length encoding is applied to the secrete image M as per the described method at beginning. This compressed data M’ is then hidden into cover image using LSB substitution technique to get the stego image S. At receiver, M’ is obtained from stego image using suitable extraction algorithm. Finally, decoding algorithm gives secrete image M from M’.

The MSB bits of secrete images (which are to be hidden) are separated and embedded in place of LSB bits of vessel image. This modified image is known as “stego image.” Later on, at the receiver side, the LSB bits of stego image are extracted

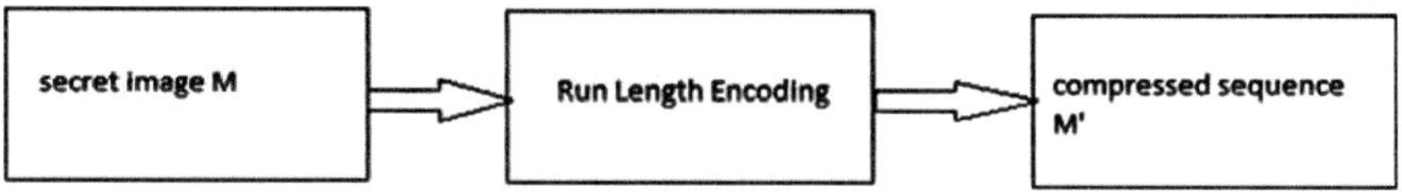

Fig. 4 RLE encoding

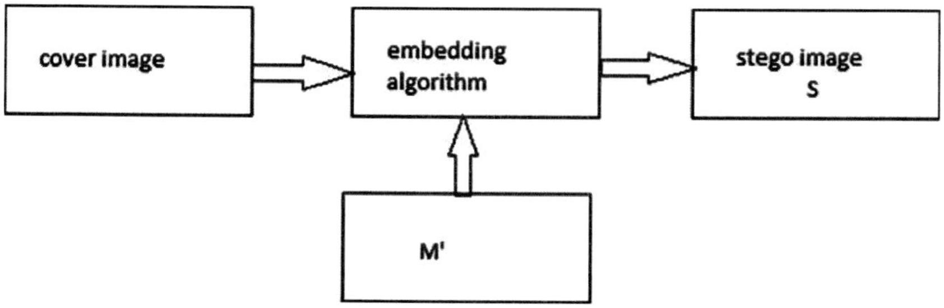

Fig. 5 Embedding procedure

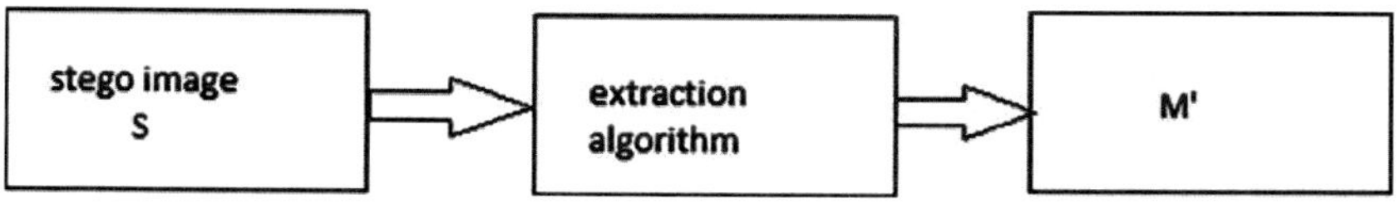

Fig. 6 Extraction procedure

Fig. 7 Decoding procedure

and skillfully managed to get back secret image. Instead of image, the text message also can be embedded and retrieved successfully using LSB + RLE substitution technique.

4 Validation of Proposed Work

4.1 MSE and PSNR

Mean square error (abbreviated as MSE) and peak signal to noise ratio (abbreviated as PSNR) are the two popular measures in image processing for quantifying distortion between original and modified or reconstructed image.

MSE gives measurement of the average of the squares of the errors. It calculates the average of squared difference between the estimated values and what is estimated.

Estimator's quality is determined by its MSE value. It is always positive. The minimum the MSE the better is the performance.

PSNR is calculated by taking the ratio of maximum possible power of a signal and the power of corrupting noise. PSNR is usually expressed in terms of the logarithmic decibel scale to accommodate very large dynamic range of signals.

$$\text{MSE} = \frac{1}{M \times N} \sum_{i=1}^{M} \sum_{j=1}^{N} \left[I(i, j) - I'(i, j) \right]^2 \tag{1}$$

$$\text{PSNR} = 10 \log_{10} \frac{255^2}{\text{MSE}} \tag{2}$$

4.2 *Moments of Distribution*

Moments are considered as additional performance metrics for data security of stego images. Moments refer to the tendency of set of values to cluster around some particular value. This may be useful to characterize that set, which is normally performed by calculating sum of integer powers of the values, known as "moments." First four moments are discussed and utilized in the work are

- *Mean*: It estimates the value around which central clustering occurs. It is the first moment of data.

$$\text{mean} = \frac{1}{N} \sum_{i=0}^{N-1} Z_i \tag{3}$$

- *Variance*: It characterizes the "width" or "variability" of dataset around *mean* value. It is the second moment of data.

$$\text{variance} = \frac{1}{N} \sum_{i=0}^{N-1} (Z_i - \mu)^2 \tag{4}$$

- *Skewness*: It characterizes the degree of asymmetry of a distribution around the mean. Positive skewness refers a distribution with an asymmetric tail extending out toward more positive z and vice versa.
  It is a non-dimensional quantity.

$$\text{skewness} = \frac{1}{N} \sum_{i=0}^{N-1} \left(\frac{Z_i - \mu}{\sigma} \right)^3 \tag{5}$$

- *Kurtosis*: It measures the relative peakedness or flatness of a distribution relative to normal distribution. A distribution is termed as "leptokurtic" if it has positive kurtosis else as "platykurtic" if it has negative kurtosis.

It has no dimension.

$$\text{kurtosis} = \left[\frac{1}{N}\sum_{i=0}^{N-1}\left(\frac{Z_i - \mu}{\sigma}\right)^4\right] - 3 \tag{6}$$

4.3 Payload Capacity

Payload capacity is determined by considering image sizes on disk. Sample calculation for flower.bmp is shown below.

In a vessel image flower.bmp (74.9 KB), total five secret images of different sizes are embedded.

Image	Size
Blood.bmp	10.8 KB
ELIZA.bmp	5.05 KB
Boat.bmp	17 KB
Lena.jpg	17 KB
Cameraman.jpg	4.07 KB
TOTAL	53.92 KB

$$\begin{aligned}\text{Payload capacity} &= \frac{(10.8\ \text{KB} + 5.05\ \text{KB} + 17\ \text{KB} + 17\ \text{KB} + 4.07\ \text{KB})}{74.9\ \text{KB}}\\ &= \frac{53.92\ \text{KB}}{74.9\ \text{KB}}\\ &= 71.9\%\end{aligned} \tag{7}$$

5 Results

The secrete images used in the work are shown below.

In this work, four vessel images of different capacity are used. Each vessel image is embedded with secret images shown in Fig. 8. Each vessel image has its own payload capacity. After embedding, we refer it as "stego image."

Table 1 lists the changes in MSE and PSNR values of stego image at each embedding stage from (1) to (last). Table 1 shows that the MSE value of stego image does

Fig. 8 Secret images

Table 1 MSE and PSNR of Stego images

Stego image	Embedded images	MSE	PSNR
barbara.bmp(76 KB) 	Blood.bmp (10.8 KB)	8.5423e−005	88.8151
	ELIZA.bmp (5.05 KB)	1.5225e−004	86.3053
	Boat.bmp (17 KB)	5.4712e−004	80.7500
	Lena.jpg (17 KB)	6.6599e−004	79.8961
	Cameraman.jpg (4.07 KB)	8.4069e−004	78.8845
flower.bmp(74.9 KB) 	Blood.bmp (10.8 KB)	3.4782e−005	92.7173
	ELIZA.bmp (5.05 KB)	7.3938e−005	89.4421
	Boat.bmp (17 KB)	3.2949e−004	82.9524
	Lena.jpg (17 KB)	3.6383e−004	82.5218
	Cameraman.jpg (4.07 KB)	4.4829e−004	81.6152
baboon.bmp(45 KB) 	Blood.bmp (10.8 KB)	5.4917e−005	90.7337
	ELIZA.bmp (5.05 KB)	1.8758e−004	85.3990
	Lena.jpg (17 KB)	2.9906e−004	83.3732
	Cameraman.jpg (4.07 KB)	4.7151e−004	81.3959
strawberry.bmp(37 KB) 	Blood.bmp (10.8 KB)	6.9528e−005	89.7092
	ELIZA.bmp (5.05 KB)	1.5158e−004	86.3243
	Lena.jpg (17 KB)	1.6558e−004	85.9408

not increase much even when all secret images are hidden in vessel image. At the same time, it is observed that PSNR value of stego image does not decrease much when all secret images are hidden in vessel image. Thus, the perception of stego image is very much similar to original image, and it is impossible to predict that the data might be hidden secretly.

Table 2 Calculation of moments

Vessel image	Before embedding				After embedding			
	Mean	Variance	Skewness	Kurtosis	Mean	Variance	Skewness	Kurtosis
barbara.bmp (76 KB)	0.4282	0.1949	0.0753	2.2068	0.4273	0.1948	0.0753	2.2075
flower.bmp (74.9 KB)	0.3326	0.2173	1.0563	3.4258	0.3321	0.2171	1.0547	3.4198
baboon.bmp (45 KB)	0.4045	0.1723	−0.1076	2.1129	0.4039	0.1723	−0.1147	2.1178
strawberry.bmp (37 KB)	0.3577	0.1812	1.1095	3.7022	0.3577	0.1807	1.1122	3.7096

Table 3 Comparison of number of bits with and without RLE

Vessel image	No. of bits required to be embedded	
	Without RLE	(Proposed method) with RLE
barbara.bmp (76 KB)	2,05,419	1,16,628
flower.bmp (74.9 KB)	2,05,419	1,16,628
baboon.bmp (45 KB)	1,56,267	81,774
strawberry.bmp (37 KB)	91,440	49,869

Table 2 lists the experimental results obtained by calculating first four statistical moments of vessel image before and after embedding. It is observed that there are insignificant changes in skewness and kurtosis values. At the same time, there is very little decrease in mean and variance of image before and after embedding. This assures security of stego image.

Table 3 details number of bits required to be embedded without RLE and with RLE (proposed method). There is significant decrease (almost 50%) in number of bits if we use RLE compression. This allows more images (data) to embed in vessel image which in turn increases the payload capacity of stego image.

Table 4 compares the payload capacity of stego images without RLE and with RLE(proposed method). It is observed that payload capacity increases from 40 to 60% if no compression techniques are used. But if we use RLE compression, payload capacity increases to 71–88%.

Table 4 Comparison of payload capacity with and without RLE

Vessel image	Payload capacity without RLE (%)	(Proposed method) payload capacity with RLE (%)
barbara.bmp (76 KB)	41.90	70.9
flower.bmp (74.9 KB)	42.55	71.9
baboon.bmp (45 KB)	61.77	82
strawberry.bmp (37 KB)	59.59	88

Table 5 Entropy and compression ratio

Secret images	Entropy	Compression ratio (%)
Lena	3.1663	52.46
ELIZA	3.0004	63.52
Boat	2.7150	70.90
cameraman	2.6692	49.21
Blood	2.5183	54.26

Table 5 lists entropy of each secret image with its individual compression ratio. Entropy is directly proportional to randomness and is inversely proportional to redundancy within dataset. In general, as entropy decreases, it is observed that compression ratio increases (for images Lena, ELIZA and Boat). But RLE compression method adopted in this literature depends on inter-pixel redundancy within image. Thus, other factors like how pixels are distributed within image, the scanning order used (row-wise, column-wise, zig-zag) also plays significant role.

6 Extracting Data

In this work, three MSB bits of secret image are embedded into LSB bits of RGB frames of vessel image, two bits per frame. Thus, each pixel of vessel image accumulates six bits of secret data. Thus, coarse information of six pixels of secret image (two pixels per frame) is hidden into each pixel of stego image. When this data is extracted, it is grouped in three bits. Each set of three bits is then padded with five zeros to make it 1 byte. From this, approximated pixel of secret image is constructed at receiver's end.

Another advantage offered by the proposed method is that it supports blind decoding while extracting the secret images. The original cover image is not required while extracting hidden data from stego image.

7 Conclusion

Instead of three-bit evaluation, we can go for four or five bit to achieve better quality of extracted image. But there is trade-off between number of evaluation bits and payload capacity. Also, more bits embedded in stego will result in more distortion.

Robustness can be boosted by encrypting RLE compressed data before embedding at the cost of increase in complexity.

The proposed method offers advantages as high embedding capacity, less complex than BPCS [3], high data security, less distortion of vessel image and blind decoding while extracting image.

Future Scope
The LSB + RLE combination tested here works successfully for text and image data. The same can be extended further to embed and extract audio data.

References

1. Kawaguchi E (2005, September) BPCS-steganography–principle and applications. In: International conference on knowledge-based and intelligent information and engineering systems. Springer, Berlin, Heidelberg, pp 289–299
2. Cox I, Miller M, Bloom J, Fridrich J, Kalker T (2007) Digital watermarking and steganography. Morgan Kaufmann
3. Anderson RJ, Petitcolas FA (1998) On the limits of steganography. IEEE J Sel Areas Commun 16(4):474–481
4. Artz D (2001) Digital steganography: hiding data within data. IEEE Internet Comput 5(3):75–80
5. Eiji Kawaguchi KN, Niimi M, Noda H, Eason RO, Sensui-cho II (1999) A concept of digital picture envelope for Internet communication. Inf Model Know Bases X 51:343
6. Joseph P, Vishnukumar S (2015, April) A study on steganographic techniques. In: 2015 global conference on communication technologies (GCCT). IEEE, pp 206–210
7. Chan CK, Cheng LM (2004) Hiding data in images by simple LSB substitution. Pattern Recog 37(3):469–474
8. Laskar SA, Hemachandran K (2012) High capacity data hiding using LSB steganography and encryption. Int J Database Manage Syst 4(6):57

Adaptive MoD Chatbot: Toward Providing Contextual Corporate Summarized Document as Suggestions and Reported Issue Ticket Routing

Shiva Prasad Nayak, Archana Rai, Kiran Vankataramanappa, Jalak Arvindkumar Pansuriya, and Joerg Singler

Abstract After advent of Internet and cloud computing technology, many of the software solutions are hosted in cloud infrastructure (a.k.a SaaS) to enable customers to pay for services as they use. Having this situation, the software non-functional requirements such as 'availability' and 'reliability' have become critical as customers expect 24 × 7 support for their hosted applications. In such scenario, chatbot or conversational AI becomes more critical to act as the first point of contact to support customer's queries to reduce multiple round trips between DevOps (Cloud infrastructure support team) to an extent possible. Currently, Team of MoD (manager on duty) is the first point of contact to support for any upcoming customer queries and if in need will be percolated down to team of developer on duty (DoD) for speedy issue resolution. In this paper, we are trying to build a novel solution of adaptive chatbot in providing contextual Q&A responses with built-in natural language template generation, building technical document corpus specific knowledge base, building document text summarization to fit into chatbot real estate, creating corporate technical document indexing to provide important document links and summarized text as response, creating support tickets for the issue reported with automated ticket routing classification to enable speedy resolution of customer queries by DoD with the techniques from information retrieval, information extraction, NLP, NLG, text summarization, text classification and ontology. This paper proposes customized

S. P. Nayak (✉) · A. Rai · K. Vankataramanappa
SAP Labs India Pvt. Ltd., 138, EPIP Zone, Whitefield, Bengaluru 560066, India
e-mail: shiva.nayak@sap.com

A. Rai
e-mail: archana.rai@sap.com

K. Vankataramanappa
e-mail: kiran.venkataramanappa@sap.com

J. A. Pansuriya · J. Singler
SAP AG, Dietmar-Hopp-Allee 16, 69190 Walldorf, Germany
e-mail: jalak.arvindkumar.pansuriya@sap.com

J. Singler
e-mail: joerg.singler@sap.com

N. N. Chiplunkar and T. Fukao (eds.), *Advances in Artificial Intelligence and Data Engineering*, Advances in Intelligent Systems and Computing 1133,
https://doi.org/10.1007/978-981-15-3514-7_50

text ranking algorithm by ensuring more weightage is given to document sentences having n-gram linked entities from ontology knowledge base.

Keywords Chatbot · Manager on Duty · Developer on Duty · N-gram linked entity · Corporate document search index · Corporate document summarization · Weighted N-gram linked entity TF-IDF sentence text ranking · Clustering contextual conversations · Natural language chatbot template generation · Issue ticket routing classifier

1 Background

The chatbot or conversational agent facilitates a natural language interface to users for submitting their queries, and underlying system provides suitable response in return. The chatbot ELIZA [1] and ALICE [2] were one of the initial chatbots known to us. In today's world, chatbots were developed across various industries which are significantly making good impact for people in society. The companion chatbot [3] developed by Endurance company for senior people and patients suffering from Alzheimer's disease, insomnobot 3000 [4] developed by Casper company for people suffering from insomnia patients, MedWhat [5] chatbot focuses in making medical diagnoses faster, easier and more transparent for both patients and physicians which are some of the well-known chatbots developed from medial industry. Some of the entertainment industry chatbots are Disney zootopia [6] for solving crimes with fictional characters, Marvel [7] guarding the galaxy with comic book crossovers. Some of the non-profit organizations chatbots are U-Report [8] from UNICEF focus on large-scale data gathering via polls for urgent social issues. Some of real estate chatbots are Roof Ai [9] that helps real estate marketers to automate interacting with potential leads and lead assignment via social media.

Modern commercial chatbots, such as those developed with Lingubot™ [10], SAP Conversational AI [11], Slack [12] technology, offer sophisticated development environments, rich in capabilities, collaborative platform to build, train, deploy and monitor intelligent conversational agents with complex goal-driven behavior.

2 Introduction

Multiple phases are involved in software development life cycle (SDLC) [13]; each phase of the software cycle produces various artifacts and deliverables required for the next phase in the software life cycle. It starts with market requirement gatherings which produce market research documents, architecture documents which are then translated into design documents. As part of development cycle, source code is produced according to the design specification. After coding and development, the testing cycle verifies the deliverables against requirements specification. The testing team follows Software Testing Life Cycle (STLC) [14]; once product quality attained to the satisfaction level, it is released to market and then follows the maintenance cycle.

The outcome in each of these software lifecycle phases evolves various corporate documents and many tools to support lifecycle of the same. The following are the well-known tools used for this purpose.

(a) Jira: To keep track of epics, user stories, to derive backlog items and to keep track of their progress made.
(b) Wiki: To depict architectural and conceptual designs. Also act as collaboration tool for topic discussions.
(c) Bug tracking: System outage and any application-specific blockers are recorded here for enabling speedy resolution by the concerned teams.
(d) BCP incidents: Business continuity plan incidents are recorded here for enabling speedy resolution by the concerned teams.
(e) Blogs: Technical information about product-specific details is recorded here.

As The document structure and contents vary between them, in addition, they are stored and retrieved from multiple repositories. Keeping track on each of these information for specific application has become mere to impossible in facilitating quicker resolutions for customer queries.

The Software as a Service (SaaS) [15, 16] model enables software applications to host in cloud infrastructure for customers to access the applications whenever they want, wherever they want, pay as they use, and with the additional service level agreement of 99.9xxx, the availability and reliability software non-functional requirement become critical. Currently, team of MoDs is expected to be the first point of contact to support and resolve any upcoming customer queries and if not will be percolated down to team of DoDs. As most part of the MoD activities will be in providing high level information related to the customer's query such as corporate technical document links, resolving the issue based on similar issue encountered earlier and if required creating necessary support ticket in the relevant system to expedite the process. This part of providing links to important corporate technical documents, identifying similar issues/resolutions provided earlier and creating necessary support ticket in the system can be automated with adaptive MoD chatbot. The MoD chatbot can act as first point of contact for customers to raise queries, and system provides contextual information as response which reduces multiple round trips, turnaround time and automatically creates issue tickets to expedite the process.

2.1 Notations Used in MoD Chatbot

(a) Let $D_W = \left\{d_W^1, d_W^2, d_W^3, \ldots, d_W^k\right\}$ be set of wiki documents and $N_w = |D_w|$ be size of wiki documents.
(b) Let $D_N = \left\{d_N^1, d_N^2, d_N^3, \ldots, d_N^k\right\}$ be set of NGP issues and $N_N = |D_N|$ be size of NGP issues.

(c) Let $D_I = \{d_I^1, d_I^2, d_I^3, \ldots, d_I^k\}$ be set of BCP incidents and $N_I = |D_I|$ be size of BCP incidents.
(d) Let $D_J = \{d_J^1, d_J^2, d_J^3, \ldots, d_J^k\}$ be set of Jira documents and $N_J = |D_J|$ be size of Jira documents.
(e) Let $D_B = \{d_B^1, d_B^2, d_B^3, \ldots, d_B^k\}$ be set of blogs and $N_B = |D_B|$ be size of blogs.
(f) Let L be n-gram linked entity from ontology knowledge base.
(g) Let T be computed tf-idf score for a given linked entity L.
(h) Let L_W^i be ith n-gram linked entity of wiki document. Similar notations for NGP issues, BCP incidents, Jira documents and blogs.
(i) Let T_W^i be tf-idf score of L_W^i. Similar notations for NGP issues, BCP incidents, Jira documents and blogs.
(j) Let U_W^i be Wikipedia/DBpedia URL of L_W^i. Similar notations for NGP issues, BCP incidents, Jira documents and blogs.
(k) Let $d_W^i = \{S_W^1, S_W^2, S_W^3, \ldots, S_W^k\}$ be set of document sentences of ith wiki document. Similar notations for NGP issues, BCP incidents, Jira documents and blogs.
(l) Let $|d_W^i|$ be sentence size of ith document in wiki corpus. Similar notations for NGP issues, BCP incidents, Jira documents and blogs.
(m) Let f_W, f_N, f_I, f_J, f_B be damping factor of wiki documents, NGP issues, BCP incidents, Jira documents and blogs, respectively.

3 MoD Chatbot Overview

At high level, Fig. 1 outlines the MoD chatbot components developed in the scope of this solution. To begin with, customers start with natural language query (chatbot interface) in knowing information about product technical details, or steps to set up an environment, or an issue blocker and so on.

I Document Corpus Knowledge base (KBs)
II Document Search Index
III Clustering Contextual Conversations and Generating Natural Language Templates
IV Document Text Classifier for Ticket Routing

Hereinafter referred as Component *I*, *II*, *III* and *IV*

3.1 Component I

Figure 2 outlines the steps to build corpus-level documents entity knowledge base

The job scheduler at time '*t*' triggers this activity in crawling the temporal changes made to corporate documents repository, performs document preprocessing, POS tagging and NER in identifying the named entities (location, organization, person).

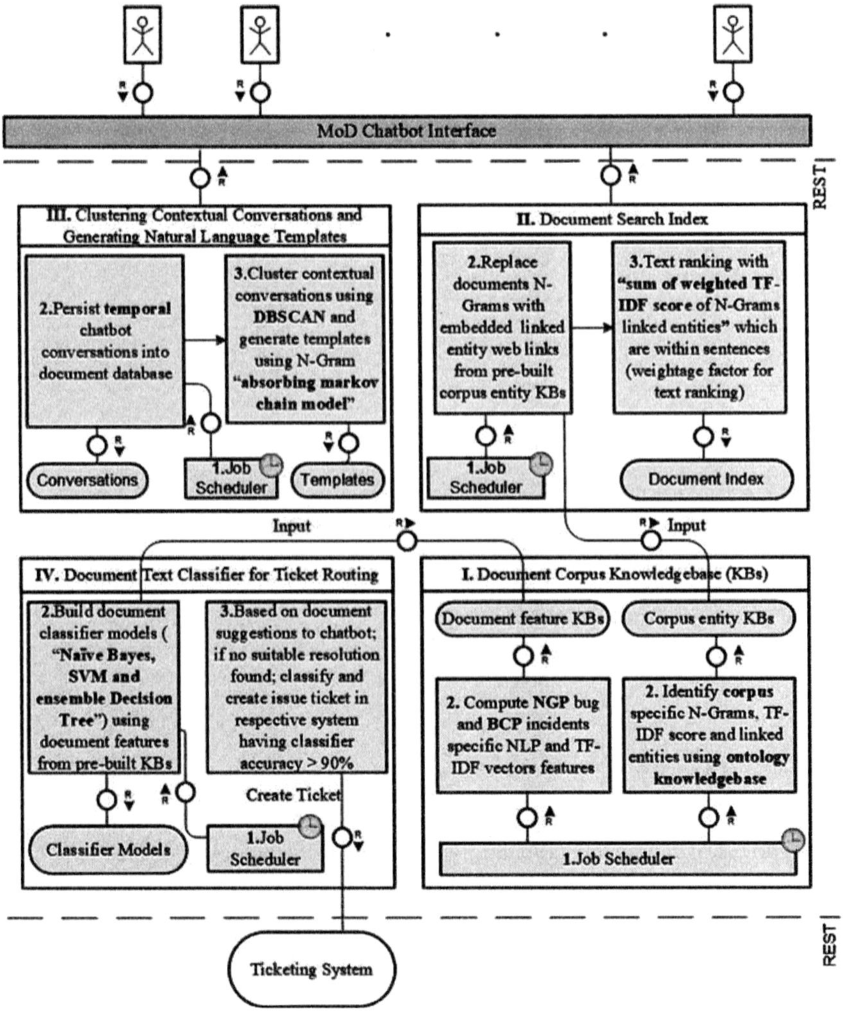

Fig. 1 MoD chatbot block diagram

The next step is to perform entity linking with Wikipedia/DBpedia ontology knowledge base and perform NED to resolve entity disambiguation by extracting dbo:abstract n-gram named entities and by looking back the availability in document corpus, more the percentage of dbo:abstract named entity availability in document corpus constitutes relevant linked entity. The next step is to compute corpus-level tf-idf score for all identified n-gram linked entities and to persist along with Wikipedia/DBpedia URL details [17–22].

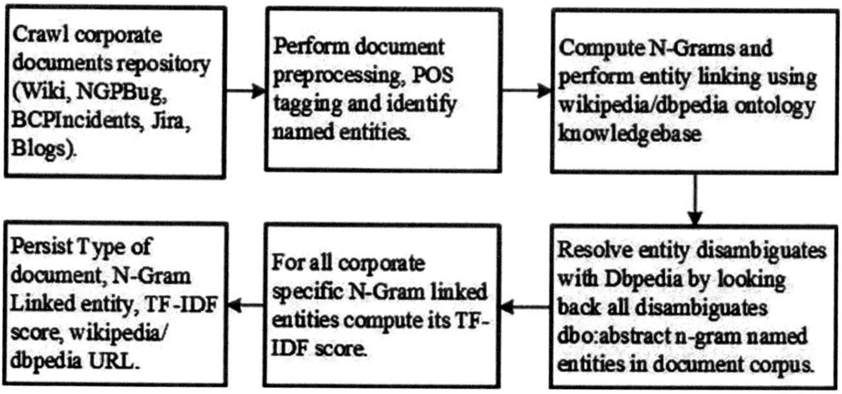

Fig. 2 Steps to build corporate document corpus knowledge base

Table 1 Corporate document corpus-level knowledge base

Document type	Linked entity	TF-IDF score	Linked entity URL
D_W	L_W^i	T_W^i	U_W^i
D_W	L_W^j	T_W^j	U_W^j
...	...	...	...
D_N	L_N^i	T_N^i	U_N^i
...	...	...	...

This information is then used for document summarization, embedding n-gram linked entities in document and building document search indexer (Table 1).

Algorithm 1: prepareCorporateDocumentCorpusKBs

```
Input: None
Output: None
foreach document types do
    foreach type specific documents do
        ngram_named_entities ← perform_NER(document);
        L ← extract_LinkedEntities(ngram_named_entities);
        U ← getDbpedia_Wikipedia_URL(L);
        foreach ngram_linked_entity from L do
            T ← T ∪ compute_TFIDF(ngram_linked_entity);
        end
        Persist(Document_Type, L, T, U);
    end
end
```

3.2 Component I

Figure 3 outlines the steps to build corpus-level document features.

The job scheduler at time '*t*' triggers this activity in crawling temporal corporate-specific issues from NGP bug and BCP incidents, perform document preprocessing, compute Tf-Idf features such as word count, average word density, word level and n-gram level tf-idf features. Extract NLP features by performing POS tagging in computing frequency distribution of nouns, verb, adjective, adverb, pronoun features and persist in to repository which are then used as features for building document ticket classifier (outlined in Component *IV*) (Table 2).

Algorithm 2: prepareCorporateIssueFeaturesKBs

Input: None
Output: None

```
foreach document types do
    foreach type specific documents do
        /* Compute NLP features                                  */
        wordcount ← extract_WordCount(document);
        avgWordDensity ← compute_AvgWordDensity(document);
        posTags ← perform_POSTagging(document);
        verbs, nouns, adjective, adverbs ←
        compute_FrequencyDistribution(posTags);
        /* Compute TFIDF features                                */
        ngramTFIDF ← compute_TFIDF(document);
        persist(NLPfeatures, TFIDFfeatures);
    end
end
```

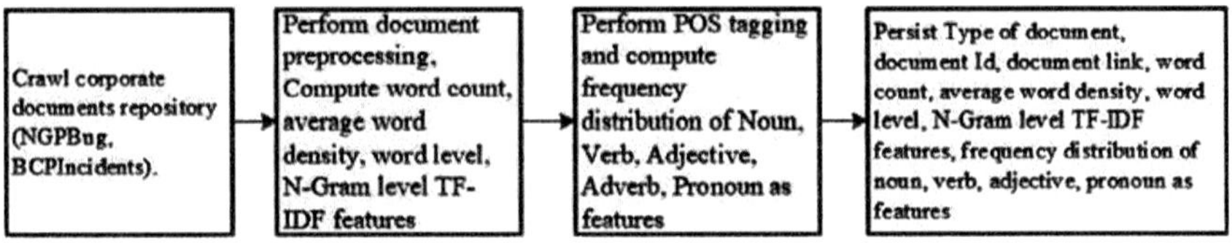

Fig. 3 Steps to build corporate document features knowledge base

Table 2 Corporate issue documents feature knowledge base

Document type	Document	NLP {features frequency distribution of noun, verb, adjective, adverb, pronoun as features}	TF-IDF features {n-gram word count, average word density, n-gram tf-idf as features}
D_N	d_N^i	$\{nlp\_features\}_N^i$	$\{tfidf\_features\}_N^i$
D_N	d_N^j	$\{nlp\_features\}_N^j$	$\{tfidf\_features\}_N^j$
...	...	...	...
D_I	d_I^i	$\{nlp\_features\}_I^i$	$\{tfidf\_features\}_I^i$
...	...	...	...

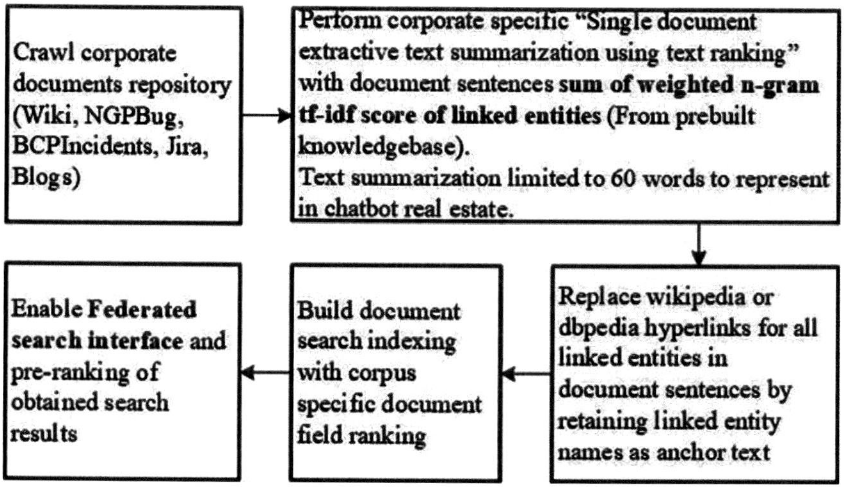

Fig. 4 Steps to build corporate document search indexing

3.3 Component II

Figure 4 outlines the steps to build corporate document search index. The job scheduler at time '*t*' triggers this activity in crawling temporal corporate-specific documents, performing document-specific 'single document extractive text summarization using text ranking' with sum of weighted tf-idf score of n-gram linked entities of sentences, replace n-gram linked entities with URL links in documents, build document-specific search indexing with document-specific fields ranking.

Build federated search index [23] to get ranked search results from document-specific search index for a given single query in chatbot.

This federated document search indexing will be used in providing contextual summarized document in chatbot (limited to 60 words to fit chatbot real estate) as suggestions.

Algorithm 3: buildDocumentSearchIndex

```
Input: noOfwords
Output: None
foreach document types do
    foreach type specific documents do
        rankedSentenceMatrix ← performDocumentTextRanking(document);
        summarizedText ←
        textSummarization(rankedSentenceMatrix, noOf words);
        toIndex ← toIndex ∪ {related_Doc_Fields, docText, summarizedText};
        performTypeSpecificSearchIndex(document_Type, toIndex)
    end
end
```

Algorithm 4: federatedSearchResults

```
Input: SearchQuery
Output: federatedSearchResults
foreach document type indexes do
    docTypeResults ←
    getSearchResults(document_Type, index, searchQuery);
    allSearchResults ← allSearchResults ∪ docTypeResults;
end
federatedSearchResults ← performPreRanking(allSearchResults);
```

Document summarization using text ranking [24, 25] was done by assigning weights to graph edges with sum of weighted n-gram linked entity tf-idf score between document sentences.

Document sentences as vertex:

$$v(d_W^i) = \left\{S_W^1, S_W^2, S_W^3, \ldots, S_W^k\right\} \tag{1}$$

Forming sentence edges when there is atleast one common n-gram linked entity between them

$$e(d_W^i) = (L_W^1 || L_W^2 ||, \ldots, || L_W^m) \in (S_W^c, S_W^k) \tag{2}$$

Sentence edge weights are computed as 'sum of weighted n-gram linked entity tf-idf score' in the order of trigram, bigram and unigram which are mutually exclusive from each other. This solution is limited to trigram and beyond which found to be a computationally intensive task.

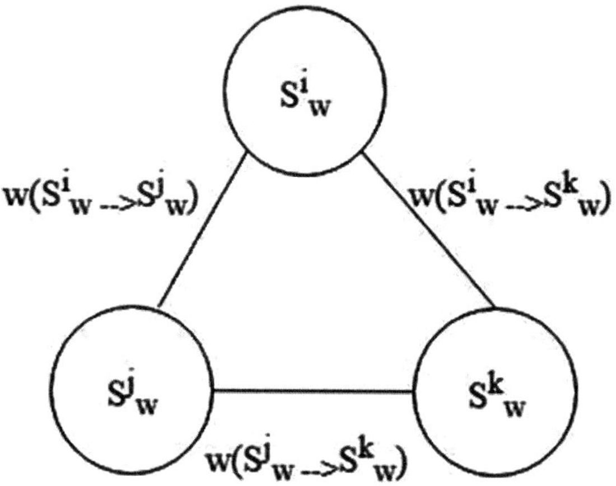

Fig. 5 Sample document sentence relation graph

$$w(S^c_W \rightarrow S^k_W) = \sum_{j=1}^{m} \frac{T^j_w}{N_W} \{\text{for} L^j_W \in (S^c_W, S^k_W) \in d^i_W\} \tag{3}$$

Figure 5 illustrates graphical representation of graph nodes (document sentences), edges (at least one common n-gram between sentences) and its weights (sum of weighted n-gram linked entity tf-idf score between sentences). Corpus-specific damping factor and document text ranking with sum of weighted n-gram linked entity tf-idf is considered as weightage.

$$\text{TR}(d^i_W) = \frac{f_W}{|d^i_W|} + (1 - f_W) \sum_{v \in \text{adj}(d^i_W)} \frac{\text{TR}(v)}{\text{deg}(v)} * w(v \rightarrow \text{adj}(v)) \tag{4}$$

Table 3 illustrates document fields used for index ranking and damping factor used for document summarization. These below damping factor ratio are heuristic in nature and after several iterations of experimentation found to be optimized value for our corporate corpus.

3.4 Component III

Figure 6 outlines the steps to build chatbot conversation template. The job scheduler at time '*t*' triggers this activity in crawling the temporal chatbot conversations from chatbot repository and then persists in to document database for further analysis. The next step is to extract conversations which users have provided low rating (ratings <3 out of 5) for our responses to user's questions, cluster responses using Revised DBSCAN [26] with Jaccard similarity [27], name and persist the dense clusters into

Table 3 Document fields for index ranking and optimized damping factors for document summarization

Document type	Document fields ranking order	Optimized damping factor
D_W	{Title, Link, Document Text, Summarized document}	$f_W = 0.73$
D_N	{NGP Bug number, Title, Link, Component Id, Description, Summarized Description}	$f_N = 0.67$
D_I	{Incident Number, Title, Link, Component Id, Description, Summarized Description}	$f_I = 0.65$
D_J	{ID, Title, Link, Description, Summarized Description}	$f_J = 0.81$
D_B	{Title, Link, contents, summarized contents}	$f_B = 0.85$

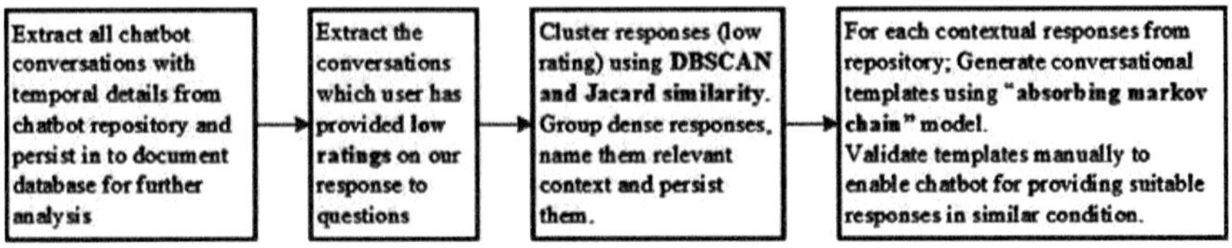

Fig. 6 Chatbot conversation template generation

relevant context (related to product architecture, product setup, technical know-how, issues and so on).

For each contextual response from repository, generate chatbot conversational templates using absorbing Markov chain model [28], validate templates manually to enable chatbot from providing suitable responses in similar condition.

Algorithm 5: clusterChatbotConversations

Input: conversationList
Output: None

foreach *chatbot conversations* **do**
  if $conversation.response\_Rating <= 3$ **then**
    $convLowR \leftarrow convLowR \cup extractHistory(conversation.conversationID)$;
  end
end
$clusters \leftarrow build\_DBSCAN\_Cluster(convLowR, eps = 5, minPts = 50)$;
$persist(cluster\_group, data\_points)$;

The DBSCAN parameters epsilon and minimum points have learned based on distance to nearest neighbor [29] and counting points neighbors histogram distribution for our corpus. Manually name the clustered data points in to relevant context.

Algorithm 6: generateChatbotTemplate

Input: contextualClusterDataPoints
Output: None
foreach *data point from contextualClusterDataPoints* **do**
    $contextGroup \leftarrow contextGroup \cup getConversationText(datapoint)$;
end
$Model \leftarrow build\_Absorbing\_Markov\_Chain\_Model(contextGroup)$;
$Templates \leftarrow generatePossibleTemplates(Model)$;
$Persist(Templates)$;

From generated templates, choose suitable contextual relevant templates and persist them in to chatbot conversational knowledge base. These templates enable chatbot for providing suitable response in similar condition.

3.5 *Component IV*

Figure 7 outlines the steps to build document corpus text classifier for ticket routing. The job scheduler at time '*t*' triggers this activity in crawling temporal corporate-specific issues from NGP bug and BCP incidents. Get document-specific NLP and tf-idf features from pre-built KBs, build text classifier using Naïve Bayes [30], SVM [31] and ensemble decision tree [32] models.

We built these multiple classifier models in identifying suitable classifier for our corpus and finally zeroed down on one among the multiple variants of decision trees models by applying bagging/boosting techniques.

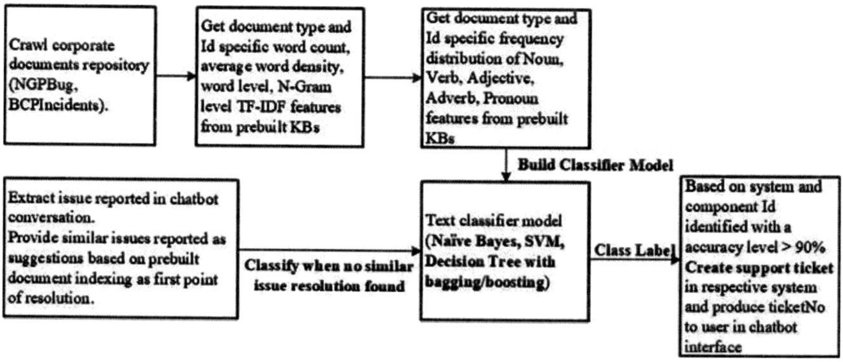

Fig. 7 Steps to build document corpus text classifier for ticket routing

When issue is reported in chatbot conversation, provide suitable resolutions available from pre-built search index, if the provided resolution found to be unsatisfactory, then the issue is classified from pre-built classifier model in knowing which component the ticket should be created.

If the classifier accuracy found to be >90%, then ticket is created automatically in respective system, otherwise, it will be logged, and after some manual analysis, the ticket will be created under respective component.

Algorithm 7: buildIssueClassifierModel

```
Input: None
Output: model
foreach corpus issue types do
    foreach type specific issues do
        classLabel = issue.documentType+' :' +issue.componentId;
        nlpFeatures ← extract_NLPFeatures_FromKBs(issue);
        tfidfFeatures ← extract_tfidfFeatures_FromKBs(issue);
        featureMatrix = featureMatrix ∪
        {classLabel, issue.description, nlpFeatures, tfidfFeatures};
    end
    model = buildClassifierModel(featureMatrix);
end
```

Algorithm 8: createIssueTicket

```
Input: conversationId
Output: ticketNo
issue ← extractIssueFromChatbot(conversationId);
searchIndex_Suggestions ←
getSuggestions_From_FederatedSearchIndex(issue);
if isSuitable(searchIndex_Suggestions) == False then
    (accuracy, clsLabel) ← classifyIssue(issue, model);
    if accuracy > 90% then
        ticketNo ←
        createTicket(clsLabel.documentType, clsLabel.componentId, issue);
    end
end
```

Fig. 8 *illustrates an outlook overview of MoD chatbot solution.*

4 Conclusion

The Software as a Service (SaaS) model enables software applications to host in cloud infrastructure for customers, and the availability, reliability software non-functional requirement become critical. Currently, team of MoDs is expected to be the first

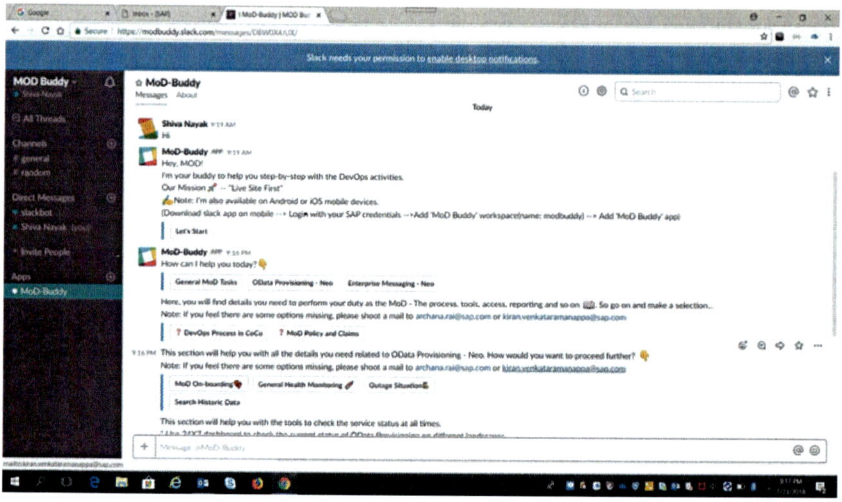

Fig. 8 MoD chatbot solution

point of contact in supporting ongoing customers interactions. In this solution, we proposed MoD chatbot being the first point of contact in answering most of the queries from customers by providing suitable contextual documents as response, identifying suitable resolution for issues based on similar resolutions available in the system, creating automated tickets in system such that DoD can takeover in providing speedy resolutions. We proposed methodology on building corpus-level and document-level knowledge base, clustering contextual low rating responses using DBSCAN, generating contextual chatbot templates using absorbing Markov chain model for low rating responses, computing classified n-gram linked entity using ontology knowledge base, computing document text ranking using sum of weighted n-gram linked entity tf-idf score as graph edge weights, embedding n-gram linked entity Web links in document text which are used in chatbot to get additional details from Wikipedia or DBpedia knowledge base, building federated document search indexing to enable personalized pre-ranking of search results, building automated ticket routing classifier from ensembled decision tree which enables in identifying the system and component id to create automated ticket for the issues reported in chatbot conversation. This solution increases the customer satisfaction index by reducing number of round trips in getting to know beforehand the documents links related to product architecture, technical know-how details, Wikipedia/DBpedia links as an additional information for the important technical keywords, important blog links containing steps to set up products, links to resolution provided earlier for a similar kind of issue, speedy resolution of new issues reported in products to an great extent.

References

1. Weizenbaum J (1966) ELIZA—a computer program for the study of natural language communications between man and machine. Commun ACM 9(1):36–45
2. Wallace R. The anatomy of ALICE. http://www.alicebot.org/anatomy.html
3. Endurance robots home page. http://endurancerobots.com/azbnmaterial/a-robot-companion-for-senior-people-and-patients-with-alzheimer-s-disease/. Last accessed 10 Feb 2018
4. Insomnobot 3000 home page. http://insomnobot3000.com/. Last accessed 15 Feb 2018
5. Medwhat bot home page. https://medwhat.com/. Last accessed 21 Feb 2018
6. Disney Zootopia home page. https://www.topbots.com/project/disney-zootopia/. Last accessed 26 Feb 2018
7. Marvel chatbot home page. https://botlist.co/collections/@sethlouey/marvel. Last accessed 2 Mar 2018
8. U-Report home page. http://unicefstories.org/tag/u-report-bot/. Last accessed 6 Mar 2018
9. Roof.AI bot home page. https://roof.ai/. Last accessed 7 Mar 2018
10. Creative Virtual. 2004–2006. CreativeVirtual.com. Web site of UK Lingubot distributor. www.creativevirtual.com. Last accessed 10 Apr 2018
11. SAP Conversational AI home page. https://www.sap.com/india/products/conversational-ai.html. Last accessed 11 Sept 2018
12. Slack home page. https://slack.com/. Last accessed 21 June 2018
13. SDLC phases home page. https://www.softwaretestinghelp.com/software-development-life-cycle-sdlc/. Last accessed 20 Jan 2018
14. Testing excellence home page. https://www.testingexcellence.com/software-development-life-cycle-sdlc-phases/. Last accessed 21 Jan 2018
15. Haojie Hang SaaS. https://www.cs.colorado.edu/~kena/classes/5828/s12/presentation-materials/dibieogheneovohanghaojie.pdf. Last accessed 5 Feb 2018
16. Deyo J (2008) Software as a Service (SaaS)
17. Nadeau D, Sekine S (2007) A survey of named entity recognition and classification. Lingvisticae Investigationes 30(1):3–26
18. Damljanovic D, Bontcheva K (2012) Named entity disambiguation using linked data. In: Proceedings of the 9th extended semantic web conference
19. Hoffart J et al (2011) Robust disambiguation of named entities in text. In: Proceedings of the conference on empirical methods in natural language processing. Association for Computational Linguistics
20. Bird S, Klein E, Loper E (2009) Natural language processing with Python: analyzing text with the natural language toolkit. O'Reilly Media Inc., Sebastopol
21. Guo Y, Pan Z, Heflin J (2005) LUBM: a benchmark for OWL knowledge base systems. Web Seman Sci Serv Agents World Wide Web 3(2–3):158–182
22. Pérez J, Arenas M, Gutierrez C (2006) Semantics and complexity of SPARQL. In: International semantic web conference. Springer, Berlin
23. Shokouhi M, Si L (2011) Federated search. Foundations and trends® in information retrieval. 5(1):1–102
24. Mihalcea R, Tarau P (2004) Textrank: bringing order into text. In: Proceedings of the 2004 conference on empirical methods in natural language processing
25. Mihalcea R (2005) Language independent extractive summarization. In: Proceedings of the ACL 2005 on interactive poster and demonstration sessions. Association for Computational Linguistics, pp 49–52
26. Tran TN, Drab K, Daszykowski M (2013) Revised DBSCAN algorithm to cluster data with dense adjacent clusters. Chemometr Intell Lab Syst 120:92–96
27. Strehl A, Ghosh J, Mooney R (2000) Impact of similarity measures on web-page clustering. In: Workshop on artificial intelligence for web search, AAAI 2000, vol 58, p 64
28. Bell MGH, Schmoecker J-D, Iida Y, Lam WHK (2002) Transit network reliability: an application of absorbing Markov chains. In: Transportation and traffic theory in the 21st century: proceedings of the 15th international symposium on transportation and traffic theory, Adelaide, Australia. Emerald Group Publishing Limited, pp 43–62

29. Rahmah N, Sitanggang IS (2016) Determination of optimal epsilon (Eps) value on dbscan algorithm to clustering data on peatland hotspots in sumatra. In: IOP conference series: earth and environmental science, vol 31, no. 1. IOP Publishing, p 012012
30. McCallum A, Nigam K (1998) A comparison of event models for naive Bayes text classification. In: AAAI-98 workshop on learning for text categorization, vol 752, no. 1, pp 41–48
31. Sun A, Lim E-P, Liu Y (2009) On strategies for imbalanced text classification using SVM: a comparative study. Decis Support Syst 48(1):191–201
32. Lei S, Weng M, Ma X, Xi L (2010) Rough set based decision tree ensemble algorithm for text classification. J Comput Inf Syst 6(1):89–95

Classification of Text Documents

Pushpa B. Patil and Dakshayani M. Ijeri

Abstract Nowadays, the text information is available in electronic form which is easily accessible to people, due to whom it is increasing day by day, but the challenging issue is to organize the information. The existing algorithm such as naïve Bayes classifier uses the maximum posterior estimation for building a classifier, but it is dependent on huge number of training samples for more accuracy and the other algorithm known as genetic algorithm that begins with an initial population which is constructed from randomly generated rules, but the accuracy of this algorithm depends on set of training examples. In this paper, we are using brute-force approach to classify the different categories of documents.

Keywords Text document · Classification · Categorization

1 Introduction

Nowadays, huge number of text documents is available in electronic form which is increasing day by day. These kinds of documents represent very large amount of information that is easily accessible. Extracting valuable information from this huge collection requires organization. Data mining helps in organizing documents automatically.

Depending on the content of documents, data mining automatically classifies documents into predefined classes. Text classification has been automated with many algorithms.

The text classification method involves some of the frequently used processes such as association rule mining, implementation of naïve Bayes classifier, genetic algorithm, and decision tree.

P. B. Patil · D. M. Ijeri (✉)
BLDEA's V. P. Dr. P. G. Halakatti College of Engineering & Technology, Vijayapur, India
e-mail: cse.ijeri@bldeacet.ac.in

P. B. Patil
e-mail: cs.pushpa@bldeacet.ac.in

N. N. Chiplunkar and T. Fukao (eds.), *Advances in Artificial Intelligence and Data Engineering*, Advances in Intelligent Systems and Computing 1133,
https://doi.org/10.1007/978-981-15-3514-7_51

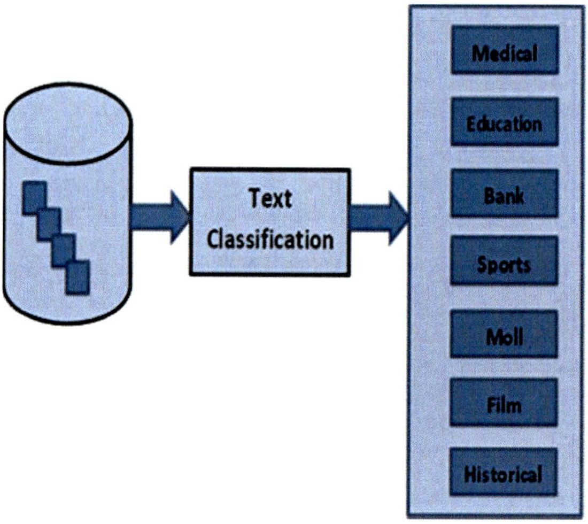

Fig. 1 Text classification example

Association rule mining identifies the connection between data items in a huge collection of data set. Taking decisions in data analysis is simplified by identifying relationships between data with respect to real-time applications.

To establish classifier, the naïve Bayes classifier utilizes the maximum posterior estimation technique. This method treats each word in a document which is unconstrained with the rest of the words in that document. Though the naïve Bayes works have been proved correctly in many studies; but to produce accurate results, it requires a large number of training data sets.

Figure 1 shows the categorization of different text documents. The database includes different kinds of documents. By using the categorization system (algorithm), we can classify those documents as different categories like sports, business, education, and science.

1.1 Problem Definition

Text categorization is the problem of grouping the related documents automatically based upon predefined categories. Examples are as follows:

- Document classification
- Web page classification
- News classification.

Document categorization problem gained a lot of importance in the few years, due to the increase in the number of digital documents.

Text categorization (TC) problem is to group the documents based upon predefined number of categories.

The paper is structured as follows:

In Sect. 2, the related work has been discussed, which covers discussion of ant clustering algorithm, naive Bayes algorithm, and grid gain framework. Section 3 covers the methodology containing the flowchart of the classification of text documents. Section 4 covers the implementation containing the algorithms used, namely document categorization method, classification method, and identify more than one document belongs to the same category. Section 5 covers the experimental results and discussions, and Sect. 6 concludes the paper.

2 Related Work

Some basic methods of text documents' classification are discussed in this work [1]. Results of its analysis are presented in the table of comparative characteristics in accordance with the list of proposed parameters.

In 2015, Vizine et al. [2] used ant clustering algorithm for classifying PDF documents automatically. The prominent application of this algorithm is Web document classification. The idea of this algorithm came into existence with reference to the behavior of some ant species.

In 2014, Harish et al. [3] proposed symbolic data analysis concepts for text document classification. Clustering of term frequency vectors is used to represent documents. Each group of the document will be treated as cluster, and each cluster is assigned with term frequency vectors which create illustrative depiction with the help of mean and standard deviation. Further, term frequency vectors are used in the form an interval valued features.

In 2016, Shathi et al. [4] proposed naive Bayes algorithm for classifying the text documents which result in the extraction of feasible information. This algorithm uses weight matrix during training text documents which is union of two techniques known as term frequency (TF) and inverse class frequency (ICF). The better and efficient performance is achieved by assigning a significant number to weighted term and added with a posteriori value during the prediction time of naive Bayes algorithm.

In 2014, Sarnovsky and Vronc [5] used boosting method on the decision tree algorithm for the classification of textual documents. Here the work is focused on the implementation of distributed boosting algorithm based on MapReduce paradigm. Grid gain framework has been implemented for distributed data processing and tested on two different data sets within their testing environment.

3 Methodology

Figure 2 shows about methodology. It contains training data sets, testing documents, word count in each document, identify data set, and display the corresponding document.

Training data set contains many numbers of documents like sports, medical, education. Testing documents containing the numbers of documents are given as input comparing of input text to training data sets and counting the words and displaying it. Depending on the number of related words present in training data set, it will say that the test document is whether sports, medical, or education related.

3.1 Training Data Set

The data set is prepared to train a model in machine learning. Significant features are considered from data set, and these features are then incorporated into a model.

Training data set acts as framework for the system in the beginning process through which further decision of new input will be decided. Supervised learning method is adopted to train the system with training data set.

Training data set includes different categories of document types and related words of each category. When we give the testing document as input, each word of

Fig. 2 Block diagram of the classification of text documents

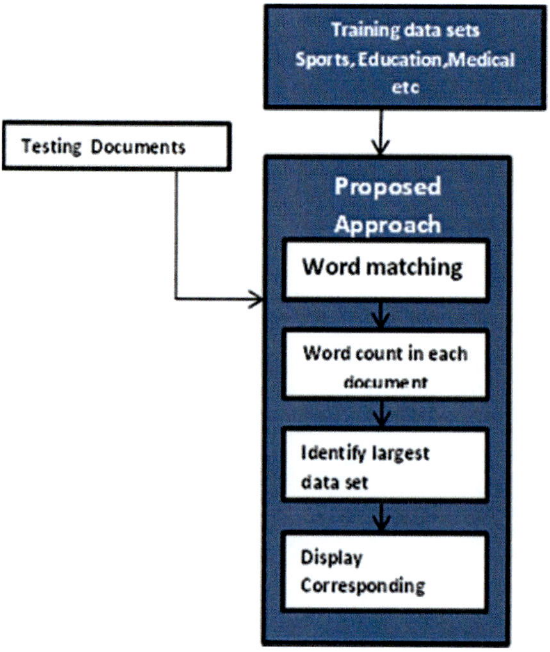

the testing document will be compared with the training data set values. If a particular word in testing document matches with any word in the training data set, then the count will get incremented.

3.2 *Testing Document*

It is very important to classify data as training data and testing data to evaluate data mining models. Once data is classified, larger percent is used for training than testing. Training and testing data are treated as similar by analysis services. The characteristics of model can be efficiently analyzed by using similar data for training and testing.

Once the model has been structured by using training data set, it can be tested by using different types of test data. It is very easy to guess the correct behavior of model due to the presence of similar attributes in testing data according to the prediction.

Testing documents are the text documents where we need to find the category of those documents by using the already existing training data sets. Each word in the testing document will be compared with all words in each category of training data sets, if that word matches with words in training data sets words, then the count will get incremented. This type of comparison is made for all of the words in testing documents.

3.3 *Word Count in Each Document (Bag of Words)*

Total number of words present in document is called as word count, and sometimes, it is necessary when a document is restricted to certain number of words.

The bag-of-words method is an elementary approach that can be used in natural language processing and information retrieval (IR). This method creates a bag (dictionary) containing count of each word from a text or document (text), whereas order of words is not considered.

The most common method used to classify documents is "bag of words" which results in count of each from document, and it can be used for training a classifier.

Each word of the testing document will be compared with each word in each category of training data sets. After the comparison is made, the next procedure is to find the words related to each kind of category. So in this step, it will find the count of words related to each category of documents.

Consider the following text information as an example to show the creation of bag of words.

- Hari wants to play cricket. Krishna wants cricket too.
- Hari also wants to play football.

With reference to these two text documents, the following is the list created:

"Hari","wants","to","play","cricket","Krishna","wants","cricket","too"
"Hari","also","wants","to","play" "football"

The dictionary consisting of count of each word will be created as bag of words as follows:

X1 = {"Hari":1,"wants":2,"to":1,"play":1,"cricket":2,"Krishna":1,"too":1};
X2 = {" Hari":1,"also":1,"wants":1,"to":1,"play":1,"football":1};

The above dictionary consists of the information in the form of key: Value pair in which key is word from example and value is count of that word. The order of occurrence is not that important.

For example

{"too":1,"Hari":1,"cricket":2,"wants":2,"play":1,"to":1} is also X1. It is also what we expect from a strict JSON object representation.

Consider another example which is the union of the above two examples.

Hari wants to play cricket. Krishna wants cricket too. Hari also wants to play football.

Its JavaScript representation will be:

X3 = {"Hari":2,"wants":3,"to":2,"play":2,"cricket":2,"Krishna":1,"too":1,"football":1, "also":1};

3.4 *Identifying Documents*

Text categorization (text classification) is the process of logically grouping of words according to predefined categories. With this approach, the documents can be logically viewed and have significant applications in real world.

In the previous step, *Algorithm 1* is calculated the words related to each kind of category. In this step, Algorithm 2 will find the category which has more related words of the given text document.

For example, if more words in particular text documents are related to medical, i.e., medical-related words—18, education-related words—5, and sports-related words—3.

Then it will find the largest one, that is medical, and it will identify the document as medical.

For example, news reports are classified as political, entertainment, weather, etc. Similarly, hospital patient records are classified according to different departments, different diseases, different surgical treatments, and even academic question paper is categorized as the main question and sub-question.

3.5 Display the Corresponding Document

In the previous step, the algorithm identifies the category in which more words of the testing document are related. In this step, the algorithm identifies the type of the category in which the document belongs and it will display the result.

Give the input as document name, compare the given document to the training data sets, and it will classify and count the number of different words present in the given input document. It will display the word count that is present in the given input document, and finally, it will display the one document name that containing many numbers of related words.

For example, training data sets contain the documents related to medical, education, sports, and bank.

Give input as testing document name: education.txt

This document contains information as follows:
Sports-related words: 4
Education-related words: 8
Bank-related words: 3
More number of words related to education: 8
So given document is: education document

The bag of words for the union of two documents is equivalent to the sum of multiplicities of each individual word from each document.

4 Implementation

In *Algorithm 1*, it will ask us to enter the document name. Then it opens that document in reading mode. Then this algorithm calls the classification method to classify the documents based on the content. The classification method counts all the words related to each category and returns the words related to each category back to this algorithm. Then this method prints the total words related to each category.

Algorithm 1: Document Categorization Method

```
Input: text document, training data set
            Output: category of the document
            1: Enter the document
            2: Open the document in read mode
            3: Call Algorithm 2 for each category
            4: for i = 1 to 7
               if sumx[i] > sumx[x]
                      assign i to max
            5: Print the total word related to each category
            6: Call Algorithm 3
```

```
7: Find the largest count
8: Print the type of the documents
9: End
```

If suppose the words related to more than one category are same, so to check that condition this method calls *Algorithm 3*. If one document has large number of related words, then it prints that corresponding document. If more than one documents has the same number of related words, then it prints all the documents name as output.

Algorithm 2: Classification Method

```
Input: Text document, training data set
           Output: largest category count
1: Read word from documents
2: for i  =0 to 40
     initialize count  = 0;
             while  !feof(fp)
             for i = 0 to 40
                          if  !isalpha(val[strlen(val)-1]) then
                               val[strlen(val)-1 = null
             ifstrcmpi(val,cat[i]) ==0 then
                   increment count
                go to step 1
 3: for i = 0 to 40
      if cmd[i]! = 0 then
                print category and count
 4: for i = 0 to 40
          add count to sum
               return sum
 5: End
```

Initialize sum to zero that is the number of words in a document. Then this algorithm reads the number of words present in each document and counts the total words present in a document as a sum and prints the category and count. Returns the sum back to called method.

Algorithm 3: Identify More Than One Document Belongs to the Same Category

```
Input: text document, training data set
          Output: Documents index value
          1: Initialize k = 0
          2: Assign max to pos[k]
                k ++
          3: for i  = 0 to 7
                if max! = i &&sumx[i] ==sumx[max] then
                         pos[k] = i
                         k ++
                         repeat step 3
                         return k;
          4: End
```

Algorithm 3 is called to identify the type of document. This method is used to identify the document in which more words are related. If the words related to a particular document are more, then it will display the corresponding document. If words related to more than one category are the same, then it will display the name of all documents in which more words are related.

5 Experimental Results and Discussion

There are seven different categories for the classification of text, namely medical, education, bank, sports, moll, film, and historical. The work has been experimented with various document samples, and it has been tested with four medical-related documents, four education-related documents, and five sports-related documents. The following are few snapshots of results.

In Fig. 3, as per document m1.txt the words related to medical are 10, education are 2, and sports, moll, film, library, and historical are 0, so maximum words are related to medical. So, it displays document as medical documents.

Figure 4 shows the classification as per the document p.txt in which the words related to medical are 10, education are 10, and sports, moll, film, library, and historical are 0, so the words present in both the medical and education category are equal. So, it displays document as both medical documents and education documents.

Table 1 describes the accuracy of different categories. There are seven documents of each medical, education, bank, sports, moll, film, and historical category. In first, second, fourth, fifth, and seventh cases, all four documents are correctly identified. So the accuracy is 100%. But in third and sixth cases, one of the bank documents is identified as two types of categories. So the accuracy is 75%.

Fig. 3 Sample result

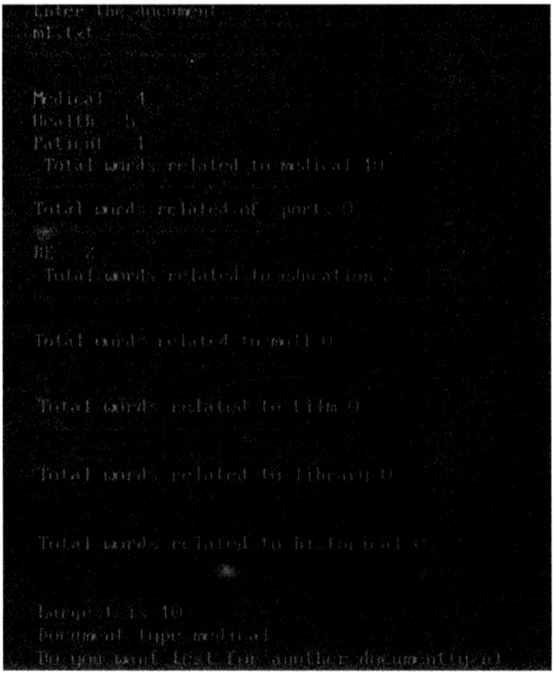

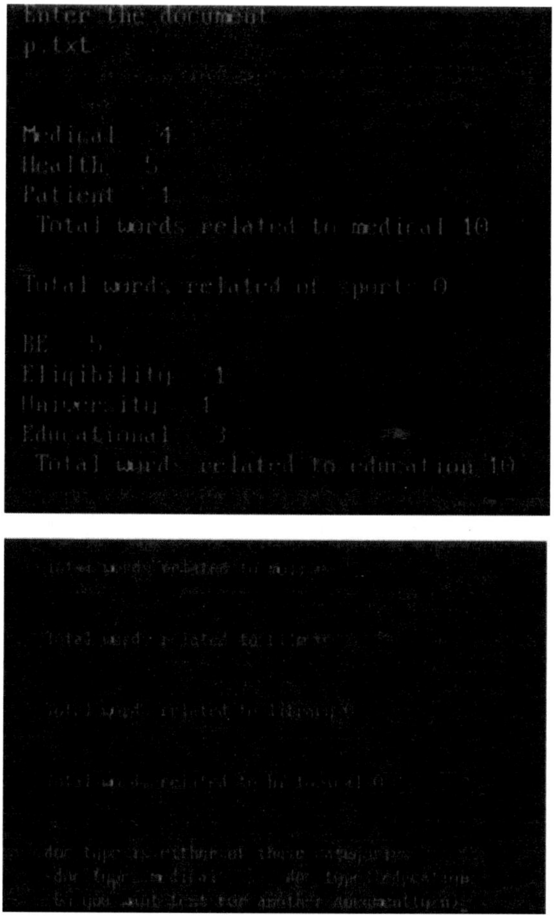

Fig. 4 Sample result for text document belongs to more than one category

Table 1 Accuracy of each category

Category	No. of testing documents	Correctly classified documents	Accuracy (%)
Medical	4	4	100
Education	4	4	100
Bank	4	3	75
Sports	4	4	100
Moll	4	4	100
Film	4	3	75
Historical	4	4	100

6 Conclusion

A large amount of text is being generated every day, and it is increasing day by day drastically. It is very difficult to extract useful information from such unstructured text. Therefore, it is important to develop an effective algorithm which automatically extracts the useful information. And for this purpose, text mining is the best approach which has gained significant attentions in recent years.

We are using the brute-force approach to classify the different categories of documents. In this approach, we are taking different kinds of documents as inputs, and based on the content, our algorithm classifies them as medical, education, sports, etc.

References

1. Golub T (2016, February 23–26) The analysis of text documents classifiers constructing methods. TCSET
2. Vizine AL, de Castro LN, Gudwin RR (2005, April 18–21) Text document classification using swarm intelligence. IEEE KIMAS
3. Harish BS, Aruna Kumar SV, Manjunath S (2014) Classifying text documents using unconventional representation. In: International conference on big data and smart computing (BIGCOMP). https://doi.org/10.1109/bigcomp.2014.6741438
4. Shathi SP, Hossain MD, Nadim M (2016, December 12–13) Enhancing performance of Naïve Bayes in text classification by introducing an extra weight using less number of training example. In: 2016 international workshop on computational intelligence. https://doi.org/10.1109/iwci.2016.7860355 (2016)
5. Sarnovsky M, Vronc M (2014, January 23–25) Distributed boosting algorithm for classification of text documents. In: 2014 IEEE 12th international symposium on applied machine intelligence and informatics (SAMI), p 201. https://doi.org/10.1109/sami.2014.6822410 (2014)

Fine-Grained Sentiment Rating of Online Reviews with Deep-RNN

Ramesh Wadawadagi and Veerappa Pagi

Abstract Increasing volume of customer reviews over the commercial Web sites have created a demand for the construction of automated content analysis systems. However, present techniques mainly focus on traditional bag-of-words (BOW) and statistical language models, ignoring semantic compositions. In contrast, deep neural networks (DNN) have exhibited greater stability in equipping on-scale sentiment prediction. Particularly, deep recursive neural networks (Deep-RNN) have been consistently used for capturing semantic compositions in natural language content when represented with structured formats (e.g., parse trees). Improved word spaces (word-embeddings) on the other hand proved to be efficient in comprehending fine-grained semantic regularities. In this paper, a fine-grained sentiment rating of online reviews based on Deep-RNN is proposed. The performance of the proposed model is evaluated through the conduction of experiments over Stanford sentiment treebank (SST) dataset. Furthermore, the effect of tuning hyper-parameters on the performance of the network is studied. The experimental results reveal that Deep-RNN exhibits better prediction accuracy compared to the traditional shallow counterparts.

Keywords Recursive neural networks · Sentiment analysis · Word-embeddings · Fine-grained sentiment rating · Deep learning

1 Overview

Online consumer reviews and ratings may influence positively or negatively any enterprise in this business world. This being the case, many online review systems offer star rating in addition to free text reviews. In spite of the fact that, different users

R. Wadawadagi (✉) · V. Pagi
Department of Computer Science and Engineering,
Basaveshwar Engineering College, Baglkot 587102, India
e-mail: rswlib@yahoo.co.in

V. Pagi
e-mail: veereshpagi@gmail.com

N. N. Chiplunkar and T. Fukao (eds.), *Advances in Artificial Intelligence and Data Engineering*, Advances in Intelligent Systems and Computing 1133,
https://doi.org/10.1007/978-981-15-3514-7_52

are inclined to different preferences associated with different aspects of a product or a service, stars, or points rating is often considered as the measure of overall opinion [1]. Inconsistency results in when the reviews are mapped to corresponding ratings, and users may struggle to obtain relevant information from a massive amount of reviews. This issue can be addressed through the construction of automated reviews rating system based on the concept of opinion polarity classification. However, identifying fine-grained opinion polarity in short sentences requires more sophisticated language models and evaluation resources. Especially, composite statements carrying opinions that involve both positive and negative phrases are even more complex for polarity detection. For example, the sentence in Fig. 1, "I just loved the resolution and picture quality of the TV, but disappointed that it does not support Bluetooth speakers" includes phrase typically associated with positive and negative aspects of the television. Furthermore, this leads to biased judgments when coarse-grained models are used. In this regard, contemporary techniques toward opinion polarity classification have not gone far beyond "BOW" model, while ignoring semantic knowledge and often operating at document level. Taking inspiration from the recent success story of Deep-RNN, this research work is an attempt toward building a fine-grained sentence and phrase-level classification model for automated rating of online reviews. Deep neural networks (DNN) are essentially composed of several hidden layers of nonlinear information processing designed to obtain higher-level hierarchies through the combination of lower level features, so as to gain each subsequent layer a more abstract and general meaning [2–4]. In particular, recursive neural networks (RNNs) are specific type of DNN designed to operate on structured input (e.g., parse trees) by means of recursive application of fixed set of weights [3, 5]. Moreover, RNN exhibits generalized form of recurrent neural networks (RcNN) with skewed tree representation [6]. In RcNN, each hidden state is outlined as a function of its previous hidden states, and hence, they are inherently deep-in-time [7]. To realize depth in space, multiple recurrent hidden units of RcNN are brought into stacked representation, similar to the way in which feed-forward networks are

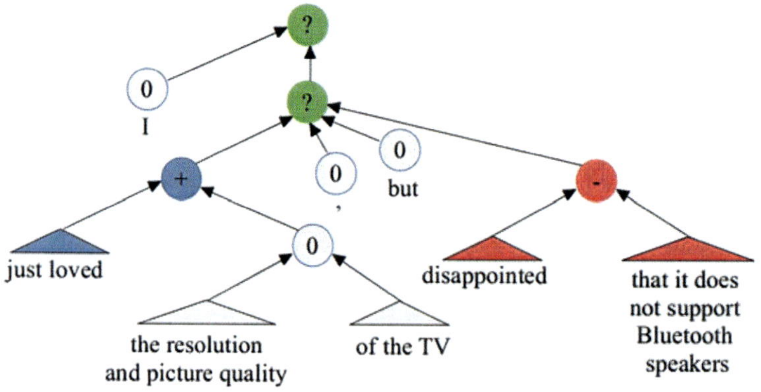

Fig. 1 Example opinion polarity detection in composite statements

stacked in conventional DNN. Unlike RcNN that are deep-in-time, RNN possesses deep-in-structure due to iterative computation of recursive connections. However, RNN often fails in modeling hierarchical representations investigated in traditional deep feed-forward neural networks (DFNN) as well as in RcNN. The missing depth parameter could be added into traditional RNN through the stacking of multiple recursive layers each on top of other. Hence, for each parse tree representation of a sentence, RNN recursively generates parent representation in a bottom-up manner through the combination of lexical terms to obtain representation for phrases, which eventually yields the sentence-level representation. Finally, sentence-level representations are applied further to obtain the classification predictions for given input sentence.

The model described in the above paragraph is basically a hybrid model that combines the features of DFNN and RNN. More technically, Deep-RNN is an extension of DFNN augmented with secondary structural processing segment inside each layer. In which case, information within the network flows in parallel through each node synchronized with each hidden layer. The weights concerning structural processing segments are distributed during forward propagation of the recursive network. On the other side, each node in the tree-structure feeds its own hidden state successively to its counterpart in the adjacent layer. Figure 2 presents the general idea of Deep-RNN, where each successive layer learns certain portion of semantic information

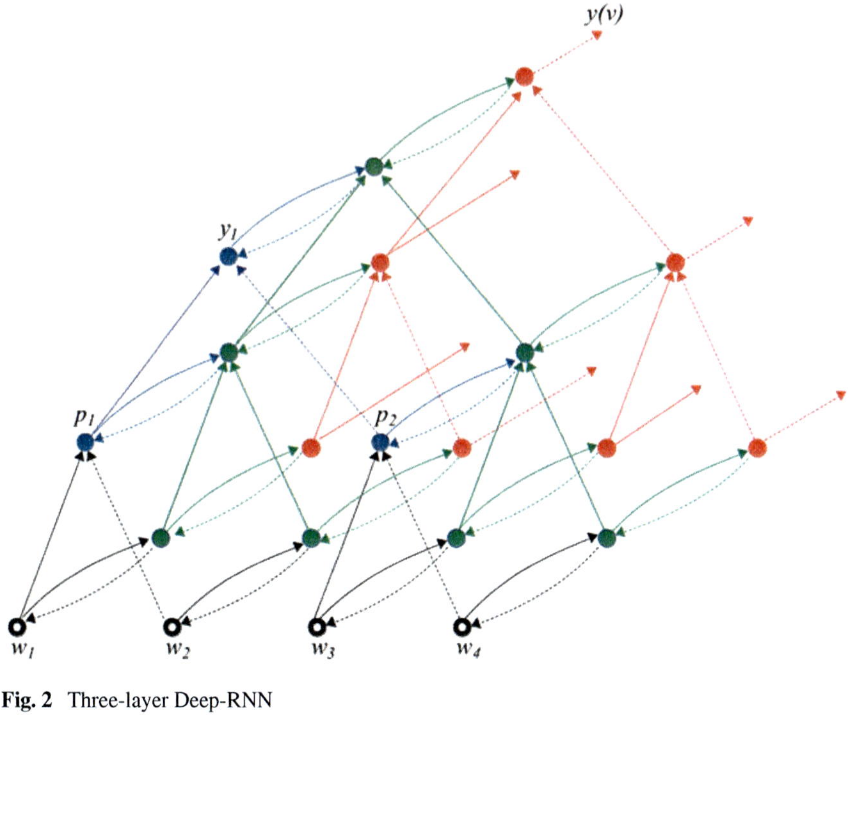

Fig. 2 Three-layer Deep-RNN

and continues toward its next layer with partial representations to achieve processing of remaining constituents. Furthermore, to estimate the performance of the proposed architecture and to carry out empirical study, the model is applied to fine-grained classification and rating of online reviews. The SST opinion dataset [8] is employed to compare the proposed work with previously published works. In SST, each node of a binary parse tree is constituted with supervised sentiment labels that help deep learning (DL), as they allow richer supervised error signal to be back-propagated over the network. The main advantage of using Deep-RNN for the task of automated rating of online reviews is that, it avoids the overhead of aspect keyword supervision. However, many methods in the literature are based on deeper analysis of aspect keywords present in the review. The remainder of the paper is organized according to the following sections. Section 2 provides a literature study on Deep-RNN used for detecting semantic compositions in natural language content. Section 3 presents background knowledge of Deep-RNN and its mathematical description. In Sect. 4, the performance of the proposed model is evaluated against SST dataset with different network parameters. Finally, Sect. 5 gives a brief summary of our research contributions, concluding remarks, and future advancements.

2 Related Works

The research on sentiment analysis (SA) has now been gradually evolved from the traditional BOW models to semantics-preserving compositions based on deep neural architectures. The following section gives a comprehensive survey on recursive network models that are successfully used to address many SA challenges. For example, in [8], a novel DNN architecture known as recursive neural tensor network (RNTN) is introduced to work on a new opinion dataset SST to capture semantic compositions in opinion content. Here, each opinion phrase represented as a parse tree embedded with word vectors is input to RNTN, that further computes representations for higher nodes (sentence representation) in the tree using the same tensor-based composition function. This hybrid model results in better prediction accuracy when applied for the task of single sentence sentiment detection. Yet another variant of recursive network known as adaptive recursive neural network (AdaRNN) used for target-dependent classification of Twitter sentiment data is also proposed [9]. AdaRNN, basically studies the context and semantic relationships among the lexical terms and adaptively propagates the sentiment of words to target classes. This model employs adaptive multi-compositionality layer in RNN which consists of more than one composition functions to adaptively propagate sentiments as distributions over these functions. Further, the research uses manually annotated datasets for empirical study to illustrate the improvement of AdaRNN over many baseline methods. Subsequently, a model based on stacked multiple recursive layers to perform fine-grained opinion classification using DRNN is presented [3]. The efficiency of DRNN is experimentally evaluated against several shallow recursive networks on SST dataset. The researchers further investigated their model qualitatively through input perturbation and studied the nearest neighboring phrases of given test samples. The work justifies toward the idea of adding depth to conventional RNN can capture different aspects of compositionality at different layers.

In [10], a framework based on RNN for the task of political ideology detection in public opinion statements is discussed. In order to learn effectively, the subsentential segments from political annotations, public opinions are crowdsourced at a phrase and sentence level. However, the model requires richer dataset for fine-grained analysis; hence, the authors developed a new political ideology dataset annotated at phrase level. The model is further studied to understand that RNN is quantitatively more efficient than current techniques that use syntactic and semantic features independently. The work also illustrates how correct the proposed model detects ideological bias in complex syntactic constituents. Yet another method to determine sentiment of an aspect of an entity using PhraseRNN that considers both dependency and constituency tree of a phrase is presented [11]. Basically, it is a hybrid model that combines the feature of RNN and AdaRNN. PhraseRNN is made to learn representations of the target aspect enriched by syntactic knowledge form the parse trees. Through the integration of dependency relations and phrases, a hierarchical structure is constructed for semantic propagation. The model outperforms several contemporary approaches when evaluated on different datasets. However in [12], a ranking framework based on RNN to rank opinion sentences for multi-document summarization is developed. This model formulates ranking task as a hierarchical regression that concurrently measures the salience of a sentence and its phrases over the parse tree. This enables learning of word-level to sentence-level compositions from reference summaries. Further, they are used to rank features automatically using hand-tailored feature vectors. Then, a hierarchical regression is applied over combined learning and raw features. Finally, ranked sentences and lexical terms are used to extract informative and unique sentences to prepare summary. Empirical study reveals that the proposed model outperforms over many contemporary summarization techniques. Another hybrid model that combines the features of recursive networks (RNN) and recurrent networks (RcNN) to address the problem of opinion classification of large movie reviews is presented [13]. To demonstrate the effectiveness of this hybrid model, a comparative study on the performance of different sentiment analysis techniques is conducted. The work demonstrates that handcrafted features (BOW with TF-IDF, bi-grams, etc.) will carry complementary discriminative information in addition to existing representations learned from neural network models. Additionally, the proposed architecture is independently experimented with RNN for sentence-level analysis, and with RcNN for passage analysis. The authors claim that this architecture is to be a more intuitive model since it imitates the way humans interpret passages. Finally, in [14] an application of recursive network for the task of Twitter sentiment analysis (TSA) is discussed. Tweet messages are generally blended with heterogeneous text content such as emoticons, shorthands, abbreviations, and many other information that challenges the task of sentiment analysis. The authors in this work tried different combinations of filtering layers with RNN to determine the suitability of network for a given dataset. Tweet messages are first preprocessed and then represented as binary dependency trees. In order to optimize the performance, network hyper-parameters are tuned and regularized with L2 regularization. Further, they also proved that balanced dataset and intermediate labels will play a major role in network training. Imbalanced dataset leads to decreased efficiency in predicting negative labels, whereas intermediate labels will take underfitting. Experimental analysis reveals that RNN with one-hidden layer a produces good result.

3 Deep-RNN: Deep Recursive Neural Networks

Recursive neural networks (RNNs) are primarily nonlinear adaptive models designed to work on structured inputs (directed positional acyclic graphs). Hence, for a given structured input, transformations are applied recursively to obtain the representations for higher-level nodes learned from its descendants. More precisely, a predetermined set of weights are recursively applied over a parse tree of an input sentence to obtain the class prediction. As a consequence, the network acquires semantic representations in such way that, each subsequent layer in a hierarchy will potentially gains a more abstract representation [7, 15]. Generally, the above methodology could be applied to any sort of structured inputs (graph structures). However, in this case the usage of RNN is restricted only to the positional binary trees (e.g., parse trees) as they are simple and efficient for representing natural language text. Thus, for a given binary parse tree of n-gram sentence initialized each leaf node with a word vector, and then RNN computes the parent representations using the following state equations.

$$x(v) = f\left(W_L \times x(\mathrm{ch}_l(v)) + W_R \times x(\mathrm{ch}_r(v)) + b\right) \tag{1}$$

where $\mathrm{ch}_l(v)$ and $\mathrm{ch}_r(v)$ are left and right descendants of v, respectively, W_L and W_R are the synaptic weights that bind left and right children to the parent v, and b is a bias vector. Further, it is noted that representations at terminal and non-terminal nodes lie in the same space, provided W_L and W_R are square matrices, and $\{\mathrm{ch}_l(v), \mathrm{ch}_r(v)\} \in W_E$ are not distinguishable from terminal and non-terminal nodes, where W_E is a $d \times V$ dimensional word-embedding matrix. The parent representations are then obtained through the recursive application of transformations to left and right subphrases in a bottom-up manner that are in same semantic space. Finally, a problem specific output layer is determined according to the following equation.

$$y(v) = g\left(W_g \times x(v) + c\right) \tag{2}$$

where W_g is the output weight matrix, c is the bias vector to the output layer, and $y(v)$ is a class prediction for the parent node v. For the task of fine-grained reviews, rating $y(v)$ is a prediction of sentiment polarity of the opinion phrase taken from the subtree at root v. Meanwhile, external errors occurred at y are back-propagated from root toward its descendants [16]. In the aforementioned model, both terminal and non-terminal nodes are assumed to be similar for classification. However, applications based on parse trees need discrimination among terminal and non-terminal nodes, because terminal and non-terminal nodes have different representations in parse trees. Hence, it is important to determine whether the incoming edge emanates from terminal or non-terminal nodes. In order to decide this, a simple parameterization of the weights W is employed. Equation 3 illustrates this idea more technically.

$$z(v) = f\left(W_L^{\mathrm{ch}_l[v]} \times z_l(\mathrm{ch}_l[v]) + W_R^{\mathrm{ch}_r[v]} \times z_r(\mathrm{ch}_r[v]) + b\right) \tag{3}$$

If $z(v) = x(v)$, then v is a leaf node, and $z(v) \in \chi$, i.e., vector space of lexical terms, otherwise $z(v) \in \varphi$, and it is vector space of phrases. Furthermore, if v is terminal node, then $W^v = W^{xz}$ otherwise $W^v = W^{zz}$. The weight matrix W^{xz} is related to transformation matrix that connects word to phrase space and W^{zz} is related to transformation matrix which connects phrase space to itself. Additionally, the dimension of W^{xz} is defined to be $|x| \times |z|$ and W^{zz} is $|z| \times |z|$, indicates that even a large pre-trained word vectors with a small number of hidden units can be used to train a network without a quadratic dependence on the word vector dimensionality $|z|$.

The above definition of RNN describes the property of depth in structure. In order to achieve depth in time and space, multiple layers of recursive networks are hierarchically stacked, so that each hidden layer yields a different representation space and exhibits a more abstract representation of the input space. More specifically, the idea of Deep-RNN is illustrated in Eq. 4.

$$z(v) = f\left(W_L^{(i)} \times z_l(\mathrm{ch}_l^{(i)}[v]) + W_R^{(i)} \times z_r(\mathrm{ch}_r^{(i)}[v]) + J^{(i)} z(v)^{(i-1)} + b\right) \tag{4}$$

Here, i refers to the index value of the layer in an array of stacked layers, $J^{(i)}$ defines the weight matrix that connects ith hidden layer to the $(i+1)$th hidden layer, and W_L, W_R, and $b(i)$ are defined as in Eq. 1 within each layer i. It is also important to note that, differentiating among terminal and non-terminal nodes is essential only for the input layer of Deep-RNN. Hence, mapping is performed using two separate J's $(i = 2)$, one among the two is for terminal nodes as $J^{xz}(i = 2)$ and another for non-terminal nodes as $J^{zz}(i = 2)$. As a result, all nodes above input layer are represented in the same space except the input layer. Finally, class prediction can be obtained through the binding of last hidden layer with an output layer using Eq. 5 as follows.

$$y(v) = g\left(W_g \times x(v)^{(i)} + c\right) \tag{5}$$

where l defines the total number of layers in the DRNN model.

4 Experimental Setup

In order to explore the features of the proposed Deep-RNN model, a series of experiments have been carried out. Beginning from the selection of dataset to the choice of activation units and network training, we examined different combinations for determining the optimal hyper-parameters of the network. The following sections present the arrangement of experiments and steps involved in the process of training and evaluating the Deep-RNN model.

4.1 Dataset Used

The efficiency of the proposed Deep-RNN model is realized through a recently published sentiment treebank dataset known as Stanford sentiment treebank (SST) dataset [8]. It is the first of its kind dataset with fully labeled parse trees designed to study the compositional effects of sentiment in subjective content. The SST dataset includes fine-grained sentiment labels for 215,154 phrases in the form of parse trees of 11,855 opinion sentences, with an average length of 19.1 words per sentence. Further, these sentiment labels are projected over an ordered integer labels that range from 0 through 4, using softmax activation function in order to formulate the supervised task as a five-class problem. This corpus is basically derived from the dataset introduced in [17] which consists of 11,855 opinion sentences captured from movie reviews. It is then parsed using Stanford parser to generate total of 215,154 phrases from those parse trees, where three human judges are being involved in annotating the phrases. The volume and coarseness of SST dataset enable the researchers to conduct complex experiments on compositional models that need structured input and supervised training. In this research work, two different instances of data samples are considered for experimentation. Firstly, the proposed Deep-RNN model is tested on the entire dataset for binary and fine-grained classification of sentiment ratings. Hence, the data samples in treebank dataset are split into train (8544), dev (1101), and test (2210) splits. Secondly, the experiment is conducted for sentiment ratings on the composite statements. Here, a dataset containing only composite statements is considered for testing. However, training and validation splits are kept unaltered. The selection of composite statements from available 2210 test samples is made manually.

4.2 Pre-trained Word Vectors

Recently, semantic word spaces (word-embeddings) have shown great success in capturing fine-grained semantic regularities for text analysis [18–20]. The intuition of word-embedding lies in the fact that words having similar meaning would take similar representations. Basically, word vectors consist of low-dimensional real values that model syntactic and semantic information of individual words. Eventually, these vectors are used as pre-trained features for many text classification tasks. The Majority of word vectors are based on distance or angle between two-word vectors as a quality measure of such representations. However, techniques based on word similarity are advantageous which measures the proximity of word vectors that are semantically identical. In this research work, we employ publicly available low-dimensional (50, 100, 300) word vectors called Glove [19], a new global log-bilinear regression model trained on Google News dataset.

4.3 Weight Initialization

Initialization of parameter weights is the most crucial step in developing any neural network model. However, the selection of correct initialization technique will significantly reduce the time required to converge when trained with gradient descent. To illustrate the benefit of selecting initialization techniques on training time, we use random parameter initialization technique and also a pre-trained model. In random initialization technique, both the compositional matrices W_L, W_R and word-embedding matrix W_E are initialized with real-valued random numbers, so that, representations for words and phrases are arbitrarily projected over the vector space without any training. The other approach is to initialize W_E with a pre-trained model Glove [19] as discussed in Sect. 4.2, which gives the word analogies of the associated word categories. Furthermore, the effect of using pre-trained model over a random initialization is studied in Sect. 5.4.

4.4 Activation Functions

In accordance with the research work reported in [21], it is advantageous to apply rectifier units for training deep architectures without pre-training step. Hence, a rectified linear unit (ReLU) $f(x) = \max\{0, x\}$ is used for hidden layers. ReLU avoids vanishing gradients problem as it returns 0 if it receives negative inputs (gradient zero) and restores its current value for positive inputs. However, for a given labeled input, the objective is to learn representations for fine-grained (five-class) sentiment polarities. To attain this, a standard softmax activation of Eq. 6 is used for the output layer that takes node's vector $z(v)$ as input and produces prediction $y(v)$.

$$g(x)_i = \frac{e^{x_i}}{\sum_{k=1}^{K} e^{x_k}} \tag{6}$$

The output of the softmax function is equivalent to a categorical probability distribution, i.e., the probability that any of the classes are true.

4.5 Regularization of Network

Overfitting is a common problem in deep neural architectures when trained with a large number of parameters. To overcome the problem of overfitting, a simple and powerful technique called dropout regularization [22] is used. In dropout regularization, certain units from neural network are selected randomly and ignored during training. In this experiment, for each hidden layer the probability of dropout rate is initially set to 1.0; further, it is tuned between 0.5 and 1.0 over the validation set. Because dropped units are shared within the network, the same units of the hidden layer are dropped at each node.

4.6 Training the Network

In order to train the proposed Deep-RNN model, we use stochastic gradient descent (SGD) with a fixed learning rate of 0.01. However, to update the training parameters a diagonal variant of AdaGrad [23] is used. The main benefit of using AdaGrad is that it automatically adapts the learning rate to the parameters and achieves smoother and quicker convergence. Weights are updated regularly after processing a mini-batch of 32 samples. Number of epochs for training is initialized to 200. To avoid bias as it happens when training the samples in ordered manner, training data is shuffled after every epoch. Finally, the recursive weight matrix W^{hh} inside a layer is initialized as $0.5I + e$, where I is the identity matrix and e is a small uniformly random noise. This justifies that, the representation of each node is approximately initialized to the average of its two descendants. Further, errors on a validation set are measured during training and stopped (early stopping) if validation error does not gets improved.

5 Results and Discussion

This section presents the experimental results of the proposed model and discusses strong baselines which use similar dataset. Further, several quantitative evaluations on the test dataset are experimented and compared with the Deep-RNN.

5.1 Baselines

In the first place, a naive Bayes (NB) classifier baseline that operates on bigram counts is considered [8]. Then, a shallow RNN that uses pre-trained word vectors and *tanh* activation [10, 21], followed by a matrix–vector RNN (MV-RNN), where each word is assigned a matrix–vector and composition is defined by matrix–vector multiplications [24]. Yet another baseline from [8], a recursive neural tensor network (RTNN) that is defined by bilinear tensor product is also used. However, the results for baselines are directly borrowed from the publications mentioned with the same ratio of train, dev, and test partitioning.

5.2 Fine-Grained and Binary Sentiment Ratings on Entire Dataset

The model is evaluated for predicting both fine-grained (five-class) and binary sentiment (negative and positive) ratings. First, the network is trained for fine-grained sentiment ratings, where a five-dimensional posterior probability output vector of a softmax layer is mapped to five-scale integer ranking (1 = *very negative*, 2 = *negative*, 3 = *neutral*, 4 = *positive*, 5 = *very positive*). However, the same five-dimensional probability vector obtained from the above step is further fused into a two-valued vector for binary prediction. The parameter values depth ($l = 3$, *number of layers*), width ($|h| = 200$, number of hidden units), and dimension ($d = 200$, *word vector dimension*) are instantly acquired from [3] as optimal values. Table 1 presents the empirical results of Deep-RNN compared with other related to previous works.

5.3 Sentiment Ratings on Composite Sentences

Identification of compositions in composite opinion statements that involve both positive and negative phrases is a challenging task. In this case study, a dataset containing only composite statements are considered for testing. However, training

Table 1 Comparative analysis of Deep-RNN with previous works

Network model	Fine-grained	Binary
SVM [25]	40.7	79.5
NB (bigram) [8]	41.9	83.1
RNN [21]	43.2	82.4
MV-RNN [24]	44.4	82.9
RTNN [8]	45.7	85.4
Deep-RNN (3, 200, 200)	**49.6**	**87.6**

and validation splits are kept unaltered. The selection of composite statements from test split (from 2210 samples) is done manually. Training parameters are tuned from the previous setup to new values as depth ($l = 4$), width ($|h| = 175$), and dimension ($d = 300$), respectively. The experiment is conducted for the evaluation of both fine-grained and binary sentiment ratings. Two different types of composite sentences have been considered for analysis. In type one, sentences that start with positive and end with negative opinions are studied. For this instance, the negation changes the overall sentiment of a sentence from positive to negative. On the other hand, in type two cases, sentences that start with negative, but end with positive are considered. In such a case, the positive portion of a sentiment changes the overall sentiment of a sentence from negative to positive. The average accuracy of both test cases discussed above is determined to be 46.5% for fine-grained and 88.4% for binary sentiment ratings.

5.4 *Effect of Tuning Hyper-parameters on Performance*

Tuning the hyper-parameters will play a critical role in scaling the performance of the network. Perhaps, initialization of weights could be one such important parameter which has significant impact on the network performance. Figure 3a, b demonstrates the impact of initializing weights with pre-trained word vectors on network performance over random initialization of weights. The experiment is being carried out for running 200 epochs. The curves on the plot illustrate the rate at which the model gets converged as the number of epochs increases. This proves that, initializing network weights with pre-trained word vectors has faster convergence than initializing with random weights.

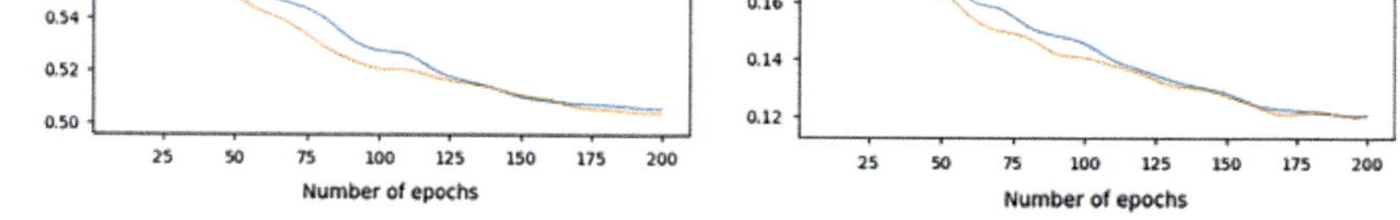

(a) Fine-grained classification (b) Binary classification

Fig. 3 Effect of tuning hyper-parameters on Deep-RNN

6 Conclusion and Future Directions

In this paper, a methodology for fine-grained rating of online reviews using Deep-RNN is proposed. The model is used for quantitative evaluation of compositionality over SST dataset and compared with many baselines. Experimental results show that Deep-RNN outperforms the baselines and accomplishes the best performance toward fine-grained sentiment rating. Furthermore, the effect of initializing weights with pre-trained word vectors on network performance over random initialization of weights is also studied. From the point of future directions, yet there is enough scope for investigating the functionalities of hidden layers and their benefits. Tuning the hyper-parameters of network model and studying its impact on the performance were always an open issue. Design and validation of task-specific deep networks require much attention from the research community.

References

1. Lee TY, BradLow ET (2011) Automated marketing research using online customer reviews. J Mark Res 48(5):881–894
2. Liu W, Wang Z, Liu X, Zeng N, Liu Y, Alsaadi FE (2017) A survey of deep neural network architectures and their applications. Neurocomputing 234:11–26
3. Irsoy O, Cardie C (2014) Deep recursive neural networks for compositionality in language. In: Advances in neural information processing systems, pp 1–4
4. Bengio Y (2009) Learning deep architectures for AI. Found Trends Mach Learn 2(1):1–127
5. Gori M, Maggini M, Sarti L (2003) A recursive neural network model for processing directed acyclic graphs with labeled edges. In: Proceedings of the international joint conference on neural networks, Jantzen Beach, Portland, Oregon
6. Elman JL (1990) Finding structure in time. Cogn Sci 14(2):179–221
7. Pascanu R, Mikolov T, Bengio Y (2013) On the difficulty of training recurrent neural networks. In: ICML-2013
8. Socher R, Perelygin A, Wu JY, Chuang J, Manning CD, Ng AY, Potts C (2013) Recursive deep models for semantic compositionality over a sentiment treebank. In: Proceedings of the conference on empirical methods in natural language processing, EMNLP '13
9. Dong L, Wei F, Tan C, Tang D, Zhou M, Xu K (2014) Adaptive recursive neural network for target-dependent twitter sentiment classification. In: Proceedings of the 52nd annual meeting of the association for computational linguistics, Baltimore, MD, USA, pp 49–54
10. Iyyer M, Enns P, Boyd-Graber J, Resnik P (2014) Political ideology detection using recursive neural networks. In: Proceedings of the 52nd annual meeting of the association for computational linguistics, Baltimore, MD, USA, pp 1113–1122
11. Thien Hai Nguyen, Kiyoaki Shirai.: PhraseRNN: Phrase Recursive Neural Network for Aspect-based Sentiment Analysis. Proceedings of the 2015 Conference on Empirical Methods in Natural Language Processing, Lisbon, Portugal, pp. 2509?2514, (2015)
12. Cao Z, Wei F, Dong L, Li S, Zhou M (2015) Ranking with recursive neural networks and its application to multi-document summarization. In: Proceedings of the twenty-ninth AAAI conference on artificial intelligence, Austin, TX, pp 2153–2159
13. Timmaraju A, Khanna V (2015) Sentiment analysis on movie reviews using recursive and recurrent neural network architectures. Semantic Scholar, pp 1–5
14. Yuan Y, Zhou Y (2015) Twitter sentiment analysis with recursive neural networks. Project report (Stanford), pp 1–8. https://cs224d.stanford.edu/reports/YuanYe.pdf

15. Van VD, Thai T, Nghiem M-Q (2017) Combining convolution and recursive neural networks for sentiment analysis. In: Proceedings of the eighth international symposium on information and communication technology. ACM
16. Poria S, Cambria E, Hazarika D, Vij P (2016) A deeper look into sarcastic tweets using deep convolutional neural networks. In: Proceedings of COLING 2016
17. Pang B, Lee L (2005) Seeing stars: exploiting class relationships for sentiment categorization with respect to rating scales. In: ACL, pp 115–124
18. Mikolov T, Sutskever I, Chen K, Corrado G, Dean J (2013) Distributed representations of words and phrases and their compositionality, pp 1–9. arXiv:1310.4546
19. Pennington J, Socher R, Manning CD (2014) GloVe: global vectors for word representation. In: Proceedings of the 2014 conference on empirical methods in natural language processing (EMNLP), pp 1532–1543
20. Liu P, Qiu X, Huang X (2015) Learning context-sensitive word embeddings with neural tensor skip-gram model. In: Proceedings of the 24th international conference on artificial intelligence (IJCAI'15), pp 1284–1290
21. Socher R, Lin CC-Y, Ng AY, Manning CD (2011) Parsing natural scenes and natural language with recursive neural networks. In: Proceedings of the 28th international conference on international conference on machine learning (ICML'11), pp 129–136
22. Melamud O, Goldberger J, Dagan I (2016) Context2vec: learning generic context embedding with bidirectional LSTM. In: Proceedings of the 20th SIGNLL conference on computational natural language learning (CoNLL), pp 51–61
23. Srivastava N (2014) Dropout: a simple way to prevent neural networks from overfitting. J Mach Learn Res 15(1):1929–1958
24. Socher R, Huval B, Manning CD, Ng AY (2012) Semantic compositionality through recursive matrix-vector spaces. In: Proceedings of the 2012 joint conference on empirical methods in natural language processing and computational natural language learning (EMNLP-CoNLL '12), pp 1201–1211
25. Kalchbrenner N, Grefenstette E, Blunsom P (2014) A convolutional neural network for modelling sentences. arXiv preprint arXiv:1404.2188

Analysis of Strategic Market Management in Light of Stochastic Processes, Recurrence Relation, Abelian Group and Expectation

Prasun Chakrabarti, Tulika Chakrabarti, Siddhant Bane, Biswajit Satpathy, Indranil SenGupta, and Jonathan Andrew Ware

Abstract This paper entails a novel approach of analysis of strategic market management based on renewal reward stochastic process. The paper has also pointed out a discovered fact that clearly infers realization of cost analysis of product revalidation in light of Brownian motion with drift. The paper indicates a rare and new concept of how can compound stochastic process be applied to sense business cost analysis. In demand–supply analysis, there lies the essence of realization of alternating renewal theory-based customer satisfaction. Furthermore, the paper also shows a novel analysis of product upgradation in light of conditional expectation and simple random walk. Facts related to recurrence relation, Abelian group and expectation indicate a non-conventional approach of business gain prediction.

Keywords Renewal reward stochastic process · Brownian motion with drift · Compound Poisson process · Alternating renewal theory · Conditional expectation · Simple random walk · Recurrence relation · Abelian group

P. Chakrabarti (✉) · B. Satpathy
Department of Business Administration, Sambalpur University, Sambalpur, Odisha 768019, India
e-mail: drprasun.cse@gmail.com

P. Chakrabarti
Techno India NJR Institute of Technology, Udaipur, Rajasthan 313003, India

T. Chakrabarti
Department of Chemistry, Sir Padampat Singhania University, Udaipur 313601, India

S. Bane
Department of Computer Science and Engineering, ITM Universe, Vadodara, Gujrat 391510, India

I. SenGupta
Department of Mathematics, North Dakota State University, NDSU Dept # 2750, Minard Hall 408E12, Fargo, ND 58108-6050, USA

J. A. Ware
University of South Wales, Pontypridd, Cymru CF37 1 DL DU, UK

N. N. Chiplunkar and T. Fukao (eds.), *Advances in Artificial Intelligence and Data Engineering*, Advances in Intelligent Systems and Computing 1133,
https://doi.org/10.1007/978-981-15-3514-7_53

1 Introduction

Strategic planning [1] entails an analysis of the factors external to any business that affect strategy. The analysis is to be sensed in a convenient way for getting optimum throughput in terms of profit. Any business is related to opportunity and threat. An efficient judgement of opportunity and threat will facilitate to realize the status of sales and profit in the present state. Statistical prediction of the trend of gain is a significant aspect in this perspective. The other output is the investigation of factors leading to strategic uncertainty regarding a business. An opportunity or a threat leads to a notable fluctuation in pattern of the sales and profit. Marketing Myopia [2] refers to the need of investigations of both opportunity and threat. An external analysis points out variation in gain estimate. The pivotal factors include investment capital, probability of market penetration, growth direction. The business trend can be investigated on the basis of apriori events and statistical trend analysis and simulation [3]-based analysis plays a pivotal role in this essence. In few cases, due to random events, strategic uncertainty [4] is to be studied and the inference has to be realized for future forecasting. The counting process $\{N(t), t \geq 0\}$ is called a renewal process [5]. A geometric Brownian motion (GBM) or exponential Brownian motion is a continuous-time stochastic process in which the logarithm of the randomly varying quantity follows a Brownian motion (also called a Wiener process) with drift. A compound Poisson process [6] is a continuous-time (random) stochastic process with jumps. The jumps arrive randomly according to a Poisson process, and the size of the jumps is also random, with a specified probability distribution. An alternating renewal process [7] models a system that, over time, alternates between two states, which we denote by 1 and 0 (so the system starts in state 1). Generically, we can imagine a device that, over time, alternates between *on* and *off* states. A popular random walk model [8] is that of a random walk on a regular lattice, where at each step the location jumps to another site according to some probability distribution. In a simple random walk [9], the location can only jump to neighboring sites of the lattice, forming a lattice path.

2 Cost Analysis Based on Analogy to Renewal Reward Process

In this section, a novel fact related to renewal reward stochastic process-based testing cost of a product has been pointed out. After careful consideration of customer needs and expectations, a product is developed and then it undergoes the verification and validation phase. The following assumptions are hereby mentioned:

(i) A business organization X outsources the verification and validation phase to another organization Y.
(ii) Y collects p number of distinct products within a time stamp t_1.

(iii) Y conducts survey of the impact of all the validated products of X, and the cycle is completed once the compiled feedback is reported to X by Y.
(iv) Testing cost of Y per product per unit time is m.
(v) Once a product is ready for market launch, an additional fixed cost c is associated for futuristic upgradation purpose.
(vi) The time between arrival times of pth and $(p + 1)$st products is T_p

The expected time of a cycle in this case is therefore pt_1, and the corresponding cost is as follows:

$$E\left[\text{cost of the cycle}\right] = (m * t_1 * p * (p - 1))/2 + c$$

The average cost incurred by $X = c(\text{average}) = (m * (p - 1))/2 + c/(p * t_1)$

The aforesaid cost analysis entails the validity of the computation based on the renewal reward stochastic process.

3 Cost Analysis Based on Analogy to Motion with Drift

A novel analogy is hereby depicted in context to realization of cost analysis of product revalidation in light of Brownian motion with drift. In due course of time due to market trend, focus on the user needs has to be noted, and accordingly, the specifications of the product have to be incorporated. If immediate action is not taken, then the existence of the product in the market tends to deteriorate with time.

The following assumptions are hereby mentioned:

(i) The impact of the product changes its state with drift coefficient β ($\beta > 0$).
(ii) At the state S_B, the product becomes useless in the market.
(iii) A cost c_1 is associated to return back the product in use in the market after incorporating additional desirable specifications as per market need. The state transition is $S_B \rightarrow S_I$ where S_I is the initial state.
(iv) The role of business analyst is crucial, and if analysis for upgradation of the product is not carried out before reaching the state S_B, then cost incurred is c_2.
(v) At state S_R, an attempt of product rectification is being done.

The average cost = $(c_2 + c_1 * (1 - p))$/(E[time to reach state S_R] where p is the probability of successful upgradation of the product. The principle that tries to rectify the product when the state is S_R has a long-run average cost of $(\beta * (c_2 + c_1 * (1 - p)))/(R - I)$, where the time period of observation is $(R - I)$.

4 Cost Analysis Based on Analogy to Compound Poisson Process

In product revalidation stage, an observation of optimum specification of a product for further incorporation plays a pivotal role. The following assumptions are hereby mentioned:

(i) After trend analysis of market impact and susceptibility, suppose additional features have to be incorporated for upgrading x products of a company.
(ii) The features are finalized at a Poisson rate α.
(iii) Costs associated for incorporation of all features for each product form a set of independent and identically distributed random variables that is independent of the feature selection process.
(iv) $c(t)$ is the total cost of incorporating all features in x products by time t.

The aforesaid exercise is to be carried out in regular intervals of time and that will facilitate in framing optimum business strategy. The total cost involved in the revalidation stage for x products is as below:

$$c(t) = \sum_{i=1}^{N(t)} c_i \text{ for } t \geq 0; \tag{1}$$

where $\{N(t),\ t \geq 0\}$ is a Poisson process and $\{c_{i,}\ i = 1, 2, \ldots.\}$is a set of costs associated with incorporating all features of each product. The aforesaid analysis suffices the validity of the principle of the compound Poisson process.

5 Analysis of Demand and Supply for a Product Based on Analogy to Alternating Renewal Theory of Stochastic Process

In case of demand–supply analysis, the supply of raw materials for incorporating features of a product is an important aspect. If there is a demand for a particular product, and due to non-availability, customers have to wait, then to some extent the impact of the product in the market is affected. Herein lies the essence of realization of renewal theory-based customer satisfaction. The following assumptions are hereby mentioned:

(i) The company sells single type of product.
(ii) Customers have placed order for the product, and the time stamps of the orders placed adhere to renewal process having non-lattice intertribal distribution D.
(iii) y is the bias or threshold value that indicates the margin beyond which there will be scarcity of product to be so.

(iv) J customers have given order and received the product, while K customers have to wait for some time after placing order as there is no supply for the moment.
(v) If inventory level after serving a customer is below bias y, then an order is placed to bring it back to p. Hence, the store uses (y, p) ordering policy.

If inventory level after serving a customer is β, the amount ordered is $(p - \beta)$ if $\beta < y$ while there will be no order placed if $\beta \geq y$. Also, $\beta(t)$ is the inventory level at time t, and it is desirable for $\lim_{t\to\infty} P\{\beta(t) \geq \beta\}$. Then, if $\beta(0) = p$, then it is evident that the supply is adequate whenever inventory level is at least ld β and non-adequate otherwise. Herein lays the validity of the principle of alternating renewal theory of stochastic process.

6 Analysis of Upgradation of a Product as Per Market Need Based on Analogy to Conditional Expectation and Simple Random Walk

Based on the customer feedback regarding a product in the upgradation stage, certain amount of time is required to complete the exercise. In the validation, if the product is found satisfactory, then it is relaunched in the market. If it is found to be upgraded further, then again it undergoes the validation phase before becoming market-ready. The following assumptions are hereby mentioned:

(i) S represents the state of a particular product to be market-ready after upgradation.
(ii) P_i denotes the ith product.
(iii) Four products P_1, P_2, P_3 and P_4 need to be re-validated. Let $t1$ unit of time P_1 is found to be satisfactory for market launch, and after $t2$ unit of time P_2 is found to be unsatisfactory and hence again

$\Delta t2$ is needed for it to be market-ready, $t3$ unit of time P_3 is found to be satisfactory for market launch and after $t4$ unit of time P_4 is found to be unsatisfactory, whereby Δ $t4$ is needed for it to be market-ready.

Now, $E\left[e^{ts}|_{p=1}\right] = e^{t1\cdot t}$. Since S represents state of product to be market-ready, hence additional time $\Delta t2$ has same distribution as S.

So, $E\left[e^{ts}|_{p=2}\right] = e^{(t2+\Delta t2)\cdot t} = e^{t2.t} E[e^{ts}]$

Similarly, $E\left[e^{ts}|_{p=3}\right] = e^{t3\cdot t}$ and $E\left[e^{ts}|_{p=4}\right] = e^{t4\cdot t} E[e^{ts}]$

Therefore, the moment generating function of the state of all products be market-ready is given by

$$E[e^{ts}] = 1/4\left(e^{t1\cdot t} + e^{t2\cdot t} E[e^{ts}] + e^{t3\cdot t} + e^{t4\cdot t} E[e^{ts}]\right) \tag{2}$$

In this context, a novel concept of symmetric simple random walk-based trials in validation phase can be pointed out. The following assumptions are hereby mentioned

(i) A product is successful in the market as per customer feedback. Let S denote the state to be of satisfactory level.
(ii) Due to stochastic arrival of a competitor, certain upgradation is desirable to maintain the satisfactory level of that particular product.
(iii) For $p > 0$, let us suppose Q represents the number of trials in the upgraded phase to state p prior to its return to initial state.

Hence,

$$Q = \sum_{n=1}^{\infty} Y_n \tag{3}$$

where $Y_n = 1$ (if a visit to state p occurs at time n and the state S is not met before n) and $Y_n = 0$ otherwise.

Based on the property of symmetric simple random walk and on the fact that the product has again become market-ready at time n after Q trials in the upgradation phase, we can denote

$$E[Q] = \sum_{n=1}^{\infty} P\{S_n > 0, S_n > S_1, \ldots S_n > S_{n-1}, S_n = p\} = 1 \tag{4}$$

7 Business Gain Forecasting Based on Recurrence Relation, Abelian Group and Expectation

As per the invent of data simulation and modeling, the significance of judgmental approaches to business forecasting plays a pivotal role. With the proliferation of software tools, the accurate analysis is possible that will definitely facilitate to come to a conclusion leading optimum business strategy. In this perspective, we propose some discovered facts leading to some directions in business thought with analogy to statistical established theories.

Conjecture: *Investigation on time phases of business loss and subsequent gain can be governed by the principle of recurrence relations and Abelian group property.*

Justification:

Based on the steady demand–supply [10] status, it is extremely important to note the initial timing instant of profit–loss analysis ($T1$), timing instant of incidence of event loss for the first time ($T2$) as well as the immediate subsequent timing instant of regaining profit overcoming the pseudo-loss ($T3$). Supervised machine learning will facilitate proper trend analysis of business growth, and accordingly, irregular pattern-based n offsets ($\Delta 1$, $\Delta 2$, $\Delta 3$ … Δn) have to be taken into consideration. For the justification purpose, let us take three iterations. Now, $T2 = T1 + \Delta 1, T3 = T2 + \Delta 2 = T1 + (\Delta 1 + \Delta 2)$. On the basis of this fact, the general equation will suffice the validity of a recurrence relation as $T(x) = T(x-1) + \alpha$, with initial condition(seed) $T3 = T1 + (\Delta 1 + \Delta 2)$ and α being the constant toward event (regaining profit as per initial measure).

In this context, we represent another interesting fact of considering the analogy of identity element to be the first timing instant of observation of any event (profit or loss). The proceeding timing instants will be of lagging nature, while the futuristic (predictive) timing instants with respect to the observed event's initial time stamp will be of leading nature. If $T_E(L_t)$ denotes the time stamp of incidence of event (profit or loss), $T_E(L_{t-1})$ as the estimate lagging time with respect to $T_E(L_t)$ and $T_E(L_{t+1})$ as the amount of leading time with respect to $T_E(L_t)$, then the entire time stamps of realizing the past, present and future events can be represented by Abelian Group (G, +) where $T_E(L_{t-1}, L_t, L_{t+1}) = \{0, \pm 1, \pm 2, \ldots \pm \infty\}$, the identity element '0' being $T_E(L_t)$. The sets representing correlation of past and future events with respect to present are $T_E(L_{t-1}, L_t) = \{(-1, 0)(-2, 0), \ldots (-\infty, 0)\}$ and $T_E(L_t, L_{t+1}) = \{(1, 0)\ (2, 0), \ldots\ (\infty, 0)\}$, thereby indicating the validity if Abelian property.

Herein lays the essence of framing a novel business strategy of estimation of effective time stamp of profit–loss observation. As per Conjecture, if a proper prediction of time stamp T_{n+1} based on previous observation timing instant T_n is done based on curve fitting, then as per the pre-specified production time phase, ample time of strategy can be formed toward optimum demand–supply and related increase in probability of profit margin. Based on arbitrary numerical data (see Table 1), graphical analysis has been performed in order to investigate the best possible curve fit, and in this context, five iterations have been taken into consideration.

From Fig. 1, it is evident that optimum prediction of third time stamp of analysis based on the second one is governed by the polynomial (degree 4) best curve fitting and the equation is

$$y = -0.0068x^4 + 0.6351x^3 - 21.285x^2 + 305.06x - 1550.9 \quad (5)$$

In the similar manner, Fig. 2 entails that optimum prediction of fourth time stamp of analysis based on the third one is governed by the polynomial (degree 4) best

Table 1 Business profit–loss analysis at irregular timing instants

Iterations	Initial time stamp of analysis $T1$	First inter-analysis time gap $\Delta 1$	Second inter-analysis time gap $\Delta 2$	Third inter-analysis time gap $\Delta 3$	Second time stamp of analysis $T2$	Third time stamp of analysis $T3$	Fourth time stamp of analysis $T4$
1	10	3	5	7	13	18	25
2	12	7	9	11	19	28	39
3	15	12	16	16	27	43	59
4	18	8	10	9	26	36	45
5	20	1	2	3	21	23	26

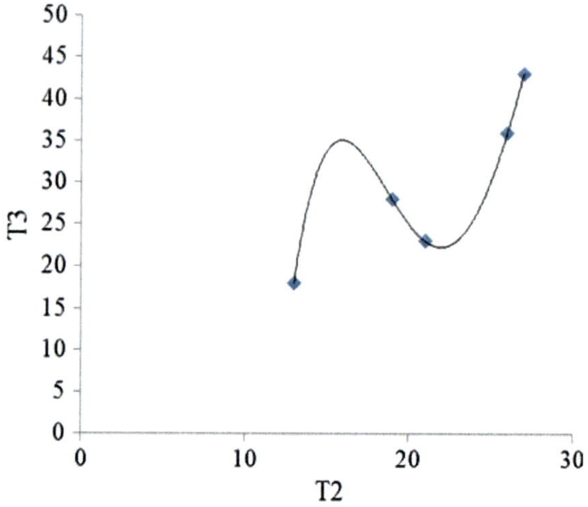

Fig. 1 Curve fitting-based prediction of third time stamp of analysis based on second one

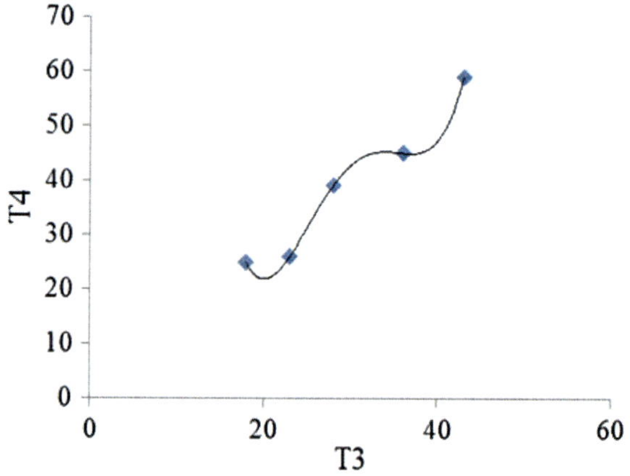

Fig. 2 Curve fitting-based prediction of fourth time stamp of analysis based on third one

curve fitting and the equation is

$$y = 0.0013x^4 - 0.1578x^3 + 6.9688x^2 - 131.04x + 909.83 \tag{6}$$

Annual investigation of the incidence of events [profit, loss] is significant in business analysis. Based on N observations (if M denotes the status of profit while N-M represents loss), the probability distribution can be computed (see Table 2). In a particular time phase (suppose in a span of ten years), the frequency of one-time/two-time/three-time .../six-time loss has to be noted, and then based on the expected value, [11] proper strategy has to be formed.

Expectation $E(F(L)) = 1(0.3) + 2(0.1) + 3(0.05) + 4(0.2) + 5(0.25) + 6(0.1) = 3.3$. The expected value is 33%, and accordingly, the related business gain strategy is to be focused in a proper direction. We hereby propose a discovered fact that based

Table 2 Discrete probability distribution of sensing the incidence of event (loss)

Frequency of event loss, $F(L)$ per year over a particular time phase	Discrete probability distribution $P(F(L))$
1	0.3
2	0.1
3	0.05
4	0.2
5	0.25
6	0.1

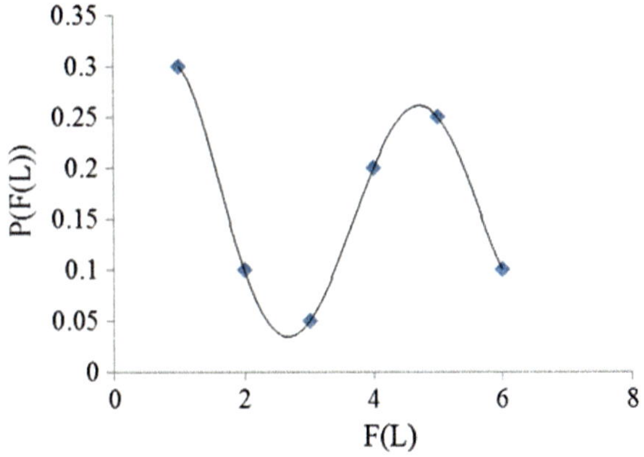

Fig. 3 Prediction of discrete probabilistic value related to event loss and frequency

on the polynomial (degree 5) curve fitting, a futuristic percentage of loss can be predicted accurately (see Fig. 3). It is governed by the following equation

$$y = 0.0046x^5 - 0.0833x^4 + 0.5437x^3 - 1.5167x^2 + 1.6517x - 0.3 \quad (7)$$

8 Conclusions

This paper throws some light on non-conventional approaches of analysis of strategic market management based on renewal reward stochastic process, Brownian motion with drift, compound stochastic process, alternating renewal theory and simple random walk. Discovered fact related to analysis using recurrence relation, Abelian group and expectation has also been cited.

References

1. Aaker DA (2005) Strategic market management. Wiley
2. Levitt T (1960, July–August) Marketing myopia. Harvard Bus Rev 45–56
3. Cobb D, Charles W, Douglas PK (1925) A theory of production. Am Econ Rev Suppl 139–165
4. Samuelson PA (1976) Economics, 10th edn, Chap. 20. McGraw-Hill Book Company, New York
5. Ross S (1996) Stochastic processes, 2nd edn. Wiley, India
6. Olofsson P (2005) Probability, statistics, and stochastic processes. Wiley
7. Giri PK, Banerjee J (1999) Introduction to statistics. Academic Publishers
8. Argon NK (2011) Alternating renewal processes. Wiley (2011)
9. Revesz P (2013) Random walk in random and non-random environments, World Scientific Publishing Co. Pte. Ltd.
10. Laudon KC, Laudon JP (2014) Management information systems: managing the digital firm. Pearson Education Inc.
11. Anderson TW (1984) An introduction to multivariate statistical analysis. Wiley, New York

Peer-to-Peer Distributed Storage Using InterPlanetary File System

A. Manoj Athreya, Ashwin A. Kumar, S. M. Nagarajath, H. L. Gururaj, V. Ravi Kumar, D. N. Sachin, and K. R. Rakesh

Abstract In today's Web, HTTP is the default protocol to transmit files. It works efficiently to move small files. But HTTP fails to implement other more efficient file distribution techniques. The InterPlanetary File System (IPFS) is a peer-to-peer globally mounted file system which addresses the challenges which can resolve data reliability, fault-tolerance, consistency, and non-repudiation issues in the current system. In this paper, we have proposed an innovative and efficient way of storing and retrieving the files on Web using IPFS; adding or uploading files is made more easier and assures high security to data. Many approaches are made where the files are stored in a distributed system but the approach using decentralized storage protocol is not used and it provides a new dimension to the application.

Keywords Blockchain · Decentralized · Distributed · Peer-to-peer network · IPFS · IPFS cluster

A. Manoj Athreya (✉) · A. A. Kumar · S. M. Nagarajath · H. L. Gururaj · V. Ravi Kumar · D. N. Sachin · K. R. Rakesh
Computer Science and Engineering, Vidyavardhaka College of Engineering, Mysuru, Karnataka, India
e-mail: manoj2375@gmail.com

A. A. Kumar
e-mail: ashwinkumar.cs@vvce.ac.in

S. M. Nagarajath
e-mail: nagarajath.cs@vvce.ac.in

H. L. Gururaj
e-mail: gururaj1711@vvce.ac.in

N. N. Chiplunkar and T. Fukao (eds.), *Advances in Artificial Intelligence and Data Engineering*, Advances in Intelligent Systems and Computing 1133,
https://doi.org/10.1007/978-981-15-3514-7_54

1 Introduction

The InterPlanetary File System (IPFS) is a peer-to-peer distributed file system that seeks to connect all computing devices with the same system of files. IPFS works similar to BitTorrent where the objects are exchanged in a single versioned controlled repository. In other words, IPFS is block storage model which provides high-throughput with content-addressed hyperlinks, where the address of the link is based on the content it holds. This block storage model is formed by a generalized type of data structure called Merkle DAG, using which versioned file systems are built. [11] IPFS works by combining a distributed hash table (DHT) and a self-certifying nature of file system which is ensured but public-key cryptography. There is no single point of failure in the IPFS, and nodes do not need to trust each other [1].

As the world is moving toward decentralization revolution, a lot of dapps are developed every day. A true dapp needs a truly decentralized storage with a decentralized content distribution network. IPFS implies the rules on the movement of data and content in its network. This file system layer offers properties such as:

- the Web pages that are truly distributed.
- the websites will have no original server to respond to the requests that can run entirely on client-side browsers [2].

1.1 Content Addressing

The method of content addressing is used in IPFS which identifies content in the network. In today's Web, the addressing is different where the same function is performed using IP addressing.

IPFS objects are the entities present in the IPFS address which contains a list of IPFS links and content data which is being addressed. The large files added to IPFS are broken into many smaller chunks of data and resolved to an array of IPFS links that points to the broken pieces of the original data. This type of addressing in the IPFS network ensures that the address specified will always result in the request to the same file [12].

Another advantage of the content addressing system is that as long as the content hosted by a node is available, it can be retrieved from the IPFS network. This eradicates the problem of broken link which exists in today's internet. IPFS does not host duplicate links or files because the links point to the same hash value which is given as the result of the addressed contents. Thus, to retrieve a file from the network, one copy of the file is enough which can be hosted by any node across the network [1, 9].

1.2 IPFS Objects

The IPFS address is resolved through the data structure called IPFS objects which has two fields: Data and links.

Data—information or content which is converted into binary data of size less than or equal to 256 kB.

Links—consist of link structures which points to other IPFS objects present in the IPFS network.

A link in the IPFS object has data fields which involves fields such as name, hash, and size [10].

Name—the name given to the link.
Hash—the value of the hash generated for the IPFS object.
Size—the IPFS object size is specified using this field [1].

1.3 Directory Structures

A directory in IPFS is an immutable globally mounted and is represented by IPFS object links which points to files or other directories.

1.4 Versioned File Systems

IPFS makes use of similar data structure that enables versioned file system used in git. The commit object in the versioned file system has one or more links with names which points to previous commits and the link which contains name object which references the file system structure which is initiated by that commit.

1.5 Distributed Hash Tables

Distributed hash tables (DHT) are considered to be a database which is distributed in a peer-to-peer network where with the help of keys associated with the values can be stored and retrieved. Without any central coordinating force, the nodes in the network balance and store the data. Distributed hash tables are fault-tolerant that enables a system to continue operating properly even if there is a failure and resilient when key/value pairs are replicated [7].

1.6 Blockchains

Nowadays, Bitcoin's technology is seeking more attention toward its second part and how the underlying technology of blockchain can be used for just more than money and transaction. There are other applications that have incorporated blockchain. Name coin is a service which represents decentralized name registration database which provides mechanism of identification of accounts and allowing users to create their own cryptocurrency. Another service called colored coins which is a more advanced application providing decentralized exchange, where a token can be traded for other tokens and provides blockchain identity. Thus, the current trends in blockchain involve building private side chain of a network or implementing a protocol on top of blockchain [2, 8].

2 Related Work

The potential solution that the blockchain companies are researching is in the field of decentralized storage. Decentralized storage networks make use of the cryptography and encryption algorithms which ensures the integrity of the data stored in the network. The projects that are trying to overcome the decentralized storage problems are listed below.

2.1 Swarm

Swarm is a decentralized protocol developed and maintained by Ethereum. It works alongside Whisper for Messaging, which is a decentralized messaging protocol. Swarm protocol provides storage for Ethereum's public record which is stored in a decentralized manner, which mainly consists of the distributed dapp code and data which also includes blockchain data [6].

2.2 BigchainDB

Improving the speed and efficiency of transactions is the main objective of the most blockchain solutions. But, BigchainDB does not attempt to improve the blockchain protocol and takes a different route which utilizes existing consensus engine called tendermint to handle the P2P communication and stores data in the NoSQL database MongoDB, through which powerful queries on the data can be made. Multiple ownership and native support on the digital assets are provided [3].

2.3 Storj

This is another decentralized storage platform powered by the Ethereum network. Storj is a platform as well as a cryptocurrency and consists of dapps or decentralized apps which are the applications run on a distributed system. The technology in Storj is similar to BitTorrent which uses bitswap protocol. When a user wants the file, the request is sent to Storj which uses DHT to locate all the links and clubs them together. Before sharing the files, the files are encrypted and the ownership is verified using the uploader's private key [5].

2.4 Siacoin

Sia is a project which was started in 2013 at HackMIT and launched in 2015. They aim to create an efficient storage solution around the globe by leveraging the free space of the hard capacity which enables a marketplace for the data, where free hard disk space can be rented through the other peers in the network. Siacoin uses a unique type of smart contract called 'file contract' where the hosts get paid for proving that they have stored a file. All data is encrypted before being uploaded to the network and only the user can decrypt them again [4].

3 Proposed Scheme

The main module of any software is the storage of data in the database. The control over the database is the major priority among many other factors and becomes the master if the control given to them. Blockchain technology involves blocks where the data is stored inside blockchain network. The copy of the data in the blocks will be received when a node joins the network. So, there is no particular master of the data in this technology.

(A) ***IPFS Primer***

Normal Uploading

To create an uploading feature, a developer needs to receive data from the browser and then store it somewhere. It could be a cloud service or other file hosting service which allows a developer to write server code to modify the image for example. There also could be multiple storage solutions. All these solutions increase the amount of bandwidth used by the application. A 1 mb upload becomes 2 mb because the server needs to upload it onto the storage solution (Fig. 1).

IPFS Uploading

For uploading to IPFS, the standalone browser can be used where data is pushed

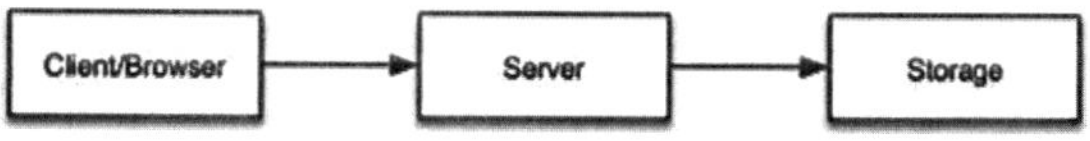

Fig. 1 Normal file uploading to server

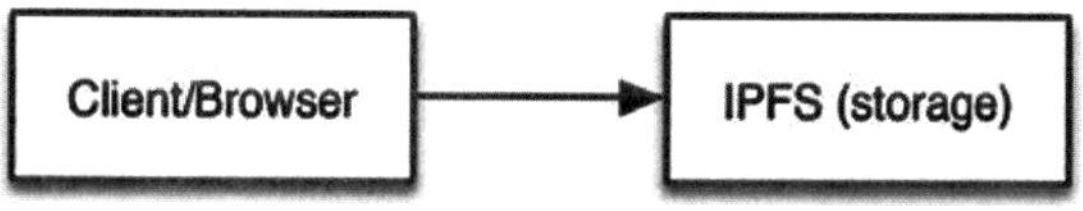

Fig. 2 IPFS file uploading

onto IPFS. Through this, the previous 2 MB upload in the client/server architecture becomes 1 mb. This ensures or saves the network costs. For the successful upload of the data to IPFS, the file reader and buffer class are used and the recent implementation of the IPFS API comes pre-packaged with these classes (Fig. 2).

Connecting to IPFS
To connect to IPFS either local or remote node, the IPFS package must be installed and the following commands enable the connection to the IPFS network,

ipfs init: The command will initialize the ipfs repository and node will be created along with key pair and a repository containing ipfs objects.

ipfs daemon: The IPFS daemon spins up local IPFS network through the browser. The daemon also provides a Web UI for the IPFS node created locally. The default URL of the Web UI is 127.0.0.1:5001/webui.

(B) ***IPFS Cluster***

Collective pinning and composition for IPFS. IPFS cluster is a piece of software that enables coordination between IPFS daemons on different hosts. IPFS cluster runs on raft consensus algorithm. IPFS cluster consists of two main applications namely:

ipfs-cluster-service: for starting your cluster peer.
ipfs-cluster-ctl: for interacting with the cluster peer through various inbuilt APIs.

There are two ways to connect peers in cluster, one setting a predefined peerset or bootstrapping nodes.

ipfs-cluster-service init: IPFS cluster peers are run with the ipfs-cluster-service command, more precisely service.json, which manages the cluster behavior, along with the peerstore file which stores the peers multi addresses. Those files can be found inside ~/.ipfs-cluster. This command will create cluster configuration $userprofile/.ipfs-cluster/service.json. This file will be used later to start a new cluster or join existing one.

ipfs-cluster-service daemon: This IPFS cluster command will start the cluster peer where raft will be initialized with the init_peerset. Peers will elect a Raft Leader and then become ready to receive data from other peers.

(C) *IPFS API*

In this paper, we are using the Python IPFS API which is a HTTP client library for IPFS.

To be able to allow cross-origin resource sharing (CORS) configuring, ipfs is necessary to return headers for CORS to work. This can be done through the following commands:

1. ipfs config --json API.HTTPHeaders.Access-Control-Allow-Methods '["PUT", "GET", "POST", "OPTIONS"]'
2. ipfs config --json API.HTTPHeaders.Access-Control-Allow-Origin '["*"]'

To be able to upload the files to the IPFS and successfully retrieve the data back, the file reader and the buffer at client side are used.

File reader allows you to read files in different formats such as readAsArrayBuffer, readAsBinaryString, readAsDataUrl, readAsText.

Buffer at the browser side is used for reading or manipulating streams of binary data. Buffer class instances are fixed-sized and have raw memory allocations which resembles to the arrays of integers. The buffer size is initialized at the time of creation and cannot be changed.

To create the application of the Python micro-Web framework, flask is used through which different routes are created for the events of uploading and viewing the files that are uploaded to the IPFS network. Python implementation of IPFS HTTP client library provides useful functions to manipulate the streams of data in the IPFS network.

The connection to the IPFS is made through the function connect(), which takes five arguments:

ipfsApi.**connect**(*host*, *port number*, *base* = *'api/v0'*, *chunk_size* = *4096*, ***defaults*)

The client or host is a TCP client for interacting with IPFS daemon. A client instance will not actually establish a connection to the daemon until at least one of its methods is called which is specified in the connect function. The connect function may throw the following exceptions if the connection to the IPFS client is not possible: Version Mismatch, Error Response, Connection Error, Protocol Error, Status Error, or Timeout Error.

To add a file or directory to the IPFS, the function add() is used which takes care of file reader and buffer.

add(*files*, *recursive* = *False*, *pattern* = '**', **args*, ***kwargs*)

Parameters
files (str)—A file path to either a file or directory.
recursive (bool)—Controls if files in subdirectories are added or not
pattern (str | list)—Single *glob* pattern or list of glob patterns and compiled regular expressions to match the names of the file paths to keep.

After adding a file, the hash value of the file is returned which can be accessed through the get() function which downloads a file, or directory of files from IPFS, and files are placed in the current working directory (Figs. 3 and 4).

The added files undergo the process of pinning where the nodes in the network pin the uploaded files or documents which ensures that the files do not vanish in the network. IPFS nodes treat the data stored like a cache, which means that there is no guarantee that the data will continue to be stored. Pinning tells IPFS network that the data is important and should not be garbage collected. Planning can be done using pin_add() which pins objects to local storage and returns a list of IPFS objects that have been pinned.

pin_add(*path*, **paths*, ***kwargs*)

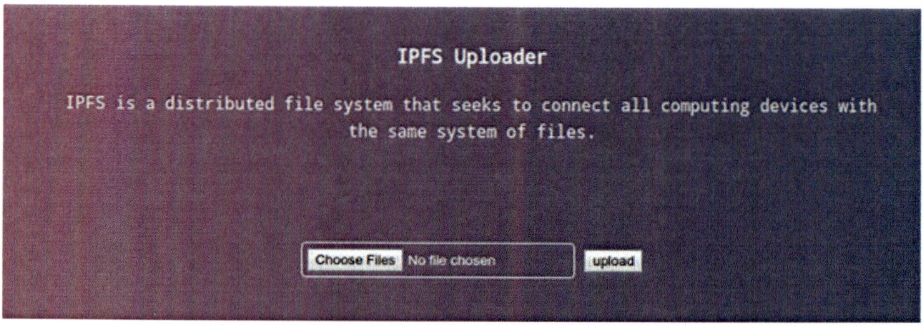

Fig. 3 Flask application to upload files to IPFS

Fig. 4 File uploaded to IPFS and its URL

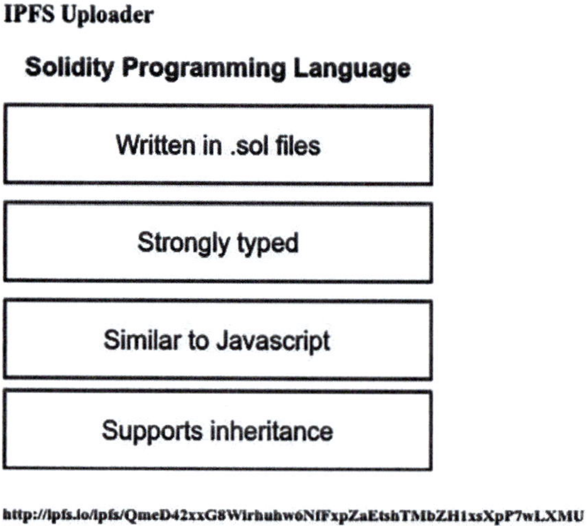

Parameters
path (str)—Path to object(s) to be pinned.
recursive (bool)—Recursively unpin the object linked to by the specified object(s)

Distributed private storage network can be set up using IPFS cluster which is a standalone application and a CLI used to allocate, replicate, and track pins across a cluster of IPFS daemons.

IPFS cluster is an orchestration tool or software to gain control of the IPFS daemons or the nodes present in the IPFS network. An IPFS cluster is created among number of nodes and the nodes share a pinset which contains CIDs which are cluster-pinned and their properties.

Cluster peers communicate using libp2p. To be able to exchange messages using libp2p, that each node contains a private key and has its own Peer ID. To ensure that the nodes in the network only exchange messages with known parties, an additional secret key is shared among all nodes in the clustered IPFS network.

4 Result Analysis

This section analyzes the model from the perspective of file size, network traffic overhead, and new node synchronization speed. The proposed model was tested using RRDtool which is an open-source, high-performance data logging and graphing for time series data, and a lightweight command-line JSON processor called 'jq'.

To evaluate the network parameters, a sample of 400 MB data is taken and two temporary nodes are connected to the IPFS network. The IPFS node on the host sends a specified amount of random data for the other node to get, and the IPFS daemon running on the other node downloads the data specified by hash reference. It will also collect benchmark data about the download, and finally, the result in the form of a graph is obtained as shown in Fig. 5.

The figure shows the download and upload rate of the nodes hosting IPFS. The file growth rate and the duplicated blocks while downloading are specified. The wantlist managed by the want manager in bitswap protocol is shown, where the wantlist objects are received by the node from the sender.

5 Conclusion

In this paper, as proposed we have used a decentralized storage, IPFS for reliable and user-friendly interactions with the IPFS network. Since the data is distributed among the IPFS nodes, it assures more security to the data as it is secured by the implementation of the hashing algorithm used in cryptography, and, in turn, the data is stored in nodes of different computers in a network. The focus is mainly on components that strengthen the functionality of Web-based distributed file storage and retrieval of stored files. The IPFS clustering service ensures a swarm of nodes which

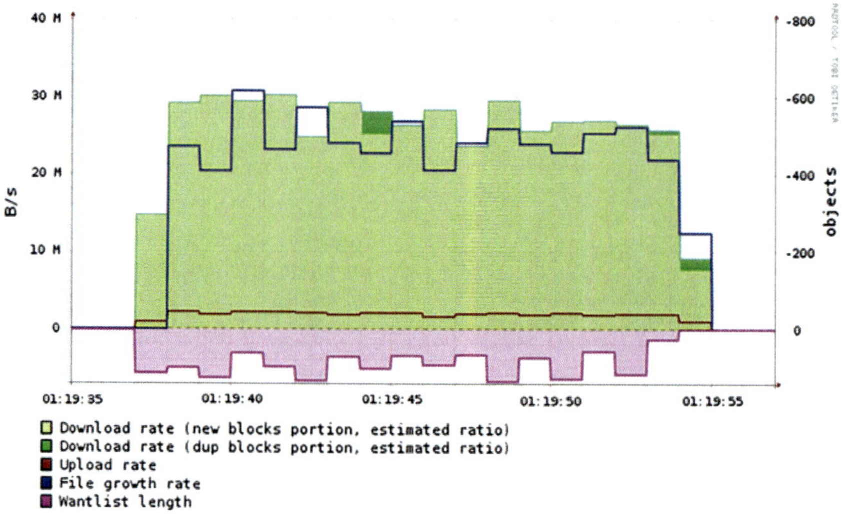

Fig. 5 Results

can be orchestrated for the collective pinning mechanism which ensures the validity of the files that are uploaded to the network. The deployment of specialized agents, such as search agents for specific applications, or the proposal of improved network topologies and protocols are some of the performance improvement opportunities to explore.

References

1. Nakamoto S (2009, May 24) Bitcoin: a peer-to-peer electronic cash system
2. Tenorio-Fornés A, Hassan S, Pavón J (2018) Open peer-to-peer systems over blockchain and IPFS: an agent oriented framework. CryBlock'18, Munich, Germany
3. BigchainDB 2.0 the Blockchain Database. Whitepaper, BigchainDB GmbH, Berlin, Germany, May 2018
4. Sambra A, Guy A, Capadisli S (2016, April 11–15) Building decentralized applications for the social web. WWW 2016 Companion, Montréal, Québec, Canada. ACM 978-1-4503-4144-8/16/04. http://dx.doi.org/10.1145/2872518.2891060
5. Storj Labs, Inc. (2018, October 30) Storj: a decentralized cloud storage network framework. In: IEEE international conference on fog and computing edge computing (ICFEC)
6. Pieper P, Lehes T (2018, April) Swarm: the blockchain for private equity. White Paper
7. Confais B, Lebre A, Parrein B (2017) An object store service for a fog/edge computing infrastructure based on IPFS and scale-out NAS. In: 2017 IEEE/ACM 1st international conference on fog and computing edge computing (ICFEC)
8. Wang S, Zhang Y, Zhang Y (2018) A blockchain based framework for data sharing with fine-grained access control in decentralized storage system. IEEE Access, 29 June 2018
9. Zheng Q, Li Y, Chen P (2018) An innovative IPFS-based storage model for blockchain. In: 2018 IEEE/WIC/ACM international conference on web intelligence (WI)

10. Alam S, Kelly M, Nelson ML (2016) Interplanetary Wayback: the permanent web archive. In: Joint conference on digital libraries (JCDL) 2016
11. Chen Y, Li H, Li K, Zhang J (2017) An improved P2P file system scheme based on IPFS and Blockchain. In: 2017 IEEE international conference on Big Data (BIGDATA)
12. Ali MS, Dolui K, Antonelli F (2017) IoT data privacy via blockchains and IPFS. IoT 2017: the seventh international conference on the Internet of Things

Knowledge Base Representation of Emails Using Ontology for Spam Filtering

V. Bindu and Ciza Thomas

Abstract With the development of the Internet, emails became the swiftest, cheapest as well as the simplest form of communication. The viability of it has been tempered by the noxious spreading of undesirable, unsolicited bulk mails called spam. Currently, to endeavor the issue of extending volumes of spam, researchers have contributed much work on spam filtering. One of the greatest challenges faced by the researchers is the novel techniques adopted by the spammers to bypass spam from the developed filters. Even the users' interests also get changed consequently resulting in the need for an adaptive filter for classification. If the emails are semantically annotated, an effective search and reasoning process can be provided. As ontologies provide semantics of data understandable for machines, better decisions can be made, thereby providing better solutions. In this paper, we illustrate an innovative method to extract knowledge from emails and represent them in ontological pattern consisting of hierarchical structures. This knowledge-based framework that represents the information artifacts of the email domain familiar to the users aids the classifier to point out the actual spam or ham.

Keywords Ontology · Knowledge base · Protégé

1 Introduction

Email is one of the basic services of the Internet used by large number of users. Prominently, as today's business transactions as well as individual activities rely on email service benefits, the users are confronted with extensive amount of emails. The manual categorization of the emails is a tedious task and hence to classify the unwanted mails from the inboxes also becomes intricate. The effective approach to

V. Bindu (✉)
Sree Chitra Thirunal College of Engineering, Thiruvananthapuram, Kerala, India
e-mail: bindunn@gmail.com

C. Thomas
College of Engineering, Thiruvananthapuram, Kerala, India

N. N. Chiplunkar and T. Fukao (eds.), *Advances in Artificial Intelligence and Data Engineering*, Advances in Intelligent Systems and Computing 1133,
https://doi.org/10.1007/978-981-15-3514-7_55

deal with this issue is to build a complete domain knowledge and logical reasoning system that hurdles the knowledge crevices between the assorted mails.

Ontologies have been acknowledged as the indispensable innovation for molding and utilizing data for the effectual management of knowledge. With the belief that building email ontology will provide domain knowledge for classification, we illustrate in this paper the detailed steps of the ontology creation methodology using domain of information artifacts from the emails. Our ontology creation aims at sharing information knowledge among users and software, analyzing domain knowledge making it more explicit for decision and building rules for feature identification for classifiers.

The main challenge in this work is to extract the information from a specific domain as email consists of multi-domain concepts. Initially, we develop the structure of email-based ontology that provides the header as well as the body part knowledge. Along with those facts, the domain name-specific information is included to extract the useful information. All these concepts along with their relations identified are represented using formal languages which provide the required semantic knowledge level structured description of the domain. These descriptions lay the foundation for building rules and calculate the associated probabilities of features required for the classifier.

The rest of this paper is structured as follows: Sect. 2 outlines prior works with the background for building ontology. Section 3 explains the art of building the ontology for representing the knowledge base for emails. Section 4 explains how to formulate the rules for ontology extraction. Section 5 describes the design environment. Section 6 portrays the inference system developed for knowledge base representation of emails. Section 7 consummates bestowing the future.

2 Background and Related Works

Ontology plays a key role in the field of any information retrieval systems. Being a field of artificial intelligence, it basically solves all the information heterogeneity problems. Ontology characterizes a common vocabulary for researchers who need to share information in an area. It incorporates machine-interpretable meanings of essential concepts in the domain and relations among them [1].

Noy and McGuinness of Stanford University [1] clarify the requirement for creating ontologies as to share regular comprehension of the structure of information among individuals or programming agents, empower reuse of domain information, make explicit domain presumptions, isolate domain knowledge from the operational knowledge and investigate domain knowledge. Creating ontology is akin to characterizing a set of information and their structure for different programs to utilize.

In the field of email classification and spam filtering using ontology, only a few research works are done. Major contributions to this area belong to Seongwook Youn and Dennis McLeod. Their research work gave immense contributions to the development of ontology for classifying emails based on the contents in the emails.

In [2], Seongwook Youn has developed an experimental system with two levels of ontology, one as global level and other as user-customized ontology. The papers [3, 5] explain an adaptive ontological method that provides efficient spam filtering. They also developed a personalized ontology spam filter to make decisions for gray emails [4]. In paper [6], the authors have presented the survey of various tools developed for instigating metadata of ontologies which can further help in developing new tools with more capabilities.

Beseiso et al. introduced [7] an exploration work that looks at the ontology extraction technique from the email systems embracing versatile pattern rules based on the extracted methods. The proposed architecture is intended to deal with the unstructured emails and the ontologies that are extricated from the email.

Dave Salmen, Bill Mandrick of Data Tactics Corporation along with the State University of New York at Buffalo have given tutorials in developing email ontology explaining the scope, domain, process of development, levels of ontology as well as the classes of email ontology.

Taghva et al. [8] built up a mail ontology to partake formally domain data with their Bayesian email classifier.

Kim et al. built up a user preference ontology framework that formally speaks to the imperative ideas and principles derived from a data mining process. They additionally gave an inference engine that uses the information to predict the user's activity on new incoming mails [9].

3 Building the Mail Profile Ontology

For realizing the knowledge base representation of emails, it is important to construct the ontologies based on the concepts obtained from different domains of the emails. The greatest challenge in designing and building the ontology for emails is that it is a multi-domain conceptual problem. The ontology developed ought to obviously portray the structure of the email, for example, the sender, the recipients of the messages, the subject and the substance body at the underlying dimension, and after that incorporate domain-specific information, so the helpful data could be separated. This design provides the advantage of clear separation of domain spaces in the ontological structure making the information extraction step and the learning process easier. Moreover, it improves the human communication process involved in conceptual modeling activities required in the mail classification.

3.1 Methodologies for Building Ontology

The methodologies that we followed have five phases that are data collection, identification of concepts and relationships within the data, designing the ontology based on the information gathered from data collection, identification of good research

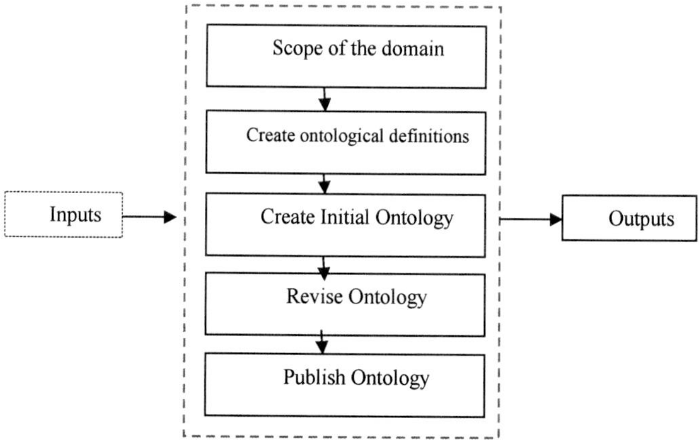

Fig. 1 Process for ontology development

tools for building ontology and evaluating the ontology. For data collection, we created a survey to a group of people to acquire the knowledge about the concepts and relationships that the user will choose for the classification. The subtleties of information data are outside the extent of this paper but have its importance in the final stage of evaluation process.

The most important process is the building and evaluation of email ontology. Building ontologies involves a repeatable process involving mainly five activities as shown in Fig. 1. In every process phase, identifying the inputs and the outputs is an important criterion. The input to the system can be subject matters of email, users' preferences, authoritative sources/definitions as well as the relevant databases. Each activities' output will serve as the input of the next activity, and the final output of the system can be Web Ontology Language (OWL) file or domain lexicon or knowledge information which can be stored as descriptions or briefings.

3.2 *Process for Ontology Development*

In our research work, we aim at representing the knowledge base of a user mailbox by creating email domain ontology. This in general involves defining the classes with subclasses (if any), properties which includes domain and range, property types for the defined objects, data types and annotations as well as the subproperties, defined instances and levels of ontology with reuse.

Scope of the domain. Defining the span of the purpose of the ontology is the initial step which is stated in the scope of the domain. This phase decides where to start, whether a modular ontology is required, what are the levels to be created as well as where to stop creating ontology. For that, we defined the objective as the validation of specific email characteristics which aids in the decision of classification of emails.

The conceptual criteria like user preferences, databases and other sources serve as the inputs while outputs will be the domain definitions as well as the list of various terms extracted from the data capable of giving knowledge. As the initial step to start, we have to design what type of ontology level has to be created. There are upper-level ontology, middle-level ontology and lower-level ontology. Upper-level ontology (ULO) includes the basic formal ontology (BFO) and relationship ontology (RO), while middle-level ontology (MLO) and lower-level ontology (LLO) include information artifact ontology (IAO), email domain ontology (EDO), contact ontology (CO), computer network ontology (CNO) and software ontology (SWO). Moreover, the approaches in the ontology can be top-down or bottom-up approach. Here, we provided combinations of both approaches as certain concepts are explicit and while other are vaguer. The email ontology general architecture that we designed is displayed in Fig. 2. The architecture consists of various dimension ontologies with structure of email, for example, the sender, the recipients of the messages, the subject and the substance body at the top dimension pursued with lower level consisting of multiple domains corresponding to each entity that is defined in the next level.

Create ontological definitions. In the second phase, based on the known taxonomies, indexes, subject matter feedbacks, etc., ontological definitions will be created as well as relations are identified. In our system, we defined email agent, email characteristics and email domain as the three prominent taxonomic hierarchies for representing the email ontology. Along with that we added the structural entities which include the domain names in the email address, role of messages entities as well as actions

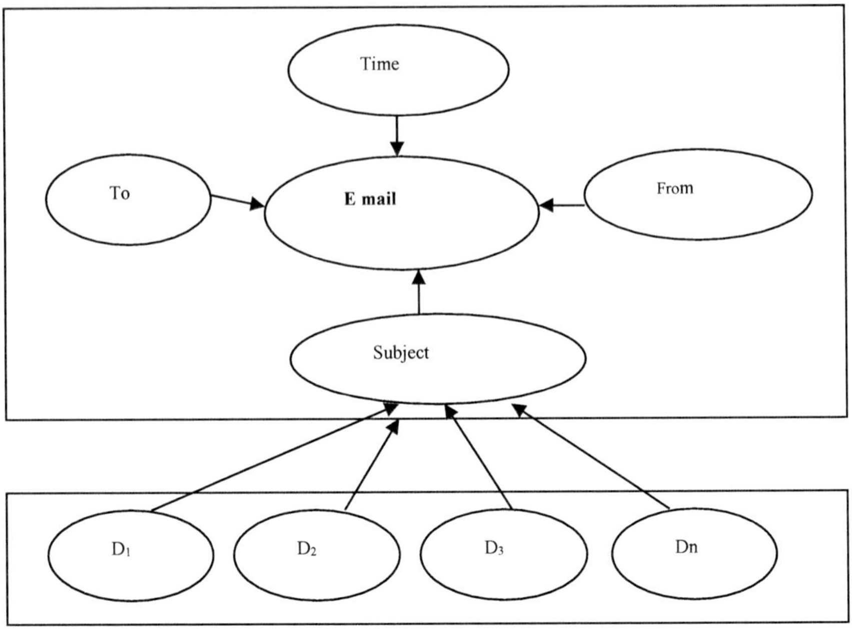

Fig. 2 General email ontology

entities. Based on the entities defined, we found the relations and events. This step is the most important step that provides the base of the knowledge extraction.
Create initial ontology. Once the entities, events and relations between them are found, then by using suitable tools or design environments, initial ontology can be created which can relate entities and events. The output obtained at this stage is the graphical representations or languages used for ontology like OWL files or briefing of subject matter that explains the knowledge base.
Revise and publish ontology. The ontology created can be revised based on suitable evaluation criteria, and the second phase can be modified further. This evaluation criterion can be defined in the first phase by providing suitable metrics. As a result, the ontology is modified, and if the metrics defined for criteria are satisfied, the ontology can be published.

4 Building Rule-Based Ontology

Ontology is a knowledge base model which inevitably involves evaluation of mathematical expressions of concepts, relationships and rules of user's interest in the area. Developing rule-based ontology for ontology extraction is ideal for extracting the knowledge which involves the following steps.

- Identify the problem space and create a mind map how to represent the concepts hierarchically.
- Discover the conceptual relationships involved in building the ontology.
- Identify a tool and configure it with suitable reasoner and annotation template.
- Using the tool, convert the concepts identified into relevant classes and subclasses.
- Properties for the concepts have to be discovered and created. Three types of properties are defined in OWL called object properties, data properties as well as annotation properties. Two individual relationships are provided by the object properties while data type properties present the relationship between one individual and data values. Annotation is data about data, i.e., metadata.
- As Semantic Web is all about giving significance to concepts, the most important step is to give importance to meaning concepts. This stage is the most important stage that provides the decision-making process.
- To describe the data about data, annotation to concepts is added.
- Using any rule-based languages available in the tool, rules can be created.
- Individual instances are made, and data type properties are allotted for every information in the indexed data.
- The reasoners provided in the tools, support on the rules is a major factor in classification. Some reasoners support rules partially while others support fully. Using suitable reasoner, classification can be invoked. If there is any variance in the ontology, then reasoner gives the reason for the inconsistency. If there is no oddity at that point, the ontology is reliable and the classification happens depending on the principles and importance that gave in the equivalent section of the class.

- Using any query language provided in the tool, queries can be issued in the ontology for fetching significant data which gives the final decision results.

5 Design Environment

In this section, we depict the tool, reasoners and the query language that is utilized as a part of the ontology creation for knowledge representation. As the tool for ontology editing, we chose Protégé an ontology and knowledge base editor produced by Stanford University. Protégé is a tool that enables the building of domain ontologies, customized data entry forms to enter data, and permits the meaning of classes, class hierarchies, variables, variable-value restrictions and the relationships between classes and the properties of these relationships.

Protégé is free and open software that can be downloaded from [10]. This tool has the additional advantage that it underpins the Web Ontology Language (OWL) [11] and enables users to: load and save OWL and Resource Description Framework (RDF) ontologies, edit and visualize classes, properties and Semantic Web Rule Language (SWRL) rules, define logical class characteristics as OWL expressions, execute reasoners such as description logic classifiers and edit OWL individuals for Semantic Web markup.

Along with the tool to determine whether the ontology developed is consistent or not, the reasoner written using the Web Ontology Language (OWL) called HermiT is available. For optimizing nominal, conjunctive query answering and incremental reasoning, another reasoner called pellet an open-source Java-based OWL 2 reasoner is also provided. These two reasoners available as plug-ins in Protégé will provide the effective reasoning in our system.

The Semantic Web Rule Language (SWRL) tab is a Protégé plug-in that provides a development environment for working with SWRL rules and SQWRL (query language for OWL) queries. For rule-based reasoning in the last step of ontology creation, we use SWRL tabs in Protégé which allows an efficient interaction between the rule engine implementation and the user. It provides a bridge between an OWL model with SWRL rules and a rule engine. This will import SWRL rules and relevant OWL classes, individuals, properties and restrictions from an OWL model; write that knowledge to a rule engine; allow the rule engine to perform inference and to assert its new knowledge insert that asserted knowledge into an OWL model.

To aid further classification of categories, SPARQL (SPARQL Protocol and RDF Query Language) query interface is likewise accessible in Protégé. By utilizing these queries, users will have the capability to retrieve and manipulate data stored in Resource Description Framework (RDF) format.

6 Research: Developing and Evaluating Inference System

In view of the steps expressed in Sect. 4, we developed an inference system that represents the knowledge base of emails using Protégé. Inference system basically relies on describing the knowledge of emails using ontology and evaluation by rule-based reasoning. The rules give good pluralistic relations with deductive learning, while ontology concentrates on the portrayal of the unary concept. Consequently, the steps involved in building and evaluating an inference system basically involve the design of ontology representation and rule-building systems.

6.1 Ontology Building

We started our research with accumulating information from emails, providing scaffold for recording the gathered information remembering the ultimate objective to outline the ontology. Based on these steps using a mind mapping tool, we created a structure of the concepts and the relationships obtained from the information database.

These concepts can be standardized following the protocols as well as the observed concepts from the emails. The major concepts that we use here are domain names, email agents, message role, email addressed to, subject of email, action carried out and email characteristics. These concepts can be further extended depending on the users' choice. Mind map of basic concepts of email domain is shown in Fig. 3.

In implementation phase, these concepts were converted into classes and subclasses under the 'Thing' as superclass in Protégé software. We created a hierarchy for this, and this implementation is shown in Fig. 4.

Relationships corresponding to these concepts based on the observation were recognized. The next step corresponds to assigning of properties to the objects, data as well as annotations. Here, the objects correspond to each of the individual classes, and the properties of object correspond to the relationship between them. In our design, we treated each of these as explicit objects and assign no relation between them.

Data type properties provide the relationship between instances of classes, RDF literals and XML Schema data types. OWL uses most of the built-in XML Schema data types like string, float, decimal, etc.

Corresponding to the classes, we have identified relationship and their properties are assigned. The important data properties along with the description and value we assigned are given in Table 1. As Semantic Web is all about giving meaning to concepts, this is the most important criteria in building ontology. This provides the foundation for classification as meaning clearly specifies how each of the classes forms the concept for classification. In order to provide an email classification using any class hierarchies present in ontology, we can give literal meanings to it by defining the data properties with value. Even combinations of data properties can be provided

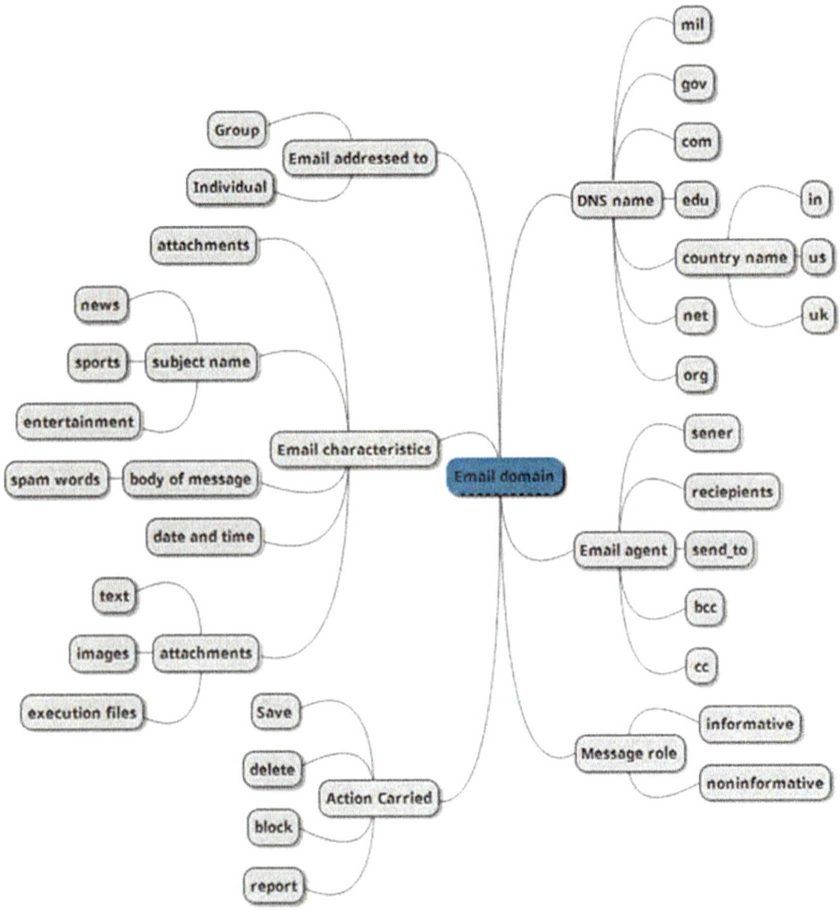

Fig. 3 Mind map showing basic concepts of email domain

which give better classifications. Association rules between various classes and their responses were identified. Using SWRL, the reasoning rules are expressed.

6.2 *Inference System Implementation*

After the ontology development and rule construction, we proceeded to content evaluation. The goal was to validate whether the ontology was illustrative of knowledge gained in the delimited scope. For this, we have to provide the data for classification. These data are added as individuals in the sample class. Using the property assertion view, the values for the data type properties are provided to these individuals. As in our system, this corresponds to different email information. We have added manually

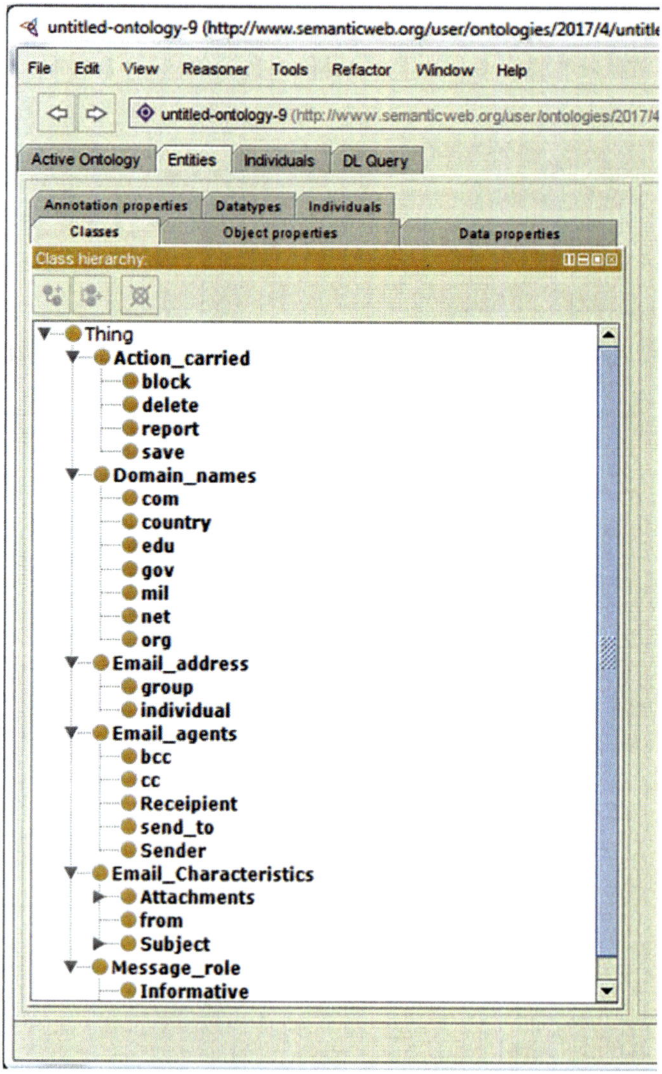

Fig. 4 Protege implementation of email domain ontology

email data along with its asserting properties to provide the data to inference system. The inference system now works on ontology-based inference and rule-based reasoning.

As for the ontological inference, we require a reasoner to test the consistency in ontology, analyze the subsumption relation, describe the concept as well as retrieve the knowledge base. In our system, we have used pellet as the reasoner. As there were no inconsistencies in the modeling and semantic description tool, the ontology was consistent and clear with no formal and semantic contradictions. The reasoner clearly

Table 1 Some data properties

Sl. No.	Data properties	Data type and description
1	Has the domain name	String, obtained from email address of the sender
2	Is addressed to	Number, obtained from the header field of email
3	Has sender name	String, obtained from email address of the sender
4	Has an attachment	Boolean, obtained from the header
5	Has done an action	Double, user input
6	Has the subject name	String, obtained from the header field of email
7	Contains the data	String, obtained from the body of email

classified the concepts based on the rules and meaning provided in the equivalent section of the class.

Our system is capable enough of retrieving concept names and individual names in the knowledge base, the retrieval of individual pairs related to their roles and the retrieval of role level. As the reasoner approves it to be consistent ontology, this can be published. The corresponding OWL file as well as the ontology graph equivalent to that of the mind map is shown in Figs. 5 and 6, respectively.

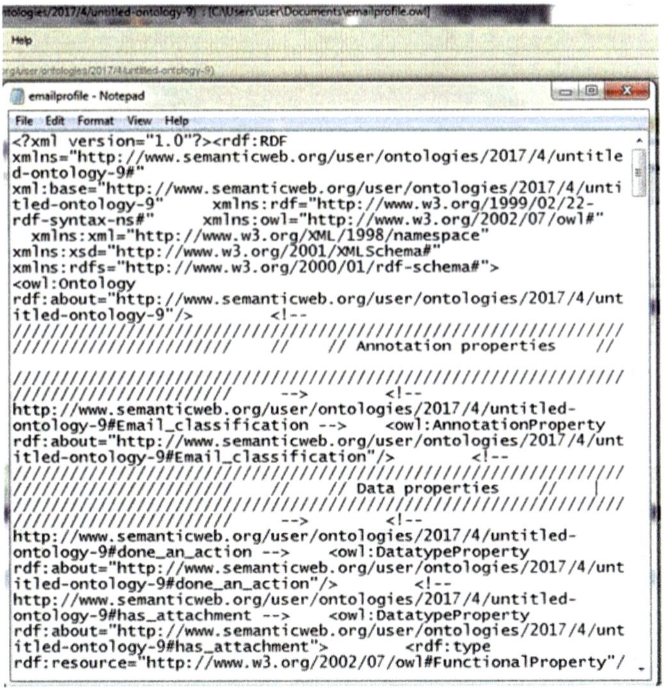


```
<?xml version="1.0"?><rdf:RDF
xmlns="http://www.semanticweb.org/user/ontologies/2017/4/untitle
d-ontology-9#"
xml:base="http://www.semanticweb.org/user/ontologies/2017/4/unti
tled-ontology-9"      xmlns:rdf="http://www.w3.org/1999/02/22-
rdf-syntax-ns#"      xmlns:owl="http://www.w3.org/2002/07/owl#"
  xmlns:xml="http://www.w3.org/XML/1998/namespace"
xmlns:xsd="http://www.w3.org/2001/XMLSchema#"
xmlns:rdfs="http://www.w3.org/2000/01/rdf-schema#">
<owl:Ontology
rdf:about="http://www.semanticweb.org/user/ontologies/2017/4/unt
itled-ontology-9"/>          <!--
////////////////////////////////////////////////////////////////
///////////////////////     //     // Annotation properties     //

////////////////////////////////////////////////////////////////
///////////////////////          -->          <!--
http://www.semanticweb.org/user/ontologies/2017/4/untitled-
ontology-9#Email_classification -->     <owl:AnnotationProperty
rdf:about="http://www.semanticweb.org/user/ontologies/2017/4/unt
itled-ontology-9#Email_classification"/>          <!--
////////////////////////////////////////////////////////////////
///////////////////////     //     // Data properties     //
////////////////////////////////////////////////////////////////
///////////////////////          -->          <!--
http://www.semanticweb.org/user/ontologies/2017/4/untitled-
ontology-9#done_an_action -->     <owl:DatatypeProperty
rdf:about="http://www.semanticweb.org/user/ontologies/2017/4/unt
itled-ontology-9#done_an_action"/>          <!--
http://www.semanticweb.org/user/ontologies/2017/4/untitled-
ontology-9#has_attachment -->     <owl:DatatypeProperty
rdf:about="http://www.semanticweb.org/user/ontologies/2017/4/unt
itled-ontology-9#has_attachment">          <rdf:type
rdf:resource="http://www.w3.org/2002/07/owl#FunctionalProperty"/
```


Fig. 5 Ontology represented as OWL file

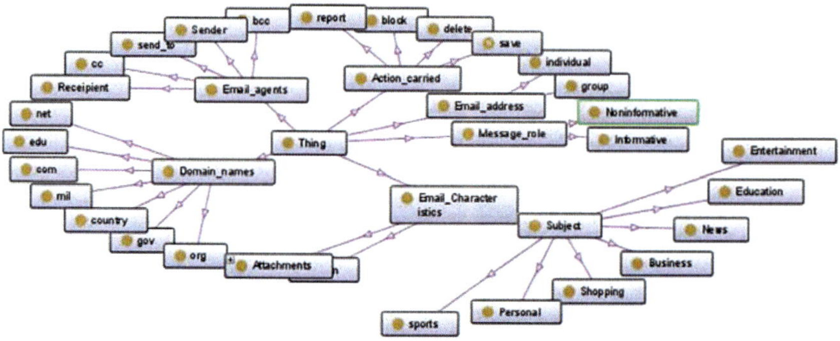

Fig. 6 Ontograph

7 Conclusions

The necessity of solutions to filter spam is increasing day by day. The widespread proliferation of spam and lack of knowledge information from emails are the key reasons in making emails difficult to tackle. Through this paper, we delineate a novel strategy in the form of ontology that represents the knowledge about emails using a domain of information artifacts familiar to the users. This knowledge acquisition framework created in Protégé helps the researchers not only to represent knowledge in the form of ontology but additionally to surmise and judge adaptively the correct choice on the new emerging mails. Consequently toward the direction of spam detection, our method of rule-based ontology is a small step forward.

For the future work, we attempt to extend the proposed framework to prepare a genuine informational collection of real email data sets utilizing this ontology and create an efficient automatic mail classification system better than the existing classifiers.

References

1. Noy NF, McGuinness DL (2001) Ontology development 101: a guide to creating your first ontology
2. Youn S (2014) SPONGY (SPam ONtoloGY): email classification using two-level dynamic ontology. Sci World J 2014
3. Youn S, McLeod D (2007) Efficient spam email filtering using adaptive ontology. In: Fourth international conference on information technology, 2007. ITNG'07. IEEE
4. Youn S, McLeod D (2009) Spam decisions on gray e-mail using personalized ontologies. In: Proceedings of the 2009 ACM symposium on applied computing. ACM
5. Youn S, McLeod D (2007) Spam email classification using an adaptive ontology. JSW 2(3):43–55
6. Youn S, McLeod D (2006) Ontology development tools for ontology-based knowledge management. Encyclopedia of E-commerce, E-government, and mobile commerce. IGI Global, Hershey, pp 858–864

7. Beseiso M, Ahmad AR, Ismail R (2012) A new architecture for email knowledge extraction. Int J Web Semant Technol 3(3):1
8. Taghva K et al (2003) Ontology-based classification of email. In: International conference on information technology: coding and computing (computers and communications), 2003. Proceedings. ITCC 2003. IEEE
9. Kim J et al (2007) Constructing a user preference ontology for anti-spam mail systems. Advances in artificial intelligence. Springer, Berlin, Heidelberg, pp 272–283
10. http://protege.stanford.edu
11. Knublauch H et al (2004) The Protégé OWL plugin: an open development environment for semantic web applications. In: International semantic web conference. Springer, Berlin, Heidelberg

Clinical Significance of Measles and Its Prediction Using Data Mining Techniques: A Systematic Review

Abhishek S. Rao, Demian Antony D'Mello, R. Anand, and Sneha Nayak

Abstract Data science techniques used in the past era for data extraction are now being replaced by data mining methods due to a lot of contemporary trends and challenges. Data mining includes implicit data extraction, which is previously unknown but potentially useful information, with the help of databases to generate new information. Medical data mining is an area of challenges since the data involves in it is imprecise, inconsistent and massive. Medical diagnostics systems are evaluated by employing large information databases, but it endures many failures to extract data from the databases. There exists no tool to discover the major relationships concerning the data. In such a case, the core knowledge of healthcare data is extracted by applying the data mining methods, thus helping in turning raw data into useful information. Measles is an immunizable disease most likely seen in infants and young children. Measles (rubeola), is an RNA virus, belongs to the *Morbillivirus* genus, and *Paramyxoviridae* family. It is one among the six major killer diseases, with accounts to 90% secondary infection rate with susceptible contacts. In this paper, the authors present a systematic review of measles cases and its mortality with the help of the taxonomical tree structure. The published research which utilizes the application of data mining techniques for early prediction of measles in terms of methods used, algorithms utilized and results obtained are evaluated. The paper provides major

A. S. Rao (✉)
Department of Information Science and Engineering, NMAM Institute of Technology, Nitte, Udupi, Karnataka 574110, India
e-mail: abhishekrao@nitte.edu.in

D. A. D'Mello
Department of Computer Science and Engineering, Canara Engineering College, Bantwal, India
e-mail: demian.antony@gmail.com

R. Anand
Department of Master of Computer Application, NMAM Institute of Technology, Nitte, Udupi, Karnataka 574110, India
e-mail: sacanand@nitte.edu.in

S. Nayak
Department of Biotechnology Engineering, NMAM Institute of Technology, Nitte, Udupi, Karnataka 574110, India
e-mail: snehanayak@nitte.edu.in

N. N. Chiplunkar and T. Fukao (eds.), *Advances in Artificial Intelligence and Data Engineering*, Advances in Intelligent Systems and Computing 1133,
https://doi.org/10.1007/978-981-15-3514-7_56

consolidation of data with respect to motive, study and methods used in the literature. The authors present the summary of the research review findings and research gaps for further study and possible application.

Keywords Data mining · Dimensionality reduction · Classification · Vaccination · Measles · Outbreak

1 Introduction

Data science is an interdisciplinary field which exploits scientific methods, processes, algorithms and systems to extract information from both structured and unstructured data. Architecture for data science is as shown below (see Fig. 1).

The major trait of data science is to understand real-time business problems. To deal with such problems, one should really need to understand the objectives of the problem identified. This gears up to the data acquisition, wherein the data has to be collected from multiple sources like web servers, logs, databases, APIs and online repositories. The core activity of data science is data modelling where machine learning techniques like KNN, decision tree, and naïve Bayes which are practically applied to the data to identify the model that fits the problem requirement to the best. Models are then trained on the training dataset and later tested to select the best performing model. The model will be deployed and maintained by testing the selected model in a pre-productive environment before deploying it in a productive environment.

Data mining (DM) refers to the implicit data extraction, which is formerly not known but potentially useful information with the help of databases to generate new information [1]. In today's era, DM has widely spread in the field of health care as it is one of the most information-intensive industries [2]. It is a known fact that knowledge, medical information and data usually keep growing on a daily basis. According to a survey, it has been estimated that a severe care hospital can generate

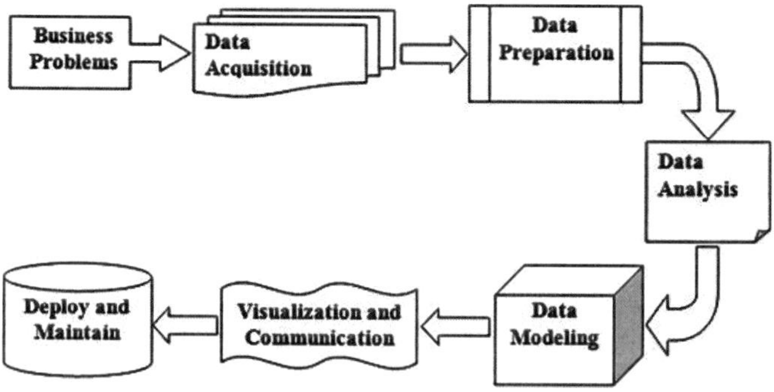

Fig. 1 An outline of data science architecture

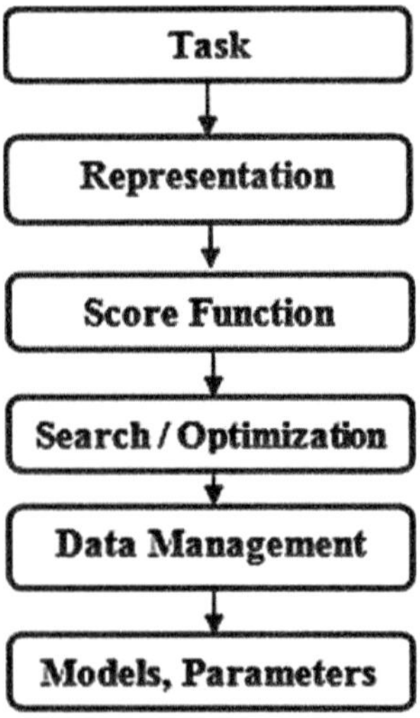

Fig. 2 Sequential steps followed for data mining algorithm

approximately 5 TB of data in a span of years [3]. The ability to use this data in order to extract quality information which is critical for health care is a crucial task. This process involves six steps namely business understanding, data understanding, data preparation, modelling, evaluation and deployment.

Steps to be followed in order to apply data mining technique for the healthcare sector are shown in Fig. 2.

The first step is the identification of the task and representing it by applying dimensionality reduction techniques in order to discover a good set of features. To measure the goodness of the model, a score function calculation should be evaluated and then the best model should be selected based on model accuracy. Data mining can also help in various healthcare-related applications like segmenting patients accurately into various groups with similar health patterns, prediction of medical diagnosis, identifying medical procedures, expenditures and utilization by analysing point of care and claim data, forecasting treatment cost, predicting duration of stay in a hospital, predicting total cost involved in patient care and many more.

Among the various data mining techniques, classification has been identified as an important problem. The ultimate goal of this technique is to precisely foresee the target class for each set of data. Classification basically involves a two-step process which includes training and testing steps. During the training step, the algorithm evaluates the data meant for learning and generally builds a classification model which would help in predicting the categorical label including unordered or discrete variables. The task of classification is to use a given set of classes for assigning a new object to a class based on the attribute values [4]. Classification techniques have been used for forecasting the treatment cost of healthcare services which has to turn out to be the main concern [2]. The importance of classification rules is to determine the class of attributes, but it does not consider the relationships of attributes [5]. The correctness of the data is checked by the model by using another dataset.

In the medical field, based on the signs and health conditions of the patient, the classification method can be used to define the diagnosis procedure and prognosis prediction [6]. In the healthcare domain, classification can be made useful using the IF-THEN prediction rule [7].

Medical data mining is an area of a challenge since the data involves in it is imprecise, inconsistent and the data is massive. Medical diagnostics systems are evaluated by employing large information databases, but it endures many failures to extract data from the database. There is no sufficient tool available to discover the major relationship concerning the data. In such a case, the core knowledge of healthcare data is extracted by applying the data mining methods thus helping in turning raw data into useful information [8]. In medical data mining, classification and prediction of data are not just a matter of accuracy but the matter of life and death. One wrong decision can have a disastrous effect on the life of patients and their families. Thus, medical data mining is considered as the decision-making framework which provides assistance for the experts to properly classify and predict the data in a quick time [9].

Although most immunizable diseases are preventable with immunization, they are not curable. Vaccination is considered to be harmless and effective. Though several obstacles to immunization compliance exist, comprising a lack of responsiveness regarding the significance of vaccines, fear of technical hitches from vaccinations and missing due dates. Vaccine-preventable diseases (VPD) for children can be successfully adhered to by complying with national immunization schedules. On the other hand, the World Health Organization (WHO) has anticipated that nearly 1.5 million children are dying every year from vaccine-preventable diseases [10], signifying that compliance to the suggested vaccinations schedule is a task for healthcare systems. The expert panel named as SAGE under the WHO has drafted few recommendations for South East Asian Regions (SEAR) stating that all children's should take two doses of measles vaccine. The chief parental obstacles to immunization include misperception and trouble in tracking vaccination schedules; missing due dates, lack of awareness about the importance of vaccines and fright of vaccinations complications and side effects. Evidence proposes that suitable vaccination coverage is linked to improved health status which in turn helps in minimizing cost. Assessing vaccine compliance helps us to estimate levels of protection all the way through the first 2 years which signifies the time during which children are most vulnerable to severe complications related to VPD [11].

Measles (rubeola), is an RNA virus, belongs to the *Morbillivirus* genus, and *Paramyxoviridae* family [12–14]. It is one among the six major killer diseases, with accounts to 90% secondary infection rate with susceptible contacts. Measles, which is an immunizable disease, is considered next in line for eradication after poliomyelitis as it is a major reason causing mortality in infants and young children of developing countries. Though measles is considered to be a childhood illness, it can affect individuals of all ages. This virus is a human pathogen and has no animal hosts for its proliferation. Transmission of this disease is mainly through direct contact or respiratory droplets. Most measles-related deaths are caused by severe obstacles including blindness, severe diarrhoea, dehydration, encephalitis, or severe respiratory contagions such as pneumonia, ear and infections. The average intermission from contact to the arrival of rash is usually 14 days (in the range of 7–18 days). Patients are generally considered to be spreadable from 4 days before the onset of rash till almost 4 days after the rash onset. Near the completion of the gestation period, patients cultivate the prodromal warning sign of high fever, cough, conjunctivitis and coryza known as the 3Cs. Even though there is no definite treatment for measles, these diseases can be very well prevented by vaccination with the available highly effective measles vaccine [15]. The infection is generally passed on by the airborne route, with a huge share of cases being self-limiting. Numerous mortality cases have been testified because of disease-associated problems [16–18].

Measles is commonly identified in infants and young children's in the population with low vaccine exposure and herd immunity as low as 10%. The measles vaccine is injectable and not easy to administer, unlike the polio vaccine. An army of a trained workforce is required for vaccine administration; therefore, one plausible solution may be the use of an aerosol technique which, with the improved delivery system can be used by field worker also. This will be very helpful in increasing the acceptability of the vaccine as it does not require injections. Maintaining exceptional immunization coverage with high-quality surveillance is highly essential. Meeting this task will be specifically critical for dealing with measles virus that will occur, as long as the virus is circulating globally [19, 20]. Another challenge is to sustain high population immunity along with exceptional immunization coverage through effective laboratory network for high-quality surveillance [21].

Severe complications are associated with measles-related mortality in children's below 5 years [12]. Diagnosis of measles can be achieved with a seropositive antibody reaction using a serological assay for measles-specific IgG or IgM titres as well as recognition of measles virus in clinical samples (e.g. blood, urine, throat swabs or nasopharyngeal secretions) by culturing [12–14]. Treatment of measles is basically by the supportive care which includes maintenance of good hydration and replacement of fluids lost through diarrhoea, intravenous rehydration may be necessary in case of severe dehydration along with which Vitamin A supplements could also be suggested.

In this paper, we have discussed various categories for measles prediction with respect to its outbreak, the role of travellers and vaccination globally. Applications of data mining and the use of various classification models for early prediction of measles along with performance metrics applied are also discussed. Summarized observation of this review along with various open issues for further research in this area is also presented.

Rest of the paper is organized as follows. Section 2 gives an overview of the literature highlighting the study, methods used and results obtained. Section 3 highlights on the data mining techniques like data preprocessing, Tabular study on feature selection consisting of datasets utilized, methods adopted and results obtained for various diseases, classification algorithms applied for early prediction and diagnosis of various diseases along with performance metrics for model validation. Section 4 highlights the summary of the review. Section 5 presents the research gaps which may help young researchers to choose the right area for further research in this direction. Section 6 concludes the paper.

2 Literature Review

This section will include a systematic review of measles cases and their mortality. It highlights various outbreak cases which occurred globally along with insight on various elimination strategies. Application of data mining for early prediction of various deadly diseases along with various classification techniques used is also highlighted which has its major implication in disease diagnosis. The overview of the literature review is shown in the form of the taxonomy tree structure as shown below (see Fig. 3).

2.1 Case Studies on Measles Outbreak

This section highlights various case studies of the measles outbreak which occurred universally. Approximately 227,245 cases of measles were reported with accounts to 18 deaths every hour as per the WHO 2012 global report [22]. Many developing countries like Syria, Afghanistan and Pakistan have encountered challenges in their health divisions due to an upsurge in outbursts with cases ranging from few thousands to lakhs resulting in momentous mortality cases [23]. Measles epidemiology over an extended period in many regions is necessary for its elimination. It also gave an insight into the fact that measles is an endemic disease which is usually transmitted locally which may lead to outbreaks in regions that have measles interruption [24]. A case study on the role of travel in the measles outbreaks even in countries like Australia where high-level immunity against measles is maintained due to non-immune travellers which may lead to a localized outbreak. Hence young children and various migrants must be given the highest priority for the prevention of measles and final elimination [25]. In California measles outbreak, the travel histories of various patients were crosschecked and it was seen that the only notable place of visit was Disney theme park located in California which attracted a lot of international travellers where measles is an endemic disease [26]. Finally, a systemic review of dengue fever and measles outburst prediction specifically on the data types, factors and sources used for forecasting was highlighted [27].

2.2 Measles Confirmed Cases and Mortality

This section gives information on various confirmed cases of measles along with a few statistical data. Worldwide, measles infects nearly 30 million children annually, and

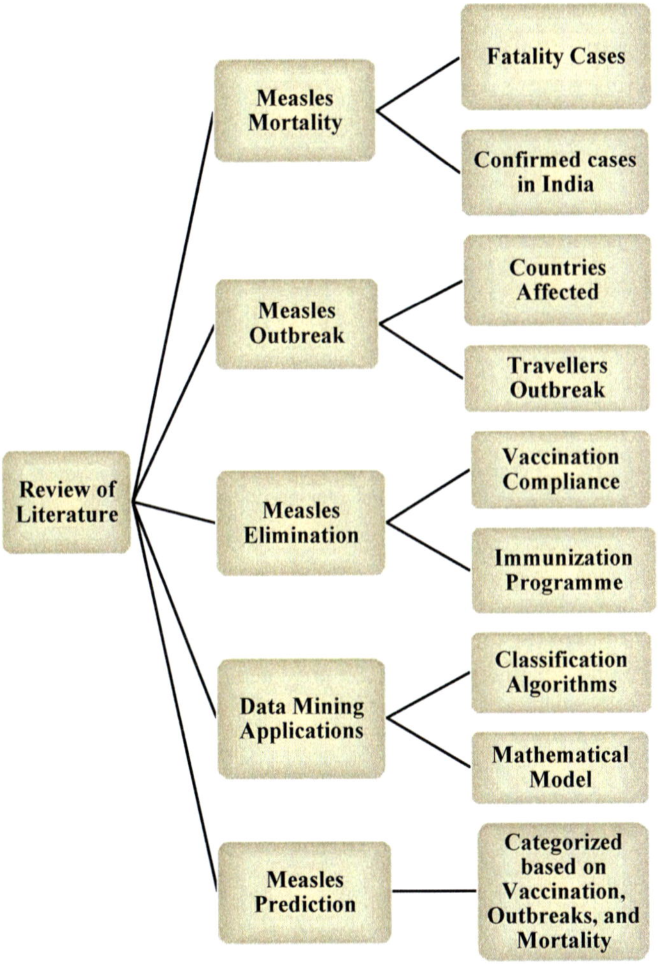

Fig. 3 Taxonomy layout of the literature review

deaths usually occur from problems related to diarrhoea, pneumonia and malnutrition. It is evident from Fig. 4 that a major increase in measles cases was seen in March and April when compared to other months in all the above said 5 years from 2014 to 2018, as the season being summer is favourable for the survival and multiplication of this deadly virus. This data was collected from the World Health Organization (WHO) records [28]. Among two-thirds of the global burden of measles demises in 2007, nearly 136,000 (range: 98,000–180,000), occurred in the South East Asian Region, with most of them occurring in India alone as per WHO. India accounted for 47% of the projected global measles death cases in 2010. The global estimates for the year 2013 suggest that nearly 0.14 million deaths were attributed to measles, which accounted for approximately 16 deaths every hour [28]. Study findings have indicated

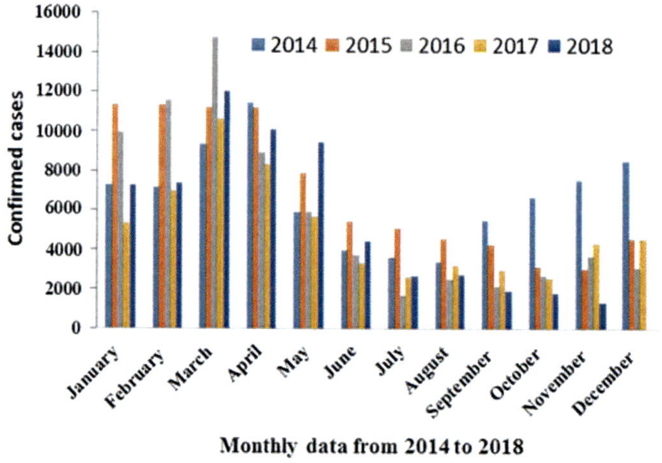

Fig. 4 Reported monthly measles cases in India for a period of 5 years

that India alone accounts to more than 50% of the global measles-associated deaths [29, 30]. In spite of the presence of these effective strategies for measles elimination, the Indian District Level Health Survey stated that only 30% of infants received the vaccine at 9 months as suggested [31]. Figure 5 depicts the measles fatality cases in India for a period of 16 years. This data was collected from various WHO sources [19, 28].

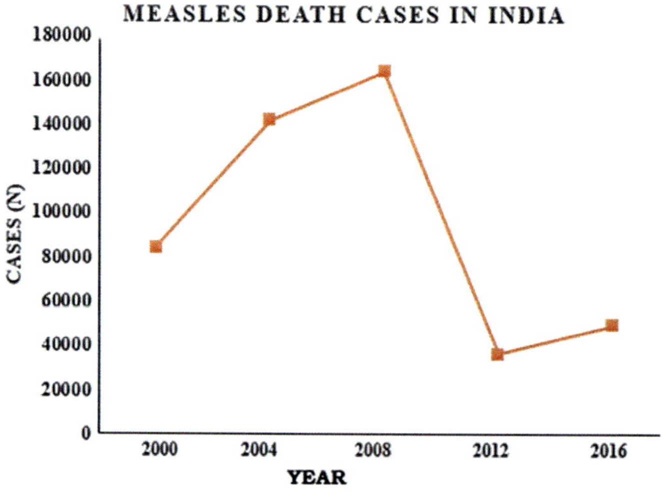

Fig. 5 Reported measles fatality cases in India from 2000 to 2016

2.3 Measles Elimination Strategies

This section highlights on few cases of a measles outbreak which occurred even though the immunization programme was in place, therefore, global elimination of measles should be targeted for as it is an epidemic disease which can spread through travellers. Vaccination compliance is a major concern in today's world for the safety of infants against various diseases. Global Vaccine Action Plan recommended by the World Health Assembly along with the countries and areas addressing an exhaustive description of measles epidemiology over a long period of time is essential to achieve measles elimination objective set by WHO Regions by 2020. Prevalence of measles in some rural and urban areas of Thailand despite the national immunization programme being well in place and spread of this deadly disease due to travellers as they were infectious on their return flights which subsequently exposed several individuals, which prompted control measures for infection prevention [31]. Various issues, strategies and challenges like population displacement, migration, wars and other social unrest which should be considered for the worldwide elimination of measles in spite of being biologically and technically feasible [32].

2.4 Application of Data Mining for Early Prediction of Measles

This section highlights various research findings where data mining techniques are applied for the early prediction of measles. Applying the sequential Monte Carlo approach of particle filtering and used a compartmental model for offering high production capacity for the measles outbreak in less vaccinated areas. Prediction results demonstrated that the model predicts the transmission pattern for measles and classifies to tell whether there will be an outbreak or not in the next month [33]. Application of a mathematical model for measles control by vaccination strongly indicated that a positive correlation exists between the contact rate with infected individuals and the spread of disease. Hence early detection and awareness about measles would help in measles transmission [34]. Data mining techniques like ANN, decision tree algorithm and naïve Bayes classifier are commonly used in developing a robust mathematical model (MM) for predicting immunizable diseases that usually affect children between 0 and 5 age group. MATLAB ANN toolbox and the statistical toolbox were used for classification and regression analysis; therefore, this approach of using MM would help in enhancing the effectiveness of routine immunization in Nigeria [35].

2.5 Methods Applied for Measles Prediction

This section of the paper categorizes various research findings into vaccination, measles outbreak as case studies and related mortality cases in the tabular format. Early measles prediction is given utmost importance in recent years as it is one among the six major killer diseases. Hence, the present model designed with the help of data mining technique will be a boon to the healthcare industry [15]. Table 1 highlights on the motive, study, methods used and the result obtained by various researchers for measles outbreak globally.

2.6 Classification Methods for Disease Diagnosis

This section of the paper provides information regarding the use of classification techniques for the early prediction of various medical datasets. Table 2 gives an overview of the use of the classification methods for the diagnosis of various diseases and the dataset used for the same along with model accuracy data. From the below table, it is evident that various classification techniques like ANN and decision tree are most commonly used for prediction of various deadly diseases where significant model accuracy is achieved. Therefore, the use of these techniques will help the duty doctor in better diagnosis of diseases.

3 Data Mining Techniques

This section gives an overview of data preprocessing and classification systems for disease diagnosis. Disease categorization and performance metrics used in model validation are also highlighted.

3.1 Data Preprocessing

Data quality is given the utmost importance before performing data preprocessing. Some of the quality-related problems like noise and outliers, duplicate data and missing values are considered. Thus considering all these qualities of data, preprocessing is indeed an important step which includes steps like aggregation, data sampling, feature selection, discretization and binarization and attributes transformation. Disease-related information is collected from various approaches like medical repositories, hospitals and health departments. Data is aggregated to consider the cumulative information like symptoms, disease description for data reduction thereby stabilizing them. Sampling techniques are applied for the stabilized data for

Table 1 Studies and methods applied by various researchers for measles prediction

Category	References	Motive	Study	Methods	Results
Vaccination	Kundrick et al. [36]	Aiming sub-national regions for vaccination promotions remains a question towards increasing global immunization coverage	The analysis was made on three metrics for ordering target areas such as 1. Vulnerable birth cohort 2. Immunization coverage 3. Effective reproductive fraction	1. 2010 measles outburst in Malawi as a case study 2. Immunization at the district and health facility catchment scale was explored 3. Effective reproductive ratio estimation was made based on the observation rate of the exponential growth of the endemic disease	Immunization coverage was below 80% in 8 districts and 354 health care facility catchments according to the survey reports
Travellers outbreak	Macintyre et al. [25]	Well-developed countries are more prune towards the non-immune travelling worker and subsequent importation leading to confined epidemics	Estimating the incidence of measles and its characteristics	1. Questionnaire 2. Variable distribution 3. Calculation of age-specific incidence rates	Visiting friends and relatives should be given the highest significance for effective precautionary strategy in order to achieve the set measles eradication status

(continued)

Table 1 (continued)

Category	References	Motive	Study	Methods	Results
	Hall et al. [37]	In spite of measles being eliminated in the US since 2000, it still continues to circulate in many countries globally which could be imported through travellers	Nasopharyngeal swab or throat specimens were used for real-time reverse transcription–polymerase chain reaction (RT-PCR) at the Minnesota Department of Health	1. Maintenance of Attendance records in child care centres and schools where individuals could have been exposed to measles 2. Minnesota immunization information connection system that usually stores electronic vaccination records was used for verification of the immunization status of each individual	High immunization coverage rates across subpopulations within communities are very much necessary to prevent the spread of this pandemic disease
	Majwala et al. [38]	2016 Measles epidemics reported in Mayuge district of eastern Uganda	1. Evaluating the risk factors for measles during the exposure period from 7 to 21 days prior to rash onset 2. Assessing vaccination efficacy	1. Reviewing medical records 2. Conducting active community case-finding 3. Calculating the vaccination coverage by means of the percentage of controls vaccinated	It was identified that children's less than 5 years were highly affected. For children's aged 9–59 months. The effectiveness of the single-dose measles vaccine was nearly 75%

(continued)

Table 1 (continued)

Category	References	Motive	Study	Methods	Results
	Poletti et al. [39]	Annual measles epidemics were observed in the South West Shoa Zone of the Oromia Region, Ethiopia	Estimation of a load of disease in the affected area considering dissimilarities in accessing health care due to travel distance from nearby hospitals	1. Development of model and disease estimation 2. Combination of model data with the special analysis in recognizing severe measles contagions and preventing deaths	A total of 1819 measles cases with 36 mortality cases were documented in the hospital
Mortality	Rathi et al. [40]	227,245 cases of measles were reported with 18 deaths per hour according to the WHO 2012 global report	Determination of place, time and distribution of an outbreak and calculation of the attack rates and vaccination effectiveness	1. Data collection by Standard measles outbreak investigation formats of WHO 2. Blood specimen collection from suspects	Proportions were matched using the Chi-square test using Fisher's-Exact correction
	Murhekar et al. [41]	Conducting a retrospective cohort study to assess measles case fatality ratio and identifying the risk aspects for measles death	Identification of the risk aspects for the measles outbreak in Bihar due to low vaccination coverage	1. House to house survey 2. Blood specimen collection from suspects for laboratory confirmation 3. Blood test for the identification of IgM antibodies 4. ELISA tests 5. Questionnaire 6. Data analysis using STATA software	As per the survey report, 3670 measles cases were found with 28 deaths. Independent risk factors and Multivariate analysis affected children's under 5 years leading to mortality

Table 2 Overview of classification methods for medical diagnosis of various diseases

Disease	Database/dataset	Method used	Classification techniques	Accuracy (%)
Chronic kidney disease	Hasheminejad kidney centre, Tehran	Statistical Symbolic	ANN Decision tree	93.85 78.44
Heart disease	UCI repository, Cleveland heart disease database, Hungarian Institute of Cardiology, Budapest, Switzerland	Statistical Symbolic	GKNN ANN Decision tree Naïve Bayes	97.92 93.83 81.11 96.50
Cataract	Cataract patients database	Symbolic	Random tree	84
Breast cancer	University of Wisconsin Hospital, SEER cancer incidence database	Symbolic Statistical	Decision tree ANN Naïve Bayes LMT Bayesian network	93.62 91.21 84.50 89.7 89.7
Lung cancer	UCI repository	Statistical	GKNN Naïve Bayes	97.92 84.14
Brain cancer	UCI repository	Statistical Symbolic	SVM Decision tree	96.36 90
Tuberculosis	UCI repository	Statistical	KNN	78
Diabetes mellitus	Pima Indians diabetes data	Symbolic	Neural network C4.5	95.30 82.6
Dengue	UCI repository	Symbolic	C5.0	80
IVF	IVF research centres in Tamil Nadu and Speciality test tube baby hospitals; Infertility Research Centre of Mother-and-Child Hospital in Shiraz, Iran	Statistical	ANN and rough set theory	91
Hepatitis C	University of California Irvine (UCI) Repository	Symbolic	Decision rule	73.20
Liver disease	University of California Irvine (UCI) Repository, Fisher's IRIS dataset	Statistical Symbolic	ANN Decision tree	67.59 69.58

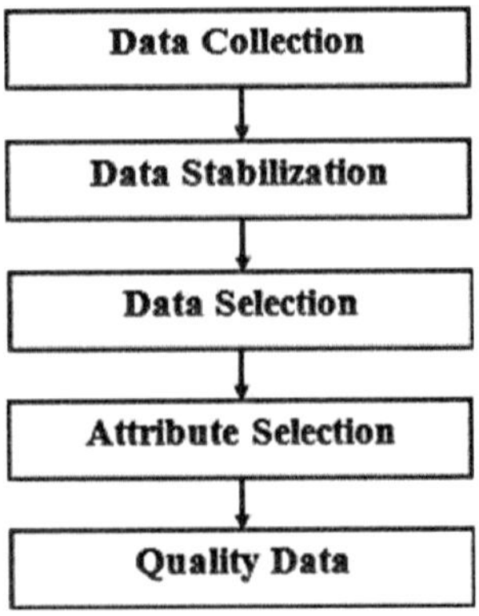

Fig. 6 Data preprocessing steps

data representation in terms of sample size. Dimensionality reduction techniques like feature selection and extraction will be practically utilized to reduce the computational load occurring during data classification. Data is finally transformed into a new set of data as per the needs of medical experts as shown in Fig. 6. Therefore, this approach will help in generating high-quality data which will help in faster disease diagnosis.

Table 3 gives details about the use of dimensionality reduction techniques for effective disease management by various researchers. It gives information regarding the disease type, the dataset utilized, methods adopted and the results obtained by the model.

3.2 Classification System

Data categorization into a number of classes is carried out by classification technique. The six most commonly used classification algorithms used for healthcare datasets are as follows:

i. **KNN (K-Nearest Neighbours)**: KNN works in a very simple way by taking into account the distance from a known data point. K neighbours will be gathered, and the majority among the data will be selected to classify the unknown data into this category.
ii. **Decision Tree**: A decision tree classically starts with a single node which subdivides into possible outcomes. Each of those outcomes leads to add nodes, which branch off into other prospects and gives a tree-like shape. They can be used to understand non-linearity and map out a mathematical algorithm which predicts the best.
iii. **Random Forest**: Random forest uses a lot of decision trees where each tree is a little different from the others. Once new data gets generated, the majority vote of the ensemble is taken to get the final results.

Table 3 Overview of dimensionality reduction techniques for disease prediction

References	Disease	Dataset	Methods	Result
Houari et al. [42]	Diabetes and thyroid	Machine learning repository	LU-decomposition (forward substitution) and Copulas	The projected model provided a scale-free depiction of dependency which was later used for detecting the repetitive values
Trabelsi et al. [43]	Diabetes, breast cancer	16 datasets with varying attributes from the UCI machine learning repository	H-Ratio	The proposed method was capable of finding out the nominal attributes which improved the performance of supervised classifiers
Almuhaideb et al. [44]	Healthcare data	25 attributes from the UCI machine learning repository	1. Feature Subset Selection Method (FSS) 2. A statistical method using the Friedman test	FSS proved to give better accuracy when more than 10 attributes were chosen for data classification
Ramaswamy [45]	In-vitro fertilization (IVF)	42 attributes from fertility clinics, research centres and hospitals in Tamil Nadu	Improvised Ant Colonized Relative Reduct Algorithm	This anticipated algorithm overperformed the existing algorithms like RR, PSO, GA, QR, ACO by improving the accuracy of the classifier which is evident from the results obtained
Peter and Somasundaram [46]	Heart	14 attributes from the UCI machine learning repository	A hybrid approach combining CFS and Bayes Theorem	The proposed hybrid method reduced the number of attributes from 13 to 4 thereby helping in getting better accuracy for the classifiers

(continued)

Table 3 (continued)

References	Disease	Dataset	Methods	Result
Rajeswari et al. [47]	Diabetes	470 instances with 28 attributes from the public healthcare centre	A hybrid model of Particle Swarm Optimization and Pulse-coupled Neural Network.	Reduction in feature attributes from 28 to 13 with minimum iterations was achieved by the proposed algorithm
Tarle and Jena [48]	Heart	UCI machine learning repository such as Switzerland, Hungarian and Cleveland	Orthogonal Local Preserving projection (OLPP)	94.93% accuracy was achieved by the proposed method for Switzerland dataset

iv. **Naïve Bayes**: Bayes theorem works on the principle of finding the likelihood of an current event based on the probability of a previously occurred event.
v. **Support Vector Machine**: Concepts of the SVM algorithm are quite simple. In SVM, a hyperplane is carefully chosen to best disperse the points in the input variable space by their class. In two dimensions, a line is visualized line whereas a hyperplane is visualized in three dimensions.
vi. **Logistic Regression**: This algorithm is also called the sigmoid function. Input values are pooled linearly using weights or coefficient values to calculate the output.

Model validation is carried out to analyse how accurately the data will be classified by the model. Evaluating the model performance based on the training set data is not acceptable in data mining because it can lead us to the problem of over-fitting. Two sorts of paradigms are used for evaluating models in data mining namely, Hold-Out and Cross-Validation. In binary classification problems, we can compare the predicted class against the actual class by a method called a confusion matrix. The most commonly used *Statistical Tests for Model Validation are* Gain and lift charts, Kolmogorov-Smirnov Chart (K-S Chart) and Receiver Operating Characteristics Curve (ROC Curve). The regression model is finally evaluated using techniques like Relative Squared Error (RSE), Root Mean Squared Error (RMSE), Relative Absolute Error (RAE) and Mean Absolute Error (MAE).

Table 4 gives an overview on the various case studies where classification models for early prediction of measles have been implemented which would be a boon to the healthcare sector as measles could be detected early thereby helping the duty doctor for quick detection of the disease for better treatment.

4 Summary of Review

This paper has given us an insight on measles and its mortality which has occurred globally. Importance of vaccination for the prevention of measles and the spread of this epidemic disease as an outbreak which occurs through travellers were highlighted. It is evident from the review that major work has been carried out in the direction of data mining for early prediction of diseases related to brain, heart, kidney, liver and many more. Use of classification models for measles prediction has been carried out in major countries like Canada, Kenya, Nigeria and the Netherlands [35, 51] for early prediction of this deadly disease. Finally, this review would help in knowledge gaining about measles, its clinical significance and use of DM techniques which would be a key parameter for fulfilling the standards set by the WHO for the eradication of this deadly disease.

Table 4 Various classification models used for measles prediction

References	Study area	Dataset	Model/algorithms used	Result obtained
Li et al. [33]	Saskatchewan, Canada	Yearly report of the Department of Public Health	Particle filtering model	The receiver operating characteristics curve of the prediction classification resulted in an AUC value of 0.893 with the minimum disagreement in the model
Fred et al. [49]	KISSI County, Kenya	Electronic health records, Saskatchewan	Deterministic model, compartmental model	Measles-related herd immunity level estimated by the model was found to be 93.75%
Idowu et al. [50]	Osun State, Nigeria	National health management information system database	The statistical-based naive Bayes technique; the decision tree-based algorithm; and artificial neural network, backpropagation network-based algorithms	Backpropagation network models were found to be better tools for predicting the occurrence of immunize-able diseases when compared to NBT and DT
Bier and Brak [51]	Dutch Bible Belt, Netherland	WHO report	Kermack Mckendrick like model	The quantitative periodic outbreak was once in 12 years. The outbreak was high in the age group of 5–14 when compared to 0–4 and 15 and above

5 Research Gap Analysis

In disease prediction, usually symptomatic cases are considered and other cases where mild illness is seen without major symptoms are neglected which is a major drawback for disease prediction since early prediction with mild illness is always better. Further, the use of machine learning repository, electronic health records, medical departments and house to house survey for vaccination history data will not only reduce the bias but also help in better model analysis and validation. The use of dynamic modelling for measles prediction uses empirical data which requires significant human involvement which leads to omitting and simplification of a few factors which does not give the real picture of the disease. Future work should focus on model adaptation into a working decision support structure which is the need of the hour. Further research on privacy issues regarding the use of DM techniques for effective healthcare management should be given the utmost importance.

6 Conclusion

Measles which is one among six major killer diseases have reported nearly 46296 confirmed cases in 2018 alone in India as per the reports of the World Health Organization. This proves that even though the goal of WHO global vaccine action plan for 2012–20 is to eradicate measles in at least five WHO regions, worldwide elimination of this deadly virus is a matter of concern which has to be addressed on a serious note. Each and every part of the world should strictly comply with the efforts for eradication of measles; in order to achieve the global termination of this disease as it is an epidemic disease. Based on travellers outbreak and measles case studies, various mathematical models were initiated for early prediction worldwide. To achieve the nation's goal of measles eradication and bring about a remarkable decrease in measles-related demises, the aim should be in designing a classifier which will bring about a momentous reduction in measles-related demises by using DM technique for effective management of this disease by early prediction thereby helping in faster diagnosis and improved health. This paper focused on making healthcare professionals aware of data mining methods like feature selection and classification techniques which could be successfully applied for huge medical databases for early prediction of disease. Hence the use of data mining in healthcare could assist the doctors and medical experts in effective decision making for disease diagnosis and prediction.

22. WHO (2013) World Health Organization 2012. Global Report Geneva, Switzerland
23. Measles Outbreak Report (2014) Measles rubella initiative, a global partnership to stop measles and rubella [Internet]. Available from: www.measlesrubellainitiative.org. Cited on Apr 2014
24. Durrheim DN, Crowcroft NS, Strebel PM (2014) Measles—the epidemiology of elimination. Vaccine 32:6880–6883. https://doi.org/10.1016/j.vaccine.2014.10.061
25. Macintyre CR, Karki S, Sheikh M, Zwar N, Heywood AE (2016) The role of travel in measles outbreaks in Australia—an enhanced surveillance study. Vaccine 34:4386–4391. https://doi.org/10.1016/j.vaccine.2016.07.023
26. Domercant JW, Guillaume FD, Marston BJ, Lowrance DW (2015) Update on progress in selected public health programs after the 2010 earthquake and cholera epidemic. MMWR Morb Mortal Wkly Rep 64:137–140
27. Nur Laila AG, Sulfeeza MD, Noor HH, Aliza AL (2017) Factors of emerging infectious disease outbreak prediction using big data analytics: a systematic literature review. In: Zulikha J, Zakaria NH (eds) Proceedings of the 6th international conference of computing and informatics. School of Computing, Sintok, pp 37–42
28. Simons E, Ferrari M, Fricks J, Wannemuehler K, Anand A, Burton A (2012) Assessment of the 2010 global measles mortality reduction goal: results from a model of surveillance data. Lancet 379:21738. https://doi.org/10.1016/S0140-6736(12)60522-4
29. Morris SK, Awasthi S, Kumar R, Shet A, Khera A, Nakhaee F (2013) Measles mortality in high and low burden districts of India: estimates from a nationally representative study of over 12,000 child deaths. Vaccine 31:4655–4661. https://doi.org/10.1016/j.vaccine.2013.07.012
30. Awofeso N, Rammohan A, Iqbal K (2013) Age-appropriate vaccination against measles and DPT-3 in India closing the gaps. BMC Publ Health 13:358. https://doi.org/10.1186/1471-2458-13-358
31. Kantele A, Valtonen K, Davidkin I, Martelius T, Vozelevskaja N, Skogberg K, Lyytikalinen O (2012) Travellers returning with measles from Thailand to Finland, April 2012: infection control measures. Euro Surveill 17:20184
32. Roy P, Priyanka V, Goel MK, Rasania SK (2015) Measles eradication: issues, strategies and challenges. J Infect Dis 46:25–28
33. Li X, Doroshenko A, Osgood ND (2018) Applying particle filtering in both aggregated and age-structured population compartmental models of pre-vaccination measles. PLOS ONE 13:e0206529. https://doi.org/10.1371/journal.pone.0206529
34. Fred MO, Sigey JK, Okello JA, Okwoyo JM, Kang'ethe GJ (2014) Mathematical modeling on the control of measles by vaccination: case study of KISII County, Kenya
35. Idowu AP, Kayode AA, Akhigbe BI, Osungbade AF, Adeosun OO (2013) Data mining techniques for predicting immunize-able diseases: Nigeria as a case study. Int J Appl Inf Syst 5:5–15. https://doi.org/10.5120/ijais12-450882
36. Kundrick A, Huang Z, Carran S, Kagoli M, Grais RF, Hurtado N, Ferrari M (2018) Sub-national variation in measles vaccine coverage and outbreak risk: a case study from a 2010 outbreak in Malawi. BMC Publ Health 18:741. https://doi.org/10.1186/s12889-018-5628-x
37. Hall V, Banerjee E, Kenyon C, Strain A, Griffith J, Como-Sabetti K, Johnson D (2017) Measles outbreak—Minnesota April–May 2017. MMWR Morb Mortal Wkly Rep 66:713–717. https://doi.org/10.15585/mmwr.mm6627a1
38. Majwala RK, Nakiire L, Kadobera D, Ario AR, Kusiima J, Atuhairwe JA, Zhu BP (2018) Measles outbreak propagated by children congregating at water collection points in Mayuge District, eastern Uganda, July–October 2016. BMC Infect Dis 18:412. https://doi.org/10.1186/s12879-018-3304-5
39. Poletti P, Parlamento S, Fayyisaa T, Feyyiss R, Lusiani M, Tsegaye A, Merler S (2018) The hidden burden of measles in Ethiopia: how distance to hospital shapes the disease mortality rate. BMC Med 16:177. https://doi.org/10.1186/s12916-018-1171-y
40. Rathi P, Narendra V, Sathiya V, Kini S, Kumar A, Sana N (2017) Measles outbreak in the adolescent population-matter of concern? J Clin Diagn Res 11:LC20–LC23. https://doi.org/10.7860/JCDR/2017/28619.10488

References

1. Osmar RZ (1999) Introduction to data mining. In: Principles of knowledge discovery in databases. CMPUT690, University of Alberta, Canada
2. Periasamy ARP, Mohan S (2017) A review on health data using data mining techniques. Int J Adv Res Comput Sci Softw Eng 7:291–297. https://doi.org/10.031206/IJCSC.2016.011
3. Shortliffe EH (1987) Computer programs to support clinical decision making. JAMA 258:61–66. https://doi.org/10.1001/jama.1987.03400170060016
4. Lashari SA, Ibrahim R, Senan N, Taujuddin NSAM (2018) Application of data mining techniques for medical data classification: a review. MATEC Web Conf 150:1–6. https://doi.org/10.1051/matecconf/201815006003
5. Periasamy ARP, Mohan S (2017) A review on health data using data mining techniques. Int J Adv Res Comput Sci Softw Eng 7:291–297. https://doi.org/10.23956/ijarcsse%2FV7I3%2F0136
6. Pushpan A, Akbar AN (2017) Data mining applications in healthcare. IOSR J Comput Eng (IOSR-JCE) 1:4–7
7. Patel S, Patel H (2016) Survey of data mining techniques used in healthcare domain. Int J Inf Sci Tech (IJIST) 6:53–60. https://doi.org/10.5121/ijist.2016.6206
8. Durairaj M, Ramasamy N (2016) A comparison of the perceptive approaches for preprocessing the data set for predicting fertility success rate. IJCTA 9:255–260
9. Doreswamy H, Salma UM (2016) A binary bat inspired algorithm for the classification of breast cancer data. Int J Soft Comput Artifi Intell Appl (IJSCAI) 5(2/3):1–21
10. Asma A, Abahussin Albarrak AI (2016) Vaccination adherence: review and proposed model. J Infect Pub Health 9:781–789. https://doi.org/10.1016/j.jiph.2016.09.006
11. Kuroski SK, Davis KL, Krishnarajah GS (2016) Completion and compliance of childhood vaccinations in the United States. Vaccine 34:87–394. https://doi.org/10.1016/j.vaccine.2015.11.011
12. Hamborsky J, Kroger A, Wolfe S (2017) Centers for disease control and prevention. In: Epidemiology and prevention of vaccine-preventable diseases, 13th edn, Supplement. Public Health Foundation, Washington, DC
13. Watson JC, Hadler SC, Dykewicz CA (1998) Measles, mumps, and rubella–vaccine use and strategies for elimination of measles, rubella, and congenital rubella syndrome and control of mumps: recommendations of the advisory committee on immunization practices (ACIP). MMWR Recomm Rep 47:1–57 [PubMed: 9639369]
14. Gershon AA (2011) Chickenpox, measles, and mumps. In: Remington JS, Klein JO, Wilson CB (eds) Infectious diseases of the fetus and newborn infant, 7th edn. Elsevier, Philadelphia, pp 661–705
15. Introduction of Measles rubella vaccine (campaign and routine immunization), National operational guidelines 2017, Ministry of health and family welfare, Government of India
16. Park K (2011) Epidemiology of communicable diseases. In: Park K (ed) Textbook of preventive and social medicine, 21st edn. Banarsidas Bhanot Publishers, Jabalpur
17. Patro BK, Shewade HD, Kathirvel S, Senjam SS, Singh MP, Ratho RK (2012) Outbreak of "modified measles" in an urban resettlement colony of North India. Indian J Publ Health 56:168–169. https://doi.org/10.4103/0019-557X.99921
18. World Health Organization (2015) Measles fact sheet N286. Available from: http://who.int/mediacentre/factsheets/fs286/en/. Cited 2 Mar 2015
19. World Health Organization (2015) Measles—fact sheet N286. Available from: http://who.int/mediacentre/factsheets/fs286/en/. Cited 20 May 2017
20. Andrus JK, de Quadros CA, Solórzano CC, Periago MR, Henderson DA (2011) Measles and rubella eradication in the Americas. Vaccine 29(S4):91–96. https://doi.org/10.1016/j.vaccine.2011.04.059
21. Ministry of Health and Family Welfare, Government of India. Measles mortality reduction India strategic plan 2005–2010, New Delhi

41. Murhekar MV, Hutin YJ, Ramakrishnan R, Ramachandran V, Biswas AK, Das PK, Gupte MD (2011) The heterogeneity of measles epidemiology in India: implications for improving control measures. J Infect Dis 204:S421–S426. https://doi.org/10.1093/infdis/jir061
42. Houari R, Bounceur A, Kechadi MT, Tari AK, Euler R (2016) Dimensionality reduction in data mining: a Copula approach. Expert Syst Appl 64:247–260. https://doi.org/10.1016/j.eswa.2016.07.041
43. Trabelsi M, Meddouri N, Maddouri M (2017) A new feature selection method for nominal classifier based on formal concept analysis. Proc Comput Sci 112:186–194. https://doi.org/10.1016/j.procs.2017.08.227
44. Almuhaideb S, Menai MEB (2016) Impact of preprocessing on medical data classification. Front Comput Sci 10:1082–1102. https://doi.org/10.1007/s11704-016-5203-5
45. Ramasamy N (2017) Feature reduction by improvised hybrid algorithm for predicting the IVF success rate. Int J Adv Res Comput Sci 8:37–40. https://doi.org/10.26483/ijarcs.v8i1.2848
46. Peter TJ, Somasundaram K (2012) Study and development of novel feature selection framework for heart disease prediction. Int J Sci Res Publ 2:1–7
47. Rajeswari S, Josephine MS, Jeyabalaraja V (2017) Dimension reduction: a PSO-PCNN optimization approach for attribute selection in the high-dimensional medical database. In: IEEE international conference on power, control, signals and instrumentation engineering (ICPCSI), 21 Sept 2017, pp 2306–2309. https://doi.org/10.1109/ICPCSI.2017.8392128
48. Tarle B, Jena S (2016) Improved artificial neural network for dimension reduction in medical data classification. In: 2016 International conference on computing communication control and automation (ICCUBEA), pp 1–6
49. Fred MO, Sigey JK, Okello JA, Okwoyo JM, Kangethe GJ (2014) Mathematical modeling on the control of measles by Vaccination: case study of KISII County, Kenya. SIJ Trans Comput Sci Eng Appl (CSEA) 2:61–69
50. Idowu AP, Kayode AA, Akhigbe BI, Osungbade AF, Adeosun OO (2013) Data mining techniques for predicting immunize-able diseases: Nigeria as a case study 1
51. Bier M, Brak B (2015) A simple model to quantitatively account for periodic outbreaks of the measles in the Dutch Bible Belt. Eur Phys J B 88:107. https://doi.org/10.1140/epjb/e2015-50621-9

A Survey on Graphical Authentication System Resisting Shoulder Surfing Attack

S. Arun Kumar, R. Ramya, R. Rashika, and R. Renu

Abstract In today's modern world, security is a major concern. To provide security, the most widely recognized authentication methods are credentials, OTP, LTP, etc, and these methods are more prone to brute-force attack, shoulder surfing attack, and dictionary attack. Shoulder surfing attack (SSA) is a data theft approach used to obtain the personal identification numbers or passwords by looking over the user's shoulder or by external recording devices and video-capturing devices. Since SSA occurs in a benevolent way, it goes unnoticed most of the time. It is one of the simple and easy methods for hackers to steal one's sensitive information. The hacker has to simply peek in while the user types in the password without any much effort involved. Therefore, this phenomenon is widely unknown to people all over the world. Textual passwords are a ubiquitous part of digital age. Web applications/mobile applications demand a strong password with at least one capital letter and a special letter. People tend to give easy passwords in order to remember them which can be easily shoulder surfed. To overcome this, graphical password techniques are used to provide a more secure password. In the graphical authentication system, the users click on target images from a challenge set for authentication. Various graphical systems have been proposed over the years which are shown to be more secure when compared to other authentication systems. In this paper, an overview of various graphical authentication systems is presented.

Keywords Shoulder surfing attack · Textual password · Graphical authentication system

S. Arun Kumar (✉) · R. Ramya · R. Rashika · R. Renu
Sapthagiri College of Engineering, Bangalore, India
e-mail: aarunkumar889@gmail.com

R. Ramya
e-mail: ramyahai01@gmail.com

R. Rashika
e-mail: rashikarajaraman@gmail.com

R. Renu
e-mail: renurudramurthybt@gmail.com

N. N. Chiplunkar and T. Fukao (eds.), *Advances in Artificial Intelligence and Data Engineering*, Advances in Intelligent Systems and Computing 1133,
https://doi.org/10.1007/978-981-15-3514-7_57

1 Introduction

Security plays a vital role in any organization. Data protection is one of the main challenges faced in any business environment. In order to protect any resources, the companies undertake various security measures. However, security has become a worldwide problem as Web sites have become an integral part of everyone's life [1]. The uncompromising security issues that have to be addressed in Web sites occur during the user authentication phase. In today's computer world, authentication is very important in order to keep the unauthorized users from accessing the protected resources. Authentication is a process that allows a device to verify the identity of a person who connects to a network resource. In order to keep the users' data private, authentication mechanism is used, wherein the user types in the username and password to access their private account. However, people performing authentication process in public results in shoulder surfing attack [2].

Shoulder surfing attack is a direct observation approach where the shoulder surfer steals the user's personal identification number (PIN) and passwords by looking over his shoulder [2, 3]. It commonly happens in public transports while the victim is commuting which involves a smart phone in almost all cases. A good example is shoulder surfing at ATMs, a crime in which a suspect watch over the victim's shoulder as he punches in his PIN number. The ATM screen asks for another transaction when the customers complete theirs. Some customers fail to notice the prompt and walk away leaving it on the screen. In this way, the thief enters the stolen PIN and pretends to be the user. But the phenomenon of shoulder surfing is not widely known [4]. Users tend to use the strategies such as hiding the device screen and shielding the device with their hand. However, by observing, one cannot get a hold with most of the victim's detailed biodata such as information about his relationships, sexual preferences, interests, hobbies, and login data. Hence, the damage shoulder surfing can cause is widely unknown [5].

Textual password approach is used tremendously all over for authentication. During the authentication phase, Web sites demand strong passwords with at least six to eight characters comprising of uppercase and lowercase alphabets, numbers, and special characters. Such passwords are believed to prevent brute-force attacks [6]. A password cannot be remembered if its strength is more. In [7] today's digital age, Web sites play a major role in one's life. People are part of an enormous amount of such Web sites with each containing the authentication phase where the user validates him by entering a password to access their private data. In order to remember all such passwords, the user tends to choose the same password for multiple Web sites which makes the password unprotected for the hackers to break. A more complex password is shoulder surfing resistant. Thus, these passwords can be easily revealed if the shoulder surfer peeks or uses video-recording devices [7].

Graphical authentication systems are used in order to overcome the disadvantages of textual password systems. Here, images are used as the password instead of a string of characters. These graphical passwords are expected to be stronger and safer than textual passwords [8]. Several studies prove that a human brain has a better ability to memorize and recollect images easily when compared to a string of characters. Since it is easy to recollect the password, the user need not choose the same password for multiple Web sites. It makes it hard for the assailants to break the password if the user prefers to use a strong graphical password. This, in turn, increases the security level during the authentication phase. A strong graphical authentication system not only safeguards the password from brute-force and dictionary attacks, but also from shoulder surfing attacks. Since shoulder surfing can create damage to the user during authentication, a strong graphical password is preferred over a textual password according to the studies conducted [9, 10].

2 Technologies for Graphical Authentication

Password-based authentication schemes have been most commonly used on many smart devices when compared to other authentication schemes. The lower complexities in implementation, computation, processing requirements, and so forth have led to the use of a password-based authentication system. Again, text-based passwords are more commonly used when compared to other existing authentication systems. However, various vulnerabilities were discovered by several cryptanalysts in text-based systems like brute-force attack, guessing attack, dictionary attack, and social engineering attack. In smart phones, the tiny screen size imposes some more constraints such as limited password length and implementation of easier authentication systems to increase performance. Moreover, the small on-screen keyboard makes typing inefficient and less precise. Consequently, the users tend to use a smaller password which makes it even more vulnerable. Since the size of smart devices is getting smaller and smaller; few authentication systems cannot be implemented in it due to its size [11].

The invention of graphical password authentication systems was triggered by the well-known limitations of textual password authentication systems. The graphical authentication systems have been generally categorized into draw metrics, loci metrics, and search metrics systems. In draw metrics-based systems, the users will have to recall and reproduce the predefined pattern on a canvas to use the system. In loci metrics-based systems, the users will have to recall and select the previously defined points in an image in order to login the system. In search metrics-based systems, the users will have to choose the predefined target images from the displayed challenge set. During the login phase, the system throws in with entirely the same images or with a few different images which were displayed to the user during the registration phase. The selection of correct target images will let the user access the system. Shoulder surfing has always been a problem in these systems because of the use of the graphical interface [12].

Many authentication systems have been evolved over the years. Today, biometric authentication system holds a prominent place as many users utilize them over textual- or graphical-based authentication systems [13]. However, one study showed that for mobile authentication, 70% of the users preferred PIN or android graphical pattern even though they are more prone to attacks. The users tend to opt for the textual-based method as they do not care about the security but the ease with which they can simply get over with the login phase. Thus, knowing this fact, the attacker will try to break into those systems which use textual- or graphical-based systems. Besides, biometrics would not be the one used for authentication if the users give more priority to the ease of use when compared to other technologies. Biometrics also lacks privacy, reliability, and security. Thus, the existence of PIN and pattern approaches is present even in the overexposure of biometrics [14].

Shoulder surfing can be done in a place jammed with people because it is comparatively easy to stand behind or beside someone and look over his shoulder for information. So, it does not need any technical skills to find the password. In [15], a dynamic pin system is used so that it becomes difficult for the attacker to break even if he observes while the user types in the password. This also requires less memory. It deals with the new method for authentication. It dynamically changes the PIN of the device. The date and time are used as a password. So, we use four digits PIN for authentication. The keylogger which is used for the finding of a password also fails to provide the attacker the ways to authenticate to the user's system. Different format of the pin is also used based on the user's preference. The user can change PIN format like it maybe h1:h2:m1:m2, m1:m2:h1:h2, h2:m1:h1:m2. The system cannot be taken down by the brute-force as the PIN changes from time to time.

Different graphical password authentication schemes are used to mark the defects of a textual password. Humans have more capacity to remember the pictures for longer duration. Since an image-based password is easier to remember, a secured graphical authentication system named pass matrix is implemented [16]. Pass matrix protects the user suffering from shoulder surfing in public places through one-time login indicator. The login indicator which is generated randomly during each phase for pass images will be unused after the session ends. A better security is provided by the login indicator in opposition to shoulder surfing attack, because a dynamic pointer is used by the user to identify the location of their password rather than selecting the password directly. In pass matrix, a part of every image is used as a password from a sequence of n images. In this, the first square is located in the first image, second square in the second image, and so on. In pass matrix, user chooses one grid from each image instead of choosing n grid in the same image. The Cued Click Point (CCP) helps the user to remember and recall their password. If the user clicks on an improper password area within the picture, the login will be failed [16].

Physical QWERTY keyboards are the most commonly accepted input device for personal and movable computing systems. This keyboard is also one of the reasons for shoulder surfing attack. The randomized keyboard then expects the user to type in something which is incorporated with an augmented reality wearable device. The user can see the keys on the randomized keyboard through augmented reality device which is commercially feasible. Different keyboard layout is made visible to the shoulder surfer, wherein he cannot deduce the actual keyboard pattern. It is important to make sure that the keystrokes done by the user cannot be easily identified by the shoulder surfer. Even if he does so, the different keyboard pattern misleads the shoulder surfer from knowing the actual password. An algorithm called individual key randomization (IKR) is used to randomize the keys on the keyboard. An algorithm called row shifting (RS) is used to shuffle the keys row-wise, whereas column shifting (CS) is used to shuffle the keys column-wise. This method overcomes the disadvantages of having a shoulder surfer peak in while the user types in the password. The above three algorithms help the user to efficiently type in the password by misleading the shoulder surfer [12].

On the basis of authentication, graphical password techniques are classified into three major types, namely draw metric, search metric, and loci metric techniques. Searchmetric technique is followed in this paper. This technique aimed at mitigating the obstacles faced in the graphical password schemes. This [17] technique consists of two phases, namely registration and an authentication phase. During the registration phase, the user is expected to enter his valid username and select images from the given set as his password. Every image is associated with three-digit code where this code has to be entered by the user to choose his image along with direction, and the same has to be remembered by the user for the entire process. During the authentication phase, the user is expected to identify the password images and the random code is associated with the images. However, for every authentication session the images will be randomized. This technique uses indirect selection such as choosing the image next to password image called the subordinate image. This subordinate image is decided based on the direction chosen initially during the registration process. The correct identification of the subordinate image for every password image from the given set leads to successful authentication else it directs the user to start the whole process again from the beginning [17].

The proposed ColorPass [18] technique follows the concept of partially observed attacker model where the user can view only the response provided by the system but not the challenges values. Here, the user chooses four pin colors. In the login procedure firstly, the user has to enter his login id and then when the system authenticates the login id, it will generate the feature table on the system that throws some challenge values in the range 1–10 to the user. The feature table can be selected depending upon the challenge values, and further, the color pin has to be selected depending upon the feature table that exactly indicates the color cell. The digit in the color cell has to be identified and submitted as a response to the given challenge by the user. The login process will be completed only after responding similarly to all the other remaining three given challenges. The response given by the user will be evaluated by the system which then the system finally decides if the user is a legitimate user or not [18].

In [19], the phase of registration, from the given various categories the user selects few images, and then in the authentication step, the user is expected to select the correct images which were used in the phase of registration. Registration phase: During this phase, the user has to enter a valid username and is required to select a minimum of 5 images and maximum of 8 images from the given categories. In this case, depending upon the length of the password the user can select only one image from each category on each page, and this process is done by typing the character not by just pointing the mouse toward the image and the same has to continue until the final category. In addition to this, the alphanumeric characters present in each image and the position of images present in each category are randomly organized. Hence, a graphical password consisting of sequence of images will be stored in the database. Authentication phase: During this phase, the user is expected to enter the registered username first and then the image categories will be retrieved by the system that is selected by the user during the phase of registration. The user chooses his image by entering alphanumeric character attached to each image. Here in each page, a fake image named NOT MY PASSWORD is automatically added to provide more security from the attacker. Any one of the images that present in each category replaces the fake image per page. Since the user knows the image selected by him in each category, the user will select the correct image; if the known image is available otherwise, a fake image will be selected in order to ignore that particular category. And also, to make the process more complex, a random category is inserted between selected categories to confuse the attacker who watches this phase of authentication [19].

Among the many shoulder surfing-resistant techniques, loci metric scheme is one among them. This CRASH technique [20] deals with loci metric scheme such as Cued Click Point. In this scheme, there are two phases such as registration phase and login phase. In the phase of registration, the user is expected to enter the credentials such as user id, username, and email id. Then further in the subsequent page, the required category and the number of images in each category will be selected by the user. Based on the selected category, images will be projected. Then an image has to be selected by the user from the given collection which will be followed by a selected image displayed in the grid format. A grid square will be selected by the user as his click point, and this process has to be repeated for certain number of images defined. After selecting the required number of predefined images, registration process gets completed. In the next phase, i.e., login phase based on the order of selection the system starts displaying all images. The user has to click on the same grid square that he clicked during phase of registration for each image. If the point clicked by the user in this phase matches with the registration phase, then the system displays the consecutive image else a wrong image may be displayed, and this process has to be repeated for certain number of images defined. If all the click points entered by the user were correct, then a random string or word will be received via email. In the subsequent page, a different set of images with completely different meaningful words will be displayed. The user has to click on the same string or word that he received via email to successfully complete the login process else the process has to repeat from the beginning. After the failure of three login attempts, the login session expires [20].

Traditional PIN-based authentication schemes are still in use in budget touch screen devices as they can be easily implemented when compared to embedded fingerprint sensor which makes the effortless login. However, this method is less resistant to shoulder surfing attack. In [21], a concept based on merging images called hybrid images is used, wherein this technology simply fools the eyes of the shoulder surfer. The core idea is on the simple observation of the variation in the distance between the screen and the user with that of the screen and the shoulder surfer. The user views the screen from the lesser distance when compared to the shoulder surfer who is at least 0.9 m away from the screen. Taking this into account, a hybrid keypad is implemented. The keypad consists of numbers with each button being the combination of two digits. The shoulder surfer is misled since the button is totally viewed differently by him with varied layouts. Consequently, the extraction of user's PIN becomes difficult. The shuffling of digits is performed in every authentication. This helps in knowing the spatial arrangement of the digits pressed. The hybrid keypad consists of two keypads. One keypad is viewed only by the user called user's keypad and the other which is visible to the shoulder surfer called shoulder surfer's keypad to confuse him. This hybrid keypad is created by using low-pass and high-pass filter parameters. This filtering helps in creating two images. The spatial frequency of both the keypads varies which differentiates the keypad layouts. The algorithm used called visibility algorithm helps to find the minimum safety distance from the user's keypad to the shoulder surfer. Therefore, a false PIN layout is created in order to protect the shoulder surfing attack [21].

This paper [22] presents a more secure pattern-key-based password authentication system where these grid points form the pattern and these grid points only point to the location of number in an integer matrix. The pattern key being the first level is followed by the secret key and then the dummy values at the last. During the registration phase, user will be given a 5×5 block grid numbered from 1 to 25. Firstly, the user types in the location number of the pattern. Then in addition to the pattern, the user registers a key for numbers 0–9. A key value ranging from "0 to 9" maps to any integers or characters or to any special characters. Followed by this, the user needs to type in the number of dummy values in this phase where dummy values precede as well as succeed the real password values. These dummy values are named as left and right dummy values. During login phase, after entering the username a next screen appears containing 5×5 grid block will appear with randomly generated numbers. The pattern choose in the registration phase has to be remembered by the user to map the key values to the selected password values along with left and right dummy values and then enter the password. If the password matches, it authenticates the user and is able to login successfully else fails in the login process [22].

The user cannot easily recall their password since he has various login ids to remember and might forget his password if he does not use them frequently. Every user chooses an easy password so that they can remember easily. So, the user tends to choose the same password login to all of his accounts which is easy for the hacker to guess. Text-based or alphanumeric passwords are difficult to remember. So, in order to overcome the disadvantages of having an alphanumeric password, graphical passwords are used. In [23], pictures are used as password as the human brain has a capacity to remember hundred images with detail. At the beginning, users are exposed to 50 images out of 70 images, wherein each character is assigned to an image. The shoulder surfer cannot easily identify which character is assigned to which image. These images will be from the random art gallery. Here, the user chooses images that are difficult to relate and are colorful. So, for every 10 characters the user selects a picture that signifies a character. The user's pass images are these 10 images. Also, the user should enter the username. The login phase takes 5 columns and 14 rows of the 70 images. The images displayed during login phase help him to recognize the password character. The account will be blocked after 3 trials. Therefore, [23] provides a secure medium for an authentication system.

3 Comparison Results

The previous works are summarized in Table 1.

Table 1 Summary of prior works

Sl. No.	Title	Approach	Pros	Cons
1	Counterfeit shoulder surfing attack using random pin [15]	Dynamic pin generation	– A good solution for shoulder surfing attack because of dynamic PIN generation – Provides more security – Safeguards from shoulder surfer	– If this method is universally accepted, then the shoulder surfer can easily deduce the password
2	Shoulder surfing-resistant graphical authentication system [16]	Pass matrix – A graphical-based password scheme	– Images are used as password which are more effective compared to textual passwords – Easy to recall and reproduce	– Long process – The user might get frustrated to undergo three phases for the password entry level – Shoulder surfer can deduce the password through concealed cameras
3	Preventing shoulder surfing using randomized augmented reality keyboards [12]	Randomized keyboard	– Additional errors during typing are identified	– The user should always wear the augmented reality devices or glasses
4	PassNeighbor: a shoulder surfing-resistant scheme [17]	Search metric style of authentication	– Shoulder surfer cannot fetch the password images by interacting with user	– The order in which the selection of password images is done as well as the direction chose has to be remembered
5	Color Combo: an authentication method against shoulder surfing attack [18]	Observable attacker model	– Shoulder surfing attack and password guessing attack can be minimized – User-friendly – Consumes less time for login process	– This system does not work for fully observable attacker model
6	Graphical password: shoulder surfing-resistant using falsification [19]	Falsification	– The false image inserted between the real images confuses a hacker while trying to capture the password	– Long process – Several steps should be undertaken during login phase
7	CRASH-Cued recall-based authentication resistant to shoulder surfing attack [20]	Locimetric-based scheme of authentication	– Balanced security and usability features – There exists downtime in both the phases	– Long process – The user might get frustrated to get over with login phase

(continued)

Table 1 (continued)

Sl. No.	Title	Approach	Pros	Cons
8	Illusion PIN: shoulder surfing-resistant authentication using hybrid images	Holographic blending	– Provides high security and can be implemented in budget touch screen devices	– It is too complicated when compared to fingerprint scanner authentication scheme – Third-party applications already use a shuffling scheme which is nearly as complex as Illusion PIN concept
9	Secure pattern-key-based password authentication system	Secured pattern – Key approach	– Highly protected password space – Impervious to shoulder surfing attack and hidden camera	– Three secured features are time-consuming – Remembering a lot of key values during authentication may frustrate the user
10	A graphical password against spyware and shoulder surfing attacks	Graphical password approach	– The images that are hard to explain – Has more password space	– Very complicated images and could not remember easily – More number of images

4 Conclusion

In this paper, the cause for shoulder surfing attack and the prevention methods is put forth. An attempt has been made to contemplate the significance of various graphical authentication systems that have been proposed over the years to overcome shoulder surfing attacks. The methods employed to overcome the disadvantages of textual passwords are presented. The system's advantages and disadvantages are presented for each paper that has been surveyed. The need for graphical authentication systems is emphasized. How textual passwords are more prone to attacks are also looked into. Furthermore, this overview will help various analysts and researchers who are keen on creating graphical authentication systems.

References

1. Gawandi M, Pate S, Snehal P, Said SK (2017) A survey on resisting shoulder surfing attack using graphical password. Int J Adv Res Comput Eng Technol (IJARCET) 6(10):1557–1561
2. Sonar AN, Suryavanshi PD, Navarkle PR, Kukre VN (2018) Survey on graphical password authentication techniques. Int Res J Eng Technol (IRJET) 05(02):26–28
3. Eiband M, Khamis M, von Zezschwitz E, Hussmann H, Alt F (2017) Understanding shoulder surfing in the wild: stories from users and observers. In: The CHI conference on human factors in computing systems (CHI 2017), At Denver, CO, USA
4. Divyapriya K, Prabhu P (2017) Image based authentication using illusion pin for shoulder surfing attack. Int J Pure Appl Math 119(7):835–840 (2018)

5. Choi D, Choi C, Su X (2017) Invisible secure keypad solution resilient against shoulder surfing attacks. In: International conference on innovative mobile and internet services in ubiquitous computing (IMIS). https://doi.org/10.1109/imis.2016.77
6. Nimbalkar P, Pachpute Y, Bansode N, Bhorde V (2017) A survey on shoulder surfing resistant graphical authentication system. Int J Sci Eng 2 (Online). ISSN 2456-3293
7. Rodda V, Kancherla GR, Bobba BR (2017) Shoulder-surfing resistant graphical password system for cloud. Int J Appl Eng Res 12:6091–6096. ISSN 0973-4562
8. Thirupathi J (2015) A comprehensive survey on graphical passwords and shoulder surfing resistant technique analysis. IJISET Int J Innov Sci Eng Technol 2(4):1130–1136
9. Chaudhar D (2015) A survey on shoulder surfing resistant text based graphical password schemes. Int J Sci Res (IJSR) 4(11) (Online). ISSN 2319-7064
10. Pawar M, Mate GS, Sharma S, Gole S, Patil S (2017) A survey paper on authentication for shoulder surfing resistance for graphical password using cued click point (CCP). Int J Adv Res Comput Commun Eng 6(1). ISO 3297:2007 Certified
11. Ranak MSAN, Azad S, Nor NNHBM, Zamil KZ (2017) Press touch code: a finger press-based screen size independent authentication scheme for smart devices. PLoS ONE 12(10):e0186940
12. Ho PF, Kam YHS, Wee MC, Chong YN, Por LY (2014) Preventing shoulder-surfing attack with the concept of concealing the password objects' information. Sci World J vol 2014:1–13
13. Su X, Wang B, Zhang X, Wang Y, Choi D (2014) User biometric information-based secure method for smart devices. Concurr Comput Pract Exp 30(3):e4150. https://doi.org/10.1002/cpe.4150
14. Davin JT, Aviv AJ, Wolf F, Kuber R (2017) Baseline measurements of shoulder surfing analysis and comparability for smartphone unlock authentication. In: CHI 2017. Denver, CO, USA
15. Yogadinesh S, Sathishkumar R, Akash L, Aakash V, Kishore Kumar K, Harichander S (2018) Counterfeit shoulder surfing attack using random pin. Int J Pure Appl Math 118:1757–1761
16. Kannadasan M, Amarnadha Reddy J, Venkat Raman K (2017) Shoulder surfing resistant graphical authentication system. Int J Sci Eng Res 8(5):194–198
17. Saeed S, Umar MS (2016) PassNeighbor: a shoulder surfing resistant scheme. In: International conference on next generation computing technologies. Dehradun
18. Anil G, John CM, Mathew PP (2016) ColorCombo: an authentication method against shoulder surfing attack. Int J Comput Sci Inf Technol Res 4(2):142–147
19. Yeung ALC, Wai BLW, Fung CH, Mughal F, Iranmanesh V (2015) Graphical password: shoulder-surfing resistant using falsification. In: 2015 9th Malaysian software engineering international conference (MySEC), Kuala Lumpur, pp 145–148. https://doi.org/10.1109/MySEC.2015.7475211
20. Sruthi PV (2015) CRASH—Cued recall authentication resistant to shoulder surfing attack. In: 2015 Online international conference on green engineering and technologies (IC-GET), Coimbatore, pp 1–4. https://doi.org/10.1109/GET.2015.7453834
21. Papadopoulos A, Nguyen T, Durmus E (2017) IllusionPIN: shoulder-surfing resistant authentication using hybrid images. IEEE Trans Inf Forensics Secur vol 12, 2875–2889
22. Zaki MH, Husain A, Umar MS, Khan MH (2017) Secure pattern-key based password authentication scheme. In: International conference on multimedia, signal processing and communication technologies (IMPACT), Aligarh, pp 171–174. https://doi.org/10.1109/MSPCT.2017.8363998
23. Darbanian E, Fard GD (2015) A graphical password against spyware and shoulder surfing attacks. In: International symposium on computer science and software engineering (CSSE), Tabriz, pp 1–6. https://doi.org/10.1109/CSICSSE.2015.7369239

Analysis of Stock Market Fluctuations Incidental to Internet Trends

Vinayaka R. Kamath, Nikhil V. Revankar, and Gowri Srinivasa

Abstract Time-series data pertaining to several related domains plays an important role as reinforcement for understanding the behaviour of that ecosystem. This paper intends to analyse one such system through the inspection of the rise/fall of stock prices for a duration in relation to buzzwords trending over the social media during the same time. We present a pipeline to elicit these insights and facilitate prediction of stock prices. In summary, the pooled data is subjected to time-series operations such as differencing transformations, smoothing and filtering to transform the data to a format amenable for further processing. Then, the trend corresponding to the keywords is illustrated by modelling their popularity using a counting-based algorithm customized for the application. This is followed by an attempt to forecast stock prices. The entire system is bundled into a software application that effectively delivers the results and visualizations in an intuitive fashion for a naive user. Throughout the pipeline, we have used a comparative paradigm that has helped us assess the proposed solution against alternatives.

Keywords Analytic tool · Data analytics · Forecasting · Stock market · Time-series · Twitter trends

1 Introduction

Data corresponding to an entity of interest which is recorded at different points in time can be classified as time-series data. The collection has high prominence because of the challenges and persistent efforts that the process of acquisition demands. The indexed points interpret many key aspects that are valuable to classify, handle and forecast the accumulated data. It is essential to model such points when the matter is of interest since an array of time-series data influences a wide range of domains like the social media presence, stock markets and the economy of a country. The stock

V. R. Kamath (✉) · N. V. Revankar · G. Srinivasa
Department of Computer Science, PES University, Bengaluru, India
e-mail: vinayakarkamath@pesu.pes.edu

N. N. Chiplunkar and T. Fukao (eds.), *Advances in Artificial Intelligence and Data Engineering*, Advances in Intelligent Systems and Computing 1133,
https://doi.org/10.1007/978-981-15-3514-7_58

market reflects the monetary strength of the participating entities. This makes it an interesting domain to indulge in, considering the influence it exerts on the goodness of the society. Stock markets can be interpreted as the public markets which aid in transfer, issue and selling of the stocks that trade on several stock exchanges or sometimes simply over-the-counter. The stocks themselves represent the fractional ownership of the corresponding company and sometimes grant special rights of the person yielding them. Stock market is the place where people offer to attain ownership of such investible assets with an intention to fund the company. Granting companies to quickly access capital from the general public, stock markets are considered crucial for the economic development.

Market sentiment is considered as the general frame of thought of the investors towards a given financial market. It amounts to the mindset of a market, particularly the psychology of the crowd involved, analogous to the price movements in the market that the crowd is involved with. The prices of the assets traded in these markets are highly driven by the sentiment of the public involved. The knowledge and predisposition of the current holders decide the volume of the stocks traded. Even though numerous factors affect the price movements, the general mood of the engaging parties is the direct consequences of these external factors. Inculcating the conception of the human sentiment can enhance the effectiveness of the model. Inspecting the correlation between the fluctuations in the stock prices and the judgment of the people can help build a strong foundation for a superior class of models that infuse the concept of market sentiment. Since it is very much impossible to survey all the participating entities, a suitable alternative that appropriately reflects the state of mind of the population has to be chosen. Social media effectively mirrors the state of mind of the people using the Internet. Any event that impacts majority of people is often actively circulated over the Internet. All these factors are in accord with an attempt to use the data from the social media for analysis of the correlation.

The rising popularity of the social media platforms attracts more people to get involved with the Internet. People are becoming more expressive with the recent advancements that surround the Internet. They do not refrain from actively voicing their opinions on online platforms. This has lead to the advancement of multitude of social media sites, almost every platform with its own variety of content. Twitter stands out from the rest of the sites because of the seriousness and authenticity of the content that gains popularity. Twitter houses over 250 million registered users out of which 50 million users sportingly share their thoughts on a regular basis. Right usage of this content can truly encapsulate the thoughts of an average citizen. Attracting celebrities and commoners, Twitter has abundance of real-time information with regards to current social trends and variety of opinions on the same incident. Behavioural economics depicts that humans are social beings, their rationality is deeply affected by the opinions of the people surrounding the group. The same goes for communities at large, and the society can be subjected to different mood states in particular. If each tweet can be considered as a concise summary of an individuals opinion or mindset towards a crucial matter, then a bundle of multiple tweets that concern the same subject is likely to express the collective opinion. By extension, it can be deduced that the variations in the economic indicators are a consequence of the changes in the sentiment of the people.

Section 2 briefs about research work that resembles our attempts and inspired this work. Section 3 describes about the data acquisition, motivation behind the work and unfolds the intention behind this work. Sections 4 and 5 explicate about the pipeline used in the analytic tool and brief about the components in the pipeline, respectively. Section 6 mitigates the experimentation procedures and the techniques used to test the validity of the model. The last section describes the insights drawn from the pipeline and concludes by unfolding the future possible enhancements.

2 Literature Survey

In [1], an ensemble model is applied on Indian stock data, related tweets and taking relevant articles into account. This model uses RNN and LSTM to forecast the stock prices. Graphical image is then passed to CNN to train the features that deal with the rise and fall of the sharp changes associated with the trend in the image. This in turn helps in recommending the suitable situations that lead to profitable investments in the given company. A hybrid artificial neural network is utilized to forecast the direction of the stock market of Japan in [2]. Genetic Algorithm (GA) leads the optimization process that aids to increase the goodness of the prediction. The hybrid GA-ANN model with TYPE 2 technical indicators is observed to lead to higher forecast accuracy. In [3], a long short-term memory (LSTM) model is built to predict stock prices. Sentiments of the investors are extracted from various forum posts and classified using a Bayesian model. This helps to understand stock market behaviour that is in accord with investor sentiments. A study involving analysis of spamming activities on Twitter based on the user activities, contents comprising the tweets and profiles of the active users is done by Sedhai and Sun [4]. Firstly, comparison of hashtags and their usage in spam and harmful tweets is done considering the orthography, count of the popular words, relative positioning and co-occurrence of the words in context. The presence of duped tweets imparts significant aid in performing context-based analysis. User-level analysis is done based on user profile information. It was observed that people who intend to attempt fraud using spams utilize the celebrated hashtags and create multiple profiles on the platform to market the favoured intentions, whereas legitimate profiles contain data in all the relevant fields.

The study in [5] involves usage of search volumes to perform the estimation and tries to find differences in the USA and China stock market by extracting data from their respective search engines. Search volumes of a particular company are exploited to inspect variations in stock prices. Various trading strategies are also proposed based on the search volume data. An automated hybrid decision tree-neuro-fuzzy system is proposed to forecast the stock market trend in [6]. The important components of the data are retrieved with the help of technical analysis, and the prominent ones are chosen using a decision tree approach. The extracted components then undergo reduction in dimensionality with the application of various techniques, and this miniature data is fed to a adaptive neuro-fuzzy system that predicts foreseeing of the stock market. Experimental results show that the proposed hybrid system has higher accuracy rate than the decision tree-based system. The study in [7] deals with real-time forecasting of stock prices based on tweets, news articles and historical data. Hybrid models are built using historical prices and sentiment analysis on Apache Spark and Hadoop HDFS to handle big data generated from news websites and social media. The architecture improves the performance as it involves parallel processing of data using Apache Spark.

3 Motivation

This paper makes an attempt to analyse the stock prices of the companies enrolled in the US stock exchanges [8] and inspect the correlation with the trend exhibited by the Internet keywords for the year 2016 with strong emphasis on delivering effective visualizations. We intended to deliver visualizations [9] that will ease the comparisons between these time-series graphs. Humble attempts at forecasting [10] the stock prices have been made at the end of the pipeline. A comparative paradigm is followed throughout all stages of the pipeline to measure the standings of the model among its counterparts. A compact and easy to use graphical user interface is designed and developed to display the plots and give independence to the user to feed the data into the pipeline.

A clean dataset [11] of full historical daily prices for almost all the companies that are registered in US stock exchanges like NYSE, NASDAQ and NYSE MKT was obtained from Kaggle, an online community for data scientists and machine learning enthusiasts. The dataset provided open, high, low and closing prices for thousands of companies along with the volume of the stocks sold since the initial public offering of the companies. The prices were adjusted for dividends and split to provide a more accurate measure of the real value of the stocks. Table 1 is a randomly chosen subset from the dataset.

The dataset offered the exchange-traded fund (ETF) prices along with the stock prices. Although the data is relevant and of importance, ETFs did not play any role

Table 1 Records from the stocks dataset—Tesla, Inc

Date	Open	High	Low	Close	Volume	OpenInt
2010-11-30	33.74	35.33	33.41	35.33	2,222,558	0
2014-02-10	189.34	199.30	189.32	196.56	12,981,696	0
2012-07-02	31.35	31.80	30.19	30.40	1,315,534	0
2016-05-13	207.78	211.20	206.70	207.61	2,401,971	0
2016-06-28	201.89	204.05	199.41	201.79	5,725,343	0

in our analysis. Considering the scale of the dataset, it was agreed that the data corresponding to the year 2016 would be used for analysis and the data from the year 2017 would be used to test the models whenever necessary. Moreover, several important events like the US Presidential elections and the Olympics occurred during the year 2016, which implied that the people had a variety of reasons to be active on social media. We believed these factors enhanced the quality of data and helped augment the resemblance of traffic on the Internet with the true intentions of the public. At first glance, it seems that multiple factors are in support of 2016 being the perfect year to inspect.

We intended to deduce the resemblance using the word counts rather than the sentiment of the text. We also ensured that these chosen words would make an attempt to reflect the general mood of the writer rather than just locution. Hashtags are the perfect case of expressing mindset or the subject of the content succinctly. Using hashtags also meant lesser effort to extract the semantics of the text but added the burden of the segmentation. It was assumed that these hashtags would summarize the content of the text in the tweet. Unfortunately, a collection of occurrences of the hashtags or something similar to our requirements was not readily available. This meant a new source which logs raw data and exposes it to the developers had to be discovered (Table 2).

www.trendogate.com offered one such page which would display the trending hashtags for a given day in various parts of the world. The required data had to be scraped from this page and stored by converting it into a suitable format. The web page provided a list of trending hashtags for a given day, which was acquired from the people of a particular state in the United States of America. Enormous scale of the data meant that the data had to be scraped in parts. We realized that scraping the data for all the days in the year meant weeks of processing time. This indicated that a sample of the data available on the website had to be nominated. It was observed that Wednesdays are usually not prone to outliers as it is right in the middle of the week. It is assumed that the mood displayed in the days of the week that are closer to the weekends is exaggerated because of the sudden change in the schedule of the subjects. It was safe to assume that our manual inspection was in accordance with the general mindset of the public. Popularly known as the weekend effect, the sudden irrational behaviour of the stock prices is a phenomenon concerning to the financial markets, it is observed that the stock returns on Mondays are substantially lower than

Table 2 Sample records from the scraped data

Regions	Date	Hashtags
Jacksonville	2016-08-24	#chatsnap
Portland	2016-05-18	#HoyLlegamosAICNE
Colorado Springs	2016-09-07	#IfMenHadPeriods
San Antonio	2016-02-17	Palomino
United States	2016-01-13	#WednesdayWisdom

those of the immediately preceding Friday. This behaviour exhibited by the stock prices agreed with our choice of the weekday and promised true values instead of getting subjected to external factors.

Even though hashtags perform a good job at summarizing the sentiment of the context [12], they try to depict the intention of the author by usage of minimal words that are chained into a single word. This augmented the need for emphasis on the words in the hashtags rather than the whole tag itself. Segmentation [13, 14] of the hashtags had to be done to map the context into several aspects of the events that the author wished to convey. Text segmentation was followed by stemming and spelling correction which aimed to bring the words to their stem form. This reduced the noise in the textual data and made sure that the data was in its purest form.

Understanding the data by exploring the variations exhibited by the patterns is necessary to design the pipeline [15]. Exploratory analysis of the data can assist in gaining a better perception of the statistical measures. It helps to detect and eliminate outliers as well as to fill in the missing data, if any. Figure 1 is one such attempt to summarize the word distribution with the help of a word cloud. Appropriate visualizations aid in attaining broader view of the dataset, which in turn imparts knowledge about the skewness, kurtosis and modality of the distribution.

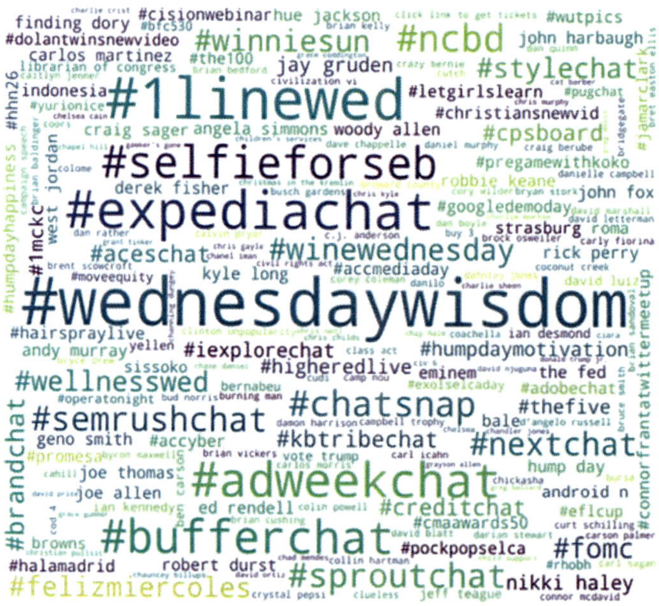

Fig. 1 Wordcloud constructed to understand frequency distribution of the hastags

4 Architecture of the System

The idea is to compare pattern extracted from the popularity of the keywords with the stock prices collected over a period of time. Figure 2 shows that the trending hashtags were associated with a state/region for every date. These hashtags were segmented into words and put back in place with the same associated features. Stop words were removed from the segmented words to filter out noise and irrelevant data. We calculated the popularity of each word on all the available dates by counting the number of states the keyword [16] was trending in. This pattern was subjected to smoothing, scaling, signal extension and data imputation to generate a continuous signal of the length equal to the company in hand. Comparing two time-series data with twining dates is very much easier than comparing two graphs of different lengths. Figure 2 throws light over the modules in the pipeline and provides a schematic representation of the connection between components.

The closing price was chosen for analysis [17] as the rest of the features had high correlation with the "Close" values. Closing prices are believed to be stable as compared to high, open and low values over a time frame and are usually the standard metric to perform analysis on. The "Date" feature is considered as the independent variable, and the "Close" is considered as the dependent variable throughout the process. It was verified that the dataset did not have any missing values, and thus, the requirement for imputation was eliminated.

Comparison of the Internet trends with the stock prices is done through the custom-designed graphical user interface. The user is free to select one company and multiple keywords to perform analysis. Stationarity of the stock prices was tested using the Dickey–Fuller test and Kwiatkowski–Phillips–Schmidt–Shin test, after which the results are displayed to the user. Based on this result, the user can choose to apply relevant filters and smoothing techniques in the subsequent stages. In the next stage, the user can choose between linear transformation, logarithmic transformation, first-order and second-order differencing. Smoothing techniques like exponential, simple exponential and moving average are available at the user's disposal. Baxter–King, Hodrick–Prescott filter and Random Walk filter can be chosen to remove the trend

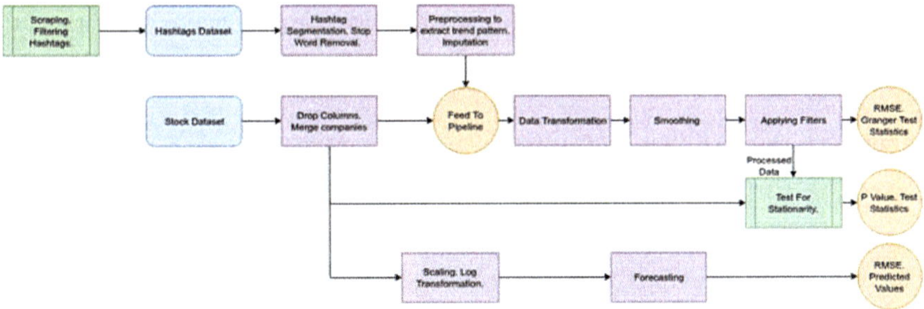

Fig. 2 Schematic of the proposed system

and seasonality. Stationarity test is performed again on the altered data to help user interpret the differences after the application of above techniques. The root mean squared error between the trend pattern and the stock data is displayed to the user. This acts as a measure of dependency between the two time-series data. Granger causality test is performed to verify whether the Internet trends play a vital role in predicting the closing prices.

The last section in the tool is dedicated specifically to forecast the values and compare multiple model for the same company. The models are trained using the data from the year 2016 and tested for the year 2017. The root mean squared error is displayed to the user to evaluate the model. User can opt to choose between regression model, LSTM, Prophet and ARIMAX to perform forecasting. The predicted values are plotted and displayed along with the training and testing data.

5 Building Blocks

The following constituents formed the stages of the pipeline. All the modules contributed at respective stages and had their own share in explaining the outcome of the analysis.

5.1 Testing Stationarity

If the statistical properties such as autocorrelation [18], expectation and variance do not change for a time series $X(t)$, it can be safely categorized as stationary. White noise follows Gaussian distribution, i.e. $N(\mu, \varphi^2)$ and does not depend on t is an instance of a perfectly stationary series.

Inspecting the varying data with a goal to comment on its stationarity is a stationarity test [19]. The test can be performed by evaluating a null hypothesis that is supporting stationarity or completed by testing for unit roots directly. It is essential to perform a check on the stationarity as many of the techniques applied on the time-series data assume the data is stationary in nature. Failing to check for stationarity can lead to erroneous results. Suitable transformations are applied on the data with an aim to convert it into a stationary series only if the results of the tests are not satisfactory.

Augmented DickeyFuller Test Augmented Dickey–Fuller Test (ADF) [20, 21] validates the null hypothesis that claims that the time-series data has a unit root. The test statistics are usually negative, which signify the opposition for the null hypothesis. Larger the magnitude of the statistic, more the disagreement with the null hypothesis. ADF closely resembles the Dickey–Fuller test, but the test is applied to an augmented model like that in Eq. 1 rather than an autoregressive model. In simple words, ADF eliminates all the structural effects like the autocorrelation in the data and proceeds to perform the Dickey–Fuller test.

$$\Delta x_t = A + B_t + \gamma x_{t-1} + \delta_1 \Delta x_{t-1} + \cdots + \delta_{p-1} \Delta x_{t-p+1} + \epsilon_t \quad (1)$$

where A is the bias, B is the coefficient on a time trend and p is the order of the lag. Unit root test is then performed considering the null hypothesis of $\gamma = 0$ against the research hypothesis of $\gamma < 0$. Starting from the higher order, the t-values of the coefficients are inspected to consider a suitable p value. The measure of the test statistic is computed and compared with the applicable critical value of the Dickey–Fuller test. The absence of unit root is confirmed only if the test statistic is less than the critical value, the null hypothesis is rejected. The statistic value is displayed to the user to exhibit the result of the test.

KPSS Test Kwiatkowski–Phillips–Schmidt–Shin (KPSS) test [22] checks the null hypothesis which presumes that the time-series is stationary and oscillates around a deterministic trend (trend-stationary) against a research hypothesis that considers the possibility of the presence of unit root. The test is performed after decomposing the series into three parts: a random walk α_t, a deterministic trend β_t and a standard error ϵ_t.

$$X_t = \alpha_t + \beta_t + \epsilon_t \quad (2)$$

Stationary data is known to have a non-varying value for the intercept or will exhibit random walk behaviour around a trend value. The decomposition helps in segregating the operations and faster validation of the criteria. The null hypothesis is framed as $H_0 : \sigma_\epsilon^2 = 0$ which indicates that the α_t is a constant. The test statistic is computed using Eq. 3 with an intention to test the null hypothesis against research hypothesis of $\sigma_\epsilon^2 < 0$.

$$\text{KPSS} = \frac{\left(T^{-2} \sum_{t=1}^{T} \hat{S}_t^2\right)}{\hat{\lambda}^2} \quad (3)$$

where $\hat{S}_t^2 = \sum_{j=1}^{t} \hat{\epsilon}_t$ denotes the residual of a regression model preformed with X_t against α_t. $\hat{\lambda}^2$ is the estimate of the long run variance of ϵ_t using $\hat{\epsilon}_t$. The test statistic is presented to the user to help him choose the optimal hyperparameters to perform analysis.

5.2 *Transformations and Differencing*

Applying transformation on the data points is the most common method to make the time-series data stationary [23]. The operations may not make the series strictly stationary sometimes, but applying such transformations always helps to increase the stationarity. Since the transformation is preceded by application of smoothing techniques and filtering of residuals, the transformations applied do not fail to enhance the efficiency of the pipeline. The following options were provided by the custom-designed tool:

1. Linear transformation.
2. First-Order Differencing.
3. Second-Order Differencing.
4. Log Transformation.

Linear transformation does not make any changes to data, i.e. $f(x) = x$ is applied on the data points. The first-order differencing mechanism takes the differences between the adjacent data points as the new data points. The new data points are given by $x_t = f(t, x_{t-1}) = x_t - x_{t-1}$. The second-order differences give the curvature of the series at a given time t. The data points are the values of these second-order differences that are obtained by using Eq. 4.

$$\begin{aligned} \Delta^2 Y_t &= \Delta(Y_t - Y_{t-1}) \\ &= \Delta Y_t - \Delta Y_{t-1} \\ &= Y_t - 2Y_{t-1} + Y_{t-2} \end{aligned} \tag{4}$$

The log transformation reduces the variability in the data. This makes the data more stationary than before. Application of $f(x) = \log(x)$ can also aid in reducing the skewness observed in the data, if any. The need for higher-order differencing is minimal as most of the series become stationary on application of second-order differencing.

5.3 *Smoothing Techniques*

With accretion of data, it can be observed that the data exhibits random variations at irregular intervals. To suppress these variations, application of smoothing techniques is essential. Usage of suitable smoothing techniques can reveal the underlying trend and aid in excavating relevant features.

Simple Exponential Smoothing With appreciable performance and simple calculation, exponential smoothing [24] is one of the widely used techniques. Exponential smoothing is very much analogous to a low-pass filter, and it helps in removal of high-frequency noises. If the original series is denoted by X_t starting at $t = 0$, the smoothed series can be obtained by using the Eq. 5.

$$\begin{aligned} S_0 &= X_0 \\ S_t &= \alpha \cdot X_{t-1} + (1 - \alpha) \cdot S_{t-1} \quad \text{where } t > 0 \end{aligned} \tag{5}$$

where α denotes the smoothing factor and agrees with $0 < \alpha < 1$. Figure 3 depicts the output obtained after the application of simple exponential smoothing for American Airlines Groups Inc.

Double Exponential Smoothing Simple exponential smoothing does not perform a good job when there is significant visible trend in the data. Involving a second

Fig. 3 Simple exponential smoothing: American Airlines Groups Inc

parameter in the process can help capture the trend patterns while reducing the noisy features in the data [25]. Present value of the series is utilized to calculate the smoothed value replacement at t by using Eq. 6.

$$\begin{aligned} X_t &= \alpha a_t + (1-\alpha)(X_{t-1} + y_{t-1}) \quad 0 \le \alpha \le 1 \\ y_t &= \gamma(X_t - X_{t-1}) + (1-\gamma)y_{t-1} \quad 0 \le \gamma \le 1 \end{aligned} \tag{6}$$

Moving Average Moving average achieves smoothing by forecasting the current data point, taking a weighted average [26] of the previous data points. This is fulfilled by sliding a large enough window across the time-series data and linearly transforming the points to forecast the current value. A simple moving average technique gives equal weightage to all the data elements in the window. Appropriate window size will help in achieving a finer curve and significantly reduce the noise in the data. Equation 7 aids in achieving the mechanism that is required. A left trailing filter is used in the pipeline. Figure 4 shows the output of a moving average operation for Amazon Inc.

$$S_t = \frac{\sum_{i \in \omega} X_i}{N} \quad \text{where } \omega = t-N+1, t-N+2, \ldots, t-1 \tag{7}$$

5.4 *Filters*

Application of filters helps in removing the seasonality and detrends the data. It acts as a high-pass filter that actively extracts the required features while isolating the high-frequency transients.

Hodrick-Prescott Filter Application of Hodrick–Prescott (HP) filter [27] assists in removing the cyclic component in the time-series data. The resulting curve is more sensitive to long-term fluctuations in the given time-series data than the short-term

Fig. 4 Moving average: Amazon.com Inc

random movements. The new series can be extracted using Eq. 8 whose parameters can be actively tuned to set the sensitivity level. HP filter believes that the time-series data consists of the trend component, a cyclical component and an error component, i.e. $y_t = \xi_t + c_t + \epsilon_t$.

$$S_t = \min_{\xi}\left(\sum_{t=1}^{T}(y_t - \xi_t)^2 + \lambda\sum_{t=2}^{T-1}[(\xi_{t+1} - \xi_t) - (\xi_t - \xi_{t-1})]^2\right) \tag{8}$$

c_t represents the cyclic component, while ϵ_t denotes the error component, ξ_t indicates the trend component in the series. λ value corresponds to the sensitivity of the trend to short-term fluctuations.

Baxter-King Filter Baxter–King (BK) Filter [28] tries to overcome the drawbacks of the HP Filter and extracts better features from the data in hand. Figure 5 helps in visualizing the effectiveness of the filter. It aims to remove the cyclic components from the data while effectively smoothing the exhibited pattern while retaining enough variance. Using a weighted moving average, the BK filter removes the cyclic components from the data. The weights are calculated with an intention to set the threshold frequencies that explain the allowable non-seasonal fluctuations of the smoothed series. The following equations assist to extract the smoother series from the original series.

$$w1 = \frac{2\cdot\pi}{w_u}; \quad w2 = \frac{2\cdot\pi}{w_l} \tag{9}$$

$$b_0 = \frac{w2 - w1}{\pi}; \quad b_j = \frac{\sin(w2\cdot j) - \sin(w1\cdot j)}{j\cdot\pi}, \quad j \geq 1 \tag{10}$$

$$\theta = b_0 + \frac{\sum_{j=1}^{K} b_j}{2\cdot K + 1}$$
$$B_j = b_j + \theta, \quad j = 0\ldots K \tag{11}$$

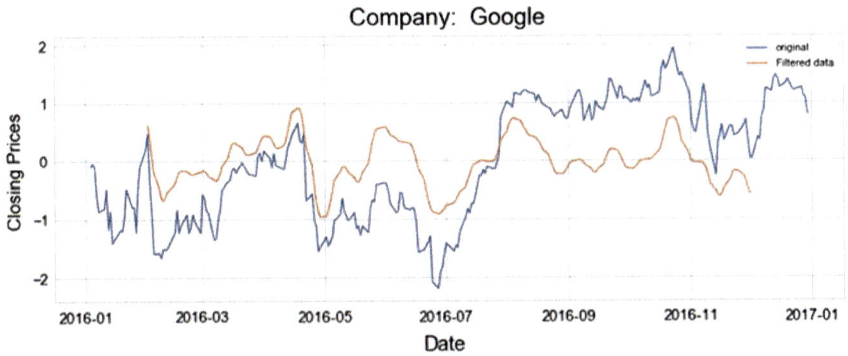

Fig. 5 Baxter-King Filter: Google Inc

The cyclical component of the series is calculated as follows:

$$S_t = Y_t \cdot B_0 + \sum_{j=1}^{K} Y_{t-j} \cdot B_j + \sum_{j=1}^{K} Y_{t+j} \cdot B_j \tag{12}$$

Christiano-Fitzgerald Random Walk Filter Christiano–Fitzgerald random walk filter (CF Filter) [29] follows the footsteps of BK filter. It attempts to formulate the process of smoothing and trend removal problem in the frequency domain. BK filter achieves smoothing by symmetric approximation with an added cost of reducing the series. Dependency on the scale factor reduces the precision of the procedure. The CF filter overcomes this by considering the whole series for computing a single data point. The new series $\hat{y}$ is computed as follows:

$$\begin{aligned} \hat{y}_t = H_0 \cdot y_t + H_1 \cdot y_{t+1} + \cdots + H_{T-1-t} \cdot y_{T-1} + \tilde{H}_{T-t} \cdot y_T \\ + H_1 \cdot y_{t-1} + \cdots + H_{t-2} \cdot y_2 + \tilde{H}_{t-1} \cdot y_1 \end{aligned} \tag{13}$$

where,

$$\begin{aligned} H_j &= \frac{\sin(iy) - \sin(ix)}{\pi i}, \quad i \geq 1 \\ H_0 &= \frac{y - x}{\pi} \\ a &= \frac{2\pi}{P_u} \\ b &= \frac{2\pi}{P_l} \end{aligned} \tag{14}$$

P_u and P_l are suitably chosen to represent the cut-off cycle length in a month. Cycles in between these thresholds are retained.

5.5 *Forecasting Models*

Using the acquired knowledge to predict outside the training range is called as forecasting. This can be done by feeding training records to the model and correcting the model parameters to fit the data. This model is used to predict the required succeeding values that are outside the training data set (Fig. 6).

LSTM Long short-term memory (LSTM) unit [30] is one among the widely used deep learning techniques that learns the long-term patterns and short-term dependencies from the data using a gating mechanism. LSTMs are special type of recurrent neural networks that are very much effective in tackling the vanishing gradient problem which often obstructs the model from learning further from the data. Figure 7 represents the sketch of the smallest cell in a LSTM unit.

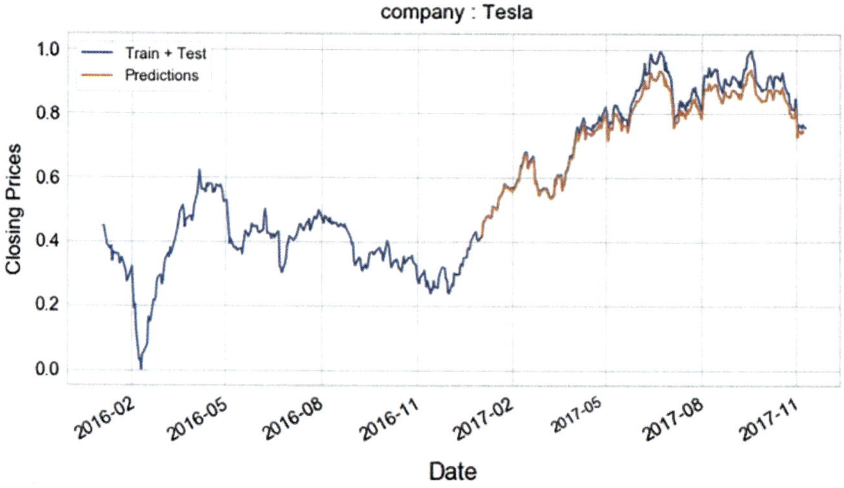

Fig. 6 Predicting Tesla stock prices using LSTM model

Fig. 7 Illustration of LSTM unit

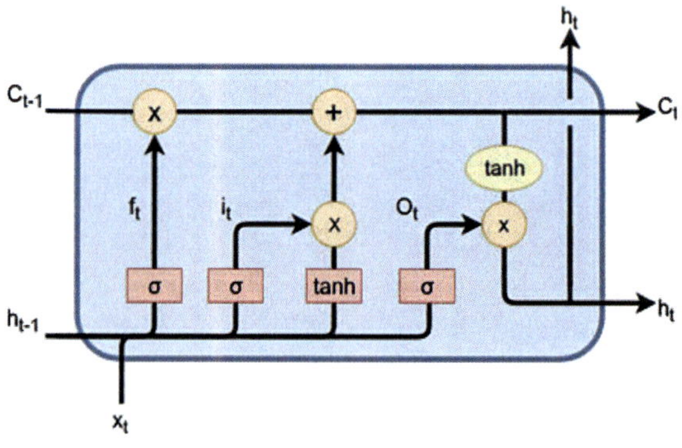

Gates are a constructive way to voluntarily let information through them. They are built using a sigmoid layer and a pointwise multiplication neuron. The output of the sigmoid decides the amount of information to let through. Conventional LSTMs consist of input gate, forget gate and output gate, each with their own functionality. The cell state is the memory of the LSTM and is the most essential component. The gates influence the amount of information that is present or flows into the cell state. The input gate decides what new information is added to the cell state, while the output of the forget gate amounts to the information that is removed from the cell state. The output of the LSTM is influenced by the output gate that depicts the amount of information present in the cell state which co-coincides with the output. The gates and the cell state can be mathematically formulated as follows:

$$\begin{aligned} i_t &= \varphi(x_t A^i + h_{t-1} B^i) \\ f_t &= \varphi(x_t A^f + h_{t-1} B^f) \\ o_t &= \varphi(x_t A^o + h_{t-1} B^o) \\ \bar{C}_t &= \tanh(x_t A^9 + h_{t-1} B^g) \\ C_t &= \varphi(f_t * C_{t-1} + i_t * \bar{C}_t) \\ h_t &= \tanh(C_t) * o_t \end{aligned} \tag{15}$$

$*$ is the elementwise multiplication operator. i, f and o are used to denote input, forget and output gates, respectively. B is the recurrent connection that connects the previous hidden layer and current hidden layer. A is the weight matrix that forms a bridge between the current hidden layer and the input. $\bar{C}$ refers to a candidate hidden state that is calculated using the previous hidden state and the current input. C represents the cell state.

The pipeline used a deep recurrent neural network with LSTM on all the hidden layers. The model used in the pipeline used a four-layer network, with 128 LSTM units on all the layers. Figure 6 provides the pattern predicted by LSTM for Tesla Inc.

Prophet Prophet is a modular regression model consisting of explicable parameters that can be tuned with very little intuition about the time series [31]. It closely resembles a generalized additive model in its approach. As in Eq. 16, the mechanism breaks the time-series data into three components: trend, holidays and seasonal constituents.

$$y(t) = t(t) + se(t) + hl(t) + \epsilon_t \tag{16}$$

The trend function is represented as $t(t)$, $se(t)$ denotes the seasonality component, while $hl(t)$ represents the effects of holidays. The error term is denoted by ϵ_t. This model is utilized to predict $y(t)$ outside the training range. The trend function is devised using a logistic growth model.

$$h(t) = \frac{X(t)}{1 + \exp(-(k + a(t)^T \psi)(t - (m + a(t)^T \gamma)))} \tag{17}$$

where X is the carrying capacity, ψ has the rate adjustments, k is the growth rate and m is the offset parameter. Seasonality is an essential component of the fitted curve, and it can be interpreted using Eq. 18.

$$s(t) = \sum_{n=1}^{N} \left(a_n \cos\left(\frac{2\pi nt}{P}\right) + b_n \sin\left(\frac{2\pi nt}{P}\right)\right) \tag{18}$$

P is the period that the time series is guessed to exhibit. Equation 18 approximates to the application of a low-pass filter, increasing N permits to fit data that changes more quickly. For each holiday component i, let D_i be the group of future and past dates for the corresponding holiday. $Z(t)$ is an attempt to conclude whether a given time t belongs to a holiday or not, where each holiday has its own parameter k. With $k \sim Normal(0, \mu^2)$, the effects of the holidays on the data can be modelled using the Eq. 19.

$$\begin{aligned} Z(t) &= [1(t \in D_1), \ldots, 1(t \in D_L)] \\ h(t) &= Z(t)k \end{aligned} \tag{19}$$

Forecasting is achieved through curve fitting, which is substantially different from the other counterparts. Figure 8 briefs about the fitted curve and the predicted values. Most of the modern forecasting models tend to note the temporal dependence of the data and continue with the forecasting. Time is employed as the regressor on the fitted curve.

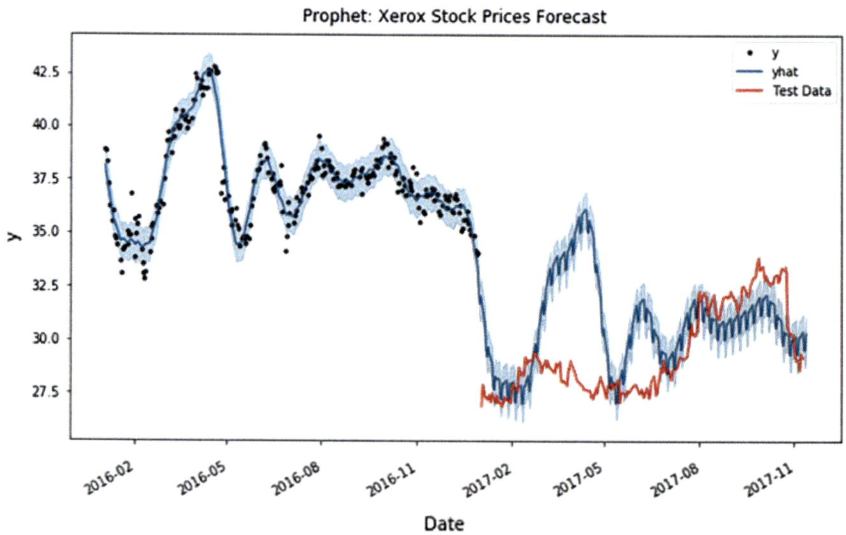

Fig. 8 Forecasting using prophet—Xerox, Corp

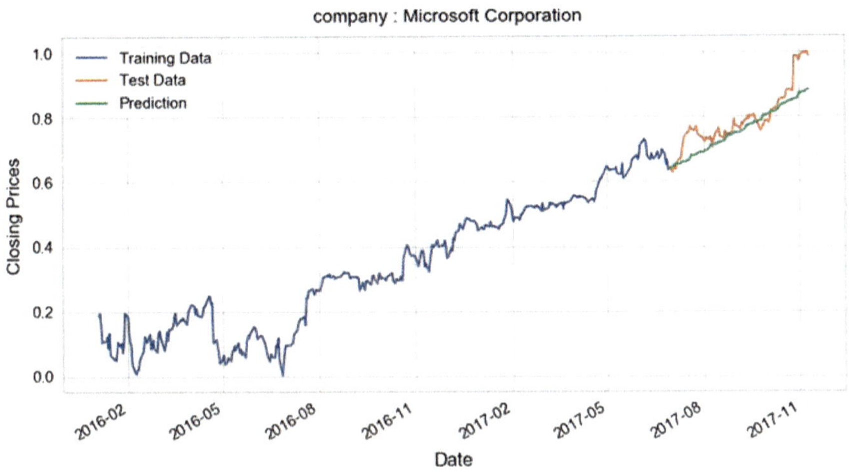

Fig. 9 Forecasting Microsoft Inc. stock prices using ARIMA

Auto ARIMA Autoregressive integrated moving average or commonly known as ARIMA is a very powerful model for basic attempts at forecasting [32] time-series data. The data preparation and parameter tuning process end up being really time consuming. It is essential to choose optimal values for p and q before the training process. This is achieved using ACF and PACF plots. Auto ARIMA eliminates the need for human intervention in deciding the parameters as the model tries out all the possible parameters and selects the best one using the ACF and PACF values. Figure 9 represents the predicted values for Microsoft. This technique closely resembles ensembling, which is proven to significantly elevate the model performance.

The model can be tuned using three parameters: the number of autoregressive terms (p), the order of differencing applied on the series (d), number of lagged forecast errors in the prediction equation (q). Using the terms from the series, we can model the equation as follows:

$$\hat{y}_t = \mu + \rho_1 y_{t-1} + \cdots + \rho_p y_{t-p} - \theta_1 e_{t-1} - \cdots - \theta_q e_{t-q} \tag{20}$$

where ρ is the slope coefficient and θ is the moving average parameter.

Linear Regression The model attempts to capture the relationship between two variables by introducing a linear entity between the dependent variable and the independent variable. The pipeline does not make an attempt to understand the measure of linear relationship between the variables, rather uses the linear regression as a prototyping mechanism to inspect the RMSE value to predict the linear relationship. Equation 21 demonstrates the model that is used to fit the data points.

$$Y = a + bX \tag{21}$$

X corresponds to the independent variable(time t) in the pipeline, while Y corresponds to the dependent variable(Stock price).

6 Experimentation

The application software was designed to have three screens, first of which allowed the user to choose the company of interest and several keywords whose resemblance he wished to inspect. The screen also displayed the test statistics from augmented Dickey–Fuller test and the KPSS test for the raw stock prices, using which the user can decide the filters and mathematical transformations to be applied in the upcoming sections.

Second screen allowed the user to choose between multiple filters, smoothing techniques and differencing transformations that were to be applied on the stock data. The test statistics of the above-mentioned stationarity tests were also displayed for the new time-series data. This was done in order to exhibit the effectiveness of the techniques that were applied. A Granger test was performed to inspect the dependency of the keyword trend pattern on the stock prices, and the test results were displayed in the second screen. The root mean squared error between the patterns was also reported to showcase the resemblance, if any. The final screen exhibits the results of forecasting, with responsiveness and privileges to choose the models. Figures 10 and 11 are snapshots from the graphical user interface.

Several smoothing and filtering techniques were applied on multiple companies to inspect and select the optimal ones. The algorithms that could capture the variations and reduce the noise were selected by visual inspection.

The pipeline was used to perform analysis on a small sample of companies. This was an attempt to develop important insights on the rise/fall of the stock prices in agreement with the popularity of the keyword. The results of the experiment in the form of root mean squared error between the two series are mentioned in Table 3.

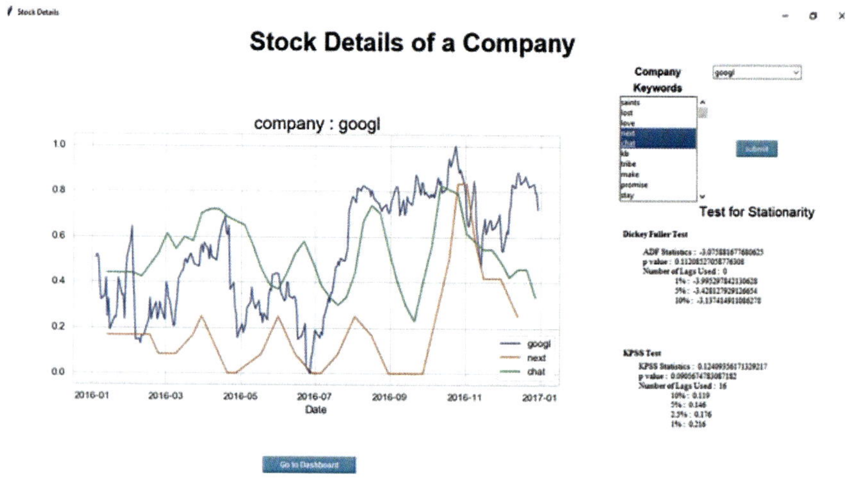

Fig. 10 Snapshot from Screen 1: GUI

Fig. 11 Snapshot from Screen 2: GUI

Table 3 RMSE values for a sample of companies with the keywords of interest

Company code	Keyword	RMSE
googl	Love	0.4458
googl	Wednesday	0.3565
fb	Chat	0.1876
aapl	Politics	0.4429
aapl	Chat	0.2348
xrx	Today	0.3884

The attempt to discover interesting findings on the correlation between the stock prices and the Internet trend resulted in a full-stack application that conveys the insights on key terms and company of interest. The tool manages to replace rapid prototyping methods with an added advantage for the user to completely utilize the data. The tool achieves its purpose to help inspect the causality of a Internet keyword with the company that is a matter of interest to the user.

Trials at forecasting the closing prices yielded very good results. The LSTM outperformed other models in most of the cases, while the Auto ARIMAX came close to the LSTM model in some cases. The results from the prophet were satisfactory on several occasions but never performed better than LSTM, and in fact, there were instances where the linear regression model performed better than the prophet.

The models were tested by predicting the closing prices including that of Google, Facebook and Apple. The root mean squared errors were calculated between the test data and the predicted values. The results are displayed in Table 4.

Table 4 Results from forecasting sampled companies

Company	Model	RMSE
Google	Linear regression	0.0984
	Auto ARIMA	0.1056
	LSTM	0.0342
	Prophet	0.3342
Facebook	Linear regression	0.0833
	Auto ARIMA	0.0985
	LSTM	0.006
	Prophet	0.2101
Apple	Linear regression	0.191
	Auto ARIMA	0.0563
	LSTM	0.0225
	Prophet	0.3110

7 Conclusion and Future Work

The tool developed in the process will help inspect the variation of the closing prices of a company that are in accordance with the trends in popularity of the keywords. The similarity in the ups and downs between the series' of interest can be analysed just by few button clicks. The freedom to choose the company and the keyword helps to focus on field of interest rather than just randomly sampling the data. It also aids in utilization of the complete dataset. This venture to analyse and visualize the data can confidently replace prototyping methods if a little more extensions are adjoined.

Ability to chose the time interval to analyse would hand over more control to the user and enhance user experience. Adding extensions where the users have the luxury to set and fine-tune the model parameters would enhance performance of the system at all stages. Forecasting using the past prices as well as historical word popularity would yield better results than traditional models that train on univariate inputs.

References

1. Hegde MS, Krishna G, Srinath R (2018) An ensemble stock predictor and recommender system. In: 2018 international conference on advances in computing, communications and informatics, ICACCI 2018. Bangalore, India, 19–22 Sept 2018, pp 1981–1985
2. Qiu M, Song Y (2016) Predicting the direction of stock market index movement using an optimized artificial neural network model. PLoS ONE 11:1–11
3. Li J, Bu H, Wu J (2017) Sentiment-aware stock market prediction: a deep learning method. In: 2017 international conference on service systems and service management, pp 1–6. https://doi.org/10.1109/ICSSSM.2017.7996306
4. Sedhai S, Sun A (2017) An analysis of 14 million tweets on hashtag-oriented spamming. J Assoc Inf Sci Technol 68(7):1638–1651

5. Jiang W (2016) "Stock market valuation using internet search volumes: US-China comparison," Summer Program for Undergraduate Research (SPUR). Available at http://repository.upenn.edu/spur/10
6. Nair BB, Dharini NM, Mohandas VP (2010) A stock market trend prediction system using a hybrid decision tree-neuro-fuzzy system. In: 2010 international conference on advances in recent technologies in communication and computing, pp 381–385
7. Seif MM, Hamed EMR, Hegazy AEFAG (2018) Stock market real time recommender model using apache spark framework. In: Hassanien AE, Tolba MF, Elhoseny M, Mostafa M (eds) The international conference on advanced machine learning technologies and applications (AMLTA2018). Springer International Publishing, Cham, pp 671–683
8. Jareo F, Negrut L (2016) Us stock market and macroeconomic factors. J Appl Bus Res 32:325–340
9. K. Simuni K (2003) "Visualization of stock market charts", proceedings of the international conference in central europe on computer graph-ics visualization and computer vision (WSCG)
10. Hegazy O, Soliman OS, Abdul Salam M (2013) A machine learning model for stock market prediction. Int J Comput Sci Telecommun 4:17–23
11. Marjanovic B (2017) Huge stock market dataset. https://www.kaggle.com/borismarjanovic/price-volume-data-for-all-us-stocks-etfs
12. Mehmood R, Maurer H, Afzal MT (2013) Knowledge discovery in hashtags. In: 2013 IEEE 9th international conference on emerging technologies (ICET), pp 1–6
13. Prince V, Labadié A (2007) Text segmentation based on document understanding for information retrieval. In: Kedad Z, Lammari N, Métais E, Meziane F, Rezgui Y (eds) Natural language processing and information systems. Springer, Berlin, pp 295–304
14. Ponte JM, Croft WB (1997) Text segmentation by topic. In: Peters C, Thanos C (eds) Research and advanced technology for digital libraries. Springer, Berlin, pp 113–125
15. Good IJ (1983) The philosophy of exploratory data analysis. Philos Sci 50(2):283–295
16. Ichinose K, Shimada K (2018) Stock market prediction using keywords from expert articles, pp 409–417
17. Sharma A, Bhuriya D, Singh U (2017) Survey of stock market prediction using machine learning approach. In: 2017 international conference of electronics, communication and aerospace technology (ICECA), vol 2, pp 506–509
18. Huitema BE (1988) Autocorrelation: 10 years of confusion. Behav Assess 10(3):253–294
19. Mitavskiy B, Cannings C (2009) Estimating the ratios of the stationary distribution values for Markov chains modeling evolutionary algorithms. Evol Comput 17(3):343–377
20. Dickey DA, Fuller WA (1979) Distribution of the estimators for autoregressive time series with a unit root. J Am Stat Assoc 74(366a):427–431
21. Sephton P (2008) Critical values of the augmented fractional dickeyfuller test. Empir Econ 35:437–450
22. Kwiatkowski D, Phillips PC, Schmidt P, Shin Y (1992) Testing the null hypothesis of stationarity against the alternative of a unit root: how sure are we that economic time series have a unit root? J Econ 54(1):159–178
23. Ogasawara E, Salles R, Porto F, Belloze K, Gonzlez Silva PH (2018) Nonstationary time series transformation methods: an experimental review. Knowl Based Syst 164:274–291
24. Anggrainingsih R, Aprianto GR, Sihwi SW (2015) Time series forecasting using exponential smoothing to predict the number of website visitor of Sebelas Maret University. In: 2015 2nd international conference on information technology, computer, and electrical engineering (ICITACEE), pp 14–19
25. Saputra ND, Aziz A, Harjito B (2016) Parameter optimization of Brown's and Holt's double exponential smoothing using golden section method for predicting Indonesian crude oil price (ICP). In: 2016 3rd international conference on information technology, computer, and electrical engineering (ICITACEE), pp 356–360
26. Hansun S (2013) A new approach of moving average method in time series analysis. In: 2013 conference on new media studies (CoNMedia), pp 1–4

27. Hodrick RJ, Prescott EC (1997) Postwar U.S. business cycles: an empirical investigation. J Money Credit Bank 29(1):1–16
28. Landon-Lane J (2002) Inverting the Hodrick-Prescott filter. Comput Econ 20(3):117–138. https://doi.org/10.1023/A:1020923129872
29. Christiano LJ, Fitzgerald TJ (2003) The band pass filter. Int Econ Rev 44(2):435–465. https://doi.org/10.1111/1468-2354.t01-1-00076
30. Hochreiter S, Schmidhuber J (1997) Long short-term memory. Neural Comput 9(8):1735–1780. https://doi.org/10.1162/neco.1997.9.8.1735
31. Taylor SJ, Letham B (2018) Forecasting at scale. Am Stat 72(1):37–45
32. Li G, Wang Y (2013) Automatic arima modeling-based data aggregation scheme in wireless sensor networks. EURASIP J Wirel Commun Netw 1:85. https://doi.org/10.1186/1687-1499-2013-85

Pseudo Random Number Generation Based on Genetic Algorithm Application

V. Pushpalatha, K. B. Sudeepa, and H. N. Mahendra

Abstract Standard computational tool like random number generator is used to create a sequence of numbers which are apparently unrelated. These numbers are used in different computations and statistics. Cloud computing is an emerging technique in all over the world to access the information via the Internet. In this paper, we proposed the hybrid model to generate a random numbers using linear-feedback shift register (LFSR) and GA to increase the key length. The hybrid model is also used to give security for the cryptographic applications. In this work, we implemented the hybrid model using the simulation tool. Modeling and simulation innovations are reasonable for assessing execution and security issues. This paper gives details about the hybrid model we proposed and the cloud simulation tool that we are using, and also we discussed about the different cloud simulation tools.

Keywords Pseudo random number · Genetic algorithm (GA) · LFSR · Cloud computing · CloudSim

1 Introduction

Nowadays, providing security to users from unauthorized user through the Internet is the major concern. It is very necessary to avoid unauthorized user from accessing confidential information or data. Cryptographic operations are used to solve the problem of confidentiality and authorization. The confidentiality helps to prevent the

V. Pushpalatha (✉) · K. B. Sudeepa
NMAM Institute of Technology, Nitte and Affiliated Visvesvaraya Technological University, Belagavi, India
e-mail: pushpav27@gmail.com

K. B. Sudeepa
e-mail: sudeepa@nitte.edu.in

H. N. Mahendra
JSS Academy of Technical Education, Bangalore and Affiliated Visvesvaraya Technological University, Belagavi, India
e-mail: mahendrahn2007@gmail.com

N. N. Chiplunkar and T. Fukao (eds.), *Advances in Artificial Intelligence and Data Engineering*, Advances in Intelligent Systems and Computing 1133,
https://doi.org/10.1007/978-981-15-3514-7_59

original data from the unauthorized users, and the authorization helps to prevent the unauthorized user from sending messages.

In this paper, the non-binary pseudo random number sequence is generated using a linear-feedback shift register (LFSR) [1]. The length of the key sequence is extended by designing hybrid model using LFSR and genetic algorithm (GA). Extending of the key length of the LFSR is the primary work in this project. To achieve more security on the cryptographic application and one time padding see [2]. The statistical tests are conducted to evaluate the randomness properties of the key sequence generated from the hybrid model. The non-binary pseudo random number which is generated using the hybrid model is applied for the cryptographic applications. The security for cryptography applications is achieved using the cloud. Here, the cloud implementation is done using simulation application, i.e., CloudSim.

The definition of cloud is defined by the National Institute of Standards and Technology (NIST) as follows. "Cloud Computing is a model for enabling convenient, on-demand network access to a shared pool of configurable computing resources (e.g., networks, servers, storage, applications, and services) that can be rapidly provisioned and released with a minimal management effort or service provider interaction" [3]. It has a capacity to strengthening, reliability, adaptability, accessibility, agility, performance, multi-occupancy, security, and maintenance [4]. Cloud computing will be a major innovation in the advancement of things to come Internet of Services, and it conveys platform, framework, and software that are accessible through membership-based administrations in a pay as-you-go model to clients [5].

1.1 Random Number Generator and Its Types

The process which generates the numbers and the random numbers which is generated is not in any sequence order is called as random number generator (RNG). The result may be the binary or non-binary sequence.

Random number generators are classified into two types.

1. True Random Number Generator (TRNG).
2. Pseudo Random Number Generator (PRNG).

1.1.1 True Random Number Generator (TRNG)

An entropy source uses true random number generator that is already available instead of inventing them. The amount of unpredictability is entropy. Low production rate is the major disadvantage of TRNG. Another important drawback of these generators is that they depend on a sort of hardware. There are mainly four TRNGs.

a. Random.org
b. Hot Bits

c. Lasers
d. Oscillators.

1.1.2 Pseudo Random Number Generator (PRNG)

The sequence of random number generator does not depend on the real word entity. To secure pseudo random number generator from the attackers, each value in the sequence can be determined, and it can be easily recalculated, if and only if the pseudo random number generators have the knowledge of initial value. To generate the random number sequence, the pseudo random number generator uses pre-calculated tables and mathematical formula tables.

2 Related Work

Fadheela Sabriet et al. have proposed a method to generate binary sequence by applying genetic algorithm. The productivity of genetic algorithm for pseudo arbitrary generation of numbers and its proficiency is upgraded by the quantity of parameters, such as mutation, initial population, size of population, and the crossover is demonstrated in this work.

Wang Yuhua et al. have discussed about evolutionary design of random numbers. They have proposed system to low-cost, high-speed, and minimum consumption of bit sequence generated based on LFSR.

Alireza Poorghanad et al. have generated a high-quality pseudo random number utilizing developmental techniques. The LFSR is used to implement the binary polynomial generator, and they have proposed generation of 128 bit string of high-speed, high-quality, and low-power random numbers using genetic algorithm.

Saiqin Long et al. have discussed about the implementation details of CloudSim and the extended CloudSim framework. The extended CloudSim framework has the center capacities of the data storage cloud, generally in two factors: (1) increasing in file stripping function on CloudSim strips into multiple blocks, (2) increasing in replica management function on CloudSim that makes the replica of the data which is stored in the CloudSim which helps to avoid data blocking.

Saleh Atiewi et al. have discussed two different cloud simulators, i.e., GreenCloud and CloudSim. The authors have done a comparative study on GreenCloud, and they concluded that compared to the CloudSim simulator, a GreenCloud has more advantages because the GreenCloud mechanism is based on NS2 network simulator, and it helps to capture the packet-level details.

3 Statistical Test

The probabilistic property describes the randomness of the generated pseudo random numbers, and the generated sequence needs to be tested. The function is designed which is used to perform operation, and it will decide whether to choose the null hypothesis or to reject. The acceptance or rejection of null hypothesis Ho will be decided by using systematical statistical rule [6].

The hypotheses in statistical testing are as follows:

Sequence is random: Null hypothesis (H_0)
Sequence is not random*:* Alternate hypothesis (H_a).

The hypothesis called H_0 (null hypothesis) confirms that the generated sequence is random, and the H_a (alternative hypothesis) confirms that the sequence is not random. Test value hypothetical testing is contrasted with a point on dispersion test is called critical value.

3.1 Chi-Square Test

The possibility of rejecting the null hypothesis by concluding false even though it is true is considered to identify the percentage of wrongly rejecting null hypothesis using statistical testing. The significance level is considered from the probability of wrong decisions in the testing. Chi-square test is invented by Karl Pearson in 1900 to test the randomness by using symbol x^2. In many of the cases to distribute various outputs of the experiment, the chi-square test is used. Assume that x is an observation, which is organized into different classes denoted by c, and each class has its lower limit and upper limit. Let us consider O_i to be the observed value of the number of observation falling into the ith category, such that $O_1 + O_2 + \cdots + O_c = x$. The probability of the observation that falls into category i is considered as p_i so that $p_1 + p_2 + \cdots p_c = 1$. Then,

$$O_i \approx xp_i$$

The statistic value for x^2 is defined as

$$x^2 = (O_1 - E_1)2/E_1 + (O_2 - E_2)2/E_2 + \cdots + (O_c - E_c)2/E_c$$

$$x_0^2 = \sum_{i=1}^{c}(O_i - E_i)2/E_i \quad (1)$$

The possibility of rejecting the null hypothesis by concluding false even though it is true is considered to identify the percentage of wrongly rejecting null hypothesis using statistical testing. The significance level is considered from the probability of wrong decisions in the testing.

Let us consider h as degree of freedom. The value should be less than the number of categories, i.e., $h = \textit{number of classes} - 1$.

3.2 Run Test

In a grouping, a run can be characterized as an arrangement of indistinguishable occasions and pursued by the distinctive occasions. The length of the run is the number of events that occur in the run, and the number of run is the total number of runs that occur in the sequence. Each run might be of different length. The run test determines the arrangements of number in a sequence to test hypothesis independence.

There are different cases of run test with respect to the number of runs and length of runs.

3.2.1 Runs up and Runs Down

The runs counted will be runs up and runs down. A keeps running up is an arrangement of numbers, each will be prevailing by an extensive number and comparable in keeps running down and the grouping will be prevailing by a more modest number. Consider that N is a number in a sequence and a is the total number of runs, the mean variance is given by

$$\mu_a = \frac{2N - 1}{3} \tag{2}$$

and

$$\sigma_a^2 = \frac{16N - 29}{90} \tag{3}$$

For $N > 20$, the distribution of a is approximately by a normal distribution. The test statistic is

$$Z_0 = \frac{a - \mu_a}{\sigma_a}$$

3.2.2 Runs Above and Below the Means

Consider b has the amount of runs, $n1$ and $n2$ be the number of observations above and below the mean, and the mean μ and variance of b for a truly independent sequence are given by

$$\mu_b = \left(\frac{2n1n2}{N}\right) + \left(\frac{1}{2}\right) \quad (4)$$

and

$$\sigma_b^2 = 2n1n2(2n1n2 - N)/N^2(N-1) \quad (5)$$

The test statistic is given by equation

$$Z_0 = \frac{b - \mu_b}{\sigma_b}$$

4 Hybrid Model of LFSR and Genetic Algorithm

This paper mainly works on extending the length of the non-binary sequence. In case of non-binary sequence, the cryptographic strength can be measured by both encryption algorithm and key generation techniques. Different types of algorithm in case of usage of maximum length sequence will give sturdy cryptographic systems. In such a manner, this paper focuses on the greatest length of non-binary sequence. Linear-feedback register is one of pseudorandom number generator (PRNG), and it generates key sequences which are periodic and limited one.

4.1 Linear-Feedback Shift Register (LFSR)

The random number of sequences is generated by using LFSR. The LFSR has N number of registers also called as stages. Each stage will show the actual status of the register at that particular time. The new status will be produced that rely upon the present status of the register by utilizing the provided feedback function f (see Fig. 1). The generated range of number will specify the modular value of the feedback function. Let m be the modular value and n be the quantity of dimensions of LFSR, and the maximal length of key sequence is given by $m^n - 1$ [1]. In the LFSR, the length is minimum; to extend the length of the feedback shift register, the genetic algorithm is used.

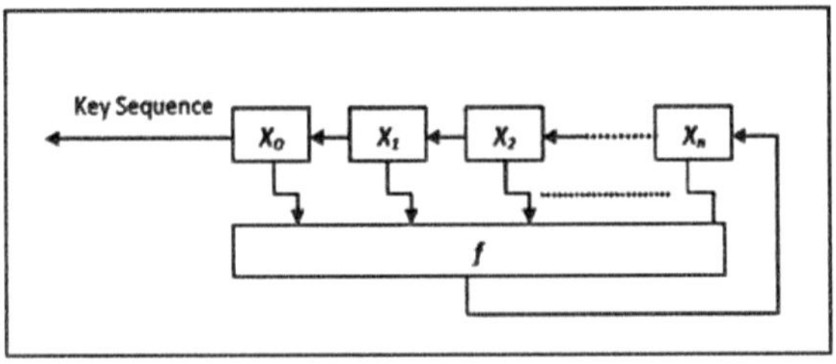

Fig. 1 Linear feedback function

4.2 Genetic Algorithm

This method of solving problem is introduced by John Holland in the year 1960s. Genetic algorithms are a kind of optimization problem, which is used to produce an optimal solution for a given problem [6].

The genetic algorithm uses three operators such as selection, crossover, and mutation. The selection operator chooses the best result and that result is used in the next generation. The crossover is the process of merging the genetic information of two individuals. The mutation operator is used to flip the values of one or more bit within the information. After all three operations are completed, the new solution will be generated. This process will be repeated until the required output produces. Figure 2 shows the workflow of genetic algorithm.

The proposed model, the pseudo random key sequence is generated from the LFSR, and from each new key, new sequence will be generated using genetic algorithm. Generation of new key using LFSR is considered as a selection operation. The GA operations carried out like mutation-1, crossover, and mutation-2 operation. The process of generating random numbers is shown in Fig. 3.

In mutation-1, two bits are considered to perform mutation operation. Let us consider 8 bits of the output from LFSR, i.e., $A_0A_1A_2A_3A_4A_5A_6A_7$, and these bits are considered as an input to the genetic algorithm process. The output of the mutation-1 operation will be $A_0\bar{A}_1A_2\bar{A}_3A_4A_5A_6A_7$ (see Fig. 4).

The output of mutation-1 operation will be input for the crossover operation. By using the crossover operator, we can merge two parent strings to get new best child string. In our work, the crossover operation is performed by shifting the least significant bit. The crossover operation output will be considered as an input to

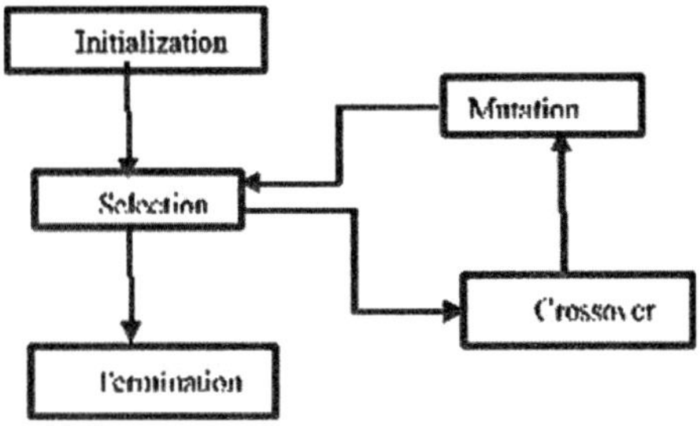

Fig. 2 Flow of genetic algorithm process

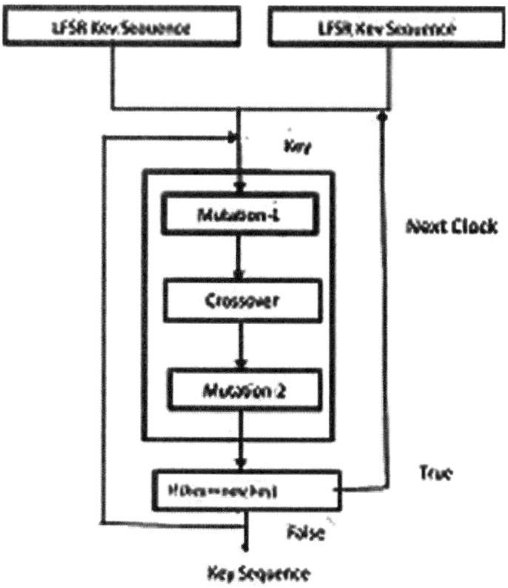

Fig. 3 Representation of genetic algorithm process using LFSR

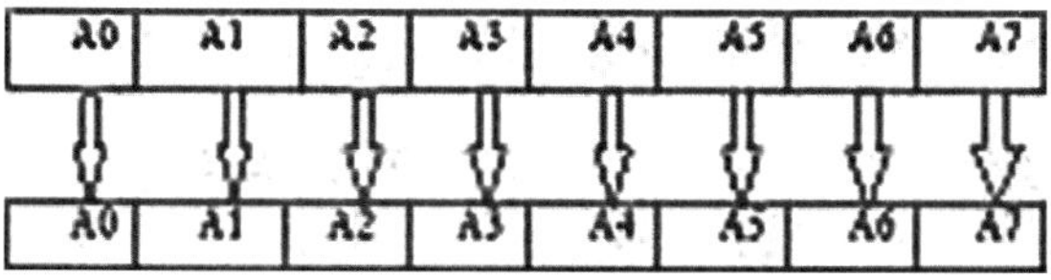

Fig. 4 Representation of mutation-1 operation

mutation-2 process. In mutation-2 operation, each bit is XORed with its previous bit, except the first bit.

5 CloudSim

The University of Melbourne, Clouds Research Center of Computer Science and Engineering Department, developed toolkit for simulation of cloud environments called as CloudSim. The CloudSim toolbox gives fundamental classes to delineating data center, virtual machines, applications, clients, computational assets, and approaches for the administrators of arranged parts of the structure (see Fig. 5) [7]. The CloudSim mainly works on SimJava. The ready-to-use environment is not provided by CloudSim for the complete execution of scenario with some specific input. To overcome from this, the user of CloudSim is required to implement their own scenario to perform the operation to get the required output by providing the input parameters. The different extensions which are given to the CloudSim are CloudAnalyst, CloudsimEx, WorkflowSim, Cloud Auction, SimpleWorkFlow, RealCloudSim, EMUSim, CloudReports, and Cloud-MIGXpress. The primary two advantages of utilizing CloudSim for execution testing comprise time adequacy and adaptability and applicability [8].

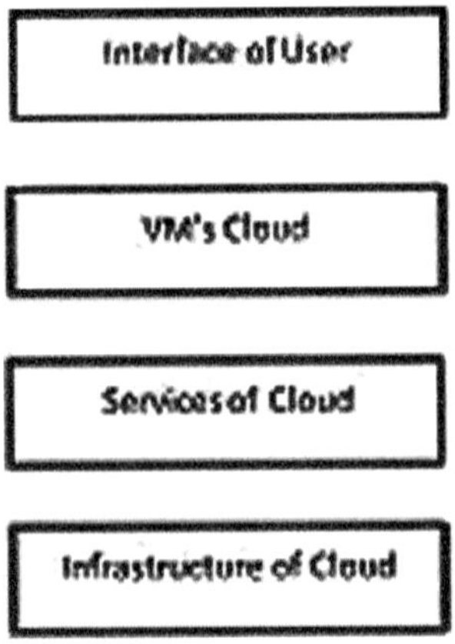

Fig. 5 CloudSim components

The other simulation technologies available are CloudAnalyst, GreenCloud, OCT, SPECI, Open Cirrus, GroudSim, Network CloudSim, iCanCloud, etc.

6 Proposed Design

The proposed hybrid model is based on non-binary LFSR generator and genetic algorithm to obtain the sequence of pseudo random number, to extend the length of key sequence which is generated by the LFSR. Each iteration output of the LFSR is considered as an input to genetic algorithm. The workflow of the LFSR and GA is shown in Fig. 3. The above sequence works on the principal in which all the elements are compared with the first input and of the genetic algorithm from LFSR; if the first element and the new key element are equal, then the sequence is terminated and the clock is given to the LFSR to generate the next input to GA (see Fig. 6). This process will repeat until the end of LFSR generation.

Let m be the sequence length of the LFSR and n be the number of random number which is generated without any repetition. For each individual input, length of a hybrid model (LHM) is the product of m and n

$$\text{LHM} = (m * n) \tag{6}$$

These procedures will be actualized utilizing simulation tool called CloudSim. The toolbox underpins framework structure and displaying of cloud segments, for example, virtual machines (VMs), provisioning procedures and data centers. The CloudSim is a Java-based toolkit, it consists of few entities which are like Java classes, and they are used to communicate with each other while simulation process is started.

The common entities in CloudSim which are used for simulation process to perform the certain actions are as follows [5]:

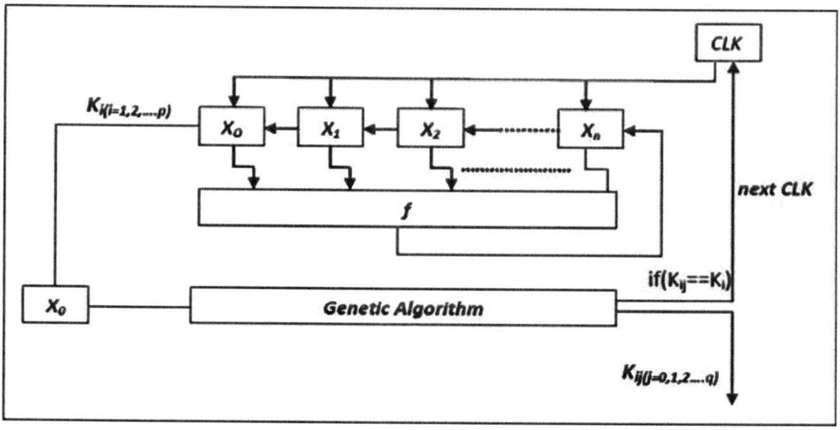

Fig. 6 General hybrid model to generate random sequence

a. **Data Center (DC)**: Data center is the center segment of the cloud framework to furnish clients with equipment and programming assets. The data center consists of more than one host.
b. **Cloud Information Service (CIS)**: The CIS is the type of a registry, and it contains all the resources that exist in the cloud.
c. **Broker**: The broker is an agent, which submits to virtual machine (VM) in a particular host and then it combines with cloudlets and applications on particular virtual machine to perform task. Once all the cloudlets are executed, the free virtual machines are destroyed.
d. **Data Center Broker**: An agent who submitted the application to the data center and who is in charge of the conveyance between data center and the virtual machine.
e. **Cloudlet**: The cloudlets are the applications in which it is run in a virtual machines.
f. **Cloudlet Generator**: The cloudlet generator will create a new cloudlet, and those cloudlets are submitted to the broker.
g. **Virtual Machine (VM)**: Virtual machine acts as a logical machine to run the applications.
h. **VM Generator**: Which produces the virtual machines in the cloud simulator and presents the new virtual machine to the broker.
i. **Host**: Physical machines where the logical machines are kept. The host will have some kind of hardware configuration (HC).
j. **VM Allocation Policy**: This is implemented by the host components as abstract class.

Figure 7 shows the CloudSim model which consists of entities like Cloud Information Service (CIS), host, data center, broker, cloudlets, and virtual machine (VM). The CIS is an entity type of a registry, it contains all the resources that exist in the cloud, and every one of the assets which are made in the cloud must get registered

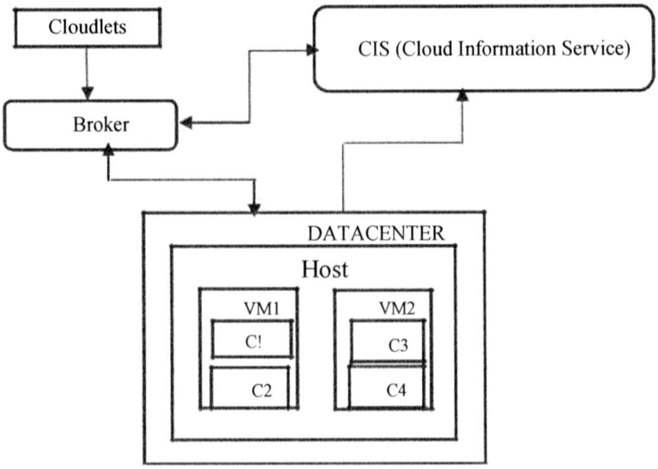

Fig. 7 Representation of CloudSim model

in the CIS previously it is utilized in cloud. The data center must have a host. The hosts are like hardware configurations which have the parameter like RAM, bandwidth, and processing elements. The host will be virtualized into many number of virtual machines. The cloudlets will run inside the virtual machines. The broker is an element of data center broker class in which it will be present in the assignment to the data center. The broker initially communicates with CIS to derive the resource information, and the CIS restores the qualities of the data center. Once the broker receives all the information of data center, it will submit the cloudlet list to the data center. Once the cloudlets are submitted to data center, it will start running in virtual machine which is resided inside the host of the data center. The virtual machine which resides inside the data center will be scheduled by the hosts, and the cloudlets will be scheduled by virtual machine which resides inside the host [5].

7 Results and Discussion

The result of pseudo random number sequence generated using the hybrid model is tested using statistical testing, and the key generated using sequence will be used for the cryptographic applications. The results of the hybrid model and cryptographic applications are discussed.

7.1 Results and Analysis of Pseudo Random Number Key Sequence

The proposed hybrid model of pseudo random number generator is used to generate the random key sequence. The hybrid model is the combination of both LFSR and genetic algorithm as shown in the figure with the initial values $X_0 = 200, X_1 = 120, X_3 = 5$. In Table 1, sequence generated from LFSR is shown in the first column which is subjected to the GA, the key generation is repeated until initial value is repeated, and that sequence of output is shown in each row of the table. The values in the last column represent the repetition of the initial values generated value from LFSR.

7.1.1 Statistical Testing

The simulation of the hybrid model is designed using LFSR and GA with the initial value, and the below tests are conducted on random number sequence to observe the randomness properties.

a. **Uniformity Test**

The uniformity of the generated key sequence is checked using this test. This test is done to comparing the distribution of the set of numbers which is generated by the simulator and the uniform distribution (see Table 2). The chi-square test is used to check the test of uniformity. This test is performed using Eq. (1).

After the satisfying all the test to check the randomness of the key sequence, then it is important to check in the cryptographic application. The next section gives the detail of the cryptographic application.

7.2 Application of Hybrid Model for Image Encryption in Cloud Environment

In the stream cipher system, encryption of an image is done using the key which is generated from pseudo random generated using the hybrid model. The generated key value is stored in the cloud environment to perform the encryption of the image. N number of cloudlets are created to perform the encryption operation. Operation of encryption algorithms are defined as follows. Let r_i be the plain text where $i = 1, 2, 3, \ldots n$. The total number elements of plain text in a stream are denoted by n, and the corresponding key element generated from the hybrid model is represented as k_i. The encryption operation is defined as follows.

Cipher text $\{c_i\} = \{r_i\} \oplus \{k_i\} \mod m$, where $i = 0, 1, 2, 3 \ldots . n$ and m modulus

Table 1 Key generated from the hybrid model

200	66	195	155	63	80	125	45	31	78	5	100	20	119	245	1	107	200
120	99	106	245	88	41	13	22	28	77	60	30	150	106	230	8	220	120
5	33	205	135	33	50	100	15	97	10	96	203	160	185	195	9	103	5
99	11	120	5	78	33	244	20	66	80	54	119	120	97	233	55	25	99
195	55	200	216	97	68	215	88	89	33	17	48	209	173	150	29	238	195
245	10	230	89	70	50	100	44	69	11	28	142	33	198	135	64	99	245
55	32	138	99	69	10	203	88	10	98	33	249	200	165	125	66	135	55
139	8	25	233	81	51	1	22	3	80	44	108	241	113	89	46	110	139
66	49	211	10	75	12	150	21	90	10	22	189	216	159	71	88	103	66
1	89	55	210	26	69	148	53	67	2	18	150	226	241	166	40	113	1

Table 2 Chi-square value of LFSR and the hybrid model

Type of PRNG	X_O^2	Critical value	Accept/reject
LFSR	3.569	11.34	Accepted
Proposed model	1.459	11.34	Accepted

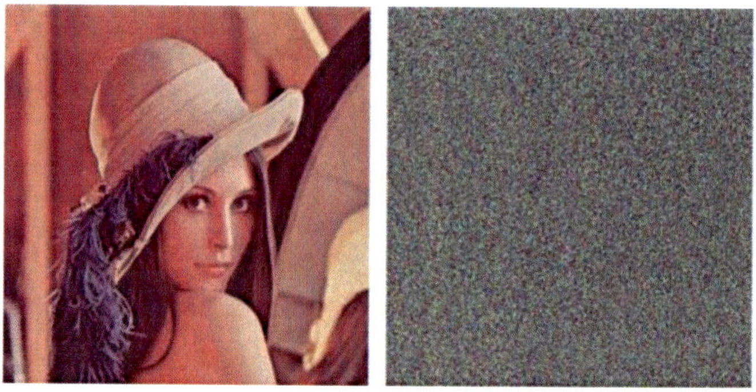

(a) (b)

Fig. 8 Encryption of an image. **a** original image, **b** encrypted image

The stream of cipher text element will be the output of the encryption.

The input image for the encryption is shown in Fig. 8a, and the encryption and decryption operation and output image of the decryption are represented in Fig. 8b.

7.2.1 Histogram Comparison of Ciphertext Image

The pixel accuracy of the cipher image and the original image will be compared to understand the firmness of the cipher system. Histogram of the original image is shown in Fig. 9, and the histogram of cipher image is shown in Fig. 10, respectively. The figure shows that the occurrence of the pixel in the cipher image is evenly distributed, it indicates that the firmness of the cipher image is more compared to the original image, and it helps to avoid the attacks.

8 Conclusion

It has been observed that the key length of the pseudo random key sequence is enhanced beyond the maximum length of LFSR. The statistical testing is performed to test that key sequence satisfies the uniformity and independence. And also the output key sequence increases the length, and it is more secured for using in cryptographic application. The cloud environment provides security for the image. Once the image

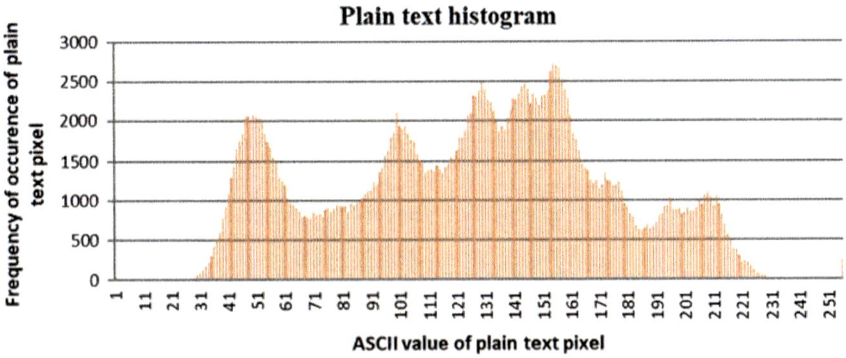

Fig. 9 Histogram of number of pixels in original shown in Fig. 8a

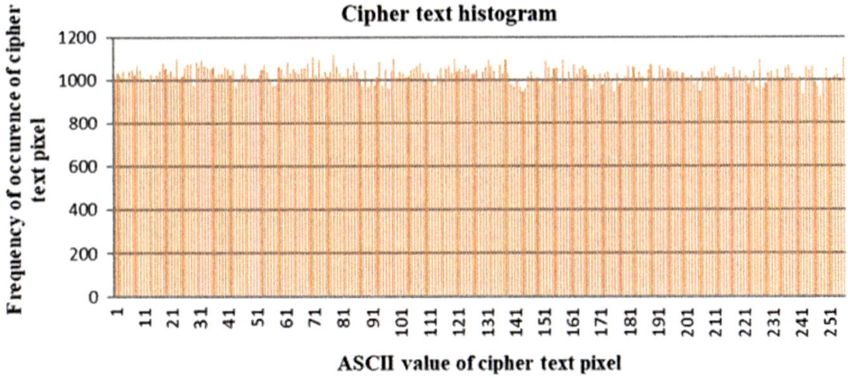

Fig. 10 Histogram of number of pixels in cipher image for the hybrid model shown in Fig. 8b

is stored in the cloud, the cloudlets will be created to perform the operation. By using the extended key length, it provides high security for encryption of the image to provide security in cloud environments. The proposed system secures the image from the attackers.

References

1. Poorghanad A, Sadr A, Kashanipour A (2008) Generating high quality pseudo random number using evolutionary methods. In: International conference on computational intelligence and security, pp 331–335. https://doi.org/10.1109/cis.2008.220
2. Singh D, Rani P, Kumar R (2013) To design a genetic algorithm for cryptography to enhance the security. Int J Innov Eng Technol (IJIET) 2(2):380–385
3. Atiewi S, Yussof S (2015) Comparison between CloudSim and green-cloud in measuring energy consumption in a cloud environment. In: 3rd international conference on advanced computer science applications and technologies, 2014, pp 9–14. https://doi.org/10.1109/acsat.2014.9

4. Pushpalatha V, Sudeepa KB, Mahendra HN (2018) A survey on security issues in cloud computing. Int J Eng Technol 7:758–761
5. Calheiros RN, Ranjan R, Beloglazov A, De Rose CA, Buyya R (2010) CloudSim: a toolkit for modeling and simulation of cloud-computing environments and evaluation of resource provisioning algorithms. Wiley Online Libr 41(1):23–50. https://doi.org/10.1002/spe.995
6. Abu-Almash FS (2016) Apply genetic algorithm for pseudo random number generator. Int J Adv Res Comput Sci Softw Eng 6(8):8–19
7. Long S, Zhao Y (2012) A toolkit for modeling and simulating cloud data storage: an extension to CloudSim. In: International conference on control engineering and communication technology, pp 597–600. https://doi.org/10.1109/iccect.2012.160
8. Malhotra R, Jain P (2013) Study and comparison of CloudSim simulators in the cloud computing. SIJ Trans Comput Sci Eng Appl (CSEA) 1(4):111–115

Analysis of an Enhanced Dual RSA Algorithm Using Pell's Equation to Hide Public Key Exponent and a Fake Modulus to Avoid Factorization Attack

K. R. Raghunandan, Rovita Robert Dsouza, N. Rakshith, Surendra Shetty, and Ganesh Aithal

Abstract Public key cryptography generates two distinct keys: one for encryption and another separate but related key for decryption. There exists only one private key for every public key to decipher the message. A variant of RSA, called the Dual RSA, has two different key pairs having separate private and public key exponents. Security of the RSA is compromised using mathematical attacks, by the factorization of '*n*'. This paper proposed an enhanced approach to Dual RSA with the help of Pell's equation and a fake modulus key. The algorithm eliminates the distribution of '*n*', whose factors compromise the RSA algorithm. Also, the solutions to Pell's equation are shared instead of the public key components. A comparative analysis is carried out with respect to RSA, Dual RSA and the proposed algorithm based on the complexity of each step of the algorithm. It is observed that the proposed system provides more security to the public key exponents and the system modulus '*n*', than RSA and Dual RSA.

Keywords Public key cryptography · Dual RSA · Pell's equation · Standard deviation · Avalanche effect

K. R. Raghunandan · R. R. Dsouza (✉) · N. Rakshith · S. Shetty
Department of Computer Science and Engineering, NMAM Institute of Technology, Affiliated to Visvesvaraya Technological University, Nitte, Udupi, Karnataka 574110, India
e-mail: rv6oasis@gmail.com

K. R. Raghunandan
e-mail: raghunandan@nitte.edu.in

N. Rakshith
e-mail: rakshithkulal017@gmail.com

S. Shetty
e-mail: hsshetty@nitte.edu.in

G. Aithal
Department of Electronics and Communications, Shri Madhwa Vadiraja Institute of Technology & Management, Affiliated to Visvesvaraya Technological University, Bantakal, Udupi, India
e-mail: ganeshaithal@gmail.com

N. N. Chiplunkar and T. Fukao (eds.), *Advances in Artificial Intelligence and Data Engineering*, Advances in Intelligent Systems and Computing 1133,
https://doi.org/10.1007/978-981-15-3514-7_60

1 Introduction

One of the greatest revolutions in cryptography is the development of public key cryptography. Rather than substituting and permuting operations, public key systems are based on mathematical functions [1]. Public key scheme involves the usage of two keys. Therefore, it is asymmetric in nature. However, the usage of two separate keys has great effects in confidentiality, distribution of keys, as well as authentication. Public key cryptography has evolved from attempts that attacked the digital signatures and the key distribution. These types of algorithms have a separate key for enciphering and a key for deciphering the messages.

RSA algorithm is used in modern computers for encrypting and decrypting the messages. It is a public key encryption algorithm. In this type of cryptosystem, the key for encryption is publicly shared and it varies from the key for decryption which is hidden. RSA procedure is very simple. Thus, it has become very popular. The RSA scheme is accepted worldwide and implemented for asymmetric encryption. It is a block cipher where the original message and encrypted message are integers between 0 and $p-1$ for some p [1].

The asymmetric property in an RSA algorithm is based on factorization of the product of large prime numbers. Based on these large prime numbers, the user creates a public key. However, the prime numbers are kept as secret. The public key is used to cipher the messages. But in current methods, if the public key has a large value, we can decode the message when the prime numbers are known. Breaking the RSA encryption is known as the RSA problem [2].

The working of RSA can be depicted in the following way:

1. Select two prime numbers, p_a and q_a, such that $p_a \neq q_a$.
2. $n = p_a q_a$ is computed, and $\Phi(n) = (p_a - 1)(q_a - 1)$ is found.
3. An integer e is chosen, where $GCD(e, \varphi(n)) = 1$ and $e \leq \varphi(n)$.
4. Integer d is found where $ed = 1 \mod \varphi(n)$.
5. Publish the public key $= (e, n)$.
6. The private key $= (d, n)$ is hidden.

Dual RSA refers to double instances of RSA sharing same public key exponent and private key exponent. It combines the instances to get a single instance of Dual RSA [3]. It has a encryption key (n_a, n_b, e) and decryption key (p_a, p_b, q_a, q_b, d). This satisfies two equations: $ed \equiv 1 \mod \varphi(n_a)$ and $ed \equiv 1 \mod \varphi(n_b)$. The equations for the keys are given by $ed = 1 + ka\,\Phi(n_a)$ and $ed = 1 + kb\,\Phi(n_b)$. Here, $k_a \Phi(n_a) = k_b \Phi(n_b)$. The idea is to construct integers k_a, k_b, k_c, such that $k_b k_c = (p_a - 1)(q_a - 1)$ and $k_a k_c = (p_b - 1)(q_b - 1)$. Here, p_a, p_b, q_a, q_b are all prime numbers.

2 Literature Survey

This particular section contains the work carried out by many researchers in the domain of RSA cryptosystem. The discussion is about the modifications of RSA in the recent past. It describes the Pell's solutions in a prime field. However, there might be a misuse of the RSA algorithm, i.e. attacks on RSA. This paper deals with elementary attacks which are explained in the following sections.

2.1 *Dual RSA*

Ibrahim [4] proposed an improvised method for RSA algorithm. In this paper, the author makes use of three prime numbers instead of two. It makes use of a, b, and c to find n, such that $n = p_a * p_b * p_c$ and $\Phi(n) = (p_a - 1)(p_b - 1)(p_c - 1)$. The value of p is chosen, such that p and $\Phi(n)$ are co-prime. Using modular arithmetic, q is found where $pq = 1 \mod \Phi(n)$. The encryption is done using $c \equiv mp \mod n$, and the message is decrypted using $m \equiv cq \mod n$.

Minni et al. [5] put forward a modified algorithm of RSA to enhance security. In this algorithm, x is shared instead of n in public key, where x is a co-prime of n. Consider that the initial prime numbers are u and v. The value of n is computed as $u * v$ and $\Phi(n) = (u - 1)(v - 1)$. A variable k_1 is computed with the conditions:

- $\sqrt{n} < k_1 < \varphi(n)$
- $GCD(k_1, \varphi(n)) = 1$
- k_1 should have short length and less Hamming distance.

The variable w is computed to substitute n. If $u > v$, let w be an integer, such that $n - u < w < n$ and $GCD(w, n) = 1$. If $u < v$, let w be an integer, such that $n - v < w < n$ and $GCD(w, n) = 1$. A variable k_2 is found, such that $k_1 k_2 \mod w = 1$. The encryption key has (k_1, w), and decryption key has (k_2, w). The message is encrypted using $ct \equiv pt^{k1} \mod w$ and decrypted as $pt = \sqrt{(ct^{k1} \mod w)}$.

K. R. Raghunandan et al. [6] put forward a variant of the RSA algorithm. In this algorithm, the author makes use of four prime numbers instead of two. The encryption key and decryption key have three components (e, f, N) and (d, g, N), where e, f, d and g are randomly chosen. The private key (d, g) completely depends on public key (e, f) which can be computed using $(d * e = 1 \mod \varphi(n))$ and $(f * g = 1 \mod \varphi(n))$. The complexity of this algorithm increases by making use of modular exponent function twice instead of once like RSA. Public keys used in the encryption side (e and f) help hacker to find private keys (d and g).

2.2 Pell's Equation

Barbeau [7] mentions about the usage of solutions modulo 11. The sum of two numbers is the ordinary sum modulo 11. Similarly, the product is the difference of the highest multiple of 11 from the ordinary product that does not exceed it. Consider a prime number t, and Z_t is a set of numbers from 0 to $t - 1$. The sum and product are defined in Z_t and replaced by the remainder after division by t. Also, $G(t, d)$ is the set of distinct solutions modulo t, of the congruence $p^2 - dq^2 \equiv 1 \mod t$.

Conrad [8] reduced the Pell's equation $x^2 - dy^2 \equiv k$, such that the solution can be obtained from $x = px_1 + dqy_1$, $y = q x_1 + p y_1$, where $p_2 - dq_2 = 1$. Pell's equation has applications in double equations (from Diophantus), rational approximations to square roots, simultaneous polygon numbers, sums of consecutive numbers, Pythagorean triangle with consecutive legs, consecutive Heronian triangles and sums of n and $n + 1$ consecutive squares.

Smith [9] defines certain theorems that can be used to determine how the Pell's equation can be solved. The Pell Class approach is not efficient to give the complete solution of Pell's equation. Most approaches do not depend on integer factorization. The unsolvability is explained by the square polynomial problem, Legendre test and modulo N tests. However, a partial criterion for solvability is given by arithmetic of solvability.

2.3 Attacks on RSA

Some elementary attacks illustrate the misuse of RSA algorithm. The aim of this proposed model is to overcome several elementary attacks that result in the factorization of the common modulus. Such types of attacks are described in the following subsections.

Elementary Attacks These types of attacks are also called mathematical attacks. Mathematical attacks focus on hitting the basic mathematical function of the algorithm. The different types of elementary attacks are described below:

Common modulus. We may wish to fix the value of modulus n instead of generating $n = pq$ for each user. The trusted central authority provides the user with a public key (n, e) and private key (n, d). Since the modulus is common, the other user can factor the modulus n using his own exponents. Thus, RSA modulus should never be shared among the entities [10].

Blinding. Let (n, e) and (n, d) be the user's public and private key, respectively. An attacker wants the signature of the user on a message $M \in Z_n^*$. However, the user denies giving his signature on M, and the attacker selects $r \in Z_n^*$, such that $M' = r^e M \mod n$ asking the user to sign this message M'. The user may provide his signature S' on M' where $S' = \left(M'\right)^d \mod n$. The attacker can now simply compute $S = \frac{S'}{r} \mod n$ to obtain the signature on the original message M.

$$Se = \frac{(S')^e}{r^e} = \frac{(M')^{ed}}{r^e} = \frac{M'}{r^e} = M \mod n$$

This technique is called blinding. Blinding is termed as an attack, but it can be used to implement anonymous digital cash [10].
Low private exponent. A user may use a small value for d for reducing the decryption time and obtain a better performance. This is because the modular exponentiation requires $\log_2 d$ time. According to Wiener [11], a small value for d may harm the cryptosystem.

Theorem 1 (M. Wiener) *Let* $n = gh$ *with* $h < g < 2h$. *Let* $d < \frac{1}{3}n^{\frac{1}{4}}$. *Given* (n, e) *with* $ed = 1 \mod \varphi(n)$, *the attacker can efficiently recover the value of* d.

Proof The proof makes use of continued fraction approximation. Since $ed = 1 \mod \varphi(n)$, consider a k, such that $ed - k\varphi(n) = 1$.

$$\left|\frac{e}{\Phi(n)} - \frac{k}{d}\right| = \frac{1}{d\Phi(n)}$$

Here, $\frac{k}{d}$ is an approximation of $\frac{e}{\Phi(n)}$. The attacker does not know $\Phi(n)$, but he may use the value of n to approximate it. Since $\varphi(n) = n - g - h + 1$ and $g + h - 1 < 3\sqrt{n}$, we have $|n - \Phi(n)| < 3\sqrt{n}$. Substituting n for $\Phi(n)$, we get

$$\left|\frac{e}{n} - \frac{k}{d}\right| = \left|\frac{ed - k\Phi(n) - kn + k\Phi(n)}{nd}\right|$$
$$= \left|\frac{1 - k(n - \Phi(n))}{nd}\right| \le \frac{3k\sqrt{n}}{nd} = \frac{3k}{d\sqrt{n}}$$

Now, $k\varphi(n) = ed - 1 < ed$. Since $e < \varphi(n)$, we have $k < d < \frac{1}{3}n^{\frac{1}{4}}$. Hence, we get $\left|\frac{e}{n} - \frac{k}{d}\right| \le \frac{1}{dn^{\frac{1}{4}}} < \frac{1}{2d^2}$. This relation is a classic approximation. It requires the computation of $\log n$ convergent of the continued fraction for $\frac{e}{n}$. Since $ed - k\varphi(n) = 1$, k and d are co-prime. Therefore, $\frac{k}{d}$ forms a reduced fraction. We can recover the value of d using this linear time algorithm [8].

Low public exponent. A small value for public exponent should be used to reduce the encryption time and signature verification time. The smallest assignment for e is 3. In order to overcome certain attacks, it is recommended to use $e = 2^{16} + 1 = 65,537$. A signature verification requires 17 multiplications when $2^{16} + 1$ value is used opposing when $e \le \varphi(n)$ is used. When we use a small value for e, we are far from a total break [10].
Hastad's broadcast attack. Let $\bar{N} = N_1 \ldots N_k$ and all g_i's are monic. Assume that the degree is d by multiplying each g_i with an appropriate power of x.

$$g(x) = \sum_{i=1}^{k} T_i g_i(x), \quad \text{where } T_i = \begin{matrix} 1 & \mod N_j,\ i = j \\ 0 & \mod N_j,\ i \ne j \end{matrix}$$

Here, T_i is known as the Chinese Remainder Coefficient, and $g(x)$ has a degree d and is monic because it is monic modulo to all N_i. Furthermore, $g(M) = 0 \mod \bar{N}$. Hastad's proof and Coppersmith's theorem are similar. However, the powers of g are not used in lattice. Thus, it obtains a weaker bound [10].

Franklin–Reiter message attack. They found an attack when two users share encrypted messages using the same modulus. Consider (n, e) as the public key of Alice and $M_1, M_2 \in Z_n^*$ be two distinct messages, such that $M_1 = f(M_2) \mod n$ where $f \in Z_n[x]$. To send the message to Rob, Jack may encrypt the message and send the ciphertext $C1$ and $C_2.C_1 = M_1^e \mod n$, M_1 is the root of the polynomial, $g_1(x) = f(x)^e - C_1 \in Z_n[x]$ and M_2 is the root of the polynomial $g_2(x) = xe - C_2 \in Z_n[x]$. A linear factor $x - M_2$ divides both the polynomials. The attacker can simply compute the gcd of g_1 and g_2 to find M_2. For $e = 3$, the gcd must be linear. The attack takes quadratic time in e for $e > 3$. Computing gcd is prohibitive for a larger value of e [10, 12].

Coppersmith's short pad attack. In this type of attack, the intruder obstructs the ciphertext and prevents it from reaching the receiver. Let $g_1(x, y) = x^e - C_1$ and $g_2(x, y) = (x + y)^e - C_2$. When $y = r_2 - r_1$, M_1 is a root of these polynomials. The root of $h(y) = \text{res}_x(g_1, g_2) \in Z_n[y]$ is $\Delta = r_2 - r_1$. The degree of h is at most e^2 and $|\Delta| < 2^m < n^{1/e^2}$. Here, Δ is a root of h modulo n. The attacker can find this root using Coppersmith's theorem. When $e = 3$, the attack is established as long as the length of the pad is one-ninth the length of the message [10].

3 Mathematical Preliminaries

Theorem 1 *According to Fermat's Little Theorem, if a and b are prime numbers and v is an integer,*

$$v^{a-1} \mod a = 1 \quad (1)$$

$$v^{b-1} \mod b = 1 \quad (2)$$

Multiply (1) *into* (2), *then we get equation* $M^{(a-1)(b-1)} \mod (a * b) = 1$. *Substitute* $n = a * b$ *and* $\varphi(n) = (a-1) * (b-1)$, *then equation will be* $M^{\varphi(n)} \mod n = 1$ *which is always equal to* $M^{k*\varphi(n)} \mod n = 1^k$. *Multiply M both sides we get* $M^{k*\varphi(n)+1} \mod n = M$.

Theorem 2 *If it is possible to replace* $k * \varphi(n) + 1$ *with the product of* x, y, z, *then* $x\,y.z = k \cdot \varphi(n) + 1$. *Now replace* $x * y * z$ *in Theorem* 1 *then* $M^{x*y*z} \mod n = M$. *Divide this equation into two parts of calculation*

$$C = \left(M^e \mod n\right)^f \mod n \quad (3)$$

$$M = C^d \mod n \quad (4)$$

Theorem 3 *Consider a non-square positive integer a. There exists a solution to $x^2 - Ry^2 = 1$. Here are some examples for different solutions to Pell's equation, by changing the value of R.*

Similarly, we can obtain the fundamental solution for different R value. By using this, we can generate a solution set.

From the Theorem 2, e and f can be referred with (x_1, y_1) and (x_2, y_2) with the equations $x_1^2 - ey_1^2 = 1$ and $x_2^2 - fy_2^2 = 1$, respectively. The values of e and f will be $\frac{(x_1^2-1)}{y_1^2}$ and $\frac{(x_2^2-1)}{y_2^2}$. Substituting this with Theorem 2, the encryption and decryption formula will be

$$C = \left(M^{(x_1^2-1)/y_1^2}\right) \mod n^{(x_2^2-1)/y_2^2} \mod n \tag{5}$$

$$M = C^d \mod n \tag{6}$$

Theorem 4 *Let $\Phi(a)$ be the count of non-negative integers $< n$ which are relatively prime to a. For $\Phi(15)$, the integers which are co-prime to 15 in the range 1–14 are 1, 2, 4, 7, 8, 11, 15, 13, 14. Therefore, $\Phi(15) = 8$. $\Phi(x)$ is always $x - 1$. This happens only when x is a prime number because a number is co-prime to any prime number.*

Theorem 5 *Modulus key n can be replaced by z, a prime number, greater than $\Phi(n)$ and less than n. Then, $\Phi(z)$ will be $z - 1$ (from Theorem 4). Now substitute z in Theorem 3,*

$$C = \left(M^{(x_1^2-1)/y_1^2}\right) \mod z^{(x_2^2-1)/y_2^2} \mod z \tag{7}$$

$$M = C^d \mod z \tag{8}$$

All of the above theorems have an application in the following sections.

4 Methodology

Many RSA variants use modulus key n in encryption and decryption, which can be factored using different factorization attacks [11] because n is a composite number. The private key component d is tracked easily when the public key component e is directly available.

4.1 Proposed Model

The proposed system uses a number z as the fake modulus, so that it is very difficult to deduce the value of n. It uses two public exponent keys that are shared indirectly with two x and y pairs which is calculated by *Pell*'s equation. Figure 1 shows the complete description of the steps involved in the algorithm. The process is divided into three phases: key generation, encryption and decryption.

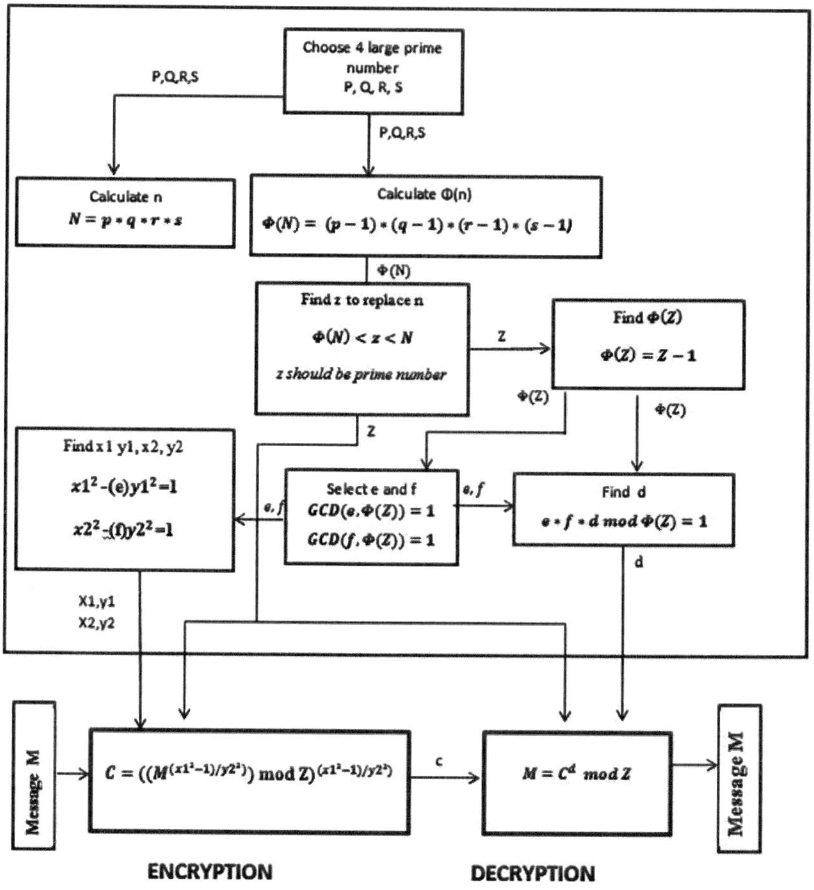

Fig. 1 Analysis of an enhanced Dual RSA algorithm using Pell's equation to hide the public key exponent and a fake modulus to avoid factorization attack

4.1.1 Key Generation

In Fig. 1, the key generation process is depicted by the following procedure:

- Select any four prime numbers p, q, r and s; n can be calculated as

$$n = p * q * r * s \tag{9}$$

- Totient function $\varphi(n)$ is calculated as,

$$\varphi(n) = (p-1) * (q-1) * (r-1) * (s-1) \tag{10}$$

- Choose a prime number z to replace n where $\varphi(n) < z < n$, and calculate

Table 1 Solutions to Pell's equation, changing the value of R

Value of R	Solution
3	(2, 1)
6	(5, 2)
10	(19, 6)
13	(649, 180)
42	(13, 2)
46	(24,335, 3588)
61	(1,766,319,049, 226,153,980)

$$\varphi(z) = z - 1 \tag{11}$$

- Find the public key exponents e and f using the equations

$$GCD(\varphi(z), e) = 1 \text{ and } GCD(\varphi(z), f) = 1 \tag{12}$$

- Find the private key component d satisfying

$$(e * f * d) \mod \varphi(n) = 1 \tag{13}$$

- Now find the (x_1, y_1) and (x_2, y_2) pairs using Theorem 3 for e and f, respectively. Table 1 gives an idea to determine the solutions to the *Pell's equation*

$$x_1^2 - ey_1^2 = 1 \quad \text{and} \quad x_2^2 - fy_2^2 = 1 \tag{14}$$

Now the key which is shared publicly is $[x_1, y_1, x_2, y_2, z]$, while the actual public key exponents e and f are hidden.

4.1.2 Encryption Process

In Fig. 1, the encryption process is depicted. The plaintext M is given as input to the encryption function. The encryption function enciphers the plaintext to find the ciphertext C using (7) from Theorem 5.

4.1.3 Decryption Process

In Fig. 1, the decryption process is depicted. This encrypted text C will be as an input to the decryption function. This input is decrypted by the receiver to recover the plaintext back using (8) from Theorem 5.

4.2 Example

Let the four prime numbers be $p = 3, q = 5, r = 11, s = 7$. Computing (9), we get $n = 1155$. Let $\Phi(n) = 480$ and the modulus key $z = 487$ using (11). Let the two public key exponents e and f be 7 and 11 which satisfy the Eq. (12). Private key exponent is calculated using (13); hence, we got $d = 101$. Now (x_1, y_1) is (8, 3), and (x_2, y_2) is (81,332, 53,989) which satisfies the formula from (14). Now public key (8, 3, 81,332, 53,989, 487) is shared to the sender by keeping private key (101, 487) secret. Let plaintext $M = 94$. The sender encrypts it to get ciphertext using the Eq. (7). Then, $C = 252$ is sent to receiver. On receiving side, the receiver obtains plaintext back using (8).

5 Result Analysis

Initially, an original image is given as input. In Fig. 2, the differences in the encryption of that particular image using normal RSA, Dual RSA and the proposed model are observed. It is also observed that the proposed system gives a better encryption of the image as compared to the other algorithms.

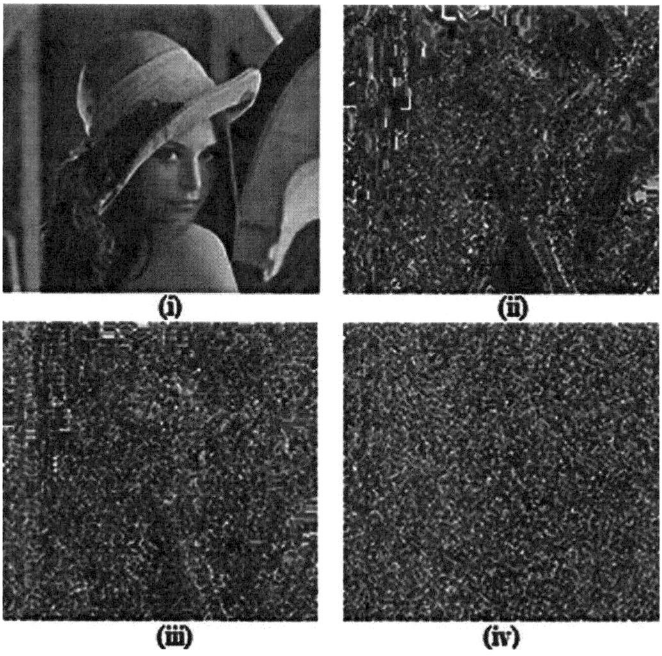

Fig. 2 **i** Original image (plaintext), **ii** encrypted using normal RSA, **iii** encrypted using Dual RSA and **iv** encrypted using proposed model

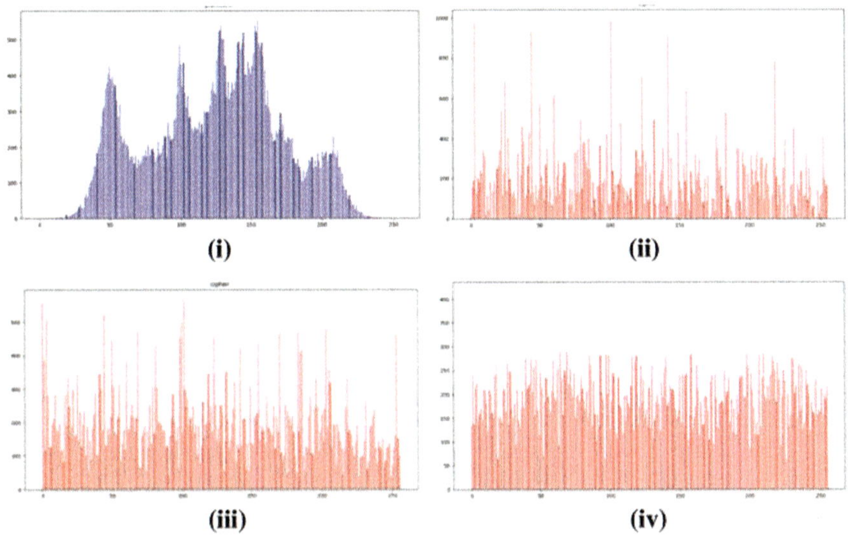

Fig. 3 Histogram graphs of **i** plaintext, **ii** normal RSA, **iii** Dual RSA and **iv** proposed model

5.1 Histogram Analysis

The pictorial representation of the probability of occurrences of the intensities with respect to the intensity values in the given image is called a histogram [13]. Figure 3 has a set of histograms plotted for the number of pixels against the frequency of occurrence of the pixel. It is seen that the proposed model gives a better histogram which has equal distribution of values.

In Fig. 3(iv), it is observed that the histogram for the proposed model has a uniform distribution and is said to be normalized. This concludes that the intruder cannot easily guess which pixel has the most frequent occurrence in the image.

5.2 Standard Deviation and Mean

The standard deviation of the occurrence of ciphertext element is compared as follows:

Let n_i be the count of occurrence of an integer i in the ciphertext, such that $0 \leq i < M$. The average number of occurrence of any integer i is given by:

$$\bar{n} = \frac{1}{M}\sum_{i=0}^{M-1} n_i \tag{15}$$

Then, standard deviation σ is given by

Table 2 Mean and standard deviation based on histogram value

	Mean value	Standard deviation
Normal RSA	197.75390625	149.19587848278516
Dual RSA	197.75390625	101.98529305738157
Proposed model	197.75390625	61.83568826939666

$$\sigma = \frac{1}{M}\sqrt{\sum_{i=0}^{M-1}(n_i - \bar{n})^2} \tag{16}$$

The average of the values in a data set is called the arithmetic mean. It can also be defined as the division of the sum of values by the total number of values [14].

In Table 2, it is observed that the proposed model gives a small value for standard deviation among the ciphertext. Thus, the conclusion is that the ciphertext is distributed equally among all the characters.

5.3 Avalanche Effect

In an avalanche effect, if there is a slight change in the input, the output will have a significant change. It is a desirable property of cryptographic algorithms, block ciphers and hash functions. In block ciphers, a slight variation in the key or plaintext causes a huge difference in the ciphertext.

Table 3 depicts the change in values of encryption and decryption from the plaintext using original public key and changed public key. Figure 4 depicts the avalanche effect when the key is changed slightly.

Figure 4 shows the variation in the ciphertext values when the key is changed from 29 to 28. Thus, it is concluded that even a small bit flip in the key gives a large difference in the ciphertext value.

Table 3 Avalanche effect

Original key = 29	Flipped key = 28
32	139
243	105
38	138
128	216
208	44
176	226

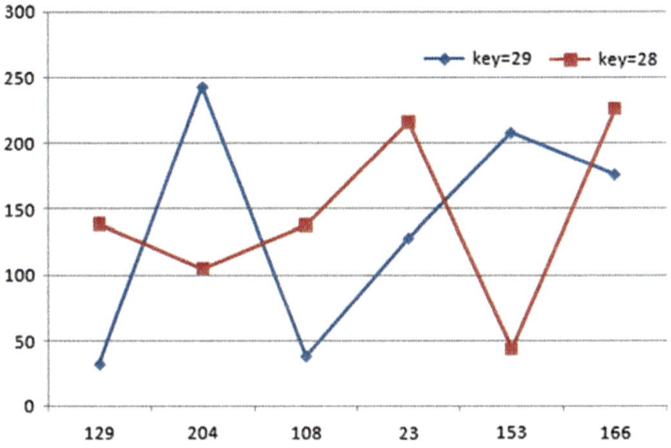

Fig. 4 Graph depicting the avalanche effect

5.4 Complexity Analysis

The computation of the amount of time taken by an algorithm depending on the length of the input is called time complexity. Similarly, space complexity is the quantification of the amount of space used. Both time complexity and space complexity depend on hardware, processors, operating system, etc. These factors are not considered while analysing the program. Only the execution time of the program is considered.

Table 4 shows the time complexity of every step in the process using RSA, Dual RSA and the proposed model. We observe that the proposed model gives the best results. Each step in the key generation has more complexity compared to RSA and Dual RSA. Due to this, it is difficult for the intruder to attack the proposed system.

Table 4 Complexity of each step in the algorithm using RSA, Dual RSA and proposed model

	RSA	Dual RSA	Proposed model
Common modulus (n)	$O(n^2)$	$O(n^3)$	$O(n^4)$
Euler's totient $\varphi(n)$	$O(n^2)$	$O(n^3)$	$O(n^4)$
Public key exponent (e)	$O(\log n)$	$O(\log n)$	$O(\log n)$
Private key exponent (d)	$O(n^2)$	$O(n^2)$	$O(n^3)$
Pell's pair (x, y)	–	–	$O(n^2)$
Fake modulus (z)	–	–	$O(n^3)$
Encryption	$O((\log n)^2)$	$O((\log n)^3)$	$O((\log n)^3)$
Decryption	$O((\log n)^2)$	$O((\log n)^3)$	$O((\log n)^2)$

Table 5 Time required to perform the attack using various attacking methods in seconds

	Normal RSA	Dual RSA	Proposed model
Fermat's factorization	0.00100016	0.003000259	955.6236587
Trail and division	0.003175933	0.257215649	1910.551775
Weiner's attack	0.001795421	0.002580976	Division by zero error

5.5 *Analysis of Attacks*

The trial division algorithm is an attack on RSA that factorize a number. Performing the trial division algorithm, one will simply check whether N is divisible by $S = 2, \ldots,$ floor(sqrt(N)). When a divisor like S is found, then $T = N/S$ is a factor. Thus, a factorization for N is found [15].

Wiener has shown that every RSA public key tuple (N, e) with $e \in Z^*_{\varphi(N)}$ that satisfies $ed - 1 = 0 \mod \varphi(N)$ for some $d < \frac{1}{3}d^{\frac{1}{4}}$ yields the factorization of $N = pq$ [16].

Fermat's factorization was proposed by a mathematician. A composite number N is written as the difference of squares: $= x^2 - y^2$. Factorization of N is obtained from this difference: Assume two non-trivial odd factors of N; s and t, such that $st = N$ and $s < t$. Thus, x and y can be found where $s = (x - y)$ and $t = (x + y)$ [15].

Table 5 shows the time required to perform the attack using particular attacking methods in terms of seconds. The value of common modulus n replaced by z used here has a length of seven bits. It can be observed that the proposed method takes more time for each of the attacks. It takes more time for the system to be cracked using Fermat's factorization and Trial Division attack [2] than the RSA and Dual RSA. But, it results in a division by zero error when the proposed system is intruded by Weiner's attack.

6 Conclusion

The RSA security depends on the factorization of a large number. This paper makes use of a fake modulus which is co-prime to the large number n. Also, the key is not shared directly between the parties. The algorithm makes use of solutions to Pell's equations as a substitute for the public key exponents. This makes it difficult for the intruder to determine the value of the public key component. Thus, the proposed algorithm eliminates the issue of mathematical attacks by providing more security to the algorithm with a slight increase in time complexity.

References

1. Stallings W (2006) Cryptography and network security, 5th edn. Pearson (2011)
2. Raghunandan KR, Shetty S, Aithal G, Rakshith N (2018) Enhanced RSA algorithm using fake modulus and fake public key exponent. In: 2018 international conference on electrical, electronics, communication, computer, and optimization techniques (ICEECCOT), Msyuru, India, pp. 755–759
3. Sun HM, Wu ME, Ting WC, Hinek MJ (2007) Dual RSA and its security analysis. IEEE Trans Inf Theory 53(8):2922–2933
4. Al-Hamami AH, Aldariseh IA (2012) Enhanced method for RSA cryptosystem algorithm. In: 2012 international conference on advanced computer science applications and technologies
5. Minni R, Sultainia K, Mishra S (2013) An algorithm to enhance security in RSA. In: 4th ICCCNT 2013, IEEE-31661
6. Raghunandan KR, Aithal G, Shetty S (2019) Comparative analysis of encryption and decryption techniques using mersenne prime numbers and phony modulus to avoid factorization attack of RSA. In: 2019 international conference on advanced mechatronic systems (ICAMechS), Kusatsu, Shiga, Japan. https://doi.org/10.1109/ICAMechS.2019.8861599
7. Barbeau EJ (2003) Pell's equation. Problem books in mathematics. Springer
8. Conrad K (2008) Applications of Pell's equation. University of Connecticut, Aug 5, 2008
9. Smith J (2009) Solvability characterizations of Pell like equations. In: Thesis on partial fulfilment of the requirements of the degree Master of Science in Mathematics, Boise State University, Aug 2009
10. Boneh D (1999) Twenty years of attacks on the RSA cryptosystem. Not AMS 46(2):203–213
11. Raghunandan KS, Shetty R, Aithal G (2017) Key generation and security analysis of text cryptography using cubic power of Pell's equation. In: 2017 international conference on intelligent computing, instrumentation and control technologies (ICICICT), pp. 1496–1500
12. Coppersmith D, Franklin M, Patarin J, Reiter M (1996) Low-exponent RSA with related messages. In: EUROCRYPT'96, vol 1070. Lecture notes in computer science. Springer, pp 1–9
13. Patel S, Goswami M (2014) Comparative analysis of histogram equalization techniques. In: International conference on contemporary computing and informatics (IC3I)
14. Raghunandan KR, Aithal G, Shetty S (2019) Secure RSA variant system to avoid factorization attack using phony modules and phony public key exponent. Int J Innov Technol Exploring Eng (IJITEE) 8(9), ISSN: 2278-3075, July 2019
15. Ambedkar BR, Gupta A, Bedi SS, Gautam P (2011) An efficient method to factorize the RSA public key encryption. In: International conference on communication systems and network technologies
16. Blomer J, May A (2004) A generalized Wiener attack on RSA. Springer, Berlin

A New Approach on Advanced Encryption Standards to Improve the Secrecy and Speed Using Nonlinear Output Feedback Mode

Dodmane Radhakrishna, Aithal Ganesh, and Shetty Surendra

Abstract The proposed cipher system is fast and secure stream cipher for software-based applications. The new cipher system is designed using output feedback mode on Advanced Encryption Standard with an extra XOR operation. It contains initial vectors (IV1 and IV2), and these IVs are updated by nonlinear functions. The increased security of the proposed cipher is due to the nonlinear functions. Along with the security, the speed of execution increased by executing Advanced Encryption Standard in only eight rounds. Various security evaluation parameters considered in the security analysis of the proposed cipher system. In addition to it, the speed of execution of the new approach analyzed with other systems with respect to the time complexity.

Keywords Symmetric-key · Advanced encryption standard · Output feedback mode · Encryption · Decryption · Plaintext and ciphertext

1 Introduction

Cryptography is an art of transforming confidential information into intangible information for the third party [1]. This transformation carried out in two ways: the symmetric-key approach and the asymmetric-key approach. In symmetric-key, only one key used to encrypt the information before transmitting at the sender side, and the same key used at the receiver to decrypt upon receiving information. In asymmetric-key, a pair of key used, in which one key used for encryption at the sender side, and

D. Radhakrishna (✉) · S. Surendra
NMAM Institute of Technology, Nitte, Karnataka 574110, India
e-mail: radhakrishna@nitte.edu.in

S. Surendra
e-mail: hsshetty@nitte.edu.in

A. Ganesh
Mangalore Institute of Technology and Engineering, MITE, Mangalore, India
e-mail: ganeshaithal@gmail.com

N. N. Chiplunkar and T. Fukao (eds.), *Advances in Artificial Intelligence and Data Engineering*, Advances in Intelligent Systems and Computing 1133,
https://doi.org/10.1007/978-981-15-3514-7_61

the other key is used for decryption at the receiver side. When compared these two standards, the asymmetric-key approaches used for exchanges of the keys or short messages, whereas the symmetric-key approaches designed to exchange large messages between parties [2]. In this work, it has decided to work with the symmetric-key approach (Advanced Encryption Standard).

In symmetric-key methods, if the information needs to be exchanged is processed in blocks (fixed-sized or various sized), then this approach is called as a block cipher [1–3]. Many standards have been proposed in the symmetric-key approach, they are Data Encryption Standards (DES), Triple DES, Advanced Encryption Standard (AES), Blowfish, etc. Out of all these standards, the Advanced Encryption Standards treated as more secure due to its reduced susceptiveness to the various attacks [1]. Therefore, it has been decided to work on Advanced Encryption Standard with respect to resistivity (as a way to increase) and time (as a way to reduce). Hence, this work centered on Advanced Encryption Standard.

Advanced Encryption Standards (AESs) are a widely accepted symmetric block cipher algorithm. This was proposed by the cryptographers, Vincent Rijmen and Joan Daemen. Later, this scheme was submitted to selection process under the name "Rijndael" [1]. In the year 2001, this new encryption standard was selected by the National Institute of Standards and Technology (NIST), the USA. Since then, AES has widely accepted and became the industry standard for almost all cryptographic applications.

The next section focuses on few symmetric-key methods and counter mode operations based on speed and secrecy of the cipher system. These studies are the basic building blocks for the new proposed cipher systems. The Sect. 3 describes the proposed system followed by the implementation section giving insight into the implementation of the proposed system. The Sect. 5 describes the test conducted on the new system followed by the conclusion and future work.

2 Literature Survey

2.1 Advanced Encryption Standard

Advanced Encryption Standard cipher system is one of the symmetric-key block cipher encryption techniques, which uses cryptographic substitution and transformation techniques.

Advanced Encryption Standard, the symmetric-key block cipher standards, was designed as repeated product ciphers of substitution and/or transformation and/or various mathematical operations. The major operation are based on the Rijndael algorithm [1, 4]. It has designed to encrypt 128-bits of the data block at a time using either by accepting the key size of 128, 192 or 256-bits. However, the size of key defines the number of rounds required for the algorithm to achieve good security. That is for 128, 192, 256-bit keys, the AES operates in 10, 12 or 14-rounds, respectively.

The plaintext/ciphertext as an input to the AES algorithm is 128-bit or 16-byte data block. The basic unit size of data for processing using the AES algorithm is bytewise [5]. All the bytes value are represented as the concatenation of its individual bit values (0 or 1) in the order from $b_0, b_1, b_2, b_3, b_4, b_5, b_6, b_7$. These bytes are represented in polynomial as follows:

$$b_7X_7 + b_6X_6 + b_5X_5 + b_4X_4 + b_3X_3 + b_2X_2 + b_1X_1 + b_0X_0 \quad (1)$$

The Advanced Encryption Standards internal operations defined to perform the operations on a two-dimensional array of bytes called the state array [1]. The state array is an arrangement of four cross four matrix consists of four rows of bytes, each containing Nb bytes, where Nb is the block length divided by 32 (four for a 128-bit key, six for a 192-bit key and eight for 256-bit key). The Advanced Encryption Standard defined to be a repeated product of substitution and transposition, termed as rounds that convert/transforms the input called as plaintext into the final output called ciphertext. Each round consists of four processing steps, which includes the one that depends on the key called AddRoundKey. The set of reverse rounds applied to convert or transform ciphertext back into the original plaintext using the same key. Figure 1 represents the sequential flow of the AES algorithm [6].

Using the Advanced Encryption Standards, the ciphertext generated offers a better security [6], and thus, Advanced Encryption Standard cipher systems have been widely accepted. But the processing time needed by this cipher systems is also more leading to increased time complexity. With these considerations, Siva Appa Rao Rapeti has suggested a new approach to reduce the time complexity under the name nonLinear feedback stream cipher system (NLFS) using the counter (CTR) mode of operation of block cipher systems [7, 8].

2.2 Counter (CTR) Mode [9, 10]

Counter mode operation presented by Whitfield Diffie and Martin Hellman in 1979. In cryptography, the block cipher could be converted to stream cipher using the counter mode of operation. The next block of key stream is generated by the successive encryption of a counter values. The counter guarantees the generation of the sequence of values that does not repeat the patterns/sequence for long time. Usually, the counter value is incremented by one.

The CTR mode of operations has designed especially to carry out encryption of blocks of data in parallel to speed up the processing. Thus, by reducing the total time, which again be the reason against the timing attacks.

In counter node, both the encryption/decryption processes are identical, which showed in Fig. 2. Each successive counter values are encrypted/decrypted and later XORed with the plaintext/ciphertext correspondingly. In CTR mode, the initial vector (IV) sometimes called nonce can be constructed using random number generators. If so, then the IV can be combined with counter values by carry free or lossless

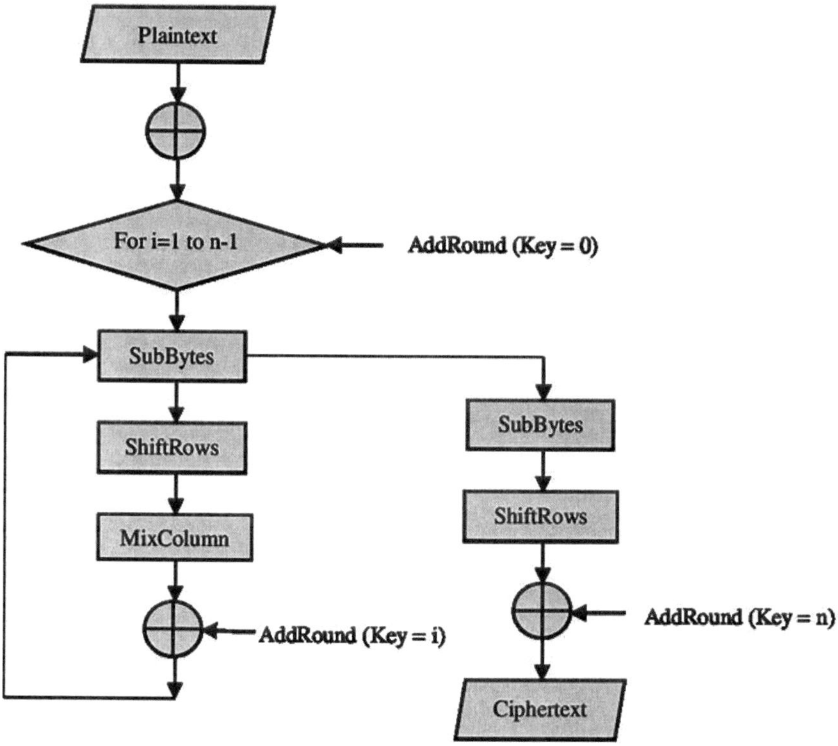

Fig. 1 General structure of advanced encryption standard algorithm

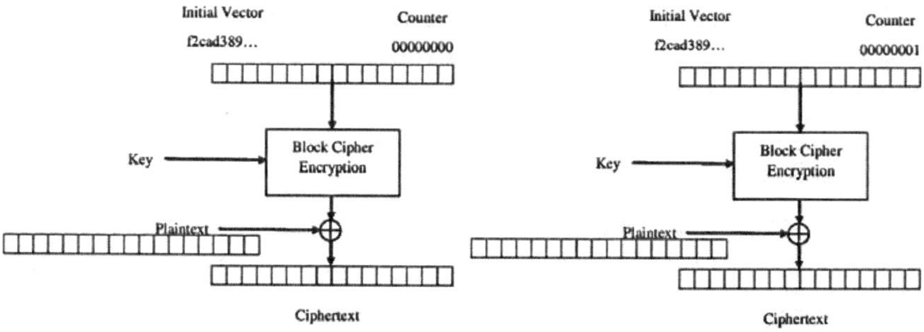

Fig. 2 Structure of counter (CTR) mode

operations such as concatenation, addition or XOR, which is then used as the actual counter as a block to be encrypted. If the IV is not random, then it will be more vulnerable to the attacks.

Figure 2 clearly depicts the possible parallel processing using counter mode operations.

2.3 Non-linear Feedback Stream Cipher System (NLFS) [11–14]

NLFS is an improvement on Advanced Encryption Standards to increase the speed of execution and hence reducing the time required to execute. However, the security of the cipher systems compromised. The various functionalities adopted in the NLFS are as follows:

1. AES secure nonlinear functions (round functions).
2. AES key generation function.
3. The S-boxes (in S-byte substitution step), which is generated using primitive-polynomial.
4. NLFS designed to have 128-bit key and 256-bit initial vector.
5. NLFS designed with two nonlinear 512-bit buffers to have the internal state of 1024-bit.

NLFS has designed in two modes. First mode of operation is the basic mode or also called as NLFS-synchronous mode. In NLFS-synchronous mode, the key stream is independent of input (plaintext and ciphertext). The second mode is self-synchronous mode. In self-synchronous mode, the key stream generation depends on cipher text. The NLFS implemented to have a single round using counter mode by maintaining the nonlinearity in the key stream generations. This leads to the reduction in the time complexity both by reducing the number of rounds of Advanced Encryption Standard and parallel execution of each block using counter mode of operations. However, the major drawback is the compromise in the security since Advanced Encryption Standard carried out as a single round.

From the above inferences, it has decided to work on reducing the time complexity but not on compromising the security. Therefore, the proposed work adopts the features from the above techniques such that there is a moderate reduction in the time complexity without compromising the security.

As stated earlier, the Advanced Encryption Standard is strong but not as fast as nonlinear feedback system cipher. The ten rounds of Advanced Encryption Standard provide enormous strength, but the speed of executions is affected. However, in nonlinear feedback system cipher, the numbers of rounds are reduced one to achieve the speed and added nonlinear feedback to increase the secrecy. This nonlinearity adds a great deal of strength, but still nonlinear feedback system cipher could not match the strength of Advanced Encryption Standard. In the proposed system, we use nonlinear feedback by restricting to the eight rounds. A value-based rotation step, which rotates the eight bits based of the value of first three bits, is merit of nonlinear feedback system cipher.

3 Methodology

The work is an outcome of the merits and demerits of both Advanced Encryption Standard and nonlinear feedback system cipher. It includes two nonlinear feedbacks to improve the strength and secrecy. The proposed method designed using Advanced Encryption Standard in output feedback mode (OFB) [12] with an extra XOR operation. But unlike OFB mode, the encryption output is XORed with second nonlinear key in the proposed method. Then, this result is XORed with the actual plaintext to generate final ciphertext. The proposed method carries out the Advanced Encryption Standard process on each block only for eight rounds. The compact proposed design shown in Fig. 3. The various stages of each round are shown in Fig. 3. The various steps of the first seven rounds are SubBytes, ShiftRows, MixColumns, value-based rotation (which is an extra step-only left shift by three-bit) and AddRoundKey. The last round is similar to Advanced Encryption Standard last round that includes SubBytes, ShiftRows and AddRoundKey.

The proposed cipher system initializes two initial vectors, namely IV1 and IV2. The nonlinearity in the IV1 and IV2 is maintained as shown in Fig. 4, which is an expanded view of the proposed nonlinear feedback cipher system architecture.

In the proposed work, initially, the two IVs are initialized using the round generator, then these vectors are updated using the nonlinear functions to increase the security. Every time, the nonlinear values updated for IV1 generated from the result of XOR operations, which carried out between the previous encrypted block (generated using the eight rounds of Advanced Encryption Standard) and the IV2. The nonlinear IV2 taken as the output of the previous encrypted block, which the same values used in the generation of the IV1, as shown in Fig. 4. These nonlinear generations of IV1 and IV2 continued until the last block of plaintext to be encrypted.

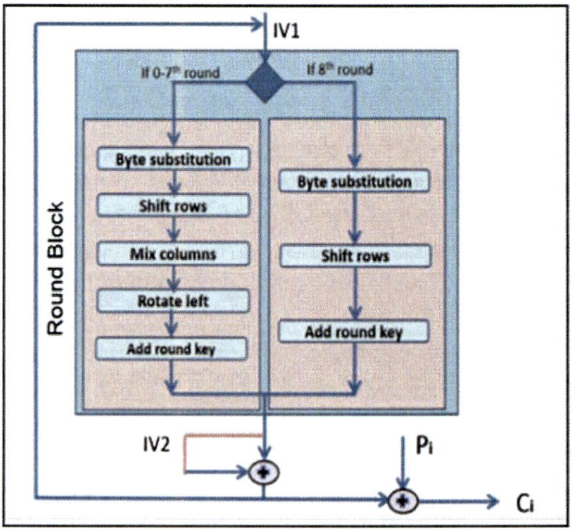

Fig. 3 Compact view of proposed nonlinear feedback cipher system architecture

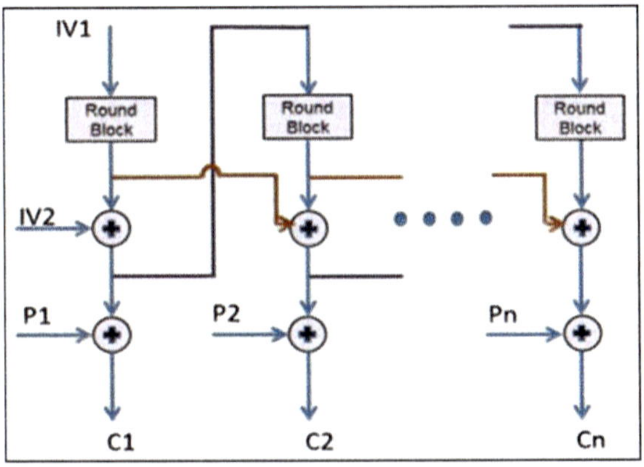

Fig. 4 Expanded view of the proposed nonlinear feedback cipher system architecture

4 Implementation

There is a lot of difference when it comes to practical implementation from the theoretical concepts. The ways followed in achieving the methodologies are as follows.

4.1 Initialization

First, need to initialize the algorithm with one 128-bits long Key and two 128-bits long initial vectors namely IV1 and IV2. The proposed system works on a fixed block size, i.e., 128-bits at a time. So, to encrypt every 128-bit of plaintext, 128-bits of key stream needed. Since the proposed cipher system is a stream cipher, it should still function perfectly if the length of plaintext block is less than 128-bits. Thus, no bit padding allowed. This case arises when the plaintext length is not a multiple of 128-bits. In such a scenario, proposed cipher system should generate only required length of key stream and hence need a length parameter which stores the length of the key stream required. The value of length variable is always 16-bytes (128-bits) unless it is the length of last block of the plaintext read. State matrix is a 4×4 column major matrix, which holds the 128-bits under execution, and all the changes done by the program reflected in the matrix. Therefore, as an initialization, state matrix is loaded with IV1. The cipher key is loaded in a key matrix, which is again a 4×4 matrix. A buffer of 128-bits kept ready to store the generated 128-bit key stream. The input file opened in read binary (rb) mode. The output file, i.e., the file in which the ciphered source data needs to be stored is opened in append mode (a).

4.2 Internal Operations

These are the operations which are identical for all input blocks unlike initialization which is done only once in the beginning.

4.3 Key Expansion

This step is implemented in coherence with its methodology. As already discussed in methodology, a total of nine subkeys are needed. These subkeys are usually called as round keys. In each step, each round key is made up of four words. Hence, it can be said that 36 words are to be generated by expanding the given 16-byte (four words) key from which four words are to be used in every single round. The first round key is the four words in cipher key itself. The next eight round keys are needed to be generated, and they are extracted by using the formulae given in methodology.

4.4 Rounds

Here, the first round key is added, i.e., for round 0, AddRoundKey (0). In this function, the state matrix is updated by XORing the contents of the state matrix with Round-Key[x], where x = round * Nb * 4 + i * Nb + j for ith row and jth column. For the next seven rounds, the functions such as SubBytes, ShiftRows, MixColumns, value-based left rotation and AddRoundKey (current round) are invoked, respectively.

4.5 The Final Round

In the last round, i.e., eighth round, the functions invoked are SubBytes, ShiftRows, value-based left rotation and AddRoundKey (eighth round), respectively [5].

It is shown that, in the last round, MixColumns step is omitted. In the end of eighth round, the content of state matrix is copied to output buffer.

4.6 Feedback

Place a copy of the contents of output buffer in the temporary buffer so that the values are preserved to give as feedback for the next set of 128-bits key stream generation. XOR the output buffer with the IV2 and keep the result in the output buffer. Now, these updated values of the output buffer are given as feedback to the next key stream generation in place of IV1. The contents in the temporary buffer are given as feedback to the next key stream generation in place of IV2.

4.7 Encryption

The input data to be encrypted is read in binary as a stream in a chunk of 128-bits (only last block is an exceptional case). The read data kept in a buffer called plaintext, and later, this is XORed with the key stream to generate the output/ciphertext as shown in Fig. 5. Hence, the corresponding cipher text is generated, later appended to the output file, and the process repeats itself for all the next plaintext blocks.

4.8 Decryption

The decryption process similar to encryption as shown in Fig. 6.

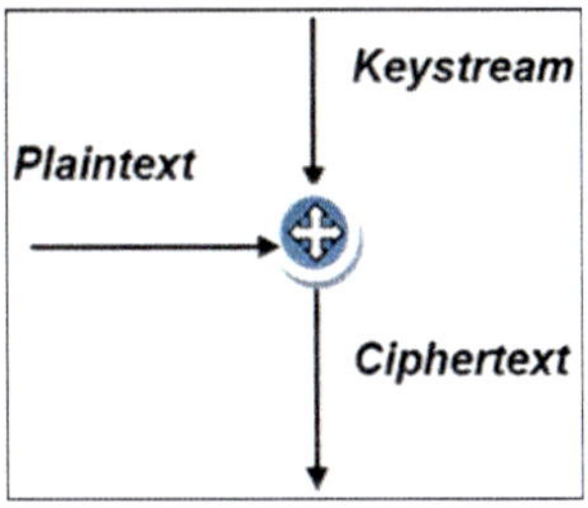

Fig. 5 Encryption in proposed system

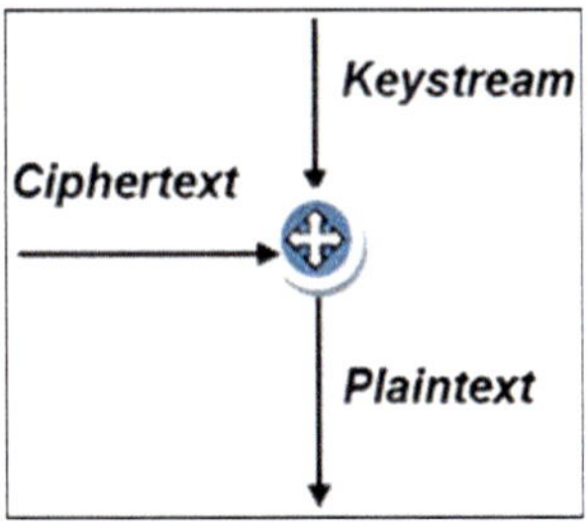

Fig. 6 Decryption in proposed system

4.9 Pseudo-Code

A brief glance on the algorithm of the proposed system is below:

The main() Function

int main()
uint8_t key[16] //Define a 16 byte cipher key
uint8_t iv1[16] //Define a 16 byte initial vector
uint8_t iv2[16] //Define another 16 byte initial vector
uint8_tinputBuffer[16] //Declare a 16 byte input buffer
uint8_toutputBuffer[16] //Declare a 16 byte output buffer
//uint8_t is a keyword used to define an unsigned integer data type of 8bits
Using FILE pointer, open source in Read Binary Mode
Using FILE pointer, open destination in Append Mode
while (not the end of the input file)
Read 16 bytes from input file in binary and load in inputBuffer
Encrypt(key, outputBuffer, length of input read, iv1)
Prepare feedback in place of iv1 and iv2 for next set of - 16byte keystream generation
tempBuffer = XOR (iv2, outputBuffer)
cipherBuffer erBuffer content to Destination file
End while
close the input file and output file
End main

The Encrypt Function

Encrypt(key, outputBuffer, length of input read, iv1)
Copy the iv1 contents to a 4x4 State Matrix
Copy key to a 4x4 Key Matrix
The given 16byte key is expanded and 9 Round keys are generated
Add first round key
for all the rounds from 1 to 7
Replace each byte of the state matrix with the corresponding value in the S box
Shift left second row by one byte, third row by 2 bytes and forth row by 3 bytes
Multiply each column by a fixed polynomial
Add current Round Key
End for
Replace each byte of the state matrix with the corresponding value in S box
Shift left second row by one byte, third row by 2 bytes and forth row by 3 bytes
Add 8th round key
Copy the contents of state matrix to outputBuffer
ReturnoutputBuffer
End Encrypt

5 Tests and Results

To evaluate the time complexity and resistivity, an experimental setup is formulated. This setup contains an Intel(R) CORE(TM) i5-7200UCPU @ 2.50GHz 2.70 GHz with 4 GB installed RAM. The evaluation results are analyzed and discussed.

Firstly, test carried out to check the randomness of the key stream and security and then have given detailed explanations about the tests, to which the proposed system was subjected. Also, the results of these tests are summarized. Same tests were conducted also on other existing designs so that the results of existing ciphers can be compared with the proposed cipher.

5.1 Repetition Test

This test verifies the reoccurrences of cipher for a corresponding input. It checks whether the same input at different bit positions will give the same cipher at that particular location.

Description

A plain text file was created with repeated content. The content of the file was "aaaaaaaaa aaaaaaaaaaa aaaaaaa aaaaaaaaaaaaa aaaaaaaaaa aaaaaaaaaa aaaaaaaa aaaaaaaaaaaa aaaaaaaaaaaaaaaaaaaa aaaaaaaaaaaaaaaaaaaa aaaaaaaa", that is for 128 times. This file was encrypted using AES-128, and the result was recorded. Same input was given to the proposed system, and output was recorded. Both results were manually examined, and the repeated occurrences were colored. In the end, the percentage of repetition was calculated for both algorithms that is Shown in Figs. 7 and 8.

Results

For the proposed system: Repetition: 29% and For AES128: Repetition: 37%.

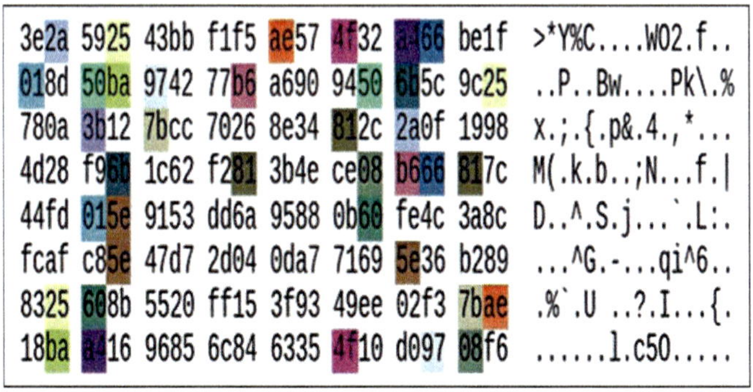

Fig. 7 Repetition test result for proposed Cipher-repetition: 29%

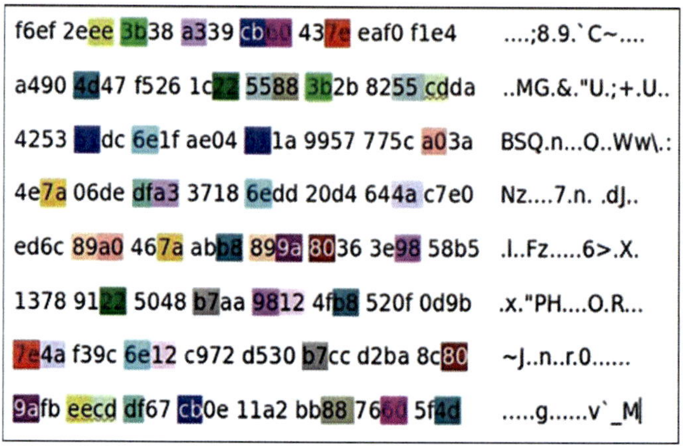

Fig. 8 Repetition test result for AES 128-bit-repetition: 37%

The test was conducted for different inputs as well. From the results, it clearly says that the proposed system has better results that is lesser the number of repetitions, better the algorithm. So can say that compared to AES 128, the proposed system has moderately better results in repetition test.

5.2 *Frequency Test*

This test verifies the equal distribution of the bits.

Description

A random input was given to the proposed system, and the cipher file was generated. Another separate program was written in C which could convert the hexadecimal cipher into its binary equivalent. This binary equivalent output was directed to a word file. Using the search engine one-bit test, two-bit test and four-bit test carried out by using the manual calculation of percentage based on the number of occurrences that is shown in Fig. 9.

Results

One-bit frequency test results should be close to 50%, and as we can see in the above table, the values are very close to the required output. Two-bit frequency test should result in an output close to 25%. The results obtained are not very close to the required output, but they are acceptable. Four-bit frequency test should result in outputs closer to 6.25%. The tables show all the values close to the optimal output except two cases which are 0000 and 1111. But still, those two outputs are acceptable. So, it can be said that the frequency test shows good results for the proposed system.

1-bit	Occurrence %
0	50.1
1	49.9

2-bit	Occurrence %
00	20.633
01	29.41
10	29.41
11	20.53

4-bit	Occurrence %
0000	4.29
0001	6.91
0010	6.48
0011	6.83
0100	6.43
0101	5.77
0110	6.72
0111	6.58
1000	6.92
1001	6.44
1010	5.83
1011	6.39
1100	6.92
1101	6.56
1110	6.82
1111	4.02

Fig. 9 Frequency test results

5.3 Efficiency Tests

There are two basic efficiency tests based on the effect of the initial vector and the key.

Impact of Key on the Key stream

This test checks how a small change in the key effects the key stream generated and hence affecting the cipher generated. The change propagation expected is very large.

Impact of IV on the Key stream

This test checks how a small change in the IV effects the key stream generated and hence affecting the cipher generated. The change propagation expected is very large.

Results

An output for a random input was generated and stored as backup. Next one-bit change was made in the key, and the output was recorded as key change. Next one-bit change was done in the IV, and the output was generated as a cipher. When all three outputs were compared, it was discovered that nothing was common, which implies, a small change in any of the two (key and IV) will propagate to be a drastic change in the output. This drastic change is shown in Fig. 10.

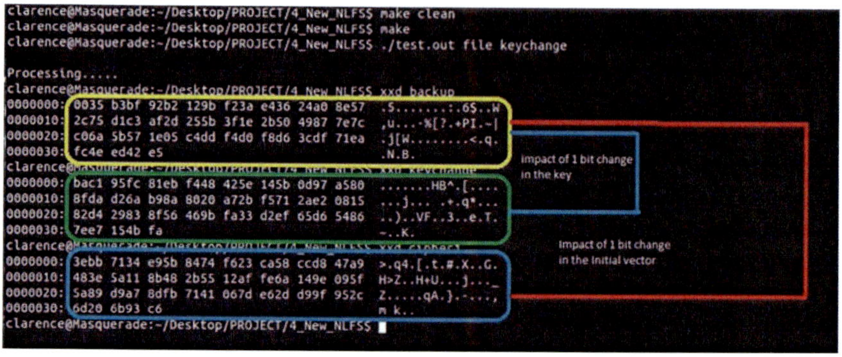

Fig. 10 Efficiency test results

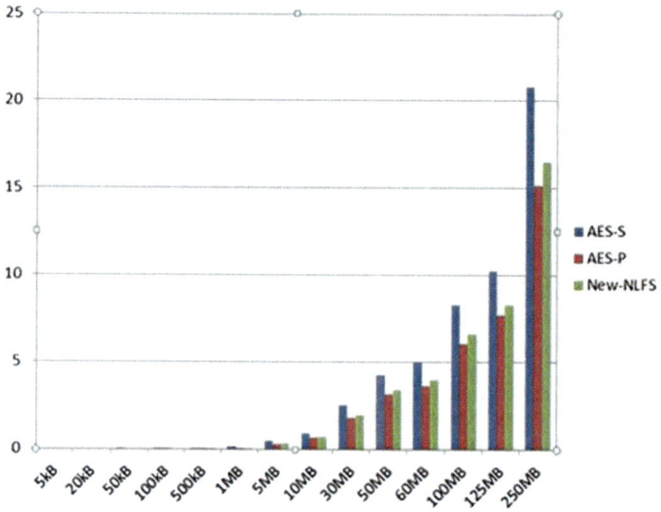

Fig. 11 Time comparison of AES128 and proposed system

5.4 *Time Tests*

In this test, we ran different sizes of input in the serial implementation of AES 128, parallelized AES128 and the proposed system cipher. The outputs in each case were recorded, which are shown in Fig. 11. From Fig. 11, it is observed that among all three algorithms, serial AES takes more time, parallel AES takes the least time, but the proposed system cipher time is intermediate. It takes just a little more time than the parallelized AES 128. Since we have a huge uprise in the security due to nonlinearity, this timing assumed acceptable.

6 Conclusion and Future Work

The proposed system cipher was designed, successfully implemented, tested and proved to be stronger than AES 128 and existing NLFS cipher. The AES 128 is stronger due to its number of internal rounds which is 10. Existing NLFS cipher reduced the number of rounds to 1 to gain speedup hence losing the strength. But existing NLFS cipher uses nonlinear feedback which can overcome the loss in strength due to the reduction in a number of rounds. The proposed system uses the concept of nonlinear feedback along with the reduction of a number of internal rounds to 8 from 10, which gives a lot of strength to the proposed system. Value-based rotation is a step included in the proposed system cipher as in existing NLFS cipher. This new step provides more nonlinearity. The key stream in proposed system cipher is independent of the plaintext, so it can be previously calculated and kept ready. Both ciphering and deciphering operations are done by using the encryption algorithm alone. The proposed system can encrypt and decrypt any type of input (audio, video, image, text, etc.). The new algorithm takes comparatively optimal time with time complexity $O(n)$ and space complexity $O(1)$.

In the future, we plan on introducing the interleaved parallelism to improve the speed. The interleaved parallelism was chosen because there is interdependency due to feedback, and it does not support normal parallelism. We also plan on reducing the amount of repetition in the output. As of now, the proposed system was compared only with AES128 and existing NLFS cipher, but in future, we plan on comparing it with more existing ciphers which are considered to be the safest and fastest.

References

1. Stallings W (2011) Cryptography and network security: principles and Practice, 5th ed. Pearson Education. ISBN 978-81-317-6166-3
2. Daemen J, Rijmen V (2002) The design of Rijndael: AES the advanced encryption standard. Springer
3. Zhang X, Parhi KK (2002) Implementation approaches for the advanced encryption standard algorithm. IEEE Circ Syst. Mag. 2:2446
4. Su CP, Lin TF, Huang CT, Wu CW (2003) A high-throughput low-cost AES processor. IEEE Commun Mag 41:8691
5. Rapeti SAA (2008) NLFS: a new non-linear feedback stream Cipher, thesis submitted in partial fulfillment of the requirements for the award of the Degree of Master of Technology in Computer Science and Engineering at Kharagpur, May 2008
6. Nat. Inst. Standards Technol. (NIST), Federal Information Processing Standards (FIPS) Publication 197, Advanced Encryption Standard, Nov 2001
7. Nat. Inst. of Standards and Technology, Advanced Encryption Standard, Fed. Information Processing Standards (FIPS) Publication 197, Nov 2001
8. Whiting D, Housley R, Ferguson N (2002) Counter with CBC-MAC (CCM)
9. Heys HM, Zhang L (2011) Pipelined statistical Cipher feedback: a new mode for high-speed self-synchronizing stream encryption. IEEE Trans Comput 60(11):1581–1595
10. McGrew D, Viega J (2005) The Galois/counter mode of operation (GCM). http://csrc.nist.gov/groups/ST/toolkit/BCM/documents/proposedmodes/gcm/gcm-spec.pdf, May 2005

11. Jung O, Ruland C (1999) Encryption with statistical self-synchronization in synchronous broadband networks. In: Proceedings of conference cryptographic hardware and embedded systems (CHES 99), pp 340–352
12. Heys HM (2003) Analysis of the statistical Cipher feedback mode of block Ciphers. IEEE Trans Comput Eng 52(1):77–92
13. Burda K (2007) Modification of OCFB mode for fast data links. Intl J Comput Sci Netw Secur 7(12):228–232
14. Maurer UM (1991) New approaches to the design of self-synchronizing stream Ciphers. In: Proceedings of conference advances in cryptology (Eurocrypt 91), pp 458–471

Cyber-Bullying Detection: A Comparative Analysis of Twitter Data

Jyothi Shetty, K. N. Chaithali, Aditi M. Shetty, B. Varsha, and V. Puthran

Abstract Cyber-bullying has become a serious issue with the increasing popularity of social media. Cyber-bullying is a form of harassment using electronic means. In this paper, we address the problem of cyber-bullying detection on Twitter data set. This proposed approach deals with text-based cyber-bullying detection. Text classification algorithms like linear regression, logistic regression, Naive Bayes, decision tree, SVM and neural networks have been widely used for classifying bullying and non-bullying tweets. The scope of these algorithms and their performances are different. Therefore, it is important to find a suitable algorithm for this data set which quickly and actually solves the problem. This paper provides a quantitative comparison of six important algorithms to detect bullying and non-bullying tweets. Comparison of these algorithms is determined in terms of precision, recall, accuracy and F1 score.

Keywords Cyber-bullying · Linear regression · Logistic regression · Naive Bayes · Decision tree · SVM · Neural networks · Feature extraction · Classification

1 Introduction

Cyber-bullying is an aggressive act that can be carried out by a single person or a group of aggressive people, using online platforms, repeatedly against a person who is not capable of defending himself. Twitter is one of the social networking service

J. Shetty (✉) · K. N. Chaithali · A. M. Shetty · B. Varsha · V. Puthran
NMAM Institute of Technology, Nitte, Karnataka 574110, India
e-mail: jyothi_shetty@nitte.edu.in

K. N. Chaithali
e-mail: chaithali9710@gmail.com

A. M. Shetty
e-mail: aditism294@gmail.com

B. Varsha
e-mail: Varsha14Vinod@gmail.com

N. N. Chiplunkar and T. Fukao (eds.), *Advances in Artificial Intelligence and Data Engineering*, Advances in Intelligent Systems and Computing 1133,
https://doi.org/10.1007/978-981-15-3514-7_62

on which users communicate using tweets. Cyber-bullying is especially present on Twitter according to the survey done by Pew Center, and Twitter users face many forms of bullying such as death threats, stalking and sexual abusive threats. Cyber-bullying has become a serious issue these days. A system, which can detect such text, would certainly be a great use.

Previous works on cyber-bullying show that machine learning is a powerful tool to analyse the bullying behaviour. In this paper, we attempt to conduct analysis on "tweets" using various machine learning algorithms. We attempt to classify the tweets whether it is bullying and non-bullying. Supervised machine learning techniques can be used to do this. Linear regression [1], logistic regression [2], Naive Bayes [3], SVM [4], decision tree [5] and neural networks [6] are few supervised learning algorithms. Linear regression is a statistical data analysis method. It is used to determine the linear relationship between a dependent variable and one or more independent variables. Logistic regression is a regression analysis method used when the dependent variable is binary. This is a predictive method for analysing. Then, Naive Bayes is a technique which uses Bayes' theorem with a naive assumption that any pair of features is independent for a given class. Decision tree algorithms are also supervised learning method, and they are nonparametric. This method can be used for both classifications and regression task. The neural network is an algorithm that endeavours to recognise the relationships in a given set of data through a process that is carried out just like how the human brain carries out processes. Neural networks do not require redesigning of output criteria for changing input since it is adaptive. SVM is one of the supervised machine learning algorithms that can be used for both classifications as well as regression challenges.

This paper focuses on these machine learning algorithms, compares the accuracy and also shows the best-suited algorithm for selected data set. The bullying data set is extracted from Twitter using Twitter API. NLTK and scikit-learn are used to perform pre-processing and algorithm execution.

2 Related Works

In the field of machine learning, many works have been done in order to construct new classifiers, and many researches are still going on to construct further new classifiers. Comparative analysis of classification techniques has been provided by many researchers. With the help of machine learning, we can detect language patterns used by bullies, and to detect cyber-bullying content automatically rules are developed [7].

A methodology was used to analyse user's behaviour along with the information contained in the tweets, with the aim to identify the most appropriate features to detect Twitter spammers in real time. They have compared many machine learning techniques such as Naïve Bayes, random forest, K-nearest neighbours and support vector machine classifiers in order to identify Twitter spammers.

A work was proposed in order to address the cyber-bullying problem based on texts, where dynamic and selective portrayals of tweets are crucial for an efficient detective approach. Comprehensive experiments on Twitter and MySpace were conducted in order to demonstrate that the proposed methods perform very well than the other text representation learning methods [8].

A model was proposed, which concurrently detects fomenter and targets of bullying along with new swear words by initiating with the aggregation of social interaction and a source catalogue of bullying pointers. The goal was set based on the regularity of participant-vocabulary. This method was tested on Ask.fm and Twitter data sets and showed that the given technique can identify new swear word list along with targets and fomenter [9].

An unsupervised approach was proposed to adopt in order to detect on the basis of growing hierarchical self-organising map cyber-bully which exists all over social media. In this approach, there are many improvised features that are used to capture syntactic and semantic communicational behaviour of capable cyber-fomenters. On testing the approach with Twitter data set, the accuracy reached 72% compared to Naïve Bayes that reached to 67%.

3 Methodology

3.1 Data Description

The data set used is in the form of comma-separated value files. The file contains tweets and their corresponding sentiments. The training data set is a csv file of type tweet_id, sentiment and tweet where the tweet_id is a unique integer identifying the tweet, the sentiment is either 1 (bullying) or 0 (non-bullying), and tweet enclosed in “”. The test data set is a csv file of type tweet_id and tweet.

3.2 Design Model

The work model of the proposed work is as shown in Fig. 1.

3.3 Pre-processing

The transformations applied to our data set prior of feeding it to an algorithm is referred to as pre-processing. Tweets contain various types of noise that can damage the performance of machine learning algorithm. Objective of this step is to make raw text ready for mining. The data cleaning methods used, in this paper are

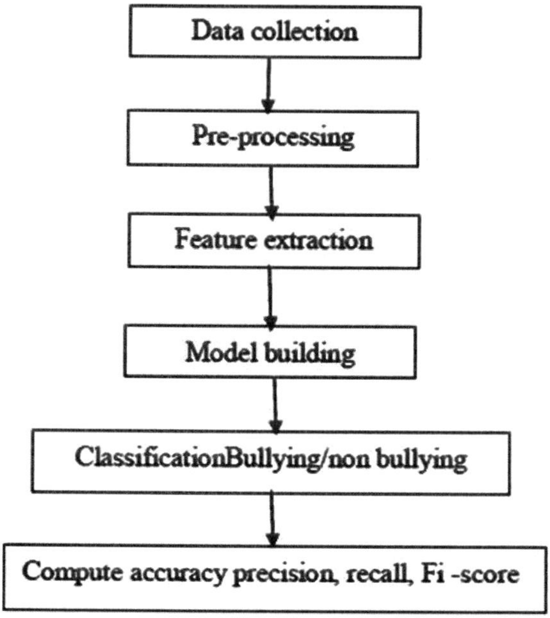

Fig. 1 Work model of proposed work

- removing twitter handler
- removing URLs.

That is we cannot analyse a tweet by reading Twitter handler or URLs. Sometimes it can lead to over-fitting as well. Other pre-processing methods followed are,

- Removing punctuation, numbers and special characters.
- Tokenization
- Removing stop words
- Stemming.

Tokenization is the process of dividing a text into a set of meaningful pieces. Each tweet will be divided into tokens. After tokenization, we remove stop words from the data set. The Natural Language Toolkit (NLTK) is a collection of libraries useful for tokenization, removal of stop words and stemming. This toolkit contains built-in stop word dictionary, and we fetch each word from the data set to check if it is not in stop word dictionary. After stop word removal, the data will be stored in a separate file.

3.4 Feature Extraction

To analyse the pre-processed Twitter data, it needs to be represented as features. In this paper, text features are constructed using Bag of Words and TF-IDF. Bag of

Fig. 2 Word cloud for bullying words

Words is the list of unique words in the text corpus. In TF-IDF, value is calculated using term frequency and inverse document frequency.

The data set which we used consists of 18,000 tweets and out of that 75% is made use for training and the rest is 25% is utilised to test the model.

Word cloud is an image which is a collection of words that are used in particular text, where size of each word represents that rate of occurrence of that particular word in the text. We use the word cloud to indicate the frequency of bullying and non-bullying word in Twitter data set as shown in Figs. 2 and 3.

4 Experiment and Results

The experiment and results are the outcomes of analysing the six different machine learning algorithms that have been used to classify the data. In this paper, standard machine learning evaluation metrics such as accuracy, recall, F1 score and precision are used for evaluation purpose. The following process is being implemented by loading the Twitter data set which is further followed by pre-processing which involves removal of stop words, tokenization and stemming. Then, the data set is split into train and test data with the percentile of 75% and 25%, respectively. The pre-processed data is represented as text features using TF-IDF and Bag of Words. They are then subjected to different machine learning algorithms to classify the tweets as bullying or non-bullying. The result obtained is used to

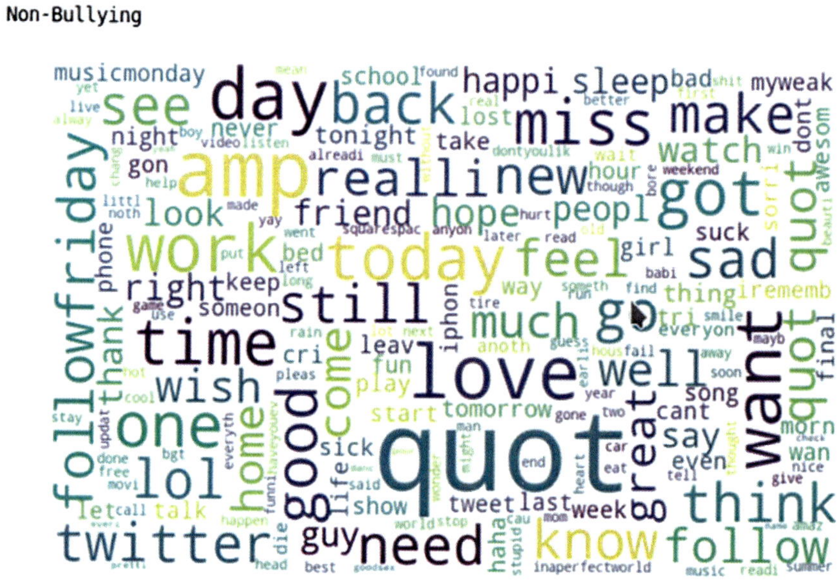

Fig. 3 Word cloud for non-bullying words

plot the confusion matrix which is used to calculate the performance measures. The model is being run using the built-in Python packages which involve scikit-learn, pandas, model_selection, numpy, pre_processing, TfidfVectorizer, CountVectorizer, confusion_matrix, matplotlib.pyplot, keras and accuracy_score.

4.1 TF-IDF for Feature Extraction

The algorithms are compared using the data available in Figs. 4, 5 and 6, and evaluation metrics are shown in Table 1. Accuracy, F1 score, precision and recall of different machine learning algorithms using TF-IDF in feature extraction are shown in Figs. 7, 8, 9 and 10, respectively.

Among the machine learning algorithms, SVM performed the best with an average accuracy 99%, followed by decision tree, neural network and logistic regression with accuracy of 98%, then linear regression with 94.5%, and the least performance was given by Naive Bayes with an average accuracy 93% using TF-IDF method.

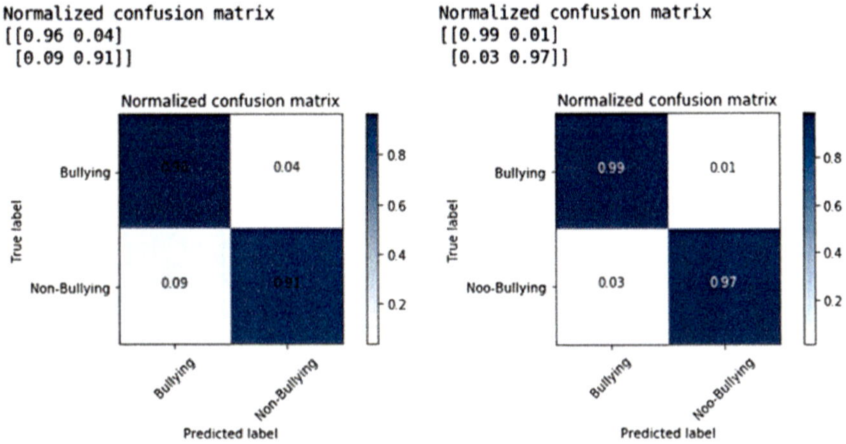

Fig. 4 Confusion matrix of Naive Bayes and logistic regression

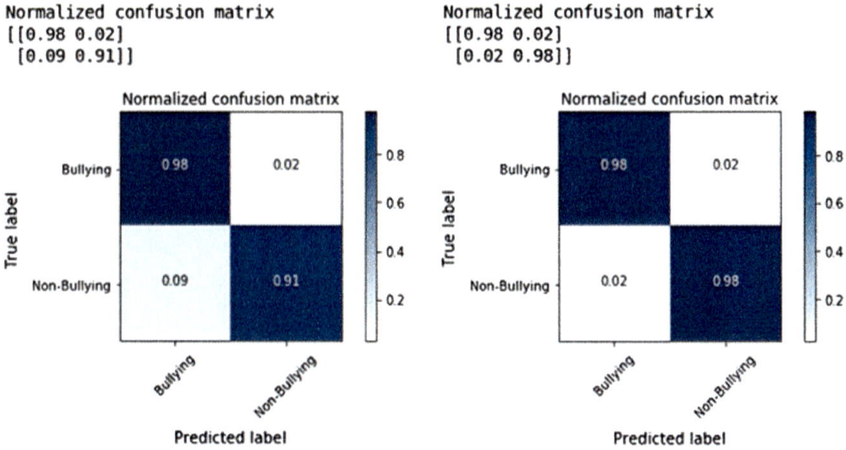

Fig. 5 Confusion matrix of linear regression and decision tree

4.2 Bag of Words for Feature Extraction

The algorithms are compared using the data available in Figs. 11, 12 and 13, and the evaluation metrics are shown in Table 2. Accuracy, F1 score, precision and recall of different machine learning algorithms using Bag of Words in feature extraction are shown in Figs. 14, 15, 16 and 17, respectively.

Among the machine learning algorithms SVM, neural network and logistic regression performed the best with a mean accuracy of 98%, succeeded by decision tree with 97% accuracy, then linear regression with 85% accuracy, and the least performance was given by Naive Bayes with an average accuracy 77% and using TF-IDF

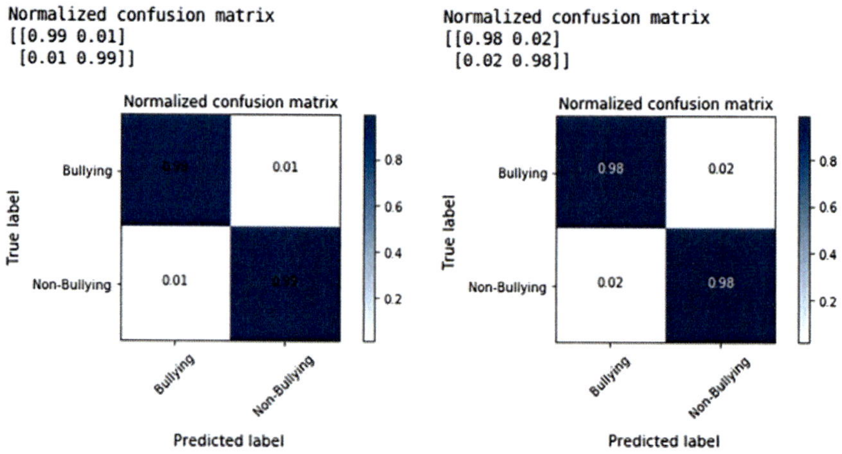

Fig. 6 Confusion matrix of SVM and neural network

Table 1 TF-IDF for feature extraction

Algorithm	Accuracy	Precision	Recall	Fi-score
Linear regression	0.945	0.916	0.98	0.947
Logistic regression	0.98	0.97	0.99	0.98
Naive Bayes	0.935	0.91	0.96	0.936
SVM	0.99	0.99	0.99	0.99
Decision tree	0.98	0.98	0.98	0.98
Neural network	0.98	0.98	0.98	0.98

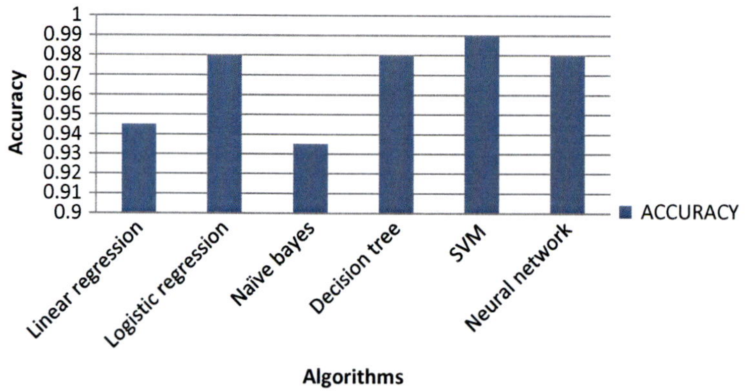

Fig. 7 Accuracy of different machine learning algorithms using TF-IDF in feature extraction

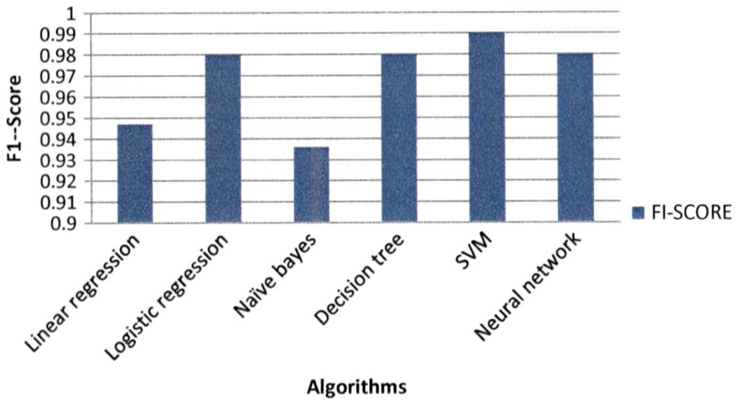

Fig. 8 F1 score of different machine learning algorithms using TF-IDF in feature extraction

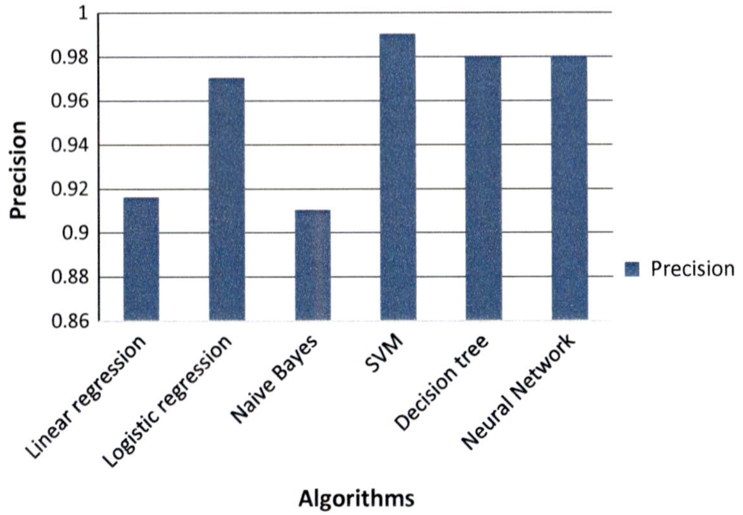

Fig. 9 Precision of different machine learning algorithms using TF-IDF in feature extraction

method. When the recall is high and precision is low, it shows that most of the bullying tweets are recognised accurately as bullying and there are few non-bullying tweets which are recognised as bullying. When precision is high and recall is low, it shows that many bullying tweets are predicted as non-bullying and less non-bullying tweets are predicted as bullying. F1 score takes in account of cases where bullying tweets are predicted as non-bullying and non-bullying tweets are predicted as bullying. F1 score is considered useful than accuracy when un-even class distribution is present in the data.

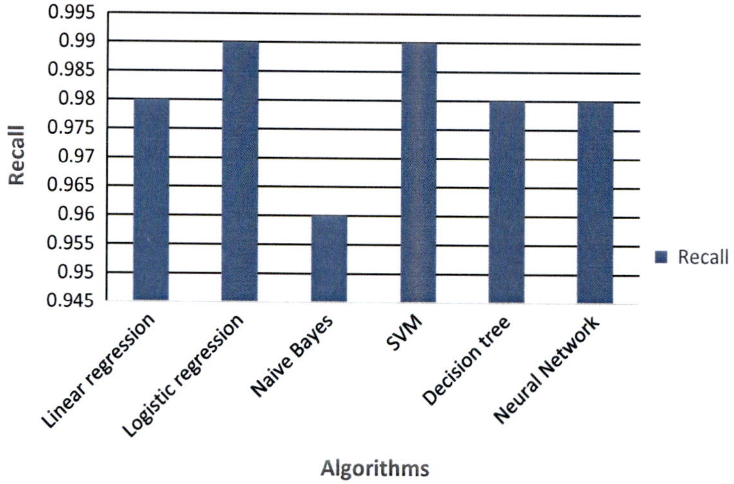

Fig. 10 Recall of different machine learning algorithms using TF-IDF in feature extraction

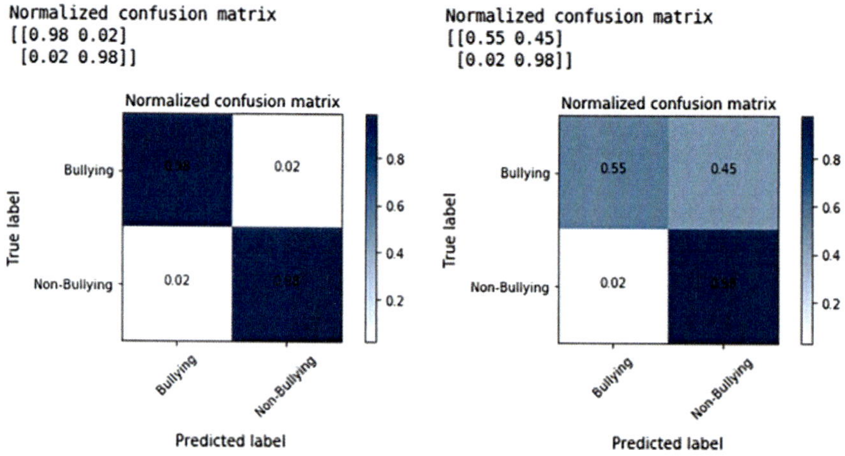

Fig. 11 Confusion matrix of SVM and Naive Bayes

4.3 Accuracy Based on Number of Tweets

Accuracy based on a number of tweets using TF-IDF and Bag of Words is shown in Figs. 18 and 19, respectively.

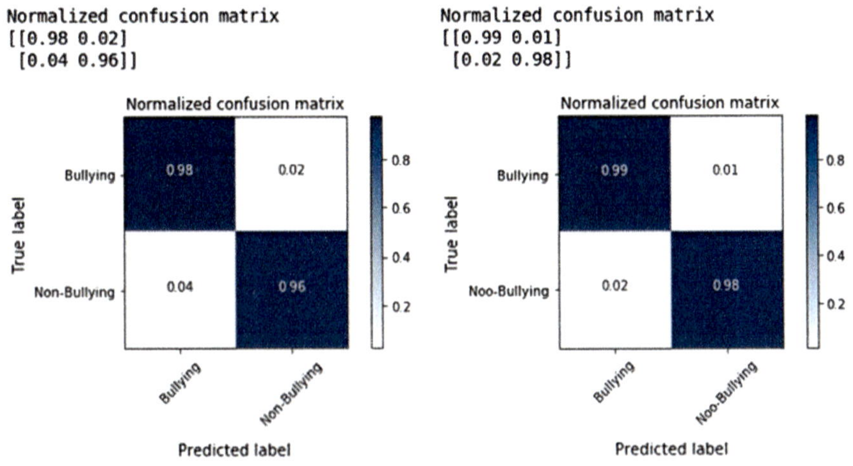

Fig. 12 Confusion matrix of decision tree and logistic regression

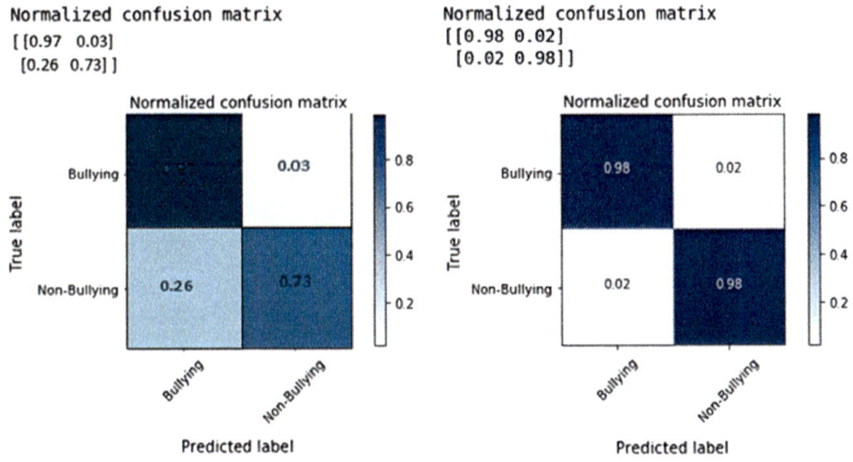

Fig. 13 Confusion matrix of linear regression and neural network

Table 2 Bag of words for feature extraction

Algorithm	Accuracy	Precision	Recall	Fi-score
Linear regression	0.85	0.79	0.97	0.87
Logistic regression	0.98	0.98	0.99	0.98
Naive Bayes	0.77	0.96	0.55	0.70
SVM	0.98	0.98	0.98	0.98
Decision tree	0.97	0.96	0.98	0.97
Neural network	0.98	0.98	0.98	0.98

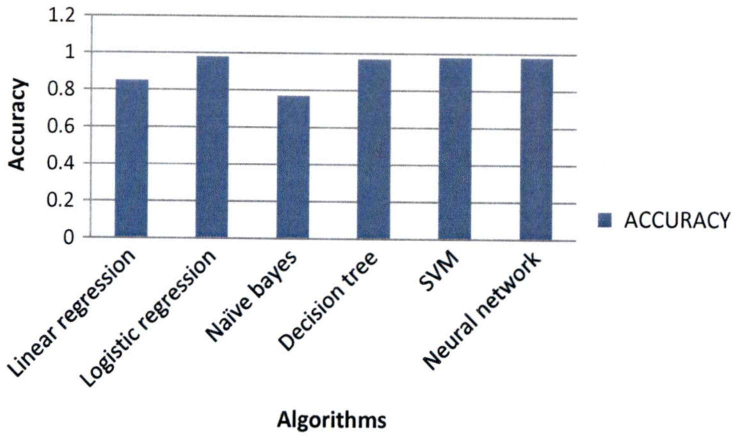

Fig. 14 Accuracy of different machine learning algorithms using Bag of Words

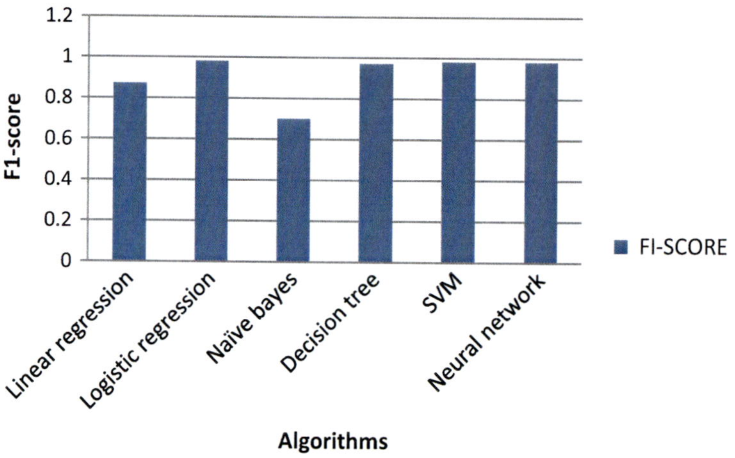

Fig. 15 F1 score of different machine learning algorithms using Bag of Words

5 Conclusion

The experimentation results prove that SVM performs better with TF-IDF as a feature extraction method and SVM performs better with Bag of Words as feature extraction method. In this paper, a qualitative analysis has been made to evaluate the performance of the six important machine learning algorithms.

This study can be extended to include ensemble classifiers to improve the performance, and also theoretical model needs to be built to explain the experimentation results.

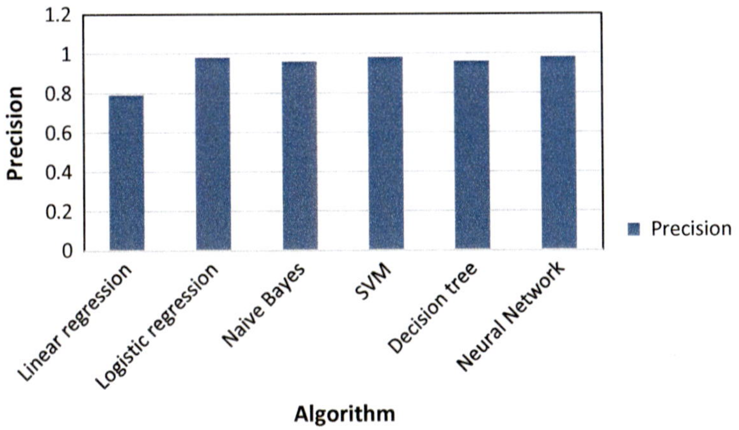

Fig. 16 Precision of different machine learning algorithms using Bag of Words

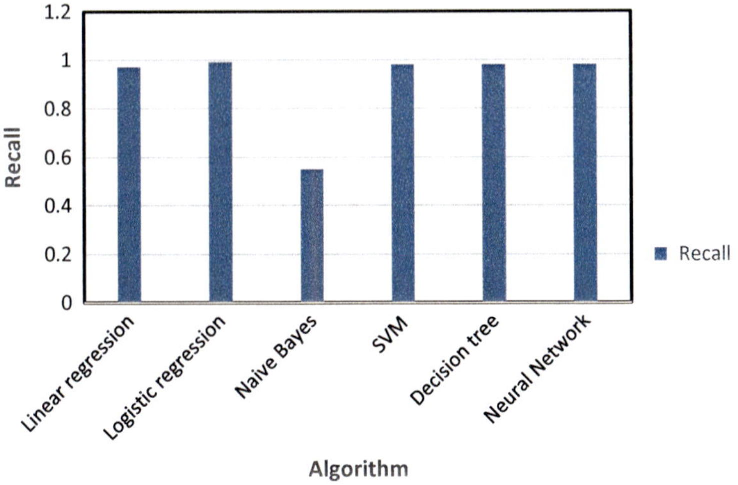

Fig. 17 Recall of different machine learning algorithms using Bag of Words

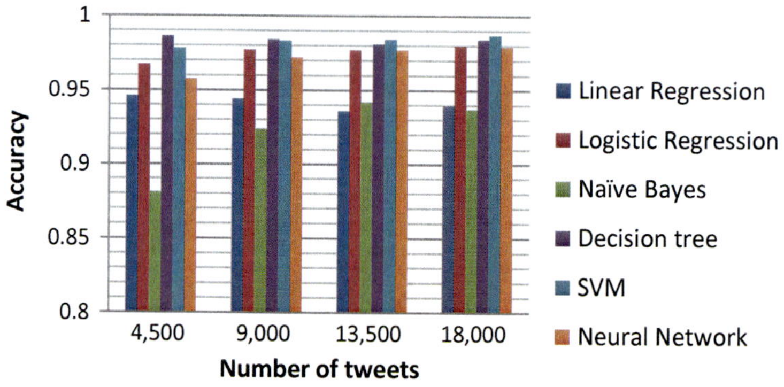

Fig. 18 Accuracy based on number of tweets using TF-IDF

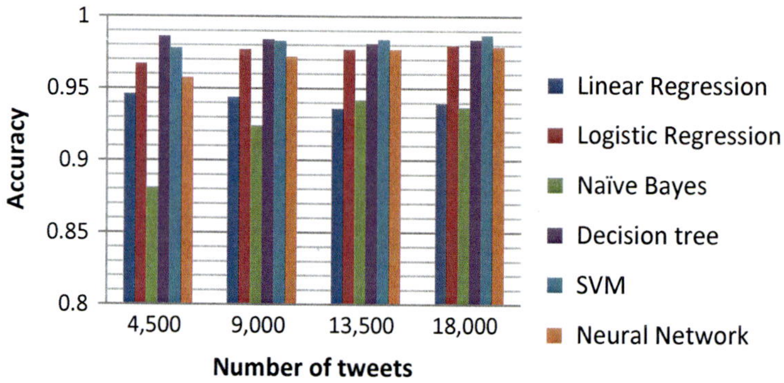

Fig. 19 Accuracy based on number of tweets using Bag of Words

References

1. Stephen P, Jaganathan S (2014) Linear regression for pattern recognition. In: International conference on green computing communication and electrical engineering (ICGCCEE). doi 10.1109/ICGCCEE.2014.692139
2. Haifley T (2002) Linear logistic regression: an introduction. In: IEEE international integrated reliability workshop final report. doi:10.1109/IRWS.2002.1194264
3. Zhang H, Li D (2007) Naive Bayes text classifier. In: 2007 IEEE international conference on granular computing. https://doi.org/10.1109/grc.2007.40
4. Awad M, Khanna R (2015) Support vector machines for classification. In: Efficient learning machines. Apress, Berkeley. doi https://doi.org/10.1007/978-1-4302-5990-9_3
5. Li L, Zhang X (2010) Study of data mining algorithm based on decision tree. In: 2010 international conference on computer design and applications. https://doi.org/10.1109/iccda.2010.5541172
6. Mishra M, Srivastava M (2014) A view of artificial neural network. In: International conference on advances in engineering & technology research. doi 10.1109/ICAETR.2014.7012785
7. Meda C, Bisio F, Gastaldo P, Zunino R (2014) A machine learning approach for twitter spammers detection. Recent Adv Electr Electron Eng. https://doi.org/10.1109/CCST.2014.6987029

8. Liang H, Sun X, Sun Y, Gao Y (2017) Text feature extraction based on deep learning: a review. EURASIP J Wirel Commun Network 2017(1):211. https://doi.org/10.1186/s13638-017-0993-1
9. Zhao R, Mao K (2017) Cyberbullying detection based on semantic-enhanced marginalized denoising auto-encoder. IEEE Trans Affect Comput 8(3):328–339. https://doi.org/10.1109/taffc.2016.2531682

An Optimal Wavelet Detailed-Coefficient Determination Using Time-Series Clustering

C. I. Johnpaul, Munaga V. N. K. Prasad, S. Nickolas, G. R. Gangadharan, and Marco Aiello

Abstract Time-series clustering is used to find out similar patterns which occur in a time-series data. Time-series sequences are characterized by many features. The trend and seasonality, for instance, are the two important features which reveal the behavior of a time-series sequence and can be used to group them. One of the common dimensionality reduction method for time-series data is the usage of wavelets. In this paper, we analyze the importance of wavelet-based time-series clustering which helps to reduce the length of time-series dimensions, in-turn reducing the time needed for the clustering task. Determining an optimal set of coefficients for wavelets is always challenging. While it plays an important role in reducing the time taken for clustering, to test/verify our claims, we analyze a renewable energy dataset which contains the periodical energy load of various generators located in Europe. The clustering of this time-series data helps to identify similar electrical nodes. We compared eight different clustering algorithms with and without using wavelet decomposition on the renewable energy data. Minibatch k-means and Brich clustering show better performance on using wavelet-based decomposition.

Keywords Time-series · Clustering · Wavelet · Detailed coefficients · Renewable energy

1 Introduction

Time-series record is the periodical observation of an attribute over a period of time. This type of data is used to represent the functioning of various entities from different domains. Stock markets, industrial production, sports, manufacturing, trans-

---

C. I. Johnpaul (✉) · S. Nickolas · G. R. Gangadharan
National Institute of Technology Tiruchirappalli (NIT-T), Tiruchirappalli, India
e-mail: johnpaulci@gmail.com

C. I. Johnpaul · M. V. N. K. Prasad
Institute for Development and Research in Banking Technology (IDRBT), Hyderabad, India

M. Aiello
University of Stuttgart, Stuttgart, Germany

N. N. Chiplunkar and T. Fukao (eds.), *Advances in Artificial Intelligence and Data Engineering*, Advances in Intelligent Systems and Computing 1133,
https://doi.org/10.1007/978-981-15-3514-7_63

portation, etc., use time-series data for obtaining useful intuitions. It can be hourly, daily, weekly, monthly, or yearly data, depending on the nature of the application [1]. The two major challenges in time-series clustering are the series dimensional length and the presence of a similar trend, respectively. The loss of significant values during the dimensionality reduction will impact the performance of clustering mechanism. Hence, a transformation is required for the time-series data to find out the significant values to perform the clustering [2]. Wavelets are one of the suitable methods to represent the time-series data to obtain the significant values. The wavelet representation of a time-series data contains two fundamental groups of values, namely detailed and approximation coefficients. The data in time domain is not suitable to represent these features which indirectly map the trend of data. One of the attractive properties of wavelet features is the ability to regenerate the original time-series with the aid of wavelet coefficients. The wavelet transformation imposes various functions on the time-series data and transforms it into the frequency domain. Hence, the trend of the series is evident from the transformed details which can be used for further learning process [3].

Usually, time-series data are unlabeled. They are more frequently used for prediction and forecasting. Hence, labeling of data is not required for such tasks. The existence of a label is required for classification of a new time-series data. Unsupervised clustering mechanisms help to aggregate similar time-series data and form the groups. In some cases, the user has to specify the number of intended groups that can be formed from the data [4]. Distance measures form the fundamental operational method of clustering. If the dimension of time-series values are larger, then distance-based clustering is not feasible as it is time consuming. Appropriate dimensionality reduction methods need to be applied for improving the performance of clustering mechanism. In most of the cases, PCA is not used in time-series data, since the order of data values in time-series is essential. The graph drawn in time domain can project only the magnitude of an attribute at a particular time. For analyzing the similar trends in time-series, a frequency domain is to be created. Wavelets perform this transformation which projects out the count of time-series segments having the same frequency. Wavelet-based dimensionality transformation follows the order of data values in the time-series [5].

Clustering of time-series data helps to identify similar time-series sequences. The clustering model is used in several applications where decision making can be performed [6]. Identification of significant data points, classification of new data elements, similar behavior detection with time, intervention analysis are some of the specific use cases. Real-time data monitoring and decision making help to control the operational devices efficiently [7]. Wavelet-based methods are much useful in real-time scenarios, where the time of execution and decision making is a critical factor. The grouping of time-series data in frequency domain has many added advantages. The decomposition of time-series data into an array of sine and cosine function values can be done in frequency domain. These values can be analyzed for observing the similarity among various time-series sequences [8].

There are various ways for performing the time-series clustering. The three fundamental methods is based on model feature and shape. The wavelet-based clustering method is a variant of the feature-based method. The feature-based method extracts relevant features from a time-series data and performs various clustering over the features. The detailed and approximation coefficients obtained through the wavelet transformation form the base features for clustering the time-series data [9]. In our proposed method, we use wavelets for the clustering of time-series data. The comparison of wavelets and non-wavelet-based clustering reveals the significance of wavelets on a time-series data. The process of determining the level of wavelet decomposition also impact the time taken for clustering process. The salient contributions of this paper are as follows.

- A wavelet decomposition-based time-series clustering approach.
- A comparative analysis of various clustering algorithms on time-series clustering.
- Significant detailed-coefficient determination of the time-series data.

The remainder of this paper is organized as follows. Section 2 describes the literature relevant to time-series clustering. Section 3 describes the clustering method and approaches. Section 4 presents a case study and the results. Section 5 presents the conclusion and future scopes of the work.

2 Time-Series Clustering

The most common clustering method takes data points in chronological order and performs clustering without any transformations. Segment-based time-series clustering divides the time-series sequences into blocks and groups the sequences based on segment similarity. A multistage clustering mechanism can be adopted to improve the performance of clustering [10]. The common goal of all clustering mechanisms is to find out similarity between the data points. A variety of similarity and distance measures is used to find out the proximity of data sequences. Combination of various distance measures is also used in time-series clustering. Apart from the closeness of the data, density of data sequences also gives rise to groups. The most effective and promising distance measure for time-series clustering is dynamic time wrapping (DTW). DTW barrycenter averaging (DBA) is widely used time-series averaging method [11].

A set of time-series sequences contains nine global characteristics or features. These features are the source of valuable information about the nature of the sequences. The majority of the clustering algorithms compute these features to reveal the behavior of time-series data. The usage of these characteristics depends on the type of clustering and decision model [12]. Sub-sequence time-series clustering discusses the similarity of internal segments in a time-series data. These sub-sequences in time-series clustering reveal interesting information about data which includes the identification of sequential patterns, periodic patterns, motifs, etc. The sub-sequence-based clustering method entails high memory usage, the possibility of unexpected failures for some parameters, and high complexity [13].

Semi-supervised clustering methods like COBRAS are also suitable for grouping time-series sequences. It is an iterative refinement process over a set of predefined clusters. Each cluster elements are known as *super instances.* A pairwise medoid is found between the super instances to establish the closeness of the sequences. Different versions of COBRAS exist based on the usage of distance measures [14]. The presence of noise, uneven values in time-series sequences affects the clustering mechanism. Omid et al. describe various steps to consider the noisy time-series sequences and transform it to a structure ready for other clustering algorithms [4]. A noise metric is also defined to evaluate the density of noise present in the time-series sequences. The time-series data which contain electrical load usage of residential electricity customers is considered for experimenting their clustering strategies. The labeled and unlabeled time-series data requires necessary preprocessing steps to apply various clustering algorithms. The feature learning from the differential features of a time-series data involves the segments and shaplets. Haishuai *et al.* discuss a new semi-supervised shaplets learning (SSSL) method to obtain necessary information from shaplets and segments of a time-series data [15].

Rebecca et al. illustrate several learning methods on the stationary time-series data. Non-stationary time-series data cannot be used for prediction and regression tasks. The transformation of non-stationary to stationary mode is essential for applying forecasting algorithms. Proper transformation methods improve the performance of forecasting of a typical non-stationary time-series sequence [16]. Time-series classification (TSC) is a supervised model in which the incoming time-series sequence is classified into appropriate group. Compressing the time-series data and using such data for creating a classification model is challenging. Wavelets are used to compress the data in a lossy strategy. Relevant methods of compression improve the performance of classification both in accuracy and time. Moreover, efficient compression methods extensively used to overcome the difficulties of storage and processing task to a great extent [17]. The decomposition of time-series data using wavelets helps in forecasting. The accuracy of such forecasting can be improved by aggregating the model coefficients using wavelets. It is based on the multi-resolution analysis and Ordinary Least Square (OLS) regression model. The method produces independent forecasts of a univariate time-series sequence and selects the forecast with the minimum error [18].

Clustering methods are often used for finding the pattern of electricity consumption. Shape-based methods provide promising results in the grouping of time-series sequences. A shapelet contains limited number of values [19]. Lulu et al. describe a shape-based method which utilizes a modified k-means clustering procedure to group the time-series sequences. An extensive real-time experiment is performed with the data obtained from the commission for energy regulation (CER) which includes the energy consumption of Irish homes [20]. Khosravi et al. studied a case study on machine learning algorithms which predicts the speed of the wind in a wind farm. The wind speed is measured in specific time intervals and fed to different machine learning techniques including neural networks to predict the wind speed. Support vector regression (SVR), multilayer feed-forward neural network (MFFNN) are used for the prediction task [21]. Modified dynamic time wrapping method is used in time-series clustering of remote sensing images. The modification to DTW is performed with the use of Canberra distance (CD). This distance measure calculation is more reliable even in the presence of obstacles like cloud. Clustering of images is done based on the maps with the help of modified distance measures on k-means algorithm [22].

Clustering of time-series data is unsupervised in nature. Wavelet-based grouping of similar generator nodes cannot explain the physical connectivity between the nodes even if they belong to the same cluster. The physical connectivity and cluster membership can give the possibilities of load sharing in generator nodes. We analyze the possibility of load sharing along with the clustering results in Sect. 4. Section 3 explains the experiments done on time-series data.

3 Proposed Work

The workflow of our proposed wavelet-based time-series clustering is shown in Fig. 1. The preprocessing steps include the determination of time scales. Time-series data is aggregated hourly, weekly, and monthly. These aggregated time-series data are transformed into detailed and approximation coefficients with the help of wavelets [23]. The first wavelet theory was proposed by Haar in 1909. A wavelet is a small wave modified with a signal. There are large number of wavelet functions which include Haar (Daubechis 1), Daubechies 2–10, Mexcan hat, Morlet, etc. [24]. Each of these functions include a transformation equation which convert the data into a wavelet representation. Consider a Haar wavelet function. A time-series sequence S of length n is subjected to a Haar wavelet transform in the time-series values into approximation and detailed coefficients according to the Eqs. (1) and (2). Every wavelet transformation has a user-defined decomposition-level parameter which restricts the number of approximation and detailed coefficients for a particular time-series. Clustering of optimal number of coefficients helps to identify the groups reasonably fast.

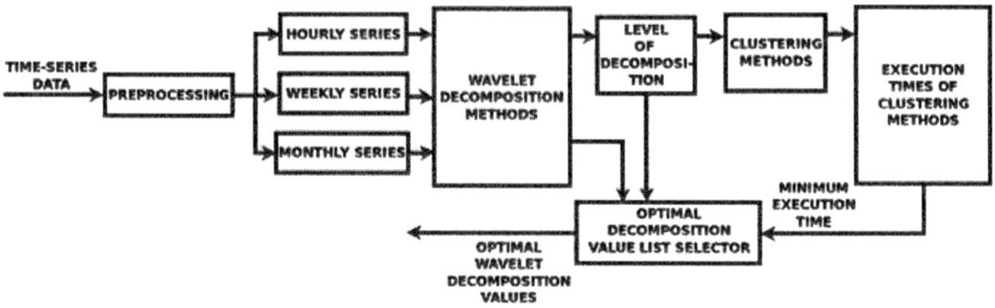

Fig. 1 Workflow of wavelet-based time-series clustering

$$\text{Approx}_{i,j} = \begin{cases} \frac{A_{i-1,2j}+A_{i-1,2j-1}}{2} & \text{if } i > 1. \\ \frac{t_{2j}+t_{2j-1}}{2}, & \text{if } i = 1. \end{cases} \quad (1)$$

$$\text{Detailed}_{i,j} = \begin{cases} \frac{A_{i-1,2j}-A_{i-1,2j-1}}{2} & \text{if } i > 1. \\ \frac{t_{2j}-t_{2j-1}}{2}, & \text{if } i = 1. \end{cases} \quad (2)$$

The approximation coefficients are denoted by $\text{Approx}_{i,j}$, where i is the level of decomposition and j is the index. The detail coefficients are represented as $\text{Detailed}_{i,j}$. t_j is the time-series sequence with the original values.

3.1 Stationarity of the Time-Series Data

Time-series data points are extremely useful when they are stationary. Stationary data points have constant statistical properties over a period of time. The statistical properties include the mean, variance, and standard deviation. If these properties are not constant, then time-series data processing will be incomplete for some applications. The performance of time-series predictions and forecasting improves when the data is stationary. Non-stationary time-series data shows a trend in the distribution of the values. The growth behavior can be increasing or decreasing. Hence, even with the help of non-stationary time-series values, it is not easy to predict the forthcoming data points. The time-series should be made stationary to proceed with prediction tasks. The methods to convert a non-stationary time-series data into a stationary include differencing, seasonal differencing, and transformation [25]. Results of various forecasting methods are discussed in Sect. 4. Broadly, there are two methods to detect the stationarity of the time-series data. They are visual and statistical methods, respectively. Visual methods capture the information from the graphs which are plotted against the time axis. The varying mean of the data is usually visible in the graph. Statistical tests include various standard methods for finding the test statistic to determine the stationarity of the data. Among the statistical methods, Augmented Dickey-Fuller Test (ADF) and Kwiatkowski–Phillips–Schmidt–Shin (KPSS) test are prominently used to determine the stationary properties a time-series data [26, 27].

3.2 Preprocessing

Time-series preprocessing entails the division of the data into hourly, weekly, and monthly taking the mean of the respective aggregate values. Stationarity tests are performed with ADF to proceed with the learning process. The results of the stationarity tests performed over the data is shown in Sect. 4. The impact of stationarity on clustering process is seldom addressed in the literature. Empirically, if there is a large variation in the stationarity of the data, it is observed that scattering of elements in the cluster is more.

3.3 Wavelet Decomposition

The decomposition of time-series values with wavelets reduces the dimension of the data [28]. The common feature of all clustering algorithms is the similarity measurement computation. Wavelet decomposition helps to reduce the time taken by the clustering methods. The similarity measures should consider all the data values if the dimensionality reduction is not performed. The wavelet coefficients are arranged in different levels. Figure 2 shows an instance of wavelet decomposition mechanism for a level 3. An instance of finding out the decomposition coefficients are as in Eqs. 1 and 2. The number of data values to be considered is determined by the decomposition level. The execution time of clustering methods varies according to the changes in the decomposition level of wavelets. Section 4 illustrates a comparative analysis with the execution time of each clustering algorithm and the level of wavelet decomposition.

3.4 Forecasting and Clustering Methods

Forecasting the time-series data is done with traditional prediction methods like naive forecasting, average and moving average forecasting, average holt-linear, average holt-winter, and SARIMA. The predicted values from each method are compared with the original values and errors are also calculated. The statistical details of the experiment are discussed in Sect. 4 with sufficient graphs. The grouping of decomposed time-series data is done with various clustering algorithms, namely affinity propagation (APC), Minibatch k-means (MBKmeans), mean shift (MSC), agglomerative clustering (AC), DBScan, spectral clustering (SC), ward clustering (WC), and birch clustering.

S	t_1	t_2	t_3	t_4	t_5	t_6	t_7	t_8
Level 1	$\mathbf{A}_{1,1}$	$\mathbf{A}_{1,2}$	$\mathbf{A}_{1,3}$	$\mathbf{A}_{1,4}$	$D_{1,1}$	$D_{1,2}$	$D_{1,3}$	$D_{1,4}$
Level 2	$\mathbf{A}_{2,1}$	$\mathbf{A}_{2,2}$	$D_{2,1}$	$D_{2,2}$				
Level 3	$\mathbf{A}_{3,1}$	$D_{3,1}$						

Fig. 2 Wavelet decomposition into approximation and detailed coefficients

4 Results and Analysis

The experiments with the proposed method is performed over a renewable energy dataset [29]. The dataset contains the details of continental European electricity system, containing the demand and renewable energy inflows for the period 2012–2014. There are 1494 buses and 2156 lines, with parameters given for 969 generators. For each bus node, signals for solar, wind, and load production is recorded in hourly basis for three years. Figure 3 shows the interconnected layout of the generators. The dataset contains the hourly load values of energy produced by each generator over two years. The results of the ADF test for the first five generator values are shown in Table 1. The critical value for 5%, 1%, and 10% of the data samples are −2.863551, −3.434902, and −2.567840, respectively. According to the ADF test, there exists two hypothesis *H0* and *H1*, respectively. If *H0* is accepted, then it signifies the presence of a unit root for the time-series data and the data is non-stationary. If *H1* is accepted, then it concludes that the time-series data is stationary.

The test static value in all the cases is less than the critical values which signifies the rejection of the null hypothesis of non-stationary time-series. Hence, the time-series data is stationary. The *p*-values of all the series are less than 0.05 which signifies the strong rejection of the null hypothesis.

Figure 4 presents the nature of time-series data for a generic generator from the dataset. The load of a typical generator node is visualized after two preprocessing modes, namely by week and three-months duration. The mean and standard deviation of the series show that the data is stationary throughout the time-period. This

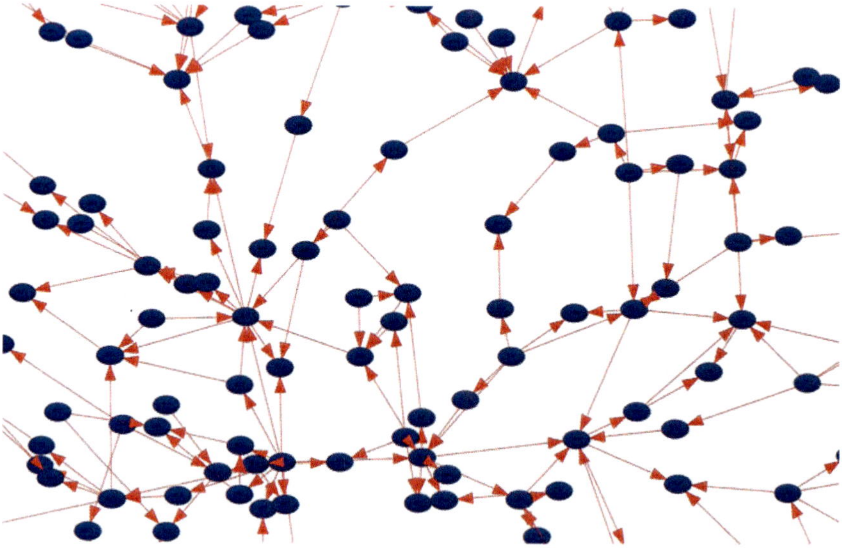

Fig. 3 Part of an interconnected network of generators

Table 1 An instance of The ADF statistic test values for the generator nodes

Node	Test static	P-value
1	−5.783485e+00	5.062056e−07
2	−6.260877e+00	4.226414e−08
3	−6.197489e+00	5.916651e−08
4	−6.537468e+00	9.529135e−09
5	−6.197504e+00	5.916192e−08

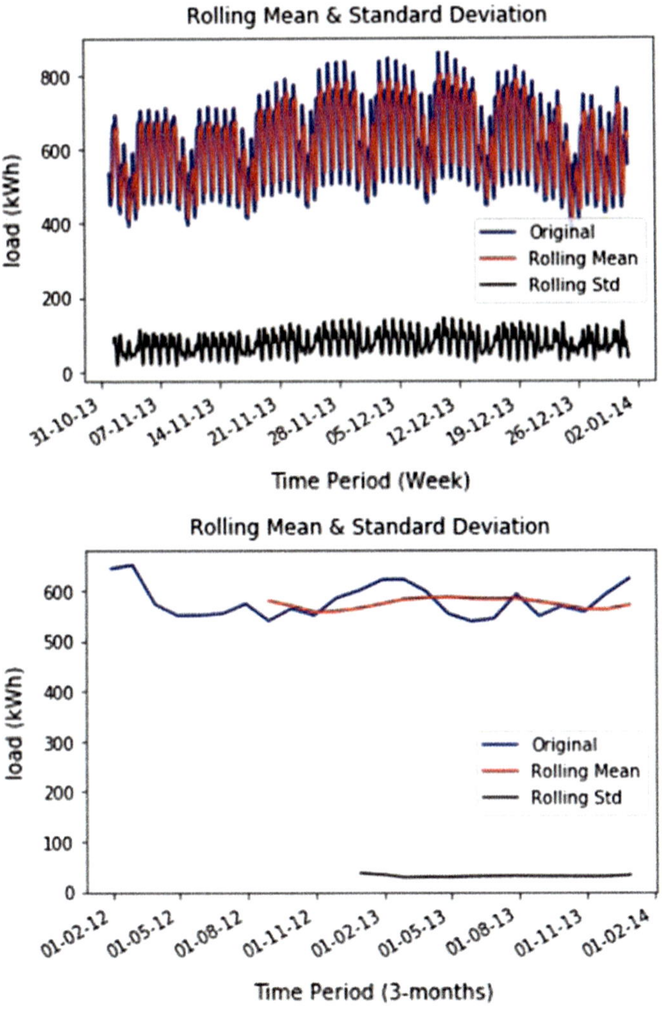

Fig. 4 Mean and standard deviation of the time-series data of a typical generator in to two preprocessing modes (week and 3-months)

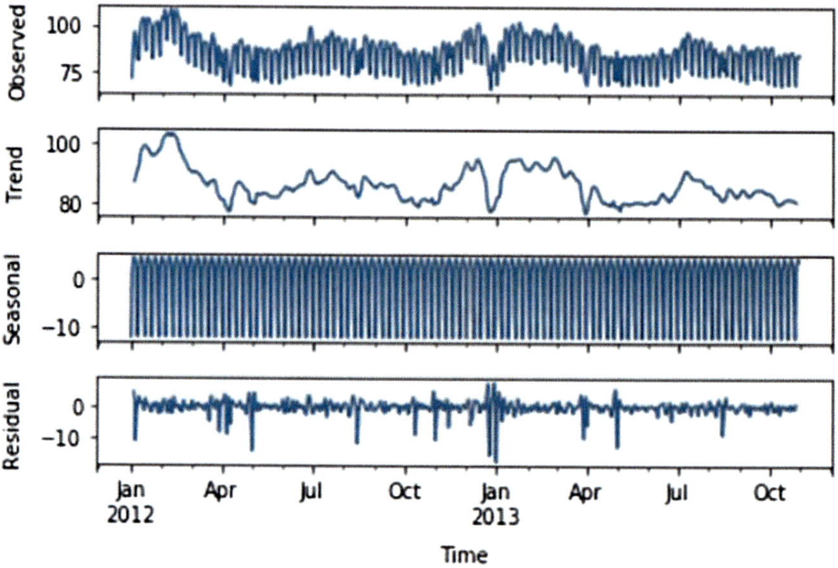

Fig. 5 Decomposed time-series data of a generator with its components

supports the ADF test results which conclude the stationarity of the series. Figure 5 presents the decomposition of the time-series data into trend, seasonal, and residual data. This decomposition shows the features of a time-series data. Figure 6 illustrates the behavior of different forecasting algorithm on time-series data. Table 2 presents the root mean square (RMS) error analysis of various forecasting algorithms. One notices that *avection rage_forecast* shows less error. One the contrary, the *Holt-linear* forecasting shows larger error. The error in the load prediis also visualized in Fig. 6. Table 3 presents the execution time of various clustering algorithms with and without wavelet transform. Wavelet transformation reduces the execution time considerably. Birch and Minibatch k-means clustering show less execution time in both the cases. Figure 7 shows the histogram representation of the execution times of various clustering algorithm. Among the algorithms, Minibatch k-means and Birch clustering have less execution time compared to others. Figure 8 presents the formation of various groups by different clustering algorithms with the execution time for wavelet and non-wavelet data. The groups formed from the wavelet are less scattered compared to non-wavelet data. Table 4 describes the detailed execution times of the clustering algorithms with varying levels of detailed coefficients. The experiments are performed with and without wavelet transformation on a fixed number of clusters. It is evident that Minibatch k-means and Birch clustering show less variance in execution times with the increase in levels. We used Daubechies (D1) function

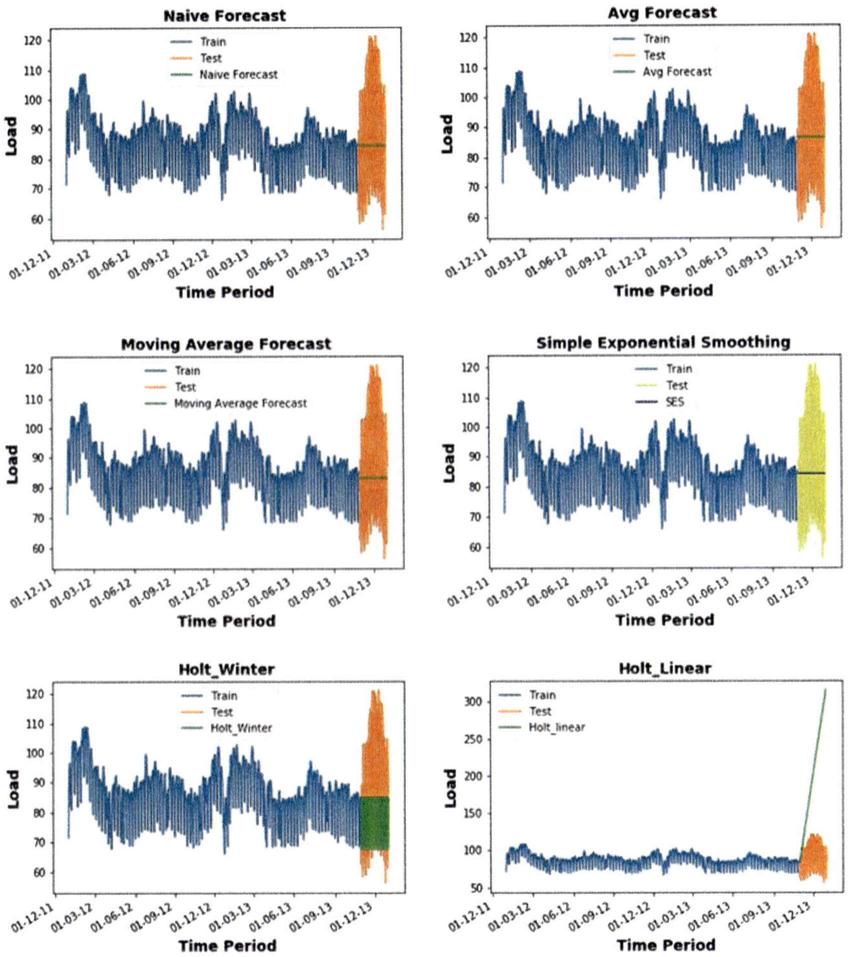

Fig. 6 Visual representation of various forecasting algorithm with their forecasting behavior

Table 2 RMSE for the clustering algorithms

Method	RMSE
Naive forecast	16.0667153131
Avg_forecast	15.6328185869
Moving_Avg	16.4654917597
SES	16.0875150375
Holt-linear	129.593625699
Holt-winter	18.7158771063
SARIMA	17.3452

Table 3 Comparison of execution times for wavelet and without wavelet clustering

Clustering method	Time (Wavelet) (s)	Time (Without wavelet) (s)
Minibatch k-means	0.01	0.05
Affinity propagation	3.27	7.41
Mean shift	0.03	0.24
Spectral clustering	1.64	1.64
Ward	0.27	0.53
Agglomerative clustering	0.34	1.01
DBSCAN	0.06	0.61
Birch	0.04	0.12

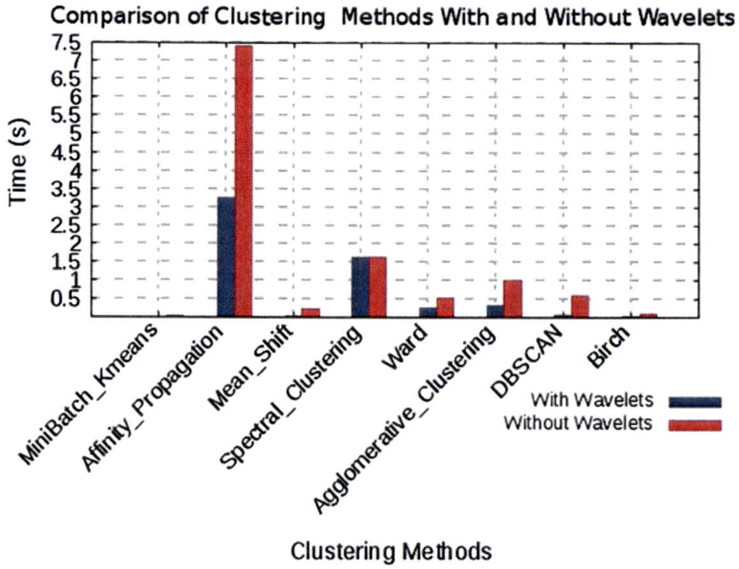

Fig. 7 Clustering of time-series with and without wavelet transformation

for wavelet decomposition to produce ten levels of detailed coefficients. According to the Minibatch k-means algorithm, level 5 shows the minimum execution time for wavelet-transformed clustering. The Birch clustering shows that the levels {1, 2, 3, 4} have the minimum execution times for wavelet-transformed clustering.

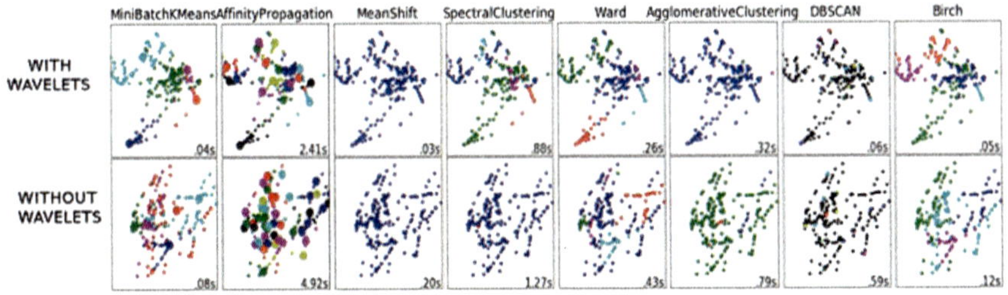

Fig. 8 An instance of clustering of time-series with and without wavelet transformation for five clusters

5 Conclusion and Future Work

The clustering of a time-series data provides a class label to various time-series sequences. The methods used for clustering mechanism is a crucial step in grouping the time-series sequences. The distance measure-based computation is an important step in such clustering methods. The time taken for clustering depends upon the number of time-series values in the sequence. There are different dimensionality reduction mechanisms existing which reduce the input size without loosing the data content. Wavelet forms a typical method of dimensionality reduction which also preserves the order of the values in the sequences. Wavelet-based grouping presents a new insight into unsupervised clustering of a time-series data. Empirically, we found that the level of wavelet decomposition and the group of algorithms that can be used for efficient clustering mechanism. We explored the possibility of various algorithms in determination of the wavelet coefficients. This knowledge of algorithms and decomposition level helps in processing new time-series values which are updated in the sequences owing to the addition of new data points. The unrelated data points can be grouped into a cluster based on their similarity in the nature of time-series values. Apart from these, the impact of new wavelet functions needs to be explored in connection with the clustering mechanism. Anomaly detection, failure detection using time-series sequences, etc., are some of the far view implications of time-series clustering.

Table 4 Comparison of execution times for wavelet (W) and without wavelet (NW) clustering for different decomposition coefficients

Level	Clustering algorithm—time (s)															
	MBKmeans		APC		MSC		SC		WC		AC		DBScan		Birch	
	W	NW	W	NW	W	NW	W	NW	W	NW	W	NW	W	NW	W	NW
0	0.02	0.06	3.22	6.40	0.02	0.20	0.62	1.22	0.10	0.40	0.08	1.06	0.01	0.73	0.03	0.13
1	0.01	0.06	5.58	7.06	0.02	0.20	1.21	1.35	0.10	0.38	0.08	0.98	0.01	0.60	0.02	0.15
2	0.02	0.07	6.00	6.54	0.02	0.20	0.38	1.17	0.11	0.38	0.09	0.98	0.01	0.60	0.02	0.15
3	0.02	0.07	3.94	6.98	0.06	0.20	0.86	1.18	0.12	0.39	0.11	0.97	0.01	0.60	0.02	0.13
4	0.02	0.06	2.87	6.68	0.06	0.20	0.45	1.50	0.13	0.38	0.15	0.95	0.01	0.60	0.02	0.15
5	0.01	0.06	3.79	6.51	0.02	0.20	0.65	1.17	0.27	0.38	0.21	0.95	0.02	0.61	0.03	0.13
6	0.02	0.09	3.38	6.62	0.03	0.20	0.73	1.18	0.21	0.38	0.34	0.96	0.03	0.59	0.03	0.13
7	0.02	0.06	2.83	6.65	0.03	0.20	1.22	1.19	0.26	0.54	0.33	0.97	0.06	0.60	0.04	0.14
8	0.03	0.06	2.94	6.45	0.05	0.21	0.80	1.19	0.44	0.38	0.40	0.95	0.09	0.59	0.04	0.11
9	0.03	0.09	3.28	6.38	0.05	0.20	0.86	1.36	0.50	0.40	0.46	1.04	0.18	0.62	0.06	0.14
10	0.05	0.10	4.03	6.62	0.10	0.20	0.92	1.18	0.68	0.39	0.65	0.96	0.27	0.60	0.10	0.14

Acknowledgements The research is supported by the Netherlands Organization for Scientific Research (NWO) in the framework of the Indo-Dutch Science Industry Collaboration Program, project NextGenSmartDC (629.002.102).

References

1. Lim BY, Wang JG, Yao Y (2018) Time-series momentum in nearly 100 years of stock returns. J Banking Fin 97:283–296
2. Ferreira LN, Zhao L (2016) Time series clustering via community detection in networks. Inf Sci 326:227–242
3. Islam MS, Pears R, Bacic B (2018) A wavelet approach for precursor pattern detection in time series. J Electr Syst Inf Technol 5(3):337–348
4. Motlagh O, Berry A, O'Neil L (2019) Clustering of residential electricity customers using load time series. Appl Energy 237:11–24
5. Bayrak G (2018) Wavelet transform-based fault detection method for hydrogen energy-based distributed generators. Int J Hydrogen Energy 43(44):20293–20308
6. Duan L, Yu F, Pedrycz W, Wang X, Yang X (2018) Time-series clustering based on linear fuzzy information granules. Appl Soft Comput 73:1053–1067
7. Bode G, Schreiber T, Baranski M, Müller D (2019) A time series clustering approach for building automation and control systems. Appl Energy 238:1337–1345
8. Teegavarapu RS (2019) Chapter 1—methods for analysis of trends and changes in hydroclimatological time-series. Elsevier, pp 1 – 89
9. Timeseries definitions. https://www.statsoft.com/textbook/time-series-analysis
10. Guijo-Rubio D, Durán-Rosal AM, Gutiérrez PA, Troncoso A, Hervás-Martínez C (2018) Time series clustering based on the characterization of segment typologies. Comput Res Repository abs/1810.11624
11. Ma R, Angryk R (2017) Distance and density clustering for time series data. In: 2017 IEEE international conference on data mining workshops (ICDMW), pp 25–32
12. Wang X, Smith K, Hyndman R (2006) Characteristic-based clustering for time series data. Data Min Knowl Discov 13:335–364
13. Zolhavarieh S, Aghabozorgi S, Teh YW (2014) A review of subsequence time series clustering. Sci World J, pp 1–19
14. Roelofsen P (2018) Time series clustering. Vrije Universiteit Amsterdam
15. Wang H, Zhang Q, Wu J, Pan S, Chen Y (2019) Time series feature learning with labeled and unlabeled data. Pattern Recogn 89:55–66
16. Salles R, Belloze K, Porto F, Gonzalez PH, Ogasawara E (2019) Nonstationary time series transformation methods: an experimental review. Knowl-Based Syst 164:274–291
17. Daoyuan L, Jacques K, Yves LT (2016) Time series classification with discrete wavelet transformed data: insights from an empirical study. In: The 28th international conference on software engineering and knowledge engineering (SEKE 2016) pp 01 – 06
18. Zhang K, Gençay R, Yazgan ME (2017) Application of wavelet decomposition in time-series forecasting. Econ Lett 158:41–46
19. Ji C, Zhao C, Liu S, Yang C, Pan L, Wu L, Meng X (2019) A fast shapelet selection algorithm for time series classification. Comput Netw 148:231–240
20. Wen L, Zhou K, Yang S (2019) A shape-based clustering method for pattern recognition of residential electricity consumption. J Clean Prod 212:475–488
21. Khosravi A, Machado L, Nunes R (2018) Time-series prediction of wind speed using machine learning algorithms: a case study osorio wind farm, Brazil. Appl Energy 224:550–566
22. Zhao Y, Lin L, Lu W, Meng Y (2016) Landsat time series clustering under modified dynamic time warping. In: 2016 4th IEEE international workshop on earth observation and remote sensing applications (EORSA), pp 62–66

23. Time series modeling. https://www.analyticsvidhya.com/blog/2015/12
24. van Berkel M (2010) Wavelets for feature detection; theoretical background, literature study. Eindhoven University of Technology
25. Stationarity and non stationarity of a time series. https://www.analyticsvidhya.com/blog/2018/09/non-stationary-time-series-python/
26. Stationarity of time series. https://machinelearningmastery.com/time-series-data-stationary-python/
27. Emerencia AC, van der Krieke L, Bos EH, de Jonge P, Petkov N, Aiello M (2016) Automating vector autoregression on electronic patient diary data. IEEE J Biomed Health Inf 20(2):631–643
28. Wavelets. https://nicolasfauchereau.github.io/climatecode/posts/wavelet-analysis-in-python/
29. Energy dataset. https://zenodo.org/record/999150

A Novel Data Hiding Technique with High Imperceptibility Using a 3-Input Majority Function and an Optimal Pixel Adjustment

P. V. Sabeen Govind, M. Y. Shiju Thomas, and M. V. Judy

Abstract Maximizing the payload capacity without losing the visual quality of the original image is a challenging problem in stegnography. This paper puts forth a data hiding technique with high imperceptibility and utilize the scope of a 3-input majority (MAJ3) function. Generally, a majority function takes an odd number of inputs and returns the majority value as its output. If the output of the majority function and the secret bits we want to hide are different, then we have defined an optimal pixel adjustment so that the receiver can extract the obscure data back. Empirical results unfold that this recommended approach increased the embedding capacity without comprising the visual quality. Complexity of the proposed algorithm is less compared to other image-based data hiding algorithms.

Keywords Cover image · Data hiding · Visual quality · Imperceptibility

1 Introduction

Information hiding is a method which camouflages secret information into a cover media. The presence of the secret information is unnoticeable with the help of the human visual system, and hence, many of them use this technique to ensure confidentiality. The two factors to assess the performance of a data hiding algorithm is its capacity to embed the amount of unknown data which is called as the payload capacity and its ability to retain the visual quality of the stego image. There is an

P. V. Sabeen Govind (✉) · M. V. Judy
Department of Computer Applications, Cochin University of Science and Technology, Kalamassery, Kerala, 682022, India
e-mail: sabingovindpv@gmail.com

M. V. Judy
e-mail: judy.nair@gmail.com

P. V. Sabeen Govind · M. Y. Shiju Thomas
Rajagiri College of Social Sciences (Autonomous), Kalamassery, Cochin, Kerala 683104, India
e-mail: shijuthomas@rajagiri.edu

N. N. Chiplunkar and T. Fukao (eds.), *Advances in Artificial Intelligence and Data Engineering*, Advances in Intelligent Systems and Computing 1133,
https://doi.org/10.1007/978-981-15-3514-7_64

implicit trade-off between these two parameters, so it is challenging to develop an algorithm with high embedding capacity without degrading the visual quality.

Besides a spatial domain [1], data hiding can be performed in a transform domain [2] as well. Spatial domain methods are computationally efficient than transform domain techniques. Histogram shifting [3] and LSB replacement [4] are the two common techniques used in the spatial domain. In the transform domain, many of the researchers utilized Fourier transform, discrete cosine transform [5], and discrete wavelet transform [6]. Transform domain techniques offer a high imperceptibility and robustness. Amin et al. [7] proposed a data hiding technique in transform domain, which is based on the effective use of DCT coefficient. In their scheme, a secret message is entrenched in quantized DCT coefficient and provides an acceptable level of image quality. Maleki et al. [8] take the advantage of mathematical distribution and modulus function, and their stego image quality is found to be satisfactory. Bassil [9] developed a stegnographic algorithm which was based on Canny edge detection.

Sun [10] contemplated a data hiding technique which stands on the effective use of Huffman encoding, they selected edge pixels for data embedding, and edge pixels are identified using the Canny edge detector. Vanmathi et al. [11] suggest a technique which is based on fuzzy logic and they also employed the edge information to conceal secret data. Fuzzy inference rules are developed to identify edge area and would detect more number of edges than a Sobel or Canny edge detector. Visual quality of their scheme is high as compared to Khamrui et al. [12], which are based on genetic algorithm. Some researchers utilized the concept of image interpolation [13] and obtained a good result. Recently, Kaw et al. [14] proposed an alternative method to interpolation called pixel repetition, and their computational complexity is less.

The paper is organized as follows. Working of the 3-input majority function and the process of embedding and extraction of data are given in Sect. 2. Implementation and analysis of data are inclined in Sect. 3, and concluding points are in Sect. 4.

2 Proposed Method

A 3-Input majority function (MAJ3) is the heart of the proposed algorithm. So, the working of the MAJ3 is given in the following subsection.

2.1 Working of the 3-Input Majority Function

Generally, a majority function takes an odd number of inputs and returns the majority value as its output. Equation (1) describes the working of the function, and Table 1 shows the various input combinations and their corresponding output.

$$\text{MAJ3}(X, Y, Z) = (X \wedge Y) \oplus (X \wedge Z) \oplus (Y \wedge Z) \quad (1)$$

Table 1 Working of a MAJ3 function

3-input MAJ function			
Inputs			Outputs
X	Y	Z	MAJ3(X, Y, Z)
0	0	0	0
0	0	1	0
0	1	0	0
0	1	1	1
1	0	0	0
1	0	1	1
1	1	0	1
1	1	1	1

where "$\wedge$" denotes bitwise AND, and "$\oplus$" denotes bitwise exclusive-OR.

2.2 *Embedding and Extraction of Data*

The embedding method using a MAJ3 function and the process of extraction is explained in this section. We are processing three cover pixels $C(i, i+1, i+2)$ and extracting the least significant bits (LSB). Majority function will take these three LSB values as input and return corresponding output based on Table 1. If the secret data SD $\subseteq$ {0,1} and the output of the MAJ3 function are equal, then we can directly copy these three pixels values as stego pixels. Otherwise, the proposed scheme changes LSB with minimum distortion. Detailed algorithm for data embedding is given below.

Step 1: Input cover image C, and secret data SD $\subseteq$ {0, 1}
Step 2: Process three cover pixels C(i, i+1, i+2) and SD[i], where SD[i] represents the secret bit we want to hide.
Step 3: Find the LSBs of C (i, i+1, i+2)
Step 4: Input these three LSB values (l_1,l_2,l_3) into MAJ3
Step 5: If MAJ3(l_1, l_2, l_3) = SD[i]
Then SI(i, i+1, i+2) = C(i, i+1, i+2), where SI(i, i+1, i+2) represents the stego pixels.
Step 6: If MAJ3(l_1,l_2,l_3) $\neq$ SD[i]

Case 1: MAJ3 (l_1, l_2, l_3) = 0 and SD[i] = 1

For the input combinations MAJ3(1,0,0), MAJ3(0,0,1), MAJ3(0,1,0), and MAJ3(0,0,0), the output becomes 0. Then, we have defined an optimal pixel adjustment shown below, so that the receiver can extract the secret data correctly. For the first three cases, we need to change only a single LSB in the cover pixel because of which the quality of the stego image is least affected. Optimal pixel adjustment in terms of MAJ function is given below.

$$\text{MAJ3}(l_1, l_2, l_3) = \text{MAJ3}(1, 0, 0), \text{Then } l_2 \text{ is changed to } 1.$$
$$\text{MAJ3}(l_1, l_2, l_3) = \text{MAJ3}(0, 0, 1), \text{Then } l_1 \text{ is changed to } 1.$$
$$\text{MAJ3}(l_1, l_2, l_3) = \text{MAJ3}(0, 1, 0), \text{Then } l_1 \text{ is changed to } 1.$$
$$\text{MAJ3}(l_1, l_2, l_3) = \text{MAJ3}(0, 0, 0), \text{Then } l_2 \text{ and } l_3 \text{ is changed to } 1$$

Case 2: $\text{MAJ3}(l_1, l_2, l_3) = 1$ and $\text{SD}[i] = 0$

For the input combinations MAJ3(1,1,0), MAJ3(0,1,1), MAJ3(1,0,1), and MAJ3(1,1,1), the output becomes 1. In this case, the following optimal pixel adjustment is used:

$$\text{MAJ3}(l_1, l_2, l_3) = \text{MAJ3}(1, 1, 0), \text{Then } l_2 \text{ is changed to } 0.$$
$$\text{MAJ3}(l_1, l_2, l_3) = \text{MAJ3}(0, 1, 1), \text{Then } l_1 \text{ is changed to } 0.$$
$$\text{MAJ3}(l_1, l_2, l_3) = \text{MAJ3}(1, 0, 1), \text{Then } l_1 \text{ is changed to } 0.$$
$$\text{MAJ3}(l_1, l_2, l_3) = \text{MAJ3}(1, 1, 1), \text{Then } l_2 \text{ and } l_3 \text{ is changed to } 0.$$

Receiver can do the reverse process and facilitate the hidden data. Pseudocodes for data embedding and extraction are given in Figs. 1 and 2, respectively. A numerical illustration is given in the next section.

```
Data_Embed (Source_Image, Sec_Data)
Start
  CI[,] = Read Source_Image;
  Stego_Image [,] = CI;
  BIT_SD [] = (Sec_Data);
  Initialize k = 0, p = 0 and q = 0;
  For all bits in BIT_SD
   If (BIT_SD [k] != MAJ3(CI[p, q],CI[p, q+1],CI[p, q+2]))
     /*Case 1:  One bit Need to Change */
    POS=FIND_POS (CI[p, q], CI [p, q+1], CI [p, q+2], Negation(BIT_SD[k]));
    Stego_Image [p, q+POS] = BIT_SD[k];
   End If
   If(BIT_SD[k] != MAJ3(CI [p, q], CI [p, q+1], CI [p, q+2]))
    /*Case 2: Two bits Need to Change */
    POS =FIND_POS (CI [p, q], CI [p, q+1], CI [p, q+2],Negation(BIT_SD[k]));
    Stego_Image [p, q+POS] = BIT_SD[k];
    End If
  Increment p, q and k respectively; /* Next Iteration */
End
```

Fig. 1 Pseudocode of data embedding

```
Data_Extraction
Input :Stego_Image
Output :Sec_Data
Start
  SI[,] = Read Stego_Image;
  Initialize k = 0, p = 0 and q = 0;
  For all bits in SI
     Bit_Sec_Data[k]= MAJ3(SI[p, q],SI[p, q+1],SI[p, q+2]);
     Increment p,q and k respectively;
  End
  Sec_Data = (Bit_Sec_Data);
  Display Sec_Data;
 End
```

Fig. 2 Pseudocode of data extraction

2.3 *Numerical Example*

A numerical example is given below and Fig. 3 shows some cover pixel values. Suppose our secret data $S = [0\ 0\ 1\ 0\ 1\ 1]$.

Take the first three pixels $C(i, i + 1, i + 2)$, here 54, 54, 55.

Find the binary equivalent of $(54)_2 = 110110$, $(54)_2 = 110110$, $(55)_2 = 110111$

Extract the LSB of these three pixels, here we have 001. Find MAJ3(0,0,1) = 0.

Output of MAJ3 function is 0 and the first secret bit we want to hide is 0. Then, we can copy these three cover pixels as stego pixels.

Move on to the next three cover pixel values 55, 55, 56. Extract the LSB and input to MAJ function, MAJ3(1,1,0) = 1. Here, output of MAJ is 1 and next secret bit we want to hide is 0. Then, change the LSB of 55 to 0, i.e., the new stego pixel is 54. This way we can embed all the secret data and the final stego image is given in Fig. 4.

Receiver acquires the stego image, and using MAJ3, he can take out the secret data back. Extracting the LSB of the first three stego pixels, here we have 0, 0, 1.

Fig. 3 Cover pixels

54	54	55	55	55	56
56	56	56	58	62	61
48	53	50	46	42	40

Fig. 4 Stego pixels

54	54	55	54	55	56
56	57	57	58	62	61
49	53	50	46	43	41

These three values are passed to MAJ3, and then he will get the first secret bit, i.e., 0. Continuing by this fashion, the devisee can extract all the obscure data.

3 Implementation and Analysis of Results

To prove the efficiency of the proposed algorithm, standard benchmarked test images of size 512 × 512 are used as shown in Fig. 5. Simulations are done using MATLAB, and secret data is randomly generated. The following five metrics are acclimated to gauge the quality of the stego image: (1) peak signal-to-noise ratio (PSNR) [2], (2) universal image quality index (Q) [15], (3) structural similarity index (SSIM) [16], (4) normalized cross-correlation (NCC), and (5) normalized absolute error (NAE) [17]. In SSIM, image distortion is assessed in terms of luminance, structure, and contrast, so that it has become a dominant method of evaluating image quality. Table 2 gives the comparison results using PSNR and Q measurements, and it is clear that the suggested modus operandi outperforms in respect of payload capacity and maintains an excellent visual quality. In Table 4, SSIM and NCC values are closer to one that indicates the similarity of cover and stego image, and NAE values are low. Table 4 concludes that the proposed technique offers high imperceptibility.

Fig. 5 Test images

Table 2 Comparison with other methods

Cover image	Khamrui et al. [12]	Shuliang's scheme [10]	Vanmathi et al. [11]	Proposed
Lena				
Capacity	80,000	38,427	71,504	87,381
PSNR	44.86	63.48	59.84	56.01
Q	0.9996	0.9999	0.9999	0.9999
Peppers				
Capacity	80,000	35,946	73,704	87,381
PSNR	44.82	63.23	59.41	55.94
Q	0.9995	0.9998	0.9998	0.9997
Elaine				
Capacity	80,000	35,852	73,704	87,381
PSNR	44.78	62.76	59.35	55.91
Q	0.9996	0.9998	0.9999	0.9998
Boat				
Capacity	80,000	37,451	72,765	87,381
PSNR	44.86	62.18	59.16	56.02
Q	0.9995	0.9999	0.9998	0.9997

Table 3 shows a statistical comparison between the original and stego image, and it is clear that calculated deviation is negligible. Figure 6 displays the histogram comparison, and both the source image and stego image have indistinguishable histograms. Payload capacity of various schemes is shown in Fig. 7, and it is evident that the recommended procedure surpasses the other methods in terms of the embedding rate. Table 4 concludes that the proposed technique offers high imperceptibility.

Table 3 A statistical comparison between original and stego image

Test image	Cover image			Stego image		
	Mean	Std. deviation	Entropy	Mean	Std. deviation	Entropy
Lena	120.9353	47.3670	7.4432	120.9367	47.3661	7.4455
Baboon	126.7271	40.7884	7.6712	126.7286	40.7902	7.6721
Peppers	117.7670	53.2044	7.5942	117.7667	53.2062	7.5951
Elaine	136.3563	46.0558	8.3201	136.3517	46.0589	8.3202
Boat	141.7101	49.1267	8.7416	141.7069	49.1289	8.7433
Goldhill	110.0652	50.9839	7.2775	110.0649	50.9831	7.2792

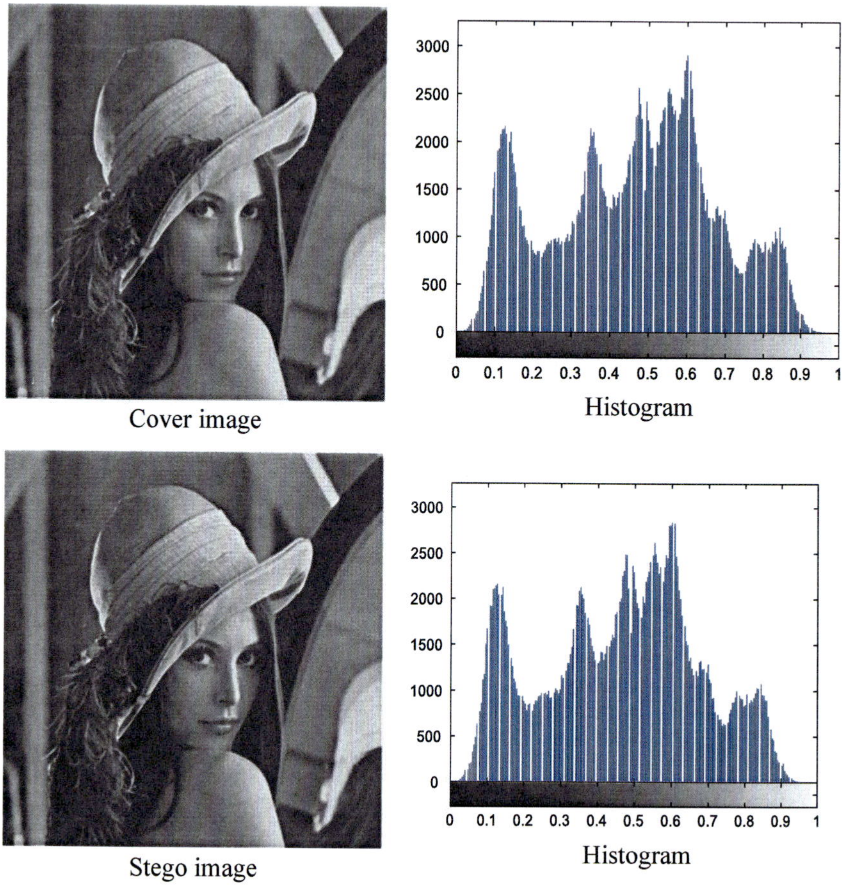

Fig. 6 Histogram comparison

Fig. 7 Payload capacity

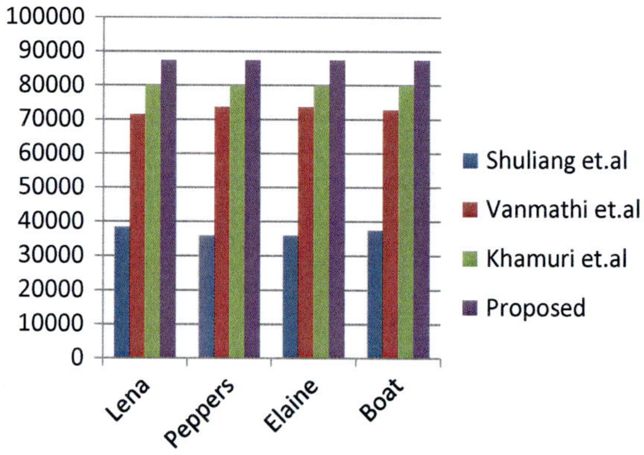

Table 4 Quality matrices

Image	PSNR (dB)	SSIM	NCC	NAE
Lena	56.01	0.9997	1.0000	0.0014
Baboon	55.97	0.9994	1.0000	0.0013
Peppers	55.94	0.9997	1.0000	0.0014
Elaine	55.91	0.9997	1.0000	0.0012
Boat	56.02	0.9996	1.0000	0.0012
Goldhill	55.89	0.9994	1.0000	0.0015

4 Conclusions

This paper proposes a data hiding technique with high imperceptibility. Working of a majority function is utilized for data embedding and extraction. Optimal pixel adjustment is performed so that the receiver can easily reconstruct the secret back. Visual quality of the proposed method is remarkable, and payload capacity is high in comparison with some recent data hiding techniques. In order to ensure more security, we can use some chaotic method before the data embedding process. Experimental results reveal that suggested method outshines the other methods. To further increase the payload capacity, we can use some adaptive strategies and hide more secret bits. In the future, we will try to define a reversible optimal pixel adjustment so that the receiver can reconstruct the cover image too.

References

1. Rasmi A, Mohanapriya M (2017) An extensive survey of data hiding techniques. Eur J Appl Sci 9(3):133–139
2. Shaik A, Thanikaiselvan V, Amitharajan R (2017) Data security through data hiding in images: a review. J Artif Intell 10:1–21
3. Chen J (2014) A PVD-based data hiding method with histogram preserving using pixel pair matching. Sig Process Image Commun 29:375–384
4. Pradan A, Sekhar KR, Swain G (2018) Digital image steganography using LSB substitution, PVD, and EMD. Math Probl Eng. https://doi.org/10.1155/2018/1804953
5. Lin C-C, Shiu P-F (2010) DCT-based reversible data hiding scheme. J Softw 5(2)
6. Xuan G, Yang C, Zhen Y, Shi YQ, Ni Z (2004) Reversible data hiding using integer wavelet transform and companding technique. Lecture notes in computer science. Springer, pp 115–124
7. Amin M, Abdullkader HM, Ibrahem HM, Sakr AS (2014) A steganographic method based on DCT and new quantization technique. Int J Netw Secur 16(4):265–270
8. Maleki N, Jalali M, Jahan MV (2014) Adaptive and non-adaptive data hiding methods for grayscale images based on modulus function. Egypt Inform J 15:115–127
9. Bassil Y (2012) Image steganography based on a parameterized canny edge detection algorithm. Int J Comput Appl 64(4):35–40
10. Sun S (2016) A novel edge based image steganography with 2k correction and Huffman encoding. Inf Process Lett 116(2):93–99
11. Vanmathi C, Prabhu S (2017) Image stegnography using fuzzy logic and chaotic for large payload and high imperceptibility. Int J Fuzzy Syst 20(2):460–473

12. Khamuri A, Mandal JK (2013) A genetic algorithm based stegnography using discrete cosine transformation. In: International conference on computational intelligence: modeling techniques and applications, Procedia Technology, vol 10, pp 105–111
13. Zhang X, Wang X (2017) High capacity data hiding based on interpolated image. Multimedia Tools Appl 76(7):9195–9218
14. Kaw J, Parah S, Sheikh J, Bhat M (2017) A new reversible stenographic technique based on pixel repetition method (PRM) and special data shifting (SDS). In: International conference on image information processing. https://doi.org/10.1109/iciip.2017.8313710
15. Wang Z, Bovik AC, Sheikh HR, Simoncelli EP (2004) Image quality assessment: from error visibility to structural similarity. IEEE Trans Image Process 13(4):600–612
16. Wang Z, Bovik AC (2002) A universal image quality index. IEEE Sig Process Lett 9:81–84
17. Parah S, Sheikh J, Akhoon J, Loan N (2018) Electronic Health record hiding in images for smart city applications: a computationally efficient and reversible information hiding technique for secure communication. Future generation computer systems. Elsevier

Designing and Testing of Data Acquisition System for Satellite Using MIL-STD-1553

B. L. Lavanya and M. N. Srinivasa

Abstract Data acquisition (DAQ) system is developed for acquiring, processing and storing the tracking the position of the satellite. Tracking the spacecraft orientation and measuring actual position of satellite are done using inertial measurement unit (IMU). An IMU consists of three accelerometers and gyroscopes. Accelerometer measures the linear acceleration, and gyroscope measures angular velocity of the object that it is mounted on. Accelerometers along with gyroscopes form an integral part of satellite's inertial guidance unit. Using DAQ system, the data can be measured from accelerometer or gyroscope or combination of both that can be used for space application. Designed DAQ system has accelerometer with Military Standard (MIL-STD) 1553 interface for data communication, to log the data without missing any of the samples at every 32 ms along with certain analog parameters in a deterministic way. Acquired acceleration data has to be converted to engineering units using numeric integration technique. This system display data in real time on the monitor and also store in a file for further off-line analysis. Hardware and software are developed using Arduino board along with MIL-STD-1553 protocol chip with serial peripheral interface (SPI), matrix laboratory (MATLAB) and Arduino IE, respectively.

Keywords Data acquisition (DAQ) system · MIL-STD-1553 · Arduino · Serial peripheral interface (SPI) · MATLAB

1 Introduction

Tracking the position of satellite and measuring actual position of satellite are done using inertial measurement unit (IMU). An IMU consists of three accelerometers

B. L. Lavanya (✉)
NMAM Institute of Technology, Nitte, India
e-mail: lavanya.bl@nitte.edu.in

M. N. Srinivasa
Laboratory for Electro-Optics Systems, ISRO, Bangalore, India

N. N. Chiplunkar and T. Fukao (eds.), *Advances in Artificial Intelligence and Data Engineering*, Advances in Intelligent Systems and Computing 1133,
https://doi.org/10.1007/978-981-15-3514-7_65

and gyroscopes [1], where accelerometer measures the linear acceleration or vibration monitoring and gyroscope measures rate or angle of rotation [2] of the object that it is mounted on. Accelerometers are the most used sensors in the vibration-based condition monitoring in the present days [3]. Micro-machined gyroscopes are used to measure rate or angle of rotation. The navigation performance is influenced by the parameters of gyro and accelerometer [4]. Microelectromechanical system (MEMS) devices have low cost, smaller size, broad frequency response and lower power dissipation. MEMS accelerometers are used to measure static and dynamic accelerations. MEMS-based inertial sensors are widely used in pedestrian navigation [4]. There are three types of MEMS accelerometers capacitive, piezoelectric and piezoresistive. Piezoresistive and piezoelectric are used for stress measurement, and capacitive accelerometers are used for displacement measurement [5]. Capacitive MEMS accelerometers devices have low cost, high degree of accuracy, low power consumption, lower temperature coefficient and high sensitivity [6], are easy to integrate, and are best suited for measuring low-frequency vibration, motion and steady-state acceleration. Silicon capacitive accelerometers are used in numerous applications such as large volume automotive accelerometers [2, 7, 8] and broad frequency response [9]. They have high sensitivity, good DC response and noise performance, low drift, low temperature sensitivity, low power dissipation and a simple structure [2, 10], and they are susceptible to electromagnetic interference (EMI). The sensor has digital output interface, which allows easier and ensures higher immunity to EMI [3].

To know the actual position of the satellite at any time instant, the actual position vector of the spacecraft with respect to some reference axis can be acquired from accelerometer, which measures its instantaneous acceleration in three orthogonal axes by double integrating. Accelerometer (MEMS technology) are used in many applications such as inertial navigation [6], inertial guidance systems, automotive safety systems (airbags and electronic suspension) [9], numerous consumer applications, robotics, machine and vibration monitoring, biomedical applications for activity monitoring [4], military, electronic stability control (ESC) [6], motion tracking, navigation, building and structural monitoring, automobile/transport, medical applications, industry and health monitoring (tremor diagnostic system) [5, 11].

In earlier days, avionics systems contain point-to-point wiring system. Military Standard (MIL-STD) 1553 was developed to reduce these complexities of wiring. MIL-STD-1553 data bus is a dual redundant, bidirectional and Manchester II encoded data bus [12, 13] which has extremely low error rate, one word fault/ten million words [14]. It widely used in application like aircraft control, avionics and heath maintenance of the satellite [12].

In satellite communication, radio frequency (RF) signals like telemetry and telecommand play important role. Telemetry provides the health of the spacecraft, which is acquired from various sensors, and telecommand is the command sent to spacecraft, which is demodulated and decoded in satellite. Data acquisition (DAQ) plays an important role in testing and calibration of sensors and also sending and receiving RF signals to and from satellite. Sensors are interfaced to satellite to measure the changes in the spacecraft orientation and actual position of the satellite, which consists of signal conditioning and analog-to-digital converter (ADC). Sensor signal contains noise, so it can directly process this signal so that noise removal is done in pre-processing, and signal conditioning circuitry manipulates a signal into a form that is suitable for input into an ADC. Once data is stored in a computer, they can be displayed on real time, analyzed or manipulated. Data can be stored in a file for off-line analysis such as converting into engineering units. Existing DAQ system is complex, so it is very difficult to debug the errors and can log data for short time. The proposed DAQ system overcomes this problem, which is simple compact design, logs a data for very long time and sends and receives telecommand and telemetry signals. A DAQ system can easily record reading (data logging) for very small time interval continuing for long time periods in computer.

Remaining paper is organized as follows. The DAQ system is presented in Sect. 2. In Sects. 3 and 4, flowchart of printed circuit board (PCB) design and PCB design of DAQ system are presented. Section 5 explains the software design. The numerical test results are presented in Sect. 6. The paper is finally concluded.

2 Data Acquisition System

The data acquisition (DAQ) system software functionality includes accessing the sensor data and providing telemetry and telecommand through MIL-STD-1553. The software graphical user interface (GUI) provides facility for data logging, ON, OFF, Up reset pulse commanding, 1553 telemetry or telecommanding and displays real time data stored in text file for converting it into engineering unit for off-line analysis. Development of DAQ system for inertial sensor with MIL-STD-1553 acquires data from accelerometer-based capacitive-sensing closed-loop accelerometer. In capacitive accelerometers, the acceleration changes the distance between the moving electrode and stationery electrode; hence, the differential capacitance gap will change, and by measuring this change, acceleration can be measured.

DAQ unit comprises of sensors, DAQ hardware and a computing device. This device can be connected to computer using a serial port, parallel port or custom interface to gather useful measurement data that can be stored in a file for off-line analysis. The accelerometer unit is interfaced through MIL-STD-1553 serial communication bus to DAQ hardware. DAQ hardware contains HI-6131 protocol IC, which provides interface between host processor and MIL-STD-1553 via four-wire serial peripheral interface (SPI) to Arduino Mega 2560 board as shown in Fig. 1. The software receives the input data from MEMS-based accelerometer sensor through 1553 interface and converts these sensor data into engineering unit, and this data is displayed on PC and stored in a file for further off-line analysis.

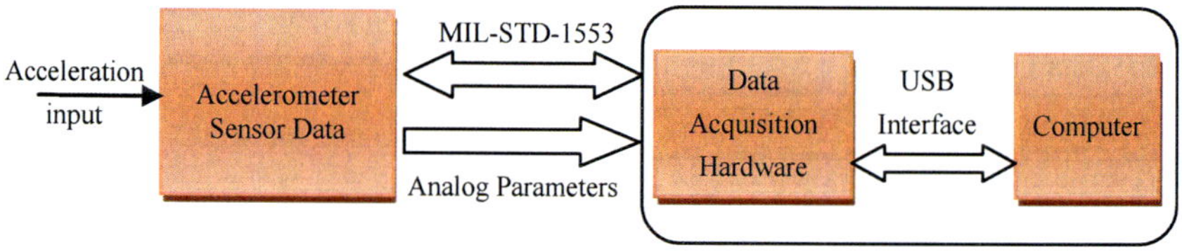

Fig. 1 Block diagram of DAQ system configuration

3 Flow Chart of PCB Design

A PCB should consist of analog signal processing circuits, RT address selection option, HI 6131 IC along with MIL-STD 1553 interface and signal processing related to ON, OFF, Up reset and sync for the sensor that has to be designed and fabricated. Flowchart of DAQ PCB is shown in Fig. 2. PCB is designed using OrCAD software.
Step 1: Hierarchical and schematic designs are basic steps involved in designing a PCB.
Step 2: Check for schematic errors. If there is error, go to Step 1. If schematic is found correct, proceed to Step 3.
Step 3: Bill of material (BOM) is generated. Component list, its style and quantity are listed. Routing is done using software. After routing, once again netlist is generated and verified.
Step 4: Errors in routing netlist is checked, if errors are found go to Step 2. If it is error-free, design files (top, bottom layer of PCB, Gerber file, etc.) for manufacturing of PCB are created. After final approval from higher authority, it is sent to manufacture along with PCB specifications.
Step 5: After PCB is manufactured, it has to be tested and each component has to be wired and tested to verify that.

4 PCB Design of DAQ System

Hierarchy gives an overview of PCB card, with input and output signals, which signal enters the PCB card and leaves the card is shown in Fig. 3. Hierarchical design has a connector, which has input signal to PCB card-like thermistor input (positive and negative), power supply, digital and analog ground, dual redundant bus, shield line (BUS A positive and negative, BUS B positive and negative), it also has output line SYNC_OUT, ON_COMD_OUT, OFF_CMD_OUT and Up_Rst_OUT. Other side of Holt console has turret (T1–T9), T1–T4 represent four lines for SPI communication (CEn, SO, SI and SCK), and T5–T9 represent input signal for sensor SYNC_IN, ARD_GND, ON_COMD_IN, OFF_CMD_IN and Up_Rst_IN. Test point (TP1–TP10) gives analog output and ARD_GND. In analog circuitry part-1 schematic, there are five thermostats used in the card for checking the temperature of the PCB card. Circuit diagram of voltage regulator LP3876ET-3.3 IC, which regulates 5 V

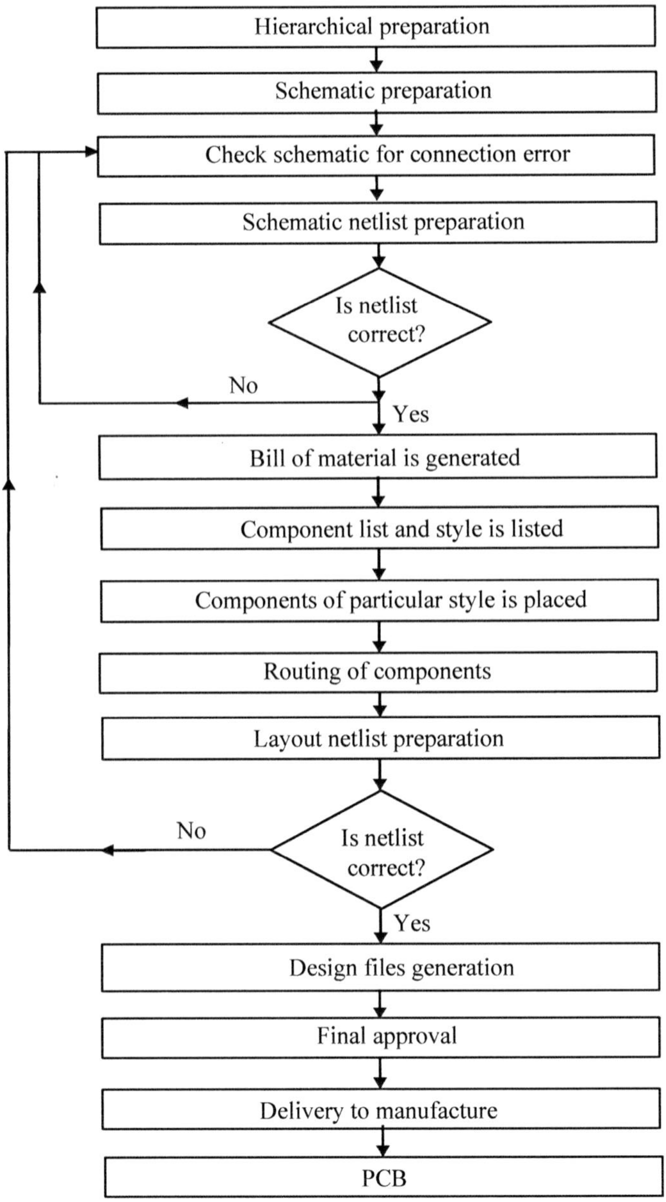

Fig. 2 Flowchart of DAQ PCB design

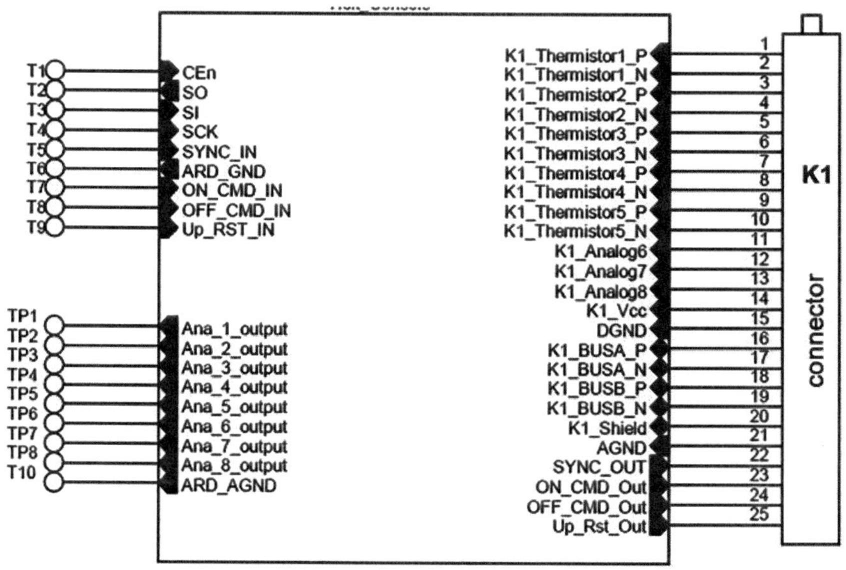

Fig. 3 Hierarchy design of PCB

input to 3.3 V as output, is shown in Fig. 4. Analog circuitry part-2 schematic is designed to handle Sync (sync1 and sync2), ON command, OFF command and Up reset pulse, and it is given to 16 bit dual supply translating trans-receiver. Sync output from trans-receiver is given to a CMOS quad differential line driver designed for data transmission.

The final top and bottom layer of the PCB design is shown in Figs. 5, 6, 7 and 8. Components are placed in both layers in order to reduce the board size. Double-sided PCB card is manufactured of dimension 17.1 cm × 10.9 cm with thickness of 0.2 cm. PCB card along with Arduino board together makes data acquisition hardware. DAQ hardware can be used for acquiring data from inertial sensor which is small size, portable and easy to use.

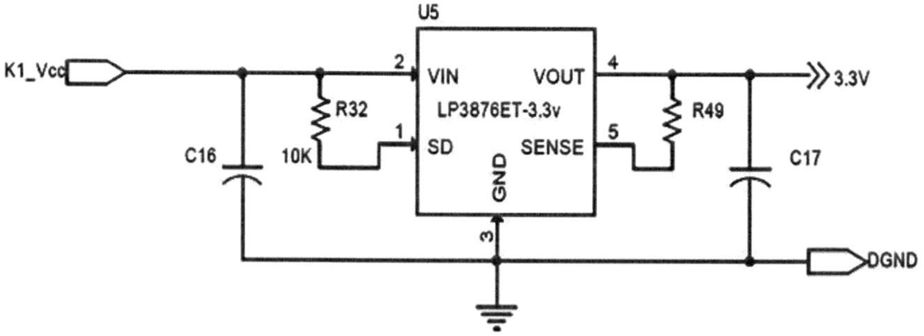

Fig. 4 Circuit diagram of voltage regulator LP3876ET-3.3 IC

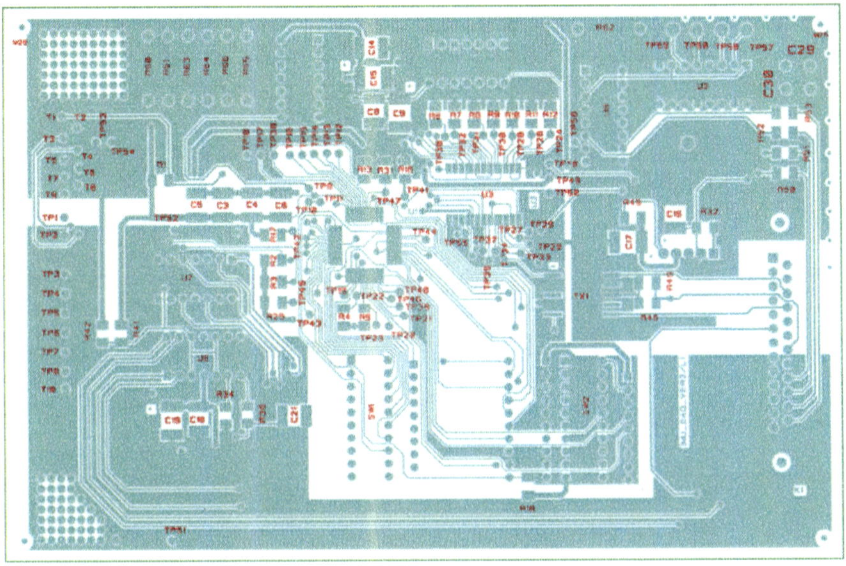

Fig. 5 Top-layer of DAQ PCB

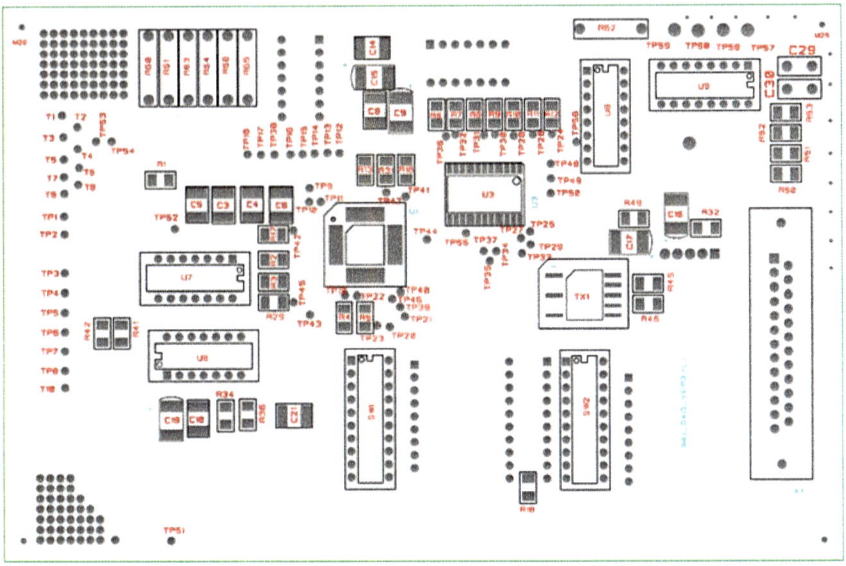

Fig. 6 Top-layer routing of DAQ PCB

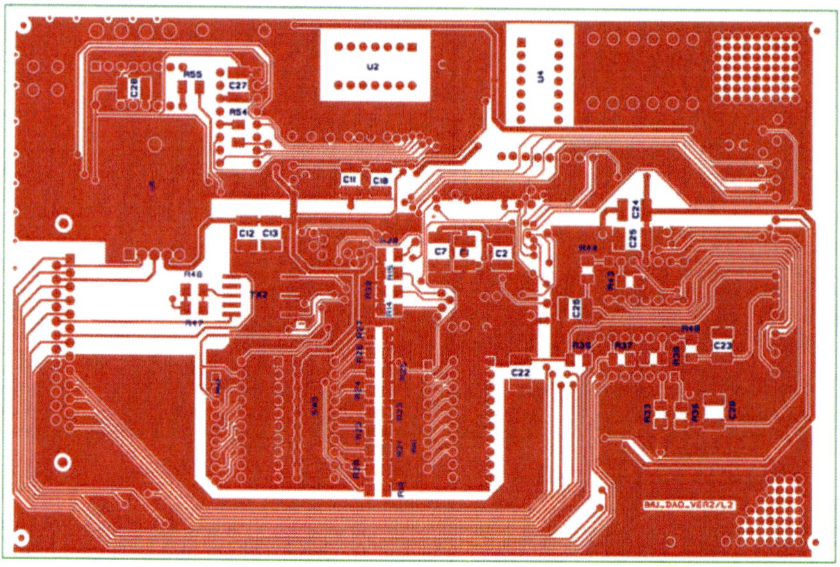

Fig. 7 Bottom-layer of DAQ PCB

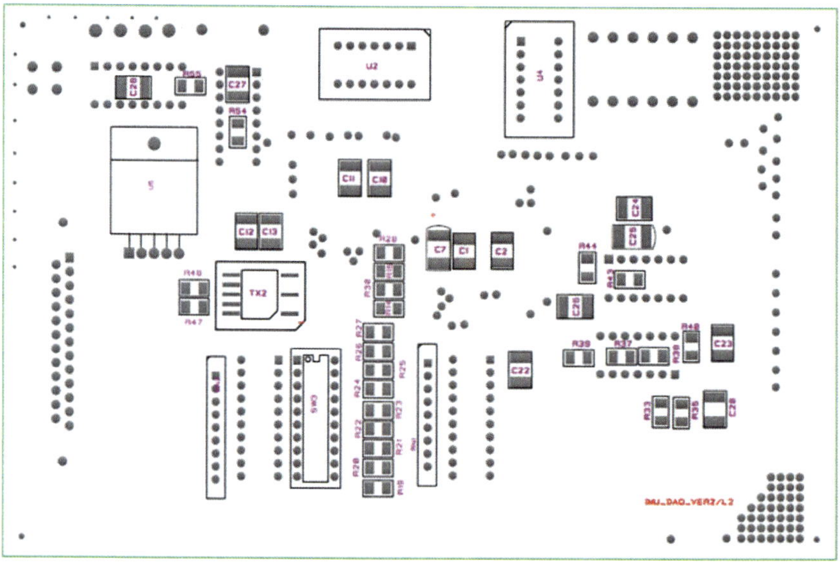

Fig. 8 Bottom-layer routing of DAQ PCB

5 Software Design

Software design consists of two parts. First part is the data logging software with GUI interface using matrix laboratory (MATLAB). Second part is for Arduino board consisting of device driver for HI6131 protocol IC along with analog parameter logging and 1553 communication and is developed using Arduino IDE.

5.1 MATLAB GUI Design

The MATLAB GUIDE tools greatly simplify the process of designing and building GUI that is layout designing and programming. GUI will have following graphical objects: connect button, read button and pulse command panel, sync panel, close button and menu bar. Connect button is used for connecting Arduino board to MATLAB using serial communication (COM port and baud rate same as Arduino IDE setting COM port and baud rate). Read button in used for reading eight analog channels data. Pulse command panel has three pulse commands: ON, OFF and uP reset button. Sync panel has two sync pulses, sync1 and sync2. Close button is for closing GUI, and menu bar is for 1553 command and telemetry command to send and view the data. After creating GUI, it is converted into .exe application using MATLAB compiler "mcc –m," macro to generate a stand-alone application.

5.2 Arduino IDE

Arduino integrated development environment (IDE) is used to initialize the memory, register and configure HI6131 IC as bus controller. Through DAQ system, sensor operation such as ON command, OFF command and uP reset command can be controlled. These commands are the pulse commands generated from Arduino board using timer for proper operation of sensor. These pulses are generated using timer present in Arduino Mega, which generate three pulses using single timer or using three different timers. Presently, three different timers (16 bit) are used to generate pulses of 64 ms pulse on digital pins of Arduino board. Analog channel A0–A7 (eight channels) is used for measuring eight analog channels. Measured data can be monitored in serial monitor (baud rate, 9600). After reading analog data, it has to be logged in file for any analysis of measured data. Data logging is very important, so data is stored in text file with respect to the data stored in a file. A device driver for HI6131 IC has to be written through SPI line, using Arduino IDE. HI6131 IC is coded in such a way that it works as BC in 1553. Memory and register initializations have to be done. After initializing memory and register, communication between BC and RT should happen.

6 Test and Results

The goal of testing is to remove or check any bugs and error in both hardware and software. Software and hardware have to be tested to ensure proper operation of DAQ system. Partial testing of the PCB card is performed using a digital multi-meter. The measurement PCB requires many surface mounts and through-hole solder joints, which is required for manual testing. The voltages of the electrical components were tested to ensure that the programming and components were functioning correctly. Software testing consists of two parts. First part is testing of developed MATLAB GUI for user interface using USB interface. Second part is testing developed of device driver for HI6131 IC along with analog parameter logging and 1553 communication which is developed using Arduino IDE. PCB card testing consists of analog signal processing related to ON, OFF and uP reset for the sensor which is generated. In software test, GUI has to be tested for various conditions. Software test has to be carried out for testing designed code, checking any error in code and removing error. During software testing, it has to be tested for some test case which is evaluated to determine if program is performing as per the requirement. Hardware is tested for verifying its output as per the design. After manufacturing of PCB, electrical test is carried out. Later, voltage regulator and supporting circuit are wired and tested, and then, operational amplifier (Op Amp) is wired. Thermistor is used for checking the temperature, and output voltage range is noted. After checking, analog part is wired along with its supporting circuits which are tested. Pulse command is generated from Arduino board, and through serial wire these pulses are given to trans-receiver. Output of trans-receiver is given to non-inverting hex buffer, to protect any damage to the signal. Sync output from trans-receiver is connected to CMOS quad differential line driver, which accepts TTL or CMOS input voltage levels and converts RS-422 output level, power down without loading down the bus.

6.1 Software Testing

The Arduino Mega 2560 can be programmed with the Arduino IDE. Arduino board is plugged into a computer's USB port. After writing code, it is compiled and uploaded to board. Eight analog channels data is monitored in serial monitor. Pulses are generated using timer present in Arduino Mega 2560 analog data which is in text format which can be saved or viewed GUI. Arduino is programmed both for transferring data and receiving data. Flowchart for software testing is shown in Fig. 9.
Step 1: GUI has different buttons for different operations. Initially, Arduino board is in disconnected state. If connect button is pressed, then Arduino is connected through USB cable of 9600 baud rate, COM port. If there is any error in connection, error message will be displayed on the screen.

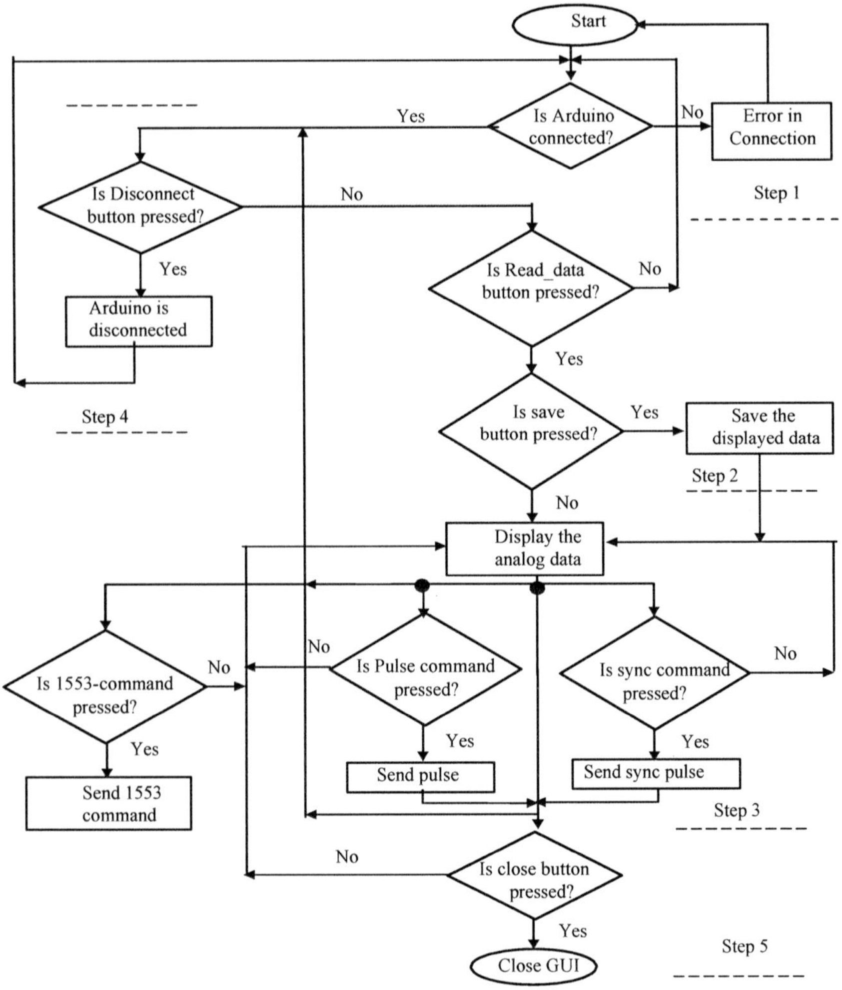

Fig. 9 Flowchart for software testing

Step 2: Read operation takes place in Step 2. When Read data button is pressed, analog data will be displayed on the screen, with 1 μs delay. If save button is pressed, it will be saved in text format and displayed in GUI.

Step 3: When pulse or sync command is pressed, pulse or sync command is generated at digital pins of Arduino. Pulse command which controls the operation of sensor and sync command is used for synchronization. 1553 telemetry command for 3 axes can be sent and value can be previewed to control the satellite operation by pressing 1553 command.

Step 4: Disconnect button status is checked every time. If disconnect button is pressed, then Arduino will be disconnected. Arduino will work only if it is in connected state.
Step 5: If close button is pressed, then GUI will be closed and Arduino status will be changed to disconnect. Close status will be confirmed before closing GUI and then goes to Step 1.

6.2 Hardware Testing

After PCB is manufactured, electrical check has been conducted for all components. Voltage regulator and its supporting circuit are wired and output is found to be 3.293 V. All analog circuits are wired and tested. 5 V Crystal oscillator 50 MHz is checked and mounted on dip IC base. Later, trans-receiver is mounted. In trans-receiver, when 1DIR is equal or low, trans-receiver will convert 5 V signal level to 3.3 V signal, and when 2DIR is high, it will convert 3.3 V signal level to 5 V signal level at 2A port, output is shown in Fig. 10.

Pulse commands are ON, OFF and uP RESET command, and these pulse commands are derived from timer of Arduino board. Output of trans-receiver is given to non-inverting hex buffer with amplitude of 5 V and 64 ms pulse which is given as input signal, and output signals are observed. Differential signalling is a method in which electrical informations are transmitted using two complementary signals. Technique of differential line signal transmission reduces electronic crosstalk, electromagnetic interference, noise emission and noise acceptance in the system. CMOS quad differential line driver IC is enabled using pin 4, EN. SYNC_TRANS is a sync output from trans-receiver which is connected to pins 1 and 7, and other inputs pins 9 and 15 are grounded. Output of this IC is positive pulse (0–5 V) at pin 2 and negative pulse (0 to −5 V) at pin 3 which are shown in Fig. 11.

Pulse commands ON CMD, OFF CMD and up RESET are generated using three different timers TIMER1, TIMER3 and TIMER4 (16 bit). These signals have single pulse of 64 ms ON period. These pulses are shown in Fig. 12. Start_sync1 and

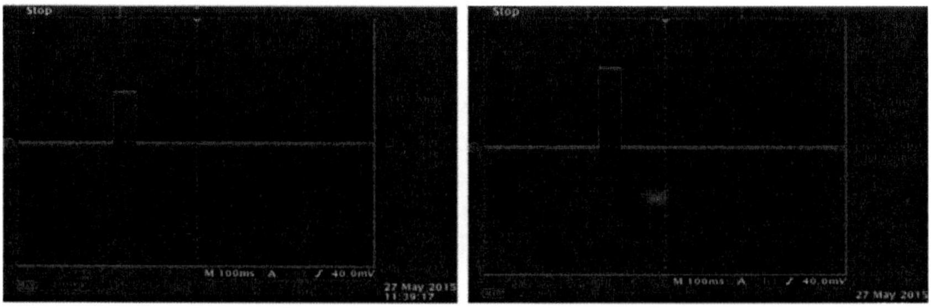

Fig. 10 Trans-receiver output when 1DIR is low and when 2DIR is high

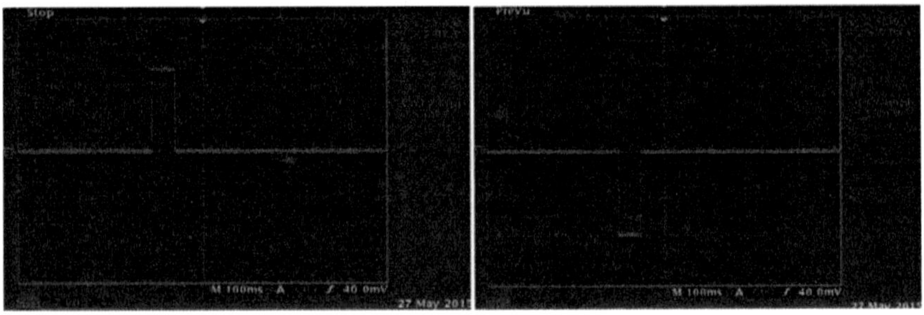

Fig. 11 Differential IC output at pin 2 and pin 3

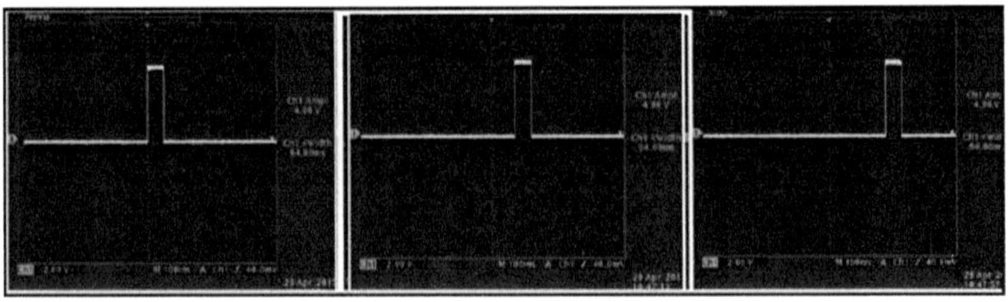

Fig. 12 ON CMD, OFF CMD and Up RESET pulse

start_sync2 are the two sync pulses, which can be generated using hardware or software. START_Sync1 is a clock of 4 ms ON pulse, 12 ms OFF pulse and a total of 16 ms period. START_Sync2 is a clock of 4 ms ON pulse, 28 ms OFF pulse and a total of 32 ms period. Sync pulse is as shown in Fig. 13.

ON CMD, OFF CMD and up RESET are the pulse command given from Arduino board for proper operation of sensor which is sent by clicking ON CMD, OFF CMD and UP RESET button. GUI of DAQ and GUI of DAQ reading data are shown in Figs. 14 and 15. In menu bar, start 1553 command and telemeter menu, save new

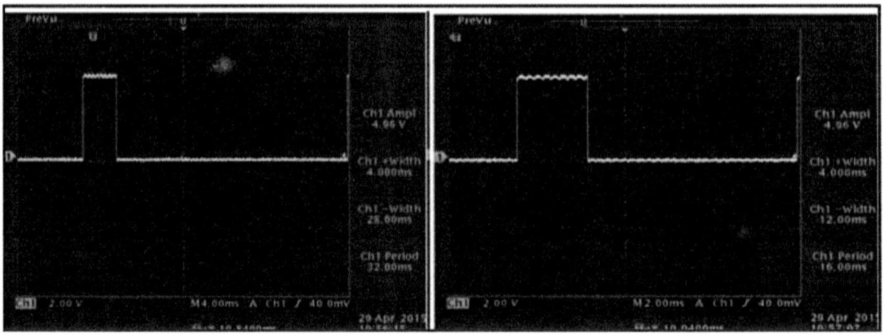

Fig. 13 Start_sync1 and Start_sync2 pulse

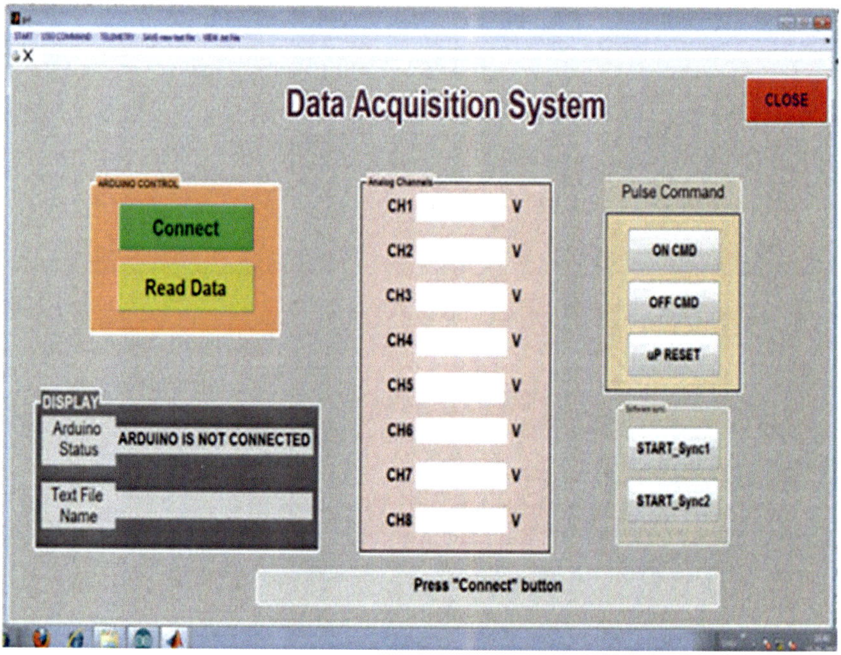

Fig. 14 GUI of DAQ system

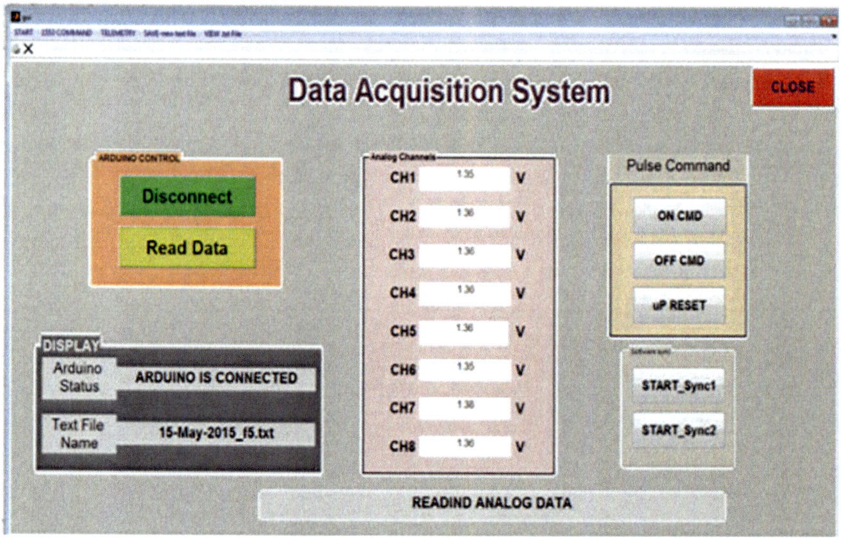

Fig. 15 GUI of DAQ reading analog data

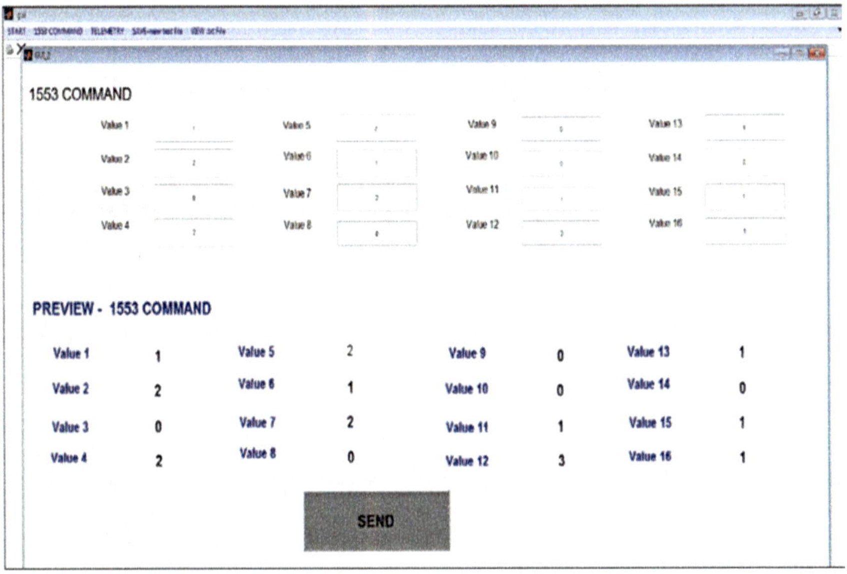

Fig. 16 GUI of DAQ for 1553-command

text file and view text file menus that are present. Telecommands are generated for sub-addresses (SA), in three axes. For BC to RT1 (Rx command) SA1, two words are used. For RT1 to BC (Tx command) SA1—23 words, SA2—13 words, SA3 and SA4—32 words are reserved, SA5—32 words and SA6—24 words. SA5 and SA6 used for board test. Using GUI, 1553 command can send and also preview the values which are entered. By clicking send button, it will ask for confirmation for sending data, if yes is pressed, data will be send, otherwise return to the window without sending data. GUI of 1553 command is shown in Fig. 16. Exit from GUI can be done, just pressing close Figure button or close button. When "CLOSE" button is pressed, it will ask for confirmation to closing of GUI.

7 Conclusion

In measurement system, data acquisition and data logging are very important. Data acquisition contains data acquisition hardware and a computing device to acquire real-time data from space and display it in monitor. The accelerometer is having MIL-STD-1553 interface of data communication for developing and a DAQ system for acquiring, processing and presentation of accelerometer sensor data. Arduino board is programmed for development of device driver for HI6131 IC along with analog parameter logging and 1553 communication. MATLAB GUI is developed, which is used for operation of sensor as per the requirements of DAQ system. All displayed sensor data is stored in text file and converted into engineering unit for off-line analysis. All software and hardware designs are tested, and results are noted. It is found that results are as per the requirement of the system. Instead of Arduino Mega 2560, other Arduino board can be used for testing DAQ system. DAQ system can be tested for both accelerometer and gyroscope. DAQ system can be designed and tested for other MIL-STD interfaces, and simulation can be carried out using different software.

References

1. Chan PY, Ripin ZM (2013) Development of wearable inertial sensors for measurement of hand arm tremors. In: IEEE international conference on smart instrumentation, measurement and applications (ICSIMA)
2. Yazdi N, Ayazi F, Najafi K (1998) Micro-machined inertial sensors. Proc IEEE 86(8)
3. Dosedel M, Havranek Z (2018) Design and performance evaluation of smart vibration sensor for industrial applications with built in MEMS accelerometers. In: 18th international conference on mechatronics mechatronika (ME)
4. Xinxi Z, Rong Z et al (2015) The performance impact evaluation on bias of gyro and accelerometer for foot mounted INS. In: IEEE 12th international conference on electronic measurement & instruments
5. Biswas S, Gogoi AK (2014) Design and simulation of piezoresistive MEMS accelerometer for the detection of pathological tremor. In: IEEE conference SOUTHEASTCON
6. Lanniel A, Alpert T (2018) Evaluation of frontend readout circuits for high performance automotive MEMS accelerometers. In: 14th conference on Ph.D. research in microelectronics and electronics (PRIME)
7. Sherman SJ, Tsang WK, Core TA (1992) A low cost monolithic accelerometer: product/technology update. In: Technical digest IEEE electron devices meeting (IEDM'92), Dec 1992, pp 160–161
8. Ristic L, Gutteridge R, Kung J (1993) A capacitive type accelerometer with self-test feature based on a double pinned poly silicon structure. In: Technical digest 7th international conference on solid-state sensors and actuators (Transducers'93), Yokohama, Japan, June 1993, pp 810–812
9. Wang L-P, Wolf RA (2003) Design, fabrication, and measurement of high-sensitivity piezoelectric micro-electromechanical systems accelerometers. J Microelectromech Syst 12(4)
10. Pak M, Fernandez FV (2017) Optimization of a MEMS accelerometer using a multi objective evolutionary algorithm. In: 14th international conference on synthesis, modeling, analysis and simulation methods and applications to circuit design (SMACD)
11. Tan TD, Anh NT (2011) Low-cost structural health monitoring scheme using MEMS-based accelerometers. In: Second international conference on intelligent systems, modelling and simulation
12. MIL-STD-1553 (1990) Designer's guide, 3rd edn. Data Device Corporation, Bohemia, New York
13. MIL-STD-1553 tutorial by Condor Engineering, Inc., USA
14. MIL-STD-1553 tutorial by AIM GmbH

Optimizing People Sourcing Through Semantic Matching of Job Description Documents and Candidate Profile Using Improved Topic Modelling Techniques

Lorick Jain, M. A. Harsha Vardhan, Ganesh Kathiresan and Ananth Narayan

Abstract People sourcing for a particular job role in any corporate venture is a painstaking, progress-impeding task, especially given the changing job market trends, quality of candidate profile and the sheer number of applicants. This paper elucidates an improvised topic modelling approach, where the semantic analysis stage is prepended with text summarization, for the procedure of people sourcing for job roles using topic modelling and machine learning techniques. The distinction between choosing the perfect candidate for the job and choosing a good candidate who is adept in certain domains but not relevant for the job and getting the candidate up to speed by providing on job training, and the downtime involved in the process is certainly a deciding factor in hiring a potential employee. The following paper aims to alleviate this issue by describing the algorithm which identifies the most suitable candidate in the applicants' pool through a novel, robust approach which uses topic modelling techniques like Latent Semantic Indexing (LSA) and Latent Dirichlet Allocation (LDA). It is empirically found that the precision of profile matching task was enhanced by using text summarization (through TextRank model) for both LSA and LDA by 21.4% and 50% respectively, and LSA outperforming LDA with regard to precision, in both with and without text summarization cases by 23.8% and 51.57%, respectively.

This work was done when the author (Lorick Jain) was in PES University
This work was done when the author (M. A. Harsha Vardhan) was in PES Institute of Technology.

L. Jain
Cisco systems Inc., Bengaluru, India
e-mail: lorick.jain@gmail.com

M. A. Harsha Vardhan
LEAP lab, Indian Institute of Science, Bengaluru, India
e-mail: harshavardhan.ma@gmail.com

G. Kathiresan (✉) · A. Narayan
PES University, Bengaluru, India
e-mail: ganesh3597@gmail.com

A. Narayan
e-mail: ananth.narayan@gmail.com

N. N. Chiplunkar and T. Fukao (eds.), *Advances in Artificial Intelligence and Data Engineering*, Advances in Intelligent Systems and Computing 1133,
https://doi.org/10.1007/978-981-15-3514-7_66

Keywords People sourcing · Topic modelling · Latent semantic analysis · Latent Dirichlet allocation · Job search · Text summarization · TextRank

1 Introduction

Hiring for IT roles, resource allocation and management at present is usually tended by human resource (HR) personnel of an IT firm or a company manually. This very task of sourcing people manually by a set of people has inherent and inevitable human bias, ignorance or prejudice involved, which might hinder the progress of a company by making the resource hired a potentially liability or burden to the company in the long run. Added to this is the risk of a large chunk of prospective employees to the company, with pertinent skills being overlooked attributing to the large number of applications along with the related documents to be processed by the HRs. This is further aggravated by diverse, multi-format, non-standardized feature of each applicant's resume or cover letters. This problem is addressed by an attempt to automate the job of HR sourcing by using a Job Description document to semantically rank candidates based on candidate work experience and contextual area of work. This would not just save time and money of the HRs and the company, but also enable them to better engage with the task of hiring worthy candidates and managing hired candidates.

One another concern about the process of hiring is the dynamism involved in the industry requirements, with new frontiers and domains of business like self-driving cars of the future, DNA genome analysis, climate change, cancer predictions, trust networks and many other fields being opened by advanced technologies like artificial intelligence, machine learning, blockchain, etc. Any broad filters involved in filtering resumes and eliminating unfit candidates are deemed to underperform in this rapid dynamism of changing industry requirements and trends. Also, such generic filters cannot match into work description provided by the candidates which has a lot of information on the relevance of the profile.

The existing practice of people sourcing involves a filter on CVs based on the grades/scores of a student, followed by sifting through educational qualifications and matching skill sets in relevant areas and thereby screening candidates for the process. As in [1], time available for recruiters when visiting a campus for a job fair or campus placement session is a major constraint in sourcing people of required skill sets into an organization. Shenoy and Aithal [1] also envisage the potential of attempts to automate the recruitment process, like this one, once the selection process is fully online and computerized. Adding to this, relevance has also been a critical problem for recruiters at companies [2]. This algorithm aims to help recruiters save time and maintain efficiency in the laborious task of resume filtering, particularly by reducing candidates who might be falsely selected as a potential employee through the current manual HR-driven selection process, and thus ensuring relevance to the organization. In other words, the algorithm mainly intends to reduce the rate of false positives in the hiring task. Ranking and selection are extremely important to the company as they try to stay away from possible false positives and good selection should be the Key Performance Indicator (KPI) for any tool that is developed [2]. Hence, the metric that will be used to judge the quality of this tool will be the number of true and false positives (the candidates who are right for the job, and the candidates who are not right for the job based on the semantic matching of their interests and area of experience to the Job Description). The potential employee may be lost if this critical problem of selection and ranking is not solved. To improve ranking results, we need to deeply understand the information and recruiter's intentions behind them.

Assumptions regarding the characteristics of what makes one job applicant better for the available position than the others are as follows:

1. Relevance of all the sentence to the Job Description and related information.
2. The influence of the CVs in extractive summarization of the entire corpus.
3. Semantic relevance of topics discussed in the CV to each other as well as the query Job Description.
4. Influence of the CV on the subtopics and the relevance of the subtopic in the corpus as well as the Job Description.

Topic modelling, the heart of this semantic matching engine, helps to analyse huge amounts of unlabelled text. Topic modelling helps distinguish between words based on its contextual clues, meanings and multiple meanings. Hence, this is a very important phase which helps to semantically generate relevant content and thereby helps in ranking the content and improve the score of documents.

2 Related Work

Document Summarization: Summarization technique is one of the areas tackled in this paper. Among the various existing techniques, graph-based approaches are explored in this paper. LexRank is a popular choice which is a stochastic graph-based method which uses eigenvector centrality in a graph representation of sentences [3]. Another popular extraction algorithm is TextRank. TextRank is a classified as a general-purpose graph-based algorithm that uses a stochastic matrix along with a damping factor.

Job recommendation is another area related to the problem tackled by this paper, as it involves similar conundrums of understanding job-related documents from both the parties, i.e. the employer and the applicant, like job description, resume, cover letter, etc. Standard multivariate job recommender systems are outperformed by deep learning variants, as in [4], which introduces a big data, clickstream-based job board recommendation system, named Deep4Job. The two-way approach involved in implementing this is as follows: an LSTM-based implementation along with a time-series encoding. The results were compared, and the LSTM approach was claimed to be better than the state-of-the-art hybrid collaborative recommendation system, Smart4Job (alluding to [5]).

Continuing on about Internet-based online recruitment, work [6] attempted to extract structured documents from applicants' resume which was needed for automatic screening of applications. Zaroor et al. [6] present a job and resume classification system called JRC. It matches resumes under certain occupational categories, by first segmenting the resume into sections, extracting knowledge out of them using NLP techniques, followed by classification by using [7], unlike our approach where we take a holistic consideration of the resume and the job description document (in both with and without text summarization case) without any segmentation into sections.

Another method [8] describes the process of job matching by constructing a skill matrix for the required job and match it with a skill matrix entered by the applicant. The patent depends mostly on the entered skill matrix which is not automatic or inferenced by any algorithm given just the resume of the person.

3 Implementation

Traditional methods involving the use of filters to sift through candidate profiles based on GPA, university/institution, previous work place are not adequate as they fail to generate semantic relationships between the job description role and the CV at hand, also by far, unable to judge the content itself. This filtering process works based on a string- or text-matching approach. Text-matching features suffer from the vocabulary gap between queries (Job Description) and document. To truly achieve meaningful results, we must use techniques of semantic relationships. The semantic gap between queries and the actual content is the main barrier for improving base relevance. The following steps are valid for the query document (Job Description document) as well.

Data Collection—The data for the following exercise was obtained from the authors' educational institution's placement cell. The data obtained was curriculum vitae documents (CVs) in the PDF format which are 420 in number. The obtained CVs were from different disciplines of engineering ranging from branches of biotechnology to mechanical, electrical and electronics to civil and computer science. These CVs were collected from an entire batch of students, whose GPA was greater than or equal to 6.0 on a scale of 10.0 (usually companies keep a filter of 8.0), and were eligible for placements. The CVs consist of performers, mediocre and low-performing students based on their overall profile strength weighed using projects, coursework, GPA, etc.

Parsing—The CVs obtained were in the form of PDFs. To generate insight from these documents and to feed it to the model in a manner acceptable by the model, parsing was done. The parsing was done using TextRact [9], an open-source Python library. The output of this phase yielded the CVs as text files, excluding special symbols and designs like lines and borders.

Pre-processing—Various stop words that are common to the CVs were removed as they do not provide relevant information to the algorithm to decide closeness to the job description (JD). The stop words removed were common words like education, skills, name of the student's educational institution and hobbies. Part-of-speech tagging was done to check the sentence structure. POS tagging can be used to understand the structure of grammar and sifting for grammatical errors. The documents are also checked for spelling errors. Essentially, a sentence must consist of a noun phrase, verb and a subject or an auxiliary verb, subject and a main verb. The sentences where this specified construct is not present are manually fixed. Lemmatization is also performed on the text obtained post-parsing the PDFs (Fig. 1).

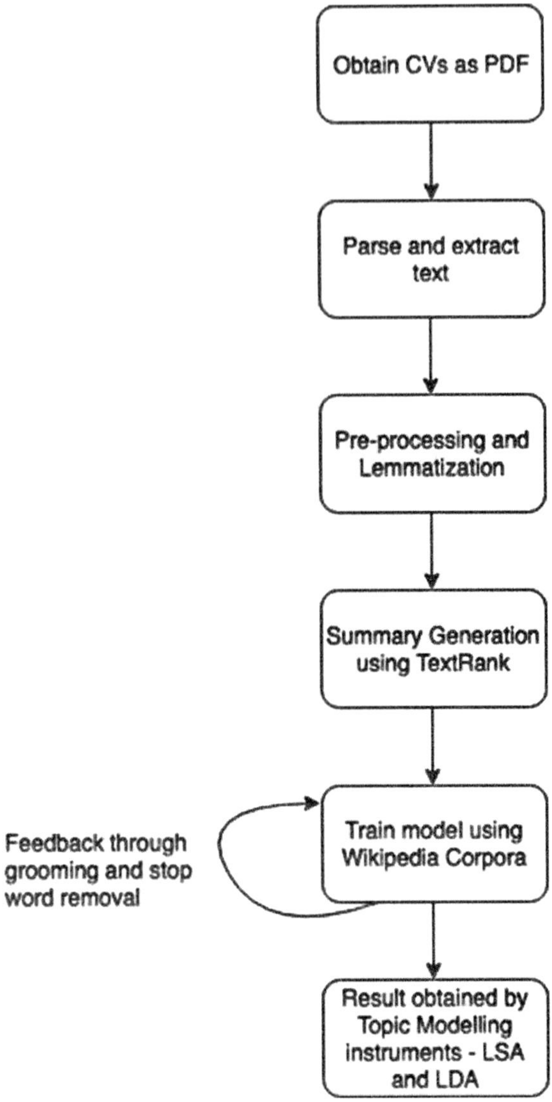

Fig. 1 Comparing LSA versus LDA with TextRank

Summary Generation Through TextRank—The document's sentences so far contain a wide range of information which might not be useful for the final result. Therefore, we reduce the number of sentences by taking the summary. The use of graph-based ranking algorithm works well in these scenarios because more relevant keywords contribute to a co-occurrence relation [10]. More technical details are regarded as more "representative" compared to personal details which have an overall less frequency in all documents combined. In order to summarize the documents, we have used TextRank algorithm in text extraction form. A combined document is tagged with original document details and sent as input to the algorithm which outputs the most important sentences. Using the tags, we split the combined documents into individual summarized resumes.

Topic Modelling—The algorithms used to match the user work experience and skills with the job description documents are Latent Semantic Analysis (LSA) and Latent Dirichlet Allocation (LDA) implemented using Gensim [11]. Latent Semantic Analysis works by creating word vectors in the LSA space and then finding the similarity between the query and the document using cosine similarity and works good on sentences as it maintains order of the words in the sentence. LDA works by creating a normal distribution of words by randomly choosing topics and then checks for P(word|Topic).P(Topic|Document); the highest score is chosen as the final topic. The summaries obtained were first subjected to a two-step process—bag-of-words model creation and tf–idf matrix generation. These methods essentially enable or help in the formation of vectors (CVs—each CV is a document vector) which can be projected into the Latent Semantic Analysis space, which is then subjected to singular value decomposition to decompose the sparse matrix. The query vector is also subjected to the same vector transformation and projected in a similar fashion into the LSA space. Then, cosine similarity is used as a similarity measure to find which document vector the query vector (Job Description) is closest to. Cosine similarity is computed in this higher-dimensional space. The matching CVs can then be obtained based on a threshold cut-off similarity value. For the purpose of this paper, the top 10 CVs were manually looked at for results, true and false positives. This LSA algorithm was compared with the results of the LDA transformation. The above procedures of parsing, tf–idf and bag-of-words model are repeated for LDA as well. The process for LDA involves generation of topics and then projecting these topics that are generated from the document vectors (CVs) into the LDA space. The same process is repeated for the query vector (Job Description document vector). The topics generated from the query vector is also projected in this space, then cosine similarity measure is computed between the topics of the query vector, and the document vectors and the top 10 highest scoring documents are retrieved. LDA involves computing cosine similarity between the topics of the job description (query) and the document vectors (CVs). Other similarity measures can be used as well, like Jaccard similarity, but for the semantic match in higher-dimensional space between text2vec models, cosine similarity is preferred [12].

Machine Learning—LSA and LDA models are first trained on the Wikipedia corpus to generate a space/matrix of English documents that serve as a base for projection of vectors of the CVs and job description documents. The space is created, and the models are then validated by projecting the above vectors in the space, following which grooming is performed and certain stop words are removed to improve performance. The number of true positives increased, and the number of false positives decreased for the LDA model. This performance improvement and validation could be performed only on the LDA model as this model is an online learning model, which ensures the online updation (updation in batches) of the matrix/space of vectors every time a new batch of documents are ingested into the model. LSA, however, is just a trained model which creates a space/matrix of Wikipedia corpora.

4 Results

The metric used to test the performance of the tool will be the number of true and false positives. The true negatives are the ones that are not of much significance as they are the candidates who are outright rejected because of poor academic performance in terms of failure rate. False negatives are not detrimental to the company as well, since there is a new batch applying for jobs every year; no scarcity in the number of applicants. False negatives also do not cost the company much in terms of their spend on employees. Hence, it is clear that the evaluation metric precision is well suited for this scenario. The number of true positives increased by an average of 11% (averaged for 10 runs), and the number of false positives decreased by 26% (averaged for 10 runs) for the LDA model.

Manual observation of results also showed that candidates with a GPA lower than the conventional filter (8.0 GPA on a scale of 10) had the skill set (as seen in the summaries, generated by TextRank, of the projects worked on by the candidate) required by the job as described in the Job description document. This result contradicts the first GPA filter used by prevailing hiring practices to remove worthy candidates on the pre-text of lower GPAs. LDA without TextRank resulted in poor results since summaries were not generated correctly, and there was not a clear identification of representative topics; however, LSA was still able to maintain satisfactory results in comparison to the application of TextRank, due to semantic matching of longer sentences between the JD and the CV. LSA helps in understanding the semantic nature of sentences, and both the JD and the CV have sentences depicting skill required and the skill present; hence, LSA is better to semantically interpret, unlike LDA which tries to generate topics and then compute similarity. LSA performs a better job in the higher-dimensional space in terms of mapping similarity as the semantic meaning is encapsulated in the higher-dimensional spaces a word vector projected in this space. Results which clearly portrayed this when LSA had better precision than LDA, in both with and without text summarization cases by 23.8% and 51.57%, respectively.

The aforementioned TextRank algorithm, when used as a pre-processing step for the topic modelling techniques, yielded a precision increase in both LSA and LDA by 21.4% and 50%, respectively (Table 1; Figs. 2 and 3).

Table 1 (True positive, false positive) pairs of the techniques

	LSA with TextRank	LSA w/o TextRank	LDA with TextRank	LDA w/o TextRank
Run 1	(9,1)	(8,2)	(7,3)	(5,5)
Run 2	(8,2)	(6,4)	(7,3)	(3,7)
Run 3	(9,1)	(8,2)	(8,2)	(5,5)
Run 4	(10,0)	(7,3)	(7,3)	(4,6)
Run 5	(9,1)	(8,2)	(7,3)	(3,7)
Run 6	(10,0)	(8,2)	(7,3)	(3,7)
Run 7	(10,0)	(8,2)	(7,3)	(3,7)
Run 8	(10,0)	(8,2)	(7,3)	(3,7)
Run 9	(10,0)	(6,4)	(7,3)	(3,7)
Run 10	(9,1)	(7, 3)	(7,3)	(3,7)
Precision	0.933	0.733	0.711	0.355

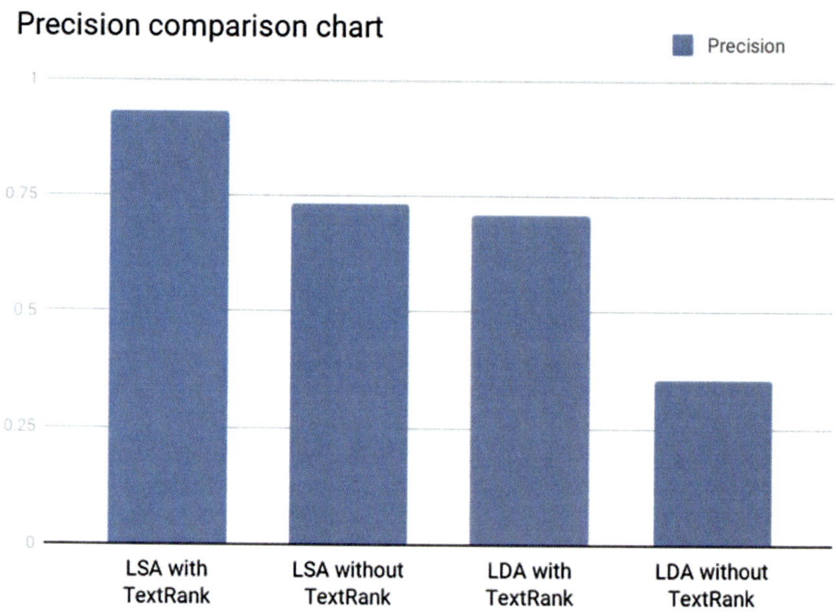

Fig. 2 Comparing precision of techniques

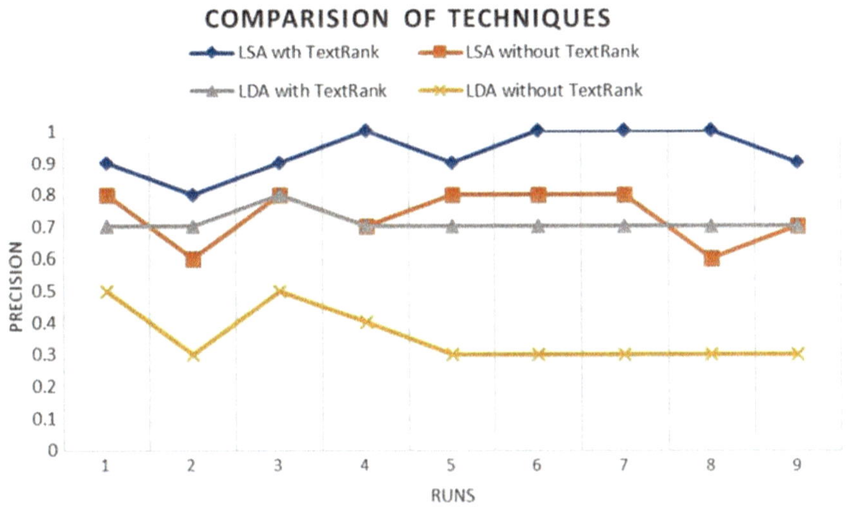

Fig. 3 Precision versus run graph

5 Conclusion

The process followed in this paper, using the methodology presented here will usher in further research to ensure candidate selection in a non-biased and automated fashion. The techniques presented in this paper show that semantic understanding by models provide better results. LSA thus performs better than LDA, in both cases of with and without the pre-processing step of text summarization using TextRank. It is also imperative to observe the novel approach used here where text summarization enhances performance of the topic modelling techniques harnessed in this study. Hence, the idea of utilizing the task of text summarization as a pre-processing step for topic modelling and semantic interpretation might be used in fields than people sourcing, thus opening up more scope for research. Various other deep learning as well as machine learning models can be employed to gain similar insight which will open new avenues of research in this field and bring about a change in the way the industry currently performs selection of candidates. In a gist, this paper is concluded with the key takeaways, namely firstly, creating a novel algorithm of improved semantic interpretation, and secondly, utilizing it for the radical and economically crucial work of people sourcing.

Acknowledgements The authors would like to thank their university's placement cell for providing data which formed the basis of the research presented above.

References

1. Shenoy V, Aithal PS (2016) Changing approaches in campus placements—a new futuristic model (June 29, 2016). Int J Sci Res Mod Educ 1(1):766–776. https://ssrn.com/abstract=2802247
2. Virtual HR. Retrieved 15 Aug 2018 from http://www.virtualhr.co.ke/top-five-recruiting-challenges-hr-professionals-face-and-how-to-overcome-them/
3. Erkan G, Radev DR (2004) Lexrank: graph-based lexical centrality as salience in text summarization. J Artif Intell Res 22:457–479
4. Benabderrahmane S, Mellouli N, Lamolle M, Mimouni N (2017). When deep neural networks meet job offers recommendation. In: Proceedings of international conference on tools with artificial intelligence. IEEE, Boston. https://doi.org/10.1109/ICTAI.2017.00044
5. Benabderrahmane S, Mellouli N, Lamolle M, Paroubek P (2017) Smart4job: a big data framework for intelligent job offers broadcasting using time series forecasting and semantic classification. Big Data Res 7:16–30. https://doi.org/10.1016/j.bdr.2016.11.001
6. Zaroor A, Maree M, Sabha M (2017) JRC: a job post and resume classification system for online recruitment. In: Proceedings of international conference on tools with artificial intelligence, IEEE, Boston. https://doi.org/10.1109/ICTAI.2017.00123
7. Kmail A, Maree M, Belkhatir M, Alhashmi S (2015) An automatic online recruitment system based on exploiting multiple semantic resources and concept-relatedness measures. In: Proceedings of the international conference on tools with artificial intelligence, IEEE, Vietri sul Mare, Italy, pp 620–627. https://doi.org/10.1109/ICTAI.2015.95
8. Defoor W (2001) Method for matching job candidates with employers. Patent No. 09/188,422, Filed Nov. 9th., 1998, Issued November 15th
9. Malmgren D (2014) textract. Retrieved 15 Dec 2018 http://textract.readthedocs.io/en/stable/
10. Mihalcea R, Tarau P (2004) Textrank: bringing order into text. In: Proceedings of the 2004 conference on empirical methods in natural language processing. http://www.aclweb.org/anthology/W04-3252
11. Řehůřek R (2009) gensim. Retrieved 15 Dec 2018 from https://radimrehurek.com/gensim/
12. Zahrotun L (2016) Comparison Jaccard similarity, cosine similarity and combined both of the data clustering with shared nearest neighbor method. Comput Eng Appl J 5(1):11–18

Mining Associations Rules Between Attribute Value Clusters

Shankar B. Naik

Abstract The approach presented in this paper clusters values of each attribute in a relational database and then finds associations between two clusters belonging to different attributes. The first experiment has shown that the approach proposed is accurate in discovering clusters and associations between them. The second experiment conducted on real dataset has discovered clusters of values of attribute 'petiole thickness,' wherein each cluster is uniquely associated with a particular species of a plant.

Keywords Association mining · Clusters · Relational database · Attribute values · Rules

1 Introduction

This paper presents an approach to cluster values of an attribute based on the values of another attribute in a relational dataset. The approach works in two phases. In the first phase, the values of an attribute are grouped together to form clusters. In the second phase, the clusters are analyzed to generate association rules. An association rule has two clusters such that each cluster belongs to two different attributes.

Consider a database containing information about user profiles, created on a social Web site, having attributes of the user. Consider the attributes 'workplace' and 'institute.' For a user, if attribute 'workplace' has value x and attribute 'institute' has value i, then institute i offers education relevant and applicable to workplace x indicating that there is the association between them. For another user, if attribute 'workplace' has value x and attribute 'institute' has value j, then institute j and the workplace x are associated but it also means that both the institutes i and j offer similar kind of education applicable to the workplace x and belong to a cluster of institutes which offer a particular kind of education. In this way, each attribute can have a set of

S. B. Naik (✉)
Government College of Arts, Science and Commerce, Sanquelim, Goa, India
e-mail: xekhar@rediffmail.com

N. N. Chiplunkar and T. Fukao (eds.), *Advances in Artificial Intelligence and Data Engineering*, Advances in Intelligent Systems and Computing 1133,
https://doi.org/10.1007/978-981-15-3514-7_67

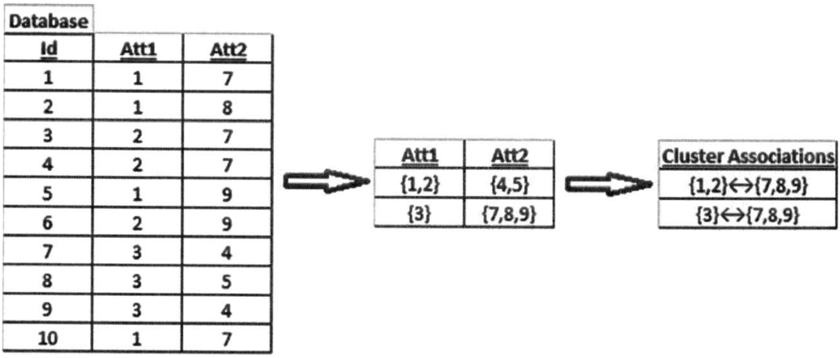

Fig. 1 Data set, clusters, and cluster associations

clusters. Two clusters belonging to two different attributes can be associated with each other if there exists a pair of values, one belonging to each cluster and both the values belong to the same user. An association between two clusters, each belonging to the attributes ‘institute’ and ‘workplace,’ implies that all the institutes present in one cluster provide education which is applicable to all the workplaces in the other cluster.

The approach presented in this paper has two steps. The first step identifies the clusters for each attribute. In the second phase, the clusters are analyzed to generate associations rules between them.

This approach can be made more intelligent by incorporating methods such as text mining and semantic analysis.

Figure 1 explains the approach presented in this paper. Consider attributes Att_1 and Att_2 such that Att_1 can take values from the set $V_1 = \{1, 2, 3\}$ and attribute Att_2 can take values from the set $V_2 = \{4, 5, 6, 7, 8, 9\}$. D is the dataset having ten records. The approach uses a data structure which is used to store the clusters generated for the attribute. The clusters in the data structure are added and updated as the elements are processed. The algorithms presented in this paper in the subsequent sections are executed on the data stored in the data structure to generate clusters. This clusters are again stored in the memory to generate association rules among them. The stored clusters are then analyzed. Clusters generated for the attribute Att_1 are $\{1, 2\}$ and $\{3\}$, while clusters generated for the attribute Att_2 are $\{4, 5\}$ and $\{7, 8, 9\}$.

The approach then generates associations $\{1, 2\} \leftrightarrow \{7, 8, 9\}$ and $\{3\} \leftrightarrow \{4, 5\}$. The details of these algorithms are presented in the subsequent sections.

The kind of study presented in this paper is a recent one. To the best of our knowledge, no such work has been done. Although, a study on network transition analysis has been performed [1, 2] which is limited only to job-related data.

2 Background and Motivation

A similar study has been presented in [3, 4]. The study in [3] was limited to the clustering of values of a single attribute. Both studies in both papers focused on a data stream environment. Whereas, the work carried out in this paper clusters values of two attributes and finds relations between two clusters of different attributes in a static database. Although, a study on network transition analysis has been performed [1, 2] which is limited only to job-related data.

3 Problem Definition

3.1 Preliminaries

Let the attributes under consideration be Att_1 and Att_2. Let $\text{Val}_1 = \{a_{11}, a_{12}, \ldots, a_{1n}\}$ be the set of values for attributes Att_1 and $\text{Val}_2 = \{a_{21}, a_{22}, \ldots, a_{2m}\}$ be the set of values for attribute Att_2.

Definition 1 Tp_l denotes a pair of values (a_{1i}, a_{2j}) such that $a_{1i} \in \text{Vl}_1$, $a_{2j} \in \text{Vl}_2$ and both a_{1i} and a_{2j} are present in the record l of the database.

Definition 2 If there exist two pairs $\text{Tp}_i = (a_{11}, a_{2i})$ and $\text{Tp}_j = (a_{12}, a_{2j})$ such that $a_{2i} = a_{2j}$ and $i \neq j$, then the values $a_{11} \in \text{Tp}_l$ and $a_{12} \in \text{Tp}_m$ are similar and are denoted as $a_{11} \sim a_{12}$.

Corollary 1 *If $a_{11} \sim a_{12}$ and $a_{12} \sim a_{13}$, then it implies that $a_{11} \sim a_{13}$, i.e., all three are similar.*

Definition 3 Let $\sim (a_{11}, a_{12})$ denote the similarity between a_{11} and a_{12}. The value of $\sim (a_{11}, a_{12})$ is number of records containing the pair Tp_i in the dataset such that $a_{11} \in \text{Tp}_i$ and $a_{12} \in \text{Tp}_i$.

Definition 4 A cluster, denoted as C, is defined as the set $C = \{a_{ji}/\text{all}\, a_{ji}\text{s are similar}\}$. All values in the cluster C belong to the attribute A_j.

Definition 5 If Tp_1 and Tp_2 contain identical sets values, then they are similar.

Definition 6 If $\text{sim}(a_{11}, a_{12}) \geq s_0$, then a_{11} and a_{12} are highly similar.

Significance of s_0 At times, a dataset can have few records which result in clustering two dissimilar values getting clustered together producing wrong clusters. Such records are a very few. The value of s_0 will prevent dissimilar values getting clustered.

Definition 7 Let Cl_{1i} be a cluster having values in Val_1, and Cl_{2j} be a cluster having values from Val_2. An association between clusters Cl_{1i} and Cl_{2j}, denoted as $\text{Cl}_{1i} \leftrightarrow \text{Cl}_{2j}$, exists if $\text{sim}(a_{1a}, a_{2b}) \geq s_0$ for some $a_{1a} \in \text{Cl}_{1i}$ and $a_{2b} \in \text{Cl}_{2j}$.

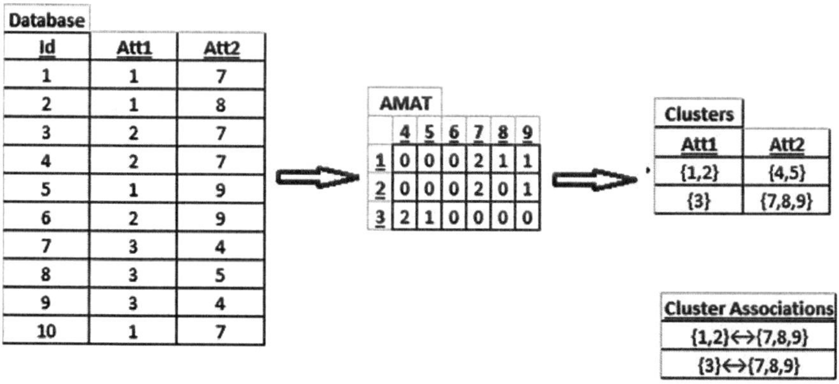

Fig. 2 Intermediate summary data structure

3.2 *Problem Statement*

Given Att_1, Att_2, Val_1, Val_2 and D, the problem is to generate the sets $\text{Cluster}_1 = \{\text{Cl}_{11}, \text{Cl}_{12}, \ldots\}$ for Attr_1 and $\text{Cluster}_2 = \{\text{Cl}_{21}, \text{Cl}_{22}, \ldots\}$ for Attr_2 and the set $\text{Association} = \{\text{Cl}_i \leftrightarrow \text{Cl}_j / \text{Cl}_i\text{s} \in \text{Cluster}_1 \text{ and } \text{Cl}_j\text{s} \in \text{Cluster}_2\}$.

4 The Approach

Data Structure The data structure is shown in Fig. 2. It has the following parts.

Matrix AMAT AMAT is a matrix of size $m * n$, where m is the size of the set Val_1, and n is the size of the set Val_2. The rows of matrix AMAT correspond to the elements in Val_1. The columns of matrix AMAT correspond to the elements in Val_2. The cell $\text{AMAT}[a_i][a_j]$ stores the count of pairs Tp such that $a_{1i} \in \text{Tp}$ and $a_{2j} \in \text{Tp}$.

Cluster Lists CList_1 and CList_2 $\text{CList}_1 = \{\text{Cl}_{11}, \text{Cl}_{12}, \ldots\}$ is the set of clusters for Attr_1, and $\text{CList}_2 = \{\text{Cl}_{21}, \text{Cl}_{22}, \ldots\}$ is the set of clusters for Attr_2.

5 The Algorithm

The algorithm presented in this chapter works in three steps which are described in the following subsections.
Step 1—Initialize_Matrix Executed in the beginning, this step scans the entire database to update the matrix AMAT. In this step, when a new pair (a_i, a_j) is read, the value of $\text{MAT}[v_i][v_j]$ is increased by one.

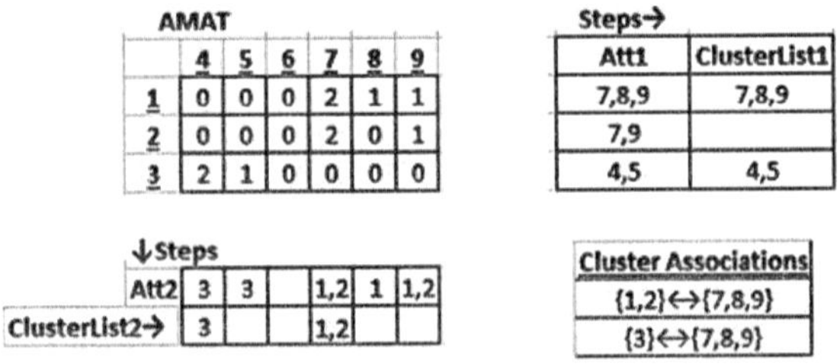

Fig. 3 Generate_Cluster step example

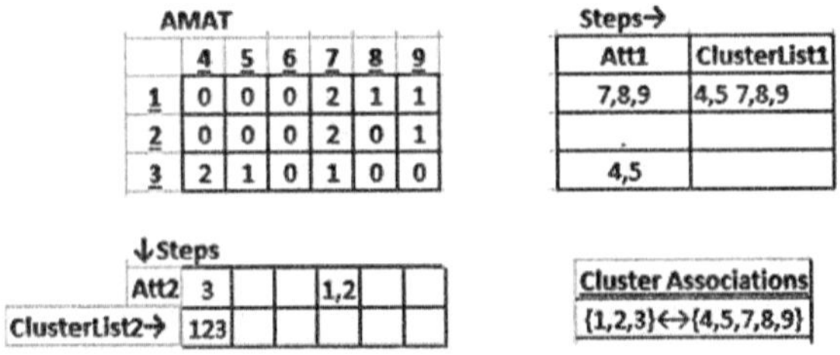

Fig. 4 Generate_Association step example

Step 2—Generate_Cluster It generates clusters from the matrix AMAT. In the beginning, it generates clusters for each row in AMAT in the following way. For a row in AMAT, it clusters the values of the columns which are having the matrix value greater than or equal to s_0 and store them in CList_1. That is, for a row v_{1i} in matrix MAT, the Generate_Cluster step will form the cluster $\{a_{2s}/\text{MAT}[a_{1i}][a_{2s}] \geq s_0\}$. Then, the clusters in CList_1 having at least one common element in them are merged together. In a similar way, clusters are obtained for other attributes. This step has a time complexity of $O(n^2)$. Figure 3 demonstrates the step.
Step 3—Generate_Association In this step, associations of the type $\text{Cl}_{1i} \leftrightarrow \text{Cl}_{2j}$ are generated between two clusters $\text{Cl}_{1i} \in \text{CList}_1$ and $\text{Cl}_{2j} \in \text{CList}_2$ if $a_1 \in \text{Cl}_{1i}$, $a_2 \in \text{Cl}_{2j}$ and $\sim (a_1, a_2) \geq s_0$. Figure 4 demonstrates this step.

The time complexity of the Generate_Association step is $O(n^2)$.

6 Experimental Analysis

The experiments were conducted on two real datasets on a system having the following parameters (Table 1).

6.1 Experiment on Iris Sapling Dataset

Dataset Description The iris sapling database is a table having information about 150 iris samples. Each record in the table stores information about 'petal-width,' 'petal-length,' 'sepal-width,' and 'sepal-length,' all in cm, of the 'species' of an iris

Table 1 System parameters

Parameters	Values
Processor	Intel Core™ i3
Memory (RAM)	3 GB
Operating System	Windows 7
Implementation Language	C++
Compiler	GNU GCC

plant. The dataset is available on the Web site [5]. The attributes considered in the experiment are 'petal-width' and 'species.' The values of attribute 'petal-width' are in the range [0.1 cm, 2.5 cm]. The values observed for attribute 'species' are 'iris-verginica,' 'iris-versicolor,' and 'iris-setosa'.

Results Experiment was conducted for different values of minimum support. The results are shown in Table 2. For minimum support of 0, there is one cluster generated for each attribute and one association rule generated. For minimum support of 1, there are two association rules generated. The 'species' cluster containing 'iris-setosa' is associated with the 'petal-width' cluster having values in the range [0.1, 0.6]. Similarly, the cluster containing 'iris-versicolor' and 'iris-virginica' is associated with the cluster containing values in the range [1.1, 2.5]. It can be observed that the widths of petals of 'iris-setosa' species are between 0.1 and 0.6 cm while widths of petals of 'iris-versicolor' and 'iris-virginica' species are between 1.1 and 2.5 cm. The parameter minimum threshold is significant. When the assigned value is 1, it means that every observation is significant. Not a single observation is ignored. The algorithm is sensitive to each observation.

The sensitivity of the algorithm is based on the value of minimum threshold. When the assigned value is 3, the number of association rules generated is three as shown in Table 2. The cluster containing 'iris-versicolor' and 'iris-virginica' is split into two clusters with each cluster associated with cluster each containing values of petal-widths within a unique range. It is observed that the species 'iris-versicolor' have their petal-widths in the range [1.1, 1.6] while the species 'iris-virginica' have their petal-widths in the range [1.8, 2.5]. The value of minimum threshold set to 3 means that a minimum of three observations are required for a value to belong to a cluster and also for a cluster to associate with another cluster to form as association rule. In other words, a observation is ignored if it occurs once or twice.

6.2 Experiments on Synthetic Dataset

Dataset Description Data for this experiment was generated using IBM Synthetic Generator [6, 7]. The datasets are generated in and stored in a text file [8–10]. Parameters of the dataset are described in Table 3.

Table 2 Clusters and associations generated for different values of minimum support

Minimum support	Petal-width clusters	Species clusters	Association rules
0	{0.2 0.4 0.3 0.1 0.5 0.6 1.4 1.5 1.3 1.6 1 1.1 1.8 1.2 1.7 2.5 1.9 2.1 2.2 2 2.4 2.3}	{Iris-setosa Iris-versicolor Iris-virginica}	{0.2 0.4 0.3 0.1 0.5 0.6 1.4 1.5 1.3 1.6 1 1.1 1.8 1.2 1.7 2.5 1.9 2.1 2.2 2 2.4 2.3} ↔ {Iris-setosa Iris-versicolor Iris-virginica}
1	{0.2 0.4 0.3 0.1 0.5 0.6} {1.4 1.5 1.3 1.6 1 1.1 1.8 1.2 1.7 2.5 1.9 2.1 2.2 2 2.4 2.3}	{Iris-setosa} {Iris-versicolor Iris-virginica}	{0.2 0.4 0.3 0.1 0.5 0.6} ↔ {Iris-setosa}, {1.4 1.5 1.3 1.6 1 1.1 1.8 1.2 1.7 2.5 1.9 2.1 2.2 2 2.4 2.3} ↔ {Iris-versicolor Iris-virginica}
3	{0.2 0.4 0.3 0.1} {1.4 1.5 1.3 1.6 1 1.1 1.2} {1.8 2.5 1.9 2.1 2.2 2 2.4 2.3}	{Iris-setosa} {Iris-versicolor} {Iris-virginica}	{0.2 0.4 0.3 0.1} ↔ {Iris-setosa}, {1.4 1.5 1.3 1.6 1 1.1 1.2} ↔ {Iris-versicolor}, {1.8 2.5 1.9 2.1 2.2 2 2.4 2.3} ↔ {Iris-virginica}

Table 3 Database parameters for cluster association generation

Parameters	Values
Number of rows	2000 K
Number of attributes values per attribute	10

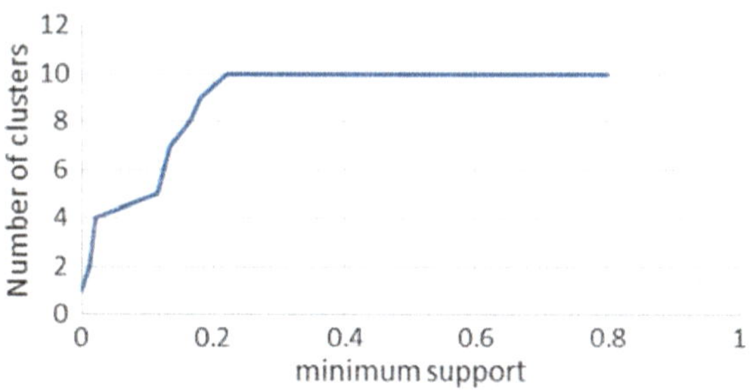

Fig. 5 Number of clusters versus minimum support

Results In this experiment, the value of s_0 was changed from 0 to 1 to observe the number of clusters generated of an attribute. It was observed that the number of clusters generated was less for lower values of s_0. However, the number of clusters generated was proportional to the value of s_0 for lower values. For $s_0 \geq 2$, the number of cluster generated remained constant. At this stage, each cluster has only one value in it (Fig. 5).

Table 4 Synthetic data experiment

Parameters	Values
Precision	Min: 0.78, Max: 0.89
Recall	Min: 0.73, Max: 0.86
Accuracy	95%

In the above experiment, the value of s_0 was changed from 0 to 1 to observe the precision, accuracy, and recall of the proposed method. The observations are shown in Table 4.

7 Conclusion

This paper presents an algorithm to cluster values of attributes in a database based on the values of another attribute in the same database. It also generates cluster association.

The proposed algorithm is implemented on two datasets. The experiment on real dataset has revealed interesting clusters of two attributes of a plant, i.e., 'petal-width' and 'species,' and associations between these clusters. The experiment on synthetic dataset proved that the algorithm is accurate enough to generate correct patterns.

The study presented in this paper was limited only to the values of an attribute. In the future, the study will incorporate methods like text mining and sentiment analysis, thus not limiting to the values only.

References

1. Cheng Y, Xie Y, Chen Z, Agrawal A, Choudhary A, Guo S (2013) Jobminer: a real-time system for mining job-related patterns from social media. In: Proceedings of the 19th ACM SIGKDD international conference on knowledge discovery and data mining. ACM, pp 1450–1453
2. Xu H, Yu Z, Yang J, Xiong H, Zhu H (2016) Talent circle detection in job transition networks. In: Proceedings of the 22nd ACM SIGKDD international conference on knowledge discovery and data mining. ACM, pp 655–664
3. Naik SB, Pawar JD (2017) Clustering attribute values in transitional data streams. In: 2017 international conference on computing, communication and automation (ICCCA). IEEE, pp 58–62
4. Naik SB, Pawar JD (2017) Mining association rules between values across attributes in data streams. In: 2017 international conference on computational intelligence in data science (ICCIDS). IEEE, pp 1–6
5. www.kaggle.com
6. https://ibm-quest-synthetic-data-generator.soft112.com
7. Agrawal R, Imieliński T, Swami A (1993) Mining association rules between sets of items in large databases. In: ACM sigmod record, vol 22. ACM, pp 207–216

8. Naik SB, Pawar JD (2015) A quick algorithm for incremental mining closed frequent itemsets over data streams. In: Proceedings of the second ACM IKDD conference on data sciences, CoDS '15. ACM, New York, NY, USA, pp 126–127
9. Naik SB, Pawar JD (2013) An efficient incremental algorithm to mine closed frequent itemsets over data streams. In: Proceedings of the 19th international conference on management of data, COMAD '13. Computer Society of India, Mumbai, India, pp 117–120
10. Naik SB, Pawar JD (2012) Finding frequent item sets from data streams with supports estimated using trends. J Inform Oper Manag 3(1):153

Machine Learning Approach to Stock Prediction and Analysis

Bhal Chandra Ram Tripathi, T. Satish Kumar, R. Krishna Prasad, and Visheshwar Pratap Singh

Abstract The stock market is becoming a highly anticipated field of analysis. The data emerging every moment in the stock market globally is in petabyte size. The data analysts are working continuously on the data generated in the stock market to make capital-based predictions. The effort is to predict the future of stocks by using the data and make the best use of the financial environment. The crucial perspective of all the analysis is the generation of the relevant data and through reliable resources. The experiment is dependent on Twitter for the generation of data through its API. The database of one million data used to make an accurate prediction of 75–79%. The use of sentimental analysis, natural language processing, and convolution neural network makes the backbone of the overall research. The benchmark algorithm named STOCKP is an attempt to touch the expected accuracy of the prediction of stock market and make the best monetary stability.

Keywords Stock market · Twitter API · Sentimental analysis · Natural language processing · Convolution neural network · Data analytics

1 Introduction

There is a continuous flow of data all around the digital-to-analog resources/devices. Platforms and benches are used by resource personalities to produce their views [1]. These views are hypothetically good ground for analysis and are enjoyed most by the

B. C. R. Tripathi (✉) · R. K. Prasad
Global Academy of Technology, Bengaluru, Karnataka, India
e-mail: bhalchandra.chandra7@gmail.com; rkp_rgp@yahoo.co.in

T. S. Kumar
BMS Institute of Technology, Bengaluru, Karnataka, India
e-mail: satish.savvy@gmail.com

V. P. Singh
Data Science and Telecommunication, Aegis School of Business, Mumbai, Maharashtra, India

N. N. Chiplunkar and T. Fukao (eds.), *Advances in Artificial Intelligence and Data Engineering*, Advances in Intelligent Systems and Computing 1133,
https://doi.org/10.1007/978-981-15-3514-7_68

data analyzers to make certain assumptions. The researchers are applying continuous attempts to make prediction against the most important aspect of economy, i.e., stock market [2].

The business organizations and global leaders use the platforms such as Twitter, Facebook, and LinkedIn to produce their views through multiple resources. The views, when analyzed and found against the trends, and sudden changes in stock market are found to be key role playing agent. The researchers and market analysts are closely using this generated data to perform certain level of future predictions. The researchers are continuously attempting to make the predictive algorithms to be over 70% accurate but it is 55–60% [3].

The use of iterative resources and traditional data analysis is lacking to solve the real-time cases of prediction. The iterative approaches utilize the algorithms to extract data continuously but are not able to meet the humongous increase in demands of the market. The attempts used are mostly requiring the human intervention, making the next model to be available for analysis late, and the results are affected by 1–2%.

The proposed work in this paper employs the automated approach and use of machine learning. The domain of machine learning is a great advantage to the research as it is allowing the CPU and GPU to work in a balanced way and produce the best-optimized result. The source of data for the analysis is from Twitter, and the implementation work uses the Twitter API and social media platforms. The data generated is well balanced with the automated approach and algorithm that is developed is capable to produce the benchmark prediction [2, 4]. The use of parallel computing code to make the best use of GPU and CPU reduces the computation of results to microseconds. The modules of machine learning and parallel computing environment provide the freedom to make the vision to reality; the modules such as sentimental analysis, natural language processing, and hash cloud smoothen the research work.

In this research, the benchmark results are produced and are compared with the results generated from existing platforms. The initial iterations and improvements are capable to match a benchmark of 55.7% but with the rigorous improvements in data and architecture, it has enhanced to 62%.

In this work, we propose a novel method to handle the feelings of individual and to establish its significance in the current trends of regional stock markets. Using the architecture to be trained with the historical data and perform the training for few weeks, the initial data set consists of one million records from different individual's views over social platforms and the reaction to those views. The architecture is then automated after rigorous iterated process.

The paper is divided into multiple sections including this section which provides the motivation of performing the research. Section 2 includes two experimental setups that were used to perform the research, Sect. 3 covers the motivation to perform the research, Sect. 4 deliberates on the Architecture of the research, and Sect. 5 concludes the research.

2 Experimental Setup

The different tools, benchmarks, and environment used in this research are mentioned in this section.

2.1 Processing Unit

There is an immense amount of inflow and outflow of data used for the research through various platforms; the use of GCC compiler with better optimization resulted in the approximate usage of resources. We have considered to work with Open64 [5], LLVM [6], ROSE [7], Phoenix [8], and GCC [9] but we moved forward with GCC. The GCC is an open-source compiler with more language supports and better optimization result handling in the current open-source environment. It is capable of core data processing with the availability of results in microseconds.

2.2 Optimization

The open-source GCC provides all-around support to optimize the processing but the use of proper flags is important user input. The initial training requires the appropriate flags to input with the data while processing. There are many flags available in GCC, which are unique in their perspective that makes it necessary to use global levels (-Os, -O1, -O2, -O3). After the use of global levels, we can make it more automated results in handling the flag. The optimization of data is an important task due to the modules needed to deploy on the optimized data will lead to a more accurate result.

2.3 Platform

The research is generic in nature, and the platform used is common to most consumers. We have surveyed and found that the developers and predictors work more on Linux environment than other generic environments. The use of cluster of Intel Core Processor with octa-core processing, which is supported by 8 GB RAM and GPU of 2 GB and having Linux support, allows making the results be approximated.

2.4 *Benchmark of Evaluation*

Certain benchmarks are set for the inputs to be processed. The benchmarks are important, as it will make the quality of processing always improve rather than varying against examples. There are certain negotiable situations, which need special attention, while evaluating. The views generated during the tense circumstances need to be taken utmost care, as the views are available few weeks before the actual tense situations. The benchmarks of predicting the stock market is dependent upon the preprocessing of the data that represents the views, and if the data is lagging in certain preconfigured benchmarks, then it is not recommendable to process further. If there are chances to repair the data generated through views, then appropriate modules are run to iterate. The stock market core algorithm is an engine, which requires preprocessed elements rather than the processing done there.

3 Motivation

The researchers in the field of the financial ecosystem are facing a large amount of data for analysis. The data generating through social platforms, live television debate broadcasting, views and other means are heterogeneous in nature and are complete indifferent in benchmark evaluation. There are about 20 stock exchanges around the world, which is self-capable of generating data of size petabytes. The stock markets are also interdependent in business; this makes it tough to analyze and make any prediction regarding the next moment of stock. There are multiple approaches to solve the business problem of stock prediction, and they are mostly acquiring the iterative and manual intervention roles [10]. There are cases of improper handling that lead to response that are considered mispredictions and have been the reason for individuals to lose faith in such predictions.

The challenge exists, and they motivate and create an all-around research in collecting the important data that gets generated in certain organizations about the views on social platforms. The experiment is set up with Tesla and Amazon to analyze their stocks by the statement of their respective officials in public domains. There has been continuous monitoring of stock markets, which trade most socks of these organizations such as of countries India, USA, China, and Switzerland and we found that there are certain criteria that these organizations make before releasing some statements in the platforms. The statements of Elon Musk on twitter were examined and its relation to different country stock exchanges and we found that the most affected market is of US stock exchange and the least is from China stock exchange. An attempt was made to collect all such previous statement data on Twitter and also their response against them in stock prices. This made us realize that there is much similarity in history of events after analyzing one million records. Figure 1 shows the graphical

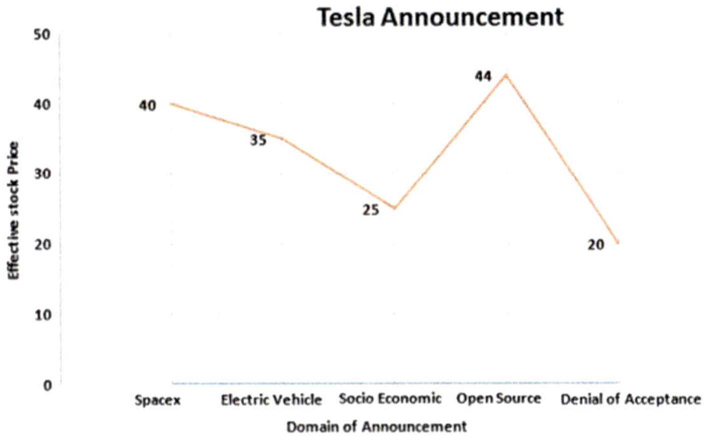

Fig. 1 Analysis of announcement by Tesla Executives on twitter and its effect on stock market

representation of views presented by Executive of Tesla on Twitter and their effect on stock market. It clearly shows that the people welfare views are affecting most of the stocks and making the stock of Tesla getting most profit.

The research is solely dependent on the development of a robust algorithm, and with the continuous attempts, it is near to touch the accuracy as expected. The algorithm employs the basic functionalities of sentimental analysis, key extraction, key communication, neural establishment, and neural optimization.

4 Architecture

The architecture of the research is simple but it requires a high processing background. In this section, we will consider the data identification, data extraction, data cleaning, core algorithm-I, and core algorithm-II phases of the architecture. The choice of Twitter site is made because of the fact that it is a social networking platform and is structured as a directed graph. Each user in twitter can choose the number of other users to follow and can be similarly followed by other users [11]. The topics of discussions in Twitter vary from sharing messages with friends, fans, family, or customers. Social media are, at an increasing rate, they are used in political campaigns [12–14] and for the decentralized coordination of political protests [15–17], are some example.

4.1 Data Identification

It is the crucial phase in the research. The need to extract the data is an important milestone but identification of appropriate data is much important than extraction. There is an availability of a surplus of data. The importance of what response against a statement by an Executive would accept for use is found. The views by an important official over twitter can lead to several million reactions but we need to extract the profiles of reactionaries. This will help to save the processing time and to extract only the responses of people whose responses are analyzed as a de facto.

The data identification phase not only saves the unnecessary data to be processes but also the resources, which used wisely, and can produce more optimized results. Figure 1 shows the analysis of that was announced by Tesla Executives on Twitter that affects the stock market.

4.2 Data Extraction and Data Cleaning

The scrappers are fed into with the results of data identification, and it makes the extraction to be particular. The research consider in case having the twitter as a platform to be considered for extraction of data then the syntax to make it possible is-

```
for tweet in tweepy.Cursor(api.search,q = "#%s"%u,tweet_mode =
'extended',
        count = n,lang = "en",since = "2019-01- 01").items():

df = df.append({'TimeStamp':tweet.created_at,'Tweet':tweet.
full_text},ignore_index = True)
```

There are multiple ways to extract with the help of identification keywords.

The data is then cleaned to remove the symbols and unnecessary characters such that it is in pure sentence form without any extras. The below snipper is a one-level cleaning and is part multilevel filtering.

```
def tidy(x):
    x = re.sub("@[\w]*", " ", x)
    x = x.lower()
    x = re.sub("(https?://[\w./]*)", " ", x)
    x = re.sub("[^a-z#]", " ", x)
    x = re.sub("#[\w]*", " ", x)
    x = ' '.join([w for w in x.split()])
return x
```

4.3 Core Algorithm

The algorithm is only the heart and soul of our research. We are developing two phases of algorithm phase-phase-II, and I, which will be making the expected results to be touched. GCC being a great optimizing compiler has supported the algorithm to work in full flourished manner [9]. The phase-I is having the modules such as sentimental analyzer, natural language processor, shift analyzer, and redundancy settle. The algorithm is still in test phase with giving the expected result of 69% accuracy. The base of both the algorithms is to create a real world on what our imaginations [8, 18].

The phase-II gets the inputs from phase-I, and it is the only phase helping to touch the expected accuracy of 75%. It has modules, which are taken from dynamic programming, and word-based analysis. Figure 2 is able to showcase the overall flow of data and appropriate analysis. The technical aspects are fixed with the above-mentioned processing capacity.

5 Conclusion

The aim of the research is to make the use of machine learning and neural network to make a reliable platform, which is accurate in financial stock predictions. We are very much confident with its infusions and beginning results; the algorithm based on basic to advance libraries has a perfect blend of imagination and reality. We are hopefully looking for better-optimized results in future enhancements by using cross-domain technology. We hope the future advancements will robust out vision.

Fig. 2 Architectural flow diagram

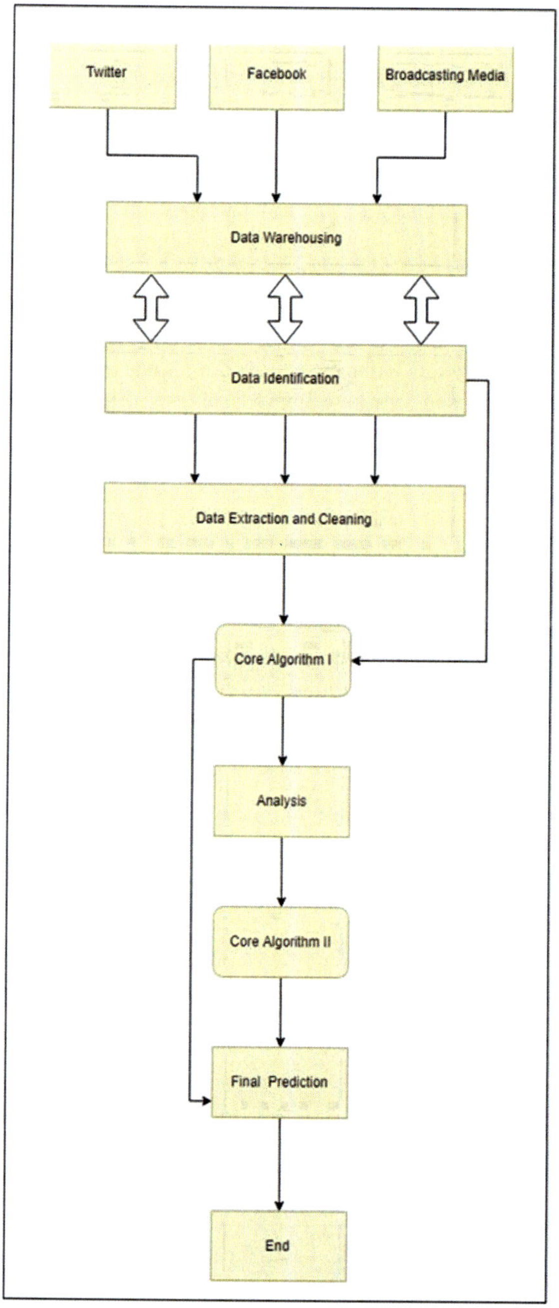

References

1. Shynkevich Y, Mcginnity TM, Coleman S et al (2014) Forecasting stock price directional movements using technical indicators: investigating window size effects on one-step-ahead forecasting. In: IEEE conference on computational intelligence for financial engineering & economics (CIFEr). https://doi.org/10.1109/cifer.2014.6924093
2. Bradley DA, Mandeville MJ (1988) Stock market prediction: the planetary barometer and how to use it. Llewellyn Publications, St. Paul, MN
3. Market reactions at the equity offerings announcement: a short window event study (2018) J Account Financ. https://doi.org/10.33423/jaf.v18i8.116
4. Determinants of stock market integration. Stock market integration. https://doi.org/10.1057/9781137381705_4
5. Open64: an open source optimizing compiler suite. http://www.open64.net
6. LLVM: the low level virtual machine compiler infrastructure. http://llvm.org
7. ROSE: an open source compiler infrastructure to build source-to-source program transformation and analysis tools. http://www.rosecompiler.org/
8. Phoenix: software optimization and analysis framework for microsoft compiler technologies. https:// connect.microsoft.com/Phoenix
9. GCC: the GNU Compiler Collection. http://gcc.gnu.org
10. Asadifar S, Kahani M (2017) Semantic association rule mining: A new approach for stock market prediction. In: 2nd conference on Swarm intelligence and evolutionary computation (CSIEC). https://doi.org/10.1109/csiec.2017.7940158
11. Angiani G, Cagnoni S, Chuzhikova N et al (2016) Flat and hierarchical classifiers for detecting emotion in tweets. AI*IA 2016 advances in artificial intelligence lecture notes in computer science 51–64. https://doi.org/10.1007/978-3-319-49130-1_5
12. Golbeck J, Grimes JM, Rogers A (2010) Twitter use by the U.S. Congress. J Am Soc Inf Sci Technol. https://doi.org/10.1002/asi.21344
13. Graham T, Jackson D, Broersma M (2014) New platform, old habits? candidates' use of Twitter during the 2010 British and Dutch general election campaigns. New Media Soc 18:765–783. https://doi.org/10.1177/1461444814546728
14. Conway BA, Kenski K, Wang D (2015) The rise of Twitter in the political campaign: searching for intermedia agenda-setting effects in the presidential primary. J Comput-Mediat Commun 20:363–380. https://doi.org/10.1111/jcc4.12124
15. Choudhary A, Hendrix W, Lee K et al (2012) Social media evolution of the Egyptian revolution. Commun ACM 55:74. https://doi.org/10.1145/2160718.2160736
16. Tan L, Ponnam S, Gillham P et al (2013) Analyzing the impact of social media on social movements. In: Proceedings of the 2013 IEEE/ACM international conference on advances in social networks analysis and mining—ASONAM 13. https://doi.org/10.1145/2492517.2500262
17. Steinert-Threlkeld ZC, Mocanu D, Vespignani A, Fowler J (2015) Online social networks and offline protest. EPJ Data Sci. https://doi.org/10.1140/epjds/s13688-015-0056-y
18. Jadhav R, Wakode MS (2017) Survey: sentiment analysis of Twitter data for stock market prediction. Ijarcce 6:558–562. https://doi.org/10.17148/ijarcce.2017.63129

A Novel Approach for Error Analysis in Classified Big Data in Health Care

S. Kavitha, Mahesh S. Nayak, and M. Hanumanthappa

Abstract The healthcare industry has observed an abundant development ensuring the growth of novel computer technologies and that moved this zone to create more medicinal information, which brought manifold arenas of research. It has encouraged scholars to relate the entire mechanical revolutions such as enormous information analysis, extrapolative analysis, instrument erudition and erudition procedures to mine valuable information as well as assist in resolutions construction. Many exertions are not only completed to manage the outburst of medicinal information but also to attain suitable data from it on the other hand. Due to the possibilities of prognostic analysis in immense information, and the usage of mechanism erudition procedures, forecasting imminent will not be a challenging chore, particularly for medication as forecasting illnesses as well as antedating the treatment suited imaginable. This paper brings a map of the development of enormous information in healthcare arena, and a knowledge procedure is related based on medicinal information. The aim is to envisage enduring kidney ailments by employing decision tree (B4.8) procedure.

Keywords Big data · Classification · Error · Healthcare · Data mining

1 Introduction

The learning coined as developmental origins of health and diseases (DOHaD) has effectively demonstrated the significance of progressive archives of entities in envisaging and/or elucidating the illnesses which an individual is undergoing. The medical

S. Kavitha (✉)
Dayananda Sagar Institutions, Bangalore, India
e-mail: s.kavitha527@gmail.com

M. S. Nayak
Research and Development Centre, Bharathiar University, Coimbatore, India
e-mail: mnayak67@yahoo.com

M. Hanumanthappa
Department of Computer Science and Applications, Bangalore University, Bangalore, India
e-mail: hanu6572@hotmail.com

N. N. Chiplunkar and T. Fukao (eds.), *Advances in Artificial Intelligence and Data Engineering*, Advances in Intelligent Systems and Computing 1133,
https://doi.org/10.1007/978-981-15-3514-7_69

arena has its unlimited role in this plenty of data due to some technical advances in the arena like cloud computing which has repositioned the assessments of maintenance outside the hospital and has prepared them accessible everywhere and at any moment [1–4], laparoscopic operation and robotic operation that substituted traditional operation [5], as well as keen families that let inmates self-care and nursing employing modest expedients, which convey outcomes on explicit functional settings. Keen applications or software are able to examine the physique symbols employing unified devices with the aim of observing [6–10], as well as health skills which assist novel approaches of biological, interactive and ecological information gathering. These contain devices which observe the occurrences with an advanced exactitude [11]. Health caution structures are extremely intricate, uneven and employ numerous data technology structures. With dealers integrating various criteria for alike or similar schemes, it is little curiosity that versatile incompetence, unused and faults in healthcare data and distribution supervision are also a usual an incidence. Subsequently, an inmate's well-being proceedings are confined in archive of inheritance structures, incompetent to be mutual with associates of the well-being civic [12–21]. They are few numerous stimuli's bringing an exertion to inspire calibration, incorporation and electronic data interchange among the numerous well-being sources.

In the recent mainly paper-based health archives domain, vital information is more often than not inaccessible at the exact interval in the lap of the medical care benefactors to authorize enhanced maintenance. It is chiefly because of the inadequacies integrated into the paper-based scheme. In an automated domain, it is really promising, if definite significant stages are taken previously, to confirm the accessibility of the accurate data at the appropriate interval. Administered knowledge is the device knowledge assignment of intrusive a purpose from categorized exercise information. We are employing administered knowledge for training set in this venture.

2 Existing System

1. N. V. Chawla, N. Blumm, D. A. Davis, N. Christakis, A. L. Barabás: Enduring ailments have been midst the key apprehensions in medical arenas since they may origin a dense problem on healthcare assets and disrupt the eminence of life. In this paper, we recommend a new context for primary valuation on enduring ailments by withdrawing consecutive hazard designs with phase pause data from analytical medical registers employing consecutive instructions withdrawing and cataloging demonstrating methods. With a comprehensive workflow, the projected outline contains four stages, namely information preprocessing, hazard design removal, organization demonstrating and post-investigation. For experiential assessment, we validate the efficiency of our projected outline with a case study on the initial valuation of COPD. Through investigational assessment on a significant countrywide medical record in Taiwan, our method cannot only originate ample consecutive risk outlines but also excerpt new designs with valuable perceptions for further medicinal study, for example, learning new indicators and

healthier cures. To the best of our information, this is the foremost effort addressing the question of removal consecutive risk designs with phase intermissions as well as cataloging models for the initial valuation of enduring ailments.
2. Felzemburghet, R. B. Reis, G. S. Ribeiro, R. D. al.: Leptospirosis is a likely dangerous ailment mainly affecting low-income people, with a projected annual frequency of 1.03 million contaminations globally. This ailment has signs often muddled with other feverish patterns, for instance, dengue fever, influenza and viral hepatitis, often creating analysis stimulating. Enlightening the correctness of initial analysis of patients with leptospirosis will upsurge the rapidity of apt antibiotic cure conveyance, and both will recover medical results for this possibly lethal ailment. The authors conducted a study of clinically and epidemiologically distinct leptospirosis cases to forecast ailment employing information removal grouping procedures. They piloted four groups of tests to assess the presentation of the procedures, evaluating their prognostic correctness of employing diverse training and assessment data sets. The JRIP algorithm attained 84% sensitivity employing a data set of only established leptospirosis cases, and a particularly of 99% employing a data set of only definite dengue cases. Thus, the method effectively projected leptospirosis cases, distinguished them from alike feverish diseases, and may signify a novel instrument to help health specialists, mainly in widespread zones for leptospirosis, quickening embattled cure and reducing ailment intensification and transience.
3. M. Skubic, M. Rantz, G. Demiris, B. K. Hensel: Current developments in wireless device systems for omnipresent well-being and movement observing structures have activated the option of speaking social requirements in keen settings through distinguishing hominid actual deeds. While the description of surge in that system needs effectual acknowledgment methods, it is also subject to doubtful conclusion-based confidentiality outbreaks. In this research work, we suggest an outline which proficiently distinguishes hominid actions in keen abodes grounded on comprehensive removal method. Besides, we suggest a method to enrich the confidentiality of the poised human recognized actions employing a revised form of micro-aggregation method. A wide authentication of our outline has been executed on standard information groups accommodating rather likely outcomes inters of exactness and privacy-utility adjustment.

Our prevailing structure employs breast tumor data sets which have two categories' frequent occurrences and no frequent occurrences. Preprocess the information and later categorizing the information employing resolution base which is to be employed as a base classifier for AdaBoost procedure with many repetitions is set to 2 and weight verge for weight clipping is set to 10 AdaBoost. M1 procedure is employed, which employs the base classifier decision stump (AdaBoost_DS) and reweighting, the number of repetitions is set on 10, and weight verge for weight clipping is set on 100. Equating the properly categorized case and correctness of grouping, AdaBoost execution of resolution stump recovers exactness. AdaBoost, short for "Adaptive Boosting," is a mechanism knowledge meta-algorithm; it can be employed in combination with various other kinds of knowledge procedures to recover their presentation. The result of the other knowledge algorithms ('weak learners') is collective into a weighted quantity which signifies the concluding result of the enhanced classifier.

3 Problem Specifications

a. Medical information models are often not well composed in their course markers.
b. Most of the prevailing catalog in approaches inclined to achieve low on a group of information that is enormously overloaded.
c. An accumulative amount of applications set up over the cloud function on information models which are enormous as well as intricate and it develops challenging to collect, mass, examine and envisage. So this brings a measurability matter.
d. Here, in equality matter and measurability matter are vanquished.

4 Limitations of Existing Algorithms

a. Merging sets, in case trees will only be advantageous if the distinct sets are exact and unlike from each other.
b. The information excavating approaches exactness differs reliant on the types of the information groups and the extent of information between the preparation and analysis clusters.
c. The collective features of the healthcare information groups are extremely unbalanced information groups, whereby the mainstream and the marginal classifier are imbalanced consequential estimate in accurate when tracked by the classifiers.
d. Remaining features of well-being information groups are the absent ideals.
e. Employing only designed information in well-being information groups.

5 Scope of Research

The proposed system is used in clinical study and it is possible to apply for real medical information investigation. The classification algorithm proposed is very useful in Bioinformatics field and drug occurrence finding.

6 Advantages for Work

- It is easy to comprehend and construe and able to manage both arithmetical and definite information, which needs petite facts planning, for likely to authenticate a model employing numerical assessments, accomplishes healthy with big information groups.
- It is vigorous, which defines that accomplishes fine even though its expectations are somewhat desecrated by the exact ideal from which the facts were produced.

7 Decision Tree for Classified

The prime usage of decision tree is in processes research investigation for scheming restricted likelihoods. Employing decision tree, decision makers can select paramount substitute and denial from root to leaf specifies exclusive class apart grounded on extensive data improvement. Decision tree is extensively employed by various researchers in healthcare sector. DT is alike to the flowchart in which each non-leaf knots signifies an examination on a specific quality and each division signifies an effect of that examination and each leaf node have a class tag. The node at the highest maximum tags in the tree is called root node. For instance, we have financial institution decision tree which is employed to determine that a person must fund the loan or not. Erecting a result for any issue does not require any kind of field information. Decision trees are a classifier which employs tree-like graph. Cataloging of patient into high risk and low risk class employing decision tree.

8 Proposed Algorithm

Input: The set of training dataset: T_{d}; A vector of attribute values: V_{d}
Label class of object: L_{d}
Output: The class of Z_{d}

Step 1: Accessing the preprocessing Health Data set
Step 2: Known data set creation
Step 3: Tanning data set creation
Step 4: Skewness based Workshop creation for Health data set
Step 5: Prepare for data normalization with all tests set
Step 6: Some tidying for tanning data set for view result with confusion matrix.

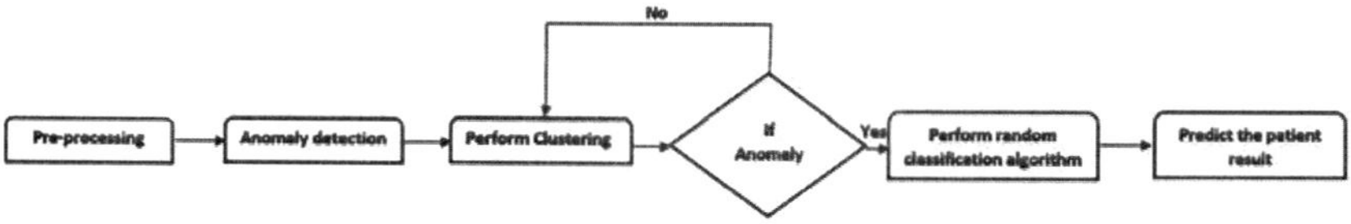

Fig. 1 Flow of the proposed method

9 Used Date Set

We employed the information groups Bangalore Kidney Foundation in Padmanabhanagar. This information contains 400 instances and 24 integer attributes, two, not enduring kidney disease.

10 Flow of Work

Figure 1 shows that after getting the unprocessed data, we first used preprocessing algorithm to get proper data set. After anomaly detection, we calculate clustering technique for classification algorithms to get proper result.

11 Experimental Results

We used our data set for getting error analysis of classified health big data set. We used all the proposed data set. Table 1 represented the performance analysis. Table 2 demonstrated error analysis. We discriminate six basic steps in the performance analysis process with data collection, data transformation and data visualization. Data collection is the progression by which data about program presentation are obtained from an executing program. Data are normally collected in a file, either during or after execution, although in some situations, it may be presented to the user in real time.

Table 1 Performance analysis

Evaluation criteria	B4.8
Correctly classified instances	476
Error	0.24
Incorrectly classified instance	7
Time to build model (s)	0.15
Accuracy	77%

Table 2 Error analysis

Root relative squared error %	18.69
Evaluation criteria	B4.8
Mean absolute error	0.07
Kappa statistic	0.93
Relative absolute error %	5.98
Root mean squared error	0.128

12 Conclusion

In order to get the maximum exactness among classifiers which is significant in medicinal analyzing with the features of information being engaged, we should enterprise a mix ideal which could decide the stated problems. The information withdrawal has played a vital part in the healthcare field, particularly in forecasting many kinds of ailments. The analysis is extensively being employed in envisaging ailments; they are widely employed in medicinal analyzing. To conclude, there is no information removal technique to decide the questions in the well-being information sets. In imminent, extra effort will be shown in this arena so to concoct cures and the lifespan of the inmate by appropriately sustaining and studying the health field information for guidelines is to enrich forecasts employing mixture representations.

References

1. Peters SG, Buntrock JD (2014) Big data and the electronic health record. Ambul Care Manag 37(3):206–210
2. Weil AR (2014) Big data in health: a new era for research and patient care. Health Aff 33(7):1110
3. Suthaharan S (2014) Big data classification: problems and challenges in network intrusion prediction with machine learning. SIGMETRICS Perform Eval Rev 41(4):70–73
4. Rao BP (2015) of the notes: brief notes on big data: A cursory look
5. Hermon R, Williams PA (2014) Big data in healthcare: what is it used for?
6. Raghupathi W, Raghupathi V (2014) Big data analytics in healthcare: promise and potential. Health Inf Sci Syst 2(1):1
7. Fang OH, Mustapha N, Sulaiman MN (2010) Integrating biological information for feature selection in microarray data classification. In: Second international conference on computer engineering and applications (ICCEA), vol 2. IEEE, pp 330–334
8. Groves P, Kayyali B, Knott D, Van Kuiken S (2013) The big data revolution in healthcare. McKinsey Quarterly, vol 2
9. Chandra S, Motwani D (2016) An approach to enhance the performance of hadoop mapreduce framework for big data. In: 2016 international conference on Micro-electronics and telecommunication engineering (ICMETE), IEEE, pp 178–182
10. Satish KR, Kavya N (2014) Big data processing with harnessing hadoop-mapreduce for optimizing analytical workloads. In: International conference on contemporary computing and informatics (IC3I), IEEE, pp 49–54
11. Mukherjee M, Paul T, Samanta D (2012) Detection of damaged paddy leaf detection using image processing. J Glob Res Comput Sci (JGRCS) 3(10): 7–10. ISSN: 2229-371X

12. Panayides AS, Pattichis CS, Pattichis MS (2016) The promise of big data technologies and challenges for image and video analytics in healthcare. In: 2016 50th Asilomar conference on signals, systems and computers, IEEE, pp 1278–1282
13. Hall MA (1999) Correlation-based feature selection for machine learning. Ph.D. dissertation, The University of Waikato
14. Paul M, Samanta D, Sanyal G (2011) Dynamic job scheduling in cloud computing based on horizontal load balancing. Int J Comput Technol Appl (IJCTA) 2(5): 1552–1556. ISSN: 2229-6093
15. Liu H, Motoda H (2007) Computational methods of feature selection. CRC Press
16. Saeys Y, Inza I, Larra˜naga P (2007) A review of feature selection techniques in bioinformatics. bioinformatics 23(19): 2507–2517
17. Samanta D, Sanyal G (2012) Development of edge detection technique for images using adaptive thresholding. Int J Inf Process (IJIP) 6(2)
18. Hall MA (2000) Correlation-based feature selection of discrete and numeric class machine learning
19. Liu H, Yu L (2005) Toward integrating feature selection algorithms for classification and clustering. IEEE Trans Knowl Data Eng 17(4):491–502
20. Samanta D, Paul M, Sanyal G (2011) Segmentation technique of SAR imagery using entropy. Int J Comput Technol Appl (IJCTA) 2(5): 1548–1551. ISSN: 2229-6093
21. Dharavath R, Singh AK (2016) Entity resolution-based jaccard similarity coefficient for heterogeneous distributed databases. In: Proceedings of the second international conference on computer and communication technologies, Springer, pp 497–507

Multi-join Query Optimization Using Modified ACO with GA

Vikas Kumar and Mantosh Biswas

Abstract Since the era of big data, enhancement of the speed of the database query is the critical problem. Fundamental part of any database management system is query optimizer. So, it should be able to choose the suitable search strategy which can give quicker response even with complex and large amount of data. The expansion of database application areas, where a query may include more than 100 joins, traditional query optimization techniques, i.e., greedy-based optimization and dynamic programming unable to find the solution but its observed from the metaheuristic algorithm like genetic algorithms (GA), particle swarm optimization (PSO) and ant colony optimization (ACO) algorithms gives optimal solution. These algorithms do not guarantee the best optimal solution, rather near best solution, which is tolerable. However, there is the limitation of slow convergence and high computation effort in GA and stagnation with premature convergence in ACO. Therefore, we proposed the hybridization of our modified ACO with GA to overcome the above limitations for multi-join query optimization and experimentally shown our proposed method on randomly created dataset gives better results over GA and ACO in terms of distance and execution times.

Keywords Multi-join query · Query execution plan · Ant colony optimization · Genetic algorithm · Database management system

1 Introduction

A query is a request for information from a database and database is a collection of data in an organized manner. A query plan is a set of steps used to retrieve data from database and the increasing demand of database management system applications day by day, user has to deal with a huge amount of data, the complexity of the query

---

V. Kumar · M. Biswas (✉)
Department of Computer Engineering, National Institute of Kurukshetra, Kurukshetra, India
e-mail: mantoshb@gmail.com

V. Kumar
e-mail: virvikas2@gmail.com

N. N. Chiplunkar and T. Fukao (eds.), *Advances in Artificial Intelligence and Data Engineering*, Advances in Intelligent Systems and Computing 1133,
https://doi.org/10.1007/978-981-15-3514-7_70

is also growing gradually. There is a need to retrieve the data from the database in a fastest possible way and query optimization plays an essential role in improving the overall performance of the database systems. The data query optimization refers to the process of producing an optimal execution plan for a given query. The objective of query optimization is to provide minimum response time and maximum throughput, but the problem of finding the best query plan is similar to traverse each node of the graph exactly once to find a minimum-cost path. Nowadays researcher proposed various data query optimization methods, i.e., Chande et al. [1] proposed genetic algorithm-based optimization where the initial population of GA generated randomly. Each individual is represented by chromosome. Chromosome represents joins between relations using the path representation approach. Then, a new population is generated based on crossover and mutation operator. Zheng [2] proposed parallel ant colony algorithms, where the colony is divided into multiple colonies equal to the computation cores for parallel solution. Yinghua et al. [3] proposed a hybrid intelligent algorithm using GA and particle swarm optimization algorithm for multi-join query optimization for smart grid data. Petković [4] explores the optimization of queries by genetic algorithms and compares it with the conventional query optimization component. Al Saedi et al. [5] show the execution time of artificial bee colony algorithm is better than particle swarm optimization algorithm for query optimization. Wilschut et al. [6] proposed the no. of execution strategies for parallel evolution of multi-join query.

Le et al. [7] reexplore the classical problem of multi-query optimization in view of RDF/SPARQL. Therefore, based on the above literature survey, the challenges [8] find during optimization are, firstly, if an optimizer is not able to explore the whole search space, the one cannot guarantee to find out the best execution plan. So, we have to find the method which can efficiently explore the whole search space. Secondly, many traditional models cannot estimate the cost correctly and the lastly important challenge is joining ordering, which control the amount of data flow. So, the efficiency of a query highly depends on order of joins. In this paper, we proposed the hybrid algorithm for join ordering problem to overcome the above challenges and details are below section.

Figure 1 shows the complete process of a query. It starts with input query in terms of SQL query, it goes to parser, and parser generates the algebra tree related to query.

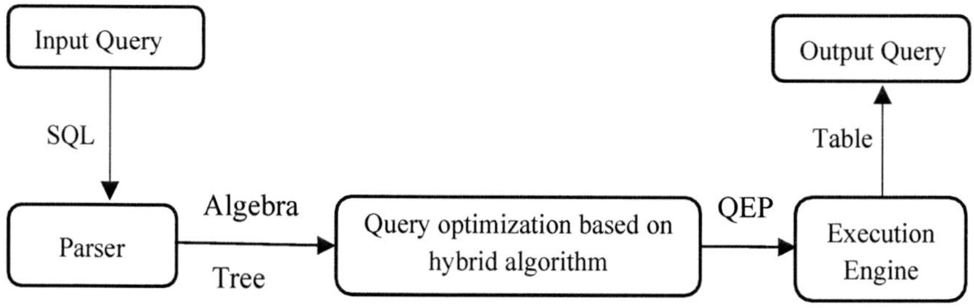

Fig. 1 Complete process of a query

Then, our algorithms select the optimum join order from multiple joins combinations. Based on join order, the query execution plan (QEP) is generated. Execution engine traverses the query execution plan and gives the appropriate retrieved data as output (Figs. 2 and 3.

Our paper is distributed in three different sections: Section 1 contains introduction; Sect. 2 shows our proposed algorithm; Sect. 3 shows the experimental result and analysis, and finally concludes the paper in conclusion.

2 Proposed Method

Multi-join query optimization is NP-hard problem [3] and simple queries among very less tables and indexes, there is always a choice for the optimal algorithm. For larger and complex queries, such as multi-joins with multiple indexes and subqueries, classical algorithms like Dijkstra's algorithm fail to find the solution. Therefore, to use the combination of metaheuristic algorithms like GA, PSO and ACO give good results. Figure 4 shows the sequential steps of our proposed method. The task of query planner is to select the best query plan from all of this possibility using our proposed method based on GA which explores the search space randomly and then evolve with time, with ACO, which use the probabilistic-based heuristic and pheromone factor for join order problem and details are as follows.

2.1 Genetic Algorithm (GA)

GA is based on Darwin's theory, "Survival of the Fittest," so only good individuals perform reproduction of the solution. Premature convergence problem of GA makes difficult to achieve the best solution.

Figure 2 shows the flowchart of genetic algorithm, in which we first create a random population, and then we find the shortest distance among the first population. We select the elitist individual based on reproduction and mutation operation and generates the next population until the stopping criteria are met.

In our proposed method, the parameters are defined by Zukhri [9] for crossover and mutation for generating new population and the best individual found by the proposed modified ACO and then evolve by GA. If in any iteration, GA individual has less distance, then we also update the pheromone using the path of that individual. Below, we define modified ant colony algorithm and then hybrid algorithm.

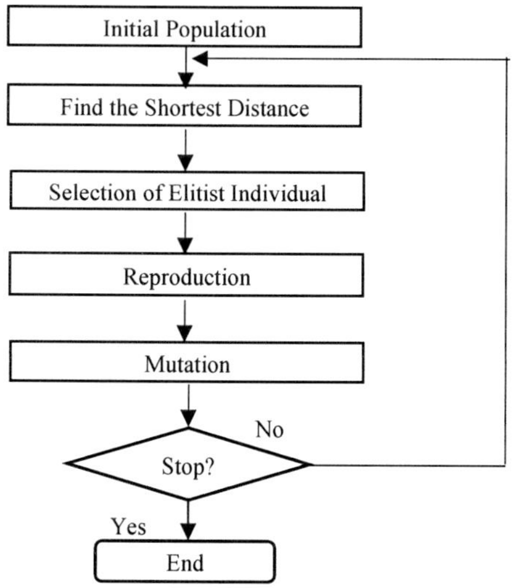

Fig. 2 Flowchart of basic genetic algorithm

2.2 *Modified Ant Colony Optimization*

Ant colony system (ACS) shortest path finding capability inspired Gianni to develop an entire new routing algorithm. The self-organizing, adaptive nature capability and positive feedback are the essential features of ACS. The ACS has been with success enforced for shortest path tour in traveling salesman problem, and it is a heuristic algorithm supported greedy approach. The biological ants move to find food source by learning and adopting pheromone with in the surroundings. After each iteration, every ant has a route based on the given transition probability (1).

2.2.1 ACO Transition Probability

$$p_{ij}^k(t) = \begin{cases} \frac{\tau_{ij}^{\alpha}(t)\eta_{ij}^{\beta}(t)}{\sum_{u \in N_i^k(t)} \tau_{iu}^{\alpha}(t)\eta_{iu}^{\beta}(t)} & \text{if } j \in N_i^k(t) \\ 0 & \text{if } j \notin N_i^k(t) \end{cases} \quad (1)$$

where τ_{ij}^{α} denotes the pheromone concentration and $\eta_{ij}^{\beta}(t)$ denotes the heuristic factor. $N_i^k(t)$ is the set of possible transition from node i. $p_{ij}^k(t)$ denotes the probability of selecting of node j from node i for ant k. The parameters are defined by Zukhri [9].

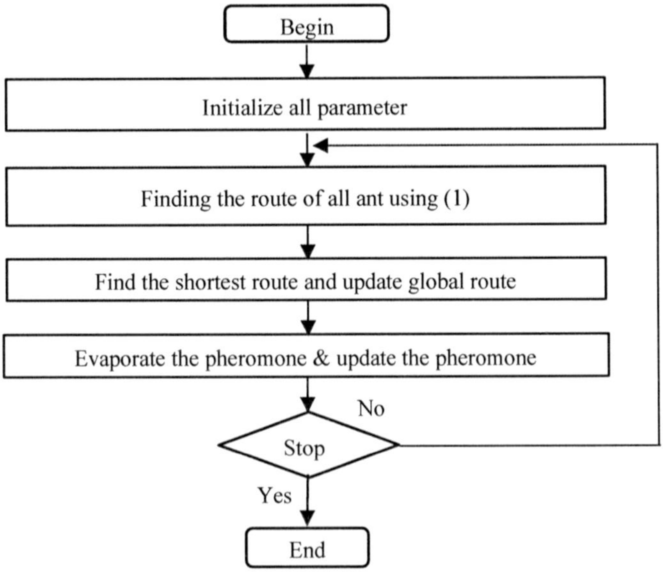

Fig. 3 Flowchart of ant colony algorithm

2.2.2 Modified ACO Transition Probability

Instead of α and β, we take only one parameter α for pheromone and $1 - \alpha$ for heuristic by below function.

$$p_{ij}^{k}(t) = \begin{cases} \frac{(\alpha\tau_{ij}(t))\cdot(1-\alpha)\eta_{ij}(t)}{\sum_{u \in N_i^k(t)}(\alpha\tau_{iu}(t))\cdot(1-\alpha)\eta_{iu}(t)} & \text{if } j \in N_i^k(t) \\ 0 & \text{if } j \notin N_i^k(t) \end{cases} \tag{2}$$

Then, we calculate the distance of each ant. We found the best path of iteration and update the global best path. After that, we evaporate the pheromone and update the pheromone matrix.

In Fig. 3, we show the ant colony algorithm flowchart, in the first step, we initialize all the parameter like no. of ant, pheromone evaporation rate, value of alpha and beta, etc. We find the route of all the ants using transition probability shown in Eq. (1). Then, we find out the best route from the iteration and update the global best route. After that, we evaporate the pheromone based on evaporation rate and update the pheromone based on no. of ant following a particular path. We iterate this process until we met the stopping criteria.

Figure 4 shows the process of our proposed hybridization method, in which we combine some parameter like no. of ant of ACO and GA population size into one parameter, no. of generation of GA and iteration of ACO into iteration. Instead of using transition probability of ACO, we use transition probability of modified ACO. We use the concept of dynamic changing the value of alpha. In the earlier

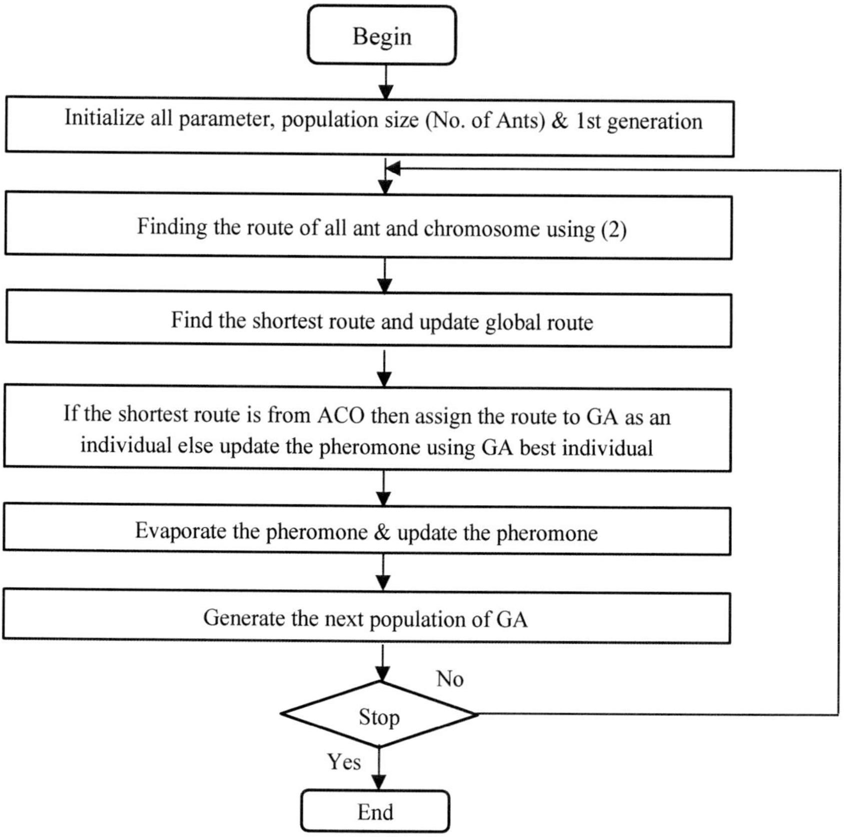

Fig. 4 Flowchart of the proposed method

stages, we take low value of alpha, by which we can exploit less and explore more. As iteration increases, we slightly increase the value of alpha so that exploitation gradually increases and exploration decreases. We also set the upper limit of alpha less 1 to give importance to both the factors. We find the shortest distance of GA as well as modified ACO. If modified ACO has less distance, then give its best route as an individual to GA to evolve more else we update the pheromone of ACO by using GA best individual. By this, there is co-operation of both methods.

3 Simulation Results

We have shown experimental results for multi-join query optimization of our proposed method in Table 1 that represent comparative distance and Table 2 represents

Table 1 Comparative distance (in km) results based on GA [1], ACO [10] and proposed

No. of iteration	GA		ACO		Proposed	
	Best	Average	Best	Average	Best	Average
10	3475	3593	396	424.2	382	417.3
20	3188	3219	351	386.2	366	379.5
30	2938	3052	370	381.7	353	379.9
50	2867	2919	349	367.9	346	365
100	2623	2754	348	363	334	355.2

Table 2 Comparative execution (in sec) results based on GA [1], ACO [10] and proposed

No. of iteration	GA	ACO	Proposed
	Time (s)	Time (s)	Time (s)
10	0.36	3.86	3.53
20	0.71	8.494	7.332
30	0.89	11.348	10.519
50	1.28	15.8788	14.773
100	2.9	37.309	34.43

execution time with two existing methods, namely genetic algorithm [1] and ant colony algorithm [10] with proposed method using our considered randomly dataset.

3.1 Experimental Parameter

The considered dataset consists of 100 join costs, here, we assume a join as a city and cost between two joins as a distance between two cities. The simulation of our proposed method is performed ten times with different no. of iterations, i.e., 10, 20, 30, 50 and 100. Implementation of the proposed method and other existing methods have been done by Python 3.14 running at Intel Cores i7-8650U CPU @ 1.90 GHz with 16 GB RAM. In our proposed method, for GA, the population size is 100, mutation probability, Pm (mutation rate) is 0.01 and no. of elite individual is 30. ACO parameters: α (influence of pheromone) is 1, β (influence of heuristic factor) is 2 and evaporation of pheromone, ρ is 0.5, and for hybrid algorithm parameters: α is 0.2, ρ is 0.5, Pm is 0.01, no. of ants/population size 100 and no. of elite individual is 30.

3.2 Result Analysis

We take 100 cities, we have dataset which contains distance between two cities. Figure 5 bar chart shows comparative distance, using data shown in Table 1, based on GA [1], ACO [10] and proposed. Figure 6 shows the computation time using data shown in Table 2 based on GA [1], ACO [10] and proposed from our experiment. Figure 5 shows, GA is worst among all, since GA is randomly generating approach, for large no. of cities, and it converges very slowly. Our proposed algorithm is slightly better than ACO in terms of distance as well as computing time. Figure 6 shows the computation time of GA is good and our proposed gives better results than ACO.

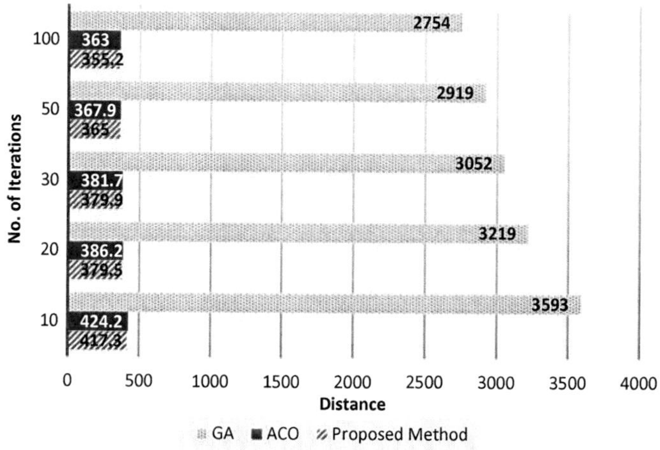

Fig. 5 Bar chart for average distance of tour considered dataset using GA [1], ACO [10] and proposed

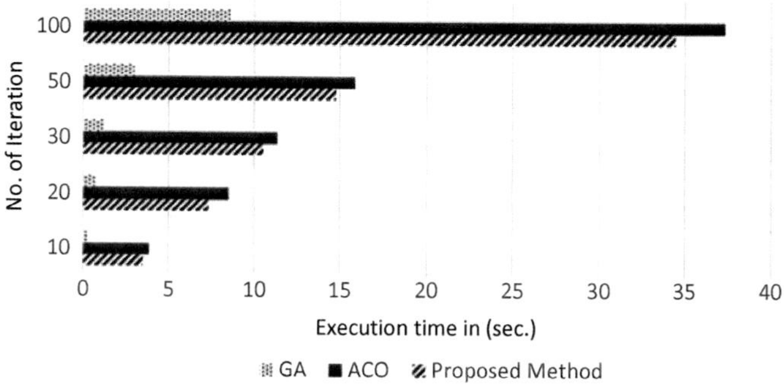

Fig. 6 Bar chart for computing time considered dataset using GA [1], ACO [10] and proposed

4 Conclusion

An optimize algorithm can improve the efficiency of queries along with reducing the query execution cost. In this paper, we have proposed a modified ACO with GA method for join ordering problem in query optimization. This proposed method focused on co-operation of both algorithm and dynamically changing exploitation and exploration of search space and it can improve accuracy and convergence rate. Simulation results show that our proposed method gives good results over GA [1] and ACO [10]. In the future, any other metaheuristic algorithms can be hybridized like artificial bee colony with GA, etc., for query optimization, which might give better result from existing algorithm.

References

1. Chande SV, Sinha M (2011) Genetic optimization for the join ordering problem of database queries. In: Annual IEEE India conference, pp 1–5
2. Zheng W, Jin X, Deng F, Mo S, Qu Y, Yang Y, Li X, Long S, Zheng C, Liu J, Xie Z (2018) Database query optimization based on parallel ant colony algorithm. In: 3rd IEEE international conference on image, vision and computing, pp 653–656
3. Yinghua H, Yanchun M, Zhang D (2015) The multi-join query optimization for smart grid data. In: 8th international conference on intelligent computation technology and automation, pp 1004–1007
4. Petković D (2010) Comparison of different solutions for solving the optimization problem of large join queries. In: Second international conference on advances in databases, knowledge, and data applications, pp 1–5
5. Al Saedi AKZ, Ghazali RB, Deris MBM (2014) An efficient multi join query optimization for DBMS using swarm intelligent approach. In: 4th world congress on information and communication technologies, pp 113–117
6. Wilschut AN, Flokstra J, Apers PM (1995) Parallel evaluation of multi-join queries. In: International conference on management of data, vol 24, No. 2, pp 115–126
7. Le W, Kementsietsidis A, Duan S, Li F (2012) Scalable multi-query optimization for SPARQL. In: IEEE 28th international conference on data engineering, pp 666–677
8. Wagh A, Nemade V (2017) Query optimization using modified ant colony algorithm. Int J Comput Appl 167(2):29–33
9. Zukhri Z, Paputungan IV (2013) A hybrid optimization algorithm based on genetic algorithm and ant colony optimization. Int J Artif Intell Appl 4(5):63–74
10. Li N, Liu N, Dong Y, Gu J (2008) Application of ant colony optimization algorithm to multi-join query optimization. Lecture notes in computer science, vol 5370, pp 189–197

The Impact of Distance Measures in *K*-Means Clustering Algorithm for Natural Color Images

P. Ganesan, B. S. Sathish, L. M. I. Leo Joseph, K. M. Subramanian, and R. Murugesan

Abstract In image processing, clustering algorithms applied to the segmentation of images. Image segmentation is the practice of clustering a complete image into many meaningful non-overlapped clusters. This is a vital step in computer vision and data analytics because the result of the segmentation process has an impact on other subsequent processes. In image processing, distance is expressed as distance in pixels or shortest path between two data points on the grid, two centers of pixels. Most clustering algorithms utilize distance measures to cluster alike data points (pixels in the case of image) into the same group, while unalike data points are clustered into different groups according to image attributes. The proposed work evaluates the efficiency of the *K*-means clustering with three distinct distance measures.

Keywords *K*-means clustering · Distance measure · Color model · CIELAB · Euclidean · City block

P. Ganesan (✉)
Department of Electronics and Communication Engineering, Vidya Jyothi Institute of Technology, Aziz Nagar, C. B. Road, Hyderabad, India
e-mail: gganeshnathan@gmail.com

B. S. Sathish
Department of Electronics and Communication Engineering, Ramachandra College of Engineering, Eluru, Andhra Pradesh, India
e-mail: subramanyamsathish@yahoo.co.in

L. M. I. Leo Joseph
Department of Electronics and Communication Engineering, S. R. Engineering College, Warangal, Telangana, India
e-mail: leojoseph@srecwarangal.ac.in

K. M. Subramanian
Department of Computer Science Engineering, Shadan College of Engineering and Technology, Hyderabad, India
e-mail: kmsubbu.phd@gmail.com

R. Murugesan
Department of Electronics and Communication Engineering, Malla Reddy College of Engineering and Technology, Hyderabad, Secunderabad, India
e-mail: rmurugesan61@gmail.com

N. N. Chiplunkar and T. Fukao (eds.), *Advances in Artificial Intelligence and Data Engineering*, Advances in Intelligent Systems and Computing 1133,
https://doi.org/10.1007/978-981-15-3514-7_71

1 Introduction

Segmentation is the most important process in computer vision and data mining. A sizable number of clustering algorithms were developed to address the segmentation problem. However, due to the complexity and variety of images, the design and modeling of effective and efficient image segmentation is still a difficult task. The term segmentation in image processing is illustrated as the procedure of separation of a test image into non-overlapped small groups or sub-images [1]. The image can be segmented based on these clusters. Image clustering is a scheme of allocating similar pixels to the same cluster [2]. The two categories of clustering can be performed as either hard or soft. In the first case, every data point is stringently allocated to any one particular cluster, whereas a particular pixel to more than one clusters in the soft clustering [3].

2 Distance Measures

In an image, a particular pixel, say '$p1$' at normal (x, y) coordinates has four immediate neighbors and four diagonal neighbors as shown in Fig. 1. Every pixel (data point) has a unique space away from (x, y). The immediate neighbor coordinates and the diagonal neighbors coordinates are clearly depicted in Fig. 1.

Distance measures are utilized to compute the minimum distance from each element of an object to the background [4, 5]. Distance is termed as the span of a line sector between points or lines [6]. In image processing, distance is expressed as the shortest distance between two pixels [7]. The Euclidean method computes the least space between two points p and q is

$$\sqrt{\left[(x-s)^2+(y-t)^2\right]} \tag{1}$$

Figure 2 illustrates the computation of distance using Euclidean distance.

Fig. 1 Image pixel and its neighbors

(x-1, y+1)	(x, y-1)	(x+1, y-1)
(x-1, y)	*P (x,y)*	(x+1, y)
(x-1, y-1)	(x, y+1)	(x+1, y+1)

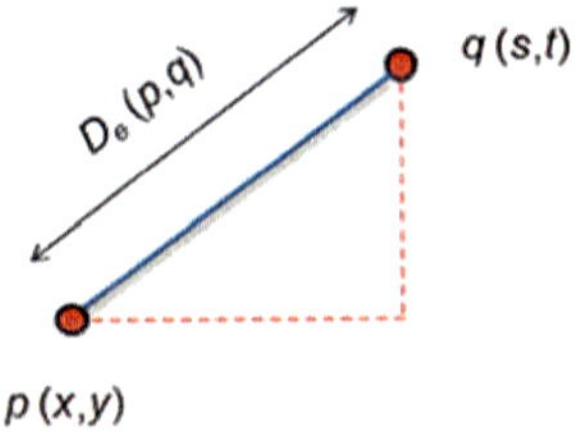

Fig. 2 Euclidean distance

The Euclidean distance of two pixels is the computation of the square root of the summation of the squares of the variation of resultant values [8]. In Euclidean squared method, there is no need of square root as shown in Eq. (2). So the segmentation or clustering is faster in Euclidean squared distance metric than normal Euclidean distance. Moreover, the clustering result is not changed in this method.

$$(x - s)^2 + (y - t)^2 \tag{2}$$

In this method, each centroid (center of the cluster) is considered as the average of the total points in that particular segment [9]. This measure is widely utilized and works very well with datasets with isolated clusters. The major drawback of this measure is more susceptible to outliers (Fig. 3).

In city block method, the distance computed as the summation of complete disparity of their corresponding points. It is also known as Manhattan distance, Manhattan length, $L1$ norm and $L1$ distance. The city block distance computes the distance between two points using Eq. (3)

$$|x - s| + |y - t| \tag{3}$$

In this, each centroid is the median of points in that cluster [9]. This measure is widely utilized in K-means clustering algorithm and work very well with datasets with isolated clusters [10]. The major drawback of this measure is more susceptible to the outliers.

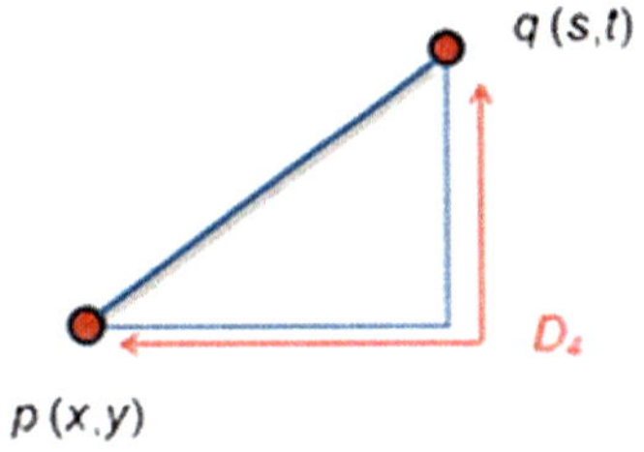

Fig. 3 City block distance

Cosine distance computes the angular distance of vectors (points). This measure is independent of vector length and invariant to rotation [11]. It is mostly used in document similarity applications. Equation (4) illustrates this measure between two vectors (*a*, *b*, *c*) and (*x*, *y*, *z*).

$$1 - \frac{ax + by + cz}{\sqrt{\mathrm{Abs}(a)^2 + \mathrm{Abs}(b)^2 + \mathrm{Abs}(c)^2}\sqrt{\mathrm{Abs}(x)^2 + \mathrm{Abs}(y)^2 + \mathrm{Abs}(z)^2}} \quad (4)$$

3 CIELAB Color Space

CIEXYZ color model is based on human perception as an alternative to traditional RGB space [12, 13]. This device-independent color space covers the gamut of perceived colors [14]. In CIEXYZ color space, three imaginary primaries were produced [15]. The conversion RGB to CIEXYZ color space is given by

$$\begin{bmatrix} X \\ Y \\ Z \end{bmatrix} = \begin{bmatrix} 0.4124 & 0.3575 & 0.1804 \\ 0.2126 & 0.7151 & 0.0721 \\ 0.0193 & 0.1191 & 0.9502 \end{bmatrix} \begin{bmatrix} \mathrm{R} \\ \mathrm{G} \\ \mathrm{B} \end{bmatrix} \quad (5)$$

In CIELAB, the intensity of the system is extracted from the luminance channel (*L*) and chrominance from *a* and *b* channels [16, 17]. The transformation equations for CIELAB are:

$$L = 116\left(\frac{Y}{Y_0}\right)^{1/3} - 16 \quad \text{if} \frac{Y}{Y_0} > 0.008856 \quad (6)$$

$$L = 903.3\left(\frac{Y}{Y_0}\right) \quad \text{if} \frac{Y}{Y_0} \leq 0.008856 \quad (7)$$

$$a = 500\left\{f\left(\frac{X}{X_0}\right) - f\left(\frac{Y}{Y_0}\right)\right\} \quad (8)$$

$$b = 500\left\{f\left(\frac{X}{X_0}\right) - f\left(\frac{Z}{Z_0}\right)\right\} \quad (9)$$

4 *K*-Means Clustering Algorithm

K-means clustering is one of the widely used cluster analysis (segmentation) in image processing and data analytics. The major intend of this algorithm is to split n number of unlabeled data (observations) into k number of clusters. This algorithm allocates every data point to any one of the cluster based on likeness. This iterative method begins with starting estimates for k number of centroids (arbitrarily produced or picked up from the dataset [18]. The basic steps for this clustering are data assignment and centroid updating. In the first step, every data point is allocated to its adjacent centroid. This is based on the distance measures computation as shown below.

$$J = \sum_{j=1}^{k} \sum_{i=1}^{n} \| x_i - c_j^2 \| \tag{10}$$

where $\left\| x_i - c_j^2 \right\|$ = space between a data point and the respective cluster centers.

C_j = set of centroids and x = data point.

In the second step, the computation of centroids is performed using the mean value of the entire data points allocated to that particular cluster. This procedure is repeated until one of the following condition (stopping criteria) is met

- Utmost amount of iterations is accomplished
- Sum of the distances is reduced
- No data points modify centroid.

5 Experimental Result and Discussion

The suggested technique is explicated based on the outcome is as follows. Figure 4 illustrates the test data to evaluate the efficiency of the proposed technique.

The end results of the proposed system with Euclidean distance for three, four, five and six clusters are depicted in Fig. 5 through Fig. 8 (Figs. 6 and 7).

The end results of the proposed system with city block distance measure for three, four, five and six clusters are depicted in Fig. 9 through Fig. 12 (Figs. 10 and 11).

The results of the proposed method with cosine distance for three, four, five and six clusters are depicted in Fig. 13 through Fig. 16 (Tables 1, 2 and 3, Figs. 14 and 15).

6 Conclusion

The proposed work discussed the impact of the distance measure in *K*-means clustering algorithm. In the proposed work, three distance measures are discussed. The *K*-means clustering utilized the distance measures to cluster alike pixels into the same group, while unalike pixels are clustered into different groups according to

Fig. 4 Test image

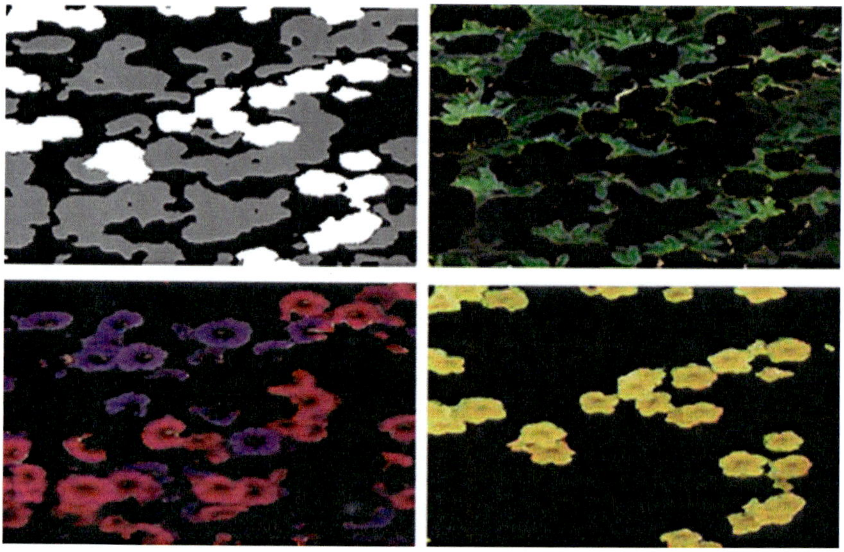

Fig. 5 Three clusters for Euclidean distance

color attribute. The test image is transformed to CIELAB color space from the traditional RGB for the competent clustering (segmentation). The transformed image is clustered into three, four, five and six clusters. From the result analysis, the work concluded that Euclidean distance measure is more efficient as compared to others.

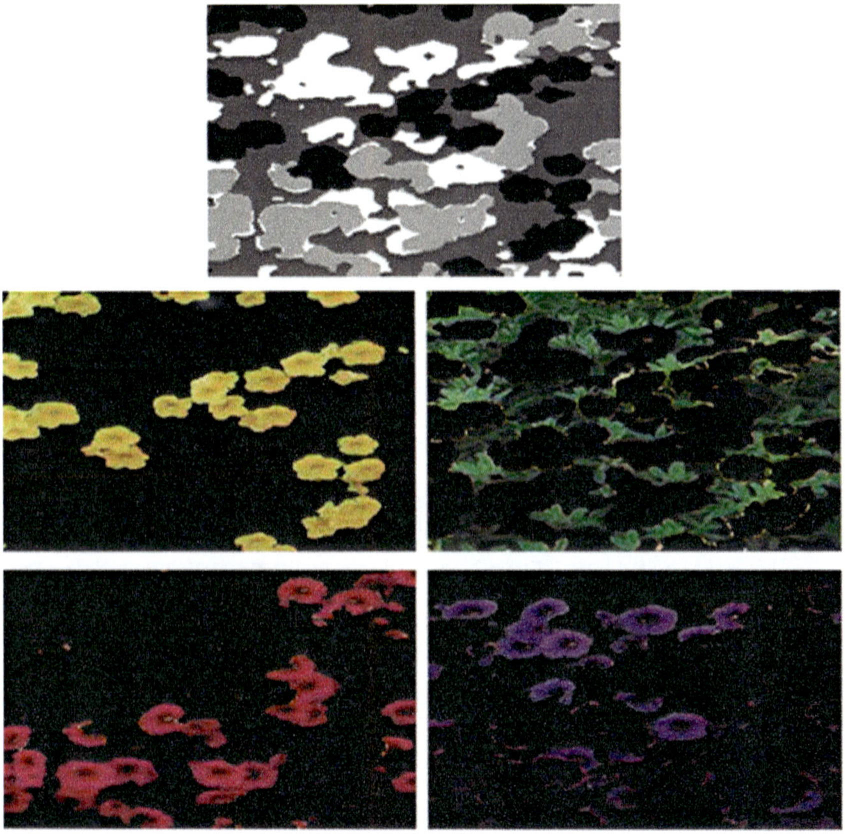

Fig. 6 Four classes for Euclidean distance

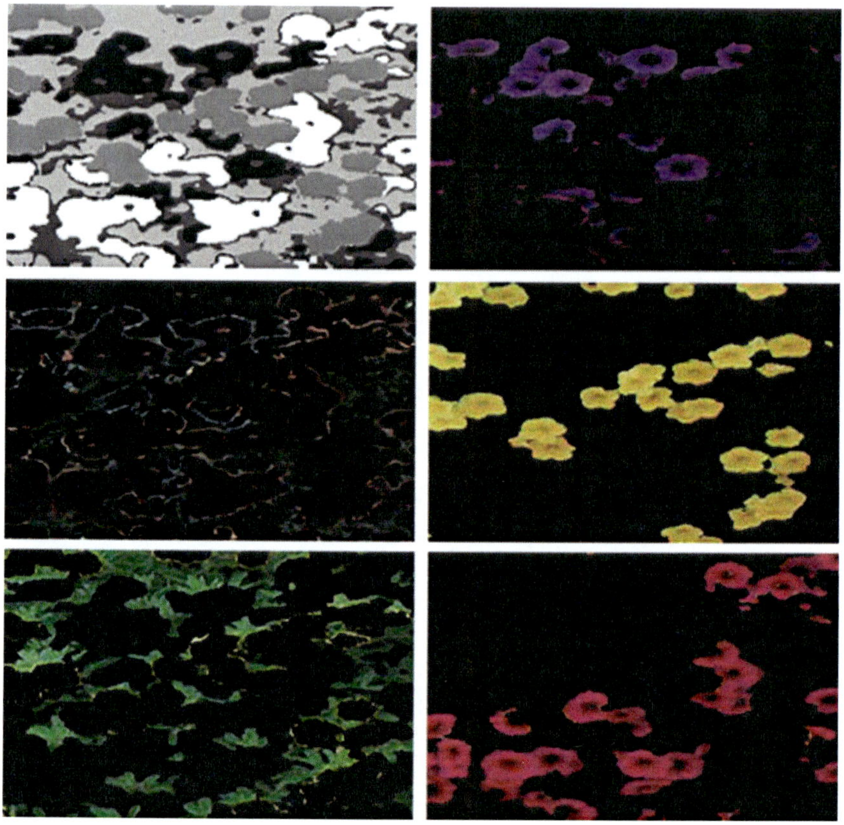

Fig. 7 Five clusters for Euclidean distance

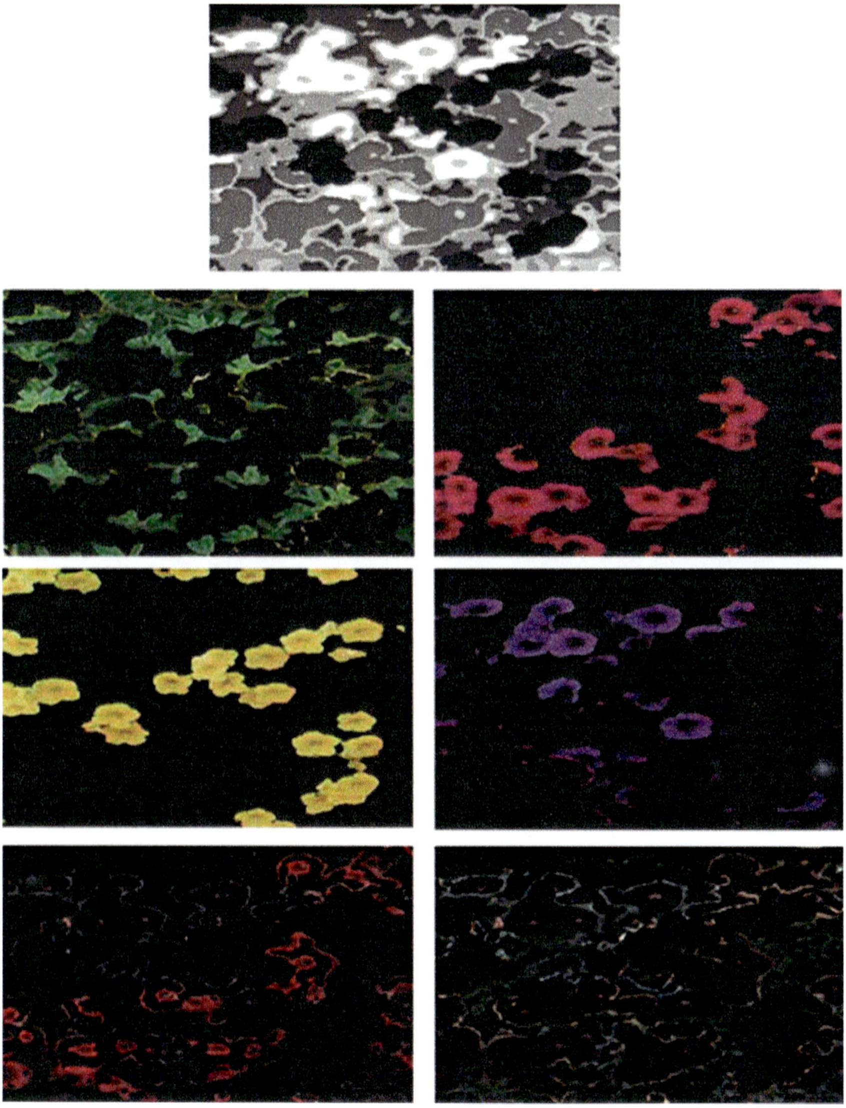

Fig. 8 Six classes for Euclidean distance

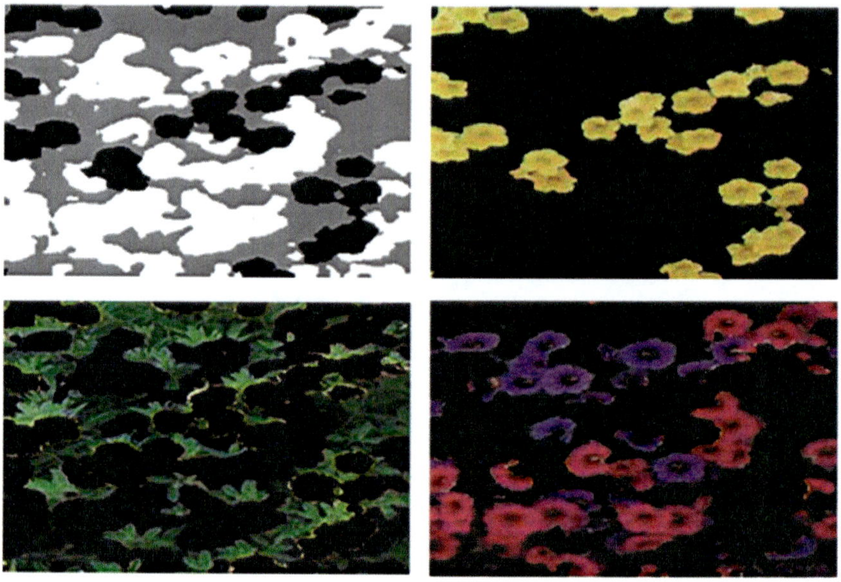

Fig. 9 Three clusters for city block distance

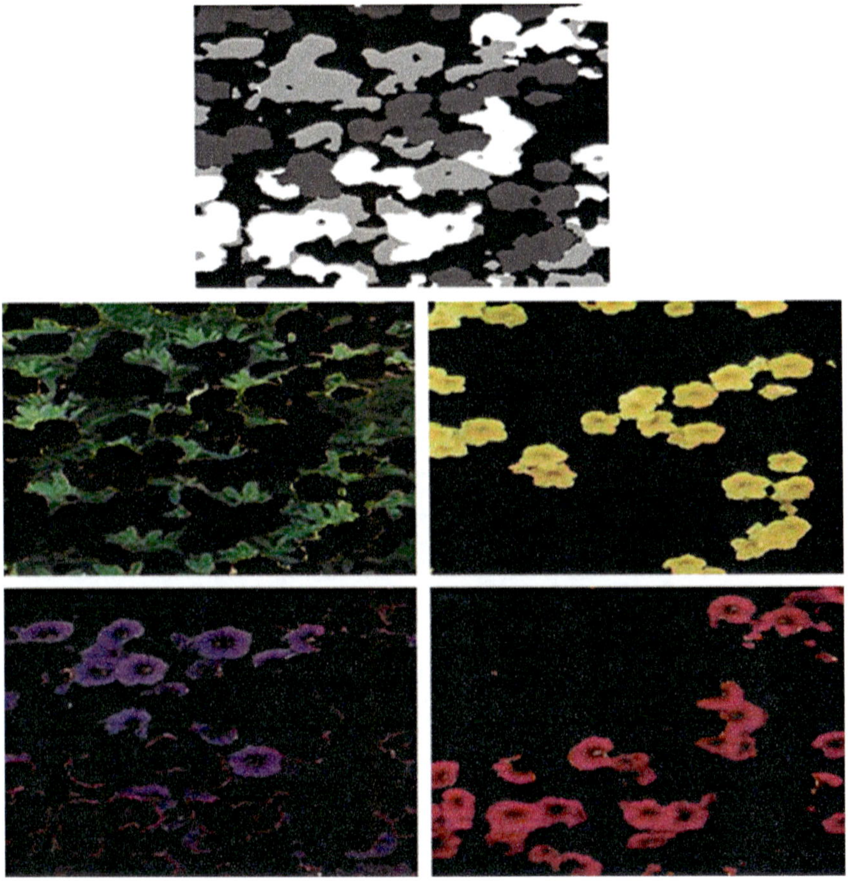

Fig. 10 Four clusters for city block distance

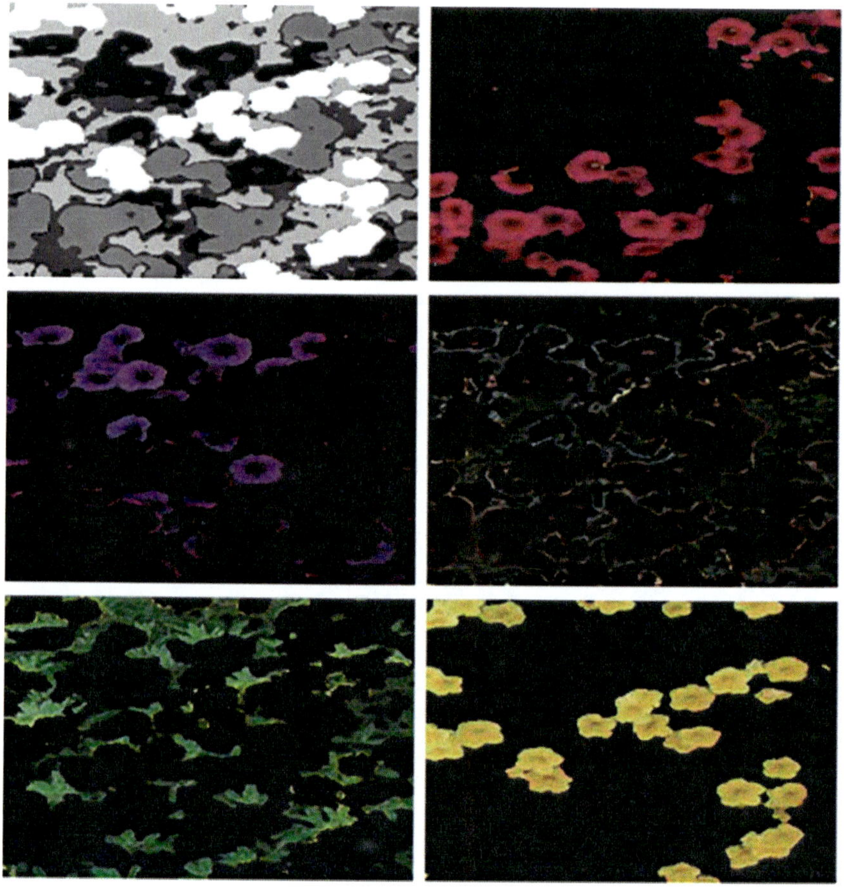

Fig. 11 Five clusters for city block distance

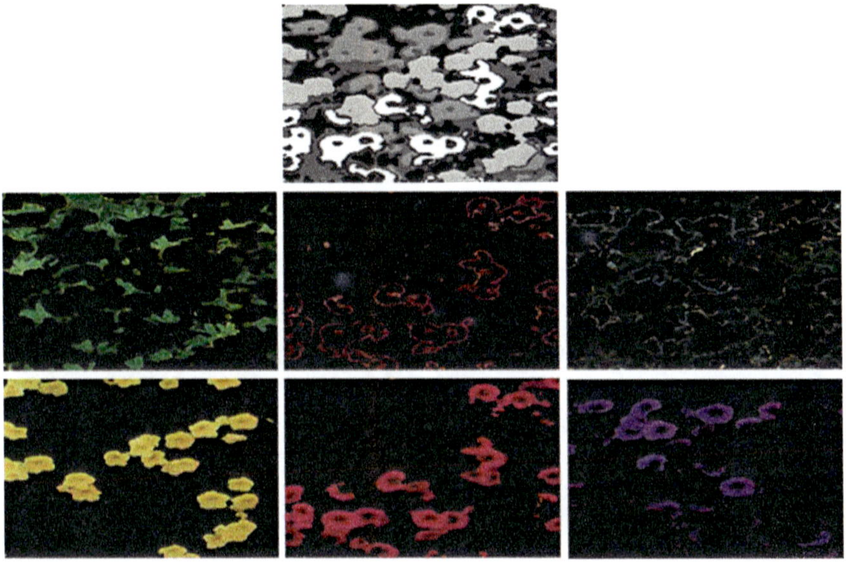

Fig. 12 Six clusters for city block distance

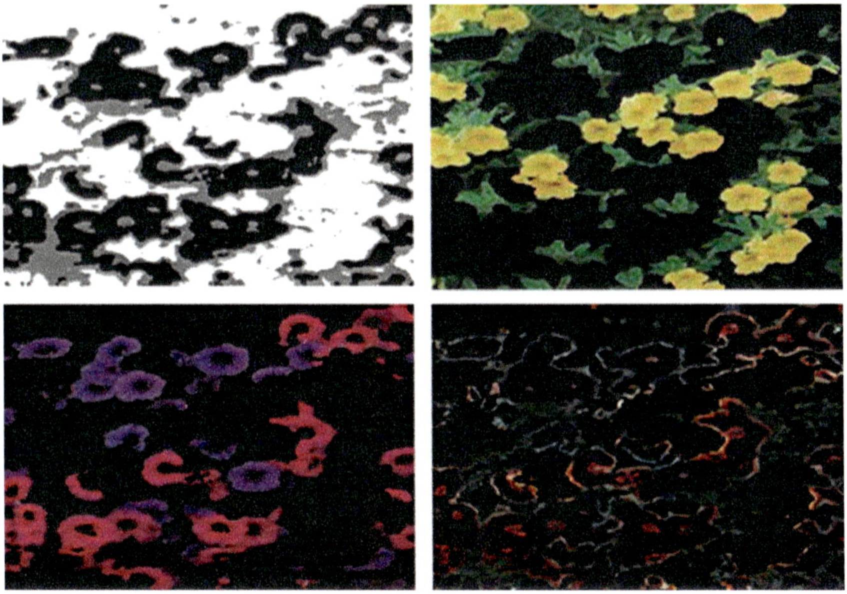

Fig. 13 Three clusters for cosine distance

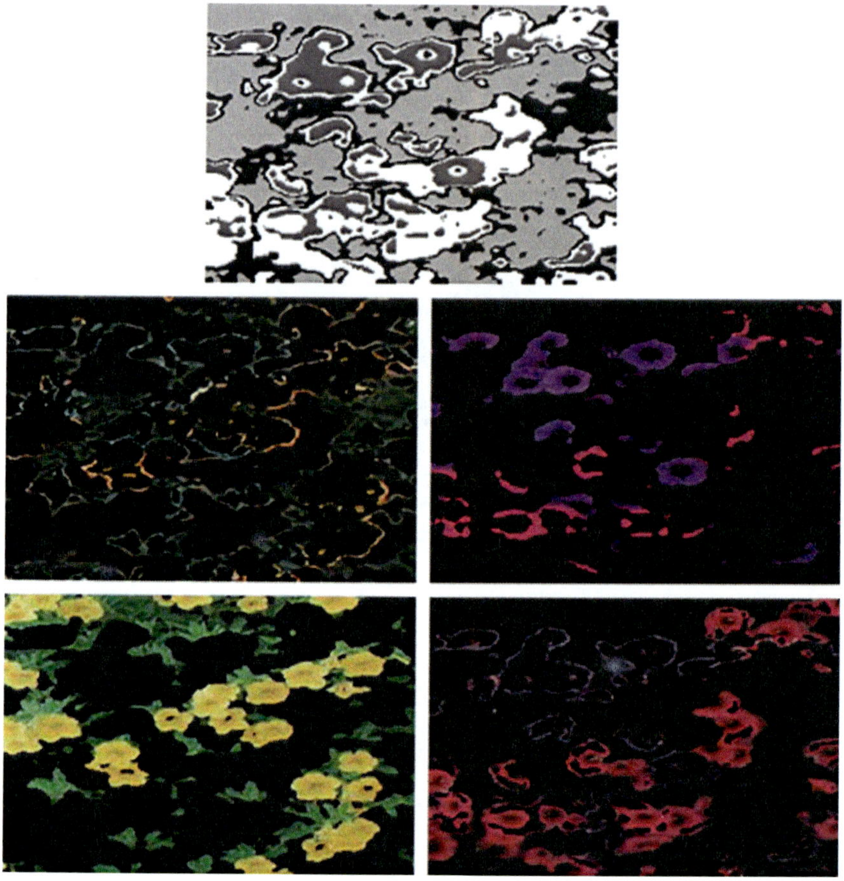

Fig. 14 Four clusters for cosine distance

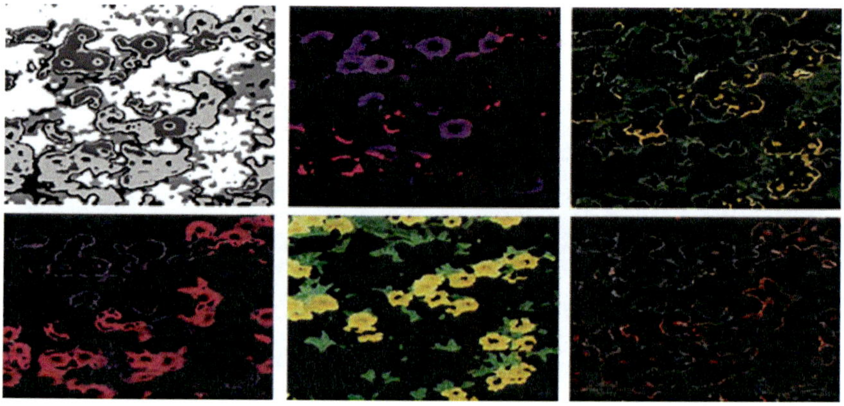

Fig. 15 Five clusters for cosine distance

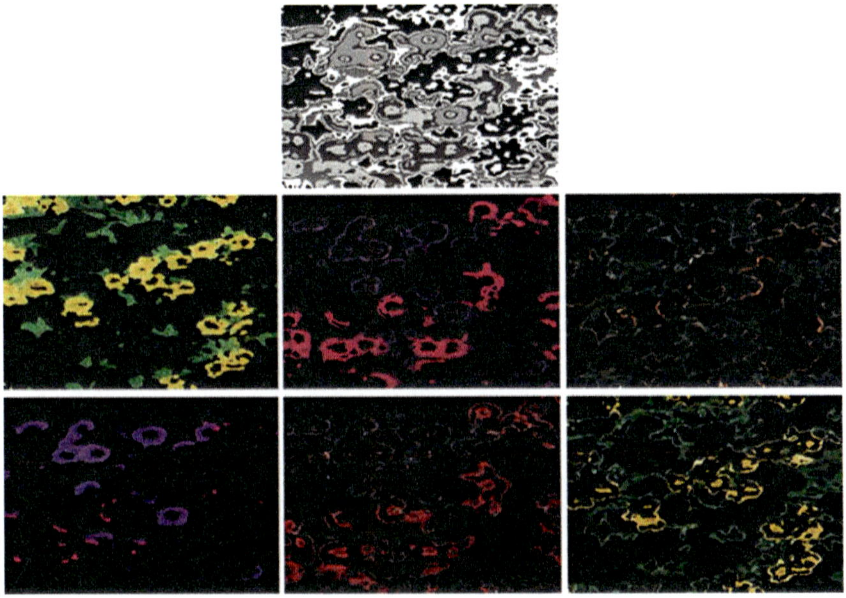

Fig. 16 Six clusters for cosine distance

Table 1 Outcome of city block distance measure

S. No	No. of clusters	Cluster centers		Sum of distance	Computational cost (in sec)
1	3	121 105 165	131 149 127	172,345 329,861 271,630	1.2348
2	4	109 127 149 177	148 199 112 142	397,803 136,136 242,554 167,744	1.7803
3	5	154 124 176 103 127	101 136 142 152 200	116,887 169,451 175,701 183,406 133,379	1.9543
4	6	102 160 122 154 128 183	153 140 137 100 200 142	165,355 79,430 159,168 106,949 129,792 85,382	2.2417

Table 2 Outcome of Euclidean distance measure

S. No.	No. of clusters	Cluster centers		Sum of distance	Computational cost (sec)
1	3	110.2319 163.4002 126.6309	147.3384 125.9793 196.6840	6,309,152.53 1,354,019.20 1,867,116.60	1.1501
2	4	126.5014 109.8447 174.0056 149.0681	196.8907 147.9837 141.4920 108.0707	1,757,504.92 5,848,318.32 2,734,709.51 3,262,663.04	1.8279
3	5	175.4438 103.6043 126.9541 154.8519 127.9244	141.2271 153.6434 197.3228 99.5023 134.1122	2,311,094.34 2,397,468.54 1,586,986.60 1,146,911.21 1,990,108.16	6.0204
4	6	127.1253 101.1720 176.8560 115.3358 139.2323 156.9625	197.4496 158.5439 141.6468 142.2818 128.0268 96.39465	1,534,503.30 1,150,701.51 1,966,011.69 1,140,228.84 1,291,064.44 690,262.24	2.9926

Table 3 Outcome of cosine distance measure

S. No	No. of clusters	Cluster centers		Sum of distance	Computational cost (sec)
1	3	0.80752 0.68679 0.54879	0.58565 0.72471 0.83464	35.6466057734419 19.5957271854142 25.6727289310815	4.3236
2	4	0.64244 0.84602 0.53691 0.76222	0.76511 0.53048 0.84279 0.64597	11.2070817507634 8.69869069206367 13.7607008516890 11.3046431111871	2.3868
3	5	0.69790 0.85683 0.60753 0.77946 0.52760	0.71525 0.51352 0.79370 0.62544 0.84888	5.13388477811076 4.90849866636880 5.11608079694687 7.09236884030025 8.15141894885545	4.0672
4	6	0.52054 0.79906 0.66110 0.86572 0.74285 0.58623	0.85334 0.60057 0.74968 0.49903 0.66884 0.80974	5.50136318801420 3.15645187279221 3.42919557063826 2.64203963064101 3.25697462510619 3.41005014890094	10.7817

References

1. Kalist V, Ganesan P, Sathish BS, Jenitha JMM (2015) Possiblistic-fuzzy *C*-means clustering approach for the segmentation of satellite images in HSL color space. Procedia Comput Sci 57:49–56
2. Sathish BS, Ganesan P, Shaik KB (2015) Color image segmentation based on genetic algorithm and histogram threshold. Int J Appl Eng Res 10(6):5205–5209
3. Sajiv G, Ganesan P (2016) Comparative study of possiblistic fuzzy *C*-means clustering based image segmentation in RGB and CIELuv color space. Int J Pharm Technol 8(1):10899–10909
4. Luo P, Xiong H, Zhan G, Wu J, Shi Z (2009) Information-theoretic distance measures for clustering validation: generalization and normalization. IEEE Trans Knowl Data Eng 21(9):1249–1262
5. Liu C, Liu J, Peng D, Wu C (2018) A general multiobjective clustering approach based on multiple distance measures. IEEE Access 6:41706–41719
6. Androutsos D, Plataniotiss KN, Venetsanopoulos AN (1998) Distance measures for color image retrieval. In: Proceedings on image processing, ICIP 98, vol l2, pp 770–774
7. Bouhmala N (2016) How good is the Euclidean distance metric for the clustering problem. In: 5th IIAI international congress on advanced applied informatics (IIAI-AAI), Kumamoto, pp 312–315
8. Malkauthekar MD (2013) Analysis of Euclidean distance and Manhattan distance measure in face recognition. In: Third international conference on computational intelligence and information technology, Mumbai, pp 503–507
9. Ganesan P, Rajini V, Kalist V, Krishna PV (2014) Comparative study of performance of distance measures in fuzzy *C* means clustering for CIELUV color images. In: International conference on control, instrumentation, communication and computational technologies (ICCICCT), Kanyakumari, pp 95–100
10. Su MC, Chou CH (2001) A modified version of the *K*-means algorithm with a distance based on cluster symmetry. IEEE Trans Pattern Anal Mach Intell 23(6):674–680
11. Modh Jigar S, Brijesh Shah, Shah Satish K (2012) A new *K*-mean color image segmentation with cosine distance for satellite images. Int J Eng Adv Technol 1(5):72–76
12. Ganesan P, Sajiv G, Leo LM (2017) CIELuv color space for identification and segmentation of disease affected plant leaves using fuzzy based approach. In: Third international conference on science technology engineering & management (ICONSTEM*)*. IEEE Press, pp 889–894
13. Ganesan P, Palanivel K, Sathish BS, Kalist V, Shaik KB (2015) Performance of fuzzy based clustering algorithms for the segmentation of satellite images—a comparative study. In: IEEE seventh national conference on computing, communication and information systems (NCCCIS). Coimbatore, pp 23–27
14. Ganesan P, Sajiv G (2017) User oriented color space for satellite image segmentation using fuzzy based techniques. In: International conference on innovations in information, embedded and communication systems (ICIIECS). Coimbatore, pp 1–6
15. Shaik KB, Ganesan P, Kalist V, Sathish BS (2015) Comparative study of skin color detection and segmentation in HSV and YCbCr color space. Procedia Comput Sci 57:41–48
16. Ganesan P, Sajiv G (2015) Unsupervised clustering of satellite images in CIELAB color space using spatial information incorporated FCM clustering method. Int J Appl Eng Research 10(20):18774–18780
17. Sajiv G, Ganesan P (2016) Comparative study of possiblistic fuzzy *C*-means clustering based image segmentation in RGB and CIELuv color space. Int J Pharm Technology 8(1):10899–10909
18. Yesilbudak M (2016) Clustering analysis of multidimensional wind speed data using *k*-means approach. In: IEEE international conference on renewable energy research and applications (ICRERA), Birmingham, pp 961–965

Designing an Adaptive Question Bank and Question Paper Generation Management System

Pankaj Dwivedi, R. Tapan Shankar, B. Meghana, H. Sushaini, B. R. Sudeep, and M. R. Pooja

Abstract This paper discusses the design and implementation of automatic question paper generation and retrieval system for the engineering domain. The system uses C#.NET for user interface design. SQL server 2017 is used for database storage. Question paper generation is accomplished using SAP crystal report 2016. The system consists of four modules, namely administrator and login module, question input module, question retrieval module, and evaluation module. The question paper is generated using a dynamic algorithm with minimal redundancy. Since the inputted question items are marked for their difficulty index, the question paper may be customized as per testing requirement, i.e., basic to advanced level of difficulty. Evaluation module generates password-protected expert validated answer keys for the objective type of questions and answer cues for the subjective type of questions. The system may duly cater to the need of an instant generation of confidential different sets of question papers having same difficulty index for the purpose of competitive engineering examinations or tests.

Keywords Assessment system · Automatic evaluation · Question bank · Course outcome

P. Dwivedi (✉)
Educational Technology Unit, Central Institute of Indian Languages, Mysuru, Karnataka, India
e-mail: pankaj.linguistics@gmail.com

R. T. Shankar · B. Meghana · H. Sushaini · B. R. Sudeep · M. R. Pooja
Department of Computer Science, Vidyavardhaka College of Engineering, Mysuru, Karnataka, India
e-mail: tapan_ramesh@yahoo.com

B. Meghana
e-mail: meghanab97@gmail.com

H. Sushaini
e-mail: sushainih1997@gmail.com

B. R. Sudeep
e-mail: sudeepxpress@gmail.com

M. R. Pooja
e-mail: pooja.mr@vvce.ac.in

N. N. Chiplunkar and T. Fukao (eds.), *Advances in Artificial Intelligence and Data Engineering*, Advances in Intelligent Systems and Computing 1133,
https://doi.org/10.1007/978-981-15-3514-7_72

1 Introduction

Tests and examinations are part of the competence and performance evaluation of learners in both formal and informal institutional setups. They are used to determine an individual's potential, suitability, and also as means to assess learning outcome. While the universities' paper setting guidelines usually takes only format of question paper into account; quality of the question paper is rarely focused upon due to time and human constraints. It is rather challenging for the teachers to cover all the aspects of the course objective and avoid duplication of questions [1]. While there is yet a general consensus about the essential role and intervention of a human-teacher in the teaching-learning process, revolution in the technology has made it possible for computers to assess and evaluate the learners' performance in many areas of studies. With universities and other educational institutions expanding their reach to students globally and largely through online and distance mode courses, there has emerged a parallel requirement for their objective, automated, fast and secure assessment process absent of much human effort and monetary burden.

A few efforts have been made for the development of questioning and evaluation systems. Products like "WebAssign" [2, 3] and "WebCT" [2, 4] provide immediate feedback on every attempted question. Physics homework system (The Andes Physics Tutor) is also an example of similar work. These question items either strictly follow categories as given in Bloom's taxonomy [5, 6] or as these categories may deem fit into in a particular area of subject into account. MILES contains a vast database of question items for Hindi, Tamil, Kannada, Telugu, and Urdu languages for assessing language proficiency among second language learners of these languages [7]. A paper generation systems using J2EE tools based on B/S architecture is proposed by Cen et al. [8]. Summary Street, a computerized tutor based on latent semantic analysis (LSA), provides support to students on content writing, summarizing, etc.; it also provides continuous feedback to students [9]. Some systems focus more on assessing performance rather than providing a mechanism for automation [10]. Research indicates that students using this technology may score significantly higher. LSA-based system, APEX, assesses the writing skills [11]. There are some other question paper generators based on RDBMS or linear searching method, and therefore question selected from database are often inefficient unless all of these questions are marked for their difficulty index and duly validated by experts.

Further, in this paper, the overall architecture of our system giving detailed working of its four modules, namely administrator verification and login module, question input module, question retrieval module, and evaluation module have been discussed. Also, advantages and future prospects have been explained from the perspective of users and implementation.

2 System Architecture

The system being discussed here broadly has four modules, namely: administrator and login module, question input module, question retrieval module, and evaluation module.

2.1 Administrator Verification and Login Module

This module primarily verifies the authenticity of the username/password of the item writer, data inputter (question inputter), and question retriever. Every item writer is assigned with a unique ID denoted as iw_id in the database. Academic and professional credentials of the item writers such as name, qualification, areas of expertise, designation, professional affiliation, and contact details are also stored as metadata in the database for the purpose of further verification and cross-reference in the future. The details of the question inputters are also stored.

The system provides access to three different users, viz. an administrator, a question inputter, and a question retriever. Item writers provide the questions to be stored in the database to the question inputters who, in turn, feed them into the database. As evident from Figs. 1 and 2, unique IDs are created and assigned to both for item writers and/or question inputters through a sign-up process. Created IDs are verified and approved by the administrator on account of their professional credentials. After the approval of IDs, question retriever and/or question inputter can log into the system. Question inputters have access to the question input module, whereas the question

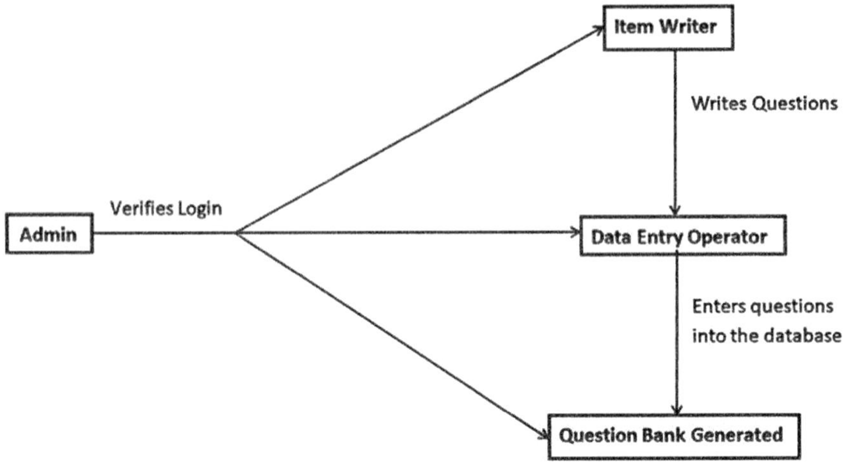

Fig. 1 Administrator verification flowchart

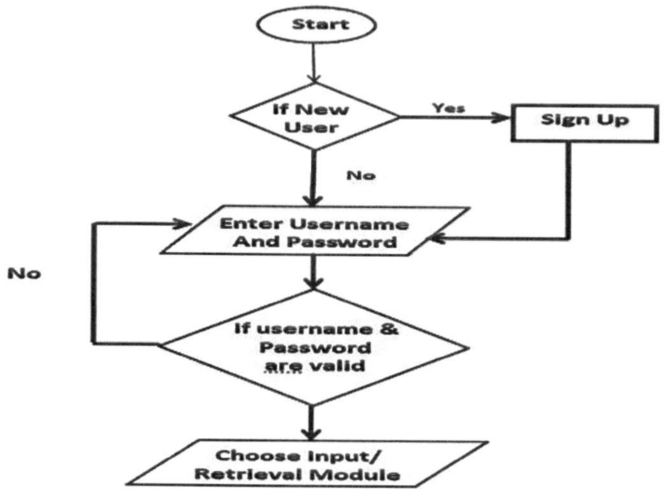

Fig. 2 User login flowchart

retrievers have access only to the question retrieval module. Unlike question inputters and question retrievers, the administrator has access to both the question input and question retrieval modules. Pattern matching and OTP-based authentication are used for the purpose of secure accesses.

2.2 *Question Input Module*

This module is the part of the system where questions are inserted into the database by the question inputters. Question items are prepared by established professionals and academicians in their respective areas. Item writers type the questions on a MS Word doc (user-friendly) format, whereas question inputters have to feed the question items into database through a uniquely designed application form. It classifies subjects/areas/content based on the seven parameters which need to be considered before entering question items into the database. These parameters are such as an instruction to the examinee, *question/item stem (with answer key/cue), scoring procedure, course outcome, item type/format used, and difficulty level.* These parameters indirectly relate to six cognitive levels as given in Bloom's taxonomy, namely *recall, understand, apply, analyze, evaluate, and create* [5]. Each subject is assigned with a unique identifier to make the identification easier. These identifiers help generation and retrieval process easier. Also, question count may be done from time to time to see the number of items under a specific subject/topic. Therefore, in case, a piece of information changes with time under a particular subject/topic, new question items may be added, deleted or modified to keep the question bank most suited for contemporary needs.

1. **Subject Identifier**
2. **Instruction to the examinee :**
3. **Question/Item Stem**
   (with answer key/cue) :
4. **Scoring Procedure : Analytical/Enumerative/Global quality**
5. **Course Outcomes :**
6. **Item Type/Format used : CA/MC/MF/RA/Mg.**
   SQ/Cn/SA/LA//PS or conversion
7. **Difficulty Level : Basic/Intermediate/Advanced**

Fig. 3 Question item format

Figure 3 represents a MS Word template/doc format of the question input format presented to the item writers. The parameter 1 indicates identifier (a specific numerical code) assigned to a particular subject/area. Question item writers usually write the name of the subject/topic under which a given question item falls within. These identifier codes are assigned by the data inputter from subject identifier tree diagram (SITD). In cases, when one question may fall within more than one subject scopes, subject identifier is assigned to the most-suited category after due validation from the experts other than the item writers. The parameter 2 provides space for the specific instructions which are to be conveyed to the examinee before attempting the question. The parameter 3 is question item stem (QIS) in which actual question item is written with its respective the answer key/cue. Question items and key/cue are fed into separate columns in database through question input form (QIF) a by data inputter.

The parameter 4 provides the question inputter with an option to specify the scoring procedure as it is defined by the item writer. In principle, there are three types of scoring procedures—analytical, enumerative, and global quality. But presently most of the questions inputted in this system follow analytical procedure for the purpose of scoring. The parameter 5 measures the course outcome. The course outcome is nothing but the main objective behind a asking a particular question. In this system, course outcome is defined to be of three types—analytical, understanding, and educational, following Bloom's taxonomy [5]. The parameter 6 is item type, i.e., objective or subjective type, and also format used for the particular inputted question. In objective category, there are five types of questions-constant alternative, multiple choice, multiple facet, rearrangement, and matching, while the subjective type of question is of five types, namely—simple question, completion, short answer, long answer, problem solving, and transformation. The parameter 7 measures the difficulty of a given question item. This system makes use of three levels of difficulty, namely basic (low, 0), intermediate (medium, 1), and advanced (high, 2) for the purpose of understanding. However, there is a scope of further refinement of difficulty layers making

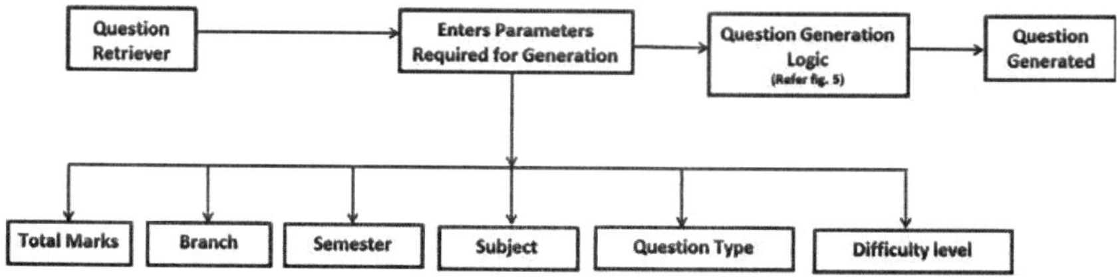

Fig. 4 Question retrieval diagram

it more suited for the different levels of education such as primary, secondary, higher secondary, graduate, and postgraduate.

2.3 *Question Retrieval Module*

This module takes in different inputs for generating the question paper. This system generates the question paper with the help of pre-request form. In the pre-request form, the requester has to fill the details of total marks, question type, branch, semester, subject/topics, difficulty, etc. These questions may further be divided into further sections or subsections as per the requester's need in line with the parameters given in Fig. 3 and as discussed in Sect. 2.2. Since the inputted question items are marked for each parameter, a question paper may be generated to cater to different types and levels of testing requirements.

Figure 4 shows the diagram of the retrieval module, while Fig. 5 shows the functional flowchart of the module. Based on the requester's requirement (given in request form), the retriever selects the various parameters given in Fig. 3, and manually entered them into question retrieval window. Probability of redundant retrieval of question items is very low. To achieve normalization, separate tables have been made to store the details of individual subjects and branches with corresponding unique identifiers. Similarly, question/answer stems and answer cues consistent with certain predefined parameters are stored in separate tables. Built-in SQL server and ASP.NET functions checksum and random, respectively, are used for the purpose of retrieval. The ID associated with each question is passed as a parameter to these functions which help retrieve a new question. All parameters are independent of one another, i.e., the selection of a parameter does not affect or limit the scope of another parameter by any manner.

2.4 *Evaluation Module*

This module helps paper evaluators to make the process of paper evaluation easier and faster. As mentioned in the design of the input module, answer keys for the objective

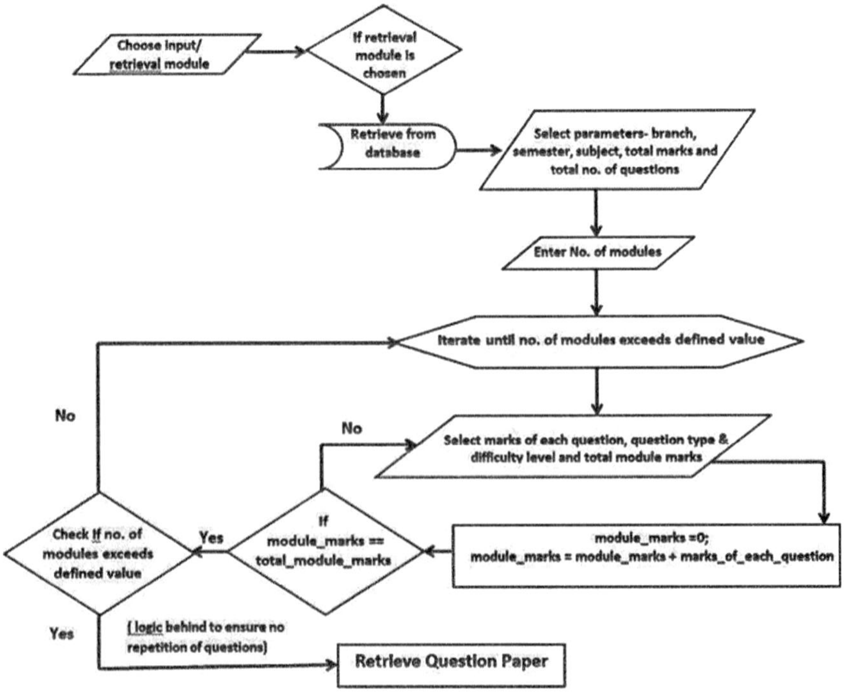

Fig. 5 Question retrieval flowchart

type of question items and answer cues for the subjective type of question items are also fed into the database along with their respective question stem. While answer keys are fixed to the objective type of question items, there can be various ways to provide an answer to a subjective type of question item. Therefore, the answer cues are written in such a way that it provides an analytical breakdown of the whole answer with marks assigned to each part of the answer. This breakdown of answers serves as a standard for paper evaluators to provide marks without any discrepancy. The answer cues provided in the evaluation module are consistent with the course outcomes of the course whose question paper is being generated. Automated evaluation is achieved using stored answer keys using simple conditional constructs. Manual evaluation of the subjective type of question items is done following stored answer cues. Since subjective question items do not have fixed answering patterns, answer cues with their analytical breakdown are fed into the system for system to recognize specific patterns in writing content. We can come to the conclusion that the evaluation module removes the prospect of bias in paper evaluation to a major extent.

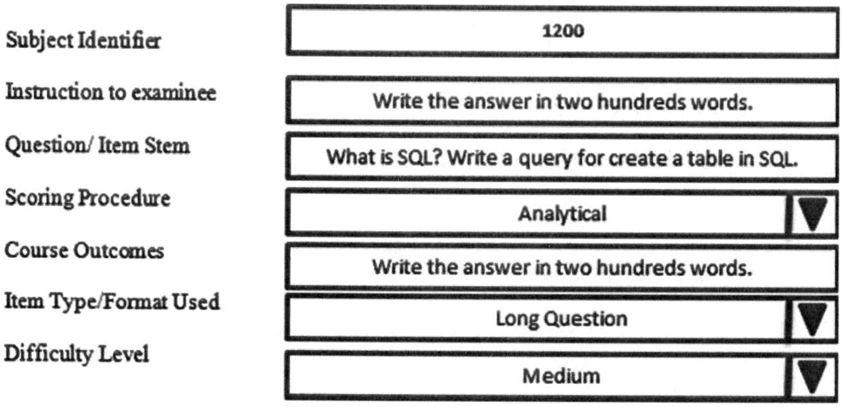

Fig. 6 Question retrieval window

3 Implementation of the System

For designing the graphical user interface (GUI), C#.NET is used. First screen comes with the options of administrator, item writer, or inputter login. Based on the type of login, second screen auto appears. For example, an item writer gets a screen as given in Fig. 3 (a MS Word template), however, for data inputter, following screen appears for entering the question items based on the seven parameters.

Once one question item is properly inputted and saved, the new question item may be selected from the item writer question format. For storing the question bank, SQL 2017 server is used, and SAP crystal report 2016 is used for generation of the question paper. The administrator has a privilege login and therefore she/he can generate the question paper and respective answer key/cues as requested by the requester (Fig. 6).

4 Advantages of the System

The main advantage of this system is that it reduces human effort as well as the needs of different resources that are used to generate a question paper. The proposed system greatly reduces the possibility of postponement or cancelation of examination in case a question paper gets leaked or is tampered with by someone. Since, generated question paper has minimal redundancy, different sets of questions paper for same subject, same difficulty index, same format, etc., may be prepared simultaneously, and therefore making it an ideal system for any competitive or entrance examination. The system is dynamic in nature, i.e., new question items can be added or modified depending on the requirements, and therefore as the passage of time, systems have the capability to become more robust. Overall, the system may duly cater to the need of instant generation of confidential different sets of questions.

5 Future Enhancements

Apart from fulfilling our basic objectives, this system may be improved in terms of the reachability and efficiency. For example, it may be enhanced to become a Web application as opposed to desktop application. The system may be customized to work in any area of engineering, science, humanities, etc.

References

1. Nalawade G, Ramesh, R (2016) Automatic generation of question paper from user entered specifications using a semantically tagged question repository. In: 18th IEEE eighth international conference on technology for education (T4E). 2–4 Dec 2016. https://doi.org/10.1109/t4e.2016.038
2. Zahorian SA, Lakdawala VK, Gonzalez OR, Starsman S, Leathrum Jr JF (2001) Question model for intelligent questioning systems in engineering education. In: 31st ASEE/IEEE frontiers in education conference. 10–13 Oct 2001. https://doi.org/10.1109/fie.2001.963871
3. WebAssign Homepage, http://www.webassign.com. Last accessed 26 Feb 2019
4. WebCT Homepage, http://www.webct.com. Last accessed 21 Feb 2019
5. Dalton E (2018) The new bloom's taxonomy, objectives, and assessments (December 3 2003). Retrieved from http://gaeacoop.org/dalton/publications/new_bloom.pdf on 2 Feb 2018
6. Forehand M (2011) Bloom's taxonomy-emerging perspectives on learning, teaching and technology. The University of Georgia
7. Dwivedi P, Rajgopal K, Srinivasan RK (2016) Multipurpose indian language evaluation system and question bank. India International Science Festival (IISF)—Young Scientists' Conclave (YSC) 8–11 Dec 2016
8. Cen G, Dong Y, Gao W et al. (2010) A implementation of an automatic examination paper generation system. Math Comput Model 52: 1339–1342. https://doi.org/10.1016/j.mcm.2009.11.010
9. Franzke, M, Kintsch, E, Caccamise, D, et al. (2005) Summary Street®: Computer Support for Comprehension and Writing. J Educ Comput Res 33(1):53–80. https://doi.org/10.2190/DH8F-QJWM-J457-FQVB
10. Anderson, J (2005) Mechanically inclined: building grammar, usage, and style into writer's workshop. Stenhouse Publishers
11. Lemaire B, Dessus P (2001) A system to assess the semantic content of student essays. J Educ Comput Res 24:305–320. https://doi.org/10.2190/G649-0R9C-C021-P6X3

Securing Media Information Using Hybrid Transposition Using Fisher Yates Algorithm and RSA Public Key Algorithm Using Pell's Cubic Equation

K. R. Raghunandan, Shirin Nivas Nireshwalya, Sharan Sudhir, M. Shreyank Bhat, and H. M. Tanvi

Abstract Information is the source of investment as well as income. Protecting information has become one of the primary objectives of security algorithms. In the field of multimedia, there is scope for extracting sensitive data without any loss of information which implies that there is no trace of an intrusion. Multimedia data must be confidential, and its integrity and identity must be preserved as it encounters authentication, data encryption, data decryption, digital watermarking and other similar activities. Providing multimedia security ensures content-based security. This paper introduces an enhanced algorithm for the transmission of images between the sender and recipient by addressing the most relevant security aspects of image transformation as well as transmission.

Keywords Multimedia security · Image transformation · Public key cryptography · Encryption · Decryption

1 Introduction

Multimedia technology has a significant impact on privacy, making data protection important, now more than ever. In this domain, the cryptographic technique has a

K. R. Raghunandan (✉) · S. N. Nireshwalya · S. Sudhir · M. S. Bhat · H. M. Tanvi
Department of Computer Science and Engineering, N. M. A. M. Institute of Technology, Affiliated to Visvesvaraya Technological University, Nitte, India
e-mail: raghunandan@nitte.edu.in

S. N. Nireshwalya
e-mail: shirinnivas@hotmail.com

S. Sudhir
e-mail: sharan.sudhir9@gmail.com

M. S. Bhat
e-mail: bhatshreyank@gmail.com

H. M. Tanvi
e-mail: tanvihm97@yahoo.com

N. N. Chiplunkar and T. Fukao (eds.), *Advances in Artificial Intelligence and Data Engineering*, Advances in Intelligent Systems and Computing 1133,
https://doi.org/10.1007/978-981-15-3514-7_73

prime role to communicate data between the sender and receiver. This paper aims at implementing an algorithm for a safe and secure transmission of digital media by discussing the RSA cryptosystem and pixel manipulation in an image. Our proposed algorithm is developed to be tested against the standard RSA implementation so as to provide high level of security in the vast domain of multimedia [1].

1.1 Multimedia

Early representations of data used printed materials or digital text. Multimedia is a form of combined digital data representations that include text, audio, images, animations, video or interactive data. Multimedia is more sophisticated than simple text and images. It is distinctively less expensive to produce, and it provides a window for user interaction with the data. It has proved to be a vital concept ever since the evolution of the Web from mainly textual layout to a graphical layout [1].

In the present world, storage of data and transmission has taken a back seat and information security is given more preference. Images are the most common type of data stored and shared widely over the Internet. They are vulnerable since most of the data stored is in unencrypted format and data transmissions between the sender and receiver usually have a weak security strategy. It is important to secure these images from intruders trying to access them illegally. An algorithm that focuses on encryption of images aims at minimizing the unlawful extraction or distortion of images, thus providing absolute security for the original image that is intended to be sent [2].

The authors, G. A. Sathish Kumar, K et al. proposed an integrated pixel scrambling with the addition of diffusion technique [IISPD] as a new approach for image encryption. The algorithm focuses on utilizing the full disorderly property of engineered map and reducing the time difficulty. The permuting address for rows and columns is calculated by the algorithm by taking the pixels that are adjacent to each other in the original image and bit xor'ing them. This is done without any information about the probability density function, thus the complexity of the algorithm is low. Additional diffusion is performed after scrambling [3].

Gururaj Maddodi et al. have demonstrated the use of neural networks and a pseudo-random sequence generator based on chaos and chaotic encryption that follows DNA rules for the storage and transfer of images. The authors have used genetic encryption for all experimental results and security analyses [4].

Authors, Paulo S. L. M. Barreto et al. introduced contextual information. In this the determination of fingerprint H_t, the hash function H is fed. Thus, a single dependency per block is required for exact containment of image modification. For detecting the block-wise rotation, block index t is inserted. If a block X'_t is altered, then hash block chaining version 1 (HBC1) will report that $X'_{(t+1) \bmod n}$ is invalid (besides X'_t itself) [5].

Lihua Tian et al. represent a new colour image segmentation algorithm called MSHC based on mean shift and hierarchical clustering algorithm. Segmented regions are formed by pre-processing the input image, which retain the advisable disconnectedness attributes of the image. Instead of the number of image pixels, the number of segmented zones is considered as the input data scale of the HC algorithm. The proximity matrix is calculated by determining the closeness between each cluster. Then, the final segmentation results are acquired by applying the ward algorithm. The MSHC algorithm is then employed on colour image and medical image segmentation [6].

Jiao J., et al.derived an original image grouping method that is based on scattered image pixel representation style. By assuming that every occurrence can be restored by the scattered linear grouping of other occurrences, the model distinguishes the graph contiguity composition and graph weights by scattered linear coefficients calculated by explaining l^{-1} minimization. In order to find the structure of the cluster, the authors make use of the spectral grouping algorithm having coefficients as graph weight matrix [7].

HyderYahya proposes a new method to encrypt and decrypt an image of a fingerprint by using a spiral method for transposition of a pixel. The first step is to use a transposition cipher to encrypt the image and then insert the same inside another image by using the LSB technique. This transposed image is sent without any revelation of sensitive data; the original image is obtained by applying the inverse LSB process which is followed by reverse encryption [8].

1.2 Cryptosystem

The most vital aspect of computer networks is the concept of data communication. This brings with it, the issues dealing with the protection of data. Data must be protected from attacks at all costs. A system that assists in data communication is called a "Cryptosystem." Cryptosystems deal with the complexities of data encryption and data decryption. While encryption works on plain text to produce cipher text, decryption works on the produced cipher text to get back the original data. This is done based on keys—public keys and private keys [9].

If both keys used are the same, then it is termed as "Symmetric Cryptography." Such systems are relatively easy to break into. To overcome this, public key cryptography has been implemented; this makes use of related key pairs. One of the pairs is called public key and is used for the encryption process, and other is called private key and is used for the decryption process. The former is disclosed while the latter is kept hidden. This ensures secure communication as only the receiver knowing private key is able to decrypt the cipher text to obtain the original data.

The main disadvantage of the symmetric key encryption is that the all the steps included should share the key used for encryption of data before decryption process. Public key cryptography is mainly used as a process to ensure data confidentiality, originality of data exchange and to store data. Most public key cryptosystems are variations of the RSA algorithm.

1.2.1 RSA Cryptosystem and its Properties

It is the first public key cryptosystem, and it is still the most secure algorithm. It has properties that deal with the complexity of factorization as well as the computation of modular powers and inverting exponentiation [9]. The RSA system uses easy large prime numbers. The whole working of RSA can be summarized as follows:

1. Two prime numbers, p and q, relatively large are selected such that they are not equal.
2. $n = p * q$ is computed, and $Z(n) = \text{lcm}\ (p - 1,\ q - 1)$ is found.
3. Integer d which is relatively prime to $Z(n)$ is picked such that $d \leq Z(n)$
4. e is computed as the multiplicative inverse of d mod $Z(n)$, keeping in mind that $e * d = 1$ mod $Z(n)$. Usually, Euclidean algorithm is extended to carry out this procedure.
5. Public key $= (e, n)$ is published.
6. Private key $= (d, n)$ is kept secret.

Balram Swamia, Ravindar Singh and Sanjay Choudhary introduce DMRJT which uses Jordan Totient function. This is similar to RSA apart from the fact that it uses two pairs of keys. One pair generates encrypted data that can only be decrypted by the corresponding pair. During encryption, the sender obtains the recipients public key. Cipher text is calculated and sent to recipient. During decryption, the recipient uses sender's private key to extract the original text from message representative. Security-wise DMRJT offers double the security that RSA has [10].

The authors WenxueTan et al., proposed a method to generate secure keys, using the concept of key security coefficient. The authors perform an extensive analysis of chosen cipher text attack with the aim of perfecting the RSA crypto scheme [11].

NaQi et al. have given a brief introduction towards the generation of keys and provided with some knowledge related to the encryption and decryption algorithm. The user input is Chinese characters which are turned into URL code, and this URL code is in turn converted to ASCSII code to encrypt–decrypt. He has proposed that to enhance RSA algorithm and to make it more secure, a large value for p and q should be chosen [12].

Considering the methods wherein data has been secured using different methods, our objective is to come up with a RSA variant providing more security to the image over the medium it is travelling, hence leading to less probability of the image getting hacked or identified by hackers.

1.3 Motivation

Our chief motivations are described as follows:

i. An image is represented as information and sent from sender to receiver. While transmission, the intruder tries to steal this information by cracking private key.

The main objective is to secure the transmission and to store or pass data in an efficient way.

ii. The information here sent is in the form of an image, and standard RSA algorithm has already been applied on image [9]. Our motive is to enhance the security feature of the image, i.e. as standard RSA has been cracked by various methods, we have come up with the concept of an RSA variant wherein the encrypted image is formed using a fake modulus that has no correlation with the modulus equation, hence improving the security of the image.

2 Mathematical Preliminaries

2.1 *Algebraic Structure*

Algebraic structure is a set of binary operations on a non-empty set S. Binary operation '$*$' is considered as a closure operation on S if it applies to all the elements of the non-empty set [13].

Consider S as a non-empty set.

Then, $\mathrm{S} \times \mathrm{S} = \{(x, y) : x \in S,\ \ y \in S)\}$.

If $f : S \times S \rightarrow S$, then f is called a binary operation on a set S.

Example On a set of natural numbers N, multiplication is considered to be a binary operation. The product of two natural numbers is likewise a natural number, i.e.

$$a + b \in N.$$

2.2 *Groups*

An algebraic structure $S = (P \odot 1)$ is said to be a group, where $\odot$ is a binary operation or group operation. The identity element of the group is 1. It is also important that the conditions mentioned below hold true [13]:

i. Associative property: $(p \odot q) \odot r = p \odot (q \odot r)$, $\odot$ is an associative operation.
ii. Identity property: $p \odot 1 = 1 \odot p = p$.
iii. Inverse property: $p \odot p^{-1} = p^{-1} \odot p = 1$.
iv. Commutative property: $p \odot q = q \odot p$, $\odot$ is a commutative operation.

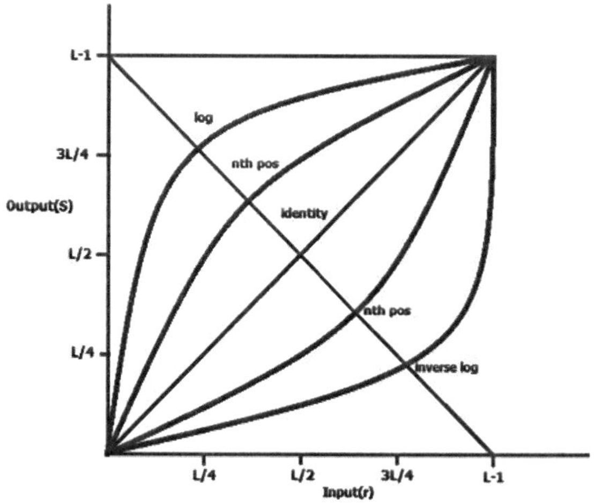

Fig. 1 Graph depicting the negation of the image pixels

2.3 *Negative Images*

Negative images are pictures where the light areas of the subject display as dark and the dark areas of the subject appear light. Consider an input image whose intensity lies in the range $(0,\ L-1)$. The negative version of the image is computed using negative transformation, which is given by,

$$A = L-1-B$$

where B is input pixel and A is output pixel. While computing image negation, each pixel is deducted from $(L-1)$. To calculate image negation of greyscale images, pixel values of original values are deducted from 255. Figure 1 represents a graph of input pixels versus output pixels [14].

2.4 *Transformation of Pixels*

Fisher-Yates Shuffling Algorithm As an element is randomly picked in the first iteration, the probability that the ith element (including the last element) goes to last position is $\frac{1}{k}$.

The possibility that ith element goes to the second last position can be proved to be $\frac{1}{k}$ by splitting it into two cases.

Scenario 1 $i = k - 1$ *(index of last element)*:

The chances of the element at the last position to be swapped at the second last position = (probability that element at last position does not remain at its original position) x (probability that the last element is swapped by selecting the index selected previously once again). The probability equation is denoted below as Eq. (1) [13].

$$\text{probability} = \left(\frac{k-1}{k}\right) x \left(\frac{1}{k-1}\right) = \left(\frac{1}{k}\right) \quad (1)$$

Scenario 2 $0 < i < n - 1$ *(index of non-last)*:

The probability of element at position '*i*' going to the second position = (probability that element at position '*i*' is not picked in previous iteration) x (probability that element at position '*i*' is picked in this iteration). The probability equation is denoted below as Eq. (2) [13].

$$\text{probability} = \left(\frac{k-1}{k}\right) x \left(\frac{1}{k-1}\right) = \left(\frac{1}{k}\right) \quad (2)$$

2.5 *The Mathematics behind RSA Cryptosystem*

Two different functions define the security of RSA, namely encryption and key generation. Prime numbers are significant in RSA because when the product of the four chosen numbers is multiplied, the resultant number can only be reduced by its primes. Thus, if a technique for factorizing is developed, the RSA cryptosystem can be easily hacked [9].

The Euler Totient Function Given a non-negative integer $n \geq 2$, the Euler Totient function '$\phi(n)$' is defined by

$$\phi(n) = Z_n = \{a,\ 0 < a < n,\ \gcd(a; n) = 1 \quad (3)$$

where set Z_n is termed as a group of units modulo n, as denoted above in Eq. (3). [9].

From Eq. (3), it is obvious that $\phi(p) = p - 1$ whenever p is prime.

Greatest Common Divisor (GCD) Assume $i,\ j \in Z$. Non-negative integer m is the GCD of i and j if:

(a) $m|i$ and $m|j$,
(b) If k is a non-negative integer that satisfies $k|i$ and $k|j$, then $k|m$.

The GCD of i and j is represented as $\gcd(i, j)$ [9].

The Fundamental Theorem of Arithmetic Given a positive integer $N \geq 2$, the prime factorization of n is written as

$$N = p_1^{a1} p_2^{a2} \ldots p_k^{ak} = \prod_{i=1}^{k} p_i^{ai} \tag{4}$$

where $p_1, p_2, \ldots, p_k$ are the prime numbers, as denoted above in Eq. (4) [9].

2.5.1 Definition of RSA

The private key 'k_1' used in RSA algorithm uses two big prime numbers that are of the same length, p and q and decrypting exponent b such that

$$\gcd(b, \phi(pq)) = \gcd(b, (p-1)(q-1)) = 1.$$

The key 'k_2' also called as the public key is constructed using $n = p * q$ and the encrypting exponent a such that:

$$ab \equiv 1 \bmod \phi(n).$$

Note that b does have an inverse modulo $\phi(n)$. The function used for encryption is as denoted below in Eq. (5):

$$e_{k_1}(w) = \left(w^a \bmod n\right) \tag{5}$$

And the function used for decryption is as denoted below in Eq. (6):

$$e_{k_2}(c) = \left(c^b \bmod n\right) \tag{6}$$

During encryption, a message block must be coded as an integer that may vary between the ranges $0 \leq w \leq n-1$. The special case of the Chinese remainder theorem mentioned below plays a significant role:

Lemma $x \equiv y \bmod n$ *if and only if both* $x \equiv y \bmod p$ and $x \equiv y \bmod q$.

Now, we follow through the RSA steps mentioned in the module 1.2.1.

In decryption, if $\gcd(w, n) = 1$, then making use of Euler's theorem for a number l, we get Eq. (7) as denoted below:

$$c^b \equiv \left(w^a\right)^b = w^{ab} = w^{1+l\varphi(n)} = \mathrm{w}\left(w^{\varphi(n)}\right)^l \equiv w \cdot 1 = w \mod n \tag{7}$$

Next, if $\gcd(w, n) \neq 1$, we can have three scenarios:

- $w = 0$, since $c^b \equiv (w^a)^b = 0^b = 0 \mod n$
- $p|wbutw \neq 0$.

Now $w = pt$ where $\gcd(q, t) = 1$.
It is obvious that $c^b \equiv w^{ab} \equiv w \mod p$.
Fermat's little theorem for a number l, we get Eq. (8)

$$w^{ab} = w^{1+l\varphi(n)} = w\left(w^{\varphi(n)}\right)^l = w\left(w^{(p-1)(q-1)}\right)^l = w\left(w^{q-1}\right)^{(p-1)} = w \cdot 1 = w \mod q \quad (8)$$

By the Lemma: $c^b \equiv w^{ab} \equiv w \mod n$.

- $q|w$ but $w \neq 0$

This can be handled just like the previous scenario.

2.6 Cubic Analogue of Pell's Equation

The equation is given by

$$a^3 + kb^3 + k^2c^3 - 3kabc = 1 \quad (9)$$

Here, we derive cubic analogue of Pell's Eq. (9) and apply it on multinomial variants analogously to the definition of the multinomial case [15].

Consider k as non-perfect cube where $k \in Z$. An equation x^3 - $k = 0$ is formed having roots α, $\alpha\phi$ and ϕ^2, where α is a real root and $\phi = (1/2) * \left(-1 + \surd - 3\right)$. We examine the numbers of the form $a + b\alpha + c\alpha^2$ where a, b and c are rational numbers. Grouping all real numbers of the form $a + b\alpha + c\alpha^2$ into a set $Q(\alpha)$ is given in Eq. (10)

$$Q(\alpha) = \left\{\pi = a + b\alpha + c\alpha^2 : a,\ b,\ c \in Q\right\} \quad (10)$$

Here, $Q(\alpha)$ defined as a field. Select two elements at random in a field to compute addition, subtraction, multiplication and division. The result obtained is a real number belonging to $Q(\alpha)$. In addition, norm function is defined below.

Definition The norm N_α is a function on $Q(\alpha)$ for every $\pi = a + b\alpha + c\alpha^2$ is defined as $N(\pi) = \left(a + b\alpha + c\alpha^2\right)\left(a + b\phi\alpha + c\phi^2\alpha^2\right)\left(a + b\phi^2\alpha + c\left(\alpha\phi^2\right)^2\right)$, where $a,\ b,\ c \in Q$ [16].

If norm is always rational, then $a,\ b$ and c are rational. This is explained while simplifying the norm. Here, $a,\ b$ and c are integers. Therefore, subset is given by using Eq. (11):

$$Z(\alpha) = \left\{\alpha = a + b\alpha + c\alpha^2 : a,\ b,\ c \in Z\right\},$$

Here, subset $Z(\alpha)$ is a ring, i.e. the addition (subtraction) and the multiplication of two component are also a component that belongs to $Z(\alpha)$. The equation yields solutions that are non-trivial units in $Z(\alpha)$.

Using Pell's equation, $a^3 + kb^3 + k^2c^3 - 3kabc = 1$, we fix the value of $a = 1$ and get the equation $kb^3 + k^2c^3 - 3kbc = 0$, dividing both sides with values of k that are positive, we simplify the equation and get $b^3 + kc^3 - 3bc = 0$ [15].

$$\begin{aligned} b^3 + kc^3 - 3bc = 0 \langle == \rangle kc^3 &= 3bc - b^3 \\ c \neq 0 \langle == \rangle k &= \frac{3bc - b^3}{c^3} \\ &= \frac{b\left(3c - 3b^2\right)}{c^3} \end{aligned} \tag{11}$$

Using Eq. (11), we can derive $k = \frac{3b}{c^2} - \frac{b^3}{c^3}$.

3 Proposed Work

To increase the security performance of the image encryption process, first we convert the original image to a negative version of itself. Then, the pixel locations of the negative image are shuffled. The next step takes place at the receiver's end, where the receiver gets the shuffled image and an algorithm is applied to get the de-scattered image which then yields the original image as shown in Fig. 2.

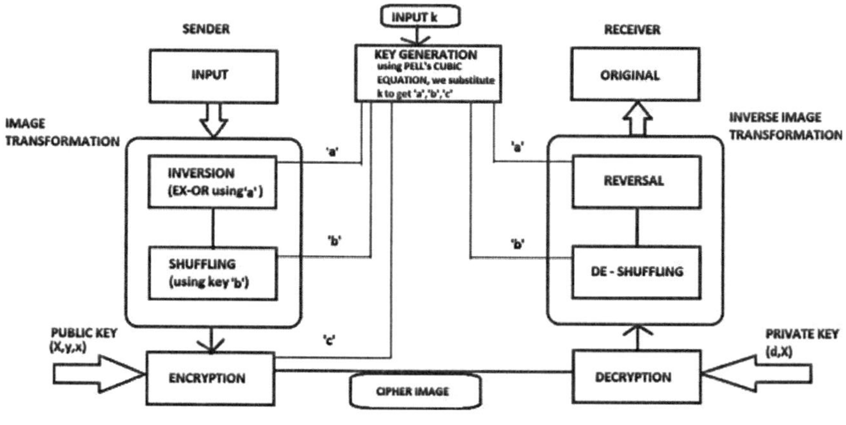

Fig. 2 Block diagram representing the processes involved

3.1 Inversion of Original Image to Negative Image

Inversion of an image is a process where light areas of an image appear dark and dark areas appear light. In a negative colour image, red areas are reversed cyan colour, green areas are reversed magenta colour, blue areas are reversed to yellow colour and vice versa. In this section, we will be converting our input image into its negative version.

Algorithm: Process of negation
Input: Image and key 'k'
Output: Inverted image
Step 1: Retrieve a list of all pixel values (R, G, B).
Step 2: Compute new (R, G, B) values (to replace the previous values) for each pixel by performing XOR operation with k.
Step 3: Assign the new (R, G, B) values to the pixel and repeat the process until all the pixels have new (R, G, B) values.
Step 4: Form the new image in negative contrast using the newly formed list of pixels as shown in Fig. 3.

Consider an image having pixel P_i, whose colour model is $(\mathrm{R}_i, \mathrm{G}_i, \mathrm{B}_i)$. i refers to the pixel in the ith location. Figure 3 represents the operation of image negation.

From the Pell's equation defined in the module 2.6, k is derived. Using the value of k, new values of $(\mathrm{R}_i, \mathrm{G}_i, \mathrm{B}_i)$ values are computed for the ith pixel by performing the XOR operation i.e.,

$$\begin{aligned} \mathrm{R}_i' &= \mathrm{R}_i \oplus k \\ \mathrm{G}_i' &= \mathrm{G}_i \oplus k \\ \mathrm{B}_i' &= \mathrm{B}_i \oplus k \end{aligned} \quad (12)$$

The above new values of (R, G, B) are stored in the pixel P_i' i.e.,

$$P_i' = \left(\mathrm{R}_i', \mathrm{G}_i', B_i'\right)$$

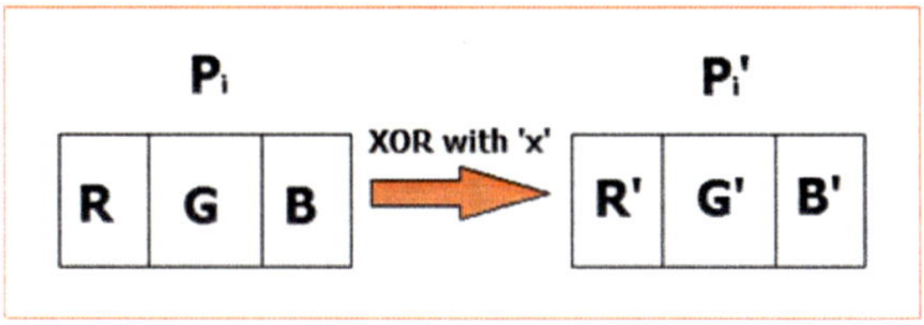

Fig. 3 Block diagram representing the negation process of a single pixel

3.2 Conversion of Negative Image to Shuffled Image

The negative image is taken into consideration, and an algorithm is applied which leads to pixel scattering. This algorithm yields a new image that is unique to the original image.

Algorithm: Process of Shuffling the pixels of the negative image
Input: Negative image and seed value '*b*'
Output: Shuffled image
Step 1: Retrieve a list of all pixel values (R, G, B).
Step 2: Calculate the total number of pixels (length).
Step 3: Create a list storing numbers from *0* to the total number of pixels (length).
Step 4: Based on seed value, shuffle the created list using the inbuilt function shuffle, which uses *Fisher-Yates* algorithm.
Step 5: Create a new list with length equal to the total number of pixels and all its elements set to *0*.
Step 6: Use elements of the shuffle list as a new location for the original pixel value.
Step 7: Using the new pixel list, form an image.

3.3 Process of Encryption and Decryption

The shuffled image is encrypted on the sender's side using a public key and forwarded to the receiver. The receiver decrypts the image using the private key given by the sender.

3.3.1 Key Generation

Using Eq. (3), the product of four prime numbers is calculated as the modulus n, i.e.,

$$n = p * q * r * s$$

Using Eq. (4), $\phi(n)$ is calculated such that,

$$\phi(n) = (p-1) * (q-1) * (r-1) * (s-1)$$

Calculate c such that

- $1 < c < \phi(n)$
- $\text{Gcd}(c, \phi(n)) = 1$

Next, x and y pairs that satisfy the equation $x^2 - cy^2 = 1$ are generated. Find d which satisfies the equation

$$(c * d) \bmod \phi(n) = 1$$

To replace n, compute X such that,

- $X > n$
- $(c * d) \bmod \phi(X) = 1$

Now, $[X, y, x]$ forms the public key, while $[d, X]$ is taken as the private key.

3.3.2 Encryption

While sending an image from sender to receiver, image is encrypted using public key pairs $[X, y, x]$. This is possible using cipher text. The cipher text C is calculated using the plain text P,

$$C = \left(P^{\frac{(x^2-1)}{y^2}} \right) \bmod X \tag{13}$$

3.3.3 Decryption

To decrypt cipher text, receiver should be knowing private key exponent $[d, X]$. The plain text P is computed to retrieve original image using cipher text C, i.e.

$$P = C^d \bmod X \tag{14}$$

3.4 De-shuffling of the Shuffled Image to Negative Image and Inversion of Negative Image to Original Image

The output of the previous step acts as the input to this step. The inverse transformation algorithm is applied to the shuffled image which gives the negative image that was shuffled back. From this de-shuffled image, the recipient obtains the original image.

Example Equation (11) is used to obtain values for variables a, b, c. We explore one such case mentioned in [3], Consider ($k = 10$, $\text{a} = 1$, $\text{b} = -5$, $\text{c} = 5$) and taking absolute values for keys a, b, c, we get $(1, 5, 5)$.

Take an image as input and EX-OR each pixels with value a = 1.
Consider a pixel having $P_i = (\text{R}_i,\ \text{G}_i,\ \text{B}_i)$ where $i = 10$.

$$\text{R}_{10} = 13$$
$$\text{G}_{10} = 9$$
$$\text{B}_{10} = 12$$

Using Eq. (12), EX-OR operation is computed, i.e.,

$$\text{R'}_{10} = 13 \oplus 10 = 3$$
$$\text{G'}_{10} = 9 \oplus 10 = 19$$
$$\text{B'}_{10} = 12 \oplus 10 = 2$$

Here, $P_{10} = (3,\ 19, 2)$
Shuffle the above pixel value with seed value $b = 5$ using *Fisher-Yates* algorithm [13].

For encryption, we use value $c = 5$ and four prime numbers that satisfy this value are used. Consider those prime numbers to be 2, 3, 5, 7 and we get,

$$n = 210,\ \phi(n) = 48,\ d = 29,\ X = 546$$

Substituting these values using Eq. (13),
for $\text{R} = 3$, $\text{G} = 19$, $\text{B} = 2$, we get cipher text C as (243, 23, 32).
For decryption, using Eq. (14),
we get back (R, G, B) = (3, 19, 2).

After de-shuffling using the same seed value in Fisher-Yates algorithm [1] and reverse inversion

$$\text{R'}_{10} = 3 \oplus 10 = 13$$
$$\text{G'}_{10} = 19 \oplus 10 = 9$$
$$\text{B'}_{10} = 2 \oplus 10 = 12$$

(R, G, B) = (13, 9, 12) are obtained. This illustrates that after decryption, resultant values are equal to the original pixel value P_{10}.

4 Results and Analysis

For quality evaluation, analysis and understanding of the level of pixel distortion in the image, we consider measures such as histogram, avalanche effect, mean and standard deviation which exhibit visual detail of image characteristics. The proposed algorithms for transformation and encryption of the image are implemented using Python 3 IDLE. We use a standard colour image, Fig. 4 (687 × 1024) retrieved from graded databases [17] as the input image throughout the application. The proposed algorithms defined above at each stage produce a new cipher image, as shown in Fig. 5 [16, 17]. The statistical representation of that image is represented below.

4.1 Histogram of Encrypted Image

A image histogram is basically a graph that illustrates the distribution of pixel intensities in an image. In Fig. 6, which shows the histogram for our encrypted image, we can observe that the pixel values are uniformly distributed. This prevents any statistical attacks and makes it difficult for the hacker to access the original image [18].

Fig. 4 Original image

Fig. 5 Encrypted image

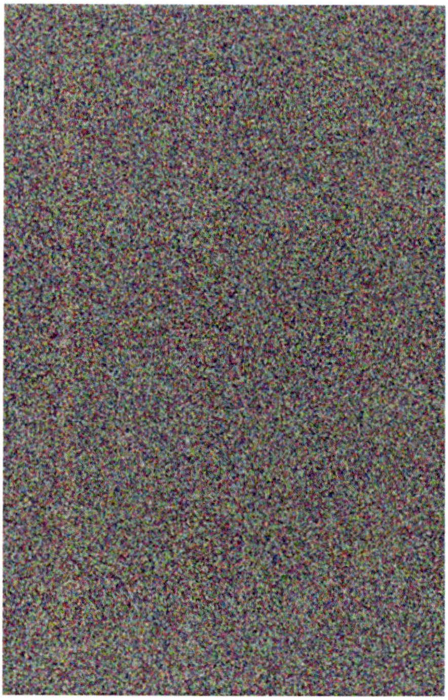

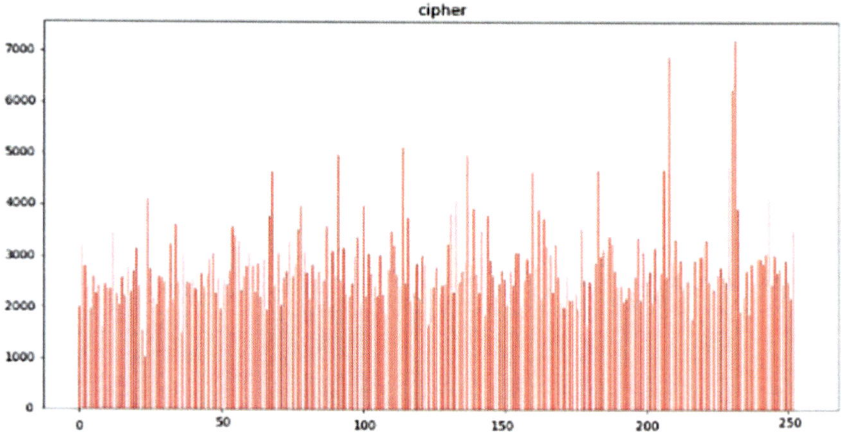

Fig. 6 Histogram representation of encrypted image

4.2 *Mean and Standard Deviation*

Mean is the set of values defined by ratio of the sum to total number of values in data.

Table 1 Tabular representation of mean and standard deviation of the image at each stage

Image type	Mean	Standard deviation
Original input image	109.434	91.485
Inverted image	209.314	44.788
Shuffled image	209.314	44.788
Encrypted image	164.451	85.850

Consider there are n number of pixels, mean of these pixels is given by

$$M = \frac{\sum xi}{\bar{n}} \tag{15}$$

where $i = 0, 1, 2, 3, 4, \ldots, n$.

Standard deviation is a measure used to compute variation occured for certain set of data given.

$$\text{SD} = \sqrt{\frac{\sum_{i=1}^{N} (C_i - M)^2}{N}} \tag{16}$$

where C_i is cipher text value of plain text i, N is number of characters, M is mean [16, 17].

Mean and standard deviation derived from Eqs. (15) and (16) are represented in Table 1.

4.3 Avalanche Effect

The avalanche effect is a characteristic property of cryptographic algorithms, wherein if there is a slight change in the input, the resulting output changes significantly. This property ensures that the message cannot be predicted easily even if the hacker employs statistical analysis method. A typical graph that depicts avalanche effect captures the distribution of the length of (principal) avalanches in the abelian sand pile model [19]. The avalanche effect is statistically represented in Fig. 7.

5 Conclusion

An enhanced RSA algorithm for securing media information using public key cryptography is proposed along with image transformation in order to double the security. The original image is transformed into a new negative shuffled image using the pixel negation and transformation which improves security and is hard to reveal by the adversary. The enhanced algorithm provides a more secure interface since four prime

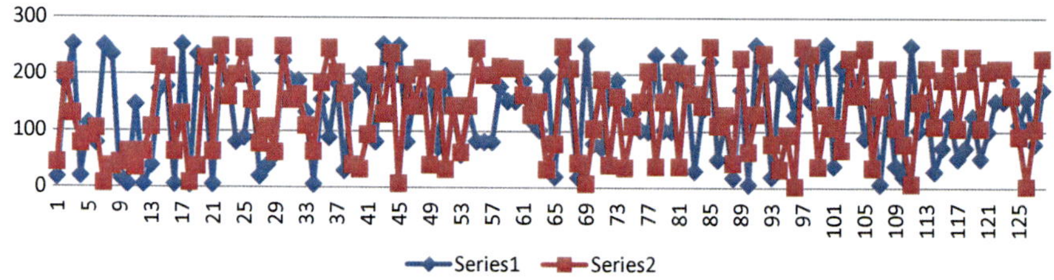

Fig. 7 Representation of the avalanche effect in encrypted image

numbers are used as opposed to only two in the standard RSA. The value obtained as the product of these four primes makes factorization much more difficult, again when compared to standard RSA. The proposed algorithm has been implemented and tested on a transformed image. Since the transformed image is already non-identifiable, it provides an additional layer of security, thus serving its purpose and proving to be useful in the safe transmission of multimedia.

References

1. Pea RD Learning through Multimedia
2. Anandakumar S Image Cryptography using RSA algorithm in network security
3. Kumar GS, Bagan KB, Vivekanand V (2011) A Novel algorithm for image encryption by integrated pixel scrambling plus diffusion [IISPD] utilizing duo chaos mapping applicability in wireless systems. Procedia Comput Sci 3:378–387. ISSN 1877-0509
4. Maddodi G, Awad A, Awad D et al. (2018) A new image encryption algorithm based on heterogeneous chaotic neural network generator and dna encoding. Multimed Tools Appl 77:24701–24725. https://doi.org/10.1007/s11042-018-5669-2
5. Barreto PSLM, Kim HY, Rijmen V (2002) Toward secure public-key blockwise fragile authentication watermarking. IEE Proc – Vis Image Sig Process 149(2):57–62
6. Tian L, Han L, Yue J (2016) Research on image segmentation based on clustering algorithm. Int J Signal Process Image Process Pattern Recogn 9(2):1–12. http://dx.doi.org/10.14257/ijsip.2016.9.2.01
7. Jiao J, Mo X, Shen C (2010) Image clustering via sparse representation. In: Boll S, Tian Q, Zhang L, Zhang Z, Chen YPP (eds) Advances in multimedia modeling, MMM 2010. Lecture Notes in Computer Science, vol 5916. Springer, Berlin, Heidelberg
8. Atown HY (2015) Hide and encryption fingerprint image by using LSB and transposition pixel by spiral method
9. Raghunandan KR, Aithal G, Shetty S (2019) Comparative analysis of encryption and decryption techniques using mersenne prime numbers and phony modulus to avoid factorization attack of RSA. In 2019 international conference on advanced mechatronic systems (ICAMechS), Kusatsu, Shiga, Japan
10. Swami B, Singh R, Choudhary S (2016) Dual Modulus RSA Based on Jordan-totient Function. Procedia Technol 24:1581–1586, ISSN 2212-0173. https://doi.org/10.1016/j.protcy.2016.05.143
11. Tan W, Wang X, Lou X, Pan M (2011) Analysis of RSA based on quantitating key security strength. Procedia Eng 15:1340–1344. ISSN 1877-7058

12. NaQi WW, Zhang J, Wang W, Zhao J, Li J, Shen P, Yin X, Xiao X, Hu J (2013) Analysis and research of the RSA algorithm. Inf Technol J 12(9):1818–1824. https://doi.org/10.3923/itj.2013.1818.1824
13. Ruohonen K (2014) Mathematical cryptology. (trans: Kangas J, Coughlan P)
14. Hussain S, Lone MM Image enhancement techniques: A review. MTech Computer Science, Department of Computer Science and IT, University of Jammu, Jammu, India
15. Truong T Cubic Pells equation. Johannes Hedberggymnasiet, Sweden
16. Shi C, Wang SY, Bhargava B (1999) MPEG video encryption in real-time using secret key cryptography. PDPTA: pp 2822–2828
17. Raghunandan KR, Aithal G, Shetty S (2019) Secure RSA variant system to avoid factorization attack using phony modules and phony public key Exponent. Int J Innovative Technol Exploring Eng (IJITEE) 8(9). ISSN: 2278-3075
18. Sneha HL Pixel intensity histogram characteristics: basics of image processing and machine vision
19. Austin D, Chambers M, Funke R, Puente LD, Keough L The Avalanche polynomial of a graph

Analysis of Tuberculosis Disease Using Association Rule Mining

Ankita Mohapatra, Sangita Khare, and Deepa Gupta

Abstract Tuberculosis (TB) is a chronic infectious disease and remains a serious explanation for death globally. Most of the infected people are from poverty-stricken communities with low healthcare infrastructure. Machine learning (ML) shows a unique way to facilitate the diagnosis of TB. ML can provide deep insights into large biomedical datasets, as well as it can uncover new biomedical and healthcare knowledge. ML techniques have to be utilized to successfully recognize TB occurrence at an early stage or re-occurrence. Association rule mining and decision tree can be used for prediction of the occurrence and re-occurrence of TB. Association rules and decision tree give a meaningful and efficient way to define and present certain dependencies among the attributes in a dataset ML algorithms help to identify, discover hidden and valuable information, and have been massively used and explored in the medical domain. The main purpose of this study is to apply ML techniques for extracting hidden patterns, which are significant to predict TB at an early stage.

Keywords TB dataset · Machine learning · Association rule mining · Decision tree

1 Introduction

The world's population is rising day by day, and along with it, there is an increase in health issues and the cost of health care. The healthcare industries are facing many challenges in identifying diseases and diagnosing them in an effective way at an early stage. As per WHO, in 2017 around 10 million people were infected with TB out of that 1.6 million people died [1]. Among developing countries—India, Indonesia,

A. Mohapatra (✉) · S. Khare · D. Gupta
Department of Computer Science and Engineering, Amrita School of Engineering, Amrita Vishwa Vidyapeetham, Bengaluru, India
e-mail: 040ankita@gmail.com

S. Khare
e-mail: k_sangita@blr.amrita.edu

D. Gupta
e-mail: g_deepa@blr.amrita.edu

N. N. Chiplunkar and T. Fukao (eds.), *Advances in Artificial Intelligence and Data Engineering*, Advances in Intelligent Systems and Computing 1133,
https://doi.org/10.1007/978-981-15-3514-7_74

Myanmar, and Bangladesh remain most severely affected countries from TB, mainly due to a lack of social awareness about health and hygiene. In today's world, where data plays an important role in industries like e-commerce, health care, retail domain, and social media, ML plays a vital role. Huge volumes of data can be analyzed to retrieve useful information and to find the relation among data. In this way, ML can provide the new era of understanding the data and give an effective solution for a given problem. ML can be used for classification and prediction of diseases. While making the diagnosis of TB in pathology, it is very difficult to identify patients infected with pulmonary TB or non-pulmonary TB. In case of pulmonary TB, small amount of bacteria is found in comparison with non-pulmonary TB.

In the medical system, there are many tests which are done to diagnose TB like a clinical report, microscopy test, chest radiography, and sputum test, etc., and still, it has some limitations like high cost, time consumption, etc. ML algorithms are used for classification, clustering, and association rule mining. In the case of diseases, it can be used to classify diseases, using algorithms like neural networks, decision tree, apriori, *k*-means, etc. In the medical dataset, there might be many attributes that are not required for the diagnosis of a particular disease, manually removing these attributes would be very challenging, but using ML techniques, such attributes can be eliminated and ranked based on their importance.

This paper is divided into the following sections, and the next section describes the literature survey where various works that are done in the past using ML techniques on TB are listed. Section 3 gives details about the dataset used in this work. In Sect. 4, the working mechanism of the algorithm used with complete architecture is explained. Section 5 shows the process of rules generation using decision tree algorithm and association rule mining. Section 6 concludes the work done in this paper.

2 Literature Survey

Many studies have been done on TB disease by experts of TB, chest physicians, or respiratory physicians and pulmonologists. In their work, they have used x-ray reports, microscopic images, and clinical data. Their main concern is basically finding the symptoms, medicines, blood free from bacteria, surgery and identifying the types of TB, but none of them have explained about the importance of ML in the diagnosis of TB at an early stage.

Zulvia et al. [2] described classification of the TB using data mining techniques; for this they collected dataset and then divided that dataset based on two factors—one is the internal factor (weight, height, age, sex), and another is the external factor (family size, family income). For classification, they applied support vector machine (SVM) algorithm for the diagnosis of TB. Wu et al. [3] described that many diseases which have similar symptoms like TB makes it very difficult to identify the actual disease. In such cases, using the random forest algorithm is useful for the classification of the diseases. Subiyanto et al. [4] demonstrated a method to diagnose respiratory infection of infants using the C4.5 algorithm with an accuracy of 81.48%. Dominic et al. [5] launched comparative research on data mining approaches with various chronic diseases. Rusdah et al. [6] showed a comparative analysis of C4.5, Naïve Bayes, backpropagation, and SVM for diagnosis of pulmonary TB. SVM technique showed better accuracy in the diagnosis of pulmonary TB. Shukla et al. [7] focused on partitions of the large medical dataset into smaller partitions using k-means clustering algorithm. Benbelkacem et al. [8] explained methods for handling large medical dataset by creating a decision tree which fetches hidden information. Karthik et al. [9] implemented clustering algorithms which identified highly infected region of TB. Rusdah et al. [10] demonstrated methods to identify TB using different data mining techniques like SVM, bagging, and random forest. There are many classes for distribution analysis of clinical data like predation of value, comparison of varied data processing techniques, prediction and structuring the unstructured clinical knowledge [11, 12]. Gupta et al. [13] focused on CMS data where the prime goal is to understand ICD9 codes by applying various ML algorithms in chronic diseases.

After going through the literature, it is seen that ML techniques would play a vital role in controlling many aspects of TB disease. Adding to that diagnosis of TB using association rule mining is not yet explored. Association rule mining is helpful for getting the co-relation between attributes. In this paper, the main aim is in-depth understanding of TB attributes, finding out the co-relation between attributes, and generating the rules for the diagnosis of TB at early stage using association rule mining and decision tree technique. For this analysis, R programming language is used. Next section gives the summary of TB dataset.

3 Data Source

Collection of data is an important step before implementation. Data was collected from a local hospital. In the dataset, information was collected regarding various tests done by patient and diagnosis report. This dataset consists of total 26 attributes taken from 1000 patients. Table 1 explains the relevant attributes used in this study.

4 Methodology

ML techniques would play an important role in describing several aspects of TB disease. In this paper, main focus is to understand TB attributes, check the co-relation between attributes and to find the important attributes that helps in the identification of TB using association rule mining and decision tree technique.

As shown in Fig. 1, overall methodology is divided into four stages. Stage 1 is the data segregation in terms of overall TB dataset and symptoms TB dataset. Overall, TB dataset contains body symptoms and pathology test data. Attributes of overall TB dataset are chest pain, sweating at night, temperature, cough more than two weeks,

Table 1 Attribute description

Dataset	Attributes	Description	Value		
Overall TB dataset	Chest pain	Pain of heart	y = high, n = normal		
	Sweating at night	Sweating during sleep times	y = high, n = normal	Symptoms	
	Temperature	Temperature is more than 100° F	y = high, n = normal		
	Cough > 2 weeks	Coughing more than 2 weeks	y = high, n = normal		
	Fever > 2 weeks	Fever more than 2 weeks	y = high, n = normal		
	Breathless	Not to get enough air into your lungs	y = high, n = normal		
	Continuous weight loss	Losing weight continuously	y = high, n = normal		
	Cough with blood	Hemoptysis	y = high, n = normal		
	Contact with TB patients	Any patient contact with TB patient	y = high, n = normal		
	Fever	Body temperature is high	y = high, n = normal		
	Blood group	Classification of blood	y = high, n = normal		
	Blood pressure	Pressure of blood, normal bp is 120/80	y = high, n = normal		
	s.creatinine	Molecules in blood, normal range 0.6–1.2 (mg)	y = high, n = normal		
	s.bilirubin	Molecules in blood, normal range 0.1–1.2 mg/dL	y = high, n = normal		

(continued)

Table 1 (continued)

Dataset	Attributes	Description	Value	
	s.alk.phosphatase	Molecules in blood, normal range 44–147(IU/L)	y = high, n = normal	
	Sodium	Molecules in blood, normal range 135–145 mEq/L	y = high, n = normal	
	Potassium	Molecules in blood, normal range 3.5–5.0 mEq/L	y = high, n = normal	
	ESR	Body infection rate, range −5 to 25	y = high, n = normal	
	Pulse	Rate of heartbeat	y = high, n = normal	
	Lymphocytes	High range white blood cells, normal range 0.5–5.0	y = high, n = normal	
	Platelets	Tiny blood cells, normal count-150,000–450,000	y = high, n = normal	
	HGB	Hemoglobin, normal range 12.0–15.5	y = high, n = normal	
	WBC	Increase the immune system of body, normal range 3.5–10	y = high, n = normal	
	HIV	Human immunodeficiency virus	y = positive, n = negative	
	RBC	High red blood cell, normal range 4.06–5.63	y = high, n = normal	
	Class level	TB is present or absent	y = presence, n = absent	

fever more than two weeks, breathless, continuous weight loss, cough with blood, contact with TB paticnts, fever, blood group, blood pressure, s.creatinine, s.bilirubin, s.alk.phosphatase, sodium, potassium, ESR, pulse, lymphocytes, platelets, HGB, WBC, HIV, RBC, and class level. Symptoms dataset contains body symptoms data, and the attributes of symptoms TB dataset are chest pain, sweating at night, temperature, cough more than two weeks, fever more than two weeks, breathless, continuous

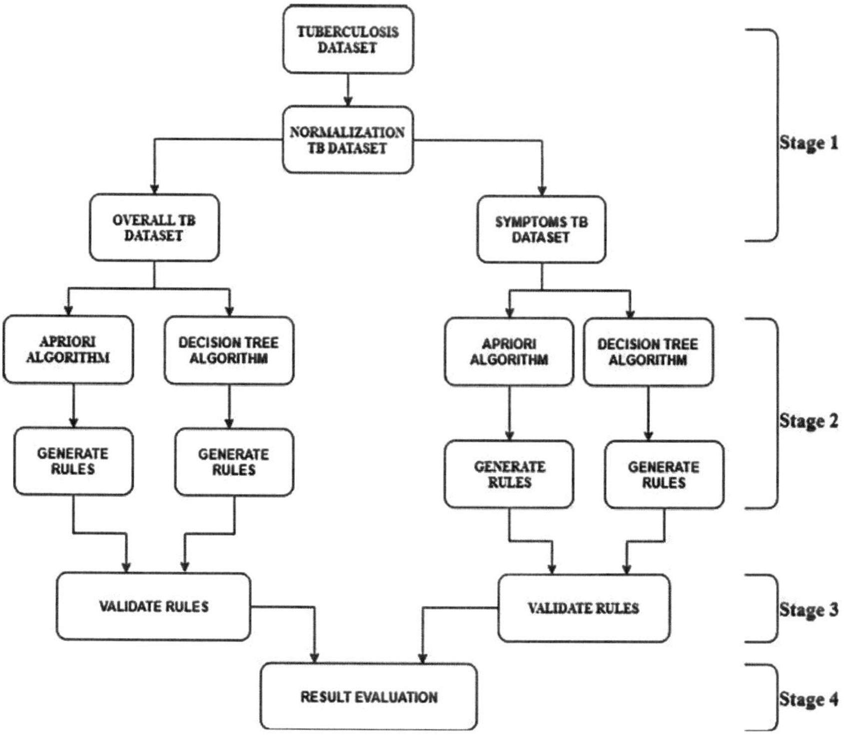

Fig. 1 Schematic diagram of the proposed approach

weight loss, cough with blood, contact with TB patients, fever, blood group, and blood pressure. Stage 2 is the rule generation process. Stage 3 is the rule validation stage which combines features from different algorithm, and stage 4 is known as result evaluation in which rules are generated. Brief description of all the stages is as follows.

4.1 Stage 1: Data Preparation Stage

In the data preparation stage, missing values are removed from the dataset. For the apriori algorithm, the dataset should be categorical data, but many attributes in the dataset are having numerical values. In normalization process, conversion of all numerical values to categorical values is done based on the norms used in medical domain. For example, HGB normal range 12.0–15.5 is defined as '*n*' and higher than that range defined as '*y*'. After analyzing the dataset, data has to be segregated into two categories: One is a symptoms TB dataset, which consists of 12 attributes out of 26 attributes, e.g., chest pain, fever more than 2 weeks, etc. Another is overall dataset, which has 26 attributes including the class level which means TB is present or absent. Here, normalization is done that will be the input for rule generation stage.

4.2 *Stage 2: Rule Generation Module*

In rule generation module, rules are generated using two algorithms: One is an apriori algorithm and another is a decision tree algorithm. In this paper, attributes => class are generated that may be in the form of classification like all factors are related to the presence of TB.

4.2.1 Apriori Algorithm

Association rule offers an indication of how the assorted attributes in a dataset are co-related. An initial process is to identify frequently occurring attributes with minimum threshold support and then finding a rule $X => Y$ with certain minimum confidence, where X and Y are frequently occurring attribute or attributes. Here, X and Y are reciprocally exclusive. Lift is outlined as the ratio of support expected if X and Y are frequently occurring that were independent [14]. Apriori algorithm is a common methodology of association rule mining that was developed by Aggarwal and Srikanth [14]. Apriori algorithm helps to generate the rules of TB disease. Rule generated can be calculated as follows: LHS => RHS [support, confidence, lift].

For the generated rule support, confidence and lift can be defined as follows: X and Y, which shows those sets of attributes which are mutually exclusive. As shown below in Eq. (1), Rule $X => Y$ means that whenever attribute set X is there in the row of the dataset, then it has attribute set Y with some probability [15].

$$\text{Rule } X => Y \tag{1}$$

The frequency of item set is known as support. In Eq. (2), support of the rule is outlined as percentage of row within the TB dataset containing both attributes in X and Y [15].

$$\text{Support}(S) = \frac{\text{frequency}(X, Y)}{N} \tag{2}$$

Equation (3) described percentage of row in each X and Y with respect to X can be defined as confidence [15].

$$\text{Confidence}(C) = \frac{\text{frequency}(X, Y)}{(X)} \tag{3}$$

Equation (4) defined as lift is the ratio of the observed support to that expected if X and Y were independent of each other [15]. In TB dataset, lift is used for correlation measurement.

$$\text{Lift} = \frac{\text{Support}}{\text{Support}(X) \times \text{Support}(Y)} \tag{4}$$

4.2.2 Decision Algorithm

A decision tree (DT) is a classification algorithm which generates tree and rules for model representation for different classes of the dataset. Decision tree is a tree structure that is sort of a flowchart. In DT, the internal node is known as attribute, and leaf node is known as class level. Attributes which were used in the decision trees follow a splitting mechanism based on entropy, information gain, and gain ratio [4].

For the calculation of decision tree in TB dataset, combination of various classes is used. Let us assume c is taken as the number of classes, and $p(S, j)$ is defined as proportion of instances used to Jth class. Hence, the entropy (S) is formulated as in Eq. (5) [15]

$$\text{Entropy}(S) = -\sum_{j=0}^{c} p(S, j) \times \log p(S, j) \tag{5}$$

Therefore, for calculating information gain over training TB dataset, T is used and it is defined as in Eq. (6) [15]

$$\text{Gain}(S, T) = \text{Entropy}(S) - \sum_{v \in \text{Values}} \frac{|T_{s,v}|}{|T_s|} \text{Entropy}(S_v) \tag{6}$$

When the value (T_s) is a set of S in T and $(T_{s,v})$ is a subset of T in attributes, S has value of accordingly, gain ratio over the attribute S is defined as in Eq. (7) [15]

$$\text{GainRatio}(S, T) = \frac{\text{Gain}(S, T)}{\text{SplitInfo}(S, T)} \tag{7}$$

Hence, splitting the decision tree in TB dataset is termed as SplitInfo, and the calculation for SplitInfo (S, T) is formulated as in Eq. (8) [15]

$$\text{SplitInfo}(S, T) = \sum_{v \in \text{Values}} \frac{|T_{s,v}|}{|T_s|} \times \log \frac{|T_{s,v}|}{|T_s|} \tag{8}$$

Once the rules are generated with two datasets like symptoms vs diagnosis of TB and overall dataset vs diagnosis of TB is over then proceeds to the validation module.

4.3 Stage 3: Validation Rules Modules

Once the rules got generated from both the algorithm, validation of the rules took place to identify common attributes using symptoms and overall dataset. Once it got completed, result evaluation stage would be giving its accurate value.

4.4 Stage 4: Result Evaluation

Using decision tree classifier, rules are evaluated for symptoms dataset and overall TB dataset. This process took the validated rules from both the dataset and gives it approximate accuracy.

5 Results and Discussion

Class attributes are two values depending on TB disease, i.e., *y* or *n*, where *n* implies the absence of TB disease and *y* is a presence of TB disease. While applying the apriori algorithm, the support with a threshold value of 30% and confidence which has a threshold value of 90% is chosen in order to get meaningful rules. These rules were filtered as class level *y*, where the main aim is to analyze the study of factors responsible for TB disease as well as symptoms attributes.

5.1 Analysis Rules Using Apriori Algorithm

Details of the analysis are based on the apriori algorithm using R programming having both symptoms and overall TB dataset.

5.1.1 Experimental Result and Rule Analysis Using Apriori Algorithm for Overall TB Dataset

Analysis of the rules is based on the apriori algorithm based on confidence 90% and to support 30%. Here, identified 11 major rules established the presence of TB disease.

Figure 2 shows the observation based on rules where there are many common attributes like the presence of breathless, high cough, sweating at night, continuous weight loss, and cough with blood reflects the infection of TB disease. Whereas the occurrence of continuous weight loss and cough with blood is the most common attributes that are occurring in many rules that shows the presence of TB. If a person

[breathless=y,fever>2weeks=y,continousweightloss=y,chestpain=y,cough withblood=y]=>[class_Level=y]
[support=0.427,confidence=0.9446903,lift=1.8097515]

[bodytemperature=y,cough>2weeks=y,fever>2weeks=y,continousweightl oss=y,HIV=y]=>[Class_Level=y]
[support=0.317,confidence=0.9256988,lift= 1.8014]

[RBC=y,breathless=y,sweatingatnight=y,fever>2weeks=y,cough=y,fever =y,chestpain=y,coughwithblood=y]=>[classlevel=y]
[support=0.309,confidence=0.9279279,lift= 1.7776]

[breathless=y,fever>2weeks=y,continousweightloss=y,chestpain=y]=> [class_Level=y]
[support=0.427,confidence=0.9446903,lift= 1.809751]

[breathless=y,cough>2weeks=y,continousweightloss=y,chestpain=y]=> [class_Level=y]
[support=0.407,confidence=0.9046703,lift= 1.909751]

[cough>2weeks=y,sweatingatnight=y,continousweightloss=y,chestpain=y]=>[classlevel=y]
[support=0.327,confidence=0.95546903,lift= 1.899751]

[cough>2weeks=y,fever>2weeks=y,continousweightloss=y,chestpain=y] => [classlevel=y]
[support=0.437,confidence=0.9546903,lift= 1.809951]

[bodytemperature=y,cough>2weeks=y,sweatingatnight=y,chestpain=y,co ugh=y,fever=y,contiousweightloss=y]=>[classlevel=y]
[support=0.353,confidence=0.9592391,lift= 1.8376]

[breathless=y,cough>2weeks=y,fever>2weeks.=y,chestpain=y,continous weightloss=y]=>[classlevel=y]
[support=0.307,confidence=0.9246988,lift= 1.7714]

[breathless=y,cough>2weeks=y,fever=y,chestpain=y,continousweightloss =y,contactTBpatients=y]=>[classlevel=y]
[support=0.329,confidence=0.9379279,lift= 1.7876]

[breathless=y,sweatingatnight=y,fever>2weeks.=y,coughwithblood=y]=> [classlevel=y] [support=0.409,confidence=0.9299209,lift= 1.7779]

Fig. 2 Generated rules for overall TB dataset

is having HIV positive, cough more than 2 weeks, fever more than 2 weeks, and continuous weight loss, it shows 92% confidence including support of 31% and lift of 1.80. If the lift value is greater than 1, then antecedent and consequent are positively correlated. There are two clinical test attributes which include RBC having more than normal range and others have HIV positive, and if it is a combination with any other symptoms, then the probability of getting infected with TB is high.

5.1.2 Experimental Result and Rule Analysis Using Apriori Algorithm for Symptoms Dataset

Considering symptoms dataset with R programming language where rules are generated based on 90% of threshold values and identified eight major rules establish the presence of TB disease.

Figure 3 shows the observation based on the above rules which includes many common attributes such as the presence of cough, continuous weight loss, cough with blood, and fever more than two weeks implies infection of TB. Presence of breathless, sweating at night, and cough with blood are common attributes which occur many times that can show the high possibility of TB infection. If a person is having cough more than 2 weeks, fever more than 2 weeks, and continuous weight loss and body temperature is high, it shows there are 94% confidence of chances of having TB disease including support of 42% and lift of 1.819.

```
[breathless=y,cough>2weeks=y,fever=y,chestpain=y,contactTBpatients=y]=>
[classlevel=y]
[support=0.309,confidence=0.9279279,lift=1.777640]

[cough>2weeks=y,sweatingatnight=y,fever>2weeks=y,continousweightloss=y]=>
[classlevel=y]
[support=0.427,confidence=0.9446903,lift=1.819751]

[breathless=y,sweatingatNight=y,fever>2weeks=y,continousweightloss=y,contact
TBpatients=n]=>[classlevel=y]
[support=0.424,confidence=0.926903,lift=1.809751]

[bodytemperature=y,breathless=y,chestpain=y]=> [classlevel=y]
[support=0.307,confidence=0.9246988,lift=1.771454]

[bodytemperature=y,breathless=y,continousweightloss=y,chestpain=y]=>
[classlevel=y]
[support=0.323,confidence=0.9446988,lift=1.871455]

[bodytemperature=y,cough>2weeks=y,chestpain=y,coughwithblood]=>
[classlevel=y]
[support=0.423,confidence=0.9346988,lift=1.874455]

[bodytemperature=y,breathless=y,fever>2weeks=y,continousweightloss=y]=>
[classlevel=y]
[support=0.417,confidence=0.9346903,lift=1.839751]

[bodytemperature=y,continousweightloss=y,cough>2weeks=y,fever>2weeks=y]=
> [classlevel=y]
[support=0.427,confidence=0.9446903,lift=1.819751]
```

Fig. 3 Generated rules for symptoms TB dataset

5.2 *Analysis Rules Using Decision Tree Algorithm*

Decision tree (DT) is the classification algorithm which generates tree and rules for model representation for different classes of dataset like overall TB dataset, symptoms dataset using R programming language.

5.2.1 Experimental Result and Rule Analysis Using Decision Tree Algorithm for Overall TB Dataset

An observation is based on decision tree which includes overall dataset that have 26 attributes and generates a tree which gives the rules. From that tree, many common attributes are extracted such as the presence of breathless, sweating at night, HIV positive, cough with blood and continuous weight loss that leads to infection of TB. Based on DT algorithm, a unique rule has been identified which says if any person came in contact with a TB infected person, then there is more chances of getting infected with TB. If an individual is having HIV positive, high RBC range, high WBC range, then the possibility of that individual infected with TB would be high.

5.2.2 Experimental Result and Rule Analysis Using Decision Tree Algorithm for Symptoms Dataset

Decision tree implementation on symptoms dataset where it generates a tree results in rules.

Figure 4 observed based on decision tree using symptoms dataset with attributes like breathless, sweating at night, cough with blood and continuous weight loss, and chest pain indicates the symptoms of TB. For validation, two datasets have been used as shown in Table 2. This defines the correctly classified instances with the accuracy of 90.05% in the overall TB dataset and 96.44% accuracy in the symptoms TB dataset.

Analyzing both the algorithm, i.e., association rule mining algorithm and decision tree are compared on the basis of two datasets, i.e., overall and symptoms dataset. Both the algorithms generate the rules using the same attributes coming from two components. Once both the algorithm executes, it generates a rule. By the validation process, it is noted that the rules generated by association rule mining algorithm and decision tree algorithm are almost the same. Important attributes which are helping to diagnosis of TB are breathless, cough, sweating at night, RBC, HIV, cough with blood, RBC, continuous weight loss, fever more than two weeks, and platelets, whereas sodium, s.creatinine, s.bilirubin, blood pressure, ESR, HGB, potassium, blood group, pulse, and s.alkposphatase are the attributes which do not add any noticeable impact to TB.

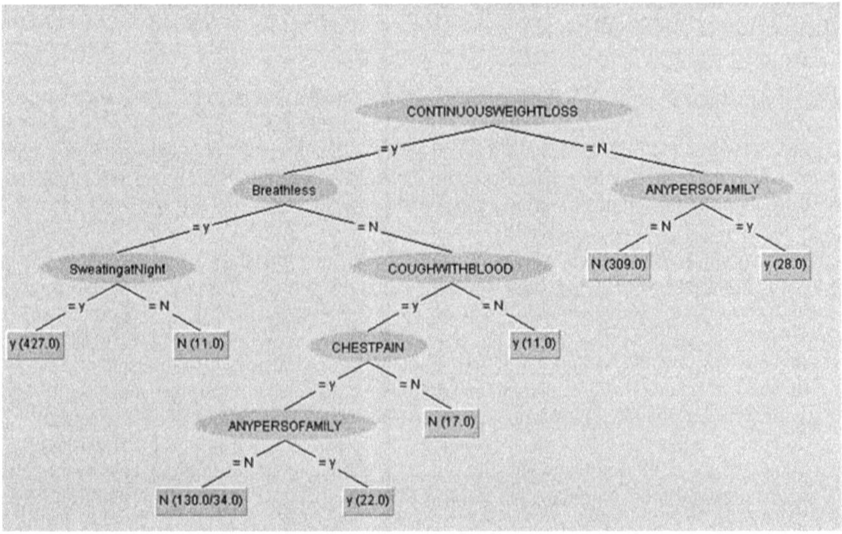

Fig. 4 Decision tree symptoms TB dataset

Table 2 Decision tree validation

Dataset	Correctly classified instances (%)
Overall TB dataset	90.05
Symptoms TB dataset	96.44

6 Conclusion

This paper shows the division of the TB data into categories through a process which elaborates the symptoms of TB in an individual using decision tree and association rule mining algorithm. By capturing the symptoms of the patients, the algorithm is applied to the dataset which generates the rules that are having high accuracy approximately 90.05% in overall TB dataset and 96.44% in symptoms dataset. It is also analyzed that using body symptoms one can predict the infection of TB in an individual. In this study, it is found that breathless, cough, sweating at night, continuous weight loss, cough with blood, HIV, fever, and chest pain are the important attributes for the diagnosis of TB. By analyzing these rules, report can be prepared which shows whether the individual is suffering from TB or not. This not only proves to be cost effective for the patients but also time saving for the doctors too.

References

1. WHO Homepage, https://www.who.int/news-room/fact-sheets/detail/TB. Last accessed 21 Feb 2019
2. Zulvia F, Kuo R, Roflin E (2017) An initial screening method for TB diseases using a multi-objective gradient evolution based support vector machine and C5.0 DECISIONTREE. In: IEEE 41st annual computer software and applications conference (COMPSAC), pp 204–209. https://doi.org/10.1109/compsac.2017.57
3. Wu Y, Wang H, Wu F (2017) Automatic classification of pulmonary TB and sarcoidosis based on random forest. In: 10th international congress on image and signal processing, biomedical engineering and informatics (CISP-BMEI), pp 1–5. https://doi.org/10.1109/cisp-bmei.2017.8302280
4. Subiyanto, Mulwinda A, Andriani D (2017) Intelligent diagnosis system for acute respiratory infection in infants. In: 3rd international conference on science in in-formation technology (ICSITech), pp 558–562. https://doi.org/10.1109/icsitech.2017. 825717
5. Dominic V, Aggarwal A, Gupta D, Khare S (2015) Investigation of chronic disease correlation using data mining techniques. In: 2nd international conference on recent advances in engineering and computational sciences (RAECS), pp 1–6. https://doi.org/10.1109/raecs.2015.7453329
6. Rusdah, Winarko E, Wardoyo R (2015) Preliminary diagnosis of pulmonary TB using ensemble method. In: 2015 international conference on data and software engineering (ICoDSE), pp 175–180. https://doi.org/10.1109/icodse.2015.7436993
7. Shukla M, Agarwal S (2014) Hybrid approach for TB data classification using optimal centroid selection based clustering. In: 2014 students conference on engineering and systems, pp 1–5. https://doi.org/10.1109/sces.2014.6880115
8. Benbelkacem S, Atmani B, Benamina M (2013) Treatment TB retrieval using decision tree. In: 2013 international conference on control, decision and information technologies (CoDIT), pp 283–288. https://doi.org/10.1109/codit.2013.6689558
9. Karthik D, Vijayarekha K, Dhivya V (2015) Identifying new regions infected by mycobacterium TB(TB): case study in Kumbakonam region, ThanjavurDistrict. Int J ChemTech Res CODEN (USA), pp 341–345(2015). IJCRGG ISSN: 0974-4290
10. Rusdah, Winarko E (2013) Review on data mining methods for TB diagnosis. In: Information systems international conference
11. Garg S, Rupal N (2015) A review on TB using data mining approaches. Int J Eng Dev Res. ISSN: 2321-9939
12. Asha T, Natarajan S, Murthy K (2012) Data mining techniques in the diagnosis of TB. www.intechopen.com. https://doi.org/10.5772/30504
13. Gupta D, Aggarwal A, Khare S (2016) A method to predict diagnostic codes for chronic diseases using ML techniques. In: Fifth IEEE international conference on computing communication and automation (ICCA), pp. 281–287. https://doi.org/10.1109/ccaa.2016.7813730
14. Khare S, Gupta D (2016) Association rule analysis in cardiovascular disease. In: Second international conference on cognitive computing and information processing (CCIP), SJCE, Mysuru, India, pp. 1–6. IEEE. https://doi.org/10.1109/ccip.2016. 7802881
15. Han J, Kamber M, Pei J (2017) Data mining: concepts and techniques, 3rd edn. Elsevier Science & Technology, US The Morgan sKaufmann Series

Scalable Two-Phase Top-Down Specification for Big Data Anonymization Using Apache Pig

Anushree Raj and Rio D'Souza

Abstract In big data applications, data privacy is one of the most concerned issues because processing of large-scale privacy-sensitive dataset often requires computation power provided by public cloud services. Sub-tree data anonymization, which is getting a good trade-off between data utility and distortion, is a popularly adopted scheme to anonymize dataset for privacy preservation. Top-down specialization (TDS) and bottom-up generalization (BUG) are two ways to fulfill sub-tree anonymization. Anonymizing dataset via generalization to satisfy certain privacy requirements such as *k*-anonymity is a widely used category of privacy-preserving techniques. It is a challenge for existing anonymization approaches to accomplish privacy preservation on privacy-sensitive large-scale dataset due to their insufficiency of scalability. Several methods focus on developing algorithms to determine the best cut of the taxonomy tree, the optimal value of *k* and the best option of anonymization technique either by top-down specification or by bottom-up generalization. This paper proposes a scalable two-phase top-down specialization (STPTDS) to anonymize large-scale dataset using Apache Pig on Hadoop.

Keywords Top-down specification · Anonymization using Pig Latin · Scalable two-phase TDS · Big data anonymization

1 Introduction

Data anonymization is the technique, wherein the information that discloses the identity is removed from dataset, so that the people who are defined by the information can remain unknown [1], i.e., sensitive data is de-identified though its format and data type are preserved. Internet makes data more reachable and hence privacy of data is of more concern.

A. Raj (✉)
St Agnes College Autonomous Mangalore, Mangalore, Karnataka, India
e-mail: anushree.raj@gmail.com

R. D'Souza
St Joseph Engineering College Mangalore, Mangalore, Karnataka, India

N. N. Chiplunkar and T. Fukao (eds.), *Advances in Artificial Intelligence and Data Engineering*, Advances in Intelligent Systems and Computing 1133,
https://doi.org/10.1007/978-981-15-3514-7_75

An example in Samarati [2] showed that linking medication records with a voter list can uniquely identify a person's name and their medical information. In response, new privacy acts and legislations are recently enforced in many countries.

Data anonymization means hiding characteristics or sensitive data in data records. The privacy of an individual can be effectively preserved even though the personal information is exposed to data users for various analysis and mining. A variety of privacy models and data anonymization approaches have been proposed and extensively studied [3–10]. However, while applying these traditional approaches to big data anonymization, it gives rise to scalability and efficiency challenges.

Anonymization methods, based on k-anonymity are broadly classified into two categories. The first category comprises of techniques that generalize data from the bottom of the taxonomy tree toward its top and are referred to as the bottom-up generalization (BUG). The second one is based on walking through the taxonomy tree from the top toward the bottom known as the top-down specialization (TDS).

Bottom-up generalization (BUG): In a k-anonymous dataset, each record is indistinguishable from at least $k - 1$ other records with respect to QID. Bottom-up generalization approach of anonymization is an iterative process starting from the lowest anonymization level [11].

Top-down specification (TDS): It is an iterative process starting from the topmost domain values in the taxonomy trees of attributes. Each round of iteration consists of three main steps, namely finding the best specialization, performing specialization and updating values of the search metric for the next round [12]. Such a process is repeated until k-anonymity is violated, to expose the maximum data utility. The goodness of a specialization is measured by a search metric. The proposed algorithm adopts the information gain per privacy loss (IGPL), a trade-off metric that considers both the privacy and information requirements, as the search metric in the proposed approach. A specialization with the highest IGPL value is regarded as the best one and selected in each round.

All k-anonymity methods identify a set of attributes known as quasi-identifiers (QID). QIDs are attributes that contain private information, where adversaries can use to unveil hidden data by linking them to related external elements [13]. Current anonymization algorithm aims to improve data privacy by generalizing the QID attributes through processes that utilize taxonomy tree, interval, or suppression. QID attributes are replaced by taxonomy tree values or intervals. The best taxonomy or interval values are named as the best cut. In TDS method, the entropy is used to calculate the highest QID score for the best cut. Other methods implement the median calculation to find out the best cut within the numerical QID attributes [14]. However, these methods do not adequately address the scalability issues. Large data size is an obstacle for these techniques, as they need to fit the data into the limited memory that is available. Processing big data may overwhelm the hardware and result in performance inefficiencies even in parallel computing environments. To overcome such inefficiencies, some techniques split the large size of data randomly into small chunks of data blocks, so they can fit into the memory for carrying out the required computations [15]. However, this resolution does not provide a real scalability solution, since it degrades the performance and increases the masking of anonymized data.

Master–slave architecture using HDFS is one of the solutions to process big data. The Hadoop framework application works in an environment that provides distributed storage and computation across clusters of computers. Hadoop consists of four types of nodes—a NameNode, DataNodes, JobTracker and TaskTracker. HDFS nodes (NameNode and DataNodes) provide a distributed file system where the JobTracker manages the jobs and TaskTrackers run tasks that perform parts of the job. Users submit MapReduce jobs to the JobTracker, which runs each of the Map and Reduce parts of the initial job in TaskTrackers and collects the final results.

The Hadoop distributed file system (HDFS) is based on the Google file system (GFS) and provides a distributed file system that is designed to run on commodity hardware.

HDFS is a block-structured file system where each file is divided into blocks of a pre-determined size. These blocks are stored across a cluster of one or several machines. Apache Hadoop HDFS architecture is shown in Fig. 1 follows a *master–slave architecture*, where a cluster comprises of a single NameNode (master node) and all the other nodes are DataNodes (slave nodes). Name node is the master node that manages the file system namespace and controls access to files by clients. It maintains and manages the DataNodes. It records the metadata of all the files stored in the cluster. It keeps a record of all the blocks in HDFS and in which nodes these blocks are located. DataNodes are the slave nodes in HDFS. DataNodes are the slave daemons or process which runs on each slave. The actual data is stored on DataNode. Secondary NameNode, it works concurrently with the primary NameNode. The secondary NameNode is one which constantly reads all the file systems and metadata from the RAM of the NameNode and writes it into the hard disk or the file system. Blocks are the continuous location on hard drive where data is stored. HDSF stores each file as blocks which are scattered throughout the Hadoop cluster. Hadoop performs only two stages: map and reduce. Developers are forced to process their

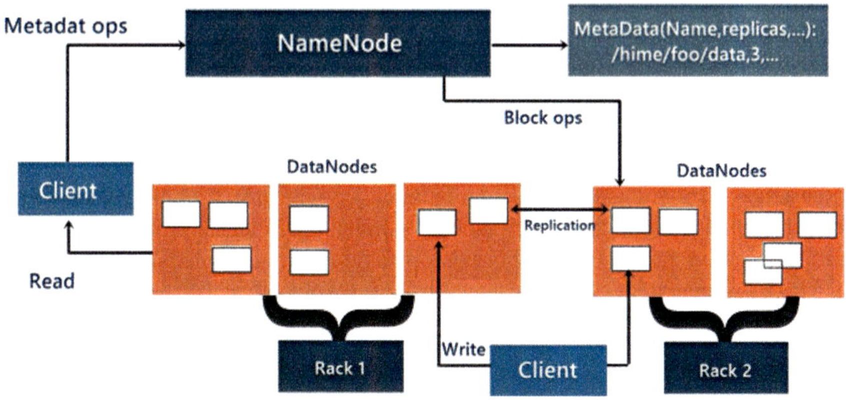

Fig. 1 HDFS architecture with master and slave

commands within one of these two stages. Based on this core structure, complex jobs need to be split into two or more jobs to fulfill the two-stage process. In Hadoop, this would be divided into two jobs of (map, shuffle) and (map, reduce) for each job. Data would be read from the disk, filtered and stored back to the disk with the first job. Again, data would be read from the disk, grouped and stored back to the disk. The MapReduce core structure consists of YARN and HDFS, and these two Hadoop native processes are used intensively in Pig. They provide reliability, performance and scalability for Pig. Pig divides jobs into small tasks, and, for each task, Pig reads data from HDFS, and returns data to HDFS once the process is completed. This in and out consumes considerable time. Multidimensional sensitivity-based anonymization method (MDSBA) was proposed for big data processing frameworks. Its core concept is applying optimized anonymization procedures and algorithms by splitting data into small tasks, so they can be parallelized among the cluster nodes. The reduce phase is expensive because it involves data partitioning, data serialization and de-serialization, data compression and disk I/O [16]. In data anonymization, the reduce phase is presented by SQL grouping commands, which causes high shuffling processes. To reduce the groupBy effect in the anonymization application, a filtration command is initiated to split data logically as per nominal values. This type of split reduces the performance degradation caused by the shuffling process. This reduces the amount of shuffling among cluster nodes during the reduce stage.

Proposed scalable two-phase TDS algorithm is designed for Hadoop ecosystems such as Spark, Hive and Pig. Traditionally, SQL language is the database dominator to manage and alter data. In Hadoop ecosystems, SQL-like programs replace traditional SQL database. Pig Latin is a Hadoop tool that manages warehouse system, by using a proprietary scripting language. Pig Latin treats data as a set of tuples, which supports tackling extensive dataset. Thereby, substantial parallelism and a slew of optimization techniques are supported. Pig provides customized support for a user-defined function (UDF), by supporting many widely used languages, such as Java and Python [17].

Pig's programming language referred to as Pig Latin is a coding approach that resides at the top of Hadoop and provides high degree of abstraction for MapReduce programming but is a procedural code not declarative. Pig Latin code can be extended through various user-defined functions that are written in Python, Java, Groovy, JavaScript and Ruby. Pig has tools for data storage, data execution and data manipulation. The proposed algorithm implements Pig scripts with UDF tool. Pig Latin scripting language is used to write the code for analyzing the data, and the compiler converts the code into the equivalent MapReduce code. Pig is a Hadoop extension that simplifies Hadoop programming by giving you a high-level data processing language while keeping Hadoop's simple scalability and reliability. Pig is architected from the ground up with support for user-defined functions [18].

2 Related Works

The concept of anonymity was proposed in [1]. Bottom-up generalization was used to achieve anonymity in Datafly system [19] and μ-Argus system [20]. LeFevre et al. [21] addressed the scalability problem of anonymization algorithms via introducing scalable decision trees and sampling techniques. Fung et al. [22–25] proposed the TDS approach that produces anonymous dataset without the data exploration problem. Wang et al. [26] adapt a bottom-up generalization approach which works iteratively to generalize the data. Identifying the best generalization is the key to climb up the hierarchy at each iteration. Li et al. [27] proposed two classification-aware data anonymization methods. It combines local value suppression and global attribute generalization. The attribute generalization is found by the data distribution, in spite of privacy requirement. Generalization levels are optimized by normalizing mutual information for preserving classification capability. BUG was proposed recently for anonymization using MapReduce [28]. The algorithms in [29, 30] explain the MRBUG driver which operates twice in intermediate and final stages. It first merges anonymization and then applies generalization. It implemented the BUG driver to leverage information gain and security. Pandilakshmi et al. [31] proposed advanced BUG. It splits data into smaller partitions and runs the MRBUG driver on a partitioned dataset. The anonymization method [32] uses a hybrid combination of BUG and TDS to anonymize data. A threshold value of k is determined by several algorithms to distinguish BUG from TDS use. Some hybrid methods were recently proposed for big data by Zhang et al. and Irudayasamy et al. [25, 28, 29]. Our research mainly focuses on the scalability and efficiency issue of TDS anonymization on large dataset.

3 Proposed Work

The paper presents a scalable two-phase top-down specialization (STPTDS) approach to generalize a table to satisfy the anonymity requirement while preserving its usefulness to classification. TDS generalizes the table by specializing it iteratively starting from the most general state. At each step, a general (i.e., parent) value is specialized into a specific (i.e., child) value for a categorical attribute, or an interval is split into two sub-intervals for a continuous attribute. This process is repeated until further specialization leads to a violation of the anonymity requirement.

TDS Algorithm

1. Initialize every value in T to the top most value.
2. Initialize Cuti to include the top most value.
3. while some $x \in \cup$Cuti is valid and beneficial do
4. Find the Best specialization from $\cup$Cuti.
5. Perform Best on T and update $\cup$Cuti.

6. Update Score(x) and validity for $x \in \cup$Cuti.
7. end while
8. Return Generalized T and ∪Cuti.

Initially, Cuti contains only the top most value for its attribute. The valid, beneficial specializations in ∪Cuti form the set of candidates to be performed next. At each iteration, it finds the candidate of the highest Score, denoted Best (Line 5), apply Best to T and update ∪Cuti (Line 6), and update Score and validity of the candidates in ∪Cuti (Line 7). The algorithm terminates when there is no more candidate in ∪Cuti, in which case it returns the generalized table together with ∪Cuti.

Our objective is to evaluate if the proposed scalable two-phase TDS preserves privacy through generalization and anonymization for a given large set of big data.

Proposed Scalable Two-Phase TDS Algorithm Using Pig

STPTDS has three components, namely data partition, anonymization level merging and data specialization.

```
READ File
FILTER by Class
GROUP by(SOCIAL, RACE, SEX)
FILTER IF Record_no > K

    Execute UDF to ungroup the Records
    STORE in SG(fully-equivalent group) Output

ELSE

    Execute UDF to ungroup the Records
    STORE in SSG(semi-equivalent group) Output

END
READ SSG
GROUP by(RACE, SEX)
FILTER IF Record_no > K

    Execute UDF to Anonymize and ungroup the Records
    STORE in SG(fully-equivalent group) Output

ELSE

    Execute UDF to ungroup the Records
    STORE in SSG(semi-equivalent group) Output

END
READ SSG
GROUP by(SEX)
FILTER IF Record_no > K

    Execute UDF to Anonymize and ungroup the Records
```

Table 1 Adult dataset and G(QID) groups used in the below experiments

All attributes	G(QID)1	G(QID)2	G(QID)3
AGE, JOB, MARITAL_STATUS, EDU, SOCIAL, RACE, SEX, POSITION, COUNTY, COUNTRY, SALARY	AGE, JOB, MARITAL_ STATUS, EDU	SOCIAL, RACE, SEX, POSITION	COUNTY, COUNTRY, SALARY
	Class: EDU	Class: POSITION	Class: SALARY

STORE in SG(fully-equivalent group) Output

ELSE

Execute UDF to ungroup and suppress the Records
STORE in SG(semi-equivalent group) Output

END

The experiment has used adult dataset, taken from UCI machine learning repository. The attributes include {AGE, JOB, MARITAL_STATUS, EDU, SOCIAL, RACE, SEX, POSITION, COUNTY, COUNTRY, SALARY}. Attributes are divided into three groups of QIDs. The first QID group is G(QID)1 = {AGE, JOB, MARITAL_STATUS, EDU}, where EDU is the class attribute. The second group is G(QID)2 = {SOCIAL, RACE, SEX, POSITION}, where POSITION is the class attribute. And finally, the third group is G(QID)3 = {COUNTY, COUNTRY, SALARY}, where SALARY is the class as shown in Table 1. For this experiment, we only anonymized group G(QID)1, by grouping QIDs for {AGE, JOB, MARITAL_STATUS} in the first stage, then {JOB, MARITAL_STATUS}, in the second stage, and finally {MARITAL_STATUS} only in the third stage. Table 1 shows the dataset attributes and G(QID) groups. This dataset is chosen since it contains a considerable number of personal attributes which could create a large number of Q_IDs. This dataset is more often used by researchers in various methods and is prominence in data anonymization. The adult dataset was enlarged up to seven different sizes by using MySQL code.

The TDS algorithm involves various iterations when calculating the best cut and scores. The algorithm reduces the size of data flowing to the UDF program. The UDFs are used to iterate a large size of arrays and it also executes almost all anonymization processes. The UDFs are implemented in different locations. It involves mainly two purpose anonymization and ungrouping. While anonymizing it implements three masking types of intervals, taxonomy tree and suppression are implemented. The aim of the algorithm is to reduce the size of data flowing to the UDF program. The ungrouping UDF reads each wrapped array, counts the number of objects and maps each array with indices. A function is defined in which it can update the array size on each wrapped array.

Experimental setup: The laboratory set up includes eight virtual machines, with one master and seven workers. Each node contains 4 core CPUs at 2.4 Hz, with a physical memory of 8 GB. Memory size on each worker starts from 16 GB, with two quad-core processors. However, the required size of memory and processor in each worker of the cluster depends on three main factors: the data size, the time required to complete the job, and the number of workers and masters within the cluster. Pig was set up on the NameNode to run the Pig Latin script. Adult data [33] was deployed for the experiments. Data was randomly enlarged for seven different sizes.

4 Results and Discussion

The proposed paper presents a scalable two-phase top-down approach, a practical and efficient algorithm for determining a generalized version of data that masks sensitive information and remains useful for modeling classification. This top-down approach is a natural and efficient structure for handling categorical and continuous attributes.

When the data size is small it is possible to avoid data split. Hence, the classification error was low. The low classification error indicates the high level of information gain. As the data size increases, the anonymization algorithms need to split large data size into small blocks. This split affects the data equivalency to a certain extent.

The proposed work investigated the processing time for various data sizes (5, 10, 15, 20, 25, 30 and 35 GB). The experiment was conducted to compare the processing times for different sizes of dataset handled by different scenarios with four workers and one master, five workers and one master, six workers and one master and seven workers and one master. More data growth showed better performance as the workers increased. Figure 2 represents the experimental results for different scenarios.

The BUG and TDS methods have also been implemented in big data anonymization. A few amendments were applied to suit the big data frameworks regarding parallelization and distribution. The core concept of k-anonymity and algorithms are applied in both cases of TDS and BUG. The comparison of some of these methods in traditional data and big data is shown in Table 2.

5 Conclusion

Big data analytics require proper anonymization tools and methods that furnish for secure procurement of information. Current anonymization methods may not be able to process large data sizes efficiently. Some of the anonymization methods are not capable of discriminating between the roles and potentially different access privilege levels of the users, resulting in the same levels of the information obscurity for their clients. To overcome some of these limitations, a scalable two-phase TDS algorithm is proposed. It investigates the scalability problem of large-scale data anonymization by TDS and proposes a highly scalable two-phase TDS approach using Pig. Dataset

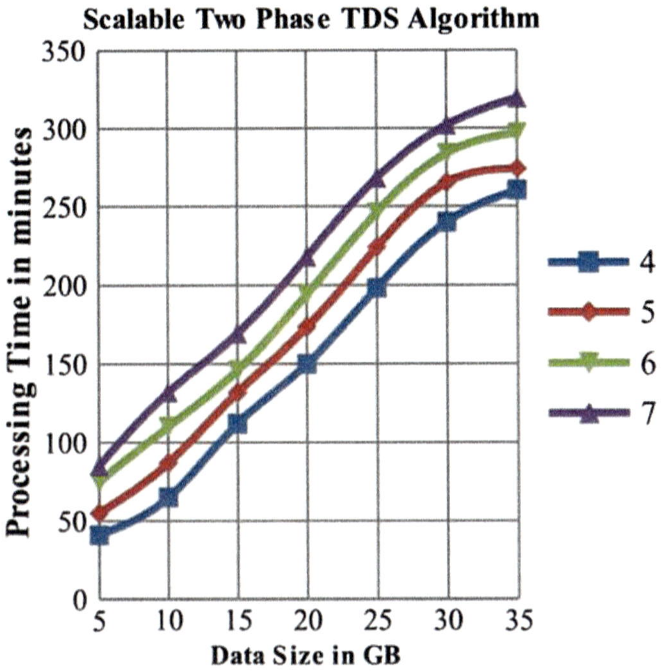

Fig. 2 Process time comparison between different scenarios with 4 workers, 5 workers, 6 workers and 7 workers. The overall performance was better while increase in workers

are partitioned and anonymized in parallel in the first phase, producing intermediate results. Then, the intermediate results are merged and further anonymized to produce consistent k-anonymous dataset in the second phase. UDFs are implemented in java to implement these two phases. The framework provides capabilities for efficient anonymization of big data, by incorporating contemporary anonymization tools, such as Pig Latin and UDF programming. The most outstanding feature of Pig programs is that their structure is open to considerable parallelization making it easy for handling large dataset.

The proposed work does not support for data streaming anonymization. Most recent big data frameworks provide complete solutions for the data stream. Hence, it is need to amend the current work to cope with the continuous data streaming. The proposed work can be further extended using other Hadoop ecosystems.

Table 2 Comparison of proposed work with already proposed works

Methods	Concepts	Strength	Limitation
Lefevre, "workload-aware anonymization techniques for large-scale dataset" [34]	Anonymization algorithms that incorporate a target class of workloads, consisting of one or more data mining tasks as well as selection predicates and the dataset much larger than main memory	High efficiency and quality data overcome problem of scalability	Problem of measuring the quality of anonymized data fails to work in the top-down specialization (TDS) approach
Iwuchukwu "*K*-anonymization as spatial indexing: toward scalable and incremental anonymization" [35]	*K*-anonymizing a dataset is similar to building a spatial index over the dataset using R-tree index-based	Achieve high efficiency and quality anonymization multidimensional generalization, high accuracy	More compaction is needed to achieve high-quality anonymization different indexing algorithms provide different issues
Dean and Ghemawat "MapReduce: simplified data processing on large clusters" [36]	MapReduce is a programming model implementation for processing and generating large dataset performs Map() and Reduce()	Allows us to handle lists of values that are too large in memory the model is easy to use	MapReduce is not suitable for a short online transactions
Fung et al. "top-down specialization for information and privacy preservation" [37]	TDS approach to produce anonymization set with data exploration problem	Improved efficiency	Failed to handle large set of data
Jiang et al. "a secure distributed framework for achieving *k*-anonymity" [38]	Distributed algorithm to anonymization for vertical partitioning data	Provides secure anonymization and integration	No scalability
Irudayasamy et al. "parallel bottom–up generalization approach for data anonymization using MapReduce for security of data in public cloud" [28]	Highly scalable parallel BUG approach using MAP reduce	Scalable and efficient	Insufficient privacy for data

(continued)

Table 2 (continued)

Methods	Concepts	Strength	Limitation
Fung, Wang "anonymizing classification data for privacy preserving" [22]	State of art approach for centralized TDS anonymization denoted as CentTDS	Scalable and efficient	Low scalability on large dataset.
Zhang et al. "a scalable two-phase top-down specialization approach for data anonymization using MapReduce on cloud" [39]	Leveraged MapReduce to anonymize large-scale data	Scalable and efficient	Privacy preserving is a challenge due to increase in large volumes of dataset
Zhanag, Wang "a hybrid approach for scalable sub-tree anonymization over big data using MapReduce on cloud" [40]	Hybrid approach for scalable sub-tree anonymization	Scalable	Large crowd effect of large data is underutilized and leads to distributed data anonymization
Proposed "a scalable two-phase TDS for big data anonymization using Apache Pig	Anonymize large-scale dataset	Scalable and efficient	Fails for data streaming

References

1. Sweeney L (2002) *k*-anonymity: a model for protecting privacy. Int J Uncertain, Fuzziness Knowl-Based Syst 10(5):557–570
2. Samarati P (2001) Protecting respondents' identities in microdata release. IEEE Trans Knowl Eng 13:1010–1027
3. Liu W, Uluagac AS, Beyah R (2014) MACA: a privacy-preserving multi-factor cloud authentication system utilizing big data. IEEE INFOCOM Workshops, pp 518523
4. Rahmani A, Amine A, Mohamed RH (2014) A multilayer evolutionary homomorphic encryption approach for privacy preserving over big data. In: Proceedings of international conference on cyber-enabled distributed computing and knowledge discovery, pp 19–26
5. Zhang X, Yang C, Nepal S, Liu C, Dou W, Chen J (2013) A MapReduce based approach of scalable multidimensional anonymization for big data privacy preservation on cloud. In: Proceedings of 3rd international conference on cloud and green computing, IEEE, pp 105–112
6. Hasan O, Habegger B, Brunie L, Bennani N, Damiani E (2013) A discussion of privacy challenges in user profiling with big data techniques: The EEXCESS use case. In: IEEE international congress on big data, pp 1–6
7. Jensen M, Kiel (2013) Challenges of privacy protection in big data analytics. In: Proceedings of international congress on big data, IEEE, pp 235238
8. Chakravorty A, Wlodarczyk T, Rong C (2013) Privacy preserving data analytics for smart homes. IEEE security and privacy workshops, pp 1–5

9. Hayashi K, Yokohama (2013) Social issues of big data and cloud: privacy, confidentiality, and public utility. In: Proceedings of 8th international conference on availability, reliability and security, pp 506–511
10. Li L, Goodchild MF, Barbara S (2013) Is privacy still an issue in the era of big data—location disclosure in spatial footprints. In: Proceedings of 21st international conference on geoinformatics, IEEE, pp 1–4
11. Wang K, Yu PS, Chakraborty S (2004) Bottom-up generalization: a data mining solution to privacy protection. USA, pp 249–256
12. Fung BCM, Wang K, Yu PS (2005) Top-down specialization for information and privacy preservation. USA, pp 205–216
13. Sun Y, Yin L, Liu L, Xin S (2014) Toward inference attacks for *k*-anonymity. Pers Ubiquit Comput 18(8):1871–1880
14. Zhang X, Yang C, Nepal S, Liu C, Dou W, Chen J (2013) A MapReduce based approach of scalable multidimensional anonymization for big data privacy preservation on cloud, pp 105–112
15. Zhang X, Yang LT, Liu C, Chen J (2014) A scalable two-phase top-down specialization approach for data anonymization using MapReduce on cloud. IEEE Trans Parallel Distrib Syst 25(2):363–373
16. http://hadoop.apache.org/hdfs/
17. Gold S (2016) Python: Python programming learn Python programming in a day-a comprehensive introduction to the basics of Python and computer programming. ACM Digital Library. https://dl.acm.org/citation.cfm?id=3055663. Accessed 20 Apr 2017
18. http://www.slideshare.net/kevinweil/hadoop-pig-and-twitter-nosql-east-2009
19. Sweeney L (1998) Datafly: a system for providing anonymity in medical data. In: International conference on database security, pp 356–381
20. Hundepool, Willenborg L (1996) μ- and τ-argus: software for statistical disclosure control. In: Third international seminar on statistical confidentiality, Bled
21. LeFevre K, DeWitt DJ, Ramakrishnan R (2008) Workload-aware anonymization techniques for large-scale data sets. ACM Trans Database Syst 33(3):1–47
22. Fung BCM, Wang K, Yu PS, Chen R (2007) Anonymizing classification data for privacy preservation. IEEE Trans Knowl Data Eng 19(5):711–725
23. Mohammed N, Fung B, Hung PCK, Lee CK (2010) Centralized and distributed anonymization for high-dimensional healthcare data. ACM Trans Knowl Discov Data 4(4):Article 18
24. Fung B, Wang K, Wang L, Hung PCK (2009) Privacy preserving data publishing for cluster analysis. Data Knowl Eng 68(6):552–575
25. Fung BCM, Wang K, Chen R, Yu PS (2010) Privacy-preserving data publishing: a survey of recent developments. ACM Comput Surv 42(4):1–53
26. Wang K, Yu PS, Chakraborty S (2004) Bottom-up generalization: a data mining solution to privacy protection. In: 4th IEEE international conference on data mining, (ICDM'04). Piscataway: IEEE
27. Li J, Liu J, Baig M, Wong RCW (2011) Information based data anonymization for classification utility. Data Knowl Eng 70(12):1030–1045
28. Irudayasamy A, Arockiam L (2015) Parallel bottom–up generalization approach for data anonymization using map reduce for security of data in public cloud Indian. J Sci Technol 8:1. https://doi.org/10.17485/ijst/2015/v8i22/79095
29. Irudayasamy A, Arockiam L (2015) Scalable multidimensional anonymization algorithm over big data using map reduce on public cloud. J Theor Appl Inf Technol 74:221–231
30. Balusamy M, Muthusundari S (2014) Data anonymization through generalization using map reduce on cloud. In: 2014 international conference on computer communication and systems, IEEE, pp 039–042. https://doi.org/10.1109/icccs.2014.7068164
31. Pandilakshmi K, Banu GR (2014) An advanced bottom up generalization approach for big data on cloud 3:1054–1059
32. Zhang X, Yang LT, Liu C, Chen J (2014) A scalable two-phase top-down specialization approach for data anonymization using MapReduce on cloud. IEEE Trans Parallel Distrib Syst. https://doi.org/10.1109/tpds.2013.48

33. Becker RK (1996) Adults data. ftp://ftp.ics.uci.edu/pub/machine-learning-databases
34. LeFevre K, DeWitt DJ, Ramakrishnan R (2008) Workload-aware anonymization techniques for large-scale datasets. ACM Trans Database Syst 33(3):1–47. Incognito: efficient full-domain *k*-anonymity. In: Proceedings of ACM SIGMOD international conference management of data (SIGMOD'05), pp 49–60
35. Iwuchukwu T, Naughton JF (2007) k-anonymization as spatial indexing: toward scalable and incremental anonymization. In: Proceedings of 33rd international conference very large data bases (VLDB'07), pp 746–757
36. Dean J, Ghemawat S (2008) Mapreduce: simplified data processing on large clusters. Comm ACM 51(1):107–113
37. Fung BCM, Wang K, Yu PS (2005) Top-down specialization for information and privacy preservation. USA. https://doi.org/10.1109/icde.2005.143
38. Jiang W, Clifton C (2006) A secure distributed framework for achieving *k*-anonymity. VLDB J 15(4):316–333
39. Zhang X, Yang LT, Liu C, Chen J (2014) A scalable two-phase top-down specialization approach for data anonymization using MapReduce on cloud. IEEE Trans Parallel Distrib Syst. https://doi.org/10.1109/tpds.2013.48
40. Zhang X, Liu C, Nepal S et al (2014) A hybrid approach for scalable sub-tree anonymization over big data using MapReduce on cloud. J Comput Syst Sci 80(5):1008–1020

Segmentation of Lip Print Images Using Clustering and Thresholding Techniques

S. Sandhya, Roshan Fernandes, S. Sapna, and Anisha P. Rodrigues

Abstract Segmentation process forms a vital component of image processing. The major objective of lip print image segmentation is to separate the image pixels into foreground pixels that contain the region of interest and background pixels that mainly consists of noise. This partition of the original lip print image into various meaningful representations makes it easy to analyze the image. There are various segmentation techniques available that can be used for certain problem statements. But in some cases, many of the existing techniques need to be combined along with our knowledge on the domain to productively solve a problem on image segmentation. In this paper, algorithms to effectively segment the original lip print image into upper and the lower lip are presented. Thresholding and clustering techniques are used to segment the lip print images. Results show that the presented techniques provide good performance and better segmentation results. It also shows that the noise pixels are effectively categorized as the background pixels while the portion consisting of the region of interest is categorized as the foreground pixels effectively.

Keywords Lip prints · Segmentation · Thresholding · Machine learning · Image processing · *k*-means clustering · Histogram thresholding · Edge detection

1 Introduction

The prime objective and the first and foremost step of image processing is to redeem the data that is primarily necessary from the original image in such a way that it would not impact the other important features of the given original image. For this to be fulfilled, enhancing of the image becomes an important prerequisite. Once the noise is removed from the given image, any other operations can be easily performed on that image.

S. Sandhya · S. Sapna (✉)
Department of Information Science and Engineering, NMAM Institute of Technology, Nitte, India
e-mail: sapna_s@nitte.edu.in

R. Fernandes · A. P. Rodrigues
Department of Computer Science and Engineering, NMAM Institute of Technology, Nitte, India

N. N. Chiplunkar and T. Fukao (eds.), *Advances in Artificial Intelligence and Data Engineering*, Advances in Intelligent Systems and Computing 1133,
https://doi.org/10.1007/978-981-15-3514-7_76

Segmentation is one among the most fundamental and significant steps in image processing. It can be termed as a pre-processing stage. In this stage, the original image is broken down into several regions mainly to determine the lines, curves or objects present in an image. Each of these regions will represent some kind of features like intensity, contrast, color, etc.

The most prominent step in lip print-based person identification system is the segmentation of the digital lip print images into upper lip and the lower lip. Lip print images usually consist of two different components. The foreground consists of upper lip and lower lip that forms the regions of interest and the background consists of the unwanted noise [1]. The foreground represents the actual image that is formed by the contact of the lip with the recording agent. This needs to be separated from the unwanted noise.

Selection of a proper segmentation algorithm is a very important task. Segmentation results are further used in the feature extraction phase. If the segmentation process does not provide proper results then the feature extraction algorithms would extract erroneous features from the noise present. This would also effect the identification phase. There are many segmentation techniques available that can be utilized for certain problem statements. But in some cases, many of the existing techniques need to be combined along with our knowledge on the domain to effectively solve a segmentation problem and also achieve better results.

Some of the existing techniques that can be used in image segmentation are edge detection-based methods, clustering methods, thresholding methods, fuzzy theory-based methods and region growing-based methods. These segmentation techniques find various applications in fields like medical image processing, pattern recognition, face detection, etc. [2].

Lip prints are the regular lines shaping ridges and depressions on the human lips. They are unique like the fingerprints. It has been affirmed that lip prints recover in the wake of encountering changes, for instance, aggravation, minor damage, etc. However, major injuries may result in some scars that can bring some alterations in the pattern.

There are relatively few freely accessible databases of lip print images. The lip prints given by The Biometric Research Center, University of Silesia and the SUT-Lips-DB database are the freely accessible lip print datasets for research purposes.

2 Related Work

Zaitoun and Aqel state that segmentation of images is an important aspect of image processing. There is no specific solution for segmentation problem. Many of the existing techniques need to be combined along with our knowledge on the domain to productively solve a problem on image segmentation. This work presents a review on many different segmentation techniques like layer-based method that deals with the shapes and block-based methods that deals with the color, pixels and various other features [3].

Bazen and Gerez presented an algorithm for fingerprint segmentation. The main aim of segmentation is to conclude what portion of the image belongs to the foreground and what part of the image belongs to the background/noise. The algorithm makes use of the features like variance, mean and coherence. These are called the pixel-based features. The accuracy obtained by using the described method was almost 93% [4].

Raju and Neelima present an histogram thresholding procedure for image segmentation. The major aim of this method was to divide an image into various segments and represent the image into an meaningful one and also simplify the analysis of the image. The proposed approach is said to be very simple and also achieves better results. It is being tested on various sets of images and also compared against different segmentation approaches. This technique proved to be the one providing better results when compared to the other approaches [5].

Bora and Gupta present an approach that uses *k*-means clustering along with edge detection methods for segmentation. Segmentation forms a vital component in image processing techniques. Clustering is an unsupervised classification technique. Research shows that clustering methods can be used for segmentation. The distance measure used in clustering is 'cosine' distance. Sobel method is used for edge detection. This approach considers only color images and provides effective segmentation results [6].

Vala and Baxi present an approach for the segmentation of digital images using Otsu's thresholding approach. The author states that the calculations involved in this approach are very simple and hence easy to implement. Otsu's algorithm based on improved histogram produces good results in case of salt and pepper noises [7].

Bhargava et al. proposed an approach that uses Otsu's approach in order to convert an image into its binary equivalent. This method calculates a global threshold value based on which the pixels are divided into foreground and background pixels, converting an image into its binary equivalent. This method finds many applications due to its simplicity [8].

Panwar et al. provide a comparative analysis on *k*-means clustering and the thresholding techniques. *k*-means clustering results in division of the given input image into k different clusters. It is also called unsupervised classification technique. Thresholding techniques are simple by nature. It involves finding a threshold and then binarizing the input image. These techniques are compared based on SNR, MSR and PSNR. Thresholding technique was found to perform better with good results [9].

Thakur and Madaan conducted a study on some the image segmentation techniques. The authors have effectively described and compared various techniques like edge-based methods, thresholding methods, fuzzy theory-based methods, clustering techniques and region growing methods. These segmentation techniques find various applications like medical image processing, pattern recognition, face detection, etc. The authors also state that there is no general solution for segmentation since it depends on various factors like pixel intensity, texture, content, domain, etc. [2].

Smacki et al. displayed a strategy for lip print identification utilizing the dynamic time warping calculation. The proposed strategy worked in two phases. The initial phase included pre-processing and second stage included extraction of features. The proposed strategy was tried on the lip prints acquired from 30 people. Agreeable outcomes were accomplished utilizing the proposed technique. Author additionally expressed that further adjustments made to the DTW calculation will result in better outcome [10].

Smacki presented a strategy for investigating the lip prints utilizing fast normalized cross-correlation system. At first, standardized images were created by disposing of the background subtleties. The images were then divided into their constituent parts: upper lip and lower lip. After the parting was done, the obtained parts were adjusted horizontally. Feature extraction was performed to extract the lines and pattern details on the lip prints. This examination was directed on 300 images. The proposed technique was tried on full lip pictures and the sub-images as well. Sub-image examination brought about high precision. However, the investigation did not take the blur and disfigured lip print images into consideration [11].

Wrobel et al. proposed a technique for lip print identification dependent on section correlation. Sections allude to the patterns or lines present on the lips. Hough transform was utilized to extricate the areas and an algorithm was concocted to look into the similarities and differences between the sections. Testing was led on 45 lip images. Error rate was determined dependent on the length of the areas removed. In some cases, where the section length was over 30, the error rate was observed to be high [12].

Bhattacharjee et al. proposed a way to use statistical model in distinguishing people based on lip prints. Accurate match and fast match algorithms were utilized. Feature vector was extracted and Euclidean distance was computed between the extracted feature vectors for verifying the lip prints. Testing was performed on 20 images. Accurate match algorithm yielded better outcome over fast match algorithm. Further improvement in precision can be accomplished by using a larger datasets [13].

Porwik and Orczyk performed recognition of lip prints and also correlation of lip prints utilizing DTW calculation and the Copeland vote counting approach. The similitudes in the lip prints were decided utilizing DTW algorithm. DTW algorithm yielded poor outcomes. So as to improve the correctness of the approach, DTW algorithm was merged with the Copeland vote counting approach where pair-wise vote counting technique was utilized to classify an obscure item. Study was directed on 120 lip print images. Author also opined that the methodology presented can be viably utilized in criminological investigations [14].

Wrobel et al. proposed a methodology for recognizing an individual dependent on the lip prints. Significance was given to the bifurcations on the lip images. The dark pixels in lip print were examined for separating the bifurcations. Bifurcation networks were contrasted to distinguish the lip prints. This examination brought about a similarity measure. Lower similarity measure demonstrated high comparability in the lip prints. The examination was performed on 120 lip print images. The best outcome was accomplished with an error rate of 23% [15].

Sharma et al. presented a person identification method using lip print images. Pre-processing was performed on the acquired images to eliminate unwanted noise. Feature extraction phase resulted in extraction of patterns for the purpose of matching the prints. Brute force algorithm was employed in pattern matching. Test was conducted on 200 images. It yielded an accuracy of 89.5%. The study did not consider the factors like scarring and changing pattern with age for the analysis [16].

A detailed study of different techniques used for segmentation of lip prints is given in [17, 18].

3 Segmentation

Research shows that there are many approaches available that can be utilized for the image segmentation purpose. In this work, thresholding and clustering methods are used for segmentation of the upper and the lower lip.

3.1 Thresholding

The least complex property that the pixels in a region/segment can share is the intensity. So a characteristic method to fragment such regions is through thresholding, the dissociation of dark and light regions.

To begin with, a threshold value is chosen and the pixels with value more than the threshold are converted to one, whereas the ones with the value less than the threshold are converted to zero. Thresholding technique converts the input grayscale image into its binary equivalent [9, 19].

If a function $F(y, z)$ is used to represent the thresholded edition of the function $G(y, z)$ at some threshold value T, thresholding can be represented as in Eq. (1).

$$F(y,z) = \begin{cases} 1 \text{ if } G(y,z) \geq T \\ 0 \quad \text{otherwise} \end{cases} \tag{1}$$

Thresholding technique always takes a color image or a grayscale image as input. The output produced is the binary equivalent of the input image that represents the segmentation results. Typically, in thresholding operation, white pixels are treated as the foreground and black pixels are treated as the background.

The important task in the thresholding operation is the selection of the threshold value. There exist various methods that can be used to select a threshold value. The user can manually select the threshold value on a trial and error basis that can be used on all the images to transform the input image into its equivalent binary representation, or any algorithms like Otsu's thresholding method can be used to automatically determine the threshold value based on the pixel distribution in the image. The histogram can be generated and the threshold value can be selected based on its peaks and valleys as well. But this method is generally difficult to implement. There are three types of thresholding algorithms. Global thresholding is used when the distribution of intensity of the foreground objects and the background noise is adequately distinct. When different threshold values are used for various regions, the operation is termed adaptive thresholding. Local thresholding is performed by dividing the images into sub-images and then choosing a threshold.

For segmentation of the lip print images, initially a threshold value is automatically determined using the Otsu's thresholding algorithm. The Otsu's algorithm is a global thresholding technique that uses the histogram of the grayscale image to determine the threshold [20]. The major objective of the Otsu's algorithm is to maximize the interclass variance and minimize the intra-class variance. Once the threshold value is chosen, this value is used to segment the original lip print image. The pixels with value more than the chosen threshold value are categorized as the foreground pixels that form the contact with the recording agent and the pixels with value lesser than that of the threshold are categorized as noise. Figure 1 shows the flowchart for segmentation using Otsu's thresholding technique.

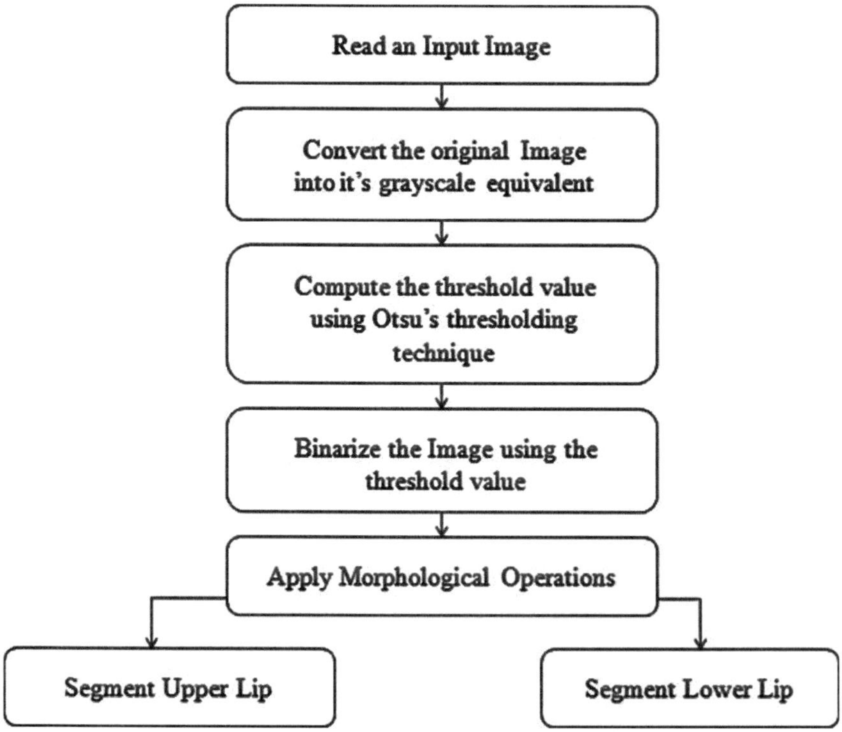

Fig. 1 Flowchart for segmentation of the lip print images using Otsu's thresholding

3.2 k-Means Clustering

Clustering is an operation in which a number of data points or objects are grouped so that the data points in a group are similar to one another in some sense when compared to the points in the other groups. These groups can be termed as clusters. It is also termed as unsupervised classification since it deals with a collection of objects that are not associated with any labels [21].

k-means clustering is one of the clustering methods that are being utilized in this work to segment the original lip print images into upper lip and lower lip. Initially, centroid values need to be specified based on which the objects are categorized into different groups. In *k*-means clustering technique, firstly, we need to select '*k*' data values as the initial cluster centers. Secondly, the separation between each of these cluster centers and each of the data points are calculated using some distance measures like Euclidean distance, Minkowski distance, etc. In this work, Euclidean distance is used as the distance measure. Based on the obtained distance, each of the data point is assigned to the cluster that is nearest to it. Averages of each and every cluster are updated and process is repeated until there is no change in the centroid value. The data points which are similar to each other grouped into a single cluster. Higher the Euclidean distance lesser the similarity and vice versa [9]. Figure 2 shows

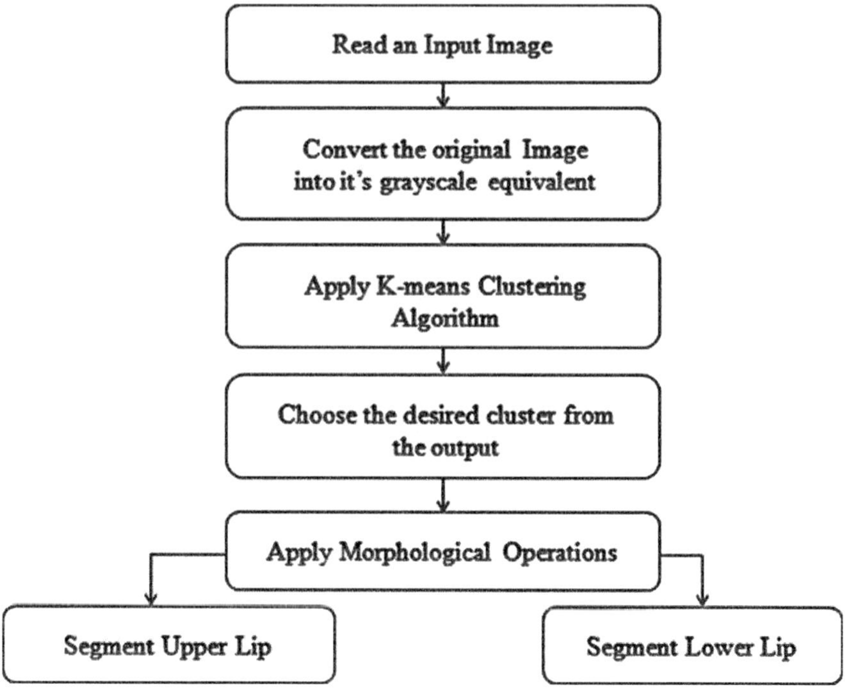

Fig. 2 Flowchart for segmentation of the lip print images using *k*-means clustering technique

the flowchart for the segmentation using k-means clustering technique. Equation (2) represents the objective function used in k-means clustering technique.

$$\text{Objective Function} \rightarrow J = \sum_{j=1}^{k} \sum_{i=1}^{n} \left\| x_i^{(j)} - c_j \right\|^2 \quad (2)$$

where k represents the number of clusters, c_j represents the centroid of cluster j and $\left\| x_i^{(j)} - c_j \right\|$ signifies the distance function. The only hitch of this technique is that the user needs to specify the number of clusters in advance.

4 Results and Discussion

This research work brings out the performance comparison of the thresholding technique and k-means clustering approach for segmenting the lip print images. The implementation was done using MATLAB. The result obtained is then used to compute the peak signal to noise ratio (PSNR) and mean squared error (MSE). SNR and MSE are utilized to analyze the performance since it has a high estimate to human impression of reconstruction quality. Higher value of PSNR and lower value of MSE show that the image quality is higher. The comparison of the segmentation algorithms using PSNR and MSE parameters represents the optimal performance in terms of human perception. Figure 3 shows the results obtained using the thresholding technique and Figs. 4, 5 and 6 show the results obtained using k-means clustering technique.

Table 1 shows the PSNR and MSE values for the results obtained using thresholding method, whereas Table 2 shows the PSNR and MSE values for the results obtained using k-means clustering technique.

Fig. 3 Segmentation using thresholding method

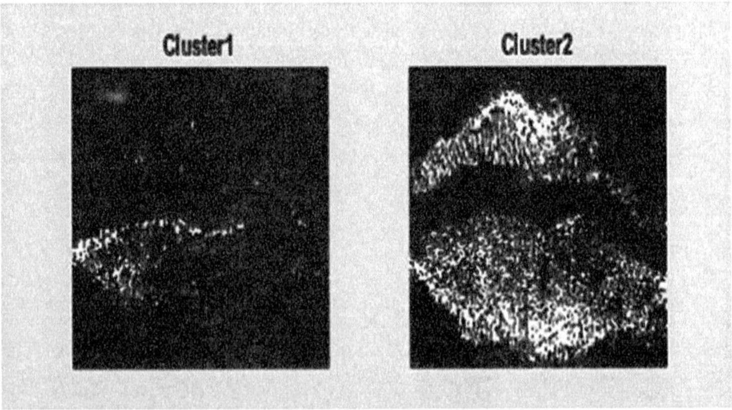

Fig. 4 Segmentation results when no. of clusters $k = 2$

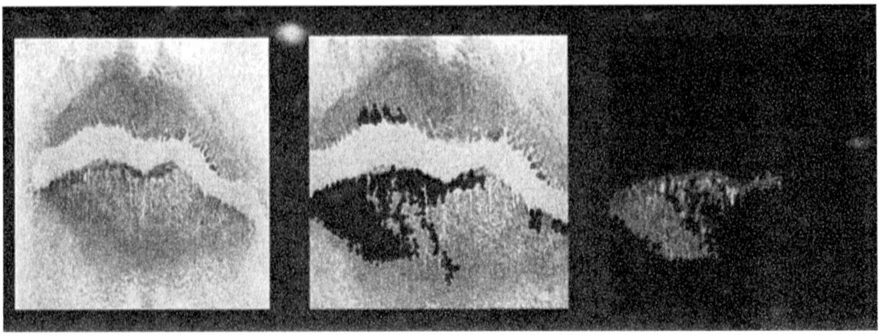

Fig. 5 Segmentation results using cluster 1

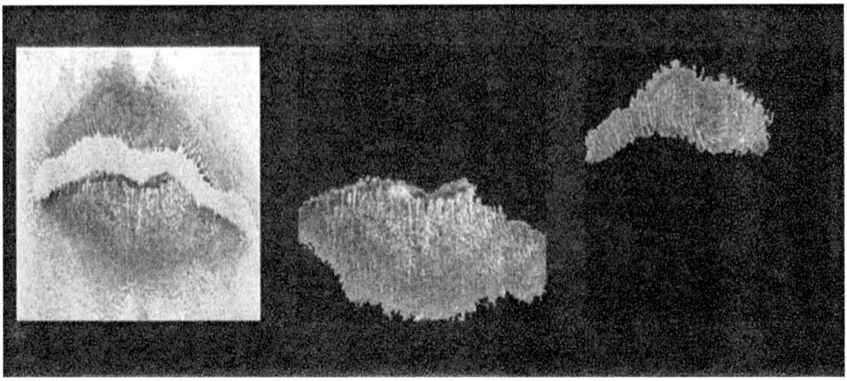

Fig. 6 Segmentation results using cluster 2

Table 1 PSNR and MSE values for segmentation results obtained using Otsu's thresholding

Sl. No	Input image	PSNR	MSE
1	Person_1.jpg	4.9691	2.0701e−04
2	Person_2.jpg	4.3100	2.4165e−04
3	Person_3.jpg	5.8277	1.4994e−04
4	Person_4.jpg	5.7956	1.7121e−04

Table 2 PSNR and MSE values for segmentation results obtained using *k*-means clustering

Sl. No	Input image	Cluster	PSNR	MSE
1	Person_1.jpg	Cluster 1	2.1946	3.9231e−04
		Cluster 2	4.5747	2.2678e−04
2	Person_2.jpg	Cluster 1	3.2254	3.0956e−04
		Cluster 2	4.2940	2.4285e−04
3	Person_3.jpg	Cluster 1	3.5706	2.8577e−04
		Cluster 2	5.6361	1.5659e−04
4	Person_4.jpg	Cluster 1	4.3231	2.7785e−04
		Cluster 2	5.3687	1.9287e−04

Figure 7 shows the graph plotted with PSNR values for input images using thresholding and *k*-means clustering techniques and Fig. 8 shows the graph plotted with MSE values for input images using thresholding and *k*-means clustering techniques.

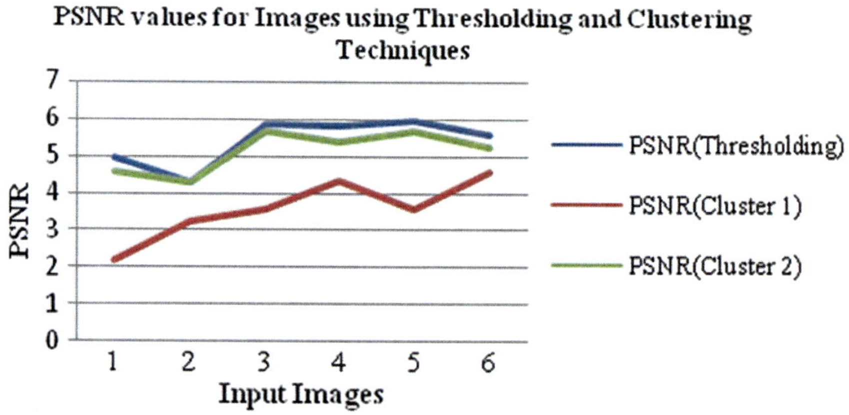

Fig. 7 PSNR values for images using thresholding and *k*-means clustering techniques

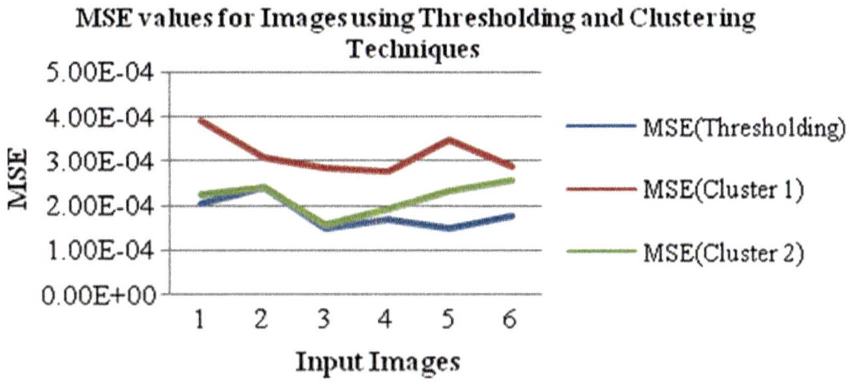

Fig. 8 MSE values for images using thresholding and *k*-means clustering techniques

5 Conclusion and Future Work

This paper presents two segmentation approaches for effectively segmenting the lip print images. The two methods used are thresholding technique and *k*-means clustering technique. The performance of the two techniques is also analyzed using PSNR and MSE methods. It can be observed that for most of the images in the dataset PSNR values are higher and MSE values are lower when thresholding technique is used when compared to the PSNR and MSE values when *k*-means clustering is used. Hence, we can conclude that thresholding technique provides fairly better results. But each of these techniques has its own share of disadvantages. The thresholding technique is dependent on the histogram of the grayscale image. Hence, a noisy image may result in improper threshold. The output of *k*-means clustering technique depends on the number of clusters. Selection of the number of clusters inappropriately may give improper results. Hence, the number of clusters needs to be selected wisely.

This work segments the upper lip and lower lip effectively from a lip print image. As a part of future work, efficient feature extraction techniques can be used to extract features from each of the segmented region and classification techniques can be applied to identify a person effectively based on the lip prints.

References

1. Choras M (2010) The lip as a biometric. Pattern Anal Appl 13(1):105–112
2. Thakur P, Madaan N (2014) A survey of image segmentation techniques. Int J Res Comput Appl Robot 2(4):158–165
3. Zaitoun NM, Aqel MJ (2015) Survey on image segmentation techniques. In: International conference on communication, management and information technology. ScienceDirect, Prague, pp 797–806
4. Bazen AM, Gerez SH (2001) Segmentation of fingerprint images. In: ProRISC workshop on circuits, systems and signal processing. Veldhoven, The Netherlands

5. Raju PDR, Neelima G (2012) Image segmentation by using histogram thresholding. Int J Comput Sci Eng Technol 2(1):776–779
6. Bora DJ, Gupta AK (2014) A novel approach towards clustering based image segmentation. Int J Emerg Sci Eng 2(11):6–10
7. Vala HJ, Baxi A (2013) A review on otsu image segmentation algorithm. Int J Adv Res Comput Eng Technol 2(2):289–387
8. Bhargava N, Kumawat A, Bhargava R (2014) Threshold and binarization for document image analysis using Otsu's algorithm. Int J Comput Trends Technol 17(5):272–275
9. Panwar P, Gopal G, Kumar R (2016) Image segmentation using *K*-means clustering and thresholding. Int Res J Eng Technol 3(5):1787–1793
10. Smacki L, Wrobel K, Porwik P (2007) Lip print recognition based on DTW algorithm. In: ACM Trans. Asian language information processing, vol. 6(2)
11. Smacki L (2013) Latent lip print identification using fast normalized cross-correlation. In: International conference on biometrics and Kansei engineering
12. Wrobel K, Doroz R, Palys M (2013) A method of lip print recognition based on sections comparison. In: International conference on biometrics and Kansei engineering
13. Bhattacharjee S, Arunkumar S, Bandyopadhyay SK (2012) Personal identification from lip-print features using a statistical model. Int J Comput Appl 55(13):30–34
14. Porwik P, Orczyk T (2012) DTW and voting-based lip print recognition system. In: International conference on computer information systems and industrial management, pp 191–202
15. Wrobel K, Doroz R, Palys M (2015) Lip print recognition method using bifurcations analysis. In: Asian conference on intelligent information and database systems. Cham, pp 72–81
16. Sharma P, Deo S, Venkateshan S, Vaish A (2011) Lip print recognition for security systems: an up-coming biometric solution. In: Proceedings of the 4th international conference on intelligent interactive multimedia systems and services. Berlin, Heidelberg, pp 347–359
17. Sandhya S, Fernandes R (2017) Lip print: an emerging biometrics technology-a review. In: IEEE international conference on computational intelligence and computing research. Coimbatore, pp 1–5
18. Chowdhury S, Biswas SB, Hazra J (2015) Lip imprint based biometric identification: a survey. Int J Sci Eng Res 6(11)
19. Khan MW (2014) A survey: image segmentation techniques. Int J Futur Comput Commun 3(2):89–93
20. Liu D, Yu J (2009) Otsu method and *K*-means. In: Ninth international conference on hybrid intelligent systems. Shenyang, China, pp 344–349
21. Lihua Tian JY (2016) Research on image segmentation based on clustering algorithm. Int J Signal Process, Image Process Pattern Recognit 9:1–12

Filtering-Based Text Sentiment Analysis for Twitter Dataset

Hiran Nandy and Rajeswari Sridhar

Abstract Text sentiment analysis for the blogs appearing in popular microblogging Web sites has become a very important area of research. Among the popular microblogging Web sites, twitter is most popular and has a limit of 140/280 characters per tweet. Due to the geopolitical and socioeconomic diversity of its users and summarized content, in this work twitter has been chosen for collecting corpus required for text sentiment analysis. Till date, two main approaches are into consideration for text sentiment analysis—lexicon-based and machine learning-based. No research work till date has focused on filtering-based text sentiment analysis that has been used here. The training data has been filtered based on six human emotions, and all their English synonyms and this filtered training dataset were used for training the classifier model. This model gives significantly better accuracy than traditional machine learning-based models on the test data.

Keywords Filtering · Text sentiment analysis · Twitter dataset · Human emotion

1 Introduction

In the recent past, microblogging has become a very effective means of communication. As the authors can post these blogs free of cost sitting from home, these blogs carry a huge amount of geopolitical and socioeconomic variance. Social networks are being used by millions of people to express their emotional state or to put forward their opinions or to just mention activities of their daily life [1]. It also gives business owners a direct means to communicate with the potential consumers and politicians a direct means to communicate with their potential voters. In this scenario, automatic sentiment classification of these blogs becomes a really important task.

H. Nandy (✉) · R. Sridhar
Department of Computer Science and Engineering, National Institute of Technology Tiruchirapalli, Tiruchirapalli, Tamil Nadu 620015, India
e-mail: hiran.tiucse@gmail.com

R. Sridhar
e-mail: srajeswari@nitt.edu

N. N. Chiplunkar and T. Fukao (eds.), *Advances in Artificial Intelligence and Data Engineering*, Advances in Intelligent Systems and Computing 1133,
https://doi.org/10.1007/978-981-15-3514-7_77

Facebook [2], Twitter [3], Pinterest [4], etc., are some of the most popular microblogging Web sites. According to a research [5], Twitter is the most popular among these microblogging Web sites. Moreover, the 140-character limit, which is later revised to 280-characters of each tweet, ensures expressions that are already in summarized form. As the number of users in twitter and number of posted tweets per day are constantly increasing, twitter can reflect the current view of society on any topic very effectively.

Because of all these practical facilities, this work has chosen twitter as its source of data corpus. As an alternative of traditional text sentiment analysis approaches, this paper proposes a novel idea of filtering-based text sentiment analysis, where all the six basic human emotions have been used to filter the training dataset. The authors got significantly better results while using this filtering-based approach over traditionally available approaches.

The arrangement of the remaining paper is as follows—Sect. 2 highlights the traditionally available approaches of text sentiment analysis. Section 3 illustrates the research methodology that this work has adopted. Performance analysis of the proposed scheme is mentioned in Sect. 4. Section 5 concludes the paper by stating the future scopes of the current work.

2 Related Work

Great reach and high population of social media have influenced many researchers to select text sentiment analysis as their area of research. Jose et al. [6] have highlighted how social media is the fastest growing and easily accessible medium for sentiment analysis because of its rapid growth and free accessibility. Tang et al. [7] have mentioned the simplicity and effectiveness that social media brings in organizations' brand-building. Lohmann et al. [8] on the contrary spotted the high risk of privacy breaching in social media. Twitter is the most popular microblogging Web site where people, regardless of their socioeconomic background and geopolitical boundary, share their day-to-day activities and express their opinions about certain topics. It has a unique constraint of 140/280 maximum characters per tweet [6, 8, 9]. Collective opinion about a certain topic can be extracted from twitter very effectively [10, 11]. According to a research [12], sentiment analysis is a natural language computing technique that can quantify the sentiments or opinions associated with a particular section of a document. A broad survey on existing approaches for information retrieval based on opinion mining has been carried out [13]. Authors have used web blogs to construct corpus of sentiment analysis and have used emoticon symbols for detecting user's mood [14]. A document can be classified into two sentiments—negative and positive [15, 16]. Two main approaches have been highlighted for text sentiment analysis—lexicon-based and machine learning-based [17–21]. A machine learning-based approach is used to find out customer's perspective by classifying their tweets into two categories—positive or negative [22]. Amolik et al. [23]

used naïve Bayes and SVM techniques to classify review tweets of upcoming Bollywood and Hollywood movies. The tweets of users across Pakistan before an election has been analyzed by [24] using already existing W-WSD, WordNet, and TextBlob sentiment analyzing tools, and SVM and naïve Bayes machine learning techniques have been used to test the obtained results. However, till now, no research work has focused on human emotion-based filtering of training data for twitter sentiment analysis which has been addressed in this work.

3 Research Methodology

The overall workflow of the proposed system is shown in Fig. 1. The following subsections details the processing done in each of the modules.

3.1 Data Collection

In this work, a pre-annotated sentiment analysis dataset from www.kaggle.com is being used. The dataset has three attributes—a. Tweet ID, b. Sentiment text, and c. Sentiment. The dataset consists of 99,989 number of tweets and each classified into two sentiments—**Positive (1)** and **Negative (0)**.

3.2 Train–Test Split

In the beginning of the work, the whole data is divided into two parts—training data and test data. 20% of the whole dataset has been considered as test data, and it has not been used for training purpose.

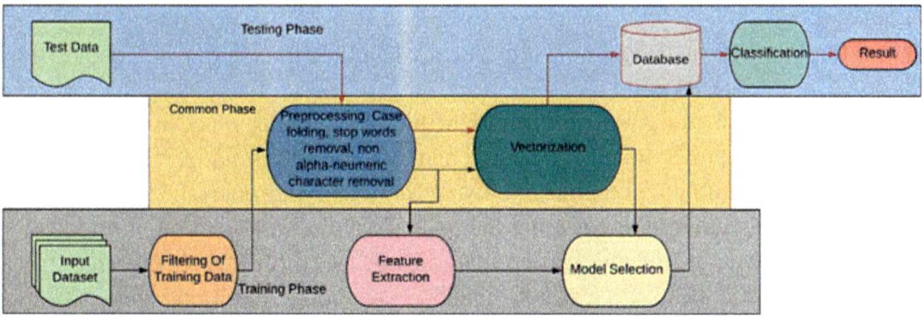

Fig. 1 Overall block diagram

3.3 Data Filtering

During the 1970s, psychologist Paul Ekman identified six basic human emotions that are experienced by all human beings universally. These emotions are—**Happiness, Sadness, Disgust, Fear, Surprise,** and **Anger**. Among these six emotions, for sentiment analysis this paper considers "Happiness" as a positive emotion and "Fear," "Sadness," "Disgust," and "Anger" as negative emotions. "Surprise" can represent both positive and negative state of emotions based on the context. The training data is filtered based on these six terms and all their synonyms in English. The original training data had 52.35% of total tweets classified as "Positive" (1) and 47.65% of the total tweets classified as "Negative" (0). After the filtering of the training data, total of 30,659 tweets remained as training data and among them 51.91% were classified as "Positive" (1) and 48.19% were "Negative" (0). Only this filtered dataset has been used for further processing.

3.4 Data Preprocessing

The filtered dataset was preprocessed with the following preprocessing techniques—
Undefined Character Removal. All the characters except alphanumeric characters (A–Z, a–z, 0–9), mentions (@), hashtags (#) were removed using regular expression. ":" and "/" characters were also kept to extract URLs later. ")", "(", "|", "-"—these four characters were kept to extract emoticon-based features.
Case Folding. For normalization purpose, the whole data is converted into lower case.
Stop Words Removal. All the stop words of the English language were removed using NLTK toolkit of python. Although "no" and "not" are considered as stop words in NLTK, they have been kept in our data as otherwise it could change the semantic meaning of a sentence completely.
Removing Repeated Characters. In all the words for three or more successive occurrences of the same character, only three occurrences have been kept using regular expressions.
Removing Small Words. All the words that are less than three letters long have been discarded.
Tokenization, Stemming, and Lemmatization. Using blank space as separator, all the sentences have been tokenized first. Then WordNet Lemmatizer is used to get the lemma of each token, and after that using Snowball Stemmer, the base forms of each token have been identified. Later the tokens have been joined into sentences.

To check the nature of the training data in this work, the frequently occurring words in positive and negative tweets using "word cloud" diagrams were given in Fig. 2a, b.

The test data was also preprocessed using these six techniques.

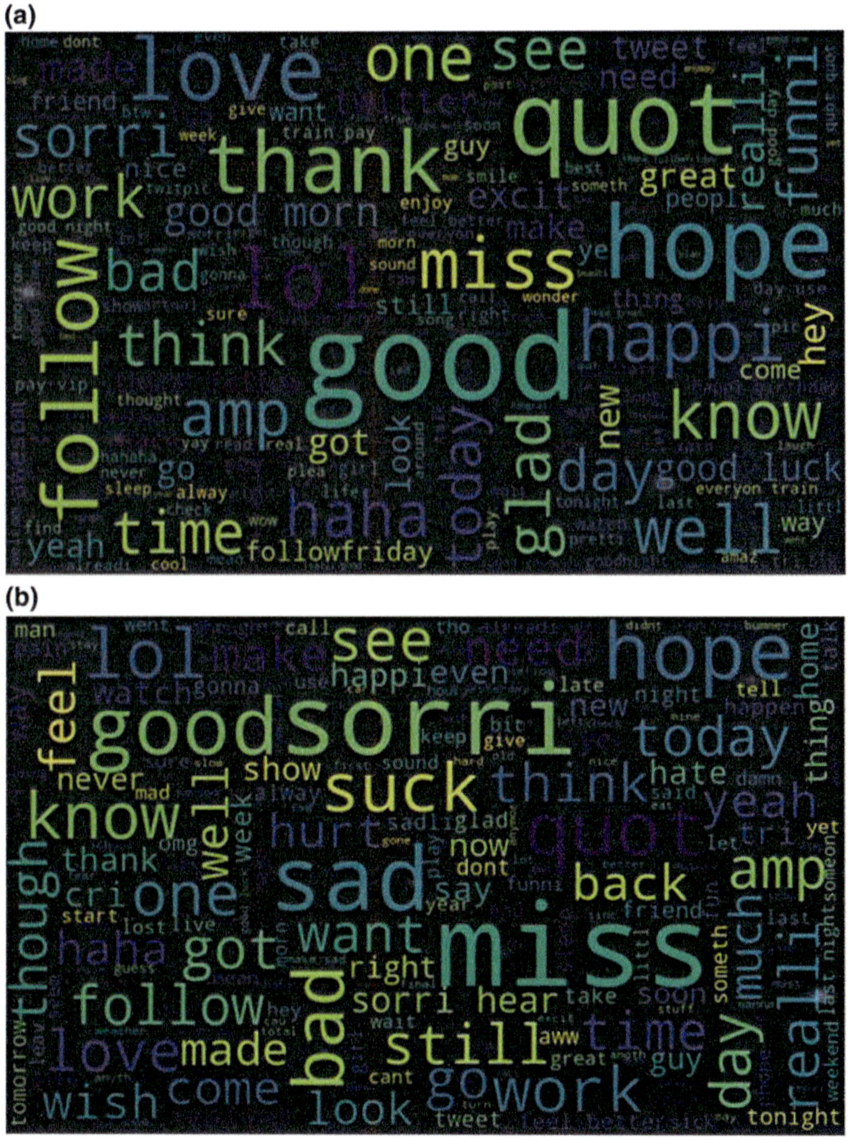

Fig. 2 **a** Positive word cloud, **b** negative word cloud

3.5 Feature Extraction

The features extracted in this work can broadly be classified into two types—1. text-based features and 2. non-textual or numeric features. The extraction technique of these features is as follows:

Text-Based Features. Two different text-featuring techniques have been used—a. Weighted bag of words and b. TF-IDF.

Weighted Bag of Words. In this text vectorization technique, each tweet is represented as a bag of its words. Frequency of each word in a document is used to create the vector. As the rarer words play the most vital role in finding document similarity, 1000 most common and 1000 rarest occurring words were collected from the dataset. For each most occurred word, its frequency has been calculated as: ceil (0.5 * (actual frequency)), and for each of the rarest words, its frequency has been calculated as (2 * actual frequency). For all other words, frequency is kept as its original frequency. While taking bigram into consideration, if any bigram contains one or more most frequent word, it is considered as a most frequent bigram. If it contains one or rarest words, it is considered as a rare bigram and if one of the words of a bigram is the most frequent word, and another is one of the rarest words, then the bigram has been considered as a normal bigram.

TF-IDF. While using TF-IDF text vectorization technique, for 1000 most common words, Laplace smoothening technique was used to distinguish between not present terms and most frequent terms.

Non-textual Features. For creating a repository of hashtags, emoticon symbols and URLs, all the tweets of last 15 days have been collected with the same terms that were used for text filtering. 5,18,669 numbers of unique tweets were extracted, among which 2,87,014 tweets came from negative keywords search and 2,31,655 tweets came from positive keywords search. After all the preprocessing of the collected tweets, top 5000 hashtags, top 50 emoticon symbols, and top 1000 URLs from both positive and negative category tweets were collected using regular expression and for positive and negative hashtags, emoticon, and URL separate repositories were created. Already available positive and negative word lists for English language tweets were taken and enhanced by adding the top 1000 positive and negative terms from the aforementioned repository. Already available swear word list for the English language was also used. After creating those repositories, non-textual features were extracted from the training data:

Normalized Average Adjective and Adverb Count. All the tweets were tokenized, and tokens were POS-tagged. After that, this feature was extracted using the formula:

$$\frac{0.5 * (\text{adjective count of the tweet} + \text{adverb count of the tweet})}{\text{length of the tweet}} \tag{1}$$

Normalized Positive or Negative Hashtags count. The hashtags of a tweet have been extracted using regular expression. Then it has been compared with the hashtag repository that was previously built. Then, count of hashtags was calculated using the following formula:

$$\begin{cases} 0, & \text{if no hashtag is found in the tweet} \\ \frac{\text{Number of positive or negative hashtages}}{\text{Total number of hashtages}}, & \text{otherwise} \end{cases} \tag{2}$$

Normalized Positive or Negative Emoticon symbol count. The emoticon symbols of a tweet have been extracted using emoticon dictionary. Then it has been compared with the hashtag repository that was already built. After that, this feature was extracted using the following formula:

$$\begin{cases} 0, & \text{if no emoticons is found in the tweet} \\ \frac{\text{Number of positive or negative emoticons}}{\text{Total number of emoticons}}, & \text{otherwise} \end{cases} \quad (3)$$

Normalized Positive or Negative URL count. The URLs of a tweet have been extracted using regular expression. Then it has been compared with the hashtag repository built earlier. This feature was also evaluated using the formula:

$$\begin{cases} 0, & \text{if no URLs is found in the tweet} \\ \frac{\text{Number of positive or negative URLs}}{\text{Total number of URLs}}, & \text{otherwise} \end{cases} \quad (4)$$

Normalized Relative Positive Word Count. Total number of positive and negative words frequency of a tweet has been calculated using the previously created positive and negative word repository. Now the normalized relative frequency has been calculated using the following formula:

$$\left\{ \frac{\text{Number of positive words} - \text{Number of negative words}}{\text{Length of the tweet}} \right. \quad (5)$$

Normalized Swear Word Count. Swear words of a tweet have been identified using the previously created swear word repository. Swear word counts have been normalized using the formula:

$$\left\{ \frac{\text{Number of swear words in a tweet}}{\text{Length of the tweet}} \right. \quad (6)$$

Normalized Stressed Word Count. All the words that have three consecutive occurrences of the same letter have been considered as a stressed word. As the stressed words represent extreme emotions, this feature has been considered. The feature is normalized using the formula:

$$\left\{ \frac{\text{Number of stressed words in a tweet}}{\text{Length of the tweet}} \right. \quad (7)$$

While extracting features, total word count of a tweet has been considered as its length. The textual and non-textual features together have been used for sentiment prediction.

3.6 *Training the Classifier*

Logistic regression, KNN, and SVM classifiers were used for the sentiment analysis purpose as these classifiers have been proven to be effective for sentiment analysis by researchers. Tenfold cross-validation technique has been used to predict the accuracy of a classifier. Logistic regression classifier was found to be most effective.

3.7 *Testing the Accuracy*

Three parameters have been used to test the accuracy of the classifier model—a > Confusion matrix, b > F-score, and c > Accuracy. Confusion matrix helped to understand true negative and false positive counts.

$$F\text{-score} = \frac{2 * \text{Precision} * \text{Recall}}{\text{Precision} + \text{Recall}} \tag{8}$$

4 Results and Analysis

As mentioned earlier, three main classification models have been considered in this work, namely logistic regression, KNN, and SVM, for their good performance in text analytics in previous researches. For text features, "Bag of Words" and "TF-IDF" models turned out to be most effective. Various scenarios were taken into consideration, and the performances of all three classifiers in those scenarios were compared.

4.1 *Filtering Versus Non-filtering*

After extracting textual and non-textual features from the training dataset, at first whole unfiltered data was used for training purpose and the accuracy of three classifiers with the test data was tested. After that, the filtered training data was used, and accuracy of all three classifier models was again checked with the unfiltered test data. The comparison of accuracy of three classifiers is shown in Fig. 3.

For all three classifiers, bigram-based bag-of-words technique has been used for text featurization. As the filtering-based approach proposed in this paper could eliminate most of the nonpolar sentences and ensured all the training data is associated with at least one human emotion, the model could achieve significantly high result than the results obtained using traditional method.

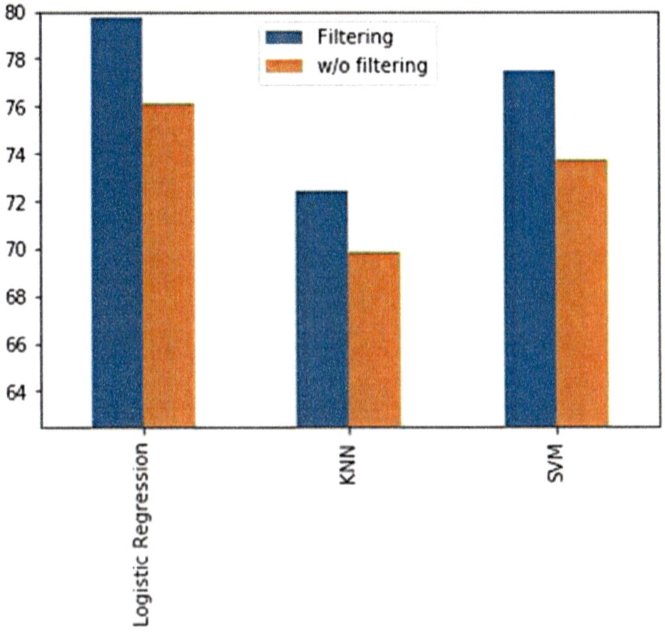

Fig. 3 Comparison of results with filtered and non-filtered training data

4.2 Importance of Non-textual Features

To check the relevance of non-textual features, the classifier models were first trained with only textual feature and their accuracies were tested. After this, seven non-textual features were concatenated with the textual feature. At first, 1000 dimensional vectors were used for text featurization. After that, to make sure the textual features do not dominate over non-textual features, principal component analysis technique was used for dimensionality reduction of textual features. 80% variation of the training data was kept and only top 44 dimensions of the textual feature that preserved 80% variation was considered. After this, total number of features became $44 + 7 = 51$. The comparison of performances of different classifier models being trained with and without non-textual features has been represented in Fig. 4.

For these comparisons also, bigram-based bag-of-words vectorization technique has been used for text featurization and the dimensionality of the text feature has been reduced using PCA technique. It is evident that inclusion of non-textual features gave a significant hike in the performance of the classification models.

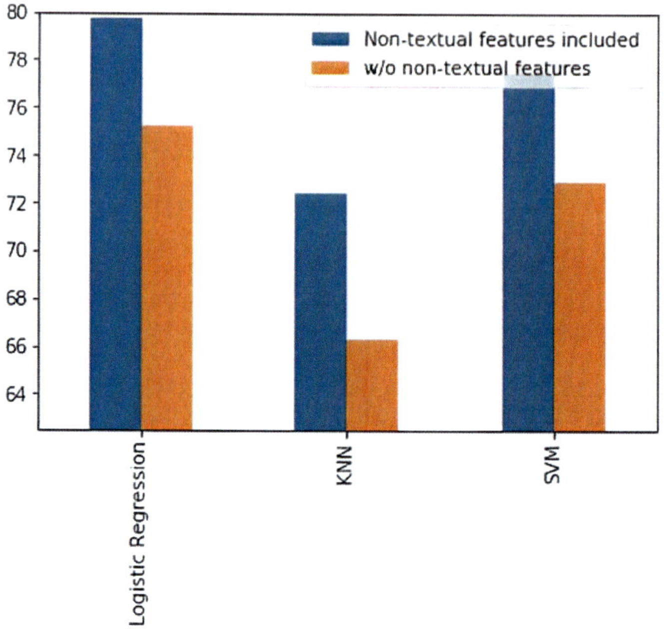

Fig. 4 Comparison of accuracies with and without non-textual features

4.3 *Comparisons of* F*-Score*

Confusion matrices were constructed from the results of the classifier models on the test data. From the confusion matrix, "precision" and "recall" were calculated. "Precision" is referred to as the fraction of relevant instances among retrieved instances, and "recall" is referred to as the fraction of the relevant instances that have been retrieved. The comparison of *F*-scores of different classifiers taking bag-of-words text vectorization technique and TF-IDF text vectorization technique is shown in Fig. 5.

In all the classifier models, bag-of-words text featurization technique has outperformed TF-IDF text featurization technique. Among all the N-gram models of bag-of-words technique, the bigram model was found out to be most effective.

5 Conclusion and Future Scope

In this era of micro-blogging Web sites, twitter sentiment analysis has become an important area of research. This work has tried to present a novel idea of filtering-based text sentiment analysis. To the best of our knowledge, this technique has not

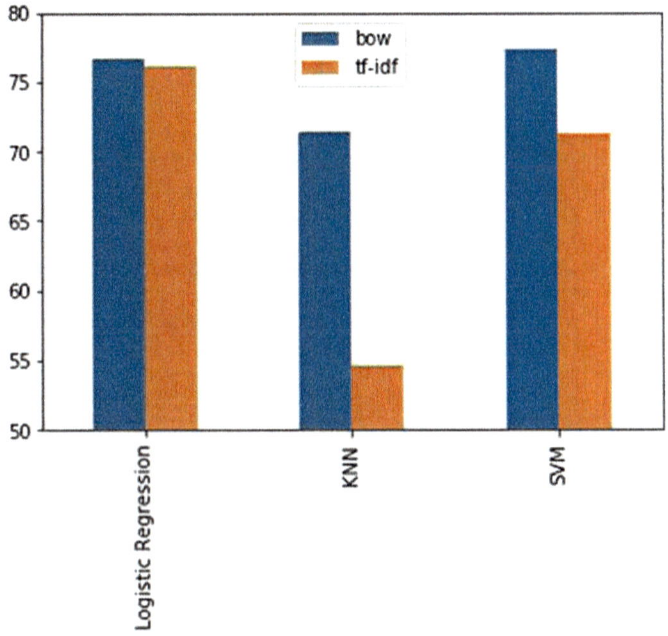

Fig. 5 Comparison of F-scores for different classifier models

been implemented before. The paper has also defined a way for creating semiautomated to automated repository of URLs, emoticons, positive and negative words, and hashtags. The future works would try to consider the media attached to a tweet also rather than considering only the text data. Moreover, the future work would also try to inspect how this filtering-based approach works in that broader scenario. The future scope of this paper also includes the identification of behavioral pattern of a twitter user from the inspection of his tweet pattern. For example, if the tweet pattern of the recent past of a suicide attempter reflects anything about his anxiety or if the tweet pattern of a student can predict his indifferent behavior in the future.

References

1. Rambocas M, Gama J (2013) Marketing research: the role of sentiment analysis. The 5th SNA-KDD Workshop'11. University of Porto
2. Facebook website. https://www.facebook.com/
3. Twitter website. https://twitter.com/
4. Pinterest Website. https://in.pinterest.com/
5. Google SEO trends. http://www.googleseotrends.com/microblogging-sites/
6. Jose A, Bhatia N, Krishna S (2012) Twitter sentiment analysis. National Institute of Technology, Calicut
7. Tang Q, Gu B, Whinston AB (2012) Content contribution in social media: the case of YouTube. In: 2nd conference of social media. Hawaii, Maui. https://doi.org/10.1109/hicss.2012.181

8. Lohmann S, Burch M, Schmauder H, Weiskopf D (2012) Visual analysis of microblog content using time-varying co-occurrence highlighting in tag clouds. In: Annual conference of VISVISUS. University of Stuttgart, Germany. http://doi.org/10.1.1.395.3268.
9. Lai P (2012) Extracting strong sentiment trend from Twitter. nlp.stanford.edu
10. Osimo D, Mureddu F (2010) Research challenge on opinion mining and sentiment analysis. In: Proceeding of the 12th conference of Fruct association. United Kingdom
11. Pak A, Paroubek P (2010) Twitter as a corpus for sentiment analysis and opinion mining. In: Proceedings of LREC Special Issue of International Journal of Computer Application. Universite de Paris-Sud, France
12. Carpenter T, Way T (2010) Tracking sentiment analysis through Twitter. ACM computer survey. Villanova University, Villanova
13. Bo Pang, Lee L (2008) Opinion mining and sentiment analysis. Found Trends Inf Retr 2(1–2):1–135. https://doi.org/10.1561/1500000011
14. Yang C, Lin KH, Chen H (2007) Emotion classification using web blog corpora. In: WI '07: proceedings of the IEEE/WIC/ACM international conference on web intelligence. Washington, DC, USA, pp 275–278. IEEE Computer Society. https://doi.org/10.1109/wi.2007.51
15. Saif H, He Y, Alani H (2011) Semantic sentiment analysis of Twitter. In: Proceeding of the workshop on information extraction and entity analytics on social media data. Knowledge Media Institute, United Kingdom. https://doi.org/10.1007/978-3-642-35176-1_32
16. Prabowo R, Thelwall M (2009) Sentiment analysis: a combined approach. International World Wide Web Conference Committee (IW3C2). University of Wolverhamption, United Kingdom. https://doi.org/10.1016/j.joi.2009.01.003
17. Taboada M, Brooke J, Tofiloski M, Voll K, Stede M (2011) Lexicon based methods for sentiment analysis. Assoc Comput Linguist. https://doi.org/10.1162/coli_a_00049
18. Annett M, Kondrak G (2009) A comparison of sentiment analysis techniques: polarizing movie blogs. In: Conference on web search and web data mining (WSDM). University of Alberia. Department of Computing Science. https://doi.org/10.1007/978-3-540-68825-9_3
19. Goncalves P, Benevenuto F, Araujo M and Cha M (2013) Comparing and combining sentiment analysis methods. In: Proceedings of the first ACM conference on online social networks—COSN'13. https://doi.org/10.1145/2512938.2512951
20. Kouloumpis E, Wilson T, Moore T (2011) Twitter sentiment analysis: the good the bad and the OMG!, vol 5. ICWSM, International AAAI
21. Sharma S (2008) Application of support vector machines for damage detection in structure. J Mach Learn Res
22. Sarlan A, Nadam C, Basri S (2014) Twitter sentiment analysis. In: International conference on information technology and multimedia (ICIMU). Putrajaya, Malaysia. https://doi.org/10.1109/icimu.2014.7066632
23. Amolik A, Jivane N, Bhandari M, Venkatesan M (2016) Twitter sentiment analysis of movie reviews using machine learning techniques. Int J Eng Technol (IJET) 7:2038–2044
24. Hasan A et al (2017) Machine learning-based sentiment analysis for Twitter accounts. Math Comput Appl 23(1):11. https://doi.org/10.3390/mca23010011

A Comparative Analysis of Clustering Quality Based on Internal Validation Indices for Dimensionally Reduced Social Media Data

Shini Renjith, A. Sreekumar, and M. Jathavedan

Abstract Almost all modern industries leverage data analytics to deal with various dimensions of their business like demand forecasting, targeted marketing, and supply chain planning. In addition to historic data, social media data has also become a prominent source of input for data analytics. The key challenges observed with social media data are its huge volume and high dimensions that need to be dealt with. Clustering is the proven strategy in data analytics to segregate the relevant data for processing and thereby reducing the impact of huge volume. Dimensionality corresponds to the diverse features of the data subject being represented. The application of dimensionality reduction techniques can help in reducing the computational intensiveness caused by the curse of dimensionality. This paper covers an experimental analysis using four popular dimensionality reduction techniques – two linear and two nonlinear approaches – to verify the impact of dimensionality reduction on cluster quality using internal clustering validation indices.

Keywords Clustering · Curse of dimensionality · Data analytics · Dimensionality reduction · Internal clustering validation indices · Social media

S. Renjith (✉) · A. Sreekumar · M. Jathavedan
Department of Computer Applications, Cochin University of Science and Technology, Kochi, Kerala 682022, India
e-mail: shinirenjith@gmail.com

A. Sreekumar
e-mail: sreekumar@cusat.ac.in

M. Jathavedan
e-mail: mjvedan@cusat.ac.in

S. Renjith
Department of Computer Science and Engineering, Mar Baselios College of Engineering and Technology, Thiruvananthapuram, Kerala 695015, India

N. N. Chiplunkar and T. Fukao (eds.), *Advances in Artificial Intelligence and Data Engineering*, Advances in Intelligent Systems and Computing 1133,
https://doi.org/10.1007/978-981-15-3514-7_78

1 Introduction

Data analytics [1, 2] is the structured process through which a given data set is inspected to extract or reveal any underlying pattern or information. Modern industries extensively rely on data analytics techniques to take informed business decisions. It can help organizations to achieve better operational efficiency, strategize marketing initiatives, carry out targeted customer services, determine optimal time to launch new products or services, detect potential customer churn so that appropriate retention actions can be performed, and most importantly enable the business to respond quickly to the changing trends and demands in the markets.

Organizations leverage data from various channels for data analytics purpose. The main sources include organization's historical performance, industry trends collected by corresponding authorities, and appropriate social media data in the form of reviews, testimonials, forums, etc. The common characteristics of all these sources are huge volume and high dimensionality. Both these aspects result in the need of very high computational power and associated increase in processing time. While clustering [3] is considered as the solution to reduce the volume of data to be processed, dimensionality reduction [4] is considered as a mechanism to reduce the data dimension, both with the aim of optimizing performance.

As part of this experimental analysis, we verified four different dimensionality reduction approaches on the same data set and measured the influence on the quality of clustering for partitioning and hierarchical clustering. We used *k*-means clustering [5–8] and agglomerative hierarchical clustering (AGNES) [9–12] as the candidate cases for partitioning and hierarchical clustering, respectively. The dimensionality reduction techniques evaluated include principal component analysis (PCA) [13, 14], independent component analysis (ICA) [15], *t*-distributed stochastic neighbor embedding (*t*-SNE) [16, 17], and locally linear embedding (LLE) [18, 19] – two linear and two nonlinear approaches.

This work is part of a series of experiments [20–22] being carried out on various aspects of clustering process performed on large and high-dimensional data sets. Section 2 of this paper provides a quick recap on the clustering algorithms and dimensionality reduction techniques covered in this empirical analysis. Section 3 briefs on the related literature, and Sect. 4 explains the research methodology adopted, infrastructure and tools leveraged, and details of data set used for the experiments. Section 5 documents our observations from this analysis, and Sect. 6 records our inferences.

2 Antecedents

2.1 *Clustering*

Clustering [3] is an unsupervised learning process through which data sets can be sliced into a set of clusters which comprises of similar entities. Based on the approach adopted, clustering algorithms are classified as partitioning, hierarchical, density-based, or model-based clustering.

k-Means Clustering k-means [5–8] is the most widely practiced partitioning clustering algorithm. It distributes the data set entities into k distinct clusters with the help of an iterative process. The iterations are carried out with the aim of achieving minimum possible value for the total within cluster variation (TWCV) for the clusters being generated. The TWCV is mathematically represented as (1).

$$\text{TWCV} = \sum_{k=1}^{K} \sum_{E_i \in C_k} (E_i - \mu_k)^2 \quad (1)$$

where K is the cluster count and E_i is an entity in cluster, C_k with centroid, μ_k.

Agglomerative Hierarchical Clustering Agglomerative hierarchical clustering (AGNES) algorithm aka agglomerative nesting [9–12] is a bottom-up clustering approach. Starting with singleton clusters for every entity in the data set, the algorithm iterates by merging the nearest clusters in subsequent steps to finally result in a single cluster with all entities from the data set. The algorithm takes irreversible steps in each iteration to form clusters, though it lacks global distribution details of the data at this time which is considered as the key challenge of this approach.

2.2 *Dimensionality Reduction*

Dimensionality reduction [4] is the statistical solution for moderating the high number of attributes of an entity without losing much information in the process. This has evolved as a key data transformation step in machine learning to bring down the need for high computational power. Reduction of dimensions is achieved by determining the set of principal attributes either through feature selection or feature projection. Feature selection is performed by choosing a subset of prevailing features using an optimization criterion, whereas in feature projection the high-dimensional vector space representing all attributes of the entity is transformed into a lower dimensional vector space.

Dimensionality reduction approaches are categorized into two types – linear and nonlinear. The most common linear approaches include PCA and ICA. There are multiple nonlinear approaches that exist like *t*-SNE, LLE and ISOMAP.

Principal Component Analysis (PCA) PCA [13, 14] is a feature projection approach for dimensionality reduction, and it extracts a new set of attributes called principal components as linear combination of original attributes. In PCA, the principal components are populated with the first new attribute explains the maximum variation. The second new attribute attempts to explain the remaining variation and so on. Typically, it is observed that more than 60% of the variation in a data set is explained by initial four principal components in PCA.

Independent Component Analysis (ICA) ICA [15] is yet another common linear dimensionality reduction technique. The key difference between ICA and PCA is that ICA looks for independent attributes (not dependent on other attributes), whereas PCA looks for uncorrelated attributes (not linearly related). Unlike PCA, ICA does not arrange or order the resulting attributes from dimension reduction process.

***t*-Distributed Stochastic Neighbor Embedding (*t*-SNE)** *t*-SNE [16, 17] is a nonlinear approach to achieve reduced number of dimensions. This approach can retain the local and global structure of the data even after performing dimensionality reduction so that nearby points in original data set will be transformed to nearby points in low-dimensional space, and the overall geometry of the data set is preserved as such. *t*-SNE considered to be one of the best candidates for data visualization problems.

Locally Linear Embedding (LLE) LLE [18, 19] preserves the topology or neighborhood structure as such in the lower dimensional space using a nonlinear approach. LLE anticipates the data set to be smooth and nonlinear and is sensitive to noises and outliers.

2.3 Internal Cluster Validation

Cluster validation approaches are used to gauge the quality of the clusters determined as the result of clustering process. In internal validation mechanism, cluster quality is measured in terms of a set of indices representing quality score, whereas external validation mechanisms compare the clustering result with an existing data. Typically, internal validation is the only option to deal with large volume of real data like the one from social media.

3 Related Works

Most of the early literature in the area of clustering covered the theoretical aspects of various clustering approaches in detail [23]. Later works started giving more emphasis to big data context [24–27]. Another set of literature, though less in volume focused on empirical analysis of clustering. Lau and King [28] experimented with information retrieval from image databases using competitive learning and rival penalized competitive learning – both are unsupervised neural network clustering algorithms. Maulik and Bandyopadhyay [29] compared performances of *k*-means, single linkage, and simulated annealing clustering approaches with the help of internal validation indices of Davies–Bouldin, Dunn, Calinski–Harabasz, and Index I.

Wei et al. [30] evaluated the performances of CLARA, CLARANS, GAC-R3, and GAC-RARw. Zhang [31] compared *k*-harmonic means (KHM), *k*-means, and expectation maximization clustering algorithms for their performance. Wang and Hamilton [32] evaluated DBSCAN and DBRS as part of their experiments.

Poonam Dutta [33] measured the performance of PAM, CLARA, CLARANS, and fuzzy *c*-means clustering algorithms with more focus on the outlier detection efficiency. Fahad et al. [34] leveraged ten different data sets to measure the clustering quality of fuzzy *c*-means, expectation maximization, BIRCH, DENCLUE, and OptiGrid. The performance of EM and the *k*-means clustering algorithms is measured and compared by Jung et al. [35] by applying logistic regression analysis. Bhatnagar et al. [36] compared *k*-means, fuzzy *c*-means, hierarchical, Gaussian mixture modeling, and self-organized map clustering techniques by measuring the performance for a specific task of grouping manufacturing firms.

Renjith et al. performed a series of experimental analyses on various aspects of clustering like comparative analysis of clustering quality for various algorithms [20], performance evaluation of clustering algorithms for varying cardinality and dimensionality [21], and impact of dimensionality reduction in clustering performance [22]. We also referred to the big volume of literature present on various dimensionality reduction techniques and their applications like [13–19, 37–42].

4 Methodology

We adopted a four-stage approach in this research as shown in Fig. 1.

4.1 Determination of Optimal Cluster Count

k-means and other partitioning clustering algorithms need the anticipated cluster count to be specified as an input to the clustering process. The optimal cluster count, *k* is highly subjective to the data set and similarity measures under consideration. This experiment made use of the R package NbClust [43] to determine optimal cluster count by computing 26 different indices supported by it. The package adopts voting method to determine and propose the optimal cluster count.

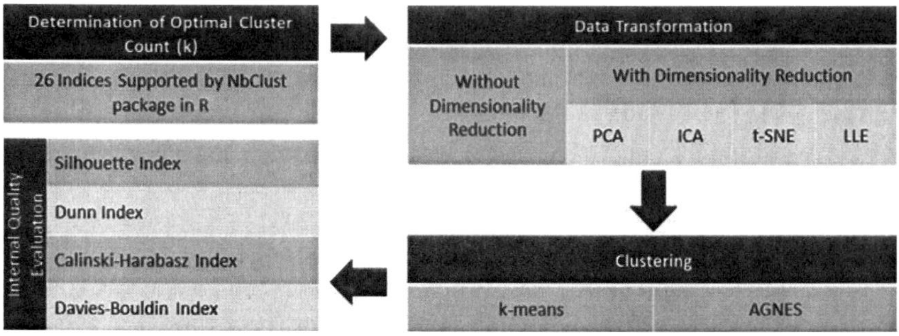

Fig. 1 Four-stage approach adopted in the empirical study

4.2 Data Transformation

As part of the experiments, we performed clustering of the high-dimensional data with and without applying dimensionality reduction techniques. The study is performed using four dimensionality reduction techniques – two each from linear (PCA and ICA) and nonlinear (*t*-SNE and LLE) categories. We leveraged frequently used implementations of these algorithms that are available as R packages for attaining the dimensionality reduction.

4.3 Clustering

We used the most popular partitioning clustering algorithm (*k*-means) and hierarchical clustering algorithm (AGNES) to verify the impact of dimensionality reduction on clustering quality. Like in the case of dimensionality reduction, we leveraged R implementations of these algorithms in our experiments.

4.4 Internal Validation of Clustering Quality

We used four different internal evaluation criteria, namely Silhouette Index [44], Dunn Index [45], Calinski–Harabasz Index [46], and Davies–Bouldin Index [47] to perform the internal validation on clustering quality. We leveraged the R package clusterCrit for measuring these indices.

Table 1 R packages leveraged in this analysis

R package used	Purpose
NbClust	To find optimal count of clusters [43]
factoextra	To pull out and represent results of multivariate data analyses
stats	Implementation of *k*-means algorithm and PCA
caret	Implementation of ICA
cluster	Implementation of AGNES algorithm
clusterCrit	Internal evaluation of clusters
lle	Implementation of LLE
Rtsne	Implementation of *t*-SNE

4.5 Tools Used

We used R [48], the open-source programming language for statistical calculations and data visualization and RStudio [49, 50], the most popular integrated development environment for R for these experiments. The specific packages used for this research are detailed in Table 1. The hardware used to conduct the experiments is Intel Core i7-6500U, 2.50 and 2.59 GHz dual-core x64-based processor with 12.00 GB RAM.

4.6 Data Set Used

We used the Jester data set 1 [51] which contains around 4.1 million anonymous ratings in the range of -10 to $+10$ corresponding to 100 jokes rated by 73,421 readers which are gathered from April 1999 to May 2003. A quick outline of the data set is made available in Table 2.

Table 2 Jester data set 1 – a quick summary

Data set description	Joke ratings from Jester online Joke recommender system
Source	http://eigentaste.berkeley.edu/dataset/
Reference	[51]
Cardinality	73,421
Dimension	100
No. of ratings	Around 4.1 million
Range	-10 to 10

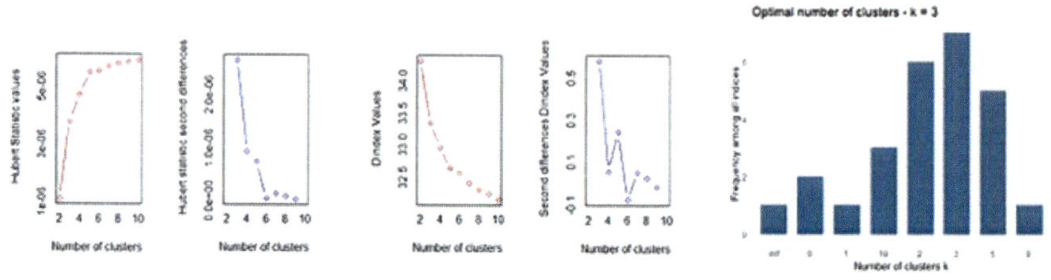

Indices	Scott	Cindex	Ball	McClain	Dindex	KL	Silhouette	Duda	PseudoT2	SDbw	Hartigan	Marriot	TrCovW	TraceW	Friedman	Beale	Ratkowsky	CH	CCC	Rubin	DB	Frey	Hubert	PtBiserial	Dunn	SDindex
Jester	-inf.	10	10	10	9	5	5	5	5	5	3	3	3	3	3	3	3	2	2	2	2	2	2	1	0	0

Fig. 2 Determination of optimal count of clusters

5 Empirical Research

5.1 Determination of Optimal Cluster Count

The optimal cluster count is decided with the help of the R package NbClust which leverages 26 different indices to decide the optimal count by applying voting method. The optimal cluster count is identified as 3 through our experiments. The detailed result from NbClust is shown in Fig. 2.

5.2 Dimensionality Reduction

Dimensionality reduction is attained by using R implementation of the corresponding algorithms. Details of the four algorithms, invoked R functions and details of corresponding packages, are shown in Table 3.

Table 3 Dimensionality reduction techniques covered in the analysis

Dimensionality reduction technique	R functions used	Applicable R package
Principal component analysis (PCA)	prcomp()	stats
Independent component analysis (ICA)	preProcess() predict()	caret
t-distributed stochastic neighbor embedding (*t*-SNE)	Rtsne()	Rtsne
Locally linear embedding (LLE)	lle()	lle

Table 4 Clustering algorithms covered in the analysis

Clustering algorithm	Clustering type	R function used for clustering	Applicable R package
k-means clustering	Partitioning	kmeans()	stats
AGNES clustering	Hierarchical	agnes()	cluster

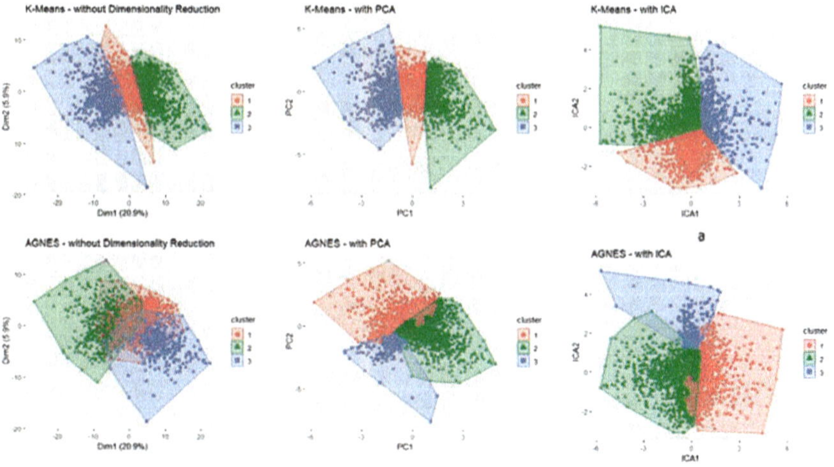

Fig. 3 Two-dimensional view of clusters formed with linear dimensionality reduction techniques when cardinality of the sample is 5000

5.3 *Clustering*

We have covered two clustering algorithms as part of this analysis – *k*-means and AGNES. Table 4 summarizes the details of the corresponding R functions and packages used for this evaluation, and Figs. 3 and 4 represent the two-dimensional plots of the resulted clusters.

5.4 *Internal Validation of Clustering Quality*

We measured the goodness of the clustering process using internal validation approaches. For our experiments, we have selected Silhouette Index [44], Dunn Index [45], Calinski–Harabasz Index [46], and Davies–Bouldin Index [47] as the internal validation indices to check resulted from cluster quality.

Figure 5 depicts the impact of four different dimensionality reduction approaches in clustering quality for *k*-means and AGNES clustering algorithms against these

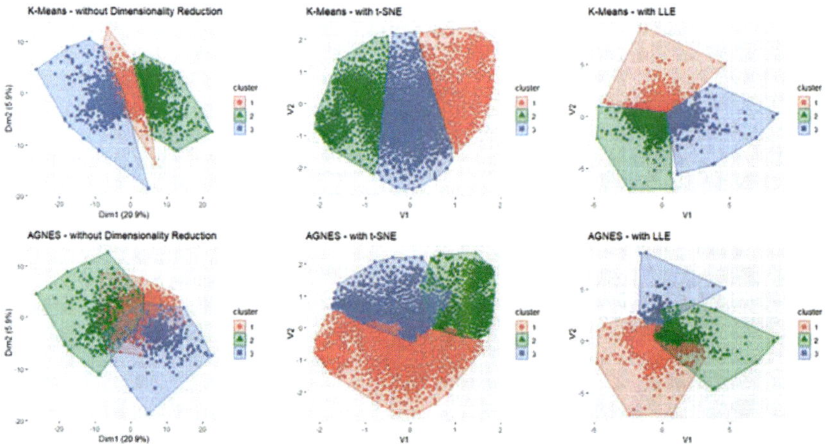

Fig. 4 Two-dimensional view of clusters formed with nonlinear dimensionality reduction techniques when cardinality of the sample is 5000

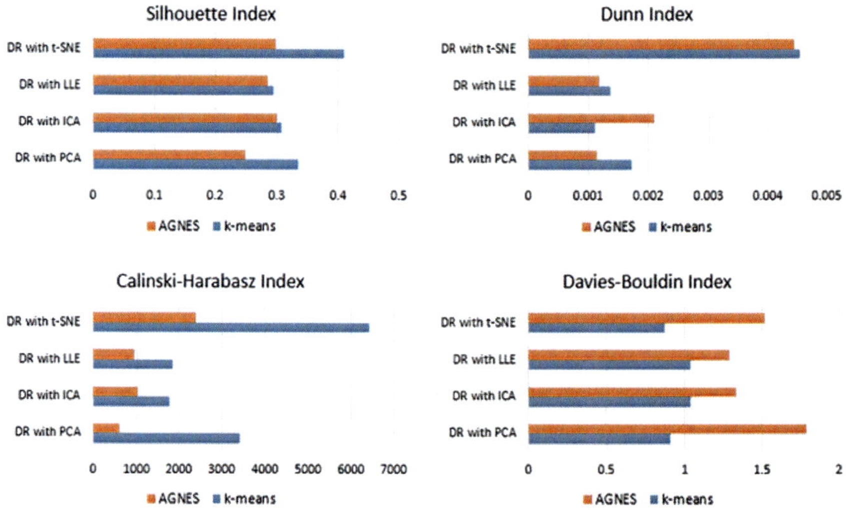

Fig. 5 Internal evaluation indices for different dimensionality reduction techniques

indices. Refer to Appendices 1 and 2 for the complete set of observations from internal evaluations conducted using the R package clusterCrit.

A high value for Silhouette index represents better clustering quality. In our experiments, *t*-SNE produced better results in the case of *k*-means clustering followed by PCA. For AGNES, *t*-SNE and ICA provided almost similar results. With Dunn index where higher values indicate more goodness, dimension reduction using *t*-SNE outperformed all other techniques both in the case of *k*-means and AGNES. A higher value represents relatively better clustering quality with Calinski–Harabasz index. Based on our observations, dimensionality reduction using *t*-SNE resulted in better quality measures for both *k*-means and AGNES. As per Davies–Bouldin index, a lower value is goodness indicator and *t*-SNE and PCA fared better with *k*-means clustering in terms of quality clustering, whereas LLE and ICA performed better for AGNES. So, the key inference is that with the data set under consideration, *t*-SNE produced better results in most of the scenarios.

6 Conclusion

Data analytics has become an inevitable part of various business operations across industries. Social media has become a predominant source of information in such scenarios and the inherent issues of high volume, and dimensionality has to be dealt with while processing such data. A combination of dimensionality reduction with clustering is the best available solution in hand. However, there are multiple algorithms available for both these operations and the user has to decide on which algorithm will be the best suited for the scenario in hand.

As part of this experimental evaluation, we verified the impact on clustering quality for two clustering algorithms in combination with four dimensionality reduction approaches. Our experiments are performed on the Jester data set 1 using two candidate clustering algorithms from partitioning and hierarchical approaches – *k*-means and AGNES. We selected two linear (PCA and ICA) and two nonlinear (*t*-SNE and LLE) dimensionality reduction techniques for verifying their influence on clustering quality by applying them prior to actual clustering activity. Our experiments proved the superiority of *t*-SNE approach for dimensionality reduction with *k*-means and AGNES for the data set under consideration.

The key callout is that the nature of the data set has a strong influence on the performance of the clustering and dimensionality reduction algorithms. So, we recommend conducting an experiment of this type on the data set under consideration to decide on the best suitable approaches – steps to be followed, tools used, specific packages required, etc., are detailed in this paper. We believe that this approach can be referenced as a simple framework for various research purposes.

Appendix 1

Complete observations from internal evaluation conducted on the *k*-means clustering results using the R package clusterCrit

Index	ClusterCrit variable	Goodness indicator	Dimensionality reduction using			
			PCA	ICA	LLE	*t*-SNE
Ball–Hall index	$ball_hall	max diff	17.20457	1.391324	1.232667	232.7926
Banfield–Raftery index	$banfeld_raftery	min	11,029.27	572.3903	626.2364	27,286.89
C index	$c_index	min	0.09553103	0.2327327	0.2312412	0.08800492
Calinski–Harabasz index	$calinski_harabasz	max	3417.288	1782.438	1847.939	6416.44
Davies–Bouldin index	$davies_bouldin	min	0.9058952	1.034618	1.039139	0.8715404
Det ratio index	$det_ratio	min diff	3.871588	3.047042	3.026433	7.844465
Dunn index	$dunn	max	0.00173042	0.00110608	0.00136502	0.00453987
Baker–Hubert Gamma index	$gamma	max	0.7785525	0.5188169	0.4677813	0.7813951
G plus index	$g_plus	min	0.05533569	0.1182925	0.1279114	0.04907176
GDI index	$gdi11	max	0.00173042	0.00110608	0.00136502	0.00453987
GDI index	$gdi12	max	0.01493041	0.01049106	0.02023069	0.02933953
GDI index	$gdi13	max	0.00511206	0.00372532	0.00717291	0.01019923
GDI index	$gdi21	max	1.21189	1.324565	1.406772	1.227989
GDI index	$gdi22	max	10.45642	12.56334	20.84945	7.936054
GDI index	$gdi23	max	3.580196	4.461175	7.392295	2.75879
GDI index	$gdi31	max	0.2752019	0.2248528	0.1981029	0.5592138
GDI index	$gdi32	max	2.374494	2.132702	2.936039	3.614
GDI index	$gdi33	max	0.8130083	0.7573111	1.04099	1.256326
GDI index	$gdi41	max	0.2410528	0.1885064	0.1703023	0.4925548
GDI index	$gdi42	max	2.079849	1.78796	2.524013	3.183207
GDI index	$gdi43	max	0.7121241	0.6348952	0.8949034	1.10657

(continued)

(continued)

Index	ClusterCrit variable	Goodness indicator	Dimensionality reduction using			
			PCA	ICA	LLE	*t*-SNE
GDI index	$gdi51	max	0.09224953	0.09543682	0.08576157	0.2129281
GDI index	$gdi52	max	0.7959463	0.9052065	1.271053	1.376078
GDI index	$gdi53	max	0.2725258	0.3214341	0.4506593	0.4783627
Ksq DetW index	$ksq_detw	max diff	7,186,235,240	73,842,109	74,344,937	3.02E + 12
Log Det ratio index	$log_det_ratio	min diff	6768.324	5570.856	5536.924	10,299.04
Log SS ratio index	$log_ss_ratio	min diff	0.3131567	−0.3377083	−0.3016196	0.9431728
McClain–Rao index	$mcclain_rao	min	0.3721302	0.5643422	0.595994	0.4384923
PBM index	$pbm	max	42.68164	0.9640561	0.6983515	1497.468
Point Biserial index	$point_biserial	max	−2.693745	−0.4353606	−0.3859055	−11.7986
Ray–Turi index	$ray_turi	min	0.2860079	0.4883412	0.4338252	0.2302405
Ratkowsky–Lance index	$ratkowsky_lance	max	0.3568949	0.3725436	0.3764579	0.4178183
Scott–Symons index	$scott_symons	min	14,296.68	−6276.843	−5815.897	47,340.83
SD index	$sd_scat	min	0.5612758	0.7125735	0.6275758	0.2424001
SD index	$sd_dis	min	0.3689349	1.142325	1.029613	0.07119309
S Dbw index	$s_dbw	min	1.177152	2.50059	4.480659	1.999409
Silhouette index	$silhouette	max	0.3342391	0.3077699	0.2943706	0.4099526
Tau index	$tau	max	0.5503895	0.3637915	0.3243153	0.5235662
Trace W index	$trace_w	max diff	56,644.87	5836.337	5748.384	1,175,143
Trace WiB index	$trace_wib	max diff	2.784339	1.556713	1.47942	5.971501
Wemmert–Gancarski index	$wemmert_gancarski	max	0.5552955	0.5115594	0.4629613	0.5244954
Xie–Beni index	$xie_beni	min	5550.067	14,184.04	6752.686	2710.222

Appendix 2

Complete observations from internal evaluation conducted on the AGNES clustering results using the R package clusterCrit

Index	ClusterCrit variable	Goodness indicator	Dimensionality reduction using			
			PCA	ICA	LLE	*t*-SNE
Ball–Hall index	$ball_hall	max diff	19.53912	1.331913	1.623174	363.1799
Banfield–Raftery index	$banfeld_raftery	min	15,293.4	1734.322	1796.481	29,752.68
C index	$c_index	min	0.3628751	0.3158354	0.3416631	0.2416017
Calinski–Harabasz index	$calinski_harabasz	max	615.9316	1015.008	944.1628	2396.612
Davies–Bouldin index	$davies_bouldin	min	1.79305	1.329694	1.286655	1.515427
Det ratio index	$det_ratio	min diff	2.089563	2.089563	2.005657	4.858279
Dunn index	$dunn	max	0.00114581	0.00209901	0.00118646	0.00444984
Baker–Hubert Gamma index	$gamma	max	0.3062376	0.3735996	0.2849121	0.4667844
G plus index	$g_plus	min	0.1733258	0.1564965	0.1784953	0.1258491
GDI index	$gdi11	max	0.00114581	0.00209901	0.00118646	0.00444984
GDI index	$gdi12	max	0.01376865	0.02173099	0.0153745	0.02884337
GDI index	$gdi13	max	0.00473243	0.00763023	0.00537622	0.00997892
GDI index	$gdi21	max	1.15714	1.296565	1.365893	0.9185128
GDI index	$gdi22	max	13.90482	13.42333	17.6997	5.953694
GDI index	$gdi23	max	4.779239	4.713223	6.189309	2.059795
GDI index	$gdi31	max	0.1850741	0.2396983	0.1761109	0.3479464
GDI index	$gdi32	max	2.223952	2.481595	2.282103	2.255348
GDI index	$gdi33	max	0.7643965	0.871342	0.7980158	0.7802811
GDI index	$gdi41	max	0.09938497	0.16374	0.1200127	0.2061969
GDI index	$gdi42	max	1.194264	1.695199	1.555164	1.336545
GDI index	$gdi43	max	0.4104816	0.5952212	0.5438165	0.4624033

(continued)

(continued)

Index	ClusterCrit variable	Goodness indicator	Dimensionality reduction using			
			PCA	ICA	LLE	*t*-SNE
GDI index	$gdi51	max	0.1013306	0.1180803	0.08362988	0.1522455
GDI index	$gdi52	max	1.217644	1.222485	1.083704	0.9868379
GDI index	$gdi53	max	0.4185175	0.4292411	0.3789543	0.3414155
Ksq DetW index	$ksq_detw	max diff	1.3315E+10	107,678,003	112,182,708	4.88E+12
Log Det ratio index	$log_det_ratio	min diff	3684.775	3684.775	3479.858	7903.421
Log SS ratio index	$log_ss_ratio	min diff	−1.40031	−0.9007937	−0.9731472	−0.0416343
McClain–Rao index	$mcclain_rao	min	0.7376162	0.6732493	0.7332948	0.6335125
PBM index	$pbm	max	7.771969	1.091036	1.552822	518.4593
Point Biserial index	$point_biserial	max	−0.8765914	−0.3172137	−0.2494482	−7.426781
Ray–Turi index	$ray_turi	min	1.920044	0.928044	1.039746	1.106356
Ratkowsky–Lance index	$ratkowsky_lance	max	0.3103157	0.3103157	0.3023539	0.4276503
Scott–Symons index	$scott_symons	min	20,174.66	−3912.774	−3847.927	48,827.03
SD index	$sd_scat	min	0.7649894	0.6738317	0.8583248	0.4549016
SD index	$sd_dis	min	0.5789202	1.703494	1.819146	0.1006037
S Dbw index	$s_dbw	min	3.363968	2.699607	2.499008	2.769472
Silhouette index	$silhouette	max	0.2485357	0.3003016	0.2853209	0.2976705
Tau index	$tau	max	0.2164711	0.2640874	0.2013074	0.3207045
Trace W index	$trace_w	max diff	107,595.6	7111.126	7257.464	2,140,160
Trace WiB index	$trace_wib	max diff	0.9718295	0.9718295	0.9111961	2.86745
Wemmert–Gancarski index	$wemmert_gancarski	max	0.1370079	0.3178442	0.3221082	0.3178927
Xie–Beni index	$xie_beni	min	14,445.39	5647.428	10,638.43	2375.582

References

1. Kohavi R, Rothleder N, Simoudis E (2002) Emerging trends in business analytics. Commun ACM 45(8):45–48. https://doi.org/10.1145/545151.545177
2. Kantardzic M (2011) Data mining: concepts, models, methods, and algorithms. Wiley
3. Cattell R (1943) The description of personality: basic traits resolved into clusters. J Abnorm Soc Psychology 38:476–506. https://doi.org/10.1037/H0054116
4. Pudil P, Novovičová J (1998) Novel methods for feature subset selection with respect to problem knowledge. In: Feature extraction, construction and selection, pp 101–116. https://doi.org/10.1007/978-1-4615-5725-8_7
5. Hartigan J, Wong M (1979) Algorithm AS 136: a *K*-means clustering algorithm. J Roy Stat Soc: Ser C (Appl Stat) 28(1):100–108. https://doi.org/10.2307/2346830
6. MacQueen J (1967) Some methods for classification and analysis of multivariate observations. In: Proceedings of the fifth Berkeley symposium on mathematical statistics and probability, vol 1, no 14. Oakland, CA, USA, pp 281–297
7. Lloyd S (1982) Least squares quantization in PCM. IEEE Trans Inf Theory 28(2):129–137. https://doi.org/10.1109/TIT.1982.1056489
8. Forgey E (1965) Cluster analysis of multivariate data: efficiency versus interpretability of classification. Biometrics 21(3):768–769
9. Kaufman L, Rousseeuw P (2009) Finding groups in data: an introduction to cluster analysis. Wiley. https://doi.org/10.1002/9780470316801. -->
10. Lukasová A (1979) Hierarchical agglomerative clustering procedure. Pattern Recogn 11(5–6):365–381. https://doi.org/10.1016/0031-3203(79)90049-9
11. Zepeda-Mendoza M, Resendis-Antonio O (2013) Hierarchical agglomerative clustering. Encycl Syst Biol 886–887. https://doi.org/10.1007/978-1-4419-9863-7_1371
12. Roux M (2018) A comparative study of divisive and agglomerative hierarchical clustering algorithms. J Classif 35(2):345–366. https://doi.org/10.1007/S00357-018-9259-9
13. Hotelling H (1933) Analysis of a complex of statistical variables into principal components. J Educ Psychol 24(6):417–441. https://doi.org/10.1037/H0071325
14. Abdi H, Williams L (2010) Principal component analysis. Wiley Interdiscip Rev: Comput Statistics 2(4):433–459. https://doi.org/10.1002/wics.101
15. Isomura T, Toyoizumi T (2016) A local learning rule for independent component analysis. Sci Rep 6. https://doi.org/10.1038/srep28073
16. Maaten L, Hinton G (2008) Visualizing data using *t*-SNE. J Mach Learn Res 9:2579–2605
17. Maaten L (2014) Accelerating *t*-SNE using tree-based algorithms. J Mach Learn Res 15:3221–3245
18. Roweis S, Saul L (2000) Nonlinear dimensionality reduction by locally linear embedding. Science 290(5500):2323–2326. https://doi.org/10.1126/science.290.5500.2323
19. Ridder D, Kouropteva O, Okun O, Pietikäinen M, Duin R (2003) Supervised locally linear embedding. Artif Neural Netw Neural Inf Process—ICANN/ICONIP 2003:333–341. https://doi.org/10.1007/3-540-44989-2_40
20. Renjith S, Sreekumar A, Jathavedan M (2018) Evaluation of partitioning clustering algorithms for processing social media data in tourism domain. In: 2018 IEEE recent advances in intelligent computational systems (RAICS). IEEE Press, Thiruvananthapuram, India, pp 127–131. https://doi.org/10.1109/raics.2018.8635080
21. Renjith S, Sreekumar A, Jathavedan M (2020) Performance evaluation of clustering algorithms for varying cardinality and dimensionality of data sets. Materials Today: Proceedings. https://doi.org/10.1016/j.matpr.2020.01.110
22. Renjith S, Sreekumar A, Jathavedan M (2019) Pragmatic evaluation of the impact of dimensionality reduction in the performance of clustering algorithms. In: Advances in electrical and computer technologies, ICAECT 2019, Lecture notes in electrical engineering. Springer, Coimbatore, India
23. Xu R, WunschII D (2005) Survey of clustering algorithms. IEEE Trans Neural Netw 16(3):645–678. https://doi.org/10.1109/TNN.2005.845141

24. Shirkhorshidi A, Aghabozorgi S, Wah T, Herawan T (2014) Big data clustering: a review. In: The 14th international conference on computational science and its applications—ICCSA 2014. Springer International Publishing, Guimaraes, Portugal, pp 707–720. https://doi.org/10.1007/978-3-319-09156-3_49
25. Sajana T, Sheela Rani C, Narayana K (2016) A survey on clustering techniques for big data mining. Indian J Sci Technol 9(3):1–12. https://doi.org/10.17485/IJST/2016/V9I3/75971
26. Ajin V, Kumar L (2016) Big data and clustering algorithms. In: 2016 international conference on research advances in integrated navigation systems (RAINS). IEEE Press, Bangalore, India, pp 101–106. https://doi.org/10.1109/rains.2016.7764405
27. Dave M, Gianey H (2016) Different clustering algorithms for big data analytics: a review. In: 2016 international conference system modeling and advancement in research trends (SMART). IEEE Press, Moradabad, India, pp 328–333. https://doi.org/10.1109/sysmart.2016.7894544
28. Lau T, King I (1998) Performance analysis of clustering algorithms for information retrieval in image databases. In: 1998 IEEE international joint conference on neural networks proceedings, IEEE world congress on computational intelligence (Cat. No.98CH36227). IEEE Press, Anchorage, AK, USA, pp 932–937. https://doi.org/10.1109/ijcnn.1998.685895
29. Maulik U, Bandyopadhyay S (2002) Performance evaluation of some clustering algorithms and validity indices. IEEE Trans Pattern Anal Mach Intell 24(12):1650–1654. https://doi.org/10.1109/TPAMI.2002.1114856
30. Wei C, Lee Y, Hsu C (2003) Empirical comparison of fast partitioning-based clustering algorithms for large data sets. Expert Syst Appl 24(4):351–363. https://doi.org/10.1016/S0957-4174(02)00185-9
31. Zhang B (2003) Comparison of the performance of center-based clustering algorithms. In: Advances in knowledge discovery and data mining, PAKDD 2003, Lecture notes in computer science, vol 2637. Springer, Seoul, Republic of Korea, pp 63–74. https://doi.org/10.1007/3-540-36175-8_7
32. Wang X, Hamilton H (2005) A comparative study of two density-based spatial clustering algorithms for very large datasets. In: Advances in artificial intelligence, AI 2005, lecture notes in computer science, vol 3501. Springer, Victoria, BC, Canada, pp 120–132. https://doi.org/10.1007/11424918_14
33. Poonam Dutta M (2012) Performance analysis of clustering methods for outlier detection. In: 2012 second international conference on advanced computing and communication technologies (ACCT 2012). IEEE Press, Rohtak, India, pp 89–95. https://doi.org/10.1109/acct.2012.84
34. Fahad A, Alshatri N, Tari Z, Alamri A, Khalil I, Zomaya A, Foufou S, Bouras A (2014) A survey of clustering algorithms for big data: taxonomy and empirical analysis. IEEE Trans Emerg Top Comput 2(3):267–279. https://doi.org/10.1109/TETC.2014.2330519
35. Jung Y, Kang M, Heo J (2014) Clustering performance comparison using *k*-means and expectation maximization algorithms. Biotechnol Biotechnol Equip 28(2):S44–S48. https://doi.org/10.1080/13102818.2014.949045
36. Bhatnagar V, Majhi R, Jena P (2017) Comparative performance evaluation of clustering algorithms for grouping manufacturing firms. Arab J Sci Eng 43(8):4071–4083. https://doi.org/10.1007/S13369-017-2788-4
37. Kohonen T (1997) Exploration of very large databases by self-organizing maps. In: International conference on neural networks (ICNN'97), vol 1. IEEE Press, Houston, TX, USA, pp PL1-PL6. https://doi.org/10.1109/icnn.1997.611622
38. Ding C, He X, Zha H, Simon H (2002) Adaptive dimension reduction for clustering high dimensional data. In: 2002 IEEE international conference on data mining. IEEE Computer Society, Maebashi City, Japan, pp 147–154. https://doi.org/10.1109/icdm.2002.1183897
39. Wang Q, Li J (2009) Combining local and global information for nonlinear dimensionality reduction. Neurocomputing 72(10–12):2235–2241. https://doi.org/10.1016/J.NEUCOM.2009.01.006
40. Araujo D, Doria Neto A, Martins A, Melo J (2011) Comparative study on dimension reduction techniques for cluster analysis of microarray data. In: The 2011 international joint conference on neural networks. IEEE Press, San Jose, CA, USA, pp 1835–1842. https://doi.org/10.1109/ijcnn.2011.6033447

41. Chui CK, Wang J (2013) Nonlinear methods for dimensionality reduction. Handb Geomath 1–46. https://doi.org/10.1007/978-3-642-27793-1_34-2
42. Song M, Yang H, Siadat S, Pechenizkiy M (2013) A comparative study of dimensionality reduction techniques to enhance trace clustering performances. Expert Syst Appl 40(9):3722–3737. https://doi.org/10.1016/J.ESWA.2012.12.078
43. Charrad M, Ghazzali N, Boiteau V, Niknafs A (2014) NbClust: an R package for determining the relevant number of clusters in a data set. J Stat Softw 61(6):1–36. https://doi.org/10.18637/JSS.V061.I06
44. Rousseeuw P (1987) Silhouettes: a graphical aid to the interpretation and validation of cluster analysis. J Comput Appl Math 20(November):53–65. https://doi.org/10.1016/0377-0427(87)90125-7
45. Dunn J (1973) A fuzzy relative of the ISODATA process and its use in detecting compact well-separated clusters. J Cybern 3(3):32–57. https://doi.org/10.1080/01969727308546046
46. Calinski T, Harabasz J (1974) A dendrite method for cluster analysis. Commun Stat Theory Methods 3(1):1–27. https://doi.org/10.1080/03610927408827101
47. Davies D, Bouldin D (1979) A cluster separation measure. IEEE Trans Pattern Anal Mach Intell PAMI 1(2):224–227. https://doi.org/10.1109/tpami.1979.4766909
48. Team RC (2009) R: a language and environment for statistical computing. R Foundation for Statistical Computing, Vienna, Austria
49. Tierney L (2012) The R statistical computing environment. Lect Notes Stat. 435–447. https://doi.org/10.1007/978-1-4614-3520-4_41
50. Racine J (2011) RStudio: a platform-independent IDE for R and Sweave. J Appl Econ 27(1):167–172. https://doi.org/10.1002/JAE.1278
51. Goldberg K, Roeder T, Gupta D, Perkins C (2001) Eigentaste: a constant time collaborative filtering algorithm. Inf Retr 4(2):133–151. https://doi.org/10.1023/A:1011419012209

Anomaly Detection for Big Data Using Efficient Techniques: A Review

Divya Jennifer D'Souza and K. R. Uday Kumar Reddy

Abstract There has been a trending impacting application in different sectors such as hospitals, banking, defense, retail and social networks, the rate at which information needs to be kept safe has been a major concern. The security being compromised at a rapid rate which is shown through the increasing rise in frauds has caused a serious impact. Anomaly detection has affected huge sectors and made its way in various applications including detect of fraudulence records, patterns, behaviors, third-party attack or intruder detection in networks and cyber-security. A great deal of research is being carried out in these areas, which uses various approaches such as statistical, data analytics, machine learning and so forth from which anomalies could be extracted. The interdependencies between data can be depicted and represented using graphs in an effective way. However, few works have focused on detecting anomalous behaviors, patterns, links and points in data represented as graphs with multi-dimension especially for detecting outliers in unstructured data. Indeed, data represented as graphs show interdependencies and should be taken into account while detecting anomalies. In this paper, a survey is carried out on the static, dynamic and machine learning approaches for detecting anomalies in the graph data structures. In addition, more focus is toward efficient graph-based techniques that can help in anomaly detection in big data transactions which are presented. Along the way, few open issues and research directions are also mentioned.

Keywords Anomaly detection · Big data · Dynamic data · Graphs · Machine learning

D. J. D'Souza (✉) · K. R. Uday Kumar Reddy
NMAM Institute of Technology, Nitte, Karnataka, India
e-mail: jenniferdsouza87@nitte.edu.in

K. R. Uday Kumar Reddy
e-mail: krudaykumar@nitte.edu.in

N. N. Chiplunkar and T. Fukao (eds.), *Advances in Artificial Intelligence and Data Engineering*, Advances in Intelligent Systems and Computing 1133,
https://doi.org/10.1007/978-981-15-3514-7_79

1 Introduction

Security is the major issue that has come into consideration in recent times. Detecting frauds in an online transaction can be done using any type of technique. Traditional approaches include using the various process of data mining such as extraction, transform, loading, predict, associate, knowledge discovery and outliers detection, in detecting fraudulent behavior. Although an extensive work [1] has been carried out on detecting frauds and outliers for static data, for structured and unstructured data, a focus is on detecting anomalies in attributed dynamic data by developing a novel/hybrid approach based on graph mining techniques as graph-based techniques are more ubiquitous and easier and faster to depict the correlation. There are several extensive studies to detect anomalies for both static and dynamic network topologies [2, 3]. It is known that in data mining, there lies a necessity for fast, robust, ethical and reliable services for detecting an anomaly. Hence, there is a need to optimize or create a hybrid graph-based anomaly detection technique that can detect anomalies on a dynamic network.

Anomaly could be named or addressed differently in various contexts. For instance in mathematical terms, numbers which do not fall into a defined mathematical model or which deviates from the normal flow could be termed as "unusual records." In financial applications transactions, customers, or any automated system that do not follow the predefined rules, could be termed as "suspicious." In health firms, abnormal cluster of cells or unusually less/more amount of any cell formations, malfunction of any organs lead to unhealthy life patterns. There are various medical terms to indicate the same. The unusual patterns or records or suspicious entities could be termed as anomalies. The study in this area to foresee such anomalies using various efficient techniques and to curb their existence if it harms the normal flow of execution is of much interest.

Let $G = (V(G), E(G))$ be a graph. Here, $V(G)$ represents the node of the graph G, and $E(G)$ represents an edge of G. Let w represent weights associated with connecting edges. Weights can take positive or negative real or integer value. Graph can be either directed or undirected. The nodes could have extra attributes associated with it and are denoted as $n_1, n_2, n_3, \ldots$ so on. Similarly, edges too may or may not have attributes and is denoted by e_1, e_2, e_3, so on if exits.

In this paper, a review of efficient graph-based techniques that can help in anomaly detection in big data transactions is presented. Section 2 explains in detail the literature survey on the static, machine learning and dynamic methods for detecting an anomaly. Section 3 focuses on the concluding remarks and open issues and gives the conclusion that could be drawn from the review.

2 Related Works

Various approaches have been studied and executed in detecting anomalies such as density- [4] and dynamic-based [5], graph-based [6], distance-based [7], clustering-based [8], statistical approaches [9], prediction-based [10] and window-based [11] methods. The recent study shows various machine learning methods [12] for detecting anomalies on high-dimensional data sets, see [4, 7, 13–16]. The techniques used in these papers can be classified into density-based [4, 13] and distance-based [14, 17] approaches. The drawback here is that it is difficult or highly complex to map the data sets to non-spatial networks or to generalize these techniques [18, 19]. To overcome these issues, we have various graph-based methods [3, 5, 6, 20–22].

Although graph-based anomaly detection has a number of advantages, it has few drawbacks [19]. The links in graphical representation can be random, and also outliers can be determined only by examining the behavior of the edges in the individual graph objects. Works related to clustering and graph streams are discussed in [2, 6, 9, 14, 20, 21, 23].

Numerous techniques for detecting outliers in a network have been explained in [5, 13, 24]. But, these techniques could be applied only on static networks [9] and do not suit well for streaming or dynamic data represented as graphs.

The dynamic and evolving nature of streaming data poses a peculiar challenge which is not addressed in graph mining techniques such as outliers' detection. Moreover, streaming data sets can be looked upon for examining and processed at most once during the computation.

In the following subsection, we discuss the literature related to anomaly detection for a statistical data set. The techniques used could be mathematical models, distance based, density based, clustering based or prediction based.

2.1 Anomaly Detection for Statistical Data Set

Barnett and Lewis [9] discuss methods which use one-dimensional data set. Statistical-based approach for detecting an anomaly is one of the oldest techniques that were used for detecting an anomaly which uses named distribution model for the data set for computing the unusual records.

Deshmukh et al. [25] explain techniques which work on one-dimensional data set. It becomes tedious to evaluate and to create a model when the dimensions grow. By applying various statistical models for the distribution of data, if the points lie under low probability such data is considered as anomalous according to Sreevidya [26]. Anomaly detection methods for statistical data sets fall under two groups: parametrized and non-parametrized techniques. According to Zhang [27] to build a model, the data set needs to be trained in case of parametric methods. But this creates a problem if the data is streaming as we cannot consider full data set. The main reason lies as statistical models take sample data to derive a conclusion. Branch et al. [28]

explain the advantages of unsupervised non-parametrized techniques for detecting unusual records. These techniques can be used in streaming data, but the drawback here would be these techniques are applicable only on single or low-dimensional dataset. It does not suit well for high-dimensional data set.

Anomaly Detection Based on Distance Consider a point or a node V(G), that is, nearest neighbors as discussed in [15, 17]. Anomaly detection based on distance can be defined as how far or near the neighboring node lies. Knorr et al. [17] also introduce novel definition based on the concept of distance with regard to a point p in data set as an outlier with respect to the parameters K and λ or less than p. Ramaswamy et al. [16, 29] redefined Knorr's explanation of distance-based anomaly detection.

As mentioned in Chugh et al. [30], any type of data whose distance or similarity measure if available could easily define distance-based anomaly detection approach on such a setup. Since this technique would not need a complete detailed understanding of the underlying application domain, this approach can be also used in streaming data because it does not rely on the assumption of any data distribution. To defend his theory, Zhang [27] states that these methods are not suitable for multi-dimensional data set because of the amount of noise it has in real-world application.

Anomaly Detection Based on Density Breunig et al. [31] explain the significance of the local outlier factor (LOF). This value could be assigned to all the data points lying in the plane based on its neighboring density. According to Papadimitriou et al. [32] instead of LOF, local correlation integral (LOCI) would be better. This is mainly due to the reason for difficulty in selecting and assigning suitable value as the LOF. Pokrajac et al. [33] propose an incremental LOF algorithm which is best fit for density-based data stream. The drawback of this technique is that it can detect any change that occurs in the stream data, but it cannot the distinction between the new normal data behaviors and anomalies. The solution was proposed by Karimian et al. [34] in his theory which dealt with the improvement of incremental local outliers' factor [4], which could make the distinction between the new data behaviors and outliers.

The comparison for density- and distance-based algorithms and methods was listed by Zhang [27] and suggested that the density-based techniques would be more effective because it contains the density of both neighbors. But drawback using this approach would be that density-based approaches do not adhere to good accuracy as dimension increases and hence will not suitable for a wider range of data set.

Anomaly Detection Based on Clusters Clustering-based methods group similar data points together to form clusters and apply various models and techniques on such data set. According to Amini and Wah [35], efficient cluster-based methods such as (D-Stream, MRStream, FlockStream, DenStream, DengrisStream, CluStream, SWClustering) give better insights and results on data sets of higher dimension.

To group streaming data, the following are the key requirements collection of data, summarizing the data, processing to eliminate irrelevant data and detecting anomalous data. Toshniwal and Toshniwal [12] propose a technique where weights are allocated to attributes according to its relevance. Chandola et al. [36] describe the main advantage of clustering techniques stating that anomaly detection technique is fast if clusters are used because the comparison of a cluster is done with every test case instead of a single data point which significantly reduces time.

Amini [37] explains one more DenStream clustering technique that consists of the following three phases. The first phase uses a transitory memory which saves previously removed or erased and unimportant data points. In the second phase, DenStream guarantees the probability of data to form clusters, thus increasing their accuracy rate.

Clustering techniques also have their own disadvantages [25]. To state few clustering techniques mainly focus on how to create clusters rather than finding anomalies. Another drawback is that most of the clustering algorithms would require specifying the number of clusters in advance [28] which cannot be applied in streaming data or real-world data set as we do not know the amplitude and size of data in advance and hence cannot create clusters in advance and then apply the processing techniques on those.

Shunye and Wang [23] described an enhanced k-means clustering technique using three steps. In the first step, dissimilarity of the matrix is constructed. Second, based on dissimilarity matrix Huffman's tree is created using the Huffman algorithm. Output obtained from Huffman's tree is the initial centroid. Third, k-means clustering technique is made suitable for k initial centroids to get k cluster outputs.

Tables 1 and 2 summarize statistical anomaly detection methods with their pros and cons and applications where these techniques are suitable.

Table 1 Statistical- and density-based methods for graph-based anomaly detection

Techniques	Article	Advantages	Disadvantages	Applications
Statistical-based anomaly detection parametric/nonparametric methods	Thakkar et al. [38], Barnett and Lewis [9], Deshmukh and Kapse [25], Sreevidya [26], Zhang [27]	Useful for single or low dimensioned data set (nonparametric)	Parametric methods cannot be used in data streams Performs poorly on high-dimensional data set Less effective Very complex	Regression models and applications which involve numerical data
Density-based anomaly detection	Zhang [27], Ramaswamy et al. [29]	Better than distance-based AD, detects data changes. Hence used for streaming data set	Complicated, expensive, cannot distinguish data behavior and anomalies	Health care

Table 2 Distance- and clustering-based methods for graph-based anomaly detection

Technique	Article	Advantages	Disadvantages	Applications
Distance-based anomaly detection	Branch et al. [28], Knorr and Ng [17], Chugh et al. [30], Ramaswamy et al. [14]	Used in data stream Easy to understand and implement Do not rely on assumed distribution to fit the data	Not effective for high-dimensional data	Wireless sensor network, breast cancer analysis
Clustering-based anomaly detection	Tellis and Souza [39], Amini and Wah [35], Thakran and Toshniwal [12]	Less complex Very effective on high-dimensional data set	"*S*" or oval-shaped clusters are difficult to identify	Intrusion detection

2.2 *Observations on Anomaly Detection for Statistical Data Set*

To Summarize, static plain graphs exploits the graph structure and finds unusual patterns to spot anomalies. It could be used on unsupervised data. Also, algorithms used in statistical approaches have some drawbacks where these techniques are exponential in nature and also not readily applicable to graph data that increases continuously in size and volume.

Static attributed graphs are graphs which have a little more information about the nodes or edges which may not be unique. These extra information or characteristics of the planar data represented as graph determine the unusual spots, edges, nodes, subgraphs and detect outliers that could be used on semi-supervised data. The drawback here is anomaly detection consumes more time. Distance-based algorithms are mainly suited when you have single or low-dimensional data sets. It does not respond well to high-dimensional data. Density-based algorithms are better compared to distance-based approaches but again do not respond well to multi-dimensional data set. Clustering-based algorithms work well with a huge amount of data set as we group them accordingly and apply the algorithms on clusters. But difficulty would be in previously assuming the number of clusters which is needed in some of the clustering-based techniques.

2.3 Machine Learning-Based Approaches

According to Tax et al. [40], constructing one class classifier such as support vector machine (SVM) is an anomaly detection technique which has gained much popularity. But it overfits a model to train the data, that is, if the total count of learned classes exceeds one, then there is a possibility that data will belong to a single class corresponding to multiple learned classes.

In [38], the authors show that with the universal class distribution, outliers' detection problem could be converted into classification problem by distributing outliers' data. Analysis that determines outliers' detection along with the problem of classification although important is needed by this algorithm which is generally unknown. Such distribution is not required by and does not give differing perspective to club classification to detect anomaly which is concluded in their works. In accordance with Kuncheva and Breiman [41, 42], the random forest is used for classification and regression which is the other version of bagging. According to their works, boosting faster performs better than bagging and boosting. In random forest technique, the random tree is treated as the base classifier. This approach is also known as ensemble learning which takes the decision tree as the base classifier.

One popular technique which is named random forest approach takes the following two parameters which could be adjusted in each node: (i) the total count of variables to be chosen which is usually fixed in all nodes; (ii) the total number of trees that are used to build the forest. This method establishes a number of individual trees that are used to determine if a given set is normal or abnormal [1].

We conducted research based on this approach by using online credit card data set from Kaggle repository to detect fraudulent transactions and below were the results and insights obtained. The data consisted of 284,807 transactions with a total of 492 fraudulent transactions.

The following observations were made as depicted in Fig. 1.

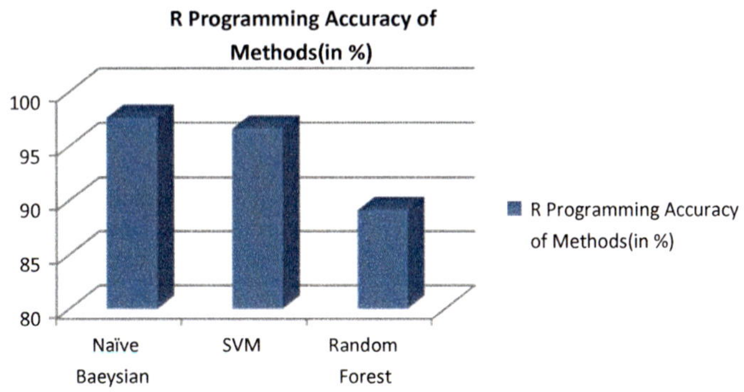

Fig. 1 Bar graph for methods Naïve Bayesian, SVM and random forest using *R* Programming

(i) Naïve Bayesian is easy to train and understand the results. Also, the model is smaller than those compared to SVM or random forest. However, it is really fragile to overfitting without any regularization assumption.
(ii) Random forests are easier to understand but are slower to train. Some objections for it are overfitting. Also, all the randomization parameters have to set such as the selection of nodes and randomization of instance variables.
(iii) SVM suits and expands to multi-dimensional data set, and it works very well for all sorts of data (unstructured, structured and semi-structured data). It has generalization in practice, and the risk of overfitting is less. However, it is tough to incorporate business logic.

2.4 *Dynamic Anomaly Detection Techniques*

Feature-Based Approaches Berlingerio et al. [43] discuss the GED, graph footprints, graph matching distance, error correcting, maximum common subgraph (MCS) and discontinuity detection in a dynamic graph. In [44], the authors show how to detect unusual changes in time-evolving or dynamic streaming data set represented as communication graphs using the concept of variance in the graph diameter. This difference or variance can be defined as the greatest of the longest shortest paths for all vertices. They also gave the concept of role extraction.

According to the authors in [1, 10], the anomalous features could be detected by extracting at least one feature from a graph instance and comparing this feature in consecutive time intervals using any relevant metric. The resulting series obtained is fed to a predictive model such as auto-regressive moving average (ARMA), and the obtained residuals are processed. The instances which exceed the given threshold are considered anomalous.

Decomposition-Based Approaches In [11, 45], the authors discussed on temporal outliers which are detected using matrix or tensor decomposition-based approaches for dynamic data streams. The singular and eigenvalues, eigenvectors are appropriately selected for interpretation.

The authors in [46] show how to find anomalous patterns in a highly traffic communication graph. This new approach works on the principle of finding similarity between cells instead of finding unusual patterns. The adjacency matrix for the network is built on this principle, and the similarity index is computed based on the destination hosts that overlap each other.

In [24], the paper is based on the concept of reconstruction of graph. The sparse graphs are taken into notice their low-rank approximations are noted at regular interval for their summaries. While reconstructing the graph if the error rate changes significantly at some time, then such graph is considered unusual.

Clustering-Based Approaches In [1], the authors pointed out the key points of their research involves around using a probabilistic model GHRG (generalized hierarchical random graphs) to detect change point in networks. This approach divides the planar graph to a set of groups which are nested and whose communications are represented as dendrograms.

The authors in [47] discuss the relationship between MDL and tensor-based approaches which tracks "important" communities over time. In [48], the authors propose an algorithm which is parameter-free and is based on the MDL approach. Their technique discovers node partitions in directed, bipartite and dynamic graphs and also monitors their progress over time to capture changes and events [49–52].

Window-Based Approaches The authors in [53] focus on the concept of graph snapshots, scan statistics for detecting an anomaly. Mongiovi et al. [45] gave the importance of NP-hard, heaviest dynamic subgraph (HDS) to solve the problem. Kashima and Idé [46] describe how to present activity of every node could be matched with its activity in the past x time ticks. It also explains how to create new node based on its role transition which is derived from the previous steps [54–56].

The main challenges of these approaches are with time and space. Can one decide on the window size and duration that needs to be considered to process the data? The other problem would be how efficiently could one segregate the noise from the real-time data when it needs to be added to the window to process?

Tables 3 and 4 summarize the various techniques for anomaly detection for dynamic data sets.

2.5 *Observations on Dynamic Anomaly Detection Techniques*

Dynamic anomaly detection approaches focus on streaming or time-evolving data. Hence, the challenge lies in the data that could be captured and processed only once. Feature-based approaches concentrate on any significant feature in the graph structure such nodes or subgraph or edges or attributes and carry out the process.

Decomposition-based approaches work on the principle of converting the graph into mathematical models such as matrix or tensors and calculate the appropriate factors such as eigenvalues and eigenvectors to detect unusual activities.

Cluster-based approaches are suitable when you have larger data set and communities with unusual high or low clusters are examined.

Window-based approaches take into consideration snapshots of a data stream for particular interval of time to process the data.

Table 3 Feature- and decomposition-based dynamic graph anomaly detection approaches

Technique	Article	Observations	Applications
Feature based	Amini [37], Yu et al. [11]	– Observe and collect a "good summary" from each snapshot of the input graph – Compare and contrast consecutive graphs using any similarity function – When the distance is greater than a manually or automatically defined threshold characterize the corresponding snapshot as anomalous	Mobile data
Decomposition based	Yu et al. [11]	– Detect temporal anomalies by relying on matrix or tensor decomposition of the time-evolving graphs – Interpret appropriately selected eigenvectors, eigenvalues or singular values	Financial/online transactions

3 Concluding Remarks and Open Issues

This paper presents a survey on techniques for anomaly detection, namely statistical, machine learning and dynamic-based approaches on plain, attributed data which could be analyzed for performing research based on the data obtained. Each of these problems is undoubtedly the basic and fundamental ones and is well known in the research community.

The different approaches listed and executed have their own pros and cons and should be applied as per the context to come up with the at most better solution for a given problem. As pointed out in the previous section, very few methods aim at detecting outliers or unusual patterns or records in streaming graphs with attributes. As per the summary on statistical-based approaches and dynamic-based approaches in Sect. 2, we get a clear insight on how these techniques have their own advantages and complexities. Hence, it could be observed that new or hybrid approaches of anomaly detection techniques are in demand and need. It is known that the data at which is being generated at an exploding rate has made way to think in new and different perspectives to tackle the big data and enhance the security.

It may be noted that techniques used on static graphs work either on attributed or plain graphs but not when one has both options available. It is recommended to work/research on approaches that can work with both, i.e., methods that are applied to

Table 4 Community- and window-based dynamic graph anomaly detection approaches

Technique	Article	Observations	Applications
Community based	Amini and Wah [35], Karimian et al. [34]	– Monitoring the changes in the whole network is not necessary – Monitor graph communities or clusters over time and report an event when there is structural or contextual change in any of them	Facebook data analysis, sentimental analysis
Window based	Fujimaki et al. [57], Mongiovi et al. [45]	– These methods rely on a time window/frame in order to spot outliers, anomalous patterns and behaviors in the input graph – A number of previous instances are used to model the "normal" behavior, and the incoming graph is compared against those in order to characterize it as normal or anomalous	Twitter, YouTube data analysis

plain graphs but can also use attribute information if available. Moreover, developing graph-based techniques is to be considered which gives better visual representation and understanding of the underlying problem?

It would be very helpful if we could induce methodologies wherein anomalies could be predicted and detected on time which could prevent lossy situations. In addition, identifying and considering the key elements is also an important factor. For example, giving an importance to setting up appropriate window size or history of data or previously deleted data and identifying attributes of nodes, edges also take significance in detecting anomalies.

From the perspective of anomaly detection, the recent discussions are focused on the current trending applications which are dynamic in nature. The discussions are focused mostly on the algorithmic computation of such problems which are evolving rapidly in a time frame (real-time applications). The summary of these techniques, their pros and cons, are also given for a better understanding to draw appropriate insight while applying these techniques. Finally, the open issues and future research are also discussed.

References

1. Akoglu L, Tong H, Koutra D (2014) Graph-based anomaly detection and description: a survey, April
2. Aggarwal C (2011) Social network data analytics. Springer
3. Aggarwal C, Xie Y, Yu P (2009) Gconnect: a connectivity index for massive disk-resident graphs. PVLDB 2(1):862–873
4. Breunig M, Kriegel HP, Ng R, Sander J (2000) LOF: identifying density-based local outliers. In: SIGMOD Conference, pp 93–104
5. Henzinger M, King V (1999) Randomized fully dynamic graph algorithms with poly logarithmic time per operation. J ACM 46(4):502–516
6. Kernighan BW, Lin S (1970) An efficient heuristic for partitioning graphs. Bell Syst Tech J 49:291–307
7. Knorr E, Ng R, Tucakov V (2000) Distance-based outliers: algorithms and applications. VLDB J. 8(3–4):237–253
8. Kim M-S, Han J (2009) A particle-and-density based evolutionary clustering method for dynamic networks. PVLDB 2(1):622–633
9. Barnett V, Lewis T (1994) Outliers in statistical data, vol 3, no. 1. Wiley, New York
10. Pincombe B (2005) Anomaly detection in time series of graphs using arma processes. Asor Bull 24(4):2–10
11. Yu Y, Zhu Y, Li S, Wan D (2014) Research article time series outlier detection based on sliding window prediction college of computer and information. Hohai University, Nanjing 210098, China Correspondence should be addressed to Yufeng Yu; hhuheiyun@126.com. Received 18 July 2014; Accepted 15 Sept 2014; Published 30 Oct
12. Thakran Y, Toshniwal D (2012) Unsupervised outlier detection in streaming data using weighted clustering. In: 2012 12th international conference on intelligent systems design and applications (ISDA), pp 947–952, IEEE
13. Aggarwal C, Yu P (2001) Outlier detection for high dimensional data. In: SIGMOD Conference, pp 37–46
14. Ramaswamy S, Rastogi R, Shim K (2000) Efficient algorithms for mining outliers from large data sets. In: SIGMOD Conference, pp 427–438
15. Knorr E, Ng R, Tucakov V (2000) Distance-based outlier: algorithms and applications. VLDB J 8(3–4):237253
16. Ramaswamy S, Rastogi R, Shim K (2000) Efficient algorithms for mining outliers from large data sets. In: ACM SIGMOD Record, vol 29, no. 2, pp 427–438
17. Knorr EM, Ng RT (1999) Finding intentional knowledge of distance-based outliers. In: Proceedings of the 25th international conference on very large data bases. Edinburgh, Scotland, pp 211–222
18. Jin W, Jiang Y, Qian W, Tung AKH (2006) Mining outliers in spatial networks. In: DASFAA Conference, pp 156–170
19. Karger DR (1994) Random sampling in cut, flow, and network design problems. In: STOC, pp 648–657
20. Aggarwal C, Wang H (2010) Managing and mining graph data. Springer
21. Aggarwal C, Zhao Y, Yu P (2010) On clustering graph streams. In: SIAM Conference on Data Mining, pp 478–489
22. Akoglu L, McGlohon M, Faloutsos C (2010) Oddball: spotting anomalies in weighted graphs. In: PAKDD Conference, pp 420–421
23. Vitter JS (1985) Random sampling with a reservoir. ACM Trans Math Softw 11(1):37–57
24. Frederickson GN (1985) Data structures for on-line updating of minimum spanning trees, with applications. SIAM J Comput 14(4):781–798
25. Deshmukh MMK, Kapse AS (2016) A survey on outlier detection technique in streaming data using data clustering approach. Int J Eng Comput Sci 5(1)
26. Sreevidya SS, (2014) A survey on outlier detection methods. Int J Comput Sci Inf Technol (IJCSIT) 5(6)

27. Zhang J (2013) Advancements of outlier detection: a survey. ICST Trans Scalable Inf Syst 13(1–3):e2, January-March 2013
28. Branch JW, et al (2006); Kadam N, Pund MA (2013) Cluster based and distance based approach for outlier detection. Int J Adv Res Comput Sci 4(2)
29. Ramaswamy S, Rastogi R, Kyuseok S (2000) Efficient algorithms for mining outliers from large data sets. In: Proceedings of ACM SIDMOD international conference on management of data
30. Chugh N, Chugh M, Agarwal A (2014) Outlier detection in streaming data a research perspective. In: International conference on parallel, distributed and grid computing, IEEE
31. Breunig MM, Kriegel HP, Ng RT (2000) LOF: identifying density-based local outliers. In: ACM conference proceedings, pp 93–104
32. Papadimitriou S, Kitagawa H, Gibbons PB, Faloutsos C (2003) Loci: Fast outlier detection using the local correlation integral. In: Proceedings 19th international conference on data engineering, pp 315–326. IEEE
33. Pokrajac D, Lazarevic A, Latecki LJ (2007) Incremental local outlier detection for data streams. In: IEEE symposium on computational intelligence and data mining, pp. 504–515. March, CIDM 2007. IEEE
34. Karimian SH, Kelarestaghi M, Hashemi S (2012) I-inclof: improved incremental local outlier detection for data streams. In: 16th CSI international symposium on artificial intelligence and signal processing (AISP), pp 023–028. IEEE
35. Amini A, Wah TY (2013) Requirements for clustering evolving data stream. In: 2nd international conference on soft computing and its applications (ICSCA'2013), 25–26 Sept 2013
36. Chandola V, Banerjee A, Kumar V (2009) Anomaly detection: a survey. ACM Comput Surv (CSUR) 41(3):15
37. Amini A (2014) An adaptive density-based method for clustering evolving data streams. Doctoral dissertation, University of Malaya
38. Thakkar P, Vala J, Prajapati V (2016) Survey on outlier detection in data stream. Int J Comput Appl 136:13–16
39. Tellis VM, D'Souza DJ (2018) Detecting anomalies in data stream using efficient techniques: a review. In: 2018 international conference on control, power, communication and computing technologies (ICCPCCT)
40. Zhou Y, Cheng H, Yu JX (2000) Graph clustering based on structural/attribute similarities. PVLDB 2(1):718–729
41. Bakar ZA, Mohemad R, Ahmad A, Deris MM (2006) A comparative study for outlier detection techniques in data mining. In: 2006 June edition IEEE Conference on cybernetics and intelligent systems, pp 1–6
42. Ramesh Kumar B, Aljinu Khadar KV (2017) A survey on outlier detection techniques in dynamic data stream. Int J Latest Eng Manag Res (IJLEMR) 2(8):22–30. ISSN: 2455-4847 www.ijlemr.com
43. Berlingerio M, Koutra D, Eliassi-Rad T, Faloutsos C (2012) A scalable approach to size-independent network similarity
44. Alguliyev RM, Aliguliyev RM, Imamverdiyev YN, Sukhostat LV (2017) An anomaly detection based on optimization. International Journal of Intelligent Systems and Applications 12:87–96. Published Online Dec 2017. In: MECS (http://www.mecs-press.org/). https://doi.org/10.5815/ijisa.2017.12.08
45. Mongiovi M, Bogdanov P, Ranca R, Singh AK, Papalexakis EE, Faloutsos C (2013) Netspot: spotting significant anomalous regions on dynamic networks. In: Proceedings of the 13th SIAM international conference on data mining (SDM). Texas-Austin, TX
46. Idé T, Kashima H (2004) Eigen space-based anomaly detection in computer systems. In: Proceedings of the 10th ACM international conference on knowledge discovery and data mining (SIGKDD). Seattle, WA, pp 440–449. ACM
47. Araujo M, Papadimitriou S, Gnnemann S, Faloutsos C, Basu P, Swami A, Papalexakis E, Koutra D (2014) Com2: fast automatic discovery of temporal (comet) communities. In: Proceedings of

the 18th Pacific-Asia conference on knowledge discovery and data mining (PAKDD). Tainan, Taiwan
48. Sun J, Faloutsos C, Papadimitriou S, Yu PS (2007) Graphscope: parameter-free mining of large time-evolving graphs. In: Proceedings of the 13th ACM International Conference on Knowledge Discovery and Data Mining (SIGKDD). San Jose, CA, pp 687–696. ACM
49. Aggarwal C (2006) On biased reservoir sampling in the presence of stream evolution. In: VLDB Conference, pp 607–618
50. Aggarwal C, Han J, Wang J, Yu P (2003) A framework for clustering evolving data streams. In: VLDB Conference, pp 81–92
51. Gao J, Liang F, Fan W, Wang C, Sun Y, Han J (2010) On community outliers and their efficient detection in information networks. In: ACM KDD Conference, pp 813–822
52. Sun Y, Yu Y, Han J (2009) Ranking-based clustering of heterogeneous information networks with star network schema. In: ACM KDD Conference, pp 797–806
53. Priebe CE, Conroy JM, Marchette DJ, Park Y (2005) Scan statistics on enron graphs. Comput Math Organ Theory 11(3):229–247, October. ISSN 1381–298X
54. Chatfield C (2004) The analysis of time series: an introduction, 6th edn. Chapman and Hall CRC
55. Moayedi HZ, Masnadi-Shirazi MA (2008) Arima model for network traffic prediction and anomaly detection. In: International symposium on information technology, vol 4, pp 1–6
56. Knorn F, Leith DJ (2008) Adaptive Kalman filtering for anomaly detection in software appliances. In: IEEE conference on computer communications workshops, pp 1–6
57. Fujimaki R, Yairi T, Machida K (2005) An anomaly detection method for spacecraft using relevance vector learning. Adv Knowl Discov Data Min 3518:785–790

Data Science and Internet of Things for Enhanced Retail Experience

Irfan Landge and Hannan Satopay

Abstract The penetration of technology is ever increasing and is encompassing more and more sectors day by day and rightly so—the advantages are huge. Specifically, the technology in consideration is the Internet of Things and data science. The implementation of these technologies is still unexplored in the field of retail. In this paper, we have discussed the importance of data in the field of data science and later explored the data that can be obtained and harnessed from the sector of retail. Many of the possible data science use cases in the field of retail have also been explored. Consequently, the architecture of the Internet of Things has been discussed and the possible use cases of the technology in the sector of retail have also been explored. The inclination is toward using these technologies for creating an enhanced retail experience for both the customers and the retail store manager by including on-the-cart billing, automated inventory management, product correlation and prediction for product replenishing duration. Finally, the realization, exploration, and visualization of the above view have been done in the implementation and the result's section of the paper. The conclusion establishes the importance of the implementation of the above technologies and the benefits it brings toward creating an enhanced retail experience.

Keywords Smart retail · Data science · Internet of Things · Predictive analysis · Inventory tracking · Automated billing

1 Introduction

Retail is a sector that can provide many avenues for technology to bloom. The technology in consideration in this paper is the Internet of Things and data science [1, 2].

I. Landge (✉)
Mumbai University, Mumbai, India
e-mail: irfan.ieee.in@gmail.com

H. Satopay
Vidyalankar Institute of Technology, Mumbai, India
e-mail: hannan.satopay@outlook.com

N. N. Chiplunkar and T. Fukao (eds.), *Advances in Artificial Intelligence and Data Engineering*, Advances in Intelligent Systems and Computing 1133,
https://doi.org/10.1007/978-981-15-3514-7_80

Retail is a people-oriented business which makes it ripe and ready for exploration through data science. Despite the massive amounts of data available, many of today's decisions are driven by human opinion and observation which leaves room for bias and error [3]. The reason is that managers are often only half right when it comes to understanding in-store problems and customer behavior. Data science can help reduce that bias by arming the managers with data insights to make tailored improvements to the retail experience [4]. At the same time, a growing need to increase customer loyalty and deliver the best in-store experience is driving the adoption of the Internet of Things (IoT) in the retail sector. The potential benefits of deploying the Internet of Things in the retail sector are positively huge [5]. The focus of this paper is in exploring these technologies. In the view of it, these technologies have been utilized in providing the following features related to the customers—automated product cost and total cart cost calculation by using the data from the cloud database to enable automated billing. This will help the customers save time in the checkout line and be well informed on the total cost of the products in their cart to stay within their budget. For the retail managers, the following features will be achieved—automated inventory management, inventory tracking, product correlation and prediction for product replenishing duration. Automated inventory management allows the inventory to be updated in real time as and when the product is being added or removed. Inventory tracking allows the managers to see the live feed of the current store inventory and attach notifications for the same. Product correlation allows for a comparison graph to be generated between the performance of each of the products belonging to the same category. This allows the analysis of selecting the product which performs better and place it appropriately on the proper shelf. Prediction for product refilling takes the historical product sale data to make a prediction of when that product needs to be replenished so that sales do not suffer due to low number of availabilities of the product.

2 Data Science

2.1 Data in Data Science

Data in today's world is no more than just bits and bytes but has become a way to drive a business forward. It all depends on what data is being considered and how it is used for business growth. The retail field is one of the most important economic drivers. Data can be technically defined as information which is saved or stored and can be retrieved for different types of analysis. The impact and influence of data are based on how it is processed in the view of the target business. The data is collected in different forms, formats, styles and from different sources. The data that has been collected can either be in a structured format or a non-structured format. The number of parameters (fields) for a dataset plays a very important role in understanding the influencing factors for the end result. Unstructured nature of the data is mainly getting generated by present sources like blog, sensors, online forms. The meaningful data is essential for applying the data science principles and performing the analysis. Precise data can make the prediction more accurate and intelligent. The huge amount of raw data is required to be converted into meaningful data for further consumption.

2.2 Data in Retail

In sales, the word retail is used related to the quantity of sale. The bulk sale is mainly considered as wholesale, and there are specific customers who are involved in such large quantity purchasing. On the other hand, the retail sale is mainly targeted toward small quantity or distributed unit-based sale. The large customer base is available for such retail sales. E-shopping is further adding to the customer base for the sector of retail. The data becomes significant in the decision making for all involved in the retail cycle which starts with the producer/manufacturer and ends with the customer. The main driving parameter is sale which depends on which products are purchased by the customer. The rate of sale, availability of stock, inventory management, offers for promoting the sale and many other factors are having a very high influencing factor toward performing analysis and making informed decisions. The digital technology is helpful in providing the customers, a friendly user interface. The digital technology is mainly used for collecting data from various sources like payment gateway, online and mobile purchase, point of sale transactions. This data that has been collected is mainly unstructured and requires to be processed using different algorithms in order to extract more knowledge and insights. Data collected by different modes and interfaces can be used for understanding customer behavior, shopping trends, popular products and so on [6]. Figure 1 depicts the data being collected from different sources and used for various analytics and personalization.

2.3 Data Science Use Cases in Retail

Depending on the requirements of the retail industry, multiple use cases can be formulated for the use of data science [7]. The most common use cases are:

1. Customer Behavior: The customer interacts with the system through multiple interface options. Mobile phone, Web sites, social media are some of the popular interfaces the customer or shopper uses while ordering or giving feedback about the product. The data about the customer responses is gathered in many formats which leads to an unstructured collection of data. This data is required to be processed for analyzing customer's sentiments and behavior.
2. In-store Experience: Although online purchase options are available, there are many customers who are interested in visiting and purchasing from the offline stores. The main difference between online and offline stores is the physical

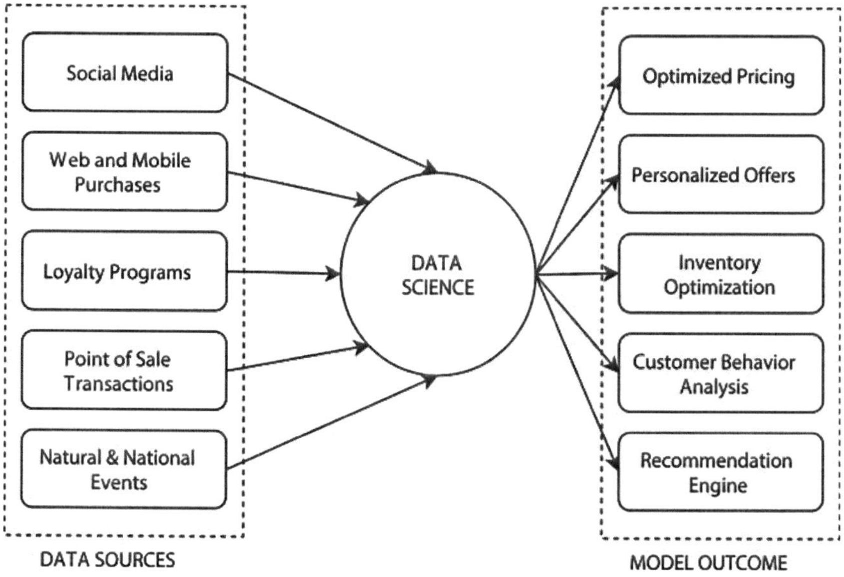

Fig. 1 Data science for the retail sector

examination of the products. In the case of an in-store purchase, the shopper gets personalized experience before purchasing the product. The data collected from the shopper's experience can be used for product placement and understanding the in-store experience of the customer.

3. Promotional Activities: The data gathered from a different point of sales can be used for understanding which products are having a fast or a slow selling rate. The promotional schemes can be worked out based on the sale data for increasing revenue. The product pricing can also be calculated from the sales data schemes.
4. Inventory Management: The sales data collected can be statistically processed to manage the inventory. The number of product units available in the stock and the need for them to be reordered can be easily worked out. The predictive analysis is used for increasing revenues with futuristic decision making.

3 The Internet of Things

3.1 The Internet of Things Architecture

Internet of Things (IoT) is becoming a more common technology for controlling devices on a network. In the Internet of Things, the network of multiple objects or elements is formed using the Internet. The unique object identity is maintained using the IP address of each element. All interconnected object can share the data or

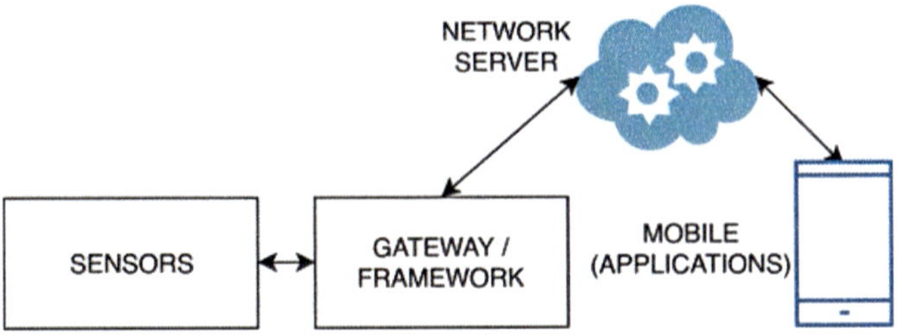

Fig. 2 Internet of things architecture

communicate with each other through the Internet. The IoT architecture consists of sensors which are used for data collection. Based on the application domain and data requirement, sensors are selected for capturing the data. The IoT architecture also consists of a gateway or framework that acts as a means for connecting the sensor to the network. The connectivity can be provided through Wi-Fi, Bluetooth or other protocols for the communication. The network server will process the collected data and instruct the devices connected with it to perform desired smart operations. The mobile applications are widely used for interacting with the systems [8]. Internet of Things architecture is shown in Fig. 2.

3.2 *Internet of Things Use Cases in Retail*

Depending on the requirement of the retail industry, multiple use cases can be formulated for the utilization of the Internet of Things [9–11]. The most common use cases in the retail industry are:

1. Smart Shelves: Inventory management has always been an expensive and tedious process for a long time. Smart shelves enabled through the Internet of Things can simplify the process by automatically monitoring the inventory and send alerts to the store manager if a certain item is running low or its expiration date is near. It can bring crucial advantages and thereby be useful in avoiding oversupply and a shortage of goods. Through tracking inventory, the operating mistakes, as well as the costs, are minimized.
2. Beacons: Beacons are useful in the retail industry through which the stores will able to collect the customer data, send personalized push notifications, increase customer loyalty and even enable in-store navigations for the customers. Using beacons, retailers can easily reach their audience and provide an engaging shopping experience.
3. Automated Checkout: With the Internet of Things, one can set up a system to read tags on each item as and when the product is being added to the shopping cart. A checkout system would then tally the items up and automatically calculate the

total cost of the products which will allow the customers to pay directly through the mobile payment interfaces [12].

4 Implementation

In the proposed implementation, a smart retail shopping experience must be created which will allow the many benefits of the Internet of Things and data science to be reflected and explored. The basic block diagram of smart retail system is shown in Fig. 3.

For the sake of convenience, the proposed system can be subdivided into three subsystem, viz. the Internet of Things subsystem, the data science subsystem and the central database subsystem. It is best if the central database is maintained on the cloud. The former two subsystems can be treated as independent of each other thereby allowing the possibility of customization based on the requirements. One common element that exists between the former two subsystems is the Central Database subsystem. It also to be noted that there is a many to one relation between the Internet of Things subsystem and the central database subsystem. It means there can be N number of unique devices connected to the Internet providing data to a single database and hence the name—central database subsystem. There is a one-to-one relation between the central database subsystem and the data science subsystem which means that the data science subsystem provides a single dashboard to handle all the data from the database subsystem.

4.1 *The Central Database Subsystem*

The technology stack that has been utilized for the central database subsystem on the cloud is Google Firebase. It is a commercial level cloud server though any other cloud services would have worked just fine. A database was created containing four categories having five products each. The data for the products was filled randomly

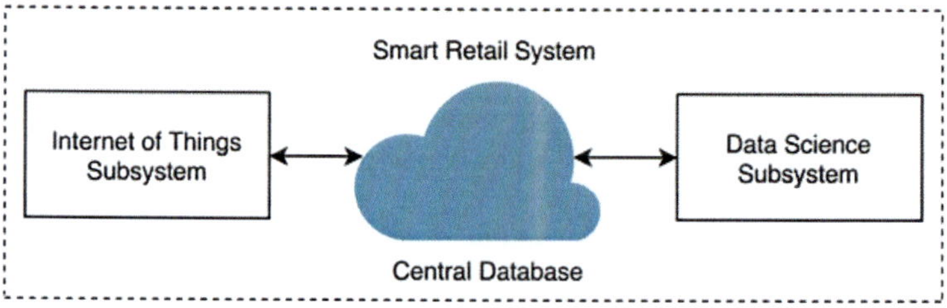

Fig. 3 Basic block diagram

for the demonstration, consumption and prototype verification purpose. During the actual implementation, the data will be updated in real time as and when the transaction takes place. The nature of this data will be dynamic; i.e., it keeps on changing on a regular interval. The second type of data is the static data, which contains the information linked with the product such as tag ID, name, price, category and original quantity. Two different data tables were created having certain static and dynamic parameters. The first table, 'Inventory Data' contains the parameters for each product category such as Product Name, Current Quantity (QCurrent), Original Quantity (QOriginal) and the Timeline. The second table, 'Tag Data' contains the Tag ID and the respective details of the product linked to it such as Product Name, Category and Price. This database allows the hardware device to obtain the data directly from the cloud instead of using the hardcoded values thereby adapting in any scenario. For example, during festive seasons, the price can automatically be adjusted in the cloud database which will be reflected during shopping. The benefits for such cloud integration are tremendous. The database in a typical key-value pair format has been illustrated in Fig. 4.

The problem with such a structured data approach is the requirement of multiple data tables which causes redundancy. The way to tackle this is to utilize non-structured approach. In such an approach, the data is placed in nested tables. It is quite useful for working with large sets of distributed data such as in our case for the sector of retail. The selection of the cloud database should be done based on the compatibility with the embedded hardware. A typical non-structured database is shown in Fig. 5. Further, during the exploration of the database, the nested functionality is clear.

Biscuit

Product Name	Qcurrent	Qoriginal	Timeline
HideNSeek	7	20	['20','19','19','15','13','10','10','9','8','7']
ParleG	10	30	['30','29','28','19','15','15','12','12','11','10']
Monaco	10	15	['15','15','15','14','13','13','13','12','12','10']
Oreo Chocolate	10	40	['40','38','37','25','22','20','15','15','11','10']
Oreo Strawberry	25	40	['40','39','37','31','31','30','27','27','27','25']

Detergent Powder

Product Name	Qcurrent	Qoriginal	Timeline
Surf Excel	9	30	['30','30','29','22','21','20','20','19','15','9']
Ariel	15	30	['30','27','26','25','20','18','16','14','13','15']
Tide	17	30	['30','29','29','25','23','20','20','19','18','17']
Patanjali	10	30	['30','29','22','15','13','10','10','9','8','10']
Rin	25	30	['30','30','29','27','27','27','26','26','25','25']

Edible Oil

Product Name	Qcurrent	Qoriginal	Timeline
Fortune	7	20	['20','19','19','15','13','10','10','9','8','7']
Samrat	15	25	['25','24','24','20','20','18','16','14','13','15']
Gemini	15	30	['30','27','26','25','20','18','16','14','13','15']
Sweekar	30	40	['40','39','39','38','38','38','37','35','33','30']
Nature Fresh	10	15	['15','15','15','14','13','13','13','12','12','10']

Shampoo

Product Name	Qcurrent	Qoriginal	Timeline
Dove	7	25	['25','19','19','15','13','10','10','9','8','7']
Head & Shoulders	10	50	['50','48','40','25','22','20','15','15','11','10']
Clinic Plus	10	30	['30','29','22','15','13','10','10','9','8','10']
Sunsilk	15	40	['40','38','37','25','22','20','17','17','16','15']
Pantene	17	30	['30','29','29','25','23','20','20','19','18','17']

Tag ID	Product Name	Category	Price
19B4A1-19B4A20	HideNSeek	Biscuit	20
19B4A21-19B4A50	ParleG	Biscuit	10
19B4A51-19B4A65	Monaco	Biscuit	15
19B4A66-19B4A105	Oreo Chocolate	Biscuit	25
19B4A106-19B4A145	Oreo Strawberry	Biscuit	25
19B4B1-19B4B30	Surf Excel	Detergent Powder	100
19B4B31-19B4B60	Ariel	Detergent Powder	110
19B4B61-19B4B90	Tide	Detergent Powder	100
19B4B91-19B4B120	Patanjali	Detergent Powder	95
19B4B121-19B4B150	Rin	Detergent Powder	105
19B4C1-19B4C20	Fortune	Edible Oil	150
19B4C21-19B4C45	Samrat	Edible Oil	160
19B4C46-19B4C75	Gemini	Edible Oil	155
19B4C76-19B4C115	Sweekar	Edible Oil	145
19B4C116-19B4C130	Nature Fresh	Edible Oil	170
19B4D1-19B4D25	Dove	Shampoo	210
19B4D26-19B4D75	Head & Shoulders	Shampoo	220
19B4D76-19B4D105	Clinic Plus	Shampoo	205
19B4D106-19B4D145	Sunsilk	Shampoo	200
19B4D146-19B4D175	Pantene	Shampoo	215

Fig. 4 Structured database **a** inventory data **b** tag data

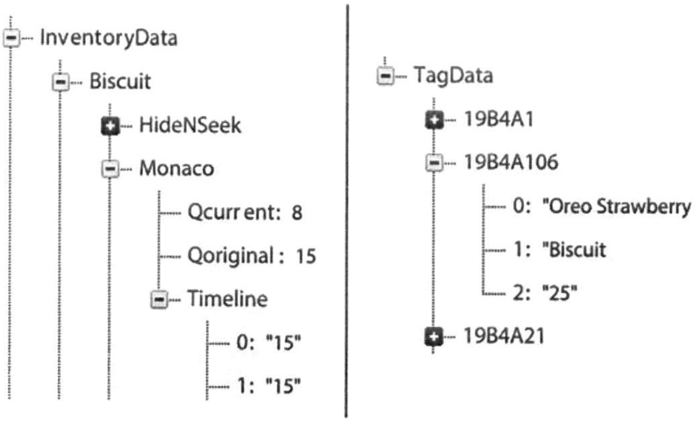

Fig. 5 Expanded view of the database **a** inventory data **b** tag data

4.2 The Internet of Things Subsystem

The technology stack considered for the implementation of the Internet of Things Subsystem is NodeMCU, MFRC522 RFID Card Read/Writer, LCD and LiPo battery system. The technology stack has been chosen by taking into consideration the ease of prototyping for concept verification, cost and functionality. The choice of technology stack that will be used commercially needs to consider the various factors such as scalability, usability, functionality, power considerations, cost, speed considerations, range and security [13]. The shopper/customer would be using a smart cart which is the Internet of Things enabled. Whenever a product is added or removed from the cart, the RFID action is fired [14]. When this happens, the tag ID attached with the product is sent to the central database and the product data stored in the central database is pulled and displayed on the screen to the shopper. The total amount in the cart is updated based on the products being added or subtracted from the same. Next based on whether the product was added or removed, the respective functions are fired which updates the current quantity in the central database. After this, based on the date, the timeline of the product is updated with the instantaneous product quantity. Finally, when the customer has finished with the shopping, the final cart value is displayed, and the payment can be done easily at the checkout counter without the hassle of billing the products . The proposed internet of things model for enhancing reatail experience is shown in Fig. 6.

4.3 The Data Science Subsystem

With the help of this data, it is possible to develop a data product in the form of a Web app. Through the Web app, the manager will be able to monitor the inventory in real

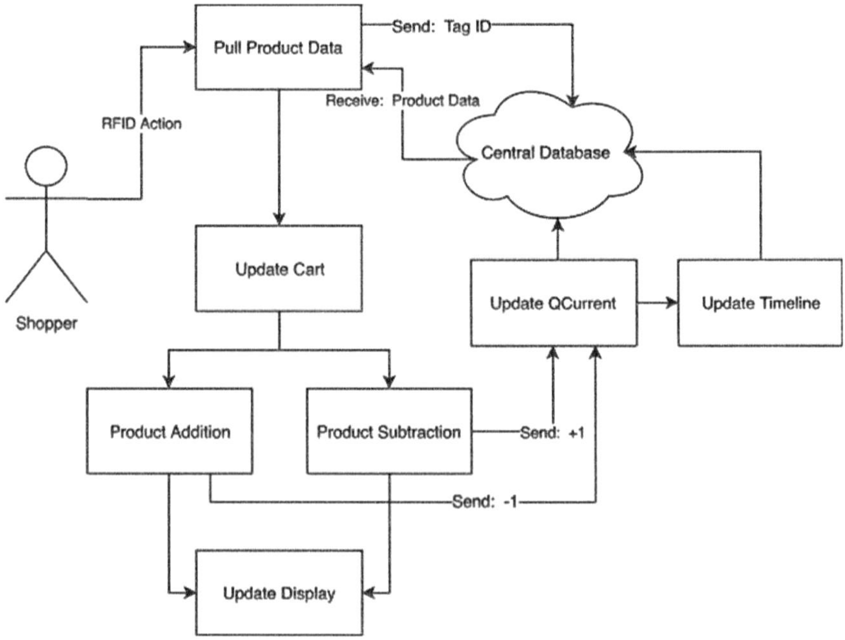

Fig. 6 Internet of things model

time. A product correlation data model has been created which helps in analyzing the performance of the products in different categories, and a prediction data model has been created which predicts the time by which the product needs to be replenished based on the historic data.

Model for Product Correlation The product correlation is created with the help of the original and the current product quantity for each product stored in the central database subsystem. To obtain the product correlation between the products of the same category, the following model can be utilized:

$$\text{Product Correlation} = \frac{\frac{\text{QOriginal}-\text{QCurrent}}{\text{QOriginal}}}{\sum_{i=0}^{n}\frac{\text{QOriginal}_i-\text{QCurrent}_i}{\text{QOriginal}_i}} * 100 \quad (1)$$

Model for Product Replenishing Duration Prediction It is possible to use alinear regression technique to develop a model for product replenishing duration prediction. The data source for such a model is obtained from the timeline data in the inventory database. Based on the historical product inventory, the model will be able to predict by which date the product needs to be replenished. We will be using the method of least square. The model can be represented mathematically as:

Given points $(x_1, y_1), (x_2, y_2), \ldots, (x_n, y_n)$, the slope and intercept of the least-squares line $y = mx + b$ can be obtained as:

$$m = \frac{n(\sum x_i y_i) - (\sum x_i)(\sum y_i)}{n \sum x_i^2 - (\sum x_i)^2} \tag{2}$$

$$b = \frac{(\sum x_i^2)(\sum y_i) - (\sum x_i)(\sum x_i y_i)}{n \sum x_i^2 - (\sum x_i)^2} \tag{3}$$

After obtaining the slope and the intercept, it will be possible to plug in the value of x (it will be zero for predicting when the product will be finished) in the equation of least-squares line to obtain the value y which will be the date when the product will finish and needs to be replenished. The threshold for replenishing can be set according to the speed by which more quantity of that product can be added, and accordingly, the adjustment can be made for the prediction.

Now by obtaining a correlation between the two variables, it is possible to analyze the model. A high value of correlation implies that there is a high linear dependence between the two variables. This value in turn can be used for performance evaluation, and the threshold for product replenishing can be adjusted based on the same. The correlation can be obtained as:

$$r = \frac{\sum x_i y_i - \frac{\sum x_i \sum y_i}{n}}{\sqrt{\left(\sum x_i^2 - \frac{(\sum x_i)^2}{n}\right)\left(\sum y_i^2 - \frac{(\sum y_i)^2}{n}\right)}} \tag{4}$$

After obtaining the value of r, the coefficient of determination can be calculated simply as $R^2 = r^2$. An R^2 between 0 and 1 indicates the extent to which the dependent variable is predictable. An R^2 of 1 means the dependent variable can be predicted without error from the independent variable. An R^2 of 0 means that the dependent variable cannot be predicted from the independent variable. It implies that greater the value of the coefficient of determination, the better is the model.

5 Results and Discussion

Retail experience can be enhanced for both the customers as well as the manager with the utilization of data science and the Internet of Things. The results are discussed with reference to the data science perspective and the Internet of Things Perspective. The Internet of Things implementation focuses on the collection of relevant data and reducing the checkout time using on-the-cart billing. The web dashboard provides the environment to run the data science model for inventory management, product performance and the replenishing duration prediction by utilizing the collected data in real time and provides a graphical outcome.

5.1 The Internet of Things Perspective

The Internet of Things comes with many benefits for the sector of retail. It is important to quantify such benefits in the form of results and learn the impact. When we look from the perspective of the Internet of Things in the sector of retail, two main benefits are being reflected from the current implementation level. One of the benefits is that the data requirements for data science are fulfilled through the utilization of the Internet of Things in the form of a Smart Cart. The second benefit that it brings is the reduction in checkout time for customers. It is an important benefit as the data gathered related to the customer's emotional behavior while waiting at the checkout for more than five minutes gives the result as shown in Fig. 7.

As can be visualized from the Venn diagram in Fig. 7, all the customers who had to wait at the counter for more than 5 min showed signs of frustration. Furthermore, close to 33% of the customers tends to abandon the cart when forced to wait at the checkout for more than 5 min. At least 50% of the customers who waited at the checkout for longer than 5 min decides to purposefully avoid the retailer in the future. A threshold limit of 2.5 min is considered after which the customers will become frustrated if there is no progress in the line. In the view of the above results, it is important to know where the time of an individual goes during the checkout process which can be obtained as

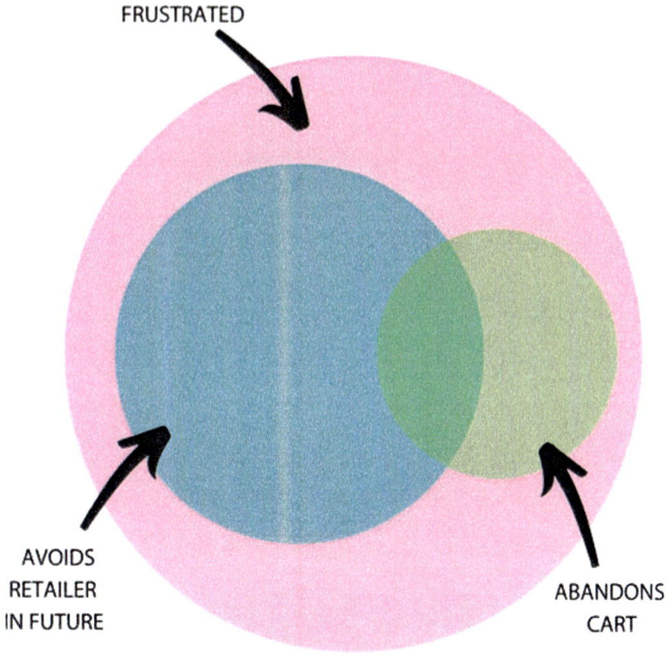

Fig. 7 Customer reaction (after waiting at checkout for over 5 min)

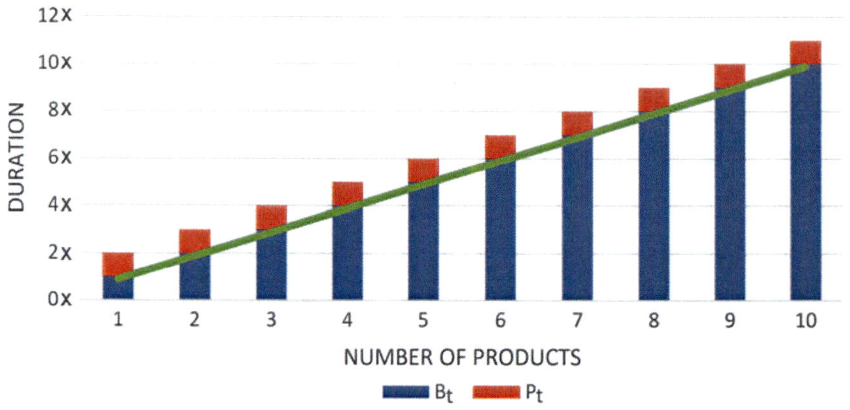

Fig. 8 Parameters influencing the checkout time

$$\text{Checkout Time}(C_t) = \text{Time to generate bill}(B_t) + \text{Time to pay bill } (P_t) \quad (5)$$

There is a linear relationship between the items in the cart and the time to scan all the products in the cart and generate the bill as can be seen from Fig. 8. It is to be noted that in Fig. 8, the term x denotes the relative time. Even though the average time to pay the bill is minimum, the time to generate the bill is time consuming and dependent on the number of items in the cart. Now by including the Internet of Things in the process, the parameter B_t is eliminated as it happens automatically on the cart itself and the only deciding factor that remains for checkout is the parameter P_t which is the time to pay the bill, which is minimal.

5.2 *The Data Science Perspective*

The model of the data science is executed, and the outcome is provided in the form a web dashboard. The web dashboard can be accessed by the retail store manager to derive benefits such as—inventory tracking, product correlation and product replenishing duration prediction.

Inventory Tracking In the web dashboard, the manager can see the live inventory data for all the products in the store, the visualization of which is shown in Fig. 9. The live inventory data is updated from the data stored in the central database which is based on the products being added/removed from the shopper's cart in real time.

When referring to the bar chart in Fig. 9, the different colors represent the different states of the product and help in obtaining a clear picture of the current state of the inventory. It is to be noted that an overlapping mechanism have been applied instead of simply stacking the different product states on each other. The extent of the color green denotes the original product quantity, the extend of the color blue denotes the

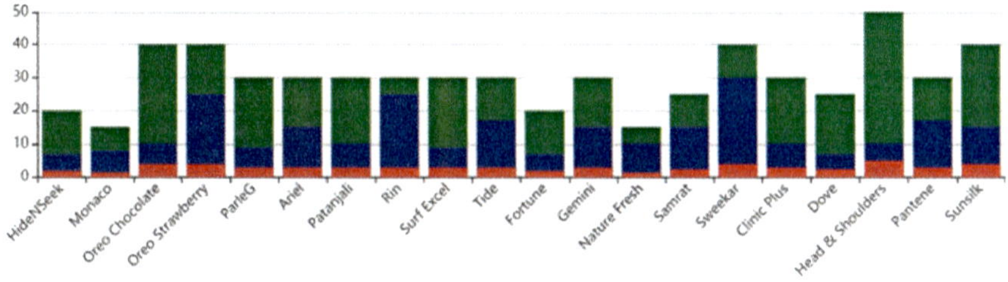

Fig. 9 Inventory tracking

current product quantity, and the extent of the color red denotes the threshold quantity for the product. The threshold quantity is calculated as 10% of the original product quantity. The visualized graph of the inventory can be represented in the form of a table and explored for two scenarios as shown in Table 1.

After going through the visualization of the inventory in Fig. 9 and consequently looking at the values in Table 1, one can easily interpret that the sale of the first product was more while the sale of the second product has been very less. One can also clearly visualize that due to the higher number of sales, the first product will reach the threshold quantity limit very soon and thus needs to be replenished accordingly. The threshold quantity limit is used to send an alert when the quantity of a product goes beyond threshold and/or the product needs to be replenished. This limit can be varied by the manager depending on the business scenario.

Product Correlation Product correlation is an important parameter as it helps in competition analysis and helps in making the decision on the placement of the products on the shelf. A relative analysis helps in gaining insights as to the demand of each product in relation to the other products in the same category. It can also be treated as a performance metrics for the products. To obtain such a correlation between the products, the original product quantity and the current product quantity of every product in the category need to be provided to the product correlation model. The model can be tested for a reference product category to understand the model behavior and the model outcome, and for that, the data from Table 2 has been utilized.

Now for calculating the performance metric, the pre-requisite equation can be obtained from Eq. 1 as

Table 1 Inventory tracking scenarios

Scenario	Original quantity	Current quantity	Threshold limit
HideNSeek	20	7	2
Rin	30	25	3

Table 2 Product quantity data

Product	Original quantity (OQ)	Current quantity (CQ)	OQ–CQ
HideNSeek	20	7	13
Monaco	15	8	7
Oreo chocolate	40	10	30
Oreo strawberry	40	25	15
ParleG	30	9	21

$$\sum_{i=0}^{n} \frac{QOriginal_i - QCurrent_i}{QOriginal_i} = \frac{13}{20} + \frac{7}{15} + \frac{30}{40} + \frac{15}{40} + \frac{21}{30} = 2.942 \quad (6)$$

Using the value obtained from Eq. 5 in Eq. 1 and plugging the product quantity data from Table 3, the product correlation as performance metric for each individual product of the category is obtained.

The performance metric in Table 3 has been graphically represented in Fig. 10.

The pie graph of the performance metrics, as seen in Fig. 10, is available in the dashboard for the manager to visualize and compare the performance of the products with respect to each other and make the business decision accordingly.

Table 3 Performance metric

Product	HideNSeek	Monaco	Oreo chocolate	Oreo strawberry	ParleG
Correlation metric (%)	22.09	15.86	25.5	12.75	23.8

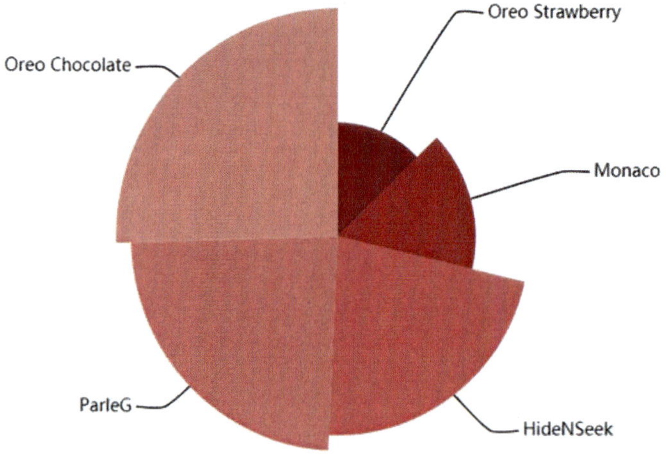

Fig. 10 Product correlation

Product Replenishing Duration Prediction The product replenishing duration is an important parameter which allows for predicting the future inventory based on the historical consumption data. The most important data that influences the predicted value is the timeline data generated for a product. A reference product has been taken into consideration to obtain the result by passing the timeline data as the parameter to the prediction model and obtain the product replenishing duration for the same. The data for the reference product is as given in Table 4.

Calculation of the parameters to obtain the equation of a line

$$m = \frac{n(\sum x_i y_i) - (\sum x_i)(\sum y_i)}{n\sum x_i^2 - (\sum x_i)^2} = \frac{10 * 686 - 55 * 158}{10 * 385 - 3025} = -2.33939394 \quad (7)$$

$$b = \frac{(\sum x_i^2)(\sum y_i) - (\sum x_i)(\sum x_i y_i)}{n\sum x_i^2 - (\sum x_i)^2} = \frac{385 * 158 - 55 * 686}{10 * 385 - 3025} = 28 \quad (8)$$

By plugging these values in the equation of a line, we get

$$y = -2.34x + 28 \quad (9)$$

Now to obtain the product replenishing duration (x), one can simply substitute y as zero (denoting that the product quantity will become zero) as

$$x = \frac{28}{2.34} = 11.9658 \approx 12 \quad (10)$$

To determine how strong the relation is between the two variables, the coefficient of determination can be determined as

$$R^2 = \left(\frac{\sum x_i y_i - \frac{\sum x_i \sum y_i}{n}}{\sqrt{\left(\sum x_i^2 - \frac{(\sum x_i)^2}{n}\right)\left(\sum y_i^2 - \frac{(\sum y_i)^2}{n}\right)}} \right)^2$$
$$= \left(\frac{686 - \frac{55*158}{10}}{\sqrt{\left(385 - \frac{3025}{10}\right)\left(2996 - \frac{24964}{10}\right)}} \right)^2 = 0.81250 \quad (11)$$

We have obtained a high value of R^2 which shows that there is a strong relationship between the two variables and so the predicted value obtained from the model is quite accurate. Further, the outcome of the model can be clearly visualized by the graph in Fig. 11 which have been constructed by feeding the Timeline data from the central database to the product replenishing duration prediction model.

Table 4 Timeline data

Date (x_i)	1	2	3	4	5	6	7	8	9	10
Quantity (y_i)	30	28	17	15	15	12	12	11	10	8
x_i^2	1	4	9	16	25	36	49	64	81	100
y_i^2	900	784	289	225	225	144	144	121	100	64
$x_i * y_i$	30	56	51	60	75	72	84	88	90	80

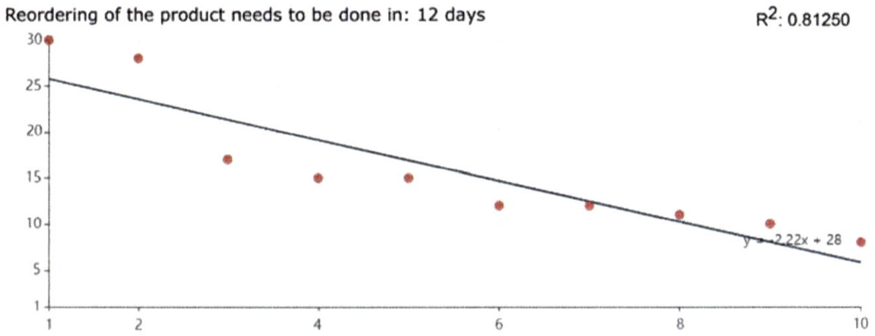

Fig. 11 Product replenishing duration prediction

The graph is shown on the manager dashboard and one can clearly visualize the historical data of the product plotted on the graph. The manager can decide as to when the product needs to be replenished by referring to the predicted replenishing duration. The coefficient of determination metric has also been given which allows the manager to quantify the interdependency of the variables and adjust the decided threshold of when to reorder the products.

6 Conclusion

The data science subsystem can work independently of the Internet of Things subsystem, but the former still requires some data on which the model can run. The data requirements can be fulfilled in many numbers of ways, but it is very clear that the utilization of the Internet of Things in the form of a Smart Cart is much superior as it allows for automatic and real-time data gathering. The time required for the bill generation adds the most delay to the checkout process which has been completely eliminated by the inclusion of the Internet of Things, and the only deciding factor that remains is the time to pay the bill which is already quite minimal. The data being obtained from the Internet of Things implementation can be harnessed by the data science model. It can provide actionable insights, prediction, tracking and management features which would have taken hours of manual labor and that too with the chances of error. By saving the valuable time and obtaining real-time insights, the proposed work is one of the key solutions toward providing an enhanced retail experience.

References

1. Ke C (2013) Maoming retail business marketing environment analysis. In: 2013 sixth international conference on business intelligence and financial engineering, pp 504–507. https://doi.org/10.1109/bife.2013.105
2. Sorace S, Pantano E, Priporas C-V, Iazzolino G (2015) The future role of digital technologies in emerging technology-based retail environments. In: 2015 8th international conference on u- and e-service, science and technology (UNESST), pp 72–76. https://doi.org/10.1109/unesst.2015.19
3. Krishna A, Deshwal P (2016) Customer service experience and satisfaction in retail stores. In: 2016 3rd international conference on computing for sustainable global development (INDIACom), pp 3719–3723
4. Liu L, Zhou B, Zou Z, et al (2018) A smart unstaffed retail shop based on artificial intelligence and IoT. In: 2018 IEEE 23rd international workshop on computer aided modeling and design of communication links and networks (CAMAD), pp 1–4. https://doi.org/10.1109/camad.2018.8514988
5. Petrescu M, Krishen AS (2018) Novel retail technologies and marketing analytics. J Mark Anal 6:69–71. https://doi.org/10.1057/s41270-018-0040-z
6. Kuriakose B, Mathai V, Baby A, Jose J (2018) Enhanced portable customer experience using community computation in offline retail. In: International conference on intelligent data communication technologies and internet of things (ICICI) 2018 lecture notes on data engineering and communications technologies, pp 15–22. https://doi.org/10.1007/978-3-030-03146-6_2
7. Semenov VP, Chernokulsky VV, Razmochaeva NV (2017) Research of artificial intelligence in the retail management problems. In: 2017 IEEE II international conference on control in technical systems (CTS), pp 333–336. https://doi.org/10.1109/ctsys.2017.8109560
8. Evans C, Bilal M (2007) Developing a WAP application for mobile retail customers. In: 2007 2nd international conference on pervasive computing and applications, pp 328–332. https://doi.org/10.1109/icpca.2007.4365463
9. Desai H, Guruvayurappan D, Merchant M, et al (2017) IoT based grocery monitoring system. In: 2017 fourteenth international conference on wireless and optical communications networks (WOCN), pp 1–4. https://doi.org/10.1109/wocn.2017.8065839
10. Dlamini NN, Johnston K (2016) The use, benefits and challenges of using the internet of things (IoT) in retail businesses: a literature review. In: 2016 international conference on advances in computing and communication engineering (ICACCE), pp 430–436. https://doi.org/10.1109/icacce.2016.8073787
11. Shankara P, Mahanta P, Arora E, Srinivasamurthy G (2015) Impact of internet of things in the retail industry. In: On the move to meaningful internet systems: OTM 2015 workshops lecture notes in computer science, pp 61–65. https://doi.org/10.1007/978-3-319-26138-6_9
12. Zheng Y, Li Y (2018) Unmanned retail's distribution strategy based on sales forecasting. In: 2018 8th international conference on logistics, informatics and service sciences (LISS), pp 1–5. https://doi.org/10.1109/liss.2018.8593273
13. Landge IA, Satopay H (2018) Secured IoT through hashing using MD5. In: 2018 fourth international conference on advances in electrical, electronics, information, communication and bio-informatics (AEEICB), pp 1–5. https://doi.org/10.1109/aeeicb.2018.8481007
14. Al-Kassab J, Thiesse F, Fleisch E (2010) O transponder, where art thou?: a case example of RFID data analytics in retail. In: 2010 internet of things (IOT), pp 1–8. https://doi.org/10.1109/iot.2010.5678446

Machine Vision

An Experimental Study on the Effect of Noise in CCITT Group 4 Compressed Document Images

A. Narayana Sukumara, Mohammed Javed, D. K. Sreekantha, P. Nagabhushan, and R. Amarnath

Abstract Consultative Committee for International Telephony and Telegraphy (CCITT) group 4 compressed binary document images support Modified Read (MR) and Modified Modified Read (MMR) compression schemes, which use three modes such as horizontal, pass, and vertical modes for encoding the data, along with run-length and modified Huffman codes. This compression scheme is widely used because of its efficiency in storage and transmission of data. Noise is a part of every data. In order to study the effect of noise in CCITT group 4 compressed binary document images, we performed the analysis of three modes, without using the decompression operation. From the analysis, it was found that noise, in general, affects the vertical mode at rapid rate compared to the other two modes. This experimental study would pave a way to perform automatic detection, analysis, and removal of noise directly in the compressed domain.

Keywords CCITT group 4 documents · Compressed binary document image · Modified Read (MR) · Modified Modified Read (MMR) · Statistical noise models · Horizontal mode · Pass mode · Vertical mode

1 Introduction

Compression is the art and science of reducing the size or amount of data without altering the information contents or its meaning. In the present era of big data,

A. Narayana Sukumara (✉) · D. K. Sreekantha
Department of Computer Science and Engineering, NMAM Institute of Technology, Nitte, Karnataka 574110, India
e-mail: nsukumara@gmail.com

M. Javed · P. Nagabhushan
Department of IT, IIIT Allahabad, Prayagraj 211015, India

R. Amarnath
Department of Studies in Computer Science, University of Mysore, Mysore, Karnataka 570009, India

N. N. Chiplunkar and T. Fukao (eds.), *Advances in Artificial Intelligence and Data Engineering*, Advances in Intelligent Systems and Computing 1133,
https://doi.org/10.1007/978-981-15-3514-7_81

compression is extensively used for reducing its size for efficient storage. Various compression algorithms are available like run-length, Huffman [6], LZW [27] etc., out of which few are lossy and few are lossless. If the compression scheme used is lossy, the information is obviously lost. In some cases, lossy schemes are preferred whereas in some lossless are preferred, that to depends on the type of information being portrayed.

Nowadays, digital devices and circuits used in transmission and storage have the capability to compress the data automatically. Over the decades researchers worked extensively on OCR and DIA, to make information extraction system more efficient than before [21, 22]. In order to store the extracted data, these systems relied heavily on one or the other compression techniques. The objectives of these systems were to recognize the text and graphics from scanned document images and to extract as much information as possible. As the systems became more efficient, data being captured and stored is more, making the data growth exponential. Considering the capabilities of the devices to compress the data immediately after acquisition, it is not a good idea to decompress for further processing, since it requires more computational resources. In order to process such data, conventional techniques use decompression, thereby increasing processing operation cost. One way to reduce the cost of processing is to directly process the data in its stored format, i.e., compressed format.

Few initial attempts were made to process such information without the use of compression, directly in its compressed format [3, 15, 16, 18]. Motivated from these initial attempts, several researchers worked extensively on processing directly in the compressed domain for various applications [8, 10–13, 20] and their attempts were successful.

Noise is generally part of every data. Noise is an *unwanted signal*, that seeps in during acquisition, transmission, compression, and storage process. Most of the time, noise is due to electronic circuit interference effects. Noise is said to corrupt the original information in a random manner. From the previous research works, we came of across several real-time noises in CCITT group 4 compressed binary document images like clutter [1], marginal [4], interfering strokes [26], etc. In this scenario, to study the behavior and effect of these real-time noises in the compressed domain without knowing their intensity or strength is highly an impossible case. Also, it is not possible to model the real-time noises. In this setting, we used an alternate approach to study the effect, with the help of known noise types such as Gaussian, salt & pepper, and speckle to which statistical models are available. These above-said noises are loosely part of compressed binary images; however, it will provide an alternate approach toward knowing in-depth, behavior of real-time noises.

There is no tangible study made in the literature with respect to noise detection and removal in CCITT group 4 compressed binary document images. In this research work, we present a study on impact of noise in compressed document images directly in the compressed domain without using the decompression. Following are the key concepts used in the analysis of noise directly in compressed domain.

2 CCITT Encoding

CCITT is one of the famous compression schemes extensively used for binary document images and supports lossless data compression. The CCITT has developed universal communication protocols that support transmission of binary images over telephone and data network lines. For encoding two levels of data CCITT supports, three different algorithms based on run-length such as Group 3 One-Dimensional (1D) i.e. Modified Huffman (MH), Group 3 Two-Dimensional (2D) i.e. Modified Read (MR) [23], and Group 4 Two-Dimensional (2D) i.e. Modified Modified Read (MMR) [24]. These algorithms are used to encode variety of file formats including TIFF and PDF. These standards are for bi-level still images i.e. binary images. CCITT Group 3 and Group 4 are the widely used standards for compression of binary images. Group 3 compression scheme is part pf ITU-T T.4 fax and Group 4 compression scheme is part of ITU-T T.6 fax.

2.1 *Modified Huffman (MH)*

In Modified Huffman, 1D left to right scan of a line, each line is encoded as a set of run lengths, each of these sets represents, runs of white or black pixels, with alternating runs of white and black runs. Each of these runs encoded as a number of bits from the Modified Huffman table, consisting of makeup and terminating codes. If run-length is 63 or less, terminating code is used. If run-length is 64 or more, upto 2560, makeup and a terminating code are used. For terminating each line, a special EOL code is used.

2.2 *Modified Read (MR) and Modified Modified Read (MMR)*

Modified Read is a 2D left to right scan of a line coding method, where coding line refers to the scan line, reference line refers to the previous one, changing element is a element whose value is different from that of the previous element, reference element is a element whose position determines the mode to be used to code, and coding mode refers to the method to code the position of each changing element along the coding line. Position of the changing element is encoded in the current coding line with respect to the position of current reference element. In this encoding, to minimize the transmission error, a Modified Huffman coded line is sent at regular intervals, commonly known as K factor which is by default 2 or otherwise 4. Following are the changing elements used, they are as follows

s_0—reference element in the coding line
s_1—next element to the right side of s_0
s_2—next element to the right side of s_1

r_1—next element to the right side of s_0 and inverse the value of s_0
r_2—next element to the right side of r_1

where s_0, s_1, s_2 are part of the coding line, where as r_1, r_2 are part of the reference line. After identifying both the lines, i.e., coding, reference, and elements i.e. s_0, s_1, s_2, r_1, r_2, various coding modes are used such as pass, vertical, and horizontal mode to code. Pass mode is used to code when the position of r_2 lies to the left of s_2. Vertical mode is used to code when the relative distance between s_1 and r_1 is less than or equal to 3. Horizontal mode is used to code whenever either of the two modes is not possible. Table 1 shows various modes and their code words (Fig. 1).

Modified Modified Read is similar to Modified Read, except the coding of first line and avoids the coding of every *K*th line. Table 2 shows how the compressed data is generated, specifying the location in terms of rows and columns for the sample binary data along with its equivalent generated code due to various modes. Total number of modes used to encode the sample data is 19, out of which first row is encoded as horizontal mode, second, third, fourth row are coded with vertical mode, at fifth row, fifth column data encoded with pass mode, rest of the two modes are vertical modes (Figs. 2, 3 and 4).

Table 1 CCITT 2D table for various modes and their code words [5]

Mode	Codeword
Pass	0001
Horizontal	001+L(s_0s_1)+L(s_1s_2)
Vertical	
s_1 below r_1	1
s_1 one to the right side of r_1	011
s_1 two to the right side of r_1	000011
s_1 three to the right side of r_1	0000011
s_1 one to the left side of r_1	010
s_1 two to the left side of r_1	000010
s_1 three to the left side of r_1	0000010
Extension	0000001xxx

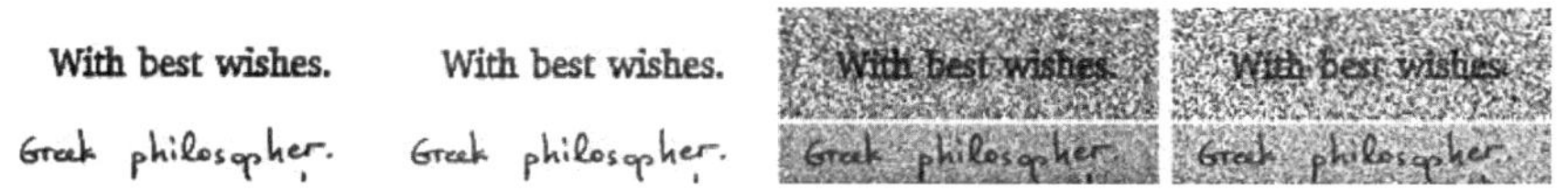

Fig. 1 Binary text documents, sample original image (First column), with 40% Gaussian noise (Second column), with 40% speckle noise (Third column) and 40% salt & pepper noise (Fourth column) used in experimental analysis

Table 2 A sample binary data (0-Black, 1-White) representing various modes with their location in (m, n) i.e. *m*th row, *n*th column and their generated code (+ sign is used for appending purpose)

Line No.	Binary data	Horizontal mode	Pass mode	Vertical mode	Final-generated code
1.	1 1 1 0 0 0 0 1 1 1 1 0 0 0 1	(1, ∗)	–	–	0011000011101110 000111
2.	1 1 1 0 0 0 0 1 1 1 1 0 0 0 1	–	–	(2, 1) + (2, 4) + (2, 8) + (2, 12) + (2, 15)	1 + 1 + 1 + 1 + 1
3.	1 1 1 0 0 0 0 1 1 1 1 0 0 0 1	–	–	(3, 1) + (3, 4) + (3, 8) + (3, 12) + (3, 15)	1 + 1 + 1 + 1 + 1
4.	1 1 0 0 0 0 0 0 1 1 1 0 0 0 1	–	–	(4, 1) + (4, 3) + (4, 9) + (4, 12) + (4, 15)	1 + 010 + 011 + 1 + 1
5.	1 1 1 1 1 1 1 1 1 1 0 0 0 0 0	–	(5, 5)	(5, 9) + (5, 11)	0001 + 010 + 011

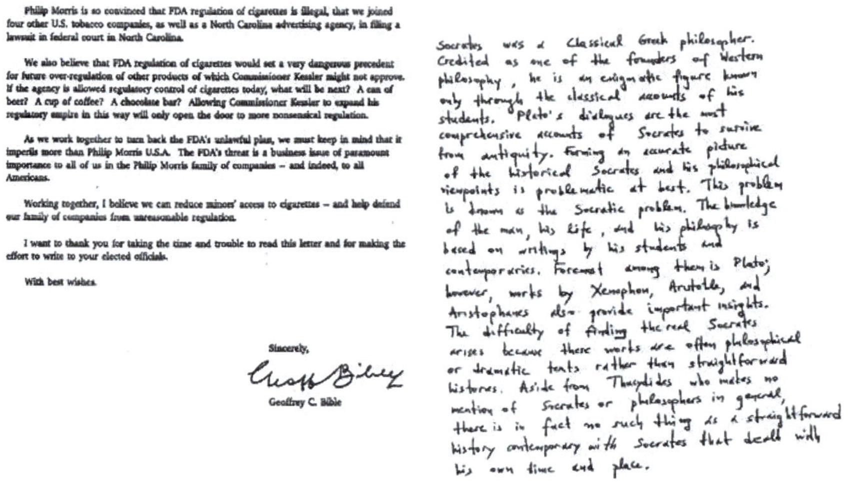

Philip Morris is so convinced that FDA regulation of cigarettes is illegal, that we joined four other U.S. tobacco companies, as well as a North Carolina advertising agency, in filing a lawsuit in federal court in North Carolina.

We also believe that FDA regulation of cigarettes would set a very dangerous precedent for future over-regulation of other products of which Commissioner Kessler might not approve. If the agency is allowed regulatory control of cigarettes today, what will be next? A can of beer? A cup of coffee? A chocolate bar? Allowing Commissioner Kessler to expand his regulatory empire in this way will only open the door to more nonsensical regulation.

As we work together to turn back the FDA's unlawful plan, we must keep in mind that it imperils more than Philip Morris U.S.A. The FDA's threat is a business issue of paramount importance to all of us in the Philip Morris family of companies – and indeed, to all Americans.

Working together, I believe we can reduce minors' access to cigarettes – and help defend our family of companies from unreasonable regulation.

I want to thank you for taking the time and trouble to read this letter and for making the effort to write to your elected officials.

With best wishes.

Sincerely,

Geoffrey C. Bible

Socrates was a Classical Greek philosopher. Credited as one of the founders of Western philosophy, he is an enigmatic figure known only through the classical accounts of his students. Plato's dialogues are the most comprehensive accounts of Socrates to survive from antiquity. Forming an accurate picture of the historical Socrates and his philosophical viewpoints is problematic at best. This problem is known as the Socratic problem. The knowledge of the man, his life, and his philosophy is based on writings by his students and contemporaries. Foremost among them is Plato; however, works by Xenophon, Aristotle, and Aristophanes also provide important insights. The difficulty of finding the real Socrates arises because these works are often philosophical or dramatic texts rather than straightforward histories. Aside from Thucydides who makes no mention of Socrates or philosophers in general, there is in fact no such thing as a straightforward history contemporary with Socrates that dealt with his own time and place.

Fig. 2 Document images from the standard dataset Tobacco800 (Printed document image), ICDAR2013 (Handwritten document image)

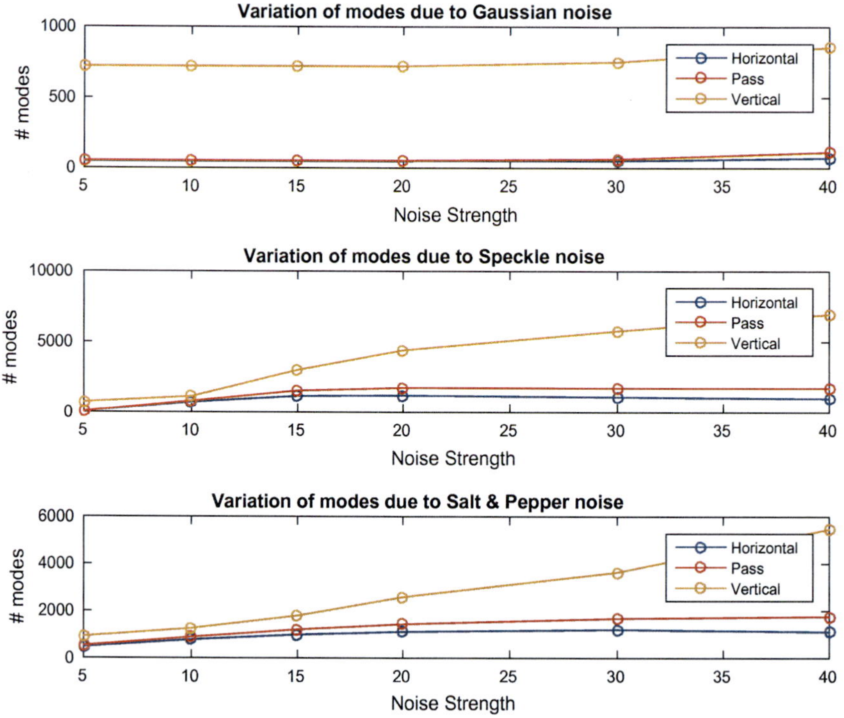

(a) Printed Document Image

Fig. 3 Variations in modes due to noise for sample binary text documents [1]

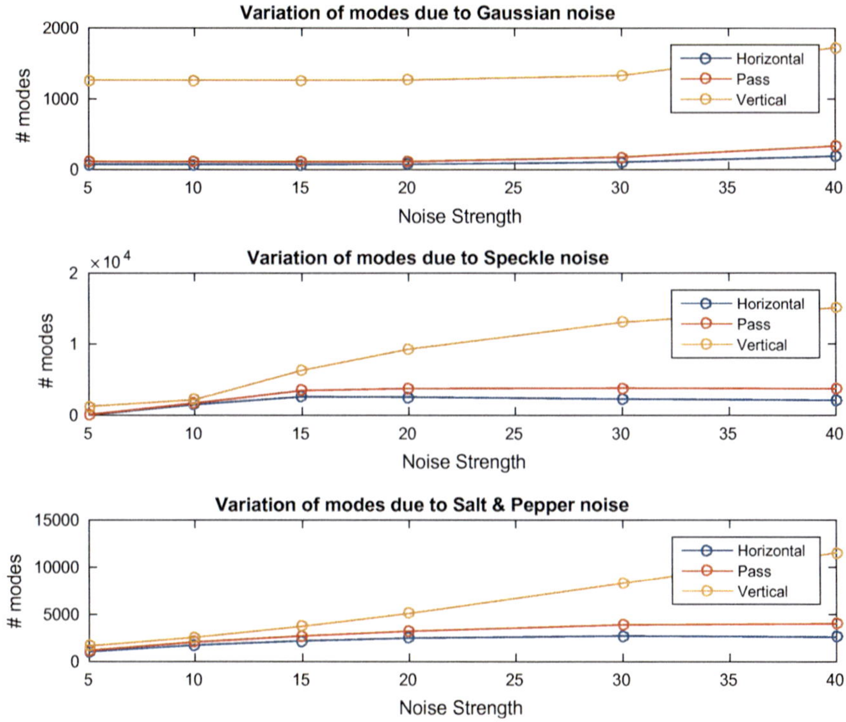

(b) Handwritten Document Image

Fig. 3 (continued)

2.3 *Statistical Noises*

In this research paper, we focus on three common types of noises with respect to CCITT group 4 compressed binary document images. They are as follows

(a) Gaussian noise
(b) Salt & Pepper noise
(c) Speckle noise.

Gaussian Noise Gaussian is a additive noise, independent of pixel, and signal intensity [2, 25] and with a probability distribution, known as Gaussian distribution. Various factors are responsible for this noise such as effects of poor illumination, electronic circuit, etc. Probability distribution function with μ mean, σ standard deviation is given by

$$p(x) = \frac{1}{\sqrt{2\pi\sigma^2}} e^{\frac{-(x-\mu^2)}{2\sigma^2}} \tag{1}$$

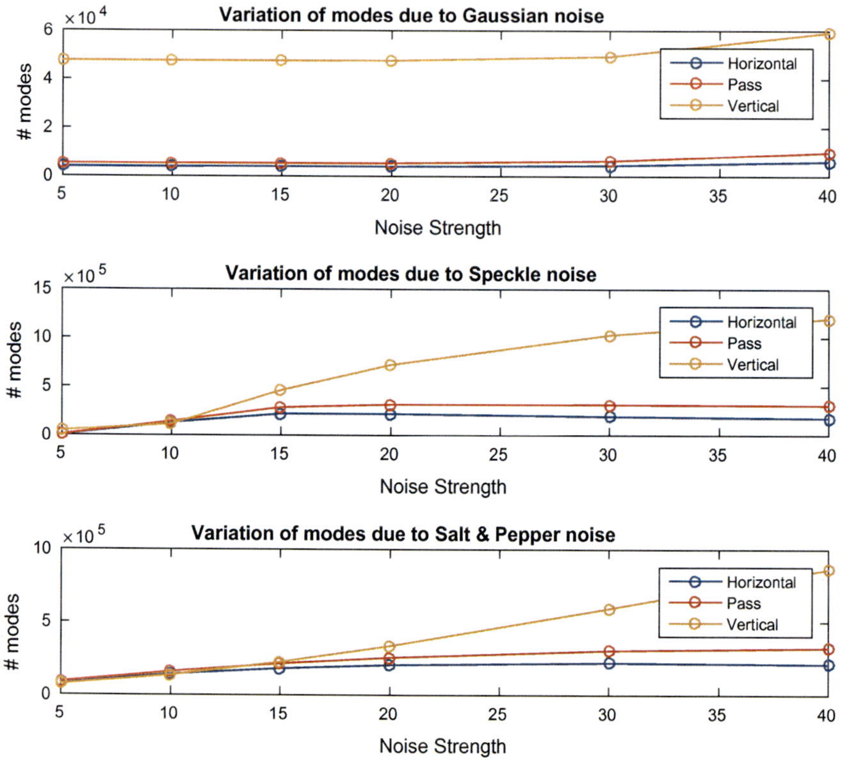

(a) Printed Document Image

Fig. 4 Variations in modes due to noise for document images from the standard dataset Tobacco800, ICDAR2013 [2]

Salt & Pepper Noise Also known as impulsive noise, appears as black and white pixels in white and black regions. This type of noise results from various data encoding, transmission, etc. [2, 25]. Probability distribution function with probabilities of intensity values p_m and p_n is given by

$$p(x) = \begin{cases} p_m & \text{for } x = m \\ p_n & \text{for } x = n \\ 0 & \text{otherwise} \end{cases} \tag{2}$$

Speckle noise Image can be degraded during acquisition and transmission processes by this multiplicative noise. This noise degrades the finer details, edges, and limits the contrast by making it difficult to detect.

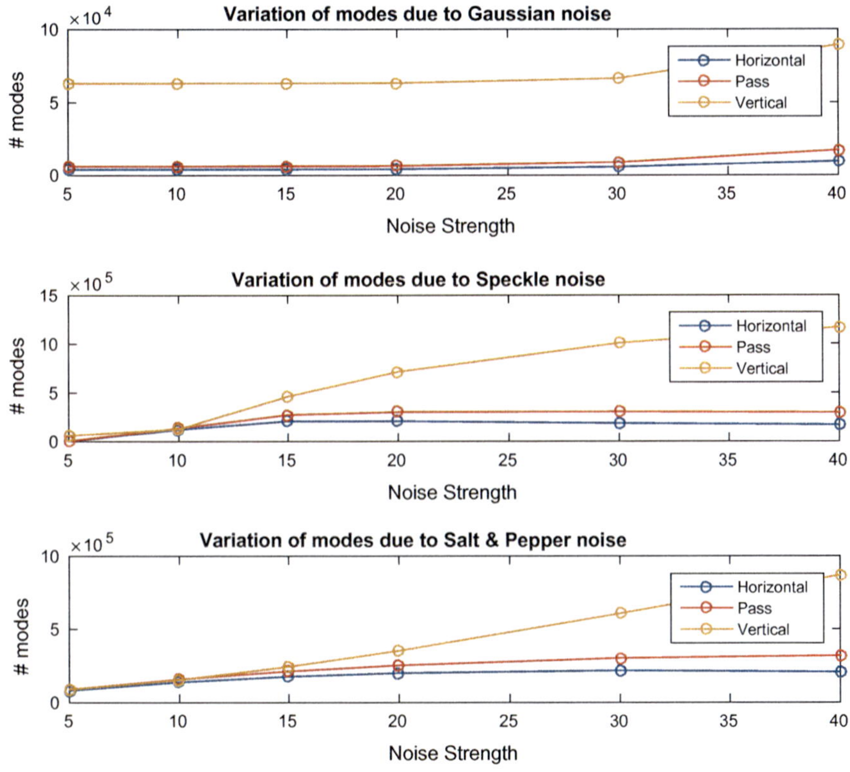

(b) Handwritten Document Image

Fig. 4 (continued)

3 Motivation and Problem Statement

The purpose of this work is to explore the possibility of noise detection, analysis, and removal directly in the compressed domain for CCITT group 4 compressed binary document images. Previous research work on CCITT group 4 compressed binary document images provides an interesting insights on word searching [16], document retrieval [15], duplicate document detection [7, 14], OCR [19], similarity measure [17], image analysis [9].

4 Experimental Analysis and Observation

For the experimental analysis, to study the behavior of noises in CCITT group 4 compressed binary document images, the standard datasets, i.e., Tobacco800 (Printed),

ICDAR2013 (Handwritten) are used. To the standard document image, we added various statistical noise, i.e. ., Gaussian, speckle, and salt & pepper noises at regular interval from 5, 10, 15, 20, 30, and 40%, for both printed and handwritten documents, and plotted the variations of modes, i.e., vertical, horizontal, and pass modes. Following are the observation made from the analysis of different modes of both MR and MMR compression schemes due to the effect of randomness of noise distribution of all noises

1. number of horizontal mode is less compared two the other two
2. number of pass mode is more compared to the horizontal mode and is less compared to the vertical mode
3. number of vertical mode is more compared to the other two.

For noises, in case of Gaussian noise, variation of all three modes observed is very low, i.e., the modes do vary at a slower rate. In case of speckle and salt & pepper noise, vertical mode increase rapidly, i.e., at higher rate, whereas for horizontal and pass mode, variation observed is very small. We can conclude from the analysis that "Number of Horizontal Mode + Number of Pass Mode < Number of Vertical Mode."

5 Conclusion

In this research work, we studied the effect of various statistical noises on the CCITT Group 4 compressed binary document images using horizontal, pass, and vertical modes. From the analysis, it is evident that, noise will have a strong impact on the vertical mode of the MR and MMR compression schemes of CCITT Group 4 compressed document images. Also to observe that sum of horizontal and pass modes is always less then the vertical mode. With the help of these new insights, a way is opened for trying de-noising operations directly in compressed document images.

References

1. Agrawal M, Doermann D (2009) Clutter noise removal in binary document images. In: 10th international conference on document analysis and recognition, 2009. ICDAR'09. IEEE, pp 556–560. https://doi.org/10.1109/ICDAR.2009.277
2. Bovik AC (2010) Handbook of image and video processing. Academic Press, Cambridge
3. Chang SF (1995) Compressed-domain techniques for image/video indexing and manipulation. In: Proceedings of international conference on image processing, 1995, vol 1. IEEE, pp 314–317. https://doi.org/10.1109/ICIP.1995.529709
4. Fan KC, Wang YK, Lay TR (2002) Marginal noise removal of document images. Pattern Recogn 35(11):2593–2611. https://doi.org/10.1109/ICDAR.2001.953806
5. Gonzalez RC, Woods RE et al (2002) Digital image processing
6. Huffman DA (1952) A method for the construction of minimum-redundancy codes. Proc IRE 40(9):1098–1101. https://doi.org/10.1109/JRPROC.1952.273898

7. Hull JJ (1997) Document matching on ccitt group 4 compressed images. In: Document recognition IV, vol 3027. International Society for Optics and Photonics, pp 82–88. https://doi.org/10.1117/12.270061
8. Javed DM, Nagabhushan P, Chaudhuri B (2015) Automatic page segmentation without decompressing the run-length compressed text documents. J Inf Process Syst 11
9. Javed M, Krishnanand S, Nagabhushan P, Chaudhuri B (2016) Visualizing ccitt group 3 and group 4 tiff documents and transforming to run-length compressed format enabling direct processing in compressed domain. Procedia Comput Sci 85:213–221. https://doi.org/10.1016/j.procs.2016.05.214
10. Javed M, Nagabhushan P, Chaudhuri BB (2014) Direct processing of document images in compressed domain. 2014 Fourth IDRBT Doctoral Colloquium. https://doi.org/10.13140/2.1.2523.9042
11. Javed M, Nagabhushan P, Chaudhuri BB (2014) Extraction of projection profile, run-histogram and entropy features straight from run-length compressed text-documents. In: Proceedings of 2nd IAPR Asian conference on pattern recognition. ACPR 2013. https://doi.org/10.1109/ACPR.2013.147
12. Javed M, Nagabhushan P, Chaudhuri BB (2016) Automatic extraction of text and non-text information directly from compressed document images. In: International conference on hybrid intelligent systems. Springer, pp 38–46. https://doi.org/10.1007/978-3-319-52941-7_5
13. Jingpeng L, Dalin J (2011) Survey on the technology of image processing based on dct compressed domain. In: 2011 International conference on multimedia technology (ICMT). IEEE, pp 786–789. https://doi.org/10.1109/ICMT.2011.6001991
14. Lee DS, Hull JJ (1999) Duplicate detection for symbolically compressed documents. In: Proceedings of the fifth international conference on document analysis and recognition, 1999. ICDAR'99. IEEE, pp. 305–308. https://doi.org/10.1109/ICDAR.1999.791785
15. Lu Y, Tan CL (2003) Document retrieval from compressed images. Pattern Recogn 36(4):987–996. https://doi.org/10.1016/S0031-3203(02)00127-9
16. Lu Y, Tan CL (2003) Word searching in ccitt group 4 compressed document images. In: Proceedings of seventh international conference on document analysis and recognition, 2003. IEEE, pp 467–471. https://doi.org/10.1109/ICDAR.2003.1227709
17. Lu Y, Tan CL, Fan L, Huang W (2001) Similarity measure for ccitt group 4 compressed document images. In: Proceedings of 2001 international conference on image processing, 2001, vol 1. IEEE, pp. 1118–1121. https://doi.org/10.1109/ICIP.2001.959247
18. Mandal MK, Idris F, Panchanathan S (1999) A critical evaluation of image and video indexing techniques in the compressed domain. Image Vis Comput 17(7):513–529. https://doi.org/10.1016/S0262-8856(98)00143-7
19. Marti U, Wymann D, Bunke H (2000) OCR on compressed images using pass modes hand hidden Markov models. In: Proceedings of IAPR workshop on document analysis systems, pp. 77–86. 10.1.1.33.6174
20. Mukhopadhyay J (2011) Image and video processing in the compressed domain. Chapman and Hall, CRC
21. Nagy G (2000) Twenty years of document image analysis in pami. IEEE Trans Pattern Anal Mach Intell 1:38–62. https://doi.org/10.1109/34.824820
22. O'Gorman L, Kasturi R (1995) Document image analysis, vol 39. IEEE Computer Society Press Los Alamitos. https://doi.org/10.1007/BF02703309
23. Recommendations-T.4: Standardization of group 3 facsimile apparatus for document transmission. Tech. rep., The International Telegraph and Telephone Consultative Committee (CCITT) (1985)
24. Recommendations-T.6: Facsimile coding schemes and coding control functions for group 4 facsimile apparatus. Tech. rep., The International Telegraph and Telephone Consultative Committee (CCITT) (1985)

25. Shapiro L (1992) Computer vision and image processing. Academic Press, Cambridge
26. Tan CL, Cao R, Shen P, Wang Q, Chee J, Chang J (2000) Removal of interfering strokes in double-sided document images. In: Fifth IEEE workshop on applications of computer vision, 2000. IEEE, pp 16–21. https://doi.org/10.1109/WACV.2000.895397
27. Welch TA (1984) Technique for high-performance data compression. Computer 52. https://doi.org/10.1109/MC.1984.1659158

A Neck-Floor Distance Analysis-Based Fall Detection System Using Deep Camera

Xiangbo Kong, Zelin Meng, Lin Meng, and Hiroyuki Tomiyama

Abstract The research content of this paper is to present a skeleton analysis-based system for senile people, which can detect the fall accident. Depth image is applied to extract the 3D joint coordinate of the senile people. In the proposed algorithm, three points on the ground are selected to get the plane equation of the ground plane. The 3D coordinate of the neck is tracked continually, and the space length from the neck to the ground plane is analyzed. If the joint of neck is close to the floor and this situation lasts for more than one min, fall is detected and alert is sent to the relatives or healthcare centers. In this study, images of living room are taken to built a data set, and the proposed method is compared with shape analysis method and deep learning methods. The proposed method gives a better result.

Keywords Fall detection · Elderly person · Skeleton analysis · Deep learning

1 Introduction

With the acceleration of aging all over the world, health care for elderly persons has become an important problem, and fall accident is an important reason of threatening their health. As reported by the WHO, these are very serious accidents which strike the senile people, and these even caused many death of senile people. Worldwide, about 424 thousand senile people lost their lives in fall accident [1]. When an elderly person lives by herself/himself and she/he has a fall accident, it is difficult to grasp this situation for the hospital or family members. Therefore, automatic system which

X. Kong (✉) · Z. Meng
Graduate School of Science and Engineering, Ritsumeikan University,
1-1-1, Nojihigashi, Kusatsu 525-8577, Japan
e-mail: xiangbo.kong@tomiyama-lab.org

L. Meng · H. Tomiyama
College of Science and Engineering, Ritsumeikan University,
1-1-1, Nojihigashi, Kusatsu 525-8577, Japan

N. N. Chiplunkar and T. Fukao (eds.), *Advances in Artificial Intelligence and Data Engineering*, Advances in Intelligent Systems and Computing 1133,
https://doi.org/10.1007/978-981-15-3514-7_82

can process the images in real time and detect the danger is very important for senile people who have a reclusive life.

Many fall detection algorithms have been proposed up to now, and these methods can be contextualized according to two tracks: wearable devices-based fall detection [2–4] and vision-based fall detection [5–8].

The approaches of [2, 3] provide wearable camera-based fall detection systems. In study [2], a novel algorithm is proposed to classify significant activities like sitting or lying, and in study [3], gradient local binary pattern feature-based classification is proposed to detect the fall, which is over-performing histogram of oriented gradients features. As the camera is worn by senile people, it monitors continually no matter indoor or outdoor. However, wearable camera is inconvenient for senile people to use, and it may cause privacy problems outdoor. The approach of [4] provides a power-efficient fall detection system using wearable sensors. In this work, the backward wrapper features are extracted with high accuracy while considering the power consumption and get a good result. The main drawback of wearable sensor-based fall detection system is that senile people are only protected when they wear the sensor. If the elderly person forgets to use the wearable sensors or charge them, these systems do not work.

Image processing-based danger detection systems solve these problems well. This is because the senile people do not need to use or charge any sensors. Study [5] proposes an external ellipse-based fall detection which analyzes the shape of the senile people's external ellipse. The approach of [6] analyzes the tangent line of the senile people's outline to detect the fall. However, outline-based fall detection methods or external ellipse-based fall detection methods cannot get a high accuracy as the features are too simple to fit various situations. The approach of [7] employs deep learning and the approach of [8] employs machine learning (oriented gradient feature is extracted, and SVM is employed to classify the features)-based shape analysis method to analyze the shape of the person, in order to detect the fall. However, the method in [7] fails in dark rooms due to the limit of RGB cameras, although weak light is a main reason of fall accidents. Thanks to the infrared-based deep camera, the system proposed in [8] performs well in weak light, but the accuracy is still not high enough.

This paper proposes a depth image-based fall detection without shape analysis. The deep learning-based 3D joint extraction is employed to get the 3D coordinate of the neck and the ankle. The fall is detected by calculating the space length from the neck to the ground plane. The rest parts are presented as follows. Section 2 gives a general description of the system proposed in this paper, and the algorithm of this system is introduced in Sect. 3. Section 4 gives an analysis of experimental results, and Sect. 5 is the conclusion.

2 System

Figure 1 is a general description of the proposed fall detection system. The deep camera captures the depth image of the room continually and extracts the 3D joint coordinate of the detected persons in real time. In the preprocessing, the processing unit records the 3D ankle coordinate three times, so three points of the ground plane are gotten (This is because the ankle is very close to the floor). The plane equation of the ground plane is calculated with these three points. When system starts, the deep camera captures the depth image and the processing unit detects the persons. When a person is detected, the 3D coordinate of the neck is recorded and the space length of the neck to the ground plane is calculated. If the space length is less than a threshold, a timer starts up. If this situation lasts over one min, an alert is sent to relatives or healthcare centers.

3 Algorithms

3.1 Human Detection

Human detection is an important preprocessing of this system. In this paper, the infrared sensor-based deep camera captures the depth image, and the features given by the following formulas of the images are used to detect the human [9]. The features

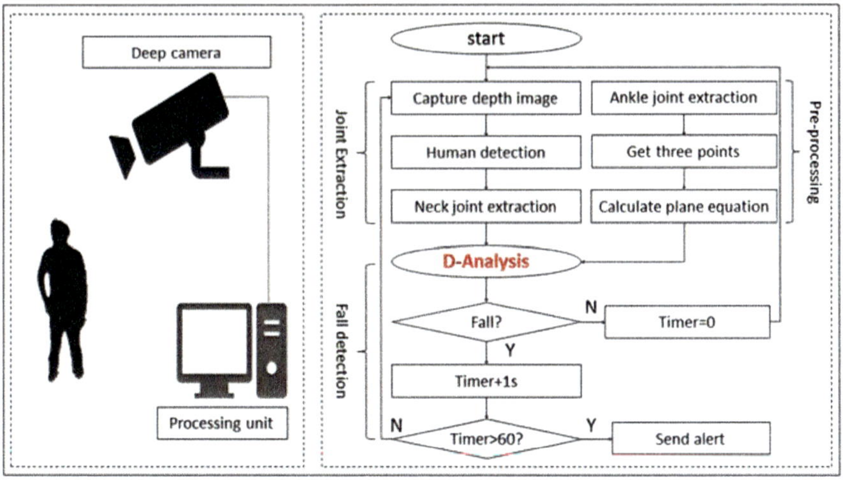

Fig. 1 General description of the system

are calculated by Formula (1). In this formula, d_{Ir} is the space length from pixel ω to the lens of the deep camera in the depth image *Ir*.

$$G_{\delta}(\mathrm{Ir}, \omega) = d_{\mathrm{Ir}}\left(\omega + \frac{m}{d_{\mathrm{Ir}}(\omega)}\right) - d_{\mathrm{Ir}}\left(\omega + \frac{n}{d_{\mathrm{Ir}}(\omega)}\right) \tag{1}$$

3.2 Plane Equation of the Floor

In order to calculate the space length from the neck to the ground plane, the plane equation of the ground plane is necessary. The deep camera captures the images of the person, and the 3D coordinate of the ankle joint is recorded three times, as shown in Fig. 2. Points $A(\chi_1, \gamma_1, \zeta_1)$, $B(\chi_2, \gamma_2, \zeta_2)$, $C(\chi_3, \gamma_3, \zeta_3)$ give the 3D position of the ankle point in different places. So, the a vector is given by Formula (2), and this vector is perpendicular to plane (3). The plane equation of the ground plane is given by Formula (3), and the space length from the neck point $N(\chi_0, \gamma_0, \zeta_0)$ to the ground

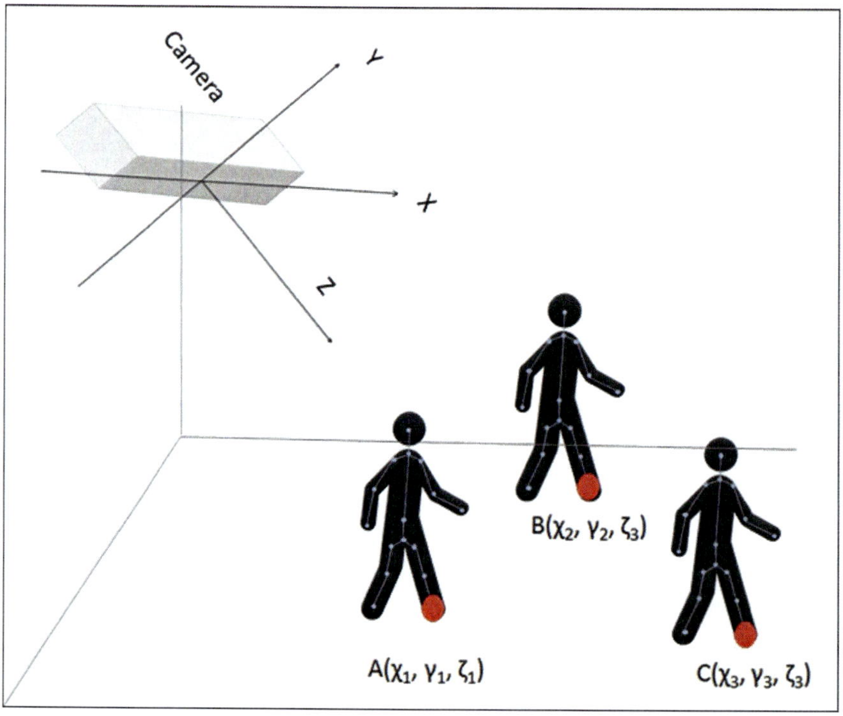

Fig. 2 Using 3D ankle joint coordinate to calculate the plane equation of the floor

plane D is given by Formulas (4)–(8). Note that these formulas are the same as the formulas for calculating plane equations.

$$\mathbf{NV} = \begin{bmatrix} \mathbf{i} & \mathbf{j} & \mathbf{k} \\ -\chi_1 + \chi_2 & -\gamma_1 + \gamma_2 & -\zeta_1 + \zeta_2 \\ -\chi_1 + \chi_3 & -\gamma_1 + \gamma_3 & -\zeta_1 + \zeta_3 \end{bmatrix} \tag{2}$$

$$\begin{bmatrix} \chi - \chi_1 & \gamma - \gamma_1 & \zeta - \zeta_1 \\ \chi_2 - \chi_1 & \gamma_2 - \gamma_1 & \zeta_2 - \zeta_1 \\ \chi_3 - \chi_1 & \gamma_3 - \gamma_1 & \zeta_3 - \zeta_1 \end{bmatrix} = 0 \tag{3}$$

$$a = \begin{bmatrix} \gamma_2 - \gamma_1 & \zeta_2 - \zeta_1 \\ \gamma_3 - \gamma_1 & \zeta_3 - \zeta_1 \end{bmatrix} \tag{4}$$

$$b = \begin{bmatrix} \chi_3 - \chi_1 & \zeta_3 - \zeta_1 \\ \chi_2 - \chi_1 & \zeta_2 - \zeta_1 \end{bmatrix} \tag{5}$$

$$c = \begin{bmatrix} \chi_2 - \chi_1 & \gamma_2 - \gamma_1 \\ \chi_3 - \chi_1 & \gamma_3 - \gamma_1 \end{bmatrix} \tag{6}$$

$$d = \chi_1 \begin{bmatrix} \gamma_3 - \gamma_1 & \zeta_3 - \zeta_1 \\ \gamma_2 - \gamma_1 & \zeta_2 - \zeta_1 \end{bmatrix} + \gamma_1 \begin{bmatrix} \chi_2 - \chi_1 & \zeta_2 - \zeta_1 \\ \chi_3 - \chi_1 & \zeta_3 - \zeta_1 \end{bmatrix} + Pz_1 \begin{bmatrix} \chi_3 - \chi_1 & \gamma_3 - \gamma_1 \\ \chi_2 - \chi_1 & \gamma_2 - \gamma_1 \end{bmatrix} \tag{7}$$

$$D = \frac{|a\chi_0 + b\gamma_0 + c\zeta_0 + d|}{\sqrt{a^2 + b^2 + c^2}} \tag{8}$$

4 Experimental Results

4.1 Data Set

An enhanced data set built in [8] is used to evaluate the experimental results, which is demonstrated in Fig. 3. In this data set, 6000 training images and 12,000 test images are included. In this data set, images of three persons are taken by cameras with two positions, and the posture contains person walks in the room, person walks with a stick, persons fall longitudinally or careening toward.

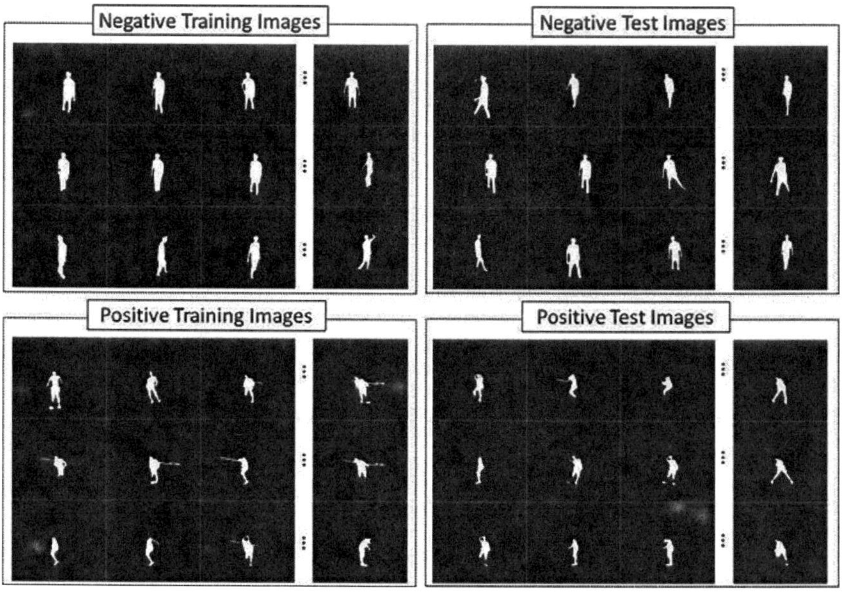

Fig. 3 Data set for evaluation

4.2 Evaluation Metrics

The method suggested by [10] is used to evaluate the experimental results. *True*(T) means detecting the state of the senile people successfully, and *false*(F) means a mis-detection. *Positive*(P) is defined as a fall state, and *negative*(N) is defined as a safe state. Formulas (9)–(12) evaluate the experimental results.

$$Sen = TrPos/(TruPos + FalNeg) \tag{9}$$

$$Spe = TruNeg/(TruNeg + FalPos) \tag{10}$$

$$Acc = (TruPos + TruNeg)/(TruPos + TruNeg + FalPos + FalNeg) \tag{11}$$

$$Err = (FalPos + FalNeg)/(TruPos + TruNeg + FalPos + FalNeg) \tag{12}$$

4.3 Experimental Results and Comparison

The fall accident is analyzed by the proposed neck-to-floor distance method, as shown in Fig. 4. The transverse axis of Fig. 4a is the frame number (approximate to

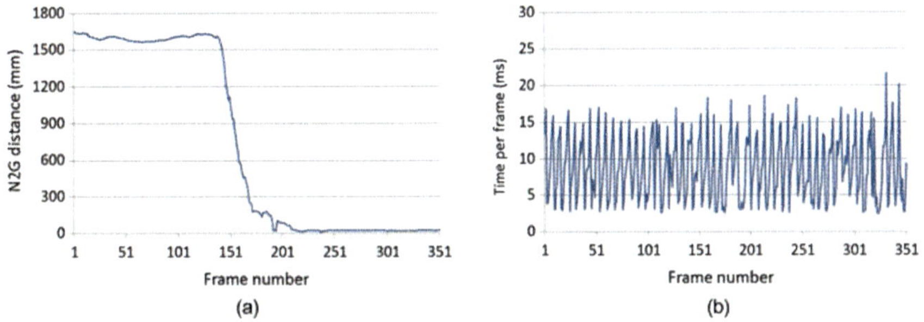

Fig. 4 Experimental result of a fall accident analyzed by the proposed method

Table 1 Comparisons

Method	Ses(%)	Spe(%)	Acc(%)	Err(%)
[5]	58.7	78.7	68.7	31.3
[8]	99.8	99.8	99.8	0.2
AlexNet	100	99.9	99.9	0.1
Proposed	**100**	**100**	**100**	**0**

the time). The vertical axis in the graph means the space length between the neck and the floor. In other words, Fig. 4a shows the change of the neck-to-floor space length with time. In this figure, when the fall accident occurs, the neck joint approaches to the ground. The space length decreases from about 1600 to about 20 mm. According to our experiments, 600 mm is set as a threshold, as other activities do not let the neck approach to the floor within 600 mm. The execution time for each frame is about 3–30ms, as illustrated in Fig. 4b.

The proposed fall detection method in this paper is compared with external ellipse analysis proposed in [5], shape analysis proposed in [8], deep learning (AlexNet), and the proposed method gives the best result, as shown in Table 1.

External ellipse analysis is a very simple algorithm which does not need complex calculation, and it performs well when an elderly person walks in the room or falls vertical to the camera direction. However, when an elderly person bends over and walks with a stick, the external ellipse analysis does not make sense. Moreover, when an elderly person falls to other directions, it is also hard for external ellipse analysis method to detect the fall, as the ellipse is very similar with standing cases. Machine learning-based [8] or deep learning-based shape analysis gives a high accuracy, but the computation is too complicated. The proposed methods give the best accuracy with a high execution speed, less than 50ms per frame with core i5 5200U.

5 Conclusion

In this paper, a neck-floor space length analysis-based low-cost, real-time, wearable device free-fall detection system is proposed. Deep camera is employed to get the 3D joint coordinate. Three coordinates of the floor are used to calculate the plane equation of the floor, and the space length between the neck and the ground is calculated in real time. If the neck is very closed to the floor and this situation lasts for over one minute, an alert is sent to relatives or healthcare centers. In this paper, the proposed method is compared with external ellipse analysis, machine learning (HOG-SVM)-based shape analysis and deep learning (AlexNet)-based shape analysis and gives a 100% accuracy with a short execution time. However, there are still problems left in this study. Although a large number of images are contained in the data set [8], many cases are not taken into consideration, for example, sitting on the chair, sleeping on the sofa and falling behind the furniture. In the experiment, we found that these situations caused many mis-detection. The construction of a complete data set concerning images with all cases consideration and the analysis of these cases are one of the future research topics.

References

1. World Health Organization. https://www.who.int/ageing/en/
2. Ozcan K, Mahabalagiri AK, Casares M, Velipasalar S (2013) Automatic fall detection and activity classification by a wearable embedded smart camera. IEEE J Emerg Sel Top Circuits Syst 3(2):125–136
3. Ozcan K, Velipasalar S (2015) Wearable camera-and accelerometer-based fall detection on portable devices. IEEE Embed Syst Lett 8(1):6–9
4. Wang C, Redmond SJ, Lu W, Stevens MC, Lord SR, Lovell NH (2017) Selecting power-efficient signal features for a low-power fall detector. IEEE Trans Biomed Eng 64(11):2729–2736
5. Gunale KG, Mukherji P (2015) Fall detection using k-nearest neighbor classification for patient monitoring. In: International conference on information processing 2015. IEEE, pp 520–524
6. Kong XB, Meng L, Tomiyama H (2017) Fall detection for elderly persons using a depth camera. In: International conference on advanced mechatronic systems 2017. IEEE, pp 269–273
7. Min W, Cui H, Rao H, Li Z, Yao L (2018) Detection of human falls on furniture using scene analysis based on deep learning and activity characteristics. IEEE Access 6:9324–9335
8. Kong XB, Meng ZL, Nojiri N, Iwahori Y, Meng L, Tomiyama H (2019) A HOG-SVM based fall detection IoT system for elderly persons using deep sensor. Procedia Comput Sci 147:276–282
9. Dalal N, Triggs B (2005) Histograms of oriented gradients for human detection. In: International conference on computer vision and pattern recognition 2005. IEEE, pp 886–893
10. Noury N, Fleury A, Rumeau P, Bourke AK, Laighin GO, Rialle V, Lundy JE (2007) Fall detection-principles and methods. In: Annual international conference of the IEEE engineering in Medicine and Biology Society 2007. IEEE, pp 1663–1666

An Introduction to Sparse Sampling on Audio Signal by Exploring Different Basis Matrices

A. Electa Alice Jayarani, Mahabaleswara Ram Bhatt, and D. D. Geetha

Abstract In this paper, we introduce the compressive sensing for audio signal acquisition. Traditional sampling algorithm collects a large amount of data and reduces the amount of data in the compression steps. In this paper, we propose to implement compressive sensing on audio signals rather than the traditional sampling for signal acquisition which gives less sampled data. The algorithm is tested on various audio signals and proved that the signals can be successfully reconstructed using l1 minimization method. The results show the proposed algorithm can be used for the high-speed application of audio signals where the analog-to-digital converter is burdened with more samples. We have also generated a basis matrix of compressive sensing using different transforms, and a comparative analysis is made. The comparative analysis is made by analyzing the signal-to-noise ratio, and it is observed that the DCT basis matrix gives a better SNR. The SNR of DCT-based compressive sensing is greater than 35 dB.

Keywords Audio signal · Compressive sensing · Basis matrix

1 Introduction

The conscious by which the sound is perceived is called a hearing. Person hearing is interpreted basically by the human auditory system (HAS) which embraces outer ear, the middle ear, and the inner ear. When the sound reaches the HAS, it will be in the form of time-varying pressure waves. Hence, the sound pressure is described as the force per unit time, and the unit is pascal (Pa). Swift changes in pressure in HAS are called sound. Also, the sound intensity is expressed as power per unit area, and its unit is W/m^2. The human sense of hearing is between 20 Hz and 20 kHz and ranging from 0 up to 120 dB in sound pressure level (SPL) [1].

A. Electa Alice Jayarani (✉) · D. D. Geetha
Reva University, Bangalore, Karnataka, India
e-mail: electalice@gmail.com

M. R. Bhatt
BMSCE, Bangalore, Karnataka, India

N. N. Chiplunkar and T. Fukao (eds.), *Advances in Artificial Intelligence and Data Engineering*, Advances in Intelligent Systems and Computing 1133,
https://doi.org/10.1007/978-981-15-3514-7_83

In traditional analog-to-digital conversion (ADC), Nyquist sampling is used for conversion of analog signals to discrete signals. Consider the analog signal as $x(t)$, the sampling is given by

$$x_s(t) = x(ns) \cdots - \alpha < n < \alpha \quad (1)$$

where s is the sampling period. The analog signal can be recovered if the sampling frequency is greater than or equal to twice the frequency of the message signal, i.e., $f_s \geq 2f_m$ [2]. Currently, there are various researches focused on compressive sensing or sparse sampling for signal acquisition. In traditional sampling, if the samples required are N, then in compressive sensing, the measurement $M \leq N$ samples per second are enough [3].

In early 2000, the author in [4] proved that the signal can be recovered from the fewer samples, and it gave a big opening of a new technique called as compressive sensing [5]. From then, a lot of research is going on compressive sensing to be used in various areas such as a camera in a phone and medical application. In [6], the author proposes compressive sensing for image, whereas in [7], the compressive sensing is used in the application of human–machine interface. The compressive sensing is used for the accurate signal acquisition of various types of data [8, 9]. In the following sections, we define compressive sensing for audio signal acquisition and reconstruction of primary signal using l1 minimization.

2 Compressive Sensing

If the majority of the components of the signal are zero, then the signal is said to be sparse in nature. The message of the audio signal in the time domain is spread all over, but when phrased in a proper basis, the coefficient that contains the message is condensed. For example, a signal $s(t), s \in R^n$ can be expressed in the basis coefficient as $S = \varphi s$ as shown in Fig. 1. We can see that when the audio signal is expressed in a transform domain, the signal is sparse in nature.

The compressive sensing is a non-uniform sampling which yields fewer samples, and the signal can be recovered using mathematical convex programming. It is expressed as

$$Y = AS \quad (2)$$

where A $A \in R^{mxn}$ is the measurement vector matrix and Y is the compressed vector $Y \in R^m$.

Considering $S \in R^n$ and the transform coefficient is k sparse, then the measurement m of the basis matrix is selected by generating a random vector uniformly. It is shown in [10] k sparse vector that x can be reconstructed from $y = Ax$ using $l1$ minimization provided

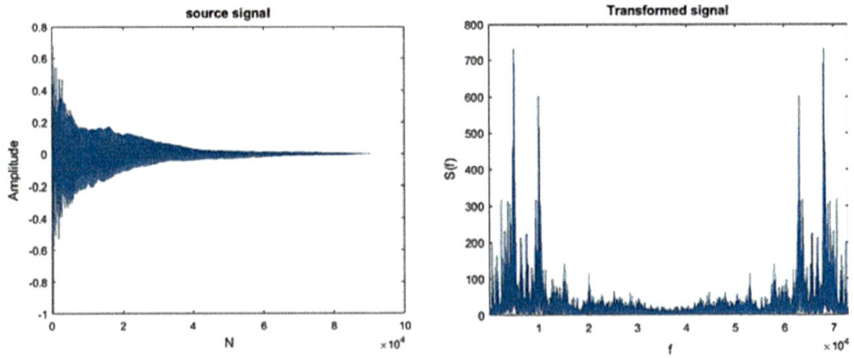

Fig. 1 Source signal in time domain and frequency domain

Table 1 Different k and m

S. no.	k	m
1.	3	≥14
2.	5	≥20
3.	7	≥25
4.	10	≥32
5.	13	≥39
6.	15	≥43
7.	20	≥51

$$m \geq Ck \ln\left(\frac{n}{k}\right) \tag{3}$$

where $C > 0$ is a universal constant independent of k, n, m. In Eq. (3), m is directly proportional to k, and hence, if the sparsity is considered small, then the measurement m can also be chosen small in comparison with n, so that the solution of an underdetermined system of linear equation is reasonable. Different sparse k signals and the corresponding m measurements are listed in Table 1.

3 Basis Matrix

3.1 Discrete Fourier Transform

The most important tool used in digital signal processing to analyze the frequency content of any signal is Fourier transform. The discrete Fourier transform is given as

$$X(K) = \sum_{n=0}^{N-1} x(n) e^{\frac{-j2\pi Kn}{N}} \tag{4}$$

And inverse DFT is given by

$$x(n) = \frac{1}{N}\sum_{K=0}^{N-1} X(K)e^{\frac{j2\pi Kn}{N}} \tag{5}$$

DFT represents a signal by a set of orthogonal sinusoidal functions, and DFT is a unitary transform.

3.2 Discrete Cosine Transform

Discrete cosine transform DCT represents a signal by a cosine function oscillating at different frequencies. DCT is given by

$$X(K) = \alpha(K)\sum_{n=0}^{N-1} x(n)\cos\left[K\left[\frac{\pi}{M}\left(n+\frac{1}{2}\right)K\right]\right] \tag{6}$$

where $K = 0, 1, \ldots, N-1$

$$\alpha(K) = \begin{cases} \sqrt{\frac{1}{N}} & \text{if } K = 0 \\ \sqrt{\frac{2}{N}} & \text{if } K \neq 0 \end{cases}$$

DCT is similar to DFT but representing only real numbers, and it is widely used in JPEG compression.

3.3 Haar Wavelet Transform

The simplest and the fastest wavelet transform is Haar transform. The Haar function is denoted as $h_k(x)$ where k is given as

$$k = 2^p + q - 1, \quad k = 0, 1, \ldots, N-1$$

where

$N = 2^n$
$0 \leq p \leq n-1, 0 \leq q \leq 2^p$
k is the order of the function.

The Haar function is defined as

$$h_0(x) \equiv h_{00}(x) = \frac{1}{\sqrt{N}}, \quad x \in [0, 1]$$

and

$$h_k(x) \equiv h_{pq}(x) = \frac{1}{\sqrt{N}} \begin{pmatrix} 2^{\frac{p}{2}} & \frac{q-1}{2^p} \le x < \frac{q-0.5}{2^p} \\ -2^{\frac{p}{2}} & \frac{q-0.5}{2^p} \le x < \frac{q}{2^p} \\ 0 & \text{otherwise} \end{pmatrix} \tag{7}$$

The amplitude and the width of the function which involves the value other than zero are given by p, and position of the nonzero value is given by q. The Haar transform matrix for the $N = 2$ is given below

$$H_2 = \frac{1}{\sqrt{2}}\begin{bmatrix} 1 & 1 \\ 1 & -1 \end{bmatrix}$$

It is observed $H = H^*$ and $H^{-1} = H^{\mathrm{T}}$; therefore,

$$H^{\mathrm{T}} H = I$$

where I is the identity matrix.

4 Proposed Algorithm

As a traditional sampling takes a large amount of samples, and in the time of compression, the data size is reduced, so it is wasting the bandwidth. To avoid this problem, we propose the following algorithm which is based on compressive sensing.

1. The original signal $s, s \in R^n$ is divided into frames of size 256 for better performance. Different frame sizes are also tested, and as it is not affecting the performance, we are not discussing all different frame sizes.
2. The original signal is converted to the transform domain using the basis matrix φ such as

$$S = \varphi s \tag{8}$$

   where
   $\varphi \in R^{nxn}$
   $S \in R^n$
3. Now, using the measurement matrix, perform non-uniform sampling which yields m samples where $m \ll n$ such as

$$Y = \theta S \tag{9}$$

where

$\theta \in R^{mxn}$

$Y \in R^{m}$

4. Recover the original audio signal using $l1$ minimization method.

5 Results

A set of 10 source audio clips are chosen for the experiment. All the clips are mono-channel with less than 60 s duration sampled with 44.1 kHz having audio data width as 8 bits. All the audio clips generated include solo musical instruments like violin, guitar, piano, flute, equinox, bass, Handel, track, Mary Song, Backstreet Boys song, Crazy Frog—Axel F, Emilie big world, and different frequency clips. For better implementation, the source signal is reduced to frames with the samples of 256 for each frame.

The compressive sensing is performed on the various audio signals with the different measurement vectors, and the reconstruction is performed as explained in the algorithm. The algorithm is implemented using MATLAB 2016 in Intel Core i5 processor. Figure 2 shows the results, and it can be observed that the original signal can be recovered even when we have less samples.

Figure 2d shows the primary signal and the compressed signal with less samples and the recovered signal. From the above results, we can conclude that the original signal can be recovered from the less sampled data. Hence, this proposed algorithm can be used in ADC for high-speed application.

5.1 *Signal-to-Noise Ratio*

Signal-to-noise ratio (SNR) can be used as an objective parameter to scale the integrity of the audio signal. SNR is defined as the measurement of similarity between the primary signal and the signal recovered by using compressive sensing. SNR is given by

$$\text{SNR} = 10\log_{10}\frac{\sum_{i=1}^{n}(s)^2}{\sum_{i=1}^{n}(s-s')^2}\text{dB} \tag{10}$$

where

s is the primary audio signal.

And s' is the recovered audio signal after sparse sampling.

Table 2 shows the SNR for the different audio signals.

(a) 1st Audio – bass

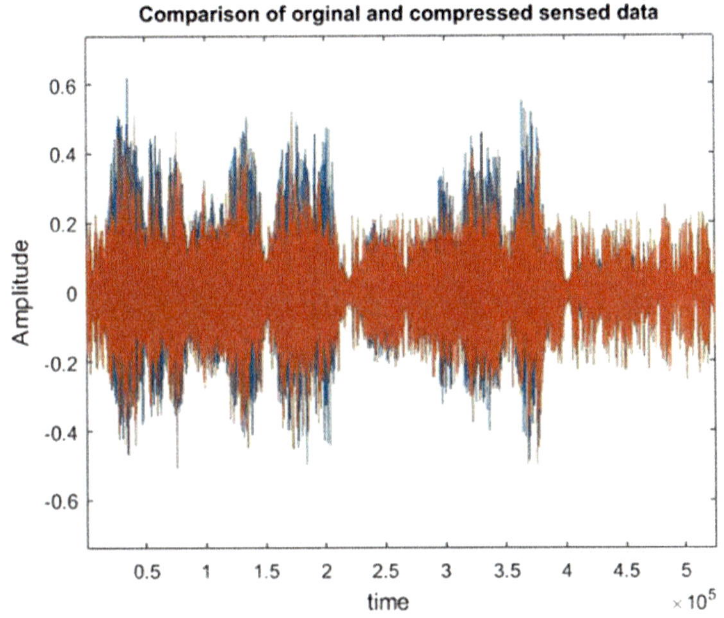

(b) 2nd Audio- handel

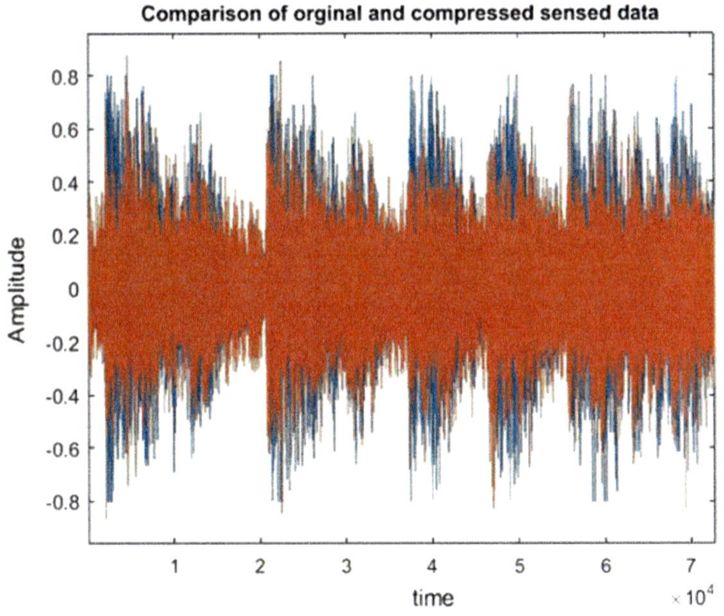

Fig. 2 Experiment results of comparison of original and compressed sensed data

(c) 3 rd Audio - piano

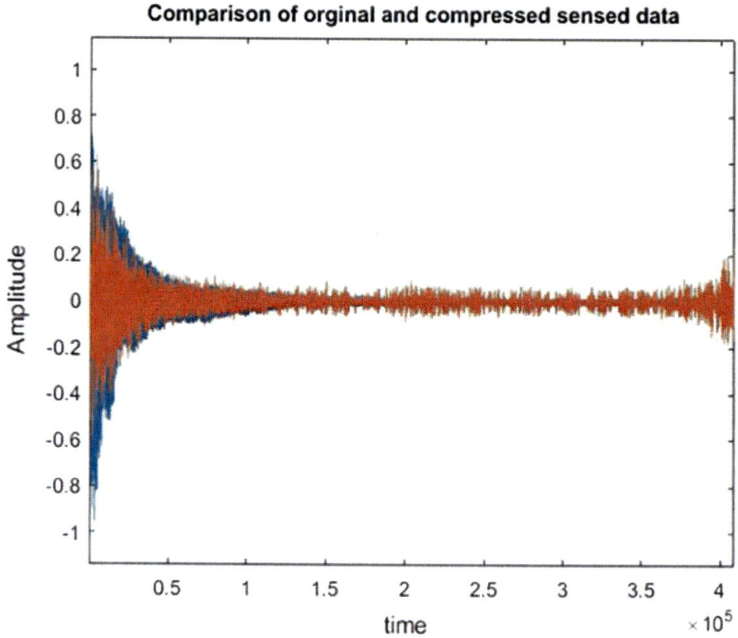

(d) 4 th Audio- Guitar

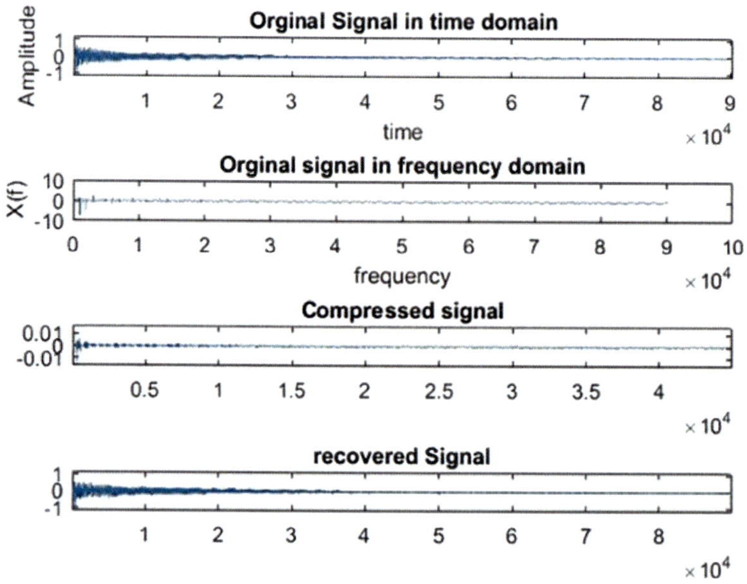

Fig. 2 (continued)

Table 2 SNR measurement

S. no.	Audio signals	Basis matrix	SNR (dB)
1.	Guitar	DFT	25.123
		DCT	44.063
		Haar	31.340
2.	Piano	DFT	22.092
		DCT	35.560
		Haar	21.220
3.	250	DFT	24.080
		DCT	36.670
		Haar	22.342
4.	440	DFT	23.636
		DCT	46.866
		Haar	28.9357
5.	Handel	DFT	24.4921
		DCT	45.3582
		Haar	29.1520

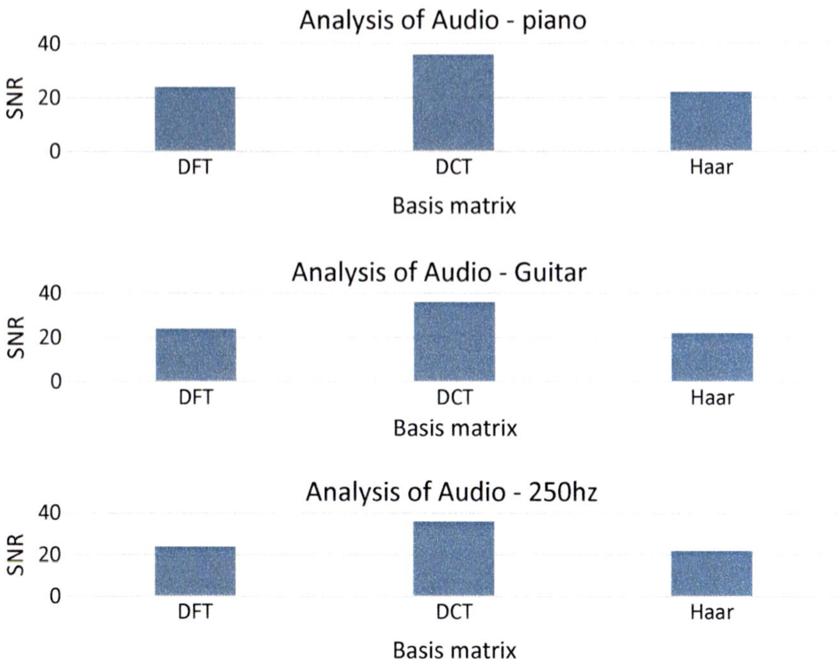

Fig. 3 SNR analysis

From Fig. 3, it can be observed that the SNR is better for the basis matrix of discrete cosine transform (DCT).

6 Conclusion

In traditional analog-to-digital converter (ADC), Nyquist sampling is used which results in N sampled data, and the data is reduced in the compression process. In this process, we proposed and implemented a sparse sampling which results in m samples where $m \ll n$. Thus, the proposed algorithm can be used in ADC and reduces the burden of ADC. The results show that the recovered signal is near to the original signal. It is also observed that the DCT basis matrix produces better result compared to the other basis matrix.

References

1. Lin WHA (2015) Audio watermark: a comprehensive foundation using MATLAB. Springer International Publishing Switzerland. https://doi.org/10.1007/978-3-319-07974-5
2. Thompson SK (2012) Sampling, 3rd edn. Wiley, New York, 446 p. https://doi.org/10.1002/9781118162934
3. Engelberg Shlomo (2012) Compressive sensing [instrumentation notes]. J IEEE Instrum Measur Mag 15(1):42–46. https://doi.org/10.1109/MIM.2012.6145261
4. Candes Emmanuel J, Romberg Justiin K, Terence Tao (2006) Stable signal recovery from incomplete and inaccurate measurements. J Commun Pure Appl Math 59(8):1207–1223. https://doi.org/10.1002/cpa.20124
5. Donoho DL (2006) Compresses sensing. J IEEE Trans Inf Theory 52(4):1289–1306. https://doi.org/10.1109/TIT.2006.871582
6. Ha PH, Lee W, Patanavijit V (2014) An introduction to compressive sensing for digital signal reconstruction and its implementation on digital image reconstruction. In: International Electrical Engineering Congress (iEECON), Chonburi, pp 1–4. https://doi.org/10.1109/ieecon.2014.6925959
7. Mantecón T, Mantecón A, del-Blanco CR, Jaureguizar F, García N (2015) Hand-gesture-based human-machine interface system using compressive sensing. In: International symposium on consumer electronics (ISCE), Madrid, pp 1–2. https://doi.org/10.1109/isce.2015.7177828
8. Bhadoria BS, Shukla U, Joshi AM (2014) Comparative analysis of basis and measurement matrices for non-speech audio signal using compressive sensing. In: IEEE international conference on computational intelligence and computing research, Coimbatore, pp 1–5. https://doi.org/10.1109/iccic.2014.7238453
9. Christensen MG, Østergaard J, Jensen SH (2009) On compressed sensing and its application to speech and audio signals. In: Conference record of the forty-third Asilomar conference on signals, systems and computers, Pacific Grove, CA, pp 356–360. https://doi.org/10.1109/acssc.2009.5469828
10. Foucart S, Rauhut H (2013) A mathematical introduction to compressive sensing. Springer, New York, 634 p. https://doi.org/10.1007/978-0-8176-4948-7

Retrieval of Facial Sketches Using Linguistic Descriptors: An Approach Based on Hierarchical Classification of Facial Attributes

S. Pallavi, M. S. Sannidhan, and Abhir Bhandary

Abstract The facial sketch of the offender may be one of the essential evidences in arresting the offender. Usually, sketch artist or software packages are used to generate the facial sketch on the basis of the eyewitness's description, which includes discrete areas of the face such as nose, mouth and eyebrows. Most of the time, eyewitness cannot describe all the facial cues of a criminal in terms of length and width. So it may be a time-consuming process for both the artist and the eyewitness. To overcome this difficulty, our work aims to make the job of the eyewitness easier by describing the criminal's facial parts in terms of natural language such as 'long nose' and 'big eyes'. To address this, we have adopted standard linguistic descriptors in English language which provides an easy human perception of facial regions. In this paper, we have trained more than 200 sketches to detect the faces and extract the specific corner points. Then, they are classified hierarchically in terms of linguistic hedges based on the Euclidean distance. Finally, the sketches are retrieved based on the score. Experimental results are obtained by using the benchmark datasets like PRIP_HDC and CUHK.

Keywords Salient facial features · Linguistic descriptors · Corner point detection · Euclidean distance · Hierarchical classification

1 Introduction

The area of facial identification system is considered to be a very prominent and powerful discipline of current biometrics. The essence of the procedure of acquisition of required data is unique by itself and it cannot be interrupted or changed when

S. Pallavi (✉) · M. S. Sannidhan · A. Bhandary
NMAM Institute of Technology, Nitte, Udupi, Karnataka 574110, India
e-mail: pallavisajangadde@gmail.com

M. S. Sannidhan
e-mail: sannidhan@nitte.edu.in

A. Bhandary
e-mail: abhirbhandary@nitte.edu.in

N. N. Chiplunkar and T. Fukao (eds.), *Advances in Artificial Intelligence and Data Engineering*, Advances in Intelligent Systems and Computing 1133,
https://doi.org/10.1007/978-981-15-3514-7_84

compared to iris recognition, finger printing and driver license identification [1]. Nowadays, capturing the way a person identifies individuals, specifically the face of an individual, is quite a confronting issue, and the maximum amount of the research work is concentrated on describing the importance of the process of identification. One of the important applications of face recognition lies with law enforcement agencies [2, 3].

The objective of the criminal investigation is to identify the suspect based on the vocal description of eyewitness. Undoubtedly, humans are more efficient in identifying others with reference to the face or belongings or particular regions seen previously [4, 5]. However, it is not possible to precisely recollect huge amount of individual faces within a very short span of time. In addition to that, computers are exclusively used in recognising faces through the adoption of greatly refined standard techniques. Moreover, digital computer systems are not able to handle the issues of change in postural factors of any object, faces captured in poor lighting conditions and other time-invariant factors. Ultimately, the systems are not able to capture the natural method of human depiction and identification of individual faces [6–8].

It is gratifying to observe that the human's manner of defining the salient regions of face in a spoken language is considered to be not so complex. One can describe the salient facial cues of suspect like mouth, nose, eyes, etc., using the linguistic descriptors. It is needless to say here that, the particular facial points may significantly have vital sense, specifically from the succeeding perspectives. Firstly, there is a possibility of aiding an eyewitness to retain memory system and also the time factor when unimportant facial cues of criminals are removed during the process of classification. Secondly, the regions including the important facial features may have turned to be covered and that may affect human perception which is difficult to describe in terms of numeric measurements [9].

Taking these into consideration, our work leads to the retrieval of criminal sketches from the collection of mugshot database established on the verbal description of salient features provided by the eyewitness. For example, 'suspect has big eyes', 'suspect has short eyebrows', etc. Here, calculation of the distance between facial regions helps in labelling the most important facial features. Finally, classifying the facial areas hierarchically make it capable of quantifying the greater granularity of data being the result of method [1, 10]. The originality of our research abides by the authentic aspects that our work supports descriptions in English language, not number-based values. Aforesaid research path may cast light on gist of the identification process, notably over the circumstances of using the inherent individual capability to determine the others.

This research article is categorised accordingly: Sect. 2 presents review conducted, next section briefs the adopted methodology for sketch retrieval and fourth section discusses the experimental outcomes and the final section with conclusion and future work.

2 Literature Survey

Karczmarek et al. [1] proposed an extended version of system to make a decision for multi-criteria based on linguistic descriptors. The authors have efficiently shown the process of describing a face with the help of linguistic descriptors for both single and group of the features. To implement this, they made use of analytical hierarchy process (AHP) where pairwise comparison is done between set of salient features and it is then realised to two levels of hierarchy in order to emphasise the importance of features. The evaluation of important features in this study is carried out by the experts coming from the different fields such as forensic and psychology. The most salient features are described in such a manner that it would eliminate the need of remembering unimportant facial cues. Finally, they introduced an entropy-based evaluation of facial features to compare with the individually estimated facial features values.

Karczmarek et al. [6] have reviewed the usage of linguistic descriptor-based face recognition as it is very difficult to recognise the facial properties with the huge dataset of images. Therefore, authors have seen that the eyewitness can easily give the description of the suspect using linguistic descriptors like 'big eyes', 'small nose', etc., which may enhance the face recognition process. The usage of linguistic descriptor may fill the gap between human description of faces and facial features. It was found that the rate of recognition using linguistic descriptors is high compared to other existing techniques.

Rahman et al. [9] presented a system to overcome the uncertainties about the facial description by onlooker. This system converts imprecise facial illustration into a whole face. Therefore, authors have introduced sketching with word (SWW) technique which imitates between sketch expert and eye witness. In SWW, computation of object includes fuzzy geometric objects, e.g., fuzzy line, triangle and parallel. All the facial components must be f-object of f-geometry. Since eyewitness has to fragment face into labels. Hence, for face recognition, the fuzzy granule has been applied. Different linguistic descriptors have been used. Proposed technique may be used for the identification of other cues related to weapons and finger-print recognition in forensic science.

Cheng et al. [11] proposed a novel learning procedure for retrieval of images based on content using analytical hierarchy process (AHP). The AHP process is implemented in such a way that it improves the issue of multi-interpretation features of an image when concentrated on various types of characteristics. Next, ranking for retrieved images, based on the similarity index, is given by combining the high and low level semantic distance. Finally, a learning of semantic mechanism for retrieval of images based on content is presented. Experimental study coveys that the proposed technique outperforms in comparison with traditional text- and content-based retrieval methods.

Gumede et al. [12] presented a system for face recognition where the most salient facial features are selected for recognition and verification without considering the face configuration. Scale-invariant feature transform (SIFT) and Speed-Up Robust

Features (SURF) feature descriptors have been used. To take out the local features from eyes, nose, mouth, etc., these descriptors rely on the facial features gradient. This paper concludes that when feature selection is applied, the performance is increased by 97%. Here, the experiment is conducted on two face databases, namely SCface and CMU-PIE and experiment results show that proposed technique achieves better results.

Happy and Routray [13] proposed a system for expression recognition to efficiently classify six universal expressions. It calculates the active patches using block-based LBP histogram and finds important areas on face for various expressions. After the selection of salient patches, system selects the expression based on the majority vote. Here, facial fiducial-point detection is carried out using CLM model and DRMF technique which helps in recognising the expression. Experiments are conducted on CK+ and JAFFE facial datasets. CK+ dataset gives an *F*-score of 94.39 and 92.22% of *F*-score on JAFFE.

Karczmarek et al. [14] examined the possible abilities of fuzzy measure to represent and evaluate the importance of the most salient facial features information. The overall review is as below. Firstly, an input image is pre-processed like cropping, scaling and enhancement. Secondly, the most important facial regions like nose, eyes, mouth, etc., are calculated in preliminary test by assigning highest accuracy rates to the selected facial areas. Thirdly, accuracies of all the region by using principal component analysis (PCA) and PCA followed by linear discriminant analysis (LDA) which gradually leads to the recognition rates. Finally, Euclidean distance is determined. *K*-fuzzy measure is constructed using the accuracies of all salient facial segments. The series of experiments are found to be conducted on AT&T and ORL. They concluded that fuzzy measure can be helpful in combining the pieces of information consisted within the facial regions.

Kumar et al. [15] proposed the describable visual attributes for verification of face and searching of images. They are simply the labels which describe appearance of images taken into consideration. This paper mainly concentrates on faces and the describable attributes, for example, gender, nose, face shape, etc. The advantage of labelling the images based on attributes is that when a new object is given, the system recognises the face without the need of further training with less number of attributes. The classifiers are built using large datasets of attributes which calculates the absence, presence and degree of attributes. They found that adding more attributes and training vigorously can help to yield better results. Finally, two new datasets are introduced with describable attributes and identities, namely FaceTracer and PubFig.

Raya et al. [16] present a system based on the Viola-Jones algorithm for detecting the face which is captured using the surveillance camera as a CCTV. Laser range finder is used to locate the moving object with respect to the input received. Once the face is detected, it clicks the image and then it sends to the server for processing. Three types of experimental tests are conducted. First test is to figure out the standard face position to be recorded and identified by the camera. Second test is to find how long (distance) the camera can capture and identify the object and the last is to check how fast system sends identified picture to server. The results have demonstrated that standard position of face is $<45^\circ$ even when there is a moving face. If the distance is more than 5 m then face cannot be captured by camera. Within one second, all captured images will be sent to server for further processing.

Jain et al. [17] proposed the face recognition system to overcome the security problems. Here, Viola-Jones algorithm is used to detect the face. This algorithm mainly has four phases; (1) Haar features (2) Integrated images (3) Adaboost algorithm and (4) Cascade classifiers. Two important dimensionality reduction techniques have been applied to the datasets, namely principal component analysis (PCA) and linear discriminant analysis (LDA). These algorithms are analysed in varying illumination level, brightness condition for varying number of images in datasets. The experiment is conducted to solve the security, safety of banking applications.

3 Methodology

The proposed methodology as depicted in Fig. 1 mainly has two stages, namely hierarchical classification stage and query-sketch stage.

In first stage, the salient facial features of inputted sketch are extracted and corner points are calculated. Linguistic descriptors are used to label the facial attributes based on the distance. In second stage, based on the eyewitness description, appropriate sketches are retrieved for the maximum score.

Detailed explanation of each stage is as follows:

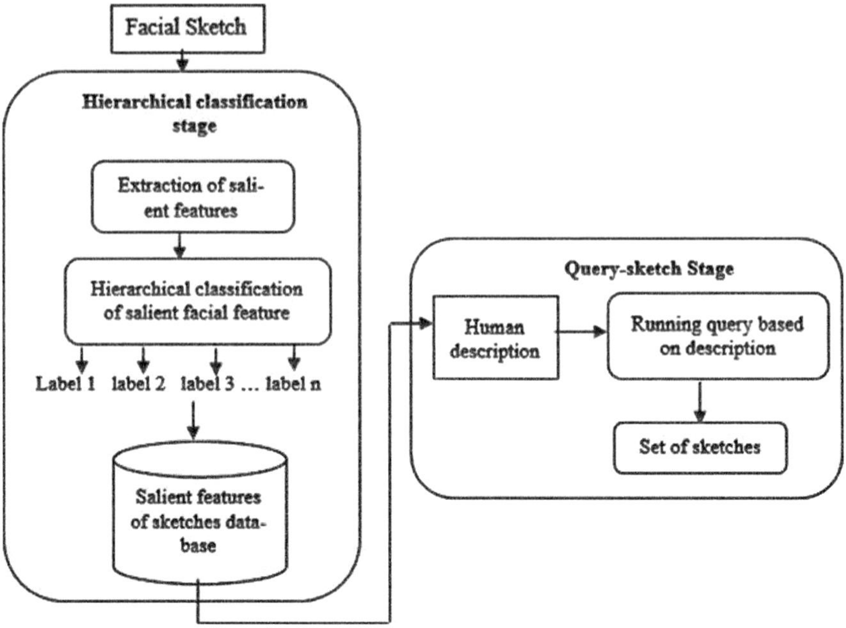

Fig. 1 Proposed system for sketch retrieval

3.1 Hierarchical Stage

3.1.1 Extraction of Salient Features

The first objective of the proposed technique is to extract the salient facial parts including nose, eyes, mouth and eyebrows. By making use of Viola-Jones algorithm, we extract the facial parts one by one. This algorithm is considered to be the first framework for the object detection which yields good recognition rate. This algorithm has three techniques for detecting the facial regions:

1. **Integral Image**:

It is a first component which lets the features used as a detector to be calculated quickly. For feature extraction, Haar features are of a rectangular type measured by an integral image. Haar features can be calculated using the mathematical formula as below [18]:

$$\sum_{1\le i\le N}\sum_{1\le j\le N} I(i,j)_{\text{white}} - I(i,j)_{\text{black}} \quad (1)$$

where I is the image, (i, j) is the location.

2. **Ada boost**:

It is a machine-learning algorithm applied over the detected face in order to have only the relevant features by eliminating the redundant features by building the strong classifiers. Strong classifier is obtained by summing the weak classifiers as below [19]:

$$F(x) = a_1 f_1(x) + a_2 f_2(x) + \cdots a_n f_n(x) \quad (2)$$

where $F(x)$ is strong classifier, $a_{1,}a_{2,}a_n$ are the co-efficient of weak classifiers. $f_1(x)$, $f_2(x)$, $f_n(x)$ are weak classifiers.

3. **Cascading:**

It is composed of stages containing strong classifiers to deduce whether the given object is face or not in a less amount of time.

In MATLAB, there is a Computer Vision Toolbox which comprises of a cascade object detector which constructs detector to efficiently detect objects. By default, the setting is made to detect the faces, but we can modify the detector to detect other facial parts like nose, mouth, eyebrows or upper body as stated in classification model. Here, Haar features-based detector builds the object, adjusted depending upon the user classification model as indicated in the XML input file.

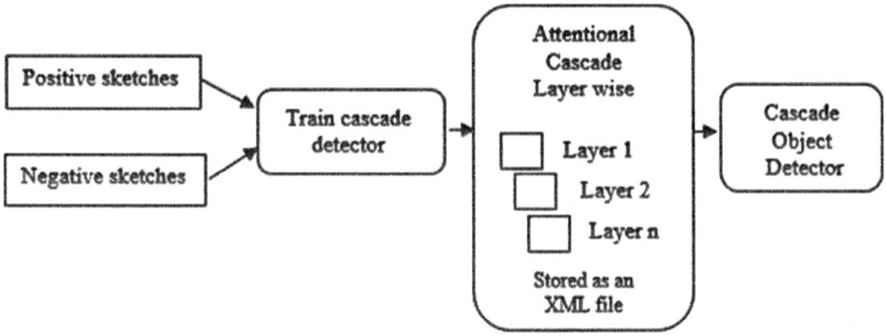

Fig. 2 Training of attentional cascade using positive and negative input samples

A. **Training the cascade detector**

Usually Computer Vision Toolbox has only a small number of pre-trained classifiers. These classifiers are not sufficient in detecting faces. There is a need for further training of cascade detectors. In our proposed work, we are training around 200 composite sketches generated from images of CUHK and PRIP-HDC databases to extract the salient facial features like mouth, nose, eyes, etc.

The training of cascade detector is performed using a group of positive samples (windows with face) and group of negative images. The set of negative images generates the negative samples. Number of cascade layers, Haar feature type and function parameters have to be specified to yield the accurate detector.

Training of cascade is done layer by layer as shown in Fig. 2.

Layer 1 training:

(1) Positive samples are calculated, which is not more than the total number of user-inputted samples.
(2) Negative samples are generated from user-inputted negative images.

Layer 2 training:

(1) Uses the output from layer 1.
(2) All positive samples are classified and removes misclassified samples as negatives.
(3) Same number of positive samples are calculated for remaining positive samples.
(4) Negative samples are generated by preparing negative images and misclassified positive samples.

Layer *n* training:

(1) Uses the output from previous layers.
(2) All positive samples are classified and removes misclassified samples as negatives.

We pass function parameters such as Training Size and False Alarm Rate which is nothing but False Positive rate accepted for each layer. True Positive Rate, number of cascade layers, positive and negative sample locations have to be set along with the type of feature used in detection. In this study, Haar-like feature is used.

B. **Salient feature detection based on the cascade training**:

In order to detect the facial parts using the detector, the name of the XML file has to be supplied in which the result of training of cascade object is saved. Once after the detection, on calling a specific method known as step, it returns a matrix BBOX of the order M X 4, where *M* corresponds to the bounding boxes consisting of the facial areas detected. Each and every row consisting of four items [*x y* width height], that corresponds to the values in pixels, the corner region of upper-left and bounding box magnitude. The succeeding steps have to be followed to use the detector acquired after training: (1) opening of the intended picture; (2) construct the object for detection using detector; (3) recognise the particular facial areas from the picture; (4) facial areas are labelled; (5) view the picture with labelled faces.

Figure 3 is the outcome obtained after the successful training of the attentional cascade object.

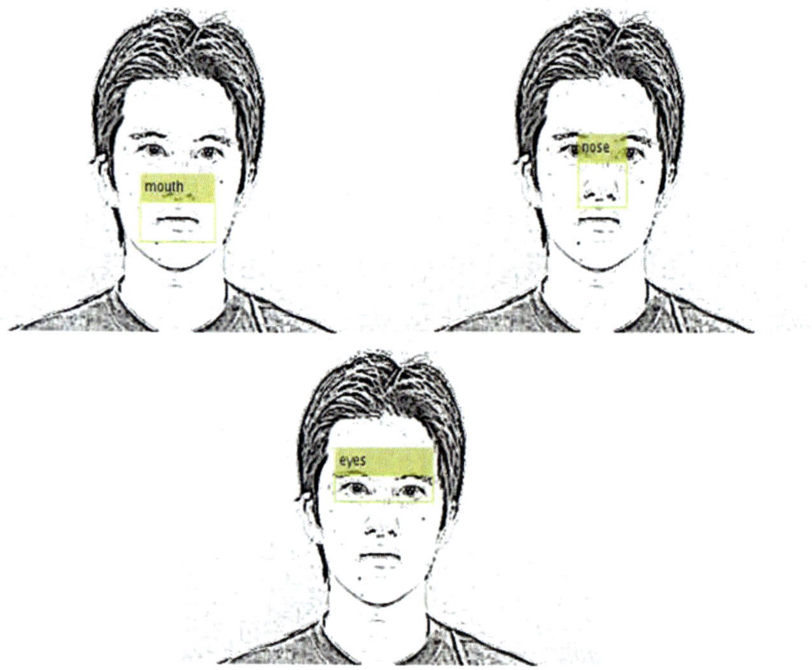

Fig. 3 Detection of mouth, nose and eyes with annotation

3.1.2 Corner Point Detection and Hierarchical Classification of Salient Features

A. **Corner Point Detection**:

As initial step, corner point detection has to be made. In this paper, we apply features from accelerated segment test (FAST) corner detector to obtain corner information from the sketch. Because FAST algorithm is 10 times faster than the other corner detector algorithms [20]. In this algorithm, corner detection is done by inspecting a circular boundary consisting of sixteen pixels around the corner point. Only those points are said as a corner point if depth of a specific amount of neighbouring pixels is more or less than the depth of the pixel in the centre by any threshold value.

As stated earlier, we extract facial regions as mentioned in aforesaid technique arrived in Sect. 3.1.1 and we detect the corner points of mouth, nose, eyes and eyebrows as the following steps.

(1) Take the extracted facial region image like nose, eyes, eyebrows and mouth.
(2) Apply the FAST corner detection algorithm on the particular image.
(3) Taking one point in bounding box as the reference, we iterate over the many detected points until the maximum intensity pixel is reached.
    (a) For mouth region, we consider 2 extreme corner points of mouth and upper and lower lips corner points.
    (b) For nose region, we consider the horizontal and vertical corner points.
    (c) For eyebrows, we consider the 2 extreme ends of eyebrows as corner points.
    (d) For eyes, we consider 2 extreme ends of eyes as corner points.
(4) Compute the distance between detected corner points using Euclidean distance [21].

$$\sqrt{\sum_{i=1}^{n}(q_i - p_i)^2} \tag{3}$$

where $i = 1 \ldots n$, number of sketches, p_i and q_i are the two corner points of respective facial region.
(5) Classifying the facial regions as linguistic descriptors based on the distance measured in step 4 (Fig. 4).

B. **Hierarchical classification:**

Here, we make use of analytical hierarchical process (AHP) to quantify the significance of the facial regions. The motive is to select the most important facial cues in the field of human recognition. Ultimately, the alternatives are assembled into the consecutive sets of features, i.e. (a) General information, (b) Eyes region, (c) Nose region and (d) Mouth region. In this study, we demonstrate three-level of hierarchy to provide the thorough estimation of importance of the individual facial cues. Figure 5 depicts the three-level hierarchy of the overall process [1, 6].

Fig. 4 Overall corner point detection of mouth, nose, eyes and eyebrows

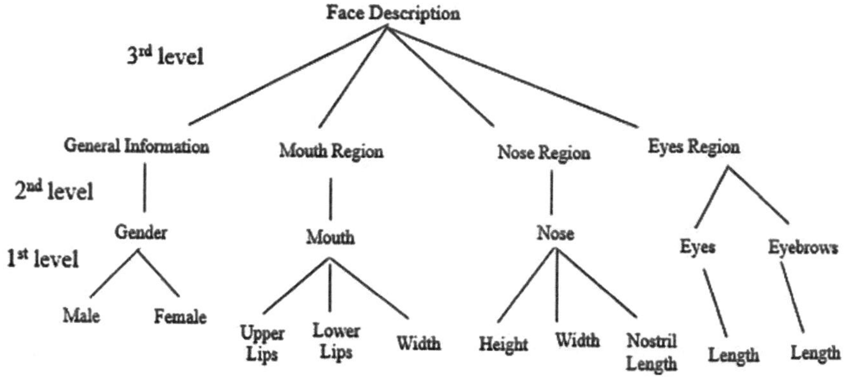

Fig. 5 Hierarchical classification of facial attributes

These atomic cues listed in the tree structure makes it easier to explain in terms of natural language. One can easily notice the particular details of the person such as shape of eyebrows, length of the eyes than normal impression with reference to a given part of face. Figure 5 depicts the three-level hierarchy of selected facial regions.

According to paper [9], the membership values of facial regions are calculated in inches using the membership function. So, in our study, the particular facial regions are labelled by mapping from the pixel value obtained to the inches using the standard pixel-inch conversion.

The equation is as below [22]:

$$\text{Inches} = \frac{\text{Pixel}}{\text{DPI}} \tag{4}$$

Table 1 Pixel to inch conversion for the respective linguistic descriptors

Label	Pixels	Inches
Mouth		
1. Width		
1.1. Narrow	20–40	0.21–0.42
1.2. Medium	41–45	0.43–0.47
1.3. Wide	46–52	0.48–0.54
2. Upper lip distance		
2.1. Thick	10–13	0.10–0.14
2.2. Thin	6–9	0.06–0.09
3. Lower lip distance		
3.1. Thick	10–12	0.10–0.13
3.2. Thin	6–9	0.06–0.09
Nose		
1. Width		
1.1. Narrow	19–21	0.20–0.22
1.2. Medium	22–23	0.23–0.24
1.3. Wide	24–26	0.25–0.27
2. Height		
2.1. Short	17–19	0.18–0.20
2.2. Medium	20–21	0.23–0.24
2.3. Long	22–24	0.23–0.25
3. Nostril size		
3.1. Big	7–9	0.07–0.09
3.2. Small	4–6	0.04–0.06
Eyes		
1. Length		
1.1. Narrow	12–17	0.13–0.18
1.2. Wide	18–20	0.19–0.21
Eyebrow		
1. Length		
1.1. Narrow	13–19	0.14–0.20
1.2. Wide	20–28	0.21–0.30

where DPI is dots per inch.

Table 1 depicted below presents the values of pixel-inches conversion performed for the labelled attributes presented in the Fig. 5 of hierarchical classification.

Table 2 Questions and their possible descriptors

Questions	Description
Describe the gender	Male, female
Describe width of the mouth	Narrow, medium, wide
Describe the upper lip distance	Thick, thin
Describe the lower lip distance	Thick, thin
Describe width of the nose	Narrow, medium, wide
Describe height of the nose	Short, medium, long
Describe the nostril size	Big, small
Describe the eyes length	Narrow, wide
Describe the eyebrow length	Narrow, wide

3.2 *Query-Sketch Stage*

3.2.1 Human Description

An eyewitness with or without the knowledge of crime gives the verbal description of a criminal. In our research work, we have considered some samples of possible queries and their expected answers that mimic the real-time interrogation of crime investigation. Table 2 presents different questionnaires samples along with the different options of answering the questions to detect a criminal.

3.2.2 Running Query Based on Description and Sketch Retrieval

Here, we run a query against description of an eyewitness. Then human description is mapped to the corresponding labelling of facial regions such as nose, eyes and eyebrows stored in the database to retrieve the matching sketches.

For a particular region of a face, if there is an exact match between the query answer and labelled region then, the score of region r is set to 1, otherwise 0. The mathematical equation for the same is depicted below.

$$\text{Score}_r = \begin{cases} 1, & \text{if match true} \\ 0, & \text{if match false} \end{cases} \tag{5}$$

Up on setting the score of individual labelled regions, the final score of a particular sketch S revealing the success rate of total number of matches for n labelled regions against the query is calculated as:

$$\text{Score}_S = \sum_{r=1}^{n} \text{Score}_r \tag{6}$$

$Scores_S$ is total score calculated for sketch S

$$\%\text{retrieval} = \frac{\text{Score}_S}{n} * 100 \tag{7}$$

%retrieval gives the retrieving percentage of the sketches. According to the theory of probability, it is stated that if %retrieval results 50% then, there is a 50–50 chances of retrieving the criminal's sketches. If %retrieval is closer to 100% then, there is an exact sketch retrieval of a criminal [23]. Hence, we have to set the threshold retrieval score as 50% which only retrieves the sketches having scores greater than or equal to 50%.

4 Experimental Results and Discussion

The proposed technique is implemented using MATLAB tool. The goal is to retrieve the sketch of a criminal out of many, based on the score obtained for maximum matching of facial cues. Here, face detection is carried out using Viola-Jones algorithm by training the cascade detector consisting of a dataset of 200 samples of sketches, out of which 100 male and 100 female sketches. The experimental studies are conducted on two benchmark datasets CUHK and PRIP-HDC. Every sketch used for the experiment is resized into uniform dimensionality of 300 * 400 for the accurate and uniform corner point detection.

4.1 *Different Questionnaire and Sketch Retrieval*

An eyewitness is queried about the description of a suspect by referring to the linguistic description Table 2. Then, the set of the maximum scored sketches are retrieved by matching the eyewitness description.

Below are some sample query statements provided by the eyewitness:

Query statement 1:

Gender is Female. She has narrow mouth. Thick upper lip. Thin lower lip. Narrow eyes. Nose height medium. Nose width is medium. Nostril size is small. Wide eyebrows.

Figure 6 depicts various sketches retrieved on executing query statement 1 based on gender bias.

Table 3 presents the value of labelled attributes corresponding to the retrieved sketches having the score value greater than or equal to 50%.

It is evident from Table 3 that many of the attribute values are very closer to the query statement given by the eyewitness even if there is no accurate match found for

Fig. 6 Retrieval of sketches having the score ≥50%

any of the sketch in the database. This provides an advantage for the eyewitness to identify an offender even if there is a slight verbal discrepancy in his/her statement.

Figure 7 depicts the plotting of the score for the sketches in the dataset based on the gender bias against the query statement 1.

From Fig. 7, it is clear that, a sketch whose score is greater than or equal to 50% is a closer match for an eyewitness description.

Query statement 2:

Gender is Male. He has medium mouth. Thin upper lip. Thin lower lip. Wide eyes. Nose height is short. Nose width is narrow. Nostril size is big. Narrow eyebrows.

Figure 8 depicts various sketches retrieved on executing query statement 2 based on gender bias.

Table 4 presents the value of labelled attributes corresponding to the retrieved sketches having the score value greater than or equal to 50%.

It is evident from Table 4 that many of the attribute values are very closer to the query statement given by the eyewitness even if there is no accurate match found for any of the sketch in the database. This provides an advantage for the eyewitness to identify an offender even if there is a slight verbal discrepancy in his/her statement.

Figure 9 depicts the plotting of the score for the sketches in the data set based on the gender bias against the query statement 2.

From Fig. 9, it is clear that, a sketch whose score is greater than or equal to 50% is a closer match for an eyewitness description.

Table 3 Description of retrieved sketches with the score

Gender	Mouth width	Upper lip distance	Lower Lip distance	Nose height	Nose width	Nostril size	Eyes length	Eyebrow length	Score (%)
Female	Narrow	Thick	Thick	Medium	Medium	Small	Narrow	Narrow	88
Female	Narrow	Thin	Thin	Short	Wide	Small	Narrow	Narrow	50
Female	Narrow	Thin	Thin	Medium	Narrow	Small	Narrow	Narrow	75
Female	Narrow	Thick	Thin	Medium	Medium	Small	Narrow	Wide	88
Female	Narrow	Thin	Thick	Medium	Narrow	Small	Narrow	Narrow	63
Female	Narrow	Thin	Thick	Medium	Medium	Small	Wide	Narrow	75
Female	Narrow	Thick	Thick	Medium	Narrow	Small	Narrow	Narrow	50

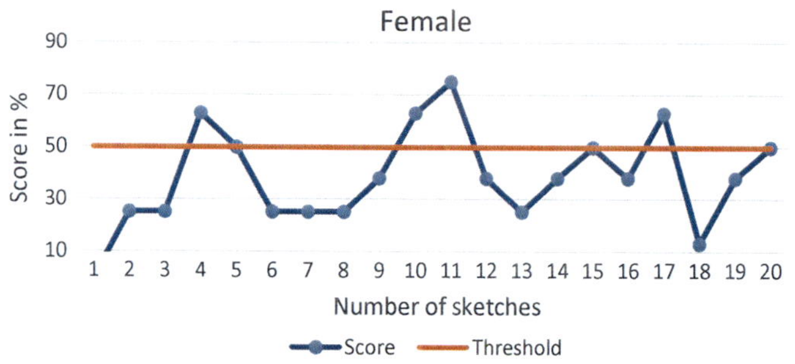

Fig. 7 Plotting of graph for the sketch retrieval for Female gender

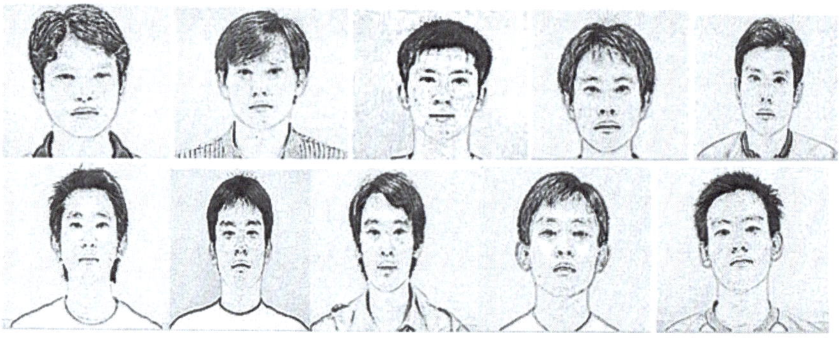

Fig. 8 Retrieval of sketches having the score ≥50%

5 Conclusion and Future Work

This research article exhibits a peculiar methodology for retrieving the sketches of offender from vocal description of an eyewitness. The approach is based on the human facial detection and extraction of corner points using FAST algorithm. Then, the hierarchical classification has been proposed in order to provide the importance to the crucial features accounted by humans in explaining the faces. The uniqueness of this research work is that only the spoken language is used to produce the better experimental results rather than the numerical measures. Finally, based on the human description, score has been calculated and then a particular threshold for score is set. Sketches having the threshold value are retrieved. The experiment is conducted on two benchmark datasets, namely CUHK and PRIP-HDC. Experimental studies have also proved that the retrieval of the sketches is possible even by the slight inaccurate depictions made by the eyewitness. Thus, the system is an attempt to provide a better result considering the theories of probability in human depiction.

In future, the work can be extended to implement an autonomous system for detecting and classifying the features automatically using the concepts of transfer learning and deep learning. There is subtle amount of work which can be carried

Table 4 Description of retrieved sketches with the score

Gender	Mouth width	Upper lip distance	Lower lip distance	Nose height	Nose width	Nostril size	Eyes length	Eyebrow length	Score (%)
Male	Medium	Thick	Thin	Medium	Narrow	Big	Narrow	Wide	75
Male	Medium	Thick	Thin	Medium	Wide	Small	Narrow	Wide	50
Male	Medium	Thick	Thin	Long	Narrow	Big	Narrow	Narrow	63
Male	Wide	Thick	Thin	Short	Medium	Big	Wide	Wide	50
Male	Medium	Thick	Thick	Medium	Narrow	Big	Narrow	Wide	63
Male	Medium	Thick	Thin	Medium	Wide	Small	Narrow	Wide	50
Male	Medium	Thick	Thick	Medium	Narrow	Big	Narrow	Wide	63
Male	Narrow	Thick	Thin	Medium	Wide	Big	Narrow	Wide	50
Male	Medium	Thick	Thin	Short	Medium	Big	Narrow	Wide	75
Male	Wide	Thick	Thin	Long	Medium	Big	Narrow	Wide	50

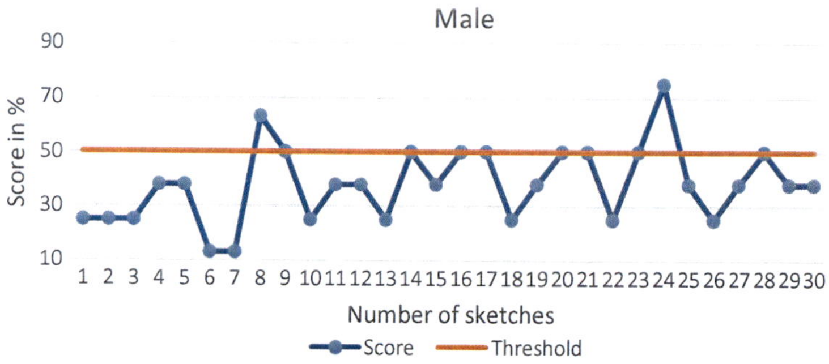

Fig. 9 Plotting of graph for the sketch retrieval for male gender

out to classify the linguistic descriptors into the local languages favourable for the eyewitness as many may not have clear understanding of the English language.

References

1. Karczmarek P, Pedrycz W, Kiersztyn A, Rutka P (2016) A study in facial features saliency in face recognition: an analytic hierarchy process approach. Soft Comput 21:7503–7517. https://doi.org/10.1007/s00500-016-2305-9
2. Sannidhan MS, Ananth Prabhu G (2016) A comprehensive review on various state-of-the-art techniques for composite sketch matching. Imp J Interdisc Res (IJIR) 2:1131–1138
3. Kokila R, Sannidhan MS, Bhandary A (2017) A study and analysis of various techniques to match sketches to Mugshot photos. In: 2017 international conference on inventive communication and computational technologies (ICICCT), pp 41–44. https://doi.org/10.1109/icicct.2017.7975243
4. Kokila R, Sannidhan MS, Bhandary A (2017) A novel approach for matching composite sketches to mugshot photos using the fusion of SIFT and SURF feature descriptor. In: 2017 international conference on advances in computing, communications and informatics (ICACCI), pp 1458–1464. https://doi.org/10.1109/icacci.2017.8126046
5. Pallavi S, Sannidhan MS, Sudeepa KB, Bhandary A (2018) A novel approach for generating composite sketches from mugshot photographs. In: 2018 international conference on advances in computing, communications and informatics (ICACCI), pp 460–465. https://doi.org/10.1109/icacci.2018.8554564
6. Karczmarek P, Kiersztyn A, Rutka P, Pedrycz W (2015) Linguistic descriptors in face recognition: a literature survey and the perspectives of future development. In: 2015 signal processing: algorithms, architectures, arrangements, and applications (SPA), pp 98–103. https://doi.org/10.1109/spa.2015.7365141
7. Pallavi S, Sannidhan MS, Bhandary A (2018) A comprehensive review on various state-of-the-art techniques for image enhancement. Int J Eng Technol 7:860–864. https://doi.org/10.14419/ijet.v7i3.34.19576
8. Achar S, SSannidhan MS, Bhandary A (2018) A comparative analysis of quality metrics between different image enhancement techniques for facial sketches. Int J Eng Technol 7:794–798. https://doi.org/10.14419/ijet.v7i3.34.19562
9. Rahman A, Beg M (2014) Face sketch recognition using sketching with words. Int J Mach Learn Cybern 6:597–605. https://doi.org/10.1007/s13042-014-0256-y

10. Saaty T, Vargas L (2001) Models, methods, concepts & applications of the analytic hierarchy process. International series in operations research & management science. https://doi.org/10.1007/978-1-4615-1665-1
11. Cheng S, Chou T, Yang C, Chang H (2005) A semantic learning for content-based image retrieval using analytical hierarchy process. Exp Syst Appl 28:495–505. https://doi.org/10.1016/j.eswa.2004.12.011
12. Gumede A, Viriri S, Gwetu M (2018) Selecting salient features from facial components for face recognition. Image and video technology, pp 63–75. https://doi.org/10.1007/978-3-319-92753-4_6
13. Happy S, Routray A (2015) Automatic facial expression recognition using features of salient facial patches. IEEE Trans Affect Comput 6:1–12. https://doi.org/10.1109/taffc.2014.2386334
14. Karczmarek P, Pedrycz W, Reformat M, Akhoundi E (2013) A study in facial regions saliency: a fuzzy measure approach. Soft Comput 18:379–391. https://doi.org/10.1007/s00500-013-1064-0
15. Kumar N, Berg A, Belhumeur P, Nayar S (2011) Describable visual attributes for face verification and image search. IEEE Trans Pattern Anal Mach Intell 33:1962–1977. https://doi.org/10.1109/tpami.2011.48
16. Raya I, Jati A, Saputra R (2017) Analysis realization of Viola-Jones method for face detection on CCTV camera based on embedded system. In: 2017 international conference on robotics, biomimetics, and intelligent computational systems (Robionetics). https://doi.org/10.1109/robionetics.2017.8203427
17. Jain U, Choudhary K, Gupta S, Pemeena Privadarsini M (2018) Analysis of face detection and recognition algorithms using Viola Jones algorithm with PCA and LDA. In: 2018 2nd international conference on trends in electronics and informatics (ICOEI). https://doi.org/10.1109/icoei.2018.8553811
18. Alionte E, Lazar C (2015) A practical implementation of face detection by using Matlab cascade object detector. In: 2015 19th international conference on system theory, control and computing (ICSTCC), pp 785–790. https://doi.org/10.1109/icstcc.2015.7321390
19. Vikram K, Padmavathi S (2017) Facial parts detection using Viola Jones algorithm. In: 2017 4th international conference on advanced computing and communication systems (ICACCS), pp 1–4. https://doi.org/10.1109/icaccs.2017.8014636
20. Jeong K, Moon H (2011) Object detection using FAST corner detector based on smartphone platforms. In: 2011 first ACIS/JNU international conference on computers, networks, systems and industrial engineering, pp 111–115. https://doi.org/10.1109/cnsi.2011.60
21. Rachid A, Said S, Bouzid M (2017) Euclidean and geodesic distance between a facial feature points in two-dimensional face recognition system. Int Arab J Inf Technol 14:565–571
22. Roer G (2020) Convert pixels to inches easily with this image size calculator|Uproer. In: Uproer. https://uproer.com/articles/image-size-calculator-px-in/
23. Tanaya K (2007) Introduction probabilistic image processing and bayesian networks. In: https://www.smapip.is.tohoku.ac.jp/~kazu/tutorial-lecture-note/ALT&DS2007Tutorial/KazuyukiTanaka.pdf

Simplified SVD Feature Construction in Multiangle Images to Identify Plant Varieties and Weed Infestation

K. Ramesh and Andrews Samraj

Abstract It is essential to identify the weed growth in plant infection at an earlier stage, which is an important procedure of precision agriculture. In this research paper, we found that the optimum methods by positioning appropriate optical sensors to acquire the most relevant images to interpret the plant, identification of weed infections in the plant bed, and accurate identification of intrusion through the images. We introduced a variant of singular value decomposition for this purpose to achieve the best possible results. The performance of this modified singular value decomposition by our work is found better than the conventional singular value decomposition-based feature extraction. Comparison of aerial and portrait images was done to identify the best choice according to the required identification.

Keywords Modified singular value decomposition · Analysis of variance · Threshold classification · Image processing

1 Introduction

The farmers of Tamilnadu state of south India are the habitual planters of vegetables and millet varieties. The outspread of organic farming and its benefits attracted these farmers toward modification in the farming techniques. This is due to the high price for the organic products which are free from chemical fertilizers and poisons pesticides. The price for the organic products is always premium and brings a great profit to the farmers. The interest in organic farming and the associated novel cultivation suggestions makes the farmers to experiment various innovative techniques into their farming activity.

K. Ramesh (✉)
Department of CS, CA & IT, Karpagam Academy of Higher Education, Coimbatore, India
e-mail: krmca86@gmail.com

A. Samraj
Department of Computer Science and Engineering, Mahendra Engineering College, Namakkal, Tamilnadu, India
e-mail: andrewsmalacca@gmail.com

N. N. Chiplunkar and T. Fukao (eds.), *Advances in Artificial Intelligence and Data Engineering*, Advances in Intelligent Systems and Computing 1133,
https://doi.org/10.1007/978-981-15-3514-7_85

The process like weed control, pest attack, intrusion detection, and expellsion are the area of interest to the farmers due to the high possibilities of automation in recent days. Conventionally, plant monitoring experiments were done by various image processing techniques. Usually, these experiments are carried out to identify the plant care requirements like watering, clearing weeds, addition of nutrients to the land, and expulsion of objects like foreign elements [1–3].

The motivation to do such automation originates from the farmers found it extremely difficult in expelling birds like parakeets, peahens and Peacocks, insects like grasshoppers, rabbits and wild hogs from the plant beds where maize of different varieties, chilli plants, beats like radish and tapioca. The major advantage of the system is to preserve the expertise and knowledge of the farmers is not replaced by the proposed system. In fact, the experience of farmers using the proposed system gets enhanced to a better level where the monitoring cannot be done using naked eyes. Various works similar to this have been reported by researchers working in this area and a continuous improvement is found in this field of work [4–6]. The focus and novelty of this proposed work can be identified in the algorithm that identifies the objects which works on a suggested variant of the regular singular value decomposition method in order to elevate the identification to give augmented result in terms of better classification [1, 7, 8].

In this proposed work, we implemented a system that identifies plants to distinguish the weeds from the crop by a combination of image processing and classification techniques to be a perfect solution to the problem [9]. Apart from this, plant identification and weed monitoring are other few associated applications also benefit the farmers in identification of intrusion and event detection. This application uses a stream of images acquired continuously by a fixed or moving cameras deployed for this purpose. The cameras can be fixed on posts erected in strategic locations or embedded into flying drones. The ultimate aim of the proposed work is to precisely identify and care individual plants in order to support highest yields in precision agriculture.

2 Methods and Materials

2.1 Image Acquisition

Acquisition of images of plants is a process in which obtaining the pictures from various angles and transforming them into a digital image are the basic steps in the image processing. The cameras used in this process are of pattern 5 MP, f/2.4, autofocus, and equipped with LED flash. Such cameras are positioned on both angles say oblique aerial as well as portrait angle, to carry out this image acquisition for different vegetable plants with and without weed infestation [10]. The images were captured throughout the plant bed with an uniform height [11]. A distance of 3 ft from the ground level is the height of the camera position for all the images used in the experiment. The characteristics of the images in the dataset are its size that is made uniformly cropping. The resolution of the images captured is 720 × 1280 pixels which makes 329 pixels per inch intensity. The preprocessing session gives other details about the picture dataset prepared by us in this work.

2.2 *Preprocessing*

Images were selected from the captured images which are bright and uniformly clear. The acquired images from multiangles say oblique aerial and portrait direction where resized by cropping to make them with same dimensions and uniform data size. The time of capturing images is at the bright sunlight to avoid gloomy or additional lighting noise. All these cropped aerial and portrait pictures are then converted into gray scale for further processing (Figs. 1 and 2).

a. Leaf of Ladies finger b. Ladies finger among weeds

Fig. 1 Gives the sample images taken from aerial angle of leaf of ladies finger and ladies finger among weed

a. Ladies Finger plant b. Ladiesfinger plant among Weeds

Fig. 2 Shows the images of ladies finger and ladies finger weed taken in portrait angle

2.3 Singular Value Decomposition

The cropped and gray scale images were taken for singular value calculations using the technique, singular value decomposition (SVD) by the computation of SVD.

$$A = U_{mxn} S_{mxn} V^{\mathrm{T}}_{mxn} \quad (1)$$

Then singular value decomposition can be enhanced by various ways to suit the application involved. Many modifications to SVD feature construction is done in the procedure for enhancing the patterns [3, 12].

After arranging the obtained singular values in descending order, the high five SVD principle component values were selected as the feature combination. This process is followed for forming the feature for all images uniformly throughout the experiments. The older method suggested for classification based on threshold values in our previous works [13] for calculating SVD features of aerial images for different plants is not followed in this proposed work. The new procedure followed is selecting the first five SVD values and adding them to form a single numerical value as the feature for all four categories of the plants ladies finger, ragi, brinjal, and the ladies finger plants that are infested by grass weed.

On continuous analysis and modifications based on our regular observation on these SVD values, we try to modify the SVD feature by simplification. This is done instead of using top five SVD values, but we simplified the features by applying techniques to modify its dimension to improve easy classification.

2.4 The Modification Proposed to SVD Simple (MSVD)-Varient

The proposed variant of SVD is developed by adding them to a single numerical value as the sum of major five SVD values (features) of every individual images used in this experiment. These SVD values are arrived after applying the regular SVD operations on every image.

$$S = \mathrm{SVD}[P_{i\,\ldots.n}] \quad (2)$$

S is the SVD matrix arrived after applying regular SVD on the experimental pictures P_i to P_n.

$$M_{\mathrm{s}} = \sum_{1}^{j} S \quad (3)$$

The modified-SVD M_s is the summation of first j values of S for that particular picture. The parameter j can be modified according to the applications. We constantly used $j = 5$ throughout the experiments.

The reason to simplify the feature vector by the proposed modification is due to the better performance and convenience we found while working on these experiments. The first five SVD values of each image from each class where added together as one single value for a single representation of a single image. This feature compression (or modification) helps us for doing easier comparisons and simpler operations. We calculated interdifference and intradifference by the distance calculation procedure where the SVD value sums of a particular class were averaged and considered as a reference feature. We pretested this variant of SVD by a popular experiment.

3 Experiments and Results

3.1 Experiment 1—Using MSVD on Industrial Sound Signals

Industrial automation requires a continuous monitoring process without disturbing the production. Tool wear monitoring is one of such important area where researchers were trying to identify the degree of tool wear from the sound waves without stopping the machining processes [7, 12]. We considered this sound wave analysis processed for testing the fitness of our proposed modified SVD before subjecting it to analyze the plant monitoring problem.

Ten sound signals of each fresh, slight wear, and severely worn-out tool's machining processes were taken for this fitness experiment. The signals were analyzed in two sets for clearer comparison and understanding of proposed technology. The ten signals with its 100 values (10 × 100) and 8 signals randomly were taken from the ten signals (8 × 100) forms the set1 and set2, respectively. These sets were subjected to regular SVD function and our proposed modified-SVD function, and the results were taken for the comparison [13, 8]. The compared results were given in Table 1.

The result shows that the MSVD performs well with much better differences in categories than the regular SVD.

Table 1 Differences among three categories of tool conditions from data set 1 and 2 using traditional SVD

Method/signals	Fresh	Slight	Severe	Differences		
				Fresh-slight	Fresh severe	Slight-sever
8 × 8—SVD	6.369	5.728	4.681	0.641	1.688	1.046
10 × 10—SVD	6.547	6.106	4.694	0.441	1.853	1.411
AVG	6.459	5.917	4.687	0.541	1.770	1.229

Table 2 Differences among three categories of tool conditions from data set 1 and 2 using MSVD

Method/signals	Fresh	Slight	Severe	Differences		
				Fresh-slight	Fresh-severe	Slight-sever
8 × 8—MSVD	8.701	8.103	5.845	0.598	2.856	2.258
10 × 10—MSVD	6.406	5.657	4.977	0.749	1.429	0.680
AVG	7.553	6.880	5.411	0.673	2.142	1.469

This is to justify the application of MSVD on the images in our plant monitoring experiments (Table 2).

3.2 Threshold Classification of Areal Images

In Fig. 3, The gap is calculated as the difference between a reference feature and the other features of the same class which is considered as intradistance. The values in calculating such distance are represented by the MSVD values

$$d(\text{ir}) = d(\text{avg}) - d(i\text{th sample}) \tag{4}$$

Whereas the interdistance was calculated as the difference between the reference MSVD values of one class with the MSVD value of an image sample from any other class.

The Euclidean distance is calculated by

$$\|q - p\| = \sqrt{(q - p) \cdot (q - p)} \tag{5}$$

Figure 3 shows the distances between classes of images by calculating the differences between MSVD features of ten images of any class given in the diagram with the MSVD features of the any ten images belonging to the other class. The comparison of all classes shown in the diagrams they are Ladies finger and Ragi, Ladies finger and ladies finger with Grass weed, Brinjal and Ragi and Brinjal and ladies finger with Grass weed.

This difference mentioned in Table 3 is found to be small and not helpful in distinguishing each class easily while comparing images. Hence, threshold augmentation by widening the gap between the classes is found necessary for a better discerning. We have created a new augmentation technique to improve the difference for the better discerning of the classes.

The first column of Table 3 is arrived by calculating the sum of first five SVD values (features) of every individual images 1–10 used in this experiment.

The second column 'Diff 1' which is difference of the very first SVD values of all 1–10 brinjal images added and averaged and subtracted from the first column.

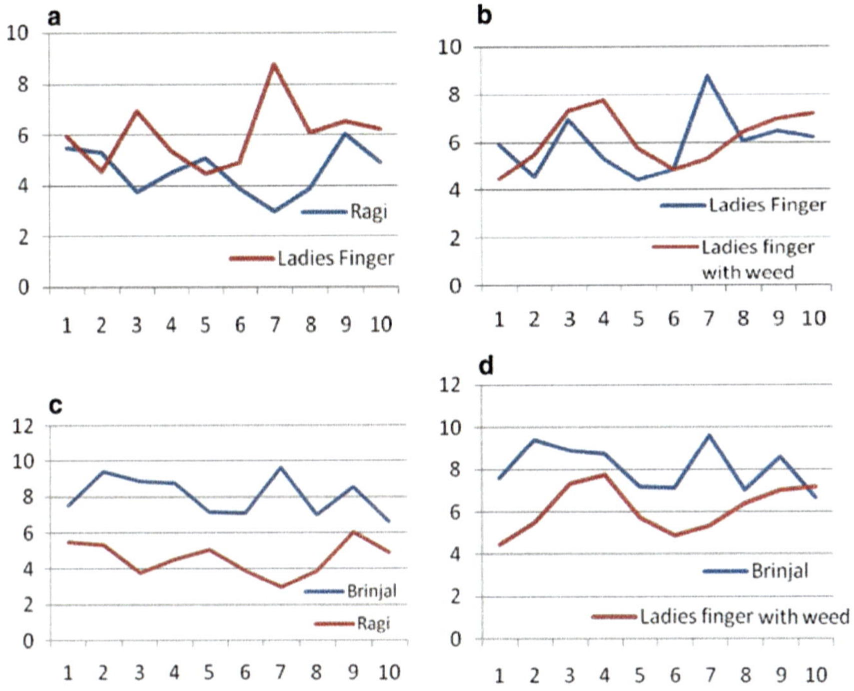

Fig. 3 **a** Distance between leaf of ragi and ladies finger. **b** Distance between leaf of ladies finger and ladies finger among weed. **c** Distance between leaf of brinjal and ragi. **d** Distance between leaf of brinjal and ladies finger among weed

Table 3 Interplant MSVD differences by first value of MSVD features

S. no	Feature difference in ladies finger and ragi	Feature difference in leaf of ladies finger and ladies finger among grass weed	Feature difference in brinjal and ragi	Feature difference in brinjal and ladies finger among grass weed
1	0.98	1.428	1.472	1.920
2	2.095	1.298	2.781	1.984
3	0.305	0.382	0.821	1.509
4	0.327	0.323	3.214	3.209
5	0.583	0.453	1.643	1.513
6	0.638	0.635	1.402	1.398
7	0.968	1.577	3.061	3.670
8	2.155	1.474	1.045	0.364
9	1.260	0.919	1.105	0.764
10	1.208	2.048	0.627	1.467
AVG	1.052	1.054	1.717	1.770

This column 3 continued the same process of all images and listed. This method is repeated with the second-largest SVD value for all ten brinjal plant images and the averages subtracted from the column 1. Column 4 ‘Diff 3’, column 5 ‘Diff 4’, and column 6 ‘Diff5’” are also created by following the same procedure with consecutive lower values.

The new augmentation technique is developed for easier decision making. The SVD features of brinjal images were taken for intradifference calculations. The differences is calculated as follows

$$X_{B0} = Y\left[Y_{ij} + Y_{ij+1} \cdots Y_{ij+k}\right] / [k+1]$$

where $i = 0, j = 0, k = 4$.

The initial values are

$$\begin{aligned} X_{B1} &= Y\left[_{i+1,j} + Y_{i+1,j+1} \cdots Y_{i+1,j+k}\right] / [k+1] \\ &\vdots \\ X_{B9} &= Y\left[Y_{i+10,j} + Y_{i+10,j+1} \cdots Y_{i+10,j+k}\right] / [k+1] \end{aligned} \quad (6)$$

Intradifference is used to calculating within the same subject image. Interdifference is used to determine and compare within and other images (Fig. 4).

3.3 *Threshold Classification of Portrait Images*

This augmentation procedure for intradifference and interdifference of portrait images of brinjal and brinjal with ragi are given in Tables 3 and 4.

The modified way of SVD feature construction is followed to form the SVD feature of brinjal images and is listed in column 1 of Table 4. The difference 1 shown in the second column of Table 4 is arrived by subtracting the sum of all top most (biggest) SVD values of the available ten images from the modified-SVD feature of the first brinjal images mentioned in the corresponding first row and column of Table 4.

$$\text{(E.g.) } 7.544 - 3.552 = 3.992$$

The same procedure for image 2 is followed and is

$$\text{(E.g.) } 9.384 - 3.552 = 5.832$$

In column 2 ‘Second diff’, the average of the second-largest SVD value of all images were subtracted from SVD feature of all brinjal images ‘Third Diff’, ‘Fourth

Fig. 4 Flow of the proposed MSVD-based system

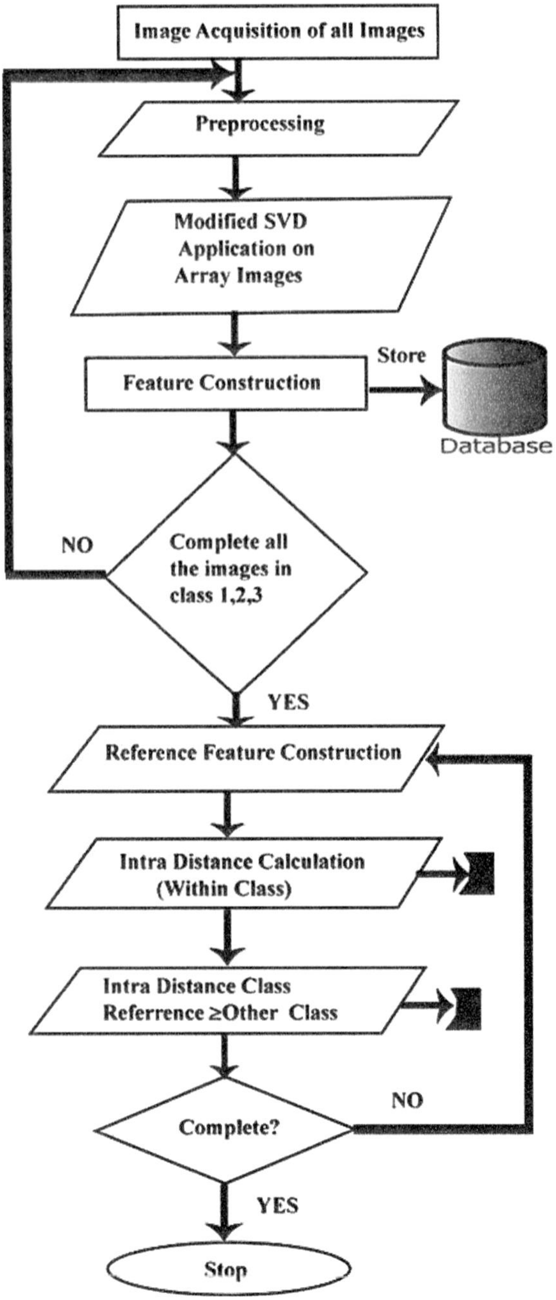

Table 4 Average differences in portrait brinjal images (intra)

MSVD features brinjal images	Top difference	Second difference	Third difference	Fourth difference	Fifth difference
7.544	5.687	6.553	6.811	6.767	6.908
9.384	7.528	8.393	8.651	8.608	8.748
8.883	7.027	7.893	8.151	8.107	8.248
8.741	6.885	7.751	8.009	7.965	8.106
7.167	5.311	6.177	6.435	6.391	6.532
7.117	5.261	6.127	6.385	6.341	6.482
9.602	7.746	8.612	8.870	8.826	8.967
6.988	5.132	5.997	6.255	6.212	6.352
8.564	6.708	7.574	7.832	7.788	7.929
6.631	4.774	5.640	5.898	5.854	5.995
AVG	6.205	7.071	7.329	7.285	7.426

Diff', 'Fifth Diff', are calculated with the third, fourth, and fifth values of the SVD in the same way.

Table 5 is formed as a comparison of features of two different plants say brinjal and ragi. Here, also the brinjal features of ten images constructed by the proposed MSVD method are taken (sum of all largest SVD value of that image) in the first column.

The second column 'Top diff' is calculated by subtracting the average of all biggest SVD values of ragi images from the brinjal feature. But the difference between

Table 5 Differences in average MSVD values of brinjal and ragi images (inter)

MSVD features brinjal images	Highest difference	Next difference	Third level difference	Fourth level difference	Fifth level difference
7.544	3.992	5.751	6.365	6.684	6.866
9.384	5.832	7.591	8.206	8.524	8.706
8.883	5.332	7.090	7.705	8.023	8.206
8.741	5.190	6.948	7.563	7.881	8.063
7.167	3.616	5.374	5.989	6.307	6.490
7.117	3.566	5.324	5.939	6.257	6.439
9.602	6.051	7.809	8.424	8.743	8.925
6.988	3.436	5.195	5.810	6.128	6.310
8.564	5.013	6.771	7.386	7.704	7.886
6.631	3.079	4.838	5.452	5.771	5.953
AVG	4.511	6.269	6.884	7.202	7.384

interfeature and intrafeature in this method is found to be very less and could not be reliable for plant identification. So, we proceeded to the next improvement [14].

3.4 *Threshold Comparison of Picture Types*

Table 6 provides the consolidated matching of classification efficiency using both the aerial and images captured in portrait angle. A simple observation noted from Table 3 provides a suggestion that the plants when they appear with flowers, and/or fruits, as well as grains are capable of giving better classification with portrait angles. Whereas plants infested with lot of weeds were found with improved differences in aerial angle images.

In the same Table 6, we combined the difference between classes of distinct plants by making an average of the differences obtained by using aerial as well as portrait images. The main objective of combining these differences arrived from the aerial and the portrait images is due to their supremacy over the other in instances. Therefore, it is noted that the aerial images are no way inferior to their counter parts that the portrait images and vice verse. This novel combination technique proposed by us is consistent and assured method of classifying images from different angles and their combinations.

Considering the differences between aerial or portrait images is not only a significant method since the difference values are higher in some cases for aerial and in some cases for portrait, while comparing the different varieties of combinations.

Table 6 Differences found in comparison and combination of aerial and portrait images

S. No.	Combination	Values of portrait differences	Values of leaves in aerial angle differences	Difference after combination
1	Brinjal and ragi	3.485	2.855	3.170
2	Brinjal and ladies finger	2.105	1.312	1.708
3	Ladies finger and ragi	1.380	1.543	1.461
4	Ladies finger and ladies finger with grass	0.190	1.582	0.886
5	Brinjal and ladies finger with grass	1.914	2.895	2.404
6	Ragi and ladies finger with grass	1.571	0.039	0.805
Average		1.774	1.704	1.739

In Table 6, we represent the combinations of various types of images formed by using averaging the differences we found while using MSVD on aerial as well as stem pictures of plants.

At the same time, we found that the differences of aerial images are higher than that of the portrait images in all plant verities calculated during intraplant images differences. But we found some exceptions in the image that are captured while there is a weed infestation. But, during the plant identification process, the interdifferences of aerial images influence the process by identifying the plants since the differences are more than the portrait images.

3.5 Superiority of MSVD Feature

The one-way ANOVA test is carried out to analyze the variance between combinations of comparisons of plants as a part of the research work.

This is to ensure the fitness of the proposed feature construction method modified singular value decomposition (MSVD), over the conventional technique followed in classical SVD.

The classical SVD takes the first '*n*' largest singular values as feature components were *n* is specified by the parameter setting. For example, if image A, produces *X* number of singular value sorted from highest value to the lowest value, $X - n$ lower values will be omitted in the feature vectors. The feature vector will be constructed as

$$F = \{f_1, f_2, f_3 \ldots f_n\} \quad \text{where } n < X \text{ and } n \in X \tag{7}$$

$\{f_{n+1}, f_{n+2}, f_{n+3} \ldots f_n\}$ will be omitted. In the case of our proposed modification in SVD feature construction, the feature is formed as a summation of the first '*n*' highest value singular values, i.e.,

$$F = \{f_1 + f_2 + f_3 + \cdots + f_n\} \quad \text{where } n \in X \tag{8}$$

Here, also, lower singular values will be omitted.

3.6 ANOVA Test

ANOVA is a power analytical tool which is helpful for comparative analysis two variances calculated in this process to give the statically significance. The ratio is of independent nature and cannot be altered by any operation.

$$S^2 = 1/n - 1 \sum \left(Y_i - \overline{Y}\right)^2 \tag{9}$$

Here, the devisers are called degree of freedom, the summation is known as sum of squares, and the result is called mean square. The one-way *ANOVA* for brinjal and ragi, ladies finger and brinjal, ladies finger and ragi, ladies finger and ladies finger with weed, brinjal and ladies finger with weed, ragi and ladies finger with weed was carried out using both the classical SVD feature and modified-SVD feature proposed in this paper.

The result is presented in Table 8. The result shows that the proposed MSVD features reflect the variance better than the conventional SVD feature.

An ANOVA test was carried out with the conventional SVD features and with the MSVD features created by the proposed method. In which the differences and similarities were found better in the calculations done with the proposed technique of modified SVD. The ANOVA test, i.e., significantly separate classes of images are given in Table 7 and significantly revival similarities with in the class even after modification images by means of weed is listed in Table 7. A suitable classifier which can identify the feature of plants from each other will be adopted in our future work.

Comparison of ladies finger and ladies finger with weed seems to be difference from all the comparisons given in Table 7. Table 8 shows the proposed method of intradifference of brinjal leaf images. The average of all top ten SVD value is subtracted from the average of first five SVD values of each sample image.

Table 7 ANOVA comparison of classical and modified-SVD features

Combinations	Conventional method 1	Conventional method 2	Proposed feature method
Ladies finger and ladies finger with weed	0.394	0.343	0.728

Table 8 Shows the proposed method of intradifference of brinjal leaf images

SVD features brinjal images	Difference	Difference	Difference	Difference	Difference
7.5441	3.992	5.751	6.365	6.684	6.866
9.3844	5.832	7.591	8.206	8.524	8.706
8.8838	5.332	7.090	7.705	8.023	8.206
8.7417	5.190	6.948	7.563	7.881	8.063
7.1678	3.616	5.374	5.989	6.307	6.490
7.1177	3.566	5.324	5.939	6.257	6.439
9.6029	6.051	7.809	8.424	8.743	8.925
6.9885	3.436	5.195	5.810	6.128	6.310
8.5647	5.013	6.771	7.386	7.704	7.886
6.6311	3.079	4.838	5.452	5.771	5.953
AVG	4.511	6.269	6.884	7.202	7.384

4 Conclusions

The proposed modified-SVD is deployed to calculate the features and the differences by means of a process specified using only the higher singular values. In this proposed method, we did calculations to form the features by adding those MSVD values from all sample images divided by the number of samples instead of conventional methods that construct the feature by selecting the first '*n*' numbers of SVD values of an image. The proposed method of constructing the feature is by calculating the average of the higher position SVD values. Subsequently, this calculated average is subtracted from the sum of first five SVD values of each sample image. In the ANOVA test, the proposed method of SVD feature confirms better clarity than the conventional features of conventional SVD.

References

1. Bali AS, Batish DR, Singh HP (2017) Phytotoxicity and weed management potential of leaf extracts of *Callistemon viminalis* against the weeds of rice. Acta Physiologiae Plantarum 25, 1–9
2. Paikekari A, Ghule V, Meshram R (2016) Weed detection using image processing. Int Res J Eng Technol 3(3):1220–1222
3. Deepa S, Hemalatha R (2015) Weed detection using image processing. Int J Res Comput Appl Robot 3(7):29–31
4. Lin Fenfang, Zhang Dongyan, Huang Yanbo (2017) Detection of corn and weed species by the combination of spectral, shape and textural features. Sustainability 9:1335
5. Choudhary J, Nayak S (2016) A survey on weed detection using image processing in agriculture. Int J Comput Sci Eng 4(6):97–100
6. Siddiqi MH, Lee S, Lee Y-K (2011) Efficient algorithm for real-time specific weed leaf classification system. J Commun Comput 891–830
7. Prakash K, Samraj A (2017) Tool flank wears estimation by simplified SVD on emitted sound signals. IEEE conference on emerging devices and smart system. IEEE, India
8. Ramesh K, Samraj A (2017) Weed growth and intrusion detection by multi angle images of vegetable plants using modified SVD. In: The 8th international conference on information technology IEEE conference publications. IEEE, Jordan
9. Symonds P, Paap A, Alameh K (2015) A real-time plant discrimination system utilizing discrete reflectance spectroscopy. Comput Electron Agric 117:57–69
10. Malemath VS, Hugar SM (2016) A new approach for weed detection in agriculture using image processing techniques. Int J Adv Sci Tech Res 3(6):356–359
11. Barrero O, Rojas D, Gonzalez C (2016) Weed detection in rice fields using aerial images and neural networks. In: XXI symposium on signal processing, images and artificial vision (STSIVA). IEEE, Bucaramanga, Colombia
12. Samraj A, Sayeed S, Raja JE (2011) Dynamic clustering estimation of tool flank wear in turning process using svd models of the emitted sound signals. Int Sch Sci Res Innov 5(8):1644–1648
13. Ramesh K, Samraj A (2017) Identification of weed growth and intrusion in plant beds by modified singular value decomposition of areal sensory images. In: IEEE conference on emerging devices and smart system. IEEE, India
14. Tits L, Keersmaecker WD, Somers B (2012) Hyperspectral shape-base dunmixing to improve intra- and interclass variability for forest and agro-ecosystem monitoring. ISPRS J Photogramm Remote Sens 74:163–174

Old Handwritten Music Symbol Recognition Using Radon and Discrete Wavelet Transform

Savitri Apparao Nawade, Rajmohan Pardeshi, Shivanand Rumma, and Mallikarjun Hangarge

Abstract Optical music symbol recognition deals with the conversion of scanned musical symbols into digital readable form. Though existing techniques have achieved considerable result for printed musical symbols, for handwritten musical symbol, there is a lot of room for research as it comes with several challenges such as degradation, skew, and non-uniformity of symbols. In this paper, we have presented discrete wavelet transform and radon transform to extract directional multi-resolution features from the input musical symbols. Further, k-NN classifier is used to classify the music symbols and achieved an encouraging accuracy of classification as 95.04% with ten-fold cross-validation.

Keywords Cross-validation · Discrete wavelet transform · Radon transform · Nearest neighbor classifier

1 Introduction

Optical music symbol recognition aims at automatic reading of music symbols by means of computer. It is easy to recognize the printed music symbols by a computer, whereas recognition of old handwritten music symbols is a difficult task due to variations in writing styles, instrument used to write the symbols, height, and width of the symbols, thickness of the symbols, mode of the writer, degradation due

S. A. Nawade (✉) · S. Rumma
Department of PG Studied and Research in Computer Science,
Gulbarga University, Kalaburagi, India
e-mail: savitri.warad@gmail.com

S. Rumma
e-mail: madhurrajmohan1@gmail.com

R. Pardeshi · M. Hangarge
Department of PG Studies and Research in Computer Science,
Karnatak College, Bidar, India
e-mail: mhangarge@yahoo.co.in

N. N. Chiplunkar and T. Fukao (eds.), *Advances in Artificial Intelligence and Data Engineering*, Advances in Intelligent Systems and Computing 1133,
https://doi.org/10.1007/978-981-15-3514-7_86

to aging of the paper, etc. All these challenges make the old handwritten music symbols recognition problem more difficult compared to simple handwriting or printed symbol recognition. This paper presents a multi-resolution feature-based old handwritten music symbols recognition system.

The remaining part of the paper is as follows. Section 2 deals with related research work; in Sect. 3, research methodology is presented, Sect. 4 contains experimental findings and discussion, and in Sect. 5, conclusion is drawn.

2 Related Work

During the last two decades, music symbol recognition has received huge attention. In this section, we have summarized the related work reported for music symbol recognition. Fornes et al. [1] introduced a system which recognizes the old handwritten music symbols which belong to 17–19 century, on applying dynamic time warping method, and compared the performance of Zernike moments with DTW and observed that Zernike moments are giving very poor results. Baro-Mas [2] presented a method which recognizes both online and offline handwritten music symbols. They have used CVC-MUSICMA dataset for experimentation which consists of single and composite music symbols. To recognize single music symbols, Blurred Shape Model (BSM) and Zoning feature extraction methods were used and applied a k-NN classifier. They have observed that Zoning of music symbols was useful for recognition of printed music sheet. They removed long ties and braces in the composite music symbols on applying staff line estimation, calculated the aspect ratio of aligned long joined components, and then, they segmented symbols by considering vertical projection of the symbol. Oh et al. [3] worked on the recognition of online handwritten music symbols by considering HOMUS database which consists of strokes and music symbols. Applied stroke classification to classify the strokes based on three features viz histograms of directed, undirected movement angles and the music symbol size. Calvo-Zaragoza et al. [4] introduced a system to recognize the music symbols both online and offline. In offline method, they have tested on their own dataset of 3800 music symbols written by 32 music composers, and for online, they have used e-pen and tablet for digitizing the music symbols. And after comparing both online and offline methods, they observed that offline method achieves higher accuracy. However, the authors have not reported the reason for getting high recognition accuracy for offline symbols compared to online music symbols. Pugin et al. [5] compared both printed and handwritten music symbol recognition by considering GAMERA and ARUSPIX and also compared the performance of all classifiers such as SVM, k-NN, and HMM and observed higher accuracy for printed music symbols. Due to the variations in scale, thickness, shape of handwritten symbols yielded lower results. Chanda et al. [6] worked on offline handwritten music symbols. They have used structure shape descriptors and texture analysis-based features. They have compared the performance of modified quadratic discriminant function (MQDF) and support vector machine (SVM) and obtained the highest accuracy for

MQDF classifier. Nawade et al. [7] worked on projection profiles to recognize the old handwritten music symbols and achieved an accuracy of 95.72.%. Nawade et al. [8] reported directional multi-resolution spatial features which recognize the old handwritten music symbols and obtained 98.16% accuracy. Miyo and Maryuyum [9] introduced an online handwritten music scores recognition system. They considered directional strokes and series of data as features of the image, and on applying SVM classifier, they achieved a higher accuracy of 97.60%. Raphael et al. [11] introduced a top model and template matching methods to recognize the isolated music symbols.

3 Proposed Methodology

In this paper, proposed a three steps recognition system of handwritten music symbols. The three steps are preprocessing, feature extraction, and classification of the symbols. Feature extraction is carried out based on combining discrete wavelet transform and radon transform features. Old handwritten music symbols are then classified using k-NN classifier. The proposed method flowchart is given in Fig. 1.

Preprocessing: In preprocessing, usually input image is binarized with Otsu method, and normalization of the image is done; after that, some morphological operations are applied on the image. But in our case, already input images are binarized and isolated ones. Therefore, we have not implemented any preprocessing algorithm here.

Feature Extraction: Feature extraction is very significant step, to compute the meaningful information from raw music symbols, we have used a frequency-based radon transform technique, to compute directional energies, these are nothing but projections (sum of the lines) along any arbitrary line on the xy plane or two dimensional spaces. It is normally represented as $g(\rho, \theta)$ and is used to reconstruct the image. When this is applied to any input image, in our case the music symbol, then it

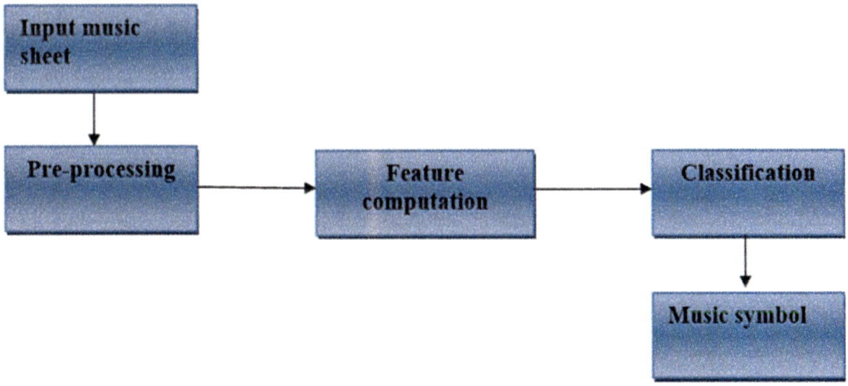

Fig. 1 Proposed method flowchart

will reconstruct the image by collecting all the back projections (lines) of the image. Normally, mathematical expression for any straight line is represented as

$$a\cos\theta + a\sin\theta = \rho \tag{1}$$

where a and b represent x and y-axis. Here, we have to collect all the lines and integrate them, by changing ρ value and keeping θ constant, this can be mathematically represented as

$$g(\rho_j, \theta_k) = \iint_{-\infty}^{\infty} f(a, b)\delta(a\cos\theta_k + b\sin\theta_k - \rho_j) \tag{2}$$

Then, θ value has to be changed and should be put in the same line, after computing all the projections, a final Radon transform equation can be represented as

$$g(\rho_j, \theta_k) = \iint_{-\infty}^{\infty} f(a, b)\delta(a\cos\theta + b\sin\theta - \rho) \tag{3}$$

After computing the directional energies by applying radon transform, these radon coefficients are fed to discrete wavelet transform and further decomposed in three sub-bands which encode the input information in three directions such as vertical, horizontal, and diagonal. Another one is approximation. By performing multi-resolution analysis, we obtained coefficients, and these coefficients are called as RWT coefficients [12]. In this method, DWT $Daubechies_9$ is used for multi-resolution analysis. We have applied four-level decomposition to radon energy coefficients which gives four DWT sub-bands at each level. In this way, we got totally $4 * 4 = 16$ RWT sub-bands.

Fixed-size feature vector is computed by calculating entropy (E) and standard deviations (S) from each RWT sub-bands. In this work, we have used totally 32 features = 16 sub-bands * 2 statistical measures. Entropy and standard deviations are presented as below:

$$E = -\sum p(r_i)\log_2 p(r_i),\ S = \sqrt{\frac{1}{N-1}\sum (K_{mn} - \overline{K})^2}$$

Where r_i is the variable of discrete radon and denotes the RWT coefficients, and $P(r_i)$ is its probability of co-occurrence, and RWT coefficients and their mean are denoted by K_{mn} and $\overline{K}$, respectively. More details on combination of radon and wavelet transform-based directional multi-resolution features are given in [10, 12]

Nearest Neighbor Classifier: To do the classification of old handwritten music symbols, here we are using a simple NN classifier. Let us assume that $B(b_i \ldots b_m)$ is used as training pattern vector, and $D(d_i \ldots d_m)$ is used as testing pattern vector. The basic process for classification is summarized in following steps:

- Calculate the city block distance between the test pattern and each of its neighbors.
- Assign the test pattern to the class of its nearest neighbor.

City block distance between two patterns is given by:

$$d_{\text{cityblock}} = \sum_{i=1}^{M} |b_i - d_i| \tag{4}$$

4 Results and Discussion

Our proposed method was focused to recognize isolated old handwritten music symbols, and dataset consist of a total of 4098 music symbols in which 2128 are clefs and 1970 are accidentals symbols such as accidentals of four types: Accidental Doub sharp, Accidental Flat, Accidental Natural, and Accidental sharp and three clef types: Clef Alto, Clef Bass, and Clef Treble. Samples from the dataset are shown in Fig. 2. We have applied k-NN classifier with the $k = 3$ on the dataset. The city block distance measure is used by the K-NN classifier to classify the symbols based on the directional multi-resolution features. The performance of proposed methodology is evaluated with ten-fold cross-validation. In Table 1, recognition accuracy of different music symbols is given. From the table, we can observe that Accidental Doub sharp has achieved the higher accuracy 99.59%, whereas Accidental sharp achieved the lower one 83.81%. For better understanding, we have given the confusion matrix in Table 2. We have compared our approach with existing method in Table 3. Our method outperforms over the existing method presented in [1].

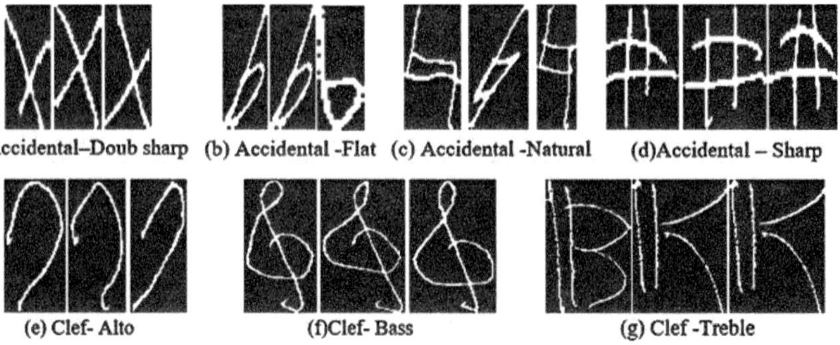

Fig. 2 Samples of old handwritten music symbols

Table 1 Recognition accuracy of music symbols of seven classes

Class	Music symbol	Recognition accuracy (%)
1	Doub Sharp_Accidental	99.59
2	Flat_Accidental	94.01
3	Natural_Accidental	93.43
4	Sharp_Accidental	83.81
5	Alto_Clef	93.95
6	Bas_Clef	98.00
7	Treble_Clef	98.53

Table 2 Confusion matrix

Class no.	Class 1	Class 2	Class 3	Class 4	Class 5	Class 6	Class 7
1	496	0	0	0	0	1	0
2	4	487	20	6	0	0	0
3	1	15	441	14	0	0	0
4	1	28	45	404	3	0	0
5	0	0	3	9	715	17	15
6	0	1	0	0	7	540	1
7	0	0	0	0	10	2	808

Table 3 Comparison with similar work

Method	Dynamic time warping	Our method
Accuracy (%)	89.55	95.04

5 Conclusion

In this study, we have employed combined radon and discrete wavelet transform on isolated old handwritten music symbols to get directional multi-resolution information of the music symbol which is used to classify the music symbols using k-NN and achieved an accuracy of 95.04%. In future, we will concentrate on recognition of combined musical symbols with different approaches. We also aim to classify the handwritten and printed music symbol sheets.

Acknowledgements We acknowledge the Alicia Fornes, Pattern Recognition and Document Analysis Group, Computer Vision Center, Universitat Autnoma de Barcelona, 08193, Bellaterra, Barcelona, Spain, for creating this dataset.

References

1. Fornes A, Llados J, Sanchez G (2007) Old handwritten musical symbol classification by a dynamic time warping based method. In: International workshop on graphics recognition. Springer, Berlin, Heidelberg, pp 51–60. https://doi.org/10.1007/978-3-540-88188-9_6
2. Baro-Mas A (2015/16) Recognition of handwritten music scores. TFGEN Enginyeria Informatica Escola Enginyeria (EE) Universitat Autonomade Barcelona (UAB)
3. Oh J, Son SJ, Lee S, Kwon J-W, Kwak N (2017) Online recognition of handwritten music symbols. Int J Doc Anal Recogn (IJDAR) 79–89. https://doi.org/10.1007/s10032-017-0281-y
4. Calvo Zaragoza J, Oncina J, Inesta JM (2013) Recognition of online handwritten music symbols. In: International workshop on machine learning and music, ECML/PkDD, Prague
5. Pugin L, Hockman J, Burgoyne JA, Fujinaga I (2008) Gamera versus Aruspix—Two optical music recognition approaches. In: ISMIR 2008 session 3COMR, alignment and annotation. http://citeseerx.ist.psu.edu/viewdoc/summary?doi=10.1.1.154.4765
6. Chanda S, Das D, Pal U, Kimura F (2014) ICFHR offline hand-written music symbol recognition. In: 14th International conference on frontiers in handwritting recognition (ICFHR). IEEE, pp 405–410. https://doi.org/10.1109/ICFHR.2014.74
7. Nawade SA, Pardeshi R, Dhawale C, Hangarge M (2018) Old handwritten music symbol recognition using the combination of foreground and background projection profiles. In: 3rd International conference on Internet of Things: smart innovation and usages (IoT-SIU). IEEE. https://doi.org/10.1109/IoT-SIU.2018.8519881
8. Nawade SA, Hangarge M, Dhawale C, Reaz MBI, Pardeshi R, Arsad N (2018) Old handwritten music symbol recognition using directional multi-resolution spatial features. In: International conference on smart computing and electronic enterprise (ICSCEE), pp 1–4. https://doi.org/10.1109/ICSCEE.2018.8538370
9. Miyao H, Maruyama M (2007) An online handwritten music symbol recognition system. IJDAR 9(1):4958. https://doi.org/10.1007/s10032-006-0026-9
10. Jadhav DV, Holambe RS (2009) Feature extraction using Radon and wavelet transforms with application to face recognition. Neurocomputing 72:1951–1959. https://doi.org/10.1016/j.neucom.2008.05.001
11. Raphael C, Wang J (2011) New Approach to optical music recognition. In: @12th International society for music information retrieval conference (ISMIR 2011)
12. Veershetty C, Pardeshi R, Hangarge M, Dhawale C (2018) Radon and wavelet transforms for handwritten script identification. In: Ambient communications and computer systems. Springer, Singapore. pp 755–765. https://doi.org/10.1007/978-981-10-7386-1_63

Gender Recognition from Face Images Using SIFT Descriptors and Trainable Features

Sneha Pai and Ramesha Shettigar

Abstract Nowadays, recognition of gender from facial image became an important problem in computer vision, security, verbal–nonverbal communication and human–computer interaction application. Recognition of gender is a challenging research problem because facial image contains many information such as gender, facial expression, age, ethnic origin in computer-aided applications, based on the facial image quality gender recognition depends. In this paper, a new gender recognition method is proposed that combines both scale-invariant feature transform (SIFT) as domain-specific approach and combination of shifted filter responses (COSFIRE) as trainable features. The proposed method will give better performance of variation in poses, different expressions and changes in illumination condition. This method tested by taking gender face recognition technology (GENDER-FERET) results shows that GENDER-FERET dataset will give better result in various illumination conditions. COSFIRE algorithm which is used for visual recognition problem will work better in given trainable character. The advantage of using SIFT is it will not get affected by changes in scale, blur, rotation, change in illumination and affine transformation.

Keywords Scale-invariant feature transform · Combination of shifted filter responses · Trainable feature · Gender recognition · Gender face recognition technology

1 Introduction

Nowadays, gender recognition is considered interesting area of research. Since face is an important part of human body, looking at face human can recognize gender, age [1], ethnicity [2], emotions and state of mind [3] but recognition using computer

S. Pai (✉) · R. Shettigar
NMAM Institute of Technology, Nitte, Udupi, Karnataka 574110, India
e-mail: snehapai188@gmail.com

R. Shettigar
e-mail: rameshshettigar@nitte.edu.in

N. N. Chiplunkar and T. Fukao (eds.), *Advances in Artificial Intelligence and Data Engineering*, Advances in Intelligent Systems and Computing 1133,
https://doi.org/10.1007/978-981-15-3514-7_87

is a difficult task because all faces consist of same specific arrangements (nose, eyes and others are in the same position) and also affected by variation in poses, usage of occlusions such as scarves, hats and glasses, and also it is affected by illumination conditions. Gender classification has many uses such as person face recognition, and human and computer interfaces, and it also improves performance of many applications. Since 1990, studies are going on to recognize gender of human faces.

Recently, many new varieties of appeal motivated researchers to work on automatic gender recognition input face from images and videos. Since gender recognition is a challenging problem [5], there is a great demand to perform face recognition and input from camera and video. Main disadvantage of this system is it requires more time because reference database has thousands of samples stored from this and finding match to input image requires more compilation time. We can overcome this problem by grouping gender based on age group and then based on properties of database comparing input image only in that group of datasets; using this method, we can reduce search space to some extent.

Figure 1 represents problems related to gender recognition systems. First row shows problems related to variation in poses, second row shows problem related to different occlusions used, and third row shows variation in illumination condition. Using face alignment algorithm [4], normalization of pose is done; this method can solve problems related to variation in poses to some extent. Based on varies surveys

Fig. 1 Main disadvantages of gender recognition systems

conducted from last few years, many methods were produced to solve gender recognition [5]. It is observed that automatic gender recognition not reached generalization capability yet.

The main motivation of our research is to recognize gender from face images by combining SIFT descriptors and trainable features. Using the proposed algorithm, recognition of multiple expressions can be done. To evaluate performance, we use GENDER-FERET database, which works better in illumination condition.

2 Literature Survey

Hatipoglu and Kose [6] proposed new gender recognition algorithm by combining two algorithms; they are speed-up robust feature (SURF)-based bogs of visual words and support vector machine algorithm. This algorithm is tested on face to see the efficiency of the proposed method. FERET dataset contains 3560 samples of frontal, left and right faces. The proposed method can identify good gender recognition on FERET database, and recognition rate of left and right is lower compared to frontal images.

Jain et al. [7] discussed how gender classification can be done using frontal facial images. The proposed system is more efficient compared to existing gender recognition system. For the experiment, the authors have taken FERET database which has 500 images where there are 250 males and 250 females. Each image feature vector in low-dimensional space is detected using independent component analysis (ICA). Experiment shows that using SVM algorithm in ICA space will achieve 96% accuracy; this is better than existing classifiers.

Nguyen and Huong [8] discussed human face which contains more discriminative feature vector, so it is a critical task to do gender classification. By using input image of the face, they were able to generate smaller version of original image and then apply LPQ operator on both images. For the extracted image, they found global facial description by combining the features of LPQ and SIFT. Then, they used SVM classifier to find gender. They used LFW and Adience databases. By using LFW and Adience databases, they obtained gender recognition accuracy rate of 96.51% and 80.5%, respectively.

Anusha et al. [9] discussed LDA, PCA and LBP algorithms; these can detect expression from face and also recognize gender from the image. Main problem of existing system is it will fail to recognize image due to low-resolution images, variation in expression, etc. In this paper, gender classification can be done using facial patches and to evaluate performance Cohn–Kande and JAFFE databases are used.

Ben Abdelkader and Griffin [10] discussed how gender classification is done using face images. Here to overcome problem such as variation in poses and illumination changes, the authors used local region analysis which extracts gender classification features by matching face of N local regions against face images of M fixed regions; this can be done using FaceIt algorithm. Dimensionality of MN-dimensional vector can be reduced using Karhunen–Loeve transform method. Here, the authors implemented holistic feature extraction method for comparison purpose. Gender classification is performed on SVM and FLD. This method is tested on cross-validation database which has 13,000 frontals as well as nearly frontal face images. They obtained gender recognition rate of 94.2%.

Akbulut et al. [11] discussed how gender recognition came into existence due to the rapid advancement in social media and mobiles. Main application of gender recognition is identifying face and expression from the face. In this paper, the authors have used LRF-ELM and CNN deep learning methods for gender classification. The test is conducted on face dataset generated for the recognition of age and gender. Gender recognition rate using ELM is 80% and CNN is 87.13%.

Ferizal et al. [12] have discussed use of PCA combined with LDA for human face recognition. In the first stage, the authors used processing technique such as resizing and equalizing the histogram, and in the second stage removing of variation in background from the image is done by adding oval masking. To improve performance of PCA + LDA method furthermore, they used 9 nearest neighbors' classification process which gives better gender recognition. Result shows that with face masking 89.70% gender recognition rate is obtained and without adding face masking 84.16% gender recognition rate is obtained.

Ting et al. [13] presented how gender recognition is done by looking at the camera; the first stage is image processing; here, they extracted many facial features from the image. Then, it is passed to the rule-based classifier. This technique is applied to static passport-like photographs and video sequences. They used database which has bald men and females wearing headscarf. Result shows by using static face images, gender recognition for females is 100% and 98% for males. For video sequences, they obtained gender recognition rate of 87.5% for females and 70% for males.

Shan [14] presented a new method in which gender recognition is done by taking real-life faces from labeled face wild database. Here, they described faces using LBP and AdaBoost method used to select discriminative features. Then, they classified gender by using SVM method along with the LBP features and obtained 94.81% gender recognition rate.

Lian and Lu [15] presented a new method called multi-view gender classification which is used to represent facial images by taking shape and texture information. In the first stage, they divide face area into small regions; using this, they extracted LBP histogram to form single vector. Based on idea of LBP, the authors proposed a new approach called multi-resolution LBP which uses SVM for classification. For the experiment, they used CAS-PEAL database by which they obtained 95.78% accuracy.

Azzopardi et al. [16] proposed a new algorithm for gender recognition based on COSFIRE filters. COSFIRE filters are trainable, and based on given prototype, automatic configuration is done. To show effectiveness of the proposed method, the authors used GENDER-FERET dataset which consists of 474 tests as well as training samples. They obtained gender recognition rate of 93.7%.

Azzopardi et al. [17] discussed that identifying of gender from face images is an important application which is affected by pose variation, illumination changes and different expressions which makes gender recognition challenging for computer systems. In this paper, gender recognition is done by combing trainable shape and color features. In particular, by means of trainable COSFIRE filters, the proposed method fuses edge and color blob-based features. For the classification, the authors used SVM classification model. Also, they used GENDER-COLORFERET dataset using which they obtained 96.4% of accuracy.

Sushama and Rajinikanth [18] discussed detection of human face from images. In this paper, face recognition is done using SIFT and DRLBP. The first stage is preprocessing, and then feature of face is extracted by using SIFT algorithm. These features can be utilized to detect face using back-propagation network (BPN). Result shows that combining SIFT and DRLBP gives more accurate result rather than using them separately.

Dileep and Danti [19] discussed that face is an important part of human body; by looking at the face, a person can recognize gender, age and identity but by using computer vision it is difficult because it is affected by different illumination conditions. In this paper, human age and gender classification are done by taking feed-forward propagation neural network algorithm. On the output of neural network classifier, multiple hierarchical decision is applied; using this, performance of system can be further improved. To test efficiency of the proposed method, they conducted experiment on benchmark database using which they obtain accuracy of 95%.

Zhao et al. [20] made use of SVM and blocking LBP method for gender recognition. In this method, face image is divided into several blocks and using LBP histogram, features are extracted and it is cascaded to form face feature vector. Using this feature vector, gender recognition is carried out using SVM. Result shows that the proposed method will give higher accuracy.

Çoban and Gokmen [21] proposed local Zernike moment method. It is a new feature extraction method for the gender recognition. Here, they combined LZW and LBP feature extraction, and for the gender classification they used SVM. To test efficiency of the proposed method, they conducted experiment on FERET and LFW databases. FERET database yields 99.57% accuracy, while LFW database yields 97.71% accuracy.

Cerkezi and Topal [22] proposed a gender recognition method based on uniform local binary pattern (LBP) algorithm. LBP descriptor is more efficient in pattern recognition applications. Using LBP, features are extracted from the image, and then face alignment is done. Then to create final feature vector, local uniform LBP histograms are combined. To evaluate the performance of the proposed system, experiment is conducted on FERET database which yields 93.57% accuracy.

Camalan and Şengul [23] discussed gender prediction using face images. LBP features are extracted from face. KNN and DA methods are used to classify gender from face images. For better performance, images are cropped, then features are extracted from image, and then classification algorithm is applied on these images. To evaluate the performance of the proposed system, they conducted experiment on FERET database which consists of 530 female and 731 male images. Using KNN and DA methods, they achieved 82.47 and 77.16% of gender recognition rate, respectively.

3 Proposed Methodology

Figure 2 represents system design of the proposed system. Preliminarily, we take input image from database and apply Viola–Jones algorithm [24]; using this, we can detect the key point from the image. Using these key points, we rotate the image horizontally and resize image to 128 × 128 pixels by applying face alignment algorithm. Then, COSFIRE algorithm is directly applied on dataset. But before applying SIFT, we have to detect key point from the image such as nose, mouth and eye. Then, we combine the output of SIFT and COSFIRE algorithm and then apply SVM classifier which classifies gender from database. We provide more information about above components in the next sections.

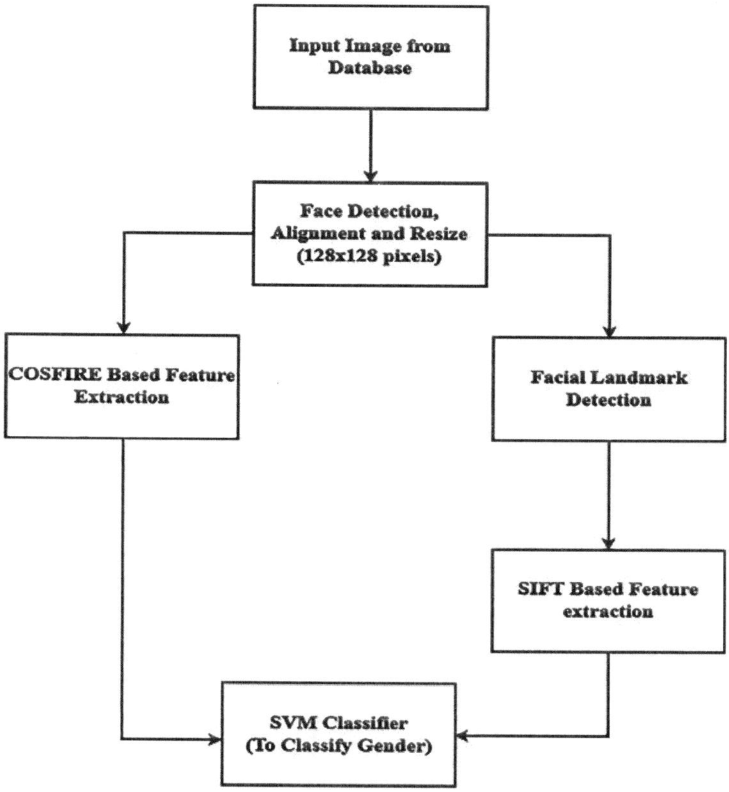

Fig. 2 System design of the proposed method

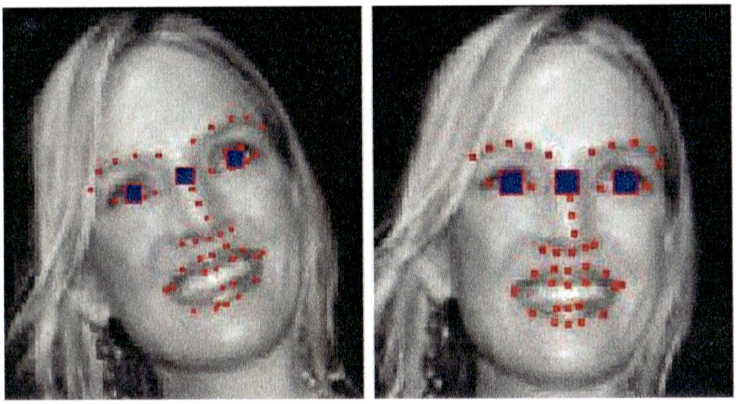

Fig. 3 Proposed face alignment algorithm

3.1 Detection of Face and Its Alignment

At the first stage for the given input image, Viola–Jones algorithm is applied [24]. It does not require any preliminary preprocessing filters. It will detect face in different scales. Later, detect key points from the face and also average location of left eye, right eye and center of left and right eye. Using these points, rotate the image horizontally.

In Fig. 3, red dots indicate key points from the face and blue dot indicates left eye, right eye and center of left and right eye; using these points, rotate the image horizontally and then resize the image to 128 × 128 pixels by cropping.

3.2 COSFIRE-Based Feature Extraction

The following sections show the brief explanation of four stages involved in COSFIRE approach. For further technical details on COSFIRE filters, we refer the reader to read [16].

3.2.1 Configuration Process of COSFIRE Filter

Configuration process of COSFIRE filter is an automatic process. In the first stage, by using different scales and orientations for particular image, we apply bank of Gabor filters and place their response. Second, at a point of interest, by using number of concentric circles around it, we can obtain local maxima Gabor responses by determining positions along these circles. To achieve maximum response for these points, we have to calculate four parameters; they are λ, θ, ρ and φ where λ is scale, θ is orientation, ρ is distance and φ is polar angle. Later to obtain local maximum

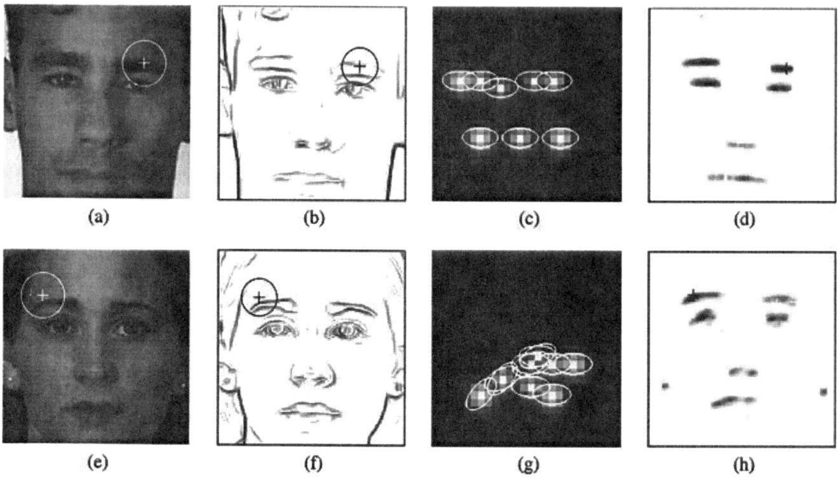

Fig. 4 Configuration process of two COSFIRE filters

Gabor responses, we combine these four parameters by using Eq. (1).

$$\text{Sf} = \{(\lambda_i, \theta_i, \rho_i, \varphi_i) \mid i \in 1 \dots n\} \tag{1}$$

Figure 4 represents configuration procedure of two COSFIRE filters. In this figure, eyebrows are considered as prototype patterns for male and female images. In the figure, (a) and (e) represent training image of size 128 × 128 pixels. (b) and (f) represent bank of Gabor filters and its corresponding inverted response. (c) and (g) represent structures of the COSFIRE filters where circle is prototype patterns selected. (d) and (h) are the inverted response map of the concerned COSFIRE filter.

3.2.2 Response of COSFIRE Filter

It has four stages. In the first stage using set Sf, we have to find unique pairs of parameters such as scale (λ) and orientations (θ) and Gabor linear filtering is applied to these parameters. The second stage is blurring. Gabor response map is blurred using Gaussian function. In the third stage by utilizing polar vector, each Gabor response which is blurred is shifted so that it supports the center. The final stage is multiplication which comprises all the Gabor responses which are blurred and shifted and are combined using geometric mean function. Figure 4d, h represents the response maps of COSFIRE filters.

3.2.3 Forming a Feature Descriptor

In this stage for each face image, the collection of COSFIRE filter are applied. By using spatial pyramids which has three levels, we can obtain COSFIRE filter responses. In level zero which has single tile, we calculate global maximum responses. Level one has 4 tiles. We have to take global maximum responses for each of the 4 tiles. Level two has 16 tiles. We have to take global maximum responses for each of the 16 tiles. For k COSFIRE filters, it has 21 tiles of spatial pyramids $(1 + 4 + 16 = 21)$. These 21 k-element feature vectors are used to train SVM classification model.

Figure 5 shows application of the two COSFIRE filters. In the figure, (b, f) represent COSFIRE filter response maps. (c, g) represent zero level which has single tile. (d, h) represent level one which has 4 tiles. (e, i) represent level two which has 16 tiles. (j) represents resulting face descriptor. For level zero, we apply COSFIRE filter 1 and COSFIRE filter 2 one time. For level one, we apply COSFIRE filter 1 and COSFIRE filter 2 four times. For level two, we apply COSFIRE filter 1 and COSFIRE filter 2 sixteen times.

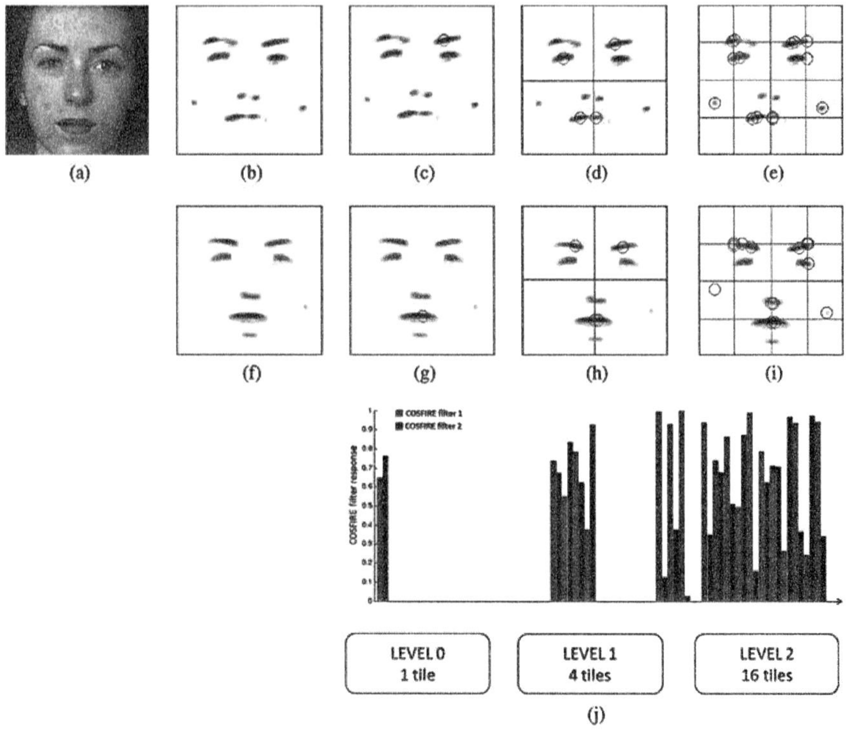

Fig. 5 Application of the two COSFIRE filters

3.3 SIFT-Based Feature Extraction Method

In this stage, key points are detected from the image. Then, eliminate key points which are not required. Then, compute the best orientation for each key point region. Finally, compute the SIFT descriptors and later apply SVM classifier.

3.4 Classification Model

Here, we combine the decision made by COSFIRE and SIFT algorithm and apply SVM classifier. Here, we used support vector machine classification model with the following chi-squared kernel Eq. (2).

$$K(x_i, y_i) = \frac{(x_i - y_i)^2}{\frac{1}{2}(x_i + y_i) + \varepsilon} \tag{2}$$

where x_i and y_i represent descriptor, i and j are training images and ε represents very small value 1.

4 Experimental Result

The following section provides information about datasets used and also experimental result for the proposed system.

4.1 Dataset

To calculate the performance of the proposed system, the experiment is conducted on GENDER-FERET [22] dataset which is created based on FERET dataset. It consists of 474 training images which have 237 male and female images each and 472 testing images which have 236 male and female images each. This database will give better result in various illumination conditions. Figure 6 shows different illumination conditions involved in GENDER-FERET dataset.

4.2 Experiments

In the following section, we analyze result of our proposed technique on GENDER-FERET dataset. First, we check result of SIFT-based and COSFIRE-based algorithms

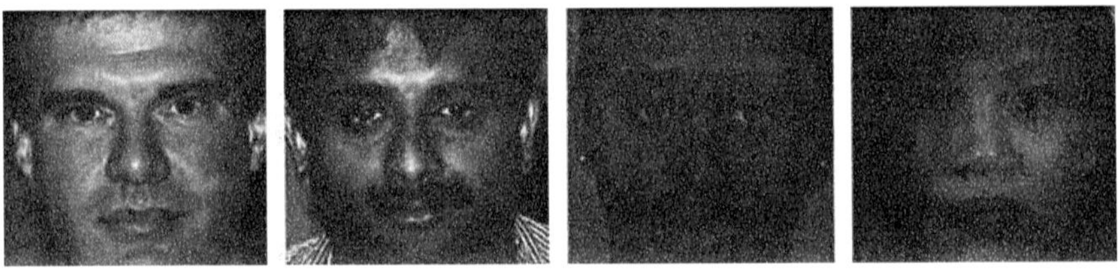

Fig. 6 A pair of images of face from GENDER-FERET database

Table 1 Results of COSFIRE-based technique on GENDER-FERET datasets

Dataset (GENDER-FERET)				Recognized image	Technique	Accuracy (%)
Training images		Test images		442	COSFIRE-based	93.64%
Male	Female	Male	Female			
237	237	236	236			

separately; then, we show result of fusing COSFIRE-based and SIFT-based classifiers. For the experiments below, we use GENDER-FERET dataset consisting of 474 training images which have 237 male and female images each and 472 testing images which have 236 male and female images each.

4.3 *Outcome of COSFIRE-Based Classifier*

For the COSFIRE technique, same procedure is used reported in [16]. COSFIRE filter has 180 COSFIRE filters where 90 male and 90 female filters are used. Table 1 represents the results of COSFIRE-based technique. Using this technique, we obtained accuracy of 93.64%. It was able to recognize 442 images out of 472 images.

4.3.1 Outcome of SIFT-Based Method

SIFT algorithm does not require configuration parameter. Table 2 shows the results

Table 2 Results of SIFT-based methods on GENDER-FERET dataset

Dataset (GENDER-FERET)				Recognized image	Technique	Accuracy (%)
Training images		Test images		447	SIFT-based	94.74%
Male	Female	Male	Female			
237	237	236	236			

Table 3 Results of fusion technique on GENDER-FERET dataset

Dataset (GENDER-FERET)				Recognized image	Fusion technique	Accuracy (%)
Training images		Test images		454	COSFIRE-SIFT-based	96.14%
Male	Female	Male	Female			
237	237	236	236			

of SIFT-based technique. Using this method, we obtained accuracy of 94.74%. It was able to recognize 447 images out of 472 images.

4.3.2 Fusing SIFT- and COSFIRE-Based Methods

In the below section, we examine the fusion result of two algorithms that are COSFIRE-based and SIFT-based classifiers. In order to make decision, we use SVM classification method. Experiment shows that the proposed method will give more accuracy. Table 3 represents the results of fusion technique. Using this method, we can obtain accuracy of 96.14%. It was able to recognize 454 images out of 472 images.

4.3.3 Comparison Between Existing and Proposed Methods

Here, we compared result of the proposed method and existing method based on GENDER-FERET dataset. Table 4 shows comparison result of the proposed and existing methods. Azzopardi et al. [25] made use of RAW LBP HOG method using which 92.6% gender recognition rate is obtained. Azzopardi et al. [16] made use of COSFIRE method using which 93.7% gender recognition rate is obtained. Azzopardi et al. [26] made use of COSFIRE SURF method using which 94.7% gender recognition rate is obtained. Finally, the proposed method is made use of fusing SIFT and COSFIRE algorithms; using this method, 96.1% gender recognition is achieved.

Table 4 Proposed method compared with existing method using GENDER-FERET dataset

Technique	Description	Accuracy (%)
Azzopardi et al. [25]	RAW LBP HOG	92.6
Azzopardi et al. [16]	COSFIRE	93.7
Azzopardi et al. [26]	COSFIRE SURF	94.7
Proposed	COSFIRE-SIFT	96.1

5 Conclusion and Future Works

In this paper, gender recognition from face images is done by fusing SIFT descriptor and trainable feature. Here, SIFT is domain approach and COSFIRE is a trainable feature. COSFIRE algorithm which is used for visual recognition problem will work better in given trainable character. Advantage of SIFT is it will not have affected changes in the scale, blur, rotation, change in illumination and affine transformation. The proposed method will give better performance in variation poses, different expressions and changes in illumination condition. The results show that combining both SIFT-based and COSFIRE-based algorithms will give better result using GENDER-FERET database, i.e., 96.14% accurate than the existing method.

Future work of our proposed method can be improved by using different datasets and applying different algorithms which can further improve gender recognition rate. Automatic face recognition is not up to the mark which can be improved by combing different algorithms. Also, using face sketch dataset can perform gender recognition.

References

1. Fu Y, Guo G, Huang TS (2010) Age synthesis and estimation via faces: a survey. IEEE Trans Pattern Anal Mach Intell 32:1955–1976. https://doi.org/10.1109/tpami.2010.36
2. Ng M, Ciaramitaro V, Anstis S et al (2006) Selectivity for the configural cues that identify the gender, ethnicity, and identity of faces in human cortex. Proc Natl Acad Sci 103:19552–19557. https://doi.org/10.1073/pnas.0605358104
3. Rolls ET, Baylis GC, Hasselmo ME (1989) The role of expression and identity in the face-selective responses of neurons in the temporal visual cortex of the monkey. Behav Brain Res 32:203–218. https://doi.org/10.1016/s0166-4328(89)80054-3
4. Murphy-Chutorian E, Trivedi MM (2009) Head pose estimation in computer vision: a survey. IEEE Trans Pattern Anal Mach Intell 31:607–626. https://doi.org/10.1109/tpami.2008.106
5. Ng C, Tay Y, Goi B (2012) Recognizing human gender in computer vision: a survey. In: Lecture notes in computer science, pp 335–346. https://doi.org/10.1007/978-3-642-32695-0_31
6. Hatipoglu B, Kose C (2017) A gender recognition system from facial images using SURF based BoW method. In: 2017 International conference on computer science and engineering (UBMK), pp 989–993. https://doi.org/10.1109/ubmk.2017.8093405
7. Jain A, Huang J, Shiaofen F (2005) Gender Identification Using Frontal Facial Images. In: 2005 IEEE international conference on multimedia and expo. https://doi.org/10.1109/icme.2005.1521613
8. Nguyen H, Huong T (2016) Unconstrained gender classification by multi-resolution LPQ and SIFT. In: 2016 3rd National foundation for science and technology development conference on information and computer science (NICS), pp 212–217. https://doi.org/10.1109/nics.2016.7725652
9. Anusha A, Jayasree J, Bhaskar A, Aneesh R (2016) Facial expression recognition and gender classification using facial patches. In: 2016 International conference on communication systems and networks (ComNet), pp 200–204. https://doi.org/10.1109/csn.2016.7824014
10. Ben Abdelkader C, Griffin P. A local region-based approach to gender classification from face images. In: 2005 IEEE computer society conference on computer vision and pattern recognition (CVPR'05)—Workshops, p 52. https://doi.org/10.1109/cvpr.2005.388

11. Akbulut Y, Sengur A, Ekici S (2017) Gender recognition from face images with deep learning. In: 2017 International artificial intelligence and data processing symposium (IDAP), pp 1–4. https://doi.org/10.1109/idap.2017.8090181
12. Ferizal R, Wibirama S, Setiawan N (2017) Gender recognition using PCA and LDA with improve preprocessing and classification technique. In: 2017 7th International annual engineering seminar (InAES), pp 1–6. https://doi.org/10.1109/inaes.2017.8068547
13. Ting C, Sheikh U, Abu-Bakar S (2010) Gender estimation based on physiological features of the face. In: 10th International conference on information science, signal processing and their applications (ISSPA 2010), pp 201–204. https://doi.org/10.1109/isspa.2010.5605542
14. Shan C (2012) Learning local binary patterns for gender classification on real-world face images. Pattern Recogn Lett 33:431–437. https://doi.org/10.1016/j.patrec.2011.05.016
15. Lian H, Lu B (2006) Multi-view gender classification using local binary patterns and support vector machines. Adv Neural Netw ISNN 2006:202–209. https://doi.org/10.1007/11760023_30
16. Azzopardi G, Greco A, Vento M (2016) Gender recognition from face images with trainable COSFIRE filters. In: 2016 13th IEEE international conference on advanced video and signal based surveillance (AVSS), pp 235–241. https://doi.org/10.1109/avss.2016.7738068
17. Azzopardi G, Foggia P, Greco A, Saggese A, Vento M (2018) Gender recognition from face images using trainable shape and color features. In: 2018 24th International conference on pattern recognition (ICPR), pp 983–1988. https://doi.org/10.1109/icpr.2018.8545771
18. Sushama M, Rajinikanth E (2018) Face recognition using DRLBP and SIFT feature extraction. In: 2018 International conference on communication and signal processing (ICCSP), pp 994–999. https://doi.org/10.1109/iccsp.2018.8524427
19. Dileep M, Danti A (2016) Multiple hierarchical decision on neural network to predict human age and gender. In: 2016 International conference on emerging trends in engineering, technology and science (ICETETS), pp 1–6. https://doi.org/10.1109/icetets.2016.7603026
20. Zhao H, Gao F, Zhang C (2012) A method for face gender recognition based on blocking-LBP and SVM. In: 2012 2nd International conference on consumer electronics, communications and networks (CECNet), pp 1527–1530. https://doi.org/10.1109/cecnet.2012.6201793
21. Coban B, Gokmen M (2014) Gender classification with local zernike moments and local binary patterns. In: 2014 22nd Signal processing and communications applications conference (SIU), pp 1475–1478. https://doi.org/10.1109/siu.2014.6830519
22. Cerkezi L, Topal C (2018) Gender recognition with uniform local binary patterns. In: 2018 26th Signal processing and communications applications conference (SIU), pp 1–4. https://doi.org/10.1109/siu.2018.8404587
23. Camalan S, Sengul G (2016) Gender prediction by using local binary pattern and K nearest neighbor and discriminant analysis classifications. In: 2016 24th Signal processing and communication application conference (SIU), pp 2161–2164. https://doi.org/10.1109/siu.2016.7496201
24. Viola P, Jones M (2004) Robust real-time face detection. Int J Comput Vis 57:137–154. https://doi.org/10.1023/b:visi.0000013087.49260.fb
25. Azzopardi G, Greco A, Vento M (2016) Gender recognition from face images using a fusion of SVM classifiers. In: Lecture notes in computer science, pp 533–538. https://doi.org/10.1007/978-3-319-41501-7_59
26. Azzopardi G, Greco A, Saggese A, Vento M (2018) Fusion of domain-specific and trainable features for gender recognition from face images. IEEE Access 6:24171–24183. https://doi.org/10.1109/access.2018.2823378

Multiscale Anisotropic Morlet Wavelet for Texture Classification of Interstitial Lung Diseases

Manas Jyoti Das and Lipi B. Mahanta

Abstract Interpretation of various lung textures found in the interstitial lung disease (ILD) is essential for the assessment of the disease, and it constitutes a vital step in diagnosis. In the present study, high-resolution computed tomography images are used in classifying healthy lung texture and five other texture patterns that mostly occur in ILDs. A texture classifier based on anisotropic Morlet wavelet has been developed, and images are analysed with the mathematical model of the Morlet wavelet in various scales, orientations and anisotropic ratios. Support vector machine (SVM) as classifier is used, and the input feature vector for the SVM is created from the quantised values of wavelet coefficients. Quantisation of the wavelet coefficients is done by calculating Rényi's entropy and estimating the parameters of generalised Gaussian distribution from the coefficient distribution. The results highlight that considering anisotropic property in the wavelet results in precision increase for a few texture patterns. The influence of scale on classification is also studied.

Keywords ILD · Lung texture · Anisotropy · Multiscale · Morlet wavelet

1 Introduction

The appearance and consistency of the surface of an object are due to its variability in light dispersion. Texture refers to such surfaces. In image processing, texture is considered as an essential feature for identifying an object [1]. The texture may be a repeated pattern or apparent repetition of few elementary patterns [2]. Texture plays an important role in classifying interstitial lung diseases (ILDs). ILD is a broader term used to refer around 150 different kinds of lung diseases. In terms of differential diagnosis in addition to medical tests, radiological imaging plays a vital part in it. High-resolution computed tomography (HRCT) is widely used to analyse the texture of a lung tissue visually. HRCT helps radiologists to identify the disease type based on the texture of the CT image [3, 4]. Figure 1 shows different lung texture

M. J. Das (✉) · L. B. Mahanta
Institute of Advanced Study in Science and Technology, Paschim Boragaon, Guwahati, India
e-mail: manas.ork@gmail.com

N. N. Chiplunkar and T. Fukao (eds.), *Advances in Artificial Intelligence and Data Engineering*, Advances in Intelligent Systems and Computing 1133,
https://doi.org/10.1007/978-981-15-3514-7_88

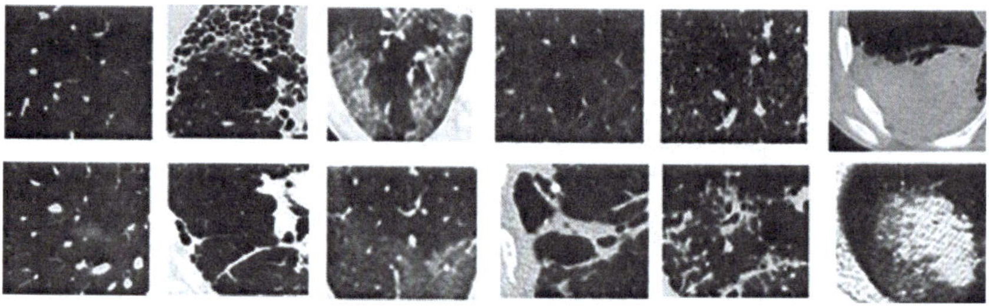

Fig. 1 Different lung texture classes; the rows show the inter-class variability, from left to right: Healthy, fibrosis, ground glass, emphysema, micronodule and consolidation

classes: healthy, ground glass, fibrosis, emphysema, micronodules and consolidation. Conventionally, or manually, lung textures are challenging to classify; as observed, inter-class variation of texture sometime may be low and intra-class variation may be high within the same class. The texture features are very subtle [5] and convey non-determinism; the feature selection algorithm needs to be good in capturing the fine texture differences.

The literature reveals many texture analysis techniques which are based on a single spatial resolution. For texture analysis, grey levels of an image are used to create histograms [6–8], co-occurrence distribution [9, 10], binary patterns in a limited neighbourhood [11] and filters [12, 13] for a single spatial resolution. Image analysis can also be done in multiple resolution or multiscale. Multiscale analysis extracts texture features from an image using wavelet transform [14, 15]; in this analysis, the energy of different frequencies present in the image can be analysed using different scales and orientations of the wavelet function. Rotation-invariant Gabor filters [16, 17] for texture analysis are also proposed; multiscale transform like Riesz transform [18] is used to create the feature vectors of a support vector machine (SVM) classifier. Wavelet with isotropic properties is used [19] as feature descriptor for lung textures.

These multiscale and rotation-invariant filters assume that there is no predominant orientation in the texture (isotropic). Since lung texture is very complex in nature, some level of orientation dependency (anisotropy) prevails [20, 21].

For classifying texture pattern, orientation information from wavelet coefficients is often coded horizontally or vertically [22] within a neighbourhood and is inadequate for anisotropy. In our implementation, phase/orientation information has been coded jointly within a neighbourhood.

In this paper, we are using Morlet transform that is anisotropic in nature. The distribution of the coefficient of Morlet transform in each scale, from fine to coarse, exhibits a histogram which shape is like a bell and has a long tail. Such a shape can be modelled with a generalised Gaussian distribution (GGD). Rényi entropy can be used to quantise anisotropy [23]; the parameters from the GGD and Rényi's entropy are used as an input feature vector for the SVM with classification score [24]. To classify different textures of the lung, SVM can be considered as an effective classifier [25].

The remainder of the paper is structured as follows. Section 2 presents the data set we used in this work, describes Morlet transform and describes the quantification

Table 1 Spread of 32 × 32 ROI blocks among six different texture patterns

Tissue type	Healthy (H)	Emphysema (E)	Ground glass (G)	Fibrosis (F)	Consolidation (C)	Micronodule (M)
No. of scans	14	10	37	51	18	28
ROIs	230	87	320	418	210	335
32 × 32 ROIs	7680	3218	7620	8251	6176	7820

methods of the wavelet coefficients by Rényi's entropy and GGD. In Sect. 3, we report and discuss our findings. Finally, Sect. 4 recalls the main contribution of the work done.

2 Method and Materials

2.1 Data Set

The number of HRCT scans (each scan corresponds to a patient) for this study is a part of internal data set collected from a National Accreditation Board for Testing and Calibration (NABL, India) radiological laboratory. In this study, 85 scans are used and a single scan may contain multiple textures; a total of 158 image stacks containing six different texture patterns are there. Experienced radiologists with 8–10 years of expertise delineate region of interest (ROI) indicating the pattern it contains. Overlapping 32 × 32 blocks are extracted from it. The thickness of each slice is 1.4 mm, and slice increment is 0.7 mm. A visual inspection is done on all of the scans, and the scans that do not contain any artefacts (viz. patient breathing movement, beam hardening, ring effect, etc.) were considered for this study. Depending on the extensiveness of the disease, it is possible to collect healthy tissue from scans that have other pathological lung tissue textures. Table 1 illustrates the spread of the texture patterns.

2.2 Morlet Transform

A hybrid isotropic polyharmonic B-spline wavelet [19] is used for texture feature classification; the study assumes that there is no dominant orientation in the texture pattern. Rotated wavelet frames [20] and contourlet transform are used to obtain results for lung texture pattern classification considering only the direction (angle) of the texture. Lung textures require information of both phase and amplitude [18]; thus, complex wavelets are suitable. Real-valued wavelets are suited for isolating

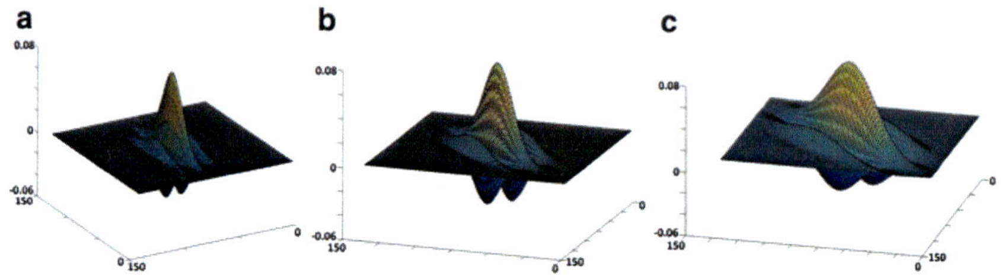

Fig. 2 Morlet wavelets at various orientations and anisotropies. **a** $\theta = 0°$, $L = 0.8$. **b** $\theta = 45°$, $L = 0.7$. **c** $\theta = 45°$, $L = 0.5$

peaks and discontinuities. Morlet wavelet is implemented since it is complex and has both the properties for frequency resolution and angle selectivity. A simple Morlet wavelet can be Gaussian function modulating a sine function. An anisotropic Morlet wavelet [26] is given by:

$$\psi(x, \theta; L) = \frac{1}{\Pi} e^{ik_0.ACx} e^{-1/2(Cx.A^T ACx)} \quad (1)$$

where $\boldsymbol{k}_0 = (0, k_0)$ is a wave vector with $\boldsymbol{k}_0 \geq 2\pi$. $\boldsymbol{C}$ is a transformation matrix which is linear and is given by $\boldsymbol{C} = \begin{bmatrix} \text{cosine}\theta & \text{sine}\theta \\ -\text{sine}\theta & \text{cosine}\theta \end{bmatrix}$, and $\boldsymbol{A}$ defines the anisotropy and is a positive definitive matrix given by $\boldsymbol{A} = \begin{bmatrix} L & 0 \\ 0 & 1 \end{bmatrix}$. The angle θ is measured with respect to the positive x-axis. The anisotropy ratio L is defined as the relationship between the length of the wavelet in theta direction and the length of the wavelet orthogonal to it.

In Fig. 2, orientation and anisotropy ratios of few Morlet are illustrated. To calculate orientation of a texture pattern $L < 1$, Morlet wavelet shows more efficient result [20].

Continuous wavelet transform (CWT) of Morlet wavelet is complex-valued coefficient; CWT of a function $f(x)$ is an integral transform with translation, rotation and dilation of the wavelet and is given by [26, 27]:

$$C_\psi(a, b, \theta; L) = \int_x f(x) \psi^*_{a,b}(x, \theta; L) \mathrm{d}x \quad (2)$$

For a scale $\boldsymbol{a}(a > 0)$, having orientation θ, location parameter $\boldsymbol{b}$ and anisotropic ratio L, $C_\psi(a, b, \theta; L)$ is the complex wavelet coefficient, ψ^* is the complex conjugate of ψ, and

$$\psi_{a,b}(x, \theta; L) = a^{-1} \psi\left(\frac{x - b}{a}, \theta; L\right) \quad (3)$$

where $\psi(x, \theta; L)$ is the Morlet wavelet defined in Eq. (1) and $1/a$ is known as the scaling parameter and keeps the energy of the wavelet proportional for every value of a. We fix the value of L and do the analysis for various (a, b, θ), where $a \, \varepsilon \, \{\bar{a}_1, \bar{a}_2, \ldots, \bar{a}_n\}$ and $\theta \, \varepsilon \, \{\bar{\theta}_1, \bar{\theta}_2, \ldots, \bar{\theta}_n\}$. Equation (2) is a convolution integral, and substituting Eq. (3) into it we get:

$$\begin{aligned} \mathcal{C}_{\psi}(a, b, \theta; L) &= a^{-1} \int_x f(x) \psi^* \left(-\frac{b - x}{a}, \theta; L \right) \mathrm{d}x \\ &= a^{-1} f(b) * \psi^* \left(-\frac{b}{a}, \theta; L \right) \end{aligned} \tag{4}$$

where * is the convolution operator. The magnitude ($\mathcal{M}$) and phase ($\mathcal{A}$) of complex Morlet coefficients ($\mathcal{C}$) are given as:

$$\mathcal{C}_{\psi}^{L}(b, a, \theta) = \left\{ \mathcal{M}^{L}(b, a, \theta), \mathcal{A}^{L}(b, a, \theta) \right\} \tag{5}$$

with

$$\mathcal{M}^{L}(b, a, \theta) = \left[\left(\mathrm{Re}\left(\mathcal{C}_{\psi}^{L}(b, a, \theta) \right) \right)^2 + \left(\mathrm{Im}\left(\mathcal{C}_{\psi}^{L}(b, a, \theta) \right) \right)^2 \right]^{1/2} \tag{6}$$

and

$$\mathcal{A}^{L}(b, a, \theta) = \mathrm{Arg}\left(\mathrm{Re}(\mathcal{C}_{\psi}^{L}(b, a, \theta) + \mathrm{Im}(\mathcal{C}_{\psi}^{L}(b, a, \theta) \right) \tag{7}$$

The phase information is jointly defined with its neighbourhood as:

$$\Phi^{L}(x, y; a, \theta) = \arctan\left(\frac{\mathcal{A}^{L}(x, y + 1; a, \theta) - \mathcal{A}^{L}(x, y; a, \theta)}{\mathcal{A}^{L}(x + 1, y; a, \theta) - \mathcal{A}^{L}(x, y; a, \theta)} \right) \tag{8}$$

2.3 *Rényi Entropy*

Rényi's entropy is a good descriptor of anisotropic texture. Morlet coefficients are characterised by a 2D histogram [28] of coefficients. Suppose the image block is of size M × N. Let $\mathcal{M}(x, y)$ be a coefficient at a pixel location (x, y). Mean magnitude value of a square neighbourhood of size $\mathbb{N}$ cantered at (x, y) is written as:

$$\mathcal{M}_{\mathrm{avg}}(x, y) = \left\lfloor \mathbb{N}^{-2} \sum_{j=-(\mathbb{N}-1)/2}^{(\mathbb{N}-1)/2} \sum_{i=-(\mathbb{N}-1)/2}^{(\mathbb{N}-1)/2} \mathcal{M}(x + i, y + j) \right\rfloor \tag{9}$$

We use a window size of 3 × 3, and to include all the rows and columns at the borders (i.e. if the dimension of the image is not a multiple of 3), we overlapped the window at borders to incorporate all the rows and columns. A paired value is created with the coefficient $\mathcal{M}(x, y)$ and the average coefficient $\mathcal{M}_{\text{avg}}(x, y)$. The normalisation is done by calculating the occurrence of the paired value and can be represented as follows:

$$\mathcal{M}_{\text{norm}}(i, j) = \frac{1}{MN} \sum_{x=0}^{M} \sum_{y=0}^{N} \delta(i, j) \tag{10}$$

where

$$\delta(i, j) = \begin{cases} 1, & M(x, y) = i \text{ and } \mathcal{M}_{\text{avg}}(x, y) = j \\ 0, & \text{Otherwise} \end{cases}$$

The normalised 2D histogram is used to find Rényi's entropy as follows:

$$\text{Ry} = \frac{1}{1-\alpha} \log \sum_{x=1}^{N} \sum_{y=1}^{M} (\mathcal{M}_{norm}(x, y))^{\alpha} \tag{11}$$

where α is the order of the entropy; we use $\alpha = 2$, and the entropy values calculated are used as a feature vector for the learning of the classifier. The feature vector is given as $F_{\text{Ry}} = \left[\text{Ry}_{1,1}, \text{Ry}_{2,1}, \ldots, \text{Ry}_{a,\theta}\right]$.

2.4 Generalised Gaussian Density

The marginal distribution of the coefficients resembles the shape of a generalised Gaussian curve; to model such a distribution, GGD [29, 30] is used. Two varying parameters can be used to approximate the coefficient of the wavelet transform as shown below:

$$p(x; \alpha, \beta) = \frac{\beta}{2\alpha\Gamma(1/\beta)} e^{-(x|/\alpha)^{\beta}} \tag{12}$$

where Γ is a gamma function and α is a scale parameter which define the width of the curve, while β is the shape parameter which defines the rate of change of the curve. The parameters α and β are estimated by maximum likelihood estimator (MLE). Let $\Theta \in \alpha, \beta$ and $X = \{x_1, x_2, \ldots, x_n\}$ be independent and identically distributed random variables; then, MLE $\hat{\Theta}$ is defined as:

The feature vectors constructed from the estimated parameters of GGD for various scales and orientations are given by: $F^{\mathcal{M}}_{GGD} = \left[\alpha^{\mathcal{M}}_{1,1}, \beta^{\mathcal{M}}_{1,1}, \ldots \alpha^{\mathcal{M}}_{a,\theta}, \beta^{\mathcal{M}}_{a,\theta}\right]$ and $F^{\Phi}_{GGD} = \left[\alpha^{\Phi}_{1,1}, \beta^{\Phi}_{1,1}, \ldots \alpha^{\Phi}_{a,\theta}, \beta^{\Phi}_{a,\theta}\right]$.

To summarise the method, the steps taken to classify the six different lung tissue patterns are as follows:

Step-1 Find the wavelet coefficients using Eq. (4) of all the image blocks in an image ROI for a number of scales, θ number of orientations and a fixed value of L.
Step-2 Complex coefficients from Step-1 are used to find out the magnitude and phase information using Eqs. (6) and (7), respectively, for each block.
Step-3 Find Rényi entropy from the magnitudes determined in Step-2 using Eqs. (9), (10) and (11) for each block.
Step-4 Phase information from Step-2 is used to find the joint phase information using Eq. (8) for each image block.
Step-5 Estimate by MLE the parameters α and β obtained from Eq. (12) for magnitude values obtained in Step-2 and joint phase information from Step-4 for each image block.
Step-6 Repeat Step-1 to Step-5 for all the ROIs.
Step-7 Train a SVM classifier with feature vector generated in Step-3 and Step-5 for all ROI.
Step-8 Repeat Step-1 to Step-7 for different values of L.

3 Results and Discussion

This section provides the details of the parameters used, classifier's set-up, and discusses the results thus obtained. In this work, a total of 85 HRCT scans are used, and the image blocks generated for the six different lung texture patterns are stated in Table 1.

3.1 Wavelet Parameters

In this study, we are considering five values of $L(0.2, 0.4, 0.6, 0.8, 1.0)$, and the number of orientations is taken in the interval $[-\pi, \pi]$ with a step size of $\pi/4$. To negate edge effect of an $M \times N$ image, we focus on a $\bar{M} \times \bar{N}$ central part of the image where $\bar{M} < M$ and $\bar{N} < N$, and we choose $\bar{M}, \bar{N}$ such that the coefficients of the wavelet transform are not affected by the edge of the image. This limits the maximum scale parameter that the wavelet can take, we call this scale $a_{\max}$, and beyond this edge effect will affect the coefficients. We also define $a_{\min}$ the minimum scale at which the wavelet is well resolved. For Morlet wavelet, $a_{\min}$ is found out to be 1 and $a_{\max}$ is found out to be 3.1 for our considered image block size. As most of the information is contained in the finest resolution, we are using $a_{\max}$ to be 2.2 and use an increment of 0.2 from $a_{\min}$.

3.2 Classification Using SVM

To check the performance of the described model, SVM as classifier is implemented. A leave-one-patient out cross-validation (CV) is considered. This CV technique is similar to a trained system trying to classify a new image block from a new scan that it has not been trained with. In contrast to the CV method used in this paper, *N*-fold cross validation ($N = 5, 10, \ldots$) when used with SVM can produce unsuitable SVM parameters when the folds are randomly drawn and number of samples for a class is small [31]. In this work, SVM with radial basis function (RBF) kernel is used. To optimise the SVM, RBF parameters γ and C are tuned, and a logarithmic grid search from 10^{-3} to 10^{3} is done for γ and [1, 10, …, 100] for C. A SVM is trained for each value of L (i.e. in total five classifiers are trained).

For a value of L, the input feature vector to the SVM consists of nine orientations, seven scale values, four GGD parameters (two from magnitude two from orientation) and one Rényi entropy value, so the size of the input feature vector is $9 \times 7 \times 5$. In Fig. 3, the impact of scale on precision is shown, and the values are taken on average for a different anisotropy ratio L. As evident from the mean value, we can deduce that, after including four scales, i.e. from $a = 1$ to $a = 1.6$, the precision of the system flattens out. This trend of flattening out after few scales may be due to the wavelet property that code much of the information in lower scales. This reduced feature vector size (i.e. $9 \times 4 \times 5$) decreases computation time. The computation time gets reduced for two sections of the method. In one section, computation time gets reduced by 38% for calculating the Morlet coefficient, and in another section computational time gets reduced by 16% for the classifier.

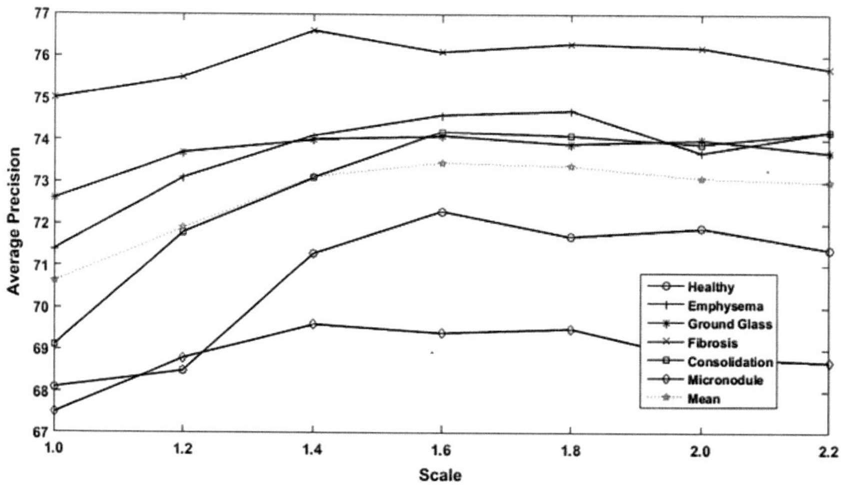

Fig. 3 Change in average precision with increasing number of scale

3.3 Impact of Isotropy on Classification

Different anisotropic ratios of the Morlet wavelet have different precision levels for the classification of lung texture as seen from Fig. 4, and the confusion matrices in the figure are computed optimally by considering four scales. To classify fibrosis, $L = 0.4$ shows the maximum precision, so fibrotic lung pattern got some degree of anisotropy in it. Precision for emphysema is maximum for $L = 0.6$; in emphysema, the airspace gets enlarged and the walls covering the airspace sometimes get hardened as evident from Fig. 1; these seemingly honeycomb-like structures can be reason for having some anisotropy.

Ground glass, consolidation, micronodule and healthy lung tissue structure do not have much of anisotropic texture in it, so they show a better precision for near isotropic Morlet wavelet. A HRCT image is shown in Fig. 5a which is composed of different texture classes; image blocks are created from it and passed through the classifiers. The class which has the highest classification score among the classifiers is taken as the class for that image block. In Fig. 5b, classified image blocks are false colour coded.

0.2	H	E	G	F	C	M
H	**68.1**	6.1	7.1	9.5	1.7	7.5
E	11.3	**71.9**	6.8	5.3	1.5	3.2
G	8.2	16.2	**70.4**	4.3	0.3	0.6
F	4.4	0.7	2.6	**79.5**	12.3	0.5
C	3.2	0.7	7.3	21.1	**66.8**	0.9
M	23	0.8	2.4	2.6	4.1	**67**

0.4	H	E	G	F	C	M
H	**70.6**	5.9	6.8	7.9	1.4	7.3
E	9.2	**76.1**	5.9	4.4	1.2	3.2
G	10.1	14.2	**71.3**	3.6	0.4	0.4
F	3.5	0.9	2.8	**81.2**	11.1	0.5
C	2.8	1.3	5.2	22.1	**68.3**	0.2
M	21.3	0.5	2.5	2.5	4	**69.2**

0.6	H	E	G	F	C	M
H	**71.3**	5.9	7	9.3	1.5	5
E	8.7	**78.3**	5.1	4.1	1	2.8
G	9.2	12.6	**74.6**	2.8	0.2	0.5
F	5.1	1.2	2	**78.8**	11.8	1.1
C	1.8	0.9	4.7	20.2	**72.3**	0.1
M	20.8	0.7	2.3	2.1	3.7	**70.4**

0.8	H	E	G	F	C	M
H	**75.3**	4.9	6.4	7.9	1.2	4.3
E	9.8	**74.1**	6.5	4.7	1.2	3.7
G	8.7	10.1	**78.4**	1.8	0.7	0.3
F	5.7	1.5	2.7	**75.6**	12.8	1.6
C	1.2	0.2	3.6	17.8	**77.1**	0.1
M	18.6	1.1	2.1	1.8	2.9	**73.5**

1.0	H	E	G	F	C	M
H	**76.1**	4.7	6.2	7.8	1.1	4.1
E	9.8	**74**	6.5	4.8	1.2	3.7
G	8.8	10	**78.2**	1.8	0.8	0.4
F	5.7	1.5	2.7	**75.4**	13.1	1.6
C	1.7	0.2	2.8	15.3	**79.8**	0.2
M	18.7	1.1	2.1	1.8	2.9	**73.4**

Fig. 4 Confusion matrix for different L values which are mentioned in the top left box. Maximum precision achieved for a particular texture pattern is underlined

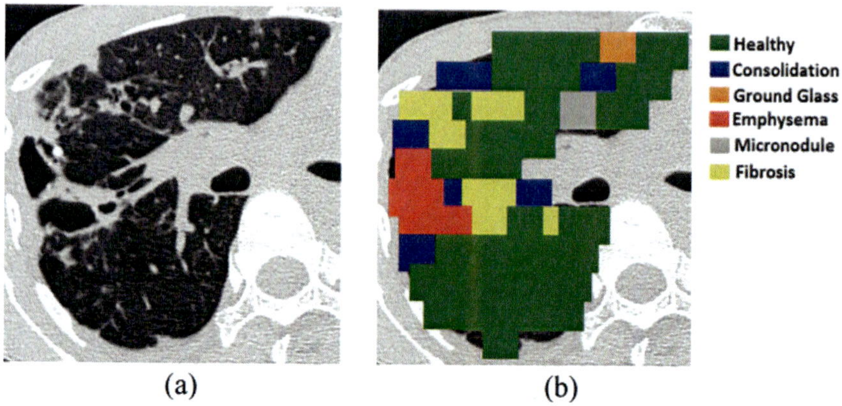

Fig. 5 **a** HRCT lung image consists of different lung texture patterns. **b** Lung pattern detected for each image block by the classifiers is colour coded

4 Conclusion

In this work, HRCT scans are used to classify most prominent lung texture classes in ILD, viz. emphysema, ground glass, consolidation, fibrosis, micronodule. An anisotropic Morlet wavelet is used in a multiscale set-up with various orientations, and the anisotropic property of the wavelet reveals that few of the lung texture patterns are composed of anisotropic texture. The results highlight that considering anisotropy and directional feature in lung texture classification can increase the overall accuracy of the system. Also, the number of scales we consider for classification has bearings on accuracy as well as computation time. The total number of scales in between $a_{\min}$ and $a_{\max}$ depends on the size of the image, this number can be large, and this large number of scales increases the feature vector size considerably. It is shown that for the developed method not all the scales are necessary to have high accuracy, and this means an optimal number of scales can be used which will reduce the computation time for the classifier and wavelet coefficient calculation.

References

1. Brodatz P (1966) Textures: a photographic album for artists and designers. Dover Publication, New York
2. Raghu PP, Yegnanarayana B (1996) Segmentation of Gabor-filtered textures using deterministic relaxation. IEEE Trans Image Process 5(12):1625–1636. https://doi.org/10.1109/83.544570
3. Fetita C, Chang-Chien KC Brillet PY, Preteux F, Grenier P (2007) Diffuse parenchymal lung diseases: 3D automated detection in MDCT. In: Ayache N, Ourselin S, Maeder A (eds) Medical image computing and computer-assisted intervention—MICCAI 2007, Lecture notes in computer science, vol 4791, pp 825–833. https://doi.org/10.1007/978-3-540-75757-3_100

4. Uppaluri R, Hoffman EA, Sonka M, Hartley PG, Hunninghake GW, McLennan G (1999) Computer recognition of regional lung disease patterns. Am J Respir Crit Care Med 160(2):648–654. https://doi.org/10.1164/ajrccm.160.2.9804094
5. Webb WR, Muller NL, Naidich DP (2001) High-resolution CT of the lung. Lippincott Williams & Wilkins, Philadelphia
6. Asherov M, Diamant I, Greenspan H (2014) Lung texture classification using bag of visual words. In: SPIE medical imaging. International society for optics and photonics, vol 9035. https://doi.org/10.1117/12.2044162
7. Depeursinge A, Racoceanu D, Iavindrasana J, Cohen G, Platon A, Poletti A, Muller H (2010) Fusing visual and clinical information for lung tissue classification in high-resolution computed tomography. Artif Intell Med 50(1):13–21. https://doi.org/10.1016/j.artmed.2010.04.006
8. Song Y, Cai W, Zhou Y, Feng DD (2013) Feature-based image patch approximation for lung tissue classification. IEEE Trans Med Imaging 32(4):797–808. https://doi.org/10.1109/TMI.2013.2241448
9. Xu Y, Sonka M, McLennan G, Guo J, Hoffman E (2006) MDCT-based 3-D texture classification of emphysema and early smoking related lung pathologies. IEEE Trans Med Imaging 25(4):464–475. https://doi.org/10.1109/TMI.2006.870889
10. Zavaletta VA, Bartholmai BJ, Robb RA (2007) Nonlinear histogram binning for quantitative analysis of lung tissue fibrosis in high-resolution CT data. In: SPIE Progress in biomedical optics and imaging, vol 6511. https://doi.org/10.1117/12.710220
11. Sorensen L, Shaker SB, de Bruijne M (2010) Quantitative analysis of pulmonary emphysema using local binary patterns. IEEE Trans Med Imaging 29(2):559–569. https://doi.org/10.1109/TMI.2009.2038575
12. Anthimopoulos M, Christodoulidis S, Christe A, Mougiakakou S (2014) Classification of interstitial lung disease patterns using local DCT features and random forest. In: 36th Annual international conference of the IEEE engineering in medicine and biology society (EMBC), pp 6040–6043. https://doi.org/10.1109/embc.2014.6945006
13. Kale M, Mukhopadhyay S, Dash JK, Garg M, Khandelwal N (2016) Differentiation of several interstitial lung disease patterns in HRCT images using support vector machine: role of databases on performance. In: SPIE medical imaging. international society for optics and photonics, vol 9785. https://doi.org/10.1117/12.2216743
14. Bovik AC, Clark M, Geisler WS (1990) Multichannel texture analysis using localized spatial filters. IEEE Trans Pattern Anal Mach Intell 12:55–73. https://doi.org/10.1109/34.41384
15. Mallat S (1989) A theory for multiresolution signal decomposition: The wavelet representation. IEEE Trans. Pattern Analysis Mach Intell 11(7):674–693. https://doi.org/10.1109/34.192463
16. Haley GM, Manjunath BS (1995) Rotation invariant texture classification using modified Gabor filters. In: Proceedings of IEEE international conference image processing, Washington DC, vol 1, pp 262–265. https://doi.org/10.1109/icip.1995.529696
17. Manthalkar R, Biswas PK, Chatterji BN (2003) Rotation invariant texture classification using even symmetric Gabor filters. Pattern Recogn Lett 24(12):2061–2068. https://doi.org/10.1016/S0167-8655(03)00043-6
18. Depeursinge A, Foncubierta–Rodriguez A, Van de Ville D, Müller H (2012) Multiscale lung texture signature learning using the Riesz transform. In: Ayache N, Delingette H, Golland P, Mori K (eds) Medical image computing and computer-assisted intervention—MICCAI 2012, Lecture notes in computer science, vol 7512, pp 517–524. https://doi.org/10.1007/978-3-642-33454-2_64
19. Depeursinge A, Van de Ville D, Platon A, Geissbuhler A, Poletti PA, Muller H (2012) Near-affine-invariant texture learning for lung tissue analysis using isotropic wavelet frames. IEEE Trans Inf Technol Biomed 16(4):665–675. https://doi.org/10.1109/TITB.2012.2198829
20. Tolouee A, Abrishami-Moghaddam H, Garnavi R, Forouzanfar M, Giti M (2008) Texture analysis in lung HRCT images. In: Digital imaging computing: techniques and application, pp 305–311. https://doi.org/10.1109/dicta.2008.27

21. Vo K, Sowmya A (2009) Directional multi-scale modeling of high resolution computed tomography (HRCT) lung images for diffuse lung disease classification. In: 13th International conference on computer analysis of images patterns, vol 5702. Springer, New York, pp 663–671. https://doi.org/10.1007/978-3-642-03767-2_81
22. Vo A, Oraintara S (2010) A study of relative phase in complex wavelet domain: property, statistics and applications in texture image retrieval and segmentation. Sig Process Image Commun 25(1):28–46. https://doi.org/10.1016/j.image.2009.09.003
23. Gabarda S, Cristóbal, G, Rodríguez P, Miravet C, Del Cura JM (2010) A new Renyi entropy-based local image descriptor for object recognition. In: Society of photo-optical instrumentation engineers, vol 7723. https://doi.org/10.1117/12.854901
24. Chang CC, Lin CJ (2011) LIBSVM: a library for support vector machines. ACM Trans Intell Syst Technol 2(3):27:1–27:27. https://doi.org/10.1145/1961189.1961199
25. Li S, Kwok JT, Zhu H, Wang Y (2003) Texture classification using the support vector machines. Pattern Recogn 36(12):2883–2893. https://doi.org/10.1016/S0031-3203(03)00219-X
26. Neupauer RM, Powell KL (2005) A fully-anisotropic Morlet wavelet to identify dominant orientations in a porous medium. Comput Geosci 31:465–471. https://doi.org/10.1016/j.cageo.2004.10.014
27. Murenzi R (1989) Wavelet transforms associated to the n-dimensional euclidean group with dilations: signal in more than one dimension. In: Combes JM, Grossmann A, Tchamitchian P (eds) Wavelets. Inverse problems and theoretical imaging. Springer, Berlin, Heidelberg. https://doi.org/10.1007/978-3-642-75988-8_22
28. Sahoo P (2004) A thresholding method based on two-dimensional Renyi's entropy. Pattern Recogn 37(6):1149–1161. https://doi.org/10.1016/j.patcog.2003.10.008
29. Sharifi K, Leon-Garcia A (1995) Estimation of shape parameter for generalized Gaussian distributions in subband decompositions of video. IEEE Trans Circ Syst Video Technol 5:52–56. https://doi.org/10.1109/76.350779
30. Wouwer GV, Scheunders P, Dyck DK (1999) Statistical texture characterization from discrete wavelet representations. IEEE Trans Image Process 8(4):592–598. https://doi.org/10.1109/83.753747
31. Batuwita R, Palade V (2013) Class imbalance learning methods for support vector machines. In: He H, Ma Y (eds) Imbalanced learning: foundations algorithms and applications. Wiley, New York. https://doi.org/10.1002/9781118646106.ch5

A Review of Intelligent Smartphone-Based Object Detection Techniques for Visually Impaired People

R. Devakunchari, Swapnil Tiwari, and Harsh Seth

Abstract Visual aids have been around for a long time to help blind people overcome the challenges and difficulties they face in everyday life. Moreover, a lot of research has been done and money has been spent to improve on these technologies. With the emergence of smart devices came the idea of real-life detection of objects. This brought about a drastic change in the market and the lives of many visually impaired people. This paper puts forward an overview of many such technologies used in the recent past years.

Keywords Speech recognition · Image detection · Visually impaired

1 Introduction

The Blind People's Association, India, released a report that put the fraction of Indian people who are visually impaired at nearly 1.9% [1]. Millions of people in this world live with some form of visual affliction that hinders them from leading normal, fruitful or even independent lives. Visual challenges induce social awkwardness and low self-confidence among them, and they are often alienated from their peers at a young age. Consequently, such people require constant care and help them going out and around, and often with daily routines. With time, some people learn to walk with sticks, but it is also just to make sure that they are not stepping on or colliding with something. The past few years have witnessed the emergence of new technologies to help visually impaired people, but most of the devices launched are expensive. The economically weak have fewer options, and fewer still are reliable.

R. Devakunchari (✉) · S. Tiwari · H. Seth
Department of Computer Science Engineering, SRM Institute of Science and Technology, Chennai, Tamil Nadu, India
e-mail: devakunchari.r@ktr.srmuniv.ac.in

S. Tiwari
e-mail: tiwariswapnil97@gmail.com

H. Seth
e-mail: harshseth242@gmail.com

N. N. Chiplunkar and T. Fukao (eds.), *Advances in Artificial Intelligence and Data Engineering*, Advances in Intelligent Systems and Computing 1133,
https://doi.org/10.1007/978-981-15-3514-7_89

Recent advances in information technology (IT), and mobile technology especially, have made it possible for people with disabilities like visual impairment to lead better, more comfortable lives using IT-based assistive technologies. Technology has, in some measure, and can, significantly, help make individuals more capable of participating in social activities and leading normal, independent lives. The domain of IT-based assistive technologies, and the range of assistance they can possibly provide, is immense. Of late, ubiquitous computing—a model of human–computer interaction designed to suit the natural human environment—has seen a rising trend. With truly ubiquitous computing, a user need not even be aware that they are using information technology for everyday tasks. Mobile assistive technologies are the result of numerous efforts to deliver assistance to disabled people and intelligently applied ubiquitous computing. Mobile phones are cheaper, portable, lightweight, and easier to operate; aids delivered via phones will easily permeate into the general population, being devoid of the stigma against the more traditional aids. The proposed model is built upon this advantage with the aim of helping the visually impaired. The application will inform the user about the environment irrespective of whether it is indoor or outdoor without having to use any sensor or chip.

The organization of the paper is as follows. Initially, in Sect. 2, we have discussed about the work done related to IT-based technologies in order to help the visually impaired people. Section 3 brings forward some of the speech recognition techniques present in the market right now. Similarly, in Sect. 4, we have emphasized on the various image recognition techniques. Section 5 is a discussion on the various techniques that we talked about. At last in Sect. 6, we have concluded the paper by analyzing the various techniques and then talking about the future work that can be done for the same.

2 Related Work

2.1 The vOICe

The vOICe [2] was developed at the University of Birmingham by Leslie Kay in 1974. The vOICe is an image to audio device which captures an image from a webcam. The capital letters in vOICe mean ‘Oh, I see!’. The image is then scanned from left to right and the feedback is given through speech. The information is dependent on the location and the brightness of the objects. This intelligent system was launched to assist the visually impaired find the objects around.

Four experiments were performed using this visuo-auditory device. In the first experiment, the users were blindfolded and equipped with the device. They had to move it around and locate the object by moving toward it. In the second one, the task was performed in a constrained environment. The third one was then based on the auditory information which they had to use to find the objects around them. The fourth one included the user's ability to discriminate objects based on categories they fell in.

2.2 The Vibe

The Vibe [3] is a device similar to the vOICe. It is also a visuo-auditory device which employs a wide angle camera which is placed in front of the user. The camera captures an image which is passed to the system. The images are then converted to appropriate grayscale and then scanned. After having scanned the image, the feedback is given out in the form of speech through the headphones. There are a few laws which determine the speech type for the feedback.

1. The grayscale images are scanned from left to right. Thus, the speech produced describes the image in the same direction.
2. If the visual signal is high, then the voice becomes acute. Also, pitch varies in accordance with the signal.
3. The sound becomes more intense as the brightness of the image increases.

2.3 Martinez et al. Approach

Recently, an approach similar to the vOICe was developed by Martinez et al. [4]. It is also an object recognition device which converts image to speech information. It has a different image processing method as compared to the vOICe. The image is scanned from the center to the left and then to the right. When the left half is scanned, the speech output is given out from the left earpiece of headphone, and when the scanning jumps to the right half, the speech is given out by the right earpiece. Figure 1 is an example of an image being scanned by the Martinez et al approach. The image scanning is done from center to the left and then in the right.

Fig. 1 Martinez et al. approach

3 Speech Recognition Systems

In order to find the suitable item for the visually impaired person, their voice is used as an input to our system. These voice signals are converted into string objects using various speech recognition techniques. Therefore, these voice signals are used as input to recognize the nearby object as per the user's requirement. Again, after the required object is ascertained by the system, a vocal feedback is provided to the user in order to make the process simpler and convenient for the visually impaired people. Figure 2 depicts the process of speech recognition. The input wave is analyzed, and then the features are extracted. These intermediate results are classified, and the resultant string is given out as the final output.

3.1 Google Speech API

Google Cloud Speech-to-Text [5, 6] enables users to convert audio to text with the help of powerful neural network models in a user-friendly API. It can recognize 120 languages and variants to support the global user base. There is an option to enable voice command and control, transcribe audio from call centers and more. It can process real-time audio streaming or a pre-recorded audio by using Google's machine learning technology. It can even identify the language it is being spoken to in.

3.2 Microsoft Cognitive Services—Bing Speech API

This API is used for converting both from speech to text and vice versa [7, 8]. The speech-to-text conversion is carried out on cloud by calling the API to recognize the input audio in real time or even a stored audio clip, which is processed, and then as per the requirement, one can retrieve partial as well as complete results. It is most commonly used to build voice triggered smart apps. It has different pricing models.

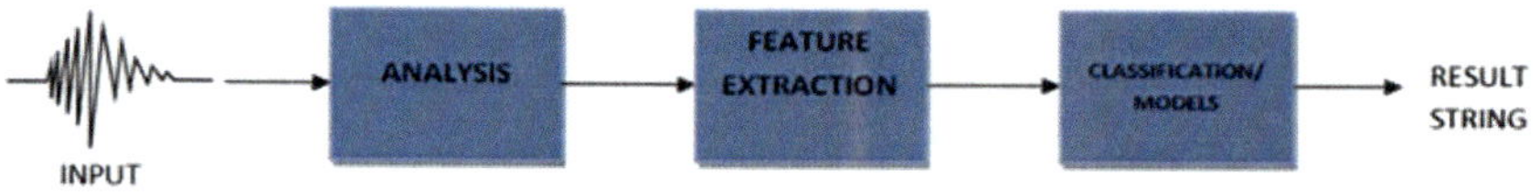

Fig. 2 Process of speech recognition

3.3 Wav2letter++

It is an open-source speech processing toolkit provided by Facebook AI Research [9]. It is written completely in the language C++, and it encompasses ArrayFire tensor library and flashlight machine learning library for maximum efficiency. It is one of the first fully convolutional toolkits which is built on top of flashlight. It facilitates scalability to multiple languages. It can work with raw speech data irrespective of the quality of the audio. This gives it an added advantage.

4 Image Recognition

4.1 Scale-Invariant Feature Transform (SIFT)

The SIFT algorithm [10] is a feature detection algorithm used in computer vision for recognizing images by describing local features in an image. Images are first transformed into feature vectors. These are invariant to image translation, scaling and rotation and also partially invariant to illumination changes and robust to local geometric distortion. To make the image recognition process scale and rotation invariant, the algorithm requires high computation power and time therefore making the process slow. This is accounted by the computation of gradients of each pixel which becomes a tedious task specially for low-powered devices.

The implementation of SIFT for features extraction facilitates the process of image recognition for the visually impaired. This technique is resilient to the movement of the objects. Unlike in OCR, when the user has to make sure that the label on the product is facing the camera which is a tedious thing for a blind person, SIFT can extract features from moving objects using the *K*-means clustering background separation technique.

Figure 3 is an example of how SIFT works. There are two pictures of the same building, one taken from the database and the other one is fed to the system to get the image recognition done. It shows the correctly matched key points encircled in green and the wrongly matched ones encircled in red.

4.2 Speeded-up Robust Features (SURF)

The SIFT algorithm as described above was summarized to be comparatively slow, and there was a need of a faster algorithm. In SURF algorithm [11], a Box Filter is used to approximate Laplacian of Gaussian with difference of gaussian for finding scale space as compared to SIFT algorithm where LAG is computed without using a Box Filter.

Fig. 3 Features extraction using SIFT [cc.gatech.edu]

SURF being a very widely used algorithm is less used now because of deep-layered models like CNNs which marginally outperform these traditional methods. The main disadvantage these types of algorithms provide is that the engineered features are based on predefined assumptions which is not reasonable for every problem unlike in deep-layered algorithms which work like human brain with neurons deeply interconnected to each other and therefore work on a relationship-based model.

Figure 4 shows a comparison of features extraction done by SIFT and SURF. The same sample photograph is given as an input to both the algorithms, and the extractions are performed. As evident from the figure, SURF is capable of matching more number of key points as compared to SIFT.

Fig. 4 Features extraction using SIFT and SURF

4.3 Convolutional Neural Network (CNN)

CNNs are the most widely used algorithm for applications such as image recognition, sound recognition, video processing, etc. CNNs are a deep neural-layered architecture model which use multilayer perceptrons designed to require minimal preprocessing. The main advantages include shift invariance, reduced size of representation, faster computation, and most importantly, the CNN models are logic based unlike traditional machine learning models which are rule based. Surprisingly, when CNNs were introduced in 2012, they reported an error rate of only 0.23% on the MNIST dataset which was a huge leap over traditional machine learning algorithms. The major problem with CNNs till date is that they require a large dataset for training to avoid the problem of overfitting.

Image recognition by CNNs has proved to be the most effective out of all the other deep learning algorithms. They can decrease image noise, features of the signals and simplify neural networks. In order to locate objects for the visually impaired, the classification problem is solely done by CNNs. The recognition speed is found to be low though.

4.4 Regions Convolutional Neural Network (R-CNN)

R-CNN is the state-of-the-art CNN-based deep learning object recognition algorithm [12]. Conventionally, in CNNs, for each input image, convolutions are performed at every pixel of the image using a sliding window mechanism. However, there may be instances where the aspect ratio or the orientation of each object might be different in the same image. Therefore, performing convolutions using different-sized filter would become computationally very expensive. Therefore, a method of R-CNN was introduced which first defines around 2 K regions in any individual image. Then, for each individual image, object classification is done using CNNs. Finally, unlikely image positions are dropped using regression.

Faster R-CNNs were used in a smartphone-based guiding system to develop a visual aid for the blind. It was deployed in two networks. The first network helps in finding a rough position of the object and the second one in determining the object type. This design helps get more accurate results but the speed turned out to be slow.

5 Discussions

The visuo-auditory devices have been in the market for a very long time now. These devices continue helping the blind and the visually impaired in the object recognition through speech information. But the accuracy of the launched products is not very high. There is no mechanism to help the visually impaired store some information like where they kept a particular object through an interactive assistant. Also, the real-time detection through an interactive assistant is not yet possible.

6 Conclusion and Future Work

The past few years have seen a drastic improvement in the technologies for building intelligent systems to help the visually impaired in many ways. Yet there is always scope for improvements, especially because of the improving technologies. The focus of this project is to improve the user experience by adding more features to the existing works and helping them live a more independent life without having to spend any extra money on the electrical devices present in the market.

The focus is to build a system for visually impaired people which not only detects objects for them in real time but also helps them find various objects of their need around them. Our key motivation is to recognize speech data and map them to clusters of pixels showcased via dynamic moving camera of smartphone. Also, the work would include an interactive environment with an assistant which could remember and remind them where their belongings are kept.

References

1. Blind People's Association India. https://www.bpaindia.org/. Last Accessed 11 Jan 2019
2. Auvray M, Hanneton S, O'Regan JK (2007) Learning to perceive with a visuo—auditory substitution system: localisation and object recognition with 'The vOICe'. Perception 36:416–430
3. Durette B, Louveton N, Alleysson D, Herault J (2008) Visuo-auditory sensory substitution for mobility assistance: testing The VIBE. In: Workshop on computer vision applications for the visually impaired, Marseille, France
4. Martinez BDC, Vergara-Villegas OO, Sanchez VGC, Jesus Ochoa Domnguez H, Maynez LO (2011) Visual perception substitution by the auditory sense. Lecture notes in computer science, vol 6783. Springer, Heidelberg, pp 522–533
5. Sak H, Senior A, Rao K, Beaufays F, Schalkwyk J (2016) Google voice search: faster and more accurate. Wayback Mach. Archived 9 Mar 2016
6. Jason K, The power of voice: a conversation with the head of google's speech technology. Tech Crunch. Last Accessed 21 July 2015
7. Speaker Identification (WhisperID). Microsoft research. Microsoft. Last Accessed on 21 Feb 2014
8. Metz C. Microsoft bing rides open source to semantic search. The Register. Last Accessed on 5 Jan 2019
9. Han KJ, Chandrashekaran A, Kim J, Lane I (2018) The CAPIO 2017 conversational speech recognition system
10. Jabnoun H, Benzarti F, Amiri H (2014) Visual substitution system for blind people based on SIFT description. IEEE
11. Chincha R, Tian Y (2011) Finding objects for blind people based on SURF features. IEEE
12. Lin BS, Lee CC, Chiang PY (2017) Simple smartphone-based guiding system for visually impaired people, MDPI. Sensors

Devakunchari Ramalingam, M.E, Ph.D. is an Assistant Professor at the Department of Computer Science Engineering, SRMIST, Chennai. She completed her research in big social data analytics at Anna University, India. Her research interests include social network analysis and big data analytics. She has published around 20 papers in national and international conferences and journals and received awards for the research contribution to society. She is active member of professional societies like ISCA, IET, IEI.

Swapnil Tiwari has completed her B.Tech in Computer Science from SRM Institute of Science and Technology, KTR. Her research interests include big data analytics and machine learning.

Harsh Seth has also completed his B.Tech in Computer Science from SRM Institute of Science and Technology, KTR. His research interests include machine learning and artificial intelligence.

Stereo Vision-Based Depth Estimation

Zelin Meng, Xiangbo Kong, Lin Meng, and Hiroyuki Tomiyama

Abstract This paper presents a lightweight depth estimation method for feature point based on binocular vision technology. First, based on the ORB feature and the brute force matching method, we obtained matching feature point pairs for two frames of image. Secondly, according to the obtained pixel coordinates of the matching point pairs, the motion of the camera can be estimated by the principle of the epipolar constraint. Finally, based on triangulation theory and camera motion, we implemented depth estimation of feature points in the input image. The experiment results indicate that our proposed approach can achieve good real-time performance, high depth estimation accuracy, and has broad application prospects in the case of constrained computing resources.

Keywords Computer vision · Feature extraction · Feature matching · Motion estimation · Depth estimation

1 Introduction

In recent years, machine vision and image processing technologies have received widespread attention from researchers. In the industrial field, machine vision has broad application prospects. For distance measurement, machine vision ranging technology based on binocular vision technology has gradually become an alternative to traditional laser ranging and ultrasonic ranging. For the technical solution of this paper, firstly, we compare and filter the current mainstream feature detection extraction methods, for instance, ORB, SIFT and SURF. We finally decide to adopt open-source and real-time ORB features. Then, we use the binary BRIEF descriptor to characterize the extracted ORB feature points. In addition, in terms of feature matching, we use the brute force matching algorithm to quickly match the feature points in the input two images. Then, based on the feature detection and extraction of the input image frame, combined with the epipolar constraint, the estimation

Z. Meng (✉) · X. Kong · L. Meng · H. Tomiyama
Ritsumeikan University, 1-1-1 Noji-Higashi, Kusatsu, Shiga 525–8577, Japan
e-mail: zelin.meng@tomiyama-lab.org

N. N. Chiplunkar and T. Fukao (eds.), *Advances in Artificial Intelligence and Data Engineering*, Advances in Intelligent Systems and Computing 1133,
https://doi.org/10.1007/978-981-15-3514-7_90

of the camera motion between frames is realized. Finally, using the principle of triangulation, depth estimation based on binocular vision technology is realized.

In the rest of this paper, we discuss feature extraction and matching approaches in Sect. 2, we describe our motion and depth estimation method in Sect. 3, then present the evaluation results, and end with the conclusions in Sect. 4.

2 Feature Description and Matching

2.1 Feature Extraction and Description

In general, for the feature description of SFIT or SURF [1, 2], not all dimensions have an actual effect in the matching. Some algorithms convert the SIFT feature descriptor into a binary code string, and then the code string uses the Hamming distance to match the feature points. This method could greatly improve the matching performance between features, because the realization of calculation of Hamming distance is convenient in modern computer architecture.

Based on those mentioned above, BRIEF descriptor [3] came into being, which provides a shortcut for calculating binary strings without having to calculate a feature descriptor like SURF does. Firstly, we smooth the input image, and then based on normal distribution, selecting a patch near the feature points. In the next step, we randomly select several point pairs in patch. Then, for every selected point pair (p, q), we make a comparison of the two points' brightness. If $I(p)$ is larger than $I(q)$, for this pair of points, the value of one of the generated binary strings is 1. If $I(p)$ is smaller than $I(q)$, the value corresponding to the binary string is 0. Between all 128 point pairs, we compare to generate 128-bit binary string. For the option of number of point pairs, we can set it to 128, 256 or 512. These three parameters are available in OpenCV, and the default parameter in OpenCV is 256. However, in our experiment, we set the parameter as 128. Once the dimension is selected, we can match these descriptors with Hamming distance.

In general, BRIEF is a very efficient way to extract descriptors of feature points. It is worth noting that for BRIEF, it is just a feature descriptor that does not provide a way to extract feature points. So, we will have to employ a feature point positioning method in the design, such as FAST, SIFT, SURF [4], etc.

In this article, we employ a new feature with local invariance—the ORB feature [5]. It can be seen from its name that ORB feature combines FAST feature and BRIEF feature descriptor. In addition, it can highly improve the performance relative to the previously proposed algorithms. ORB is a highly efficient alternative of algorithms like SIFT or SURF. One more thing to be mentioned that ORB is open source, while the other two are both patented. In other words, if we employ either of them in the design for commercial use, purchasing a license is needed.

The ORB feature includes the detection method of FAST feature and BRIEF descriptors. The ORB feature achieves better than the originals. First of all, it utilizes

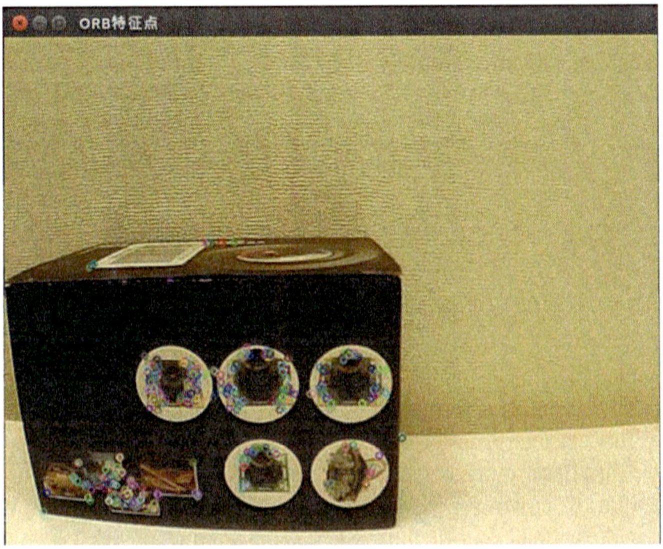

Fig. 1 ORB-based feature detection and extraction

FAST feature method to search feature points. After that it adopts Harris corner point measurement method to screen the N feature points which have largest Harris corner response value from the group of detected feature points. In terms of our experiment result of ORB feature extraction, it is shown in Fig. 1.

We know that FAST feature points are not scale invariant, so we can achieve scale invariance by constructing Gaussian pyramids and then detecting corner points on each layer of pyramid images. Then, for local invariance, we still have a problem that is not solved, i.e., the FAST feature points do not have direction.

In the paper of ORB, a gray-scale centroid of mass method is proposed to solve this problem. The so-called gray-scale centroid of mass refers to the center of the image block gray value which is regarded as weight. This approach has been employed aiming at representing a direction. For one image block B, we describe the moments in regards to block B as:

$$m_{pq} = \sum_{x,y \in B} x^p y^q I(x, y), p, q = \{0, 1\} \quad (1)$$

where $I(x, y)$ indicates gray-scale value of point (x, y). Thus, the image's centroid of mass is described as the following:

$$C = \left(\frac{m_{10}}{m_{00}}, \frac{m_{01}}{m_{00}} \right) \quad (2)$$

So, the direction of each FAST feature point refers to the angle between its block's geometric center and centroid of mass:

$$\theta = \arctan\left(\frac{m_{01}}{m_{10}}\right) \quad (3)$$

2.2 Feature Matching

After extracting the image features, feature matching solves the problem of data association between images in the data set, that is, to determine the corresponding relationship between the currently seen landmarks and the previously seen landmarks. By matching the descriptors between the image I_1 and image I_2 accurately, we can reduce a lot of burden for the subsequent motion estimation and optimization operations. In the actual experiment [6–8], we consider the images of two moments. Assume that we have obtained feature points $X_m, m = 1, 2, \ldots M$ in the first frame and feature points $X_n, n = 1, 2, \ldots N$ in the second frame. The simplest and most direct way is brute force matcher. That is, for each feature point X_m, and all X_n, the distance of descriptors is measured, and then the nearest one is ranked as the matching point. The descriptor distance represents the similarity between the two features. In practical application, different distance measure norms can also be taken. For binary BRIEF descriptors, we often use Hamming distance as a measure—the Hamming distance between two binary strings, referring to the number of different digits. In other words, longer Hamming distance indicates lower similarity. In Fig. 2, matching results of the ORB feature extracted from input images are shown. In addition, the screened good matching results have been shown in Fig. 3.

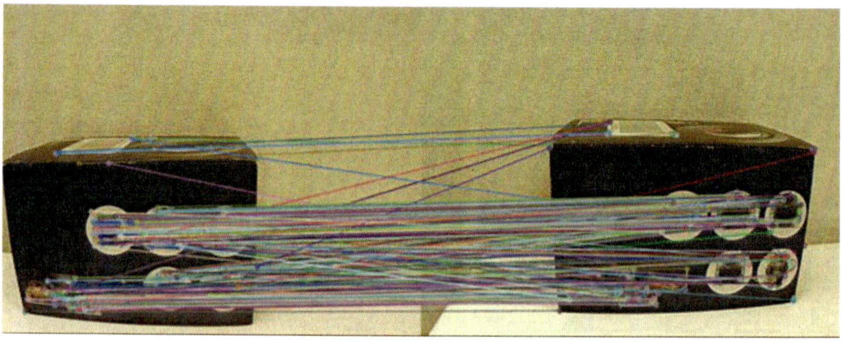

Fig. 2 Brute force matching for feature points

Fig. 3 Screening results of good match

3 Motion and Depth Estimation

3.1 Epipolar Constraint

Suppose we get a pair of matching feature points from the two images, p_1 and p_2, as shown in Fig. 4. If there are a number of such matching points, the camera's motion between the two frames is possible being recovered by the correspondence of these two-dimensional image points. The geometric relationship of the corresponding matching points in two frames of image is demonstrated in Fig. 4.

In the above figure, we want to find the motion between the two frames of image I_1, I_2, let the motion of the first frame to the second frame be R, t. The two camera centers are O_1 and O_2, respectively. There is a feature point p_1 in image I_1, which correspond to feature point p_2 in image I_2. Both of them are obtained by feature

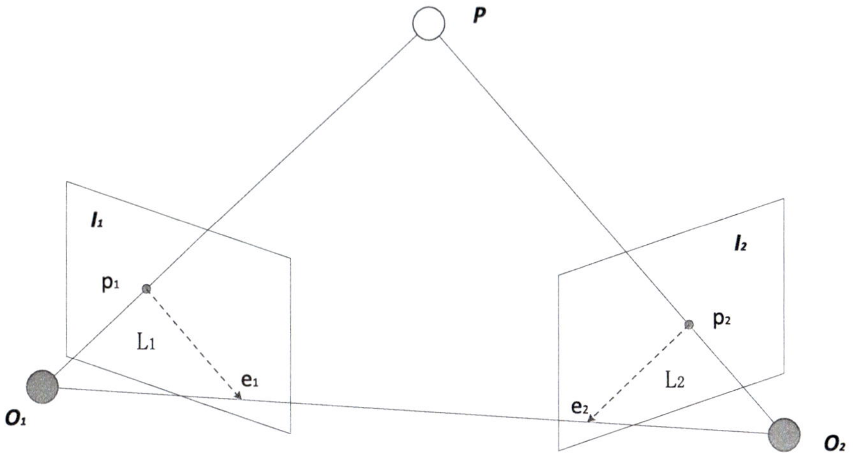

Fig. 4 Scheme of epipolar constraint

matching. In the case of correct matching, the two are the projections of the same spatial point on the two imaging planes. Habitually, we use the related terms of epipolar constraints to describe the geometric relationship of the matching pairs of p_1 and p_2. First, the connection $O_1 p_1$ and the connection $O_2 p_2$ intersect at point P in three-dimensional space. At this time, the three points O_1, O_2, and P can determine a plane. That is the so-called epipolar plane. The intersection of the $O_1 O_2$ line and the image plane I_1 and I_2 are e_1 and e_2. Intersection points e_1 and e_2 are called epipoles, and $O_1 O_2$ is called baseline. The intersection line L_1 and L_2 between the epipolar plane and the two image planes I_1 to I_2 are called epipolar line.

From the first frame image, the ray $O_1 p_1$ is the spatial position where a certain pixel may appear, because all points on the ray are projected to the same pixel. Assuming that the P position is not known, from the second frame image, the connection $e_2 p_2$ (the epipolar line among the image I_2) is the position of the projection where P may appear, that is, the projection of the ray $O_1 p_1$ in the second camera. Since the projection position of the p_2 pixel is determined by feature point matching, the spatial position of$P$ and the motion of the camera can be inferred.

Considering camera coordinate frame of the first image, let spatial position of P be $P = [X, Y, Z]^T$, p_1 and p_2 are the pixel coordinates of the matching point pair. x_1 and x_2 are the coordinates of the matching point pair in the imaging coordinate system. According to the epipolar constraint, the following equation holds:

$$p_2^T K^{-T} t^\wedge R K^{-1} p_1 = 0 \tag{4}$$

The geometric meaning of the epipolar constraint is that O_1, P, O_2 are coplanar. Both the translation and the rotation are included in the epipolar constraint. In general, the middle part is described as two matrices: the fundamental matrix F and the essential matrix E.

$$E = t^\wedge R, F = K^{-T} E K^{-1}, x_2^T E x_1 = p_2^T F \tag{5}$$

The epipolar constraint gives a succinct representation of the spatial positional relationship of two matching points. Therefore, the camera pose estimation problem becomes the following two steps:

1. Find the essential matrix E based on the pixel position of the matching point pair.
2. Decompose the E matrix and find the rotation matrix R and t, which is the desired motion of camera.

3.2 *Triangulation*

After getting the motion of the camera, we use triangulation to estimate the depth information of the pixel. The principle of triangulation is shown in Fig. 5. Similar to the epipolar constraint, considering the images I_1 and I_2, with the left image as the

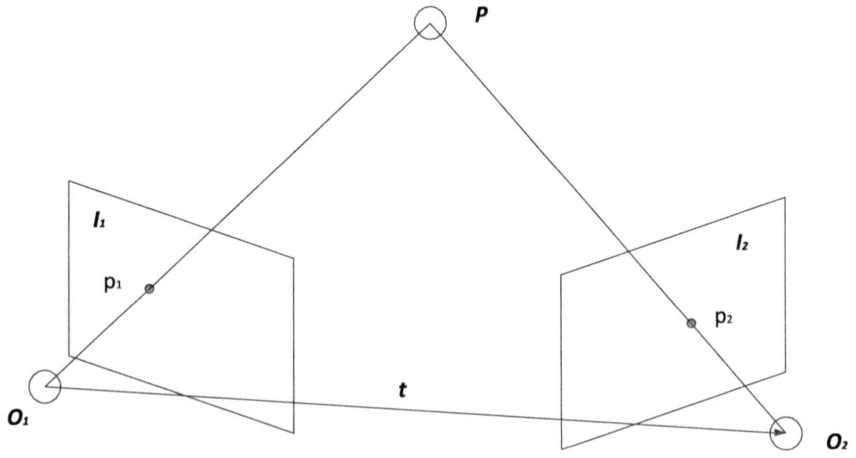

Fig. 5 Scheme of triangulation

reference and the motion of right image is described by matrix R and t. Similarly, O_1 and O_2 are the camera's optical center. There are feature points p_1 among the I_1 images and feature points p_2 among the I_2 images. Let the straight line O_1p_1 and the straight line O_2p_2 intersect in the scene at the point P, that is, the position of the map point corresponding to the two feature points in the three-dimensional scene. According to the epipolar geometry, x_1 and x_2 are the imaging plane coordinates of two feature points, and the following relationship is satisfied:

$$s_1x_1 = s_2Rx_2 + t \tag{6}$$

Since we already know the motions R and t of the camera, the amount to be determined is the pixel depths s_1 and s_2. According to the above formula, we use the method of solving one by one to solve the unknowns s_1 and s_2. For any pair of matching point pairs, we first multiply the left side of the above formula by x_1^T, then the left side of the above equation is 0, and the right side of the equation becomes the equation with the unknown quantity s_2. It is known that rotation matrix R and translation matrix t which representing camera's motion can solve the depth of feature point p_2 which is on the image I_2. After that return the value of s_2 to the above equation, and the depth s_1 of the feature point p_1 on image I_1 can be solved. Table 1 shows the estimation error and measurement accuracy of the proposed approach.

4 Conclusions

This paper proposes a depth estimation solution based on binocular stereo vision with good real-time and accuracy. Based on good real-time and accuracy, the depth

Table 1 Estimation error and measurement accuracy

Pixel coordinate	Reprojection pixel coordinate	Reprojection error	Actual distance (cm)	Measured distance (cm)	Measurement accuracy (%)
0.165	0.164	0.001	15.00	14.78	98.6
0.165	0.164	0.001	20.00	19.62	98.1
0.165	0.164	0.001	30.00	29.45	98.2

estimation scheme proposed in this paper has broad application prospects in industrial application environments. In addition, after many experimental evaluations, the depth estimation solution proposed in this paper is feasible. Regarding future research, we will make optimization so as to reduce the reprojection error. Optimization could make the proposed depth estimation approach achieve better accuracy performance.

References

1. Rosin PL (1999) Measuring corner properties. Comput Vis Image Underst 73:291–307
2. Lowe DG (2004) Distinctive image features from scale-invariant keypoints. Int J Comput Vision 60:91–110
3. Michael C, Vincent L, Christoph S, Pascal F (2010) BRIEF: binary robust independent elementary features. European conference on computer vision. Springer, Heidelberg, pp 778–792
4. Bay H, Tuytelaars T (2006) Surf: speeded up robust features. Computer vision ECCV. Springer, Heidelberg, pp 404–417
5. Rublee E, Rabaud V (2011) Orb: an efficient alternative to sift or surf. In: 2011 IEEE international conference on computer vision. IEEE Press, pp 2564–2571
6. Meng Z, Kong X, Meng L, Tomiyama H (2018) Coordinate transformations for binocular vision systems. In: International symposium on advanced technologies and applications in the internet of things
7. Meng Z, Kong X, Meng L, Tomiyama H (2018) Distance measurement for pose estimation based on binocular vision technology. In: Taiwan and Japan conference on circuits and systems
8. Meng Z, Kong X, Meng L, Tomiyama H (2018) Distance measurement and camera calibration based on binocular vision technology. In: 2018 international conference on advanced mechatronic systems. IEEE Press, pp 342–347

A Dynamic Programming Algorithm for Energy-Aware Routing of Delivery Drones

Yusuke Funabashi, Atsuya Shibata, Shunsuke Negoro, Ittetsu Taniguchi, and Hiroyuki Tomiyama

Abstract Drones draw increasing attention as vehicles for home delivery services. Energy consumption is one of the most critical problems in delivery drones due to the limited capacity of batteries. This paper studies a routing problem for energy minimization of delivery drones. This paper formally defines energy minimizing vehicle routing problem (EMVRP) and proposes a dynamic programming algorithm to efficiently solve the problem. Experiments show the effectiveness of the proposed algorithm in terms of both quality of results and algorithm runtime.

Keywords EMVRP · Dynamic programming · Drone delivery

1 Introduction

Unmanned aerial vehicles (UAVs) or drones are collecting a lot of attention these days for hobby and business purposes. Especially, drones are considered as promising vehicles for home delivery services since they do not suffer from road traffic congestion [1, 2]. Another advantages of delivery drones against automotive trucks include low CO_2 emission and low labor cost.

Y. Funabashi (✉) · A. Shibata · S. Negoro · H. Tomiyama
Graduate School of Science and Engineering, Ritsumeikan University,
1-1-1 Nojihigashi, Kusatsu 525–8577, Japan
e-mail: yusuke.funabashi@tomiyama-lab.org

S. Negoro
e-mail: shunsuke.negoro@tomiyama-lab.org

H. Tomiyama
e-mail: ht@fc.ritsumei.ac.jp

I. Taniguchi
Graduate School of Information Science and Technology, Osaka University,
1–5 Yamadaoka, Suita, Osaka 565–0871, Japan
e-mail: i-tanigu@ist.osaka-u.ac.jp

N. N. Chiplunkar and T. Fukao (eds.), *Advances in Artificial Intelligence and Data Engineering*, Advances in Intelligent Systems and Computing 1133,
https://doi.org/10.1007/978-981-15-3514-7_91

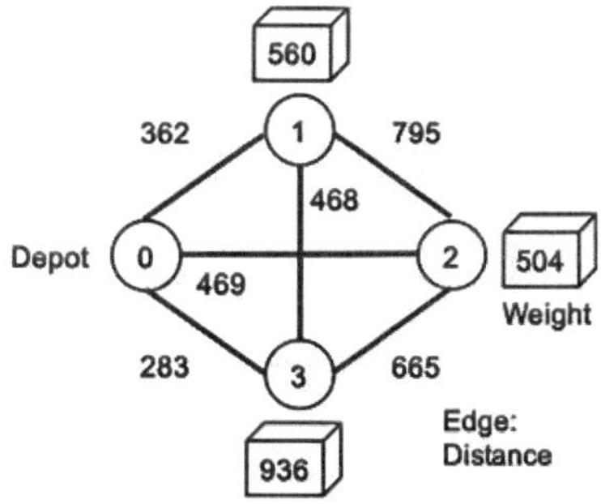

Fig. 1 An example problem

This paper addresses a kind of *vehicle routing problem (VRP)* for delivery drones. Given a set of items to deliver, the problem asks an optimal route which starts from a depot, delivers all of the items to customers, and comes back to the depot. The problem is similar to the traditional *traveling salesman problem (TSP)* which asks the shortest route in distance. On the other hand, our problem asks the route with the minimum energy consumption, which is called an *energy minimizing vehicle routing problem (EMVRP)* [3]. Energy consumption is much more critical in delivery drones than in delivery trucks since battery capacity of drones is very limited.

Let us consider an example shown in Fig. 1. The node labeled "0" denotes a depot, and the other three nodes denote customers (i.e., shipping destinations). The numbers in the boxes represents the weight of the items to deliver, and the numbers on the edges represent the distance between the two places. The optimal route for TSP is shown in Fig. 2a. The total distance of the route is 1964 ($= 362 + 468 + 665 + 469$). However, this route may not be optimal for EMVRP. The energy consumption of drones depends not only on the flight distance but also on the total weight of loaded items. The heavier the load is the higher the energy consumption is. This implies that heavy items should be delivered as early as possible. The optimal route for EMVRP is shown in Fig. 2b.[1] The total distance of the EMVRP route is 2015 ($= 283 + 468 + 795 + 469$), which is longer than the TSP-optimal route. Thus, the shortest route is not always same as the minimum-energy route.

This paper proposes a dynamic programming algorithm for EMVRP. The proposed algorithm efficiently finds exactly optimal routes in terms of energy consumption.

The rest of this paper is organized as follows. Section 2 surveys related work. Section 3 describes the routing problem for delivery drones, and Sect. 4 proposes a dynamic programming algorithm for the problem. Section 5 presents evaluation and finally Sect. 6 concludes this paper.

[1] Actually, the optimal EMVRP route depends on the drone.

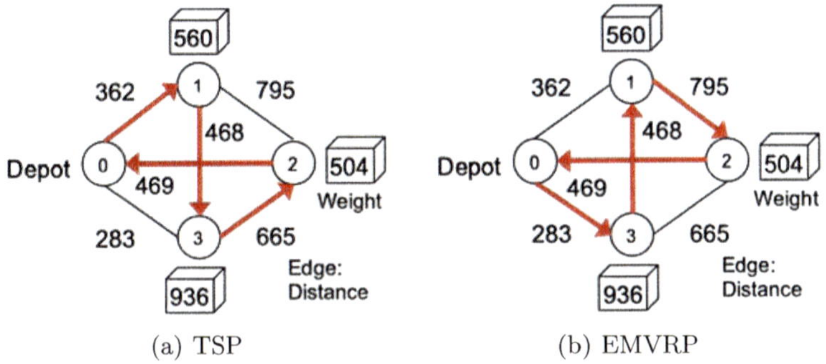

Fig. 2 Optimal routes

2 Related Work

The traveling salesman problem (TSP) is a classic problem, asking the shortest route which visits all customers and comes back to the origin. TSP is known to be an NP-hard problem, and a large number of heuristic, meta-heuristic, and exact algorithms have been developed so far. One of the simplest yet efficient heuristic algorithms is the nearest-neighbor (NN) algorithm, which selects the nearest customer one by one until all customers are visited. Exact algorithms for TSP include dynamic programming algorithms developed by Bellman [4] and by Held and Karp [5].

Vehicle routing problems (VRPs) extend the TSP for delivery purposes. A given set of items are delivered to customers by vehicles such as trucks. There exist various VRPs with different constraints and objectives. Examples of the constraints include the number and capacity of vehicles, scheduling of vehicles and drivers, and so on.

Energy minimizing vehicle routing problems (EMVRPs) further extend VRPs for energy minimization [3]. The energy for travel from a point to another is defined as a function of the distance between the two points and the weight of the vehicle and carrying items. Then, EMVRP asks the route which visits all customers with the minimum energy. In general, EMVRPs as well as VRPs are NP-hard. In [3], Kara formally defined EMVRP as an integer-programming problem. The energy is assumed to be proportional to the product of the distance and the weight. In [6], Negoro proposed a weight-first algorithm for EMVRP, which delivers the lightest item one by one. Their results show that the weight-first algorithm is worse than the classic NN algorithm unfortunately. In [7], Dorling developed a more detailed model for energy consumption of multimotor drones, and also proposed a simulated-annealing algorithm for EMVRP. In [8], Wang extended EMVRP for heterogeneous vehicles and presented an integer-programming formulation. Then, in [9], Wang proposed a genetic algorithm for the EMVRP for heterogeneous vehicles.

To the best of our knowledge, existing algorithms for EMVRP are heuristic or meta-heuristic ones, which do not guarantee the optimality of obtained solutions. This is the first paper which proposes a fast yet exact algorithm for EMVRP.

3 The Routing Problem for Delivery Drones

This section defines the EMVRP addressed in this paper.

We are given N items to deliver. Without loss of generality, no two items are to be delivered to the same customer. In other words, multiple items to the same customer are packed into a single package in advance. Then, the number of customers is also N. The items are numbered from 1 to N. The customer where item i $(1 \leq i \leq N)$ is to be delivered is called customer i or point i. The depot is numbered 0 as shown in Fig. 1.

This paper assumes that the items are delivered in a single trip by a drone. All of the items are uploaded onto the drone at the depot, and the drone starts a trip. If the total weight of the items exceeds the capacity of the drone, the items need to be partitioned into groups before routing and delivery, and how to partition the items is out of the scope of this paper.

Let $w(i)$ denote the weight of item i and $d(i1, i2)$ denote the distance between points $i1$ and $i2$. Also, let $x(j)$ denote the jth visited customer, which is the decision variable of the routing problem. Since a route starts and ends at the depot, we define:

$$x(0) = x(N+1) = 0 \tag{1}$$

Also, all of the customers are visited once, which is formally defined as follows:

$$1 \leq x(j) \leq N \quad (1 \leq j \leq N) \tag{2}$$

$$x(j1) \neq x(j2) \quad (1 \leq j1, j2 \leq N, j1 \neq j2) \tag{3}$$

The classic TSP asks the shortest route in distance, and its objective function is defined as:

$$\text{minimize} \quad \sum_{j=0}^{N} d(x(j), x(j+1)) \tag{4}$$

On the other hand, the goal of our EMVRP is energy minimization. Let $e(w, d)$ denote the energy consumption of the delivery drone, which is a function of payload weight w and flight distance d. Function $e(w, d)$ depends on the drone, and is assumed to be given. For example, in [3], $e(w, d)$ is defined as

$$e(w, d) \propto d \times (w + W_{\text{drone}}) \tag{5}$$

where W_{drone} is the weight of the drone itself. It should be noted that this work is not restricted to Formula (5).

Let $W(j)$ denote the total payload when the drone leaves jth visited point. When the drone starts a trip, all of the items are loaded. Therefore, the following formula holds.

$$W(0) = \sum_{i=1}^{N} w(i) \tag{6}$$

When the drone makes the jth stop ($1 \leq j \leq N$) at point $x(j)$, an item of weight $w(x(j))$ is unloaded. Therefore, the total payload when the drone leaves point $x(j)$ is defined as:

$$W(j) = W(j-1) - w(x(j)) \tag{7}$$

Then, the objective function of our EMVRP is defined as follows:

$$\text{minimize} \quad E = \sum_{j=0}^{N} e(W(j), d(x(j), x(j+1))) \tag{8}$$

The EMVRP addressed in this paper is formally defined as follows. Given w, d, and e, find x which minimizes the objective function (8) while meeting the constraints (1)–(3), (6) and (7).

4 Dynamic Programming Algorithm

This section proposes an exact algorithm for the EMVRP defined in the previous section. The proposed algorithm is based on dynamic programming [5].

4.1 Principles

In general, dynamic programming (DP) is an approach to mathematical optimization problems. DP divides a given problem into smaller sub-problems in a recursive manner. Then, using the optimal solutions of the sub-problems, DP finds the optimal solution for the original problem. DP runs efficiently by avoiding recomputation of similar sub-problems, where the similar sub-problems denote the sub-problems whose optimal solutions are the same with each other. In the design of DP algorithms, it is crucial to derive a recurrance relation between an original problem and sub-problems.

Let $\mathbb{S}$ denote a set of customers who are already visited, and let i be the last-visited customer in $\mathbb{S}$. We call a pair $(\mathbb{S}, i)$ as a *state*. Obviously, the initial state is $(\{0\}, 0)$. Then, we define a problem asking the minimum energy consumption $E(\mathbb{S}, i)$ for delivery from the initial state to state $(\mathbb{S}, i)$. Now, we can derive a recurrance formula to calculate $E(\mathbb{S}, i)$ as follows.

$$E(\mathbb{S}, i) = \max\{\ E(\mathbb{S}\backslash i, i') + e(W'(\bar{\mathbb{S}}) + w(i), d(i', i)) \\ \mid i' \in \mathbb{S}\backslash i\ \} \tag{9}$$

Algorithm 1 Dynamic Programming for EMVRP

```
W_init ← ∑W
for next_customer ∈ 𝕍 do
    dp[1 << (next_customer − 1)][next_customer] ← energy(depot to next_customer with W_init)
    Weight[1 << (next_customer − 1)] ← (W_init − W_next_customer)
end for

for state in (2^N − 1) do
    for next_customer ∈ 𝕍 do
        if next_customer has not been visited yet then
            for prev_customer ∈ 𝕍 do
                dp[state|(1 << (next_customer − 1))][next_customer] ←
                    min(dp[state][prev_customer] + energy(prev_customer to next_customer with Weight[state]),
                        dp[state|(1 << (next_customer − 1))][next_customer])
                Weight[state|(1 << (next_customer − 1))] ← Weight[state] − W_next_customer
            end for
        end if
    end for
end for

min_cost ← INFINITE
for prev_customer ∈ 𝕍 do
    min_cost ← min(dp[2^N − 1][prev_customer] + energy(pre_customer to depot without payload), min_cost)
end for
```

Recall that i is the latest customer in $\mathbb{S}$. In the formula, i' denotes the second-latest customer. $E(\mathbb{S} \backslash i, i')$ is the minimum energy consumption for flying from the depot to i', and $e(W'(\bar{\mathbb{S}}) + w(i), d(i', i))$ is the energy consumption for flying from i' to i. $W'(\bar{\mathbb{S}})$ denotes the total weight of items which are not yet delivered, which is formulated as:

$$W'(\bar{\mathbb{S}}) = \sum_{k \notin \mathbb{S}} w(k) \tag{10}$$

In Formula (9), it should be noted that, when departing from i', item i is still loaded on the drone. Therefore, $w(i)$ is added to $W'(\bar{\mathbb{S}})$. Also, it is obvious that the energy consumption at the initial state, i.e., before leaving the depot, is zero.

$$E(\{0\}, 0) = 0 \tag{11}$$

The original routing problem asks the minimum energy consumption when the drone departs from the depot, visits all of N destinations, and comes back to the depot. Formally, the original problem asks:

$$E(\{0, 1, 2, \ldots, N, 0\}, 0) \tag{12}$$

The problem in Expressoin (12) is recursively partitioned into sub-problems according to Formula (9), reaching Formula (11), and then, the optimal route with the minimum energy consumption is obtained.

4.2 Algorithm

Based on the principles in Sect. 4a, a dynamic programming algorithm is developed in this work. The pseudo code of our algorithm is outlined in Algorithm 1.

W_{init} is the total weight of all items to be delivered, and $\mathbb{V}$ is a set of customers. $state$ represents a set of customers who are already visited. Actually, $state$ is a bit-vector of length N, where N is the number of customers. If customer i is visited, the $(i-1)$th bit is set. $dp[state][customer]$ is a two-dimensional array which stores the energy consumption, and it corresponds to $E(\mathbb{S}, i)$ in the previous section. For example, $dp['0011'][2]$ stores the energy consumption when the drone already visited customers 1 and 2 and the drone is now at customer 2. Lines 2–4 calculates the energy consumption from the depot to the first customer. Then, Lines 7–18 travel all of remaining customers, and finally, in Lines 21–23, the drone comes back to the depot and the minimum energy consumption is calculated. Lines 7–18 are the main part of the DP algorithm. Instead of recursive procedure calls, the algorithm calculates the energy with three-level nested loops. The computational complexity of our DP algorithm is $O(2^N \times N^2)$, which is much faster than an exhaustive search of $O(N!)$.

5 Evaluation

The effectiveness of our proposed DP algorithm is evaluated through experiments. Our DP algorithm as well as several existing algorithms are implemented in Python and are compared in terms of the runtime of the algorithms and the quality of solutions (i.e., the energy consumption of the obtained routes).

5.1 Experimental Setup

Five routing algorithms shown below are compared in the experiments.

- *DP-TSP:* A dynamic programming algorithm for TSP. It finds the shortest route without considering energy.
- *NN-TSP:* The nearest-neighbor algorithm for TSP. It selects the nearest customer one after another.

- *NN-EM:* A simple heuristic algorithm for EMVRP. It selects the minimum-energy neighbor one after another. It is similar to NN-TSP, but the selection criteria are not distance but energy.
- *BF-EM:* A brute-force algorithm for EMVRP. It exhaustively explores all possible routes (i.e., $N!$ routes) to find the energy-minimum one.
- *DP-EM:* Our DP algorithm proposed in this paper.

In order to calculate the energy consumption of delivery drones, the energy model in Equation (5), which is presented in [3], is used. Based on the power measurement results presented in [10], we set W_{drone} is 297.5 and the maximum payload is 48. A total of 80 instances of EMVRP are randomly generated, where the number of customers ranges from 5 to 20. For each number of customers, there are five problem instances. The experiments are conducted on Intel Core i5 processor.

5.2 Results

Figure 3 shows the results on the runtime of the five routing algorithms. The complexities of our NN-TSP and NN-EM implementations are $O(N^2)$, and their runtime is less than 1 millisecond in any cases. The complexity of BF-EM is $O(N!)$, and BF-EM fails to find optimal solutions within an hour for problems with more than 12 customers. The complexity of DP-TSP and DP-EM are exponential, still they are practical for delivery to 20 customers.

Figure 4 shows the results on energy consumption obtained by the five algorithms, where the results are normalized to DP-TSP. As mentioned above, there exist five problems for each number of customers. In the figure, the average of the five results is depicted. The results show that the two heuristic algorithms, NN-TSP and NN-EM,

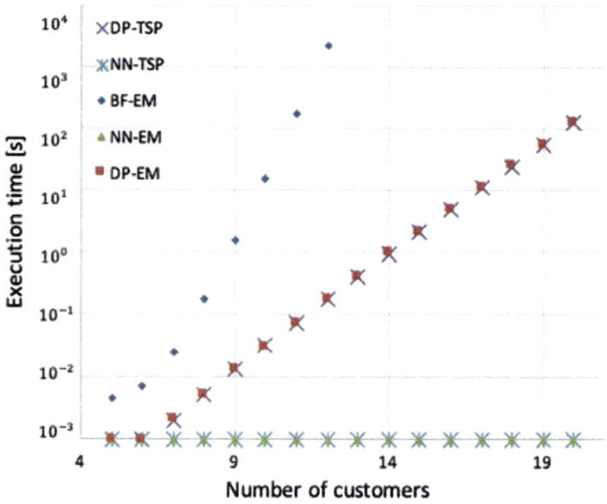

Fig. 3 Runtime of the routing algorithms

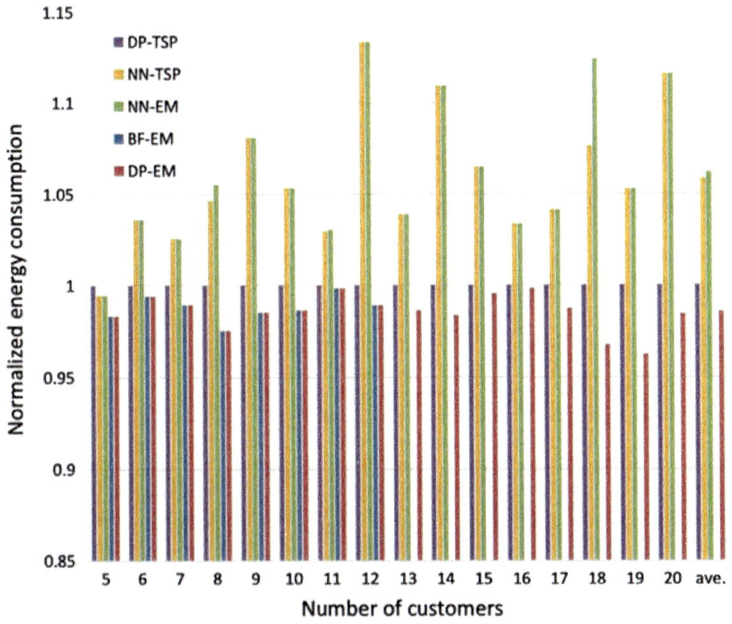

Fig. 4 Energy consumption of the obtained routes

are not effective. As long as BF-EM finds solutions, it is confirmed that our DP-EM algorithm successfully finds the same solutions as BF-EM does. Also, it is observed that DP-TSP is not optimal in terms of energy consumption. DP-EM outperforms DP-TSP by 1.5% on average.

6 Conclusions

Drones are expected to be popular vehicles for home delivery services in near future. In this paper, we have proposed a dynamic programming algorithm for an energy minimizing routing problem for delivery drones. Experimental results show that the proposed algorithm efficiently finds optimal routes.

The routing problem addressed in this paper is static one, meaning that a set of items to deliver is fixed. In the future, we plan to address dynamic problems where delivery orders arrive at the depot next from next over time.

References

1. Amazon Prime Air. http://www.amazon.com/primeair
2. Project Wing. https://x.company/projects/wing/
3. Kara I, Kara BY, Yetis MK (2007) Energy minimizing vehicle routing problem. In: International conference on combinatorial optimization and applications, Springer, pp 62–71
4. Bellman R (1962) Dynamic programming treatment of the travelling salesman problem. J ACM 9:61–63
5. Held M, Karp RM (1962) A dynamic programming approach to sequencing problems. J Soc Ind Appl Math 10:196–210
6. Negoro S, Taniguchi I, Tomiyama H (2016) Fundamental analysis of low energy path routing for delivery quadcopters. In: International technical conference on circuits/systems, computers and communications 2016
7. Dorling K, Heinrichs J, Messier GG, Magierowski S (2017) Vehicle routing problems for drone delivery. IEEE Trans Syst Man Cybern Syst 47:70–85
8. Wang S, Liu X (2016) Energy minimization vehicle routing problem with heterogeneous vehicles. In: International conference on service systems and service management 2016
9. Wang S, Wu Y (2017) A genetic algorithm for energy minimization vehicle routing problem. In: International conferemce on service systems and service management 2017
10. Negoro S, Sukezane D, Shibata A, Maekawa K, Taniguchi I, Tomiyama H (2018) Measurement and modeling of quadcopter energy with ROS. In: Workshop on synthesis and system integration of mixed information technologies 2018

Qualitative Approach of Empirical Mode Decomposition-Based Texture Analysis for Assessing and Classifying the Severity of Alzheimer's Disease in Brain MRI Images

K. V. Sudheesh and L. Basavaraj

Abstract Medical image processing systems are widely adopted in several real-time diagnostic systems due to their significant nature of information extraction and processing which helps to predict the early stages of illness. Alzheimer's disease (AD) is one of the most chronic and challenging diseases in the medical diagnostic field. This disease is responsible for neurodegenerative brain disorder and attacks the brain cells and nerves that result in affecting the brain functionality and finally cause dementia. In this work, the prime focus is on the early prediction of Alzheimer's disease using image processing-based machine learning techniques. Erstwhile, extensive studies are researched using pathological- and MRI-based systems which show the issues caused due to the brain's white matter damage. Nevertheless, these studies do not provide that how white matter damage is associated with the AD and its classification at multiple stages. Conferring to the proposed approach, an improved feature extraction technique is introduced by combining empirical mode decomposition and gray-level co-occurrence matrix (GLCM). In order to abstract robust features, several image preprocessing steps are applied such as image enhancement, and later feature extraction is applied followed by the classification where multiple classifiers such as KNN, decision tree classifier, RBF, and support vector machine classification are castoff to assess the performance of feature extraction technique. Projected methodology obtains promising performance for the classification of various stages of AD and consequently can be employed for real-time application for early prediction of AD. An extensive experimental study is carried out for the anticipated approach that is implemented on OASIS brain imaging dataset. Experimental study shows that support vector machine, KNN, decision tree, and RBF classification techniques achieve that the accuracy as 39.06%, 80.20%, 98.43%, and 77.08%, respectively. Combination of the proposed feature extraction technique and decision tree classification scheme achieves a promising classification performance.

K. V. Sudheesh (✉) · L. Basavaraj
Department of Electronics and Communication Engineering, ATME College of Engineering, Mysuru, Karnataka, India
e-mail: sudheesh.vvce@gmail.com

L. Basavaraj
e-mail: basavaraj.atme@gmail.com

N. N. Chiplunkar and T. Fukao (eds.), *Advances in Artificial Intelligence and Data Engineering*, Advances in Intelligent Systems and Computing 1133,
https://doi.org/10.1007/978-981-15-3514-7_92

Keywords Alzheimer's disease · Co-occurrence matrix · Decision tree classifier · Dementia · Empirical mode decomposition · T1 weighted MRI

1 Introduction

Recent growth in technology has led to the dramatic growth in several systems such as medical imaging and computer vision systems. This advancement in medical imaging systems by means of computer vision application helps to offer better diagnosis solutions and healthcare systems. In this field of medical healthcare system development, brain monitoring and identification of brain dementia are considered as a crucial task for doctors and clinicians [1]. Normally, the main cause of brain tumor is an abnormal increase in human brain cells, which can have a serious impact on the human body; as a result, an early prediction is highly recommended by doctors which can be useful to prevent the growth of disorder brain cells. However, brain abnormality effect may not be similar for each person, and also, it can vary based on the diagnostic sessions. Several types of brain disorders are diagnosed and identified which have different impacts on the human body. The brain tumors are primarily classified as benign and malignant tumors. Benign tumors do not contain any cancerous cell and are homogeneous in the structure which can be monitored by minor surgical eradication or radiological operations. Commonly, these types of cell swelling have less growth, whereas malignant cells grow in the form of heterogeneous cells of cancer, and these are chronic in nature which can be treated by using chemotherapy and radiotherapy. Several techniques have been developed to detect these types of brain diseases which can be useful for early stage of abnormality identification such as way of computed tomography and magnetic resonance imaging [2]. MRI techniques stand widely adopted for identifying the abnormal growth of cells' location and size but complete information extraction becomes a challenging task, thereby another painful process is presented which is known as biopsy. Biopsy is a painful technique and cannot be performed for every patient; consequently, there is a need to develop a significant technique for abnormality identification for its diagnosis in the early stage. Brain imaging system is a well-known process of clinical information extraction from the patient's medication assessment which can be used for further diagnosis purposes. Moreover, MRI systems can extract the clinical information from multidimensional images which can be used for disease monitoring after recognition and localization of affected area [3]; to complete this system of brain dementia identification and classification, various stages are included, such as data processing, information extraction, and representation. Moreover, this information on medical image data is represented in the higher level of abstraction to achieve the relevant clinical knowledge which can be helpful for decision making in the diagnostic system. In this field, effective data management, data processing, visualization, and efficient data analysis process can improve the computational performance by utilizing high-speed processors, image display units, and software programs. Recently, the growth in medical field has led toward the growth of size of medical imaging database, wherein the implementation of these techniques becomes a very complex task; hence, automated tools of medical image analysis have attracted researchers due to its advantages in clinical diagnosis system [4].

Based on MRI techniques, several methods have been proposed for AD identification and classification. In this field of the medical image, computer vision-based pattern learning plays an important role in automated MRI image analysis [5]. Furthermore, these techniques of automated brain image analysis can be used for identifying various brain-related diseases such as Alzheimer's, tumor, cerebrovascular, and inflammatory. This becomes an important research area for the medical and industrial fields for automated brain disorder's exposure and classification. A novel approach is presented for swelling cell detection and classification [6]. In order to do this, texture feature extraction is performed. Mainly two texture features are extracted where piecewise triangular prism surface area (PTPSA) procedure is used intended for first fractal texture feature mining and the second feature includes fractional Brownian motion. Later, these features are fused to formulate an efficient model of feature pattern. Machine learning-based techniques are also adopted widely for abnormality identification and classification. For classification, mainly image segmentation and support vector-based schemes have been used [7]. In this field, the most promising techniques for revealing of irregular cells are known as SVM, artificial neural network [8], fuzzy c-means [9], and expectation maximization (EM) methods, and techniques are widely adopted for detection and classification of brain syndromes. Support vector machine uses statistical parameter learning methodology and performs classification into the available classes in the dataset. This scheme has been widely used for classification purpose such as image recognition and biomedical applications. In this process of brain aberration classification, feature selection process plays an important role [10].

However, a huge amount of works have been carried out in this field of brain irregularities detection but still, computational complexity and robustness remain challenging tasks for automated brain MRI analysis tools. Generally, brain syndromes are heterogeneous in nature due to irregular shape, texture, color, and position which may lead to the complexity to perform the AD identification and classification. These processes are based on feature extraction and classification progression. Since support vector machine shows a significant performance, feature assortment and processing are very crucial task which can help to increase the performance of system. In order to increase performance, study introduces a new approach for feature extraction where empirical mode decomposition (EMD) and GLCM feature extraction techniques are combined together. Prior to this, a technique for image enhancement is also implemented for robust feature extraction.

2 Literature Review

This section provides a brief discussion in this area of brain Alzheimer's identification and classification using computer vision techniques. As discussed before, the performance of computer vision-based pattern learning scheme depends on the various stages, for instance, preprocessing, feature extraction, feature selection, and classification [11]. Image preprocessing is a very important stage in image processing.

At this stage, multiple steps are applied to enrich the image quality aimed at better dispensation such as gray conversion, image filtering, and image enhancement. In computer vision-based machine learning techniques, image preprocessing plays an important role followed by image segmentation and classification [12]. Generally, captured brain images contain brain tissues surrounded by the skull. These tissues can be removed applying segmentation scheme on captured brain image. For better analysis, image normalization is considered as an important technique [13]. Specifically, in MR image acquisition, multiple scanners are used during image capturing process which causes intensity variation and degrades the image analysis stage. Hence, histogram normalization technique is presented to overcome this matter. The comprehensive process is supported out into two stages as intensity scaling where higher intensity signal is scaled to lower intensity and another stage is histogram normalization where histogram ranges are adjusted. Most of the brain disease identification techniques require image segmentation process for a clear view of dementia and its region. For this type of application, combined feature extraction techniques show significant performance for achieving high accuracy performance. This also focuses on the skull removal from brain tissues, and later texture feature extraction model is presented [14]. Later, some techniques are introduced to improve the performance by considering a hybrid model for brain affected cell segmentation [15] where the complete process of classification is categorized into two techniques: the first technique is based on the symmetry computation models where symmetry and active contours of MRI are considered for computations. According to this process, active contour model is implemented on a given filtered image. Later, a simple computation process is applied to improve the process of segmentation where the differences of two consecutive images are computed and represented in the binary form, and later this binary form of the image is mapped on the original image to extract the segmented region. Accurate segmentation of brain disease is a very crucial task for researchers using automated process. Segmentation can help to enhance the image analysis and provide the number of information comparatively. On the other hand, machine learning process such as deep neural network-based segmentation scheme is also used for brain segmentation. This is a completely automated process for the segmentation of brain cells which is implemented for glioblastomas (both low and high grades) MRIs [16].

As mentioned, these types of disorders are completely random in nature and can appear at any place with any shape and size and contrast which is completely unpredictable. In order to cope up with this, a new approach of segmentation for MR images using deep neural networks which contain various layers to speed up the processing along with this two-phase training methodology is adopted. Similar to other machine learning practices such as deep neural network and convolution neural network-based scheme are also used for MRI segmentation [8]. Unlike previous work, this convolutional neural network (CNN) model takes kernels for computation.

These small kernels help to develop a bigger architecture of the neural network and overcome the effect of neural weights in the network.

However, brain swelling has various subclasses such as astrocytoma (AS), infancy tumor—medulloblastoma (MED), glioblastoma multiform (GBM), normal regions (NR), meningioma (MEN), and tributary tumor-metastatic (MET); therefore, a multi-class classification scheme can be used for identifying these classes of tumor in any given brain MRI image [17]. This process uses active contour model for region of interest (ROI) extraction, multiple intensity, and textured feature points are also obtained. Further, dimensionality reduction is implemented using principal component analysis (PCA). As per the classification analysis, artificial intelligence scheme is implemented to achieve multi-class classification performance. In this computer vision process, image quality is an important aspect which is responsible for the performance of irregularity classification. Some recent studies revealed that the intensity of the brain and normal tissue is not similar which requires multispectral image analysis by identifying the dementia slices using an automated process [18]. Based on this concept of intensity, variation between brain and tissue various schemes has been developed. In this field of segmentation, texture parameter extraction and conditional random field techniques are utilized which are further classified by using support vector machine [19]. For T1-weighted MRI, the 3D convolutional approach is also discussed recently [20]. Proper training is provided to the complete database which helps to achieve the contrast-enhanced scans. This process is applicable for the MRI data which contains four channels as contrast-enhanced, non-enhanced, T1-weighted, T2-weighted, and FLAIR contrasts. In order to classify the abnormality, support vector machine scheme is widely studied and adapted for medical brain MRI image classification. Wavelet-energy parameters are considered for feature descriptor computation, and the performance of support vector machine is optimized using biogeography-based optimization (BBO) scheme to achieve the robust performance [21].

This analogue describes that various recent techniques have been discussed in this field of brain Alzheimer's detection and classification using computer vision schemes. In computer vision techniques, machine learning or pattern recognition methods are widely adopted to achieve the accurate prediction and detection of the brain diseases. However, computational complexity and accuracy of abnormality detection are a challenging task for researchers. Moreover, accuracy of system can be enhanced by appropriate and robust feature extraction techniques. For this reason, there is a need to develop a robust model for accurate AD identification and classification for the medical diagnostic system.

3 Materials and Methodology

This section presents a complete description of the proposed methodology for the classification of AD multiple stages using MRI imaging systems. The complete system architecture is shown in Fig. 1.

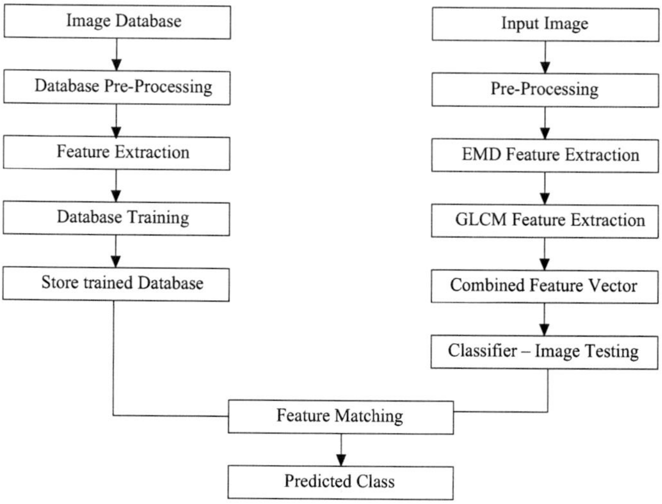

Fig. 1 System architecture

The complete process is categorized as follows: (a) During image acquisition, medical images may suffer from low-illumination and contrast issues which may lead to the inappropriate analysis and faulty diagnostic recommendations by clinicians. In order to deal with this issue, an improved approach for contrast enhancement in medical imaging for MRI images is aimed at the study. (b) The proposed approach is based on the pattern learning or machine learning approaches; accordingly, feature selection scheme is required to achieve better classification. Hence, an approach is put forth for optimal feature selection. For robust feature modeling, gray-level co-occurrence matrix (GLCM)-based featured pixels are incorporated. (c) In image processing, data transformation-based techniques play an important role to obtain the frequency domain feature by decomposing input image. In this effort, transformation-based scheme for feature abstraction is proposed. Later these features are combined into one vector to formulate the feature vector. And finally, (d) a multi-class classifier scheme is used for learning the patterns to perform the training and testing in order to identify the multiple stages of AD.

3.1 Image Enhancement

Here, a new approach is described for image enhancement using improved histogram equalization technique where histogram of input image is divided into two sub-histograms. In this work, the main aim is to preserve the image brightness and edge information during processing the low-contrast MRI image.

Let the input image be *I*, and its histogram of *k* gray level is given as:

$$\mathcal{H}(k) = n_k \quad \text{for} \quad k = 0, 1, \ldots L \tag{1}$$

n_k denotes occurrence of gray stages, and total number of gray levels are denoted by L. In this, probability density function of input image histogram is expressed as:

$$p(k) = \frac{\mathcal{H}(k)}{P} \quad \text{for} \quad k = 0, 1, \ldots, L-1 \tag{2}$$

where P represents total pixels in image which depends on the image size. In this phase, the main focus is to achieve threshold of input image which is computed by means of a threshold parameter t. At this stage, histogram with gray-level range $[0, L-1]$ is partitioned into multiple segments as upper $\left(\mathcal{H}_{\text{upper}}\right)$ and lower histograms $(\mathcal{H}_{\text{low}})$ with the histogram range $[t, L-1]$ and $[0, t-1]$, respectively, for upper and lower histograms. PDF of each histogram is formulated as follows:

$$p_{\text{low}}(k) = \frac{\mathcal{H}_{\text{low}}(k)}{n_{\text{low}}} \quad \text{for} \quad k = 0, 1, \ldots, t-1 \tag{3}$$

$$p_{\text{upper}}(k) = \frac{\mathcal{H}_{\text{upper}}(k)}{n_{\text{upper}}} \quad \text{for} \quad k = \tau, t+1, \ldots, L-1 \tag{4}$$

n_{low} and n_{upper} signify total number of pixels in lower and upper histograms. In this process of image quality enhancement, the focal point is on histogram bins of input image. Here, histogram partition is executed and is divided into two sections as lower and upper histograms. Higher frequency histograms are considered as an important part when compared with the lower frequency histograms. The complete image is deformed into multiple bins based on the intensity and histogram variations. However, these bins also contain image information, and thereupon an approach to improve is put forward the low-frequency histogram bins. In order to do this task, histograms are modified by using logarithmic computation on lower and upper histogram PDFs which are computed as follows:

$$\text{Updated}_{P_{\text{low}}}(i) = \log[p_{\text{low}}(i) + 1] \tag{5}$$

$$\text{Updated}_{P_{\text{upper}}}(i) = \log\left[p_{\text{upper}}(i) + 1\right] \tag{6}$$

In this process, combined histogram enhancement is performed by taking help of cumulative density function $c(k)$ which is denoted as:

$$c(k) = \sum_{i=0}^{k} p(i) \quad \text{for} \quad k = 0, 1, \ldots., L-1 \tag{7}$$

This function is mapped into k gray levels using transform function defined as:

$$f(k) = X_0 + (X_{L-1}.c(k) - X_0 \cdot c(k)) \quad (8)$$

where X_0 denotes minimum gray level and X_{L-1} denotes maximum gray level for the input image. With the help of (8), input image can be plotted into previously discussed dynamic range $[X_0, X_{L-1}]$ and the histogram variations can be noticed. At this stage, to balance the histogram, updated histogram values are clipped and redistributed equally before processing for normalization. Based on this, the enhanced image can be obtained as:

$$f(k) = \begin{cases} X_0 + (X_{t-1} - X_0). \sum_{i=0}^{k} \text{Update}_{P_{\text{low}}}(i) & \text{for } k = 0, 1, \ldots, t-1 \\ X_t + (X_{L-1} - X_t). \sum_{i=0}^{k} \text{Update}_{P_{\text{upper}}}(i) & \text{for } k = t, t+1, \ldots L-1 \end{cases} \quad (9)$$

3.2 Feature Extraction

Previous segment deals with image contrast enhancement technique for refining the enactment of image analysis. In this subsection, it is analyzed about the feature extraction modeling for brain MRI medical imaging system. In addition to medical image, texture analysis can provide better information and helps to improve the clinical diagnosis process [22]. Thus in this paper, a texture feature extraction model for MRI imaging applications is processed. However, in this field of pattern recognition, spatial-space frequency computation has gained attention due to its significant nature of pattern learning. Moreover, this can be enhanced by applying a multi-resolution frequency decomposition scheme for considered input signal. Medical images contain nonlinear and non-stationary information; thereupon empirical mode decomposition technique can be considered for implementation. Further, it is improved by the empirical mode decomposition process using adaptive texture analysis modeling. In compliance with the proposed approach, empirical mode decomposition (EMD) technique is used where several intrinsic mode functions (IMF) are computed by decomposing the original input signal and later processed to Hilbert transform resulting in amplitude and frequency modulated frequencies. Hilbert transform is applied on computed IMFs, and three important features are extracted which are (a) spectral entropy feature, (b) spectral magnitude, peak frequency, and (c) spectral energy. As this method is direct, data-dependent and adaptive in nature do not require linearity and stationary conditions for signal evaluation. It is implicit that any signal is self-possessed of a sequence of diverse intrinsic oscillation modes, and the EMD can be castoff as a process that brings out this decomposition of the inbound signal obsessed by its dissimilar intrinsic mode function (IMF). This can help to decompose any complex signals into various smaller IMFs and can obtain Hilbert transform. However, signal decomposition process is based on the characteristics of time and data. During IMF computation, two assumptions need to be satisfied that

(a) in the entire data, total number of extrema and zero-crossing parameters also need to consider whether these are equal or different from each other (b) computed mean value of local minima and maxima must be equal to zero.

Let the input signal is given as $a(t)$ wherein local maxima and minima points are computed. Once all local maxima points are computed, it is connected by using cubic spline curves as upper envelope as $a_u(t)$, and similarly, the local minima points are connected from lower envelope as $a_l(t)$. After connecting all the local extrema points, the mean of these envelopes is computed as:

$$m_1(t) = \frac{a_u(t) + a_l(t)}{2} \tag{10}$$

Using this, first IMF is computed as $I_1(t) = a(t) - m_1(t)$. This process is also known as shifting process. In next stage, second IMF is computed. In order to reckon this, first IMF is assumed as signal and its upper, lower, and mean values are computed and used for second IMF computation as $I_1 - m_{11} = h_{11}$. This process of IMF computation is repeated until the total IMF is obtained as:

$$I_{1(k-1)} - m_{1k} = I_{1k} \tag{11}$$

Along with the IMF computation, standard deviation between two successive IMFs is also computed, and this process is repeated until the standard deviation is less than 0.3. This standard deviation computation can be expressed as:

$$SD = \frac{\sum_{t=0}^{T} \left| I_{1(k-1)}(t) - I_{1k}(t) \right|}{I_{1(k-1)}^2(t)} \tag{12}$$

At this stage, original input signal can be represented by combining the IMF and its residue in the linear combining process, given as:

$$a(t) = \sum_{m=1}^{M} I_m(t) + r_M(t) \tag{13}$$

where, M denotes total number of IMFs, h_m denotes the mth and IMF r denotes residue which can be computed by estimating the last IMF as $c_1 = I_{1k}$ and residue can be expressed as $r_1 = a(t) - c_1$. Later, these IMFs are processed through Hilbert transform computation, and pre-envelope of IMF is calculated as follows:

$$z(t) = I(t) + jI_H(t) \tag{14}$$

where $I_H(t)$ denotes the Hilbert transform of signal $h(t)$ expressed as:

$$h_H(t) = I(t) * \frac{1}{\pi t} \tag{15}$$

where * denotes convolutional operator. With the help of Eqs. (15) and (14), it can be represented as:

$$z(t) = A(t)e^{j\varphi(t)} \tag{16}$$

In the next phase of proposed approach, amplitude of input signal and instantaneous phase are computed as $A(t)$ and $\varphi(t)$, respectively, which is given as:

$$\begin{aligned} A(t) &= \sqrt{I^2(t) + I_H^2(t)} \\ \varphi(t) &= \arctan\left[\frac{I_H(t)}{I(t)}\right] \end{aligned} \tag{17}$$

In the next, the pre-envelope data of IMF is rectified as $c(t)$ and the instantaneous frequency is computed as:

$$\omega(t) = \frac{d\varphi(t)}{dt} \tag{18}$$

Here, Fourier transform for Hilbert transform needs to be computed to achieve the auto-regressive spectrum sensing model of the input signal. Based on this, the power spectrum of nth order is computed as:

$$P_{XX}(f) = \frac{1}{\left|1 + \sum_{l=1}^{n} a_n e^{-j2fl}\right|^2} . \hat{\varepsilon}_n \tag{19}$$

where ε_p is the total error which is computed by estimating the forward and backward prediction error. As examined before that three features are extracted known as spectral peak, entropy, and energy. Power spectrum is used for peak estimation where sample-by-sample frequency range is considered for estimation. Spectral entropy features are computed as follows:

$$SP = -\sum_{f=0}^{f_s/2} \bar{P}_{XX}(f) \log\left[\bar{P}_{XX}(f)\right] \tag{20}$$

where $\bar{P}_{XX}(f)$ denotes normalized power spectral density and f_s denotes the sampling frequency. Similarly, spectral energy is given as:

$$SE = \frac{1}{N} \sum_{f=0}^{f_s/2} \bar{P}_{XX}(f) \tag{21}$$

Table 1 GLCM normalized features

Feature	Mathematical formula
Entropy	$\sum_x \sum_y P(x, y) \log P(x, y)$
Contrast	$\sum_x \sum_y (x - y)^2 P(x, y)$
Energy	$\sum_x \sum_y p^2(x, y)$
Homogeneity	$\sum_x \sum_y \frac{P(x,y)}{1+\lvert x-y \rvert}$

where N denotes total number of auto-regressive coefficients. Along with these feature extraction model, GLCM features are also computed and incorporated with Hilbert transform features to formulate a robust feature vector. Here, it is reviewed on the basis of texture feature extraction process [23] which has been adopted widely for various classification systems. Gray-level co-occurrence matrix (GLCM) features are aimed for texture analysis of any given input image. Any image is constructed by using some certain pixel values with an intensity or a gray value of image. According to GLCM, combination and occurrence of gray level in any of image are analyzed and tabulated. This tabulation process is used for texture feature analysis. Normalized co-occurrence matrix feature are shown in Table 1.

GLCM features are considered as a matrix wherein the distribution is based on the angular and distance relation between pixels. Commonly in image, pixels vary at every position which can be used for identifying various texture information using GLCM. With the help of GLCM, several features can be computed using these matrices which are classified as visual texture features, statistics feature, and information theory feature and information correlation measurement. Once the complete feature vector is formed, multi-class classifier is used to obtain the classification performance. In this research, multiple classes of MRI diseases have been considered; thereupon multi-class support vector machine classifier is aimed at classification performance analysis and comparison.

3.3 Classification Techniques

This section provides a brief deliberation about various classifiers which have been used in this study such as support vector machine, decision tree classifier, KNN classifier, and RBF classification technique.

Support Vector Machine. Here, the examination is based on the support vector machine classification scheme. SVM adapts maximization margin criteria which help to obtain the optimal hyperplane between the binary classes. As conceded that, a training feature vector f is given as $f_i \in \mathcal{V}^d, i = 1, \ldots, l$ for two classes where each class label vector is $y \in \{1, -1\}^l$. Support vector machine approach obtains the solution by solving the optimization problem given as:

$$\begin{gathered}\min_{w \in \mathcal{H},\, b \in \mathcal{V},\, \gamma_i} \frac{1}{2} w w^T + C \sum_{i=1}^{l} \gamma_i \\ \text{s.t } y_i\big(w^t \varphi(x_i) + b\big) \geq 1 - \gamma_i, \\ \gamma_i \geq 0,\, i = 1, \ldots l\end{gathered} \tag{22}$$

where w denotes weight vector, C denotes regularization constant, φ is a function used for mapping the data, and $\mathcal{H}$ denotes suitable feature space which is used for formulating decision surface. This can be solved for binary class problem; but as per the occurrence, it is deemed that a multi-class classification problem which is a more complex task to perform the classification. Rather than creating multiple binary classifiers, multi-class classifier is adopted which divides the k class problem into single objectives for training the support vectors and maximizes the margin of each class. According to this process, labeled training set is denoted as $\{(x_1, y_1) \ldots, (x_l, y_l)\}$ of cardinality l where $x_i \in \mathcal{V}^d$ and $y_l \in \{1, \ldots k\}$ which can be scrutinized as follows:

$$\begin{gathered}\min_{w_m \in \mathcal{H},\, b \in \mathcal{V}^{l\times k},\, \xi \in \mathcal{R}_i^{l\times k}} \frac{1}{2} w_m w_m^T + C \sum_{i=1}^{l} \sum_{i \neq y_i} \xi_{i,t} \\ \text{s.t. } w_{y_i}^T \varphi(x_i) + b_{y_i} \geq w_t^T \varphi(x_i) + b_t + 2 - \gamma_{i,t} \\ \gamma_i \geq 0,\, i = 1, \ldots l,\, t \in \{1, \ldots, k\} \backslash y_i\end{gathered} \tag{23}$$

Based on this, the resulting decision function is expressed as:

$$\arg\max_m \mathcal{F}_m(x) = \arg\max_m \big(w_m^T \varphi(x) + b_m\big) \tag{24}$$

KNN Classification. In this subsection, the KNN classification scheme is described briefly. According to this process, test objects are evaluated based on the closest or more similar training sample in the feature space, whereas entity is classified built on the mainstream votes of its neighbors, wherein the nearest neighbor is determined using metric function. KNN algorithm is shown in Table 2.

Decision Tree Classification. This subsection deals with decision tree classification approach. Various decision tree algorithms have been presented which are based on the top-down and recursive materialistic search technique to achieve best decision tree for a given problem. According to this technique, a single attribute is selected at a single shot from the total accessible attributes as a node in tree and decedent of this node creates a legal value for this attribute. Below given Table 3 shows a pseudo-code process of decision tree classifier.

RBF Classification. Radial basis function (RBF) classification is technique of machine learning which is built on the neural network learning. In this work, classification approach for medical image classification is utilized. RBF algorithm is shown in Table 4.

Table 2 KNN classifier

Input: k number of the nearest neighbor classifier, training sample T, set samples D **Output**: labels of the test sample
Step 1: split feature set into training and testing feature vectors Step 2: initialize the predicted labels with empty matrix as $L = [.]$ Step 3: for each d in D and t in T do Step 4: Neighbor $(d) = \{\}$ Step 5: if $\lvert\text{Neighbor}(d)\rvert < k$ then Step 6: given neighbors are as $\text{Neighbors}(d) = \text{closest}_{\text{set}(d,t)} \cup \text{Neighbors}(d)$ Step 7: end if Step 8: if $\text{Neighbors}(d) = k$ then Step 9: break Step 10: $L = \text{test}_{\text{class}(\text{Neighbors}(d))} \cup L$ Step 11: end for Step 12: performance measurement

Table 3 Decision tree classifier

Input: Feature attributes (F), training class (T), testing class, and target labels (Tr) **Output**: decision tree, prediction, and performance
Step 1: initialize the root node for tree formulation Step 2: evaluate attribute list, and if attributes of (T) and target attributes are the same as t_i Step 3: return a tree with single node along with its attributes Step 4: if (F) is empty (no attribute) Step 5: then a tree with single node with most common target value from Tr, otherwise Step 6: select A attribute from the F and classify based on entropy-based measurement Step 7: set this attribute for root and compute the legal value Step 8: for all permissible value of A, perform Step 9: add a branch below root which corresponds to $A = v_i$ Step 10: let T be a subset of A which contains $A = v_i$ Step 11: if T is empty, do Step 12: add a leaf node in below branch by the target value with utmost common value in target TR else Step 13: add sub-tree learned using Decision_tree(T, Target, Feature) Step 14: return root with the decision tree

Table 4 RBF classifier

Input: labeled training pattern, number of iteration, RBF centers, regularization constant, spreading factor **Output**: optimized network for classification
Step 1: apply k-means clustering model to find the initial closest neighbors Step 2: optima weight computation as $w = \left(G^t G + 2\frac{\lambda}{l} I\right)^{-1} G^t y$ Step 3: gradient computation as $\frac{\delta}{\delta\mu_k} E$ and $\frac{\delta}{\delta\sigma_k} E$ with optimal weight and gradient vector Step 4: estimate the conjugate direction for vector $\bar{v}$ Step 5: apply line search and minimize the δ in direction of $\bar{v}$ Step 6: Re-compute the optimal weight Step 7: store the updated vectors in the optimized RBF net Step 8: classification performance analysis

4 Experimental Results and Comparison

This section presents a complete experimental analysis using proposed image retrieval approach. Looking forward to validate the proposed method performance, a simulation study is presented by considering OASIS medical imaging database. This complete experiment is presented using MATLAB 2013 running on windows platform with Intel i3 processor.

4.1 Database Description

A brief description of Open Access Series of Imaging Studies (OASIS) is presented here [24]. This database repository contains two-type of database of brain MRI known as: (a) cross-sectional MRI and (b) longitudinal MRI wherein both databases are considered as young, middle age, demented, and non-demented older adults for the acquisition of database.

- Cross-sectional MRI Data: In this set, cross-sectional brain MRI database of 416 subjects is considered between ages of 18–96. For every user, 3–4 T1 weighted MRIs are obtained. In this, both men and women participated in database acquisition. 100 users are identified with slight to sensible Alzheimer's disease, and these users remain found to be above the age of 60.
- Longitudinal MRI Data: This database contains MRI images of 150 users who are aged from 60 to 96. Total 373 sessions of MRI imaging are considered for image acquisition. For all user, 3–4 discrete T1-weighted MRI shots are acquired for both right-handed men and women. In this, 72 images are considered as non-demented, 64 images are considered as demented, and 51 images are considered with mild-to-moderate Alzheimer's disease. Sample images of OASIS database are shown in Fig. 2.

In order to identify the clinical diagnosis status of each user, Clinical Dementia Rating (CDR) analysis is utilized in this database. CDR is a computational process which considers six domains of each user such as memory, judgment, orientation, problem-solving, functionality, home, hobbies, and personal care. [24]. A global

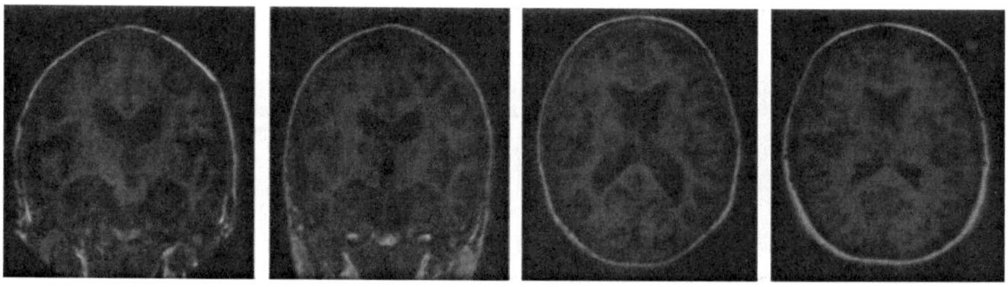

Fig. 2 OASIS database sample images

Table 5 Data acquisition details

Sequence	MP-RAGE
TE (msec) (echo time)	4.0
TR (msec) (repetition time)	9.7
TI (msec) (inversion time)	20
TD (msec) (delay time)	200
Flip angle	10
Orientation of slice	Sagittal
Thickness of slice (mm)	1.25
Total slice number	128
Image resolution	256 × 256 (1 × 1 mm)

Table 6 Data acquisition details

Disease class	No. of images	Model 1		Model 2		Model 3	
		Training (80%)	Testing (20%)	Training (70%)	Testing (30%)	Training (60%)	Testing (40%)
Class-1 (non-demented)	84	67	17	58	26	50	34
Class-2 (demented)	64	51	13	44	20	38	26
Class-3(mild–moderate)	50	40	10	35	15	30	20

CDR rating is generated wherein 0 rating indicates no dementia, 0.5 denotes very mild dementia, 1 denotes mild, 2 denotes moderate, and 3 denotes severe dementia, respectively. Dataset acquisition details are shown in Table 5.

Consequently, to demonstrate the robust performance of proposed approach, an experimental study of different ratios is deliberated for database training and testing is examined. For the study, total 198 images from longitudinal MRI database are considered. Herein, 84 images are considered as non-demented, 64 images are considered as demented, and 50 images are considered with mild-to-moderate Alzheimer's disease. Database division used for training/testing and creating different test cases as Model 1, Model 2, and Model 3 for the experiment is shown in Table 6.

4.2 *Performance Measurement*

Performance of advised approach is restrained by computing average retrieval precision, average precision, average recall, and average reclamation rate. Precision for query image (I_q) is expressed as:

$$\Pr(I_q, n) = \frac{1}{n} \sum_{i=1}^{|\text{Total Images}|} \left|\psi\left(f(I_i), f\left(I_q\right)\right)\right| R(I_i, I_1) \leq n \tag{25}$$

n is number of total retrieved images, $f(i)$ denotes category of retrieved image, R is the rank of image aimed at query image amid entire database. Similarly, recall can be computed as:

$$\Pr(I_q, n) = \frac{1}{N_G} \sum_{i=1}^{|\text{Total Images}|} \left|\psi\left(f(I_i), f(I_q)\right)\right| R(I_i, I_1) \leq n \tag{26}$$

N_G denotes total number of top matches and

$$\psi\left(f(I_i), f(I_q)\right) = \begin{cases} 1, & f(I_i) = f(I_q) \\ 0 & \text{otherwise} \end{cases} \tag{27}$$

Based on these computations, average retrieval and precision rates can be computed as:

$$\text{Avg}_{\text{Pr}} = \frac{1}{|DB|} \sum_{i=1}^{|DB|} P(I_i, n) \tag{28}$$

$$\text{Avg}_{\text{RR}} = \frac{1}{|DB|} \sum_{i=1}^{|DB|} P(I_i, n) n \leq N_G \tag{29}$$

Statistical Performance Parameter. In this section, the performance of support vector machine, decision tree, KNN, and RBF classifiers are implemented. In order to evaluate the performance, certain elements are computed for the classification accuracy, precision, recall, specificity and sensitivity parameters with the assistance of confusion matrix. Table 7 shows the recital of SVM classifier in footings of classification accuracy and other statistical parameters for every class.

According to this experiment, proposed feature extraction model is combined with support vector machine which achieves the accuracy of 39.06% with average precision, recall, specificity, and sensitivity as 39.03%, 41.62%, 56.25%, and 41.62%, respectively.

Table 7 SVM classifier performance

SVM performance			
Classification accuracy—SVM	39.06		
	Class-1	Class-2	Class-3
Precision	0.3972	0.3333	0.4406
Recall	0.4328	0.4444	0.3714
Specificity	0.5111	0.5789	0.5975
Sensitivity	0.4328	0.4444	0.3714

Similarly, the significant KNN classifier is presented whose performance is as shown in Table 8. KNN classification approach obtains 80.20% classification accuracy with improved statistical performance by reducing the false-negative classifications. In this experiment, average performance of precision, recall, specificity, and sensitivity is obtained as 77.07%, 74.37%, 87.14%, and 74.37%, respectively.

Table 9 shows the performance of decision tree classifier. Decision tree classifier obtains overall accuracy of 98.43% which is comparatively better. Furthermore, it achieves desired enactment in rapport with average precision, recall, specificity, and sensitivity as 97.91%, 98.76%, 99.13%, and 98.76%, respectively.

Later, RBF classifier is also utilized for Alzheimer's disease classification. The experimental performance of RBF classifier is shown in Table 10. RBF classification scheme achieves the overall accuracy of 77.08% along with average precision, recall, specificity, and sensitivity as 70.91%, 70.38%, 83.93%, and 70.38%, respectively. Finally, decision tree classification-based performance analysis is shown in Table 9 and shows best performance in relations to classification when compared with other classification models. Figure 3 shows the various preprocessing stages which are implemented in this work. The input image corresponding to each category is shown in Fig. 3a. This stage includes histogram enhancement, image dilation, image eroding, and image decomposition steps are implemented in section (b), (c), (d), (e), and (f), respectively.

Figure 4 presents the classification accuracy performance comparison wherein all classifiers are used to compare. Decision tree shows better performance when compared with other classifiers. As discussed before, support vector machine, KNN,

Table 8 KNN classifier performance

KNN performance			
Classification accuracy—KNN	80.20		
	Class-1	Class-2	Class-3
Precision	0.8421	0.7826	0.6875
Recall	0.9142	0.7058	0.6111
Specificity	0.7631	0.9218	0.9295
Sensitivity	0.9142	0.7058	0.6111

Table 9 Desicion tree classifier performance

Decision tree classifier performance			
Classification accuracy—DT	98.43		
	Class-1	Class-2	Class-3
Precision	0.9909	0.9811	0.9655
Recall	0.9819	0.9811	1
Specificity	0.9876	0.9927	0.9938
Sensitivity	0.9819	0.9811	1

Table 10 RBF classifier performance

RBF performance			
Classification accuracy—RBF	77.08		
	Class-1	Class-2	Class-3
Precision	0.8455	0.6153	0.6666
Recall	0.8595	0.5853	0.6666
Specificity	0.6984	0.8920	0.9275
Sensitivity	0.8595	0.5853	0.6666

RBF, and decision tree classifier are used and compared the performance in relations to classification exactness. With the help of this study, classification accuracy rate is obtained as 39.06%, 80.20%, 98.43%, and 77.08% for SVM, KNN, DT, and RBF classifiers, respectively.

Figure 5 depicts the comparative analysis in terms of precision for each class by using SVM, KNN, DT, and RBF classifiers. The study shows that the corresponding average precision is obtained as 39.03%, 77.07%, 97.91%, and 70.91% for SVM, KNN, DT, and RBF classifiers, respectively. Experimental study shows that the decision tree achieves better performance for each considered class. Similarly, recall performance is compared based on the different classifiers by considering various classes.

This performance in terms of recall is shown in Fig. 6. In compliance with the experiment results, the average recall performance is obtained by SVM is 41.62%, KNN is 74.37%, decision tree is 98.76%, and RBF is 70.38%. It is shown clearly that the combination of proposed approach with decision tree obtains better results in terms of recall.

Finally, the existent specificity and sensitivity performance comparison for all the classifier in the considered classes were shown in Fig. 7. The experimental analysis shows that the average specificity rates obtained by each classifier are 56.25%, 87.14%, 99.13%, and 83.93%, respectively, and average specificity rates obtained by each classifier are 41.62%, 74.37%, 98.76%, and 70.38%, respectively. Decision tree classifier outclasses in terms of average specificity and average sensitivity.

4.3 Comparative Analysis

Performance Measurement for Model 1

In the experiment, a comparative study for Model 1 where training and testing databases are divided in the ratio of 80–20% is presented. This experimental analysis shows classification accuracy performance separately for each classifiers and each class distinctly. At first, GLCM-based features are extracted and classification

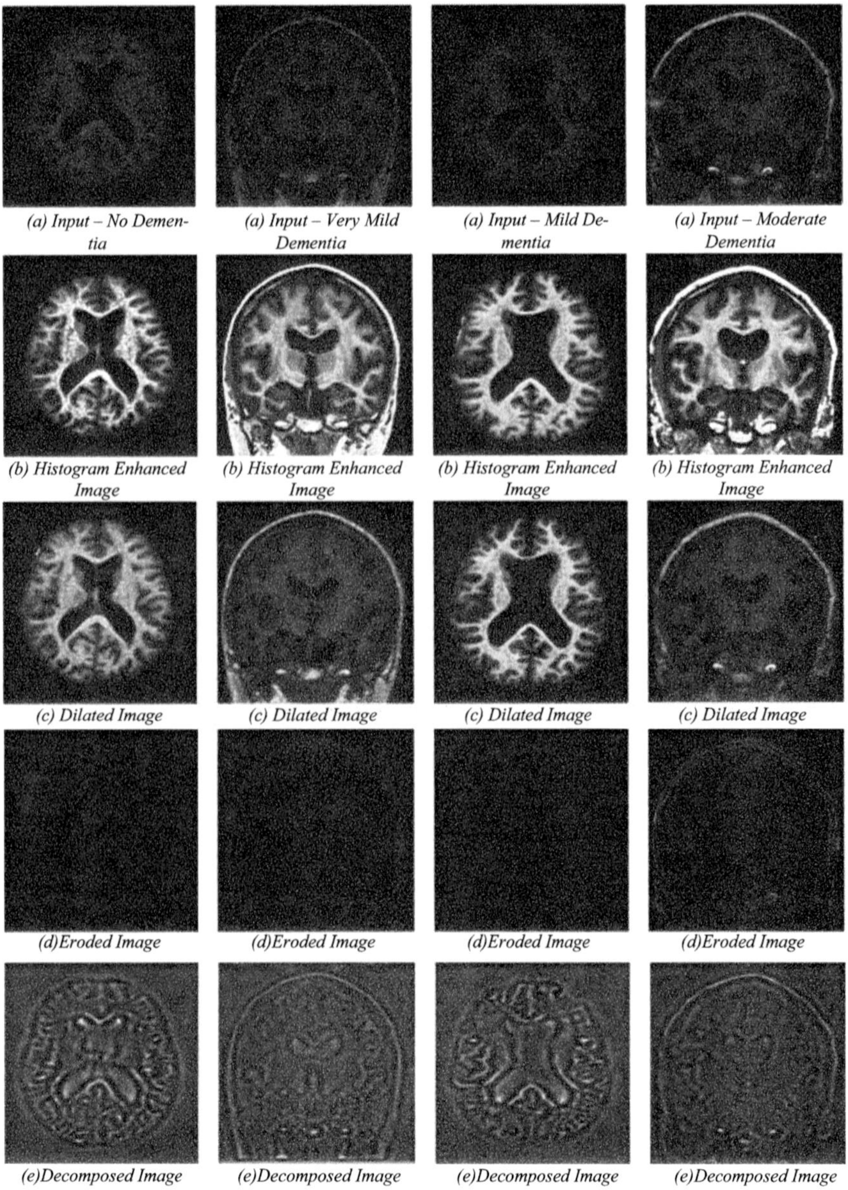

Fig. 3 Preprocessing stage for each category of images

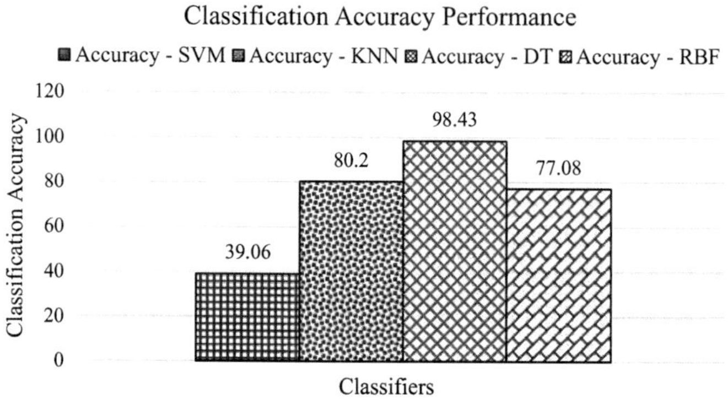

Fig. 4 Classification accuracy performance

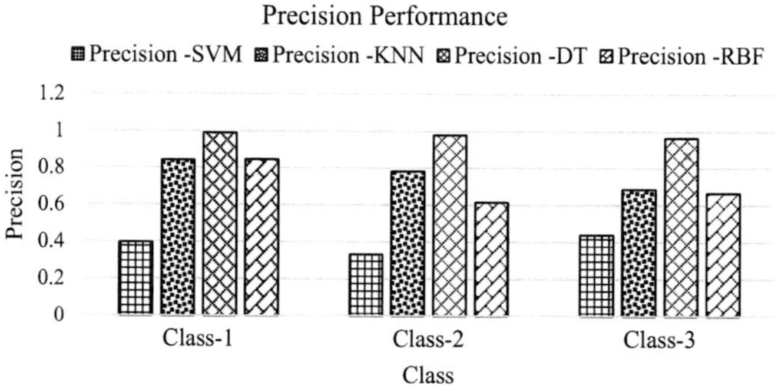

Fig. 5 Precision comparison performance

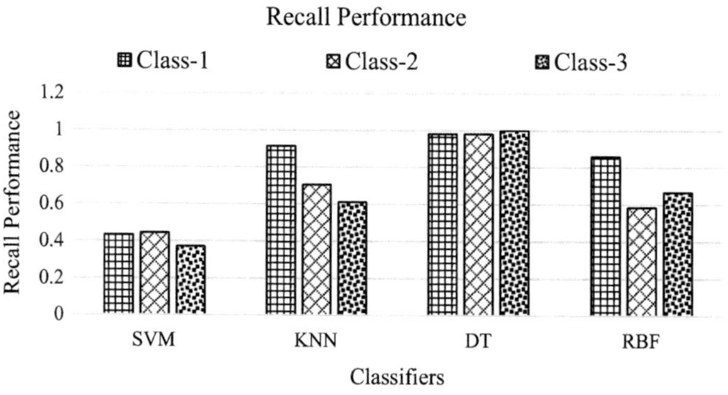

Fig. 6 Recall comparison performance

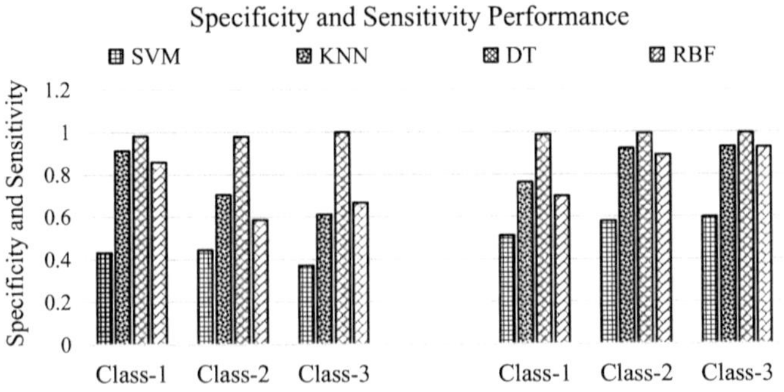

Fig. 7 Specificity and sensitivity performance

accuracy is measured for each class and average accuracy is obtained. Similar task is performed for other classifiers such as KNN, RBF, and DT. Table 11 shows a comparative analysis using four different types of the classifier which are combined with the three different types of feature extraction techniques. This experiment shows that the GLCM feature extraction techniques are not able to achieve the desired performance when the image is complex in nature. GLCM provides the classification accuracy of 48.20%, 71.49%, 77.61%, and 81.28% for SVM, RBF, KNN, and decision tree classifier, respectively. In another study, the EMD feature extraction provides the classification accuracy of 51.87%, 73.01%, 82.77%, and 89.02% for SVM, RBF, KNN, and decision tree classifier, respectively. Correspondingly, the anticipated scheme which combines the GLCM and EMD features is also evaluated which gives the classification accuracy of 60.42%, 75.58%,85.20%, and 93.34% for SVM, RBF, KNN, and decision tree classifier, respectively.

Table 11 Comparitive analysis of performance for Model 1

Classifier	Feature extraction	Class 1	Class 2	Class 3	Avg. Accuracy
SVM classifier	GLCM	62.07	36.89	45.66	48.20
	EMD	56.32	46.97	52.33	51.87
	Combined	63.37	49.68	68.22	60.42
RBF classifier	GLCM	72.3	69.88	72.29	71.49
	EMD	71.2	78.63	69.22	73.01
	Combined	78.21	75.27	73.28	75.58
KNN classifier	GLCM	79.03	78.17	75.63	77.61
	EMD	80.03	79.23	89.06	82.77
	Combined	86.31	86.08	83.22	85.20
Decision tree classifier	GLCM	82.11	75.36	86.38	81.28
	EMD	89.66	92.21	85.21	89.02
	Combined	92.31	92.08	95.63	93.34

Performance Measurement for Model 2

Similarly, the classification performance for the Model 2 where experimental data is divided into 70–30% ratio is presented. Furthermore, in this experiment also the similar procedure for performance measurement as specified in Model 1 experiment is considered.

In compliance with experiment, Table 12 shows the exploration of SVM, RBF, KNN, and decision tree classifiers that are used for the classification performance measurement of Model 2. Initially, GLCM feature extraction process is applied, and the obtained performances are 43.68%, 64.85%, 74.41%, and 77.90% for the aforementioned classifiers, respectively. Likewise, EMD feature extraction process is applied, and classification performances obtained are 47.94%, 69.25%, 76.06%, and 87.77%, respectively, for each classifier. Finally, the projected combined feature extraction scheme is applied which gives the promising classification performance as 53.30%, 73.62%, 81.73%, and 89.69%, respectively, for each classifiers.

Performance Measurement for Model 3

In this experiment, the database is divided in the form of 60%–40% ratio for training and testing, respectively, and the performance is measured for all the classifier with different feature extraction techniques in terms of classification accuracy.

In the investigation of the performance of Model 3, the experiment uses similar configurations where GLCM feature extraction technique obtains classification performance as 42.83%, 59.39%, 68.88%, and 75.51% using SVM, RBF, KNN, and decision tree classifier, respectively. Next, EMD features are considered, and classification performance of 45.70%, 65.33%, 71.98%, and 84.66 is obtained using SVM, RBF, KNN, and decision tree classifier, respectively. Finally, both features are pooled together, and the classification performance obtained as 50.43%, 71.13%, 79%, and

Table 12 Comparitive analysis of performance for Model 2

Classifier	Feature extraction	Class 1	Class 2	Class 3	Avg. accuracy
SVM classifier	GLCM	56.39	33.58	41.08	43.68
	EMD	54.21	43.22	46.39	47.94
	Combined	54.3	42.39	63.22	53.30
RBF classifier	GLCM	68.13	61.2	65.22	64.85
	EMD	69.31	80.23	58.21	69.25
	Combined	75.38	74.22	71.28	73.62
KNN classifier	GLCM	76.39	75.66	71.2	74.41
	EMD	75.68	71.28	81.22	76.06
	Combined	85.63	80.23	79.33	81.73
Decision tree classifier	GLCM	76.22	73.28	84.22	77.90
	EMD	85.69	86.39	91.23	87.77
	Combined	89.63	88.21	91.23	89.69

Table 13 Comparitive analysis of performance for Model 3

Classifier	Feature extraction	Class 1	Class 2	Class 3	Avg. accuracy
SVM classifier	GLCM	52.39	36.88	39.22	42.83
	EMD	52.2	39.68	45.23	45.70
	Combined	51.22	40.38	59.69	50.43
RBF classifier	GLCM	61.28	55.68	61.22	59.39
	EMD	65.1	75.69	55.22	65.33
	Combined	74.23	71.09	68.09	71.13
KNN classifier	GLCM	71.22	70.21	65.22	68.88
	EMD	71.25	69.03	75.66	71.98
	Combined	84.22	78.56	74.24	79.00
Decision tree classifier	GLCM	74.22	71.29	81.03	75.51
	EMD	84.22	81.23	88.54	84.66
	Combined	87.65	85.29	86.30	86.41

86.41% for SVM, RBF, KNN, and decision tree classifiers, respectively, as shown in Table 13.

Performance Measurement Analysis for Proposed and Existing Methodologies

A boundless volume of research has been supported for the precise identification of Alzheimer's in the latest ages, and various methodologies have been projected for this persistence. In this section, some of the best competitive research that has been put forward in this zone in recent existences is compared. The comparative study of existing work and the proposed work is given in following Table 14.

5 Conclusion and Future Scope

In this work, the consistent is focused on Alzheimer's disease detection and classification using machine learning schemes. According to machine learning process, feature or attributes are extracted for the entire image database which is pre-labeled with the stage of Alzheimer's based on the CDR score. Prior to that, image preprocessing technique is included which helps to improve the contrast and quality of image. Looking forward to further enhance the performance of system, the empirical mode decomposition features are later combined with the GLCM features and a feature vector is formulated. Empirical mode decomposition feature extraction includes various spectral feature extractions such as spectral entropy feature, spectral magnitude, peak frequency, and spectral energy. These features are trained and tested using various classifiers such as support vector machine, KNN, decision tree, and RBF. According to experimental study, various parameters are examined and compared with the performance of different classifiers. In this study, decision tree, SVM, RBF, and KNN

Table 14 Comparitive study of proposed method v/s existing methodologies

References	Year	Database	Techniques used	Accuracy
Ali et al. [25]	2015	Brain web (DS1), BRATS (DS2)	Morphological pyramid with FCM clustering technique	96%
Abdel Maksoud et al. [26]	2015	Brain (DS1), (DS2), (DS3)	K-means clustering integrated with fuzzy C-means system	96.83%
Selvathi and Vanmathi [27]	2018	OASIS	Convolution neural network	94.86%
Khajehnejad et al. [28]	2017	OASIS	Semi-supervised manifold learning using voxel-based morphometry with PCA	93.86%
Previtali et al. [29]	2017	OASIS	Oriented FAST and rotated BRIEF with SVM	77%
Yadav et al. [30]	2018	OASIS	Minimal redundancy maximum relevance with classifiers SVM, KNN, and naive Bayesian	95% 95% 90%
Islam and Zhang [31]	2017	OASIS	Deep CNN model	73.45%
Islam and Zhang [32]	2018	OASIS	Ensemble system of deep CNN	93.18%
Chyzhyk et al. [33]	2012	OASIS	Lattice independent component analysis (LICA) and the kernel transformation	74.25%
Savio et al. [34]	2011	OASIS	Deformation-based features with linear SVM	84%
Siddiqui et al. [35]	2017	OASIS	Discrete wavelet transforms and PCA with random forest classifier	95.70%
Proposed method	2019	OASIS	Empirical mode decomposition and gray-level co-occurrence matrix (GLCM) with decision tree classification	98.43%

classification schemes are used to put forth the classification performance as 98.43%, 39.06%, 77.08%, and 80.20% in terms of classification accuracy. This work presents a detailed experimental analysis which shows that combination of proposed feature extraction model, and decision tree classifier outperforms the performance accuracy when compared with other classifiers. There is always a scope for improvement of the overall performance of the classification technique; in this regard, in future it is planned to achieve the greater accuracy of the classification process by composition of two or three classifier models which ensembles the features to identify the different classes of AD to enhance the efficiency of the whole system. Also, it is under experiment to use another database known as Alzheimer's Disease Neuroimaging Initiative (ADNI) in the future to detect the severity of Alzheimer's disease with the aid of deep learning technique and convolution neural network classification.

References

1. Lei G, Lei Z, Youxi Wu, Li Ying, Guizhi Xu, Qingxin Y (2011) Tumor detection in MR images using one-class immune feature weighted SVMs. IEEE Trans Magn 47(10):3849–3852. https://doi.org/10.1109/TMAG.2011.2158520
2. Anitha V, Murugavalli S (2016) Brain tumour classification using two-tier classifier with adaptive segmentation technique. IET Computer Vision 10(1):9–17. doi: https://doi.org/10.1049/iet-cvi.2014.0193
3. Wong KP, Bergsneider M, Glenn TC, Kepe V, Barrio JR, Hovda DA, Vespa PM, Huang SC (2016) A semi-automated workflow solution for multimodal neuroimaging: application to patients with traumatic brain injury. Brain Inform 3(1):1–15. https://doi.org/10.1007/s40708-015-0026-y
4. Xuan C, Binh PN, Chee-Kong C, Sim-Heng O (2017) Reworking multilabel brain tumor segmentation: an automated framework using structured kernel sparse representation. IEEE Syst Man Cybern Mag 3(2):18–22. https://doi.org/10.1109/MSMC.2017.2664158
5. Esses SJ, Lu X, Zhao T, Shanbhogue K, Dane B, Bruno M, Chandarana H (2017) Automated image quality evaluation of T2-weighted liver MRI utilizing deep learning architecture. J Magn Reson Imaging. https://doi.org/10.1002/jmri.25779
6. Khan MI, Jing Z, Mohammad AI, Robert JO (2009) Fractal-based brain tumor detection in multimodal MRI. Appl Math Comput 207(1):23–41. https://doi.org/10.1016/j.amc.2007.10.063
7. Nilesh BB, Arun KR, Har PT (2017) Image analysis for MRI based brain tumor detection and feature extraction using biologically inspired BWT and SVM. Int J Biomed Imaging. https://doi.org/10.1155/2017/9749108
8. Sérgio P, Adriano P, Victor A, Carlos AS (2016) Brain tumor segmentation using convolutional neural networks in MRI images. IEEE Trans Med Imaging 35(5):1240–1251. https://doi.org/10.1109/TMI.2016.2538465
9. Seyed HD, Alan WL (2017) Spatial possibilistic fuzzy C-Mean segmentation algorithm integrated with brain mid-sagittal surface information. Int J Fuzzy Syst 19(2):591–605. https://doi.org/10.1007/s40815-016-0247-0
10. Jothi G, Hannah IH (2016) Hybrid Tolerance Rough Set-Firefly based supervised feature selection for MRI brain tumor image classification. Appl Soft Comput 46:639–651. https://doi.org/10.1016/j.asoc.2016.03.014
11. Olasimbo AA, Sharifah M, Wan A, Salman Y, Saif M (2015) Soft biometric: gender recognition from unconstrained face images using local feature descriptor. JICT 111–122. doi: arXiv:1702.02537

12. Jude HD, Anitha J (2012) Image pre-processing and feature extraction techniques for magnetic resonance brain image analysis. In: International conference on computer applications for communication, networking, and digital contents, pp 349–356. doi: https://doi.org/10.1007/978-3-642-35594-3_47
13. Sun X, Shi L, Luo Y, Yang W, Li H, Liang P, Li K, Mok VCT, Chu WCW (2015) Wang D (2015) Histogram-based normalization technique on human brain magnetic resonance images from different acquisitions. Biomed Eng Online 14(1):73. https://doi.org/10.1186/s12938-01,5-0064-y
14. Mohd SMR, Tanzila S, Fatima N, Afraz ZS (2014) 3D texture features mining for MRI brain tumor identification. 3D Res 5(1):3. doi: https://doi.org/10.1007/s13319-013-0003-2
15. Mubbashar S, Jawad HK, Kalim Q (2014) A Hybrid approach of using symmetry technique for brain tumor segmentation. Comput Math Methods Med. https://doi.org/10.1155/2014/712783
16. Mohammad H, Axel D, David WF, Antoine B, Aaron C, Yoshua B, Chris P, Pierre MJ, Hugo L (2017) Brain tumor segmentation with deep neural networks. Med Image Anal 35:18–31. https://doi.org/10.1016/j.media.2016.05.004
17. Jainy S, Vinod K, Indra G, Niranjan K, Chirag KA (2013) Segmentation, feature extraction, and multiclass brain tumor classification. J Digit Imaging 26(6):1141–1150. doi: https://dx.doi.org/10.1007%2Fs10278-013-9600-0
18. Nooshin N, Mirosalv K (2015) Brain tumors detection and segmentation in MR images: Gabor wavelet vs. statistical features. Comput Electr Eng 45:286–301. https://doi.org/10.1016/j.compeleceng.2015.02.007
19. Stefan B, Lutz-P. N, Mauricio R (2011) Fully automatic segmentation of brain tumor images using support vector machine classification in combination with hierarchical conditional random field regularization. In: International conference on medical image computing and computer-assisted intervention, pp 354–361. doi: https://doi.org/10.1007/978-3-642-23626-6_44
20. Kleesiek J, Urban G, Hubert A, Schwarz D, Maier-Hein K, Bendszus M, Biller A (2016) Deep MRI brain extraction: a 3D convolutional neural network for skull stripping. NeuroImage 129:460–469. https://doi.org/10.1016/j.neuroimage.2016.01.024
21. Yang G, Zhang Y, Yang J, Ji G, Dong Z, Wang S, Feng C, Wang Q (2016) Automated classification of brain images using wavelet-energy and biogeography-based optimization. Multimed Tools Appl 75(23):15601–15617. https://doi.org/10.1007/s11042-015-2649-7
22. Tristan G, Johan M, Isabelle M (2004) Texture based medical image indexing and retrieval: application to cardiac imaging. In: Proceedings of the 6th ACM SIGMM international workshop on multimedia information retrieval, pp 135–142. doi: https://doi.org/10.1145/1026711.1026734
23. Haralick RM (1979) Statistical and structural approaches to texture. Proc IEEE 67(5):786–804. https://doi.org/10.1109/PROC.1979.11328
24. Marcus DS, Wang TH, Parker J, Csernansky JG, Morris JC, Buckner RL (2007) Open Access Series of Imaging Studies (OASIS): cross-sectional MRI data in young, middle aged, non-demented, and demented older adults. J Cogn Neurosci 19(9):1498–1507. https://doi.org/10.1162/jocn.2007.19.9.1498
25. Ali H, Mohammed E, Eman E-D, Ahmed A (2015) Multi-resolution MRI brain image segmentation based on morphological pyramid and fuzzy c-mean clustering. Arab J Sci Eng 40(11):3173–3185. https://doi.org/10.1007/s13369-015-1791-x
26. Eman AM, Mohammed E, Rashid Al A (2015) Brain tumor segmentation based on a hybrid clustering technique. Egypt Inform J 16(1):71–81. https://doi.org/10.1016/j.eij.2015.01.003
27. Selvathi D, Vanmathi T (2018) Brain region segmentation using convolutional neural network. In: 4th international conference on electrical energy systems (ICEES). IEEE
28. Khajehnejad M, Saatlou FH, Mohammadzade H (2017) Alzheimer's disease early diagnosis using manifold-based semi-supervised learning. Brain Sci 7(8). doi: https://doi.org/10.3390/brainsci7080109. (Open Access)
29. Previtali F, Bertolazzi P, Felici G, Weitschek E (2017) A novel method and software for automatically classifying Alzheimer's disease patients by magnetic resonance imaging analysis. Comput Methods Programs Biomed 143:89–95. https://doi.org/10.1016/j.cmpb.2017.03.006

30. Rishi Y, Ankit G, Ravi BM (2018) Classification of alzheimer using FMRI data and brain network, construction opportunities for mobile IT (COMIT-2018). Comput Sci Inf Technol. https://doi.org/10.5121/csit.2018.80609
31. Jyoti I, Yanqing Z (2017) A novel deep learning based multi-class classification method for Alzheimer's disease detection using brain MRI data. In: Brain informatics. BI. Lecture Notes in Computer Science, vol 10654. Springer, Cham. doi: https://doi.org/10.1007/978-3-319-70772-3_20
32. Jyoti I, Yanqing Z (2018) Brain MRI analysis for Alzheimer's disease diagnosis using an ensemble system of deep convolutional neural networks. Brain Inf 5:2. https://doi.org/10.1186/s40708-018-0080-3
33. Darya C, Manuel G, Alexandre S, Josu M (2012) Hybrid dendritic computing with kernel-LICA applied to Alzheimer's disease detection in MRI. Neurocomputing 75:72–77. https://doi.org/10.1016/j.neucom.2011.02.024
34. Alexandre S, Manuel G, Jorge V (2011) Deformation based features for Alzheimer's disease detection with linear SVM. In: International conference on hybrid artificial intelligence systems, Wroclaw, Poland, 23–25 May 2011. Springer, Berlin, Heidelberg, Germany, pp 336–343. doi: https://doi.org/10.1007/978-3-642-21222-2_41
35. Muhammad FS, Ghulam M, Ahmed WR, Liyana S (2017) Multi-class disease classification in brain MRIs using a computer-aided diagnostic system. Symmetry 9(3). https://doi.org/10.3390/sym9030037. (Open Access)

Facial Image Indexing Using Locally Extracted Sparse Vectors

Vinayaka R. Kamath, M. Varun, and S. Aswath

Abstract Amidst all biological characteristics in a human, face emerged to be the most common biometric identifier used by humans due to its distinctive and prominent features. Moreover, the collection of facial images is a friendly, direct and a non-intruding activity, which makes it easily acceptable in nature by the users because of its non-infringement identification technology. We intend to take complete advantage of these developments and propose a novel methodology to achieve identification of the subjects by examining the facial images of the prospects. In the proposed methodology, preprocessed images are passed onto logically adaptive regression kernel for coherent facial feature extraction. The patterns recorded from these feature vectors are condensed using LDA to reduce the computational load. These sparse vectors are quantized and aggregated using VLAD with an intention to classify the descriptors in the later stages of the pipeline. Classification is achieved using CAT Boost and multi-layered perceptron to demonstrate the results using a comparative paradigm. The proposed system has been tested on benchmark datasets such as Grimace, Faces95 and Faces96. Evaluation of these datasets has been done considering the precision, recall and F1-score with an intention to perceive the best one among the proposed alternatives.

Keywords Face recognition · Feature aggregation · Image classification · Image processing · LARK

1 Introduction

Face recognition is said to be achieved when an individual's face is identified from a set of test images. The facial features are either captured from a video which is a set of frames containing an image or a still image. These facial features are captured using mathematical transformations and are stored in the form of faceprints. These

V. R. Kamath (✉) · M. Varun · S. Aswath
Department of Computer Science, PES University, Bengaluru, India
e-mail: vinayakarkamath@pesu.pes.edu

N. N. Chiplunkar and T. Fukao (eds.), *Advances in Artificial Intelligence and Data Engineering*, Advances in Intelligent Systems and Computing 1133,
https://doi.org/10.1007/978-981-15-3514-7_93

faceprints are captured and then stored in a repository. Each of these faceprints is assigned a class corresponding to the individual's face and then passed on to the predictor models for training. After the training phase is complete, the testing phase begins. In the testing phase, the facial recognition algorithm is run on the test dataset which is completely different from the training dataset and then the accuracy is computed. If a face present in the training dataset does not match with the test dataset then the algorithm assigns the closest match from the training dataset or the repository. Applications of facial recognition include its use cases in the phones as a login method for the user. The phone stores a user's faceprint and this faceprint is compared with the users log and the phone is unlocked. In order to tag individuals present in the photographs, Facebook uses an advanced facial recognition software. To find out the identity of an unknown individual, these facial recognition software's return the individuals credentials like the name, social security number and the date of birth. This technology is also used in criminal investigation to find out suspects for a given crime.

Facial recognition poses significant challenges. The complexity of the solution must increase with the wildness of the subjects. The environment in which the images are captured also imparts serious variances to the recorded image. The changes in illumination, head scale, pose of the test subject and facial expressions make the task of identification difficult.

The paper proposes a stacked pipeline to accomplish face recognition with an intent to perform well in conditions where the frontal view is subjected to different illumination, tilt and differences in expressions. Section 2 briefs about the components used in the pipeline. It explains functioning of the stages and provides the necessary background required to understand the system. Section 3 explains about the experimentation procedures as well as the datasets that were used to test the model. Last section describes about the findings and inferences drawn from the experimentation and concludes by suggesting possible future enhancements.

2 Stages of the Pipeline

This section explains the constituents of the system and imparts information about the modules and how they are chained together to form an efficient pipeline. Figure 1 highlights all the stages of the pipeline.

2.1 *Image Preprocessing*

2.1.1 Conversion to Grayscale

This preprocessing technique [1] applies to images that have 3 channels. Since each pixel contains 3 channel values, the first channel corresponds to red, second channel

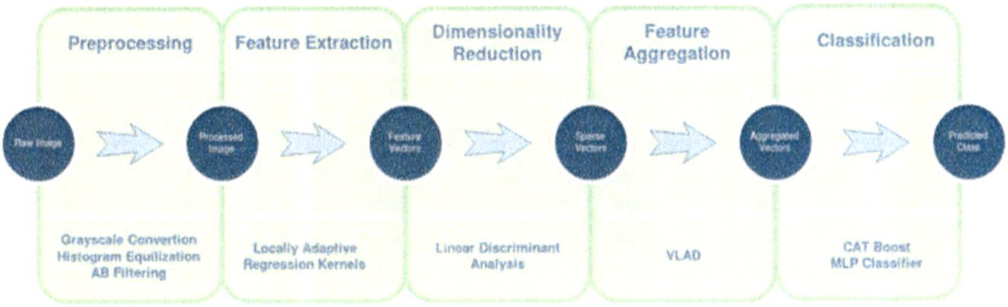

Fig. 1 Illustration of the proposed system

for the green and the third channel indicates the intensity of blue. Each pixel needs to be mapped to exactly one grayscale value [2]. This can be accomplished by three methods. In the lightness method, colours which are the most significant and least significant are taken into consideration. The average of these channels is computed as follows.

$$G(x) = \frac{(\text{Max}(\text{Red}, \text{Green}, \text{Blue}) + \text{Min}(\text{Red}, \text{Green}, \text{Blue}))}{2} \tag{1}$$

In the average method, the average value of the three channels is calculated as follows:

$$G(x) = \frac{(\text{Red} + \text{Green} + \text{Blue})}{3} \tag{2}$$

In the luminosity method [3], the weighted average of the three channels R, G, B is computed. Since humans have a quicker response to green colour than the two other colours (red and blue), green colour is assigned the highest weight. Luminosity is calculated as follows.

$$G(x) = 0.21 \cdot \text{R} + 0.72 \cdot G + 0.07 \cdot \text{B} \tag{3}$$

2.1.2 Histogram Equalization

Considering an image f represented by a $m * n$ order matrix, each pixel in a matrix has its intensity values varying in the range $[0, L - 1]$ where 0 denotes the lower bound and $L - 1$ denotes the upper bound of the matrix. Usually, in a grayscale image, the number of possible intensity values represented by L is 256.

Suppose we have an image with nonuniform distribution of intensities [4], this image is perceived as an image that has low contrast. In an attempt to improve the contrast of the image, we apply a transformation function which can convert the nonuniform distribution of intensities in the probability histogram to an uniform distribution of intensities [5]. In order to normalize the histogram of f we need to make sure that

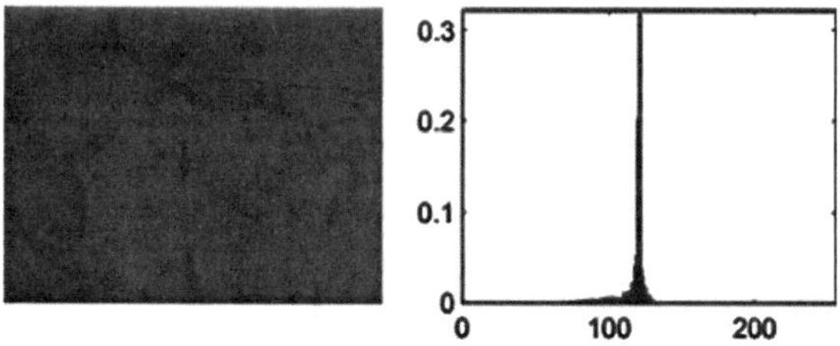

Fig. 2 Before histogram equalization

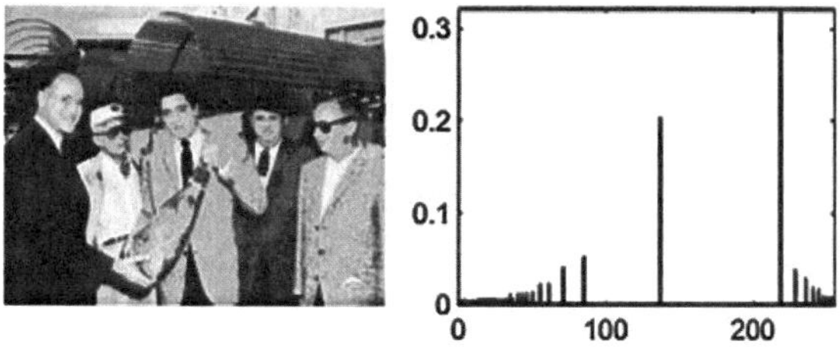

Fig. 3 After histogram equalization

$$P_n = \frac{\text{(Number of pixels with intensity } n)}{\text{(Total number of pixels)}}$$

This can be further enhanced by computing $g_{i,j}$ using Eq. (4).

$$T(k) = g_{i,j} = \text{floor}((L-1)\sum_{n=0}^{f_{i,j}} p_n) \tag{4}$$

where p_n represents the probability of occurence of n and $n \in [0, L-1]$. Figure 2 represents the histogram before equalization while Fig. 3 shows the histogram after equalization.

2.1.3 Adaptive Bilateral Filtering

Adaptive bilateral filtering [6, 7] not only helps in reducing the noise, but also enhances the sharpness of the image. Beyond retaining the overall functionality of a bilateral filter, it has sustained two modifications. Of which, the first one is an offset ζ, which is introduced in ABF to control the range filter and the later one(σ_d) is used to control the adaptiveness of offset and the width of the range filter. Conventional bilateral filter can be generated by keeping $\zeta = 0$ and a constant value of σ_d.

The response of the shift-invariant ABF at (m_0, n_0) at an impluse is given by Eq. (5).

$$h(m_0, n_0; m, n) = \begin{cases} r_{m_0,n_0}^{-1} \exp(-\frac{(m-m_0)^2+(n-n_0)^2}{2\cdot\sigma_d^2})\times \\ \exp(-\frac{(g[m,n]-g[m_0,n_0]-\zeta[m_0,n_0])^2}{2\cdot\sigma_r^2}), & [m,n] \in \Omega_{m_0,n_0} \\ 0, & \text{else} \end{cases} \quad (5)$$

The fixed dependent offset ζ is the key to the restoration of slope in ABF, which is accomplished by the transformation of images local histogram. This is done inorder to avoid the inefficient process of observing edge profiles and locating edge normals. Thereby, this amalgamation of σ_d and ζ helps in local adaptive reinforcement of the bilateral filtering technique and transforming itself into a very efficacious algorithm that proved to be proficient in sharpening as well as smoothing. On restoration of edge slope by altering the histogram of edges, ABF introduces a new technique of slope restoration. It provides a robust and reliable solution for slope restoration that is proven to be very efficient in its implementation by its capability to work well for a variety of images.

2.2 *Locally Adaptive Regression Kernels*

High-definition screens have become available but the quality of image and the noises present in the images hinder the effectiveness of the screen. This can be resolved by performing computations on the image such as denoising, deblurring and image space time upscaling. This method estimates the pixel or point by providing a classical kernel regression where the prediction of this construction is nonparametric in nature. The data model is illustrated by kernel regression [8] construction as follows:

$$y_i = z(x_i) + \epsilon_i \qquad x_i \in \omega, i = 1 \ldots S \quad (6)$$

Here, y_i represents a sample which is denoised and is calculated at $x_i = [x_i^1, x_i^2]^{\mathrm{T}}$. $Z(x)$ denotes the regression function. ϵ_i is the zero mean noise which is independently and identically distributed. P contains the total number of specimens within a window ω which is arbitrary about x and is the position of interest. It is easy to determine estimates which are pointwise in nature.

$$\begin{aligned} z(x_i) &\approx z(x) + (\Delta z(x))^{\mathrm{T}}(x_i - x) \\ &\quad + \frac{1}{2!}(x_i - x)^{\mathrm{T}} Hz(x)(x_i - x) + \cdots \\ &\approx \beta_0 + \beta_1^{\mathrm{T}}(x_i - x) \\ &\quad + \beta_2^{\mathrm{T}}\mathrm{vech}((x_i - x)(x_i - x)^{\mathrm{T}}) + \cdots \end{aligned} \quad (7)$$

H signifies the Hessian operator, vech signifies the half-vectorization operator which performs lexicographic ordering on a symmetric matrix (the lower triangular portion of this matrix) into a vector which is column stacked. The values β_1 and β_2 can be defined as follows.

$$\beta_1 = \left[\frac{\partial z(x)}{\partial x_1}, \frac{\partial z(x)}{\partial x_2}\right]^T \tag{8}$$

$$\beta_2 = \frac{1}{2}\left[\frac{\partial^2 z(x)}{\partial x_1^2}, \frac{\partial^2 z(x)}{\partial x_2^2}\right]^T \tag{9}$$

The vech operator can be illustrated using Eq. (10) and (11):

$$\text{vech}\left(\begin{bmatrix} a & b \\ b & d \end{bmatrix}\right) = \begin{bmatrix} a & b & d \end{bmatrix}^T \tag{10}$$

$$\text{vech}\left(\begin{bmatrix} x & y & z \\ y & e & f \\ z & f & i \end{bmatrix}\right) = \begin{bmatrix} x & y & z & f & e & i \end{bmatrix}^T \tag{11}$$

The kernel regression function is computed as:

$$K(C_i, x_i, x) = exp{-}(x_i - x)C_l(x_i - x) \tag{12}$$

with,

$$C_i = \sum_{K \in \omega} \begin{bmatrix} {z_{x_1}}^2(x_k) & z_{x_1}(x_k)z_{x_2}(x_k) \\ z_{x_1}(x_k)z_{x_2}(x_k) & {z_{x_2}}^2(x_k) \end{bmatrix} \tag{13}$$

These feature vectors play an important role in defining the descriptors and keypoints for an image. The overlapping patches are removed and featurized LARKS [9–11] are obtained. Key points are extracted from preprocessed images by changing the level of window size, sensitivity and smoothness. By varying these parameters we obtain different versions of vectorized LARKS on images as demonstrated in Fig. 4.

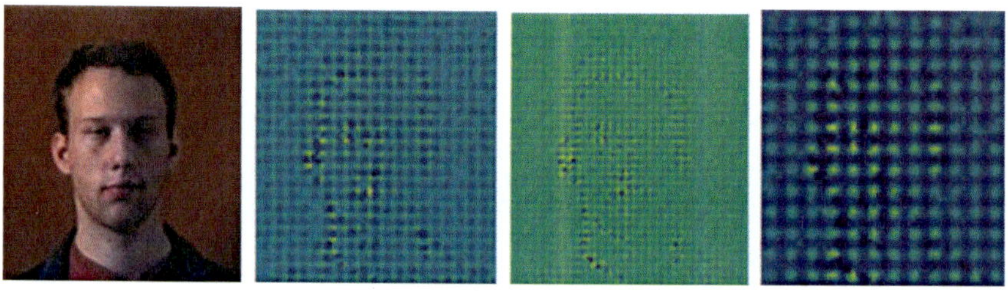

Fig. 4 Visualization of feature vectors in comparison with the original image

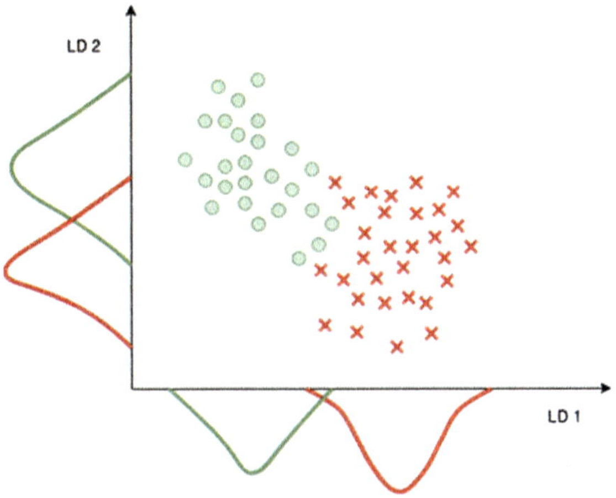

Fig. 5 Maximizing the component axis for class separation

2.3 *Linear Discriminant Analysis(LDA)*

Although it is considered to be analogous to the very popular and widely used principal component analysis [12] which aims to find component axes with an intention to elevate the variance of our data [13], to add on this the technique focuses on the axes that retains maximum discrimination between multiple classes in hand. The main intent is to project a dataset onto a dimensional space that is smaller than the current one and still maintain good class-separability, primarily to circumvent overfitting as well as to reduce the computational costs. Figure 5 provides intuition behind linear discriminant analysis.

Dimensionality reduction is performed by implementing a series of steps in the algorithm given below:

1. Computation of x-dimensional mean vectors for various classes from the dataset.
2. Intra-class scatter matrix and Inter-class scatter matrices are computed.
3. Eigenvectors and the corresponding eigenvalues for the previously calculated scatter matrices are calculated.
4. Sorting the eigenvectors in descending order by eigenvalues. Choose the eigen vectors based on their corresponding eigenvalues. Pick the k largest ones inorder to form a $d \times k$ dimensional matrix W.
5. Deployment of this $d \times k$ eigenvector matrix for the transformation of the samples onto a new subspace. This is achieved by $Y = X \times W$, where X is a $n \times d$-dimensional matrix depicting n samples, and Y refer to the transformed $n \times k$-dimensional samples in the new smaller subspace.

2.4 Vector of Locally Aggregated Descriptors(VLAD)

After the extraction of the features is carried out from the images, it is necessary to pool them. These feature vectors obtained have significant dimensionality, this dimensionality makes it difficult to perform required computation. Information regarding the image like colour, position and intensity details of neighbouring pixels are contained in these feature vectors. The task of classification does not necessarily need so many details of an image. We can safely ignore these features present in these vectors. By not dropping out features it becomes tedious and hence the time taken for the machine learning algorithm to train in these features increases. The need of VLAD becomes paramount in order to quantize these vectors. In order to achieve quantization, VLAD [14, 15] uses a fischer kernel [16] which is nonprobabilistic method and uses a codebook calculated by Kmeans++ algorithm [17]. A descriptor x_t is associated with a codebook that contains visual words that are closest to the descriptors.

Codewords are represented by $\mu_1, \mu_2 \ldots, \mu_k$. The set of local feature descriptors is given by $I = (x_1, x_2, \ldots, x_n)$, q_i^k depicts the extent to which x_i represents a data vector that is related to the cluster μ_k. With constraints $q_i^k > 0$ and $\sum^k q_i^k = 1$, the feature vector x is encoded using Eq. (14).

$$v_k = \sum_{i=1}^{N} q_i^k (x_i - \mu_k) \tag{14}$$

The residuals are stacked to obtain the vector $\phi(\hat{I})$, which is given by:

$$\phi(\hat{I}) = \begin{bmatrix} \vdots \\ v_k \\ \vdots \end{bmatrix}$$

As a next step, component wise mass normalization is performed by dividing individual vector v_k with the sum of features related to it, i.e $\sum^k q_i^k$.

2.5 Classifiers

2.5.1 CAT Boost

CAT Boost [18] follows a specialized version of gradient boost. It maintains ordered features while supporting categorical features. A categorical feature can only take finite values for a possible instance which are the labels of the categories. One of the largest challenges of machine learning is to handle categorical data. CAT Boost

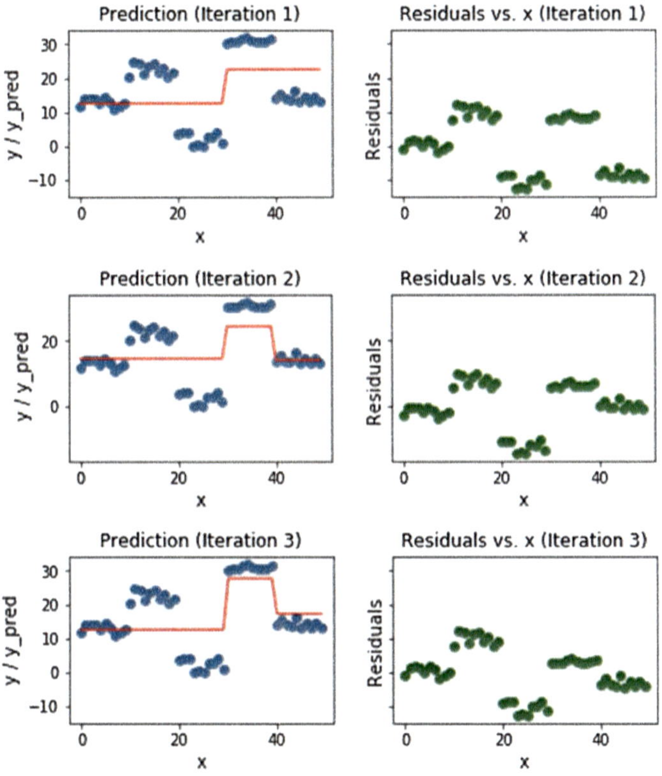

Fig. 6 Gradient boosting mechanism

has evolved to be very proficient whilst dealing with these multiple categories like audio, text, image as well as historical data.

The term Boost is derived from an algorithm known as gradient boosting [19]. Gradient boosting can be used for regression and classification that uses a group of weaker prediction models known as decision trees [20]. Figure 6 visually runs through the gradient boosting mechanism. The loss function implemented in this boosting algorithm is a variant of mean squared error loss, which is defined as follows:

$$\text{LOSS} = \text{MSE} = \sum (y_i - y_i^p)^2 \quad (15)$$

where y_i is the ith target value and Y_i^p is the ith prediction. A simple linear regressor like a decision tree is fitted on the training data. Then the error residuals are computed by subtracting the actual value and the predicted value. Another model is fitted on error residuals that uses the same input variable. Then the residual which is predicted is added to the prediction made earlier by the model. This is performed using the following equations:

$$e_1 = y - y_{\text{pred1}}$$
$$y_{\text{pred2}} = y_{\text{pred1}} + e^1_{\text{pred}} \tag{16}$$

e_1 is the error residual for the first prediction, y is the actual value of the target variable, y_{pred1} is the first prediction while e^1_{pred} is the prediction when error residuals are treated as target variables, y_{pred2} is the second prediction.

Cat Boost assigns indices to categorical columns which is encoded into one hot encoding format. Columns where unique number of categories go beyond the one hot max size, an efficient method alike to mean encoding is used to minimize overfitting. Cat Boost does this by obtaining a permutation of the input observations in an arbitrary fashion. Then it proceeds onto converting the value of labels from categorical or floating point to an integer. Categorical feature values are converted to numeric values by using the equations given below:

$$\text{avg\_target} = \frac{\text{countInClass} + \text{prior}}{\text{totalCount} + 1} \tag{17}$$

countInClass represents the frequency of occurrence of the label value which was equal to 1 for objects consisting of current categorical feature value. In Eq. (17), prior represents the preliminary value in the numerator, totalCount is the total count of objects where its categorical feature value is similar to the current one. Equation (17) can be represented as below:

$$\frac{\sum_{j=1}^{p=1} [x_{\sigma,k} = x_{\sigma_p,k}] Y_{\sigma_j} + \alpha \cdot P}{\sum_{j=1}^{p=1} [x_{\sigma,k} = x_{\sigma_p,k}] + \alpha} \tag{18}$$

2.5.2 Multi Layered Perceptron

Multi-layered perceptron(MLP) [21] can be visualized as a logistic regression classifier [22] where a trained nonlinear transformation is applied to transform the input. This transformation projects the input data onto a region which makes it linearly separable. The input layer, hidden layer and the output layer are the necessary components of MLP classifier. Figure 7 shows the arrangement of perceptrons for a multi-layered perceptron classifier.

A single hidden layer in an MLP represents a function $f : R^D -> R^L$, D being the dimensions of the input vector, L being the dimension of the output vector f(x), where f(x) can be represented by an Eq. (19).

$$f(x) = G(b^{(2)} + W^{(2)} \cdot (\delta(b^{(2)} + W^{(2)}x))) \tag{19}$$

Here, $b^{(1)}$ and $b^{(2)}$ represent the bias vectors and $W^{(1)}$, $W^{(2)}$ are the weight matrices while G and δ are the activation functions. The output of the hidden layers can be computed using Eq. (20).

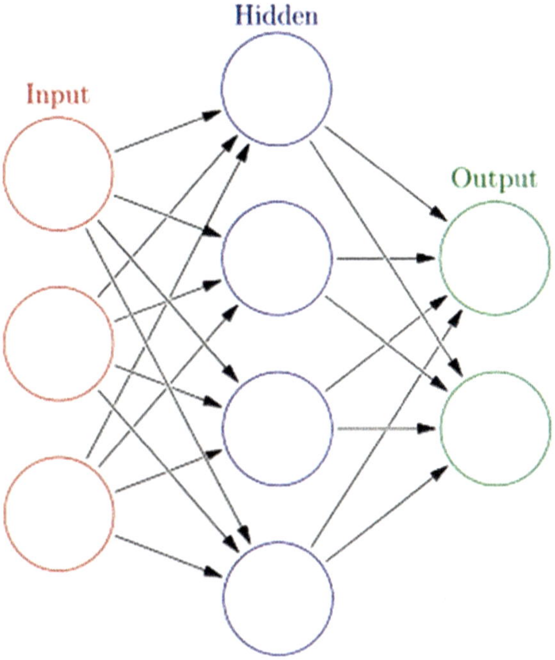

Fig. 7 Underlying architecture of a multi layered perceptron

$$h(x) = \Phi(x) = \delta(b^{(1)} + W^{(1)}x) \tag{20}$$

The weight matrix which establishes a connection between the input vector and the hidden layer is given by $W^{(1)} \in R^{D*D_h}\}$. Every column entry in $W_i^{(1)}$ are the weights running from the input units towards the ith hidden unit. The output vector can be computed using the following equation.

$$o(x) = G(b^{(2)} + W^{(2)}h(x)) \tag{21}$$

This can be scaled to include multiple hidden layers and the equations can be extended to deeper networks. Stochastic gradient descent [23] is used to optimize the model to perfectly fit the data. The collection of parameters that is utilized to optimize the model is given by $\theta = W^{(1)}, b^{(1)}, W^{(2)}, b^{(2)} \ldots W^{(i)}, b^{(i)}$. Backpropagation [24] is achieved by computing the gradients, i.e $\frac{\partial \Phi}{\partial \theta}$.

3 Experimentation

3.1 Datasets

3.1.1 Grimace

Designed and conserved by Dr. Libor Spacek, grimace consists of images of 18 individuals which offer facial expression variations for candidates belonging to both male as well as female categories. For every individual, there are 20 images with a resolution of 180 × 200 pixels. Background is kept identical across all images with minor variations in the head scale. The hairstyle of these individuals is almost similar with noticeable changes in the illumination. Figure 8 contains few of the images from the collection.

3.1.2 Faces95

The dataset is composed of a total of 72 unique individuals. For each individual, 20 images were obtained by requesting him or her to re-approach the camera once an image was captured. It offers a significant amount of variation in the head scale, and

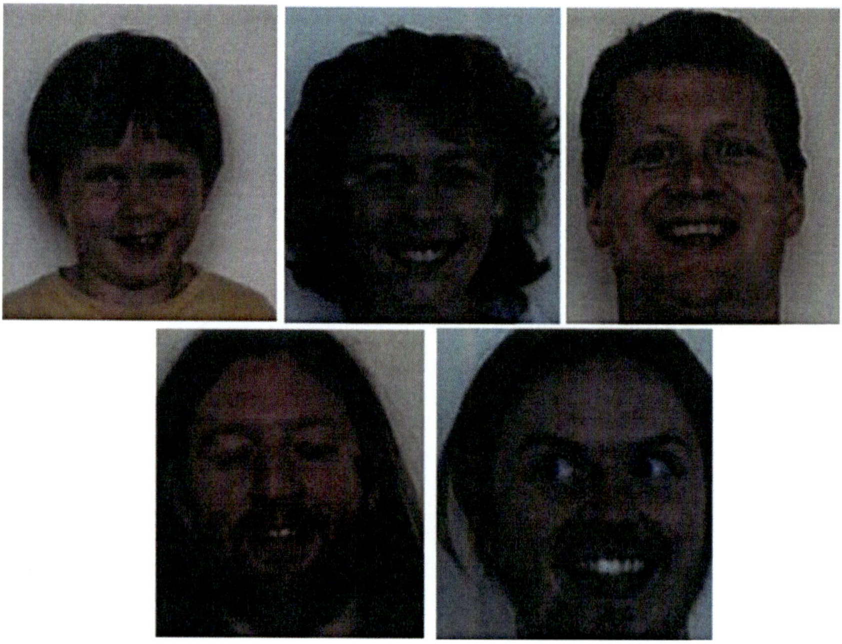

Fig. 8 Samples from grimace dataset

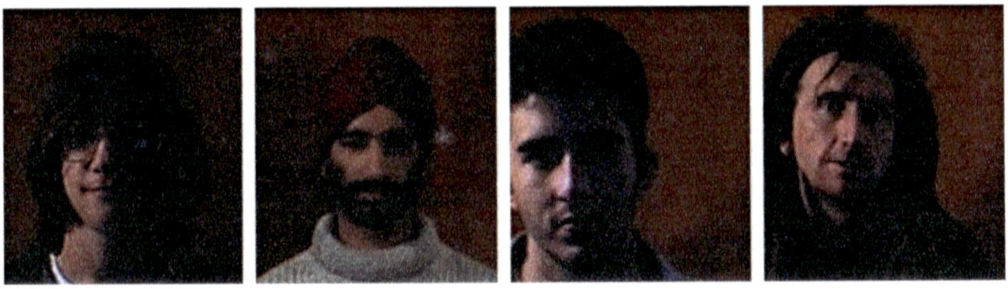

Fig. 9 Randomly chosen samples from Faces95

the variations due to shadows are very less. This often causes difference in the red background. The image resolution is of the order 180 × 200. Variations exhibited by the dataset are depicted in Fig. 9.

3.1.3 Faces96

Faces96 consists of a total of 152 individuals; for every individual, 20 images were captured. In order to impart significant head variations of every individual he is asked to take one step forward approaching the camera. A 0.5 second gap is introduced in between, while capturing successive frames. Image resolution is of the dimension 196 × 196 pixels. There is large amount of head scale variation with minor head turn, tilt and slanting. There is some amount of translation of the faces in images. As the subject approaches the camera, it results in significant lighting changes because of the artificial lighting which is part of the arrangement. Challenges offered by the data are conveyed by the Fig. 10. No variation in hairstyle of an individual is exhibited by the records.

3.2 Results

Precision measures the percentage of positive observations which were measured correctly out of the total positive observations which were predicted. Recall indicates the percentage of positive observations that were predicted correctly out of the observations present in the actual class. The weighted average of precision and recall that considers both false positives and false negatives is called F1-score.

The performance measures of two models CAT Boost and multi-layered perceptron classifier were measured on three benchmark datasets Grimace, Faces95 and Faces96. The performance was measured on these datasets with and without the application of LDA to achieve reduction in dimensionality. For the dataset Grimace, the overall F1-score is larger in case of MLP than CAT Boost which concludes that MLP performs better than CAT Boost on the Grimace dataset. For Faces95, the overall F1-score of CAT Boost with or without LDA is greater than MLP. Hence,

Fig. 10 Variations exhibited by Faces96 dataset

Table 1 Measures of correctness monitored from experimentation

		Precision		Recall		F1-score	
		CATBoost	MLP	CATBoost	MLP	CATBoost	MLP
Grimace dataset	Without LDA	0.88	0.92	0.9	0.94	0.88	0.93
	With LDA	0.85	0.91	0.83	0.92	0.83	0.91
Faces95	Without LDA	0.92	0.92	0.94	0.95	0.93	0.93
	With LDA	0.91	0.89	0.92	0.93	0.91	0.9
Faces96	Without LDA	0.93	0.9	0.9	0.89	0.91	0.89
	With LDA	0.89	0.88	0.87	0.87	0.88	0.87

CAT Boost performs better in this case. From Faces96 dataset, the overall F1-score for CAT Boost is greater than MLP. We can conclude that CAT Boost is a better model than MLP on this dataset. Table 1 helps in thorough analysis of the results and provides justification for our conclusions.

We can also observe that the overall difference in the F1-score after and before the application of the LDA for all the three datasets, for both the models is less. Hence, we can conclude that dimensionality reduction preserves variance. By taking LDA into account, we are improving the time complexity to a great extent which means that the overall time taken to perform the computation is significantly reduced.

It is safe to conclude that CAT Boost performed better on Faces95 and Faces96 dataset while MLP classifier is a better algorithm for handling expressions, i.e Grimace dataset.

4 Conclusion and Future Work

The model performed satisfactorily considering the substandard hardware, that was used to train and test the model. The recorded metrics indicate that the proposed system is production ready with accuracy close to that expected by the real-world applications. The pipeline performed well, even under the circumstances when there were drastic changes in the key parameters that defined that image. The model was tested for scale invariance, minor illumination variations and severe expression changes, and the results strongly depict the capabilities of the model.

The model shows potential for enhancements in the feature extraction stages. Feature extraction takes up most of the computational time in the pipeline. This seems to be major hurdle to train the model on the commodity hardware. Suitable alternatives that run efficiently on lightweight hardware can enhance the model performance on the hardware of our target and help us move a step closer in designing the perfect lightweight model.

References

1. P Prabhu (2016) Digital image processing techniques -a survey. Golden research thoughts 5(11)
2. Chandran S (2010) Color image to grayscale image conversion. pp 196 – 199, https://doi.org/10.1109/ICCEA.2010.192
3. Bala R, Braun KM (2004) Color-to-grayscale conversion to maintain discriminability. Proc SPIE 5293: https://doi.org/10.1117/12.532192
4. Zhihong W, Xiaohong X (2011) Study on histogram equalization. In: 2011 2nd international symposium on intelligence information processing and trusted computing, pp 177–179, https://doi.org/10.1109/IPTC.2011.52
5. Pizer SM, Amburn EP, Austin JD, Cromartie R, Geselowitz A, Greer T, ter Haar Romeny B, Zimmerman JB, Zuiderveld K (1987) Adaptive histogram equalization and its variations. Comput Vis Graph Image Process 39(3):355–368
6. Zhang B, Allebach PJ (2008) Adaptive bilateral filter for sharpness enhancement and noise removal. Image Proces IEEE Trans 17:664–678. https://doi.org/10.1109/TIP.2008.919949
7. Aleksic M, Smirnov M, Goma S (2006) Novel bilateral filter approach: image noise reduction with sharpening—Art. no. 60690f. In: Proceedings of SPIE—the international society for optical engineering. https://doi.org/10.1117/12.643880
8. Cheng PE (1990) Applications of kernel regression estimation:survey. Commun Stat Theory Methods 19(11):4103–4134. https://doi.org/10.1080/03610929008830431
9. Aleksic M, Smirnov M, Goma S (2018) Sparse locally adaptive regression kernel for face verification. Procedia Comput Sci 132:890–899. https://doi.org/10.1016/j.procs.2018.05.101. In: International conference on computational intelligence and data science

10. Seo HJ, Milanfar P (2010) Training-free, generic object detection using locally adaptive regression kernels. IEEE Trans Pattern Anal Mach Intell 32(9):1688–1704. https://doi.org/10.1109/TPAMI.2009.153
11. Seo HJ, Milanfar P (2011) Face verification using the lark representation. IEEE Trans Inf Forensics Secur 6(4):1275–1286
12. Tharwat A (2016) Principal component analysis—a tutorial. Int J Appl Pattern Recognit 3:197. https://doi.org/10.1504/IJAPR.2016.10000630
13. Tharwat A, Gaber T, Ibrahim A, Hassanien AE (2017) Linear discriminant analysis: a detailed tutorial. AI Commun 30:169–190. https://doi.org/10.3233/AIC-170729
14. Amato G, Bolettieri P, Falchi F, Gennaro C (2013) Large scale image retrieval using vector of locally aggregated descriptors. Similarity search appl. Springer, Berlin, Heidelberg, pp 245–256
15. Abbas A, Deligiannis N, Andreopoulos Y (2015) Vectors of locally aggregated centers for compact video representation. https://doi.org/10.1109/ICME.2015.7177501
16. Van Der Maaten L (2011) Learning discriminative fisher kernels. ICML 11:217–224
17. Arthur D, Vassilvitskii S (2007) k-means++: the advantages of careful seeding. In: Proceedings of the 18th annual ACM-SIAM symposium on Discrete algorithms, society for Industrial and applied mathematics, pp 1027–1035
18. Dorogush AV, Ershov V, Gulin A (2018) Catboost: gradient boosting with categorical features support. arXiv preprint arXiv:181011363
19. Natekin A, Knoll A (2013) Gradient boosting machines, a tutorial. Front Neurorobotics 7:21. https://doi.org/10.3389/fnbot.2013.00021
20. Sharma H, Kumar S (2016) A survey on decision tree algorithms of classification in data mining. Int J Sci Res (IJSR)
21. Panchal G, Ganatra A, Kosta Y, Panchal D (2011) Behaviour analysis of multilayer perceptrons with multiple hidden neurons and hidden layers. Int J Comput Theory Eng 3(2):332–337
22. Peng J, Lee KL, Ingersoll MG (2002) An introduction to logistic regression analysis and reporting. J Edu Res 96:3–14. https://doi.org/10.1080/00220670209598786
23. Bottou L (2010) Large-scale machine learning with stochastic gradient descent. In: Proceedings of COMPSTAT'2010, Springer, pp 177–186
24. Kishore R, Kaur T (2012) Backpropagation algorithm: an artificial neural network approach for pattern recognition. Int J Sci Eng Res 3(6):3

Ambient Intelligence

Smart Agro-Ecological Zoning for Crop Suggestion and Prediction Using Machine Learning: An Comprehensive Review

R. Chetan, D. V. Ashoka, and B. V. Ajay Prakash

Abstract Crop production in agriculture depends on many factors such as climate, geography, biological, economical, historical, political, socioeconomic and agro-ecological zoning. Intelligent agro-ecological zoning is at the forefront of this, the main aim is accurately suggesting and prediction of crops that ensure more production of it. This paper is a review for reassessing the research work on the agricultural crop suggestion and prediction with relevance to machine learning techniques to find research gap and to provide future research direction.

Keywords Agro-ecological zone · Machine learning · Crop suggestion · Prediction

1 Introduction

India is an agrarian country and its majority of the economy is based upon crop production. In India, the major backbone of business is agriculture. India ranks second in the world in related sectors, such as agricultural production, agriculture, forestry and fisheries which shows 17.32% in GDP growth [1]. In India, still, agriculture plays a vital role in many socioeconomic frameworks due to its broadest economic background. With refernce to the Wikipedia statistics, the number of suicide cases of

R. Chetan (✉)
Department of Computer Science and Engineering, JSS Academy of Technical Education, VTU, Bengaluru, India
e-mail: chetan.dhananjaya@gmail.com

D. V. Ashoka
Department of Information Science and Engineering, JSS Academy of Technical Education, VTU, Bengaluru, India
e-mail: dr.ashok_research@hotmail.com

B. V. Ajay Prakash
Department of Computer Science and Engineering, SJB Institute of Technology, VTU, Bengaluru, India
e-mail: ajayprakas@gmail.com

N. N. Chiplunkar and T. Fukao (eds.), *Advances in Artificial Intelligence and Data Engineering*, Advances in Intelligent Systems and Computing 1133,
https://doi.org/10.1007/978-981-15-3514-7_94

300,000 farmers in India from 1998 to 2018 are reported. Therefore, to eliminate this problem, a new approach is needed this will provide crop suggestion and prediction using agro-ecological zoning.

Agro-ecological zoning (AEZ) [2] is defined by landform, climate and soil characteristics. The zones indicate which areas are climatically suitable for different crops. The zone details depend on the data processing facility as well as the scale of the study. The crop production also depends on the AEZ. Therefore, it is important to develop simple, effective and case-specific tools that make available the up-to-date AEZ knowledge for the recommendation of crops.

Several papers have been shown in existing literature, such as Mishra et al. [3] detailed review paper on machine learning techniques in agricultural crop production till 2016. Justin et al. [4] studied the use of AEZ for increasing the yield potential. Ahamed et al. [5] studied the data mining techniques used for the recommendation of crops in different districts in Bangladesh. Rakesh kumar et al. [6] studied the crop selection method using machine learning techniques. Jain et al. [7] studied the various environmental factors for crop selection using machine learning. These review papers gave a detailed overview of data sources, machine learning techniques, methodology and models used by the previous works from 1992 to 2016 for crop selection and prediction and outlined the challenges for future work.

2 Review Process

The following steps were taken to carry out a comprehensive review process:

Step 1: Determine the review requirements.
Step 2: Frame research questions to collect review data
Step 3: Identifying the web resources.
Step 4: Data acquisition from different Web sources.
Step 5: Review of papers.
Step 6: Reporting the reviewed results.

Some of the review requirements we have considered are what are the machine learning techniques used so far, different datasets used for applying the machine learning techniques, factors considered for crop suggestion and prediction and what are the important features selection techniques used so far.

After the review requirements, we have come across different research questions which are discussed in the next section. For each research question, the following online resources like IEEE Xplore, Springer Link and so on are used for data acquisition.

As a prior study documented by Mishra [2], the research between 2013 and the end of 2016 was used as an index of papers from 1992 to 2016. Models, parameters, machine learning techniques and measurements made by performance metrics, datasets and feature extraction techniques were extracted in a spreadsheet to generate graphs and tables.

3 Research Questions

To conduct a review on crop suggestion and prediction techniques, the following research questions were framed are as follows:

RQ1: Which datasets are commonly used?
RQ2: Which techniques are used for feature selection?
RQ3: Which factors are used for crop suggestion and prediction?
RQ4: Which machine learning techniques were used in agriculture?
RQ5: What are the metrics used in crop suggestion and prediction for evaluating performance?

Based on the above research questions, let us consider one by one and highlight the issues which help us for further future work.

RQ1: Which datasets are commonly used?

The dataset is important for crop suggestion and prediction. The commonly used datasets considered are from various sources like Indian government digital initiative [8], Bangladesh Agricultural Research Institute (BARI) [9], USA [10, 11], Europe [12] and UCI machine learning datasets [13] as shown in Table 1.

RQ2: Which techniques are used for feature selection?

There are three purposes for feature selection. Improve performance, provide effective predictions and gain a better understanding of the underlying process. Appropriate features give better prediction accuracy. The regression techniques, filtering methods, consistency-based feature selection methods and random forest and other data mining algorithms have been used for feature selection [3, 4, 5, 6, 7]. Some of the features commonly used are climate variables [15, 16, 17] like max and min temperature, rainfall, soil variables like ph, type of soil and area.

RQ3: Which factors are used for crop suggestion and prediction?

Factors that have been used for the suggestion and prediction of agricultural crops include weather conditions (temperature, clouds, rainfall, humidity, etc.), the type of soil (sandy, silt, clay, peat, salt water, etc.), soil composition (PH value, phosphate, nitrogen, potassium, etc.), harvesting methods and areas (e.g., hills, riverbeds and deep areas). Methods used for crop selection refers to selecting of crops over a specific season according to different environmental and economic factors in order

Table 1 Table shows different datasets used for crop suggestion

Sl. No.	Example dataset type	References
1	Simulated dataset	[4, 14]
2	Web site extracted data	[8]
3	Agricultural research institute	[5, 9, 11, 12]
4	UCI machine learning datasets	[13]

to obtain maximum profit. The factors considered are precipitation, market price, average temperature, demand, etc. The government has fixed the minimum support prices (MSP) for the crops which also influences the farmers for crop production. The crop selection also depends on the sowing period and harvesting period which will be different for different crops. The agro-ecological zone is also an important factor for crop suggestion and prediction which is not considered in any of the work.

RQ4: Which machine learning techniques were used in agriculture?

The major application of machine learning so far used in agriculture is for predicting crop yield [18, 19, 20, 21]. The graph showed in Fig. 1 shows the inference of where machine learning is used in different applications.

From the survey, we can see that most of the algorithms like decision tree [22], regression [23], markov model [24, 25, 26], k-means [27] and neural networks [28] are used for predicting the yield of different crops. For crop selection, no such techniques have been proposed and only fuzzy modeling [29, 30, 31] is used. Still, the machine learning algorithms like gradient boosted tree (GBT), random forest (RF) and regularized greedy forest (RGF) algorithms have not considered. This can be visualized in Fig. 2.

Table 2 shows the different machine learning models used in the agricul-

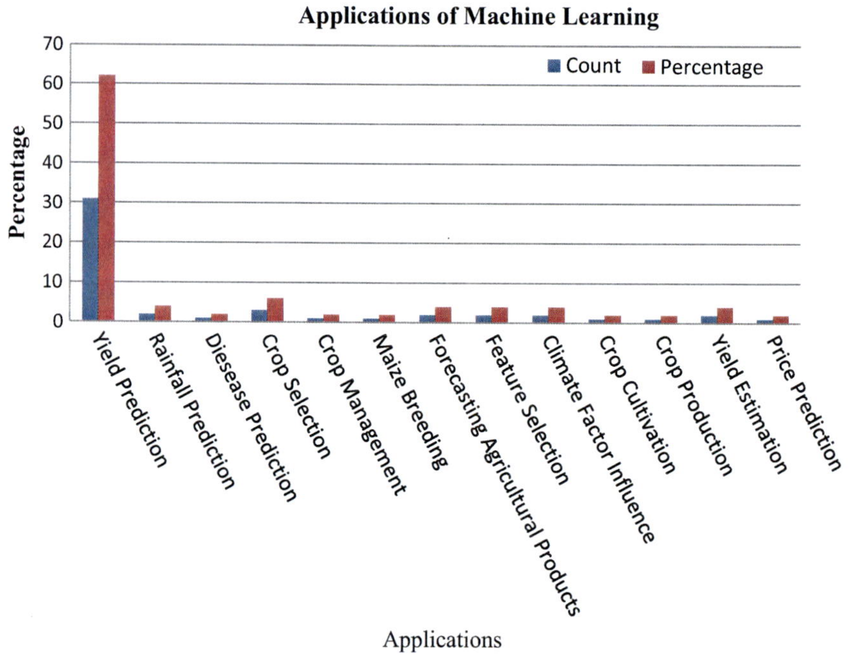

Fig. 1 Graph showing application of machine learning techniques in agriculture

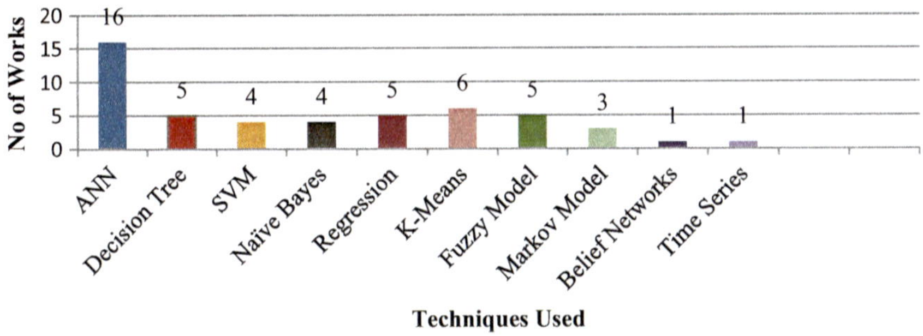

Fig. 2 Graph showing machine learning techniques used so far

Table 2 Table shows the machine learning models used in agricultural domain

Sl. No.	ML model	Domain
1	SVM	Crop yield prediction [32], quality of crops, livestock production
2	Bayesian network	Animal welfare
3	ANN [33, 34, 35, 36]	Crop management, detection of weeds and diseases, crop yield prediction

tural domains such as crop yield prediction, livestock production, animal welfare, detection of diseases in plants and quality of crops.

RQ5: What are the metrics used in crop suggestion and prediction for evaluating performance?

The root mean square error (RMSE) and mean absolute percentage error are the best generally used performance metrics. RMSE variants such as normalized RMSE as well as RMSE with cost are also used. Performance measures such as equal coefficient (EC), mean absolute relative error, R2 squared, mean relative and mean square error and variance absolute percentage error are majorly used. But some of the performance metrics such as accuracy, recall, F1, FP rate and sensitivity can also be used.

4 Future Guidelines

The comprehensive review findings on crop suggestion and prediction are as follows:

- The machine learning techniques applied on agricultural production are majorly used for crop yield prediction and less work has been done on crop suggestion. The algorithms like random forest, greedy random forest and gradient boost decision tree algorithms are not explored efficiently.

- The factors like agro-ecological zones and minimum support prices have not been considered for crop suggestion and yield prediction which becomes one of the main dependent factors.
- Datasets should include the minimum support price provided for each crops and AEZ are required.
- New methodology has to be proposed for crop suggestion and prediction, we have to combine both in the single framework because a lot of work has been done separately for crop suggestion and prediction.

It is recommended to adopt commonly used performance measurements such as MAPE, recall and RMSE. This is because researchers can determine the efficiency of the model as well as other research models.

5 Conclusion

In this paper, we have made a sincere effort for providing a comprehensive review on crop suggestion and prediction using machine learning algorithms with AEZ are documented and aggregated the finding in them. Initially, the requirements for the review such as what are the datasets to be considered, feature selection algorithms to be used, machine learning techniques have to applied were finalized. Based on the review requirements, we were able to frame the research questions. The research questions framed were really helpful in finding what the different sources of dataset to be considered, which feature selection techniques need to be applied, which machine learning techniques are used in crop suggestion and prediction. Finally, how the performance metrics such as RMSE, MAPE and recall can be considered for evaluation of the model that we are going to build. Highlighting the techniques that were not considered which helps us for our future research work.

References

1. Literature available at URL: http://statisticstimes.com/economy/sectorwise-gdp-contribution-of-india.php
2. Literature available at URL: http://wgbis.ces.iisc.ernet.in/energy/HC270799/LM/SUSLUP/FAO/SB73AEZ.PDF
3. Mishra S, Mishra D, Santra GH (2016) Applications of machine learning techniques in agricultural crop production: a review paper. Indian J Sci Technol 9(38):1–14
4. Van Wart J, van Bussel LG, Wolf J, Licker R, Grassini P, Nelson A, Boogaard H, Gerber J, Mueller ND, Claessens L, van Ittersum MK (2013) Use of agro-climatic zones to upscale simulated crop yield potential. Field Crop Res 143:44–55
5. Ahamed AM, Mahmood NT, Hossain N, Kabir MT, Das K, Rahman F, Rahman RM (2015) Applying data mining techniques to predict annual yield of major crop and recommended

planting different crops in different districts in Bangladesh. In: IEEE/ACIS 16th international conference on software engineering, artificial intelligence, networking and parallel/distributed computing (SNPD) 2015, pp 1–6. https://doi.org/10.1109/snpd.2015.7176185
6. Kumar R, Singh MP, Kumar P, Singh JP (2015) Crop selection method to maximize crop yield rate using machine learning technique. In: IEEE international conference on smart technologies and management for computing, communication, controls, energy and materials (ICSTM), pp. 138–145. 6–8 May 2015. https://doi.org/10.1109/icstm.2015.7225403
7. Jain N, Kumar A, Garud S, Pradhan V, Kulkarni P (2017) Crop selection method based on various environmental factors using machine learning. Int Res J Eng Technol (IRJET) 4(2):1530–1533
8. https://data.gov.in/sector/crops
9. https://www.barcapps.gov.bd./dbs/index.php
10. House CC (1979) Forecasting corn yields: a comparison study using Missouri data. Statistical Research Division, United States Department of Agriculture 17(16):189–200
11. https://catalog.data.gov/dataset?_metadata_type_limit=0&q=Crop+Yield
12. https://data.gov.ie/dataset/area-yield-and-production-of-crops
13. https://archive.ics.uci.edu/ml/datasets.html
14. Chakravarti A, Joshi N, Panjiar H (2015) Rainfall runoff analysis using artificial neural network. Indian J Sci Technol 8(14):1–7
15. Oludhe O, Christopher C. (2002) Deterministic and probabilistic prediction approaches in season to inter-annual climate forecasting
16. Priya SRK, Suresh KK (2009) A study on pre-harvest forecast of sugarcane yield using climatic variables. Stat Appl 8(2):1–8
17. Kumar AVTV, Rajini Kanth R (2013) A data mining approach for the estimation of climate change on the jowar crop yield in India. Int J Emerg Sci Eng (IJESE) 2(2):16–20
18. Shibayama M (1991) Estimating grain yield of maturing rice canopies using high spectral resolution reflectance measurements. Remote Sens Environ 36(1):45–53
19. Gu Y, James W, McNicol M (1994) An application of belief networks to future crop production. In: IEEE conference on artificial intelligence for applications, San Antonia, p 305–309
20. Stathakis D, Savin I, Negre T (1994) Neuro-fuzzy modeling for crop yield prediction. Int Arch Photogramm Remote Sens Spat Inf Sci 34:1–4
21. Wilcox A (2000) Factors affecting the yield of winter cereals in crop margins. J Agric Sci 135(4):335–346
22. Veenadhari S, Mishra B, Singh CD (2011) Soybean productivity modelling using decision tree algorithms. Int J Comput Appl 27(7):975–8887
23. Prasad PR, Begum SA (2013) Regression and neural networks models for prediction of crop production. Int J Sci Eng Res 4(9):98–108
24. Matis JH, Birkett T, Boudreaux D (1989) An application of the Markov chain approach to forecasting cotton yields from surveys. Agric Syst 29(4):357–370
25. Jain RC, Ramasubramalliall V (1998) Forecasting of crop yields using second order Markov chains. J Indian Soc Agric Stat 51:61–72
26. Osman J, Inglada J, Dejoux JF (2015) Assessment of a Markov logic model of crop rotations for early crop mapping, Elsevier. Comput Electron Agric 113:234–243
27. Utkarsha P, Narkhede N, Adhiya KP (2014) Evaluation of modified K-means clustering algorithm in crop prediction. Int J Adv Comput Res 4(3):1
28. Chen C, Mcnairn H (2006) A neural network integrated approach for rice crop monitoring. Int J Remote Sens 27(7):1367–1393
29. Petridis V, Kaburlasos VG (2003) FINk NN: a fuzzy interval number k-nearest neighbour classifier for prediction of sugar production from populations of samples. J Indian Soc Agric Stat 4:17–37
30. Salleh MNM (2012) A fuzzy modelling of decision support system for crop selection. In: IEEE symposium on industrial electronics and applications (ISIEA2012), Bandung, Indonesia, pp 17–22

31. Papageorgiou EI, Aggelopoulou KD, Gemtos TA, Nanos GD (2013) Yield prediction in apples using fuzzy cognitive map learning approach. Comput Electron Agric 91:19–21
32. Hong-Ying L, Yan-Lin H, Yong-Juan Y, Hui-Ming Z (2012) Crop yield forecasted model based on time series techniques. J Northeast Agric Univ (Engl Ed) 19(1):73–77
33. Monisha Kaul M, Robert L, Hill H, Walthall C (2005) Artificial neural networks for corn and Soybean yield prediction. Agric Syst 85(1):1–18
34. Co HC, Boosarawongse R (2007) Forecasting Thailand's rice export: statistical techniques vs. artificial neural networks. Comput Ind Eng 53(4):610–627
35. Gorni G, Augusto A (2008) The application of neural networks in the modeling of plate rolling processes. Miner Metals Mater Soc JOM 49(4):1–4
36. Dahikar MSS, Rode SV (2014) Agricultural crop yield prediction using artificial neural network approach. Int J Innov Res Electr Electron Instrum Control Eng (IJIREEICE). 2(1):1–4

Preparedness in the Aftermath of a Natural Disaster Using Multihop Ad hoc Networks—Drone-Based Approach

Getzi Jeba Leelipushpam Paulraj, Immanuel Johnraja Jebadurai, and J. Jebaveerasingh

Abstract During natural disasters such as flood and earthquakes, the communication infrastructure collapses leading to loss of connectivity of people. This may slag the process of rescue team reaching the spot for disaster recovery process. Multihop ad hoc networks play as an alternative providing connectivity among the people in disaster and rescue team. The information about the collapsed infrastructure is received from the Internet service providers. The choppers with small drones are sent to those locations. The drones form an ad hoc network and scan the disaster-hit location. The report is in turn sent to the chopper getting relief materials for the people in the disaster location. The proposed technique is simulated using Node MCU and Bluetooth module. Simulation results show that the proposed technique outperforms the existing techniques in terms of vital performance metrics.

Keywords Multihop ad hoc networks · Disaster recovery management · Choppers · Drones · Internet service providers

1 Introduction

Multihop ad hoc networks is an wireless ad hoc networks that deliver its tasks by taking two or more hops from source to destination. They have many vital applications that include wildlife monitoring, natural disaster recovery management, agricultural monitoring and many more [1]. Natural disaster recovery management is one of the applications of multihop ad hoc networks. During natural disasters such as flood and earthquakes, the primary disaster caused is the failure of communications infrastructure. This lack of communication disconnects people from rescue team that leads

G. J. L. Paulraj (✉) · I. J. Jebadurai · J. Jebaveerasingh
Karunya Institute of Technology and Sciences, Coimbatore, India
e-mail: getzi@karunya.edu

I. J. Jebadurai
e-mail: immanueljohnraja@gmail.com

J. Jebaveerasingh
e-mail: jebaveerasingh@karunya.edu

N. N. Chiplunkar and T. Fukao (eds.), *Advances in Artificial Intelligence and Data Engineering*, Advances in Intelligent Systems and Computing 1133,
https://doi.org/10.1007/978-981-15-3514-7_95

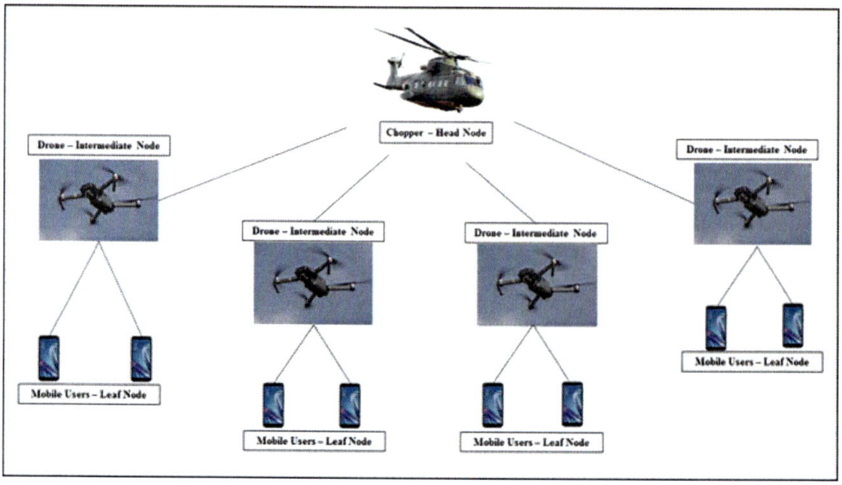

Fig. 1 Block diagram of multihop ad hoc network in disaster recovery management

even to loss of life. This paper proposes a rescue system using multihop ad hoc network after any natural disaster.

The failed infrastructures are identified using the input from Internet service providers (ISP). Choppers are sent to the identified locations. These choppers act as a head node of the network. The choppers are equipped with two or three drones which form the next level of nodes in the network. The users can for wireless network with the drones using their mobile devices. The emergency rescue app installed in the mobile devices enable the users to request essential services, viz. food, water and medical services. The block diagram of the proposed technique is shown in Fig. 1.

Section 2 discusses the various techniques available for disaster management in the literatures. Section 3 presents the proposed technique. Section 4 explains the implementation details and Sect. 5 concludes the paper and suggest future works.

2 Related Works

Infrastructure failure is a crucial failure that delays the recovery process is any disaster-hit region. Failure of infrastructure disconnects people and hinders the basic amenities to reach them in time. Many literatures have proposed various recovery techniques to overcome the infrastructure failures. In [2], the authors discuss Internet of things (IoT)-based task technology fit approach for disaster management. This approach discusses how IoT technology could offer enhanced rescue operation by information fitting. A new disaster mobility model has been proposed in [3]. The model is evaluated in terms of mobility, distribution of nodes and node density comparing the proposed disaster mobility model with random waypoint model. An optimal broadcasting mechanism for disaster scenario is proposed in [4]. Relative distance between nodes and node density parameters are the factors considered for the selection of broadcast schemes. However, using the evolutionary approach this technique selects the best parameters for the broadcast of information in disaster-hit area. A decentralized communication scheme has been proposed in [5]. The broadcasting scheme has been formulated as multi-objective optimization problem. Based on the topology characteristics, a broadcasting scheme has been proposed to minimize delay, number of retransmission and reachability. Techniques proposed in [3–5] have adopted mobile ad hoc networks for disaster recovery. Similarly, [6] surveys various disaster management networks using delay-tolerant network, wireless sensor networks and wireless mesh networks.

Solving communication discontinuity is a major challenge in disaster-hit region [7]. This paper resolves the issue by proposing a drone-based approach. The three modules, viz. chopper module, drone module and user module coordinate among themselves to offer Internet access, connectivity and issue of basic amenities. This approach not only solves the connectivity issue, but also enables the user to quickly get the basic amenities for the disaster-hit region.

3 Proposed Technique

Multihop ad hoc disaster management system (MADMS) enables the rescue team to cover the affected areas where infrastructure-based communication fails. The proposed disaster management system consists of modules, viz. application module, drone module and the chopper module. The chopper module forms the head node and identifies the infrastructure damaged locations. It identifies the location with the help of ISPs. The chopper install drone as intermediate nodes which forms wireless hotspots for the people in disaster-hit region. Users with mobile phones discover suitable drone and form wireless network installing the predefined application. Through the predefined application, the users request various vital services, viz. food, water and medical emergency. The detailed functioning of each module is explained as follows.

3.1 Chopper Module

The chopper module measures the signal strength of the mobile tower at a region with the help of the ISPs. If the signal strength is less than the required level, the chopper connects forms an ad hoc network with the other choppers connected to the nearest mobile tower.

By forming this ad hoc network, every chopper act as a wireless access point. The block diagram of the chopper module is given in Fig. 2

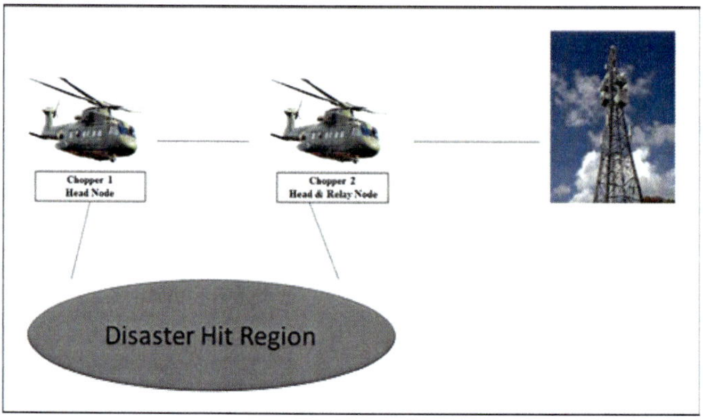

Fig. 2 Chopper module

3.2 *Drone Module*

The drone module receives Internet access from chopper module and acts as wireless access point to the user applications. The drone also has essential amenities required by the user such as food packets, water bottles and medical kit. The drones behave as intermediate nodes relaying between the user module and chopper module.

3.3 *User Module*

The user's mobile terminal connects to the drone module to access the Internet connectivity and basic amenities. The sequence diagram for acquiring Internet connectivity is explained in Fig. 3. The mobile user sends request to the drone module requesting for IP address. The drone module connects with the mobile user providing the IP address. The drone module provides the URL for downloading the amenity request application developed for android. The user module downloads the app.

Through the app, the user module can request for basic amenities such as food, water and cloths. The drone modules connect with the chopper module to collect the requested amenities and deliver it to the users.

4 Implementation Details

The proposed framework has been implemented using Internet of things (IoT) controller boards and smartphone. The chopper module is represented using Node MCU IoT board. It is an IoT platform that includes firmware which runs on the ESP8266

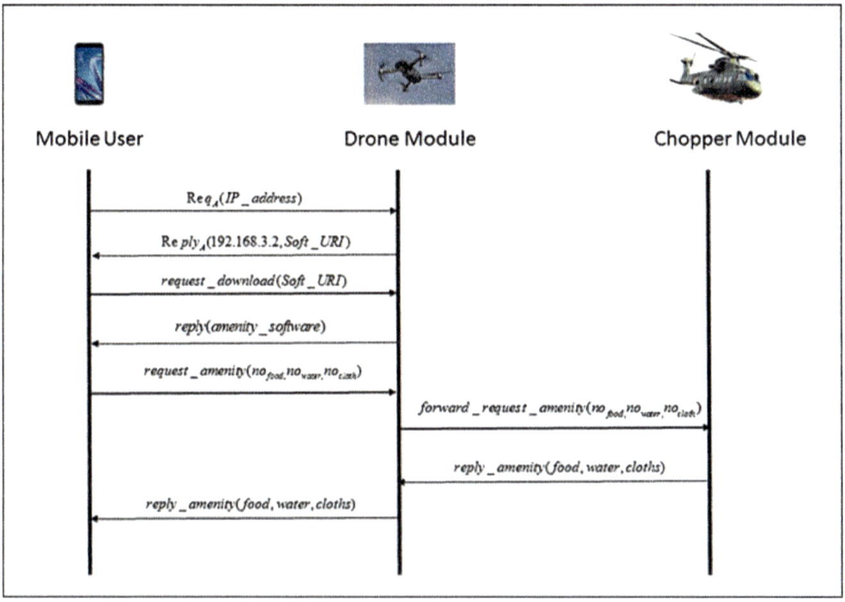

Fig. 3 Interaction between user, drone and chopper modules—sequence diagram

WiFi. The drone modules are represented by Bluetooth module. Two Bluetooth modules have been connected to the Node MCU controller board. The smartphones are used to represent the user module. The implementation block diagram is depicted in Fig. 4.

The user module connects to the Bluetooth module and forms a private network. The private network enables the smartphones to install an application. The application enables the users to request basic amenities. The request is forwarded to the Node MCU board which in turn sends a reply to the user module through the Bluetooth module. The user can visualize the reply in the installed application.

5 Conclusion

Rapid evolution of IoT, mobile technology enables to quickly solve the connectivity problem in disaster-hit region. It also enables people to receive their amenities. A drone-based approach has been proposed to solve the connectivity issue in disaster-hit region. The chopper module connects with nearest ISPs to continue the connectivity. The drones are deployed over the disaster-hit region which acts as a relay node offering Internet connectivity to the users in disaster-hit region. The users can download a mobile application through which they request their basic amenities. The amenities are issued to the users immediately by analyzing their location. The proposed technique has been simulated using Node MCU and Bluetooth module to analyze

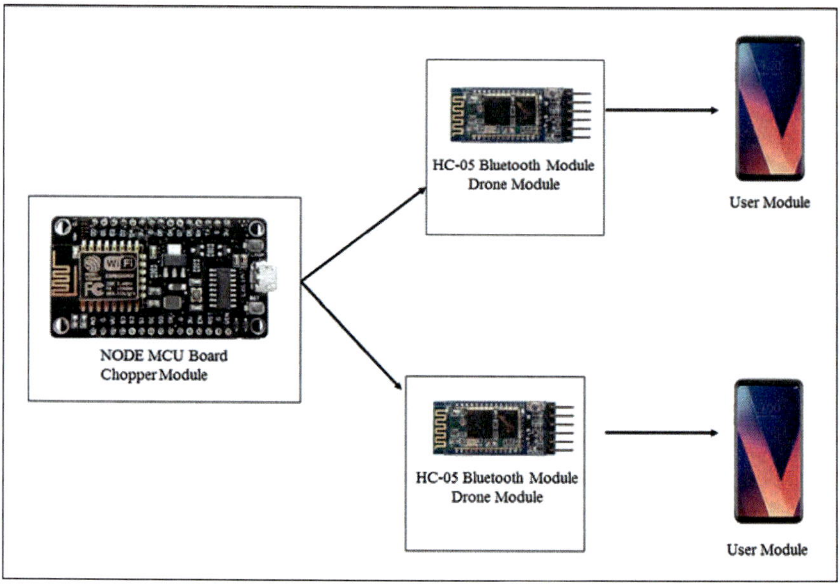

Fig. 4 Implementation block diagram

the implementation feasibility. As a future direction, the proposed technique can be implemented on real time.

References

1. Wang F, Yuan H (2010) Challenges of the sensor web for disaster management. Int J Digit Earth 3(3):260–279
2. Yang L, Yang SH, Plotnick L (2013) How the internet of things technology enhances emergency response operations. Technol Forecast Soc Chang 80(9):1854–1867
3. Aschenbruck N, Gerhards-Padilla W, Martini P (2009) Modeling mobility in disaster area scenarios. Perform Eval 66 (12):773–790
4. Reina DG, Toral SL, Leon-Coca JM, Barrero F, Bessis N, Asimakopoulou E (2013) An evolutionary computational approach for optimizing broadcasting in disaster response scenarios. In: 2013 seventh international conference on innovative mobile and internet services in ubiquitous computing, pp 94–100. IEEE
5. Reina DG, León-Coca JM, Toral SL, Asimakopoulou E, Barrero F, Norrington P, Bessis N (2014) Multi-objective performance optimization of a probabilistic similarity/dissimilarity-based broadcasting scheme for mobile ad hoc networks in disaster response scenarios. Soft Comput 18(9):1745–1756
6. Reina DG, Coca JML, Askalani M, Toral SL, Barrero F, Asimakopoulou E, Sotiriadis S, Bessis N (2014) A survey on ad hoc networks for disaster scenarios. In: 2014 international conference on intelligent networking and collaborative systems, pp 433–438. IEEE
7. Tran Q, Shibata Y, Borcea C, Yamada S (2016) Ad hoc networks on-site configuration of disaster recovery access networks made easy. Ad Hoc Netw 40:46–60

An IoT-Based Congestion Control Framework for Intelligent Traffic Management System

Md. Ashifuddin Mondal and Zeenat Rehena

Abstract The concept of smart city helps to improve the quality of urban life of the citizens while keeping in mind the environmental impacts. Smart and sustainable transportation system is one of the major contributors in order to make the city smart. Major cities around the world face enormous vehicular growth due to socioeconomic growth and rural to urban migration of the people. These results in high traffic congestion on road, road accidents, delay and have an adverse environmental impact, thus effecting smooth mobility of the citizens. Hence, traffic management authorities face difficulties to manage and reduce traffic congestion, road accidents and air pollution. In order to overcome the above-mentioned challenges, this paper proposes a framework for managing road traffic congestion in intelligent traffic management system which utilizes the available infrastructures and resources in an optimum way. The proposed framework comprises of four different modules, namely data collection module, data storage module, data processing module and business application module.

Keywords IoT · Traffic congestion · Smart city · Intelligent traffic management system · Crowdsourcing

1 Introduction

Nowadays, there is a trend of huge migration of citizens from rural to urban area throughout the world. It has been forecasted that more than 60% of the population will live in urban areas by 2030 [1]. Hence, the traditional approach of city management

Md. A. Mondal (✉)
Department of Computer Science and Engineering,
Narula Institute of Technology, Kolkata 700109, India
e-mail: ashifuddin.mondal@nit.ac.in

Z. Rehena
Department of Computer Science and Engineering,
Aliah University, New Town, Kolkata 700160, India
e-mail: zeenatrehena@yahoo.co.in

N. N. Chiplunkar and T. Fukao (eds.), *Advances in Artificial Intelligence and Data Engineering*, Advances in Intelligent Systems and Computing 1133,
https://doi.org/10.1007/978-981-15-3514-7_96

is not sufficient to serve the citizens in a better way. The concept of smart city can address the challenges related to the management of different city components like transport, energy, health, buildings, etc. [2]. Intelligent transportation system is one of the important components of smart city which has a major impact on the smooth mobility of citizens [3]. Number of vehicles is increasing enormously day by day due to socioeconomic growth and rural to urban migration of the people. The number of registered vehicle in India has increased from 0.3 million in 1951 to 142 million in 2011 according to [4]. But the capacity of existing road infrastructure is not sufficient to cope with the increasing number of vehicles. This leads to heavy traffic congestion on the roads of urban area which further results in road accidents, air pollution and loss of valuable time of the citizens. One solution to avoid the traffic congestion is to expand or build new infrastructure (road, flyover, etc.) so that it can accommodate the large number of vehicles. But this is not a feasible solution due to different factors, mainly lack of physical space to construct new roads over the existing one and lack of financial support [5].

With the help of the information and communication technology (ICT), the above-mentioned problems can be addressed efficiently. IoT is an emerging paradigm in which daily life objects equipped with sensors which sense the environmental condition and share the sensed information with each other based on standard communication protocol. Hence, the concept of IoT [6] can be used in an efficient way to manage the traffic congestion and its related problems. The sensors can be deployed over the road network or can be attached to the traffic infrastructure (traffic signal, car) for capturing traffic information and transmit the data to the central server for further processing. In turn, the central server can control traffic infrastructure (traffic signal) depending on the present road traffic condition.

On the other hand, the traditional traffic management system is not able to manage the traffic on road network of a city in an efficient way as it lacks in real-time traffic data collection technique, smart data processing technique as well as proper dissemination of traffic information to the citizens. It mainly relies on human resource (traffic sergeant) for traffic data collection, traffic monitoring and traffic management. Thus, the traditional traffic management system needs to deploy a large number of human resources to manage traffic in a large geographical area. Hence, it is not scalable enough. This leads to enormous traffic congestion on the road network in the major cities. Thus, an intelligent traffic management system needs to be designed so that traffic congestion can be reduced and managed efficiently.

The aim of this paper is to propose an IoT-based system for traffic congestion control in intelligent traffic management system. The system uses WSN [7] technology for traffic data collection. Different types of sensors, inductive loop and camera, etc., need to be deployed on the road network to collect real-time traffic data. The proposed system processes the sensed data for example no. of vehicles, speed of vehicles with the help of different techniques such as artificial neural network [8], k-means algorithm [9], etc., to infer knowledge and make a proper decision toward the traffic congestion control on road. The system uses different business applications, like smart parking management system [10–13], suggesting the best alternate route to assist the citizens so that they can take proper decision. Also, the proposed system helps the transport authority to monitor traffic status of different road segments using business applications like road traffic congestion state identification system [14]. It allows the transport authority to update the traffic rule and regulation based on traffic status on the road. The proposed system automatically controls the traffic control infrastructure (traffic light) based on current traffic status on the road segments.

The rest of the paper is organized as follows: Sect. 2 highlights the detailed review of the literature survey. The proposed system is presented in Sect. 3. Section 4 describes the workflow of the system. Section 5 evaluates the proposed system. Finally, we conclude the paper in Sect. 6.

2 Related Work

A lot of different research works have been done in different applications of intelligent traffic management system (ITMS). Also, several works are found in the literature that discusses the mitigation of congestion control in ITMS.

In [7], authors presented different phases of traffic management system and gave up-to-date review on various technologies which are used in different phases of traffic management system. They also studied about the security threats related to traffic management system.

In another study [15], authors proposed various methods of traffic data acquisition, preprocessing and integration of data. They have also discussed data fusion phase. The authors described data preprocessing and data cleaning concepts and explained two data cleaning techniques: space-based similarity search and time-based similarity search.

In [3], authors proposed a framework for context-aware intelligent traffic management system in smart cities. They discussed various challenges of existing traffic management system.

In [16], authors reviewed the current trends in intelligent transportation systems (ITS) and they discussed different components of ITS. They also discussed emerging technologies toward connected vehicle-infrastructure-pedestrian (VIP) environment.

In another study [4], authors presented detailed classification of ITS and discussed different components of ITS. They discussed different issues and challenges of ITS in India.

Though all the above-mentioned works discuss the framework of intelligent transportation system and different data collection techniques in detail, but they did not focus on congestion control system. This work discusses various components of congestion control framework for ITMS and explains the detailed workflow of the system. Also, it explains the different data collection techniques and compares their merits and demerits. It also explains the concept of crowdsourcing as another way of traffic data collection.

In the next section, we are going to propose an IoT-based congestion control system for intelligent traffic management system.

3 Proposed Framework

To resolve the key issues like air pollution, traffic congestion and road accidents due to increase in number of vehicles, this paper proposes an IoT-based congestion control framework for ITMS which integrates a wide range of components like information sensing, communication, data processing, information dissemination and existing traffic control infrastructure. The framework has four main components, namely (i) data collection module, (ii) data transmission module, (iii) data processing module and (iv) business application module. The proposed framework is depicted in Fig. 1.

1. **Data Collection Module**: ITMS require high-quality real-time traffic data for proper decision making. Real-time traffic data needs to be collected from roads and is sent to the remote data processing module. The proposed system considers heterogeneous data source (such as sensors, GPS embedded vehicles, citizens) for collecting traffic data. Traffic data acquisition can be done by the following three techniques: In-road sensors, floating car data and crowdsourcing.

   (a) **In-Road Fixed Sensors**: Traditionally, inductive loop detectors [15, 16] were used to acquire traffic data from road. This type of detector detects the vehicle based on the induced current generated when it was passing over the detector. Pneumatic tube [16] can also be used to collect traffic data. It detects the vehicle based on change of pressure inside the tube. Both the detectors collect traffic volume and speed of vehicles from the road. But these detectors are

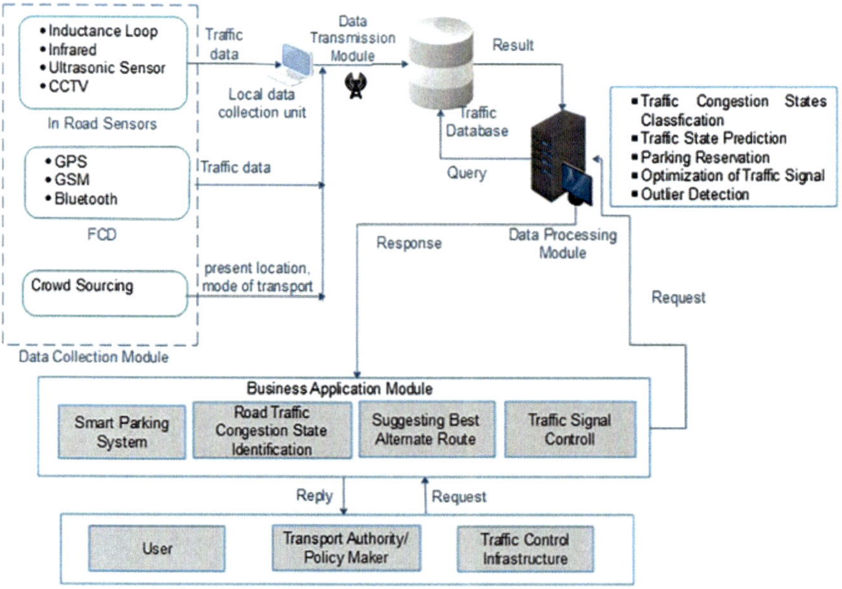

Fig. 1 Proposed congestion control framework of ITMS

rarely used as it hampers the traffic during installation process when they are deployed under the road segment. Thus, the better solution is to use RFID technology, video camera or infrared to acquire real-time traffic data like traffic volume, speed and types of vehicles.

(b) **Floating Car Data**: In this method, real-time traffic data is acquired by locating the vehicle with the help of mobile tower or GPS system over the road network. For that every vehicle should be equipped with GPS system or mobile phones. In this method, traffic information like vehicle speed, location of vehicle and direction of movement are sent continuously to the remote processing server by locating the vehicles in multiple positions in a different point of time using mobile tower or GPS.

(c) **Crowdsourcing**: Crowdsourcing is a problem solving model where crowd or group engaged together to achieve a common goal [17]. Facebook or Twitter platform can be used to collect crowdsource traffic data. Apart from that Google Distance Matrix API is also used to get real-time traffic information (estimated travel time) from Google server. Nowadays, most of the citizens uses different types of Google services (Google Map, Gmail, etc.) in their smartphone while they are traveling in the city and Google tracks the location, movement of citizens with the help of these Google services. Then, Google uses deep learning algorithms to infer traffic information (estimated travel time).

2. **Data Transmission Module**: This module helps to communicate the traffic information collected from road segments to the remote data processing module for processing of these data. Traffic data may be transmitted over wired or wireless mode (3G/4G, WiFi).
3. **Data Processing Module**: Data processing module is a very important component of ITMS as this module process the traffic data collected from road segments to infer meaningful knowledge/status about traffic. Different types of algorithms are being used like: MCDA, artificial neural network, support vector machine, Bayesian networks and other machine learning techniques to process the data. Artificial neural network (ANN)-based approach can be used in the field of smart mobility as they have knowledge building capability from given input-output data set. Like ANN, the support vector machine is another AI-based approach that can be used for solving traffic problem like incident detection. Traffic congestion state classification, traffic state prediction, traffic signal optimization and parking reservation can be done with the help of these algorithms running on remote data processing server.
4. **Business Application Module**: ITMS have different business application modules which are used by citizens and transport authority. Smart parking system [18] suggesting the best alternate route are some of the business application that can be used by citizens from their handheld devices.

4 Workflow of the System

Figure 2 depicts the workflow of the proposed system. It collects the road traffic data from heterogeneous data sources. Data sources include different types of sensors (inductive loop, infrared, RFID), camera, crowdsource which collects and transmits the data to the ITMS and store the traffic data in the database. The data processing module process the traffic data using different types of algorithms to generate meaningful knowledge. Different types of applications (smart parking system suggesting best alternate route) running in ITMS uses that knowledge to provide services to the user. The data processing module of the proposed framework uses machine learning techniques (artificial neural networks and k-means clustering) to analyze the traffic data and produce useful information (identification of congestion status, grouping of road segments based on traffic density and average speed) which will help transport authority toward decision and policymaking regarding smooth traffic management.

Figure 3 shows the sequence diagram which depicts the interaction between objects of the proposed system in sequential order with respect to the users or citizens. Traffic database module request for traffic data to the heterogeneous data sources deployed over road network and in turn they send the real-time traffic data. The above-mentioned process is repetitive one. The users can make a request with their handheld devices to the remote data processing module for specific service. The remote data processing module process the user's request by executing application-specific algorithm based on real-time traffic data and returns back the result to the user.

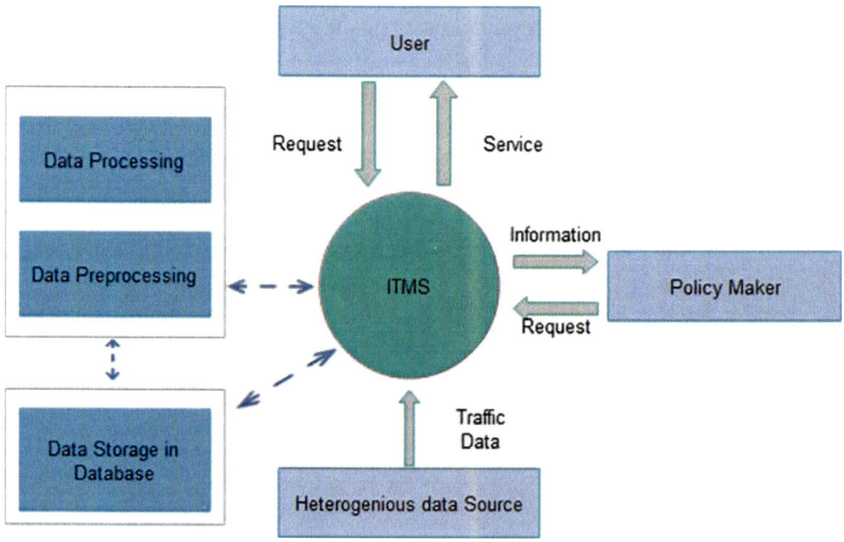

Fig. 2 Workflow of the proposed system

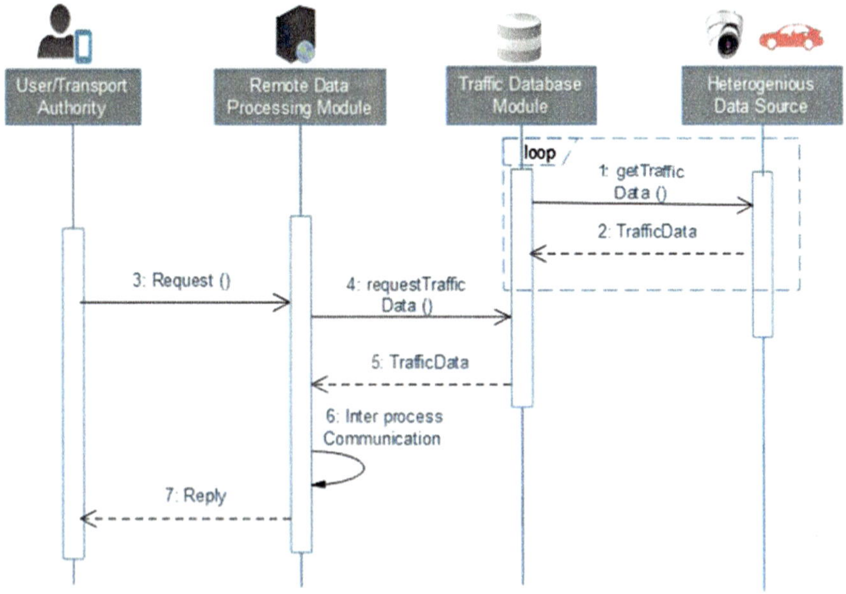

Fig. 3 Sequence diagram from user's perspective

Figure 4 shows the interaction among the objects of the proposed system with respect to the traffic control infrastructure. The proposed system manages the traffic control infrastructure based on real-time traffic data without human intervention. The data sources send the traffic data from the road network to the database. Remote data processing module process the real-time data and controls the traffic infrastructure as per context of the current situation. For example, traffic signal can be controlled automatically by the proposed system (ITMS) based on the current context of the road.

5 Evaluation of the Framework

The proposed system gathers information about road network environment in real time and adapts its behavior. The decision is made on the current data of the road network. The data includes information about road traffic like the density of vehicles in a road segment, average speed of vehicles, traffic flow and weather condition in a particular point of time. Different types of in-road fixed sensors, floating car data (GPS embedded vehicle) and crowdsource data are the potential sources of this traffic information and the performance comparison of these sources is shown in Table 1. The data processing module acts according to the predefined rule or algorithm and/or through machine intelligence based on collected traffic data. For example, the cycle

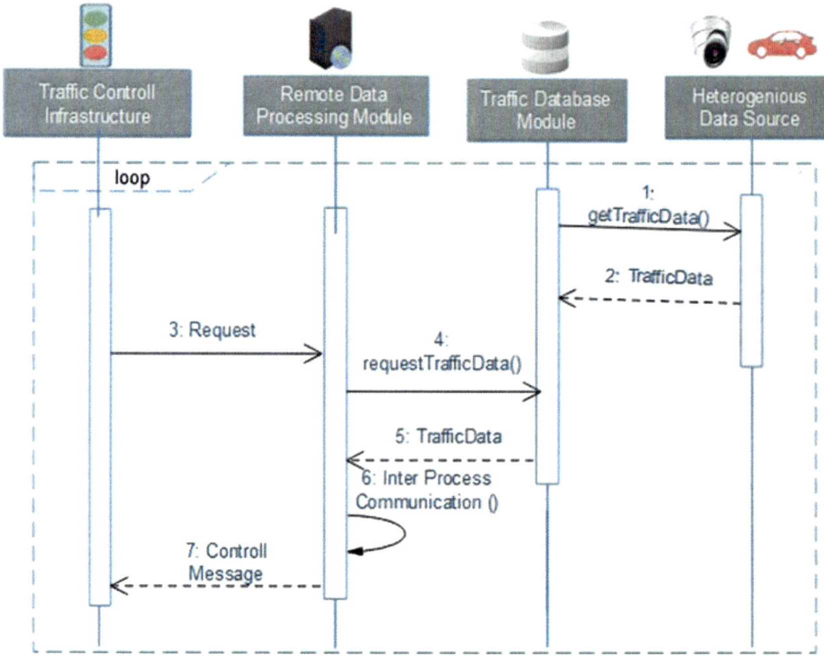

Fig. 4 Sequence diagram from traffic control infrastructure's perspective

length of traffic signal is automatically adjusted depending on the traffic status of the road intersection. The citizens can also take the service of ITMS from their handheld mobile devices. Car owner can search for parking space within a city.

As far as real-time traffic data collection is concerned, the proposed system gives importance on crowdsourcing. The paper implements the concept of crowdsourcing to acquire real-time traffic data using Google Map Distance Matrix API. In recent times, almost all citizens have their smartphones and use many Google services (Google Map, Gmail). When citizens travel along the road network, Google tracks every user by locating (longitude, latitude) them with the help of Google services and monitors the direction of movement, speed of the movement, position of the user with respect to time. The Goggle traffic database stores this information. The paper uses Python language to fetch the real-time traffic data with the help of Distance Matrix API provides by Google. Figure 5 shows the user interface to collect real-time traffic data.

Transport authority has to mention source and destination address from where he/she wants to monitor the traffic. Also, the time from when he/she wants to monitor the traffic data needs to be mentioned. The field $Interval$ indicates the time gap in seconds between two successive data collection. The collected data then stored in database as shown in Fig. 6.

Table 1 Comparison of different data collection techniques

Properties	In-road fixed sensor	FCD	Crowdsourcing
Installation cost	Expensive as sensors needs to be installed in road segments	Not expensive as it uses GPS system embedded in the vehicle to collect data	Not expensive as it uses hand held mobile devices to collect data
Maintenance cost	High maintenance cost	Low maintenance cost	Low maintenance cost
Coverage	Limited coverage (as sensors installed only in major road segments)	Wide coverage	Wide coverage
Performance	Performance affected in bad weather condition	Performs well in all-weather condition	Performs well in all-weather condition

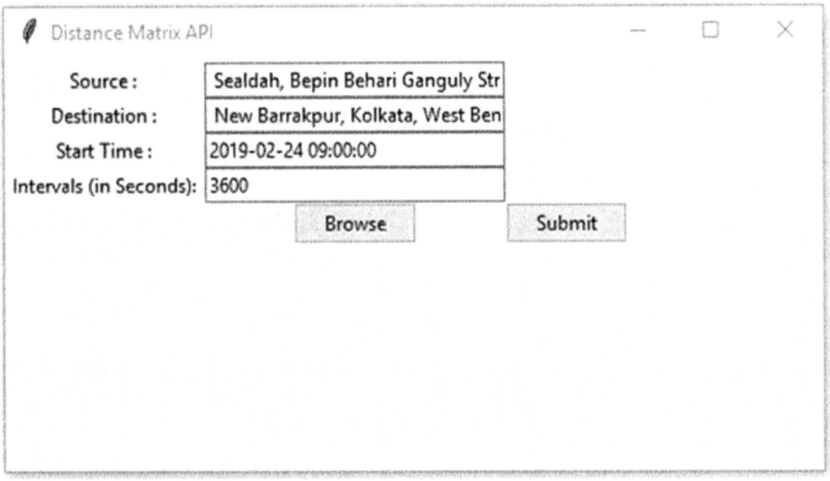

Fig. 5 Interface of data collection module (crowdsourcing)

6 Conclusion

This paper presents an IoT-based congestion control framework for intelligent traffic management system. The aim is to reduce traffic congestion, adverse effects on environment due to congestion, road accidents while satisfying the ever-increasing mobility needs of citizens. The proposed system tries to use the available traffic resources in an optimum way for satisfying citizens' mobility need. The data processing module of the system acts according to the predefined algorithms and/or through machine intelligence based on collected traffic data. The proposed system includes several business applications like smart parking system, suggesting the best

Origin : Sealdah, Sealdah, Bepin Behari Ganguly Street, Raja Bazar, Kolkata, West Bengal 700014, India

Destination : New Barrackpore, New Barrakpur, Kolkata, West Bengal 700131, India

Sr. No.	Distance		Duration		Duration in Traffic		Date-Time
1	20.1 km	20112	52 mins	3140	49 mins	2928	2019-02-24 09:00:00
2	20.1 km	20112	52 mins	3140	51 mins	3061	2019-02-24 10:00:00
3	20.1 km	20112	52 mins	3140	52 mins	3114	2019-02-24 11:00:00
4	20.1 km	20112	52 mins	3140	54 mins	3220	2019-02-24 12:00:00
5	20.1 km	20112	52 mins	3140	52 mins	3145	2019-02-24 13:00:00
6	20.1 km	20112	52 mins	3140	50 mins	2996	2019-02-24 14:00:00
7	20.1 km	20112	52 mins	3140	49 mins	2957	2019-02-24 15:00:00
8	20.1 km	20112	52 mins	3140	51 mins	3051	2019-02-24 16:00:00
9	20.1 km	20112	52 mins	3140	54 mins	3253	2019-02-24 17:00:00
10	20.1 km	20112	52 mins	3140	59 mins	3525	2019-02-24 18:00:00
11	20.1 km	20112	52 mins	3140	57 mins	3443	2019-02-24 19:00:00
12	20.1 km	20112	52 mins	3140	54 mins	3245	2019-02-24 20:00:00

Fig. 6 Crowdsource traffic database

alternate route, traffic signal optimization and traffic congestion state identification. The citizens can get the services of the proposed framework from their handheld mobile device for smooth mobility. It reduces traffic congestion and minimizes loss of valuable time of citizens by disseminating proper traffic information to the citizens' handheld devices. Also, it assists the policymaker in decision making related to traffic rule and regulation based on statistical report generated by the system. The proposed system can be used as a prototype model by the government agencies or private sectors to improve the existing traffic management system.

References

1. Petrolo R, Loscr V, Mitton N (2014) Towards a smart city based on cloud of things. In: Proceedings of the 2014 ACM international workshop on Wireless and mobile technologies for smart cities—WiMobCity 14. ACM Press, New York, USA, pp 61–66. https://doi.org/10.1145/2633661.2633667
2. Gaur A, Scotney B, Parr G, McClean S (2015) Smart city architecture and its applications based on IoT. In: The 5th International symposium on internet of ubiquitous and pervasive things, vol 52. Elsevier, pp 1089–1094. https://doi.org/10.1016/j.procs.2015.05.122
3. Rehena Z, Janseen M (2018) Towards a framework for context-aware intelligent traffic management system in smart cities. In: AW4City 2018 enhancing citizen centricity with web applications. ACM, France, pp 893-898. https://doi.org/10.1145/3184558.3191514
4. Rawal T, Devadas V (2015) Intelligent transportation system in India—A review. J Dev Manage Commun II(3)
5. Shah N, Kumar S, Bastani F, Yen I-L (2012) Optimization models for assessing the peak capacity utilization of intelligent transportation systems. Eur J Oper Res 239–251. https://doi.org/10.1016/j.ejor.2011.07.032 (Elsevier)
6. Atzori L, Iera A, Morabito G (2010) The Internet of Things: a survey. Comput Netw 2787–2805. https://doi.org/10.1016/j.comnet.2010.05.010 (Elsevier)
7. Djahel S, Doolan R, Muntean G-M, Murphy J (2015) A communications oriented perspective on trafic management systems for smart cities: challenges and innovative approaches. In:

IEEE communication surveys and tutorials, vol 17, no 1, First quarter. https://doi.org/10.1109/COMST.2014.2339817
8. Cheng T, Wen P, Li Y (2016) Research status of artificial neural network and its application assumption in aviation. In: IEEE 12th international conference on computational intelligence and security
9. Na S, Xumin L, Yong G (2010) Research on K-means clustering algorithm. In: IEEE 3rd international symposium on intelligent information technology and security informatics
10. Pham TN, Tsai M-F, Nguyen DB, Dow C-R, Deng D-J (2015) A cloud-based smart-parking system based on Internet-of-Things technologies. In: Special section on emerging cloud-based wireless communications and networks, vol 3. IEEE Open Access, pp 1581–1591. https://doi.org/10.1109/ACCESS.2015.2477299
11. Ji Z, Gonchev I, O'Droma M, Zhao L, Zhang X (2014) A cloud-based car parking middleware for IoT-based smart cities: design and implementation. Sensors 22372–22393. https://doi.org/10.3390/s141222372
12. Kianpisheh A, Mustaffa N, Limtrairut P, Keikhosrokiani P (2012) Smart Parking System (SPS) architecture using ultrasonic detector. Int J Softw Eng Its Appl 6(3)
13. Geng Y, Cassandras CG (2011) A new smart parking system based on optimal resource allocation and reservations. In: 14th International IEEE conference on intelligent transportation systems, Washington DC, USA. https://doi.org/10.1109/ITSC.2011.6082832
14. Thianniwet T, Phosaard S, P-Atikom W (2009) Classification of road traffic congestion levels from GPS data using a decision tree algorithm and sliding windows. In: Proceedings of the world congress on engineering. Springer. UK. https://doi.org/10.1007/978-90-481-8776-8_23
15. Lopes J, Bento J, Huang E, Autonious C, Ben-Akiva M (2010) Traffic and mobility data collection for real-time applications. In: 13th International IEEE annual conference on intelligent transportation systems, Madeira Island, Portugal
16. Sumalee A, Ho HW (2018) Smarter and more connected: future intelligent transportation system. In: IATSS Res 67–71. https://doi.org/10.1016/j.iatssr.2018.05.005 (ScienceDirect)
17. Chatzimilioudis G, Konstantinidis A, Laoudias C, Zeinalipour-Yazti D (2012) Crowdsourcing with smartphones. IEEE Internet Comput 16(5):3644. https://doi.org/10.1109/MIC.2012.70
18. Rehena Z, Mondal MA, Janssen M (2018) A multiple-criteria algorithm for smart parking: making fair and preferred parking reservations in smart cities. In: Proceedings of the 19th annual international conference on digital government research: governance in the data age, DG.0 2018, Delft, The Netherlands, pp 40:1–40:9. https://doi.org/10.1145/3209281.3209318

Link Prediction on Social Attribute Network Using Lévy Flight Firefly Optimization

P. Srilatha, R. Manjula, and C. Pavan Kumar

Abstract The problem of link prediction largely depends on the topological information. Social attribute network model is employed where the nodes represent both the social nodes and also the attribute nodes. The edge represents the interaction between the social nodes, the interaction between the social node and the attribute nodes but not the interaction between the attribute nodes. In this paper, firefly optimization algorithm using Lévy search is employed to predict links. The proposed algorithm accuracy is measured in terms of *AUC* and *precision* and compared with similar methods in literature. From the experimental results, it is evident that Lévy walk outperforms over other existing algorithms. From the results, we also infer that exploiting the Lévy search firefly algorithm taking shorter jumps has improved the accuracy of the link prediction algorithm over the methods that take longer jumps based on random walks.

Keywords Link prediction · Social attribute network · Lévy flight · Firefly

---

P. Srilatha · R. Manjula
School of Computer Science and Engineering, Vellore Institute of Technology, Vellore, Tamil Nadu 632014, India
e-mail: sreelatha.pulipati@gmail.com

R. Manjula
e-mail: rmanjula@vit.ac.in
URL: http://www.vit.ac.in

P. Srilatha · C. Pavan Kumar (✉)
Department of Computer Science and Engineering Anurag Group of Institutions, Hyderabad, Telangana 632014, India
e-mail: pavankumarc@ieee.org
URL: http://www.anurag.edu.in/

N. N. Chiplunkar and T. Fukao (eds.), *Advances in Artificial Intelligence and Data Engineering*, Advances in Intelligent Systems and Computing 1133,
https://doi.org/10.1007/978-981-15-3514-7_97

1 Introduction

Social network is a popular form of representing and understanding relationships among entities involved [10]. For example, online social networks used to communicate can be used to capture interaction among people and their likes and dislikes. Such information will be useful in developing new knowledge. Each individual represented as node in the social network will have attributes providing information about that individual. Understanding the nature of nodes based on their interaction with the other nodes and their attributes has become an interesting area of research in the recent past mainly to understand the evolution of network and to predict links that may occur in future. The link prediction algorithms consider a snapshot of graph structure at a time t and aim to predict the edges that will emerge in the time interval between t and a future time instance t' [10, 11, 14, 21] but will not consider the attributes of nodes involved in social network. Many real-world networks like Facebook, Google++ and Twitter contain rich information like person name, circles, group membership, interests and occupation as attributes associated with each node. Such additional information can be used to improve the accuracy of link prediction [9, 18]. But these methods improve the accuracy of the link prediction only if the attribute information is completely known.

Yin et al. [27, 28] proposed social attribute network (SAN) that uses node attribute along with node topology in social network to recommend links. In [5, 22, 26], the authors propose methods to address the problem of attributes incompleteness by using social influence and infer the importance of attribute information and homophily. However, few algorithms suffer with the scalability issues. Authors in [3, 5, 16] address the scalability issues, but these approaches do not fully used the node attribute information. Gong et al. [5] proposed the use of social attribute network(SAN) model to predict the links and also to infer the attribute information in a scalable way. Recently, to address the scalability issues nature-inspired meta-heuristic algorithms like particle swarm optimization (PSO), genetic algorithms (GA) and ant colony optimization (ACO) were employed. In these algorithms, natural species uses the exhaustive search to move over the entire search space. Naruchitparames et al. [15] use the genetic algorithm to recommend the friends based on the network structure. Bliss et al. [2] construct the linear equation by considering the combination of both node-based similarity indexes separated by a path length 2 and the unknown weights. Unknown weights are optimized using the genetic operators like mutation and crossover. But the authors do not consider the paths separated by a path length more than 2, and the social attributes play an significant role in the task of link prediction. Akbari et al. [1] consider the social features along with the network structure. Sherkat et al. [20] proposed ant colony optimization for solving the link prediction problem. But the algorithm evaluates the same node several times, and the algorithm converges when a large amount of pheromone is deposited on the edge which makes the ants impossible to search the entire parameter space. Majority of the nature-inspired meta-heuristic methods proposed in literature suffered with quick convergence; scalability of the similarity indexes is used in framing the linear equations and has not used to solve multimodal optimization problems. Inspired by the natural behaviour of the species in the world like food-finding behaviour of the ants and flashing behaviour of the fireflies, firefly optimization algorithm was developed by Yang [25] in the year 2008. The effectiveness of the algorithm comes from the communication of these fireflies, i.e. how effectively they communicate with each other which in turn depends on the flashing pattern that is directly proportional to the intensity of the flashes. Firefly algorithm is successfully used to solve the job scheduling problems [13], design optimization [24] and variable optimization problems [4]. Firefly algorithm has some advantages such as distributed computing, controlled randomization and efficient local search over the other meta-heuristic algorithms. However, the randomization typically used normal or Gaussian distribution and Lévy distribution. In our paper, we solved link prediction problem using the Lévy flight firefly optimization method. In this paper, fireflies are employed to travel over the 2-D grid constructed using the given nodes in the network. Intensity values of the fireflies are directly proportional to the surface function that gets constructed dynamically when the links are predicted. The firefly with lesser brightness attracts towards the brighter firefly, i.e. the firefly attracts towards the edge that is having higher link score. Link score is constructed using the common neighbours and friend groups. Common neighbours are computed between two nodes by sharing at least one common social node or with only common attribute node. Common neighbour [12] method is given as

$$\text{neighbourhood}(x, y) = |\Gamma(a)| \cap |\Gamma(b)| \tag{1}$$

where $|\Gamma(a)$ and $|\Gamma(b)|$ represent set of neighbours of nodes a and b, respectively. After one iteration, the new link matrix is populated and the light intensities of the fireflies are recalculated. Finally, after maximum number of iterations, link matrix that contains predicted links along with link scores is populated. We use *AUC* and *precision* as evaluation metrics to evaluate the accuracy of the algorithm with other link prediction algorithms. The rest of the paper is organized as follows: Section 2 discussed the problem formulation and the evaluation methods. Section 3 presents the framework for Lévy flight-based link prediction on social attribute network. Section 4 presents experimental set-up needed. Section 5 shows the results of the proposed algorithm and the comparative study with the other algorithms. Section 6 draws the conclusion.

2 Problem Definition and Performance Metrics

In our problem, we consider graph $G(V, E)$ to represent a network where E indicates the interactions between the vertices or nodes V. In addition to the graph G, we also considered the node attributes. For example, in Facebook, nodes are users and edges represent the friendship relationship between the users and the user attributes include school, occupation, location, hometown and gender which can be derived from the profiles of the users. Each attribute can have multiple possible values, e.g. for a node, an employer can have companies like Intel, Yahoo and Google as attribute values. We considered each possible value as a separate binary attribute. Thus, for a given graph G and distinct M attributes, an attribute matrix A is created and is denoted by $A = [\mathbf{v}_1, \mathbf{v}_2, \mathbf{v}_3, \ldots, \mathbf{v}_M]$, where $\mathbf{v}_i$ is the vector that contains the M distinct attributes for the node i. The entry value in the vector takes three possible values. The value is 1 if the attribute has some known value (positive attribute), the value is -1 if the attribute has no known values (negative attribute), and the value is 0 if the attribute is unknown (missing attribute). Given the attribute matrix A, a network graph G and M distinct attributes, an augmented graph $G_A(V_A, E_A)$ is constructed by attaching M additional attribute nodes to network graph G. For each node p in G with positive attribute a, an undirected link (or edge) is constructed between node p and attribute node a. The network that is obtained we called as social attribute network (SAN) model. In SAN, the interactions between the social nodes are represented by E, the interactions between social node and the attribute nodes are represented by E_{Att}, and thus, the set E_A is denoted by $E_A = E \cup E_{\text{Att}}$. Similarly, the set of social nodes is represented by V, the set of attribute nodes is represented by V_{Att}, and thus, the set V_A is denoted by $V_A = V \cup V_{\text{Att}}$ and the graph G_A is represented by an adjacency matrix of order $(N + M) * (N + M)$. Thus, link prediction is to find the set of links form in network over a period of time which will be in the set of non-existing links $U - E$. Set U denotes universal set containing all possible $\frac{|V|(|V|-1)}{2}$ edges, where $|V|$ is the number of nodes in graph G. The purpose of our method is to find out the edges between the two nodes which are having higher probability to form a link, i.e. the firefly optimization algorithm which is employed to find the non-existing links between the nodes tends to attract the edge that has higher probability to form a link. The fireflies are attracted to the edge based on the proximity score between the nodes connecting that edge. In our work, the proximity scores are computed dynamically. And to test the accuracy of our proposed algorithm, the existing links in E are randomly divided into two parts. The training set, E_{Train}, is considered as known information, while the testing set E_{Test} is used for testing, and no information from the testing set is used for prediction. And we consider $E_{\text{Train}} \cup E_{\text{Test}} = E$ and $E_{\text{Train}} \cap E_{\text{Test}} = \phi$. Then, the algorithm makes use of training set and gives link scores of all non-observed links from the set $U - E_{\text{Train}}$.

To quantify the prediction accuracy, the following standard metrics are calculated.

1. *AUC*: *AUC* scores are calculated by comparing scores of each testing link with the scores of non-existing links and are denoted by

$$AUC = \frac{n' + 0.5n''}{n} \tag{2}$$

where n is the total number of independent comparisons, n is the number of times the testing link having a higher score than the non-existing links, n'' is the number of times the testing link, and the non-existing link has the same score.

2. *Precision*: Given the ranking of the non-observed links, the precision is defined as ratio of L_r links that are found right to the total number of found links , L and is defined as

$$Precision = \frac{L_r}{L} \tag{3}$$

Clearly, higher precision values means higher accuracy.

3 Framework for Link Prediction Based on Lévy Flight Firefly Optimization

3.1 Basic Idea of the Algorithm

For the graph $G_A = (V_A, E_A)$, a 2-D discrete grid is constructed by locating the vertices of the graph to the coordinates of the graph. In our algorithm, fireflies are employed to randomly walk on the coordinates of the grid which denote the transitions. Algorithm starts by placing a population of fireflies randomly on the coordinates of the grid. We set the objective function proportional to link value found so far. Initially, when the contour surface is unknown and no links are found, fireflies travel more randomly and faster using Lévy walk spreading out and covering more of the entire parameter space. The firefly at node v_i attracts the next edge (v_i, v_j) with the attractive parameter β. Once a few links have been known, there will be a local maximum at that point on the grid allowing the fireflies to draw attention on those areas instead of flying quasi-randomly around the search space. The algorithm reconstructs the surface function with the new links found so far, and the intensities of the fireflies are recalculated which are directly proportional to the surface function. This procedure continues until certain number of iterations. Finally, the link matrix is populated with predicted links along with the link scores. To maintain the integrity of the algorithm and to control the behaviour of the fireflies when the algorithm finds more than ten links, we use only ten highest rank links to construct the surface function rather than using all the predicted links. Further, fireflies move to search for other potential links. Assumptions made in the firefly algorithm based on firefly behaviour are as follows:

1. All flies are unisex, so that one firefly will be attracted to other fireflies regardless of their gender.
2. Attractiveness of the fireflies is directly proportional to the brightness, i.e. lesser brightness firefly moves towards the brighter one, and they both decrease as the distance increases.
3. Brightness of the firefly calculated with respect to objective function is used to solve the problem.

3.2 Parameter Initialization

The following parameters are considered for firefly algorithm as suggested by Yang himself in arriving at optimal solution.

Parameter values are considered as suggested by Yang himself [19, 25], $\beta_0 = 1$, $\Gamma = 1$, $\alpha = 1$ and coarseness = 1. The algorithm terminates at defined value of maximum number of iterations, N_c.

3.3 Movement of the Fireflies

Movement of fireflies is based on two strategies. One is information-based movement which uses the current firefly information to generate new firefly information, and the other strategy is random movement which is employed to explore the search space to find good solutions. The random movement of the firefly is controlled by Lévy flights. Lévy flight is a random walk in which step size is drawn from the Lévy distribution. In our paper, we have considered the Lévy flights or walks because they are more efficient than Brownian random walks in exploring the larger search space because Lévy walks take more extreme steps along with the smaller steps, whereas Brownian random walks take more smaller steps and larger steps occur infrequently. Since the social networks have too many long links, thus exploring the entire network using Lévy walks is more efficient than Brownian random walks. Thus, in each iteration firefly at node v_i moves to the brighter firefly according to the attraction parameter β and it is defined as

$$\beta = \beta_0 \exp^{-\gamma r^2} \tag{4}$$

where γ is the scale parameter and r is the Euclidean distance between the firefly i and the firefly j and is given by

$$r_{ij} = \sqrt{(x_i - x_j)^2 + (y_i - y_j)^2} \tag{5}$$

The brightness or intensity value of the fireflies approximated is defined as

$$\beta = \beta_0 \exp^{-\gamma r^2} \approx \frac{\beta}{1 + \gamma r^2} \tag{6}$$

Thus, the movement of firefly by Lévy flight is given by

$$x_i = x_i + \frac{\beta_0}{1 + \gamma r_{ij}^2}(x_j - x_i) + \alpha L(s, \gamma, \mu) \tag{7}$$

Since, $L(s, \gamma, \mu) = \sqrt{\frac{\gamma}{2\pi}} \frac{1}{s^{\frac{3}{2}}}$, Eq. (7) reduces to Eq. (8)

$$x_i = x_i + \frac{\beta_0}{1 + \gamma r_{ij}^2}(x_j - x_i) + \alpha \sqrt{\frac{\gamma}{2\pi}} \frac{1}{s^{\frac{3}{2}}} \tag{8}$$

3.4 *The Fitness Broadcast*

Fireflies carry an intensity value whose quantities encode the fitness of their locations. These allow the fireflies to glow at an intensity directly proportional to the surface function or the objective function. In our paper, fitness function is constructed dynamically by using the link matrix which contains the link scores. Link scores are computed by considering the mutual friends and also the friend groups.

$$F(x, y) \propto \frac{1}{r^2} \tag{9}$$

where r is the Euclidean distance between the nodes.

3.5 *Termination Condition and Outputs*

The algorithm terminates at defined value of maximum number of iterations on largest connected component N_c of data set considered. Finally, the algorithm outputs a link matrix that contains all the predicted links. The *AUC* and *Precision* metrics are used to predict links or proximity score.

Similarity score computation based on proposed model is given in Algorithm 1.

Algorithm 1 Firefly-based Social Attribute Network Link Prediction algorithm

Input: Augmented network Graph G_A, Intensity I of the fireflies, N: Number of iterations, r_d: Neighbourhood range, α, β_0, γ.
Output: Score matrix S.

Parameter initialization:
Fireflies will be randomly distributed in parameter space on nodes of G_A;
Each firefly will have its intensity I;
while $t < N$ **do**
  while $i = n$ **do**
    m = computeneighborhood(i)
    while $j = m$ **do**
      if $(I_i < I_j)$ **then**
        Move firefly i towards j using Lévy walk;
      end if
    end while
  end while
  Compute link strength and update score matrix
  Update intensity I
end while

4 Experimental Set-up and Results of Proposed Algorithm

The proposed Algorithm 1 is implemented in MATLAB, and the performance is measured in terms of *AUC* and *precision*. A 2-D grid is considered on which fireflies are made to move based on Algorithm 1. Co-authorship data set network is considered which is a network that contains interactions between the others. For each author, terms in the papers are considered as attributes of the authors. An edge is formed if both the authors have cited the same paper. Data set includes computer science conference (ACM), International Conference on Machine Learning (ICML) and International World Wide Web (WWW) from digital bibliography library project (DBLP) [23]. *AUC* and *precision* values of proposed Algorithm 1 are given in Table 1.

5 Comparison with Other Methods in Literature

Proposed Algorithm 1 in this paper is compared with other existing methods such as common neighbour (CN) [12], Salton Index [17], Jaccard Index [6], Leicht–Holme–Newman Index-1 (LHN-1) [8] and Katz [7] Indices in terms of *AUC* and *precision*. Comparison of *AUC* of proposed Algorithm 1 is given in Table 2. Comparison of *precision* values of proposed Algorithm 1 is given in Table 3. From Tables 2 and 3, we can infer that proposed Algorithm 1 performs better than existing algorithms in literature.

Table 1 *AUC* and *precision* values of proposed Algorithm 1

Data sets	*AUC*	*Precision*
ACM	0.9455	0.718
ICML	0.9579	0.6072
WWW	0.9692	0.8100

Table 2 Comparison of AUC with other methods

	ACM	ICML	WWW
CN [12]	0.8635	0.9236	0.9505
Salton [17]	0.7932	0.9238	0.9417
Jaccard [6]	0.4223	0.5223	0.4426
LHN-1 [8]	0.8552	0.9234	0.9504
Katz [7]	0.8219	0.9126	0.9492
Proposed method	0.9455	0.9579	0.9692

Table 3 Comparison of precision with other methods

	ACM	ICML	WWW
CN [12]	0.7355	0.5418	0.6995
Salton [17]	0.24	0.4121	0.7531
Jaccard [6]	0	0.0042	0
LHN-1 [8]	0.2893	0.1862	0.4200
Katz [7]	0.1818	0.1160	0.4600
Proposed method	0.718	0.6072	0.8100

6 Conclusion

Link prediction problem inspired by the nature of firefly algorithm with Lévy flight is proposed in this paper. Node attribute information is considered along with node topology that helps in predict links better as compared to the traditional topological link prediction algorithms. Proposed algorithm was evaluated in terms of *AUC* and *precision*. From the experimental values, we infer that the proposed method outperforms the existing methods in literature. Further, it is interesting to explore additional node attributes that help to improve the accuracy of link prediction (Figs. 1 and 2).

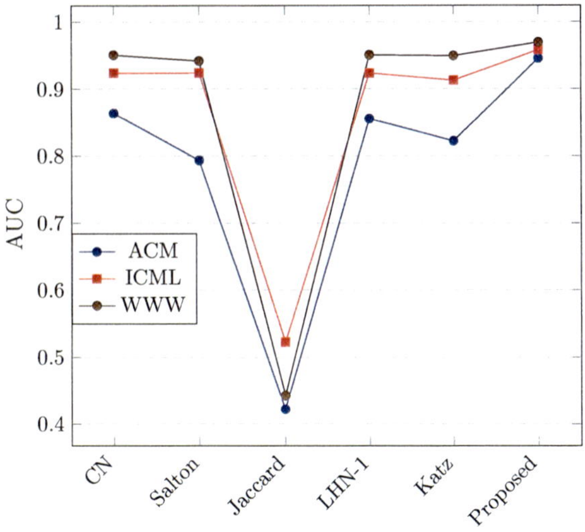

Fig. 1 Comparison of *AUC* with other methods

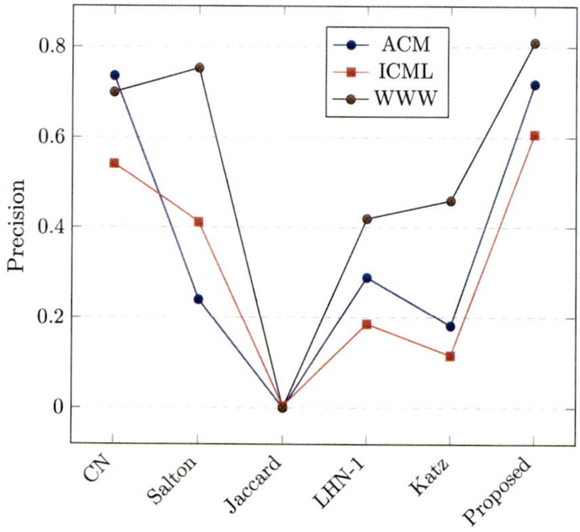

Fig. 2 Comparison of *Precision* with other methods

References

1. Akbari F, Tajfar AH, Nejad AF (2013) Graph-based friend recommendation in social networks using artificial bee colony. In: 2013 IEEE 11th international conference on dependable, autonomic and secure computing (DASC). IEEE, pp 464–468
2. Bliss CA, Frank MR, Danforth CM, Dodds PS (2014) An evolutionary algorithm approach to link prediction in dynamic social networks. J Comput Sci 5(5):750–764
3. Chen B, Chen L, Li B (2016) A fast algorithm for predicting links to nodes of interest. Inf Sci 329:552–567
4. Gandomi AH, Yang XS, Alavi AH (2011) Mixed variable structural optimization using firefly algorithm. Comput Struct 89(23–24):2325–2336
5. Gong NZ, Talwalkar A, Mackey L, Huang L, Shin ECR, Stefanov E, Shi ER, Song D (2014) Joint link prediction and attribute inference using a social-attribute network. ACM Trans Intell Syst Technol (TIST) 5(2):27
6. Jaccard P (1901) Étude comparative de la distribution florale dans une portion des Alpes et des Jura. Bulletin del la Société Vaudoise des Sciences Naturelles 37:547–579
7. Katz L (1953) A new status index derived from sociometric analysis. Psychometrika 18(1):39–43
8. Leicht E, Holme P, Newman ME (2006) Vertex similarity in networks. Phys Rev E 73(2):026120
9. Li J, Cheng K, Wu L, Liu H (2018) Streaming link prediction on dynamic attributed networks. In: Proceedings of the eleventh ACM international conference on web search and data mining. ACM, pp 369–377
10. Liben-Nowell D, Kleinberg J (2007) The link-prediction problem for social networks. J Am Soc Inf Sci Technol 58(7):1019–1031
11. Liu W, Lü L (2010) Link prediction based on local random walk. EPL (Europhys Lett) 89(5):58007
12. Lorrain F, White HC (1971) Structural equivalence of individuals in social networks. J Math Sociol 1(1):49–80
13. Marichelvam MK, Prabaharan T, Yang XS (2014) A discrete firefly algorithm for the multi-objective hybrid flowshop scheduling problems. IEEE Trans Evol Comput 18(2):301–305
14. Murata T, Moriyasu S (2008) Link prediction based on structural properties of online social networks. New Gener Comput 26(3):245–257

15. Naruchitparames J, Güneş MH, Louis SJ (2011) Friend recommendations in social networks using genetic algorithms and network topology. In: 2011 IEEE congress on evolutionary computation (CEC). IEEE, pp 2207–2214
16. Papadimitriou A, Symeonidis P, Manolopoulos Y (2012) Fast and accurate link prediction in social networking systems. J Syst Softw 85(9):2119–2132
17. Salton G, McGill MJ (1986) Introduction to modern information retrieval. McGraw-Hill, Inc
18. Scellato S, Noulas A, Mascolo C (2011) Exploiting place features in link prediction on location-based social networks. In: Proceedings of the 17th ACM SIGKDD international conference on Knowledge discovery and data mining. ACM, pp 1046–1054
19. Senthilnath J, Omkar S, Mani V (2011) Clustering using firefly algorithm: performance study. Swarm Evol Comput 1(3):164–171
20. Sherkat E, Rahgozar M, Asadpour M (2015) Structural link prediction based on ant colony approach in social networks. Physica A: Stat Mech Appl 419:80–94
21. Wang P, Xu B, Wu Y, Zhou X (2015) Link prediction in social networks: the state-of-the-art. Sci China Inf Sci 58(1):1–38
22. Yang C, Zhong L, Li LJ, Jie L (2017) Bi-directional joint inference for user links and attributes on large social graphs. In: Proceedings of the 26th international conference on world wide web companion. International World Wide Web Conferences Steering Committee, pp 564–573
23. Yang J, Leskovec J (2015) Defining and evaluating network communities based on ground-truth. Knowl Inf Syst 42(1):181–213
24. Yang XS (2010a) Firefly algorithm, stochastic test functions and design optimisation. Int J Bio-Inspired Comput 2(2):78–84
25. Yang XS (2010b) Nature-inspired metaheuristic algorithms. Luniver Press
26. Yang Y, Zhang J, Zhu X, Tian L (2018) Link prediction via significant influence. Physica A: Sta Mech Appl 492:1523–1530
27. Yin Z, Gupta M, Weninger T, Han J (2010a) Linkrec: a unified framework for link recommendation with user attributes and graph structure. In: Proceedings of the 19th international conference on World wide web. ACM, pp 1211–1212
28. Yin Z, Gupta M, Weninger T, Han J (2010b) A unified framework for link recommendation using random walks. In: 2010 International conference on advances in social networks analysis and mining (ASONAM). IEEE, pp 152–159

Secure and Energy-Efficient Data Transmission

H. V. Chaitra and G. K. RaviKumar

Abstract The cheap availability of sensor and bandwidth has led to growth of Internet-of-Things (IoT)-based application such as healthcare, military domain adopting wireless sensor network (WSN). These applications generate massive amount of data which requires strict QoS (i.e., low latency, high secure routing and energy efficiency). For provisioning QoS requirement, number of cluster-based routing model has been presented in recent times. However, these models incur energy overhead among cluster head. Thus, for improving energy efficiency, it is important to improve cluster selection algorithm. Further, for providing secure communication, the existing model is designed using public key cryptography such as RSA and Diffie Hellman. As a result, it incurs communication overhead and increases packet processing delay. Further, very limited work is done using elliptical curve cryptography (ECC). For overcoming research issues and problems, this work presents a secure and energy-efficient data transmission (SEEDT) model. The SEEDT model presents an energy-efficient cluster head (CH) selection algorithm using multi-objective parameter by enhancing imperialist competitive algorithm (ICA) and employing ECC for providing secure routing. The SEEDT model attains significant performance in terms of communication overhead, packet processing time and lifetime.

Keywords WSN · Clustering · Cryptography · Evolutionary computing

1 Introduction

Wireless sensor networks (WSNs) have attained wide adoption across various industries and organizations such as hospital, military and so on. The rapid growth and adoption are due to low-cost availability of senor device and ease of deployment [1].

H. V. Chaitra (✉)
Nitte Meenakshi Institute of Technology, Bangalore, India
e-mail: chaitrahvgowda2005@yahoo.com

G. K. RaviKumar
BGSIT, Mandya, India
e-mail: ravikumargk@yahoo.com

N. N. Chiplunkar and T. Fukao (eds.), *Advances in Artificial Intelligence and Data Engineering*, Advances in Intelligent Systems and Computing 1133,
https://doi.org/10.1007/978-981-15-3514-7_98

However, it involves distinctive security issues and problems that cannot be met utilizing traditional security methods [2]. In specific, public key-based cryptographic technique offers superior strategy for provisioning security to high-density wireless sensor networks. However, these models incur memory and energy overhead. Further, they are very slow. On the other side, the symmetrical-based cryptography method offers better performance in terms of energy efficiency and speed. However, these methods require efficient key establishment mechanism between communicating devices [3, 4]. In standard security design, an efficient key establishing method must offer secure and well-connected topography, that is, there exists secure communication path (perhaps multihop-based communication) among every contending sensor device permitting exchange of control and data packets, while meeting general limitation or constraint of wireless sensor networks.

Sunil Kumar and Shankar [5, 6] conducted extensive survey and identified problem in attaining secure and energy-efficient data transmission model for wireless senor network. Further, Rahayu et al. [7] conducted extensive survey on security mechanism on clustering based design. The survey shows that there is a need for new design that attains a noble trade-off among energy efficiency and security prerequisite of dynamic WSN-based application such as wearable computing device, tactical Internet and Internet-of-Things [8–10, 12–14]. Recently, number of energy-efficient routing design has been presented such as fuzzy-based [15, 16], hop-based [17, 18], evolutionary computing-based [19–23]. However, these models incur energy overhead between CH which is nearer to the sink; they are not suitable for optimizing multi-objective optimization. Further, none of these approaches have considered security provisioning. For provisioning security and privacy, number of approaches adopting cryptography has been presented [24–27]. Laoudias et al. [24] discussed the need for provisioning security for tracking, navigation and localization of sensor device for sensor network. Further, for provisioning security and privacy, Eletreby and Yagan [25] assumed that due to randomness nature of sensor device, there exist connectivity issues among sensor device. As a result, sharing keys among communicating device is challenging under homogenous network. Ting et al. [26] proposed signcryption scheme which consists of two stages such as offline and the online signcrypt stages. They showed that identity-based cryptography (IBC) setup for transmitting packet to a gateway in a public key infrastructure (PKI) can reduce heavy verification loads on low-power sensor devices. Further, Rahman and El-Khatib [27] presented a secure time synchronization protocol for WSN using bilinear pairing. Their model reduced communication overhead over the existing approach due to adoption of ECC rather than conventional PKI. However, to efficiently adopt elliptical curve cryptography for WSN, it is important to validate the public key. Failing to validate the public key will lead WSN susceptible to eavesdropping attack by intermediate node. Public key validation mechanism needs a public key infrastructure, to distribute and modify (attribute revocation) certificates. Further, the sensor device must be able to keep, exchange and validate the obtained certificates [28]. Distribution of certificate, certificate exchange, certificate validation and certificate storage process will lead to computation, communication and storage overheads. Thus, it is not efficient to provision such method to wireless sensor networks [29].

For overcoming research challenges, this manuscript presented a secure and energy-efficient data transmission (SEEDT) model for WSN. For attaining, firstly, this paper presents the SEEDT model which presents energy-efficient cluster head selection algorithm using multi-objective parameter by enhancing imperialist competitive algorithm (ICA) and employing ECC for providing secure routing. Since our model uses ECC [30, 31] message, computation overhead is reduced when compared to conventional model such as Diffie Hellman and RSA-based approaches. Hence, the SEEDT can offer confidentiality of packet information with minimal overhead of computation and maintenance.

1.1 Contribution of Research Work

The research contribution for work is as follows:

- Presenting a novel enhanced ICA for better cluster head selection.
- The SEEDT model reduces communication overhead, runtime and packet processing time or latency considering security provisioning using ECC.
- The SEEDT model attains better lifetime performance than the existing model considering security provisioning using ECC.

The experiment analysis is presented in Sect. 3. In the last section, conclusion along with future research direction is discussed.

2 Proposed Secure and Energy-Efficient Transmission Design for WSN

This section presents secure and energy proficient transmission design for cluster-based WSN. Firstly, this section proposed an ICA using multi-objective parameter for attaining energy-efficient clustering design for WSN. Secondly, for provisioning secure communication over WSN, this work used ECC-based security design. The architecture of SEEDT is shown in Fig. 1.

2.1 Multi-objective Optimization Problem Description for Energy-Efficient Cluster-Based Routing Design

For cluster head selection, multi-objective parameter is considered using ICA, i.e., this work considers the location and residual energy of wireless sensor nodes $\mathcal{K}$ as multi-objective parameter for CH $\mathcal{D}$ selection using ICA. This model is self-possessed of setup phase (SP) and transmission phase (TP). In SP, the cluster member

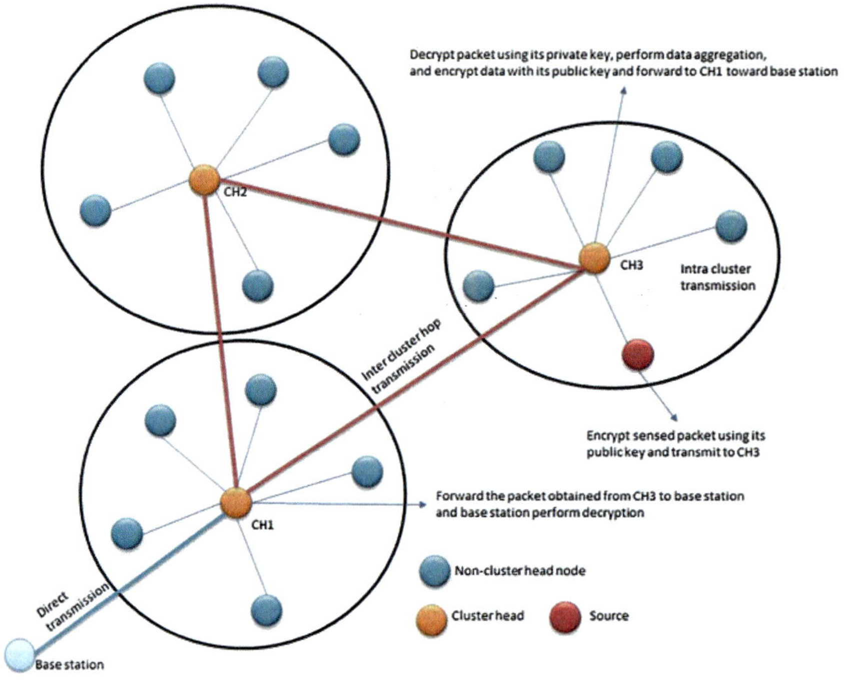

Fig. 1 Architecture of SEEDT model

and the cluster head, route between the base station and clusters are identified. In TP, the CH gathers and performs data fusion among CM. Post-aggregating data, the CH transmits data to the ideal hop sensor device (HSD)/CH till it reaches the sink/base station (BS).

The optimization problem for selecting cluster head can be described as follows

$$O_{\mathcal{D}} = \gamma * T_h^{\mathcal{D}} + (1 - \gamma) * T_m^{\mathcal{D}} \tag{1}$$

where γ depicts constant parameter for performing cost optimization, $T_h^{\mathcal{D}}$ depicts ratio of CH average residual energy over CM sensor nodes, $T_m^{\mathcal{D}}$ depicts ratio of average distance among non-CH sensor device and the BS to the average distance among the CH and the BS. For solving Eq. (1), it involves the following optimization problems (OP), where γ is a constant which depicts the impact of $T_h^{\mathcal{D}}$ and $T_m^{\mathcal{D}}$ in computing cost optimization parameter $O_{\mathcal{D}}$. Similarly, the cost function (CF) for choosing ideal HSD for carrying out inter-clustering communication (ICC) is described using the following equation

$$O_{\mathbb{D}} = \varphi * T_h^{\mathbb{D}} + (1 - \varphi) * T_m^{\mathbb{D}} \tag{2}$$

φ depicts constant parameter that described the effect of $T_h^{\mathbb{D}}$ and $T_m^{\mathbb{D}}$ in estimating parameter $O_{\mathbb{D}}$.

The CH selection optimization problem is resolved by applying enhanced imperialist competitive algorithm. First, set the parameter and initialize the OP. Along with, set country size and initialize it as M. Each country i has a position vector (PV) $g_a = \left[g_{a1}, g_{a2}, g_{a3}, \ldots, g_{aj}\right]$ and velocity vector (VV) $w_a = [w_{a1}, w_{a2}, w_{a3}, \ldots, w_{ad}]$ where j describes the dimensionality problems and a depicts positive parameter (PP) indexing country in a colony. The PV and VV are utilized to depict the present condition. Then, the model evaluates the fitness of every country. Every country evaluates its ideality using Eqs. (1) and (2). In mean time, every country keeps it local ideal strategy (LIS) $R_a = \left[r_{a1}, r_{a2}, r_{a3}, \ldots, l_{lj}\right]$ by itself and global ideal strategy (GIS) $R_l = \left[r_{l1}, r_{l2}, r_{l3}, \ldots, l_{aj}\right]$ attained by any countries within a colony. Post that it evaluates and identifies local and global ideal location using which the imperialist is added. Then, by optimizing the location and velocity, in every communicating round, there is a variation in velocity of every country toward local ideal locations and global ideal locations. The location of each and every country is reorganized using the following equations

$$g_{ab}^{\mathbb{t}+1} = g_{ab}^{\mathbb{t}} + w_{ab}^{\mathbb{t}+1}, \tag{3}$$

The velocity of each and every country is reorganized using the following equation

$$w_{ab}^{\mathbb{t}+1} = \mathbb{v}w_{ab}^{\mathbb{t}} + \mathbb{w}_1\mathbb{z}_1(r_{ab}^{\mathbb{t}} - g_{ab}^{\mathbb{t}}) + \mathbb{w}_2\mathbb{z}_2(r_{lb}^{\mathbb{t}} - g_{lb}^{\mathbb{t}}) \tag{4}$$

where the notation of g_{ab}, r_{ab} and r_{ab} is similar to w_{ab}. w_{ab} is the bth dimensionality size of ath country velocity. In general, it is bound to be between $\left[w_{\downarrow}, w_{\uparrow}\right]$ to avoid set of colony from going out of boundary condition (i.e., beyond the search space). The accelerator parameter $\mathbb{w}_1$and $\mathbb{w}_2$is measured using evolutionary conditions (EC). Coefficient $\mathbb{z}_1$and $\mathbb{z}_2$are randomly computed parameter which ranges within [0, 1] for jth dimensionality size and $\mathbb{v}$depicts inertia weight. The weights $\mathbb{V}$ play a significant role in governing impact of velocity of a particular country of the current one. Preceding process is carried out for bringing trade-off among local and global search. The weight updation is optimized or altered to avoid the enhanced ICA method in getting stuck in local optima which can be described using the following equation

$$\mathbb{V} = \left(\frac{\mathbb{I}_{\uparrow} - \mathbb{I}_C}{\mathbb{I}_{\uparrow}}\right) * \left(\mathbb{V}_{\uparrow} - \mathbb{V}_{\downarrow}\right) + \mathbb{V}_{\downarrow} \tag{5}$$

where $\mathbb{I}_{\uparrow}$ depicts max iteration size considered, $\mathbb{I}_C$ depicts current iteration size, $\mathbb{V}_{\downarrow}$ depicts minimal inertial weight and $\mathbb{V}_{\uparrow}$ depicts max inertial weight. The current ideal optimization solution is computed once the termination statement is met.

2.2 System Model for SEEDT Design for WSN

This section describes the detail of asymmetric-based ECC model for SEEDT model for cluster-based WSN. The sensor network is composed of set of cluster head $\mathcal{D}$, cluster member device K and set of hop device $\mathbb{D}$. The sensor device senses the data, encrypts using public key and transmits to the nearby CH; then, the CH decrypts the data using own private keys and aggregates the data obtained from its member. This process incurs slight energy overhead among CH. However, it eliminates redundant packet. Thus, it aids in reducing packet size. As a result, energy induced to transmit per bit of data is reduced significantly. Further, data aggregation is carried out by CH, CH then encrypts using its public key and transmits data to hop device to reach the base station. The hop device just routes packet and does not perform any encryption or decryption task. Finally, the base station decrypts the data. The traditional or existing asymmetric cryptography design such as RSA, Diffie Hellman, etc., incurs higher computation overhead affecting overall network lifetime performance of WSN. For overcoming, the SEEDT uses ECC for cluster-based routing model for WSN.

The elliptic curve (EC) of asymmetric-based security model is described in Fig. 2.

From Fig. 2, it can be said a curve that intersects two axes using feature of equation that construct from points where the line intersects these axes. Another point on the curve is obtained when a point on the curve is multiplied with a number. But, even though the actual point and the outcome are also identified, however, it is difficult to identify which parameter is utilized to carry out the product operations. Also, it is easy to perform the elliptic curve equation computation and increasingly hard to reverse. The elliptic curve is established using, $y^2 = x^3 + ax + b$.

The terms which are used, $A \rightarrow$ Point on the curve, $E \rightarrow$ Elliptic Curve and $i \rightarrow$ Maximum limit (prime number). For attainting scalable key distribution mechanism, key is distributed among cluster head. The CH is responsible for distributing keys among its cluster member. The SEEDT model is composed of the following phases such as key generation, key distribution, encryption and decryption. **Key Generation**: In key generation process, it constructs public (PBK) and private keys (PVK) for

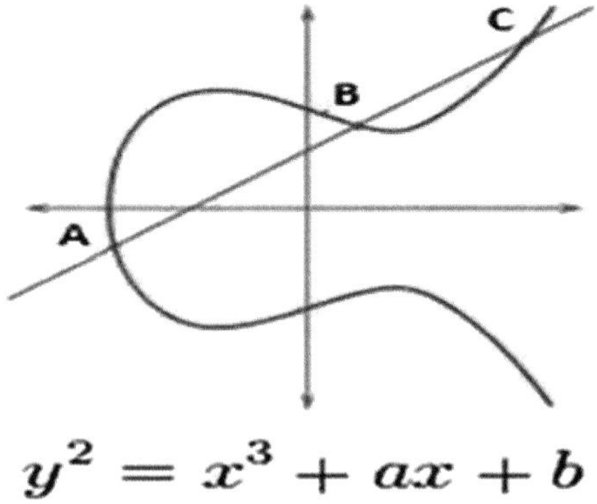

Fig. 2 Elliptical curve cryptography

performing encryption and decryption operation. Encryption process is carried out using transmitting sensor device public key. Similarly, the decryption process is carried out using receiving sensor device private key. This work chooses a parameter n which is an array of i. Public key is established using the following equation: $\mathcal{B} = n \times \mathcal{A}$, where n is arbitrarily selected in array of (1 to $i - 1$), $\mathcal{A}$ is the point on EC, $\mathcal{B}$ depicts PBK and n depicts PVK.

Encryption: The sensor device/cluster head will perform encryption operation on packet using its PBK. This work considers the packet that needs to be communicated or transmitted which is s. Then, packet s must be described on EC. This work assumes s which posses the point $\mathcal{S}$ on the EC E. Now, choose q arbitrarily in an array of $[1 - (i - 1)]$. Post that two cipher data are constructed which are depicted as T_1 and T_2. The cipher data is represented as follows $T_1 = q \times \mathcal{A}$ and $T_2 = \mathcal{S} + q \times \mathcal{B}$. Thus, lastly T_1 and T_2 have been communicated or transmitted (i.e., the point with public key is sent). **Decryption**: The cluster head/base station decrypts the data by multiplying the first component of the received point by the secret key n and subtracting it from the second component. The SEEDT model reduces communication overhead among sensor device and enhances lifetime performance which is experimentally proven in the next section below.

3 Experimental Analysis

The experiment is conducted on Window 10 OS, 64-bit I5 class processor, with 8 GB RAM. The experiment is conducted using SENSORIA simulator [32]. The simulator, the proposed SEEDT, LEACH and EEHC is designed using C# programing language using DotNet framework 4.6. Firstly, we carried out experiment analysis for assessing performance of the proposed SEEDT over EEHC in terms of packet processing time (PPT) and communication overhead (CO). Further, simulation is conducted to validate network lifetime performance by the proposed SEEDT over EEHC and LEACH routing model. The simulation parameters utilized for carrying out experiment evaluation are described as follows. The size of wireless sensor network is fixed to 100 m * 100 m; the network density size is set as 500, 600 and 700, and one base station is considered which is placed outside sensing range of sensor device. Each sensor device is given 0.1 J of energy. The sensing and transmission range of sensor device is set 3 m and 5 m, respectively. The radio, idle and amplification energy dissipation are set to 50 nj/bit, 50 nj/bit and 100 pJ/bit/m^2, respectively. The packet processing delay and bandwidth are set to 0.1 ms and 10,000 bit/s, respectively. Lastly, the transmission speed and packet length are set to 200 bit/s and 2000 bits, respectively. Here, each member node performs sensing activity 0.1 s, and sensing information size is of 2000 bits of data. Post-sensing activity, the sensor node performs encryption using 512 bit key and transmits the packet to its corresponding sensor device. The simulation process is repeated till all the sensor node in network dies.

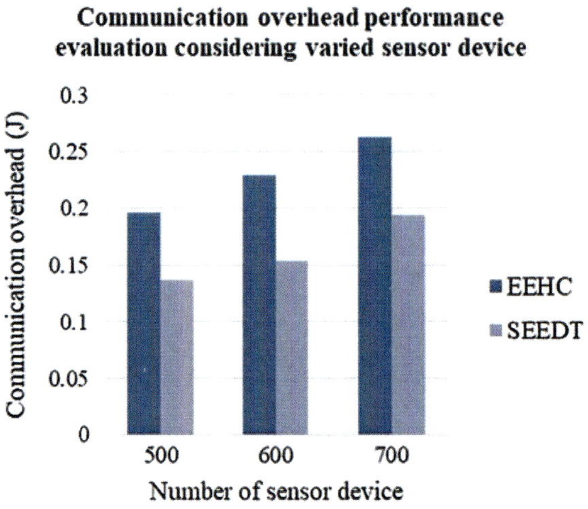

Fig. 3 Communication overhead performance evaluation for varied sensor device

3.1 Communication Overhead Performance

This section evaluated performance evaluation of communication overhead incurred by SEEDT over EEHC. The communication is computed as an energy induced in transmitting packets in control channel for attaining secure communication among sensor device toward sink. The communication overhead incurred by SEEDT over EEHC considering varied sensor device is graphically shown in Fig. 3. The outcome shows that SEEDT reduces communication overhead by 30.51%, 33.37% and 26.59% over EEHC considering 500, 600 and 700 sensor device, respectively. From Fig. 3, communication overhead increases with increase in sensor device for both protocols and linearly sharply for EEHC. This is due to that as node is increased, the member size of cluster also increases. An average CO minimization of 29.96% is attained by SEEDT over EEHC considering varied sensor device. From the result, SEEDT attained significant performance when compared with the existing model [27].

3.2 Packet Processing Time Performance Evaluation for Varied Sensor Device

This section evaluated performance evaluation of packet processing time incurred by SEEDT over EEHC. The packet processing time is computed as a time taken to securely transmit data from source to destination which composed of phases such key generation and management, encrypting and decrypting. The packet processing time taken by SEEDT over EEHC considering varied sensor device is graphically shown in Fig. 4. The outcome shows that SEEDT reduces packet processing time by

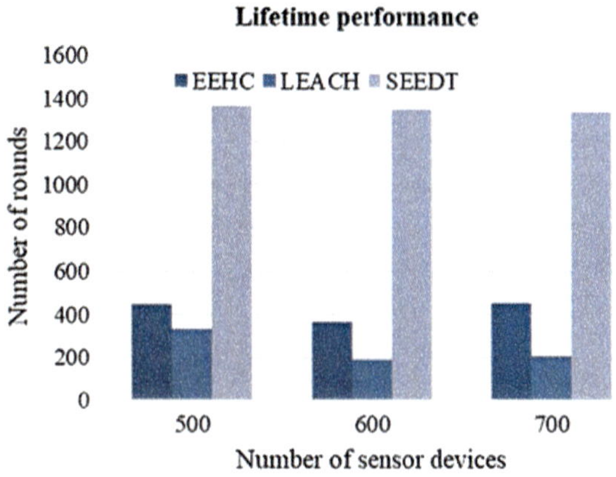

Fig. 4 Packet processing time performance evaluation for varied sensor device

24.75%, 25.58% and 34.74% over EEHC considering 500, 600 and 700 sensor device, respectively. From Fig. 4, it can be seen that the packet processing time increases with increase in sensor device for both protocols and linearly sharply for EEHC. This is due to that as node is increased, the member size of cluster also increases. An average packet processing time reduction of 28.94% is attained by SEEDT over EEHC considering varied sensor device. Further, experiment is conducted to evaluate running time. The existing model [26] attained an average running time of 9 ms, and the proposed SEEDT attained an average running time of 3.6 ms. The overall result attained shows efficiency of SEEDT model.

3.3 Lifetime Performance Evaluation Considering Varied Sensor Device

This section evaluated performance evaluation of lifetime attained by SEEDT model over EEHC [33]. The lifetime performance attained by SEEDT over EEHC considering varied sensor device is graphically shown in Fig. 5.

The outcome shows SEEDT improves lifetime performance by 75.47%, 85.68 and 84.368% over LEACH considering 500, 600 and 700 sensor device, respectively. Similarly, SEEDT improves lifetime performance by 67.2%, 73.07% and 66.12% over EEHC considering 500, 600 and 700 sensor devices, respectively. An average network lifetime enhancement of 81.84% is achieved by SEEDT over LEACH for varied sensor device. Similarly, an average network lifetime enhancement of 68.79% is achieved by SEEDT over LEACH and EEHC, respectively, considering varied sensor device. The overall result attained shows lifetime efficiency of the proposed SEEDT model.

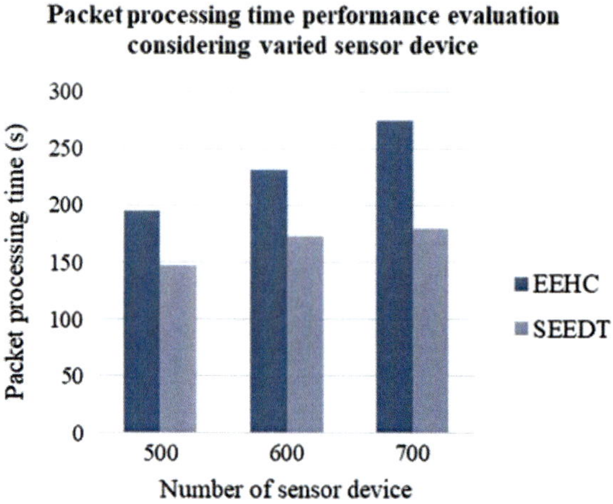

Fig. 5 Network lifetime performance under different network density sizes

4 Conclusion

This work presented an enhanced ICA using multi-objective parameter for selecting CH. Further, the WSN suffers from number of security issues which requires to bring a good trade-off between secure and energy efficacy. For attaining, this work presented secure and energy-efficient data transmission model using ECC. The SEEDT model reduces communication overhead among cluster head. From the result obtained, it can be seen that the SEEDT model reduces communication overhead by 289.96% and reduces packet processing time 28.94%, over EEHC. Further, it improves lifetime performance by 81.84% and 68.79% over LEACH and EEHC, respectively. The overall result attained by SEEDT shows robust and scalable performance. Future work would consider to minimize storage overhead among CH as the key is stored with CH. Thus, eliminating this work led CH to accomodate a greater number of cluster member. For building such model, a novel security with efficient session management is required.

References

1. Akyildiz I, Su W, Sankarasubramaniam Y, Cayirci E (2002) A survey on sensor networks. IEEE Commun Mag 40(8):102–114
2. Wang Y, Attebury G, Ramamurthy B (2006) A survey of security issues in wireless sensor networks. IEEE Commun Surv Tutor 8(2):2–23
3. Eschenauer L, Gligor VD (2002) A key-management scheme for distributed sensor networks. In: Proceeding of ACM CCS, pp 41–47
4. Chan H, Perrig A, Song D (2003) Random key predistribution schemes for sensor networks. In: Proceeding of IEEE S&P, pp 197–213
5. Sunil Kumar KN, Shankar S (2016) Security issues in wireless sensor network—a review. https://ieeexplore.ieee.org/stamp/stamp.jsp?arnumber=7784988

6. Sunil Kumar KN, Shankar S (2017) A review on security and privacy issues in wireless sensor networks. In: 2017 2nd IEEE international conference on recent trends in electronics, information & communication technology (RTEICT), Bangalore, pp 1979–1984
7. Rahayu TM, Lee S, Lee H (2014) Survey on LEACH-based security protocols. In: 16th international conference on advanced communication technology, Pyeongchang, pp 304–309
8. Cirani S, Picone M (2015) Wearable computing for the Internet of Things. IT Prof 17(5):35–41
9. Zanella A, Bui N, Castellani A, Vangelista L, Zorzi M (2014) Internet of Things for smart cities. IEEE Internet Things J 1(1):22–32
10. Simsek M, Aijaz A, Dohler M, Sachs J (2016) 5G-enabled tactile Internet. IEEE J Sel Areas Commun 34(3):460–473
11. Elijah O, Rahman TA, Orikumhi I, Leow CY, Hindia MN (2018) An overview of Internet of Things (IoT) and data analytics in agriculture: benefits and challenges. IEEE Internet Things J. https://doi.org/10.1109/jiot.2018.2844296
12. Qiu T, Chen N, Li K, Atiquzzaman M, Zhao W (2018) How can heterogeneous Internet of Things build our future: a survey. IEEE Commun Surv Tutor 20(3):2011–2027. https://doi.org/10.1109/COMST.2018.2803740
13. Sisinni E, Saifullah A, Han S, Jennehag U, Gidlund M (2018) Industrial Internet of Things: challenges, opportunities, and directions. IEEE Trans Ind Inform. https://doi.org/10.1109/tii.2018.2852491
14. Nayak P, Devulapalli A (2016) A fuzzy logic-based clustering algorithm for WSN to extend the network lifetime. IEEE Sens J 16(1):137–144
15. Nayak P, Vathasavai B (2017) Energy efficient clustering algorithm for multi-hop wireless sensor network using type-2 fuzzy logic. IEEE Sens J 17(14):4492–4499
16. Rani S, Ahmed SH, Talwar R, Malhotra J (2017) Can sensors collect big data? An energy efficient big data gathering algorithm for WSN. IEEE Trans Ind Inform 99:1
17. Sarma HKD, Mall R, Kar A (2016) E2R2: energy-efficient and reliable routing for mobile wireless sensor networks. IEEE Syst J 10(2):604–616
18. Tsai CW, Hong TP, Shiu GN (2016) Metaheuristics for the lifetime of WSN: a review. IEEE Sens J 16(9):2812–2831
19. Parsapoor M, Bilstrup U (2014) An imperialist competitive algorithm for interference-aware cluster-heads selection in ad hoc networks. In: 2014 IEEE 28th international conference on advanced information networking and applications, Victoria, BC, pp 41–48
20. Chen CH, Chen WH (2016) United-based imperialist competitive algorithm for compensatory neural fuzzy systems. IEEE Trans Syst Man Cybern Syst 46(9):1180–1189
21. Weile DS, Michielssen E (1997) Genetic algorithm optimization applied to electromagnetics: a review. IEEE Trans Antennas Propag 45(3):343–353
22. Prasad J, Souradeep T (2012) Cosmological parameter estimation using particle swarm optimization (PSO). Phys Rev D 85(12). Art ID 123008
23. Laoudias C, Moreira A, Kim S, Lee S, Wirola L, Fischione C (2018) A survey of enabling technologies for network localization, tracking, and navigation. IEEE Commun Surv Tutor. https://doi.org/10.1109/comst.2018.2855063
24. Eletreby R, Yagan O (2018) Connectivity of wireless sensor networks secured by heterogeneous key predistribution under an on/off channel model. IEEE Trans Control Netw Syst. https://doi.org/10.1109/tcns.2018.2808141
25. Ting P, Tsai J, Wu T (2018) Signcryption method suitable for low-power IoT devices in a wireless sensor network. IEEE Syst J 12(3):2385–2394
26. Rahman M, El-Khatib K (2018) Secure time synchronization for wireless sensor networks based on bilinear pairing functions. IEEE Trans Parallel Distrib Syst. https://doi.org/10.1109/tpds.2010.94
27. Oliveira LB, Dahab R (2006) Pairing-based cryptography for sensor networks. In: 5th IEEE international symposium on network computing and applications (NCA'06), Cambridge, MA, USA (fast abstract)
28. Du W, Wang R, Ning P (2005) An efficient scheme for authenticating public keys in sensor networks. In: Proceedings of the 6th ACM international symposium on mobile ad hoc networking and computing, Urbana-Champaign, IL, USA, pp 58–67

29. Joppe WB, Halderman JA, Heninger N, Moore J, Naehrig M, Wustrow E (2014) Elliptic curve cryptography in practice
30. Brow E (2010) Elliptic curve cryptography
31. Al-Karaki JN, Al-Mashaqbeh GA (2007) SENSORIA: a new simulation platform for wireless sensor networks. In: 2007 international conference on sensor technologies and applications (SENSORCOMM 2007), Valencia, pp 424–429
32. Chaitra HV, Ravikumar GK (2016) A secure and energy efficient cluster optimization by using hierarchial clustering technique. In: 2016 3rd international conference on devices, circuits and systems (ICDCS), Coimbatore, pp 93–97

A Non-cooperative Game Theoretic Approach for Resource Allocation in D2D Communication

Tanya Shrivastava, Sudhakar Pandey, Pavan Kumar Mishra, and Shrish Verma

Abstract Device-to-device (D2D) communication is believed to be a creative innovation for 5G. It makes reuse of the resources of cellular users and provides high data rate. The reusability of resources generates interference which reduces the higher data rate flow. Therefore, an efficient resource allocation scheme is required which provides higher data rate. Thus, in this paper, we propose a non-cooperative game theoretic approach for resource allocation in D2D communication. In this approach, the D2D users act as players and the channel allocation is considered as a strategy which uses signal-to-interference-plus-noise-ratio as a parameter for payoff matrix. We also elaborate on the utility function for our game which is based on Shannon's capacity. First, we have formulated the game between the D2D users and formed the payoff matrix. Second, the Nash equilibrium of the game is determined for an efficient channel allocation. This allocation ensures win-win condition of the players playing the game. Additionally, we have also given the proof of the existence of the Nash equilibrium in our game which signifies that the game contains a stable outcome and at least one Nash equilibrium. The results of simulation represent the viability and efficiency of our scheme.

Keywords Device-to-device · Game theory · Nash equilibrium · Resource allocation · Signal-to-interference-plus-noise-ratio · Shannon's capacity

---

T. Shrivastava · S. Pandey · P. K. Mishra (✉) · S. Verma
Department of Information Technology, National Institute of Technology Raipur, Raipur, Chhattisgarh 492010, India
e-mail: pavanmishra.it@nitrr.ac.in

T. Shrivastava
e-mail: tanyashrivastava151@gmail.com

S. Pandey
e-mail: spandey.it@nitrr.ac.in

S. Verma
e-mail: shrishverma@nitrr.ac.in

N. N. Chiplunkar and T. Fukao (eds.), *Advances in Artificial Intelligence and Data Engineering*, Advances in Intelligent Systems and Computing 1133,
https://doi.org/10.1007/978-981-15-3514-7_99

1 Introduction

With an exponential growth in number of device-to-device (D2D) users and proximity-based services, there is a great rise in the demand for high data rate and bandwidth. However, the existing cellular system, i.e., 4G has limited resources and it is not efficient to handle and resolve this upsurge [1]. Therefore, an updated cellular system is required which will be able to handle an enormous amount of data and will provide high data rate. The researchers and telecom companies have come up with the new cellular system, i.e., the fifth-generation cellular system (5G). The 5G will provide 1000 times more data rate and reduce the power consumption up to 10 times in comparison to 4G system. In order to achieve this objective of 5G, several new schemes and features have been proposed by various researchers.

An innovative feature of 5G technology is D2D communication [2, 3]. D2D communication allows the mobile devices in close proximity to communicate with each other without communicating to the base station [4]. In [5, 6], D2D has been proposed as an underlaying infrastructure of cellular network that considers the vicinity of communicating devices as an important factor. In a cellular network, the D2D users share the channel with the cellular users. This sharing results in co-channel interference among the users of the same channel [7]. The unwanted interfering signals present within the network results in degradation of performance of communication taking place in the networks [8, 9]. Therefore, optimal assignment of channels is a salient and back-breaking task [10, 11]. In [12], the authors proposed an approach that generates local awareness about the interference occurring amid the D2D devices and the cellular users.

The base station (BS) is responsible for allocating the channels to the demanding users by a channel allocation scheme. Consequently, the decision of the base station for the resource allocation plays a major role. These decisions are taken in a constrained environment: containing set rules and outcome. Various resource allocation schemes have been proposed in [7, 10]. But all these schemes do not focus on real-time decision making. Various factors come into play when it comes to real-time decision making. All these factors cannot be evaluated at the run time. So, we need a predefined method containing each possible scenario.

Thus, we found that game theoretic approaches deal with evaluating the result based upon the runtime conditions and thereby focuses on real-time decision making. Game theory is helpful to evaluate multiple scenarios before making any hard decisions [13]. This method either chooses the maximum profit or, say, the minimum loss attached with a strategy. While utilized in variety of disciplines, game theory is most primarily used as a tool within the study of economics [14] but it is not only limited to this. Song et al. in [15] present the applications of various models of game theory to study resource allocation taking place in D2D communication.

The game theoretic approaches for resource allocation have been illustrated in [16]. Yin et al. guaranteed the fairness between D2D and cellular users and formulated a two-stage Stackelberg game. In [17], the behavior of cooperative D2D devices for mode selection within the transmission is modelled by a coalition game that is distributed in nature. Although many researchers have worked on game theoretic resource allocation approaches, almost all of them assume that the base station is already aware about the channel specifications and standards and thus, they can do the channel allocation efficiently. These assumptions are not necessarily correct always. Therefore, a game theoretic resource allocation scheme is required which can allocate the resource without knowing the channel standards. Thus, we have found an efficient way to embed game theory in D2D communication which provides a solution to this problem by calculating the channel quality using SINR as a payoff.

We propose a game which is a non-cooperative resource allocation game which justifies the condition of Nash equilibrium and thus resolves the issues of dependency by evaluating every scenario possible [18]. This game selects all the D2D users as players and evaluates the SINR values for each strategy. Next step involves looking for a Nash equilibrium giving a channel assignment such that each D2D user is responding in the best possible way to the strategies of other D2D user. This results in win-win situation for all the players. This is a solution point where no D2D user can improve its SINR by using some other channel. Using game theoretic approach allows us to determine the outcomes of various channel allocation scenarios in parallel. This methodology precomputes all the cases possible thus, saves the bandwidth.

The rest of the paper is organized as follows. We present the network model in Sect. 2. Section 3 describes the game model while the proposed approach is illustrated in Sect. 4. Section 5 gives the proof that the proposed game is a Nash equilibrium. Section 6 discusses the performance evaluation including the simulation settings and results.

2 Network Model

We have considered a single-cell scenario within the network that comprises of base station, pair of D2D users, and the cellular users as depicted in Fig. 1. CU_1 and CU_2 are the cellular users while $D2D_{t1}$ and $D2D_{t2}$ are the transmitters communicating to the D2D receivers $D2D_{r1}$ and $D2D_{r2}$ as shown by solid blue arrows. The cellular users have been randomly pre-allocated the channels (shown with dashed blue lines).

All the D2D users are waiting for the base station to assign the channels to them. This channel assignment is shown by dashed black arrows. All the cellular users can be allocated a single channel and every single cellular user is allowed to share their channel with at least one D2D users. The number of channels present in the network can be equivalent to the number of cellular users or more. This is done so that each cellular user is assigned a channel. When the channel sharing is done, users on the same channel experience interference problems. To minimize this interference problem, we give an approach based on game theory which is described in game model.

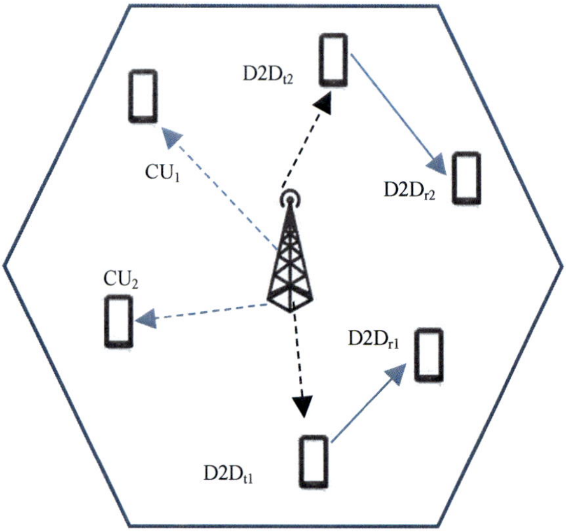

Fig. 1 Network model

3 Game Model

In this section, we illustrate a game model that comprises of multiple players and multiple strategies. The number of strategies depends upon the number of players taking part in the game. For n player game, there will be n strategies. Players play with these strategies in order to optimize their efficiency. The game results in the allocation keeping in mind that no player has any better allocation than the resulted one. Thereby forming a win-win situation for all the players.

The game model proposed in this paper is a non-cooperative form of game consisting of players, strategies, payoff function, and payoff matrix. The players play the game with their strategies in such a way that the final outcome is a Nash equilibrium. Therefore, the best respond of player to the strategies of other players. In our game, the D2D users are the players. Allocation of channels by calculating SINR is considered as the strategy. That means, if there are n channels, the players will have n strategies, i.e., assignment of channel 1, 2, 3, …, n by calculating SINR for each case. This SINR is calculated as:

$$\text{SINR}_{B_{ij}} = \frac{P_i * G_{ir}}{\sigma^2 + \sum_{j \neq i}^{n} G_{jr} * P_j} \tag{1}$$

where G_{ir} is the gain obtained by the channel between the receiver r and transmitter i, P_i is the transmission power of user i in uplink mode, σ^2 denotes variance of Gaussian noise power obtained by white additive noise, n denotes the number of total users sharing same channel. Utility function is given by Shannon's capacity which is expressed as:

Table 1 Payoff matrix

		User 2			
		Channel 1		Channel 2	
User 1	Channel 1	$\text{SINR}_{\text{c1u1}}$	$\text{SINR}_{\text{c1u2}}$	$\text{SINR}_{\text{c1u1}}$	$\text{SINR}_{\text{c2u2}}$
	Channel 2	$\text{SINR}_{\text{c2u1}}$	$\text{SINR}_{\text{c1u2}}$	$\text{SINR}_{\text{c2u1}}$	$\text{SINR}_{\text{c2u2}}$

Here, c1 and c2 denote channel 1 and channel 2, respectively, while u1 and u2 denote user 1 and user 2

$$C = B * \log_2(1 + \text{SINR}) \tag{2}$$

Here, C denotes Shannon's capacity limit for the given channel and gives the capacity of the channel in bits/second and is maximum, the channel bandwidth is given by B and is expressed in Hertz, SINR is Signal-to-interference-plus-noise-ratio calculated as illustrated in Eq. (1). Table 1 shows a layout of a two-player payoff matrix. Here, user 1 and user 2 are the two players playing the game with the strategies to acquire either channel one or channel two. Since there are two channels, there are two strategies. Denotes the value of SINR obtained when a user gets channel i and the other user is already allocated channel j. This SINR value is then used to calculate C as illustrated in Eq. (2).

4 Proposed Approach

This section elaborates on the design of non-cooperative game which is proposed in this paper. The base station considers the D2D users as the players. The proposed approaches comprise in three steps. In first step, SINR will be calculated. In the second step, a Nash game and find Nash equilibrium are formulated, and in third step, calculation of throughput and allocation of channel.

Step I: SINR Calculation—The base station determines the distance and thus calculates the interference to evaluate SINR according to Eq. (1). This SINR is determined for each and every possibility of channel sharing by the user.

Step II: Formulation of Nash game—A non-cooperative game is formulated as stated in payoff matrix, i.e., Table 1. It deals with calculating the Nash equilibrium from payoff matrix or evaluating a position or state from the payoff matrix such that no player wants to change it's state further. Thereby, ensuring a win-win situation for each player.

Step III: Calculation of throughput and channel allocation—The throughput is calculated according to Eq. (2) and then the Nash giving maximum throughput is considered as the final outcome. The channels are allocated accordingly. For example, if the Nash equilibrium giving maximum throughput is (i, j) in the payoff matrix, then the assignment of channel i is done to user 1 and the assignment of channel j is made to user 2.

Algorithm 1 states the input and output of the proposed resource allocation scheme. Step I is implemented in (i)–(iii). Step II is implemented from (iv) to (vi) steps, while Step III is implemented from (vii) to (ix) steps in Algorithm 1. The randperm(n) function in Algorithm 1 is required to allocate the channels randomly to the cellular users.

Algorithm 1. A non-cooperative game design for resource allocation

Input: Number of D2D users (m), total number of channels (n), and power of transmission of D2D and cellular users.

Output: The most optimal channel allocation with a win-win outcome.

i. Initially allot the channels randomly to the cellular users.
   Cu_ch = randperm(n)
ii. Calculate the distances between the cellular users and D2D users.
   Dist_DD (distance between all the D2D transmitters and receivers),
   Dist_CD (distance between cellular user and the D2D receivers)
iii. SINR Calculation

```
for p = 1 to n do
        for q = 1 to n do
                SINR(p, q)//According to eq. 1
        end
end
```

iv. Apply game on the calculated SINR values to calculate the optimal assignment which is a Nash equilibrium
v. Find the best response br_i for each player where $i \in [1, m]$.
vi. Find the $Nash_i$ where $i \in [1, k]$, k denotes the total number of possible Nash equilibria.
vii. Calculate the throughput of all possible $Nash_i$ $i \in [1, k]$.
viii. Compare all the throughputs.
   $Throughput(Nash_i) > Throughput(Nash_j)$ $\forall$ $i, j \in [1, k]$ and $i \neq j$.
ix. Select the best Nash with maximum throughput and allocate channel according to the selected Nash.

5 Proving Nash Equilibrium

Definition 1 If each S_i is a compact, non-empty, and convex subset of a metric space, and each $u_i(s_1, s_2, \ldots, s_N)$ is quasi-concave and continuous in $(s_1, s_2, \ldots, s_N)$, then $G = (S_i, u_i)$, $i = 1$ to N possesses at least one pure strategy Nash equilibrium.

Definition 2 A function $f(x)$ is convex if and only if $f''(x) \geq 0$, i.e., a function twice differentiable w.r.t. x is concave if it is greater than or equals to 0.

We have $S = \{s_1, s_2, \ldots, s_N\}$ as a non-empty set of strategies. These strategies s_i range from $i = 1$ to N where N denotes the number of channels. These strategies are based on the calculation of SINR given by Eq. (1).

Lemma 1 *Differentiating Eq. (1) w.r.t.* G_{ir}*.*

We assume constant power of the transmitter and constant noise level. We get $S' =$ *constant, again differentiating it w.r.t.* G_{ir}*, we get* $S'' = 0$*.*

Lemma 2 *Differentiating Eq. (1) w.r.t.* G_{ir}*.*

We assume constant power of the transmitter and constant noise level. We get, $S' = \frac{-p}{(q+rG_{ir})^2}$ *where p, q, r are constants. On Double differentiating, we get* $S' = \frac{p}{(q+rG_{ir})^3} \geq 0$.
From Lemmas 1 and 2, we get that S is non-empty, convex set of strategies.

Definition 3 A function $f(x)$ is concave if and only if $f''(x) \leq 0$, i.e., a function twice differentiable w.r.t. x is concave if it is less than or equals to 0.

We have our utility function given by $U(x)$ as $U(x) = B \log_2(1 + x)$ where x denotes the SINR value. Here, bandwidth B is assumed constant and the SINR value is variable. So, we differentiate the above equation with respect to x. We get,

$$U'(x) = \frac{B}{1+x}$$

Again differentiating $U'(x)$ we get, $U''(x) = \frac{-B}{(1+x)^2} \leq 0$.
This shows that the utility function is concave w.r.t. x.

Lemma 3 *Every concave function is also quasi-concave function but vice versa is not true.*

According to this lemma, since U(x) is concave U(x) is also a quasi-concave function.

Lemma 4 *log x is always continuous for x > 0.*

We have Eq. (2) where B is constant and log(1 + x) is always continuous for x > 0. As SINR is considered as x and SINR = 0 has no practical significance. So, SINR > 0, which means x > 0. Thus, the utility function U(x) is continuous in $\{s_1, s_2, \ldots, s_N\}$ *and quasi-concave.*

As the set of strategy is convex as well as non-empty and the function of utility is quasi-concave, there must be the existence of Nash equilibrium. Hence, Nash equilibrium exists.

6 Performance Evaluation

6.1 Simulation Settings

In this simulation, we consider four users, i.e., the two cellular users and two pair of D2D users. All the cellular users have been randomly pre-allocated the channel. Since

Table 2 Table of constant and assumed values

Parameters	Values
Power of transmission of cellular users	45 db
Power of transmission of D2D users	24 db
Noise	−174 dbm/Hz
Number of users in simulation	4
Number of iterations for simulation	5

every cellular user is required to get allocated a channel so the number of channels must be greater than or equals to 2. Here, we consider two channels. The game is formulated between the D2D users for the efficient allocation of these channels. The performance of the proposed approach is analyzed in terms of the following metrics:

1. Data rate is evaluated in terms of SINR.
2. Total data rate (throughput) received by all the users on sharing same channel.
3. Throughput of the channels is analyzed for each Nash equilibria and the final throughput is the maximum of all the throughput obtained from Nash equilibria (Table 2).

6.2 Simulation Results

This section involves evaluating the performance of the proposed resource allocation approach. Figure 2 describes the SINR values obtained in each case as stated in Table 1. We can see that case 2 (user 1 gets channel 1 and user 2 gets channel 2) and case 3 (user 1 gets channel 2 and user 2 gets channel 1) give the highest SINR values. These are the cases that are in Nash equilibrium. Our proposed approach compares

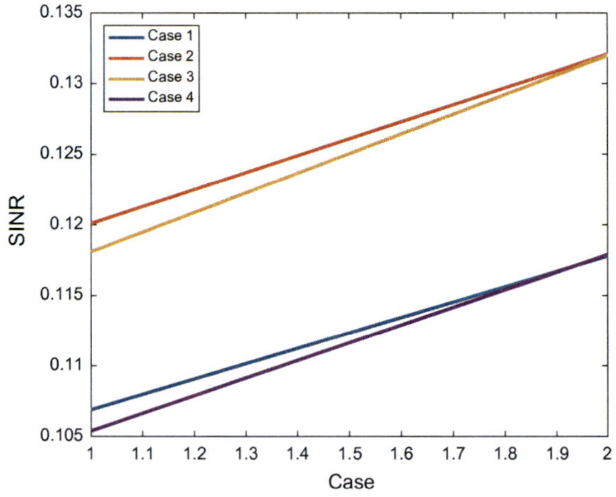

Fig. 2 SINR computation in all cases

the throughput of both these Nash equilibria and assigns the channel according to the Nash giving highest throughput. Figure 3 illustrates the SINR experienced by each user, i.e., D2D user 1 and D2D user 2 per iteration. Here, we simulate this for five iterations. The rise and fall in the SINR values are according to the distance between the cellular and D2D users. From Fig. 3, we can observe that user 1 has high SINR in comparison to user 2 for iterations 2, 3, and 4. This is because, for iterations 2, 3, and 4, user 1 has close proximity to the base station.

Similarly, Fig. 4 describes the SINR of each channel, i.e., channel 1 and channel 2 for five iterations. The rise and fall in the graph are according to the distances of the users in each iteration. In Fig. 4, we see that SINR of both the channels is comparatively lesser for iteration 3. This is because the D2D transmitter using the channels is distant from the receivers. Figure 5 depicts the total data rate. The total

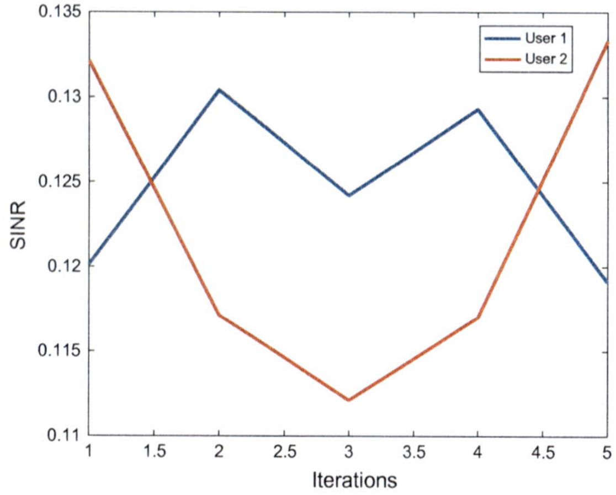

Fig. 3 SINR received by each user per iteration

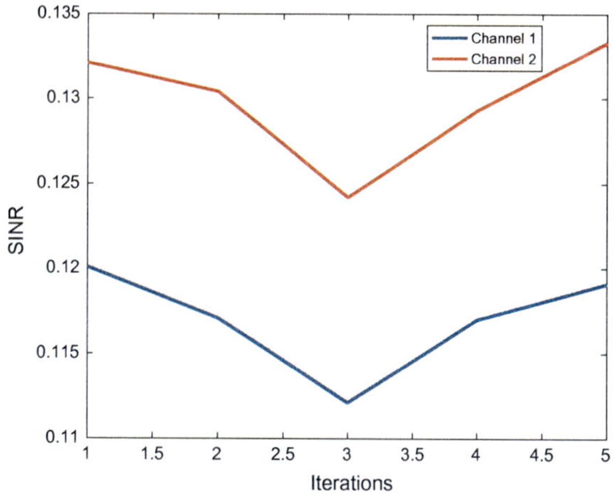

Fig. 4 SINR of channel per iteration

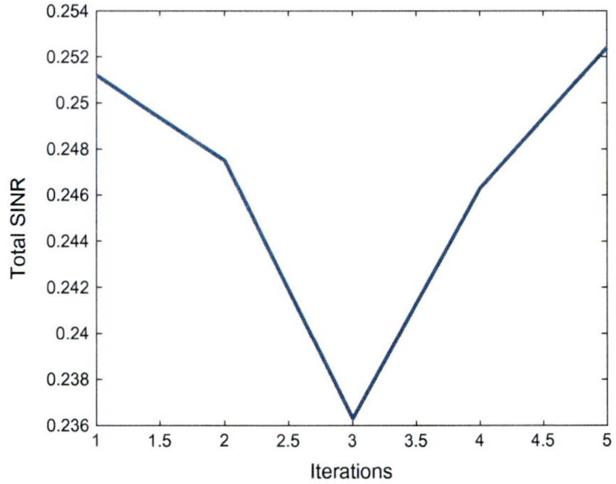

Fig. 5 Total data rate

data rate is described as the summation of SINR experienced on each channel, i.e., total SINR experienced on the channels. This total data rate varies for each iteration as the distance between the users varies which intends the SINR to vary.

Figure 6 illustrates the throughput obtained by different Nash equilibria. In our case, there are two Nash equilibria. The throughput obtained by both of these Nash is plotted against each iteration. Throughput 1 corresponds to Nash 1, while Throughput 2 corresponds to Nash 2. These throughputs for each Nash are shown for different iterations in Fig. 6. The fall at iteration 3 is because of the close proximity of the users on the same channel. This increases interference, thereby reducing the throughput. Figure 7 illustrates the Nash throughput, i.e., this is the maximum of the throughputs obtained by both the Nash, i.e., max (Throughput by Nash 1, Throughput by Nash 2). This may vary for each iteration, i.e., for some iterations, we can get maximum throughput by Nash 1, while other iterations have maximum throughput by Nash 2.

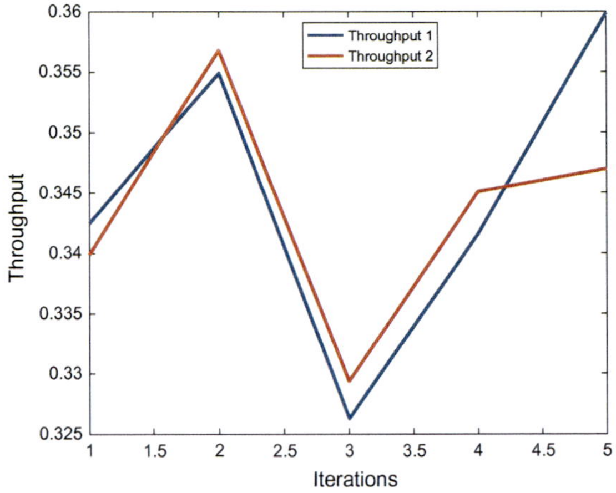

Fig. 6 Throughput obtained by different Nash equilibria per iteration

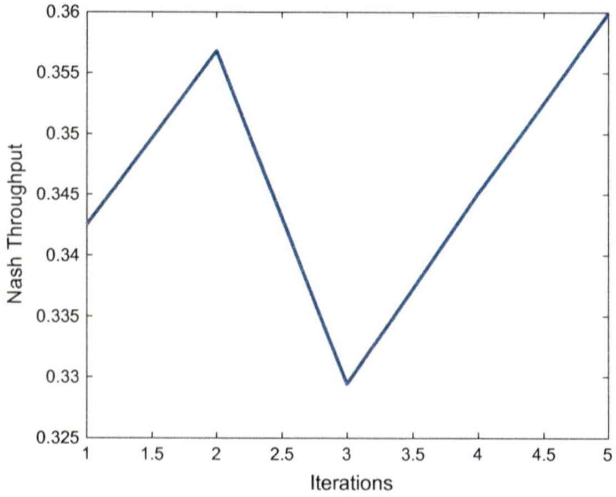

Fig. 7 Nash throughput

In Fig. 6, we see that at iteration 3, throughput 2 is higher than throughput 1. So Fig. 7 gives throughput 2 as the final throughput.

7 Conclusion

D2D communication is an important technology in 5G networks that has great potential to provide high data rates and efficient resource allocation. We proposed a non-cooperative game for resource allocation for D2D users. This approach guaranteed the win-win situation for all the players playing the game and thereby reduced the selfish behavior of the players. We have also proved that the Nash equilibrium exists for the proposed game which ensures that there must be certain stable allocation or solution to the game as at least one Nash equilibrium exists. The simulation results show that the scheme proposes an efficient win-win solution that guarantees high data rate for all the players. The future prospects and the further extension of this work are to formulate a utility function that includes pricing in the game model.

References

1. Wang CX et al (2014) Cellular architecture and key technologies for 5G wireless communication networks. IEEE Commun Mag 52(2):122–130
2. Ansari RI et al (2018) 5G D2D networks: techniques, challenges, and future prospects. IEEE Syst J 12(4):3970–3984
3. Cao Y, Jiang T, Wang C (2015) Cooperative device-to-device communications in cellular networks. IEEE Wirel Commun 22(3):124–129
4. Liu J, Kato N, Ma J, Kadowaki N (2015) Device-to-device communication in LTE-advanced networks: a survey. IEEE Commun Surv Tutor 17(4):1923–1940

5. Doppler K, Rinne MP, Jänis P, Ribeiro C, Hugl K (2009) Device-to-device communications; functional prospects for LTE-advanced networks. In: 2009 IEEE international conference on communications workshops, Dresden, pp 1–6
6. Janis P et al (2009) Device-to-device communication underlaying cellular communications systems. Int J Commun Netw Syst Sci
7. Hoang TD, Le LB, Le-Ngoc T (2016) Energy-efficient resource allocation for D2D communications in cellular networks. IEEE Trans Veh Technol
8. Safdar GA, Ur-Rehman M, Muhammad M, Imran MA, Tafazolli R (2016) Interference mitigation in D2D communication underlaying LTE-A network. IEEE Access 4:7967–7987
9. Kaufman B, Lilleberg J, Aazhang B (2013) Spectrum sharing scheme between cellular users and ad-hoc device-to-device users. IEEE Trans Wirel Commun 12(3):1038–1049
10. Zhao W, Wang S (2015) Resource sharing scheme for device-to-device communication underlaying cellular networks. IEEE Trans Commun 63(12):4838–4848
11. Alkurd R, Shubair RM, Abualhaol I (2014) Survey on device-to-device communications: challenges and design issues. In: 2014 IEEE 12th international new circuits and systems conference (NEWCAS), Trois-Rivieres, QC, pp 361–364
12. Jänis P, Koivunen V, Ribeiro Ć, Korhonen J, Doppler K, Hugl K (2009) Interference-aware resource allocation for device-to-device radio underlaying cellular networks. In: IEEE vehicular technology conference
13. Osborne MJ (2004) An introduction to game theory, vol 3. Oxford University Press, New York
14. Rabin M (1993) Incorporating fairness into game theory and economics. Am Econ Rev 1281–1302
15. Song L, Niyato D, Han Z, Hossain E (2014) Game-theoretic resource allocation methods for device-to-device communication. IEEE Wirel Commun 21(3):136–144
16. Katsinis G, Tsiropoulou EE, Papavassiliou S (2014) A game theoretic approach to the power control in D2D communications underlay cellular networks. In: 2014 IEEE 19th international workshop on computer aided modeling and design of communication links and networks, CAMAD, pp 208–212
17. Akkarajitsakul K, Phunchongharn P, Hossain E, Bhargava VK (2012) Mode selection for energy-efficient D2D communications in LTE-advanced networks: a coalitional game approach. In: 2012 IEEE international conference on communication systems (ICCS), Singapore, pp 488–492
18. van Damme E (2014) Non-cooperative games

IoT-Based Nursery Management System

Mahendra S. Naik, Sreekantha Desai, K. V. S. S. S. S. Sairam, and S. N. Chaitra

Abstract IoT is one of the developing technologies in recent years. Internet of things is a huge network, which provides human-to-human (H2H), machine-to-machine (M2M), and human-to-machine (H2M) communications. The IoT is still developing technology and has many research scopes. The IoT has varieties of applications, which makes human's life easier. These IoT applications reduce human interventions and give the results more accurate. The IoT applications can be used in every domain. One such application is the maintenance of nursery based on the environment and weather conditions. It is more efficient and cost-effective. The aim of the proposed method is to design an automated model for maintaining the useful weather condition for the good growth of the tiny plants. The maintenance of nursery without human intervention using the IoT and cloud is the key ambition of the paper. The tiny plants are grown in the nursery, and hence, the health of the plant plays an important role. The nursery, tissue culture areas are providing farmers good yielding plants. To get good quality plants maintaining various environmental parameters like temperature, humidity, and certain gasses is highly important. The model has been developed using Arduino Uno board, sensors, and Raspberry Pi model B. ThingSpeak, an open IoT platform with MATLAB analytics, is used for data collection and analysis. The simulation result gives the weather and environment conditions of the nursery on an hourly basis.

Keywords IoT · Nursery management · Growth · Humidity · Sensors

1 Introduction

The plant nursery business is one of the demanding and profitable businesses. It is also the fastest growing agro-industry all over the world. The nursery is a type of

---

M. S. Naik (✉)
PES Institute of Technology and Management, Shivamogga, India
e-mail: mahendrasnaik@gmail.com

S. Desai · K. V. S. S. S. S. Sairam · S. N. Chaitra
NMAM Institute of Technology, Nitte, Karkala, India

N. N. Chiplunkar and T. Fukao (eds.), *Advances in Artificial Intelligence and Data Engineering*, Advances in Intelligent Systems and Computing 1133,
https://doi.org/10.1007/978-981-15-3514-7_100

agriculture where plants are grown, nurtured, and sold to the home garden or commercial purpose. In nursery, the plants are grown to usable sizes and their quality and health are maintained. Nurseries are of different types; they are retail nurseries, which sell plants to the general public, and wholesale nurseries, which sell plants only for other nurseries. The public get plants from nurseries for gardens, and nurseries also provide plants for agriculture, forestry, etc. Nurseries grow a different variety of plants, like shade plants, rock garden plants, fruit bearable plants, and commercial plants. The different plants require different environment and treatment. The major role is to maintain the perfect health and quality of the plant in the nursery; this requires the large workforce, and hence, the cost of labor increases. Adopting labors to maintain the nursery is not cost-effective. The growth and quality of the plants require a very sophisticated environment. The traditional technique used in nursery affects the production rate. By adapting conventional farming techniques, the growth rate and quality in India have seen a double-figured rate. Hence, an automated and continuous monitoring system is required to maintain the physiological parameters like temperature, humidity, and light intensity in the nursery. The greenhouse temperature, weather conditions, and the proper environment can be provided for the plants using IoT. The automated system can be developed with the help of the IoT. Internet of things (IoT) is a network of things (devices) in which devices communicate and interact with each other anywhere and anytime via the Internet. Kevin Ashton, the innovator, describes the Internet of things as the network connecting objects in the physical world to the Internet. The benefit of using IoT includes portability, easy access, lower cost, lower power requirement, and reduced space. The IoT is still the developing technology. The IoT is a network of everything. The communication will be between all the devices which are connected. Hence, the routing of the data and processing of the data should be maintained. The networks are heterogeneous and the compatibility between the networks should be achieved. The research fields regarding IoT can find in all the areas.

In this proposed method, Raspberry Pi is used to develop the system for controlling and surveillance purpose. This work focuses on providing an automated system using modern technology like IoT, low-cost commodity hardware, and open-source software that is likely to solve many problems faced by the nursery industry. The designed system will sense any physiological changes within the mentioned area and notifies the concerned persons through messages. An application program is used. When the sensed data crosses the threshold values, suitable operation is performed to regulate the nursery environment remotely.

This paper is organized as follows. Section 2 briefs the overview of the methods used in IoT-based agriculture and farm maintenance. Section 3 explains the problem faced by the farmers and about the proposed work. The developed model description and its working are explained in Sect. 4. Section 5 explains the algorithms and flowchart used. Section 6 discusses the result and conclusion.

2 Literature Survey

The wireless sensor network which forms a huge network in the IoT generates a large amount of data. The proposed method developed using Arduino and Raspberry Pi in a cost-efficient method is scalable both in the terms of sensor and sensor nodes. Overall system architecture and the design of hardware and software components are presented. Some sample deployment and measurement results are also presented to demonstrate the usefulness of the system.

The authors of the paper explain the design and research views of intelligent agriculture system. Yan-e and Bing [1, 2] the authors introduce the concept of managing the agricultural farm by extracting the agricultural information and designed the system based on IoT [1]. This paper gives an implementation example of a system in agricultural production. Paper [2] explains the research on the agriculture intelligent system based on IoT. This paper presents an intelligent monitoring platform framework and system structure for facility agriculture ecosystem based on IoT. Here, the authors explain about the division of work in four layers. When the monitoring work divides among four layers, the cite results in a better ecosystem and better crop growth [2].

The study and application of the IoT technology in agriculture by Zhao et al. describes the management of greenhouse with the help of sensors and explains how to collect real-time data and to send the data through the Web or by SMS [3]. The authors of the paper have developed an IoT-based monitoring system to analyze crop environment and design the method to improve the efficiency of decision making by analyzing harvest statistics [4]. Authors developed the decision support system to forecast agricultural production using IoT sensors [4].

The authors of paper [5] designed the system, which includes an intelligent frequency conversion irrigation function, automatic control function of greenhouse combined with the sensor nodes, wireless transmission network, and sensor configuration, data collection system. The system developed in greenhouse practical application has received a good effect in Tianjin. This realized that the real-time data automatic acquisition of greenhouse environment parameters and biological information, the farmer achieved good economic and ecological benefits and the great significance to the development of modern agricultural information-based and intelligent.

In the paper [6], the researchers explain about the wireless sensor applications, and for many of the problems, the WSN will be feasible solutions by developing applications. The Raspberry Pi is a low-cost development kit, which can be programmable as a computer.

The systems are developed more efficiently along with the WSN. Raspberry Pi works as a base station which connects the sensor nodes via ZigBee protocol in the wireless sensor network, collects sensors' data from different sensors, and supplies multi-clients services including data display [7].

In paper [7, 8], authors explain how Arduino and Raspberry Pi are used in farm management and about the sensors of IoT. The authors have developed a design model to control the environmental conditions of a farm. The comparison analysis of with designed model and without a designed model is also given. According to the results, the designed system could monitor surrounding weather conditions including humidity, temperature, climate quality, and also the filter fan switch control in the farm. The developed system is very useful for the farmers as they can control the farm from anywhere and anytime, which in turn results in cost reduction, human intervention, and increases the product management and good yields [8]. The authors tell about the working scenario and concepts of IoT. The authors explain the data extraction method and computing methodology Internet of Things work primarily with the sensors. The sensors are connected to machines and everyday things. The sensors collect the required data from machines, things, or from the contexts that are used to analyze the scenario. Clouds are used to store this data and can be used whenever required. The objects and devices that are connected to the Internet can be constantly monitored [9].

3 Proposed Model

The traditional farming adapted requires human involvement, because of which the labor cost and human errors increase. This paper focuses on developing an automated system with no or minimum human intervention. The designed model aims to provide a solution to develop the IoT-based nursery management system in cost-effective and with open-source software. The nursery monitoring system is developed using Raspberry Pi, and Arduino boards and various sensors are used for the same. The Google Cloud Messaging service is used. The complete framework is very useful because the user can access and regulate the system by the remote place with the help of handheld digital devices. As said earlier by human intervention prone to many mistakes, the growth rate of plants in the nursery may reduce. The proposed system reduces human errors and ensures good growth of plants. The proposed system maintains the nursery system with least human intervention and the developed system is very cost-effective. The production and growth rate will increase by incorporating new methods. Since the labor cost in the traditional system is very high nowadays, the development of this system helps in cost-effective and quality nursery can be developed, and the same can be adopted for smart agriculture. This paper gives the model to maintain the nursery or a farm so the labor cost is reduced and yield can be increased. An extensive system was built using Raspberry Pi and Arduino Uno board for interfacing with different sensors, and IoT technique for cloud interfacing and analysis. The system was methodically investigated for various environmental parameters like temperature, humidity, air quality, etc., associated with effective nursery management. The proposed model also controls these parameters to the desired level effectively.

4 Methodology and Design

The sensors that are required to monitor the nursery and the physical parameters are connected to the Arduino board. The Arduino board continuously monitors the nursery environment and extracts the data periodically. The extracted data is then transferred to Raspberry Pi which is an interface to the Arduino board using USB; hence, the need for the logical controller is avoided. Raspberry Pi works as a server interface and as a bearer to a smartphone user or a terminal over the Internet implemented using Google Cloud Messaging (GCM) Service. The Raspberry Pi acts as an impetus to the actuator, which later regulates the various defined parameters in a nursery to the desired level. ThingSpeak, an open IoT platform with MATLAB analytics, is used for data collection and analysis. This open IoT platform is used to collect the data from various sensors used in the system. Later, the collected data is processed and various comparative analyses with the previous data are done.

Figure 1 shows the proposed model in which all the sensors are connected to Arduino Uno. The data extracted from the Arduino Uno is transmitted to the Raspberry Pi through USB; the Raspberry Pi is a small credit card-sized computer, which is configured to send the data to the cloud. Meanwhile, when data crosses the threshold level, the message is sent to his/her digital handheld device. The user can regulate the parameters which are high/low by switching on a fan, increasing the intensity of light or reducing the intensity of light, and switching on the water pump if moisture in the soil is reduced. Hence, the observing and regulating of the parameters are possible by the developed model, which helps to increase the growth rate of plants and save the labor cost.

4.1 Sensors Used

4.1.1 Temperature and Humidity Sensor (DHT11 Sensor)

Temperature and humidity are the two parameters considered in the IoT nursery management system. The DHT11 sensor is used as one of the sensors to measure the temperature and humidity in the development of the prototype. DHT11 is a digital sensor which measures the environmental temperature and humidity both. It measures the contextual temperature and humidity by exclusive digital signal acquisition technique. Excellent stability and high reliability can be obtained. The temperature is measured with the help of an NTC thermistor or negative temperature coefficient thermistor. These thermistors are usually made with semiconductors, ceramics, and polymers. The resistance of the device is inversely proportional with temperature and follows a hyperbolic curve. Temperature using NTC often found out the Steinhart–Hart equation.

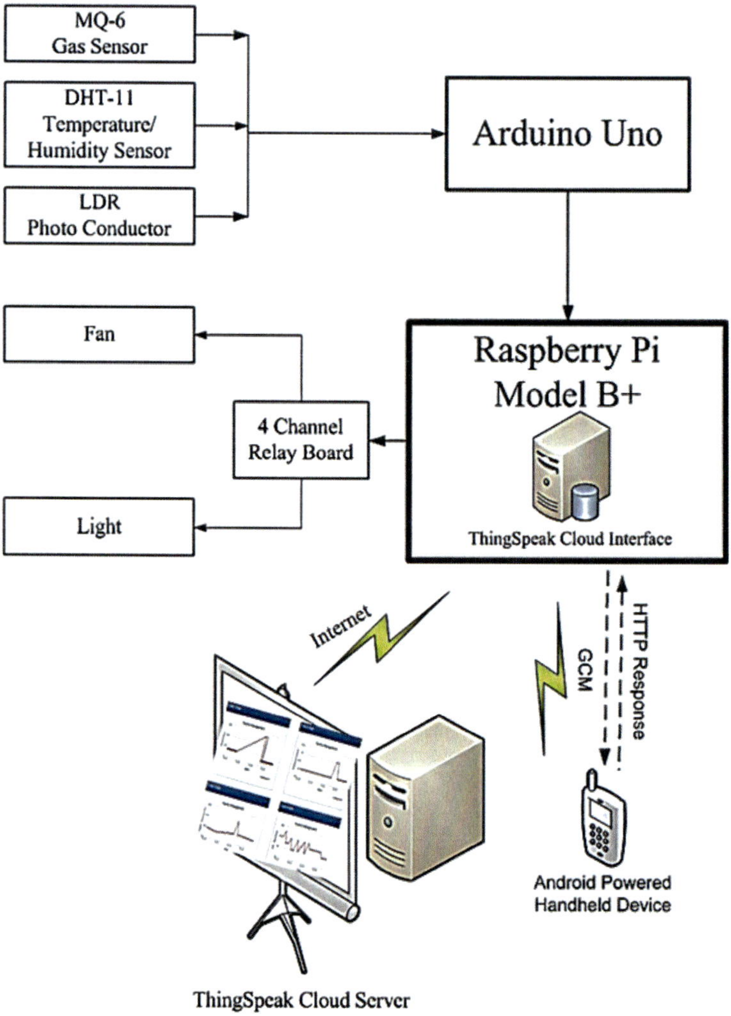

Fig. 1 Block diagram of IoT-based nursery management system

4.1.2 Gas Sensor (MQ-6 Sensor)

The MQ-6 is a gas sensor which detects the presence of harmful gasses in the environment. The gas sensor can be used with any of the sensors and an automatic system can be developed to shut down the system when the harmful gas leaks in the surroundings. An alarm can be given for the controlling system or in the home, and hence, any disaster can be avoided. The gas sensor detects combustible gasses like methane, propane, flammable gasses, and toxic gasses, which are harmful to human and organic life. In this proposed system, MQ-6 gas sensor is used in the prototype to detect the presence of any harmful gasses which shortens the growth of the plants in the nursery.

4.1.3 Photoconductor

The intensity of the light is one of the important parameters for the healthy growth of the plants in the nursery. Light-dependent resistor (LDR) is used to measure the intensity of the light. The LDR is a type of resistor whose resistance is inversely dependent on the amount of light intensity. It is basically a photoelectric cell that works on the electrical and optical phenomenon. The photoconductor is a passive component which is basically a resistor whose resistance value decreases with the decrease in the intensity of light.

4.2 Communication Process

4.2.1 Server Side

Transferring data from Arduino Uno to Raspberry Pi. The data transfer is done by Python language in Raspberry Pi to get the data through USB by serial connection.
Computing the data. After transferring data, computation of the fetched data is again done by the help of Python language in Raspberry Pi to extract the exact data required to store. The data received is a string. The required data can be extracted by using a split function.
Storing the data. This can be done by MySQL language in Raspberry Pi to store the data. The data which is stored in the database as a data repository helps in sending the data to the user mobile.
Activating relay. This can be done by Python language in Raspberry Pi to send a signal to relay from Raspberry Pi to relay which activates the devices connected to it.

4.2.2 Client Side

Getting notification message to smartphone. The changes of any parameter in the field are noted to the sender by message, which helps in collecting information about the current values that lead to taking the command.
Selecting/proceeding command to the server. Selecting a specific button in the app to send proper command toward the server.

5 Results and Interpretation

Using the android emulator in the personal computer, the developed system (GCM-based nursery management system) is tested on the Android-based mobile devices and on the Web page. Experiments were conducted under various conditions like different values of temperature, gas, etc. Many samples for every environmental condition are taken from various places for training the system. Some samples other than training samples are also used. Samples of a few snapshots of the system working on the emulator and on the mobile device are presented along with the detailed performance analysis of the developed system that is given here. In the result, the four different parameters like temperature, air quality, humidity, and air intensity are measured for different timings and the graph is shown if all the parameters go above or below the threshold level and these values are stored in the loud and compared with the previous values. At the same time, when the different parameters which are measured go above or below the threshold level, the system intimates the user by sending them the message through Google Cloud Messenger (GCM). After receiving the message, the user can control the parameters by switching lights or fans on by GCM from the place wherever he is (Figs. 2, 3, 4 and 5).

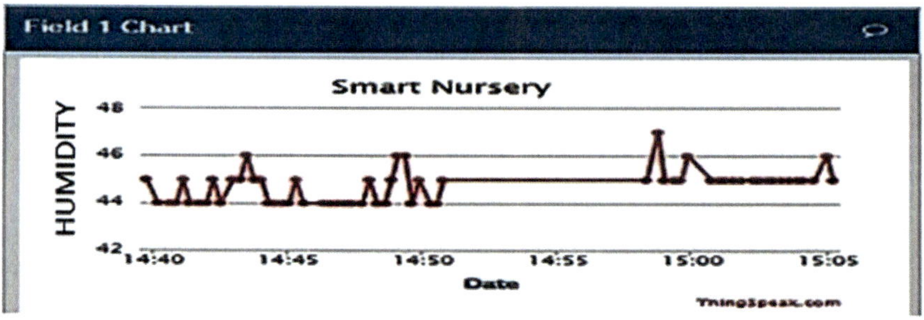

Fig. 2 Graph showing humidity variation

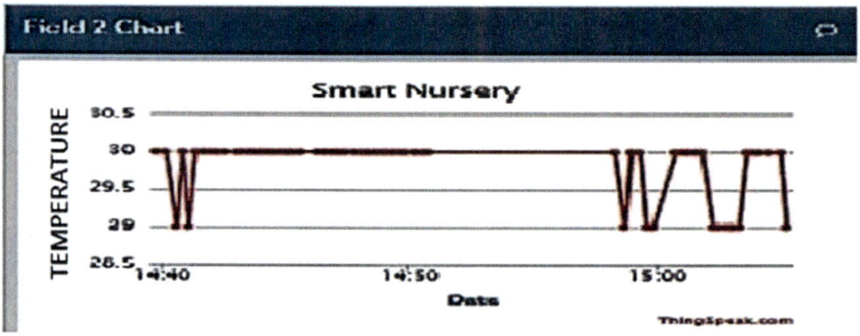

Fig. 3 Graph showing temperature variation

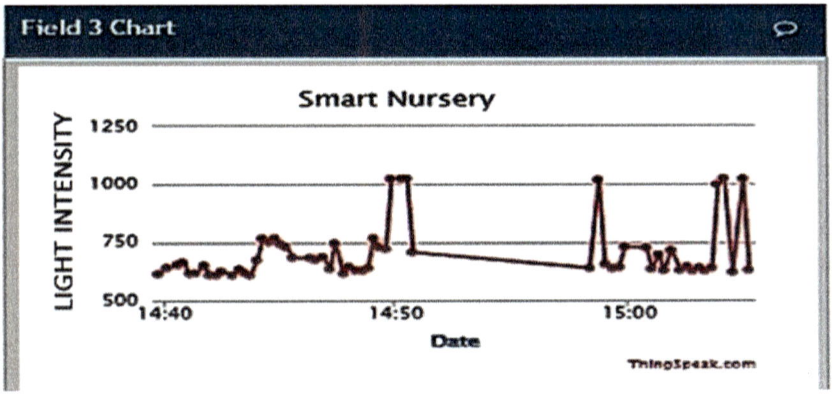

Fig. 4 Graph showing light intensity variation

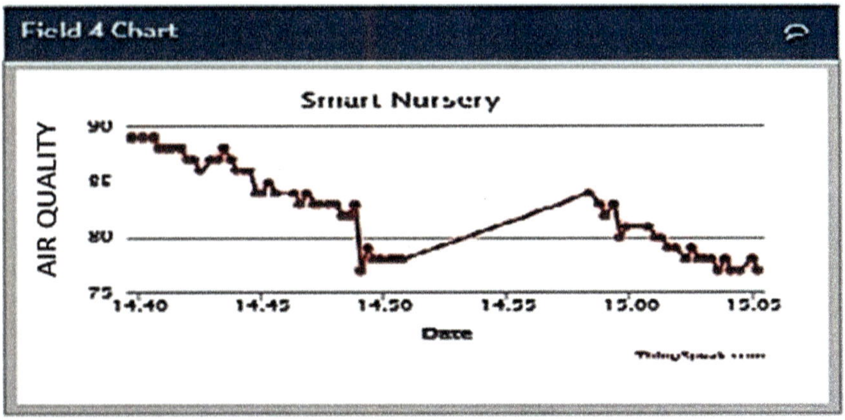

Fig. 5 Graph showing air quality variation

6 Conclusions and Future Work

The proposed system continuously monitors the physical conditions in and around the nursery and effectively regulates the surrounding environment in the permissible limit to yield the maximum production with reduced human intervention. Hence, the designed monitoring system can take the nursery business toward profit by producing healthy and quality products. In addition, the system can be extended to monitor various other environmental parameters by simply plugging up a sensor to the system. The system can be upgraded to a multiuser control environment with security. This system can be integrated with other ascents of framing like feed management, inventory management, and accounting.

References

1. Yan-e D (2011) Design of intelligent agriculture management information system based on IoT. In: 2011 international conference on intelligent computation technology and automation (ICICTA), vol 1. IEEE, pp 1045–1049. https://doi.org/10.1109/icicta.2011.262
2. Bing F (2012) Research on the agriculture intelligent system based on IoT. In: Proceedings of the 2012 international conference on image analysis and signal processing, Hangzhou, China, vol 911, p 14
3. Zhao J-C, Zhang J-F, Feng Y, Guo J-X (2010) The study and application of IoT technology in agriculture. In: 2010 3rd IEEE international conference on computer science and information technology (ICCSIT), vol 2. IEEE, pp 462–465. https://doi.org/10.1109/iccsit.2010.5565120
4. Lee M, Hwang J, Yoe H (2013) Agricultural production system based on IoT. In: 2013 IEEE 16th international conference on computational science and engineering (CSE). IEEE, pp 833–837. https://doi.org/10.1109/cse.2013.126
5. Guo T, Zhong W (2015) Design and implementation of the span greenhouse agriculture Internet of Things system. In: 2015 international conference on fluid power and mechatronics (FPM). IEEE, pp 398–401. https://doi.org/10.1109/fpm.2015.7337148
6. Nikhade SG (2015) Wireless sensor network system using Raspberry Pi and ZigBee for environmental monitoring applications. In: 2015 international conference on smart technologies and management for computing, communication, controls, energy, and materials (ICSTM). IEEE, pp 376–381. https://doi.org/10.1109/icstm.2015.7225445
7. Jindarat S, Wuttidittachotti P (2015) Smart farm monitoring using Raspberry Pi and Arduino. In: 2015 international conference on computer, communications, and control technology (I4CT). IEEE, pp 284–288. https://doi.org/10.1109/i4ct.2015.7219582
8. Abdul-Rahman AI, Graves CA (2016) Internet of Things application using tethered MSP430 to Thingspeak cloud. In: 2016 IEEE symposium on service-oriented system engineering (SOSE). IEEE, pp 352–357. https://doi.org/10.1109/sose.2016.42
9. Ferdoush S, Li X (2014) Wireless sensor network system design using Raspberry Pi and Arduino for environmental monitoring applications. Procedia Comput Sci 34:103–110. https://doi.org/10.1016/j.procs.2014.07.059

Shortest Path Discovery for Area Coverage (SPDAC) Using Prediction-Based Clustering in WSN

C. N. Abhilash, S. H. Manjula, R. Tanuja, and K. R. Venugopal

Abstract Area coverage has been one among the major limitations in wireless sensor networks (WSNs). The main drawback in WSN is their limited lifetime. Hence, this has become the popular topic in WSNs recent research trend. Optimization of area traversed paths is one of the key factors in the proposal of area coverage approaches in WSN. This work presents the shortest path discovery for area coverage (SPDAC) approach to optimize the path trajectory of mobile relay node (MRN) to solve area coverage problems through the strategic deployment of sensors. Also, we extend the algorithm to reduce static node communication overhead by introducing a new scheme prediction-based clustering protocol for energy consumption (PCP-EC). These two approaches together will help in extending lifetime of the network. Simulations are carried out using network simulator tool to analyze the reduction in number of static nodes interactions and other quality of services.

Keywords Area coverage · Energy consumption · Prediction clustering

1 Introduction

In WSNs, data collection detected by the nodes that are organized in the area of sensing is one among the most important tasks [1]. Typically, this collection of data mainly relies on wireless transmission between the sink node and sensor nodes that suffer from different situations. For example, wireless communications that

C. N. Abhilash (✉) · S. H. Manjula · R. Tanuja · K. R. Venugopal
Department of Computer Science and Engineering, UVCE Bangalore University, Bengaluru, Karnataka, India
e-mail: abhilashcn83@gmail.com

S. H. Manjula
e-mail: shmanjula@gmail.com

R. Tanuja
e-mail: r_tanuja@yahoo.com

K. R. Venugopal
e-mail: venugopalkr@gmail.com

N. N. Chiplunkar and T. Fukao (eds.), *Advances in Artificial Intelligence and Data Engineering*, Advances in Intelligent Systems and Computing 1133,
https://doi.org/10.1007/978-981-15-3514-7_101

especially includes long-distant ones may consume the restricted on-board energy supplied to sensors extremely with respect to the exponent true path loss. Also for short distant, multi-hop communications are approved, as the aggregation of data is near the base station, sensors near the base station still require more power than neighboring nodes as it incurs heavier sizes of traffic is transmitted by them that points to a lesser overall lifetime of the network. Modification has performed in the literature [2–5], but the highly inherent and unstable consumption of energy still has been a major challenge. Next approach in data aggregation from sensor networks makes use often, and controlling the mobility of few sensors [6–8], termed as mobile relay nodes (MRNs) throughout this paper.

The mobility problem is increased along the area specified to enhance the network performance; also the solutions achievable are always not much poorer than those found in a sub-region of a lesser dimension [9]. Though, the collection of data by MRNs in WSN [10, 11] poses its specific tasks as well. Because of relatively less speed of MRNs as compared to acoustic or electromagnetic waves, data gathered may incur a higher delay than multi-hop progressing, whereas the previous is feasible with more cost for energy used by the sensors.

Huge information collected with delay not merely damages the rightness of the information it also may affect in the overflow in buffer of the sensors. The delay, primarily measured through mobility and MRNs scheduling, i.e., the way they cross along the detecting region and as it gathers information by each node, has been the central focus of the research trend.

This work tackles this issue from a novel perspective. Initially, an advanced optimization method is used to shrink the tour distance of MRNs, and hence the traverse time with the hypothesis of an endless traverse speed, increases by joining the positions collectively with a nearby source of data and later passing and swapping (i.e., JPS) specific spots. The source of data is obtained either by the conventional nodes in the network, or by the dominant cluster (DC) in tiered-structured network. Next depending over the wireless faithful communication characteristics, a multi-rate transmission model is implemented that permits MRNs to gather data with a lesser proportion by traversing more distance.

As WSNs comprises of abundant less powered sensors that have the ability for detecting, computing and connecting with each other nodes. These sensors witness the event at various positions in the region, work together and transmit the processed information to sink. Hence, sensor network is exceptionally vital in virtual system (VS) for detecting and forwarding data to intricate physical domain at lesser cost [1]. Though, WSN with restricted and non-revitalized source of energy; maintaining the proficiency of sensor energy has become a serious drawback in forming the topology that disturbs the sensor lifetime with more damage. Therefore, how to reduce energy depletion and maximize the network lifespan is the central focus when new protocol is designed for sensor network. Providentially, it has been the major objectives of topological control [2, 3]. Moreover, technology in topology control is categorized into twofold. One is the power controlling, and the other as the tiered level of topological control.

For the tiered level of topology control [12], normally there can exist in four ways, i.e., cluster techniques [4–6], coupled dominating set [7, 8, 13–15] and spanning-tree techniques [9–11]. By topological control, typically we can get easy topology for a given network, reserving connecting links [16, 17] and the coverage [18–20]. Additionally, the diameter of the topology obtained cannot be increased further [21]. Clustering has proved to be a key method to reduce the energy depletion also extends the lifetime of network. In cluster formation, sensors are convened into clusters; every cluster, leader is elected by single sensor node, called as dominant cluster (DC), and sensors under them is noted as cluster associate (CAs). Every CA records the factors linked to its location and later transmits to respective DCs. As the information arrives by every CA, the DCs cumulatively gather data and transfers to sink.

On one side, as DCs are liable for accepting and accumulating the information by its CAs, transmits the accumulated information to identified endpoints, the energy depletion in DC is very high in relation to CAs. So, to solve this dispute, selecting a suitable cluster head will be the key point while proposing a cluster protocol. Further, if DCs transmits the accumulated data to the sink directly, then long-distant communication takes more energy and that leads to the impulsive expiry of the DCs. Hence, proposing a global cluster with multi-hop routing structure helps to accelerate the data being a main purpose for clustered protocol.

This work also aims on the energy in diverse scenario of WSNs in which sensors are positioned uniformly that presents an advanced approach called prediction-based clustering protocol for energy consumption (PCP-EC) where a different cluster-head selection algorithm is proposed and depends on the in-built predicted energy depletion rate of sensor nodes as the factor to contend to become a DC. Hence, the global energy depletion is optimized by computing the in-built energy depletion. In order to reduce the energy depletion further in DCs, a novel routing inter-cluster structure algorithm is proposed depends on the in-built energy ratio of sensor nodes. This work also provides strong numerical computation toward choosing optimal cluster range and minimizes the energy depletion of the total network that proves to be highly precise and realistic based on hypothetical analysis and simulation trials.

2 Related Work

Xu et al. [1] examine the occasion gathering issue by utilizing the portability of the sink hub and the spatial-worldly relationship of the occasion, for expanding the system lifetime with an ensured occasion accumulation rate. We initially demonstrate the issue as sensor choice issue and that can be tackled in time interval allotted, if worldwide learning of occasions will be accessible as there are no speed requirements on versatile destination. So likewise dissects the outline of a doable development course for portable sink to limit the speed prerequisites for a reasonable framework. An online plan is then proposed to unwind the presumption about worldwide learning of occasions and we demonstrate that the normal occasion accumulation rate can be ensured in principle. Through thorough reenactment on genuine follow information, we show that the system lifetime can be fundamentally broadened, contrasting with some different plans.

Zhuang et al. [2] presents the vitality utilization models that are not depending on the normal separate computation, yet utilizing the probabilistic separation circulations for various sensor hub areas. They are pointing at choosing the ideal matrix estimate by utilizing these more precise models, rather than proposing new effective directing plans, which is out of the extent of this paper. In spite of the fact that these models have higher calculation many-sided quality than those utilizing the normal separation, they are approved by our numerical and reenactment comes about. We have likewise affirmed that it is essential what is more, beneficial to pay the additional exertion: the ideal lattice estimates what is more, the insignificant vitality utilization is both altogether different from those essentially utilizing the normal separation.

This paper also presents few energy consumption scenarios that includes the energy depletion in WSN is recorded with the models for probable distance calculation that was treated as very challenging in the previous works. Also, illustrates how to utilize those models in a commutative manner along with the variable fitting model, by which it is possible to adjust the grid dimension and reduce energy depletion precisely. Third is to utilize those models to prove how to advance the energy rate in WSN by using variable-size dimensions. Zhou et al. [4] present an access-controlled protocol created on elliptical curved cryptography (ECC) for WSN. The entrance conventional control achieves node confirmation and key foundation for new nodes. Not the same as traditional verification strategies in view of the hub character, our entrance control convention incorporates both the hub personality and the hub bootstrapping time into the validation methodology. Consequently, our entrance control convention cannot just recognize the personality of every hub yet in addition separate between old hubs and new hubs. Furthermore, each new hub can build up imparted keys to its neighbors amid the hub validation system. Contrasted and traditional sensor arrange security arrangements, our entrance control convention can shield against most all around perceived assaults in sensor organizes, and accomplish better calculation and correspondence execution because of the more effective calculations in light of ECC than those in view of RSA.

Liu et al. [8] present a broad and reasonable grained review on clustered routing algorithms proposed in the previous study for sensor network. The benefits and goals of clustering are outlined for sensor networks, and design a new taxonomy for WSN clustered routing approaches depending on widespread and thorough attributes of cluster. In specific, systematically a few prominent clustered routing algorithms in WSN are analyzed and compared with various approaches related to taxonomy defined and with important metrics. Yu et al. [13] discuss the topological control as a vital issue in WSN. Considering intrinsic features of flatness, tiered-level structure could achieve the better scalable factor and throughput of WSN. To resolve this issue, a virtual backbone network can be constructed by using a coupled dominating set (CDS) for wireless sensor network [22]. In preceding years, efficient and fast construction of CDS has been the core research challenge in tiered level of control. In this work, a broad survey is done related to CDS and correlated disputes with several models and with definite applications.

3 Proposed System

This paper presents a novel shortest path discovery for area coverage (SPDAC) using prediction-based clustering protocol for energy consumption (PCP-EC) that includes:

- SPDAC approach is used to enhance the trail path of mobile relay node (MRN).
- SPDAC used to solve area coverage problems through the strategic deployment of sensors.
- Presents a new prediction-based clustering protocol for energy consumption.
- PCP-EC approach is used to reduce static node communication overhead.

Combining both approaches together will help in extending lifetime of the network. Simulations are carried out using network simulator tool to analyze the reduction in number of static nodes interactions and other quality of services. The implementation of the approaches is as stated:

3.1 Discover the TSP Tour Distance

For construction of the traveling salesman problem with occurrence between source S and destination D is quite direct. TSP occurrence can be solved by using present estimation approaches or whenever likely to get its precise optimal resolution using present TSP methods, e.g., the Concorde. Representing the TSP resultant tour as Ttsp, an agenda to accumulate information by each of the sensors is found by Ttsp. Though the scheduling visit might eliminate by attaining the overall ideal result in certain scenarios, but it could minimize the locality search, thereby the computational difficulty guarantees greatly to achieve better performance. By adopting this approach, its value is being legalized in [18–20]. Evidently, the absence of data rate restrictions, Ttsp is been always achievable. Finally, the join, pass and swap operations are conducted based on this visit schedule designed by TSP, using the static transmission range d.

3.2 Join-Pass Data Accumulation Spots by Welzl's Algorithm

Huge information accumulated tasks can join if the corresponding nodes are covered by the disk-radius is not greater than d, MRN can complete the tasks assigned near distinct accumulation site. Naturally, few accumulation spots emerge as short tour distance. Progressively by this process, the JPS scheme as shown in Pseudo-code 1, further decreases the number of data accumulation spots that the MRN must visit along the Ttsp found above, and by accepting a decisional form of Welzl's algorithm to join the accumulation spots with proximity sensor nodes onto a fresh information accumulation spot when necessary.

Join-Pass Swap (JPS) Algorithm

Pseudo-code 1: *(S: the group of sensor nodes; d: communication distance).*

1: Tjoin $\leftarrow \Phi$

2: find the TSP tour distance for S and D

3: Ttsp $= (l_0 + l_1 + l_2 \ldots + l_n)$

4: forall l_i $(i = 1, 2, 3 \ldots, n - 1)$ *do*

5: compute the highest $j(i \le j \le n)$ *by the Welzl decisional algorithm, where every positions in* $\{l_0 + l_1 + \ldots l_j\}$ *could be enclosed by a encircling disk with circumradius greater than d, with midpoint ci;*

6: end for

7: compute the JoinSet(i) with the highest cardinality, p $\leftarrow$ |*JoinSet (i)*|*;*

8: while p > 1 do

9: forall lj ϵ *JoinSet (i) do*

10: delete JoinSet(j); end for

11: replace them with ci in Ttsp

12: forall lj still inside Ttsp do

13: update JoinSet(j) end for

14: repeat step 7

15: end while

16: Tjoin $\leftarrow$ *Ttsp*

17: return Tjoin

The Welzl's algorithm [22] calculates the minimum encircling disk with limited set of nodes over the region in a specific time frame and yields circum-radius with midpoint of the encircling disk. This algorithm is considered differently, to identify information accumulation spot to encompass more sensors as likely inside a circum-radius of *d*. The Welzl's algorithm for taking decision is shown in Pseudo-code 2 that yields the minimum encircling disk for a specified subset of sensors if the circum-radius is not greater than *d*, else false otherwise.

Decision Tree Algorithm

Pseudo-code 2: for Welzl's Decisional

1: circumradius $\leftarrow \infty$

2: (circumradius, midpoint) = Welzl (S′)

3: if circumradius > d, then

4: circumradius is false;

5: then

6: yield circumradius and midpoint

7: endif

3.3 Cluster Setup Phase

The development of cluster stage is divided into three parts where first one the sensor node in-built energy depletion prediction stage, where time period is $T1$; next is the cluster-head contention stage, where time period is $T2$; and last is the cluster formation stage, where time period is $T3$.

3.3.1 Prediction-Based Energy Depletion Stage

During the selection of DCs, PCP-EC requires predicting the in-built energy depletion ratio for a sensor and finds it to detect the capability to become a CA. Every sensor node then transmits *Node_Msg* in the radius of R_a using both the values: the node ID along with its current energy E_{cur}. In similar interval of time, it gets the *Node_Msg* through neighboring nodes, then each sensor node calculates its average displacement to the neighboring sensors depending on the received strength of signal; generally, every node can expect and calculate its in-built energy depletion rate.

$$\text{ratio (Si)} = \frac{\sum_{i=1}^{n+1} E_{dep}(\text{Si})}{\sum_{i=1}^{n+1} E_{pres}(\text{Si})} \tag{1}$$

$$\text{ratio (Si)} = \{6(n+1)\}E_{elec} + (n+1)\left(E_{sen} + E_{dep}\right) + \left\{(21+n)\varepsilon f s R_a^2\right\} / \sum_{i=1}^{n+1} E_{pres}(\text{Si}) \tag{2}$$

where $E_{pres}(\text{si})$ represents the present sensor node energy si, the $E_{dep}(\text{si})$ represents the depletion of sensor node energy si and the amount of n nodes in the radius of R_a of si.

3.3.2 Cluster-Head Contention Stage

Once $T1$ terminates, PCP-EC initiates the cluster-head contention stage. For each sensor si, at this stage, if accepted no DC_Msg is transmitted when t_i time expires, it sends a DC_Msg in the radius of R_a to promote that it will become DC. An arbitrary value of Vr is presented to decrease the possibility that both nodes broadcast DC_Msgs at the similar period. Further, every node in the network only transmits the DC_Msg in the radius of R_a. So the possibility is that more sensor nodes in the neighboring contending region can have low expectation for equal time constraint.

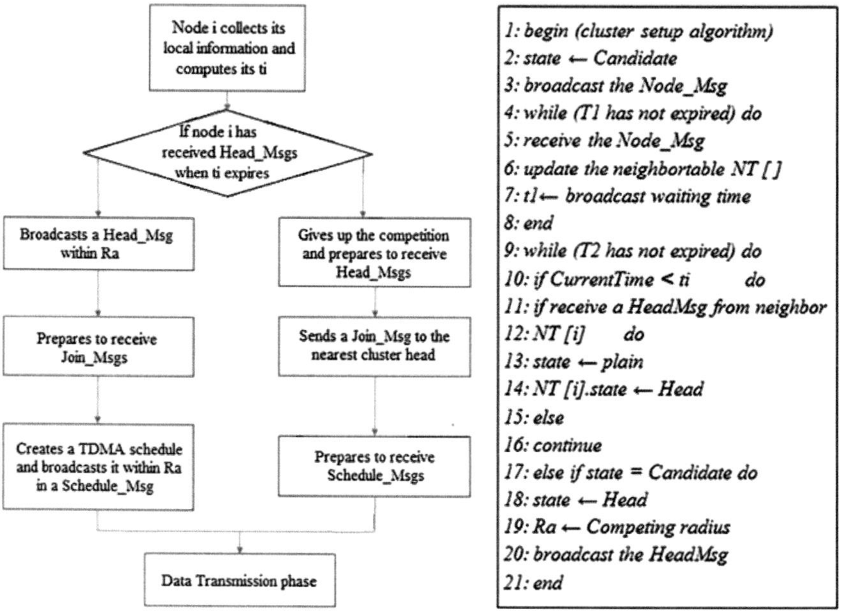

Fig. 1 Clustering development and formation stage

3.3.3 Clustering Formation Stage

The last sub-phase of the cluster development stage is the formation stage. Each sensor node selects the adjacent DC and sends a Join_Msg that holds node ID and the present energy E_{pres}. Every DC produces a time division list as per the accepted Join_Msg and transmits the list scheduled to all the member clusters by transmitting a Schedule_Msg. Every clustered node comprises of the set of nodes under the region specified near the CH. Figure 1 describes the procedure for the cluster development stage.

4 Performance Analysis

To evaluate the performance of the planned approach, consider a 400 × 400 scarce sensing region with 50–100 nodes distributed uniformly in random fashion. The amount of information to be transmitted through the MRN is 4 KB for every sensor nodes, and the continual speed for MRN is 1.1 m/s.

4.1 Node Energy and Network Lifetime

Network lifetime will check for network period in heterogeneous eventualities is greater than the homogenized eventualities. Further explanation is on PCP-EC that takes the native energy depletion magnitude relation of sensor nodes under consideration once choosing the CHs and routing nodes that takes complete benefit of the higher-energy source with cheap cost sensors under diverse circumstances; hence, the CH nominated per area unit forever the best, and therefore the lifespan of network period will be extended. Thus, PCP-EC is much suitable for each the diverse and normalized eventualities.

4.2 Simulation Parameters

Table 1 represents the values of network simulation parameters.

Table 1 Network simulation parameters

Parameter	Values
Region or boundary	400 m $\times$ 400 m
Total nodes	50–100
Total source nodes	4–12
Sensing time per channel	5 ms
Channel switching time	80 μs

4.3 Energy Consumption

The results showed in Fig. 2 represent the comparison of energy consumption. Here, the clustering-based scheme is compared with non-clustering-based JPS scheme shows that the prediction-based clustering improvises the network lifespan and saves the energy of a system. All source nodes route their energy to their CH and only CH will participate in transferring the energy to the MRN by sharing the tour path. Hence, the tour length of MRN can be minimized and also covering all the locations of the area in network.

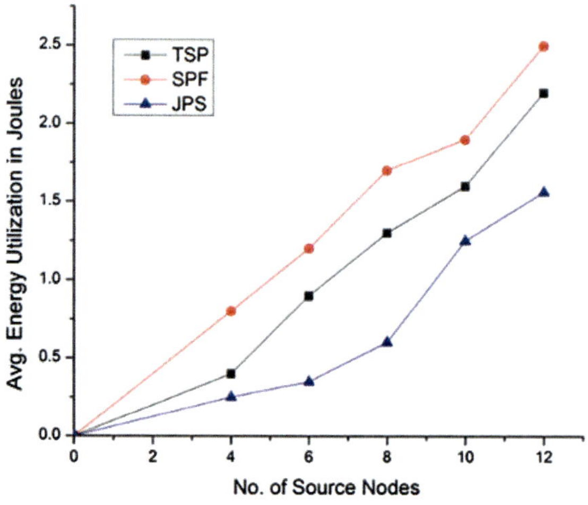

Fig. 2 Average energy consumption of clustered with JPS algorithm

4.4 Transmission Time

Figure 3 shows the average time taken by the network to transfer the collected information from source nodes to MRN. In this graph, source nodes nearby will find the optimal point to inform MRN to reach that point to collect from each of those nearby source nodes. Finding of such tour points will increase if nodes are spread across entire network. With clustering, all such source nodes send their information to DC node. DC node will aggregate the information and share its position to MRN. With this approach, the number of tour points for MRN will be reduced to the number of clusters. This will in turn reduce the total time taken to traverse the entire network and collect information from all source nodes.

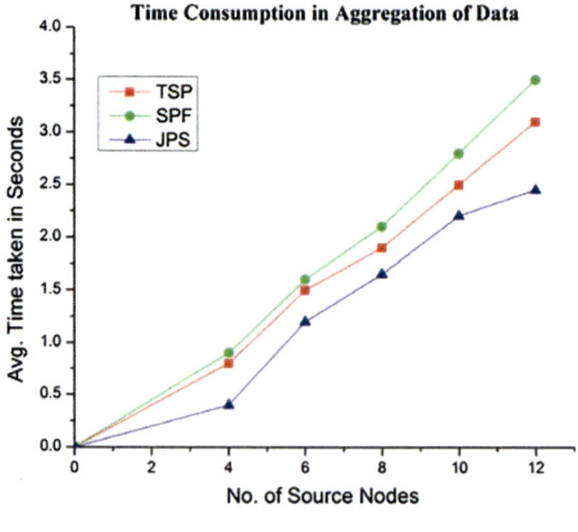

Fig. 3 Average time taken of clustered with JPS algorithm

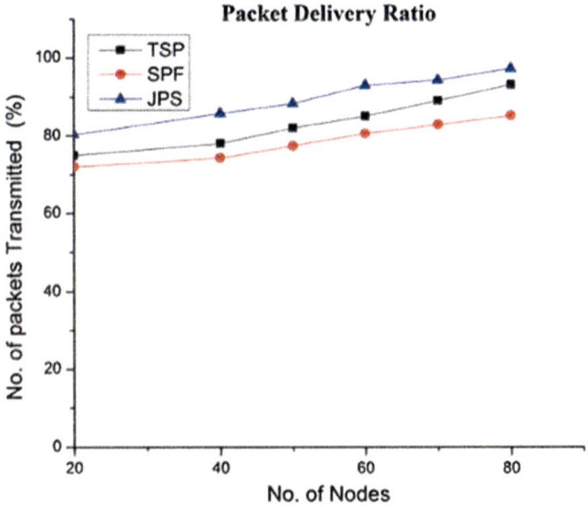

Fig. 4 Rate of successful packets delivered

CHs are the direct participants for information transfer with MRN in the proposed JPS algorithm. Selection of the CHs is always with high energy and it is also based on well condition of existence in the network. Hence, the failure of packets delivery is very minimal. Figure 4 shows the packet delivery ratio of the proposed algorithm JPS against the TSP and SPF.

As all nodes will participate in existing algorithms for information transfer to MRN, there will be high chances of packet loss either due to MRN movement from the communication range or due to different transmission range of sender nodes. Number of participants to deliver packets to MRN is fixed and is the sum of clusters in the area specified of a network. Therefore, information being captured is only from CH inside the network. CH will take the information from their source nodes and aggregates it as obtained in Fig. 5. As the number of senders fixed, the transmission channel can also be fixed for long time and collect all the information while CH is sending. This leads to more information collection by MRN per each CH. Whereas in SPF and TSP, each source node participates directly with MRN, and there will be chances that packet collision occurs within them. Hence, the information collection rate will be higher in JPS compared to the other two algorithms.

5 Conclusion

Area coverage is a key topic of study in WSN where the collection of information from all source nodes is one of key factor. Energy is a major parameter that has to be carefully optimized to cover all area traversed paths. Hence, the shortest path approach is defined to optimize the tour length of the MRN. Also, this trajectory path is further extended to prediction-based clustering scheme to reduce the path of MRN even further. Results show that area coverage can be achieved with better energy efficiency with join-pass-swap (JPS) algorithm.

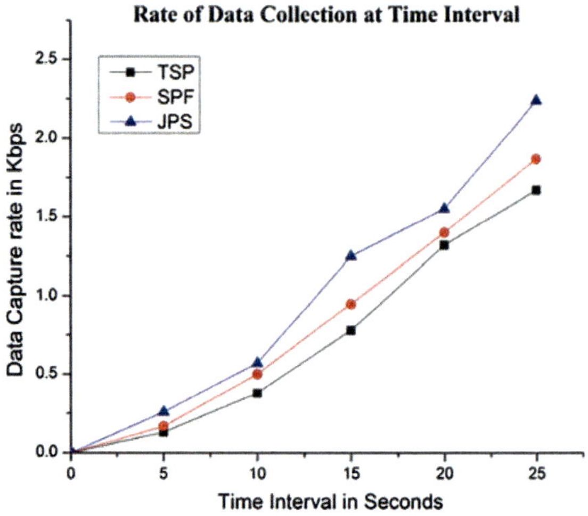

Fig. 5 Rate of data captured at regular interval

References

1. Xu X, Luo J, Zhang Q (2010) Delay tolerant event collection in sensor networks with mobile sink. In: Proceedings of IEEE INFOCOM
2. Zhuang Y, Pan J, Cai L (2010) Minimizing energy consumption with probabilistic distance models in wireless sensor networks. In: Proceedings of IEEE INFOCOM
3. Panichpapiboon S, Ferrari G, Tonguz O (2006) Optimal transmit power in wireless sensor networks. IEEE Trans Mob Comput 5(10):1432–1447
4. Zhou Y, Zhang Y, Fang Y (2007) Access control in wireless sensor networks. J Ad Hoc Netw 5(1):3–13
5. Zeng W, Sarkar R, Luo F, Gu X, Gao J (2010) Resilient routing for sensor networks using hyperbolic embedding of universal covering space. In: Proceedings of IEEE INFOCOM
6. Luo J, Hubaux J (2005) Joint mobility and routing for lifetime elongation in wireless sensor networks. In: Proceedings of IEEE INFOCOM
7. Wang W, Srinivasan V, Chua K (2008) Extending the lifetime of wireless sensor networks through mobile relays. IEEE/ACM Trans Netw 16(5):1108–1120
8. Liu X (2012) A survey on clustering routing protocols in wireless sensor networks. Sens J 12:11113–11153
9. Luo J, Huabux J (2010) Joint sink mobility and routing to maximize the lifetime of wireless sensor networks: the case of constrained mobility. IEEE/ACM Trans Netw 18(3):871–884
10. Li J, Cheng S, Gao H, Cai Z (2014) Approximate physical world reconstruction algorithms in sensor networks. IEEE Trans Parallel Distrib Syst 25:3099–3110
11. Labrador MA, Wightman PM (2009) Topology control in wireless sensor networks: with a companion simulation tool for teaching and research. Springer Science and Business Media, New York, USA
12. Santi P (2005) Topology control in wireless ad hoc and sensor networks. Wiley, Chichester, UK
13. Yu J, Wang N, Wang G, Yu D (2013) Connected dominating sets in wireless ad hoc and sensor networks. Compr Surv Comput Commun 36:121–134
14. Underwater Robots Track Oil and Ocean Life (2013) http://spectrum.ieee.org/podcast/robotics/industrial-robots/underwater-robots-track-oil-and-ocean-life
15. Todd M (2007) A different approach to sensor networking for SHM: remote powering and interrogation with unmanned aerial vehicles. In: Proceedings of sixth international workshop structural health monitoring (IWSHM)

16. Naeimi S, Ghafghazi H, Chow C, Ishii H (2012) A survey on the taxonomy of cluster-based routing protocols for homogeneous wireless sensor networks. Sens J 12:7350–7409
17. Guo L, Ai C, Wang X, Cai Z, Li Y (2009) Real time clustering of sensory data in wireless sensor networks. In: Proceedings of the performance computing and communications conference (IPCCC), Scottsdale, AZ, USA
18. Du D, Wan P (2013) Connected dominating set: theory and applications. Springer Science and Business Media, New York, NY, USA
19. Cheng X, Huang X, Li D, Wu W, Du D (2003) A polynomial-time approximation scheme for minimum connected dominating set in ad hoc wireless networks. J Netw 42:202–208
20. Li Y, Wu Y, Ai C, Beyah R (2012) On the construction of k-connected m-dominating sets in wireless networks. J Combin Optim 23:118–139
21. Xiong N, Huang X, Cheng H, Wan Z (2013) Energy-efficient algorithm for broadcasting in ad hoc wireless sensor networks. Sens J 13:4922–4946
22. Ding M, Cheng X, Xue G (2003) Aggregation tree construction in sensor networks. In: Proceedings of the 2003 IEEE 58th vehicular technology conference, Orlando, FL, USA, pp 2168–2172
23. Li Z, Li M, Wang J, Cao Z (2011) Ubiquitous data collection for mobile users in wireless sensor networks. In: Proceedings of IEEE INFOCOM

Smart Mirror Using Raspberry Pi for Intrusion Detection and Human Monitoring

Raju A. Nadaf and Vasudha Bonal

Abstract The Smart Mirror is a system in which the normal mirror is converted to behave like a smart device. Proposed Smart Mirror is designed using Raspberry Pi 3 model and by using a touch-enabled screen. The designed system is capable of acting like a regular mirror in case of normal mode of operation, and it acts like a Smart Mirror in a triggered mode of operation. The Smart Mirror thus designed is an interactive system which is capable of operating in three modes. The system can be operated interactively by accepting one of the type of "command mode," namely voice-based commands, touch- and mobile-based controls. The system is designed to display weather, temperature and latest news on the mirror. The system is primarily designed for the purpose of human monitoring and also intrusion detection system. The proposed design is thought of a bundle of package, which not only just displays information over screen, but also can be used for providing security. The system is built using hardware units like Raspberry Pi 3 model, microphone, touch screen, mobile device, camera and passive infrared sensor (PIR) sensors and programming with Python. The intrusion detection is done using simple frame difference approach, and human monitoring is implemented using Yolo machine learning technique with OpenCV.

Keywords Smart Mirror · Raspberry Pi · Intrusion detection · human monitoring

---

R. A. Nadaf (✉) · V. Bonal
Department of Computer Science and Engineering,
Basaveshwar Engineering College, Bagalkot, India
e-mail: raj.enggs@gmail.com

V. Bonal
e-mail: vasudha_125@rediffmail.com

N. N. Chiplunkar and T. Fukao (eds.), *Advances in Artificial Intelligence and Data Engineering*, Advances in Intelligent Systems and Computing 1133,
https://doi.org/10.1007/978-981-15-3514-7_102

1 Introduction

In the present world, the advancement in technology has made almost every equipment as smart device. Ranging from household things to most advance electronic gadgets, almost all are becoming smart. The smart characteristics of present devices may be due to artificial intelligence (AI), sensors and prediction systems, etc. So having this thought in mind, we are proposing a Smart Mirror which is not only capable of displaying customized information on the display screen but also act smartly and provide security when needed. Usually, mirrors are used for grooming up by people. At an average per day, at least 28 min are being spent for grooming up in front of the mirror, by a person. The basic idea is to make use of this time to keep a person updated with latest news, weather, date, time, calendar and other updates. Usually in present busy life, we do not get explicit time to check out for news and other updates. Above all, there is no availability of bundle of package that displays all such information in a common screen. Further, the system proposed is interactive in nature; hence, the user even while grooming up can give voice commands to get required and related information on screen, keeping his/her hands free. There are related products available in market, but the main difference lies in the usability of the product. The available products are mostly passive in nature with little interactivity implemented in them. The primary task of such a system is to display information on the mirror, and they accept either voice commands or touch commands or commands from mobile. The screens are being designed using a liquid crystal display (LCD) or light-emitting diode (LED) monitor along with the two-way acrylic sheet and a mirror. The design proposed is capable of working with voice commands, touch commands as well as mobile commands. The proposed work not only acts as means of providing information but also provides security and monitoring. The system can be used for intrusion detection. As soon as the human activity/motion/movement is being detected by the PIR sensor, the camera connected to Raspberry Pi is activated. Image is being captured by the camera connected to Raspberry Pi. Image processing technique such as simple frame difference approach is used to detect the intrusion. The intrusion will be communicated to the owner of the mirror through alert message along with the picture of an intruder. The system can also be used for human monitoring. Nowadays, the families are getting smaller in size. Hence, it becomes difficult to monitor children, elders and patients. Because of busy schedule and working parents, it becomes difficult task to monitor kids after they return from school. The designed system is capable of monitoring human within a coverage range of camera. If the human moves out of the sight of the camera, then an alert message will be sent to the owner of the Smart Mirror. Whenever the owner of Smart Mirror is in front of Smart Mirror, voice commands can be used to push the Smart Mirror either in intrusion detection mode or human monitoring mode explicitly, and mobile commands can be used for the same task whenever the owner is not in front of mirror. The schematic appearance and diagram of Smart Mirror are given in Figs. 1 and 2, respectively.

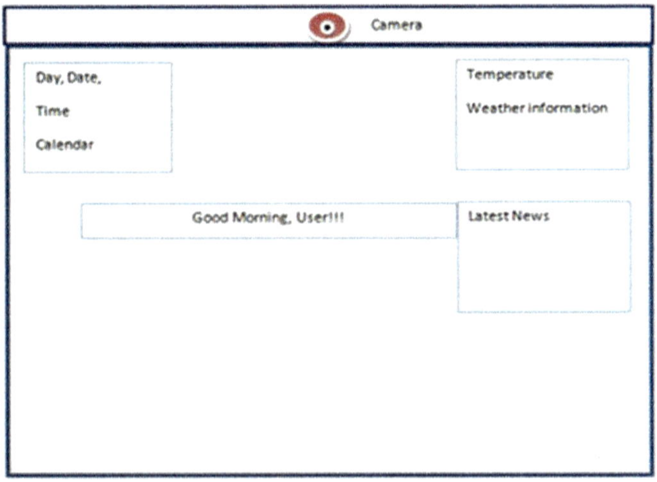

Fig. 1 Schematic appearance diagram of Smart Mirror

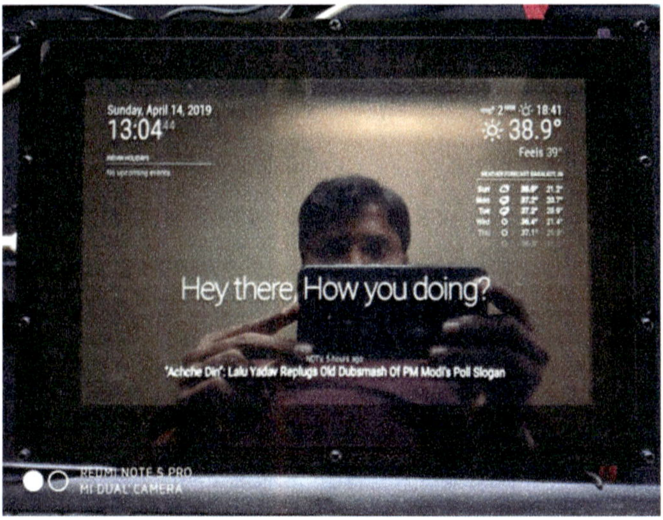

Fig. 2 Smart Mirror

2 Related Work

Several related works have already been taken up in this area. The Smart Mirror proposed and implemented so far are having variety in terms of hardware and mode of operation. Intelligent mirror which is capable of accepting voice command via the microphone and has been built with Raspberry Pi microcontroller, LED monitor and acrylic mirror displays the weather, time and location information on the screen

[1]. Smart Mirror built with Raspberry Pi and multi-control unit (MCU) can display weather and latest news updates on the screen. Humidity and temperature sensors are incorporated. IoT is implemented using cloud [2]. Futuristic multimedia-based Smart Mirrors are designed which accept voice commands. The design is based on the concepts of artificial intelligence. System alerts about weather and suggests user according to the weather. For example, if it is cold and cloudy day, we can see a message on the mirror saying "Please wear jacket today" [3]. Some of the Smart Mirrors are having Webpage-based interface and are customizable. These are operated using voice commands and make use of APIs of various Web sites. The proposed mirror makes use of Google Assistant and stores user details in the database [4]. Some mirrors designed can be used as weight and fitness trackers. The authentication is provided using face recognition. GPS navigation, Bluetooth connectivity and wireless communication are added features. Sonus technology is used for improved communication [5]. Some mirrors designed use Hermione 1.0, which is an extension of magic mirror. The platform provides the user with easy installation of a Smart Mirror for domestic use. System can be used as a home assistant and is voice-based [6]. Other systems proposed the work in two modes, viz. normal mode and smart mode. System is developed using Python and JavaScript programming tool such as Node.js. It is a voice-command-based Smart Mirror [7]. Some Smart Mirrors implemented with Raspberry and SMT32F030CT8T6 microcontroller as core controlling chips. These are voice-enabled, and special speech synthesis module is implemented using SYN6288 chip [8]. Smart Mirror is designed for theft detection in a home environment. PIR sensors used human motion detection, and the camera captures information and stores in drop box. DHT22 chip is used for theft detection, and VNC viewer is used for mobile control [9]. A comparative study of Smart Mirrors is given, and a voice-based Smart Mirror is proposed. It is AI-based system which supports human gestures and face detection. Machine learning techniques are used for making system more responsive [10]. Health monitoring Smart Mirrors are designed to detect health issues. System makes use of posture analysis algorithm (PAA) to analyze postures of human to find any changes in postures over a period of time [11]. Commercial and home-usage Smart Mirrors are designed to capture real-time data on the screen. System is voice-based and makes uses of Ambient Artificial Intelligence (AmI) technique [12]. Multi-user Smart Mirrors are designed as commercial products which are based on RFID access of employees. The device has a personalized user interface [13]. The systems proposed and implemented till now are based on different algorithms and hardware components and are primarily designed as systems which provide a means of information. The proposed system is different from the existing ones as it works interactively by providing security in real-time situation. In the next section, the issues and challenges of the proposed system are explained.

3 Issues and Challenges

The proposed model is designed for voice commands, touch commands and mobile commands. The device is primarily designed for intrusion detection and human monitoring. All these features are implemented using Raspberry Pi. The Raspberry is a microcontroller device with limited performance capacity. The synchronization of all these features into Smart Mirror is one of the challenges. Power issues, delay in content delivery and corrupting of SD card are the common technical issues with the Raspberry Pi device. Providing accuracy in intrusion detection and human identification are true challenges in a real-time scenario. In case of dynamic background in the image processing, the intensity of challenge will rise high. The non-technical issues include cost and the durability of the hardware devices. Use of the freeware and open-source software can reduce the cost of the system. The interfacing of the hardware and software is actual challenge. The proper knowledge of hardware devices and sensors is mandatory for providing better and sophisticated solutions. The processing video frames using Yolo need smart programming, because the Raspberry Pi is a small device with limited processing capabilities.

4 Proposed Model

The block diagram for the proposed model shown in Fig. 3 consists of Raspberry Pi as heart of the system. The mirror state is a synchronization unit. It is software component which is mainly responsible for synchronizing all the components connected to the Raspberry Pi. All the commands that are issued to the Raspberry Pi are passed through the mirror state. The primary task is to check whether the command can be executed at that moment or not. If already one command is under execution and the second command is being issued, then the mirror state decides whether the second issued command can be executed or not.

The striking features of the system are intrusion detection and human monitoring. The block diagram for the same is depicted in Fig. 4.

As soon as an intrusion is confirmed, then an alert message will be sent to the owner of the mirror to his/her registered mobile along with the picture of the intruder. In case of human monitoring, if the person under monitoring moves out of the camera sight, then the message will be communicated to owner of the Smart Mirror through an alert message. The system accepts voice commands, touch commands and can be controlled through mobile device. The PIR sensors are used for human movement detection, and in case of human movements, the Raspberry Pi needs to trigger the camera. Then, the intrusion is actually detected through simple frame difference approach using image processing. In case of human monitoring, the machine learning techniques are used. The Yolo with OpenCV technique for human detection is used whenever the person under monitoring moves out of the camera sight. Such a human

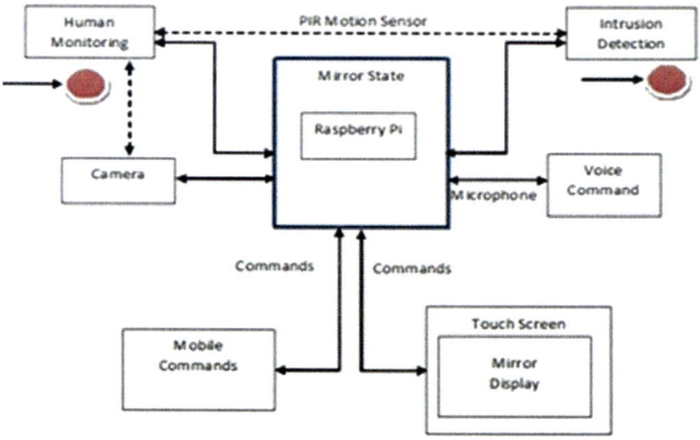

Fig. 3 Schematic appearance diagram of Smart Mirror

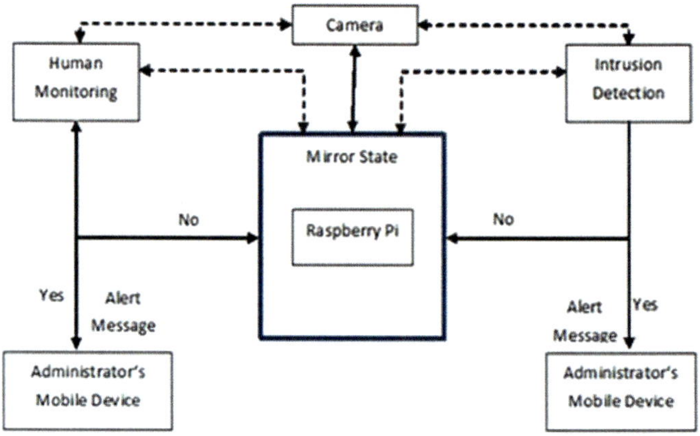

Fig. 4 Block diagram for intrusion detection and human monitoring

monitoring can be useful to monitor children, elders, patients, prisoners admitted to hospital, workers in gold shop and Prisoners in jail etc.

5 Design and Implementation

The design and implementation of the proposed Smart Mirror requires both hardware and software components. The Raspberry Pi is the main component of the system. The configuration details of the same are shown in Fig. 5.

Fig. 5 Raspberry Pi 3

5.1 *Raspberry Pi 3 Model*

The Raspberry Pi 3 model is having a CPU Quad-core processor with 64-bit ARM Cortex A53 with a speed of 1.2 GHz, has a GPU with 400 MHz and has a video-core IV multimedia, with an internal memory of 1GB with a speed of 900 MHz. It has four USB ports, a HDMI and 3.5mm audio jack. It has Ethernet of 10/100 Mbps and wireless WAN 802.11n.

5.2 *Camera*

Camera is the most important component after Raspberry. The camera is required to capture the video that is needed for intrusion detection and human monitoring. The configuration details of the same are shown below. The camera is a high-quality camera with 8 megapixel clarity and has high resolutions, which is specially designed for add-on board for Raspberry Pi. Camera is capable of taking 3280 × 2464 pixel static images. The camera is shown in Fig. 6.

Fig. 6 Raspberry Pi camera

Fig. 7 Raspberry Pi-compatible microphone

5.3 Microphone

The microphone is required to give voice commands to the Smart Mirror. The sensitivity ranges from -47 ± 4 db and has sensitivity reduce -3 db 1.5 V 3. The working voltage is 4.5 V with a frequency response of 100–16 kHz. The cable length can be approximately 2 m. The microphone is shown in Fig. 7.

5.4 Raspberry-Compatible Touch Screen

This is the display unit of Smart Unit. The navigation can be done using touch over the screen. Raspberry Pi 3 screen IPS—SunFounder 10.1 in. HDMI IPS LCD monitor is having a resolution of 1280×800 and 10.1 in. sized screen which is a IPS LCD having dimension $9.8 \times 3 \times 7.24$ in.

5.5 PIR Sensors

Motion detection is done by making use of PIR sensors. The presence of human beings can be detected by the passive infrared (PIR) sensors in a given proximity. The range of the sensor may be from 5 to 10 m.

5.6 Mobile Device

Mobile device is needed to control the Smart Mirror from a distant location also. The mobile can be any Android-based mobile with 4G Internet facility and Wi-Fi capability. The VNC client should be installed in the mobile. The mobile can be used to push the system either for intrusion detection or human monitoring.

5.7 SD Card

This is the storage unit of Raspberry Pi. A micro SD memory card of Class 10 quality having storage capacity of 32GB is required.

5.8 Softwares

Python: Python is an interpreted high-level language, which is general-purpose programming language. The software is required for coding the proposed system.

Raspbian OS: It is the Operating System (OS) required for Raspberry Pi. It is an OS which is available for free to download and use. It is based on Debian Linux. It works efficiently with the Raspberry Pi system.

Yolo with OpenCV: It is used for human detection in human monitoring. YOLO (You Only Look Once) is a method for object detection. It is the algorithm behind how the code is going to detect objects in the image.

VNC viewer: It enables the mobile device to control the Smart Mirror even from distant location. One computer or mobile can be remotely controlled by another computer using VNC. Input to the VNC viewer can be the keyboard and either mouse or touch events to VNC server. In response, the VNC viewer receives updates to the screen . It is possible to see the desktop of the Raspberry Pi inside a window on our computer or mobile device. It is possible to control it through Raspberry Pi itself.

The system consists of four modules, namely Login Module, Input Module, Intrusion Detection Module and human monitoring Module.

5.9 Login Module

The following figure shows the Login Module. As soon as the system boots up, the authentication process is taken up. It is a username- and password-driven authentication. A successful authentication allows the user to access the Smart Mirror for command execution and navigation. Figure 8 shows the Input Module.

5.10 Input Module

Figure 9 shows the input validation unit. The system accepts input through this module. The Input Module of the system is shown in Fig. 10.

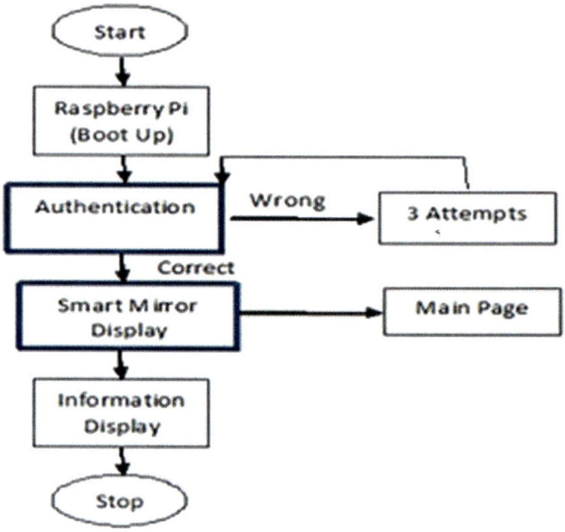

Fig. 8 Login module

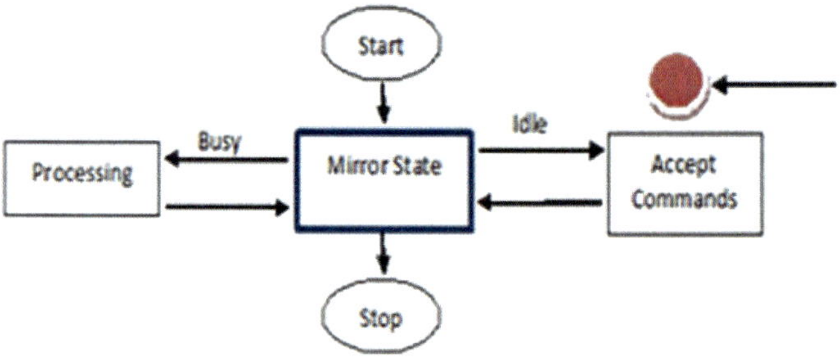

Fig. 9 Input validation unit

Input validation Unit is used to validate the input command. If Raspberry Pi is busy in processing previous command, then this module prompts required message to the user.

5.11 Intrusion Detection

The voice command or mobile command is required to push the system in intrusion mode of operation. The intrusion can be detected using a camera fitted on the Smart Mirror. The PIR sensors are used to detect motion. As soon as the motion is sensed by PIR, the mirror state activates Raspberry Pi which in turn triggers the camera. Now camera takes up the video. The video needs to be converted into frames for

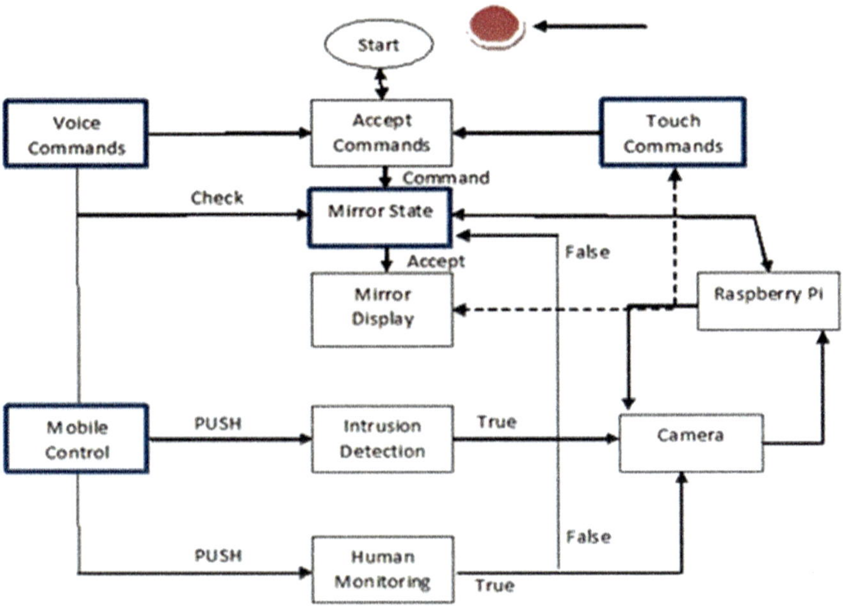

Fig. 10 Input module

processing. As soon as the system starts up, the first frame of camera video will be taken as reference (background) frame. The subsequent frames of the camera can be treated as the foreground (processing) frames. When PIR sensor activates the mirror state and Raspberry, the foreground frames are now compared with the reference frames. After subtraction from the reference frame, if there is any addition of new component or object into the foreground frame, we can detect the presence of intrusion. The simple frame subtraction approach is used for intrusion detection. Figure 11 shows the details.

The frame difference technique can be used to create reference frame dynamically by comparing adjacent frames. Any new changes in the background image are combined with reference frame to form new reference image. Now this new reference image is taken as background image and compared with the foreground image (image under processing) for the intrusion detection; Fig. 12 shows details.

The effective method of intrusion can be produced by dynamically changing the reference frames as per the deployment conditions. This is useful when some part of reference frame also involves moving objects. Once the intrusion is confirmed, the owner of the Smart Mirror will receive an alert message on his/her registered mobile along with the photograph of intruder. The formulae for frame difference calculation are as below in Fig. 13.

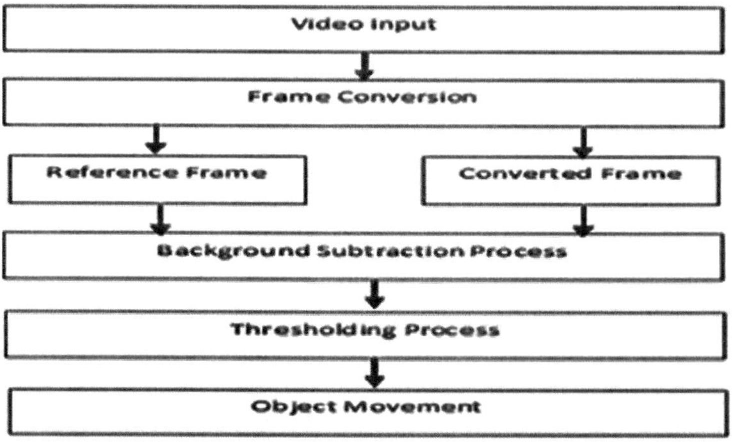

Fig. 11 Background subtraction technique for intrusion detection

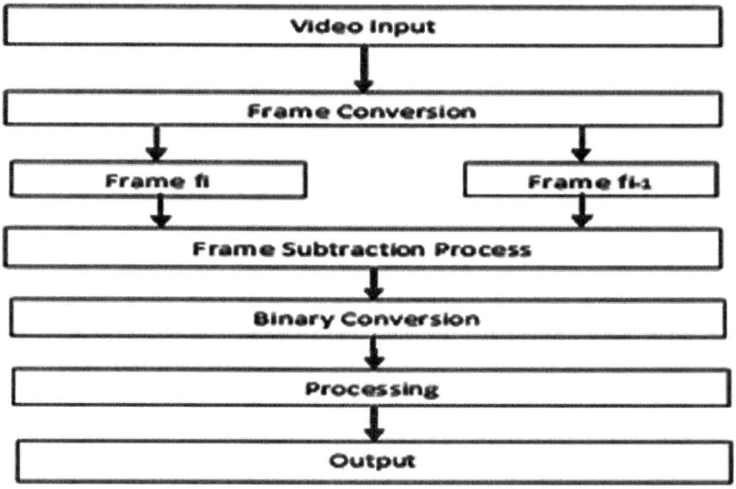

Fig. 12 Frame difference technique with binary conversion for intrusion detection

Fig. 13 Frame difference calculation formulae

$$D_{i-5} = |f_i - f_{i-5}|$$

$$D_{i+5} = |f_i - f_{i+5}|$$

$$MOV(x,y) = |RBI(x,y) - f_i(x,y)|$$

$$D(x,y) = \begin{cases} 1 & target \quad MOV(x,y) > T \\ 0 & background \quad MOV(x,y) \leq T \end{cases}$$

DK is differential. It is the difference between kth frame image fk with the k-1th frame image fk-1.

5.12 Human Monitoring

The most basic work here is to identify the human in the frames of a running video. Human detection is the first part of the Human Monitoring process. The voice command or mobile command is required to push the system in Human Monitoring mode of operation. The Human Monitoring can be done by using the Yolo with the OpenCV. Python coding is also being used. When the human under observation moves out of the camera sight, an alert message will be sent to the owner of the Smart Mirror on his/her registered mobile number. YOLO is said to be a fully convolutional network (FCN) since convolutional layers have been used here. There can be approximately 75 convolutional layers used along with skip connections and upsampling layers. No pooling is used, and to down sample the feature of maps, the convolutional layer with a stride of 2 is used. Being a FCN also YOLO is not dependant on the size of the input image. The network has to down sample the image by a factor equals to the values called stride of the network. For example, if the stride of the network is 32, then an input image of size 416 × 416 will yield an output of size 13 × 13. The classifier/regressor accepts the features that are learned by the convolutional layers. This process does the actual detection prediction. The actual prediction in YOLO is done by a convolutional layer which uses 1 × 1 convolutions. The output produced itself is a feature map. Due to the use of the 1 × 1 convolutions, the size of the prediction map will be absolutely same as the size of the feature map. In prediction map, each cell will predict a fixed number of bounding boxes in case of YOLO v3. The following Fig. 14 shows the bounding box equations.

b_x and b_y are the x, y center coordinates, b_w and b_h are width and height of the prediction, respectively. The terms t_x, t_y, t_w and t_h are the network outputs. The terms c_x and c_y are the top-left coordinates of the grid. The terms p_w and p_h are anchor dimensions for the box.

The actual object detection process can be shown in Fig. 15.

Figure 16 gives the details of the object detection using YOLO technique.

Fig. 14 Bounding box equations

$$b_x = \sigma(t_x) + c_x$$
$$b_y = \sigma(t_y) + c_y$$
$$b_w = p_w e^{t_w}$$
$$b_h = p_h e^{t_h}$$

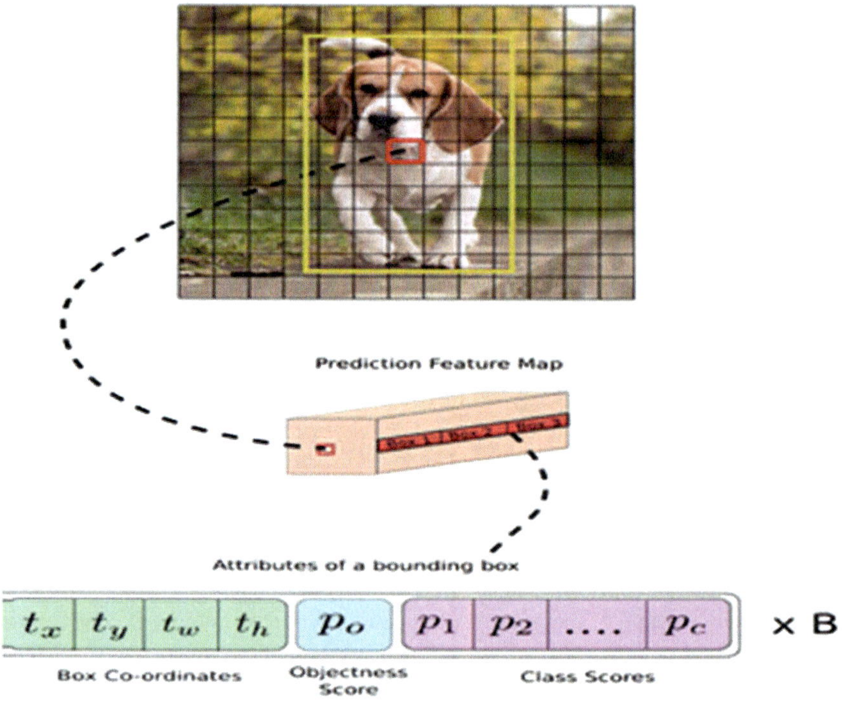

Fig. 15 Actual detection prediction process

Fig. 16 Object detection process by YOLO technique

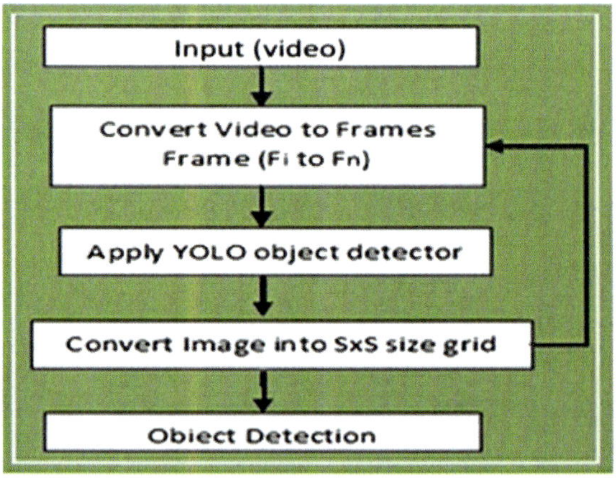

6 Conclusion and Future Work

The main theme of the proposed work is to design a product bundled with maximum possible features. The system is not just devised as a means of information provider but also an interactive system which can actively be used for intrusion detection. The system can be used as security system. Human motoring is of prime importance where we need to monitor a person. There is no absolute need of personal monitoring. Smart Mirror can be used for the same. The system can be extended as a commercial product. There is a scope for future work in this proposed system by adding artificial intelligence. The same mirror can be extended to control the home appliances and lighting. Hence, we can be able to control home appliances, even when we are getting ready for the day. In order to increase the level of security, face detection can be used for authentication.

References

1. Kumbhar PY, Mulla A, Kanagi P, Sha R (2018) Smart mirror using Raspberry PI. Int J Res Emerging Sci Techno 5(4)
2. Jagdish AP, Sonal TS, Sangaleharshada D, Dokhale A (2018) A review paper design and development of a smart mirror using Raspberry Pi. Int J Eng Sci Invent (IJESI) 7(4): 40–43 (Ver. I)
3. Kiran SR, Kakarla NB, Naik BP (2018) Implementation of home automation system using smart mirror. Int J Innov Res Comput Commun Eng 6(3)
4. Ajayan J, Santhosh Kumar P, Saravanan S, Sivadharini S, Sophia R (2018) Development of smart mirror using Raspberry-Pi 3 for interactive multimedia. In: 12th International conference on recent innovations in science and management (ICRISEM)
5. Divyashree KJ, Dr. Vijaya PA, Awasthi N (2018) Design and implementation of smart mirror as a personal assistant using Raspberry PI. Int J Innov Res Comput Commun Eng 6(3)
6. Assudani M, Kazi AS, Sherke PO, Dwivedi SV, Shaikh ZS (2018) Hermione 1.0—A voice based home assistant system. In: National conference on advances in engineering and applied science (NCAEAS)
7. Kamineni BT, Sundari PA, Suparna K, Krishna Nayak R (2018) Using Raspberry Pi to design smart mirror applications. IJETST 05(04):6585–6589
8. Sun Y, Geng L, Dan K (2018) Design of smart mirror based on Raspberry Pi. JETST 05(04):6585–6589
9. Lakshmi NM, Chandana MS (2018) IoT based smart mirror using Raspberry Pi. Int J Eng Res Technol (IJERT) 6(13)
10. Mittal1 DK, V Verma, Rastogi R (2017) A comparative study and new model for smart mirror. Int J Sci Res Res Pap Comput Sci Eng 5(6):58–61
11. Cvetkoska B, Marina N, Bogatinoska DC, Mitreski1 Z (2017) Smart mirror e-health assistant—Posture analyze algorithm. IEEE EUROCON
12. Khanna V, Vardhan Y, Nair D, Pannu P (2017) Design and development of a smart mirror using Raspberry PI. Int J Electr Electron Data Commun 5(1). ISSN 2320-2084
13. Gomez-Carmona O, Casado-Mansilla D (2017) SmiWork: an interactive smart mirror platform for workplace health promotion. Int J Electr Electron Data Commun 5(1). ISSN: 2320-2084

A Home Security Camera System Based on Cloud and SNS

Takuya Egashira, Lin Meng, and Hiroyuki Tomiyama

Abstract Security network cameras are widely deployed not only in business offices and public facilities but also in general households. Most of commercial network camera systems come with dedicated application software to monitor and control the camera system. This paper presents a home security camera system based on cloud and SNS. Unlike most of the commercial network camera systems, our security camera system does not require dedicated application software to monitor or control the camera system. Instead, we use twitter as the user interface. Also, we take advantage of the Amazon Web Service (AWS) cloud system for AI-based face authentication.

Keywords Twitter · AWS · Face authentication

1 Introduction

Obsolete security cameras simply record videos in their local storage, and humans watch the recorded videos later when necessary. Due to the advances in IoT technology, recent security cameras are connected to the Internet, and recorded videos are stored on cloud. Humans can watch camera images in real time or watch the recorded videos from a distance with PCs or mobile devices such as smartphones [1–3]. More recently, due to the advances in the computer vision and AI technologies, security cameras are able to automatically detect suspicious humans or unusual conditions in real time [4–8]. When detected, alerts are sent to security companies or the owners of the cameras through the Internet or other communication networks.

T. Egashira (✉)
Graduate School of Science and Engineering, Ritsumeikan University, 1-1-1, Nojihigashi, 525-8577 Kusatsu, Japan
e-mail: takuya.egashira@tomiyama-lab.org

L. Meng · H. Tomiyama
College of Science and Engineering, Ritsumeikan University, 1-1-1, Nojihigashi, 525-8577 Kusatsu, Japan

N. N. Chiplunkar and T. Fukao (eds.), *Advances in Artificial Intelligence and Data Engineering*, Advances in Intelligent Systems and Computing 1133,
https://doi.org/10.1007/978-981-15-3514-7_103

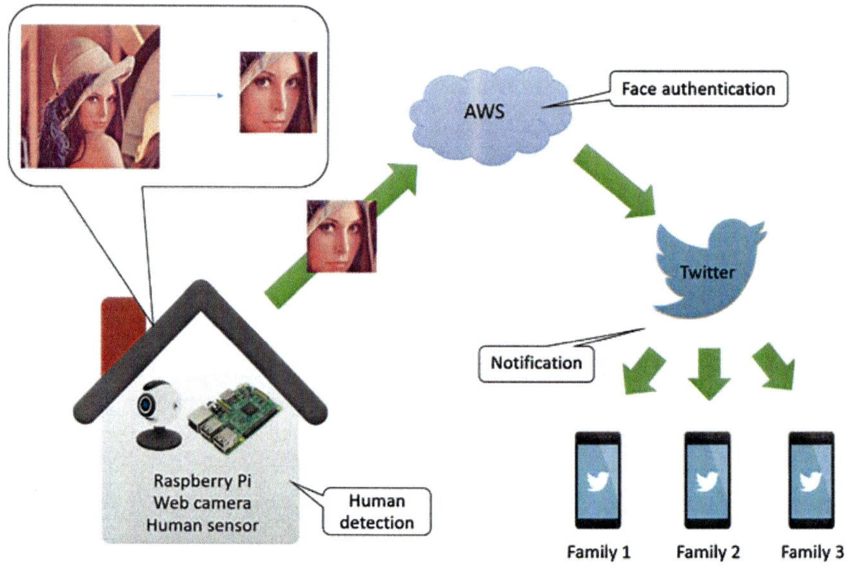

Fig. 1 Overview of our home security camera system

Most of IoT-based security camera systems rely on application software to monitor the videos and control the cameras. Users have to install the application software on their smartphones. If the software is provided for a particular kind of devices only, for example, Android devices only, the users of the camera are limited to Android users. On the other hand, it is time consuming to develop the application software for a variety of devices such as Android, iPhone, and Windows PCs.

In this work, we have developed a home security camera system based on cloud and SNS. Specifically, our system uses twitter for user interface. Twitter is one of the most popular SNSs in the world, and it is available on various devices such as Android and iPhone devices. Also, twitter is available on PCs with Internet browsers. This twitter-based interface significantly improves the user convenience. Another important feature of our security camera system is that our system uses Amazon Web Service (AWS). Our system takes advantage of sophisticated AI-based face recognition offered by AWS.

This paper is organized as follows. Section 2 outlines our home security camera system, and Sect. 3 describes its details. Section 4 concludes this paper with a summary.

2 System Overview

The overview of our home security camera system is shown in Fig. 1. The camera system consists of a camera device, AWS, Twitter, and user devices such as smartphones.

The camera device consists of a Raspberry Pi3 computer board, a human sensor, and a Web camera. The camera device is assumed to be placed in a house toward the entrance door. When a person enters the door, the human sensor detects the person and the camera takes pictures of the person. The face of the person is trimmed and sent to AWS. Then, face authentication is performed on AWS, and its result is notified to family members via twitter. Registration of new family members is also performed via twitter.

3 System Details

This section describes more details of our home security camera system.

3.1 Human Detection on the Camera Device

As briefly described in the previous section, the camera device consists of a Raspberry Pi3 board, a human sensor, and an affordable Web camera. The human sensor and the camera are connected to the Raspberry Pi board via GPIO and USB port, respectively. The camera device is assumed to be placed inside a house, and the Web camera is pointed to the entrance door.

When a person comes in from the door, the human sensor detects the person. Then, the camera convectively takes ten pictures at 0.5 s intervals, as shown in Fig. 2. We take ten pictures since the face of the person often fails to be captured clearly. For the

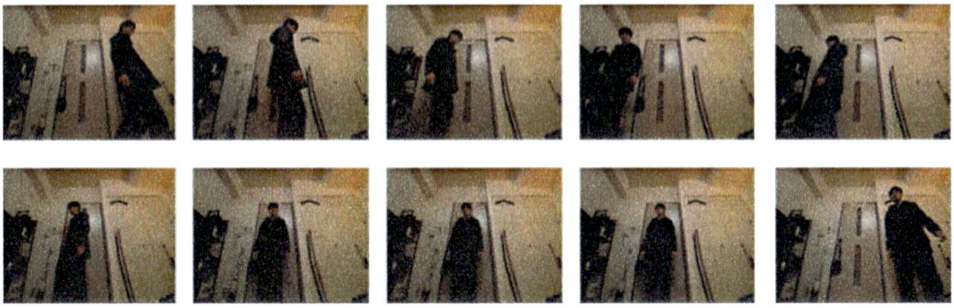

Fig. 2 Pictures taken by the camera

Fig. 3 Trimming the picture

ten pictures, our program on the Raspberry Pi3 board detects faces. We use the Haar-like feature classifier in the OpenCV library for face detection. If multiple persons are in a picture, multiple faces are detected. The detected faces are trimmed in order to reduce the size of the images as shown in Fig. 3. Then, the image files are sent from the Raspberry Pi board to AWS. Due to the trimming process, the network traffic is reduced.

3.2 Face Authentication on Cloud

In our home security camera system, we use AWS as a cloud platform. AWS provides not only computing and storage resources but also a rich set of libraries for engineers to write high-quality programs easily. Specifically, we use Amazon S3, Lambda, and Rekognition.

Amazon Simple Storage Service (Amazon S3) is a storage service on AWS. Face pictures sent from our camera device are stored in S3. Then, the face pictures are compared with preregistered face images of family members, which are also stored in S3. Our face authentication program is built upon Amazon Rekognition. Rekognition is a set of libraries for image and video analysis based on deep learning AI technology. AWS Lambda is an event-driven computing platform on AWS. We developed a Lambda program which executes our face authentication program when a new picture is stored on S3. Our face authentication program judges whether the newly arrived face image matches any of preregistered faces or not. If yes, our program returns the likelihood between the new image and the matched image.

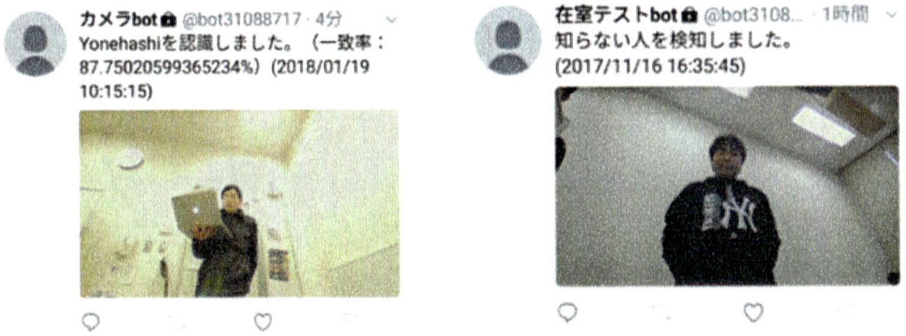

(a) Detection of preregistered person: The message says that "Yonehashi" is recognized with likelihood 87.75%.

(b) Detection of unregistered person: This message says that unknown person is detected.

Fig. 4 Notification messages on twitter **a** detection of preregistered person: The message says that "Yonehashi" is recognized with likelihood 87.75%. **b** Detection of unregistered person: This message says that unknown person is detected

3.3 Notification on Twitter

Our security camera system uses twitter to notify family members of the fact that a human comes into their house. There are a couple of advantages of using twitter. One advantage is that our system does not force users to install dedicated application software. Twitter is one of the most popular SNSs and many people already installed it on their devices and are familiar with using it. Twitter works on various devices and platforms including Android smartphones, iOS smartphones, personal computers, and so on. Another advantage is that all of the family members can receive the information at the same time.

When our face authentication program on AWS judges that the face in the received picture is a preregistered person, the program posts a message like the one in Fig. 4a. The posted message shows the name of the person and the likelihood value together with the picture. When our face authentication program judges that the face in the picture does not match any preregistered person, the program posts a warning message on twitter, showing that an unknown person is detected, as shown in Fig. 4b.

3.4 Registration on Twitter

In our security camera system, twitter is used not only for notification from AWS to family members but also for registration of new members on AWS. The registration process is shown in Fig. 5. A new member comes into the house for the first time, the person is judged as an unknown person by our security camera system, and an alert message is posted to the current members on twitter. When one of the current

Fig. 5 Registration of new member

members shares the tweet with a direct message to the Bot account (i.e., the account of the Raspberry Pi-based camera device), the member receives a reply from the Bot, asking the name of the person. Then, the member enters the name of the person; the person is newly registered into our face authentication program on AWS. Unknown person is detected.

4 Summary

This paper described a home security camera system which we developed. Our security camera system employs a world-popular SNS, twitter, as a user interface. Therefore, our system is available on various devices including Android smartphones, iOS smartphones, Windows PC, and Linux PC. Users do not have to install any application software on their devices in order to monitor and control the camera system. Also, our system takes advantage of a powerful and sophisticated cloud platform, Amazon Web Service, for AI-based face authentication.

As a future plan, we will extend our system for other SNSs such as Line, Weibo, and Facebook, since these SNSs are more popular than twitter in some regions.

References

1. Ramaswamy P (2016) IoT smart parking system for reducing greenhouse gas emission. In: International conference on recent trends in information technology 2016. IEEE, pp 1–6
2. Patchava V, Kandala HB, Babu PR (2015) A smart home automation technique with raspberry pi using IoT. In: International conference on smart sensors and systems 2015. IEEE, pp 1–4
3. Purbaya S, Sudiharto DW, Wijiutomo CW (2017) Design and implementation of surveillance embedded IP camera with improved image quality using gamma correction for surveillance camera. In: International conference on science and technology 2015, IEEE
4. Parmar CM, Gupta P, Bharadwaj KS, Belur SS (2018) Smart work-assisting gear. In: 4th IEEE world forum on internet of things 2018. IEEE, pp 724–728
5. Saba A, Nagarathna (2017) IoT based energy efficient security system. In: 3rd international conference on applied and theoretical computing and communication technology 2017. IEEE, pp 132–136
6. Lee HR, Lin CH, Kim WJ (2016) Development of an IoT-based visitor detection system. In: International SoC design conference 2016. IEEE, pp 281–282
7. Othman NA, Aydin I (2017) A new IoT combined body detection of people by using computer vision for security application. In: 9th international conference on computational intelligence and communication networks 2017. IEEE, pp 108–112
8. Quadri SAI, Sathish P (2017) IoT based home automation and surveillance system. In: International conference on intelligent computing and control systems 2017

Design, Calibration, and Experimental Study of Low-Cost Resistivity-Based Soil Moisture Sensor for Detecting Moisture at Different Depths of a Soil

S. Sunil Kumar, Ganesh Aithal, and P. Venkatramana Bhat

Abstract Soil moisture is one of the important abiotic factors for the proper growth of plant and scheduled irrigation in agriculture. Soil moisture stress would lead to the improper growth of crop and reduction in yield. Nowadays, most of the scheduled irrigation systems are depending on the compact soil moisture sensor to detect the soil moisture level and actuate the irrigation accordingly. Even though many soil moisture sensors are already available, the limitation of the existing soil moisture sensor poses further challenges for improvement in soil moisture sensor. In this work, one of the limitations of existing soil moisture sensor is unable to detect the soil moisture level at different granularities of soil considered and proposed an improvement. In the proposed work, the multiple parallel probes at different depths of soil are used to detect the soil moisture at different vertical profiles of soil using the basic soil properties of variation in soil resistivity due to soil moisture. The empirical study of proposed work showcases the behavioral characteristics of variation in soil resistivity due to moisture and relationship between change in soil resistivity of upper and lower probes and its possible use to detect the presence of soil moisture at different depths of soil.

Keywords Soil moisture · Resistivity · Vertical profile · NI WSN 3202

1 Introduction

Agriculture has been a backbone for any nation and primary source of food, which is at high risk due to the shortage of water for irrigation [1–3]. The shortage of freshwater for agriculture crops leads to the emergence of smart technologies on scheduled irrigation. In scheduled irrigation, crop will be provided with the required amount of water only when needed, hence avoids the excess supply of water to crop and

S. Sunil Kumar (✉) · P. Venkatramana Bhat
Mangalore Institute of Technology and Engineering, Moodabidri, India
e-mail: sunil@mite.ac.in

G. Aithal
Shri Madhwa Vadiraja Institute of Technology & Management, Bantakal, India

N. N. Chiplunkar and T. Fukao (eds.), *Advances in Artificial Intelligence and Data Engineering*, Advances in Intelligent Systems and Computing 1133,
https://doi.org/10.1007/978-981-15-3514-7_104

save water. In the early scheduled irrigation system, farmers started to use sprinkler system to supply water to crop. The sprinkler system running around the plant keeps soil of plants wet along with non-plants soil area. Due to the supply of water to the soil of non-plant area, it leads to waste of water and causes the germination of weeds along with normal crop plant. As a solution to the problem of sprinkle system, drip irrigation system came into existence, in which pipe with a small hole pointed to plant soil area is placed and watered in terms of drip irrigation. Drip irrigation avoids the situation of supply of water to non-plant soil area as like in sprinkle system. However, drip irrigation solves the issues related to sprinkle system of keeping only plant soil area wet but lacks intelligence in determining the required amount of water leads to the excess supply of water to plant [4–8]. The resolution to the problem of drip irrigation system towards adding intelligence on determining required amount of water for plant or crop depends on efficient, low-cost smart soil moisture-sensing technologies. Most of the existing smart scheduled irrigation methods largely depend on efficiently detecting the current soil moisture level of plant or crop and supply the water accordingly. Soil moisture has been an important abiotic factor and poses the urgent need of proper maintenance of soil moisture level towards minimizing the moisture stress effects on crops. However, existing methods for detecting soil moisture level were works well in laboratory and small-scale deployment in fields. Some of the limitations of existing soil moisture sensors lead to further research on building low-cost, reliable, and robust soil moisture sensors. This work proposes one such work towards building a low-cost resistivity-based soil moisture sensor and testifying it on the real field for experimental studies.

2 Related Work

Lot of research works were carried out in soil moisture-sensing technology, and many novel methods were proposed [9, 10]. Advancement in soil moisture technology could be the solution for various challenges of remote monitoring of soil moisture level and scheduled irrigation applications. Almost all sensing technology depends on variation in physical properties and its observation for determining the respective parameter of interest. Even soil moisture-sensing technology depends on the variation on physical characteristics such as dialectic constant of soil, resistivity of soil, temperature coefficient of soil, and others for determining the soil moisture level. Some of the researchers proposed methods based on variation in soil dielectric constant due to soil moisture [9–14]. Some of the works used the concept of time-domain reflectometry, which uses electromagnetic waves for determining the soil moisture content. The minimum dependency on soil puts the TDM approach in situ experiment, where repetitive calibration is not required [11–13]. Thermal imaging is an emerging technology to determine the soil moisture level using thermography properties of soil [15]. The drawback of variation on soil due to the soil composition puts thermography approach under further research. Another soil property, which is used for determining the soil moisture, is soil resistivity. As soil resistivity depends on the soil water content poses a scope for determining the soil moisture level based on variation in soil resistivity [16]. The limitation of soil resistance variance due to other parameters such as salinity, compactness, and fertilizer affects the accuracy of soil moisture sensor based on soil resistivity property. The low cost for design and fabrication of soil moisture sensor based on soil resistivity makes the continuous in situ use of resistivity-based soil moisture sensor [17]. Even though many soil moisture sensors are available, most of the sensors are lack in multi-point measurement of soil moisture. The multi-point measurement of soil moisture to determine the vertical profile of soil moisture in real time has been progressive research [18, 19]. Even most of the existing sensors lack in accuracy, and need for repeated calibration leads to continuous research on low-cost efficient multi-point soil moisture sensing [20, 21]. This work showcases one such work, towards designing a low-cost multi-point measurement of soil moisture in real time.

3 Proposed Work (Materials and Methods)

The proposed sensor works based on the phenomena of variation in the soil resistivity with respect to water together with soil or moisture levels in the soil. The variation in resistivity between parallel probes due to moisture content is utilized to determine the soil moisture level. In the proposed sensor, there are three electrodes (termed as probe) length of 4 cm and each is placed horizontally below to each other as shown in Fig. 1. Three electrodes are made up of copper foil sheets. These are connected to a fixed insulating stand. It facilitates to determine soil moisture content at upper surface or lower surface more specifically since the three probes divide the soil into two sections. The upper probe and lower probe are connected to (+5 V) positive terminal of the source, and the middle probe is connected to negative terminal. The upper probe and lower probes' voltage output is considered as sensor output in voltage, which calibrates the moisture level.

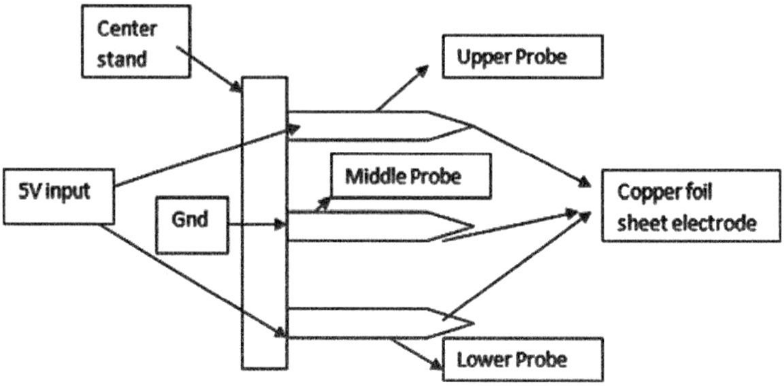

Fig. 1 Block diagram of design of soil moisture sensor

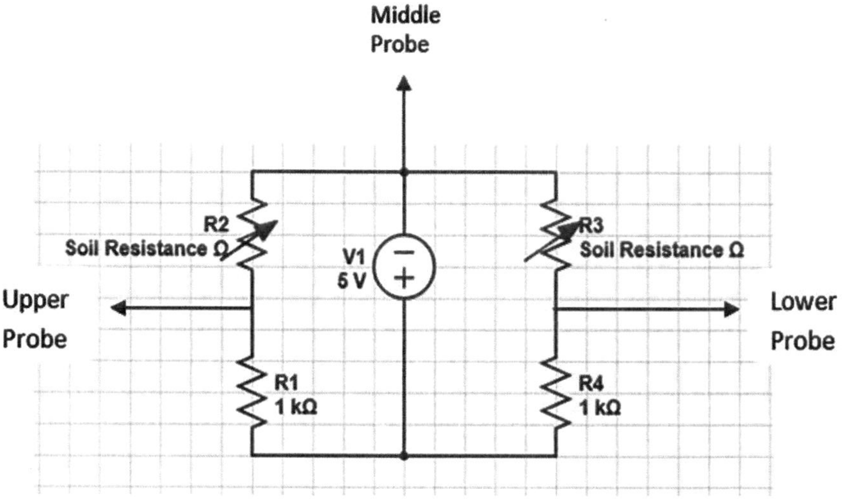

Fig. 2 Circuit layout of the proposed resistivity sensor

Figure 2 shows the circuit of the sensor with two probes. These two probes are upper, which is buried inside the soil just two inches or five centimeter from the top. The second one is lower probe which is 20 cm below the first one since, as water is poured inside the soil, the resistance of the soil decreases. To limit the current in the soil and probe, the paper introduces a current-limiting resistor in the circuit. This current limiting resistance of 10 Ω is connected in series with the soil resistance that is the upper or lower probes. Further, it is connected to the positive end of the supply. The negative end of the supply is connected to middle probe so as to complete the circuits. This allows us to find the moisture level in terms of voltage output for different levels of the soil. The output is taken from the junction of the current-limiting resistance and the soil resistance.

Figure 2 shows the sensor probe with V_{out} as output. R_1 is the resistance, which limits the current flow in the circuit. R_2 is the variable soil resistance, voltage across this, which is measured as V_{out} and is transmitted to the server. Let "I" be the current in amps flowing through the circuit. The voltage across R_2 is measured as $V = I \times R_2$. i.e.,

$$V_{\text{out}} = I \times R_2 \tag{1}$$

$$\text{Since } I = \frac{V_{\text{in}}}{R_1 + R_2} \tag{2}$$

By substituting Eq. (2) in Eq. (1)

$$V_{\text{out}} = V_{\text{in}} \times \frac{R_2}{R_1 + R_2}$$

where R_2 and R_3 are soil resistance, and V_{in} is voltage of the source.

The Voltage V_{out} across R_2 to found, so as to calculate the soil moisture between upper probe and middle probe, that is

$$V_{out} = V_{in} \times \frac{R_2}{R_1 + R_2}$$

The V_{out} across R_3 to determine the soil moisture between lower probe and middle probe is

$$V_{out} = V_{in} \times \frac{R_3}{R_3 + R_4}$$

The analog voltage output of sensor probes is used to determine the soil moisture level. The average of sensor voltage output of upper and lower probes is plotted against soil weight as the water is poured in. The polynomial model is fitted for plotted graph of average sensor voltage V_{out} versus weight of the pot. The polynomial equation determines the weight of pot based on the average sensor voltage V_{out}.

The polynomial equation determined weight is used in gravimetric method to determine the soil moisture level. The approach of the gravity matrices method to determine the soil moisture level is as follows.

Let GWC = Gravimetric water content, M_w = Mass of water, M_s = Mass of soil, W_{st} = Wet soil weight (including pot weight), D_{st} = Dry soil weight (including pot weight), D_s = Dry soil weight (excluding pot weight), P_t = Pot weight, W = Percentage of water in gravimetric water content in units (g g^{-1}) and then

$$\text{GWC}\,(w) = \frac{M_w}{M_s} \tag{3}$$

$$M_w = W_{st} - D_{st} \tag{4}$$

$$M_s = D_{st} - P_t \tag{5}$$

$$W_{st} = W_s + P_t \tag{6}$$

$$D_{st} = D_s + P_t \tag{7}$$

4 Experimental Setup and Study

In the work, two experiments are conducted to find the soil conditions. The first one is indoor and the second one is outdoor experiments. In both conditions, 1.5 kg of sandy soil is taken, dried in a hot-air oven for 24 h to get nil moisture in the soil. Empty weight of the pot is measured as P_t Kg, and then pot is filled with the dried sandy soil which is weighed again as D_{st}. The difference between these shows the net dry soil weight D_s Kg. After getting the dry soil weight, based on European method [22], the following steps are performed to get water-holding capacity of soil.

a. Water is poured into the pot till the excess water drains out from the bottom of the pot.
b. After excess water drains out from the pot completely, wet soil weight together with the pot is measured.
c. $(W_{st} - D_{st})$ gives the water-holding capacity of the soil sample.

After getting the soil's water-holding capacity, the soil is dried again by hot-air oven for the indoor and outdoor experiments, again for 24 h.

Indoor experiment: Once the soil sample is ready for the experiment, the following steps are carried out to test the proposed sensor.

1. Empty and soil filled pots are measured for its weights.
2. National Instruments Wireless Sensor Network Testbed comprised of NI WSN 3202 programmable node, and NI WSN 9791 Gateway Node is used for collecting real-time data during experiment.
3. The pot is hanged to the weight sensor.
4. Soil moisture sensor is immersed in the dry soil such that upper probe is just 5 cm below the top layer of the soil.
5. Pot is placed in chamber with two incandescent bulb of 40w each, on the top of the pot.
6. Soil moisture sensor output pins are connected NI WSN 3202 Node for collecting data in real time.
7. Weight sensor output pin is connected to NI WSN 3202 Node for real-time data collection about change in weight value due to moisture.
8. Water is poured to soil gently, from top surface until excess water drain out from the bottom.
9. NI WSN 9791 Gateway programmed to continuously monitor and log all the changes in weight and moisture sensor voltage output in the period of inducing moisture and during evaporation of water.

Outdoor Experiment

The above same steps are repeated to carry out the outdoor experiment for testing the soil moisture sensor, except the fourth step. In case of the fourth step, the experiment opens up for the sun.

5 Experimental Results and Discussion

The figure shows the upper and lower probes' voltage output during the first 15 min of period. Figure 3 discusses the soil moisture at different depths of soil. The voltage

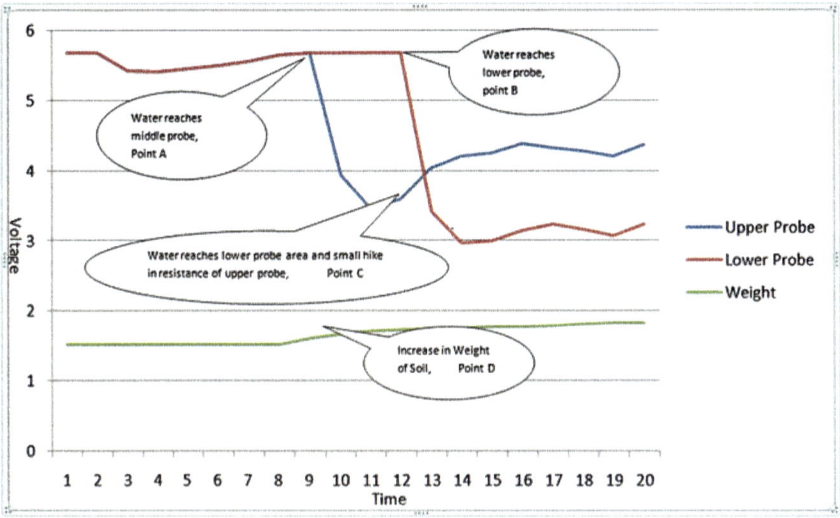

Fig. 3 Sensor output during watering stage

between upper probe and the ground point gives the resistivity of the soil between upper probe and the ground point, similarly the middle (ground) and lower probes. This technique is used for measuring the moisture level in different depths of the soil.

As water is poured in, it penetrates down and approaches the middle probe. As it approaches middle probe, soil resistance starts decreasing; hence, V_{out} of the upper probe decreases. This is demonstrated in the figure from point A to point C. As water crosses the middle probe and approaches the lower probe, it causes an increase in conductivity between lower and middle probes. Hence, soil resistivity between lower probe and middle probe starts decreasing also $V_{out,}$ as shown in point B. Since the middle probe acts as ground for both the probes and both supplies for the probes are emerging from the same source, which makes the changes in the resistance between upper and middle probes, when water reaches the lower probe. That is, the resistance between upper and middle probes increases as the water reaches the lower probe, which is shown in point C.

Figure 4 shows the upper probe and lower probe voltage output for complete experiment. Since the water is poured from the top surface, the upper probe voltage output drops earlier than lower probe voltage output. As the water approaches the lower probe area, which causes the voltage drops across lower probe as shown in Fig. 4, the graph shows the gradual increase in upper probe and lower probe voltage V_{out} during the soil water evaporation stage for complete experiment.

As part of the comparison, the proposed sensor reading to gravimetric water content, the following procedure is worked out. During the experiment, the proposed weight sensor inserted in the pot is continually monitored. Figure 4 shows the increase in weight as the water is poured into the pot and the gradual decrease in the weight

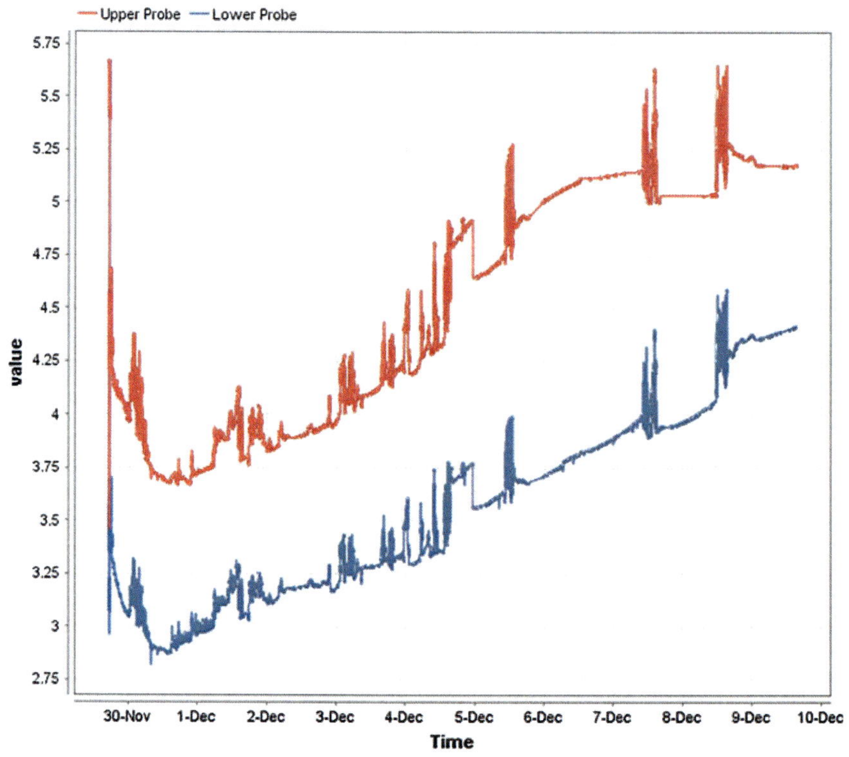

Fig. 4 Upper and lower probes' output after watering or during evaporation process

of pot during the evaporation. The point A shows the initial dry soil weight. Point B shows the weight of sample pot at saturated water-holding capacity of soil. The point B onwards to point C shows the behavior of decrease in weight as water evaporates from soil (Fig. 5).

Upper probe and lower probe output is specifically used to determine the presence of water in the pot below the upper and above the lower probes of the soil, respectively. Figure 6 shows the plot of data between average of upper probe and lower probe values to the weight of the pot. The weight decreases as the average soil resistance increases.

The data of correlation between average of upper and lower probes' sensor output V_{out} to weight of sample soil is used for regression analysis and to determine the calibration equation. The regression analysis of data provides a polynomial model and its fitted equation as follows.

$$Y = \text{intercept} + B1x^1 + B2x^2 + B3x^3 + B4x^4 \quad (8)$$

where B1 = −24.51036, B2 = 8.76999, B3 = −1.40355, B4 = 0.08424.

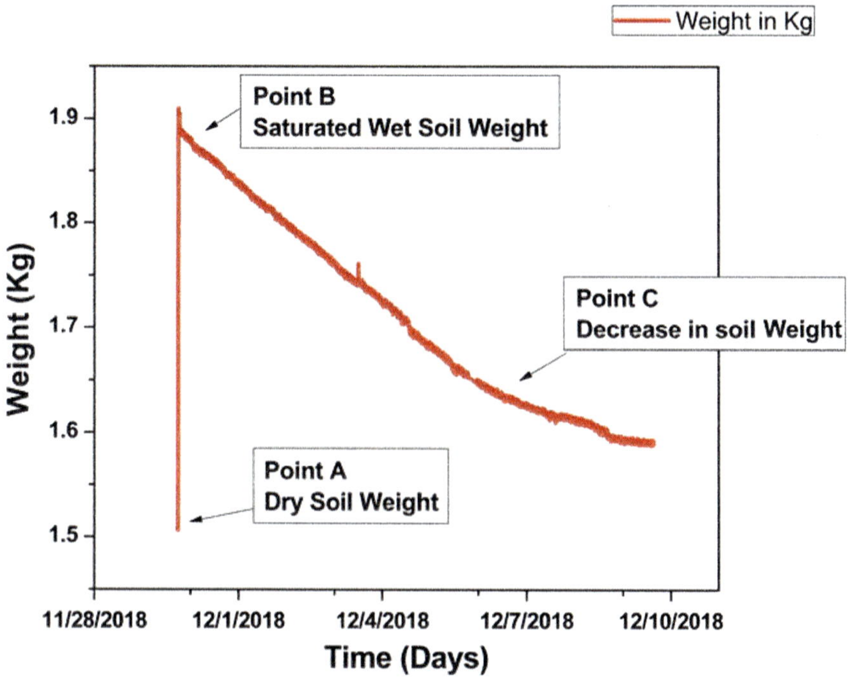

Fig. 5 Weight sensor value after watering

Hence, the approximate weight of sample pot is determined using the known average of sensor probe readings. Once the weight is determined, gravimetric water content equation is used to determine the water content in soil.

6 Calibration and Testing

Initially, the direct method of gravimetric approach is utilized to determine the weight of sample soil pot under different levels of moisture percentage on soil. As per the GWC equation, the water content of selected sample at saturation point of water-holding capacity is 26.6% where saturation level of water is 400 g and total weight of the soil is 1500 g, from Eq. (3), i.e., GWC (w) = 400 g/1500 calculated as 26.6%. Since total water-holding capacity is 400 g, it is considered as 100%. Water content percentage with respect to water-holding capacity of soil is mapped to moisture level in the range 0–100% as shown in columns 2 and 3 of Table 1. Selected soil sample is dried after determining the water-holding capacity of soil for further study of GWC water content at various moisture levels by repeating the same. The table shows the gravimetric water content at different moisture levels and corresponding weight.

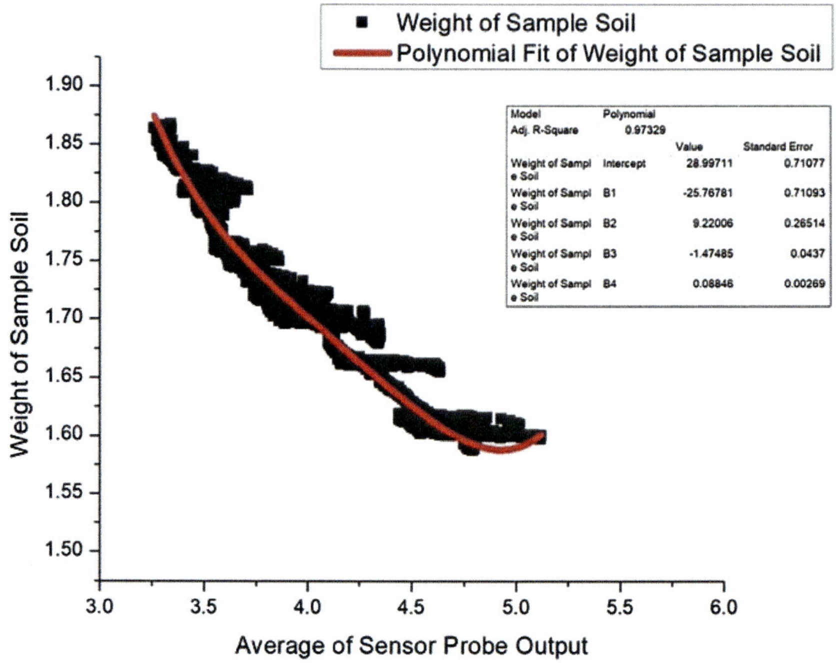

Fig. 6 Correlation between averages of sensor output to weight of soil

Table 1 Mapping of soil moisture percentage to water-holding capacity with weights

Water content percentage w.r.t. water-holding capacity (%)	Weight of soil	Moisture level with respect to 100% mapping (%)
3.3	1550	12.5
6.6	1600	25
13.6	1700	50
26.6	1900	100

The sensor probe output voltage is mapped to respective weight based on the polynomial Equation (8) of the average of sensor probe outputs v/s weight of the pot. The predicted weight based on the model is used to compute the water content in the soil, based on the gravimetric Equations (3)–(7).

The following experimental test cases are carried out to verify the accuracy of the modeled equation to determine the moisture level of soil. The preliminary steps for conducting test experiment are as follows.

1. The same soil is used for conducting the test case experiment after drying every time.
2. 12.5, 25, 50, 100% of moisture are induced to the dry soil as per Table 1

Table 2 Comparison of direct method v/s soil moisture output

Direct method induces moisture level (%)	Weight of soil under direct method	Model fitted weight value	Model calibrated moisture level (%)	Amount of water in ml
3.3	1550	1548	3.2	50
6.6	1600	1594	6.2	100
13.6	1700	1695	13.0	200
26.6	1900	1890	26.0	400

3. To make each sample of 12.5, 25, 50, 100% moisture level, corresponding amount of water is added as per Table 1 to make the expected moisture level.
4. The readings of the moisture and weight sensors are taken for each experiment

After these preliminary steps, the values of weights and of the corresponding upper and lower probes are collected. By using upper and lower probes' values, the weight of the pot is predicted based on Eq. (8). This value is compared with the weight sensor value, and error is evaluated. Table 2 shows the details of the same

In the above calibration, both upper and lower probes' output voltages are considered together as average of output voltage to determine the moisture level of soil.

The two conditions were used based on output voltage of upper and lower probes for determining the vertical profile of soil. The conditions are as follows

Let, Upper Probe $= U_p$, Lower Probe $= L_p$,
Upper Probe Voltage $= U_{PV}$, Lower Probe Voltage $= L_{PV}$,
Middle Probe $= M_P$.

Condition 1 If $U_{pv} < L_{pv,}$ Then
Presence of water content between U_p and M_p depth level of soil.

Condition 2 If $U_{pv} > L_{pv,}$ Then
Presence of water content between M_p and L_p depth level of soil.

Proof for Condition 1 As per Ohm's Law,

$$I = \frac{V}{R} \tag{9}$$

Hence for the above circuit, current I at middle probe would be

$$I = I_1 + I_2 \tag{10}$$

Using Eq (9), determine the current in upper probe and lower probe circuits is as follows

$$I_1 = \frac{V_1}{R_x} \tag{11}$$

$$I_2 = \frac{V_2}{R_y} \tag{12}$$

whereas resistance from upper probe to middle probe circuit is

$$R_x = R_1 + R_2 \tag{13}$$

whereas resistance from lower probe to middle probe circuit is

$$R_y = R_3 + R_4 \tag{14}$$

By considering R_2 and $R_{4,}$ it is a variable soil resistance
Let us consider,

R_2 = soil resistance between upper probe and middle probe areas
R_4 = soil resistance between lower probe and middle probe areas
$\mathrm{Max}_{\_\mathrm{Upper\_Resist}}$ = Maximum soil resistance value between upper probe and middle probe areas
$\mathrm{Min}_{\_\mathrm{Upper\_Resist}}$ = Minimum soil resistance value between upper probe and middle probe areas
$\mathrm{Max}_{\_\mathrm{Lower\_Resist}}$ = maximum soil resistance between lower probe and middle probe areas
$\mathrm{Max}_{\_\mathrm{Lower\_Resist}}$ = maximum soil resistance between lower probe and middle probe areas.

The value of R_2 ranges between $\mathrm{Max}_{\_\mathrm{Upper\_Resist}}$ and $\mathrm{Min}_{\_\mathrm{Upper\_Resist}}$ as shown below

$$\mathrm{Min}_{\mathrm{Upper}_{\mathrm{Resist}}} \leq R_2 \leq \mathrm{Max}_{\mathrm{Upper}_{\mathrm{Resist}}} \tag{15}$$

$$\mathrm{Min\_Lower}_{\mathrm{Resist}} \leq R_2 \leq \mathrm{Max\_Lower}_{\mathrm{Resist}} \tag{16}$$

As water poured, there will be negative charge in R_2 as follows

$$-\Delta R_2 = \left(\mathrm{Max}_{\mathrm{Upper}_{\mathrm{Resist}}} - P\right) \geq \mathrm{Min}_{\mathrm{Upper}_{\mathrm{Resist}}} \tag{17}$$

where P = Reduction component of soil resistance as water is added.
Since R_2 is part of R_x, applying Eq. (17) in Eq. (13)

$$\downarrow R_x = R_1 + (-\Delta R_2) \tag{18}$$

Using Eq. (18) in Eq. (11)

$$\uparrow I_1 = \frac{V_1}{\downarrow R_x} \tag{19}$$

Hence, rewriting Eq. (19) gives

$$\downarrow V_1 = \uparrow I_1 * \downarrow R_x \tag{20}$$

This proves that, as soil resistance between upper probe and middle probes decreases and causes decrease in upper probe voltage output.

Proof for Condition 2 Proof for Condition 2 is a continuation of Proof of Condition 1 as follows

$$-\Delta R_4 = (\text{Max\_Lower}_{\text{Resist}} - q) \geq \text{Min\_Lower}_{\text{Resist}} \tag{21}$$

where q = Reduction component of soil resistance as water distributes between middle and lower probe areas.

As represented in Eq. (21), there will be a negative charge in R_4 as water crosses the middle probe area and reaches the lower probe.

Since R_4 is part of R_y, applying Eq. (21) in Eq. (14).

$$\downarrow R_y = R_1 + (-\Delta R_2) \tag{22}$$

Using Eq. (22) in Eq (12)

$$\uparrow I_2 = \frac{V_1}{\downarrow R_y} \tag{23}$$

Applying Eq. (23) in Eq. (4) proves the second condition, and reason for hike in upper probe voltage as water reaches the lower probe.

Test case experiment results in Fig. 1 shows the impact of the above two conditions towards determining the moisture at different horizontal granular points in soil. Figure 1 shows the presence of moisture at upper surface only point A, during which upper probe voltage is lesser than lower probe voltage. Similarly, point C upper voltage greater than lower probe voltage as water approaches lower probe. In this way, the proposed low-cost soil moisture sensor detects the moisture at different depths of a soil.

7 Conclusion

In this paper, the experimental study of low-cost resistivity-based soil moisture sensor is carried out and results are discussed. The low-cost resistivity-based soil moisture sensor is utilized for multi-point measurement of soil moisture, and experimental results are discussed. The results are adequate to the indoor experiment, but lack the accuracy for outdoor experiments; due to temperature, salinity and humidity were affecting the soil resistance. The results are adequate to determine the soil moisture at different depths of soil and compared with the standard capacitive sensors available in the market.

Acknowledgements Authors like to sincerely acknowledge Vision Group on Science and Technology (VGST), Department. of IT, BT, and S&T, Government of Karnataka, Bangalore, for sanctioning research grant for the Department of Computer Science and Engineering, Mangalore Institute of Technology and Engineering, Moodabidri, under "KFIST = L2" grant schema vide GRD No. 339 dated.

References

1. Coping with Water Scarcity (2006) UN-water thematic initiatives. A strategic issue and priority for system-wide action. Available online: http://www.unwater.org/publications/coping-water-scarcity/. Accessed on 5 Sept 2017
2. AghaKouchak A, Cheng L, Mazdiyasni O, Farahmand A (2014) Global warming and changes in risk of concurrent climate extremes: insights from the 2014 California drought. Geophys Res Lett 41:8847–8852
3. Hanak E, Mount J, Chappelle C, Lund J, Medellín-Azuara J, Moyle P (2017) Public policy institute of California, water policy center. What if California's drought continues? Available online: http://www.ppic.org/publication/what-if-californias-drought-continues/. Accessed on 5 Sept 2017
4. Thompson TL, Pang HC, Li YY (2009) The potential contribution of subsurface drip irrigation to water-saving. Agric Sci China 8:850–854
5. Bernstein L, Francois LE (1973) Comparisons of drip, furrow, and sprinkler irrigation. Soil Sci 115:73–86
6. Camp CR (1998) Subsurface drip irrigation: A review. Trans. ASAE 41:1353–1367
7. Tagar A, Chandio FA, Mari IA, Wagan B (2012) Comparative study of drip and furrow irrigation methods at Farmer's field in Umarkot. World Acad Sci Eng Technol 69:863–867
8. Keeratiurai P (2013) Comparison of drip and sprinkler irrigation system for the cultivation plants vertically. J Agric Biol Sci 8:740–744
9. Walker JP, Willgoose GR, Kalma JD (2004) In situ measurement of soil moisture: a comparison of techniques. J Hydrol 293:85–99
10. Susha Lekshmi SU, Singh DN, Baghini MSA (2014) critical review of soil moisture measurement. Measurement 54:92–105
11. Topp GC (1980) Electromagnetic determination of soil water content: measurements in coaxial transmission lines. Water Resour Res 16:574–582
12. Mojid MA, Cho H (2004) Evaluation of the time-domain reflectometry (TDR)—measured composite dielectric constant of root-mixed soils for estimating soil-water content and root density. J Hydrol 295:263–275
13. Rao BH, Singh DN (2011) Moisture content determination by TDR and capacitance techniques: a comparative study. Int J Earth Sci Eng 4:132–137
14. Mioko A et al (2014) Using capacitance sensors for the continuous measurement of the water content in the litter layer of forest soil. Appl Env Soil Sci 2014
15. Hsu Wei-Ling, Chang Kuan-Tsung (2019) Cross-estimation of soil moisture using thermal infrared images with different resolutions. Sens Mater 31(2):387–398

16. Tan WY et al (2019) Newly calibrated analytical models for soil moisture content and pH value by low-cost YL-69 hygrometer sensor. Measurement 134(2019):166–178
17. Saleh M, Elhajj IH, Asmar D, Bashour I, Kidess S (2016) Experimental evaluation of low-cost resistive soil moisture sensors. 2016 IEEE international multidisciplinary conference on engineering technology (IMCET). Beirut, pp 179–184. https://doi.org/10.1109/imcet.2016.7777448
18. Yan H, Tang ZJ, Xing Z, Gao DN, Hong HX (2016) Design of soil moisture distribution sensor based on high-frequency capacitance. Int J Agric Biol Eng 9(3):122–129
19. Saeed IA et al (2019) Development of a low-cost multi-depth real-time soil moisture sensor using time division multiplexing approach. IEEE Access 7:19688–19697. https://doi.org/10.1109/ACCESS.2019.2893680
20. Chow L et al (2009) Field performance of nine soil water content sensors on a sandy loam soil in New Brunswick, Maritime Region, Canada. Sensors 9(11):9398–9413
21. Matula S, Báťková K, Legese W (2016) Laboratory performance of five selected soil moisture sensors applying factory and own calibration equations for two soil media of different bulk density and salinity levels. Sensors 16(11):1912
22. Water holding capacity determination techniques? Available online at https://www.agvise.com/educational-articles/water-holding-capacity/

An IoT-Based Predictive Analytics for Estimation of Rainfall for Irrigation

H. Shalini and C. V. Aravinda

Abstract Agribusiness is the foundation of Indian Economy. Its prosperity depends overwhelmingly on the climatic parameters. Occasions like erratic atmosphere are outside human capacity to control. Water assumes a noteworthy element in the development of a product (crop). If there is lacking water supply, chances of product disaster are more. Agriculturists fall into commitments since they have to go up against an uncommon yield's effectiveness in view of deficient water supply and other atmospheric conditions which increase the peril of their benefit and the high expense of living. Along these lines, it winds up critical to anticipate the measure of rainfall utilized for irrigation. This will guarantee the budgetary use of water. This research paper presents a study and experimentation of predictive analytics to predict the amount of rainfall for irrigation. Predictive analytics include the extraction procedure of valuable data from the informational collection given by the client and foresee critical highlights or patterns. Prediction process is performed using supervised machine learning techniques. Multiple linear regression, k-nearest neighbour, decision tree, and random forest techniques are considered for building the predictive model, where these models are evaluated using root mean squared error. Root mean squared error obtained for multiple linear regression, k-nearest neighbour, decision tree, and random forest are 0.165, 0.103, 0.094, 0.083, respectively. Evidently, random forest shows less root mean squared error compared to other models and is considered for the prediction process. Also, an IoT-based weather station has been built to retrieve the real-time data from a sample area.

Keywords Agriculture · Internet of things · Weather · Multiple linear regression · K-nearest neighbour · Decision tree · Random forest

H. Shalini (✉) · C. V. Aravinda
NMAM Institute of Technology, Nitte, Karkala, India
e-mail: shasep05@gmail.com

C. V. Aravinda
e-mail: aravinda.cv@nitte.edu.in

N. N. Chiplunkar and T. Fukao (eds.), *Advances in Artificial Intelligence and Data Engineering*, Advances in Intelligent Systems and Computing 1133,
https://doi.org/10.1007/978-981-15-3514-7_105

1 Introduction

Climate is a standout amongst the most constraining elements to the agribusiness. Climate influences how crops develop as well as the coordination's around planting, collecting, and transportation. By incorporating climate forecast models into product planting and gathering and transportation, better choices can be made ahead of time of harvest misfortunes because of climate hazards. Climate information and examination on field-by-field or zone-by-zone premise enable agriculturists to settle on educated choices during the time to boost nourishment generation, to limit natural effect, and diminish working expenses.

Machine learning (ML) is a subset of artificial intelligence (AI). ML provides the ability to automatically learn from past experience without being explicitly programmed to the system. In supervised learning, every occurrence of a training set is made out of various information characteristics and expected yield. For every input, a row of information is associated with an expected yield value. Two forms a value can take: discrete or continuous. Labels are associated with the inputs that are the reason this process of learning is called supervised. In unsupervised learning, every occurrence of the training dataset is not associated with an expected yield value. This kind of learning mainly detects patterns based on the attributes of input data. Reinforcement ML is about collaboration between two components nature and the learning specialist. The learning specialist uses two components, in particular, exploration and exploitation. When learning specialist follows up on trial and error basis it is known as exploration. When it acts upon the knowledge gained from the environment it is known as exploitation. This environment remunerates the learning specialist for the right actions. Utilizing the rewards acquired, the leaning specialist enhances its environment information to choose the following action.

This research paper presents different supervised ML techniques for the prediction of rainfall in a sample area using the weather dataset. Multiple linear regression (MLR), k-nearest neighbour (KNN), decision tree, and random forest are some of the supervised ML techniques which are considered for building the prediction model. MLR, KNN, decision tree, and random forest are regression models. Based on the evaluation of these models one of the best models is considered for constructing the prediction model for predicting the rainfall.

Also, an Internet of things (IoT)-based implementation is carried out for retrieving real-time weather data. An IoT is a network that consists of devices that are able to communicate with each other over the network. The hardware design consists of a DHT11 sensor, BMP280 sensor, and rain sensor to collect temperature, humidity, air pressure, and rainfall measurements, respectively. The data is collected in CSV file format.

Root mean squared error (RMSE) is considered for evaluating the ML models. The RMSE is a measure of the differences between values anticipated by a model and the true values observed. RMSE is applied for regression models.

Section 2 discusses the related work to the research project. Section 3 defines the objectives of the research project. Section 4 elaborates on how the objectives are implemented. Section 5 shows the results obtained during the experimentation. Section 6 concludes the research work.

2 Related Work

An asp.net MVC web application developed which is accessible to the farmers so that they can interact with the government for any kind of crop loss. Farmers are supposed to enter the details of their land like which crop has been chosen for growth, expected cost, etc., on the website. In case of any crop tragedy occurs, farmers should upload the details of the crop loss like the images or videos of the land. This will ensure that farmers will get the lost cost from the government [1]. A monitoring system has been developed in which plants can be monitored through a smartphone. Also, controlled water supply is provided to the crops for irrigation for effective utilization of water. Sensors are utilized to monitor the water availability to crops. Based on that the water is provided for crop growth [2].

An approach towards performing predictive analysis of weather parameters accurately using time series has been done [3]. Also, nonlinear autoregressive networks have been used for prediction of numerical values. The dataset considered for predictive analysis is from Manaus city from 1970 to 2015 years. Evaluation of the model has been done considering mean squared error (MSE), RMSE, and normalized root mean squared error (NRMSE) [4]. The fuzzy logic algorithm for developing a weather prediction system has been discussed. The data for the weather prediction has been collected from weather service provider. The weather prediction system using fuzzy logic algorithm will provide information weather plant should be watered. This will ensure the effective use of water in the field of agriculture [5]. Data analysis on the weather dataset from the meteorological centre of Bengaluru district has been performed. ML techniques like K-means and hierarchical clustering have been considered to extract weather parameters patterns. The results obtained will be useful for decision-making in the field of agriculture for crop growth [6].

3 Objectives

1. Design of an IoT-based weather station is to capture weather parameters like temperature, humidity, air pressure, and rainfall of a sample area and saves the data in the cloud.
2. Perform predictive analysis on the historical data collected.

4 Methodology

4.1 Architecture

Figure 1 refers to the hardware setup of the weather station. Arduino Uno is the micro-controller that will capture all the sensor values. Sensors used in the setup include DHT11, BMP280, and rain sensor. DHT11 sensor captures the temperature and humidity values, BMP280 captures air pressure values, and rain sensor captures the digital value if rain was detected. ESP8266 WiFi module gives the ability to upload data collected from sensors to cloud.

Arduino IDE software helps to write and upload code onto the Arduino board. Once the code is uploaded serial monitor displays the sensor values. ThingSpeak is an IoT platform that lets us analyse and visualize data [7]. A private channel has been created to capture each sensor values. Each parameter is displayed in different graphs. Data collected in the cloud can be extracted in CSV format and use it for predictive analysis.

4.1.1 Data Collection

Data is required to carry out any predictive analysis. Data collected includes weather parameters like temperature, humidity, air pressure, rainfall, etc. Weather parameters values are collected for the year 2018 and 2017 of the location Mangalore. The reference site from which data has been taken is https://www.worldweatheronline.com/mangalore-weather-history/karnataka/in.aspx. Access to historical weather data of any location can be extracted from this website. The data collected is saved in

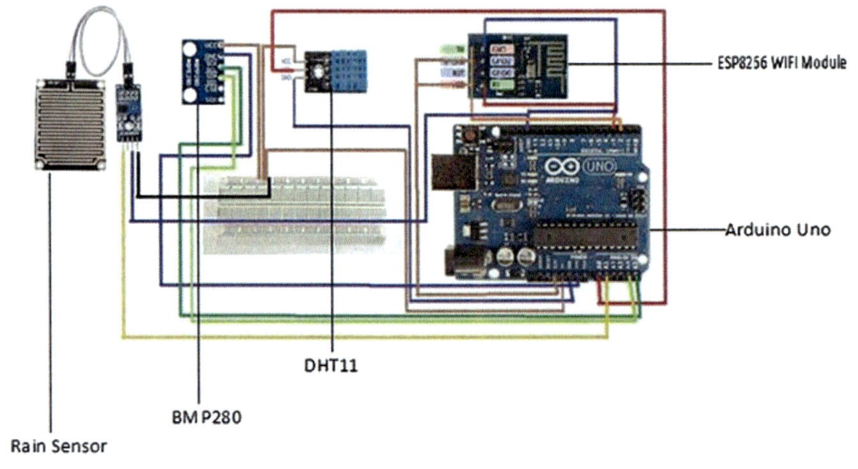

Fig. 1 Hardware setup of weather station

the CSV file format along with other parameters like DayOfYear, MonthOfYear, etc. This CSV file is taken as input for further analysis. Also, data collected from IoT can be added to this CSV file. Table 1 shows sample weather data.

4.1.2 Experimental Setup

Following connections have to be made to complete the design shown in Fig. 1. Table 2 shows the connections between DHT11 sensor and Arduino Uno board. Table 3 shows the connections between BMP280 sensor and Arduino board. Table 4 shows the connections between rain sensor and Arduino Uno board. Table 5 shows the connections between ESP8266 WiFi module and Arduino Uno board.

4.1.3 Procedure

1. Once the connections have been made connect Arduino board to laptop using USB cable.
2. Open Arduino IDE and start building the code. Sketch has two functions setup() and loop().
3. In setup() specify all the initializations, and in loop() specify code that will extract values from the sensors using libraries.
4. Once the code is completed, upload it on the Arduino board using upload button in Arduino IDE.
5. Open the serial monitor to check if it is displaying correct values of sensors.
6. ESP8266 WiFi module will connect to the internet and upload data to ThingSpeak channel.
7. ThingSpeak channel will display data collected from all sensors in a graph.
8. MATLAB visualization is possible for each of the channel.
9. Data stored can be exported as XML, JSON, CSV file.

Data is exported in CSV format and utilized for predictive analysis.

4.2 Data Analysis

Data analysis is a procedure of investigating, cleansing, changing, and displaying information with the objective of finding valuable data, informing conclusions, and supporting decision-making. Dataset is of unstructured form which has to be pre-processed to capture important attributes. Analysis will be performed on the cleansed dataset for determining the rainfall.

ML will help in predictive analysis of weather. The main idea behind ML is that the machine is trained with the experience data so that they will learn from their experience and predict the future event. Statistical methods are utilized to enable

Table 1 Weather dataset sample

Date	MaxTemp	MinTemp	Wind (km/h)	Pressure (mb)	Precipitation (mm)	Humidity (%)	DayOfYear	MonthOfYear	Year
01-01-2017	36	23	10	1012	0	58	1	1	2017
02-01-2017	34	24	9	1013	0	69	2	1	2017
03-01-2017	37	23	9	1013	0	62	3	1	2017
04-01-2017	37	23	8	1012	0	61	4	1	2017
05-01-2017	36	22	9	1012	0	66	5	1	2017

Table 2 DHT11 sensor to Arduino board connections

DHT11 sensor	Arduino Uno
VCC	5 V
GND	GND
Data pin	A0

Table 3 BMP280 sensor to Arduino board connection

BMP280 sensor	Arduino Uno
VCC	5 V
GND	GND
SCL	A5
SDI	A4

Table 4 Rain sensor to Arduino board connection

Rain sensor	Arduino Uno
VCC	5 V
GND	GND
A0 (data pin)	A1

Table 5 ESP8266 WiFi module to Arduino board connection

ESP8266	Arduino Uno
VCC	3 V
GND	GND
CHPD	3 V
RX	~3
TX	2

machines to improve with experience. Techniques of ML are trained using the dataset to develop a model. The model is designed and it is trained with various inputs in order to predict the result for test dataset. The accuracy of this prediction is evaluated. ML algorithm is deployed only when the accuracy of the prediction is acceptable. Otherwise, the ML algorithm is trained again with same dataset. ML is categorized into three different types supervised, unsupervised, and reinforcement. This paper presents supervised ML techniques for building predictive models. MLR, KNN, decision tree, and random forest techniques have been considered for building the predictive model.

4.2.1 Tools and Languages

The tool used here is Jupyter notebook and Spyder IDE, and the language used is Python.

4.2.2 ML Techniques

MLR, KNN, decision tree, and random forest techniques have been considered for comparison and selecting the best model for predictive analytics.

1. **MLR**—MLR endeavours to show the connection between at least two distinct feature and a reliant feature by fitting a straight condition to discovered information. Each estimation of the distinct feature x is related to an estimation of the reliant feature y. The equation of multivariate linear regression is as follows:

$$Y = \alpha + \beta_1 x^{(1)} + \beta_2 x^{(2)} + \cdots + \beta_n x^{(n)} \tag{1}$$

   Y—reliant variable
   X—distinct variable (can be more than one).
2. **KNN**—KNN is a basic algorithm that stores every single accessible case and anticipate the numerical target dependent on a distance measure (e.g., distance measures like Euclidean, Manhattan, Minkowski, etc.). The result of KNN regression is the average of the numerical target values. Optimal value for K has to be selected by inspecting the data. Equations for the distance functions are as follows:
   Euclidean distance function:

$$\sqrt{\sum_{i=1}^{k}(x_i - y_i)^2} \tag{2}$$

   Manhattan distance function:

$$\sum_{i=1}^{k}|Xi - Yi| \tag{3}$$

   Minkowski distance function:

$$\left(\sum_{i=1}^{k}(|x_i - y_i|)^q\right)^{\frac{1}{q}} \tag{4}$$

3. **Decision Tree**—Decision tree regression uses a decision tree (as a predictive model). It has a tree-like structure with root node on top of the tree and leaf nodes at the bottom of the tree. The main objective of decision tree is to build a model to anticipate the value of a target variable by learning simple decision rules inferred from the data features. Based on the requirement of the experimentation, the target of the model can either take set of discrete values which will be referred to as classification tree or continuous values referred to as regression tree.

4. **Random Forest**—Random forest regression is an ensemble ML method for classification or regression works by utilizing multiple decision tree, and output is the average of all the decision tree output.

4.2.3 Model Evaluation Measure

For evaluating the constructed model, RMSE is calculated. The RMSE is a measure of the differences between predicted value and actual value. Its mathematical equation is as follows:

$$\text{RMSE} = \sqrt{}(\sum_{i=1}^{k}(p_i - a_i)^2/k) \quad (5)$$

p_i predicted value
a_i actual value
k number of samples.

5 Results and Discussion

5.1 Weather Station Results

Figure 2 displays all the sensor values in the serial monitor. Weather parameters captured are uploaded to ThingSpeak cloud using channel API write key. Figure 3 shows each weather parameter value being displayed separately in the graph. Field 1 chart displays temperature value versus time. Field 2 chart displays humidity value versus time. Field 3 chart displays air pressure value versus time. Field 4 chart displays rain value versus time. Each highlighted point in the graph represents a value. Data can be downloaded in CSV format and use it for further analysis.

5.2 Data Analysis Results

5.2.1 Prediction of MaxTemp Feature Using Random Forest

Graphical representation of actual and predicted values of MaxTemp for the January month using random forest technique is shown in Fig. 4.

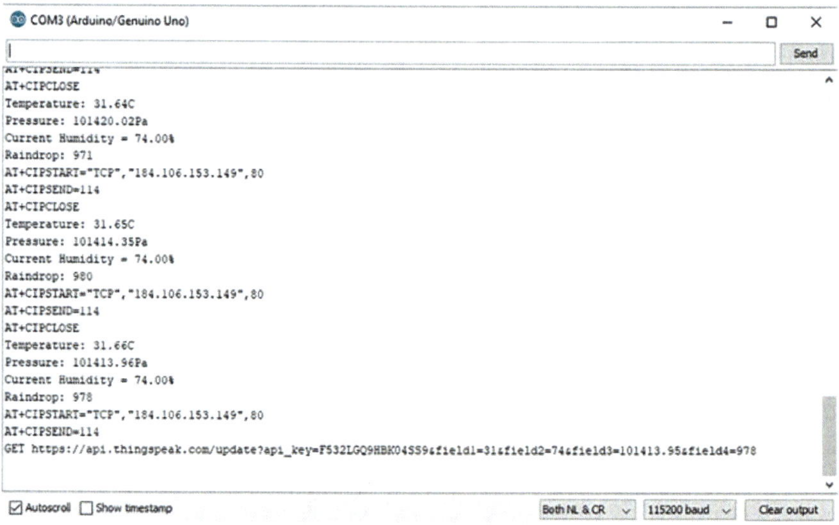

Fig. 2 Sensor values displayed on serial monitor of Arduino IDE

Fig. 3 Sensor values updated in ThingSpeak cloud. Field 1 chart represents the temperature values captured by DHT11 sensor. Field 2 chart represents the humidity values captured by DHT11 sensor. Field 3 chart represents air pressure value captured by BMP280 sensor. Field 4 chart represents rain values captured by rain sensor

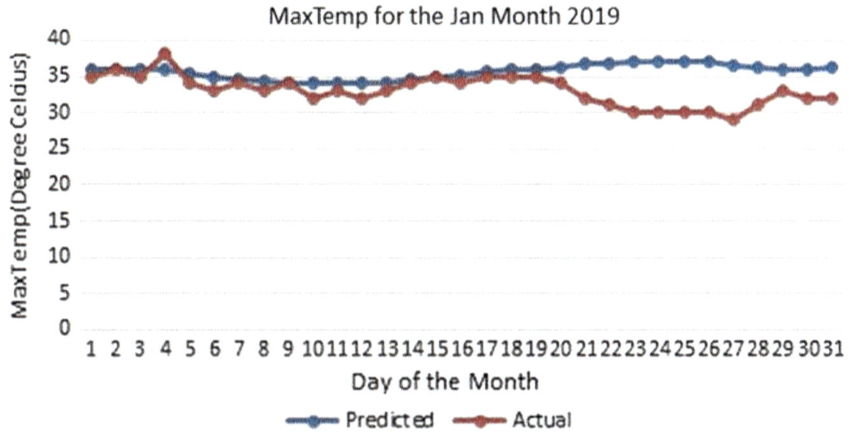

Fig. 4 Graphical representation of actual and predicted values of MaxTemp using random forest

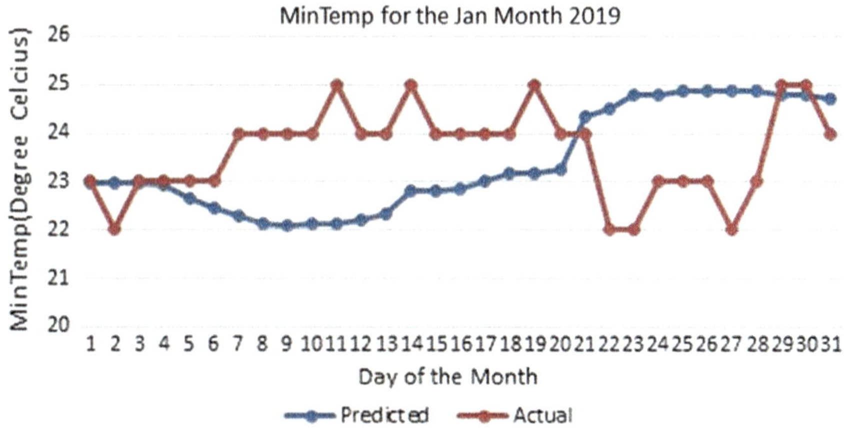

Fig. 5 Graphical representation of actual and predicted values of MinTemp using random forest

5.2.2 Prediction of MinTemp Feature Using Random Forest

Graphical representation of actual and predicted values of MinTemp for the January month using random forest technique is shown in Fig. 5.

5.2.3 Prediction of Wind Feature Using Random Forest

Graphical representation of actual and predicted values of wind for the January month using random forest technique is shown in Fig. 6.

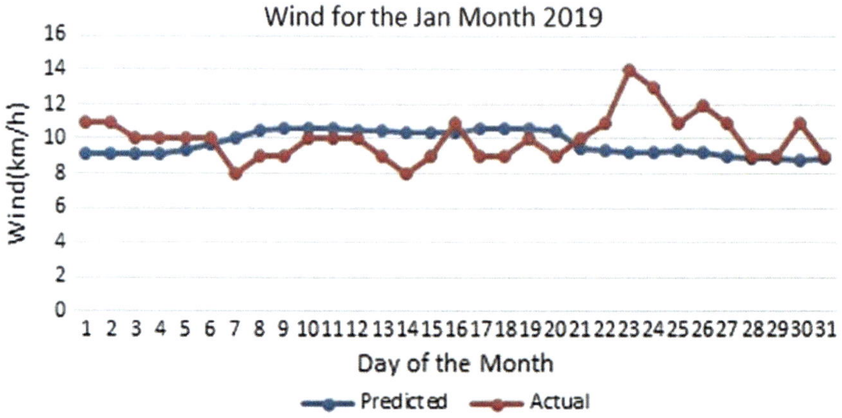

Fig. 6 Graphical representation of actual and predicted values of wind using random forest

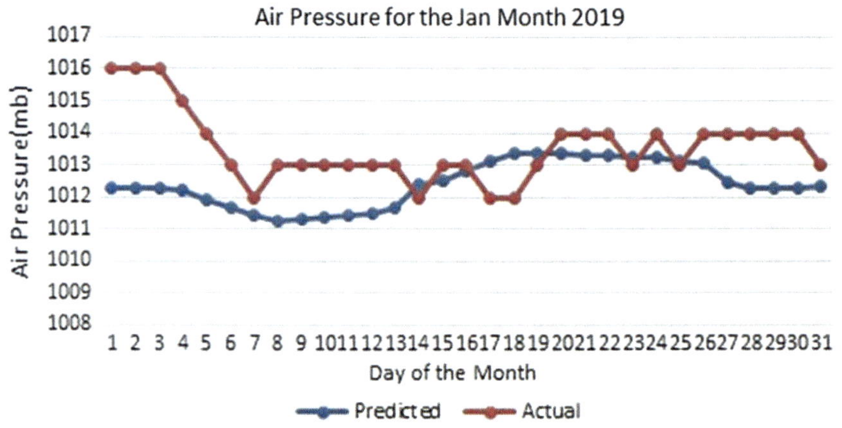

Fig. 7 Graphical representation of actual and predicted values of air pressure using random forest

5.2.4 Prediction of Air Pressure Feature Using Random Forest

Graphical representation of actual and predicted values of air pressure for the January month using random forest technique is shown in Fig. 7.

5.2.5 Prediction of Precipitation Feature Using Random Forest

Graphical representation of actual and predicted values of precipitation for the January month using random forest technique is shown in Fig. 8.

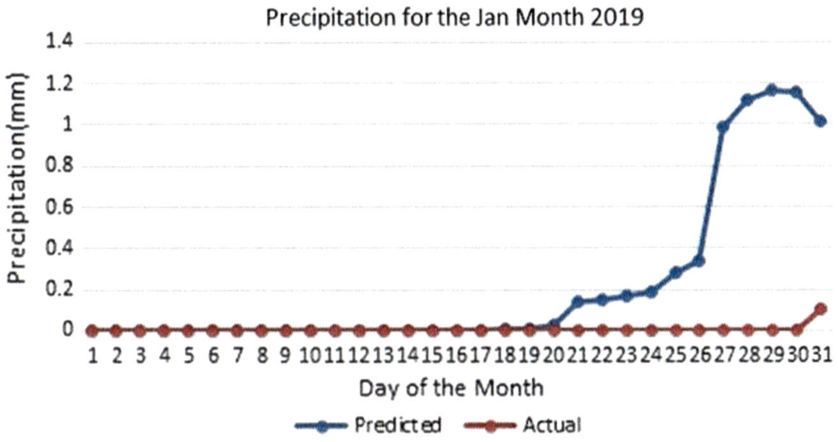

Fig. 8 Graphical representation of actual and predicted values of precipitation using random forest

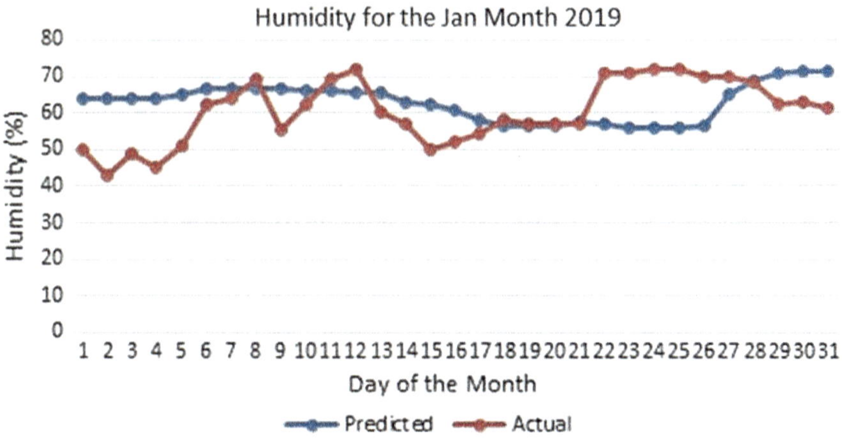

Fig. 9 Graphical representation of actual and predicted values of humidity using random forest

5.2.6 Prediction of Humidity Feature Using Random Forest

Graphical representation of actual and predicted values of humidity for the January month using random forest technique is shown in Fig. 9.

5.2.7 RMSE Comparison

RMSE is calculated for each of the ML models considered (refer Table 6). Random forest gives better results compared to other ML models because it is an ensembled method. Ensemble means a group of simple trees performing together. Results

Table 6 RMSE comparison for ML models

ML models	RMSE calculated
Multivariate linear regression	0.165
KNN	0.103
Decision tree	0.094
Random forest	0.083

achieved from group of trees are better than a single tree. Random forest shows better stability as they display less overfitting to the dataset. Predictive analysis can be carried out using random forest as the base model.

6 Conclusion

This research paper presents an IoT prototype design for capturing the weather parameters in real time as well as a comparison between few supervised ML models like MLR, KNN, decision tree, and random forest to choose the best model for predictive analysis of weather parameters for estimation of rainfall for irrigation. RMSE obtained for MLR, KNN, decision tree, and random forest are 0.165, 0.103, 0.094, 0.083, respectively. Random forest model shows the best accuracy value and hence can be considered to be the better model for the given dataset. Weather parameters are predicted for the year 2019 based on the historical weather data.

References

1. Yaganteeswarudu A, VishnuVardhan Y (2017) Software appication to prevent suicides of farmers with asp.net MVC. In: 2017 7th international conference on cloud computing, data science & engineering-confluence, Jan 2017. IEEE, pp 543–546. https://doi.org/10.1109/confluence.2017.7943210
2. Vaishali S, Suraj S, Vignesh G, Dhivya S, Udhayakumar S (2017) Mobile integrated smart irrigation management and monitoring system using IOT. In: 2017 international conference on communication and signal processing (ICCSP), Apr 2017. IEEE, pp 2164–2167. https://doi.org/10.1109/iccsp.2017.8286792
3. Sodha D, Saha G (2016) Crop management of agricultural products using time series analysis. In: 2016 IEEE international conference on recent trends in electronics, information & communication technology (RTEICT), May 2016. IEEE, pp 1456–1460. https://doi.org/10.1109/rteict.2016.7808073
4. Saha G, Chauhan N (2017) Numerical weather prediction using nonlinear auto regressive network for the Manaus region, Brazil. In: 2017 innovations in power and advanced computing technologies (i-PACT), Apr 2017. IEEE, pp 1–4. https://doi.org/10.1109/ipact.2017.8245061
5. Kurniawan AP, Jati AN, Azmi F (2017) Weather prediction based on fuzzy logic algorithm for supporting general farming automation system. In: 2017 5th international conference on instrumentation, control, and automation (ICA), Aug 2017. IEEE, pp 152–157. https://doi.org/10.1109/ica.2017.8068431

6. Shobha N, Asha T (2017) Monitoring weather based meteorological data: clustering approach for analysis. In: 2017 international conference on innovative mechanisms for industry applications (ICIMIA), Feb 2017. IEEE, pp 75–81. https://doi.org/10.1109/icimia.2017.7975575
7. ThingSpeak. https://thingspeak.com/

Smart Watering System Using MQTT Protocol in IoT

Mukambikeshwari and Asmita Poojary

Abstract Agriculture is the main source of income in the developing country like India, where life of 60% of the population depends on agriculture. As per the survey report says, 3515 farmers optioned suicide from April 2013 to November 2017. This is because usually farmers follow traditional technique for cropping that leads to many problems like improper growth of crop, lack of knowledge in using fertilizers, and improper usage of water. Ultimately, farmers take extreme steps to end up their life, not being able to face the financial issues. This problem can be solved by building an Internet of Things solution for agriculture. In our system, different sensors like soil moisture sensor, temperature sensor, and humidity sensor are positioned in fields, from where the data is collected and sent to Raspberry Pi 3. Here, the MQTT protocol acts as publisher, subscriber, and broker. The protocol receives the data from Raspberry Pi and sends it to the web application which can be operated in smartphone too. Hence, the farmer would be able to get accurate status on the condition of the field and can thereby make necessary changes if required.

Keywords MQTT · Raspberry Pi · IoT · Broker

1 Introduction

The Internet of Things (IoT) has the ability to change the world we live in; smart homes, automatic cars, etc., are all a result of the IoT equation. Notwithstanding, the utilization of innovation like IoT [1, 2] in agriculture could have the best effect.

Technologies developed based on smart farming using IoT will help agriculturist to increase the crop yield and in turn increase the fertility level of the soil [3, 4].

Smart farming is more efficient as compared to other traditional methods, where the project is structured for monitoring water supply to the crop using sensors like

Mukambikeshwari (✉) · Asmita Poojary
NMAM Institute of Technology, Nitte, Udupi 574110, Karnataka, India
e-mail: mukambika333@gmail.com

Asmita Poojary
e-mail: asmitapoojari@nitte.edu.in

N. N. Chiplunkar and T. Fukao (eds.), *Advances in Artificial Intelligence and Data Engineering*, Advances in Intelligent Systems and Computing 1133,
https://doi.org/10.1007/978-981-15-3514-7_106

humidity, temperature, moisture [5] and automated irrigation system. The farmers can get information sitting in any places [6].

The applications developed under smart farming using IoT not only target traditional systems and huge agriculture area operations, but also help uplift other slowly growing innovations in agriculture like biological farming, family farming and increase the highly transparent farming. Applying IoT technology [7–10] in the farming field not only increases crop fertility, but it also helps to solve major problem [11] like water scarcity by using the application in an efficient way.

2 Related Work

In [1], the project a robot was designed to handle various sensors like soil moisture, humidity, UV, thermo-hydro, etc. In this, the robot keeps moving in the certain path to keep track of foreign obstacle moving around. The solar technology is used for power supply. All the recorded data is stored inside the Zigbee protocol, which is programmed accordingly.

In [2], an application is designed for two seasons one for summer and one for rainy. During summer season, the soil moisture is detected through hygrometer. If it goes beyond the threshold, the warning message is sent to the user's phone saying that there is water scarcity in the field. User can give the command to switch on the motor. At the time of rainy season, radar-level sensor is used which will send a warning message to the user to remove the execess water in the crop field.

In [3], the aim of the proposed system is to change the traditional technique used in the previous proposals. Here, the soil moisture sensor activates the water sprinkler when water requirement raised in the field. The soil PH sensor will detect the fertility of the soil over a period of time, and the stored value is sent to the farmer to reduce or increase the amount of fertilizer to be used next time. This proposal was applicable only to the rice cropping.

In [4], the UAV and ground sensors are put in the field where UAV is used to spray the chemical in the field, and it is designed in such a way that the spraying should not happen to any other owner fields. Hence, to avoid the UAV keeps communicating with the WSN, the two modules used to stimulate communication between UAV and WSN are behavioural module and chemical dispersion module.

In [5], an algorithm is used to monitor the water extent. The field is installed with solar panel for power supply. The plants are observed to detect the disease by placing wireless cameras in the field. The image taken is processed with image processing technique. The developed system is less power consuming and lower in cost.

In [6] developed smart agriculture monitor system using wireless sensor network on IoT, monitoring the field through video information and collecting the data based on humidity and temperature. So better communication can achieve and reduce power consumption. System is workable in outdoor as well as green house.

In [7] developed smart agriculture monitor system using wireless sensor network on IoT, monitoring the field through video information and collecting the data based on humidity and temperature. So better communication can achieve and reduce power consumption. System is workable in outdoor as well as green house.

In [8] proposed a technique to solve the irrigation problem, and sensors dipped in the field will send the signals to the microcontroller for evaluation of the water quantity. Microcontroller handles the pump and servo motor to supply water to the crop.

In [9] designed system where android mobile application collects data of different sensors, which notifies the user for inputting proper fertilizer and watering to the field on time. Graph generated in the application shows the soil fertility level. Application and device installed in the field interact each other through cloud-based station.

In [10], Automated irrigation system which is developed for the contry Philippine, where Gizmo moisture sensor and temperature sensor dipped in the field, data is processed by the Arduino-compatible microcontroller. LCD shows the value of the moisture. Arduino Uno is coded to control the water valve.

3 Proposed Model

Flow diagram in Fig. 1 shows the working phenomenon of the entire system. When the machine switches to start mode, water level of soil is read by designated sensor [12, 13]. When the moisture level goes down, the sensor automatically sends the signal to the main machine to switch on motor. Once after required water is supplied to the crop, the motor goes to off mode. The protocol used for communication purpose is the MQTT protocol [14–16].

3.1 Message Queue Telemetry Transport Protocol

MQTT has three parameters—publisher, subscriber, and broker [17]. This protocol is very simple and lightweight messaging protocol and reads the data in an efficient way. This protocol is basically designed for devices like low bandwidth and high latency or for the unreliable networks. It is designed to decrease network bandwidth, device requirement and also trying to ensure the reliability, which gives the degree of guarantee in delivery of data.

In MQTT, a publisher publishes messages on a topics subscribed by subscriber shown in Fig. 2. Every subscriber must subscribe the topic to view the messages. There is no direct connection between the publisher and subscriber, but the communication takes place through the broker. There are many MQTT brokers available. Mosquito broker uses iot.eclipse.org server which is an open-source MQTT broker that runs on Windows and Linux. Libraries need to be getting installed to use this protocol. Various types of authentications and data security mechanisms are carried out by using MQTT protocol.

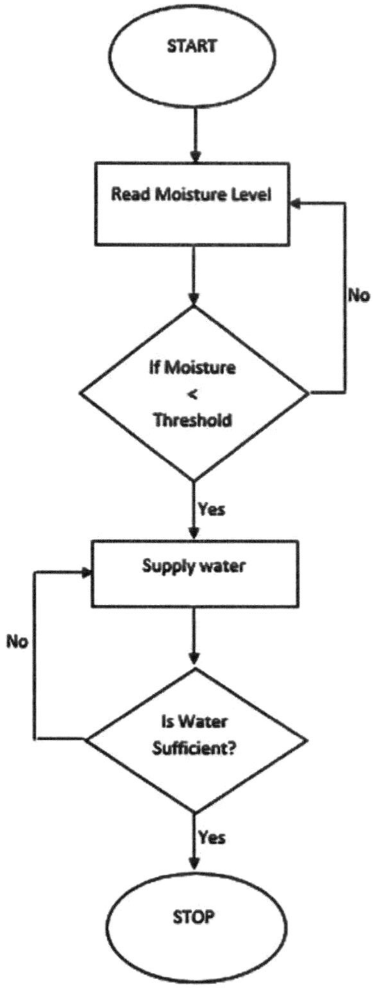

Fig. 1 Flow diagram of the model

In this system, Raspberry Pi 3 board is connected with digital soil moisture sensor, sprinkler unit, humidity/temperature sensor and motion sensor over respective GPIO pin, shown in Fig. 3.

3.2 *Raspberry Pi 3*

The Raspberry Pi 3 is connected to different sensors like moisture sensor, temperature sensor, and PIR sensors. These sensors read the data through GPIO pins of kit, after

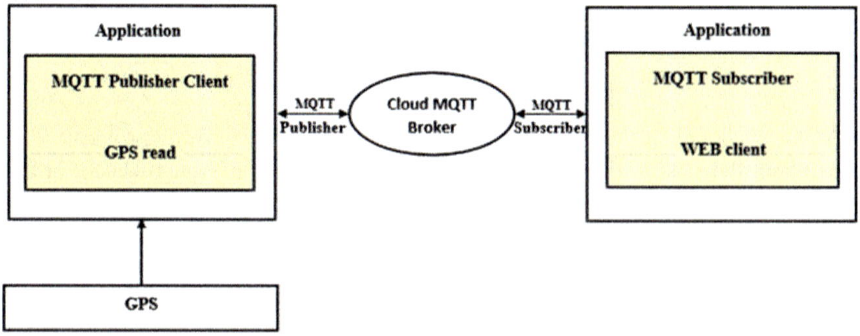

Fig. 2 MQTT protocol communication

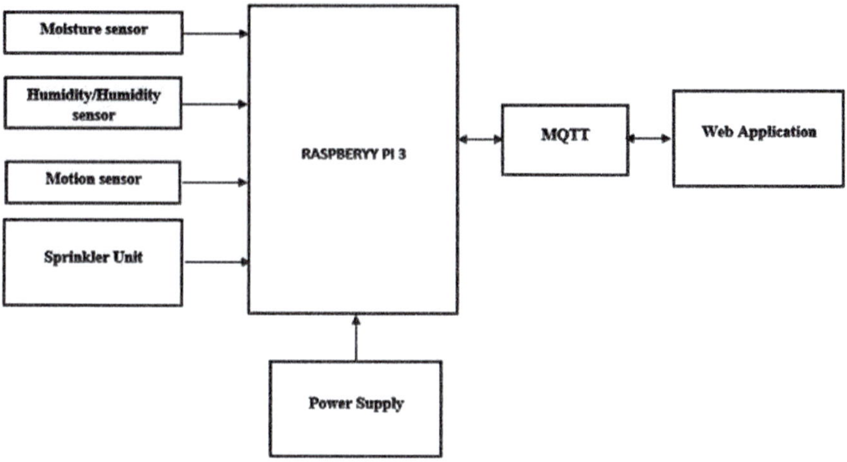

Fig. 3 Block diagram of the system

which data is passed to web server via the protocol named message queue telemetry transport to web application. The system is connected to a Wi-Fi modem [18–20] for the signal.

Digital soil moisture sensor is used to observe the water content in the soil using dielectric permittivity [21] by simply inserting sensor into the soil to be tested. The value of the water content of the soil is generated in percentage also can be 5 V for high moisture and 0 V for low.

The passive infrared sensor (PIR sensor) acts as motion sensor. This sensor detects the movements by light emitting from the object. This can be placed in sensitive areas. This will detect the movement of any objects, animals, and humans. Humidity/temperature sensor [22, 23] is a single sensor which detects the soil temperature and humidity level. This can also be used to detect air temperature/humidity. This is taken as supplementary with moisture sensor for more accurate ground data.

3.3 Soil Moisture Sensor and Sprinkler Unit

The soil moisture sensor and the sprinkler unit are the two units where sensor monitors the water level and sprinkler works as per the instruction given by control unit. The water quantity in the soil and estimates the adequate amount of water needed for the crops/plants [24].

The soil moisture sensor and sprinkler units will be connected to Raspberry Pi 3 board over GPIO pin, respectively.

The application running in the board will get the soil moisture level from the sensor [25] connected over GPIO21. It then decides whether to indicate the sprinkler based on the present moisture level of the soil. If the moisture level of the soil attempt lowers than the threshold, then application indicates the sprinkler unit to flow the water. It also monitors the water flow to the soil simultaneously. When enough water is supplied to the soil, the application indicates the sprinkler to stop water sprinkling to the soil. Thus, these units will achieve the water flow control and also provide enough water to the crops on time when it is required [15].

3.4 Motion Sensor Unit

Motion sensor continuously monitors the field. This electronic sensor detects the movement any object where it occupies. Motion of any moving object is detected by the sensor and signals immediately sent to machine to alert the user. At the application side, buzzer gets activated when the foreign object is found in the field. This will help famers to avoid visiting the farm in the midnight and also improve the productivity [26–28].

3.5 Data Monitor/Control Unit

All the sensor data received in the RPI3 board will be sent to control unit, which is basically a web/android application. In this application, all sensor data will be displayed. RPI3 board will be connected to the Internet using Wi-Fi. It uses MQTT protocol to send the data to the control unit. RPI3 board acts as a publisher MQTT client, which sends the data to any MQTT broker/cloud.

This data will be sent to the control unit, which is subscriber MQTT client, by the MQTT broker hosted on the cloud.

There are two modes of operation in this model of approach,

1. Automatic
2. Manual.

Fig. 4 System model

In automatic mode, all the decisions will be taken by the RPI3 unit itself based on the sensor inputs. For example, if the water level goes below certain threshold, it immediately indicates the soil moisture unit to supply the water.

In manual mode, based on the data received from the RPI3, user can operate any units remotely. If the application indicates the soil moisture level is too low, then user can just press the button present in the application to allow the sprinkler unit to start supply water. Thus, this module helps to closely monitor and control the water level based on humidity/temperature level, and it supplies enough water to the soil based on the requirement without wasting. Also it can be remotely operated from any place and can be controlled using just an application on the mobile phone.

In this project, the whole work is carried using Raspberry Pi 3, shown in Fig. 4. Various sensors are connected to Raspberry Pi through jumping wires. The complete output of system can be observed in web application shown in Fig. 4. Automatic water supply is carried out using the motor. 5 V DC power supply is required to Raspberry Pi through battery. Front-end design is done using HTML and JavaScript and backend coding using C programming. Accurate variation in sensor output can be observed.

4 Results and Discussion

Initially, the user must register to cloud ID (iot.eclipse.org) and port number (80) to get connected to cloud. Once after the user logs in the device, ID must be provided to get access the device. After Raspberry Pi starts sending the data to the UI through the cloud, the values of different sensors are displayed as shown in Fig. 5.

Here, the sensor values are displayed in web application. The device ID must be given to start the particular devices. Multiple devices can be used for different field. The device ID can be given by putting slash when multiple devices used. The status

Fig. 5 User interface of the proposed system

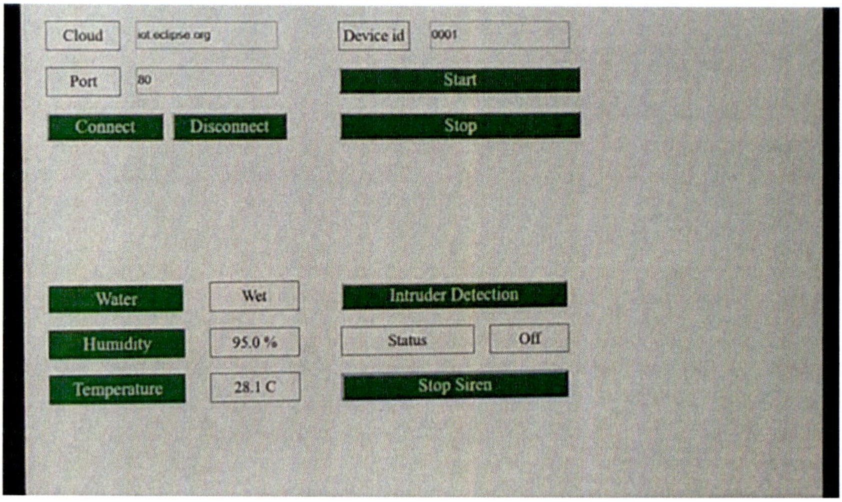

Fig. 6 Status of soil, humidity, temperature, intruder detection

of soil sensor, humidity sensor and temperature sensor is known. Intruder detection is alerted by motion sensor. At the application end, the buzzer alerts the user, shown in Fig. 6.

5 Conclusion and Future Scope

In this work, the developed system can help farmers in an automated irrigation system by measuring the soil moisture, temperature, humidity level of the farm. By using this irrigation system, farmer can avoid watering to the crop with manual intervention. The primary application of this project is to reduce water scarcity and stress of visiting the farm in the night time to check the water requirement. Also it is simple, cost-effective and less hardware required system, which helps farmer in easy installation and maintenance which provides the accurate data of sensors on time. Development of android application ++n is more helpful to the farmer, and also GSM can be used for Wi-Fi unavailability.

References

1. Krishna KL, Silver O, Malende WF, Anuradha K (2017) Internet of Things application for implementation of smart agriculture system. In: International conference on I-SMAC (IoT in social, mobile, analytics and cloud) (I-SMAC 2017)
2. Hari Ram VV, Vishal H, Dr. Dhanalakshmi S, Vidya PM (2015) Regulation of water in agriculture field using Internet of Things. In: 2015 IEEE international conference on technological innovations in ICT for agriculture and rural development (TIAR 2015)
3. Vijayakumar S, Rosario JN (2011) Preliminary design for crop monitoring involving water and fertilizer conservation using wireless sensor networks. In: 2011 IEEE 3rd international conference on communication software and networks (ICCSN)
4. Costa FG, Ueyama J, Braun T, Pessin G, Osório FS, Vargas PA (2012) The use of unmanned aerial vehicles and wireless sensor network in agricultural applications. IGARSS 2012
5. Nisha G, Megala J (2014) Wireless sensor network based automated irrigation and crop field monitoring system. In: 2014 Sixth international conference on advanced computing (IcoAC)
6. Bhanu B, Rao R, Ramesh JVN, Hussain MA (2014) Agriculture field monitoring and analysis using wireless sensor networks for improving crop production. In: 2014 Eleventh international conference on wireless and optical communications networks (WOCN)
7. Parameswaran G, Sivaprasath K (2016) Arduino based smart drip irrigation system using Internet of Things. Int J Eng Sci Comput
8. Hlaing CS, Zaw SM (2017) Plant diseases recognition for smart farming using model-based statistical features. In: 2017 IEEE 6th global conference on consumer electronics (GCCE), 24–27 Oct 2017
9. Gumaste SS, Prof. Kadam AJ (2016) Future weather prediction using genetic algorithm and FFT for smart farming. In: 2016 International conference on computing communication control and automation (ICCUBEA)
10. Ghosh S, Sayyed S, Wani K, Mhatre M (2016) Smart irrigation: a smart drip irrigation system using cloud, android and data mining. In: 2016 IEEE international conference on advances in electronics, communication and computer technology (ICAECCT). Rajarshi Shahu College of Engineering, Pune, India. 2–3 Dec 2016
11. Sarangdhar AA, Prof. Dr. Pawar VR (2017) Machine learning regression technique for cotton leaf disease detection and controlling using IoT. In: International conference on electronics, communication and aerospace technology (ICECA 2017)
12. Patil SS, Thorat SA (2016) Early detection of grapes diseases using machine learning and IoT. In: 2016 Second international conference on cognitive computing and information processing (CCIP)
13. Mat I, Kassim MRM, Harun AN, Yusoff IM (2016) IoT in precision agriculture applications using wireless moisture sensor network. In: 2016 IEEE conference on open systems (ICOS), 10–12 Oct 2016, Langkawi, Malaysia
14. Padalalu P, Mahajan S, Dabir K, Mitkar S, Javale D (2017) Smart water dripping system for agriculture/farming. In: 2017 2nd international conference for convergence in technology (I2CT)
15. Abaya S, De Vega L, Garcia J, Maniaul M, Redondo CA (2017) A self-activating irrigation technology designed for a smart and futuristic farming. In: 2017 International conference on circuits, devices and systems

16. Shi Y, Wang Z, Wang X, Zhang S (2015) Internet of Things application to monitoring plant disease and insect pests. In: International conference on applied science and engineering innovation (ASEI 2015)
17. Kodali RK, Sarjerao BS (2017) A low cost smart irrigation system using MQTT protocol. 978-1-5090-6255-3/17/$31.00 c 2017 IEEE
18. Changl H, Zhou N, Zhaol X, Cao Q, Tanl M, Zhang Y (2014) A new agriculture monitoring system based on WSNs. In: ICSP 2014 proceedings
19. Marimuthu R, Suresh A, Alamelu M, Kanagaraj S (2017) Design and development of a persuasive technology method to encourage smart farming. In: 2017 IEEE region 10 humanitarian technology conference (R10-HTC), 21–23 Dec 2017, Dhaka, Bangladesh
20. Sisyanto REN, Suhardi, Kurniawan NB (2017) Hydroponic smart farming using cyber physical social system with telegram messenger. In: 2017 International conference on Information technology systems and innovation (ICITSI)
21. Kodali RK, Jain V, Karagwal S. IoT based smart greenhouse
22. Jeong YJ, An KE, Lee SW, Seo D (2018) Improved durability of soil humidity sensor for agricultural IoT environments. In: 2018 IEEE international conference on consumer electronics (ICCE)
23. Rekha P, Ramesh MV, Rangan VP (2017) High yield groundnut agronomy: An IoT based precision farming framework. 978-1-5090-6046-7/17/$31.00 ©2017 IEEE
24. Awate A, Deshmankar D (2015) Fruit disease detection using color, texture analysis and ANN. 978-1-4673-7910-6/15/$31.00_c 2015 IEEE
25. Akyildiz IF, Su W, Sankarasubramaniam Y, Cayirci E (2002) A survey on sensor networks. IEEE Communications Magazine. Aug 2002
26. Nishwinpal K, Kumar KVG (2016) Design and implementation of citrus classification architecture on FPGA, IOSR. J VLSI Sig 6(3):18–25 (Ver. II)
27. Malavade VN, Akulwar PK (2016) Role of IoT in agriculture. In: National conference on changing technology and rural development, CTRD 2k16
28. Minbo L, Zhu Z, Guangyu C. Information service system of agriculture IoT

Internet of Things (IoT) Enabling Technologies and Applications—A Study

D. K. Sreekantha, Ashok Koujalagi, T. M. Girish and K. V. S. S. S. S. Sairam

Abstract Internet of things (IoT) is a worldwide network of the autonomous real-world and digital internet enabled things connecting and communicating with each other seamlessly. The main objective of IoT concept is to transform traditional devices to intelligent and autonomous units, which can be managed from anywhere and anytime over the Internet. IoT technologies are predicted to improve the standards of life, and at the same, there are many issues to be researched intensively. IoT enables the people and machines to discover and interact with many devices like sensors, actuators, service triggers and other things connected to the web. The recent technological innovations in the internet, cloud, big data and smartphones leveraged the design of smart factories, smart homes and smart cities. IoT market research predicted that rapid developments in IoT applications will dramatically increase the number of sensors and actuators deployed worldwide. This chapter reviewed the literature from papers from high impact journals on IoT enabling technologies, solutions and their deployments in smart environments.

Keywords IoT · Cloud computing · Wireless sensor networks (WSN) · Smart ecosystems · Wearable devices and sensors

1 Introduction

International Telecommunication Standards Board (ITU-T) defined Internet of things (IoT) as worldwide network for serving knowledge community. The sophisticated services are provided over interconnected real and digital things (ITU-T, 2012). IoT interconnects the physical and virtual world [1], and the intelligent components are embedded into various physical objects [2]. The world has benefited significantly by deploying internet-enabled solutions [3] in recent years. IoT solutions

D. K. Sreekantha (✉) · K. V. S. S. S. S. Sairam
NMAM Institute of Technology, Nitte, Karkala, Karnataka 574110 , India
e-mail: sreekantha@nitte.edu.in

A. Koujalagi · T. M. Girish
Basaveshwar Science College, Bagalkot, Karnataka, India

N. N. Chiplunkar and T. Fukao (eds.), *Advances in Artificial Intelligence and Data Engineering*, Advances in Intelligent Systems and Computing 1133,
https://doi.org/10.1007/978-981-15-3514-7_107

are having tremendous prospects to implement smart operations. Intelligent communication between interconnected devices will create and share large amounts of information enabling timely decision-making [4]. The authors demonstrated the IoT security prototype system implemented using the Arduino board. [1]. IoT should also ensure a secure device-to-device and device-to-human communication. The authors explored the opportunities and challenges for building a successful blockchain and secure IoT applications. Investments in the development of IoT solutions have been strongly supported by governments, interest groups, product companies and research institutes. IoT-European Research Cluster has focused research on context- awareness during the period 2015–2020 [4]. The implementation of blockchain and IoT technology in the public systems would quickly interconnect citizens, governments and industries [5]. Recent developments on the cloud, big data, and embedded systems have triggered the rapid growth of the IoT solutions.

2 Emerging Technologies for IoT Ecosystems Development

2.1 Sensor Technologies

IoT interconnects different devices with communication, sensing, signalling and networking capabilities. There are numerous multimedia and publishing medium apps that are having a strong influence on social life. These audio, visual and publishing medium archives are accumulating a wealth of resources. The focus is on the IoT solutions required to understand the context and subject information stored in these archives. Authors discussed the existing tools, applications, techniques and models addressing the various issues to make IoT a reality and to offer a strong appetite for IoT [6]. There are many devices to improve reliability, privacy, and safety in IoT networks. Governments and industries are interested in implementing IoT solutions for surveillance. The food industry has introduced wireless sensor networks (WSN) and RFID technologies for the modern system of surveillance, monitoring and verifying to ensure the quality of food supplied. The pervasive (P-IoT) is a new venture to determine the exact details of physical exercise. Monitoring applications of IoT can provide effective systems for gathering essential knowledge on the situation for developing security measures. IntelliSurv is a solution that identifies illegal events in different ways. This established system coordinates with the sources of emotion with sensor management in the Roman theoretical team. This system is based on the model of the belief–desire–intention (BDI), by establishing an independent decision and funding agency.

2.2 Wireless Sensor Networks (WSN)

IoT is a well-known communication system that interconnects several millions of devices. WSN gathers information and plays an important role in the implementation of many mobile-based solutions. Most of the services of WSN are linked and location-oriented. WSN is an essential part of some open IoT structures and designed to support some specific applications. Recent developments in browser technology, design techniques, and communication systems have helped in the design of smart IoT applications. This study has predicted a steady increase in the applications of WSN with multiple application sets sharing the same WSN infrastructure. Energy harvesting algorithms deal with conserving the power in wireless devices. The power consumption is a very significant factor in determining the service quality, and efficiency of wireless devices [7]. The solutions to improve the reliability, safety and privateness in IoT domain have been explored by long-term evolution (LTE) technologies.

The battery pack from several antenna batteries transmits wireless transmissions to charge the IoT devices. The biodiversity computing paradigm demonstrates the ability to present the high-quality results in communicating messages through WSN and adhoc transport networks [8]. The movement of the sensor nodes in a geographic region enables us to monitor the data transmission and energy status via distance nodes. A comprehensive overview of existing traditional and biological shell protocols was carried out. Biodegradable mobile routing mechanisms override the traditional circuits in the future. This will reduce congestion, delay, and complexity of computing required for IoT services. IoT paradigm leverages to combine thousands of million devices, several products and consultancy anywhere, anytime using different applications, and mobile apps [9].

2.3 Cloud Computing

World Wide Web is leveraging the increased interaction of people with the physical and virtual worlds. The analysis of operational data gathered during the previous years in business and industry helps in understanding the trends. This trend analysis empowers intelligent decision-making. To store and manage previous years' data, one requires large storage devices and data management systems. The need for analyzing big data triggered the growth of the cloud computing paradigm. Today, cloud database enables online storage and extraction of a large quantity of data instantly. The storage and retrieval of big data is a bigger problem in the IoT domain. IoT devices have resource constraints for processing, storing and low power at the edge. IoT technology has radically changed the data communication and device management in many software implementations [3]. The increased interconnected system size with billions of nodes creates challenges for data communication and processing, and in many cases, customer needs data history. This paper compares the domain descriptions of 26 different IoT platforms, which depend on ten specific programs.

Authors discovered that visualization and cloud-based research are inadequate with current scenarios. The authors discussed the idea of moving to the vast cloud providing opportunities and solutions that can communicate with the environment, facilitating the evolution of innovative services [10]. Widespread clouds are connected to mobile products, various sensors, and actuators to provide computing and storage resources on-demand with quality of service assurance. This document concentrates on the deployment of SAaaS-based basic features. The presence and availability of sensing and actuation as service is gradually increasing over the internet.

Cloud computing is a prototype for accessing requests for services to a configurable resource common pools (such as servers, applications, software, computers, networks and, storage) which can be provided at ease [11]. Regional IoT systems enable us to interconnect something around to access any content in a comfortable way using customizable websites and firmware (SaaS). Hence, the cloud plays the role of an interface to access the IoT device network.

The software that communicates with sensors has special specifications for storing large data, huge processing ability for dynamic data processing and high-speed networks for video and audio streaming. C-DAC regional survey research contributes to the creation of a cloud service of HPC for all research teams of different sizes [3]. The various problems associated with sensor identification as a service in the scientific clouds are discussed. The authors suggested many approaches to integrate physical devices to the Internet [12]. The accelerated growth of IoT solutions is restricted by constraints on computing, storage resources, concerns regarding confidentiality and safety at the edge nodes.

The cost-effective and reliable solutions that collectively address safety, efficiency, privacy, and workability are necessary for the sustainable development of IoT [13]. The authors suggested solutions for recognition of the human context for controlling the devices using regional sources of IoT. The additive production technique (AM) has specific abilities to provide an innovative way to quickly develop the products [14]. The emerging cloud paradigm facilitates accessing multiple sources at minimal cost and effort. A novice cloud framework that interfaces not only with solid material sources such as 3D printers but also knowledge sources such as test data should be designed. The edge computing is a technique to transfer a portion of the program from the cloud to edge devices for better interaction with them. The edge computing technique implementation in IoT investigates the power utilization patterns in IoT products [15]. The experimental set-up estimates the utilization of computing and storage resources at the edge devices. This study helps in understanding the impact on the lifetime of controlled devices driven by batteries. Authors have compared the features, advantages and disadvantages of cloud, and edge computing paradigms.

This paper proposes a prototype for offering sensing as a service over the cloud. Authors illustrated the concept of sensing as a service over the cloud through some software tools such as augmented reality for environment and agriculture observation studies. IoT achievements using RFID technologies are in early stage. The big data and sensing services can be offered through cloud in a cost-effective way to revolutionize the IoT in the consumer market. The common man can access these sensing services through their smart devices on daily basis. The cloud computing, big data and blockchain technologies are expected to bring a big revolution in IoT implementation. Authors have explored the latest studies on Industrial IoT (IIOT) domain. This paper introduces the context for SOA models and essential techniques which may be applied to crucial industry processes of IoT [16]. The authors investigated the research issues and upcoming trends related to IoT. The significant contribution from the author's study is in identifying future research problems in the domain of industrial IoT to promote further research.

2.4 IoT Middleware

IoT middleware is a software layer used in the implementation of IoT applications. IoT middleware is a computer program designed to interface the IoT devices and applications.

The authors presented the requirements of an IoT middleware through an IoT solution [17]. This IoT solution is designed for dynamic estimation of alcohol level in the blood through data sensed through smart watches. Authors have carried out an extensive literature survey of technical solutions designed for IoT middleware. IoT middleware manages the diversity in IoT nodes and takes care of critical elements of structure, flexibility and safety aspects of an IoT system. Authors have analysed the prominent IoT middleware designs from customer-centric, cloud-oriented designs, lightweight, user-based designs, and heavyweight service-oriented frameworks. The assessments of IoT middleware performance were carried out on real-life context and discovered that investigation about features IoT middleware is necessary. This investigation has considered the features such as autonomy, scalable service registering, detection, configuration, diversity and cross-operability middlewares of IoT.

The comparative study of four IoT middleware solutions such as Open IoT, UBIWARE CHOReOS and LinkSmart was carried out [11]. The review of literature on operating systems exclusively developed for IoT domain reveals the broad architecture, characteristics preferred in custom-made operating system (OS) [18]. IoT solutions have some exceptional characteristics in comparison with traditional computing products. Hence, the OS has to be developed as per the unique specifications of IoT products and targeted applications. The authors have investigated for designing the perfect OS for IoT. The subject debated in this paper enables us to control the smart real-time IoT systems and discussed a general prototype for IoT operating system. The paper [19] inspects the operational variances among several IoT and enterprise protocols to find the important technical problems that need to be considered while designing facility virtualization in IoT domains.

The authors studied various IoT and enterprise protocols with an emphasis for increasing assistance, and virtualization in IoT domains. This paper [20] discusses an architectural solution that permits end-users in developing IoT services by concentrating on domain logical problems using smartphones. Authors pointed out that confidence in the technology, real-world IoT implementation of solutions, consistency in operations, customer satisfaction will have actual influence on success of company's products and hence the profits. The exploration of a frameworks for authenticating mobile objects in smart locations in the IoT domain with four-layer architecture was carried out [21]. The objective of this framework is to focus on various IoT issues such as heterogeneity, scalability and mobility. The study results enabled authors to the design the objectives for IoT deployment in the smart city. IoT integrates device-oriented, internet-oriented and semantic-oriented approaches into one paradigm [21]. The things-oriented approach presents the facts that every device in IoT domain may be observed through ubiquitous techniques. The semantic-oriented approach points to large quantity of different types of data gathered from sensors. This data has to be analysed to understand the hidden meaning and purpose of the data using semantic methods. IoT leverages us to access the current status of devices and resources to the gather information on energy utilization in manufacturing [22]. This information enables us to monitor and control the power consumption patterns in a manufacturing process.

The authors proposed a real-time energy efficiency optimization method (REEOM) for monitoring and control of excessive power consumption in the manufacturing process. Multilevel event model and multifaceted event processing techniques are applied to attain real-time energy-linked key performance indicators (e-KPIs). This paper discussed a use case with a non-dominant genetic algorithm II applied to plan or replan the production schedule in an efficient way to save energy based on dynamic e-KPIs. The future roadmap for factories recognizes the most important problems fed today in production line today.

The emerging complications in production procedures are influencing the decision-making process [23]. This research paper proposes a framework based on IoT solutions for effective process management by decreasing the time required for taking decisions. This objective is achieved by vigorously creating context awareness on the top of current production processes. This architecture is evaluated on day-to-day production situations by executing system for real estate industry for competent project administration. This reference architecture aims at enabling project managers with the deepest context understanding of the project. This investigation work, therefore, directs the track for smarter mission administration solutions enable supervisors to competently design, screen, observe and regulate building. This tool empowers mangers to deliver the project on time, with in the budget and satisfactory value.

IoT solutions have significant features for smart transportation based on industry 4.0 [24]. Smart transportation applies IoT hybrid techniques such as wireless sensor network (WSN), radio frequency Iidentification (RFID), cloud computing and analysis of big data to improve the tracking and making decisions regarding logistics instantaneously with great precision and adaptability. The transportation activities are structured into an ontology scheme based on a framework having four levels of services. IBM Company pays emphasis on novelties in integrating data using IoT and the processing information, typically relevant in a diverse industrialized area. The UPS concentrates on processing of big IoT data to enhance LSP administration. The advantages and problems in smart device-to-device (D2D) interactions need to achieve the demands of IoT [25]. Several businesses and standardization bodies have publicized keen interest in putting D2D into practice in wireless networks. D2D approach leverages centralized administration and efficient energy utilization with traffic redirecting.

This paper discusses a real-time network developed and executed for humanoids [26]. Monitoring of networked entities, their condition, alarming, messaging, emailing and filing is a significant job to ensure reliable communication of data. The currently available conventions are compared with network planning procedures of RTNET. The authors found that RTNET organizes the data exchange of the five network interaction objects on the priority to satisfy actual demands for the restricted bandwidth of the LAN. This paper focuses on logical and technical aspects for reliability, design, improvement, and security of IoT systems [62].

The objective of this study is to improve technical ideologies and procedures for upholding intelligent regulations in IoT solutions for efficient policymaking. The physical features are significant for IoT expansion. The IoT system's operations shall be software-driven controlled from remote devices practically from every point. The development of a smart automatic system to gather thermal power plant data was presented [27]. This smart system enables simulation and trend analysis of boiler performance. The boilers are interconnected by a sensors networks driven by an IoT software in a distant place. This paper also proposes an idea of a virtual knob facility for an IoT device that can regulate the conditions of the furnace from the distant area.

2.5 *IOT Embedded Systems*

The authors suggest the development of cost-effective, power savvy and tiny image recording devices for use in IoT solutions [28]. These devices collect image data, carry out processing and transmit the image data through a 2.4 GHz Wi-Fi links. This device uses a cost-effective high-throughput STM32F7 microcontroller that can capture and transmit images at maximum speed allowing the idle core to carry out image manipulations. Authors discussed the development and importance of IoT solutions in everyday life, the general framework, popular conventions, many probable solutions, future trends and application of MQTT protocol implemented in Arduino platform [5].

The dynamical selection of route for the mobile devices in a diverse wireless mesh topology network with forwarding and the backward technique was applied to discover the occurrence of holes in interconnectivity in the system [29]. An IoT-based vigilance system for an apple orchards was created using WSN [30]. This system uses CC2530 chip and ZigBee technology for wireless transmission and distributed sensors to precisely capture and transmit environmental conditions in apple orchard. This system senses the environmental data such as infrared, temperature, lighting and humidity parameters from apple orchard. This monitoring system has a wider coverage area, higher speed, and stable data transmission, consumes less power and has higher practical value.

This paper describes the safety, and privacy challenges faced in IoT products and services delivery [31]. The authors have listed the safety problems, attacks, dangers and susceptibilities encountered in product or service delivery management. Authors discussed the Apple iPhone product delivery process for illustration and presented the loopholes and possibilities for attacks. The IoT products have restricted resources that demand a special routing protocol for transmitting the sensed data from one end to another end [32].

2.6 IoT Security Technologies

This paper describes a summary of the problems, faith and confidentiality factors that are the main bottlenecks in IoT solutions implementation and growth [33]. Globally, the IoT solutions are deployed in various businesses and industries.

The CERP-IoT of European Commission has been leading the research on IoT technologies. Multinational corporations are motivating the IoT research and promoting curiosity in IoT solutions. Smart partnerships across the world are essential to realize the optimistic prospects presented by IoT.

This situation calls for immediate technical breakthrough results and good authorities and standardization procedures to ensure safe and secure IoT solutions [34]. The smart organizations comprise very large number of smart devices interconnected together in the global system known as the IoT. Reliable interactions among IoT devices are critical for the safety and security in IoT [63]. Security can be enforced by authentication in large number of interconnected objects in IoT network. This paper discusses numerous facilities and configuration methods that can be useful in the smart domain based on the systematic literature review (SLR) technique [35]. SLR classifies the present research methods based on the IoT service configuration. These research techniques are matched with one another based on some technical features. The technical features are system accuracy, useful properties, quality of service, procedures and present working environment in domain-specific operational methods.

Authors discussed the pros and cons of each designated method and explored some clues for resolving their limitations. This paper discusses some important case studies with trustfulness and privacy mechanisms in QoS-aware service configuration methods with higher safety exposure. Author's study gave insights into the standards and solutions giving the assurance to integrate various discrete WSNs [36]. Authors proposed an architecture to match the old and fresh installations (non-IP based) preserving the probability to transmigrate to an all IP platforms subsequently. This suggested architecture is presently under testing in the real estate mechanization. The semantic model development for various component parts for an IoT system was explored by the authors [37].

2.7 Smart Technologies

Smart city. "The British Standards Institution has stated Smart City is a term denoting the effective integration of physical, digital and human systems in the built environment to deliver a sustainable, prosperous and inclusive future for its citizens". "IEEE says smart city brings together technology, government, and society to enable the following characteristics-smart economy, smart mobility, smart environment, smart people, smart living, and smart governance". The implementation of smart solutions such as smart devices, smartphones, smart meters, and hardware has leveraged the deployment of smart city solutions. Smart cities comprise of diverse smart entities such as smart homes, smart streets, and smart grids. The semantic notation leverages this model data depicted as connected data and can also be related to actual data on the internet as specific linked open data. The researchers of cloud computing and IoT networks are forming a team to collectively address the big data issues [11]. This paper studied an urban IoT system planned to assist the vision of the smart city [38].

The objective is to take advantage of cutting edge networking technologies rendering value-added services for the management of the smart city. The authors have carried out a complete study of the empowering applications, protocols and frameworks for smart city IoT. The authors studied the technical solutions, and optimum policies implemented in the smart city project of Padova island city. This paper presents the original analytical model for IoT networks being attacked [39]. The G-network is an appropriate concept to model the security threats by taking counter-attacks to account. This model is fully discussed in detail including the major operation efficiency parameters. The results obtained by conducting experiments on this model confirm the strength of the model and its satisfactory performance. The authors proposed the sharing of bicycles using IoT to leverage the transport system in the city to maintain more greenery, conserving energy, preventing pollution and supporting sustainability in China [40].

The paper pointed out the significant challenges that narrowband (NB)-IoT is addressing today and probable guidelines to overcome them [41]. Authors proposed an advanced evolution in NB-IoT technologies to drive the packets by interconnecting a large number of entities. This paper identified a range of network designs for smart city solutions and compared the equivalent characteristics for preserving QoS assurances [42]. Authors also presented participatory sensing, its associated network architecture, and QoS issues. This novel network model explores the human role in data understanding, collection, and transfer in a smart activity cycle. This network model empowers users to participate actively in utilizing smart city services and give feedback for enhancing the quality of services and improvement in IoT design architecture. Authors have studied the safety, and the privacy flaws of the Edimax plug system [43]. This paper highlighted the security flaws plug system of Edimax and advised the IoT product companies to improve the safety features in their products. This paper discovered numerous security susceptibilities, unsafe protocols used communication, absence of device identification and feeble passwords. Authors studied the impact of four types of attacks like firmware attack, brute-force attack, device scanning attack and device spoofing attack.

Smart city intends to enhance the people's life standards by using the data gathered from diverse IoT systems deployed citywide [44]. Authors presented multi-source sensor system called PortoLivingLab that enables IoT systems to integrate the services of citywide sensors working on four factors weather conditions, environment pollution, transportation and citizen movement [45]. The smart city has deployed 600 + vehicles networked 19 types of mounted sensors on them. A smart city has also established a reconfigurable crowd-sensing platform and organized several crowd-sensing drives with exceeding 600 members. The data is gathered in a shared back-end server and used for data analysis of vital activities in a smart city. The authors described the design components of PortoLivingLab, emphasizing the IoT tools and tests applied. The developing nation such as India has abundant opportunities to utilize emerging tools to build smart cities [46]. The increased migration and population growth in cities demand efficient utilization of these available resources. IoT is promising to meet these demands. The fieldwork on probable architecture design smart city based on IoT technologies was carried out in the paper [45].

The overgrowing population leads to a greater burden for scarce utilities such as clean drinking water, roads, dwellings and traffic management [13, 42, 47]. The government and other organizations are showing keen interest in efficient utilization of water. This may be accomplished by the constant administration of the distribution of water. Rapid growth in the world's population, urban concentration and crowding has raised swift growth in traffic-related and environmental pollution [48]. This environmental pollution has a definite harmful impact on social health, reduces the ambiance and overall liveability. Designing and implementing efficient approaches to control population growth and urban concentration is at the utmost importance.

Smart city project prototypes provide IoT-based integrated decision support systems to monitor the pollution in the environment using many diverse data sources. The effective pollution management policies would really transform the cities into smart cities by enabling the residents a contented life. A regular health surveying framework is particularly a vital one in a smart city [49]. Data uncertainty has been ubiquitous because of the growth of IoT systems deployed for gathering data in dam safety monitoring software [50]. Efficient Top-k treating of uncertain data is a necessity for dam safety monitoring. The objectives are to decrease the energy consumption and to decrease the time to respond to the query in IoT solutions. The technical details for developing 5G networks in line with the demanding IoT requirements are necessary for realizing the benefits of a connected living [51].

This paper also defines boundaries of legacy systems to specify the characteristics of IoT requests which are part of 3GPP standardization. 5G wireless is experiencing the most important evolution. Now, the cellular standards are summing methods to enhance the network capacity and quality. The legacy networks are falling short of satisfying IoT requirements. Smart healthcare solutions are designing the devices using IoT and wireless sensor systems for patients to monitor their health from their home and gather information about everyday activities [55].

Smart home. The smart home has an IoT integrated system consisting of sensors, actuators and processing devices for communicating, computing, controlling and providing visualization for the benefit of inmates in multiple ways. Energy efficiency, assisted living and activity monitoring are some of the fundamental features. The word smart home in IoT means growing occurrence of sensors and actuators installed in their home [52]. Every city is facing the challenge of solid waste management in an eco-friendly way. The garbage is also increasing along with the population. This large indisposed collection of garbage is causing environmental pollution, hygienic conditions of the area and also leading to disease out breaks. IoT solutions can be applied to efficiently handle solid waste [53]. The authors surveyed extensively the literature on solid waste management using IoT. The study discussed solutions to many issues such as sensing, analyzing, collecting, manipulating the waste data and producing output results for efficient treatment of solid waste. This system enables tracking the quantity of waste, waste bin locations, loading, lost and missing bins. IoT leverages track the level of the waste in each waste bin and proposes the optimal path for quickly gathering of waste from these bins using a semi- or fully automated approach [4].

IoT solutions are also developed to discover the fire in industry, home and office and alert the authorities to extinct the fire. The growing ubiquity of RFID, wireless sensor network and mobiles empowers to design smart IoT applications. Authors described the design, communication, safety and privacy problems in a smart city IoT solution [54]. Authors presented a comparative study of IoT designs that focuses on vital factors like integrity, QoS, reliability, confidentiality, etc. by describing the components and the protocols in application layer essential to manage data transmission. This paper described the implementation of many data security designs by using techniques like RFID, ZigBee, Wi-Fi and Bluetooth applied in the smart city.

Tsunami detection and alerting are very important for preventing the loss of lives and property [55]. This paper explored the literature on tsunami discovery and vigilant system, challenges and algorithms implemented using IoT technologies. This paper presented a stochastic semantic model that symbolizes data and facts from IoT devices and leverages rational thinking over vagueness without loss of meaning [56]. This system is executed as an addition to the Human Behaviour Monitoring and Support (HBMS).

Smart health care. The opportunities and challenges in building robust IoT solutions in the healthcare domain were highlighted [57]. Authors considered the numerous security requirements and tests that reveal safety hazards. The authors surveyed healthcare problems related to elderly people and kids staying in orphanage houses. These people may get consistent health checkups. The objective is to design a device with multiple sensing components in hospitals and interconnecting hospitals. Active and Assisted Living (AAL) is a study domain for caring aged persons in distress from remembrance deficiency in performing day-to-day actions [55, 56]. A favourable method to provide assistance is by using memory aids according to support requirements of the person to cope up with daily activities. This information can be gathered using IoT applications suitably designed and deployed in cloud space to meet the specific person's requirements and may be used whenever the need arises.

Smart agriculture. Precision agriculture implements IoT solutions integrated with wireless sensor networks for monitoring the parameters such as pest activities, temperature, water level, humidity, soil PH and soil nutrition levels in the field [58]. IoT enables farmers to remotely monitor their crop and farm machinery through smartphones. This paper presented some typical applications in agriculture using WSN monitoring, IoT and cloud computing as the supporting platform. The yield of the crop will be reduced because of reasons such as water leakages in water distribution systems, inefficient field activities, growing of crops requiring more water, incorrect times and soils. IoT solutions transform the traditional pumps, boosters and lighting into smart devices. IoT leverages efficient and cost-effective solutions to carry out crop monitoring, utilization of pesticides, fertilizers depending on crop health, soil health and pest control. IoT solutions deploy connected wireless sensor devices and cameras in the field to capture the data and transmit it to farmers for appropriate decision-making.

Table 1 shows the comparative study of features such as application areas.

Figure 1 shows the summary of IoT domain emerging technologies, applications and research challenges surveyed in this paper.

Table 1 Summary of features collected from the survey articles

Ref Id	Tittle	Application areas	Methodology	Highlights of results
[6]	A survey: Internet of things (IoT) technologies, applications and challenges	Smart sensors, communication technologies and Internet protocols	Novel sensor management (SM) framework for pervasive IoT acoustic surveillance	Improving robustness, security and privacy in IoT
[8]	Hierarchical architecture and protocol for mobile object authentication in the context of IoT smart cities	Ontology for human localization sensor	Support vector machines (SVM) and linear discriminate analysis (LDA) classifiers	Facilitate the location-aware sensor search
[47]	Human localization sensor ontology enabling OWL 2 DL-based search for user's location-aware sensors in the IoT	Human localization sensors using identified localization patterns	Semantic sensor web paradigm	Substantial knowledge in sensor specifications, localization methods, human location context and their interrelations
[17]	IoT middleware: a survey on issues and enabling technologies	Economic analysis and pricing models for data collection and wireless communication in IoT	Adaptive and robust designs	Efficient task allocation and security for long service time and low maintenance cost
[41]	Narrowband Internet of things: evolution, technologies and open issues	Wireless power transfer	Long-term evolution	Implementation and application of wireless power transfer

(continued)

Table 1 (continued)

Ref Id	Tittle	Application areas	Methodology	Highlights of results
	Distributed wireless power transfer system for internet of things devices	Distributed wireless power transfer system	Spatially distributed power beacons with a single antenna	Supplying power to IoT devices
[59]	Classical and bio-inspired mobility in sensor networks for IoT applications	Biologically inspired mobile routing mechanisms surpasses	Bio-inspired computing paradigms	Reduces congestion and computational complexity
[9]	Heterogeneous wireless network for IoT applications	Advanced IEEE 802.15.1 (BLE) and IEEE 802.11ah (HaLow) based on receive signal strength as threshold parameter for vertical handover	Heterogeneous wireless network scenario	Connect billions of devices and services at anytime
[3]	A survey on design and analysis of robust IoT architecture	Cloud platforms	Direct memory access controller (DMA)	The reliability of the system interfacing can be improved
[10]	Sensing and actuation as a service: a new development for clouds	Cloud of sensors and actuators	Pervasive cloud	Sensing and actuation as a service (SAaaS) paradigm
[11]	Cloud computing for Internet of things and sensing-based applications	Real-time processing of the data	Customized portals and in built applications (SaaS)	Shared pool of configurable resources
[12]	Towards the web of things: web mashups for embedded devices	Online web applications	Edge computing	Cost-effective, reliable solutions
[15]	Impact of edge computing paradigm on energy consumption in IoT, computing paradigm on energy	Power utilization patterns in IoT products	Edge computing technique implementation in IoT	Determined the lifetime of controlled devices driven by batteries
[14]	IoT-enabled cloud-based additive manufacturing platform to support rapid product development	Innovative way to quickly develop the product	Additive production technique	Emerging cloud paradigm facilitates to access multiple sources with minimal cost and effort

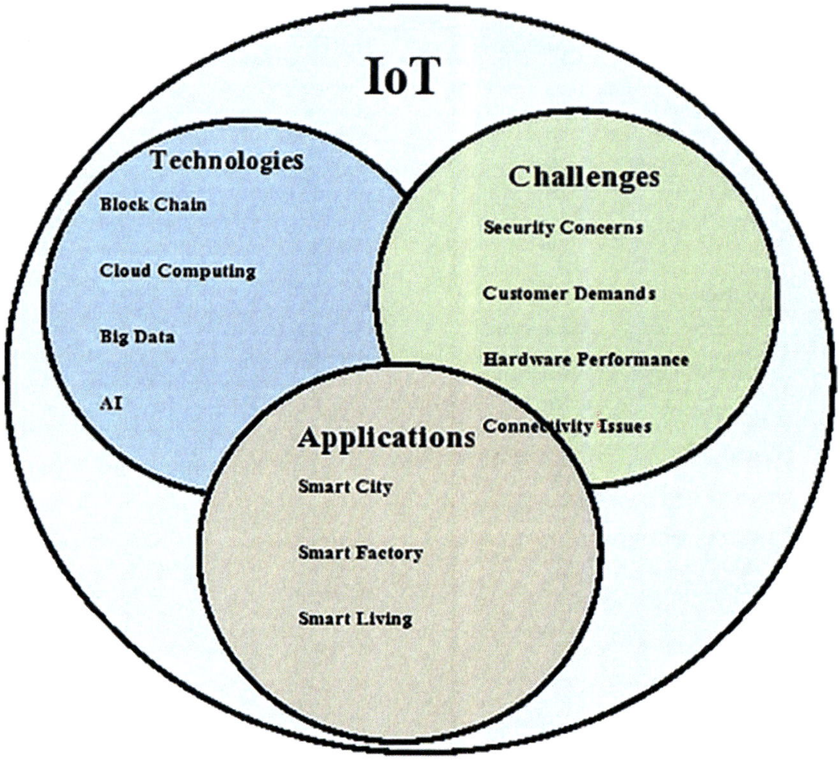

Fig. 1 IoT domain, emerging technologies, applications and research challenges

References

1. Kraijak S, Tuwanut P (2015) A survey on IoT architectures, protocols, applications, security, privacy, real-world, implementation and future trends. https://doi.org/10.1049/cp.2015.0714
2. Reyna A, Martín C, Chen J, Soler E, Díaz M (2018) On blockchain and its integration with IoT. Challenges and opportunities. Futur Gener Comput Syst 173–190. locate/fgcs. doi:/https://doi.org/10.1016/j.future.2018.05.046
3. Dhanalaxmi B, Naidu GA (2017) A survey on design and analysis of robust IoT architecture. In: International conference on innovative mechanisms for industry applications (ICIMIA 2017), IEEE, pp 375–378. https://doi.org/10.1109/icimia.2017.7975639
4. Vijayalakshmi SR, Muruganand S (2017) In: International conference on I-SMAC (IoT in social, mobile, analytics and cloud) (I-SMAC 2017), IEEE, pp 703–708. https://doi.org/10.1109/i-smac.2017.8058270
5. Luong NC, Hoang DT, Wang P, Niyato D, Kim DI, Han Z (2016) Data collection and wireless communication in Internet of Things (IoT) using economic analysis and pricing models: a survey. IEEE Commun Surv Tutor 18(4). https://doi.org/10.1109/comst.2016.2582841
6. Shah SH, Yaqoob I (2016) A survey: Internet of Things (IoT) technologies, applications and challenges. In: 4th IEEE international conference on smart energy grid engineering. IEEE Smart Energy Grid Eng (SEGE). https://doi.org/10.1109/sege.2016.7589556

7. Tiwari RA (2017) A survey on lifetime elongation of wireless sensor network using energy efficiency algorithms. In: International Conference On I-SMAC (IoT in social, mobile, analytics and cloud) (I-SMAC), IEEE, pp 181–185. https://doi.org/10.1109/i-smac.2017.8058334
8. Saadeh M, Sleit A, Sabri KE, Almobaideen W (2018) Hierarchical architecture and protocol for mobile object authentication in the context of IoT smart cities. J Netw Comput Appl. https://doi.org/10.1016/j.jnca.2018.07.009
9. Salwe SS, Naik KK (2017) Heterogeneous wireless network for IoT applications. IETE Tech Rev. https://doi.org/10.1080/02564602.2017.1400412
10. Distefano S, Merlino G, Puliafito A (2012) Sensing and actuation as a service: a new development for clouds. IEEE 11th international symposium on network computing and applications, cloud computing is among the hottest trends in ICT. https://doi.org/10.1109/nca.2012.38
11. Rao BP, Saluia P, Sharma N, Mittal A, Sharma SV (2012) Cloud computing for Internet of Things and sensing based applications. In: 6th international conference on sensing technology (ICST), pp 374–379. https://doi.org/10.1109/icsenst.2012.6461705
12. Guinard D, Trifa V (2009) Towards the web of things: web Mashups for embedded devices. Madrid, Spain in proceedings of WWW, Apr 2009
13. Siegel JE, Kumar S, Sarma SE (2017) The future Internet of Things: secure, efficient, and model-based. IEEE Internet Things J. https://doi.org/10.1109/jiot.2017.2755620
14. Wang Y, Lin Y, Zhong RY, Xu X (2018) IoT-enabled cloud based additive manufacturing platform to support rapid product development. Int J Prod Res. https://doi.org/10.1080/00207543.2018.1516905
15. Mocnej J, Miškuf M, Papcun P, Zolotová (2018) Impact of edge computing paradigm on energy consumption in IoT, computing paradigm on energy. Papers OnLine 51(6):162–167. https://doi.org/10.1016/j.ifacol.2018.07.147
16. Da Xu L, He W, Li S (2014) Internet of Things in industries: a survey. IEEE Trans Ind Inform. 10(4):2240. https://doi.org/10.1109/tii.2014.2300753
17. Ngu AH, Gutierrez M, Metsis V, Nepal S, Sheng QZ (2017) IoT middleware: a survey on issues and enabling technologies. IEEE Internet Things J 4(1). https://doi.org/10.1109/jiot.2016.2615180
18. Gaur P, Tahiliani MP (2015) Operating systems for IoT devices: a critical. In: 2015 IEEE region 10 symposium, pp 33–36. https://doi.org/10.1109/tensymp.2015.17
19. Farahmandpour Z, Versteeg S, Han J, Kameswaran A (2017) Service virtualization of Internet-of-Things devices: techniques and challenges. In: 2017 IEEE/ACM 3rd international workshop on rapid continuous software engineering (RCoSE). https://doi.org/10.1109/rcose.2017.4
20. Mansanet I, Torres V, Valderas P, Pelechano V (2015) IoT compositions by and for the crowd. JCIS
21. Cerullo G, Mazzeo G, Papale G, Ragucci B, Sgaglione L, IoT and sensor net works security. https://doi.org/10.1016/b978-0-12-811373-8.00004-5
22. Wang W, Yang H, Zhang Y, Xu J (2018) IoT enabled real-time energy efficiency optimization method for energy-intensive manufacturing enterprises. Int J Comput Integr Manuf 31(4–5):362–379. https://doi.org/10.1080/0951192x.2017.1337929
23. Ghimire S, Luis-Ferreira F, Nodehi T, Jardim-Goncalves R (2017) IoT based situational awareness framework for real-time project management. Int J Comput Integr Manuf 30(1):74–83. https://doi.org/10.1080/0951192x.2015.1130242
24. Trappey AJ, Trappey CV, Fan CY, Hsu AP, Li XK, Lee IJ (2017) IoT patent roadmap for smart logistic service provision in the context of Industry 4.0. J Chin Inst Eng 40(7):593–602. doi:https://doi.org/10.1080/02533839.2017.1362325
25. Pawar P, Trivedi A (2018) Device-to-device communication based IoT system: benefits and challenges. IETE Tech Rev. https://doi.org/10.1080/02564602.2018.1476191
26. Huang HP, Yan JL, Huang TH, Huang MB (2017) IoT-based networking for humanoid robots. J Chin Inst Eng 40(7):603–613. https://doi.org/10.1080/02533839.2017.1372224
27. Verma NK, Kumar D, Kumar I, Ashok A (2018) Automation of boiler process at thermal power plant using sensors and IoT. J Stat Manag Syst 21(4):675–683. https://doi.org/10.1080/09720510.2018.1475078

28. Tresanchez M, Pujol A, Pallejà T, Martínez D, Clotet E, Palacín J (2018) A proposal of low-cost and low-power embedded wireless image sensor node for IoT applications. In: 15th international conference on mobile systems and pervasive computing-MobiSPC-2018. Procedia Comput Sci 99–106, Science Direct. https://doi.org/10.1016/j.procs.2018.07.149
29. Kapoor C, Singh H, Laxmi V (2018) A survey on energy efficient routing for delay minimization in IoT networks. In: International conference on intelligent circuits and systems, pp 320–323, IEEE. https://doi.org/10.1109/icics.2018.00072
30. Meng X, Cong W, Liang H, Li J (2018) Design and implementation of apple orchard monitoring system based on wireless sensor network. In: Proceedings of IEEE, international conference on mechatronics and automation, Changchun, China, 5–8 Aug 2018. https://doi.org/10.1109/icma.2018.8484350
31. Omitola T, Wills G (2018) Towards mapping the security challenges of the Internet of Things (IoT) supply chain. In: 22nd international conference on knowledge-based and intelligent information and engineering systems. Procedia Comput Sci 126:441–450. https://doi.org/10.1016/j.procs.2018.07.278
32. Kabilan K, Bhalaji N, Selvaraj C, Kumaar M, Karthikeyan PT (2018) Performance analysis of IoT protocol under different mobility models. Comput Electr Eng 154–168. https://doi.org/10.1016/j.compeleceng.2018.09.007
33. Coetzee L, Eksteen J (2011) The Internet of Things—promise for the future? An introduction. In: Cunningham P, Cunningham M (eds) IST-Africa conference proceedings. IIMC International Information Management Corporation. INSPEC Accession Number: 12441774. ISBN: 978-1-905824-26
34. Grabovica M, Popić S, Pezer D, Knežević V (2016) Security measures of enabling technologies in Internet of Things (IoT): a survey. https://doi.org/10.1109/zinc.2016.7513647
35. Asghari P, Rahmani AM, Javadi HH (2018) Service composition approaches in IoT: a systematic review. J Netw Comput Appl. https://doi.org/10.1016/j.jnca.2018.07.013
36. Mainetti L, Patrono L, Vilei A (2011) Evolution of wireless sensor networks towards the Internet of Things: a survey. In: SoftCOM 2011, 19th international conference on software, telecommunications and computer networks, pp 1–6, http://ieeexplore.ieee.org/stamp/stamp.jsp?tp=&arnumber=6064380&isnumber=6064354
37. De S, Barnaghi P, Bauer M, Meissner S, Service modelling for the Internet of Things. In: Proceedings of the federated conference on computer science and information systems, pp 949–955. ISBN 978-83-60810-39-2
38. Palade A, Cabrera C, White G, Razzaque MA, Clarke S (2017) Middleware for Internet of Things: a quantitative evaluation in small scale. IEEE. https://doi.org/10.1007/s4.860-018-0055-4
39. Sarigiannidis P, Karapistoli E, Economides AA (2017) Novel, analytic framework for modeling security attacks in Internet of Things (IoT) infrastructures. IEEE Internet of Things J 4(6). https://doi.org/10.1109/jiot.2017.2719623
40. Liu L (2018) IoT and a sustainable city. In: 5th international conference on energy and environment research, ICEER. Energy Procedia 153:342–346. https://doi.org/10.1016/j.landurbplan.2010.07.009
41. Xu J, Yao J, Wang L, Ming Z, Wu K, Chen L, Narrowband Internet of Things: evolutions, technologies and open issues. IEEE Internet Things J. https://doi.org/10.1109/jiot.2017.2783374
42. Jin J, Gubbi J, Luo T, Palaniswami M (2012) Network architecture and QoS issues in the Internet of Things for a smart city. In: International symposium on communications and information technologies (ISCIT), IEEE 956. https://doi.org/10.1109/iscit.2012.6381043
43. Ling Z, Luo J, Xu Y, Gao C, Wu K, Fu X (2017) Security vulnerabilities of Internet of Things: a case study of the smart plug system. IEEE Internet Things J 4(6):1899–1909. https://doi.org/10.1109/jiot.2017.2707465
44. Santos PM, Rodrigues JG, Cruz SB, Lourenço T, d'Orey PM, Luis Y, Rocha C, Sousa S, Crisóstomo S, Queirós C, Sargento S (2018) PortoLivingLab.: an IoT-based sensing platform for smart cities. IEEE Internet Things J. https://doi.org/10.1109/jiot.2018.2791522

45. Linger RC, Hevner AR (2018) Flow semantics for intellectual control in IoT systems. J Decis Syst 27(2):63–77. https://doi.org/10.1080/12460125.2018.1529973
46. Vijai P, Sivakumar PB (2016) Design of IoT systems and analytics in the context of smart city initiatives in India. In: 2nd international conference on intelligent computing, communication and convergence. Procedia Comput Sci 92:583–588. https://doi.org/10.1016/j.procs.2016.07.386
47. Chaochaisit W, Bessho M, Koshizuka N, Sakamura K (2016) Human localization sensor ontology enabling OWL 2 DL-based search for user's location-aware sensors in the IoT. IEEE 10th international conference on semantic computing. https://doi.org/10.1109/icsc.2016.31
48. Miles A, Zaslavsky A, Browne C (2018) IoT-based decision support system for monitoring and mitigating atmospheric pollution in smart cities. J Decis Syst 27(sup1):56–67. https://doi.org/10.1080/12460125.2018.1468696
49. Parthasarathy P, Vivekanandan S (2018) A typical IoT architecture-based regular monitoring of arthritis disease using time wrapping algorithm. Int J Comput Appl. https://doi.org/10.1080/1206212x.2018.1457471
50. Mao Y, Zhong H, Chen H, Li X (2017) Two-phase PTTopk query processing algorithm for uncertain IOT data in dam safety monitoring. Intell Autom Soft Comput 23(4):581–588. https://doi.org/10.1080/10798587.2017.1316070
51. Agiwal M, Saxena N, Roy A (2018) Towards connected living: 5G enabled Internet of Things (IoT). IETE Tech Rev. https://doi.org/10.1080/02564602.2018.1444516
52. Bedi G, Venayagamoorthy GK, Singh R (2016) Internet of Things (IoT) sensors for smart home electric energy usage management. IEEE. https://doi.org/10.1109/ICIAFS.2016.7946568
53. Fallavi KN, Kumar VR, Chaithra BM (2017) Smart waste management using Internet of Things: a survey. In: International conference on I-SMAC (IoT in Social, Mobile, Analytics and Cloud (I-SMAC 2017). IEEE, pp 60–64. https://doi.org/10.1109/i-smac.2017.8058247
54. Datta P, Sharma B (2017) A survey on IoT architectures, protocols, security and smart city based applications. In: 8th ICCCNT 2017, IIT Delhi, July 3–5. https://doi.org/10.1109/icccnt.2017.8203943
55. Gomathi RM, Krishna GH, Brumancia E, Dhas YM (2018) A survey on IoT technologies, evolution and architecture. In: 2nd international conference on computer, communication, and signal processing (ICCCSP). https://doi.org/10.1109/icccsp.2018.8452820
56. Lunardi GM, Al Machot F, Shekhovtsov VA, Maran V, Machado GM, Machado A, Mayr HC, de Oliveira JP (2018) IoT-based human action prediction and support, IoT-based human action prediction and support. Internet Things 3(4):52–68, https://doi.org/10.1016/j.iot.2018.09.007
57. Thakar AT, Pandya S (2017) Survey of IoT enables healthcare devices. In: Proceedings of the IEEE international conference on computing methodologies and communication (ICCMC), pp 1087–1090, IEEE. ISBN 978-1-5090-4890-8/17/$31.00
58. Mekala MS, Viswanathan P (2017) A survey: smart agriculture IoT with cloud computing. 978-1-5386-1716-8/17/$31.00 ©2017 IEEE, https://doi.org/10.1109/icmdcs.2017.8211551
59. Hamidouche R, Aliouat Z, Gueroui AM, Ari AAA, Louail L (2018) Classical and bio-inspired mobility in sensor networks for IoT applications. J Netw Comput Appl. https://doi.org/10.1016/j.jnca.2018.07.010
60. Perera C, Zaslavsky A, Christen P, Georgakopoulos D (2014) Context aware computing for the Internet of Things: a survey. IEEE Commun Surv Tutor 16(1). https://doi.org/10.1109/surv.2013.042313.00197
61. Haller S, Karnouskos S, Schroth C (2009) The Internet of Things in an enterprise context. In: FIS 2008, LNCS 5468, pp 14–28. https://doi.org/10.1007/978-3-642-00985-3_2
62. Chatterjee S, Byun J, Dutta K, Pedersen RU, Pottathil A, Xie H (2018) Designing an Internet-of-Things (IoT) and sensor-based in-home monitoring system for assisting diabetes patients: iterative learning from two case studies. Eur J Inf Syst. https://doi.org/10.1080/0960085x.2018.1485619
63. Basu SS, Tripathy S (2018) Securing multicast group communication in IoT-enabled systems. IETE Tech Rev. https://doi.org/10.1080/02564602.2017.1407681

64. Nguyen TD, Chiem TP, Duong TH, Le DH, Nguyen XA (2017) On improving agricultural IoT management process for fault detection. J Inf Telecommun 1(3):208–223. https://doi.org/10.1080/24751839.2017.1347393
65. Chaudhuri A (2016) Cyber threat mitigation of wireless sensor nodes for secured, trustworthy IoT services. EDPACS 54(1):1–14. https://doi.org/10.1080/07366981.2016.1181416

Evaluation of Standard Models of Content Placement in Cloud-Based Content Delivery Network

Suman Jayakumar, S. Prakash and C. B. Akki

Abstract The process on the ecosystem of the computing environment of today's world includes both static and dynamic contents. The availability of these content is essential for many critical applications in reliable as well as on real-time applications. The conventional content delivery network (CDN) creases its capacity in many ways to be synchronous with modern and advance applications; therefore, the trend of shifting and setting up CDN on the cloud is adopted due to many advantages of clouds. This paper analyses many standard computational model's architectures for cloud-based content delivery networks (CCDN). The proposed study contributes to discuss the essential characteristics of the content placement problems in relation to the cloud ecosystem for highlighting all forms of problems associated with it. The study also contributes to highlight a possible solution to circumvent such a significant research problem associated with CCDN.

Keywords Content delivery network · Cloud · Caching · Content placement process · Replication · Latency

1 Introduction

The content delivery network (CDN) came into existence in 1986 with an objective to deliver the contents of the Web, because the Web design did not support the

S. Jayakumar (✉)
Visvesvaraya Technological University, Belagavi, India
e-mail: jayakumarsuman@gmail.com

S. Prakash (✉)
Department of Computer Science and Engineering, East Point College of Engineering and Technology, Bengaluru, India
e-mail: Prakash.hospet@gmail.com

C. B. Akki (✉)
Department of Computer Science and Engineering, Indian Institute of Information Technology, Dharwad, India
e-mail: akki.channappa@gmail.com

N. N. Chiplunkar and T. Fukao (eds.), *Advances in Artificial Intelligence and Data Engineering*, Advances in Intelligent Systems and Computing 1133,
https://doi.org/10.1007/978-981-15-3514-7_108

transmission of the Web components over a larger distance [1]. From then, there is an exponential growth of CDNs and as well as increase in the content owners who adopt CDN for their content distribution to the clients aiming for the best possible user experience. The modern Web applications include various e-commerce platforms and many of the personalized services where many contents are generated by the user dynamically in the form of HTML, XML, text, images, and videos [2]. The advancement into video codecs, vision systems, and other supporting mechanisms has opened up a new channel for the video content distribution which is expected to grow from 64 to 80% of overall network by 2019, that is, just in five years of span [3]. Existing smart applications of Internet of things (IoT) are also known for dissemination massive contents like video files that are dependent on using push-based content forwarding strategy as well as the generation of sources of such contents from streams of visual sensors to the upstream devices [4]. A typical CDN is utilized by Web servers to store contents in constraint environment of the network. The CDN overcomes the bottleneck of caching proxies maintained by ISP using dynamic contents sidebar. The architecture of CDN is complicated due to a varied type of network nodes which are collaborated in a typical way to deliver the content. Figure 1 shows the core component of CDN and Fig. 2 shows the typical CDN architecture [5].

The surrogate servers are also known as a non-origin server or the servers which cache the contents of the origin server. Another component is router, which distributes the content requested by the client to a best-suited surrogate server, through network elements and the server status logs in as a meta-information. The meta-information is maintained at account and audit components.

The world leader of CDN service provider includes Akamai (70% market stake), Adero, Digital Island, Inktomi, etc. The CDN designs are not available into the public domain, whereas one of the works by Zhang [6] has proposed an open CDN architecture, which brings canonical to the fact that in order to provide large-scale and economic models of content delivery and distribution, the cloud infrastructure is to be used to setup CDN [6, 7]. A cloud-based CDN, i.e., CCDN included significant

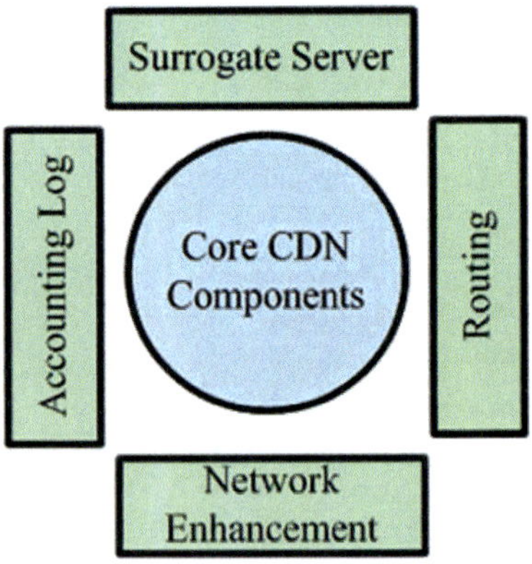

Fig. 1 Core component of CDN

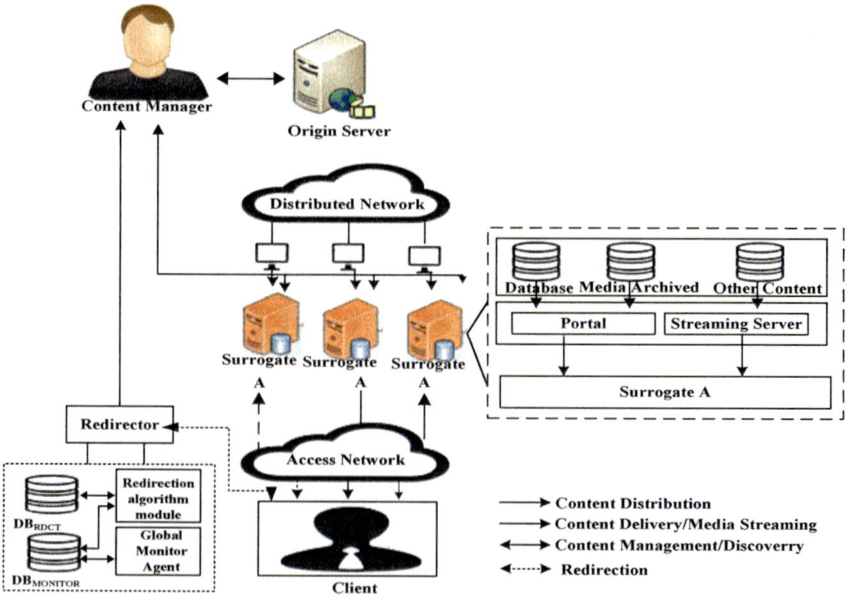

Fig. 2 Typical CDN architecture for single user

benefits like (1) scalability: due to cloud component like load balancer and elastic up and downscaling capabilities and (2) abstraction: complexities of CDN infrastructure is hidden from the user and content owners. (3) Quality of service (QoS)-based management and most importantly and (4) open standard [8]. Section 2 describes a few of the existing cloud-based CDNs followed by a discussion of beneficial points of existing approaches of CCDN as well as their pitfalls. Section 3 discusses the problems evolved models of content placement. The briefing of a feasible solution to address the pitfalls associated with existing approaches is discussed in Sect. 4. The conclusion of paper and contribution are discussed in Sect. 5

2 Existing Cloud-Based CDNs

Cloud-based CDN (CCDN) offers enhanced content distribution over the Internet with respect to media and contents over the specific Web site. The existing types of CCDNs are as follows:

- **Limelight Orchestrate**: This form of CCDN offers analytical operation essentially meant for the business purpose [9]. It offers different forms of characteristics, e.g., content control, cloud storage, traffic orientation, security, and content delivery over mobile devices, etc.

- **MetaCDN**: This form of CCDN offers content delivery management for real-time multimedia contents and static contents only [10]. This form of CCDN uses a public cloud storage system, e.g., Limelight, Amazon S3, etc., that offers only fundamental operations. Different forms of supported core components are QoS monitor, MetaCDN manager, Database redirector, Allocator, etc.
- **RackSpace**: This form of CCDN offers on-demand usage model for its storage as well as its channel capacity over the network. It harnesses the applications of Akamai CDN for controlling the latency during accessing contents. The storage operation is carried out using OpenStack, and the complete operation of content management is carried out by using time-to-live timers by RackSpace [11].
- **MediaWise**: It is a unique form of CCDN that permits a registered user to act as a provider of CDN. It offers a higher set of flexibility by allowing customers to obtain any CDN with the supportability of content creation dynamically [12].
- **Amazon CloudFront**: It is another frequently used CCDN system in the global commercial market that offers content delivery services with Amazon elastic cloud [13]. Both dynamic and static contents are supported over live streaming by this CCDN. However, Amazon CloudFront offers smaller coverage of only 32 CDN compared to 219 CDN offered by RackSpace.
- **COMODIN**: It is also known as cooperative media-on-demand on the Internet that is meant for streaming massive contents over the Internet using a collaborative network system and IP multicast [14].

Apart from the above-mentioned existing CCDN, there are few others too, e.g., CoDaaS [15] and CoDeeN [16]. CoDaaS is meant for distributing contents generated in cost-effective manner while CoDeeN is used for academic purpose exercised in Princeton University.

3 Advantages of Existing CCDN and Its Limitations

It is already known that the cloud environment significantly enhances the content delivery management of the CDN system. One of the beneficial points of CCDN is its capability of data processing over the World Wide Web as well as cost-effective storage mechanism followed by distribution functionality. Another potential advantage of CCDN is that it provides the on-demand modeling in order to render cost effectiveness where the user is permitted to initiate and terminate the services on the basis of the amount they pay without much complexity in terms and condition. Finally, CCDN offers potential service migration, interoperability, and diversified application supportability, which although it offers benefits but is shrouded by problems yet.

The potential limitations of CCDN of the existing system are as follows, viz. (1) dynamic content management, (2) developing content, (3) diversification of content, (4) ownership of CCDN, (5) personalization of CCDN, (6) cost-effective modeling, (7) security, (8) developing and working with hybrid clouds, (9) CCDN monitoring,

(10) quality of service, (11) demand prediction, (12) cloud Selection, (13) ubiquitousness, (14) flexible content management, etc. All of these above factors are highly essential to be addressed, without which CCDN has to also suffer from various other limitations associated with networking, e.g., redirection of the request, load balancing, local caching, and network proximity. An in-depth study of all the existing approaches toward CCDN shows that it is prominently designed using one cloud platform and does not offer much flexibility of content distribution dynamically. Also, in reality, there is a lot of contradiction between what is demanded by CCDN and what is discussed in existing research models. A suitable example to cite this issue is that the majority of the existing research model does not have a cost model as they are constructed over a single cloud system. A contradiction is without a standard cost model, and the existing system offers information about cost factor without even introducing the model validity of the cost factor. Another essential thing required for effective performance of CCDN is to achieve a superior content placement operation, which is again a challenge.

Therefore, in order to overcome the limitation associated with an effective content management, the following are the demand gaps, viz. (i) not much focused on supportability of HTTP/2 protocol that offers faster-processing speed compared to existing practices on HTTP/1.1, (ii) needs extensive security feature with lesser latency, (iii) should support optimization of high-end file system, e.g., images, scripts, video, etc., (iv) should offer an effective browser caching mechanism where existing approaches use just the default one, (v) needs an automated configuration system in order to process WebSockets traffic to the user's origin server, (vi) need highly efficient algorithm that is capable of identifying and mitigating bottleneck condition, and (vii) it also demands a precise network routing system that is highly optimized that always targets to reduce latency in order to achieve better quality of experience. At the same time, the existing products of CCDN is also featured by more issues, e.g., (i) existing tools are unofficially reported to encounter problems with caching times, (ii) distribution problems, (iii) defective and complex configuration problems, (iv) browser capability issue, (v) error-prone logs, (vi) dysfunctional URL, etc. Therefore, the existing CCDN is associated with both advantages as well as the limitation.

4 Content Placement Problem in CCDN

At present, there are many strategies for improving the content placement operation over CCDN. Caching is one of the essential operations for improving content placement, especially for larger content. Usage of the flash memory system is reported to offer better performance, and its cost factor can also be controlled using error-tolerant design [17]. Such a design model adaptively offers usage of error-correction code so that tolerance limit can be improved while performing content placement. An approximation approach toward controlling the stream could further set right the tolerance limit of in-network caching system that can permit better performance for transcoding. However, such problems could be well addressed if an efficient server is placed on the correct location. Basically, the content placement problem in CCDN deals with the determination of a specific file that is required to be kept under specifically selected cache. This is a problem specifically in highly distributed system like a cloud where storage requirement needs to be optimized along with increasing the number of request that is required to be processed by that cache.

However, it has been seen that routing control system acts as an impediment to server placement process that minimizes the chances of candidate sites to be properly configured [18]. The presented architecture consists of network space block that acts as an input for cross-space optimization followed by analyzing the outcome of the relationship curve. The network space consists of measured data of latency, domain knowledge associated with the cost of the site, and preferences while cross-space optimization consists of coordinated space and physical space. This model discussed that the inclusion of provisioning has potential influence over server placement. However, the interoperability issue is not discussed. Such a problem was found to be addressed using the edge caching mechanism integrated with the scheduling process [19]. A statistical model was developed considering different sizes of cache, the popularity of video for improving backhaul scheduling. The study deals with videos of the constant bit rate for developing a unique cache policy in order to overcome the scheduling problems. However, the problem associated with content placement and provisioning of resources is yet open-ended, whereas some recent joint models claim of overcoming this using resource constraint-based approach [20]. This study has introduced a specific model for controlling the network delay along with bandwidth in CCDN system that is mathematically expressed as follows,

$$\arg_{\min}\left(\sum_{i\in u, j\in c} p_j^b d_i(0)x_{ij} + \sum_{j\in C}(p_j^s s + p_j^b \hat{b}_j)z_j + \sum_{i\in u, j\in C} p^q q_{ij}|D_{ij} - D^U| + \sum_{i\in U} p_{0i}^b d_i(0)x_{io}\right)$$

The above expression represents an objective function that is constructed using the following inputs, e.g., the price of single storage and bandwidth over the cloud site as well as channel capacity between the cloud site and root cloud site. The researchers [21] have provided a model that is capable of allocating the client site dynamically depending upon the variations of the rates of user demands. An efficient content placement is highly essential for developing a better form of CDN slice that offers multiple functionalities right from transcoders to caches. In this regards, virtual network function assists in constructing the bridge between the network demands and content demands. Therefore, the platform-based approach (Fig. 3) considering CDN as a service is proven to be potential for construction such as slices of CDN that are distributed over different domains of the cloud. In recent time, a set of solutions are developed for reducing the cost associated with content placement, offering better experiences, and maintaining a balance between better user experience and cost [22]. This study has also performed the mathematical modeling using a standard decision-making approach, e.g., game theory to validate the presented solution. A similar problem is also reported to be solved using integer linear programming [23]. Such

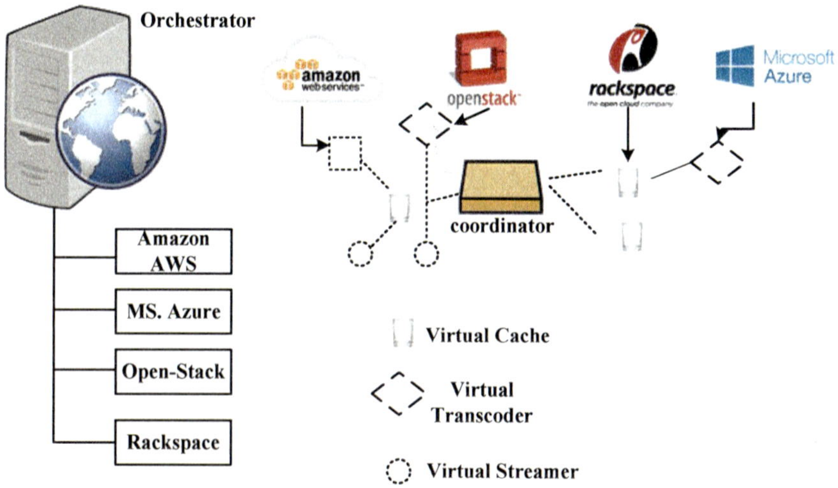

Fig. 3 CCDN in the form of service-based process

approaches tend to explore better VNFs in addition to their position information in order to control the cost and increase the quality of service. However, such approaches are carried out without even evaluating the capabilities of the provisioning system. Therefore, the service-level analytical modeling has been carried out to assess the content placement system as well as to check out its pricing strategy [24]. Such strategic solution is claimed to enhance the resource management system as well as scheduling operation for services.

The caching is one of the essential operations required for addressing an efficient content placement problem. At present, fewer models are available to focus on improving caching considering all the necessary resources in CDN. However, there are certain models where the differential caching system is presented in order to minimize the cumulative rental cost using greedy-based approach [2]. This model is also capable of performing dynamic fine-tuning of the content placement with respect to the demanded route map. This model not only controls cost but also increase the capability of processing maximized user demands in traffic. Apart from this, there are also models of content placement that emphasized on virtualized environment [25], signifying caching as dominant solution in error-prone networks [26], replication of contents and routing request [27], *k*-center problem modeling for optimization [28], traffic management for bottleneck [29], adaptive provisioning [30], and graph partitioning [31].

The recent model has been presented for addressing the cache positioning issue where it was found that the cross-layer-based framework design could offer a better formulation of the heuristic algorithm [32]. The performance of such a model can be further enhanced if the incoming traffic pattern is known or mapped. The current work toward using time and space factors for extracting the explicit pattern of the request has investigated this problem using standard dataset [33]. However, it is just a theoretical validation of such a model and does not offer comprehensive evidence of validation performance. Improvement of such theoretical model could be done if the analysis is carried out on identifying the rate of acceptance/rejection of the request, in which case, it is feasible to control. The recent heuristic-based model is claimed to reduce such rate of rejection with addressing the allocation of resource problem in order to improve the content placement operation. The operation of such an approach can be further enhanced by using coded caching process as the majority of the existing approaches are found to use uncoded schemes. A model using coded caching scheme is more justifiable for constructing an efficient placement delivery array [34]. Table 1 represents various problems addressed by different models.

Table 1 Summary of different model contribution

Problem addressed	Strategy adopted	Strength	Weakness
Storage overhead	Approximation, control stream [17]	Reduce overhead	Does not specify scalability
Server placement	Architecture-based provisioning, decision-based [18]	Minimize inter-domain traffic	Less supportability of interoperability
Concurrent request processing	Scheduling with edge caching [19]	Caching policy	Does not address traffic variability
Variation in demand rates	Allocation algorithm of bandwidth and cloud site [20]	Reduces cost	No benchmarking
Cost, quality of experience	Mathematical model, game theory [21]	Reduce cost	No benchmarking
VNF placement	Integer linear programming [22]	Reduce cost	Specific to proactive placement only
Resource management	Network function virtualization [23]	Better service deployment	Does not consider traffic variability
Dynamic caching	Iterative/greedy [24]	Maximize demands	Does not address traffic uncertainty, constraint
Cache placement	Cross-layer, integer linear programming [31]	Ensure better communication performance	Narrowed research scope
Optimizing resource allocation	Search-based optimization [33]	Control outage of request	High iterative structure
Cache placement	Coded caching [34]	Simplified mathematical model	Focused on the centralized system

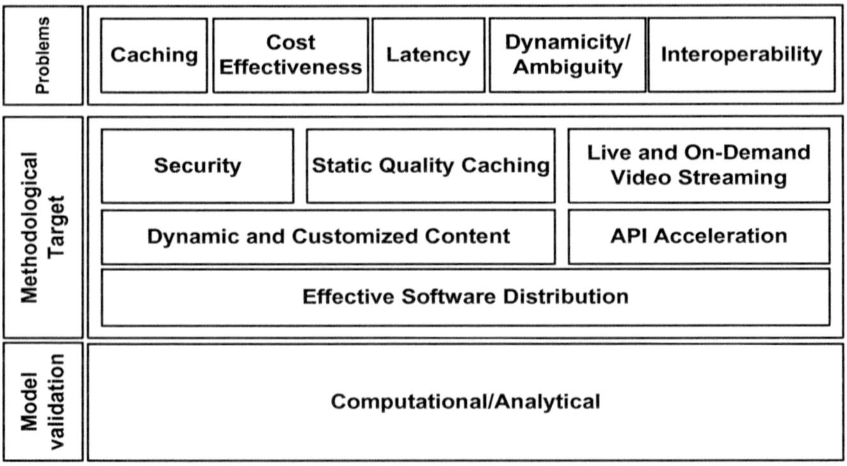

Fig. 4 Prospective architecture of content placement in CCDN

5 Prospective Architecture

After reviewing the existing system as well as our prior work [35], it is potentially felt that there is a need to develop a significant CCDN framework with faster response time for its service delivery. A closer look at the existing system will show that they have both advantages as well as the limitation and the approaches used are more or less highly symptomatic. Such solutions cannot rectify extensive factors causing problems in content placement operation in CCDN.

Hence, the core agenda of the architecture construction will be (i) to save time, (ii) to offer content privacy, (iii) geo-targeting, (iv) quick content delivery, (v) less expensive, etc. Apart from this, the secondary target for developing a novel CCDN architecture will be to achieve (i) network integration, (ii) great performance, (iii) highly programmable CDN system, (iv) cost effectiveness, and (v) higher integration supportability. The design principle should offer (i) uniform quality of caching, (ii) highly secure environment, (iii) supportability of personalization/customization of contents, (iv) effective software distribution, and (v) API acceleration. The architecture (Fig. 4) should offer a comprehensive way of addressing the existing problems that are yet to be addressed in CCDN.

6 Conclusion

The proposed study has contributed to highlight specific information associated with problems in CCDN and highlights that content placement problems, which is one of the essential operations over CDN, is not emphasized practically. Some of the essential problems are, viz. (i) the presence of distributed network makes the decision of content placement strategy very much difficult owing to presence of diversified range of data, (ii) the presence of dynamic data without any form of prioritization from the source of generation also leads to lag behind when it comes to identify a specific contents to be placed in specific storage unit, and (iii) lack of standard and effective content placement model in existing system with respect to cloud ecosystem. This paper contributes to extract the strength and weakness of existing research models toward content management in CCDN and finally highlights a tentative prospective architecture that could be considered for any future modeling of CCDN.

References

1. Vakali A, Pallis G (2003) Content delivery networks: status and trends. IEEE Int Comput 7(6):68–74
2. Siracusano G et al (2017) Re-designing dynamic content delivery in the light of a virtualized infrastructure. IEEE J Sel Areas Commun 35(11):2574–2585
3. Marković DR, Gavrovska AM, Reljin IS (2016) 4K video traffic analysis using seasonal autoregressive model for traffic prediction. In: 2016 24th telecommunications forum (TELFOR). Belgrade, pp 1–4
4. Moustafa H, Schooler EM, McCarthy J (2017) Reverse CDN in fog computing: the lifecycle of video data in connected and autonomous vehicles. In: 2017 IEEE fog world congress (FWC). Santa Clara, CA, pp 1–5
5. Fortino G, Russo W, Mastroianni C, Palau CE, Esteve M (2007) CDN-supported collaborative media streaming control. IEEE MultiMedia 14(2):60–71
6. Zhang Z (2014) Feel free to cache: towards an open CDN architecture for cloud-based content distribution. In: 2014 international conference on collaboration technologies and systems (CTS). Minneapolis, MN, pp 488–490
7. Ling L, Xiaozhen M, Yulan H (2013) CDN cloud: a novel scheme for combining CDN and cloud computing. In: Proceedings of 2013 2nd international conference on measurement, information and control. Harbin, pp 687–690
8. Wang M, Jayaraman PP, Ranjan R, Mitra K, Zhang M, Li E, Khan S, Pathan M, Georgeakopoulos D (2015) An overview of cloud based content delivery networks: research dimensions and state-of-the-art. In: Transactions on large-scale data-and knowledge-centered systems XX 2015. Springer, Berlin, pp 131–158
9. Limelight Orchestrate, https://www.limelight.com/orchestrate-platform/. Retrieved on 30 Oct 2018
10. MetaCDN, http://www.metacdn.com/. Retrieved on 30 Oct 2018
11. RackSpace, https://www.rackspace.com/. Retrieved on 30 Oct 2018
12. MediaWise, https://www.poynter.org/mediawise. Retrieved on 30 Oct 2018
13. Amazon CloudFront, https://aws.amazon.com. Retrieved on 30 Oct 2018
14. COMODIN, http://www.spanishcentral.com/translate/comod%C3%ADn. Retrieved on 30 Oct 2018
15. Jin Y, Wen Y, Shi G, Wang G, Vasilakos AV (2012) CoDaaS: an experimental cloud-centric content delivery platform for user-generated contents. In: 2012 international conference on computing, networking and communications (ICNC), Maui, HI, pp 934–938
16. Codeen, https://medlineplus.gov/druginfo/meds/a682065.html. Retrieved on 30 Oct 2018
17. Zhang X, Xiong D, Zhao K, Chen CW, Zhang T (2018) Realizing low-cost flash memory based video caching in content delivery systems. IEEE Trans Circuits Syst Video Technol 28(4):984–996

18. Yin H, Zhang X, Zhao S, Luo Y, Tian C, Sekar V (2017) Tradeoffs between cost and performance for CDN provisioning based on coordinate transformation. IEEE Trans Multimedia 19(11):2583–2596
19. Ahlehagh H, Dey S (2014) Video-aware scheduling and caching in the radio access network. IEEE/ACM Trans Netw 22(5):1444–1462
20. Haghighi AA, Shah Heydari S, Shahbazpanahi S (2018) Dynamic QoS-aware resource assignment in cloud-based content-delivery networks. IEEE Access 6:2298–2309
21. Benkacem I, Taleb T, Bagaa M, Flinck H (2018) Optimal VNFs placement in CDN slicing over multi-cloud environment. IEEE J Sel Areas Commun 36(3):616–627
22. Dieye M et al (2018) CPVNF: cost-efficient proactive VNF placement and chaining for value-added services in content delivery networks. IEEE Trans Netw Serv Manage 15(2):774–786
23. Frangoudis PA, Yala L, Ksentini A (2017) CDN-as-a-service provision over a telecom operator's cloud. IEEE Trans Netw Serv Manage 14(3):702–716
24. Hu M, Luo J, Wang Y, Veeravalli B (2014) Practical resource provisioning and caching with dynamic resilience for cloud-based content distribution networks. IEEE Trans Parallel Distrib Syst 25(8):2169–2179
25. Sengupta A, Tandon R, Simeone O (2017) Fog-aided wireless networks for content delivery: fundamental latency tradeoffs. IEEE Trans Inf Theory 63(10):6650–6678
26. Hu H, Wen Y, Chua T, Huang J, Zhu W, Li X (2016) Joint content replication and request routing for social video distribution over cloud CDN: a community clustering method. IEEE Trans Circuits Syst Video Technol 26(7):1320–1333
27. Jin Y, Wen Y, Guan K (2016) Toward cost-efficient content placement in media cloud: modeling and analysis. IEEE Trans Multimedia 18(5):807–819
28. Liu J, Yang Q, Simon G (2018) Congestion avoidance and load balancing in content placement and request redirection for mobile CDN. IEEE/ACM Trans Netw 26(2):851–863
29. Kuo W, Lin Y (2016) Resource-saving file management scheme for online video provisioning on content delivery networks. IEEE Trans Comput 65(6):1910–1920
30. Papagianni C, Leivadeas A, Papavassiliou S (2013) A cloud-oriented content delivery network paradigm: modeling and assessment. IEEE Trans Dependable Secure Comput 10(5):287–300
31. Sung J, Kim M, Lim K, Rhee JK (2016) Efficient cache placement strategy in two-tier wireless content delivery network. IEEE Trans Multimedia 18(6):1163–1174
32. Ma G, Wang Z, Zhang M, Ye J, Chen M, Zhu W (2017) Understanding performance of edge content caching for mobile video streaming. IEEE J Sel Areas Commun 35(5):1076–1089
33. Tran TD, Le LB (2018) Joint resource allocation and content caching in virtualized content-centric wireless networks. IEEE Access 6:11329–11341
34. Yan Q, Cheng M, Tang X, Chen Q (2017) On the placement delivery array design for centralized coded caching scheme. IEEE Trans Inf Theory 63(9):5821–5833
35. Jayakumar S, Prakash S, Akki CB (2018) An investigational study and analysis of cloud-based content delivery network: perspectives. Int J Adv Comput Sci Appl (IJACSA) 9(10):307–314. https://doi.org/10.14569/IJACSA.2018.091037

IoT-Based Data Storage for Cloud Computing Applications

Ankita Shukla, Priyatam Reddy Somagattu, Vishal Krishna Singh, and Mala Kalra

Abstract Growing ubiquity of radio frequency identification (RFID) and wireless sensors devices has provided the Internet of things (IoT) with a new promising avenue to build powerful applications and industrial applications. Common physical objects can be connected to the Internet world using the RFID and can enable monitoring by a single system. Storing and processing the data in the cloud platform has become a major challenge in such a system. IoT data has three traditional properties of volume, variety and velocity. Hence to address these challenges in the IoT framework, this paper puts forward a storage framework for IoT data that enables efficient storage of huge and substantial IoT data integrating the structured as well as non-structured data. The proposed framework can combine multiple databases and is extendable to Hadoop to store multiple varieties of data collected by RFID readers and sensors. Some components in the framework are further tested to process huge amount of data using a distributed file repository which can handle massive unstructured files efficiently. Further, a prototype is also developed to prove the proposed framework's effectiveness.

Keywords Internet of things · Database management model · Query adapting method

A. Shukla (✉) · M. Kalra
Department of Computer Science and Engineering, National Institute of Technical Teachers Training and Research, Chandigarh, India
e-mail: ankitashukla.nitttr@gmail.com

M. Kalra
e-mail: Malakalra2004@gmail.com

P. R. Somagattu
Department of Information and Technology, Indian Institute of Information Technology Allahabad, Prayagraj, Uttar Pradesh, India
e-mail: priyatamsomagattu@gmail.com

V. K. Singh
Department of Information and Technology, Indian Institute of Information Technology Lucknow, Lucknow, Uttar Pradesh, India
e-mail: vashukrishna@gmail.com

N. N. Chiplunkar and T. Fukao (eds.), *Advances in Artificial Intelligence and Data Engineering*, Advances in Intelligent Systems and Computing 1133,
https://doi.org/10.1007/978-981-15-3514-7_109

1 Introduction

Industrial developments are hugely influenced by the development of IoT as a new realm in communication technology [1, 2]. IoT is an umbrella word for physical objects and their representations that are uniquely identifiable in the Internet-like structure. Kevin Ashton was the first who used the term "Internet of Things" in the year 1999 which later became famous through the Auto-ID Center. Prerequisites for IoT are often seen to be radio frequency identification (RFID) tags, sensing nodes, mobile phones and actuators [3–5]. IoT uses unique addressing schemes, to connect physical objects, through an Internet-like structure to achieve common goals by leveraging interaction and cooperation among connected objects [6–9]. However, before IoT comes to final shape, few challenges must be tackled. For example, a series of challenges remain during transit and collection of data by the devices monitoring the physical world. Data is generated at high speed through RFID and sensors, and hence, data processing is required to have high throughput. Moreover, IoT data storage solution should be able to store huge amount of data generated as well as should support horizontal scaling efficiently. Furthermore, the data collected can be structured as well as unstructured and through multiple sources, and hence, data collection provisions and components should be able to support heterogeneity in terms of data generation and collection [10, 11]. For the stated challenges, a storage platform is required which can store and manage a high volume of structured and unstructured data. In the proposed framework, a database management model is used to handle structured data, while the file repository is leveraged to manage unstructured data. Hypertext Transfer Protocol (HTTP) provides an interface for applications using data from storage which is based on the proposed framework [12]. For the same purpose a RESTful service provision is proposed to support HTTP in the current framework.

2 Architecture

Any data management framework which meets IoT and cloud service requirement should be able to handle not only high volume of data but also high variety of data, because the data is generated from variety of devices like RFID readers, thermometers monitors etc. and that to at high velocity. Collected data may vary in parameters like data types, data structures and accessing methods. Hence, any one umbrella method cannot handle such heterogeneous data. Along with this, high volume and high-velocity aspects of data require IoT data management systems to have high throughput for greater efficiency. Figure 1 shows the architecture of the framework proposed, which is divided into several modules described as follows:

File repository: Major purpose of the file repository in the proposed framework is to deal with unstructured data. For this, it leverages a well-known file system for the distributed environment, Hadoop Distributed File System (HDFS). A version

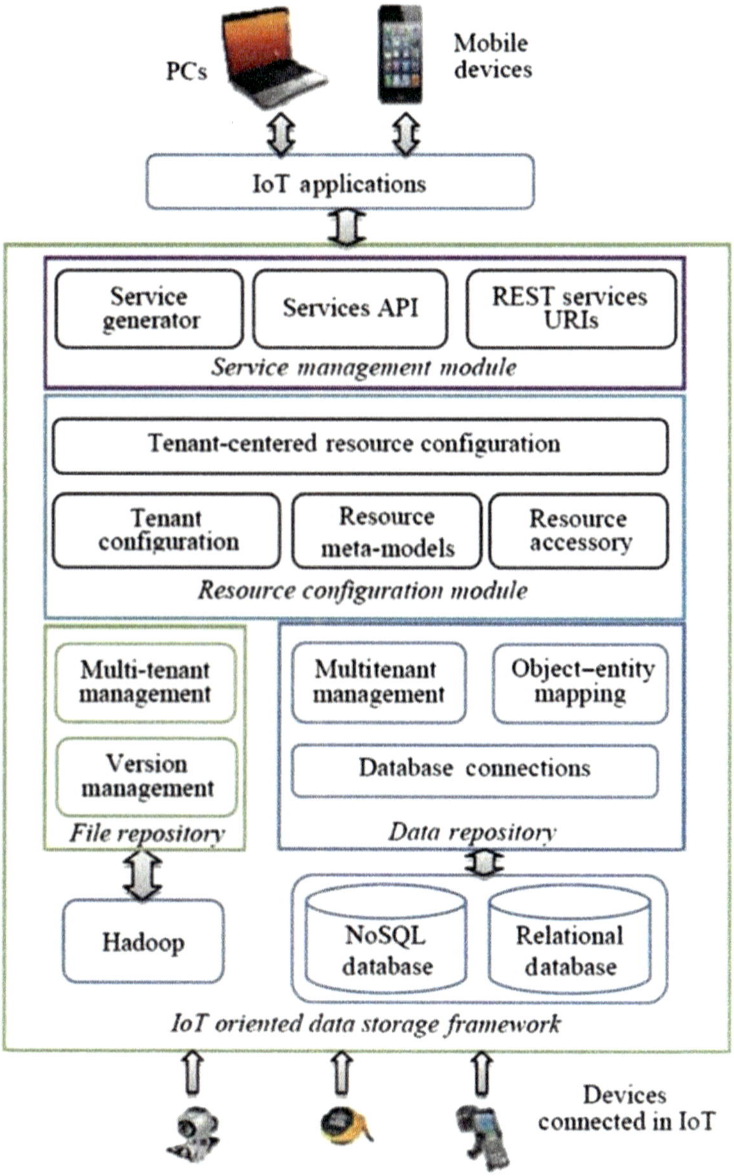

Fig. 1 Proposed framework

manager and a multitenant manager are also used to handle versioned file systems and isolating the data of tenant. Over this mechanism, a file processor deals with the management of small files.

Database module: For managing structured data it utilizes both i.e., relational database and NoSQL database. It combines multiple databases. It facilitates API creation and unified entity mapping of objects in order to hide data differences during interfacing and implementation. Thus, data access and application migration are made convenient by the database module.

Service module: This plays an important role in automatic generation of RESTful service. Firstly, from the configuration it extracts metadata and then maps the extracted data to the stored data entities like file systems, databases and repositories. All the mapping is done, based on extracted metadata for generating the RESTful service.

Resource configuration module: This module is based on a predefined metamodel and supports data management, both static and dynamic. Based on the requirement of multitenant, it configures data resources and related services. Using this module, load balancing can be done by leveraging data disposing mechanisms, thereby implementing the isolated preferences.

3 Approach

This section details the proposed approach, and data storage framework implementation is discussed in terms of: (1) database management model; (2) file repository model; (3) resource configuration; and (4) RESTful service generation.

A. **Flowchart of the approach**: Various steps like acquisition and storage of data, resource configuration, data utility, etc. on the basis of the process of disposing data define the proposed methodology. Figure 2 shows the flowchart of the approach proposed.
B. **Database management model**: Access interface unification and multiple database combination are the key tasks of database management. For this purpose, query adapting and object entity mapping are major techniques used in the proposed framework. Furthermore, the presented framework also blends multitenant data isolation method.

Object–Entity Mapping Its basic task is to integrate real physical world and virtual cyber world through object entity mapping, thereby allowing the developers to manage data in databases and also providing flexibility to process the real world objects. Database abstraction through structures of data collected is used because there are different types of databases. With respect to data storage archetype, there are four categories of mainstream NoSQL databases, given as follows: (i) key-value store, (ii) document store, (iii) column store and (iv) graph store. Despite the differences in these storage models, we locate and map a record structure containing

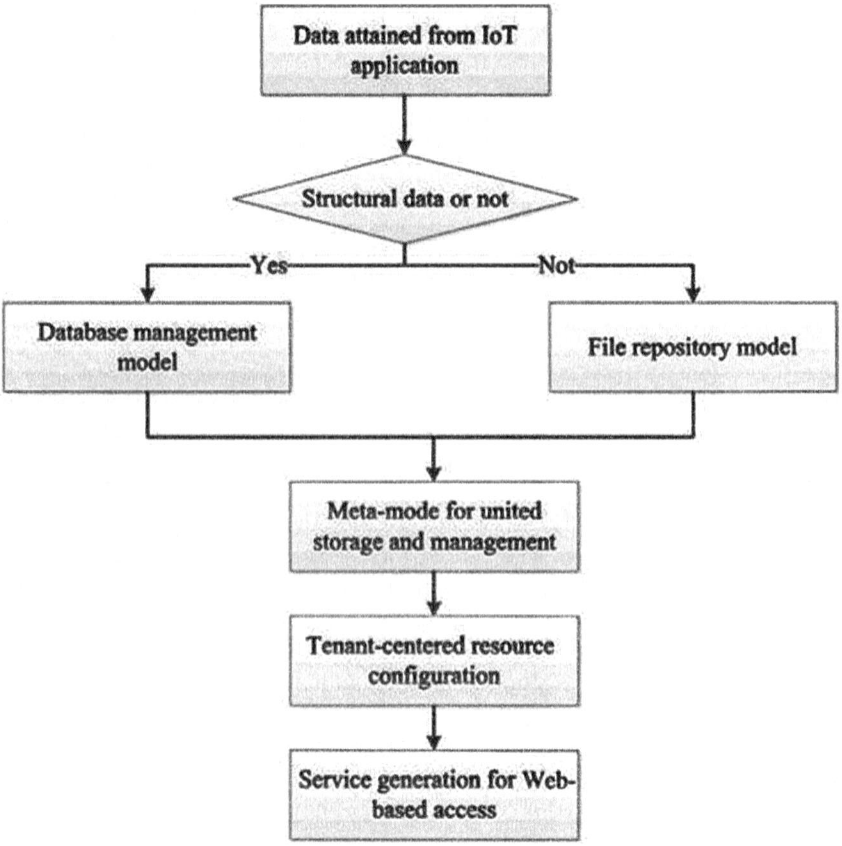

Fig. 2 Flowchart of the proposed approach

the specific properties to an object. A data entity is mapped from a collection of series of records with similar properties, for example, a table in relational databases. Moreover, in a key-value storage model a pair of key value could be considered as a record property. Similarly, a collection of key-value pairs could be taken as a series of groups from a data collection and also as the properties of a record. Moreover, the object–entities mapping requires the maintenance of the relations between the entities. In general, NoSQL databases do not support foreign key constraints; hence, the proposed framework includes this feature. Whole problem finally boils down to the method by which foreign key can be stored. The solution given in the proposed framework is based on Pedro's design with changes made according to the framework's requirements. In NoSQL databases, "table" denotes structure and is equivalent to a relational database's "table" similar to "collection" in MongoDB. To circumvent the necessary "JOIN" operation, the foreign key of each record is stored as a single property and not in the extra table. For one-to-one and many-to-one relations, only one value is stored as the foreign key property, while for relations such as many-to-many and one-to-many, the normalization of foreign key property is done for storing a set of values. Moreover, value set operations should be implemented if the database does not provide, like "add" (a known value is added to the defined set; if the set already contains the values, then no action will be taken), "contains" (a Boolean value is returned indicating whether the given value is present in the set or not), "remove" (a known value is removed from the set in case the value is already present), etc.

Query Adapting Method Queries directly generated through unified APIs cannot be accepted by heterogeneous databases. Hence, a translation is required to generate queries so that databases can accept them. This is achieved by a group of adapters. Thus, for relational databases the procedure of adapting leads to translation of unified queries into SQL queries. ORM frameworks have already implemented this. This brings forth the new challenges of implanting the adapters like: few functionalities of relational databases are not supported by NoSQL databases.

As a result, interoperability becomes a major issue during the translation of unified queries. Hence, the direct translation of operations is not a good idea because they will not be supported by target databases, and thus, operations should be implemented outside the databases. Since NoSQL databases do not support the value restrictions, the implementation of most functions is straightforward. In the proposed framework, these functionalities have been implemented by filtration of result record sets (RS) before returning it to accessor or requestor. Still the complexity of implementing the join operation persists, because the relational database systems support join operations while NoSQL databases do not have API for them since NoSQL's architecture and design is such that it does not handle join operations efficiently. Above all this, there is still the issue of seamless joining operation of entities stored in different kinds of databases.

4 Case Study and Discussion

4.1 Case Study

Hereafter, the paper discusses a case study by taking into account the process of logistics delivery as an example and demonstrates the working of the proposed framework. In the logistics business, a huge number of logistics orders and transactions are monitored using the technologies based on IoT environment such as RFID tags, sensor nodes and cameras. The data collected is first preprocessed at the terminals. The preprocessed data is then transferred to the logistics management application. This application is grounded on the framework for data storage. Multiple locations store all the types of data under different kinds of databases like MySQL database, MongoDB and file repository so that data storage and access performance are improved (as shown in Fig. 3). Proposed framework takes this logistics data (both stored and

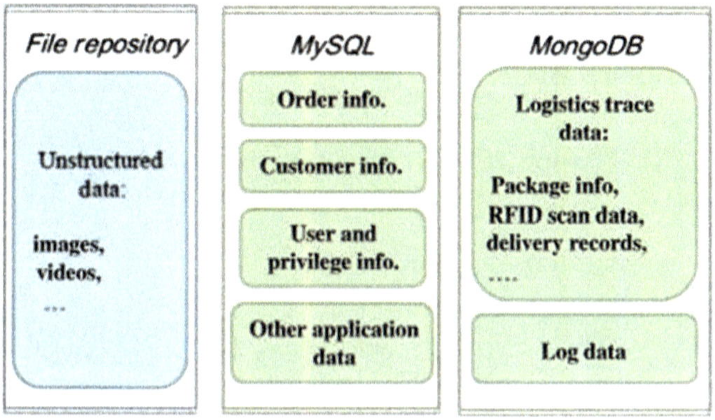

Fig. 3 Data storage and access performance

accessed data) to show the frameworks support in data management during the run-time. Storing logistics data: During the operation of logistics system, the delivery of each package and transaction is monitored by multiple devices. Hence, rapid data generation takes place. Since the transactions are multiple in numbers, it also leads to a high volume of data generation. This high-volume data generated by different devices can be stored in the proposed storage framework shown in Fig. 4.

The generated data is first preprocessed at terminals, and thus, filtered data is transferred ahead to the logistics management system so that it is in an acceptable format. In the next step, the storage of traced data in the framework is done by the logistics management system. Here, data is separated in structured and unstructured categories which are stored in databases and file repositories, respectively. Further, metadata is also taken into context, and accordingly, the structured database is stored separately into different databases based on the metadata in the configuration.

4.2 Data Resource Configuration and Storage

Storing the related data gathered from distributed sources in the defined platform makes the management and configuration of the information from different tenants easy, as depicted in Fig. 5. The IoT data files stored in the file repository are organized in two dimensions. Using device EPC generating data and the data generation timestamp, every file can be located precisely. When an EPC is provided, all the files containing the data generated by that device with respect to the EPC will be listed by the file repository.

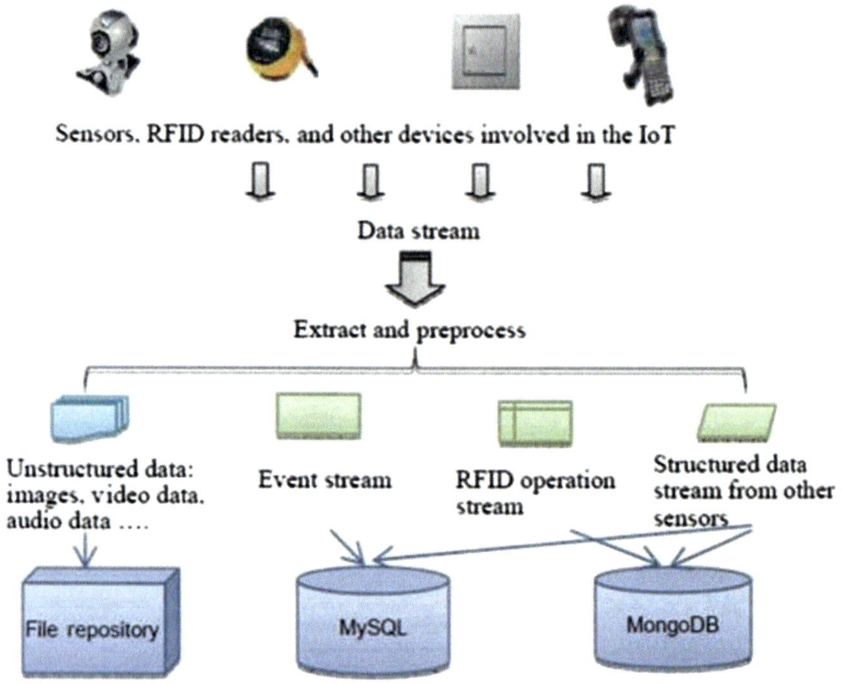

Fig. 4 Process of storing logistics

5 Conclusions and Future Work

Industrial application sector is hugely impacted by both IoT and cloud technologies. Thus, an efficient cloud-based data storage framework can increase the efficiency of IoT in terms of data storage and management services. A series of challenges exist when it comes to data storage and access like: high volume of data, data with different data types, data generating at high speed, complex data management requirements, etc. The paper has proposed a list of challenges and feasible solutions for them. Database management models have been used for structured data which extends to multiple databases and helps in developing API for unified accessing and processing of collected structured data. Similarly, file repositories based on HDFS have been used for unstructured data to implement multitenant data and version management isolation. RESTful service generating mechanisms have been used to provide cross-platform and remote data access using the HTTP interface. The merger of cloud and IoT technology in data management is likely to be used in multiple new applications. Currently in the proposed framework, limited adapters have been implemented. In the future, the authors expect to optimize the framework's performance as new avenues of applications emerge. With the future perspective, we will add on more adaptors for various NoSQL databases.

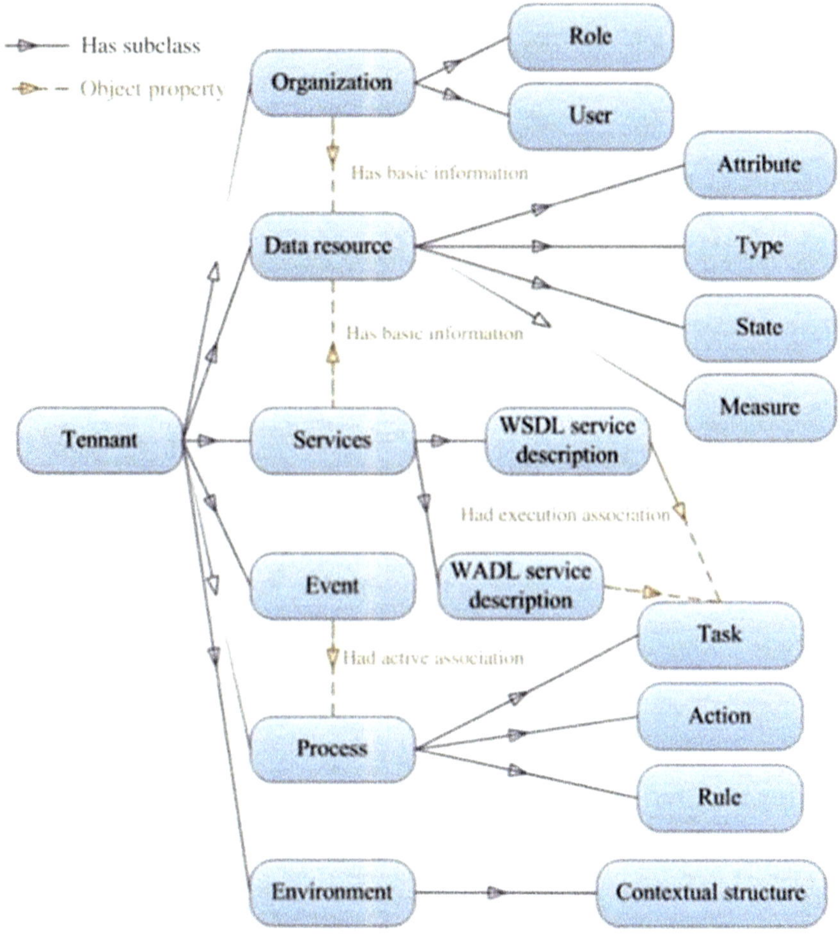

Fig. 5 Meta-model for resource configuration

References

1. Atzori L, Iera A, Morabito G (2010) The internet of things: a survey. Comput Netw 54:2787–2805. https://doi.org/10.1016/j.comnet.2010.05.010
2. Tan L (2010) Future internet: the internet of things. In: 2010 3rd international conference on advanced computer theory and engineering, pp V5-376, V5-380. https://doi.org/10.1109/icacte.2010.5579543
3. Schweppe H, Zimmermann A, Grill D (2010) Flexible on-board stream processing for automotive sensor data. IEEE Trans Ind Inform 6:81–92. https://doi.org/10.1109/TII.2009.2037145
4. Li S, Da XuL, Wang X (2013) Compressed sensing signal and data acquisition in wireless sensor networks and internet of things. IEEE Trans Ind Inform 9:2177–2186. https://doi.org/10.1109/TII.2012.2189222

5. Li Y, Li S, Song Q et al (2014) Fast and robust data association using posterior based approximate joint compatibility test. IEEE Trans Ind Inform 10:331–339. https://doi.org/10.1109/TII.2013.2271506
6. Wang L, Da XuL, Bi Z, Xu Y (2014) Data cleaning for RFID and WSN integration. IEEE Trans Ind Inform 10:408–418. https://doi.org/10.1109/TII.2013.2250510
7. Cattell R (2011) Scalable SQL and NoSQL data stores. ACM SIGMOD Rec 39:12. https://doi.org/10.1145/1978915.1978919
8. Guo J, Da XuL, Xiao G, Gong Z (2012) Improving multilingual semantic interoperation in cross-organizational enterprise systems through concept disambiguation. IEEE Trans Ind Inform 8:647–658. https://doi.org/10.1109/TII.2012.2188899
9. Shvachko K, Kuang H, Radia S, Chansler R (2010) The hadoop distributed file System. In: MSST '10 proceedings of the 2010 IEEE 26th symposium on mass storage systems and technologies (MSST), pp 1–10
10. Xu Y, Kostamaa P, Gao L (2010) Integrating hadoop and parallel DBMs. ACM 969. https://doi.org/10.1145/1807167.1807272
11. Cur O, Hecht R, Duc C Le, Lamolle M (2012) Data integration over NoSQL stores using access path based mappings. Lect Notes Comput Sci 481–495. https://doi.org/10.1007/978-3-642-23088-2_36
12. Atzeni P, Bugiotti F, Rossi L (2012) SOS (save our systems): a uniform programming interface for non-relational systems. ACM 582–585. https://doi.org/10.1145/2247596.2247671

IoT-Based Heart Rate Monitoring System

Jagadevi N. Kalshetty, P. Melwin Varghese, K. Karthik, Randhir Raj, and Nitin Yadav

Abstract Technology has changed our world in all-round manner, starting from the food processing to our daily lifestyle and in our daily chores. Heart being the most important vital organ of the body needs special focus as most of its irregularities, if not all, are life threatening. The most common heart disease involves coronary heart disease, heart arrhythmia, and tachycardia. Through this project, we are presenting a new way of monitoring and detecting heart attack and various other diseases. The system uses principle of photophelthysmology which involves measuring the variation in blood volume in tissue using light and sensor that it contains. To pump blood through body, heart first contracts and forces blood that it has received through veins into arteries after mixing it with oxygenated blood. Then, it takes rest of few millisecond and pump blood back again thus creating a cycle. This variation is recorded by heart rate sensor and shown as output on mobile application. Each person has his own high and low value which he can set in application by default after getting professional consultation from specialist.

Keywords Photophelthysmology · Heartbeat · Stethoscope · ECG

1 Introduction

Already among the death of human life in world, most of it is contributed by heart diseases. Major reason that is said to contribute to this can be classified among human factors and medical reasons. Human factor includes people eating lots of carbohydrate and less balance diet continuously for years along with less or no exercise. Medical reason includes the people who are born with such heart defects or conditions. As the person gets older, his/her heart gets weaker and hence requires routine checkup. Our product is targeting this part of population who wants to keep track of their heart health but cannot afford to go to doctor frequently due to their busy schedule and daily life.

J. N. Kalshetty (✉) · P. M. Varghese · K. Karthik · R. Raj · N. Yadav
NitteMeenakshi Institute of Technology, Bangalore, Karnataka, India
e-mail: jagadevi.n.kalshetty@nmit.ac.in

N. N. Chiplunkar and T. Fukao (eds.), *Advances in Artificial Intelligence and Data Engineering*, Advances in Intelligent Systems and Computing 1133,
https://doi.org/10.1007/978-981-15-3514-7_110

The heart rate of a healthy adult varies from 68–76 (bpm) and that of babies is around 120 beats per minute (bpm), while older people have heart rates at around 90 bpm [1]. There are a large number of heart diseases. However, most occurring of them are coronary heart disease, heart failure, arrhythmia, and high blood pressure to name a few.

Coronary heart disease is basically a group of heart conditions that leads to nausea, breathlessness, sweating, etc. It is caused due to deposition of plague (cholesterol) in the inner walls of coronary arteries. A trial fibrillation is a condition in which upper chamber of heart beats at fast and irregular rate of up to 300–400 bpm [2]. In a trial, flutter heart will be steady at around 300 bpm [3].

With advancement in field of mobile technology and android operating system, it is now much easier to send data from remote location to global severs and database. Global expansion in knowledge and usage of IoT in building of new product which is more user-friendly is great helping hand in connecting these different components which work on different ecosystem into single product. In this project, we have used knowledge of IoT to interconnect sensor unit, transmission unit, processing unit, and display unit to create complete product.

The products available in market include ECG, wrist band, stethoscope, etc. ECG is highly sophisticated machine that has wide range of feature, and however, it requires specialist to work on it and is not so cost-effective for general public. Wrist band, on other hand, is not health-specific and just lets us to know his/her heart rate. Heart disease is told to be a result of most number of deaths annually than any other disease. Every year 17.4 million people die of cardiovascular disease [4]. Global healthcare market is said to be 372 billion business by 2020 [5] and has huge potential to explore.

2 Related Background

Traditionally, stethoscope and electrocardiogram machine are used by doctors to measure person's heartbeat. Commercial product includes smart watches.

To measure heartbeat through stethoscope, doctor needs to place disc-shaped resonator near the left side of a person's chest near his/her heart and then measure number of times heart beats in a minute. However, problem with stethoscope is it does not tell heartbeats in real time. A person has to wear it and wait patiently to know his heartbeats. Also, it is not user-friendly as the person has to wait and count his heartbeats making it almost useless in case of emergency [6].

Another device used by doctors to keep record of person's beats per minute is ECG machine. ECG works by recording electrical activity produced by heart with the help of electrodes placed over chest. ECG is highly accurate, and in fact it is the most accurate machine and gives great detailed knowledge of person's health. However, it is costly and bulky which restrict its use by general public. Moreover, it requires technological expertise to operate, making it only limited to professionals and in hospitals [7].

With introduction of cheaper and easy to integrate micro-electro-mechanical (MEM) sensors, we can simply replace the existing system with smaller and easy to use newer one. Commercial smart watches use MEM sensors and are accurate to some extent. However, they do not provide complete ecosystem of safety of person which we have achieved by integrating user and doctor relatively through the same system.

3 Proposed Method

The system will have three Android applications which run concurrently for proper working of the heartbeat system application.

The application is as follows:

- Patient app
- Ambulance app
- Monitor app

Patient app This is the main part of the heartbeat monitoring system. In this, the patient or the person suffering from heart disease has to be registered. She/he should provide the details along with the problem he/she is suffering from and the contact of the dearest one to whom we can contact during heart attack. When a person is getting heart attack, the sensor after sensing the attack sends the data to the cloud (firebase: provided by Google) which is received by the other app (the ambulance app), and the app on receiving the data from the cloud locates the person who is suffering from attack and ambulance after receiving the person admits him/her to the nearby hospital for the further treatment. Meanwhile, this app also sends the message to the dearest one and provides a track on the ambulance for the family member of the patient.

Ambulance app The person who is operating the application will get the notification regarding the patient's heart attack, and this app will locate the person and admit him/her to the nearby location. Once the person is admitted into the hospital, the dearest one will receive the notification regarding their ward, that he/she been admitted to the particular hospital.

Monitor app This app is basically created for the family member of the patient to let them know about user's health. This app provides the live data of the patient's pulse rate and his/her health status. It also provides the ambulance path if called upon.

Figure 1 shows the block diagram of the proposed system. In order to detect the heartbeat, we have used invent to INVIT_11 pulse sensor [8]. It measures the variation in blood volume in the given part of body. It consists of infrared light-emitting diode that focuses IR rays on the part of body and photodiode that senses amount of light received back. The amount of light reflected back is inversely proportional to the

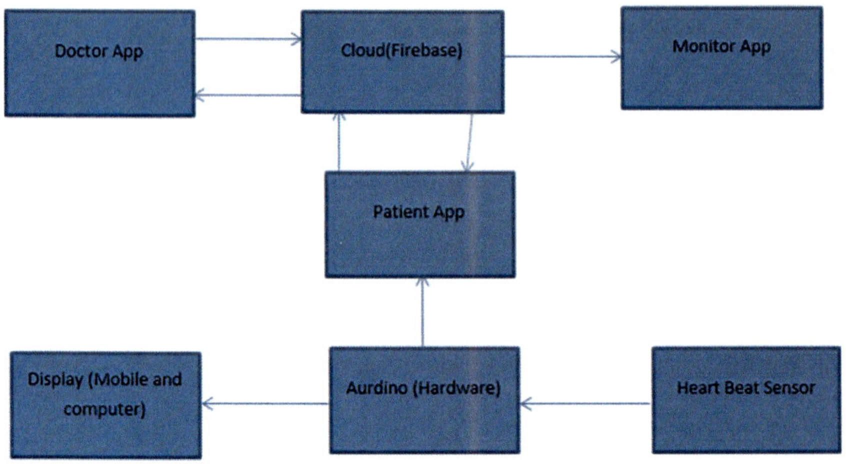

Fig. 1 Block diagram

volume of the blood. So, if a person's heart is beating heavily, then it indicates that the volume of blood in that tissue is high hence light reflected back is also less. This signal is then amplified and sent to Arduino unit and can be viewed on computer or any other display.

After doing necessary processing on the data received from the Arduino, this data is sent to the user's mobile phone that needs to run on Android platform. After receiving data from the Arduino, heartbeat reading and user's location will be sent to the cloud database. In this project, we have used firebase for this purpose.

Firebase will also keep log of all the reading of the person and location and at the same time. Firebase will also provide authentication for all the users along with other array of features. Doctor app will send back location of ambulance which user can see on his app. Monitor app which will be used by the patient's relative or family member can watch the location and reading of user in real time.

Hardware requirement:-

- Pulse sensor
- Arduino Uno board
- Android smartphone
- Bluetooth module

Arduino is an open-source platform which can be used to develop various projects [9]. Arduino Uno is Arduino board with ATmega 328P microcontroller in it.

- It has 20 pins, 6 of which are analog input pin and 14 are digital input/output pin. Along with all this, it has power USB port so that Arduino board can be powered with the help of USB cable from any electronic device.
- It has power jack to power board with AC power supply directly by connecting it with barrel jack.

Fig. 2 Pulse sensor

- It has voltage regulator that stabilizes AC current before sending it to other parts of board, powers LED to indicate if board is connected to power supply correctly.
- It has quartz crystal oscillator with frequency of 16 MHz to find out time whenever required.

Figure 2 shows the INVIT_11 pulse sensor which is used to measure heart rate. It has two different types of diode. First is light-emitting diode that transmits IR light and second is photodiode that receive the reflected light. The amount of light absorbed by tissue is directly proportional to blood it contain, hence proportional to heart rate.

Figure 3 shows the Bluetooth module HC-05 used in this proposed system [10]. It is designed for serial wireless connection. It is based on the concept of master–slave, and in default configuration, it works as slave. Master module can be any device that wants wireless connection, and in our project, it is mobile phone. It has state pin which indicates about connection; when Bluetooth module is not connected to any device, it sends signal to low state and LED flash frequently otherwise at the delay of 2 s indicating it has been connected to some device.

Data received by Arduino board from pulse sensor is sent to Android smartphone through Bluetooth module HC-05 where phone is acting as master. Android application running of user's mobile will show all the details of his/her heartbeats. Also, Android smartphone is also used as GPS module as it is used to send user's location to database server and to doctor in case of emergency to receive ambulance.

Software requirement:-

- Arduino compiler
- Android Studio
- Android 6.0 and above based smartphone

Fig. 3 Bluetooth module

Arduino compiler is used to compile Arduino code that is used to compute the heartbeat and plot graph on serial plotter to better understand person's heart condition through the graph.

Android Studio is official app-building tool provided by Android. It contains array of features such as instant run which pace up editing, building, and running cycle of app, performance profiler that tells the amount of CPU and memory required by developed in real time. It is single tool to develop app in all three platforms namely mobile, table, and TV. We have used Android Studio to develop the monitor app, patient app, and doctor app. We also used geofire feature to locate the position of user and ambulance.

Firebase is a backhand platform that can be used by anyone for developing web, Android, and IOS application. It has feature such as schemaless database in which data is stored as JSON tree with multiple nodes. It offers feature to store data offline even when there is no availability of Internet connection.

4 Simulation and Results

Figure 4 shows login page of relative app. Relative (Relative is the one who is the guardian of the patient) have to enter username and password along with name of patient they want to know current condition of that patient. Relative can also keep track of that patient's location.

Figure 5 shows login page of doctor app. Doctor will have to enter his user name, password, and name, and under personal note, it has to specify the field he is qualified into signup. After signing up, he will get notification in case of emergency.

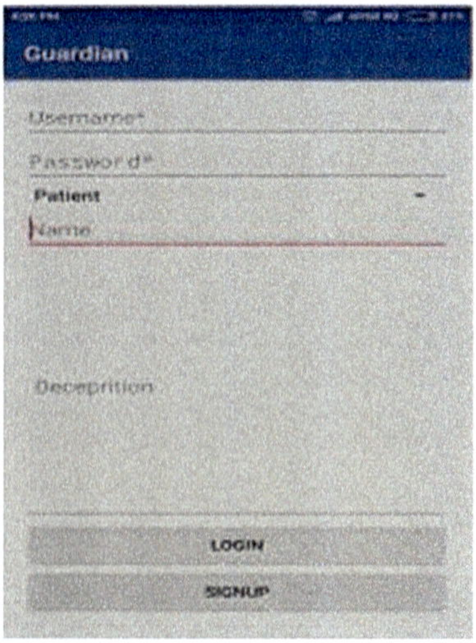

Fig. 4 Login page of relative app

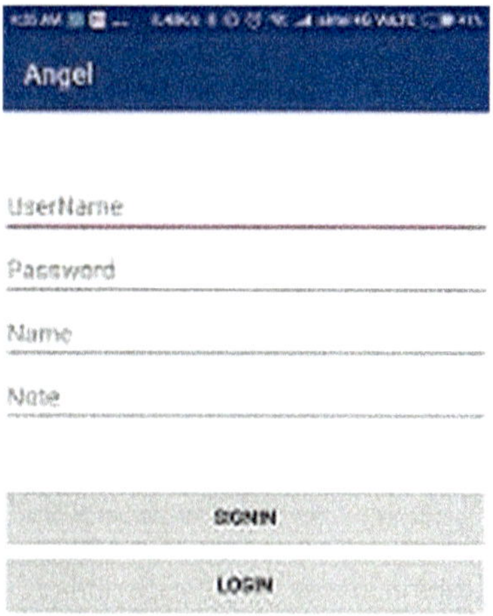

Fig. 5 Login page of doctor app

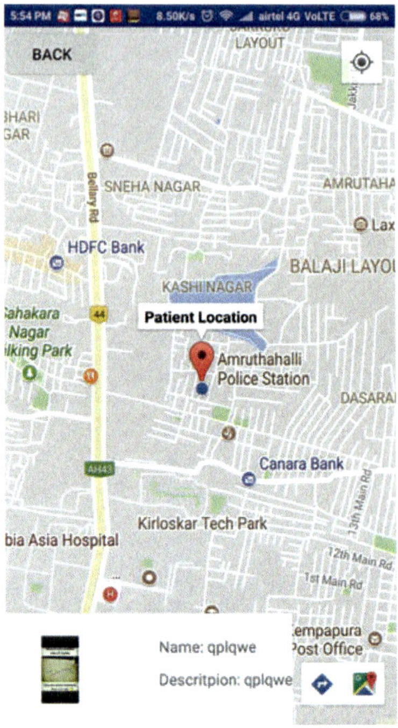

Fig. 6 Patient location in map

Figure 6 shows patient's location in map. User's (i.e., patient) location will be fetched once the emergency button is triggered, and this location can be seen in doctor and relative app.

Figure 7 shows the shortest path for ambulance from doctor's location to the patient location. Shortest path will be given based on traffic condition and other factor. This will help to reach the patience on time and provide his services on time.

Figure 8 shows the home page of patient app. Patient can check the ambulance location and nearest doctor and can trigger emergency on click of button.

Figure 9 shows heartbeat of a person on serial plotter. The variation in heart rate is calculated based on amount of blood in that part of body at that moment. This will help the doctor to read blood pressure values of the patient.

5 Conclusion

We have developed a system which is able to detect any fluctuation in heartbeat by tracking heartbeat of a user in coming future, and as the technology gets advance, more features will be added to the system. This paper focuses on the heart rate monitoring and alert which are able to monitor the heartbeat rate condition of the patient. The system determines the heartbeat rate per minute and then sends notification

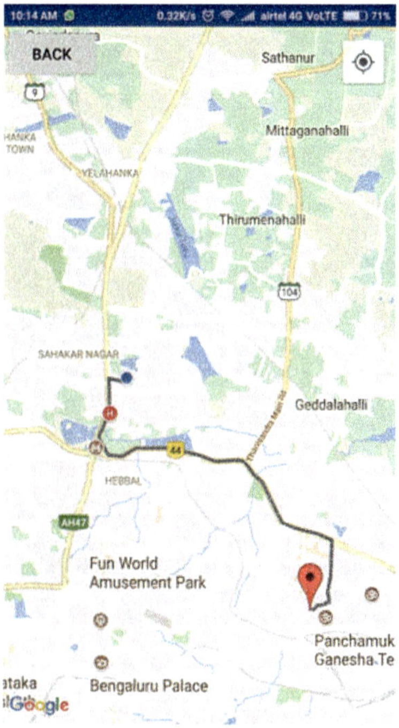

Fig. 7 Shortest path for ambulance from doctor's location to the patient location

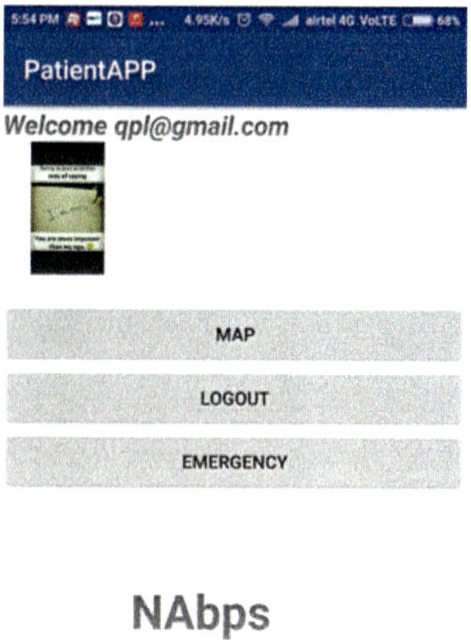

Fig. 8 home page of patient app

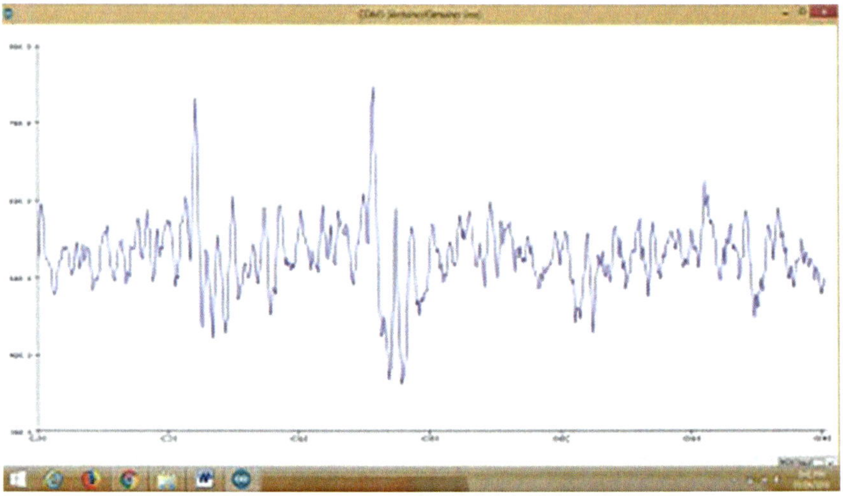

Fig. 9 Heartbeat of a person on serial plotter

and alert to the mobile phone. It is portable and cost-effective. It is a very efficient system and is very easy to handle, thus provides great flexibility and serves as a great improvement over other conventional monitoring and alert systems. Wireless components and mobile technologies are key components that would help enable patients suffering from chronic heart diseases to live in their own homes and lead their normal life, while at the same time being monitored for any cardiac events and can be effectively managed. This will not only serve to reduce the burden on the resources of the healthcare center but would also improve the quality of healthcare sector.

Addition of new sensor—new sensor such as accelerometer sensor can be integrated to Arduino, which helps in counting the number of steps and calories burnt. If the person is running fast, then number of steps count will increase at the same time hence heart rate will also increase. Such case of increase in heartbeat with increase in workout is completely normal. Hence, we can prevent false alarm with accelerometer sensor. Also, no. of calories burnt will give wider view with respect to persons health.

Diseases classifier—with addition of more accurate sensor and better algorithm, we can provide functionality of classifying disease based on frequency and average of heartbeats per minutes; this feature will provide user with wider spectrum of health care.

References

1. General heart rate https://www.mayoclinic.org/healthy-lifestyle/fitness/expert-answers/heart-rate/
2. About Atrial fibrillation http://www.heart.org/HEARTORG/Conditions/Arrhythmia/AboutArrhythmia/
3. About Atrial flutter https://en.wikipedia.org/wiki/Atrial_flutter
4. Death due to CVD http://www.who.int/mediacentre/factsheets/fs317/en/
5. Health care market value https://economictimes.indiatimes.com/industry/healthcare/biotech/healthcare/indian-healthcare-market-to-hit-372-bn-by-2022/
6. Stethoscope working https://science.howstuffworks.com/innovation/everyday-innovations/stethoscopes.htm
7. ECG machine working https://imotions.com/blog/what-is-ecg/
8. Pulse sensor unit http://invent.module143.com/pulse-sensor-how-to-use-it/
9. Arduinouno https://en.wikipedia.org/wiki/Arduino_Uno
10. Bluetooth module http://www.electronicwings.com/sensors-modules/bluetooth-module-hc-05-